Biological Wastewater Treatment

Principles, Modelling and Design

2nd edition

Biological Wastewater Treatment

Principles, Modelling and Design

2nd edition

Guanghao Chen
Mark C.M. van Loosdrecht
George A. Ekama
Damir Brdjanovic

PUBLISHING

Published by: **IWA Publishing**
Republic - Export Building, Units 1.04 & 1.05
1 Clove Crescent
London E14 2BA, UK
T: +44 (0) 20 7654 5500
F: +44 (0) 20 7654 5555
E: publications@iwap.co.uk
I: www.iwapublishing.com

First published 2020

British Library Cataloguing in Publication Data
A CIP catalogue record for this book is available from the British Library

Library of Congress Cataloguing in Publication Data
A catalogue record for this book is available from the Library of Congress

Cover design: Damir Brdjanovic
Cover image: Royal HaskoningDHV
Graphic design: Synopsis d.o.o.
English language editor: Claire Taylor
Nomenclature editor: Feixiang Zan

ISBN: 9781789060355 (Hardback)
ISBN: 9781789064186 (Paperback)
ISBN: 9781789060362 (eBook)

Editors: Guanghao Chen, Mark C.M. van Loosdrecht, George A. Ekama and Damir Brdjanovic

Authors:
Adrian Oehmen
Antonio M. Martins
Carlos M. Lopez-Vazquez
Christine M. Hooijmans
David G. Weissbrodt
Damir Brdjanovic
Diego Rosso
Di Wu
Eberhard Morgenroth
Edward J.H. van Dijk
Ernest R. Blatchley III
Eveline I.P. Volcke
Fangang Meng
Guanghao Chen
George A. Ekama
Grietje Zeeman
Gustaf Olsson
Hector A. Garcia
Ho Kwong Chui
Hui Lu
Imre Takács
Jiao Zhang
Jules B. van Lier
Kang Xiao
Kim H. Sørensen
Kimberly Solon
Mario Pronk
Manel Garrido-Baserba
Mark C. Wentzel
Mark C.M. van Loosdrecht
Michael K. Stenstrom
Michele Laureni
Mogens Henze
Nidal Mahmoud
Pernille Ingildsen
Sebastiaan C.F. Meijer
Tianwei Hao
Xia Huang
Yves Comeau

This eBook was made Open Access in June 2023.

Doi: 10.2166/9781789060362

Authors

Wastewater treatment development
Guanghao Chen
Mark C.M. van Loosdrecht
George A. Ekama
Damir Brdjanovic

Basic microbiology and metabolism
David G. Weissbrodt
Michele Laureni
Mark C.M. van Loosdrecht
Yves Comeau

Wastewater characterization
Eveline I.P. Volcke
Kimberly Solon
Yves Comeau
Mogens Henze

Organic matter removal
George A. Ekama
Mark C. Wentzel†

Nitrogen removal
George A. Ekama
Mark C. Wentzel†
Mark C.M. van Loosdrecht

Enhanced biological phosphorus removal
Carlos M. Lopez-Vazquez
Mark C. Wentzel†
Yves Comeau
George A. Ekama
Mark C.M. van Loosdrecht
Damir Brdjanovic
Adrian Oehmen

Innovative sulphur-based wastewater Treatment
Hui Lu
Di Wu
Tianwei Hao
Ho Kwong Chui
George A. Ekama
Mark C.M. van Loosdrecht
Guanghao Chen

Wastewater disinfection
Ernest R. Blatchley III

Aeration and mixing
Diego Rosso
Michael K. Stenstrom
Manel Garrido-Baserba

Bulking sludge
Mark C.M. van Loosdrecht
Antonio M. Martins
George A. Ekama

Aerobic granular sludge
Mario Pronk
Edward J.H. van Dijk
Mark C.M. van Loosdrecht

Final settling
Imre Takács
George A. Ekama

Membrane bioreactors
Xia Huang
Fangang Meng
Kang Xiao
Hector A. Garcia
Jiao Zhang

Modelling activated sludge processes
Mark C.M. Van Loosdrecht
George A. Ekama
Carlos M. Lopez Vazquez
Sebastiaan C.F. Meijer
Christine M. Hooijmans
Damir Brdjanovic

Process control
Gustaf Olsson
Pernille Ingildsen

Anaerobic wastewater treatment
Jules B. van Lier
Grietje Zeeman
Nidal Mahmoud

Modelling biofilms
Eberhard Morgenroth

Biofilm reactors
Kim Helleshøj Sørensen
Eberhard Morgenroth

Editors

Prof. Guanghao Chen, PhD

Guanghao Chen, a PhD graduate of Kyoto University, is currently a Chair Professor of Civil and Environmental Engineering, a director of the Water Technology Centre, and Head of the Hong Kong Branch of the Chinese National Engineering Research Centre in the water engineering field at the Hong Kong University of Science and Technology. His research interests include sulphur-cycle-based biological treatment process development, seawater for toilet flushing, a total water system with seawater for toilet flushing and cooling, greywater reuse, seawater-catalysed urine phosphorus recovery, and modelling of in-sewer sulphide control. He pioneered the invention and development of the SANI® process that helps to close up the water loop between fresh and brackish/saline water supplies. In 2015 he was elected an IWA Distinguished Fellow and between 2010-2018 he received three IWA awards as well as the Hong Kong Green Innovation Gold Award 2018. Prof. Chen has supervised 30 PhD and more than 20 MPhil students so far.

Prof. Mark C.M. van Loosdrecht, PhD

Mark C.M. van Loosdrecht is a scientist recognised for his significant contributions to the study of reducing energy consumption and the footprint of wastewater treatment plants through the patented and award-winning technologies Sharon®, Anammox® and Nereda®. His main work focuses on the use of microbial communities within the environmental process-engineering field, with a special emphasis on nutrient removal, biofilm and biofouling. His research interests include granular sludge systems, microbial storage polymers, wastewater treatment, gas treatment, soil treatment, microbial conversion of inorganic compounds, production of chemicals from waste, and modelling. He is a full professor and Group Leader of Environmental Biotechnology at TU Delft in the Netherlands. He is also a Fellow of the Royal Dutch Academy of Arts and Sciences (KNAW), as well as the Netherlands (AcTI), USA (NAE) and Chinese (CAE) academies of engineering. Professor van Loosdrecht has won the Stockholm and Singapore water prizes and was awarded an honorary doctorate by ETH in Zurich and Ghent University. Apart from his other achievements, he has published over 800 papers, supervised 85 PhD students so far and is an honorary professor at the University of Queensland. He was the Editor-in-Chief of Water Research and is Advisor to IWA Publishing.

Prof. George A. Ekama, PhD

George Ekama lives by a simple research credo: locally inspired, globally relevant. Ekama has widely published with more than 150 papers on wastewater treatment in top international journals. He has remained at the forefront of developments in wastewater treatment, primarily through a strong research group and has supervised 40 MSc and 25 PhD students. Focusing on municipal and industrial wastewater treatment, Ekama's work has covered a broad range of areas, from biological nitrogen and phosphorus removal, activated sludge system modelling, biological sulphate reduction, anaerobic digestion, plant-wide modelling of whole wastewater treatment plants, and investigating alternatives to desalination for augmenting urban water supply, such as seawater toilet flushing and source separation of urine. With his postgraduate students, he has twice won the Water Institute of Southern Africa's (WISA) Umgeni Award, and the WISA Piet Vosloo Memorial Prize He is a senior fellow of WISA and a fellow of the Royal Society of South Africa, UCT and the South African Academy of Engineers. Furthermore, in 2012 he won the IWA Project Innovation Award, in 2013 he was awarded The Order of Mapungubwe in Silver, and in 2017 he was recognized as an IWA Distinguished Fellow.

Prof. Damir Brdjanovic, PhD

Damir Brdjanovic is Professor of Sanitary Engineering at IHE Delft Institute for Water Education and Endowed Professor at Delft University of Technology. Areas of his expertise include pro-poor and emergency sanitation, faecal sludge management, urban drainage, and wastewater treatment. He is a pioneer in the practical application of models in wastewater treatment practice in developing countries. He is co-inventor of the Shit Killer® device for excreta management in emergencies, the award-winning eSOS® Smart Toilet and the medical toilet MEDiLOO®, with funding by the Bill & Melinda Gates Foundation (BMGF). He has initiated the development and implementation of innovative didactic approaches and novel educational products (including e-learning) at IHE Delft. Brdjanovic is co-founder and director of the Global Sanitation Graduate School. In addition to dozen of PhD students, in excess of 150 MSc students have graduated under his supervision so far. Prof. Brdjanovic has a sound publication record, is co-initiator of the IWA Journal of Water, Sanitation and Hygiene for Development, and is the initiator, author and editor of seven books in the wastewater treatment and sanitation field. In 2015 he became an IWA Fellow and in 2018 received the IWA Publishing Award.

Sponsors of the Open Access edition

中持股份
CSD Water Service

There are only a few classic works on wastewater treatment, and one of them is " *Biological Wastewater Treatment: Principles, Modelling, and Design*" published by IWA Publishing. The book revolves around the most fundamental and important aspects of biological wastewater treatment, showcasing a careful balance between theory and practice. Each chapter in the book stands on its own, yet the entire book is seamlessly integrated. Every chapter is written by world-renowned experts in the relevant field, making it a distinctive feature of this book.

Since its first edition in 2008, this book has become the best-selling publication of IWA Publishing and has received widespread international acclaim. This reflects the shift in wastewater treatment from traditional experience-based methods to the integration and application of multidisciplinary knowledge, including mechanisms, chemistry, microbiology, physics, bioprocess engineering, and mathematical modeling. The second edition of this book is led by Professor Chen Guanghao, a renowned Chinese scientist from the Hong Kong University of Science and Technology, along with Mark C.M. van Loosdrecht, George A. Ekama, and Damir Brdjanovic. Building upon the framework of the first edition, the second edition includes new chapters while revising and expanding the content of existing chapters. The addition of new authors to the original team ensures that the second edition adequately reflects the developments in knowledge and practice over the past decade since the publication of the first edition.

As the world becomes increasingly interconnected, the dissemination of knowledge is no longer confined to traditional means. This book acknowledges the transformative power of technology and the global need for accessible information. Therefore, the availability of free electronic versions is a significant aspect of its contribution. By embracing this digital age, the book aims to reach a wider audience, transcending geographical barriers and fostering a culture of open-access knowledge sharing. This initiative is a testament to the authors' commitment to advancing the field and empowering individuals across the globe.

CSD Water Service is proud to be a sponsor of the open access edition of this book. As a company dedicated to wastewater treatment and technological innovation, we recognize the critical role that wastewater treatment plays in preserving our precious resources and protecting our ecosystems. By supporting the dissemination of knowledge and the advancement of wastewater treatment practices, we strive to contribute to a cleaner and more sustainable world.

CSD Water Service

中持股份
CSD Water Service

CREATING SAFE COMFORTABLE AND SUSTAINABLE ENVIRONMENT

Established in 2009 and listed on the A-share market in 2017, CSD Water Service Co., Ltd (SH603903) is an innovative comprehensive environmental service provider with the mission of "creating a safe, comfortable, and sustainable environment". Adhering to the development principles of "technology leading, customer experience, and cooperation partners", we are committed to becoming a leading enterprise in environmental technology and have been awarded the "Top 10 Influential Enterprises in China's Water Industry" for four consecutive years.

In 2020, CSD Water Service Co., Ltd carried out a mixed reform with Yangtze River Ecological and Environmental Protection Group Co., Ltd, which is the implementing entity of the Yangtze River protection work in Three Gorges Group. After the mixed reform, Yangtze River Ecological and Environmental Protection Group became the largest shareholder of CSD Water Service Co., Ltd. As one of the main technological innovation platforms of the Yangtze River Ecological and Environmental Protection Group, CSD Water Service Co., Ltd has embarked on a career journey taking the Yangtze River Protection as an opportunity and is committed to fulfilling its more profound social responsibility and ecological vision.

Core technology system

Future oriented concept WRRF construction and operation technology system

Comprehensive organic waste treatment and utilization technology system

Industrial Park Wastewater Treatment Technology System

New Engineering Technology Direction

Key technologies for controlling rainwater overflow pollution

Key technologies and intelligent management for pipeline detection and repair

复洁环保

Clean Environment Operation

In recent years, with the continuous improvement of water treatment requirements, the emergence of new pollutants, and the urgent need for carbon neutrality and carbon emission reduction, water science and technology have developed rapidly. Since the first edition of *"Biological Wastewater Treatment: Principles, Modelling and Design"* was published, it has been used as a key textbook by many internationally famous research institutions and universities for more than ten years, and has been widely praised by the industry. The second edition of this book is led by four internationally renowned experts in the field of water treatment, and completed by dozens of experts and scholars from all over the world, which embodies the high-quality work results of all authors in this field for more than 10 years, reflects the latest progress in theoretical research and technical application of biological sewage treatment, and will effectively promote the development and progress of sewage treatment science and technology.

As one of the senior practitioners in the sewage and sludge treatment industry, I am honored to have the opportunity to write the foreword to this book on behalf of Shanghai CEO Environmental Protection Technology Co., Ltd. CEO has long been committed to the research and application of advanced technology and high-end equipment in the field of energy conservation and low carbon. Since its establishment more than 20 years ago, relying on the core technology with independent intellectual property rights and guided by green, low-carbon and recycling principle, the company has completed a large number of high-benchmark cases in the fields of sludge treatment, odor purification, sewage treatment and energy utilization of urban and industrial sewage treatment plants, which has effectively promoted the pollution and carbon reduction WWTP.

Facing the needs of global climate change and sustainable development, we believe that with the publication of this book, with the joint efforts of industry colleagues, we are expected to further fully explore the resource and energy characteristics of sewage and sludge. Through the research and application of advanced sewage treatment theory and technology, sewage treatment and sludge output paths and processes will continue to innovate. Our vision is to comprehensively improve the technical level of material and energy recycling of sewage treatment plants, and help sewage treatment plants realize the transformation from a large energy and material consumer to a treasure trove of energy resources as soon as we can. This is our common dream, and let's work hard to pursue it together!

Wenjun Huang
Chairman of Shanghai CEO Environmental
Protection Technology Co., Ltd.

SUKE
苏科环保

Water is the source of all things, the destination of all living things, and an important prerequisite for the survival and prosperity of human beings and the entire biosphere. Due to world population growth, increased water consumption, and climate change, water scarcity is becoming a serious problem in many countries. In the case of limited resources, sewage treatment plants not only need to solve the problem of pollutant removal, but also make good use of sewage to re-enter the circulation supply is also a necessary measure. With the continuous in-depth research and practice of scientific researchers and industry practitioners, sewage treatment technology is flourishing and diversified. However, the problem of emerging pollutants and the improvement of water quality discharge standards have brought more challenges to wastewater treatment technology. At the same time, the demand for circular economy and carbon neutrality under the background of the new era has also brought great changes to the sewage treatment industry.

As an enterprise focusing on technological innovation in the field of water environment, SUKE is committed to becoming a global leader in filtration and separation technology solutions. Adhering to the corporate mission of "providing cost-saving, worry-free and efficient solutions for water environment improvement, water resource utilization and water safety", we promote water treatment technology innovation. Membrane bioreactor (MBR), as a mainstream technology in sewage treatment, has become one of the research hotspots in the industry due to its high treatment efficiency, good effluent quality, and small footprint. But it also has high energy consumption high cost difficult operation and maintenance and other shortcomings. Aiming at the above pain points of traditional MBR technology, SUKE has developed a new type of energy-saving and consumption-reducing MBR technology, SUKE-LEP®-NA reciprocating membrane bioreactor technology, that supports the key project of the national key research and development plan "international cooperation in scientific and technological innovation between governments". SUKE-LEP®-NA overcomes the technical pain points of MBR by innovatively adopting reciprocating membrane mechanical vibrating instead of traditional aeration scrubbing, and provides the advantages of saving equipment, long life, low energy consumption, high nitrogen removal efficiency, and long cleaning cycle.

Collaborative innovation has become a trend in the world today. We believe this book can provide a better communication platform for scholars, and we also hope that this book can provide guidance for relevant practitioners, lay a solid foundation for realizing green water and lush mountains, and jointly promote the high-quality development of water environment. The support for the publication of "Wastewater Biological Treatment - Principles, Modeling and Design" reflects our support for technological innovation in the global water environment field and our confidence in promoting technological transformation of the industry. With the accelerated evolution of a new round of technological revolution and industrial transformation around the world, our goal is to realize the "dual carbon" vision of sustainable development of natural resources through technological innovation and build a community of shared future for mankind.

Dr. Song Canhui
Chairman of Suke Environmental Protection
Technology Co., Ltd

The Woo Wheelock Green Fund was established in 1994 and is Hong Kong's first business sector funding to provide support for environmental protection through research. Its affiliate, Wheelock Properties Group, marks its centenary in 2023. Founded and rooted in Hong Kong, Wheelock Properties has been part of the city's architectural heritage, developing sustainable residential and commercial properties that have shaped and supported Hong Kong's progress through the decades. The Group's signature projects endeavour to enhance the built environment through a deep respect of the natural environment and the earth's resources. Both the Fund and Wheelock Properties Group share the same sustainable vision – to continuously explore innovative solutions to address complex challenges in sustainable development.

We are proud to be a sponsor of *The Biological Wastewater Treatment Principles, Modelling and Design (2nd edition)*. This book aligns with our commitment to foster educated communities, particularly among the younger generation. We acknowledge the pivotal role wastewater treatment plays in safeguarding and conserving water sources, managing wastage, and contributing to sustainable urban development. We trust that this understanding will inspire necessary action towards mitigating climate change and help to build climate resilience.

We invite you to join the journey to a sustainable and resilient future.

Woo Wheelock Green Fund

Sponsors of the 2nd edition

中国水环境集团
China Water Environment

In Shanghai, China, there is a plant where people are exercising and children are playing, while wastewater treatment is invisible underground. 'Not In My Back Yard' is changing to 'Close To My Yard'. The government regards it as the most beautiful and environmentally friendly wastewater treatment plant in China. These years, we have also built such plants in Beijing, the capital of China, and in Sichuan province, the hometown of giant panda. I am the builder and operator of these projects. It is now more than a decade since the first edition of this book was published. Since then, the biological technology of wastewater treatment has changed significantly around the world. I am very happy to witness the progress from only treatment to reuse and recovery.

According to the Millennium Development Goals and the 2030 Agenda for Sustainable Development, economic growth and clear water, are two of the most pressing challenges. We hardly notice natural resources such as air, water, soil and blue sky when we have them. But we won't be able to survive without them. Industrialization has created material wealth never seen before, but it has also inflicted irreparable damage to the environment. As president Xi said, clear waters and green mountains are as good as mountains of gold and silver. We must maintain harmony between man and nature and pursue sustainable development. In the battle against environmental pollution, wastewater treatment plays a pivotal role. In the current global Covid-19 pandemic, underground wastewater treatment plants are the final barrier to protect the water environment.

This book is a widely used teaching resource for universities and a reference for engineering companies around the world including China. During my career, I and my colleagues have also benefitted a lot from the book. New advances in technology continue to occur and, as a result, this new edition includes numerous advances and represents the current state-of-the-art information. I deeply agree with the concept and technology mentioned in this book.

While resource-oriented wastewater treatment is gradually becoming the theme of the time for the global wastewater sector, China is also actively exploring its own suitable path. With the vision of turning the WWTP from a site of pollutant removal into a plant for energy, water and fertilizer and an integrated part of urban ecology. These years, China Water Environment Group, collaborating with the top experts, institutes and authorities, has constructed and operates several plants aiming to realize the quadruple goal of sustainable water supply, energy saving operation, resource recovery and environmental harmony, by integrating various innovative designs and leading-edge technologies. All this demonstrates that we are moving in the right direction. Of course, challenges still exist. We should face these problems and tackle them with all the stakeholders. We are confident that we will have more achievements when the next edition of this book is published.

侯锋

Feng Hou
Chairman of China Water Environment Group
Professor, IWA member

Water breeds life. Despite the changes of seasons and time, water is consistently essential for life. On the back of social progressing and technological upgrading, humankind has revitalized the circulation and supply of water. With the deepening of people's cognition and understanding of wastewater treatment gradually, the relevant technology has also continuously achieved important breakthroughs. However, wastewater compositions have inevitably become more complex. This has continuously brought new challenges to the wastewater treatment industry.

Biological treatment has become one of the most striking strategies with high expectations in the industry, as it uses microorganisms to convert wastewater into clean water through natural processes. Biological wastewater treatment seems easy to understand. However, considering that it integrates disciplines such as biology and biochemistry, there are still many complicated areas that we have not yet completely understood. This offers scientific researchers and industry practitioners enormous space to further tap into these technologies and apply varying theories.

As a leading enterprise in the ecological and environmental sector of China, Everbright International adheres to its corporate mission of being "Devoted to Ecology and Environment for a Beautiful World", devoted to transforming the wastes and contaminated water into clean one. In terms of wastewater treatment, the company has adopted biological treatment technologies for some of its world-class wastewater treatment projects, playing a key role in promoting the implementation of such technologies. The support of the publication of Biological Wastewater Treatment - Principles, Modeling and Design (2nd edition) showcases the company's full support for water treatment technology and R&D of practical actions, as well as expressed our confidence in the prospect of biological wastewater treatment technologies.

As the old Chinese saying goes, "the sea contains every river, and the greatness contains everything". We believe the book will be a platform collecting various voices from different experts and scholars, bringing in more constructive thoughts and inspirations for the research and practice of biological wastewater treatment. We hope this book will bring enlightening theoretical and practical basis as a guidance for implementation, establishing a solid foundation for the upgrading and application of R&D in the future industry, in order to better protect water, the essence of life of human beings.

Dr. Wang Tianyi
CEO of China Everbright International Limited
Chairman of China Everbright Water Limited
Chairman of China Everbright Greentech Limited

ATAL Engineering Group
安樂工程集團

It has been over a decade since the publication of the first edition of the Biological Wastewater Treatment Book in 2008. Shortly after that, the Hong Kong Cross Harbor Race, an annual swimming race in the world renowned Victoria Harbor first held in 1906 and suspended since 1979 due to poor water quality, was resumed. The successful implementation of the Harbor Area Treatment Scheme, the largest ever sewerage infrastructure project in Hong Kong, significantly improves water quality in the Harbor and enables the Race to resume. ATAL Engineering Limited, a pioneer in environmental engineering in Hong Kong, Macau and Mainland China, contributed significantly to the Scheme and won top awards for this project. In fact, ATAL has earned a reputation for providing state-of-the-art environmental technologies and product applications. Working at the cutting edge of environmental engineering for decades, we have built hundreds of water/wastewater treatment plants, and keep on embracing new challenges in this field.

Besides international metropolises like Hong Kong, there are also increasing public concerns on pollution in developing regions/countries where economic development has taken precedence over the environment. As inadequately treated sewage is a major contributing factor to environmental degradation, the important role of wastewater treatment in environmental protection and improvement is self-evident. While biological treatment is still the dominant wastewater treatment approach, new challenges include more stringent requirements on pollutants removal, especially phosphorus and nitrogen. Meanwhile, expectations for energy-neutral or even energy-positive wastewater treatment plants are growing. To cater for this trend, new energy saving bio-treatment processes such as anammox, aerobic granular sludge, etc. are expected to be implemented in full scale projects gradually. Our ANAMMOX project in side stream proved its energy saving features, and its application in main stream would be even more rewarding energy-wise. In this regard, readers may find the information in this book rather useful.

The second edition of the book continues to provide details on the theoretical aspects of biological treatment, and further addresses the new challenges and latest technological developments. As a technical reference book, it provides young professionals in the field of pollution control and environmental protection a deeper understanding of biological treatment technologies and emerging contaminants removal methods. Hopefully, it will help undergraduate and postgraduate students to establish basic theoretical knowledge, and build up the capacity for design and modelling. Our design, build and operation experiences in actual projects could serve as the basis to feedback and improve on future design and modelling work, as well a reference source of real-world practical engineering information. Together with fellow authors, who are all eminent water quality experts, we put together our joint efforts into meeting new challenges, developing innovative technologies and building a better environment for this and future generations on this planet, our one and only homeland.

Raymond H.M. Chan
Managing Director
ATAL Engineering Ltd.

Preface

This second edition of Biological Wastewater Treatment fully illustrates the advantage of transnational activities in order to improve the development of a topic.

The first edition was due to a significant international collaboration which I am very proud to have been part of. The book was made as support for an online course, but had the ambition to be a general textbook. This has been proven since the book has been used worldwide and translated into Spanish, Arabic, Chinese, Russian and Korean, and has become the ultimate international textbook within the topic area.

Water science and technology is developing very fast. In the 12 years since the first edition a lot of new knowledge has been developed and new scientific specialities are needed in order to handle the increasing complexity of wastewater treatment. It is a pleasure for me that younger specialists have joined the editor group of highly renowned old bigwigs for this 2nd edition.

The first edition was made when I was close to retirement and was one of my last projects. I decided very early in my professional life that I would step aside and get out of the wastewater business at a not too old age, which I did in 2011.

Good luck with the new edition but do not forget that there is a wonderful world outside the fence of the wastewater treatment plant.

Mogens Henze
Professor emeritus
Environmental Engineering
Technical University of Denmark

About the book

The first edition of this book was published in 2008 and it went on to become IWA Publishing's bestseller. Clearly there was a need for it because over the twenty years prior to 2008, the knowledge and understanding of wastewater treatment had advanced extensively and moved away from empirically-based approaches to a fundamental first-principles approach based on chemistry, microbiology, physical and bioprocess engineering, mathematics and modelling. However the quantity, complexity and diversity of these new developments was overwhelming for young water professionals, particularly in developing countries without readily available access to advanced-level tertiary education courses in wastewater treatment. For a whole new generation of young scientists and engineers entering the wastewater treatment profession, this book assembled and integrated the postgraduate course material of a dozen or so professors from research groups around the world who have made significant contributions to the advances in wastewater treatment. This material had matured to the degree that it had been codified into mathematical models for simulation with computers. The first edition of the book offered, that upon completion of an in-depth study of its contents, the modern approach of modelling and simulation in wastewater treatment plant design and operation could be embraced with deeper insight, advanced knowledge and greater confidence, be it activated sludge, biological nitrogen and phosphorus removal, secondary settling tanks, or biofilm systems.

However, the advances and developments in wastewater treatment have accelerated over the past 12 years since publication of the first edition. While all the chapters of the first edition have been updated to accommodate these advances and developments, some, such as granular sludge, membrane bioreactors, sulphur conversion-based bioprocesses and biofilm reactors which were new in 2008, have matured into new industry approaches and are also now included in this second edition. The target readership of this second edition remains the young water professionals, who will still be active in the field of protecting our precious water resources long after the aging professors who are leading some of these advances have retired. The authors, all still active in the field, are aware that cleaning dirty water has become more complex but that it is even more urgent now than 12 years ago, and offer this second edition to help the young water professionals engage with the scientific and bioprocess engineering principles of wastewater treatment science and technology with deeper insight, advanced knowledge and greater confidence built on stronger competence.

The Editors

Table of contents

doi: 10.2166/9781789060362_0001

1

Wastewater treatment development

Guanghao Chen, Mark C.M. van Loosdrecht, George A. Ekama and Damir Brdjanovic

1.1 GLOBAL DRIVERS FOR SANITATION

Sanitation was voted the greatest medical milestone since 1840 in a British Medical Journal poll in 2007 (Ferriman, 2007). This confirms the absolute importance of sanitation in achieving and maintaining good public health. In many industrialized countries, wastewater is transported safely away from the households. However, proper sewage treatment is not always in place, in particular in many developing countries where there is far less sanitation coverage in comparison with water supply. The need for proper sanitation was made explicit in the United Nations Sustainable Development Goals (SDGs). This was further accentuated by announcing the sustainable development goal number 6 – Safe Water and Sanitation which states that clean and accessible water is essential for all people in the world in view of the fact that millions of people die from diseases associated with inadequate water supply, sanitation and hygiene (WASH). Sanitation plays a central role in achieving this goal. The UN World Water Development Report 2019 calls for access to safe water and sanitation for all as essential for eradicating poverty, building peaceful and prosperous societies, and ensuring that 'no one is left behind' on the road towards sustainable development. However, despite significant efforts, progress on sanitation improvement is still slow and lagging behind. The world needs to pay attention to the call to start implementing proper sanitation solutions for all. What is important in this is to not only connect people to sanitation solutions, but also to make this connection last in an environmentally sustainable way. Sewer systems and wastewater treatment plants have proven to be very efficient in conveying and removing pathogens, organic pollutants and nutrients but they require proper operation and maintenance, and a good understanding of the processes involved.

1.2 HISTORY OF WASTEWATER TREATMENT

Wastewater treatment development was the most visible in the 20th century. Sewage has for a long time been considered a potential health risk and nuisance in urban agglomerations. However, the fertiliser value of human excreta was already recognized in very early days. In China from ancient times, *e.g.* the Xihan Dynasty (202 BC), up until recently (the 1970s), the vast majority of agricultural land was fertilized by human faeces from latrines. The ancient civilisations in the Indus valley (already in 2000 BC), the Euphrates region and Greece used public latrines which drained into sewers conveying the sewage and stormwater to a collection basin outside the city; from there, brick-lined conduits took the wastewater to agricultural fields which used the wastewater for irrigation and to fertilise crops and orchards. The sewers were periodically flushed with wastewater.

The Romans took this system further: in about 800 BC, they constructed the *Cloaca Maxima*. Initially, this central sewer system was used to drain the marsh upon which Rome was later built. By 100 AD, the system was almost complete and connections had been made to some houses. Water was supplied by an aqueduct system which carried sewage from the public baths and latrines to the sewers beneath the city and finally into the Tiber. The streets were regularly washed with water from the aqueduct system and the waste washed into the sewers.

This system worked very well because it could count on an effective government and the protection of a powerful army to maintain the far-reaching aqueducts. However, when the Roman Empire collapsed, their sanitary approach collapsed with it as well. The period between 450 and 1750 AD is therefore known as the 'Sanitary Dark Ages' (Wolfe, 1999). During this period the main form of waste disposal was simply to dispose of it in the streets, often by emptying buckets from second-storey windows. In around 1800 a collection system appeared in many cities, instigated by the city dwellers who did not want to put up with the smell any longer. It was also welcomed by the farmers around the city who found good use for this 'humanure'. In Amsterdam, a cart drove through the streets and buckets could be emptied into it. The cart was ironically named after a brand of Eau de cologne known at that time: the Boldoot cart. However, spilling during transportation and emptying of the buckets was unavoidable, and the olfactory burden on the citizens did not decrease much. By then, plans arose for a general sewer system but high investment costs and uncertainty over flushing and maintenance of the sewers put the fast implementation on hold.

In around 1900 Mr. Liernur came up with a solution. He developed a plan for separate collection of toilet water and of grey and storm water. Toilet water was to be collected through a vacuum sewer called the Liernur pneumatic sewerage system. This system found use in several European towns (Figure 1.1).

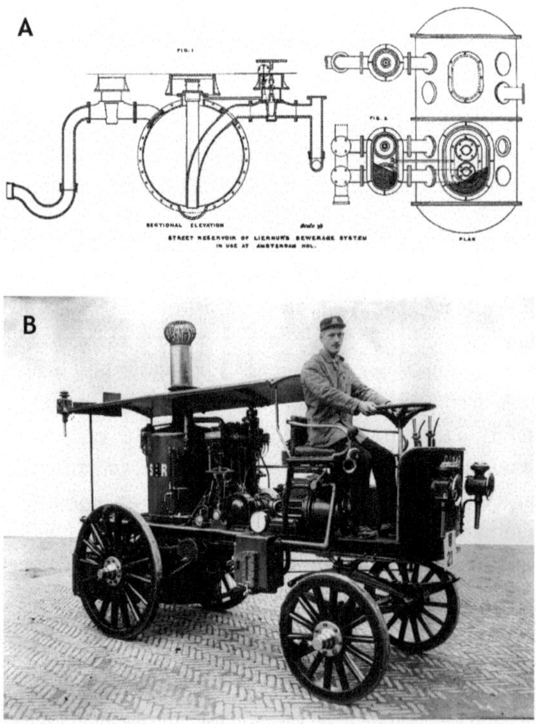

Figure 1.1 The Liernur vacuum sewer system (A) and the vehicle used for collection and transport of waste (B) (photos: Van Lohuizen, 2006).

The collected sewage did not undergo any treatment. Instead, it was spread out over land as a fertilizer. However, water-logging became a major problem, and the continuous expansion of the cities made it more difficult to find sufficient land nearby. The idea that there might be better ways, using 'organisms', gradually began to emerge (Cooper, 2001).

In the United States and the United Kingdom, organisms already found their way as applied water cleaners in what were termed biological filters: biofilms on rocks in the river bed. One of the earliest biological filters, in Salford near Manchester in the UK, stems from 1893. In the US the first filter was installed in 1901, in Madison, Wisconsin. Between 1895 and 1920 many were installed in the UK to treat sewage from towns and cities. This rapid application had a negative effect upon the later implementation of the activated sludge process in the UK after it was invented in 1913: investment money had already been spent on the biological filters.

The activated sludge process was discovered in the UK: experiments on treating sewage in a draw-and-fill reactor (the precursor to today's sequencing batch reactor) produced a highly treated effluent. Believing that the sludge had been activated, in a manner similar to activated carbon, the process was named 'activated sludge' (Ardern and Lockett, 1914).

During the first half of the 20th century, the river into which the wastewater was discharged was considered an integral part of the treatment process. The reason why 5 days is used in the biochemical oxygen demand (BOD) test is because 5 days was the longest time water spent in the rivers of the UK before it reached the sea. The book Stream Sanitation by Phelps (1944) uses mathematical modelling to calculate the maximum organic load from the oxygen sag curve to prevent the dissolved oxygen (DO) concentration falling below a minimum value at a point downstream of the wastewater discharge point. However, with the rapid growth of cities, it was soon realized that rivers could not cope with the ever increasing organic loads. As a response, the requirements increased for wastewater treatment to achieve better removal efficiencies. To reduce the oxygen demand in the river and to eliminate the toxic effect of ammonia on aquatic species, the requirement for nitrification was introduced. This led to the construction of many low-loaded trickling filter plants for organic removal and nitrification in the USA, Europe and South Africa. Anaerobic digestion was usually included in the trickling filter plants to treat the primary and trickling filter sludge produced. The discharge of nitrate from these plants was believed to be good because it provided a barrier against anaerobic conditions in the rivers and lakes. However, the trickling filters did not always nitrify very well - particularly in the winter - due to the requirement of high organic removal efficiency prior to efficient nitrogen removal.

In the second half of the 20th century a new problem in surface water emerged: that of eutrophication. Eutrophication stands for the explosive growth of algae and other water plants due to the fertilizing effect of the nitrogen (N) and phosphorus (P) discharged into rivers. In the 1960s it became clear that nitrogen and phosphorus also needed to be removed from wastewater to limit eutrophication. This inspired intensive research programs and during the 1960s the fields of bacteriology and bioenergetics were applied to wastewater treatment. By applying Monod (1950) kinetics from the field of bacteriology, Downing *et al.* (1964) showed that nitrification depends on the maximum specific growth rate of autotrophic nitrifying organisms which is slow compared with that of heterotrophic organisms. For the full-scale plant, this means that the sludge age has to be long enough to achieve consistently low effluent ammonia concentrations. So successful was the use of Monod kinetics in wastewater treatment that it is still used today in all simulation models for wastewater treatment, not only to model nitrification but also many other biological reactions. From bioenergetics, which was developed to a very advanced level by McCarty (1964), it was realized that the nitrate produced by nitrification could also be used by some heterotrophic bacteria instead of oxygen and converted into nitrogen gas. This insight led to the nitrification-denitrification activated sludge system, in which parts of the reactor were not aerated to induce denitrification. With all this new knowledge put successfully into practice, the suspended medium activated sludge system became the preferred wastewater treatment system. The post-denitrification system, in which the non-aerated (anoxic) reactor follows the aerobic reactor, was developed by Wuhrmann (1964) in Switzerland. To increase the denitrification rate in the anoxic reactor, methanol was dosed to supply the organics for the denitrification process. Because of the low nitrogen effluent values achieved with this method, this practice was widely adopted in the USA. However, methanol addition costs money, and it is rather contradictory to add organics to wastewater after first removing them. The pre-denitrification system developed by Ludzack and Ettinger (1962) formed a logical next step. In South Africa in 1972, Barnard combined the post- and pre-denitrification reactors and introduced recycle flows to control the nitrate entering the pre-denitrification reactor in the 4-stage Bardenpho system (Barnard, 1973). With this development, nitrogen removal activated sludge systems became increasingly common.

A different line of development was initiated by the work of Pasveer (1959) who progressed based on the work of Ardern and Lockett. They originally designed a fill-and-draw process. Pasveer was focusing on an economical system. The ditch system he developed was based on using one treatment unit only. There was no primary settler, no secondary settler, no digester, and so forth. In the fill-and-draw process with continuous feeding, simultaneous nitrification and denitrification occurred. The simplicity and low costs led to widespread use. Out of the Pasveer ditch system the continuous operated oxidation ditch systems evolved, based on the same principle but with a separate clarifier.

However, to control eutrophication, solely nitrogen removal was not sufficient. Phosphorus, mainly in the form of ortho-phosphate from detergents and human waste, also needed to be removed because in many ecosystems phosphorus proved to be the main enabling element for eutrophication. Unlike nitrogen, phosphorus can only be removed by converting to a solid phase. Phosphorus removal by chemical precipitation followed by tertiary filtration appeared during the 1970s. In regions where water is scarce however, like the south-western states of the USA, South Africa and Australia, indirect reuse of surface water was already high and chemical phosphate removal would cause a rapid increase in surface water salinity. Apart from the fact that salinity reduces agricultural use of surface water, its greater impact is on the durability of the water distribution system. To mitigate these impacts, water policy in South Africa in the late 1960s and early 1970s was aimed at full wastewater reclamation for redistribution to avoid both eutrophication and salination of surface water – if the high cost of chemical phosphate removal was going to be incurred, then the water may as well be reclaimed completely and returned to the distribution system rather than the environment (Bolitho, 1975; Van Vuuren, 1975).

Figure 1.2 The first (pilot) application of the Pasveer ditch system (at Voorschoten, The Netherlands in 1954). The plant capacity was 400 P.E. and 40 m³/h at dry weather flow (photo: Van Lohuizen, 2006).

Biological phosphate removal is a unique biological process that was discovered by accident. The first indication of biological phosphate removal occurring in a wastewater treatment process was described by Srinath *et al.* (1959) from India. They observed that sludge from a certain treatment plant exhibited excessive (more than needed for cell growth) phosphate uptake when aerated. It was shown that the phosphate uptake was a biological process (inhibition by toxic substances, oxygen requirement). Later, this process, referred to as enhanced biological phosphate removal (EBPR), was noticed in other (plug flow) wastewater treatment plants. The first designed processes (the PhoStrip® process) for biological phosphate removal still arose from a time when the mechanism behind the process was unknown (Levin and Shapiro, 1965). In the early 1970s due to an increased demand for nitrate removal as well as for energy savings (the 1970s energy crisis), at several places worldwide it was discovered that biological phosphate removal could relatively easily be stimulated. For example in 1974, while optimizing nitrogen removal at the Alexandria activated sludge plant by switching aerators off at the influent end of the plant, Nicholls (1975) noted low effluent phosphorus (and nitrate) concentrations. He found very high phosphate concentrations in the sludge blanket which had settled to the floor of the reactor and into which the influent wastewater descended due to a higher density than the clear supernatant. Barnard (1976) developed the Phoredox principle for biological excess phosphate removal, which introduced anaerobic and aerobic cycling in the activated sludge system. EBPR is now an established technology which opened the opportunity for phosphate removal and recovery without increasing salinity so that treated effluents could be returned to the environment or efficiently reused. As so often happens, new developments are found by accident and the understanding of how they work follows afterwards. It took many years of research in South Africa, Canada and Europe to fully understand and control the process and today there are still several facets about it that are not clear. However, not fully understanding the underlying principles has never stopped engineers and scientists from building and operating wastewater treatment plants.

The energy crisis in the 1970s associated with an increased demand for industrial wastewater treatment shifted attention from aerobic wastewater treatment to anaerobic wastewater treatment. The slow growth rate of methane-producing bacteria had always been a limitation on the process development. For the concentrated and warmer industrial wastewaters, this was less of a problem and certainly the development of the upflow anaerobic sludge blanket reactors (UASB) by Lettinga and co-workers (Lettinga *et al.*, 1980) meant a breakthrough for anaerobic treatment. Not only was this technology feasible for industrial wastewater treatment but also anaerobic treatment of low-strength municipal wastewater in tropical regions of South America, Africa and Asia could be efficiently introduced.

After a century of constructing wastewater treatment plants, many treatment plants that were initially built outside the urban area had become engulfed by residential areas. Expansion of plants became a problem and the engineers started to find more compact treatment options. Moreover, industry started to treat its own wastewater, and for industry, land use is even more critical than for *e.g.* municipalities. One successful line of development was going back to the original biofilm-based trickling reactors. A whole range of new processes was developed (biological aerated filters, fluid-bed reactors, suspension reactors, biorotors, granular sludge processes or moving-bed reactors and membrane bioreactors) which overcame the original problems of the trickling filter process.

The development of these reactors originated from the 1970s. Another development initiated in this period only became widely introduced from the late 1990s: the activated sludge process with membrane separation instead of settlers. The breakthrough for the membrane bioreactors (MBR) was made by Yamamoto and his co-workers (Yamamoto *et al.*, 1989) by integrating the hollow fibre membrane module inside the aeration bioreactor. In the early 21th century another compact technology eliminating clarifiers was developed based on granular sludge. By recognizing the basic principles behind granular growth morphology it was possible to develop aerobic granular sludge technology (the Nereda® process) which allows for a more energy-efficient and compact nutrient removal process (Pronk *et al.,* 2015).

With ever increasing effluent demands, the need arose to upgrade treatment plants instead of building new plants. Around the turn of the last century, this led to the development of a range of new processes to be integrated in existing treatment plants. The problem tackled especially by these processes is the very high nitrogen and phosphate release during anaerobic digestion of waste activated sludge, which were traditionally recycled to the activated sludge process. Apart from struvite precipitation problems, this also results in high nutrient recycling and higher effluent nitrogen and phosphate concentrations from the activated sludge system when the dewatering liquor is recycled back to the influent. Research into this problem has led to many innovations in dewatering liquor treatment. In the Netherlands, processes have been developed such as the Single reactor system for High activity Ammonium Removal Over Nitrite (SHARON®), ANaerobic AMMonia OXidation (ANAMMOX®) and processes for improved nitrogen removal and mineral crystallization processes for phosphorus precipitation for phosphorus recovery and reuse (Crystalactor®). Especially the anammox process has been developed in an array of commercial process technologies.

In the last decade there has been far more attention paid to resource recovery. Water is an obvious resource to be recovered, just as biogas. An array of new possibilities is opening up that will change the planning and design of wastewater treatment facilities in the near future (Guest *et al.,* 2009; Kehrein *et al.,* 2020). Examples of recoverable resources are cellulose, hydrogen, heat, polyhydroxyalkanoates, phosphates, proteins, extracellular polymers, etc. (Van Loosdrecht and Brdjanovic, 2014).

An important aspect of wastewater plant operation has always been its controllability. This concerns direct process control as well as indirect control of *e.g.* sludge settleability or biofilm growth. Process control has been a limiting factor from the start. Ardern and Lockett (1914) as well as Pasveer tried to minimize costs by applying fill-and-draw cycles where settling would occur in the treatment plant. In fact the Nereda® technology is also based on this principle. This requires process automation. Since the early 1970s instrumentation, control and automation (ICA) has attracted the attention of the water and wastewater industry. Since then the technical development of new processes, sensor and instrumentation technology, computer performance, communication technology and the Internet of Things, detection methods, control theory and artificial intelligence has made good progresses in early warning, monitoring and operating of wastewater treatment plants. This improvement in ICA technology made over the last few decades meant that process control became sufficiently reliable and now sequencing batch reactors are increasingly being used again (*e.g.* Nereda® technology). The increasing effluent demands, combined with a demand to save and recover resources, is pushing the need for increased process control. Although mathematical models were already developed in the early days of wastewater treatment processes, they only became in widespread use with the introduction of low-cost personal computers and the presentation of a unified activated sludge model (Henze *et al.,* 1987).

The control of sludge properties has always been a point of concern as well. Filamentous sludge and foaming caused by specific bacterial groups has always been important. Control of filamentous bacteria by the application of selector systems (Chudoba, 1973) has been successful in many cases. Nevertheless, the filamentous organism *Microthrix parvicella* is still giving regular problems in nutrient removal processes. Despite much research, which has certainly helped to obtain a better understanding of the causes and control of filamentous bulking, it is still not clearly understood to the point where the sludge settleability is quantitatively predictable for different activated sludge systems. This means that larger secondary settling tanks have to be built to cater for possible periods of poorer sludge settleability. However, in recent years the understanding of biofilm and sludge morphology has significantly increased and seems to have come together. One outcome of these theoretical developments is the introduction of aerobic granular sludge systems which can be seen as the other extreme of filamentous sludge or as a particular form of the biofilm process (Beun *et al.,* 1999).

Another major concern is wastewater and sludge disinfection and final sludge disposal in an environmentally sustainable way. The fact that wastewater contains pathogenic organisms was the reason for the start of large-scale sewerage systems and wastewater treatment plants 150 years ago. This was more or less forgotten until the middle of the 20th century when disinfection of effluents came into use. This was partly given up due to the carcinogenic compounds created during chlorination of wastewater, but recently in several areas disinfection has become an issue again, using filters, UV and ozonation. With the advance of wastewater recovery and the drive to more individually based wastewater treatment processes, disinfection is now receiving renewed attention. Final sludge disposal was originally a health risk issue because of the risk of spreading pathogens. Nowadays sludge disposal to agricultural lands is becoming more and more limiting (also as food safety standards are tending to increase) and the handling of sludge becomes more and more important. In particular sludge dewaterability and dewatering and how to minimize the problem is a strong research focus. In recent years sludge pretreatment for improving sludge dewaterability has been attracting the wastewater industry. If dewatering could be efficiently performed then sludge incineration could be used as a means to recover the energy and resource (*e.g.* phosphorous) enclosed in the sludge.

The demands on the wastewater system are continuously increasing, with nowadays an increased attention on emerging micropollutants that might accumulate in the water cycle or affect natural ecosystems. Water shortage will lead to the further development and implementation of technologies for water reclamation and reuse. Water reuse is not only limited to water-scarce regions; in water-rich areas such as Western Europe, local regulations and demands can make it economically profitable to use wastewater effluent instead of natural water to produce water for industry. All these developments take time and after more than a century of separate development, wastewater treatment and drinking water treatment are growing closer to each other. A successful example comes from Hong Kong where seawater has been supplied for toilet flushing since 1958. Nowadays it covers more than 80% of the population (7 million inhabitants), saving 750,000 tons of freshwater per day at limited energy consumption compared with seawater desalination by reverse osmosis (Chen *et al.,* 2012; Van Loosdrecht *et al.*, 2012). This dual water supply practice has enabled development of the sulphur-based wastewater treatment process (*i.e.* the SANI® process, Wang *et al.,* 2009) with the local wastewater board.

Finally, and by no means least, a major problem in wastewater collection and treatment is the training and education of a new generation of engineers and scientists to design new and to retrofit old wastewater treatment plants and of operators to run them to achieve the limits of the technologies and processes developed to date. This is particularly pertinent in developing countries where political and economic uncertainty results in skill losses to the developed countries. With the development of the technology over the past 40 years the domain of the profession has expanded from a civil engineering activity to a more process engineering and microbiology-based activity. In many universities separate environmental engineering curricula have been developed to bridge both disciplines. Today, all these processes and their technologies are mixed to create complex treatment systems where the use of principles and models are needed in order to handle the full complexity of the applications. Thus today we have a complexity of wastewater treatment as never seen before. This can be confusing and the attempts of numerous companies to market their own processes and technologies is adding to the confusion. This second edition has updated and/or revised most of the chapters of the first edition which was published in 2008 and includes new developments such as aerobic granular sludge and sulfur-based wastewater treatment. All these processes and technologies rely on the same basic processes, and as has been said: *'the bacteria have no idea of the shape of the reactor or the name of the technology, it simply denitrifies if there is nitrate, carbon source and no oxygen.'*

REFERENCES

Ardern E. and Lockett W.T. (1914). Experiments on the oxidation of sewage without the aid of filters. *Journal of the Society of Chemical Industry*, 33, 523.

Barnard J.L. (1973). Biological denitrification. *Water Pollut. Control*, 72, 705-720.

Barnard J.L. (1976). A review of biological phosphorus removal in the activated sludge process. *Water SA,* 2(3), 136-144.

Beun J.J., Hendriks A, Van Loosdrecht M.C.M., Morgenroth M., Wilderer P.A. and Heijnen J.J. (1999). Aerobic granulation in a sequencing batch reactor. *Water Research,* 33(10), 2283-2290.

Bolitho V.N. (1975). Economic aspects of wastewater treatment in South Africa. *Water SA,* 1(3), 118-120.

Chen G.H., Chui H.K., Wong C.L., Tang D.T.W., Lu H., Jiang F. and Van Loosdrecht M.C.M. (2012). An Innovative Triple Water Supply System and a Novel SANI® Process to Alleviate Water Shortage and Pollution Problem for Water-scarce Coastal Areas in China. *Journal of Water Sustainability*, 2(2), 121–129.

Chudoba J., Grau P. and Ottova V. (1973). Control of activated sludge filamentous bulking. II. selection of microorganisms by means of a selector. *Water Research*, 7(10), 1389-1406.

Cooper P.F. (2001). Historical aspects of wastewater treatment. In Decentralized sanitation and reuse: concepts, systems and Implementation. Edited by Lens P., Zeeman G., and Lettinga G., IWA Publishing, London, UK, 11-38.

Downing A.L., Painter H.A. and Knowles G. (1964). Nitrification in the activated sludge process. *J. Proc. Inst. Sewage Purif.,* 64(2), 130-158.

Ferriman, A. BMJ readers choose the "sanitary revolution" as greatest medical advance since 1840 (2007) *BMJ,* 334 (111), doi:10.1136/bmj.39097.611806.DB.

Guest J.S., Skerlos S.J., Barnard J.L., Beck M.B., Daigger G.T., Hilger H., Jackson S.J., Karvazy K., Kelly L., Macpherson L., Mihelcic J.R., Pramanik A., Raskin L., Van Loosdrecht M.C.M., Yeh D. and Love N.G.A (2009). A new planning and design paradigm to achieve sustainable resource recovery from wastewater. *Environmental Science and Technology,* 43(16), 6126-6130.

Henze M., Grady C.P.L. Jr., Gujer W., Marais G.v.R. and Matsuo T. (1987). Activated Sludge Model No. 1, IAWPRC Scientific and Technical Report No. 1, London, UK.

Kehrein P., Van Loosdrecht M.C.M., Osseweijer P., Dewulf J., Garfi M. and Duque J.A.P. (2020). A critical review of resource recovery from municipal wastewater treatment plants – market supply potentials, technologies and bottlenecks. *Environmental Science: Water Research & Technology.* DOI: 10.1039/C9EW00905A.

Lettinga G., Van Velsen A.F.M., Hobma S.W., De Zeeuw, W. and Klapwijk A. (1980). Use of the upflow sludge blanket (USB) reactor concept for biological wastewater treatment, especially for anaerobic treatment. *Biotechnology and Bioengineering,* 22, 699-734.

Levin G.V. and Shapiro J. (1965). Metabolic uptake of phosphorus by wastewater organisms. *Water Pollution Control Federation Journal,* 37, 800-821.

Ludzack F.J. and Ettinger M.B. (1962). Controlling operation to minimize activated sludge effluent nitrogen. *Water Pollution Control Federation Journal*, 34, 920-931.

McCarty P.L. (1964). Thermodynamics of biological synthesis and growth. Procs. 2nd International Conference on Water Pollution Control, 2, 169-199.

Monod J. (1950). Technique of continuous culture – theory and application. *Annales de l'Institut Pasteur*, 79, 167.

Nicholls H.A. (1975). Full scale experimentation on the new Johannesburg extended aeration plants. *Water SA,* 1(3), 121-132.

Pasveer A. (1959). A contribution to the development in activated sludge treatment. *J. Proc. Inst. Sew. Purif.,* 4, 436.

Phelps E.B. (1944). Stream Sanitation. John Wiley and Sons Inc., New York, USA.

Pronk M., De Kreuk M.K., De Bruin B., Kamminga P., Kleerebezem R.V. and Van Loosdrecht M.C.M. (2015). Full scale performance of the aerobic granular sludge process for sewage treatment. *Water Research*, 84, 207-217.

Srinath E.G., Sastry C.A. and Pillai S.C. (1959). Rapid removal of phosphorus from sewage by activated sludge. *Experientia*, 15(9), 339-340.

Van Bemmelen J.M. (1868). Het stelsel Liernur - Voor den afvoer der faecale stoffen in de steden, *De Economist,* 17(1), 401-440.

Van Loosdrecht M.C.M. and Brdjanovic D. (2014). Anticipating the next century of wastewater treatment. *Science*, 344(6191):1452-1453.

Van Loosdrecht M.C.M., Brdjanovic D., Chui H.K. and Chen. G.H. (2012). A source for toilet flushing and for cooling, sewage treatment benefits, and phosphorus recovery: direct use of seawater in an age of rapid urbanisation. *Water 21*, 14(5), 17-20.

Van Vuuren L.R.J. (1975). Water reclamation – quality targets and economic considerations. *Water SA,* 1(3), 133-143.

Wang J., Lu H., Chen G.H., Lau G.N., Tsang W.L. and Van Loosdrecht M.C.M. (2009). A novel sulfate reduction, autotrophic denitrification, nitrification integrated (SANI) process for saline wastewater treatment. *Water Research.* 43(9), 2363-2372.

Wolfe P. (1999). World of Water 2000 – The past, present and future. *Water World/Water and Wastewater International Supplement to PennWell Magazines*, Tulsa, OH, USA, pp.167.

Wuhrmann K. (1964). Hauptwirkugen und Wechselwirkugen einiger Betriebsparameter Belebtschlammsystem: Ergibnisse mehrjäriger grossversuche. *Schweigerische Zeitschrift für Hydrologie*, XXVI(2) 218.

Yamamoto K., Hiasa M., Mahmood T. and Matsuo T. (1989). Direct solid-liquid separation using hollow fiber membrane in an activated sludge aeration tank. *Water Science and Technology*, 21, 43-54.

Figure 1.3 An example of a state-of-the-art wastewater treatment plant: Sha Tin Sewage Treatment Works in Hong Kong (photo: Drainage Services Department, HK).

doi: 10.2166/9781789060362_0011

2

Basic microbiology and metabolism

David G. Weissbrodt, Michele Laureni, Mark C.M. van Loosdrecht and Yves Comeau

2.1 INTRODUCTION

Starting with the basics of general microbiology on cells, microorganisms, metabolisms and growth, this chapter provides essential knowledge on modern analytical methods to characterize microorganisms and metabolic functions. It covers the main steps to understand microbial growth and metabolisms under the principles of stoichiometry, bioenergetics, thermodynamics, and kinetics. Fundamentals of microbial growth are key to deriving quantitative information in order to design processes and develop mathematical models to predict their performance. Traditional wastewater treatment design is based on empirical rules. For a good design and development of new processes, understanding microbial ecology is important. Baas-Becking (1934) stated back at the beginning of the 20th century that '*Everything is everywhere, but the environment selects*'. It is the task of an environmental engineer to design the process in such a way that the desired microorganisms are indeed selected and can perform their tasks optimally. Operating environmental biotechnology systems relates to an efficient management of the microbial resource, by implementing microbial community engineering principles. The design and control of biological processes in wastewater treatment plants (WWTPs) or water resource recovery facilities (WRRFs) link to '*What are microorganisms and how do they grow?*'.

2.1.1 Microorganisms in biological wastewater treatment

Conversions in biological nutrient removal (BNR) processes are based on the natural cycling of elements (C, N, O, P and S) by microorganisms. The pollutants that need to be removed in wastewater treatment processes are described in Chapter 3. Microbial conversions are used to convert these pollutants into compounds that are easily separable from water, mainly gaseous compounds (methane, carbon dioxide or nitrogen gas) or solid compounds (mainly biomass). The extracellular polymers produced by the microorganisms aid in flocculating suspended cells and incorporating them into activated sludge flocs, granules, or biofilm matrices. The microbial processes used in wastewater treatment also occur in natural systems such as rivers and lakes. The 'self-cleaning' capacity of these natural systems is often limited by physical restrictions such as aeration, the amount of biomass present, and mixing. The treatment technology aims to make an engineered ecosystem where these physical limitations are minimised due to biomass retention, aeration, and proper mixing. These processes are described in other chapters in this book.

2.1.2 Microbial growth as a functional unit

Here we focus on the basic aspects of microbial growth, how it can be monitored and studied, and how it can be mathematically described. It is simply based on how microorganisms derive their cellular energy from chemical redox reactions or alternatively light (catabolism) and how they implement this energy to maintain their cell (maintenance) and build new biomass from the nutrients available (anabolism). Microbial growth systems relate to a trigonal combination of elements, electrons, and energy. On this basis, microbial growth can be formulated as a chemical reaction characterized by stoichiometry (yields), kinetics (rates), and thermodynamics (Gibbs free energy). Essentially any redox reaction that has a negative Gibbs energy can be considered a potential catabolic reaction for energy generation for microbial growth. It is this versatility in converting chemical compounds that is exploited in wastewater treatment.

2.1.3 Microbial community engineering

Microbial community engineering is applied to select the guilds of natural microorganisms that can convert waste compounds into harmless compounds or new resources. The establishment of microbial niches is driven by microbial ecology. A good combination of ecological engineering principles can help to predict the selection and conversion of, and competition between, microorganisms as the basis for process design (Kleerebezem and Van Loosdrecht, 2007; Lawson *et al.*, 2019). Microorganisms are retained in the treatment process as flocs, granules, and/or biofilms. Selective pressures are generated by acting on *e.g.* the availability of nutrients, electron donors (hereafter referred to as e-donors), most often organic matter, and electron acceptors (e-acceptors) such as dissolved oxygen (O_2) or oxidized nitrogen compounds (NO_x) such as nitrite (NO_2^-) or nitrate (NO_3^-). These variables are manipulated by WWTP designers and operators to set the conditions necessary to sustain microbial growth and conversions. Additional environmental boundaries can impact bioprocesses, *e.g.* pH, temperature, and light availability. By engineering the system, limitations in nutrients, aeration and amount of biomass can be overcome.

Design and operation of WWTPs link to the management of hydraulic (HRT) and sludge (SRT, or sludge age) retention times. This drives the selection of microorganisms based on growth rates and yields of biomass production on substrates. Uncoupling the HRT (several hours) from SRT (several days) is intended to maintain microorganisms in the tanks. Microbes that can cope with the applied SRT will be selected; the others will be washed out of the process. Affinity for substrates plays a limited role in wastewater systems. Conversions in flocs and biofilms and the mass transfer of granules are limited because of diffusional resistances. Besides deterministic factors that can be controlled by engineers, other types of stochastic or probabilistic factors can impact the presence of microorganisms in activated sludge. These comprise dynamics in chemical and biological compositions of wastewater and in environmental conditions.

2.1.4 Analytical methods for microbial ecology

Environmental biotechnology systems are driven by complex communities of microbial populations that metabolize the major nutrients. Metabolizing populations are accompanied by a diversity of flanking populations whose functionalities remain to be discovered. Continuous analytical advances in microscopy, microbiology, molecular biology, system (micro)biology, and ecophysiology are helping to examine microbial communities. Their individual lineages can now be elucidated at high resolution from phylogenies to genomes and expressed phenotypes.

These methods enable the investigation of sludge, moving from a 'black box' to a more 'grey box' and potentially to 'white box' understanding (Section 2.3.1).

The degree of resolution to target on the characterization of biomasses depends on the research question and scope of investigation from science to engineering. New-generation methods are frequently sold as promises to solve process issues. Often, microbial and functional diversities seem studied at a sensitivity level which is far too high to be relevant for process operation. Within the digitalization and artificial intelligence hub, efforts are being made to render high-throughput data informational and to link them with bioengineering concepts. Analytics are crucial to collecting information and knowledge from a biosystem. This is driving new strategies that integrate engineering and analytical concepts to manage the microbial resource.

2.1.5 Mathematical models of microbial growth

Mathematical models are efficient to aggregate knowledge and predict microbial growth processes involved in complex biomasses. Quantitative models from the series of Activated Sludge Models (ASM1-3), Anaerobic Digestion Models (ADM), or other biofilm and granular sludge models (chapters 14-17) are built on the metabolic understanding of microbial guilds that take part in the processes. Biomasses are modelled as assemblies of *e.g.* ordinary heterotrophic organisms (OHOs), autotrophic nitrifying organisms (ANOs), denitrifying heterotrophic organisms (DHOs), anoxic ammonium-oxidizing (anammox) organisms (AMOs), polyphosphate-accumulating organisms (PAOs), glycogen-accumulating organisms (GAOs), among other fermentative and methanogenic organisms. Process models are built on growth stoichiometry, bioenergetics, thermodynamics, and kinetics. Models are not only essential to understanding growth, conversions and interactions, but also to aid in process design.

2.2 BASIC ASPECTS OF MICROBIOLOGY AND METABOLISM

Microbial cells are the central functional units of biological wastewater treatment processes. It is important to grasp the essence of what they are composed of, how they metabolically function to sustain their growth, and how they regulate their metabolism to cope with ever-changing environmental conditions. This section covers the (*i*) domains of microorganisms that have evolved and regulate biogeochemical cycles on Earth, (*ii*) cell structures and components, (*iii*) trophic groups of microorganisms that compose the microbial diversity in nature and WWTPs, (*iv*) main metabolisms and nutritional and energetic requirements, as well as (*v*) environmental factors that force them to regulate their metabolism.

2.2.1 Prokaryotes, eukaryotes and viruses

Life is classified by the three domains of archaea, bacteria (eubacteria), and eukarya. These domains differentiated and evolved from a last universal common ancestor (LUCA) or concestor, *ca.* 3.3-3.8 billion years ago, as displayed by their genetic information (Figure 2.1).

The generic term 'microorganisms' comprises unicellular types of prokaryotes and eukaryotes. Prokaryotes are unicellular organisms which include archaea and eubacteria. Archaea are the primary ancestors from which bacteria and cyanobacteria differentiated. From both the microbiological and engineering points of view, prokaryotes are considered as bacteria since their structure, components, and metabolic functioning are very similar. Eukarya are commonly referred to as eukaryotes, which include unicellular organisms (protozoa, algae and yeast) and multicellular organisms (metazoa, fungi, plants and animals). Prokaryotes and eukaryotes primarily differ in their cell structure. DNA is located in the cytoplasm of prokaryotes and in a nucleus (Grk. *karyon*) in eukaryotes. Besides microorganisms, viruses and (bacterio)phages exist in microbial communities.

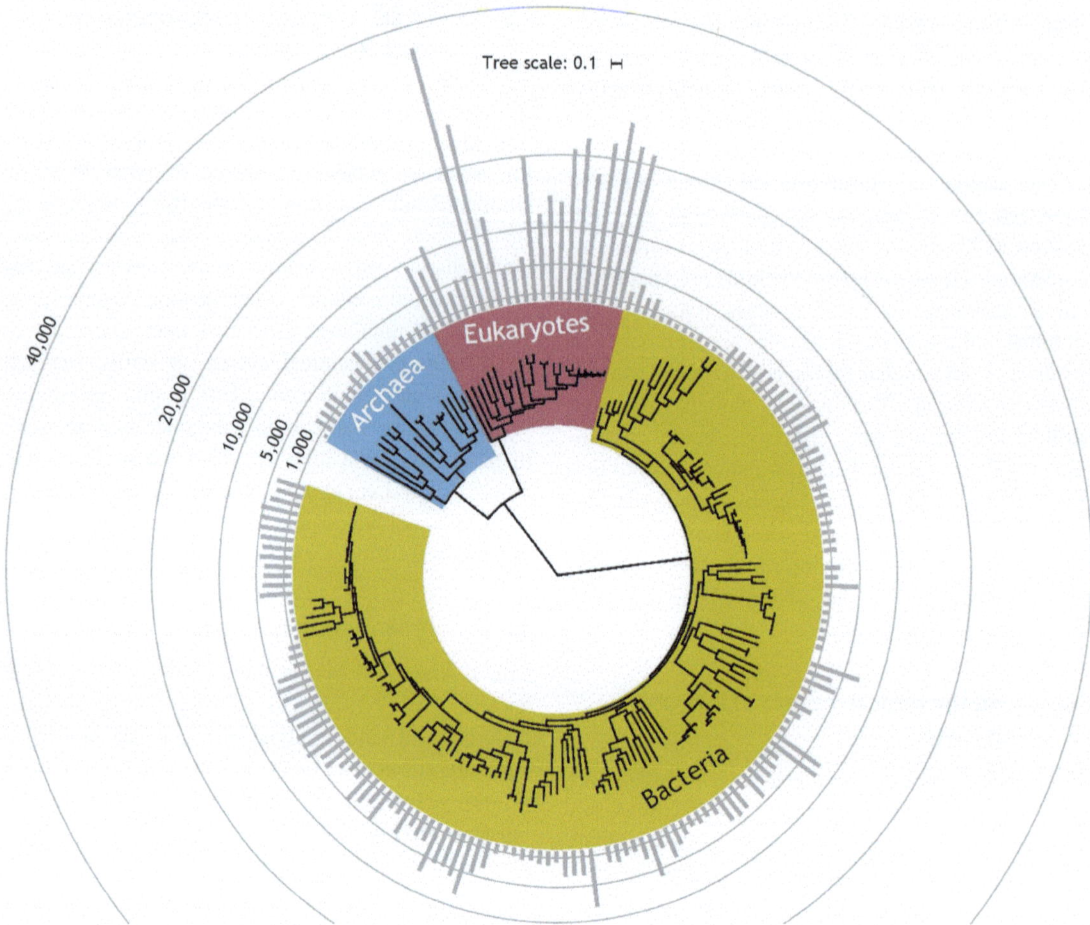

Figure 2.1 Tree of (cellular) life delineating the domains of archaea (blue), bacteria (yellow), and eukarya (purple). The domain of bacteria displays a wide diversity of microorganisms. The genome size of the different types of cells is displayed by the bar chart on a scale of 1,000-40,000 of non-transposable element-related protein-coding genes present on their DNA. Bacteria and archaea exhibit small genomes that can rapidly evolve via *e.g.* horizontal gene transfer. Viruses and phages are not displayed on the tree. Their positioning is under active debate.

Viruses and phages are pieces of genetic information (RNA or DNA) packed in a protein capsule. They transfer this DNA or RNA into host living cells (bacteria for phages) where it is multiplied. Although viruses are usually not considered living organisms as such, they are sometimes included in the definition of microorganisms. Via the infection process, phages shape communities, often by impacting the predominant populations. Viruses are classed within archaeoviruses, bacterioviruses and eukaryoviruses, based on their cellular host targets. Viruses and phages are generally not displayed in the tree of (cellular) life. In short, they are polyphyletic and have several evolutionary origins. Viral lineages do not share a single common origin such as LUCA for cellular microorganisms, nor do they share characteristics with cells. However, there is debate on where to position viruses, which cannot be dismissed as non-living material, in the tree of life (Brussow, 2009; Moreira and Lopez-Garcia, 2009). Besides the omission of viruses, the horizontal gene transfer between microorganisms challenges the tree concept in the unifying theory for biology. This chapter, to make it simple but bearing this in mind, will restrict itself to the description of cellular life when referring to microorganisms.

Besides suspended solids from the influent, the biomass of WWTPs contains selected prokaryotic and eukaryotic microorganisms as well as metazoans such as nematodes or oligochaete worms. These, such as protozoa, predate on the microbial biomass and thereby influence the composition of microbial communities and the structure of biofilms (Jürgens *et al.*, 1997; Klein *et al.*, 2016). The morphology of the various groups of organisms can be observed by microscopy (Figure 2.2).

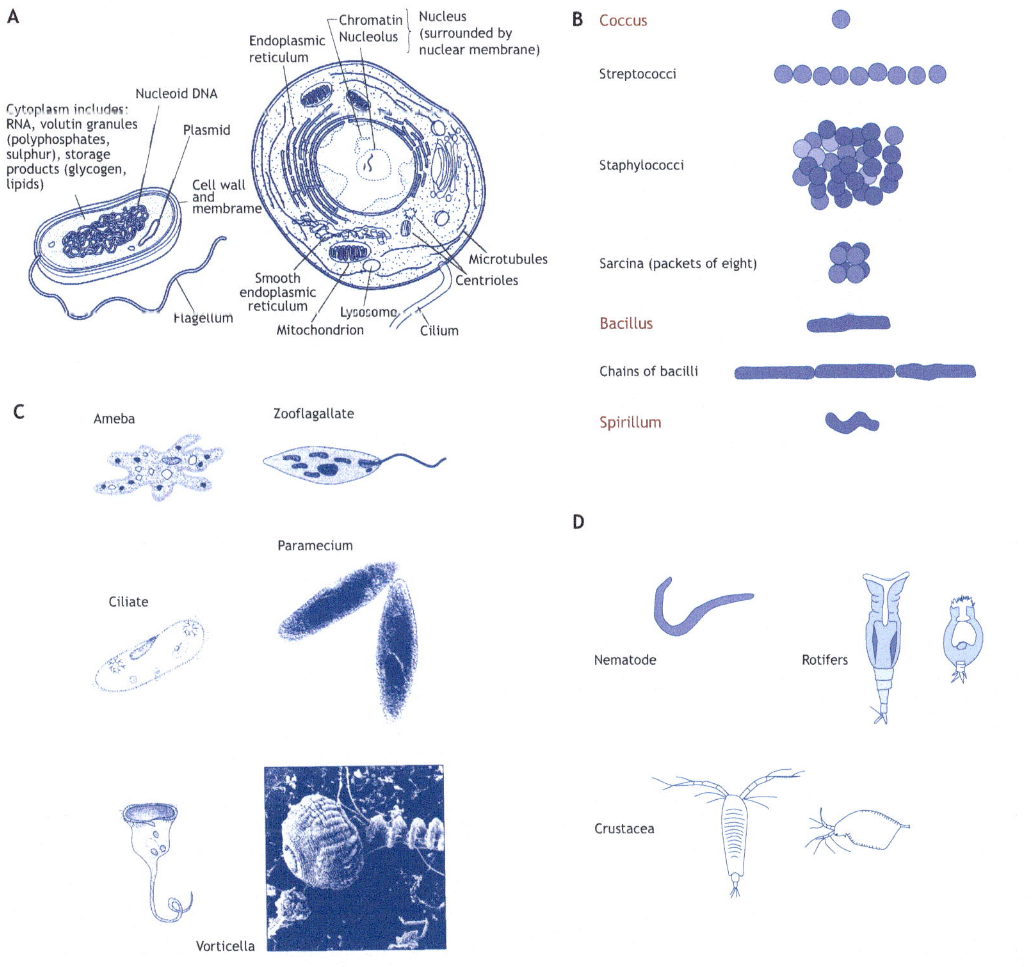

Figure 2.2 Prokaryotes and eukaryotes of activated sludge: A) structure of prokaryotic (0.5 to 5 µm) and eukaryotic (5 to 100 µm) cells; B) morphology of bacteria; C) morphology of protozoa; and D) multicellular organisms. Figures taken from Comeau *et al.* (2008), in Metcalf and Eddy (2014) and in Rittman and McCarty (2001) (adapted).

Wastewater treatment started as a sanitation technology. Sewage contains a whole range of pathogens which are partially removed in the wastewater treatment plant, either by removal by *e.g.* protozoa or by incorporating them in the sludge flocs. Many of these pathogens can therefore be found in the excess sludge or the effluent of WWTPs. Chapter 8 deals with pathogen removal from sewage.

2.2.2 Cell structure and components

2.2.2.1 Cell structures of prokaryotes and eukaryotes

Both prokaryotes and eukaryotes are composed of a plasma membrane (or cell membrane) that physically delineates the cell from its environment. The membrane is composed of a bilayer of mainly phospholipids. It allows the creation of both pH and charge gradients which are an essential aspect of energy generation by cells. While the external environment can be highly variable, cells maintain a stable intracellular chemical environment to sustain their metabolism.

Prokaryotes are simple cells with a membrane separating the cell interior (cytoplasm) from the environment. The cytoplasm contains the genetic material (nucleic acids), ribosomes, proteins, and storage polymers. Some bacteria can have a more complex cellular structure. Anammox bacteria enclose a separate compartment called anammoxosome in which energy is produced. Prokaryotic cells are protected by a cell wall outside the cell membrane. The wall is composed of a peptidoglycan layer made from polysaccharide chains cross-linked by unusual peptides containing D-amino acids. This complex is often covered by a capsule or slime layer of extracellular polymeric substances (EPS). There are broadly speaking two different types of cell wall in bacteria, called Gram-positive and Gram-negative. The names originate from the reaction of cells to the Gram stain, a test long-employed for the classification of bacterial species. Gram-positive bacteria have multiple layers of peptidoglycans, giving them a thick and strong cell wall. Gram-negative bacteria have a thin peptidoglycan layer and harbour a lipopolysaccharidic outer membrane that is absent in Gram-positive bacteria. Pili or fibrils are present at the cell surface; they can drive attachment to surfaces, and even exchange of electrons (which can be of use in bioelectrochemical systems). Cells can be motile due to the presence of flagella or pili. The cells of archaea are similar to bacteria although they have different membrane lipids and lack the peptidoglycan cell wall. Their cell walls are more diverse and to date have been less studied.

Eukaryotes are composed of a more evolved cell structure with several compartments separated from the cytoplasm by membranes. These compartments are typically a nucleus that contains the DNA, mitochondria for energy generation, a Golgi apparatus for secretion and intracellular transport, and, with phototrophic organisms, chloroplasts which convert light into cellular energy.

2.2.2.2 Elemental composition of biomass

Standard cellular compositions are considered to be independent of prokaryotes or eukaryotes. All cells are made from the same types of polymeric building materials. The four classes of cellular macromolecules include nucleic acids (DNA, RNA), proteins, polysaccharides, and lipids (Figure 2.3). These polymers form the organic or 'volatile fraction' of biomass. The four elements C, H, O, and N constitute the majority of their mass. The elemental composition of biomass is relatively universal across microorganisms. It can be expressed as a 4-element empirical chemical formula by $C_1H_{1.8}O_{0.5}N_{0.2}$ (Heijnen and Kleerebezem, 2010) that can be used when calculating the stoichiometry of microbial conversions (Section 2.5).

The non-volatile or inorganic fraction of biomass relates to P and mineral constituents such as magnesium (Mg), potassium (K), and calcium (Ca) that end up in the ash fraction after calcination of biomass. In literature, the term 'inerts' is widely used. However, this term describes both inorganic and unbiodegradable organics. The total suspended solids

(TSS) in activated sludge is the sum of volatile (VSS) and inorganic suspended solids (ISS) (Table 2.1). According to minerals that are incorporated in biomass, the elemental composition of biomass can be extended to the following 8-element formula: '$C_1H_{1.8}O_{0.5}N_{0.2}P_{0.03}S_{0.02}Mg_{0.01}K_{0.008}$…'.

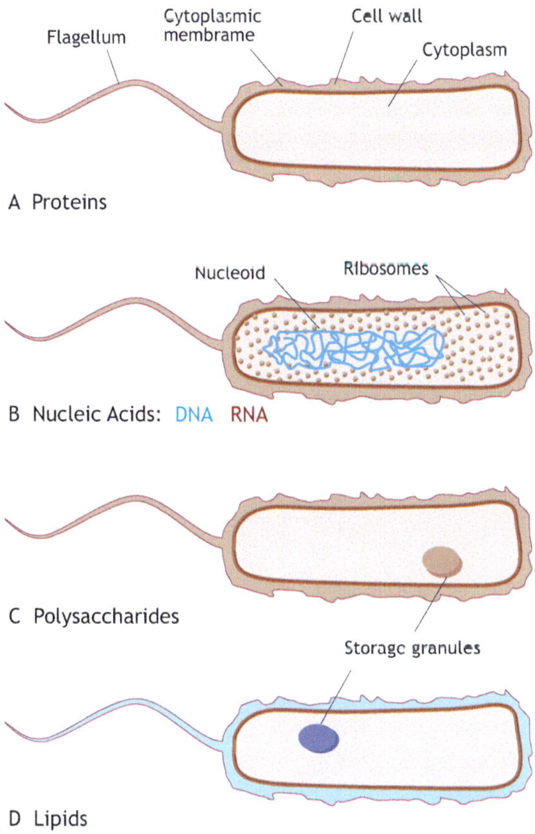

Figure 2.3 Macroconstituents of bacterial cells: proteins, nucleic acids, polysaccharides, and lipids are the main cellular macromolecules. Proteins are found in the flagellum, the cytoplasmic membrane, the cell wall and the cytoplasm. Nucleic acids (DNA and RNA) are found in the nucleoid and ribosomes (RNA is displayed as dark red spots in the light-red ribosome dots). Polysaccharides are found in the cell wall and sometimes in storage granules. Lipids are found in the cytoplasmic membrane, in the cell wall, and in storage granules. Source: Comeau *et al.* (2008) in Madigan *et al.* (2018) (adapted).

Table 2.1 Chemical composition of microbial cells.

Major cellular constituent	Typical mass fractions (% m/m)	
Organic 'volatile' fraction: C, H, O and N elements		
Proteins	30-60%	
Carbohydrates	5-30%	
Lipids	5-10%	
Nucleic acids[a)]	5-15%	(RNA)
	1%	(DNA)
Inorganic mineral 'ash' fraction: P, Mg, K, Ca, etc. elements		
Inorganic compounds	5-10% (OHOs) up to 40% (PAOs)[b)]	
As cell elements	%TSS	%VSS[c)]
Organic (VSS)	93.0	53.1
Carbon	50.0	28.3
Oxygen	22.0	12.4
Nitrogen	12.0	6.2
Hydrogen	9.0	
Inorganic (ISS)	7.0	
Phosphorus	2.0	
Sulphur	1.0	
Potassium	1.0	
Sodium	1.0	
Calcium	0.5	
Magnesium	0.5	
Chlorine	0.5	
Iron	0.2	
Other trace elements	0.3	

[a)] Nucleic acids are composed of 'volatile' C, H, O, N and 'non-volatile' P elements. DNA is composed of nucleotides based on adenine (A; $C_{10}H_{12}O_5N_5P$), cytosine (C; $C_{10}H_{12}O_6N_5P$), guanine (G; $C_9H_{12}O_6N_3P$), and thymine (T; $C_{10}H_{13}O_7N_2P$). RNA comprises nucleotides based on A, C, G, and uracil (U; $C_9H_{11}O_8N_2P$). The percentage of P is 9-11% of the masses of nucleotides. [b)] In the presence of terminal e-acceptors, polyphosphate-accumulating organisms (PAOs) grow and store phosphorus intracellularly as inorganic polyphosphate (Comeau *et al.*, 1986). Their ash fraction can reach 40% of the total solids. In the absence of terminal e-acceptors, PAOs use their polyphosphate pool to generate energy for storage and polymerization of volatile fatty acids (VFAs). When sampled and analysed under these conditions, their ash fraction will be lower and close to one of the ordinary heterotrophic organisms (OHOs). c) For an empirical formula for cells of $C_5H_7O_2N$.

In literature, one can find different expressions of elemental compositions of biomass. Although similar, compositions can be slightly variable, so if elemental measurements are available they should be used.

Biotechnologists generally compute growth models by normalizing to carbon fluxes using $C_1H_{1.8}O_{0.5}N_{0.2}$ and C-mol units for production of biomass and bioproducts from well-defined synthetic carbon feedstocks. Environmental engineers mainly compute balances based on global analyses of chemical oxygen demand (COD-based mass units) to cope with the complex compositions of wastewater, waste, and green streams. One common form used in their literature is $C_5H_7O_2N$ (corresponding in C-mol form to $C_1H_{1.4}O_{0.4}N_{0.2}$). Table 2.2 provides information on the COD content of biomasses and common organic compounds in wastewater. Microbiologists often measure biomass as protein content (units as mg proteins). Since proteins are roughly 50% of the cellular biomass, the total VSS can be estimated by doubling the protein amount.

Table 2.2 Chemical composition and theoretical chemical oxygen demand (thCOD) of microbial cells and organic substances (adapted from Comeau *et al.*, 2008).

Compound	Chemical formula	Weight (VSS)		
		CHON (g/mol)	thCOD (g/mol)	COD/VSS (g/g)
Biomass				
	$C_1H_{1.8}O_{0.5}N_{0.2}$ [a)]	24.6	33.6	1.366
	$C_1H_{1.8}O_{0.5}N_{0.2}P_{0.03}S_{0.02}$ [b)]	26.2	35.8	1.366
	$C_5H_7O_2N$	113	160	1.42
	$C_5H_7O_2NP_{1/12}$	113	160	1.42
	$C_{60}H_{87}O_{23}N_{12}P$	1,343	1,960	1.46
	$C_6H_{7.7}O_{2.3}N$	131	193	1.48
	$C_{18}H_{19}O_9N$	393	560	1.42
	$C_{41.3}H_{64.6}O_{18.8}N_{7.04}$	960	1,369	1.43
	$C_4H_6O_2$	86	144	1.67
Organic substance				
Casein	$C_8H_{12}O_3N_2$	184	256	1.39
Average organics	$C_{18}H_{19}O_9N$	393	560	1.42
Carbohydrates	$C_{10}H_{18}O_9$	282	320	1.13
Fats, oils	$C_8H_6O_2$	134	272	2.03
Oils: oleic acid	$C_{18}H_{34}O_2$	254	880	3.46
Protein	$C_{14}H_{12}O_7N_2$	320	384	1.20
Glucose	$C_6H_{12}O_6$	180	192	1.07
Formate	CH_2O_2	46	16	0.35
Acetate	$C_2H_4O_2$	60	64	1.07
Propionate	$C_3H_6O_2$	74	112	1.51
Butyrate	$C_4H_8O_2$	88	160	1.82
Methane	CH_4	16	64	4.00
Hydrogen	H_2	2	16	8.00

[a)] Four-element C-mol formula of the 'volatile' fraction of biomass used by Heijnen and Kleerebezem (2010). The degree of reduction (γ; see Section 2.4.3) of the biomass is 4.2 mol electrons per C-mole biomass, relating to a thCOD of 33.6 g COD per mole biomass. The multiplication factor between the degree of reduction and thCOD is 8 g COD per mole electrons. [b)] Six-element extended formula comprising P and S as additional elements.

Elemental compositions can be measured for organic matter and different types of biomasses. Sewage organic matter displays an elemental composition of *e.g.* $C_{11}H_{18}O_5$ (or $C_1H_{1.6}O_{0.45}$). Phototrophic organisms maximize their growth yield using light energy to assimilate a high fraction of nutrients. For instance, purple phototrophic bacteria ($C_1H_{1.8}O_{0.38}N_{0.18}$) assimilate up to 100:7:2 gCOD, gN, gP, respectively (*vs.* 100:5:1 for activated sludge). Eukaryotic algae can take the form of $C_{106}H_{263}O_{110}N_{16}$ (or $C_1H_{2.5}O_{1.04}N_{0.15}$). When measuring elemental compositions, one should bear in mind whether cells enclose intracellular storage polymers. Polyphosphate or glycogen impact elemental measurements. One should consider the elemental composition of 'active cellular biomass' without these storage compounds.

2.2.2.3 Cellular macromolecules

Nucleic acids constitute the monomers of the genetic information-containing molecules in the cell (DNA and RNA). The sequences of strands of nucleic acids are by convention given in the 5'-to-3' direction of their synthesis through the action of polymerases. The upstream 5'-end corresponds to the position of the phosphate group on the fifth carbon of the sugar ring. The downstream 3'-end relates to the unmodified hydroxyl substituent on the third carbon. DNA harbours a double strand of nucleic acids; RNA a single strand of nucleic acids. The genome of prokaryotes and eukaryotes is stored in DNA. Viruses and phages store genetic information in either DNA or RNA, depending on their types.

Deoxyribonucleic acid (DNA) stores the genetic information of cells in sequences of millions of nucleotides composed of adenine (A), cytosine (C), guanine (G) or thymine (T), a pentose sugar (deoxyribose), and an orthophosphate group. DNA is a double helix formed by two strands of polynucleotides linked by hydrogen bonds between pairs of nucleobases (*i.e.* base pairs). These base pairs are formed by A-T and C-G linkages. The sugar-phosphate entities provide the backbone structure of DNA. The total amount of genes coded by nucleotides of a cell is called a genome. In prokaryotes, the genomic DNA is dispersed in the cytosol as single amorphous, circular chromosome (nucleoid). DNA can take other forms such as mobile genetic elements that are exchanged across microbial populations. Horizontal gene transfer is an important aspect underlying antimicrobial resistance. In eukaryotes, DNA is contained in the nucleus, arranged as multiple linear chromosomes.

Ribonucleic acids (RNA) are single strands of hundreds of nucleotides with a complex 3D structure. RNA is composed of the same nucleotides as DNA except that thymine is replaced by uracil (U). At the forefront of cell regulation, the nucleotide code contained in genes at DNA level is transcribed into RNA sequences. RNA takes different forms. Messenger RNA (mRNA) carries the genetic information translated from DNA to the ribosomes to synthesise proteins. Transfer RNA (tRNA) supplies amino acids to ribosomes. Ribosomal RNA (rRNA) links amino acids to produce peptides and proteins.

Proteins are polymers of amino acids. They can confer structure (*e.g.* in the EPS) or are the basic catalysts (*i.e.* enzymes) for biochemical conversions in cells. Enzymes can be located intracellularly in the cytoplasm, on the cell membrane either facing the intracellular or extracellular environment or cross the membrane, or excreted in the extracellular medium.

Lipids (Barák and Muchova, 2013) and polysaccharides (Misra *et al.*, 2015) are involved in different functions of microbial cells. Structural phospholipids and glycoconjugates are primary components of cell walls and membranes. Lipids are polymers of fatty acids. Lipidic membranes are hydrophobic films that physically separate the aqueous cytosol from the outer aqueous environment. Microorganisms can tune the lipidic and carbohydrate compositions of their membranes to change the membrane's fluidity and permeability depending on conditions. This allows cell division and/or sporulation.

The cell wall is a 3D matrix of complex sugars and their derivatives mixed with glycoproteins. Gram-negative organisms synthesize peptidoglycan to provide mechanical strength and shape to the cells. Gram-positive organisms harbour a thicker layer of peptidoglycan and teichoic acid, since they are lacking an outer membrane. Oligosaccharides present in the periplasm are used in osmoregulation. Lipopolysaccharides bind to extracellular cations to further stabilize the structure of the outer membrane and to increase its permeability. They anchor the polysaccharides in the outer membrane system to better stabilise the outer wall structure.

2.2.2.4 Intracellular storage biopolymers

Intracellular biopolymers are crucial to sustain growth in dynamic environments such as wastewater treatment processes. They maintain a more constant intracellular metabolism by buffering the temporal availability and limitation of substrates and nutrients in the environment. Carbon storage is widespread across microorganisms in natural and wastewater environments. Under feast-famine regimes or light-dark diel cycles, microorganisms that store resources possess a competitive advantage.

Important bacterial storage polymers include poly-β-hydroxyalkanoate (PHAs) polyesters formed by storage and polymerization of volatile fatty acids (VFAs), and polysugars such as glycogen, lipids, sulphur, and polyphosphate (Figure 2.4). These are energy, carbon, and phosphorus reserves for microbial cells. Intracellular pools of glycogen are involved in GAOs and PAOs such as described in Chapter 6 on phosphorus removal. Glycogen is used as energy and reducing power to store VFAs present in the bulk as PHAs. The synthesis of glycogen is relatively simple and conserved across microorganisms, by primarily involving glycogen synthases.

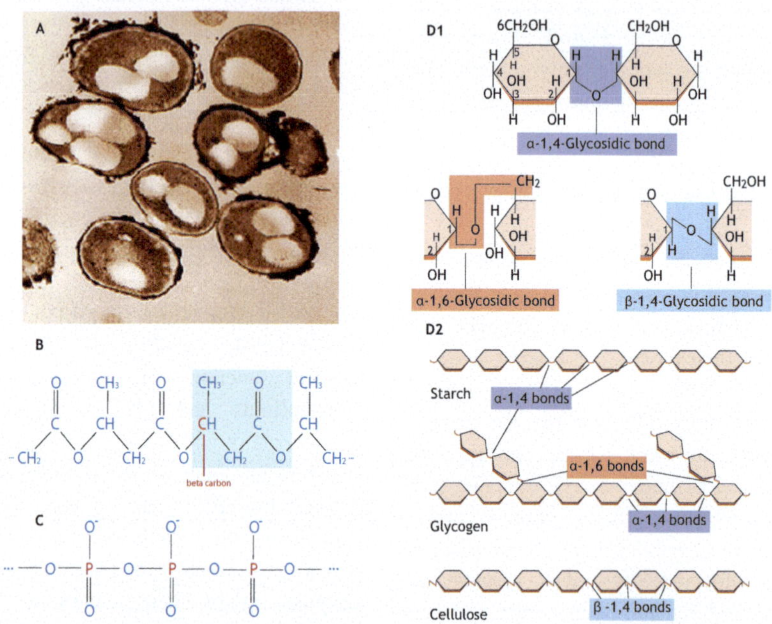

Figure 2.4 Intracellular storage polymers: A) intracellular polyesters of poly-β-hydroxyalkanoates (PHAs). Cell size is approximately 1 μm (photo: M.C.M. van Loosdrecht); B) chemical structure of poly-β-hydroxybutyrate (PHB). In poly-β-hydroxyvalerate (PHV), the CH_3 group is replaced by a CH_2CH_3 group. PHB and PHV are the two most common PHAs; C) chemical structure of polyphosphate. Polyphosphates are polymers of phosphate molecules and are stabilized by cations (e.g. Ca^{2+}, Mg^{2+}, K^{+}) interacting with charged oxygen (-O-) molecule; D) chemical structure of polysaccharides. D1 = differences in the glycosidic bonds in the position of linkage between glucose molecules and in geometry (α and β) and D2 = structure of starch, glycogen (a bacterial storage polymer) and cellulose. Source: Comeau *et al.* (2008) in Madigan *et al.* (2018) (adapted).

Polyphosphates are linear chains of phosphates whose negative charge is stabilised by cations such as K^+ and Mg^{2+}. The energy-rich phosphate-ester bond is the same as in the universal energy carrier molecule of cells (ATP; Section 2.2.3) which contains a chain of 3 phosphates. In most bacteria, polyphosphate is used as a phosphate reserve and only a limited (but important for wastewater treatment) group of bacteria use it as an energy storage compound. It is synthesized via polyphosphate kinases. These can work reversibly to cleave polyphosphate in parallel to exo- and endopolyphosphatases.

Under light/dark cycles and feast/famine regimes, microoalgae store starch or lipids that can be used to produce biofuels. Bacteria such as *Microthrix* can store lipids mainly when supplied with lipids as growth substrate.

2.2.2.5 Extracellular polymeric substances (EPS) and biofilms

Microorganisms produce a large diversity of EPSs as functional exopolymers. EPSs form the capsule (or glycocalyx) around the cell wall. They form a sticky layer at the cell surface to enable the attachment of cells and the development of biofilms on surfaces. In nature, biofilms are a prevalent growth mode. This is in sharp contrast to traditional microbiology where usually single planktonic cells are studied. EPSs can be excreted in the bulk liquid phase surrounding cells to increase the viscosity of the local environment.

Numerous microorganisms synthesize capsular polysaccharides and mucus. Mucoid pathogens use slime layers as a virulence factor for adhesion to targeted tissues. EPSs are composed of a diversity of exopolysaccharides besides proteins, extracellular DNA, and other inorganic components. Complex pathways, that are coded on very different operons of the genomic DNA, are involved in exopolymerizations. Figure 2.5 displays cells embedded in a matrix of glycoconjugates detected by fluorescent lectin binding analysis (FLBA).

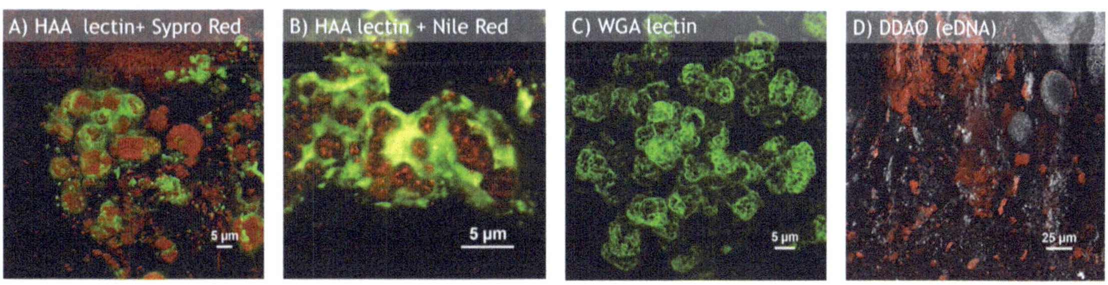

Figure 2.5 Detection of microbial cells embedded in EPS matrices in crysosections of BNR granular sludge by fluorescent lectin binding analysis (FLBA) combined with confocal laser scanning microscopy (CLSM). Glycoconjugates are stained with the green-fluorescent lectins of *Helix aspersa* (HAA) and wheat germ agglutinin (WGA) (in green). A) Microorganisms are stained with Sypro 60 that binds to proteins (in red) and B) with Nile Red that binds to lipophilic inclusions (in red); C) detection of the glycoconjugates surrounding cell clusters. D) Detection of extracellular DNA (eDNA) in the granular biofilm matrix with DDAO staining. Source: Weissbrodt *et al.*, 2013.

EPSs provide the 'glue' of cellular assemblies in the matrix of activated sludge flocs, biofilms, and granules. *Zoogloea* organisms are known to produce EPSs and are described as floc formers. However, the composition of EPSs is as yet not very well characterised (Seviour *et al.*, 2019). Under C/N imbalances, an overproduction of EPSs in activated sludge can lead to unfavourable viscous bulking phenomena during which flocs lose their density because of water entrapment in the matrix of EPSs. Immobilisation of cells by EPSs is essential in nature to prevent their wash out from their ecological niche. Biofilm carriers and granules are used in WWTPs to uncouple the SRT from the HRT, and to maintain a high biomass concentration in order to reach sufficiently large volumetric conversion rates. A detailed description of biofilms is given in chapters 17-18.

2.2.3 Metabolism and regulation

2.2.3.1 Breakdown of polymeric substrates and biosynthesis of biomass macromolecules

Microbial metabolisms consist of hydrolysing complex polymers (namely proteins, polysaccharides, and lipids) from the environment and wastewater. Extracellular enzymes break them down into simple monomers that can be taken up and used by cells for energy generation and growth (Figure 2.6). Proteins are degraded into amino acids, polysaccharides into monosaccharides, and lipids (or fats) into glycerol and fatty acids. Biomass synthesis consists of using C, N, P, and S elements provided by nutrients to reconstruct essential monomers and to re-polymerize them into functional macromolecules that form the cell. The energy required for the construction of complex molecules and cell structure is derived from redox reactions which are at the basis of life on earth, or from light.

2.2.3.2 Dissimilation and assimilation of substrates: catabolism and anabolism

Growth is the sum of catabolism and anabolism. The cellular metabolism involves the catabolism to generate energy by dissimilating substrates and the anabolism to use this energy to synthesise biomass constituents by assimilation of substrates and nutrients (Figure 2.6).

Anabolic reactions are relatively standard across all microorganisms (unity in biochemistry). Anabolism requires cellular energy supply delivered by the catabolism to synthesize cellular macromolecules. All the catabolic steps involved in the breakdown of substrates yield intermediate compounds that can be used as building blocks for the anabolic production of proteins, carbohydrates, lipids, and nucleic acids.

Catabolic reactions give the major diversity in microorganisms, based on the variety of molecules that can be used as e-donors and e-acceptors. They form the basis of most environmental biotechnological processes. Catabolism is the oxidation of an organic (*e.g.* sugars, VFAs) or inorganic (*e.g.* ammonium NH_4^+, H_2, H_2S) e-donor combined with the reduction of an e-acceptor (*e.g.* O_2, NO_x^-). The chemical energy contained in e-donors is transformed into a usable form for the microorganism. A membrane process is involved in converting the biochemical (redox) energy into adenosine triphosphate (ATP), the intracellular chemical energy transporter. ATP is used to drive cellular maintenance and growth. Catabolic reactions run via three main pathways: aerobic and anaerobic respirations, and fermentation.

Respiration pathways involve glycolysis, the Krebs cycle (or citric acid cycle, tricarboxylic acids cycle (TCA)), and the electron transport chain. Carbohydrate monomers are converted through glycolysis into pyruvate as the central intermediate. This oxidative process leads to the formation of ATP by substrate-level phosphorylation and of electrons carried by nicotinamide adenine dinucleotide (NADH). The dissimilation of organic compounds goes from pyruvate to acetyl-CoA to intermediate compounds of the Krebs cycle. Across glycolysis and the Krebs cycle, electrons are liberated and captured by NAD^+ reduced into NADH. NADH delivers the electrons to the electron transport chain until release on the terminal e-acceptor (*e.g.* O_2). This cellular respiration process activates the oxidative phosphorylation of adenosine diphosphate (ADP) into ATP. Both the electron transport chain and ATP biosynthesis occur in the cell membrane. These two processes are combined in a phenomenon called chemiosmosis. Three protein complexes are involved in the electron transport chain. They are proton pumps which generate a gradient of protons across the cell membrane. This proton gradient generates a proton-motive force which drives the ATP synthase.

Aerobic respiration pathways involve O_2 as terminal e-acceptor (*i.e.* respiration with air). Maximum three moles of ATP are produced per mole NADH oxidised at the electron transport chain. This is also expressed as a phosphate/oxygen (P/O) ratio of three moles of ATP produced per half mole of O_2 used. In reality the P/O ratio is lower, between 1.5-2 moles ATP per half mole of O_2. Overall, the aerobic respiration of one mole glucose theoretically leads to 38 moles of ATP via:

1. Glycolysis [2 ATP + 2 NADH (*i.e.* 6 ATP) = <u>8 ATP</u>]: two moles of ATP are produced by substrate-level phosphorylation via glucose oxidation into pyruvate. Two moles of NADH are formed and lead to six moles of ATP by oxidative phosphorylation in the e-transport chain.
2. Preparatory step [2 NADH (*i.e.* 6 ATP) = <u>6 ATP</u>]: pyruvate is transformed into acetyl-CoA. Two moles of NAD are produced, yielding six moles of ATP by oxidative phosphorylation.
3. Krebs cycle [2 GTP + 6 NADH (*i.e.* 18 ATP) + 2 FADH (*i.e.* 4 ATP) = <u>24 ATP</u>]: two moles of GTP (ATP equivalents) are produced by oxidation of succinyl-CoA to succinic acid. The six moles of NADH and two moles of FADH formed lead to production of 18 and four moles ATP, respectively, by oxidative phosphorylation.

In anaerobic respiration pathways, an inorganic (*e.g.* nitrite, nitrate, sulphate, iron (III)) or organic (*e.g.* tetrachloroethene) terminal e-acceptor is used instead of O_2 (*i.e.* respiration without air). Anaerobic respirations of glucose generate less ATP (< 38 moles ATP per mol glucose) than aerobic ones.

A special form of catabolism is represented by fermentation pathways. In this case, organic compounds are partly oxidised and partly reduced. This is exemplified by glucose fermentation into ethanol and CO_2. Fermentations are mainly based on glycolysis and often directly coupled to ATP generation by substrate-level phosphorylation. This leads to a relocation of electrons into an organic end-product such as VFAs, ethanol, lactate or other products depending on fermentation types. About twenty-times less ATP (2 moles ATP per mole glucose) is formed by fermentation than by aerobic respiration of glucose. In the conversion of glucose to acetate, two moles of NADH are produced. These cannot be further processed into ATP because of the absence of an external e-acceptor. NADH is regenerated either by producing H_2 or converting acetate in a more reduced product such as ethanol or butyrate.

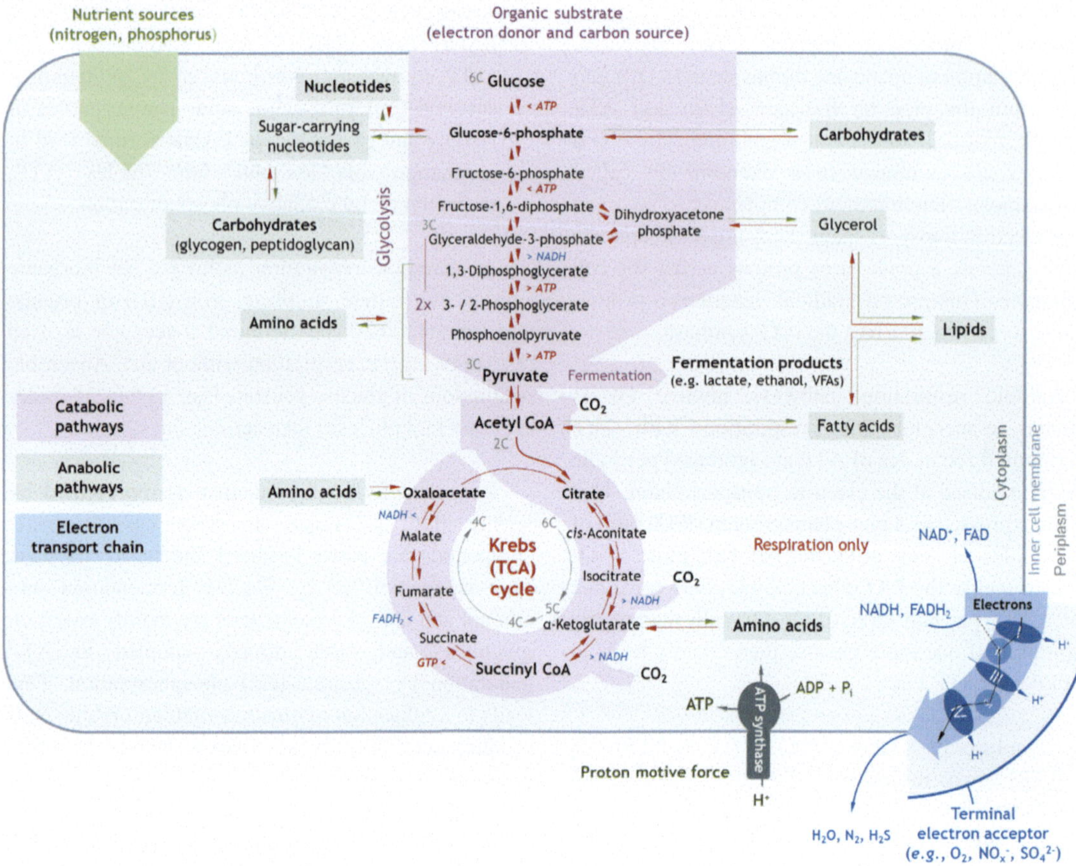

Figure 2.6 Combination of catabolic (pink) and anabolic (green) pathways for the breakdown of substrates (red arrows) and biosynthesis of biomass components (green arrows). Respiratory catabolisms involve glycolysis, the Krebs cycle, and electron transport chain. Fermentative catabolisms involve only glycolysis. The illustration is primarily given for the dissimilation of glucose. Glucose (6 carbons), after substrate-level phosphorylation, is split into two 3-carbon pyruvate molecules. In fermentations, electrons end up in fermentation product. In the case of respirations, pyruvate is further converted to acetyl CoA (2 carbons). When entering the Krebs cycle, acetyl CoA reacts with oxaloacetate (4 carbons) to form citrate (6 carbons) while releasing the co-enzyme A (CoA). Citrate is then split into 4-carbon carboxylate intermediates by losing step-wise carbon atoms with the concomitant release of O_2. Catabolism yields the production of cellular energy as ATP that will be supplied to sustain cellular maintenance and to drive anabolic reactions. In the different parts of the catabolism, ATP is either consumed (in the first steps of glycolysis) or produced (in the last steps of glycolysis, and in the Krebs cycle as GTP). Electrons are liberated as NADH in glycolysis, but mainly in the Krebs cycle for respiring microorganisms. Respirations link to the involvement of the electron transport chain: electrons are delivered by electron carriers such as NADH and $FADH_2$ and processed step-wise across a series of five membrane proteins. Three of them export protons released by the electron carriers, generating a proton motive force to drive the oxidative phosphorylation of ADP into ATP under the action of the spinning ATP synthase. The various intermediates produced in the catabolism can be used in anabolic reactions to form cellular constituents such as amino acids (for synthesis of proteins), carbohydrates, lipids, and nucleotides (for synthesis of nucleic acids) (image was built on information provided in http://classes.midlandstech.edu/).

2.2.3.3 Metabolic regulation in microbial cells: ATP, NADH, and NADPH

More closely to the basic definition of microbial growth, metabolisms are regulated by the ratios of ATP/ADP and of NADH/NAD$^+$ in the cells. The ATP/ADP ratio (ATP ⇆ ADP + P_i, $\Delta G = -30.6$ kJ/mol) influences the rate at which the ATP-ase is running. At lower ATP-ase activity, the proton motive force increases which then slows down the electron transport chain. The adenylate energy charge (AEC; Eq. 2.0) of a cell relates to the concentrations of adenine nucleotides, namely ATP, ADP and AMP (adenosine tri-/di-/mono-phosphate, respectively). The adenylate kinase catalyses their interconversion: ATP + AMP ⇆ 2 ADP. If AEC = 1, all the adenine nucleotide pool is in the form of ATP. If AEC = 0.5 all the adenine pool is in the form of ADP. If AEC = 0, all the adenine pool is in the form of AMP; the cell is 'discharged'. *Escherichia coli* has a typical AEC of 0.8 during exponential growth, which decreases to 0.5 during the stationary phase, and further <0.5 in the decay phase. Living cells harbour mostly ATP, while dying cells mostly AMP.

$$\text{AEC } (0-1) = \frac{[\text{ATP}] + 0.5 \cdot [\text{ADP}]}{[\text{ATP}] + [\text{AMP}] + [\text{ADP}]} \qquad (2.0)$$

The Krebs cycle is regulated at different levels. Three main regulatory levels involve the pyruvate, isocitrate, and α-ketoglutarate dehydrogenases, respectively.

Pyruvate oxidation into acetyl-CoA is driven by residual levels of ADP and ATP. A higher presence of ADP drives the conversion of pyruvate into acetyl CoA to fuel the Krebs cycle. A higher presence of ATP and NADH deactivates the Krebs cycle. Isocitrate and α-ketoglutarate processing steps relate to the production of NADH and release of CO_2. Depending on the NADH/NAD$^+$ ratio, the cell will process more or less pyruvate through the Krebs cycle. In short, the activity of the pyruvate dehydrogenase enzyme is activated by pyruvate and ADP, and inhibited by Acetyl CoA, NADH and ATP. NADH produced in the catabolism will be used first for cellular maintenance purposes, prior to being transformed as nicotinamide adenine dinucleotide phosphate (NADPH) for the anabolism. This goes via the pentose phosphate pathway that runs in parallel to glycolysis for anabolic purposes. The reaction NADH + NADP$^+$ ⇆ NADPH + NAD$^+$ enables the cell to regulate the residual NADH levels by shifting the equilibrium to supply electrons for either maintenance (catabolic; use of NADH) or anabolic purposes (use of NADPH). NADH and NADPH are therefore key compartmentalized pools of electrons to maintain the redox homeostasis of the cell (Xiao *et al.*, 2018). The redox stress is reflected by changes in NADH/NAD$^+$ and NADPH/NADP$^+$ ratios.

2.2.3.4 Molecular regulation in microbial cells: DNA, RNA, proteins and metabolites

For growth to take place, bacteria must be able to replicate their genetic material and carry out chemical transformations which allow the synthesis of all constituents from various precursors and energy (Figure 2.7). Chemical transformations are catalysed by enzymes (*i.e.* activated proteins). The synthesis of any protein requires its genetic expression. The first step is the transcription of DNA into RNA. RNA is translated into a protein that is then processed to make it functional. With its constituents replicated, a bacterial cell can divide into two daughter cells. In the genome of single cells, the information contained in the genes present in the DNA provides the metabolic potential of a microorganism. Genome-scale or genome-based models can be drawn to predict metabolisms. During gene expression, the genetic information is transcribed from genes into mRNA (transcriptome). After gene expression, mRNA carries the genetic information to the ribosomes for the synthesis of proteins. The genetic code supplied by mRNA is translated into proteins (the proteome). tRNA supplies amino acids to the ribosomes. rRNA binds amino acid residues into peptides and proteins. Post-translational activation of proteins leads to the formation of enzymes (enzymome). Enzyme activity depends on concentrations of metabolites (metabolome) involved in biochemical reactions. Metabolic conversions result in a flux of metabolites (fluxome) inside the cells.

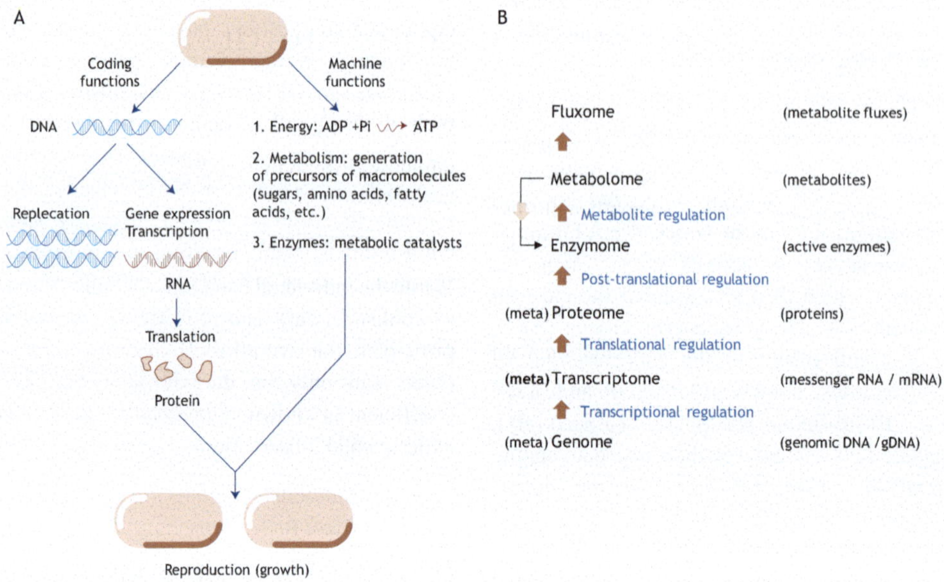

Figure 2.7 Metabolic and molecular regulation in microbial cells: A) microbial growth requires both coding and machine functions to be operational. DNA serves for replication and gene expression, first by transcription of DNA into RNA, then translation of RNA into proteins. Source: Comeau *et al.* (2008) in Madigan *et al.* (2018) (adapted); B) from DNA to metabolites: multiple levels in the molecular biology and metabolic regulation of microbial cells.

A microbial community is composed of numerous microorganisms, each of which is in a particular metabolic state. Metabolic regulation happens in individual populations of cells that mutually impact each other either by *e.g.* producing a metabolite consumed by the next organism or producing an inhibitory compound that stops the metabolism of other organisms. The overall metabolic performance of a microbial community relies on a balance across the complex metabolic network formed by the diversity of microorganisms. Across a microbial community, the different populations exhibit different regulation levels. The overall regulation of the microbiome performance is the resultant of the regulation levels of all microorganisms involved. We speak about metagenome, metatranscriptome, metaproteome, metaenzymome, (meta-)metabolome, and metafluxome to describe the amount of informational molecules carried and expressed by the microbial populations present in the microbiome. The prefix 'meta' (Grk. μετά-) and suffix '-ome' (Grk. -ωμα) describe an assembly of features in a mass of information, respectively. Nonetheless, in the field of community system microbiology, we aim to extract the regulatory information from single lineages prior to re-assembling individual information back to microbiome level. The genomes of each organism are the functional units of a microbiome, from which the genetic expression is transcribed, translated, activated, and results into actual conversions, regulated at different levels.

2.2.4 Trophic groups and metabolic diversity

Systematics can be applied to describe the trophic groups of microorganisms present in nature. The distinctions are made based on materials (chemical e-donors and e-acceptors, or photons) used by microorganisms to generate their energy via catabolism and on the type of carbon source (C source) they use in their anabolism. These features are displayed by the adjectives and prefixes used in the naming of trophic groups. Microbial diversity is primarily driven by catabolisms and the diversity of e-donor/e-acceptor associations that can be achieved in nature. Anabolisms display fine-scale differences (heterotrophs *vs.* autotrophs) for the synthesis of biomasses. However, anabolic reactions are relatively well conserved across microbial populations.

2.2.4.1 Trophic structure in microbiology and links to environmental engineering

Energy generation is differentiated by the use of either chemicals (chemotrophs) or light photons (phototrophs). Energy-generating chemical redox reactions can involve either organic (organotrophs) or inorganic (lithotrophs) e-donors. Autotrophs use inorganic C sources: they can produce their own organic resource by fixing CO_2. Heterotrophs are organically not self-sufficient and use external organic C sources for biomass synthesis. The prefixes chemo-/photo-, organo-/litho-, and auto-/hetero- can be combined to theoretically describe the catabolic and anabolic levels of any trophic group present in an environment (Figure 2.8).

The e-acceptors can range from O_2 in aerobic respiration to inorganics in anaerobic respiration. In fermentation, no terminal e-acceptor is available; electrons are relocated onto an organic catabolic product.

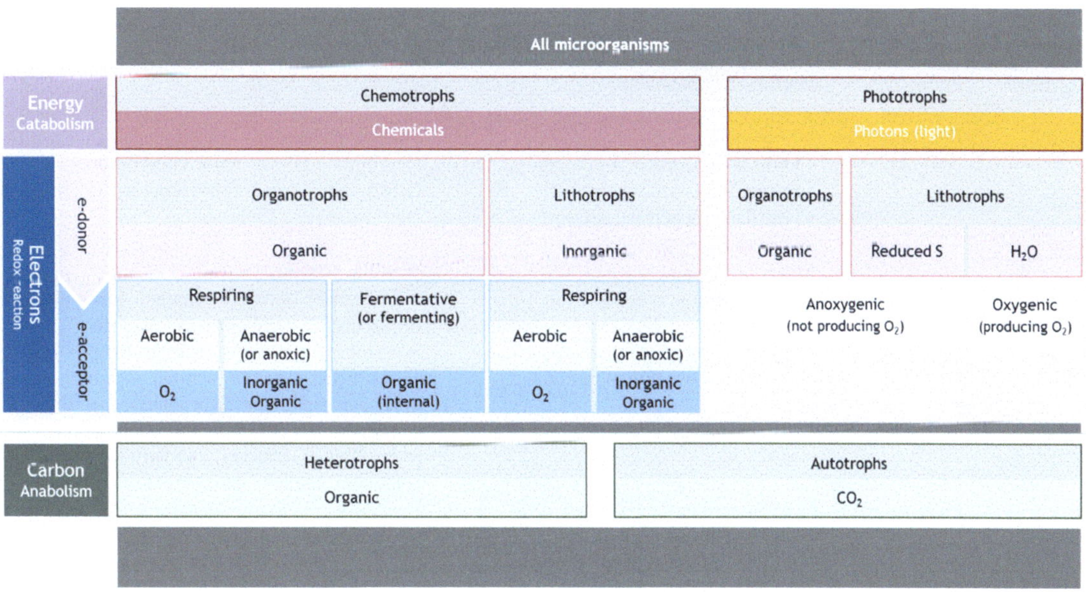

Figure 2.8 Systematics in trophic groups of microorganisms by using a combination of prefixes coding for the (*i*) source of energy originating from either a chemical redox reaction (chemo-) or light (photo-); (*ii*) source of electrons originating from either an organic (organo-) or inorganic (litho-) e-donor; (*iii*) source of carbon originating either from CO_2 (auto-) or an organic compound (hetero-). The terms can be complemented with an adjective characterizing the type of respirations or fermentations conducted by the microorganisms. Besides dissolved oxygen (aerobic respiration), microorganisms can perform anaerobic respirations by using inorganic (*e.g.* nitrite, nitrate) or organic compounds (*e.g.* tetrachloroethene) as terminal e-acceptors. In fermentations, electrons are relocated from pyruvate to end up in organic compounds such as ethanol, lactate, and VFAs, as well as dihydrogen as an inorganic fermentation product. Anaerobic respirations characterized by microbiologists are described in environmental engineering as happening under anoxic conditions. Phototrophs harbour a diversity of purple to green microorganisms. In phototrophs, light photons deliver energy to energize the electrons supplied by an e-donor that is either water by oxygenic phototrophs (the splitting of water to deliver electrons to the cell yields O_2 formation) or that is an inorganic (*e.g.* H_2S) or organic (*e.g.* acetate) compound by anoxygenic phototrophs. While it is common to classify micro-organisms according to one primary conversion, one should bear in mind that microorganisms are versatile in nature and can exhibit different types of metabolisms depending on the environmental conditions. Mixotrophy is also widespread.

Environmental engineering harbours slight differences in contrast to microbiological terms. Aerobic conditions involve O_2 as the terminal e-acceptor. Anoxic conditions (without O_2) involve inorganics such as oxidized nitrogen as nitrite and nitrate as terminal e-acceptors. Anaerobic conditions relate to the absence of any terminal e-acceptors other than carbon dioxide. Anammox organisms have been elucidated by microbiologists and are referred to as 'anaerobic' ammonium oxidizers, duly referring to their anaerobic respiration process. Engineers would refer to 'anoxic' ammonium oxidation. From the microbiological perspective we look inside the cell, from the physiology of the cell. From the engineering perspective we look at conditions surrounding the cells and the external supply of e-acceptors. As frequently experienced, terms are intermixed between fields. It is important to check and align terminologies when describing redox conditions in biological systems (Table 2.3). One method is simply by using the dictionary: oxic/anoxic (etym. 'presence/absence of O_2'), aerobic/anaerobic (etym. 'presence/absence of air'). However, it is important to keep the microbiological or engineering context in mind when reading literature.

Table 2.3 Microbiology and environmental engineering semantics (aerobic, anaerobic, oxic, and anoxic) used to describe relations of microorganisms and processes in the presence or absence of terminal e-acceptors.

Adjective	Etymology	Microbiology semantic	Engineering semantic
Aerobic	Presence of air	Aerobic respiration – O_2 used as terminal e-acceptor to respire the e-donor substrate. – Nitrifiers are aerobic respiring chemolithoautotrophs.	Aerobic process – O_2 supplied as e-acceptor via aeration. – Nitrification is an aerobic process.
Anaerobic	Absence of air	Anaerobic respiration – Other terminal e-acceptor than O_2 used to respire the e-donor substrate. – Anammox organisms are anaerobic respiring chemolithoautotrophs.	Anaerobic process – No external e-acceptor supplied. No aeration, no nitrite/nitrate supply or recirculation. – Anaerobic digestion is an anaerobic process. This process is neither aerated nor supplied with other e-acceptors. In the complexity of the ecosystem though, different anaerobic respiring (*e.g.* sulphate reducers) and fermentative (*e.g.* acidogens, methanogens) organisms coexist.
Oxic	Presence of O_2	Environment in the presence of O_2 – Seasonal mixing in lakes is important for vertical transport of oxygen across the water column.	– Used as an alternative to aerobic: full BNR can be performed in an anaerobic-anoxic-oxic (A2/O) process in a succession of anaerobic, anoxic and oxic tanks.
Anoxic	Absence of O_2	Environment in the absence of O_2 – Eutrophication unfavourably leads to anoxic surface water environments.	Anoxic process – Other terminal e-acceptor supplied in the system via *e.g.* nitrite or nitrate recirculation loops. – Denitrification and anammox are anoxic processes.
Fermentative/ fermenting		Fermentation – No external terminal e-acceptor. Electrons are relocated onto an internal organic compound produced in the metabolism. – Methanogens are fermentative chemoorganoheterotrophs.	Fermentation process – For an efficient EBPR, particulate organic matter should be hydrolysed into soluble substrates that are fermented into VFAs prior to uptake by PAOs and GAOs in the anaerobic tank of the A2/O process. – Lignocellulosic sugars can be efficiently converted into VFAs via mixed-culture fermentation.

The need for, and tolerance or sensitivity to, O_2 varies widely among aerobes. Aerobes use O_2 at different levels: obligate aerobes need it; microaerophiles require it in low levels; and facultative aerobes can function in its absence. In aerobes, enzymes for O_2 reduction are always induced. In contrast, denitrifiers are facultative aerobes and harbour constitutive enzymes for O_2 reduction. Their enzymes for nitrite or nitrate reduction need to be induced, a condition that requires the absence of O_2. All denitrifying bacteria can also use O_2, their catabolic processes being relatively similar. Sulphate reducers cannot use O_2, their catabolic process being very different from aerobic respiration. Anaerobes do not use O_2: aerotolerant microorganisms can tolerate it; obligate (strict) anaerobes do not since poisoned by the oxidant O_2.

Phototrophs are oxygenic when using water as e-donor: by splitting H_2O they release O_2. Anoxygenic phototrophs use another inorganic (*e.g.* H_2 or H_2S) or organic (*e.g.* acetate) e-donor. Phototrophs generate their energy from light. However, an e-donor is always needed for growth. Photons and electrons are two different entities. Phototrophs use photons to energize electrons supplied by the e-donor, in contrast to a redox reaction by chemotrophs. The energized electrons are processed in an electron transport chain by (non-)cyclic photophosphorylations. Photosystems harbour light-harvesting units that transfer photons to a reaction centre where the electrons get energized prior to transfer via the quinone pool to cytochromes and ferredoxin. Phototrophs comprise a battery of pigments to capture photons at different wavelengths. Their microbial diversity can be examined by analysing pigments. Like plant cells, cyanobacteria and eukaryotic algae use chlorophyll a (Chl a). Bacteriochlorophylls are used by purple (Bchl a and b) and green (Bchl c, d and e) sulphur/non-sulphur bacteria, as well as heliobacteria (Bchl g). Phototrophs contain other accessory pigments. Carotenoids confer photoprotection against photooxidation of cell constituents, providing immunity against endogenous photosensitization, and act against oxygen radicals. Carotenoids absorb blue light (450-550 nm) and transfer this energy to Bchl. Phycobilins are used by cyanobacteria and red algae to enhance the absorption of green, yellow, orange, and red lights that are less efficiently absorbed by Chl a.

Microorganisms seldom operate on one single metabolism. Metabolic combination in single cells is possible. This is called mixotrophy, which is a relatively broad term. It relates to parallel involvement of different metabolisms by a microorganism. Anabolism always involves an organic compound or CO_2 as its C source. Mixotrophs can maximize organic matter usage to synthesise biomass while deriving energy from inorganics. An aerobic respiring chemoorganoheterotroph both dissimilates and assimilates fractions of an organic compound. By becoming a chemolithoheterotroph, it can generate its energy from inorganics and more organic matter is available to anabolise biomass. The ultimate mixotroph can maximize growth by dissimilating an inorganic substrate, assimilating organic matter, and fixing CO_2 to supplement growth. Phototrophs can combine photoorganoheterotrophy and photolithoautotrophy to maximize carbon use from organics and CO_2 and/or to recycle via the Calvin cycle the CO_2 released from processing organics.

Metabolic versatility is widespread, enabling growth under ever-changing conditions. In a pure culture, individual cells of a population can harbour different metabolic states, depending on the gradients of nutrients and photons across reactor geometries and bioaggregates. Purple non-sulphur bacteria (PNSB) are models of hyper-versatility with photoorganoheterotrophy, photolithoautotrophy, dark fermentation, photofermentation, and chemotrophies in single cells.

In addition, microorganisms can be delineated between prototrophs that are able to produce essential compounds needed for growth and auxotrophs that depend on other organisms to supply their essential resource. This is less studied in wastewater environments, while there is growing interest in mixed-culture fermentation processes.

2.2.4.2 Illustration of microbial trophic groups

Combinations of respirations or fermentation with photo-/chemo-, litho-/organo-, and auto-/hetero-trophies (Figure 2.8) are illustrated here for chemotrophs and phototrophs.

Obligate aerobic respiring chemoorgano-heterotrophs grow by dissimilating an organic e-donor by respiration with O_2 only and synthesizing biomass from an organic C source that can be the same as the e-donor. OHOs use acetate as their e-donor/C source and O_2 as their e-acceptor. *Anaerobic respiring (or 'anoxic') chemoorganoheterotrophs* catabolize an organic e-donor by respiration with an inorganic e-acceptor such as nitrite or nitrate and anabolize an organic C source (which can be the e-donor). DHOs use acetate and respire it with NO_2^- or NO_3^-. *Fermentative chemoorganoheterotrophs* grow by fermenting an organic e-donor whose electrons end up in an organic compound, and using the same C source for anabolism. Acidogenic fermenters convert carbohydrates into VFAs as the end catabolic product. *Aerobic respiring chemolithoautotrophs* respire an inorganic e-donor with O_2 as the terminal e-acceptor and capture dissolved CO_2 (or its bicarbonate form at pH 7) to produce biomass. Nitrifiers fix CO_2 and respire the e-donors NH_4^+ or NO_2^- with O_2 as the e-acceptor. *Anaerobic-strict respiring (or 'anoxic strict') chemolithoautotrophs* respire an inorganic e-donor with an inorganic e-acceptor in the absence of O_2 and assimilate CO_2. Anammox bacteria fix CO_2 and respire NH_4^+ with NO_2^- or NO as e-acceptors. *Chemoorganoautotrophs* (organic e-donor, inorganic C source) such as anaerobic methanotrophic archaea and *chemolithoheterotrophs* (inorganic e-donor, organic C source) such as some oceanic extremophiles can be semantically constructed but are less widespread in nature. Their exerted metabolism should rely on thermodynamic and biochemical constrains. While mixotrophy is possible, organics are often used for catabolism and anabolism, making organic/inorganic associations less common between e-donors and C sources.

Oxygenic photolithoautotrophs such as cyanobacteria photo-energize electrons derived from water and fix CO_2. *Anoxygenic photolithoautotrophs* such as purple sulphur bacteria (PSB) dissimilate an inorganic e-donor other than water such as H_2S, energize the electrons with photons, and assimilate CO_2. *Anoxygenic photoorganoheterotrophs* such as PNSB obtain electrons from organics (*e.g.* acetate), energize them with photons, and use this substrate for anabolism. *Photolithoheterotrophs* can energize electrons from an inorganic e-donor such as H_2 using light and build biomass from organics such as acetate. *Photoorganoautotrophs* are seemingly less reported; photons would energize electrons supplied by organics, while CO_2 would be used in anabolism.

2.2.4.3 Predominant guilds of microorganisms involved in BNR from wastewater

Common guilds of microorganisms present in BNR, anaerobic digestion, and nutrient recovery systems are summarized in Table 2.4.

BNR WWTPs are designed in sequences of redox conditions achieved either in space across different tanks in flow-through processes or in time in sequencing batch reactors. Redox conditions involve alternating anaerobic (no terminal e-acceptor), anoxic (NO_x^- as e-acceptors), and aerobic (O_2 as e-acceptor) phases. These phases are combined in the presence or absence of organic matter as an e-donor and C source. Biofilms and granules comprise different redox biovolumes across gradients of e-donors and e-acceptors in the biofilm depth. This delineates niches for the proliferation of microorganisms of interest (Figure 2.9).

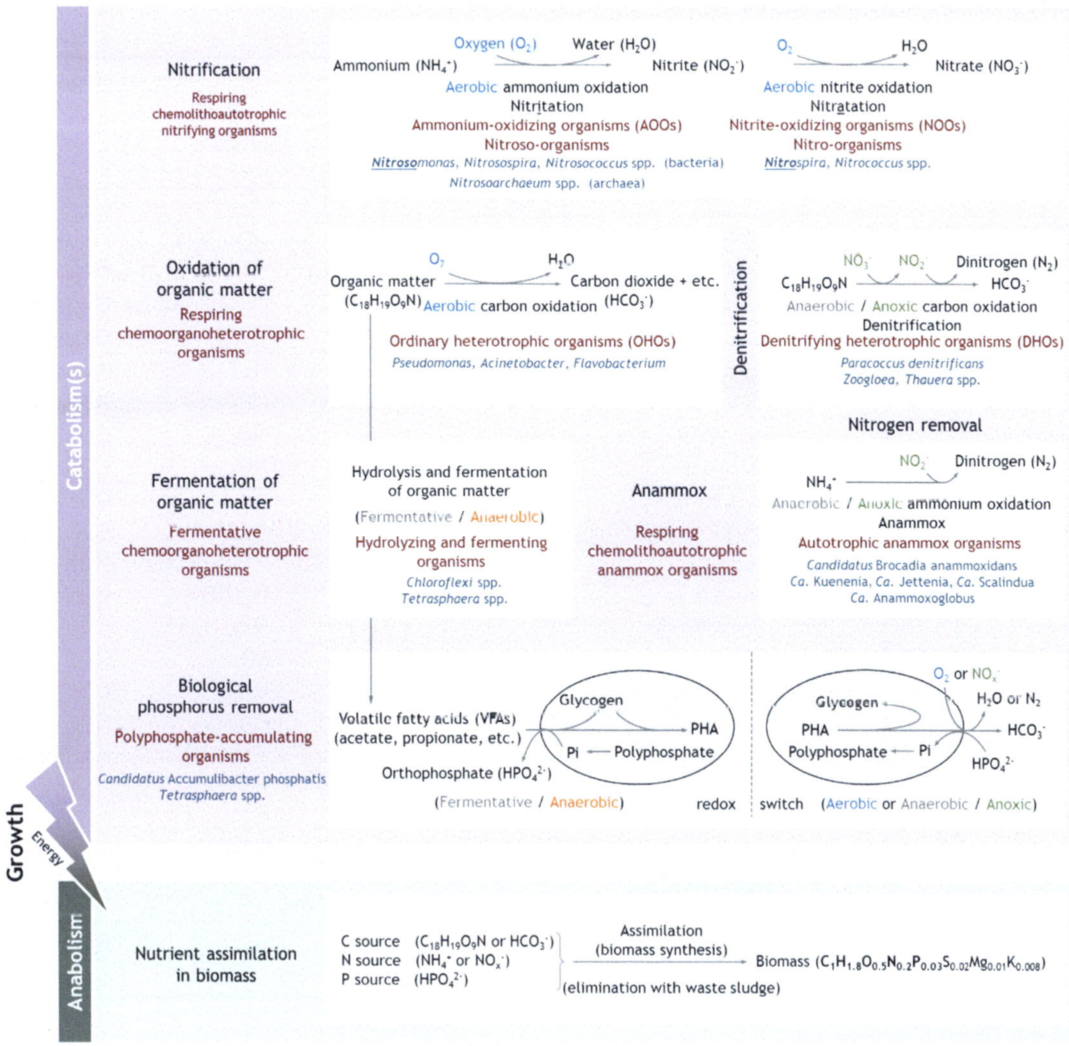

Figure 2.9 From microbial conversions to microbial guilds and microbial populations in BNR WWTPs. Involvement of C, N and P nutrients in the main catabolic and anabolic processes of bacterial populations present in activated sludge. In the conversions, the coloured terms (aerobic, anoxic, and anaerobic) are given in the engineering lexicon. The corresponding microbiological terms are given in grey (anaerobic *vs.* anoxic, fermentative *vs.* anaerobic). Based on Weissbrodt and Holliger, 2013.

Table 2.4 Trophic classification of microorganisms in WWTPs and WRRFs (adapted from Rittmann and McCarty, 2001; Metcalf and Eddy, 2003).

	Energy source				Carbon source
	e-donor		e-acceptor	Typical products	
Trophic group	Microbial guild	Type of e-donor			
Chemotrophs					
Chemoorganoheterotrophs	Aerobic heterotrophs	Organic	O_2	CO_2, H_2O	Organic
	Denitrifiers	Organic	NO_3^-, NO_2^-	N_2, CO_2, H_2O	Organic
	Fermenting organisms	Organic	Organic	Organic: VFAs	Organic
	Iron reducers	Organic	Fe (III)	Fe (II)	Organic
	Sulphate reducers	Acetate	SO_4^{2-}	H_2S	Acetate
	Acetoclastic methanogens	Acetate	Acetate	CH_4	Acetate
Chemolithoautotrophs	Nitrifiers: AOB	NH_4^+	O_2	NO_2^-	CO_2
	Nitrifiers: NOB	NO_2^-	O_2	NO_3^-	CO_2
	Anammox bacteria	NH_4^+	NO_2^-	N_2	CO_2
	Denitrifiers	H_2	NO_3^-, NO_2^-	N_2, H_2O	CO_2
	Denitrifiers	S	NO_3^-, NO_2^-	N_2, SO_4^{2-}, H_2O	CO_2
	Iron oxidizers	Fe (II)	O_2	Fe (III)	CO_2
	Sulphate reducers	H_2	SO_4^{2-}	H_2S, H_2O	CO_2
	Sulphate oxidizers	H_2S, S^0, $S_2O_3^{2-}$	O_2	SO_4^{2-}	CO_2
	Aerobic hydrogenotrophs	H_2	O_2	H_2O	CO_2
	Hydrogenotrophic methanogens	H_2	CO_2	CH_4	CO_2
Phototrophs					
Photolithoautotrophs	Microalgae, cyanobacteria	H_2O	CO_2	O_2	CO_2
	Green sulphur, green filamentous bacteria, purple sulphur bacteria	H_2S	CO_2	S (0)	CO_2
Photoorganoheterotrophs	Purple non-sulphur bacteria	Organics, acetate	CO_2, H^+	H_2	Organics, acetate

PAOs and GAOs store VFAs under anaerobic conditions and grow by respiring either O_2 or NO_x^-. Nitrifiers are composed of aerobic ammonium-oxidizing (AOOs; '*Nitroso-*' organisms, *i.e.* presented to involve a nitroso compound like HNO or NO as biochemical intermediate) that respire NH_4^+ into NO_2^- and aerobic nitrite-oxidizing (NOOs; '*Nitro-*' organisms, *i.e.* involving a nitro compound like HNO_2 as biochemical intermediate) organisms that respire NO_2^- into NO_3^-, with O_2 as the e-acceptor. AOOs and NOOs are primarily autotrophic using CO_2 as the C source, while some such as *Nitrospira* are versatile and can metabolize organics. Microorganisms are indeed always more versatile than phenotypically classified. A third nitrifying guild of complete ammonium oxidizers (comammox) has recently been elucidated, involving full aerobic respiration of NH_4^+ into NO_3^-. DHOs respire electrons from organic matter with NO_x^- as the e-acceptors. Chemolithoautotrophic denitrifiers can respire electrons from H_2S using NO_x. Anammox organisms are autotrophic and respire NH_4^+ with NO_2^- or NO. Some anammox populations are thought to convert organics. In flocs, granules and biofilms, BNR organisms are accompanied by a diversity of side populations (hydrolysers, fermenters, and grazing populations).

Phototrophs are always detected at some levels in wastewater environments. As soon as light is available they can grow independently. Several phototrophs are versatile and can grow chemotrophically as well. In the resource recovery context, phototrophic processes are gaining interest. Microalgae and cyanobacteria are 'sustainable' CO_2 sinks for carbon capture mediated by light. Phototrophs such as PNSB are used to grow a biomass with high protein content that can be valorised as a food complement when grown on a waste stream from a food industry.

Conceptual models of microbial ecosystems of BNR flocculent and granular activated sludges are excellent bases for functional analyses (Nielsen *et al.*, 2010; Weissbrodt *et al.*, 2014) (Figure 2.13). The basic metabolism of the main guilds can easily be outlined by combining the e-donor and e-acceptor for catabolism, and the C and N sources for anabolism (Figure 2.10). Simple outlines disentangle the connections of the main metabolizing populations.

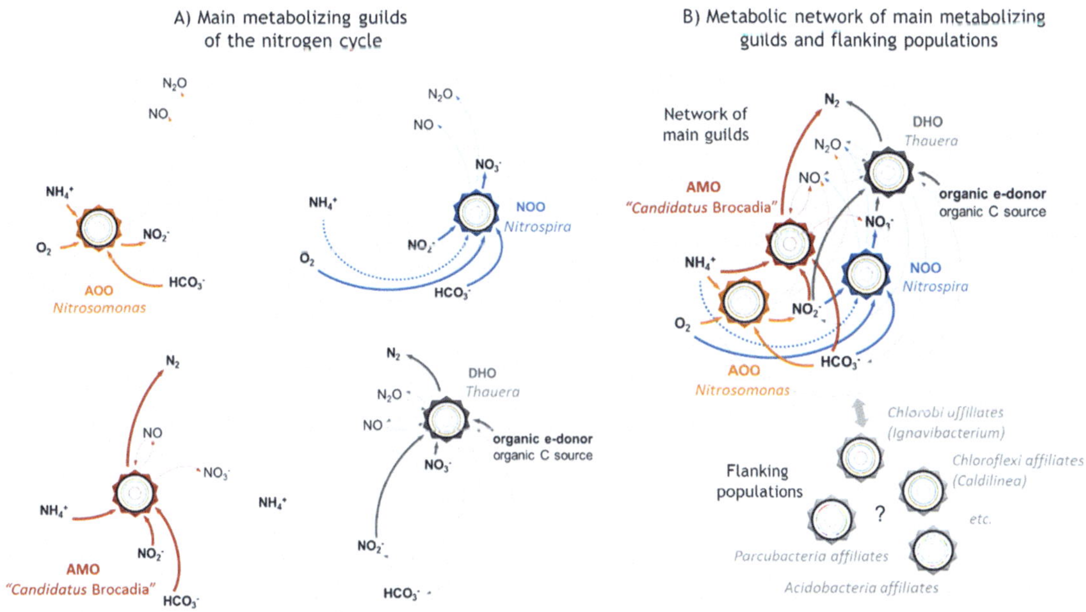

Figure 2.10 An easy way to outline the growth system of microbial guilds. A) This illustrates typical guilds of the nitrogen cycle, namely aerobic ammonium-oxidizing organisms (AOOs), aerobic nitrite-oxidizing organisms (NOOs), anammox organisms (AMOs), and denitrifying heterotrophic organisms (DHOs). The growth system of these main metabolizing guilds can be analysed individually for their C source, N source, e-donor, and e-acceptor; B) besides the main metabolizing guilds, microbial communities of activated sludge harbour a diversity of flanking populations, that have been postulated to 'connect' the microbiome. However, factual measurements for elucidating connectivity are needed. These individual pieces can then be assembled at microbiome level to delineate metabolic interactions. The gear wheels represent the microorganisms. The internal circles are the assembled genomes of these microorganisms. Source: based on Weissbrodt *et al.* (2020).

2.2.5 Microbial physiology and environmental gradients

2.2.5.1 Environmental factors

Microorganisms display physiological ranges of environmental conditions for optimal growth. Chemical, biological, and physical factors impact on conversions, maintenance, growth, and selection. Depending on the trophic groups, the major factors affecting growth relate to the availability of e-donors and e-acceptors, nutrients (C, N, P and S), mineral sources, trace metals for enzymes activities, and light. In mathematical growth models, limitation is characterized by saturation functions (Monod terms) such as $S_{S,i}/(K_{S,i} + S_{S,i})$. Under certain circumstances, microorganisms can face multiple limitations simultaneously. Microbes can tune their metabolism to the multiple limitations, by *e.g.* modulating their affinity for substrates.

Other factors involve temperature, pressure, pH, ionic strength, activity, alkalinity, and salinity (*i.e.* osmotic pressure). Temperature impacts the growth rate of microorganisms. The optimal range of temperature for each group is relatively narrow. Psychrophiles grow from *ca.* -12 °C (lowest growth temperature measured so far) to 15 °C, mesophiles between 15-40 °C, thermophiles between 40-70 °C, and hyperthermophiles between 70-110 °C. Microbes operating at a high temperature generally grow faster. By increasing temperature, a gradual increase in growth rate is observed until an abrupt drop is observed due to thermal denaturation of proteins. Different relations can describe the temperature dependence. In chemistry, the Arrhenius' law describes the exponential dependency of the rate constant (k) to the ratio of energy activation of reaction to temperature (T) of the system: $k = A \cdot e^{-Ea/(R \cdot T)}$, with the pre-exponential factor A and ideal gas constant R. In biological wastewater treatment, rates are expressed with a temperature coefficient θ, *e.g.* for biomass-specific growth rate q_X (or μ historically): $q_X(T) = q_X(T_{ref}) \cdot \theta^{(T-Tref)}$. The Q_{10} temperature coefficient measures the rate of change under an increase of 10 °C in a biological system: $Q_{10} = (R_2/R_1)^{10/(T2-T1)}$. As a rule of thumb, an increase of 10 °C leads to 2-3 times higher conversion and growth rates.

pH measures H^+ activity. pH gradients across the cell plasma membrane drive cellular bioenergetics and growth rates. Proton-motive forces are used by respiring organisms to generate ATP by oxidative phosphorylation. In metabolic models, the coefficient α expresses the dependency of a growth yield to pH. It represents the ATP needed to transport the substrate across the cell membrane. The pH dependency of a biological rate can be expressed with a Monod term such as $10^{-pH}/(K_{pH} + 10^{-pH})$. Biological rates can be dependent on both temperature and pH, and on combinations of these two parameters, such as: $q_i(T, pH) = q_i(T_{ref}, pH_{ref}) \cdot \theta^{(T-Tref)} \cdot 10^{-pH}/(K_{pH} + 10^{-pH})$ (Lopez-Vazquez *et al.*, 2009; Weissbrodt *et al.*, 2017).

Photons drive phototrophic processes across visible (VIS, 400-750 nm) and near infrared (NIR, 780 nm to 3 μm) ranges of light. Growth has seldom been reported at lower energy wavelengths in mid (MIR, 3-50 μm) or far (FIR, 50-1,000 μm) infrared. FIR is used in photomedicine to activate human cells. Short wavelengths (*i.e.* high energies) can be inhibitory or lethal to both chemotrophic and phototrophic microorganisms. The bactericidal effect of ultraviolet (UV, 10-400 nm) radiation is used to deactivate microorganisms. The effect of light on biological rates of phototrophs can be modelled in different ways (Schoener *et al.*, 2019). Both the light irradiance and wavelength have to be taken into account when addressing the impacts of light on microbial growth. A simple approximation is to use Monod terms such as $I/(K_I + I)$ with light intensity I (W/m^2 or $J/s.m^2$). Following Lambert-Beer's law of light absorbance, the residual light intensity in the biological system will decrease proportionally to growth and cell concentration.

Chemical factors can activate, limit or inhibit growth. Inhibition can be reversible or irreversible. We speak of toxicity or lethality when leading to cell death. Depending on the concentration, the same element, ion or substance can either promote growth or be inhibitory and toxic. (Eco)toxicity tests deliver dose-response curves by contacting microorganisms, algal, insect or mammalian cells to increasing concentrations of chemicals. EC50, IC50 and LC50 values express chemical concentrations that lead to half-maximum effective, inhibitory and lethal response, respectively. There are various inhibition terms in literature. They can be approximated as inverse Monod terms such as $K_{I,i} / (K_{I,i} + S_{I,i})$ with $S_{I,i}$ being the concentration of dissolved inhibitory compound and $K_{I,i}$ the half-saturation inhibitory constant.

Biological agents can impact growth. Antimicrobials are produced by an organism to deactivate a competitor. Populations that develop resistances can withstand the antimicrobial stress. Pathogenicity, infection by viruses, and predation by protozoans such as ciliates and by metazoans such as nematodes and worms are selection pressures that shape microbial communities in wastewater biocenoses.

2.2.5.2 Microbial niche establishment across gradient systems

Natural and engineered biological systems are characterized by gradients of environmental factors. Microbial diversity is shaped by redox conditions and energy metabolism, and finely tuned by nutrients that microbes assimilate. Microbial niches establish along gradients of light, gradients of e-donors and e-acceptors, gradients of C, N, P and S sources, and gradients of other minerals and trace metals that they can scavenge from substrates. Microbial niche differentiation in aquatic systems arise from interfacial (gas-liquid and liquid-solid) and transport phenomena (advection, convection and diffusion) that macroscopically shape the environment. Microorganisms shape their microenvironment via their conversions. A microbial niche will establish close to another one that supplies an e-donor or e-acceptor, C, N, P or S source, and/or other micronutrients. Diffusional limitations across the depth of sediment, biofilms and granule matrices contribute to the segregation of microbial guilds. Substrates are consumed, converted, and diffuse further step-wise across the microbial food chain.

Microbial community engineering and process design relies on microbial resource management across flows and gradients. In a flow-through WWTP, substrate and redox conditions are engineered in separate tanks to select and activate the microbial guilds necessary to perform the depollution of organic matter, and nitrogenous and phosphorous compounds. Successions of anaerobic, anoxic and aerobic conditions are necessary for full BNR. Different BNR process configuration are shown in Chapter 6, Figure 6.20. Anaerobic and anoxic tanks are mixed, whereas aeration transfers O_2 into the liquid phase of aerobic tanks. Internal recirculations of wastewater flows are included to provide e-acceptors such as NO_x^- which are produced in the nitrification tank back to the pre-denitrification tank for their reduction. For this reduction, influent organic matter serves as the e-donor and C source. Anaerobic zones should be deprived of terminal e-acceptors to stimulate the pre-fermentation of organics and uptake of VFAs by PAOs and GAOs. Biofilm and granular sludge systems used for process intensification thrive on gradients. Simultaneous BNR can be achieved in the biofilm matrix. More details on biofilm systems are developed in chapters 17-18.

2.3 MICROBIAL ECOLOGY AND ECOPHYSIOLOGY METHODS

Following on from early developments in microbiology by Koch and Hesse in Germany, the pioneers of the 'Delft School of Microbiology' in the Netherlands made key contributions in virology and microbial ecology (Beijerinck), biogeochemistry (Baas-Becking), biochemistry (Kluyver), microbiology (Van Niel), and microbial biotechnology (Kuenen and Heijnen). The integration of their principles is at the root of breakthroughs made in ecological engineering for the development of environmental biotechnologies for the biological treatment of waste and wastewater, recovery of resources, and public health sanitation.

2.3.1 Black to grey and white-box analysis of microbiomes

Over the last 100 years of activated sludge, the wastewater engineering sector has developed in parallel analytical methods from microscopy, microbiology, microbial ecology, biochemistry, ecophysiology, molecular biology, and system microbiology.

Starting from the early microscopy observations of 'animalcules' in fluids made by Van Leuwenhoek in the 17th century, analytical biosciences have made uninterrupted progress. We now benefit from a range of methods to fingerprint microbial communities, detect and characterize microorganisms (*i.e.* morphotypes, phylotypes, genotypes), and examine their metabolisms (*i.e.* phenotypes). Step by step, new methods have made it possible to 'switch light on sludge' and to move from a black-box description to more grey-box analysis of microbial communities, which combined with mathematical modelling will enable more white-box quantitative predictions on metabolisms and interactions inside microbial networks. Beyond the names of guilds and populations of mixed cultures, their metabolic functions are required to design and control processes, and predict the overall performance. Ecophysiological methods are useful to examine metabolic phenotypes.

Microbial ecology has moved forward to community system microbiology and environmental system biology. The latest 'omics' methods are providing high resolution and throughput to characterize pure cultures and complex communities, microbial lineages and their metabolic traits ranging from DNA (meta-genome), RNA (meta-transcriptome), proteins (meta-proteome) to metabolites (meta-metabolome), and their fluxes (meta-fluxome). While resolution and throughput have dramatically increased, big datasets have led to new challenges to manage, extract and visualize the key information. Wet-lab procedures are going hand in hand with dry-lab workflows of bioinformatics and biostatistics.

Several calls have been made over the last few decades to bridge microbial ecology and environmental biotechnology, but twenty years later, this need remains the *status quo*. The fields of process engineering and environmental life sciences have specialized completely independently together with expansions in analytical and computational power. Novel methods and workflows can drive new research questions, while the choice of method relies on the investigation target set at the outset. Multilateral confrontation and integration of microbiology, biotechnology and engineering expectations and principles is needed.

2.3.2 Informational molecules from microorganisms

The four predominant types of cellular macromolecules (nucleic acids, proteins, polysaccharides and lipids) are the key informational molecules that compose biomass (Section 2.2.2). They are useful to elucidate phylogenies and the functional traits of microbial populations.

DNA contains the genetic code of life. The nucleotide sequences code for genes in the genomes of microorganisms. Gene sequences allow for identification of microbial lineages and prediction of their functional potential. A multitude of methods can be applied to DNA templates. From an isolate, one can extract DNA, and sequence and analyse its genome. Out of the DNA pool of a microbial community, one can sequence the metagenome, and further bin and annotate single-lineage genomes.

RNA (primarily the mRNA) comprises the information on transcribed genes. Gene expression can be captured by RT-qPCR (see below) or high-throughput transcriptomic analyses. The (meta-) transcriptome of a microorganism or a community informs on what genetic potentials were expressed by the population(s) involved.

Proteins harbour the information that was translated from RNA. Sequencing the (meta-) proteome of an organism or a microbiome is useful to gain both phylogenetic and functional information. From the biochemical viewpoint, the presence of a protein is not sufficient to validate the metabolic function of an organism. Proteins should be activated into enzymes that catalyse biochemical conversions. Enzymatic tests help to check for protein activity. Still, an *ex-situ* activity test on a protein extract can lead to different results than a metabolic test performed with the cells (*i.e.* protein is involved under physiological cellular conditions).

Carbohydrates contain an enormous amount of functional information in the glycome of microorganisms. Sugars, phosphorylated sugars, aminosugars, and glycoproteins are functional molecules of both the intracellular and extracellular spaces. EPSs are functional polymers specific to microorganisms. However, access to glycomic information is rather difficult because of the multitude of molecules and polymers involved.

Lipids comprise information on microorganisms and their functions. Phospholipid fatty acids (PLFA) analyses can be used to characterize the composition of cell membranes. The chemical composition spectrum of PLFAs can be used to differentiate between microbial populations and their phenotypes, notably across bacterial and eukaryotic domains.

Organic and inorganic intracellular storage compounds such as PHAs, polysugars, polyphosphate, and intracellular sulphur globules are specific to the metabolism of certain microbial guilds. These compounds can be targeted by microscopy with and without stainings, from phase-contrast to epifluorescence, laser-scanning, and scanning/transmission electron microscopies. Different extraction methods can be used for a quantitative analysis of their monomers using physicochemical separations and detections.

Substrate labelling with heavy isotopes (*e.g.* ^{13}C, ^{15}N, ^{32}P) and radiolabels (*e.g.* ^{14}C) are used in ecophysiology methods to track the assimilation of substrates into cellular macromolecules and storage polymers by microorganisms in communities.

DNA, RNA, proteins, and other structural components of biomass are highly-ordered biopolymers present in either the cytoplasm or the cell membrane. Although widely used, their extractions into analytes representative of the original biological systems are always subjected to biases, making the results barely quantitative. Bearing this in mind, harmonization of molecular workflows will help comparison between results.

2.3.3 Classifications of microorganisms: morphotypes, phenotypes, and genotypes

Different types of taxonomic and functional classifications of microorganisms from morphotypes to phenotypes and genotypes have been attempted across the decades and analytical revolutions. In early eras of microbiology, microorganisms were primarily classified by visual observation via microscopy. Microscopy enabled the development of microbial classifications based on the shape of microorganisms (morphotypes). Although very useful for a rapid identification of undesired filamentous bacteria in activated sludge, morphotypic classifications seldom succeeded in decomposing phylogenies at high resolution. Biochemistry allowed Kluyver and others to classify microorganisms along metabolisms (phenotypes). Again, the classification resolution remained at a higher level. Molecular biology methods based on nucleic acids can differentiate microbial lineages at high resolution. Sequencing is the key to deciphering the nucleotide code of gene fragments, genomes, and metagenomes.

For all classification methods, a good database is needed for comparison of morphotypes, phenotypes, and sequences of nucleic acids and proteins. This is often the bottleneck in the elucidation of rich and diverse microbial communities. A combination of methods is needed for an accurate examination of microbial populations. Microscopy should always be used first to visually inspect microbiological samples. Then, more advanced and costly methods can be used by following a clear analytical strategy.

2.3.3.1 rRNA genes for taxonomic classification at high resolution

Amplification by PCR (see below) and sequencing of amplicons of bacterial 16S or eukaryal 18S rRNA genes propelled the 'Woesean reformation' (after Woese, University of Illinois, USA) for an accurate elucidation of phylotypes. Ribosomal RNA (rRNA) genes are 'universal' to microorganisms. These genes provide code for the production of rRNA that is necessary to link amino acids into peptides and proteins in the ribosomes. Thus, the DNA of every microbe contains rRNA genes, from one to multiple (up to 16) copies. The rRNA genes are composed of stem structures that are highly conserved across microorganisms and of hypervariable loop structures that mutate and evolve faster than stems. The sequences of hypervariable regions of 16S and 18S rRNA genes are used to phylogenetically differentiate between populations. rRNA-based methods drive systematics in phylogenetic trees of life (Figure 2.1).

These analytical methods have benefitted from Sanger sequencing of full rRNA gene amplicons of single populations or clones retrieved from microbial communities. The latest generation platforms allow for a high-throughput sequencing of pools of amplicons. This analytical paradigm shift went from (*i*) long and high-quality sequences (necessary for an accurate classification) but low throughput to (*ii*) high throughput and resolution across microbiomes but short and low-quality sequencing reads. New molecular workflows can target the rRNA *per se* (not the gene), by direct full-length sequencing of the rRNA pool of environmental samples.

2.3.3.2 Taxonomic classification and levels

Taxonomic classifications of microorganisms range from domain/kingdom to phylum, class, order, family, genus, species, and strain. The length and quality of sequences lead to different degrees of resolution in classifications; the longer and higher quality the sequence, the more accurate and resolved the classification. Naming of populations goes with systematics. Clear semantic, writing and formatting rules have been defined internationally. Phylogenies are often revised by international committees. For instance, the class of *Betaproteobacteria* which had existed for a long time has recently been combined within the class of *Gammaproteobacteria*. The genus and specie of a population that has been isolated are now written with full italic Latin terms such as the genus *Zoogloea* and specie *Zoogloea ramigera*. The genus and specie of a non-isolated population are not italicized but accompanied by a candidate term in italic. An example is the candidate genus "*Candidatus* Accumulibacter" and candidate species "*Ca.* Accumulibacter phosphatis"; the isolation of this model PAO has been attempted on different occasions over the last 20 years, without success. The selection conditions involving the alternation of anaerobic/aerobic conditions make it difficult to isolate "*Ca.* Accumulibacter" on agar plates.

Besides taxonomic levels, the terms 'population', 'guild' and 'community' are useful in environmental microbiology vocabulary. A population is a set of microbial cells of the same lineage (*e.g.* populations of *Zoogloea* or "*Ca.* Accumulibacter"). A guild is a set of different microbial populations that share the same function (*e.g.* OHOs, DHOs or ANOs guilds). A microbial community (*e.g.* in activated sludge) contains different populations and guilds.

Functional genes can be targeted to describe and classify microorganisms along their metabolic and cellular potentials. Phylogenetic trees can be constructed based on the information on functional genes that have been amplified by PCR and sequenced. This approach is used to differentiate between clades of microbial populations within a lineage, such as "*Ca.* Accumulibacter" clades using the polyphosphate kinase (ppk1) gene.

Full genome sequencing is used to identify and classify isolates with high resolution, to derive new species and strains, to characterize metabolic potentials, and to predict metabolic pathways.

Full metagenome sequencing followed by genome-centric metagenomics (*i.e.* binning of genomes of single lineages out of metagenomes)

(Albertsen *et al.*, 2013) has propelled a fine-scale elucidation of taxa and their full functional potential. Phylogenetic trees from populations of wastewater environments are generally made with rRNA genes or functional genes, and can now be built by comparison of genomes retrieved from metagenomes against databases of reference genomes (Rubio-Rincón *et al.*, 2019). The obtaining of genomes from populations of a community drives the development of metabolic models to predict their distributed metabolisms and interactions and to reassemble them at microbiome scale.

Proteomics and metaproteomics help to taxonomically and functionally classify microbial populations from isolates to communities based on the identified protein and peptide sequences. It provides an additional hint on the expressed functionality of the identified populations. Protein-based analyses may be considered as a rough proxy for activity.

2.3.4 Culture-dependent vs. culture-Independent methods

The analysis of microbiomes can be made via culture-dependent methods or culture-independent methods. Since the work of Koch, Hesse and colleagues on public-health-related pathogens in the 19th century, agar plates have been used to isolate colonies of microorganisms. Obtaining pure cultures of isolates has long been considered as a gold standard to characterize microorganisms, their physiology, and their metabolisms. However, only a small fraction of microorganisms is actually cultivable in pure cultures.

Enrichment cultures have furthered the field to describe the metabolism of targeted predominant populations. Beijerinck developed early principles of enrichment cultures to design a cultivation medium to the needs of a microorganism. In other terms, this can be formulated as '*Make sure that only organisms with desired properties are able to flourish in the culture.*' (Beijerinck, 1921; Van Niel, 1949; La Rivière, 1997). Combined with the postulate that '*Everything is everywhere but the environment selects.*' (Baas-Becking, 1934; De Wit and Bouvier, 2006), these principles underlie the design of WWTP processes under conditions that select for the guilds of microorganisms involved in C, N and P conversions. Since less than 25% of microbes from activated sludge can be isolated, wet-lab experiments in environmental biotechnology research thrive on enrichment cultures.

Since the Woesean reformation, molecular characterizations of mixed cultures have avoided the need for pure cultures, by extracting and analysing the pool of targeted informational molecules from complex biological samples. They are called culture-independent methods. Molecular biology methods are efficiently complemented with microscopy to visualize and localize microorganisms and with ecophysiology methods to elucidate the actual metabolic traits of these microorganisms in consortia.

2.3.4.1 Analysing taxa and functions: choosing the right method(s)

More than 35 microbial ecology methods are available to analyse taxa and their metabolisms. The key resides in the investigation question underlying the method choice from fundamental or applied microbiology perspectives. Methods involve microscopy, isolation, basic molecular biology, community fingerprinting, advanced sequencing, and ecophysiology techniques. A set of classic and modern methods often used in the field of microbial ecology and water engineering is given in Figure 2.11, followed by a description of methods commonly used in the examination of wastewater treatment systems. More details can be obtained in the sections on 'Microscopy' (Nielsen *et al.*, 2016) and 'Molecular Methods' (Karst *et al.*, 2016) of the IWA open-access eBook and videos of Experimental Methods in Wastewater Treatment [1] (Van Loosdrecht *et al.*, 2016).

[1] https://experimentalmethods.org/

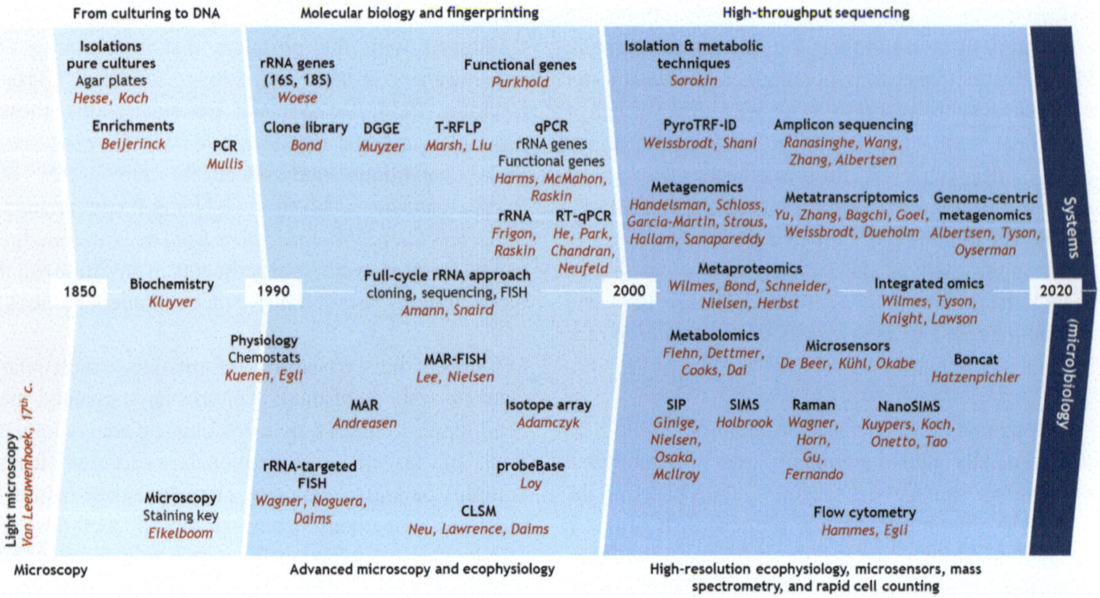

Figure 2.11 Overview of 35+ analytical methods and targets to investigate microbial ecology questions in environmental and wastewater systems. The importance is not only about the method but on the specific research question that drives the method choice. Microbial ecology involves a set of culture-based and culture-independent methods to target phylogenies and functionalities. Microscopy techniques are paramount to visualize what types of microorganisms are present in a biological sample. A simple phase-contrast microscopy measurement should always be conducted first prior to any other analysis. Molecular biology methods are widespread with the use of universal phylogenetic markers (*e.g.* bacterial 16S rRNA gene and eukaryal 18S rRNA gene) and functional genetic markers (*e.g.* targeting functional genes coding for enzymes involved in catabolic or anabolic pathways). Fingerprinting methods that evolved from *e.g.* DGGE and T-RFLP to amplicon sequencing, enable the characterization of microbial community compositions. High-throughput analyses provide high resolution on (meta-) genomes and transcriptomes (via sequencing), and on proteomes and metabolomes (via mass spectrometry). The techniques are displayed in black. Names from representative authors are given in red. Source: based on Nielsen and McMahon, 2014.

2.3.5 Microscopy, isolation, and counting methods

2.3.5.1 Visual inspection by microscopy techniques

The examination of a biological sample and underlying microorganisms should always start by visual inspection from an overall description at the bench prior to light or phase-contrast microscopy (Figure 2.12) with and without sample homogenization. Microscopy provides a first hint on the shape of bioaggregates and the predominant morphotypes of microorganisms. It differentiates between *e.g.* prokaryotes and eukaryotes, specific morphotypes such as filamentous bacteria, chemotrophs and phototrophs (pigmentation), and microorganisms with storage polymers. The application of stainings and fluorophores that target macromolecules in the samples can reveal specific features under light or fluorescence microscopy, depending on the nature of the stains used. This can be helpful *e.g.* to detect special inclusions in the cells such as storage polymers and lipids.

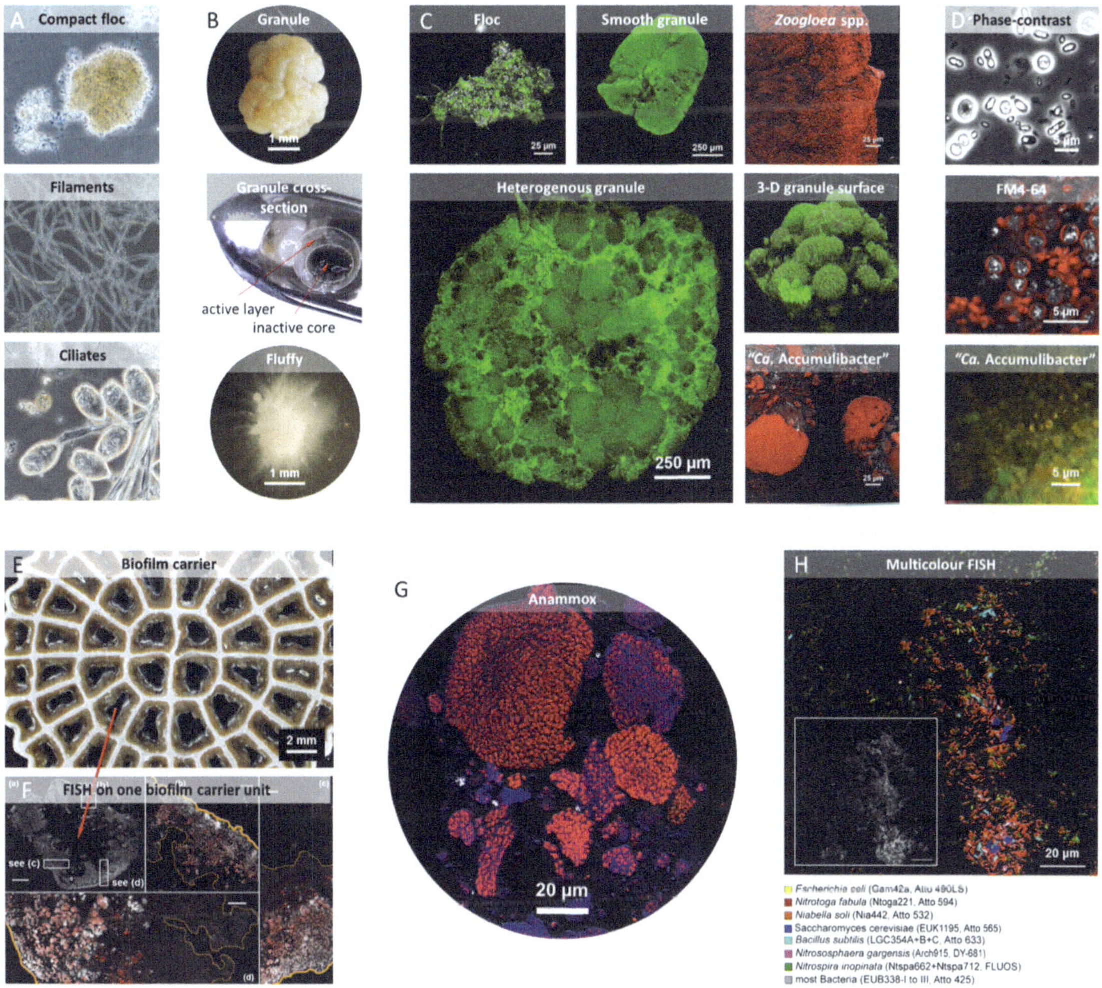

Figure 2.12 Visual inspection of biological samples from wastewater environments. A) Light microscopy observations of compact activated sludge flocs, of filamentous bulking, and of a colony of protozoa (images: D. Brdjanovic and Eikelboom (2000); scale bars not available); b) light microscopy observation of a dense mature BNR granule with cauliflower-like biofilm structure (B, top), of a granule cross section with active white outer sphere and inactive internal black core (B, middle), and of a slow-settling fluffy granules (B, bottom; aggregate is 1 mm diameter). Images by Weissbrodt (2018) (B, top), Weissbrodt, Lochmatter and Gonzalez-Gil (B, middle), and Weissbrodt *et al.*, 2012a (B, bottom); C) CLSM examinations of a floc, a smooth granule with homogenously dispersed cells dominated by fast-growing Zoogloea spp., and a heterogeneous granule of dense microcolonies dominated by the slower-growing PAO "*Ca.* Accumulibacter". Images: Weissbrodt *et al.* (2013); D) resolution at cell level with phase-contrast microscopy of PAO and GAO cells with intracellular storage products (image: N. Gubser), with CLSM and FM4-64 staining to label membrane lipids of cells filled in with inorganic storage products, *i.e.* polyphosphate in PAOs (image: Weissbrodt *et al.*, 2013), and with FISH and epifluorescent microscopy to detect "*Ca.* Accumulibacter"-like PAOs (image: L. Bittencourt Guimarães); E) light microscopy observation of a biofilm carrier sampled from a reactor operated for nitrogen removal from concentrated side stream by partial nitritation and anammox (image: Weissbrodt *et al.*, 2020). FISH-CLSM analysis to localize anammox populations; F) in a cryosection of one space unit of a biofilm carrier (scale bars of sub-images: (a) 500 µm, (b-d) 50 µm; image: Laureni *et al.*, 2019); G) in disrupted biofilm (image: Wells *et al.*, 2017); H) multicolour FISH and CLSM examination of a defined mixture of seven archaeal, bacterial, and yeast isolates (image: M. Lukumbuzya *et al.*, 2019).

2.3.5.2 Isolations and pure cultures

Microbiologists have been trying to enrich, isolate, cultivate, and characterize microorganisms of interest in pure culture collections for many years. Obtaining a pure culture of the targeted organism via a streak-plate on agar petri dishes, agar deep tubes, or serial dilution techniques followed by pour-plating or spread-plating is useful for accurately describing morphotype(s), phenotypes and genotypes. These techniques involve an agar matrix composed of all the nutrients necessary to sustain the growth of microcolonies of populations of interest. This has been the method of choice to derive new species and strains. These colonies of isolates can be picked and cultivated in liquid pure cultures for *e.g.* physiological and biochemical testing.

However, isolations of environmental microorganisms remain challenging; only a low fraction of microorganisms can be cultivated with these methods. Culturabilities range from max. 0.001% from seawater to 0.30% in soil and 25% in activated sludge samples. However, growth regimes in nature are much more complex than batch regimes applied in agar plates of liquid cultures. Microorganisms are permanently subjected to gradients and variations in environmental conditions. This limits the efficiency of selecting and isolating batch cultures in the lab. Medium compositions are also different from the aqueous compositions of the natural living media. The richness of C-containing media can also impact the diversity of microorganisms selected. Some organisms are auxotrophs and require external supply of essential resources either from a well-defined medium or from the activity of other organisms.

2.3.5.3 Advanced microscopy and rapid cell counting

Hybridization of the rRNA (*per se*) can help to detect, identify, semi-quantify, and locate microbial populations in a biological specimen (Figure 2.12). Bacterial 16S rRNA or eukaryal 18S rRNA targeted fluorescence *in-situ* hybridization (FISH) (Nielsen *et al.*, 2009) is a molecular biology method that consists of hybridizing the 16S or 18S rRNA of microbial populations using an oligonucleotide probe (*i.e.* a short strand of nucleic acids of typically *ca.* 20 nt) mounted with a fluorophore. Oligonucleotide FISH probes can be designed to tune the hybridization specificity across phylogenies. It can be made relatively unspecific to *e.g.* class level or very specific at genus level. A probe design is made by aligning rRNA sequences from known microorganisms targets, and finding a consensual sequence. Probebase (Greuter *et al.*, 2015) is a database of oligonucleotide FISH probes[2]. With FISH, known populations or guilds of microorganisms are targeted. It is important to know in advance what populations can be present. Fingerprinting the microbial community at the outset using amplicon sequencing (Section 2.3.7) can be helpful to identify the predominant populations and their relative abundances prior to choosing or designing a new FISH probe.

After sampling, microbial cells are fixed in the dark in 4% paraformaldehyde (PFA) followed by ethanol:phosphate-buffer-saline (EtOH:PBS 1:1) for Gram– organisms or directly in EtOH:PBS 1:1 for Gram+ organisms otherwise PFA would rigidify the membranes of Gram+ organisms making them impermeable to FISH probes. Cells are deposited on a lysine-coated microscopy slide and made permeable at a specific chemical stringency with formamide and in combination with heat. The fluorescent oligonucleotide FISH probe is dropped onto the specimen. It is transferred intracellularly and binds to the complementary strand of rRNA if it is present in the specimen. Cells hybridized with the FISH probe emit a fluorescent signal under the epifluorescence microscope (EFM), after removing and washing out the unbound markers. FISH can provide semi-quantitative (qFISH) information on the relative abundances of microbial populations. With dispersed cells, this is done by comparing the number of florescent hybridized cells with the total number of cells present in the sample (either unstained under

[2] http://probebase.csb.univie.ac.at/

phase contrast, or stained with another fluorophore or fluorescent FISH probe). With biofilms and granule cross sections, the semi-quantitation can be computed by the ratio of (un)coloured areas.

However, as with any molecular method, FISH can contain biases. Cell fixation, cell binding to the microscopy slide, cell permeability, probe size and intracellular transfer rate, hybridization efficiency, fluorophores and fluorescent signal intensity, and human factors can all impact on the analytical result. Analysts often introduce bias by primarily looking at where there is fluorescence on the specimen. Thus, in general, relative abundances calculated with FISH are over-represented. The 'dark information' of the specimen is part of the scientific dataset and should also be taken into account. Therefore, accuracy and confidence in FISH-EFM measurements can be enhanced by recording a randomized digital sampling of 10-30 spots on the specimen. However, flocs, granules, and biofilms can be challenging to de-aggregate, and particles of different biovolumes can cause issues in measurements.

Flow cytometry (FCM) is an elegant and fast capillary method to detect, count, and measure the metrics of microbial cells in a biological sample. However, the sample should be de-aggregated into a suspension of single cells (*e.g.* using a Potter-Elvejehm homogenizer) for an accurate and representative analysis. In FCM, single cells pass one by one through a capillary prior to laser detection. Cells can be detected based on size, shape, cellular constituents (*e.g.* low or high level of nucleic acid content), and stainings or FISH probings applied. It is an interesting alternative to EFM, provided that the sample can be homogenized, which is often the drawback with flocs, granules and biofilms. Nevertheless, FCM is very efficient at analysing the biological stability of low-concentration samples such as river water, groundwater, source water and drinking water. In the drinking water field, FCM is recognized by regulatory offices as a standard alternative to heterotrophic cell counting (Hammes and Egli, 2010).

Confocal laser-scanning microscopy (CLSM) examines biological specimens at high resolution. This advanced microscopy technique zooms in to record a full floc structure of *ca.* 50-100 μm and to dig in the architecture of biofilm and granule cross sections. The specimen can be prepared by physical sectioning of a biofilm or granule sample with a cryotome. The CLSM examination involves optical sectioning of the specimen by penetration of the laser at different depths (max. 100 μm). The 3D structure can be re-composed by a z-stack of the digital images. CLSM can be combined with different fluorescent stainings to visualize biological features and FISH probes to localize populations of interest in floc or biofilm structures. For FISH, CLSM can help to remediate issues in semi-quantitation when working with bioaggregates by enabling calculations of biovolumes.

Scanning (SEM) and transmission (TEM) electron microscopies provide high-resolution and detailed analysis of cellular structures and components of microbial cells. Their combination with energy-dispersive X-ray (EDX) spectroscopy can be used to detect elemental compositions in the specimen. Different types of fixations are needed for SEM, TEM, and environmental SEM (ESEM). Besides biochemical phenotypes, electron microscopy (EM) analysis is required to describe a new species.

2.3.6 Molecular biology and fingerprinting methods

2.3.6.1 Extraction of nucleic acids and proteins

One important step that affects the analytical outputs of molecular biology measurements relates to the extraction of nucleic acids and proteins, prior to their analysis. Nowadays extractions are mostly performed using manufactured kits. A diversity of kits and principles can be used to extract, isolate, and purify these macromolecules. The final result of the molecular analysis is impacted by the extraction method used initially (Albertsen *et al.*, 2015; Rocha and Manaia, 2020). For reproducibility and comparison purposes, it is preferable to process all the samples from the same series with the same kits and methods. A comparison of kit performances can be performed at the beginning when starting a new project. Outputs of molecular methods that do not use the same workflows and from different groups and papers can seldom be compared. Efforts are needed for more systematics in molecular biology via *e.g.* round robin tests (or ring tests). The MIDAS database and field guide to the microbes of activated sludge and anaerobic digesters[3] provides baseline information for sampling, wet-lab processing, and dry-lab analysis of microbial community compositions (McIlroy *et al.*, 2015; Nierychlo *et al.*, 2019). However, adaptations to protocols are always possible, and need duly reporting.

2.3.6.2 Polymerase chain reactions

A polymerase chain reaction (PCR) is a molecular biology technique that is used to amplify the nucleotide sequence of a fragment of a gene present on the DNA of one or multiple organisms. PCR can be performed to detect the presence or absence of a gene and of an organism carrying this gene in a microbial community. PCR can target 16S/18S rRNA genes or functional genes. The oligonucleotide sequences of the forward and reverse primers are designed accordingly.

PCR uses a set of two oligonucleotide primers, known as forward and reverse primers, of *ca.* 20 nucleotide base pairs (bp) each to target the specific double-strand DNA region of interest on the selected gene. A high-temperature-resistant polymerase enzyme is used to amplify this delineated region by adding nucleotides one-by-one to the sequence chain of the PCR product (known as the amplicon). Amplification works via sequences of temperature cycles and replication reactions performed in a thermocycler. PCR programs involve initial denaturation of the DNA template to open the double strand (*e.g.* 5-10 min at 95 °C) followed by *ca.* 30 cycles of (*i*) denaturation (*e.g.* 1 min at 95 °C), (*ii*) annealing of the two forward and reverse primers on the two strands (*e.g.* 45 s at 50-60 °C) and (*iii*) elongation of the amplicon sequence by the polymerase (*e.g.* 2 min at 70-75 °C), followed by a final elongation of the amplicons (*e.g.* 10 min at 70-75 °C). The parameter values depend on the genes and organisms targeted, and can be found in the methods section in articles. The chemical compositions of the reaction mixtures, temperatures and phase lengths of the PCR are very specific to the nucleotide composition of the genes of interest. A gene fragment with a higher content of G-C nucleotides bonds (known as the GC content) would need more intense denaturation to open them.

Depending on the type of subsequent fingerprinting or sequencing analysis, a PCR is conducted with labelled primers. PCR products are checked on an agarose gel. A band characteristic of the length of the targeted gene sequence should be detected. If the PCR was not specific, a mixture of bands of different length is obtained, necessitating that the primers and PCR reaction conditions are optimized further. When a band is lacking, this indicates that the PCR did not succeed (either the PCR is not specific or the DNA of the targeted organism is absent from the sample pool).

Quantitative (real-time) PCR (*i.e.* qPCR) provides additional power to semi-quantify the relative abundance of this gene and/or microorganism. Fluorescence is emitted every time that a PCR product is formed. From cycle to cycle, an exponential production of amplicons is detected by the fluorescence detector. However, as with any molecular study, although called 'quantitative PCR', qPCR outputs from different studies are affected by biases all across the workflows which makes them very difficult to compare (Rocha and Manaia, 2020).

Reverse transcription (RT) and qPCR (referred to as RT-qPCR) targets the RNA. The pool of RNA is extracted from the biological sample. mRNA sequences can be reverse-transcribed into complementary DNA (cDNA) prior to qPCR. This

[3] https://www.midasfieldguide.org/

provides hints on the targeted gene expression and transcriptional regulation in cells as a first proxy of microbial 'activity'.

2.3.6.3 Clone libraries

Before the advent of high-throughput sequencing, the composition of amplicons obtained by PCR was obtained by generating a clone library and sequencing the gene fragments present in each clone. In this method, after purification, the pool of PCR products is contacted to cells of *Escherichia coli*. Each fragment is ligated to a plasmid vector and transformed into one *E. coli* cell by heat shock. The pool of transformed *E. coli* cells is incubated overnight at 37 °C in a liquid culture for multiplication of the cells. The cell mixture is streak plated on agar plates prior to a second incubation in order to grow colonies of transformants. Around 100-500 colonies of clones that each contain one type of gene fragment are picked up one by one for PCR and Sanger sequencing of PCR products (*i.e.* amplicons). The number of reads of a specific sequence relates to the relative abundance of gene fragments present in the pool of amplicons. Cloning sequencing is used to identify the predominant microbial populations in the process of interest. It also provides long, high-quality sequences of the gene fragments.

2.3.6.4 Community fingerprinting

The relative abundance of gene fragments obtained after PCR from the pool of genomic DNA (gDNA) of the microbial community can be profiled by *e.g.* denaturing gradient gel electrophoresis (DGGE), automated ribosomal intergenic spacer analysis (ARISA), amplified ribosomal DNA restriction analysis (ARDRA), or terminal-restriction fragment length polymorphism (T-RFLP). A PCR is performed on the 16S or 18S rRNA genes at the outset if the objective is to profile the microbial populations. These techniques can also be used to address the diversity of a pool of fragments of a functional gene across the community. These classic methods have been widely used since the nineties in order to obtain the fingerprint of microbial communities. The output of these methods is primarily a set of bands of lengths relating to the gene fragments amplified and/or restricted. The diversity of bands gives a proxy of the diversity of operational taxonomic units (OTUs) present in a microbial community. Taxonomic classification of the gene fragments can be obtained by cloning-sequencing or combination with amplicon sequencing. In the technological transition from classic fingerprinting to pyrosequencing, the PyroTRF-ID pipeline was developed to relate terminal restriction fragments to phylotypes by comparison of T-RFLP and amplicon sequencing profiles (Weissbrodt *et al.*, 2012b). The profiles obtained by these different techniques are represented as relative abundances of OTUs. They can be described in terms of richness (*i.e.* number of different OTUs present in the profile) and diversity (*i.e.* number and dispersion of relative abundances of OTUs in the profile). Fingerprinting methods are efficient to rapidly check in *ca.* 1-2 days *e.g.* the enrichment level of targeted OTUs in a bioreactor or to detect a drastic switch in OTU predominance in the event of a process disturbance.

2.3.6.5 Modern amplicon sequencing

New developments in sequencing have enabled the pool of amplicons to be directly sequenced. Amplicon sequencing drastically improves the depth of libraries of OTUs or amplicon sequence variants (ASVs) obtained. It enables an efficient characterization of microbial community compositions by obtaining the identities of microorganisms and their relative abundances.

Amplicon sequencing consists of amplifying by PCR a targeted gene fragment from the gDNA extracted from a biological sample and sequencing the pool of amplicons obtained. This method is typically applied to bacterial 16S and eukaryal 18S rRNA gene amplicon sequencings. Sample-tagged primer sets are used to amplify hypervariable regions of these 'universal' genes from the pool of gDNA obtained from the microbial community. The tagged amplicons (*ca.* 300-600 bp) are purified, quantified, pooled, and sequenced with high throughput with new generation benchtop platforms such as MiSeq (Illumina), Ion Torrent, PacBio, Oxford Nanopore, among others.

A

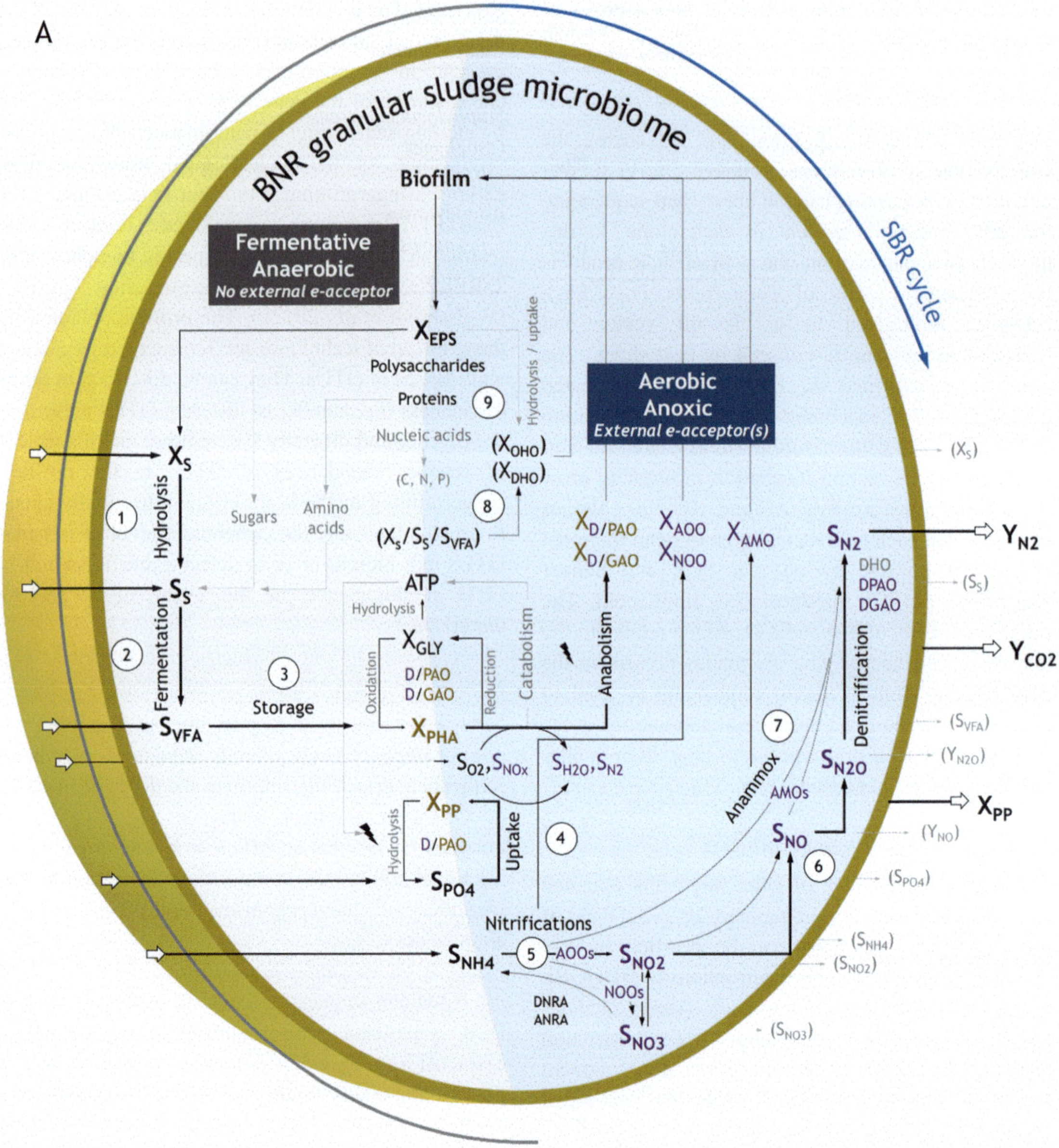
BNR granular sludge microbiome
SBR cycle
Biofilm
Fermentative
Anaerobic
No external e-acceptor
Aerobic
Anoxic
External e-acceptor(s)
X_EPS
Polysaccharides
Proteins
Nucleic acids
(C, N, P)
Hydrolysis / uptake
(X_OHO)
(X_DHO)
(X_S/S_S/S_VFA)
X_S
S_S
S_VFA
Hydrolysis
Fermentation
Sugars
Amino acids
Storage
ATP
Hydrolysis
X_GLY
D/PAO
D/GAO
Oxidation
Reduction
Catabolism
X_PHA
S_O2, S_NOx
S_H2O, S_N2
X_PP
D/PAO
S_PO4
Hydrolysis
Uptake
Anabolism
X_D/PAO
X_D/GAO
X_AOO
X_NOO
X_AMO
Anammox
AMOs
Nitrifications
S_NH4
AOOs
S_NO2
NOOs
S_NO3
DNRA
ANRA
S_NO
S_N2O
S_N2
DHO
DPAO
DGAO
Denitrification
Y_N2
Y_CO2
X_PP
(X_S)
(S_S)
(S_VFA)
(Y_N2O)
(Y_NO)
(S_PO4)
(S_NH4)
(S_NO2)
(S_NO3)
1
2
3
4
5
6
7
8
9

B Microbial guilds and lineages

1 Hydrolysers
(aerobic or anoxic)
Cytophaga
Chloroflexi
Saccharibacteria (TM7)
Sphingobacteriales
Tetrasphaera

2 Fermenting organisms
Tetrasphaera

3 PAO
Rhododcyclaceae
"*Ca*. Accumulibacter"
Dechloromonas
Tetrasphaera

GAO
Xanthomonadaceae
"*Ca*. Competibacter"
Rhodospirillaceae
Defluviicoccus

4 PAO
Rhododcyclaceae
"*Ca*. Accumulibacter"
Dechloromonas
Tetrasphaera

5 AOB
Nitrosomonas

NOB
Nitrospira
Nitrococcus

6 DPAO
Rhodocyclaceae
"*Ca*. Accumulibacter"
Dechloromonas

DGAO
Xanthomonadaceae
"*Ca*. Competibacter"
Rhodospirillaceae
Defluviicoccus

DHO*
Rhodocyclaceae
Zoogloea
Thauera
Methyloversatilis
Xanthomonadaceae
"*Ca*. Competibacter"
Thermomonas
Pseudoxanthomonas
Devosia
Tetrasphaera
Rhizobiales
Aminobacter
Bradyrhizobium
Comamonadaceae
Acidovorax
Sphingomonadaceae
Hyphomicrobiaceae
Rhodobacteraceae

7 AMO
Brocadiaceae
"*Ca*. Brocadia"
"*Ca*. Kuenenia"

8 Filamentous organisms
Sphaerotilus
Leptothrix

Zoogloeal organisms
Zoogloea
Thauera
Rhizobiales

9 Hydrolysers
(aerobic or anoxic)
Cytophaga
Chloroflexi
Saccharibacteria (TM7)
Sphingobacteriales
Rhodospirillaceae
Tetrasphaera

Figure 2.13 The Weissbrodt conceptual model of the BNR granular sludge ecosystem highlighting the main C, P and N conversions (A, left) together with the microbial guild involved and their underlying bacterial populations (B, right). High resolution on the detection of populations present in the microbiome is revealed by 16S rRNA gene amplicon sequencing. Functional traits can be attributed to lineages according to literature reports. The conceptual ecosystem model forms a strong basis to target further ecophysiological analyses. A granular sludge process is efficiently operated for full BNR in a sequencing batch reactor (SBR) that alternates sequences of an anaerobic C-feast phase and an aerobic or anoxic starvation phase in the presence of dissolved oxygen (O_2) and/or oxidized nitrogens (NO_x^-) as terminal e-acceptors. The biofilm matrix of granules induces diffusional limitations, resulting in gradients of substrate and redox conditions in the depth, enabling simultaneous nitrification and denitrification processes. Under *anaerobic conditions*, the particulate organic matter (X_S) is hydrolysed into a soluble substrate (S_S) and fermented into volatile fatty acids (S_{VFA}) by hydrolysing and fermenting organisms prior to intracellular storage of VFAs as poly-β-hydroxyalkanoates (X_{PHA}) by polyphosphate-accumulating (PAOs) and glycogen-accumulating (GAOs) organisms. PAOs and GAOs involve the degradation of glycogen (X_{GLY}) as the source of ATP and NADH for the polymerisation of PHAs. PAOs hydrolyse polyphosphate (X_{PP}) as an extra source of ATP for the active intracellular transport of acetate, resulting in the release of orthophosphates in the medium (S_{PO4}). Under aerobic and/or anoxic conditions, PAOs and GAOs grow on the intracellular pool of carbon and electrons of PHAs. Both guilds replenish their stocks of glycogen, and PAOs their stock of polyphosphate by removing orthophosphates from the wastewater. PAOs and GAOs can grow by coupling their catabolism to denitrification, becoming denitrifying PAOs (DPAOs) and GAOs (DGAOs). Oxidized nitrogens (S_{NOx} like S_{NO2} or S_{NO3}) are supplied by the oxidation of ammonium by nitrifiers via two guilds of aerobic ammonium-oxidizing (AOOs) and nitrite-oxidizing (NOOs) organisms. Some comammox populations of the genus *Nitrospira* in the guild of NOOs can fully oxidize ammonium (S_{NH4}) into nitrate (S_{NO3}). The guild of denitrifying heterotrophic organisms (DHOs) is surprisingly vast in BNR granular sludge. One puzzling question remains: what carbon and electron source do all these populations thrive on if not storing VFAs in the anaerobic phase? Anammox populations can also establish their niche in the vicinity of granules. Figure: based on Weissbrodt *et al.* (2014) and Winkler *et al.* (2018).

Datasets of 10k-100k read counts are typically obtained per sample depending on the required sequencing depth. A set of 10 k-30 k reads is sufficient when aiming at profiles of predominant populations. Increasing the sequencing depth per sample by pooling fewer samples on the same sequencing flow cell provides higher resolution to describe microbial diversity with low-abundance populations. Datasets are processed using bioinformatics routines and visualized and analysed using numerical ecology (Weissbrodt *et al.*, 2014; Albertsen *et al.*, 2015), using *e.g.* packages with the R software. Microbial community profiles are represented as stacked bar charts or heatmaps to reveal their compositions and the dynamics of the populations involved. With short reads of 300-600 bp, one can obtain taxonomic classifications from the phylum to class, order, family, and genus levels. With the latest developments of using sequencers with long-read outputs, one can now achieve sequencing of the full-length rRNAs (*per se*) of populations present in a microbial community. This considerably increases the accuracy and resolution of taxonomic classification of phylotypes beyond the genus to species levels.

The resolution obtained on microbial lineages and microbial community compositions delivers useful information to build conceptual models of an activated sludge ecosystem. Figure 2.13 exemplifies a conceptual ecosystem model built for BNR granular sludge based on the detailed phylogenetic information obtained by amplicon sequencing and the functional features retrieved from literature for all the predominant microorganisms detected.

2.3.7 High-throughput 'omic' methods

Microbiomes can be characterized at high resolution and throughput by analysis of informational macromolecules. Sequencing and mass spectrometry measure the diversity of these materials across a microbial community. Integrated 'meta-omic' techniques (Narayanasamy *et al.*, 2015) can target multiple levels from DNA to RNA, proteins and metabolites in the regulation of metabolisms across the complex microbial networks of microbial communities (Figure 2.7).

At DNA level, sequencing the metagenome of a sludge generates the full genetic information across all the microorganisms present in the microbial community. System microbiology aims further at obtaining single-lineage genomic information beyond metagenomes. Genome-centric metagenomics is the art of binning the genomes of single microbial populations out of the sludge metagenome. The genome of an organism is the essential unit that gives the full genetic potential of the microorganism of interest. Metagenome-assembled genomes (MAGs) are very important references prior to any subsequent RNA sequencing or protein mass spectrometry. They are also useful to map metabolic pathways prior to analyses of metabolites and their fluxes during microbial conversions. An example of a MAG is given in Figure 2.14 for "*Candidatus* Accumulibacter delftensis" that was retrieved from the metagenome of an EBPR sludge.

At RNA level, metatranscriptomics aims at sequencing the pool of mRNA that has been expressed by the microbial community under a set of operational conditions. Metatranscriptomes can be mapped against MAGs to highlight which genetic functions of a (set of) microorganism(s) have been transcribed from DNA to RNA. This provides information about the transcriptional regulation of these organisms.

At protein level, metaproteomics aims at sequencing the amino acid sequence of proteins and peptides present in the cells of the microbial community. The processing of a metaproteome makes efficient use of MAGs to identify proteins and their affiliation to key microorganisms. The activation of proteins as enzymes should be demonstrated for more insight into post-translational activations of conversions in microorganisms; detection of phosphorylated proteins can be of assistance in this.

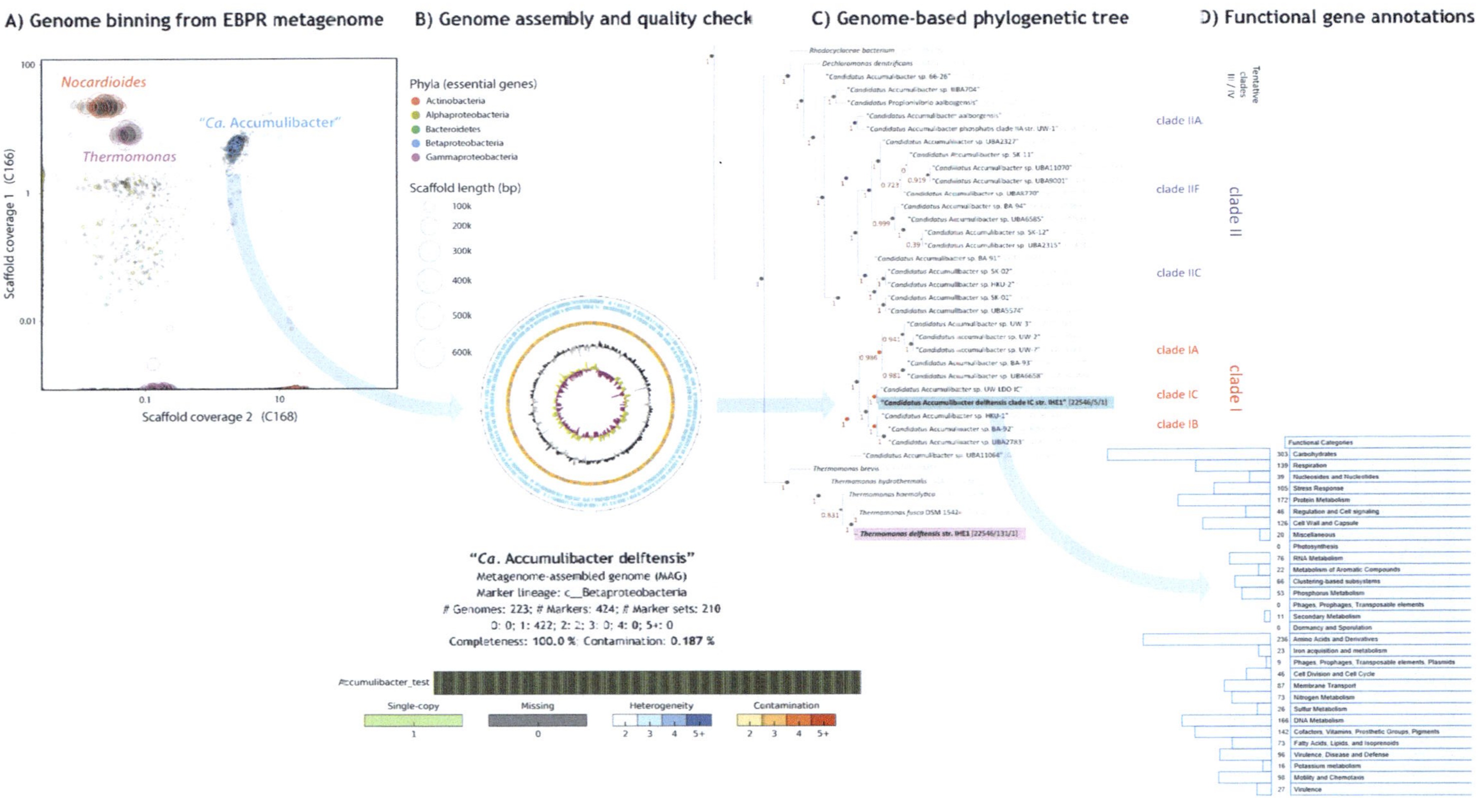

Figure 2.14 Genome-centric metagenomic analysis of "*Candidatus* Accumulibacter delftensis": A) near-complete single-lineage genomes were recovered from three different microbial populations affiliating with the "*Ca.* Accumulibacter", *Thermomonas*, and *Nocardiodes* genera by differential coverage binning from the metagenome of an EPR sludge; B) the metagenome assembled genome (MAG) of "*Ca.* Accumulibacter" displayed high levels of completeness and low contamination; C) a genome-based phylogenetic tree was constructed by mapping the "*Ca.* Accumulibacter" genome of interest with other MAGs available in public databases. It affiliated with the "*Ca.* Accumulibacter" clade IC. It was significantly different in composition to other MAGs and was name "*Ca.* Accumulibacter delftensis"; D) functional annotations of the genetic coding regions of the MAG. Source: modified from Rubio-Rincón *et al.* (2019).

The activation of a metabolic pathway leads to the production of metabolites across the network of biochemical reactions catalysed by the enzymes. The pool of extracellular and intracellular metabolites can be analysed by tandem liquid chromatography and mass spectrometry at different time points of a bioreactor subjected to a disturbance. The fluxes of metabolites can further be quantified as rates. Metabolomics and fluxomics analyses track the concentrations of metabolites and their fluxes during regulation. These analyses are often performed using *e.g.* ^{13}C or ^{15}N-labelled substrates to track transient changes in metabolic states in the microbial community in a bench-scale reactor. Metabolic flux analyses are combined with mathematical modelling in order to develop quantitative predictions of metabolic performances, from single microorganisms to networks of microbial populations in a microbial community.

Multi-omic methods can be extended with glycomics and lipidomics to measure the highly complex diversity of carbohydrate and lipidic functional molecules in microbiomes. Analyses of mobilomes can detect the set of mobile genetic elements and antimicrobial resistances that are horizontally transferred in a community. Analyses of viromes measure viruses and bacteriophages and their impacts in sludge.

2.3.8 Ecophysiology methods

Ecophysiology methods are required to examine and validate the metabolic functionalities highlighted as potentials using DNA, RNA, and protein-based molecular approaches. They make use of labelled substrates to track their conversion and assimilation into cells (Musat *et al.*, 2012; Nielsen *et al.*, 2016).

Stable isotope probing (SIP) consists of tracing the assimilation of a substrate labelled with non-radioactive stable heavy isotopes (*e.g.* ^{13}C *vs.* ^{12}C, ^{15}N *vs.* ^{14}N) into macromolecules such as DNA, RNA, proteins, carbohydrates or lipids. After incubations of sludge samples with labelled substrates, the pool of macromolecules of interest is extracted. The heavy fractions of the DNA or RNA pools are separated by ultracentrifugation prior to molecular analysis with one of the methods described above. Amplicon sequencing will lead to identifying the microorganisms that have integrated the heavy isotopes in their nucleic acids. Mass spectrometry analysis will help to identify proteins that have incorporated the heavy elements and their taxonomic affiliation.

FISH (Section 2.5.3) can be combined with microautoradiography (MAR), Raman spectroscopy, or secondary ion mass spectrometry (SIMS) to localize microbial populations and simultaneously identify which cells have actively assimilated an *e.g.* ^{14}C, ^{3}H, $^{32/33}P$, or ^{35}S radio-labelled compound (MAR, Raman, SIMS) or a stable isotope compound (Raman, SIMS). FISH-MAR is used to elucidate whether anammox populations could store organics under mainstream conditions (Laureni *et al.*, 2015). Nanoscale SIMS (NanoSIMS) provides nanoscale resolution to detect the metabolic state of single cells in a sample. These techniques apply to biomass samples that are rapidly chemically fixed after sampling to fix the metabolic state of the cells at the sampling time points. NanoSIMS is used to track carbon conversion in an anammox consortium (Tao *et al.*, 2019). However, these techniques relate to exponentially increasing costs. Their implementation requires the formulation of research questions on microbial metabolisms at the outset.

2.3.9 From microbial ecology analyses to microbial community engineering

From an engineering point of view, microbiological and molecular biology methods form a toolbox toward an essential list of microbial information for environmental bioprocess design, monitoring, and control. The underlying questions of system microbiology aim at unravelling the following: *Which populations are present/active? What are they doing? Where are they doing it? Which conditions trigger selection and activation? What are the consequences of a process disturbance on microbial selection and metabolisms? What is the feed-back impact of the population and metabolic shifts in the community on the process performance?*

If aiming at translation into the engineering context, the big datasets generated by omics should move from pure description into the development of theoretical concepts that can sustain the development of microbial community engineering strategies. WWTPs are not designed based on sequencing but primarily on stoichiometry and kinetics of microbial growth. Microbial ecology sets out the principles to engineer microbial communities. The thermodynamics of microbial growth sets the boundaries and spontaneity of microbial conversions. Biochemistry tells whether a microorganism harbours the necessary enzymatic machinery to perform the conversion. Stoichiometry gives information about the ratios of nutrients, e-donors and e-acceptors needed to sustain growth of the microbial guilds of interest. It relates to operating expenditures (OPEX), for instance the amount of O_2 that should be transferred into the liquid phase of an aeration tank. Kinetics provides information about how fast a microbial conversion will be. It relates to capital expenditure (CAPEX), such as the volumes of activated sludge tanks, *i.e.* the sizing of the plant and amount of concrete and steel needed to build the installation.

In addition to knowing about the microorganisms and metabolisms targeted (either at the outset or revealed after analytical measurements), developing a rational mathematical description of their growth is essential for understanding, prediction, and design.

2.4 MICROBIAL GROWTH BASICS

2.4.1 Microbial growth

Microbial metabolism comprises all the biochemical reactions involved in the growth and maintenance of living cells. Microbial growth is driven by thermodynamically favourable, energy-yielding redox reactions (catabolism) which are bioenergetically coupled to the synthesis of biomass (anabolism) and growth-independent, yet indispensable, processes (termed maintenance) (Figure 2.15) (Heijnen and Van Dijken, 1992; Heijnen, 1994). In the catabolic reaction, energy is generated via the transfer of electrons from an e-donor to an e-acceptor. In the anabolic reaction, this energy is used for the synthesis of cellular components from a C source, an N source, and other nutrients.

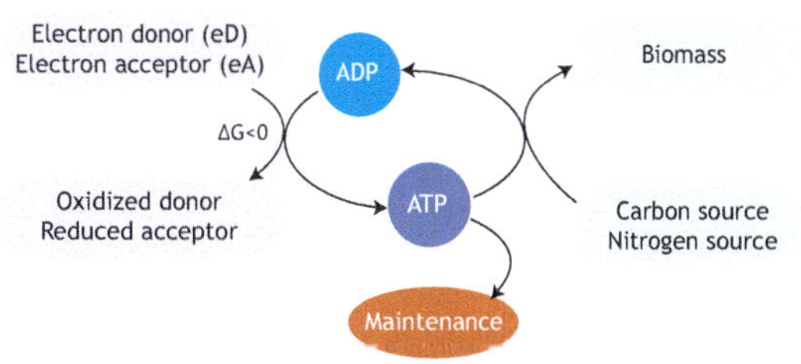

Figure 2.15 Bioenergetic coupling of the energy-yielding catabolic process with anabolic biomass synthesis and growth-independent maintenance processes. The chemical energy released in the catabolism is transferred to the energy-rich compound adenosine triphosphate (ATP). Source: adapted from Heijnen and Kleerebezem (2010).

The C sources for biosynthesis are either organic or inorganic and the anabolic processes are highly conserved (*i.e.* similar) in all bacteria. The energy sources are of three types: organic, inorganic and light. The catabolic processes can vary widely between different microbial groups. It is the catabolism, and in particular the variety of possible combinations of e-donors and e-acceptors for energy generation, that shapes the immense microbial diversity found in nature. The specific source of energy, electrons and carbon are used for the main classification of microorganisms (Figure 2.8). Microorganisms can be further classified based on the e-acceptor (*e.g.* aerobic for O_2) or metabolic product (*e.g.* acetogenic for acetate producers).

The general stoichiometry of any microbial growth reaction can be written in the form of an equation as follows in Eq. 2.1.

$$-Y_{eD/X}\cdot\text{e-donor}-Y_{C_S/X}\cdot\text{C-source}-Y_{N_S/X}\cdot\text{N-source}$$
$$-Y_{eA/X}\cdot\text{e-acceptor}+1\cdot\text{biomass}\pm Y_{H_2O/X}\cdot H_2O$$
$$\pm Y_{HCO_3^-/X}\cdot HCO_3^-\pm Y_{H^+/X}\cdot H^+ + \ldots = 0 \quad (2.1)$$

where the stoichiometric coefficients ($Y_{i/X}$) are expressed per mole of biomass produced (mol_i/mol_X).

The values of all stoichiometric coefficients are positive, and the sign preceding them derives from elemental and charge mass balances (Section 2.4.3). It is worth noting that, while we chose to define $Y_{X/X}$ as 1 mol_X/mol_X, any stoichiometric coefficient can be set to unity, with the other yield coefficients being expressed relative to this component. In Eq. 2.1 all the yield coefficients can be derived from elemental balances (C, H, O and N balances) once one of the yield coefficients is known. Traditionally this is the substrate needed to produce a unit of biomass. When organic C sources are used, the e-donor and C source generally coincide (*e.g.* chemoorganoheterotrophic metabolism).

The following sub-sections present the methodology to derive the anabolic and catabolic reactions, and how to combine them based on one measured or estimated stoichiometric coefficient to obtain the overall metabolic equation. First, the main pathways of energy generation are summarized, and the methodology to derive the stoichiometry of any redox reaction and calculate the generated energy is presented.

2.4.2 Bacterial bioenergetics

The two main pathways of biological energy generation are glycolysis and the TCA cycle (Section 2.2.3). Glycolysis partially oxidizes glucose (a sugar) to pyruvate and to acetyl-CoA. The acetyl-CoA is then fully oxidised to CO_2 in the TCA cycle (Figure 2.16). The chemical energy released in these conversions is transferred to the energy-rich compound adenosine triphosphate (ATP) and electrons are transferred to the oxidized form of the coenzyme nicotinamide dinucleotide (NAD^+) that becomes reduced to NADH.

In the presence of an e-acceptor, such as O_2 or NO_x^- such as nitrate (NO_3^-) or nitrite (NO_2^-), the NADH is regenerated back to NAD^+ by a transfer of the electrons via the electron transport chain to the e-acceptor. In parallel to this electron transport process, protons are transported across the cell membrane to the outside of the cell. The resulting pH and charge gradient create a proton motive force which is used for the transport of various compounds across the cell membrane but mainly for ATP production by the ATP-ase enzyme. During ATP generation, protons are transported back to the cell interior. Thus, there are three key central metabolites in bacterial bioenergetics: acetyl-CoA, ATP and NADH. The intracellular level of these compounds acts as a powerful regulator of the metabolism of bacteria.

Importantly, in the absence of an external e-acceptor, the cell cannot regenerate the NADH produced by glycolysis, and the TCA cycle will not function to oxidise the substrate further than pyruvate and acetyl-CoA. Under these conditions, organic carbon itself can be used as an e-acceptor and be reduced with the NADH generated in the glycolysis into products such as acetate and propionate. This process of internal electron transfer is termed fermentation.

Interestingly, most reductive biosynthetic pathways in the anabolism rely on a coenzyme that is equivalent to yet distinct from NADH, namely NADPH. NADPH is produced in the pentose phosphate pathway together with ribose-5-phosphate, a precursor of DNA and RNA. Although NADH can be converted into NADPH and vice versa, this can only happen at the expense of ATP, highlighting how cells maintain separate energy and growth systems.

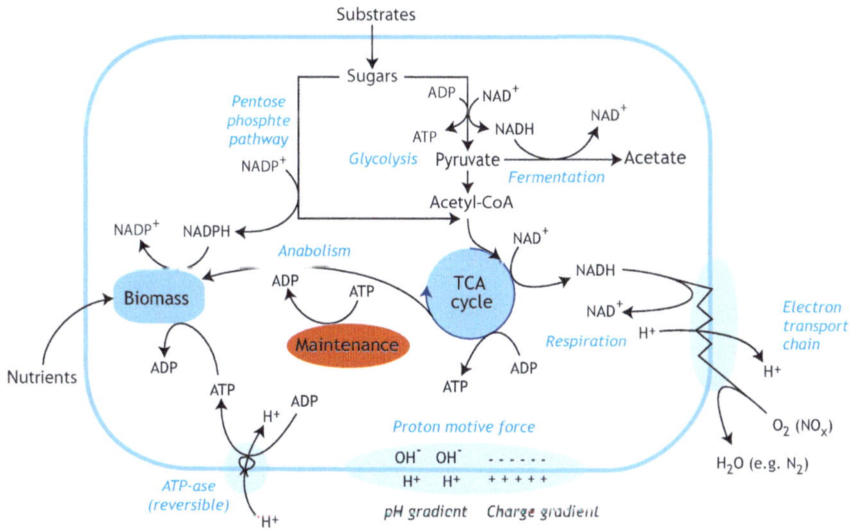

Figure 2.16 Schematic overview of bacterial bioenergetics, highlighting the central metabolites, acetyl-CoA, ATP, NADH, and NADPH, and the key catabolic and anabolic processes. Source: adapted from Comeau *et al.* (2008).

2.4.3 Redox reactions

Chemotrophic, light-independent growth is driven and sustained by the energy (as Gibbs free energy) generated in oxidation-reduction (redox) reactions. Redox reactions involve the transfer of electrons from a reduced e-donor (that is being oxidised) to an oxidized e-acceptor (that is being reduced). The stoichiometry of any redox reaction, here defined as an equation, is derived by identifying the e-donor (D) and e-acceptor (A) half-reactions:

$$\begin{aligned} D&: \ -1\cdot eD + Y^{D}_{e/eD}\cdot e^{-} \pm \ ... = 0 \\ A&: \ -1\cdot eA - Y^{A}_{e/eA}\cdot e^{-} \pm \ ... = 0 \end{aligned} \tag{2.2}$$

where, by definition, the two half-reactions are commonly expressed per mole of consumed e-donor or e-acceptor (*e.g.* $Y^{D}_{e/eD}$ in mol_e/mol_{eD}). Once the e-donor and e-acceptor are defined, the values of all the stoichiometric coefficients and the sign preceding them can be derived from elemental and charge balances. First, the elemental balance of the central atom of the e-donor and e-acceptor (*e.g.* C, N, and Fe) needs to be closed. Next, the elemental balances for O, H, and the charge are closed in subsequent steps to give the coefficients for H_2O, H^+ and e^-. For a more extensive discussion and examples of mass balance implementation in a spreadsheet, the reader is referred to Kleerebezem and Van Loosdrecht (2010).

Once the half-reactions are balanced, the complete redox reaction is defined as follows:

$$\lambda_D\cdot D+\lambda_A\cdot A=0 \tag{2.3}$$

where the multiplication factors λ_D and λ_A can be expressed as a function of known stoichiometric coefficients by defining two independent equations. First, electron neutrality, *i.e.* no net production or consumption of electrons in the complete reaction, is assumed (Eq. 2.4).

$$\lambda_D\cdot Y^{D}_{e/eD}-\lambda_A\cdot Y^{A}_{e/eA}=0 \tag{2.4}$$

Second, the value of one stoichiometric coefficient of the complete reaction is chosen. Usually, the reaction is expressed per mole of e-donor consumed as follows:

$$\lambda_D \cdot Y^D_{eD/eD} + \lambda_A \cdot Y^A_{eD/eA} = 1 \tag{2.5}$$

From eqs. 2.4 and 2.5, Eq. 2.3 can be rewritten as follows:

$$\frac{Y^A_{e/eA} \cdot D + Y^D_{e/eD} \cdot A}{Y^A_{e/eA} \cdot Y^D_{eD/eA} + Y^D_{e/eD} \cdot Y^A_{eD/eA}} = 0 \tag{2.6}$$

where by definition $Y^D_{eD/eD}$ = 1 mol_{eD}/mol_{eD}. Also, when the e-donor and e-acceptor do not coincide (*i.e.* eD ≠ eA), the e-donor is not involved in the e-acceptor half-reaction (*i.e.* $Y^A_{eD/eA}$ = 0), therefore Eq. 2.6 becomes:

$$D + \frac{Y^D_{e/eD}}{Y^A_{e/eA}} \cdot A = 0 \tag{2.7}$$

In this equation, the stoichiometric coefficients of A are normalized to balance out the electrons in the overall reaction. In contrast, when the e-donor and e-acceptor coincide (eD = eA), *e.g.* the catabolic reaction in fermentative organisms, $Y^A_{eD/eA} = Y^D_{eD/eD}$ = 1 mol_{eD}/mol_{eD} and Eq. 2.6 becomes:

$$\frac{Y^A_{e/eA} \cdot D + Y^D_{e/eD} \cdot A}{Y^A_{e/eA} + Y^D_{e/eD}} = 0 \tag{2.8}$$

Example 2.1 Aerobic glucose respiration

Under aerobic conditions, glucose is oxidised to bicarbonate and thus serves as e-donor ($C_6H_{12}O_6 / HCO_3^-$). Oxygen is the e-acceptor and is reduced to water (O_2 / H_2O). Based on the defined e-donor and e-acceptor couples, the following balanced half-reactions can be written:

$$D: -C_6H_{12}O_6 - 12\ H_2O + 6\ HCO_3^- + 30\ H^+ + 24\ e^- = 0$$
$$A: -O_2 - 4\ H^+ - 4\ e^- + 2\ H_2O = 0$$

and the overall reaction can be derived from Eq. 2.7,

$$-C_6H_{12}O_6 - 6\ O_2 + 6\ HCO_3^- + 6\ H^+ = 0$$

considering that $Y^D_{e/eD} = 24\ mol_e / mol_{eD}$ and $Y^A_{e/eA} = 4\ mol_e / mol_{eA}$.

Please note that in all examples the values of the stoichiometric coefficients are rounded.

Example 2.2 Glucose fermentation to ethanol

In absence of an external e-acceptor, glucose serves both as e-donor ($C_6H_{12}O_6 / HCO_3^-$) and as e-acceptor being reduced to ethanol ($C_6H_{12}O_6 / C_2H_6O$). The following balanced half-reactions can be written:

$$D: -C_6H_{12}O_6 - 12\ H_2O + 6\ HCO_3^- + 30\ H^+ + 24\ e^- = 0$$
$$A: -C_6H_{12}O_6 - 12\ H^+ - 12\ e^- + 3\ C_2H_6O + 3\ H_2O = 0$$

and the overall fermentative reaction can be derived from Eq. 2.8,

$$-C_6H_{12}O_6 - 2\ H_2O + 2\ C_2H_6O + 2\ HCO_3^- + 2\ H^+ = 0$$

considering that $Y^D_{e/eD} = 24\ mol_e / mol_{eD}$ and $Y^A_{e/eA} = 12\ mol_e / mol_{eA}$.

Example 2.3 Degree of reduction (γ) and chemical oxygen demand (COD)

In a redox reaction, electrons are conserved. Since most biological reactions are essentially redox balances, the electron balance is a valuable extra balance to derive the stoichiometry. To this end, the degree of reduction (γ) has been introduced in biotechnology. Essentially the amount of electrons in a chemical compound relative to reference chemical compounds is calculated. In general, the reference compounds chosen are present in most biological conversions (*e.g.* CO_2, H^+, H_2O, NH_4^+, for the elements C, H, O and N). Ammonium is taken as the reference because the valence of the N atom is the same as for N in biomass. For the other chemicals the degree of reduction is then calculated as the amount of electrons released or consumed in the redox half-reaction of the compound towards the defined reference compounds. In Example 2.1 the degree of reduction of a mole of glucose equals 24 moles of electrons and the degree of reduction of a mole of O_2 equals –4 moles of electrons. For carbon compounds, the degree of reduction is often given per carbon atom which for glucose would be 4 moles of electrons per C-mol. Note that the latter can be readily obtained.

There is a direct link between the degree of reduction and the chemical oxygen demand (COD) of a compound. Also, the COD measurement 'counts' the number of electrons released upon full oxidation to CO_2. Considering that a mole of O_2 can accept 4 moles of electrons (Example 2.1) and that a mole of O_2 has a molar weight of 32 g/mol, the COD of a mole of electrons is equivalent to 32/4 = 8 g_{COD}/mol_{e-}. In the case of Example 2.1, a mole of glucose has a COD of 24·8 = 192 g_{COD}/mol_{glc} (or 1.066 g_{COD}/g_{glc}). For O_2, the COD content is –4·8= –32 g_{COD}/mol_{O2} (–1 g_{COD}/g_{O2}). Since the COD is directly related to the amount of electrons, and electrons are a conserved moiety, the COD balance is also conserved. For example, taking the overall reaction of Example 2.1, the COD balance would be: (–192)·1 – (–32)·6 – 0·6 – 0·6 = 0.

Note that, since calculated based on the individual elements and the charge present in compounds, the γ balance or COD balance do not provide an extra independent conservation equation to solve the stoichiometry. It backs up one of the other elemental conservation equations. However, it is wise to use the γ balance or COD balance to track and check the electron flow and balances across reactions.

2.4.4 Thermodynamics basics

There are two fundamental laws in thermodynamics. The first law states that energy is conserved *i.e.* one can make an energy balance over a chemical reaction. According to the second law, spontaneous processes tend towards thermodynamic equilibrium and are associated with an increase in entropy *i.e.* a degree of disorder. Microbial cells are highly organized, low entropy systems and thus require energy to be synthetized from simple growth substrates such as O_2 and glucose, as well as a constant energy input to keep the complexity. The catabolic reaction provides this energy. In the absence of external catabolic substrates, microorganisms die and the loss of cell organization results in an overall entropy increase in line with the second law of thermodynamics.

For any given reaction, the available energy that can be used to carry out work is expressed by the Gibbs free energy change (ΔG), in kilojoule per mole of one reactant (kJ/mol_i). Energetically favourable reactions, termed exergonic, have a $\Delta G < 0$ (*i.e.* the energy of the products is less than the energy of the reactants) and can occur spontaneously. The catabolism is an exergonic reaction. Conversely, if $\Delta G > 0$ the reaction is called endergonic and requires an external input of energy to occur (*e.g.* anabolism). Once the stoichiometry of a reaction is defined, the associated ΔG can be calculated from tabulated values of the Gibbs energy of formation (G_f^0; Table 2.5). By analogy, tabulated values for the enthalpy of formation (H_f^0) are used for the calculation of the enthalpy change of a reaction (ΔH in kJ/mol_i), namely the heat that is released in exothermic reactions ($\Delta H < 0$) or absorbed in endothermic reactions ($\Delta H > 0$).

The standard change in Gibbs free energy (ΔG^0) of any reaction can be calculated as follows:

$$\Delta G^0 = \sum_i^n Y_i \cdot G_{f,i}^0 - \sum_j^m Y_j \cdot G_{f,j}^0 \tag{2.9}$$

where n and m are the number of products and substrates, respectively, and Y_i and Y_j the corresponding stoichiometric coefficients in the overall reaction. Traditionally, for the calculation of G_f^0 of a compound, the thermodynamic reference ($G_f^0 = 0$) is the elemental state and standard phase, namely C (s), H_2 (g), O_2 (g), N_2 (g), S (s) and P (s) for the main biologically-relevant elements. ΔG^0 is defined assuming the activities of all the reagents is 1. In diluted aqueous systems this corresponds to a concentration of 1 M (mol/l) for dissolved compounds and a partial pressure of 1 atm for gaseous compounds. Also, 298 K is the assumed standard temperature (T_S).

To reflect environmental conditions, ΔG^0 can be corrected for the actual concentrations. This is especially relevant for the proton concentration since the thermodynamic reference ($1\ mol_{H^+}/l$) is very different from the usual environmental concentration $10^{-7}\ mol_{H^+}/l$. The Van't Hoff equation is used to correct the value of the Gibbs free energy to the environmental pH:

$$\Delta G^{01} = \Delta G^0 \pm R \cdot T \cdot Y_{H^+} \cdot \ln\left(\frac{10^{-7}\,mol_{H^+}/l}{1\ mol_{H^+}/l}\right) \quad (2.10)$$

where R is the universal gas constant ($8.3145 \cdot 10^{-3}$ kJ/mol.K)[4], and Y_{H^+} is the stoichiometric coefficient of protons in the overall redox reaction, and the actual concentration is divided by the thermodynamic reference. The sign of the correction term is positive if protons are produced and negative if they are consumed. The concentrations of the other compounds in the equation usually deviate less from the standard conditions and are thus less often corrected for.

However, when the actual conditions strongly differ from the standard ones, and the reaction is close to equilibrium (*i.e.* ΔG^{01} close to zero; see examples 2.4 and 2.5), it is important to account for the actual activities of the reactants. The actual free energy change (ΔG^1) is calculated as follows:

$$\Delta G^1 = \Delta G^0 + R \cdot T \cdot \left(\begin{aligned} &\sum_i^n Y_i \cdot \ln\left(\frac{C_i}{1\ mol_i/l \text{ or } 1\ atm_i}\right) \\ -&\sum_j^m Y_j \cdot \ln\left(\frac{C_j}{1\ mol_j/l \text{ or } 1\ atm_j}\right) \end{aligned} \right) \quad (2.11)$$

where the activity of product i (C_i) and the activity of substrate j (C_j), are divided by the corresponding standard activity. In dilute aqueous systems, the activity of dissolved species equals their molar concentration and the activity of gaseous species corresponds to their partial pressure in atmospheres. Water and solids have activities of one.

Finally, the impact of the actual temperature (T) on ΔG^0 can be calculated based on the Gibbs-Helmholtz equation:

$$\Delta G^0 = \Delta G^0_{T_S} \cdot \frac{T}{T_S} + \Delta H^0_{T_S} \cdot \frac{T_S - T}{T_S} \quad (2.12)$$

where $\Delta H^0_{T_S}$ represents the enthalpy change of the redox reaction calculated from tabulated values of the enthalpy of formation (H_f^0; Table 2.5), and Ts is the standard temperature of 25 °C. H_f^0 can be considered constant over the ranges of concentrations and temperatures relevant for biological systems.

Example 2.4 Aerobic ferrous iron oxidation

As we will learn later on, iron is a poor e-donor but its oxidation is possible under aerobic conditions as oxygen is a strong e-acceptor. Specifically, ferrous iron serves as an e-donor being oxidized to ferric iron (Fe^{2+}/Fe^{3+}), while oxygen is reduced to water (O_2/H_2O). The two half-reactions, written according to Eq. 2.2 and combined with Eq. 2.7, result in the following overall reaction:

$$-Fe^{2+} - 0.25\ O_2 - H^+ + Fe^{3+} + 0.5\ H_2O = 0$$

The standard Gibbs free energy change of the reaction ΔG^0 can be calculated from Eq. 2.9 and Table 2.5 to be $-44.3\ kJ/mol_{Fe^{2+}}$. However, at the commonly assumed neutral pH (Eq. 2.10), the reaction runs close to thermodynamic equilibrium ($\Delta G^{01} = -4.4\ kJ/mol_{Fe^{2+}}$). It becomes favourable from a thermodynamic standpoint only under acidic conditions *e.g.* $\Delta G^1 = -38.6\ kJ/mol_{Fe^{2+}}$ at pH 1.

---

[4] kJ mol^{-1} K^{-1}

Table 2.5 Standard Gibbs free energy (G_f^o) and enthalpy (H_f^o) of formation for the organic and inorganic compounds used in the examples; values from Kleerebezem and Van Loosdrecht (2010) and references therein. The reader is referred to the free supporting material from Kleerebezem and Van Loosdrecht (2010) for a more complete list of compounds.

Element	Compound	Name	G_f^0 (kJ/mol)	H_f^0 (kJ/mol)
	e^{-1}	Electron	0.0	0.0
H	H^+	Proton	0.0	0.0
	H_2	Hydrogen	0.0	0.0
O	O_2	Oxygen	0.0	0.0
	H_2O	Water	–237.2	–285.8
N	N_2	Dinitrogen	0.0	0.0
	NH_4^+	Ammonium	–79.4	–133.3
	NO_2^-	Nitrite	–32.2	–107.0
	NO_3^-	Nitrate	–111.3	–173.0
S	HS^-	Bisulphite	12.1	–17.6
	SO_4^{2-}	Sulphate	–744.6	–909.6
Fe	Fe^{2+}	Ferrous iron	–78.9	–89.1
	Fe^{3+}	Ferric iron	–4.6	–48.5
C	CO_2	Carbon dioxide	–394.4	–394.1
	CHO_3^-	Bicarbonate	–586.9	–692.0
	CH_4	Methane	–50.8	–74.8
	$C_2H_3O_2^-$	Acetate	–369.4	–485.8
	C_2H_6O	Ethanol	–181.8	–288.3
	$C_3H_5O_2^-$	Propionate	–361.1	–510.4
	$C_6H_{12}O_6$	Glucose	–917.2	–1264.2
Biomass	$CH_{1.8}O_{0.5}N_{0.2}$	Biomass	-67.0	–91.0

Example 2.5 Interspecies hydrogen transfer

Molecular hydrogen (H_2) is a key intermediate in anaerobic microbial communities. A common example is the syntrophic growth of anaerobic ethanol oxidizers and hydrogenotrophic methanogens. Ethanol oxidizers use ethanol as an e-donor and produce acetate ($C_2H_6O/C_2H_3O_2^-$), while protons serve as e-acceptors and hydrogen is generated (H^+/H_2):

$$-C_2H_6O - H_2O + C_2H_3O_2^- + 2\,H_2 + H^+ = 0$$

Under standard conditions and pH 7 the reaction is unfavourable: $\Delta G^{01} = 9.6\ \text{kJ/mol}_{C_2H_6O}$. However, it becomes thermodynamically possible if the hydrogen partial pressure is kept sufficiently low ($\Delta G^1 = -\ 36.0\ \text{kJmol}_{C_2H_6O}$ at 0.0001 atm). Thus, ethanol oxidizers rely on the hydrogen consumption by hydrogenotrophic methanogens. Methanogens use hydrogen as an e-donor (H^+/H_2) and bicarbonate as an e-acceptor (HCO_3^- / CH_4):

$$-4\,H_2 - HCO_3 - H^+ + CH_4 + 3\,H_2O = 0$$

This second reaction is favourable at pH 7 and under standard hydrogen partial pressure of 1 atm: $\Delta G^{01} = -135.5\ \text{kJ/mol}_{HCO_3^-}$. At hydrogen partial pressures that allow for the growth of ethanol oxidizers, the reaction becomes less favourable, yet still possible ($\Delta G^1 = -\ 44.2\ \text{kJ/mol}_{HCO_3^-}$ at 0.0001 atm). Thus, as hydrogenotrophic methanogens rely on hydrogen production, syntrophic growth occurs at low hydrogen partial pressures favourable for both populations.

It is worth noting that the change in redox potential (ΔE in volt, V) is often used instead of the Gibbs free energy change (ΔG) to estimate the feasibility of a given reaction. The redox potential (E) measures the tendency of an element or compound to donate or accept electrons from a reference electrode. The standard redox potential (E^0) is defined relative to the standard hydrogen electrode, and assumes by definition the value of 0.00 V ($E^{01} = -0.414$ V at pH 7). Conventionally, redox potentials are expressed for half-reactions written as reductions, *e.g.* H^+/H_2 for the reference redox couple. Positive values characterize strong e-acceptors and negative potentials strong e-donors. ΔE^{01} is calculated as the redox potential of the electron-accepting couple minus the redox potential of the electron-donating couple at pH 7, with an equation equivalent to Eq. 2.10. Importantly, the redox potential and Gibbs free energy of the considered half-reaction are fully equivalent according to the Nernst equation:

$$\Delta G^0 = -n \cdot F \cdot E^0 \tag{2.13}$$

where n is the number of electrons donated and F is the Faraday constant (96.485 kJ/mol.V)[5].

2.5 STOICHIOMETRY OF MICROBIAL GROWTH

2.5.1 Anabolism

Anabolism is the overall process encompassing the synthesis of all biomass constituents, such as proteins, lipids, carbohydrates, DNA and RNA. Despite the immense variety of microorganisms, the content and elemental composition of microbial cells is similar for all cells. For the derivation of the anabolic reaction, the empirical formula $CH_{1.8}O_{0.5}N_{0.2}$ can be used for the organic fraction of biomass (Esener *et al.*, 1983; Heijnen and Kleerebezem, 2010). This molecular composition of biomass can be directly related to the commonly measured VSS by the molar weight of 24.6 g_{VSS} per mole of biomass. In some cases, a more accurate elemental analysis might be required, and inorganic elements such as P, Mg, K and other trace elements such as Zn, Mn, Mo, Se, Co, Cu and Ni are included (section 2.2.2.2).

The stoichiometry of the anabolism, as any other redox reaction, can be derived from the sum of an e-donor and e-acceptor half-reaction. However, in this case it is not *a priori* known whether the half-reaction of biomass production yields or consumes electrons. Once the biomass composition is defined, the carbon (C_s) and nitrogen (N_s) sources need to be identified to define the half-reaction for biomass production (An^*); to this end:

$$An^*: \quad -Y^{An^*}_{C_S/X} \cdot C_S - Y^{An^*}_{N_S/X} \cdot N_S + C_1H_{1.8}O_{0.5}N_{0.2} \pm Y^{An^*}_{H_2O/X} \cdot H_2O \pm Y^{An^*}_{H^+/X} \cdot H^+ \pm Y^{An^*}_{e/X} \cdot e^- = 0 \tag{2.14}$$

where the stoichiometric coefficients ($Y^{An^*}_{i/X}$) are defined per mole of biomass produced (*i.e.* $Y^{An^*}_{X/X}$ = 1 mol_X/mol_X). The carbon and nitrogen sources are always consumed, while the sign preceding the other reactants derives from elemental and charge mass balances. If more comprehensive biomass compositions are used, Eq. 2.14 is easily generalized to include any nutrient (*e.g.* P) by providing the corresponding source. Importantly, the sign preceding $Y^{An^*}_{e/X}$, *i.e.* the production or consumption of electrons in An^*, depends on the oxidation state of the carbon and nitrogen source relative to that of C and N in biomass, and determines the need for a complementary e-donor or e-acceptor half-reaction. For instance, if the carbon/nitrogen source is more oxidized than biomass, electrons are needed to reduce them (*i.e.* $-Y^{An^*}_{e/X}$ as for CO_2 in Example 2.6) and thus an e-donor (D) is required. Conversely, an e-acceptor (A) is required for carbon/nitrogen sources that are

[5] kJ mol^{-1} V^{-1}

more reduced than biomass (*i.e.* $Y_{e/X}^{An^*}$ as for propionate in Example 2.7). The overall anabolic reaction can be obtained as described above, and is by definition expressed per mole of biomass ($Y_{X/X}^{An}$ = 1 mol_X/mol_X):

$$An: \begin{cases} An^* + \dfrac{Y_{e/X}^{An^*}}{Y_{e/eD}^{D}} \cdot D = 0, \text{ for oxidised substrates} \\ An^* + \dfrac{Y_{e/X}^{An^*}}{Y_{e/eA}^{A}} \cdot A = 0, \text{ for reduced substrates} \end{cases} \quad (2.15)$$

In the case of chemoorganoheterotrophic growth, the same organic substrate usually serves as both e-donor and C source (Example 2.7). In the case of chemolithoautotrophic growth, a separate e-donor is required for the reduction of CO_2 to biomass and, as we will see, this does not necessarily coincide with the e-donor in the catabolic reaction (examples 2.6 and 2.12). In general, NH_4^+ is the preferred N source as it has the same oxidation state of biomass. Ammonium is also in the same chemical state as amino acids that compose proteins *i.e.* the main components of biomass. However, more oxidised nitrogen forms (*e.g.* NO_3^-) can be also used at the cost of a higher e-donor consumption (Example 2.8).

Example 2.6 Chemolithoautotrophic growth and anammox

By definition, autotrophs use HCO_3^- as a C source. If NH_4^+ is available as N sources, the biomass synthesis half-reaction can be written as follows:

$$-HCO_3^- - 0.2\ NH_4^+ - 5\ H^+ - 4.2\ e^-$$
$$+CH_{1.8}O_{0.5}N_{0.2} + 2.5\ H_2O = 0$$

As HCO_3^- is more oxidised than biomass, electrons are consumed and an e-donor is required. While different inorganic e-donors can be used (*e.g.* Fe^{2+}/Fe^{3+}, H_2/H^+), we will here consider the case of anammox bacteria that use nitrite (NO_3^-/NO_3^-) as e-donor:

$$-NO_2^- - H_2O + NO_3^- + 2\ H^+ + 2\ e^- = 0$$

Based on Eq. 2.15, the full anabolic reaction of anammox bacteria becomes:

$$-HCO_3^- - 0.2\ NH_4^+ - 2.1\ NO_2^- - 0.8\ H^+$$
$$+CH_{1.8}O_{0.5}N_{0.2} + 2.1\ NO_3^- + 0.4\ H_2O = 0$$

with a $\Delta G_{An}^{01} = 274.8\ kJ/mol_X$. It is here worth noting that the e-donor in anabolism does not necessarily coincide with the catabolic e-donor. For example, ammonium and not nitrite is the e-donor in the anammox catabolic reaction.

Example 2.7 Reduced carbon source

The biomass synthesis half-reaction with propionate ($C_3H_5O_2^-$) as the C source and NH_4^+ as N source can be written as follows:

$$-0.33\ C_3H_5O_2^- - 0.2\ NH_4^+ + CH_{1.8}O_{0.5}N_{0.2}$$
$$+0.16\ H_2O + 0.33\ H^+ + 0.46\ e^- = 0$$

As $C_3H_5O_2^-$ is more reduced than biomass, electrons are released and thus an e-acceptor is required. If oxygen is available (O_2/H_2O), the complete anabolic reaction becomes Eq. 2.15:

$$-0.33\ C_3H_5O_2^- - 0.2\ NH_4^+ - 0.12\ O_2 - 0.13\ H^+$$
$$+CH_{1.8}O_{0.5}N_{0.2} + 0.4\ H_2O = 0$$

with an associated ΔG_{An}^{01} of $-20.3\ kJ/mol_X$.

Example 2.8 Oxidised nitrogen source

In the case of chemoorganoheterotrophic growth on glucose, glucose serves both as the C source and e-donor in the anabolic reaction ($C_6H_{12}O_6/CO_2$). If NH_4^+ is the N source, the complete anabolic reaction becomes:

$$-0.18\ C_6H_{12}O_6 - 0.2\ NH_4^+ + CH_{1.8}O_{0.5}N_{0.2}$$
$$+0.05\ HCO_3^- + 0.25\ H^+ + 0.4\ H_2O = 0$$

If instead NO_3^- is used as the N source, the anabolic reaction changes as follows:

$$-0.24\ C_6H_{12}O_6 - 0.2\ NO_3^- + CH_{1.8}O_{0.5}N_{0.2}$$
$$+0.45\ HCO_3^- + 0.25\ H^+ + 0.2\ H_2O = 0$$

where the 40% increase in glucose consumption per unit of produced biomass is due to the additional requirement of electrons to reduce NO_3^- to NH_4^+.

In general, the actual Gibbs energy change of the anabolism (ΔG^{01}_{An}) is in most cases slightly positive or sometimes slightly negative (examples 2.6 and 2.7). It is however important to note that the calculated ΔG^{01}_{An} does not account for the biochemical complexity of biological systems (*i.e.* the decrease in entropy upon biomass synthesis), and represents only the chemical energy present in the biomass. The decrease in entropy upon converting simple molecules into the complexity of a cell is an irreversible process that always needs an external input of energy (or in thermodynamic terms requires work or energy dissipation) which is provided by the catabolism. One can compare this to constructing a wooden cabin. The energy content in the wood does not change during the construction of the lodge, but there is work performed associated to the decrease in entropy when a pile of wood is converted into a cabin.

2.5.2 Catabolism

The catabolism is the energy-generating reaction in microbial metabolism (Figure 2.15). In principle, any exergonic redox reaction ($\Delta G<0$) can serve as a catabolic reaction, and the immense variety of possible combinations of e-donors and e-acceptors yielding Gibbs free energy shapes the microbial diversity across all ecosystems. Once the e-donor and e-acceptor half-reactions are defined, the catabolic reaction is derived from Eq. 2.7 or 2.8 as follows:

$$\text{Cat}: \begin{cases} D + \dfrac{Y^D_{e/eD}}{Y^A_{e/eA}} \cdot A = 0 & , \text{ if } eD \neq eA \\ \dfrac{Y^A_{e/eA} \cdot D + Y^D_{e/eD} \cdot A}{Y^A_{e/eA} + Y^D_{e/eD}} = 0, & \text{if } eD = eA \end{cases} \qquad (2.16)$$

where, by definition $Y^{Cat}_{eD/eD}$ = 1 mol_{eD}/mol_{eD}. Interestingly, depending on the e-donor/e-acceptor couple, the available ΔG can vary by up to two orders of magnitude (examples 2.9-2.11).

The average Gibbs free energy change per electron for microbiologically-relevant elements with a variable oxidation state can be calculated, and gives insight into their role as e-acceptor or e-donor (Figure 2.17) (Kleerebezem and Van Loosdrecht, 2010).

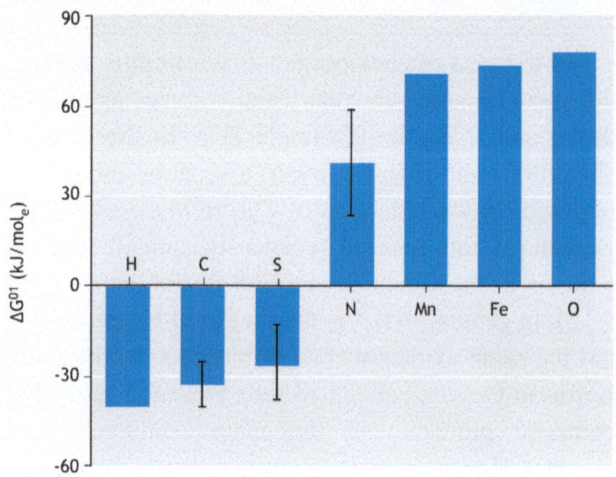

Figure 2.17 Average Gibbs free energy change per mole of electrons for microbiologically-relevant elements with a variable oxidation state obtained considering the full oxidation of each considered compound. Standard deviations are presented for atoms found in numerous oxidation states. Source: based on Kleerebezem and Van Loosdrecht, 2010.

The values are obtained from the half-reaction describing the full oxidation of each considered compound to its most oxidized state *i.e.* the e-donor half-reaction. The average and standard deviation are presented for atoms found in numerous oxidation states (C, N and S). For the other compounds, values are calculated considering the main oxidation state found in nature. C, S and H-based compounds can in general be regarded as strong e-donors ($\Delta G^{01}_D < 0$) and are thus weak e-acceptors (the reverse reaction has $\Delta G^{01}_A > 0$). Conversely, N, Fe, Mn and O_2 compounds are strong e-acceptors ($\Delta G^{01}_D > 0$) and weak e-donors.

The combination of strong e-donors and e-acceptors will always result in thermodynamically-favourable reactions. Reactions based on weak e-donor and e-acceptor couples will be thermodynamically unfavourable. Combining a strong e-donor and weak e-acceptor, or *vice-versa*, will result in a reaction whose thermodynamics (energy production or consumption) will depend on the specific compounds and environmental conditions (examples 2.4 and 2.5).

Example 2.9 Anaerobic respiration of glucose

Glucose is a strong e-donor (Figure 2.17). In the absence of oxygen, glucose can be oxidised with nitrate (NO_3^- / N_2) as the e-acceptor. This process is termed denitrification, and proceeds according to the following stoichiometry:

$$-C_6H_{12}O_6 - 4.8\ NO_3^- + 2.4\ N_2 + 6\ HCO_3^- \\ +2.4\ H_2O + 1.2\ H^+ = 0$$

with a $\Delta G_{Cat}^{01} = -\ 2{,}687.1\ kJ/mol_{C_6H_{12}O_6}$. This value is comparable to the $\Delta G_{Cat}^{01} = -\ 2{,}843.8\ kJ/mol_{C_6H_{12}O_6}$ associated with aerobic glucose oxidation (Example 2.1), as oxygen and nitrate are both strong e-acceptors. Differently, if *e.g.* sulphate is used as the e-acceptor (SO_4^{2-} / HS^-):

$$-C_6H_{12}O_6 - 3\ SO_4^{2-} + 3\ HS^- + 6\ HCO_3^- + 3\ H^+ = 0$$

less energy is gained per oxidized glucose ($\Delta G_{Cat}^{01} = -\ 453.9\ kJ/mol_{C_6H_{12}O_6}$) since sulphate is a weak e-acceptor (Figure 2.17).

Ultimately, if no external e-acceptor is available (anaerobic conditions), glucose can serve as both e-donor and e-acceptor (Example 2.2), with a further decrease in the available Gibbs free energy ($\Delta G_{Cat}^{01} = -\ 225.7\ kJ/mol_{C_6H_{12}O_6}$).

The heat released during glucose oxidation with nitrate ($\Delta H^0 = -\ 2{,}743.3\ kJ/mol_{C_6H_{12}O_6}$ or $-15.2\ kJ/g_{C_6H_{12}O_6}$) is calculated with an equation equivalent to Eq. 2.9 and values from Table 2.5. Considering the specific heat capacity of water ($4.18\ kJ/kg_{H_2O}.°C$), the anoxic oxidation of one gram of glucose with nitrate would result in a 3.6 °C temperature increase of 1 kg of water.

Example 2.10 Aerobic and anoxic oxidations of ammonium

Ammonium is a poor e-donor, yet its oxidation is successfully catalysed in the presence of strong e-acceptors such as oxygen (O_2/H_2O) or nitrite (NO_2^- / N_2). Under aerobic conditions, ammonium is oxidised to nitrite in a process termed nitrification:

$$-NH_4^+ - 1.5\ O_2 + NO_2^- + 2\ H^+ + H_2O = 0$$

with a $\Delta G_{Cat}^{01} = -\ 269.8\ kJ/mol_{NH_4^+}$. Under anoxic conditions, ammonium oxidation to dinitrogen gas provides the catabolic energy for anammox bacteria:

$$-NH_4^+ - NO_2^- + N_2 + 2\ H_2O = 0$$

with a $\Delta G_{Cat}^{01} = -\ 362.8\ kJ/mol_{NH_4^+}$.

Example 2.11 Acetoclastic methanogenesis

Acetate serves both as e-donor ($C_2H_3O_2^- / HCO_3^-$) and e-acceptor ($C_2H_3O_2^- / CH_4$) in the anaerobic production of methane. Based on Eq. 2.16 (eD = eA), the following catabolic reaction is derived:

$$-C_2H_3O_2^- - H_2O + CH_4 + HCO_3^- = 0$$

with a ΔG_{Cat}^{01} of ($\Delta G_{Cat}^{01} = -\ 31.2\ kJ/mol_{C_2H_3O_2}$), which is close to thermodynamic equilibrium.

2.5.3 Metabolism

While the individual equations for anabolism (An) and catabolism (Cat) can be derived from chemical principles, the overall metabolic or growth equation (Met) can only be derived if a biological yield coefficient is known. Typically, this is derived from experimental observations. Measured yields can be retrieved from literature reports but one has to consider the conditions under which the yields were measured.

In general terms, the stoichiometry of the overall metabolism (Eq. 2.1) results from the energy coupling of the anabolic (Eq. 2.15) and catabolic (Eq. 2.16) reactions as follows:

$$\text{Met}: \ \lambda_{Cat} \cdot \text{Cat} + \lambda_{An} \cdot \text{An} = 0 \tag{2.17}$$

where the multiplication factor for the anabolism (λ_{An} in molx/molx) is usually assumed to express the whole metabolism per mole biomass produced ($Y_{X/X}^{Met}$ = 1 molx/molx). As a result, λ_{Cat} (mol$_{eD}$/molx) can be interpreted as the number of times the catabolism needs to run to generate the Gibbs energy required to generate one mole of biomass in the anabolic reaction. The determination of λ_{Cat}, and thus of the complete metabolic equation, requires the quantification of one stoichiometric coefficient linking catabolism and anabolism. In principle, λ_{Cat} can be estimated based on the rates of production or consumption of any two reactants participating in both the catabolism and anabolism; see Kleerebezem and Van Loosdrecht (2010) for a more detailed discussion. Commonly, λ_{Cat} is determined based on the biomass yield ($Y_{X/eD}$), defined as biomass produced (X) per e-donor consumed (eD) as follows:

$$Y_{X/eD} = \frac{\lambda_{Cat} \cdot Y_{X/eD}^{Cat} + \lambda_{An} \cdot Y_{X/X}^{An}}{\lambda_{Cat} \cdot Y_{eD/eD}^{Cat} + \lambda_{An} \cdot Y_{eD/X}^{An}} \tag{2.18}$$

where by definition $Y_{X/eD}$ is the inverse of the stoichiometric coefficient of the e-donor in the overall metabolic reaction (Eq. 2.1, *i.e.* $1/Y_{eD/X}$). By definition, the biomass is not involved in the catabolic reaction (*i.e.* $Y_{X/eD}^{Cat} = 0$), and both the anabolism and overall metabolism are expressed per mole of biomass produced (*i.e.* $Y_{X/X}^{An} = \lambda_{An} = 1$ molx/molx). Also, in the metabolic reaction the term substrate (S) is often used instead of e-donor (eD), and this notation will be adopted hereafter. Thus, Eq. 2.18 becomes:

$$Y_{X/S} = \frac{1}{\lambda_{Cat} \cdot Y_{S/S}^{Cat} + Y_{S/X}^{An}} \tag{2.19}$$

where the substrate (*i.e.* e-donor in catabolism) does not necessarily coincide with the e-donor in the anabolism (Example 2.12), but it needs to be involved in both reactions.

Ultimately, if the biomass yield is determined experimentally (*i.e.* observed, $Y_{X/S}^{obs}$), Eq. 2.19 can be rearranged to obtain λ_{Cat} as follows:

$$\lambda_{Cat} = \frac{\dfrac{1}{Y_{X/S}^{obs}} - Y_{S/X}^{An}}{Y_{S/S}^{Cat}} \tag{2.20}$$

It is important to note that the measured $Y_{X/S}^{obs}$ is not constant, but it depends on the biomass-specific growth rate at which the culture is grown and already accounts for the maintenance. This will be discussed in Section 2.6.

Example 2.12 Anammox metabolism

Anoxic ammonium oxidizing (anammox) bacteria oxidize ammonium to dinitrogen gas with nitrite as the e-acceptor in the catabolism (Cat in Example 2.10), and obtain the electrons to reduce inorganic bicarbonate to biomass from the oxidation of nitrite in the anabolism (An in Example 2.6):

$$\text{Cat}: \ -NH_4^+ - NO_2^- + N_2 + 2\,H_2O = 0$$

$$\text{An}: \ -HCO_3^- - 0.2\,NH_4^+ - 2.1\,NO_2^- - 0.8\,H^+ + CH_{1.8}O_{0.5}N_{0.2} + 2.1\,NO_3^- + 0.4\,H_2O = 0$$

Here, ammonium serves as e-donor in the catabolism and as N source in the anabolism, while nitrite is the e-acceptor in the catabolism and the e-donor in the anabolism. As ammonium and nitrite are involved in both catabolism and anabolism, the growth yield on any of them can be used to obtain the overall metabolism. Here the growth yield on ammonium is used.

The experimentally-determined biomass yield of anammox $(Y^{obs}_{X/S} = Y^{obs}_{X/NH_4^+})$ equals $0.071\ mol_X/mol_{NH_4^+}$ (Lotti *et al.*, 2014). Based on Eq. 2.20 and the defined stoichiometry $(Y^{An}_{NH_4^+/X} = 0.2\ mol_{NH_4^+}/mol_X$ and $Y^{Cat}_{NH_4^+/NH_4^+} = 1\ mol_{NH_4^+}/mol_{NH_4^+}))$, λ_{Cat} becomes $13.9\ mol_{NH_4^+}/mol_X$. The complete anammox metabolism can thus be calculated from Eq. 2.17:

$$-14.1\ NH_4^+ - 16.0\ NO_2^- - HCO_3^- - 0.8\ H^+$$
$$+13.9\ N_2 + 2.1\ NO_3^- + CH_{1.8}O_{0.5}N_{0.2} + 28.2\ H_2O = 0$$

with a $\Delta G^{01} = -\ 335.8\ kJ/mol_{NH_4^+}$. This value is obtained by dividing the Gibbs free energy of the growth reaction ΔG^{01} $(-4{,}730.0\ kJ/mol_X)$ by the stoichiometric coefficient for ammonium (14.1 mol_{NH4}/mol_X). Normalizing ΔG^{01} by the substrate is informative to compare growth efficiencies on different e-donors.

2.5.4 Generalized method to estimate the maximum biomass yield

Since the coupling between anabolism and catabolism depends on an energetic efficiency, several authors have tried to develop a thermodynamic-based prediction of yield coefficient (Heijnen, 1994). Here we introduce the methodology proposed by Heijnen and Van Dijken (1992), and further developed by Heijnen (1994) and Heijnen and Kleerebezem (2010). This method is based on the need for Gibbs energy dissipation (ΔG^{01}_{Diss} in kJ/mol_X) in the conversion of simple molecules into the complexity of a cell. The Gibbs energy dissipation reflects the specific growth efficiency (*e.g.* lower for more efficient metabolisms) and equals, with the opposite sign, the Gibbs free energy change of the overall metabolic reaction. In analogy to Eq. 2.17, the Gibbs free energy balance can be written as follows:

$$\text{Met}:\ \lambda^*_{Cat} \cdot \Delta G^{01}_{Cat} + \lambda^*_{An} \cdot \Delta G^{01}_{An} = -\Delta G^{01}_{Diss} \qquad (2.21)$$

Considering that by definition $\lambda^*_{An} = 1\ mol_X/mol_X$, this equation can be rewritten as:

$$\lambda^*_{Cat} = \frac{\Delta G^{01}_{An} + \Delta G^{01}_{Diss}}{-\Delta G^{01}_{Cat}} \qquad (2.22)$$

Based on a large inventory of experimental data, Heijnen and Van Dijken (1992) observed that the Gibbs energy dissipation (ΔG^{01}_{Diss}) depends only on the anabolic e-donor and C source, with the type of microorganism and e-acceptor having little or no impact.

For heterotrophic growth on organic carbon, ΔG^{01}_{Diss} is a function of the number of carbon atoms in the carbon substrate (NoC_{C_S} in C-mol/mol$_{Cs}$) and the number of electrons per carbon atom released upon full oxidation of the carbon substrate to CO_2 (γ_{C_S} in mol$_e$/C-mol$_{Cs}$):

$$\Delta G^{01}_{Diss} = 200 + 18 \cdot \left(6 - NoC_{C_S}\right)^{1.8} + \exp\left\{\left(\left(3.8 - \gamma_{C_S}\right)^2\right)^{0.16} \cdot \left(3.6 + 0.4 \cdot NoC_{C_S}\right)\right\} \qquad (2.23)$$

This method has been derived from experimental data for NoC_{C_S} from 1 (*e.g.* formate, CO_2) to 6 (*e.g.* glucose), while covering γ_{C_S} values between 0 (CO_2) and 8 (CH_4). Based on Eq. 2.23, the ΔG^{01}_{Diss} for heterotrophs ranges between almost 200 and 1,000 kJ/mol_X (Figure 2.18), with an estimated error of 30% for any arbitrary C source (Heijnen and Van Dijken, 1992).

For autotrophic growth (*i.e.* CO_2 as C source), ΔG^{01}_{Diss} depends on the e-donor in the anabolic reaction. Electrons donated by strong e-donors, such as H_2/H^+ and HS^-/SO_4^{2-} (Figure 2.17), have enough energy to reduce inorganic carbon to biomass $(\Delta G^{01}_{An} < 0)$. In this case, based on Eq. 2.23, ΔG^{01}_{Diss} equals 1,000 kJ/mol_X. Conversely, for weak e-donors, such as Fe^{2+}/Fe^{3+} and NH_4^+/NO_2^-, $\Delta G^{01}_{An} > 0$ and the energy content of the electrons need to be increased via membrane gradients (*i.e.* reverse electron transport - RET) at the expense of Gibbs energy. This results in a significantly higher ΔG^{01}_{Diss} of 3,500 kJ/molx.

The relation in Figure 2.18 illustrates that the more a compound resembles a central intermediate of the metabolism *e.g.* pyruvate, the less energy dissipation (biochemical work) is needed to produce a cell. In this respect, it is useful to recall that biomass building blocks typically contain 4-5 carbons with a γ_{C_S} of 4.2 moles of electrons (Eq. 2.13).

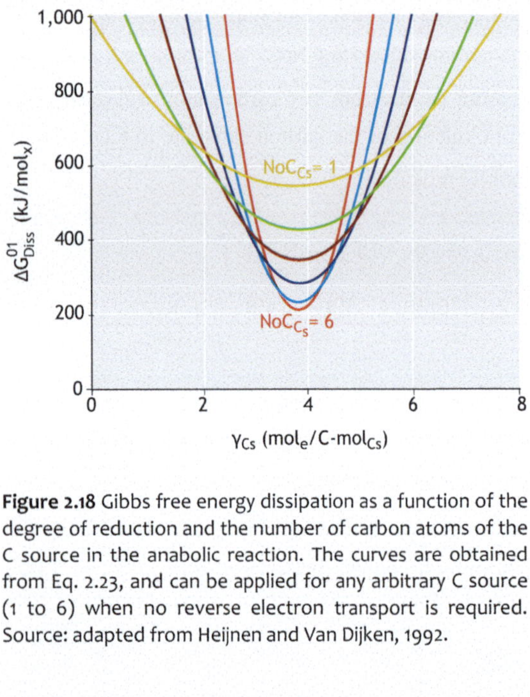

Figure 2.18 Gibbs free energy dissipation as a function of the degree of reduction and the number of carbon atoms of the C source in the anabolic reaction. The curves are obtained from Eq. 2.23, and can be applied for any arbitrary C source (1 to 6) when no reverse electron transport is required. Source: adapted from Heijnen and Van Dijken, 1992.

Therefore, for simple C sources, energy needs to be invested in the formation of new carbon bonds. Similarly, if the C source is more reduced or oxidised than biomass, more oxidation or reduction reactions are required. With $NoC_{C_S} = 6$ and $\gamma_{C_S} = 4$, glucose is an almost ideal C source as reflected in the small amount of energy dissipation required for biomass production (236.1 kJ/molx). Also, for poor e-donors in autotrophic anabolism, reverse electron transport is needed contributing to a higher ΔG^{01}_{Diss} (Heijnen and Van Dijken, 1992).

Ultimately, by substituting λ^*_{Cat} (Eq. 2.22) in Eq. 2.19, the maximum biomass yield on substrate is obtained:

$$Y^{max}_{X/S} = \frac{1}{\lambda^*_{Cat} \cdot Y^{Cat}_{S/S} + Y^{An}_{S/X}} \tag{2.24}$$

Here it is important to realize that $Y^{max}_{X/S}$ represents the maximum amount of biomass that could be produced per substrate consumed in the absence of maintenance, and is thus intrinsically higher than the experimentally determined $Y^{obs}_{X/S}$. Also, while the method allows for an estimation of the biomass yields with an error of 13% in the range 0.01-0.8 molx/mols, specific biochemistry differences might result in higher variations (see examples in Heijnen and Van Dijken, 1992).

Example 2.13 Anammox metabolism (continued)

The estimated maximum biomass yield and derived overall metabolic stoichiometry for anammox are compared with those derived experimentally (Example 2.12). Anammox bacteria are autotrophs and require reverse electron transport because ammonium is a weak e-donor, thus $\Delta G^{01}_{Diss} = 3{,}500 \text{ kJ/mol}_X$. Also, we have already calculated ΔG^{01}_{An} of 274.8 kJ/mol$_X$ and a ΔG^{01}_{Cat} of −362.8 kJ/mol$_{NH_4^+}$ in examples 2.6 and 2.10. Based on Eq. 2.22, λ^*_{Cat} becomes 10.5 mol$_{NH_4^+}$/mol$_X$, and the overall metabolism can be derived:

$$-10.7\ NH_4^+ - 12.6\ NO_2^- - HCO_3^- - 0.8\ H^+$$
$$+10.5\ N_2 + 2.1\ NO_3^- + CH_{1.8}O_{0.5}N_{0.2} + 21.4\ H_2O = 0$$

with an associated $\Delta G^{01} = -\ 327.3$ kJ/mol$_{NH_4^+}$. This value is obtained by dividing ΔG^{01} (= – ΔG_{Diss}^{01}, in kJ/molx) by the ammonium stoichiometric coefficient.

The theoretical stoichiometry and the resulting maximum yield ($Y^{max}_{X/NH_4^+} = 0.094\ mol_X/mol_{NH_4^+}$) values are consistent with those determined experimentally (Lotti *et al.*, 2014).

Example 2.14 Aerobic metabolism on glucose

Under aerobic conditions, heterotrophic microorganisms can use glucose as the catabolic e-donor (Example 2.1) and C source in the anabolic reaction (Example 2.8):

$$\text{Cat}: \ -C_6H_{12}O_6 - 6\ O_2 + 6\ HCO_3^- + 6\ H^+ = 0$$

$$\text{An}: \ -0.18\ C_6H_{12}O_6 - 0.2\ NH_4^+ + CH_{1.8}O_{0.5}N_{0.2}$$
$$+0.05\ HCO_3^- + 0.25\ H^+ + 0.4\ H_2O = 0$$

One can calculate $\Delta G^{01}_{Cat} = -2{,}843.8\ kJ/mol_{C_6H_{12}O_6}$ and $\Delta G^{01}_{An} = -24.8\ kJ/mol_X$ from Eq. 2.10 (and Table 2.5), and $\Delta G^{01}_{Diss} = 236.1\ kJ/mol_X$ based on Eq. 2.23. Accordingly, λ^*_{Cat} equals $0.07\ mol_{C_6H_{12}O_6}/mol_X$ (Eq. 2.22) and the following overall metabolism can be derived:

$$-0.25\ C_6H_{12}O_6 - 0.2\ NH_4^+ - 0.45\ O_2$$
$$+CH_{1.8}O_{0.5}N_{0.2} + 0.5\ HCO_3^- + 0.7\ H^+ + 0.4\ H_2O = 0$$

with an associated $\Delta G^{01} = -946.8\ kJ/mol_{C_6H_{12}O_6}$ expressed per mole of glucose substrate, and a $Y^{max}_{X/C_6H_{12}O_6} = 0.67\ mol_X/\text{C-}mol_{C_6H_{12}O_6}$ consistent with reported experimental values.

2.6 KINETICS OF MICROBIAL GROWTH

The previous section presented a framework for the stoichiometric description of microbial growth and this section describes growth kinetics. The combination of stoichiometry and kinetics gives a full description of microbial growth that can be further implemented in bioreactor models and serves as the basis for process design.

2.6.1 Substrate consumption rate: the Herbert-Pirt relation

It has been observed that with increasing growth rate, the yield of biomass on substrate S (typically the e-donor) also increases. This is explained by the dual use of the substrate (Figure 2.15). Microorganisms primarily consume the substrate for (*i*) growth-independent processes (maintenance, needed to maintain the complex cell system) and (*ii*) biomass production (growth). In order to describe this behaviour, the Herbert-Pirt relation is generally used. The Herbert-Pirt equation defines the relation between the biomass-specific substrate consumption rate (q_S, in $mol_S/h.mol_X$)[6] and the growth and maintenance processes as follows:

$$q_S = Y^{max}_{S/X} \cdot \mu + Y^{Cat}_{S/S} \cdot m_S \qquad (2.25)$$

where, μ ($mol_X/h.mol_X = 1/h$) is the biomass-specific growth rate, and m_S ($mol_S/h.mol_X$) is the biomass-specific substrate consumption rate for maintenance, namely the amount of substrate catabolized to provide the Gibbs energy for growth-independent (maintenance) processes. $Y^{max}_{S/X}$ is the stoichiometric coefficient of the substrate in the metabolic reaction determined by the Gibbs energy dissipation method (Section 2.5.4), which by definition is the reciprocal of the theoretical maximum biomass yield ($Y^{max}_{X/S} = 1/\ Y^{max}_{S/X}$ in mol_X/mol_S) as defined in Eq. 2.24. $Y^{Cat}_{S/S}$ is the stoichiometric coefficient of the substrate (e-donor) in the catabolic reaction Eq. 2.16, namely the reaction providing the energy for all metabolic processes, including maintenance. Considering that by definition $Y^{Cat}_{S/S} = 1\ mol_S/mol_S$, Eq. 2.25 becomes:

$$q_S = \frac{1}{Y^{max}_{X/S}} \cdot \mu + m_S \qquad (2.26)$$

In other words, Eq. 2.26 reflects the fact that the substrate is simultaneously consumed in two independent processes: the metabolic reaction (Eq.

---

[6] $mol_S\ h^{-1}\ mol_X^{-1}$

2.21), with process rate μ, and in the reaction generating the energy for maintenance, namely the catabolism (Eq. 2.16), with process rate m_S (Figure 2.19A). Furthermore, Eq. 2.26 can be rewritten as follows:

$$\frac{1}{Y_{X/S}^{obs}} = \frac{1}{Y_{X/S}^{max}} + \frac{m_S}{\mu} \tag{2.27}$$

considering that the observed biomass yield can also be defined as $Y_{X/S}^{obs} = \mu/q_S$. From Eq. 2.27 it becomes clear that the actual amount of biomass produced per substrate consumed ($Y_{X/S}^{obs}$) decreases at decreasing μ, as the relative contribution of the Gibbs energy for maintenance increases (Figure 2.19A and C). Conversely, the relative contribution of substrate consumption for maintenance decreases at high μ, and $Y_{X/S}^{obs}$ approaches the theoretical $Y_{X/S}^{max}$ (Figure 2.19C).

From a design perspective, knowing the process stoichiometry enables the amounts of substrates and products involved to be estimated. As discussed in Section 2.5, the actual stoichiometry can be derived based on the biomass yield on the substrate ($Y_{X/S}^{obs}$, Eq. 2.20). However, as this yield is a function of the growth rate (Eq. 2.27; Figure 2.19C), the maximum yield and the maintenance coefficients need to be identified to derive the actual process stoichiometry for the specific conditions of interest. This can be done experimentally directly from Eq. 2.27 by measuring $Y_{X/S}^{obs}$ at different μ in a chemostat (where μ is equivalent to the imposed dilution rate).

If no experimental data are available, $Y_{X/S}^{max}$ is derived as previously described (Eq. 2.24), while a realistic value for m_S can be derived from the Arrhenius-type correlation proposed for the biomass-specific Gibbs energy consumption for maintenance (m_G, in kJ/h.mol$_X$)[7] (Tijhuis *et al.*, 1993):

$$m_G = 4.5 \cdot \exp\left\{-\frac{69}{R} \cdot \left(\frac{1}{T} - \frac{1}{298}\right)\right\} \tag{2.28}$$

where R is the universal gas constant ($8.3145 \cdot 10^{-3}$ kJ/mol.K), and the value 69 corresponds to the associated activation energy (kJ/mol). Interestingly, Eq. 2.28 was derived based on an extensive inventory of experimental data comprising different types of microorganisms, growth conditions, e-donors (organic and inorganic) and e-acceptors, and temperatures (Tijhuis *et al.*, 1993). However, only the temperature seems to play a major role in m_G, suggesting that the energy requirements for maintenance are a more universal metabolic trait. This observation is consistent with the large similarity in biomass composition of different organisms probably resulting in similar maintenance requirements (Section 2.5.4). The reported accuracy of Eq. 2.28 is ± 40% in the temperature range 5-75 °C. From the stoichiometry of the catabolic reaction, m_S can be obtained as follows:

$$\frac{m_s}{Y_{S/S}^{Cat}} = -\frac{m_G}{\Delta G_{Cat}^{01}} \tag{2.29}$$

where by definition $Y_{S/S}^{Cat} = 1$ mol$_S$/mol$_S$, and ΔG_{Cat}^{01} is the change in Gibbs energy of the catabolic reaction (kJ/mol$_S$) as estimated in Section 2.5.2. From Eq. 2.29, the more energy is generated in the catabolic reaction (*i.e.* more negative ΔG_{Cat}^{01}), the lower the amount of substrate is required for maintenance. Any other m_i can be derived from the corresponding $Y_{i/S}^{Cat}$ catabolic reaction stoichiometry.

---

[7] kJ h^{-1} mol$_X^{-1}$

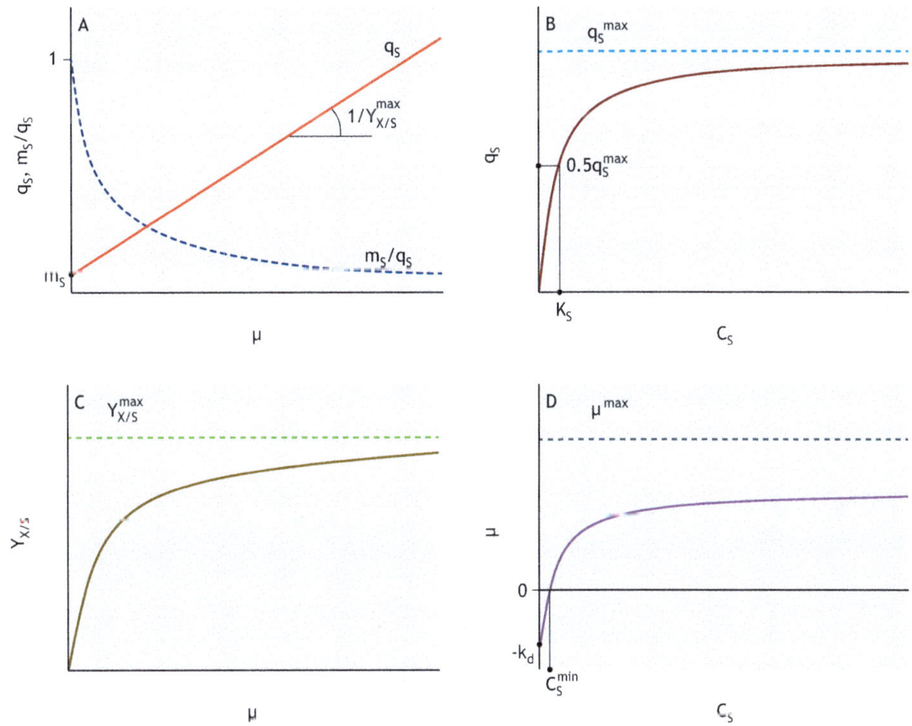

Figure 2.19 A) Biomass-specific substrate uptake rate (q_S) and biomass-specific substrate uptake rate for maintenance (m_S) normalized by q_S as a function of the biomass-specific growth rate (μ). B) Biomass-specific substrate uptake rate as a function of the limiting substrate concertation (C_S). C) Observed growth yield as a function of the biomass-specific growth rate. D) Biomass-specific growth rate as a function of the limiting substrate concertation.

2.6.2 Substrate consumption rate: saturation kinetics

It has also been observed that the biomass-specific substrate consumption rate (q_S) is a function of the actual substrate concentration (C_S), and this behaviour is adequately represented by the following hyperbolic equation:

$$q_S = q_S^{max} \cdot \frac{C_S}{C_S + K_S} \tag{2.30}$$

where q_S^{max} is the maximum biomass-specific substrate uptake rate (mol$_S$/h.mol$_X$), and K_S is the affinity constant for the substrate (mol$_S$/l), where the substrate is assumed to be the single compound limiting the growth in the system. From Eq. 2.30, it is clear that q_S approaches q_S^{max} for high C_S, while it decreases at decreasing C_S and reaches a value of $0.5 \cdot q_S^{max}$ at $C_S = K_S$ (Figure 2.19B).

The value of the affinity constant cannot be estimated on bioenergetic grounds and needs to be estimated experimentally or taken from the literature. Note that the cellular affinity coefficient is less relevant in environmental biotechnology due to the application of flocculent and biofilm systems where mass transfer dominates and apparent K_S values are used (Chapter 14).

Conversely, Heijnen (1999) hypothesized that the value of the maximum biomass-specific substrate uptake rate (q_S^{max}) is restricted by a kinetic limitation in one metabolic process. To this end, the author proposed that the rate of Gibbs energy generation in the catabolism is limited by a maximum catabolic rate of electron transport, q_e^{max} (mol$_e$/h.mol$_X$). Conceptually, this is best understood in the case of a distinct e-donor and e-acceptor, where electrons are transferred from the e-donor (substrate) to the terminal e-acceptor via the membrane-associated electron transport chain (Figure 2.15). In these conversions, a maximum value for the electron transport through the electron transport chain is assumed. This assumption is supported by the observation of *e.g.* a constant maximum oxygen uptake rate ($q_{O_2}^{max}$) in *E. coli* when grown aerobically on different substrates (Anderson and Von Meyenburg, 1980).

An Arrhenius-like correlation, analogous to Eq. 2.28, has been proposed for the maximum biomass-specific electron-transfer rate in the electron transport chain (q_e^{max} in mol$_e$/h.mol$_X$):

$$q_e^{max} = 3 \cdot \exp\left\{-\frac{69}{R}\cdot\left(\frac{1}{T}-\frac{1}{298}\right)\right\} \tag{2.31}$$

where R is the universal gas constant ($8.3145 \cdot 10^{-3}$ kJ/mol.K).

Based on Eq. 2.31, for any given catabolic reaction, the catabolic Gibbs energy generated per unit biomass q_G^{max} (kJ/h.mol$_X$) can be obtained as follows:

$$q_G^{max} = q_e^{max} \cdot \frac{\Delta G_{Cat}^{01}}{\gamma_S^*} \tag{2.32}$$

where ΔG_{Cat}^{01} is the actual Gibbs energy change of the catabolic reaction (kJ/mol$_S$) and γ_S^* the number of electrons per substrate transferred in the catabolism (mol$_e$/mol$_S$). Specifically, if an external e-acceptor is available and the e-donor is an organic substrate, $\gamma_S^* = NoC_{C_S} \cdot \gamma_{C_S}$, where NoC_{C_S} is the number of C per mole of substrate and γ_{C_S} the degree of reduction of the substrate (mol$_e$/C-mol$_S$). In the case of inorganic e-donors or fermentations, γ_S^* defines the actual number of electrons involved in the catabolic reaction per mole substrate (Heijnen and Kleerebezem, 2010).

In an analogy to Eq. 2.25, the Gibbs energy made available in the catabolism is spent by definition for maintenance and growth purposes. By considering the generalized form of Eq. 2.25:

$$q_i = Y_{i/X}^{max} \cdot \mu + Y_{i/S}^{Cat} \cdot m_S \tag{2.33}$$

and realizing that in terms of Gibbs energy, $Y_{i/X}^{max}$ = $Y_{G/X}^{max} = -\Delta G_{Diss}^{01}$ (kJ/mol$_X$), $Y_{i/S}^{Cat} = \Delta G_{Cat}^{01}$ (kJ/C-mol$_S$), and $m_s = -Y_{S/S}^{Cat} \cdot \left(m_G / \Delta G_{Cat}^{01}\right)$ (Eq. 2.29), the specific Gibbs energy generated in the catabolic reaction under maximum growth conditions (kJ/h. mol$_X$) is:

$$q_G^{max} = -\Delta G_{Diss}^{01} \cdot \mu^{max} - m_G \tag{2.34}$$

A reasonable estimation of the maximum biomass-specific growth rate (μ^{max}) can now be calculated from eqs. 2.32 and 2.34, also including the temperature effect from eqs. 2.28 and 2.31, as follows:

$$\mu^{max} = \frac{q_e^{max} \cdot \frac{\Delta G_{Cat}^{01}}{\gamma_S^*} + m_G}{-\Delta G_{Diss}^{01}} = \frac{3 \cdot \frac{\Delta G_{Cat}^{01}}{\gamma_S^*} + 4.5}{-\Delta G_{Diss}^{01}} \cdot \exp\left\{-\frac{69}{R}\cdot\left(\frac{1}{T}-\frac{1}{298}\right)\right\} \tag{2.35}$$

Ultimately, q_S^{max} can be estimated from eqs. 2.33 and 2.35.

To conclude, for every reaction there is only one independent substrate conversion rate, *e.g.* q_S^{max} , and all the other rates are directly derived from it based on the process stoichiometry. Depending on the context, conversion rates are expressed as absolute rates

(R_S, in mol_S/h), volumetric rates (r_S, in mol_S/h.l)[8], or biomass-specific rates (q_S, in mol_S/h.mol_X). However, only q_S can be used to compare the physiological results among different studies, since it reflects the actual metabolic state of cells. Conversely, volumetric rates depend on the biomass concentration in the reactor.

Example 2.15 Aerobic glucose respiration

Based on the values calculated in Example 2.14, and considering that for glucose NoC_{C_S} equals $6\ C\text{-}mol/mol_{C_6H_{12}O_6}$ and $\gamma_S = 4\ mol_e/C\text{-}mol_{C_6H_{12}O_6}$, the maximum biomass-specific growth rate (μ^{max}) of heterotrophic microorganisms growing on glucose at 25 °C can be estimated in 1.5 1/h (Eq. 2.35).

Example 2.16 Aerobic ammonium oxidation

In the oxidation of ammonium to nitrite, six electrons are transferred ($\gamma_S^* = 6\ mol_e/mol_{NH_4-N}$) and the associated Gibbs free energy change is estimated in Example 2.10. Also, as aerobic ammonium oxidizers are autotrophs and ammonium is a poor e-donor, ΔG^{01}_{Diss} equals 3,500 kJ/mol_X. The estimated μ^{max} at 25 °C is 0.04 1/h.

Example 2.17 Anaerobic (or anoxic) ammonium oxidation

Anammox bacteria (examples 2.12 and 2.13) oxidise ammonium to dinitrogen gas ($\gamma_S^* = 3\ mol_e/mol_{NH_4-N}$) and require reverse electron transport ($\Delta G^{01}_{Diss} = 3{,}500\ kJ/mol_X$). Considering the optimal anammox growth temperature of 30 °C, and the ΔG^{01}_{Cat} calculated in Example 2.12, a maximum biomass-specific growth rate of 0.16 1/h is estimated (Eq. 2.35). In this case, this is almost one order of magnitude higher that the μ^{max} observed so far under optimal conditions (Lotti *et al.*, 2015; Zhang *et al.*, 2017), and the difference can possibly be attributed to the unique membrane properties (much higher density) of anammox bacteria resulting in a slower operation of the electron transport chain.

[8] $mol_S\ h^{-1}\ l^{-1}$

2.6.3 Outlook

In conclusion, this section on the mathematical formulation of microbial growth demonstrates that, despite the existing enormous microbial diversity, any microbial growth system can be described by one parameter defining the overall growth stoichiometry - the maximum biomass yield ($Y^{max}_{X/S}$) - and three kinetic parameters, namely the maximum biomass-specific substrate (e-donor) uptake rate (q_S^{max}), the maintenance coefficient (m_S), and the affinity constant for the growth-limiting substrate (K_S).

All the other parameters typically used for the description of a microbial system can be expressed as a function of these four parameters, and be subsequently implemented in bioreactor models. In particular, the actual biomass-specific growth rate, μ, can be directly derived from eqs. 2.26 and 2.30 as follows:

$$\mu = Y^{max}_{X/S} \cdot \left(q_S^{max} \cdot \frac{C_S}{C_S + K_S} - m_S \right) \tag{2.36}$$

From this equation, it is clear that for low C_S values the actual q_S can become lower than m_S, resulting in negative biomass-specific growth rates *i.e.* the biomass concentration will decrease. The minimum substrate concentration at which growth can be sustained (C_S^{min}) is estimated by imposing Eq. 2.36 equal to zero (Figure 2.19D).

It is also worth highlighting how the proposed stoichiometric and kinetic framework relates to commonly adopted modelling approaches in the wastewater treatment field (Chapter 14). Maintenance is commonly implemented as endogenous respiration, namely biomass itself is catabolized for maintenance purposes in the absence of external substrate. Mathematically, endogenous respiration is implemented as a first-order process in biomass with the decay coefficient (k_d), and is fully equivalent to the maintenance concept described here as is evident by rewriting Eq. 2.36 (Figure 2.19D):

$$\mu = \left(Y_{X/S}^{max} \cdot q_S^{max}\right) \cdot \frac{C_S}{C_S + K_S} - \left(Y_{X/S}^{max} \cdot m_S\right) = \\ = \mu^{max} \cdot \frac{C_S}{C_S + K_S} - k_d \tag{2.37}$$

Actually, the Herbert-Pirt relation is based on the work by Herbert (1975) (endogenous decay) and Pirt (1982) (maintenance), and since both concepts result in an equivalent mathematical expression it is referred as the Herbert-Pirt relation. Alternatively, biomass decay is modelled as a separate process coupled to the production of dead biomass, a fraction of which serves as new substrate. In this case, the decay process substitutes the maintenance or endogenous respiration concept.

Here it is also important to stress that in most environmental engineering applications, the estimated Ks values can be significantly higher than the intrinsic (biological) affinity because diffusion processes in flocs and biofilms further limit substrate availability. Ks is thus typically referred to as the apparent substrate affinity.

REFERENCES

Albertsen M., Hugenholtz P., Skarshewski A., Nielsen K.L., Tyson G.W. and Nielsen P.H. (2013). Genome sequences of rare, uncultured bacteria obtained by differential coverage binning of multiple metagenomes. *Nature Biotechnology* 31, 533–538.

Albertsen M., Karst S.M., Ziegler A.S., Kirkegaard R.H. and Nielsen P.H. (2015). Back to Basics - The influence of DNA extraction and primer choice on phylogenetic analysis of activated sludge communities. *PloS One,* 10(7), e0132783.

Andersen K.B. and Von Meyenburg K. (1980). Are growth rates of *Escherichia coli* in batch cultures limited by respiration? *Journal of Bacteriology,* 144(1), 114-23.

Baas-Becking L.G.M. (1934). *Geobiologie of Inleiding Tot de Milieukunde.* Van Stockum and Zoon, The Hague, Netherlands.

Barák I. and Muchová K. (2013). The Role of Lipid Domains in Bacterial Cell Processes. *International Journal of Molecular Sciences,* 14(2), 4050-65.

Beijerinck M.W. (1921). *Verzamelde geschriften.* Delftsch Hoogeschoolfonds.

Brussow H. (2009). The not so universal tree of life or the place of viruses in the living world. *Philosophical Transactions of the Royal Society of London. Series B: Biological Sciences,* 364(1527), 2263-74.

Comeau Y. (2008). Microbial Metabolism. In: *Biological wastewater treatment: Principles, Modelling and Design,* Henze M., Van Loosdrecht M.C.M., Ekama G.A. and Brdjanovic D. (eds.), IWA Publishing, London, pp. 19-32.

Comeau Y., Hall K.J., Hancock R.E.W. and Oldham W.K. (1986). Biochemical model for enhanced biological phosphorus removal. *Water Research,* 20(12), 1511-21.

De Wit R. and Bouvier T. (2006). 'Everything is everywhere, but, the environment selects'; what did Baas-Becking and Beijerinck really say? *Environmental Microbiology,* 8(4), 755-8.

Eikelboom D.H. (2000). *Process Control of Activated Sludge Plants by Microscopic Investigation.* IWA Publishing, London, UK.

Esener A.A., Roels J.A. and Kossen N.W. (1983). Theory and applications of unstructured growth models: Kinetic and energetic aspects. *Biotechnology and Bioengineering,* 25(12), 2803-41.

Greuter D., Loy A., Horn M. and Rattei T. (2015). probeBase - an online resource for rRNA-targeted oligonucleotide probes and primers: new features 2016. *Nucleic Acids Research,* 44(D1), D586-D9.

Hammes F. and Egli T. (2010). Cytometric methods for measuring bacteria in water: advantages, pitfalls and application. *Analytical and Bioanalytical Chemistry,* 397(3), 1083-1095.

Heijnen J.J. (1994). Thermodynamics of microbial growth and its implications for process design. *Trends in Biotechnology,* 12(12), 483-92.

Heijnen J.J. (1999). Bioenergetics of Microbial Growth. In: *Encyclopedia of Bioprocess Technology: fermentation, biocatalysis and bioseparation* Flickinger M.C. and Drew S.W. (eds.), John Wiley and Sons, Chichester, UK, pp. 267-291.

Heijnen J.J. and Kleerebezem R. (2010). Bioenergetics of Microbial Growth. In: *Encyclopedia of Industrial Biotechnology*, pp. 1-66.

Heijnen J.J. and Van Dijken J.P. (1992). In search of a thermodynamic description of biomass yields for the chemotrophic growth of microorganisms. *Biotechnology and Bioengineering,* 39(8), 833-858.

Herbert D. (1975). Stoichiometric aspects of microbial growth. In: *Continuous culture 6: application and new fields,* Dean A.C.R. (ed.), Ellis Horwood, Chichester, UK, pp. 1-30.

Jürgens K., Arndt H. and Zimmermann H. (1997). Impact of metazoan and protozoan grazers on bacterial biomass distribution in microcosm experiments. *Aquatic Microbial Ecology,* 12, 131-8.

Karst S.M., Albertsen M., Kirkegaard R.H., Dueholm M.S. and Nielsen P.H. (2016). Molecular methods. In: *Experimental Methods in Wastewater Treatment,* Van Loosdrecht M.C.M., Nielsen P.H., Lopez-Vazquez C.M. and Brdjanovic D. (eds.), IWA Publishing, London, pp. 285-323.

Kleerebezem R. and Van Loosdrecht M.C.M. (2007). Mixed culture biotechnology for bioenergy production. *Current Opinion in Biotechnology,* 18(3), 207-12.

Kleerebezem R. and Van Loosdrecht M.C.M. (2010). A Generalized Method for Thermodynamic State Analysis of Environmental Systems. *Critical Reviews in Environmental Science and Technology,* 40(1), 1-54.

Klein T., Zihlmann D., Derlon N., Isaacson C., Szivák I., Weissbrodt D.G. and Pronk W. (2016). Biological control of biofilms on membranes by metazoans. *Water Research,* 88, 20-9.

La Rivière J.W.M. (1997). The Delft School of Microbiology in historical perspective. *Antonie van Leeuwenhoek,* 71, 3-13.

Laureni M., Weissbrodt D.G., Szivak I., Robin O., Nielsen J.L., Morgenroth E. and Joss A. (2015). Activity and growth of anammox biomass on aerobically pre-treated municipal wastewater. *Water Research,* 80, 325-36.

Laureni M., Weissbrodt D.G., Villez K., Robin O., De Jonge N., Rosenthal A., Wells G., Nielsen J.L., Morgenroth E. and Joss A. (2019). Biomass segregation between biofilm and flocs improves the control of nitrite-oxidizing bacteria in mainstream partial nitritation and anammox processes. *Water Research,* 154, 104-16.

Lawson C.E., Harcombe W.R., Hatzenpichler R., Lindemann S.R., Löffler F.E., O'Malley M.A., García Martín H., Pfleger B.F., Raskin L., Venturelli O.S., Weissbrodt D.G., Noguera D.R. and McMahon K.D. (2019). Common principles and best practices for engineering microbiomes. *Nature Reviews Microbiology,* 17, 725-41.

Lopez-Vazquez C.M., Oehmen A., Hooijmans C.M., Brdjanovic D., Gijzen H.J., Yuan Z. and Van Loosdrecht M.C.M. (2009). Modeling the PAO-GAO competition: Effects of carbon source, pH and temperature. *Water Research,* 43(2), 450-62.

Lotti T., Kleerebezem R., Abelleira-Pereira J.M., Abbas B. and Van Loosdrecht M.C.M. (2015). Faster through training: The anammox case. *Water Research,* 81, 261-8.

Lotti T., Kleerebezem R., Lubello C. and Van Loosdrecht M.C.M. (2014). Physiological and kinetic characterization of a suspended cell anammox culture. *Water Research,* 60, 1-14.

Lukumbuzya M., Schmid M., Pjevac P. and Daims H. (2019). A Multicolor Fluorescence *in situ* Hybridization Approach Using an Extended Set of Fluorophores to Visualize Microorganisms. *Frontiers in Microbiology,* 10(1383).

Madigan M.T., Bender K.S., Buckley D.H., Stahl D.A. and Sattley W.M. (2018). *Brock Biology of Microorganisms.* Pearson, New York, USA.

McIlroy S.J., Saunders A.M., Albertsen M., Nierychlo M., McIlroy B., Hansen A.A., Karst S.M., Nielsen J.L. and Nielsen P.H. (2015). MiDAS: the field guide to the microbes of activated sludge. *Database,* 2015, bav062.

Metcalf and Eddy AECOM (2014). *Wastewater Engineering: Treatment and Resource Recovery.* McGraw-Hill, New York, USA.

Misra S., Sharma V. and Srivastava A.K. (2015). Bacterial Polysaccharides: An Overview. In: *Polysaccharides: Bioactivity and Biotechnology,* Ramawat K.G. and Mérillon J.-M. (eds.). Springer International Publishing, Cham, pp. 81-108.

Moreira D. and Lopez-Garcia P. (2009). Ten reasons to exclude viruses from the tree of life. *Nature Reviews: Microbiology,* 7(4), 306-11.

Musat N., Foster R., Vagner T., Adam B. and Kuypers M.M. (2012). Detecting metabolic activities in single cells, with emphasis on nanoSIMS. *FEMS Microbiology Reviews,* 36(2), 486-511.

Narayanasamy S., Muller E.E., Sheik A.R. and Wilmes P. (2015). Integrated omics for the identification of key functionalities in biological wastewater treatment microbial communities. *Microbial Biotechnology,* 8(3), 363-8.

Nielsen J.L., Seviour R.J. and Nielsen P.H. (2016). Microscopy. In: *Experimental Methods in Wastewater Treatment* by Van Loosdrecht M.C.M., Nielsen P.H., Lopez-Vazquez C.M. and Brdjanovic D. (eds.), IWA Publishing, London, pp. 263-284.

Nielsen P.H., Daims H. and Lemmer H. (2009). *FISH Handbook for Biological Wastewater Treatment - Identification and quantification of microorganisms in activated sludge and biofilms by FISH.* IWA Publishing, London, UK.

Nielsen P.H. and McMahon K.D. (2014). Microbiology and microbial ecology of the activated sludge process. In: *Activated Sludge - 100 Years and Counting,* Jenkins D. and Wanner J. (eds.), IWA Publishing, London, UK, p. 464.

Nielsen P.H., Mielczarek A.T., Kragelund C., Nielsen J.L., Saunders A.M., Kong Y., Hansen A.A. and Vollertsen J. (2010). A conceptual ecosystem model of microbial communities in enhanced biological phosphorus removal plants. *Water Research,* 44(17), 5070-88.

Nierychlo M., Andersen K.S., Xu Y., Green N., Albertsen M., Dueholm M.S. and Nielsen P.H. (2019). Species-level microbiome composition of activated sludge - introducing the MiDAS 3 ecosystem-specific reference database and taxonomy. *bioRxiv*, 842393.

Pirt S.J. (1982). Maintenance energy: a general model for energy-limited and energy-sufficient growth. *Archives of Microbiology,* 133(4), 300-2.

Rittmann B.E. and McCarty P.J. (2001). *Environmental Biotechnology - Principles and Applications*. McGraw-Hill, New York, USA.

Rocha J. and Manaia C. M. (2020). Cell-based internal standard for qPCR determinations of antibiotic resistance indicators in environmental water samples. *Ecological Indicators*, 106194.

Rubio-Rincon F.J., Weissbrodt D.G., Lopez-Vazquez C.M., Welles L., Abbas B., Albertsen M., Nielsen P.H., Van Loosdrecht M.C.M. and Brdjanovic D. (2019). "*Candidatus* Accumulibacter delftensis": A clade IC novel polyphosphate-accumulating organism without denitrifying activity on nitrate. *Water Research,* 161, 136-51.

Seviour T., Derlon N., Dueholm M.S., Flemming H.-C., Girbal-Neuhauser E., Horn H., Kjelleberg S., Van Loosdrecht M.C.M., Lotti T., Malpei M.F., Nerenberg R., Neu T.R., Paul E., Yu H. and Lin Y. (2019). Extracellular polymeric substances of biofilms: Suffering from an identity crisis. *Water Research,* 151, 1-7.

Shoener B.D., Schramm S.M., Beline F., Bernard O., Martinez C., Plosz B.G., Snowling S., Steyer J.P., Valverde-Perez B., Wagner D. and Guest J.S. (2019). Microalgae and cyanobacteria modeling in water resource recovery facilities: A critical review. *Water Research,* 2, 100024.

Tao Y., Huang X., Gao D., Wang X., Chen C., Liang H. and Van Loosdrecht M.C.M. (2019). NanoSIMS reveals unusual enrichment of acetate and propionate by an anammox consortium dominated by Jettenia asiatica. *Water Research,* 159, 223-32.

Tijhuis L., Van Loosdrecht M.C.M. and Heijnen J.J. (1993). A thermodynamically based correlation for maintenance Gibbs energy requirements in aerobic and anaerobic chemotrophic growth. *Biotechnology and Bioengineering,* 42(4), 509-19.

Van Niel C.B. (1949). The 'Delft School' and the Rise of General Microbiology. *Bacteriological Reviews,* 13(3), 161-74.

Weissbrodt D.G. and Holliger C. (2013). Intensification du traitement biologique des eaux usées : Technologie à boues granulaires et Gestion des ressources bactériennes. *Bulletin de l'ARPEA - Journal Romand de l'Environnement,* 256, 12-22.

Weissbrodt D.G., Holliger C. and Morgenroth E. (2017). Modeling hydraulic transport and anaerobic uptake by PAOs and GAOs during wastewater feeding in EBPR granular sludge reactors *Biotechnology and Bioengineering,* 114(8), 1688-702.

Weissbrodt D.G., Lochmatter S., Ebrahimi S., Rossi P., Maillard J. and Holliger C. (2012a). Bacterial selection during the formation of early-stage aerobic granules in wastewater treatment systems operated under wash-out dynamics. *Frontiers in Microbiology,* 3, 332.

Weissbrodt D.G., Shani N., Sinclair L., Lefebvre G., Rossi P., Maillard J., Rougemont J. and Holliger C. (2012b). PyroTRF-ID: a novel bioinformatics methodology for the affiliation of terminal-restriction fragments using 16S rRNA gene pyrosequencing data. *BMC Microbiology,* 12, 306.

Weissbrodt D.G., Neu T.R., Kuhlicke U., Rappaz Y. and Holliger C. (2013). Assessment of bacterial and structural dynamics in aerobic granular biofilms. *Frontiers in Microbiology,* 4, 175.

Weissbrodt D.G., Shani N. and Holliger C. (2014). Linking bacterial population dynamics and nutrient removal in the granular sludge biofilm ecosystem engineered for wastewater treatment. *FEMS Microbiology Ecology,* 88(3), 579-95.

Weissbrodt D.G., Wells G.F., Laureni M., Agrawal S., Goel R., Russo G., Men Y., Johnson D.R., Christensson M., Lackner S., Joss A., Nielsen J.L., Bürgmann H. and Morgenroth E. (2020). Systems Microbiology and Engineering of Aerobic-Anaerobic Ammonium Oxidation. *ChemRxiv*, Preprint.

Weissbrodt D.G. (2018). StaRRE - Stations de récupération des ressources de l'eau: Intensification, bioraffinage et valorisation. *Aqua & Gas*, 1, 20-24.

Wells G.F., Shi Y., Laureni M., Rosenthal A., Szivak I., Weissbrodt D.G., Joss A., Buergmann H., Johnson D. R. and Morgenroth E. (2017). Comparing the Resistance, Resilience, and Stability of Replicate Moving Bed Biofilm and Suspended Growth Combined Nitritation-Anammox Reactors. *Environmental Science and Technology,* 51(9), 5108-17.

Winkler M.K.H., Meunier C., Henriet O., Mahillon J., Suarez-Ojeda M.E., Del Moro G., De Sanctis M., Di Iaconi C. and Weissbrodt D.G. (2018). An integrative review of granular sludge for the biological removal of nutrients and of recalcitrant organic matter from wastewater. *Chemical Engineering Journal,* 336, 489-502.

Xiao W., Wang R.-S., Handy D.E. and Loscalzo J. (2018). NAD(H) and NADP(H) Redox Couples and Cellular Energy Metabolism. *Antioxidants and Redox Signaling,* 28(3), 251-72.

Zhang L., Narita Y., Gao L., Ali M., Oshiki M. and Okabe S. (2017). Maximum specific growth rate of anammox bacteria revisited. *Water Research,* 116, 296-303.

NOMENCLATURE

Symbol	Description and unit	Unit	Unit
A	Arrhenius constant[a)]	a)	a)
C_i	Concentration of compound i	mol_i/l	$mol_i\ l^{-1}$
ΔG	Gibbs free energy change per mole of compound i	kJ/mol_i	$kJ\ mol_i^{-1}$
ΔH	Enthalpy change per mole of compound i	kJ/mol_i	$kJ\ mol_i^{-1}$
γ_i	Degree of reduction of compound i	mol_e/mol_i	$mol_e\ mol_i^{-1}$
k	Rate constant	b)	b)
K_i	Affinity or half-saturation constant for compound i	mol_i/l	$mol_i\ l^{-1}$
$K_{I,i}$	Half-saturation inhibitory constant	mol/l	$mol\ l^{-1}$
I	Light intensity	W/m or J/s.m	$W\ m^{-1}$ or $J\ s^{-1}\ m^{-1}$
λ_{An}	Multiplication factor for anabolism	mol_X/mol_X	$mol_X\ mol_X^{-1}$
λ_{Cat}	Multiplication factor for catabolism	mol_{eD}/mol_X	$mol_{eD}\ mol_X^{-1}$
μ	Biomass-specific rate of microbial growth	$mol_X/h.mol_X$	$mol_X\ h^{-1}\ mol_X^{-1}$
m_i	Biomass-specific rate of conversion of compound i for maintenance	$mol_i/h.mol_X$	$mol_i\ h^{-1}\ mol_X^{-1}$
m_G	Biomass-specific rate of Gibbs free energy consumption for maintenance	$kJ/h.mol_X$	$kJ\ h^{-1}\ mol_X^{-1}$
NoC	Number of carbon atoms	$C\text{-}mol/mol_i$	$C\text{-}mol\ mol_i^{-1}$
R_i	Absolute (or total) rate of conversion of compound i	mol_i/h	$mol_i\ h^{-1}$
r_i	Volumetric rate of conversion of compound i *i.e.* total rate divided by the volume of reactor	$mol_i/h.l$	$mol_i\ h^{-1}\ l^{-1}$
q_i	Biomass-specific rate of conversion of compound i *i.e.* total rate divided by the mass of cells of the population of interest in the reactor	$mol_i/h.mol_X$	$mol_i\ h^{-1}\ mol_X^{-1}$
q_G	Biomass-specific rate of Gibbs free energy production or consumption	$kJ/h.mol_X$	$kJ\ h^{-1}\ mol_X^{-1}$
Q_{10}	Temperature coefficient that measures the rate of change under an increase of 10 °C	-	-
$S_{I,i}$	Concentration of dissolved inhibitory compound	mol/l	$mol\ l^{-1}$
T	Temperature	°C or K	°C or K
θ	Temperature coefficient	-	-
$Y_{i/j}$	Stoichiometric coefficient or yield of conversion of material i per conversion of material j	mol_i/mol_j	$mol_i\ mol_j^{-1}$

[a)] Depends on reaction order *i.e.* on units of k; [b)] depends on reaction order.

Abbreviation	
A	Acceptor
An	Anabolism
Bchl	Bacteriochlorophyll
anammox	microbiol.: Anaerobic ammonium oxidation; eng.: Anoxic ammonium oxidation
Cat	Catabolism
Chl	Chlorophyll
comammox	Complete ammonium oxidation into nitrate
D	Donor
Diss	Dissipated
eA	Electron acceptor (e-acceptor)
eD	Electron donor (e-donor)
S	Substrate
X	Biomass

Acronym	
ADM	Anaerobic digestion model
ADP	Adenosine diphosphate
AEC	Adenylate energy charge
AMO	Anammox organism
AMP	Adenosine monophosphate
ANO	Autotrophic nitrifying organism
AOA	Ammonium-oxidizing archaea
AOB	Ammonium-oxidizing bacteria
AOO	Ammonium-oxidizing organism
ARDRA	Amplified ribosomal DNA restriction analysis
ARISA	Automated ribosomal intergenic spacer analysis
ASM	Activated sludge model
ASV	Amplicon sequence variant
ATP	Adenosine triphosphate
BNR	Biological nutrient removal
BONCAT	Bioorthogonal non-canonical amino acid tagging
CAPEX	Capital expenditure
cDNA	Complementary DNA
CoA	Coenzyme A
CLSM	Confocal laser scanning microscopy
COD	Chemical oxygen demand

DGGE	Denaturing gradient gel electrophoresis
DNA	Deoxyribonucleic acid
DHO	Denitrifying heterotrophic organism
EBPR	Enhanced biological phosphorus removal
EFM	Epifluorescence microscopy
EPS	Extracellular polymeric substance
ETC	Electron transport chain
FADH/$FADH_2$	Flavin adenine dinucleotide (oxidized/reduced form)
FCM	Flow cytometry
FISH	Fluorescence in-situ hybridization
FLBA	Fluorescent lectin binding analysis
FSS	Fixed suspended solids
GAO	Glycogen-accumulating organism
gDNA	Genomic DNA
GTP	Guanosine-5'-triphosphate
HAA	Helix Aspersa lectin
HRT	Hydraulic retention time
ISS	Inorganic suspended solids
LUCA	Last universal common ancestor
MAR	Microautoradiography
MIDAS	Microbial database for activated sludge
NAD^+/NADH	Nicotinamide adenine dinucleotide (oxidized and reduced forms)
$NADP^+$/NADPH	NAD phosphate (oxidized and reduced forms)
NanoSIMS	Nanoscale SIMS
NIR	Near-infrared
NOO	Nitrite-oxidizing organism
NO_x^-	Oxidized nitrogen compound
OHO	Ordinary heterotrophic organism
OPEX	Operating expenditure
OTU	Operational taxonomic unit
PAO	Polyphosphate-accumulating organism
PCR	Polymerase chain reaction
PHA	Poly-β-hydroxyalkanoate
PHB	poly-β-hydroxybutyrate
PHV	Poly-β-hydroxyvalerate
PLFA	Phospholipid fatty acid
PNSB	Purple non-sulphur bacteria
PSB	Purple sulphur bacteria
PyroTRF-ID	Identification of T-RFs by pyrosequencing data

qFISH	Quantitative FISH
qPCR	Quantitative PCR
RET	Reverse electron transfer
RNA	Ribonucleic acid
mRNA	Messenger RNA
rRNA	Ribosomal RNA
tRNA	Transfer RNA
RT-qPCR	Reverse transcription (RT) and qPCR
SIMS	Secondary ion mass spectrometry
SIP	Stable isotope probing
SRB	Sulphate-reducing bacteria
SRT	Sludge retention time
TCA	Tricarboxylic acid
T-RF	Terminal-restriction fragment
T-RFLP	Terminal-restriction fragment length polymorphism
TSS	Total suspended solids
VFA	Volatile fatty acid
VSS	Volatile suspended solids
WWTP	Wastewater treatment plant

doi: 10.2166/9781789060362_0077

3 Wastewater characteristics

Eveline I.P. Volcke, Kimberly Solon, Yves Comeau and Mogens Henze

3.1 WASTEWATER TYPES AND THEIR CHARACTERISTICS

3.1.1 Sources of wastewater

Water is consumed during various types of human activities. The required water quality depends on its purpose. This water 'fit-for-purpose' concept constitutes the basis of several water reuse schemes, minimizing the overall water consumption. Nevertheless, the production of wastewater which is finally discharged into sewers (a centralized sanitation system) or treated on-site (decentralized wastewater treatment) remains unavoidable.

The terms 'municipal wastewater' or 'sewage' refer to wastewater which is transported through sewers to a (centralized) wastewater treatment plant. It comprises wastewater from households, institutions and industries, as well as groundwater which inevitably infiltrates into the sewers through cracks and/or leaks.

The design of the sewer system affects the wastewater composition. What are referred to as separate sewer systems have a second, separate pipe to collect the stormwater from the surface run-off, while in combined sewer systems the stormwater is collected together with wastewater from households, institutions and industries, in a single pipe, resulting in dilution and higher volumes of sewage. Combined sewers are obviously cheaper but entail the risk of overloading wastewater treatment plants during long periods of heavy rainfall, resulting in a part of the wastewater being discharged into local water bodies, with no or insufficient treatment. In the worst case, even flooding from sewers can take place. For this reason, most newly built or renovated sewer systems are separate sewer systems. Combined sewer systems remain in place in old urban areas, in developing countries, and where the sewer system is difficult to access.

As well as the wastewater in the sewerage, a wastewater treatment plant also needs to handle some external inputs. The amount of external inputs is typically higher for larger plants. Examples of such

Table 3.1 Wastewater types, as determined by their origin.

Wastewater from society	Wastewater generated internally in treatment plants
• Domestic wastewater (*i.e.* from households)	• Thickener supernatant
• Wastewater from institutions (*e.g.* hospitals, schools)	• Digester supernatant
• Industrial wastewater	• Reject water from sludge dewatering
• Infiltration into sewers	• Drainage water from sludge drying beds
• Stormwater	• Filter-washing water
• Landfill leachate	• Equipment-cleaning water
• Septic tank wastewater/sludge	

external inputs are landfill leachate and septic tank wastewater. When no decentralized treatment is available, these streams are brought to a centralized wastewater treatment plant. For leachate, dedicated transport can be organized, or it can be released into a sewer near the landfill. Septic tank wastewater or sludge is typically loaded into the wastewater treatment plant by trucks. Furthermore, some specific wastewater streams are also generated internally in wastewater treatment plants, such as thickener supernatant, digester supernatant, reject water from sludge dewatering, drainage water from sludge drying beds, filter-washing water and equipment-cleaning water.

Table 3.1 summarizes the various wastewater types. The effect of external inputs and internally generated wastewater treatment streams on the wastewater treatment plant operation is discussed further in this chapter, based on their composition (Section 3.11).

3.1.2 General overview of wastewater constituents

The constituents in wastewater can be divided into main categories according to Table 3.2. The contribution of these constituents can vary considerably.

Table 3.2 Constituents present in domestic wastewater (based on Henze *et al.*, 2002).

Wastewater constituents	Specification (examples)	Hazard
Microorganisms	Pathogenic bacteria, virus and worm eggs	Risk when bathing and eating shellfish
Biodegradable organic carbonaceous materials		Oxygen depletion in rivers, lakes and fjords, in their turn resulting in fish death, odours
Nutrients	Nitrogen and phosphorus components	Eutrophication, oxygen depletion, toxic effect
Sulphur components	Hydrogen sulphide, sulphate	Odour, toxic effect, corrosion
Cellulose	Mainly originating from toilet paper	Increased sludge production
Micropollutants	Biocides, pesticides, solvents, personal care products, pharmaceuticals	Toxic effect, bioaccumulation
Metals	Hg, Pb, Cd, Cr, Cu, Ni	Toxic effect, bioaccumulation
Other organic materials	Detergents, pesticides, fat, oil and grease, colouring, solvents, phenols, cyanide	Toxic effect, aesthetic inconveniences, bioaccumulation in the food chain
Other inorganic materials	Acids, for example hydrogen sulphide, bases	Corrosion, toxic effect
Thermal energy	Discharge of water streams with different temperatures	Changing living conditions for flora and fauna
Radioactivity		Toxic effect, accumulation

The following sections elaborate on the importance, the characterization and the fractionation of various wastewater constituents (sections 3.3-3.10). For a better understanding of the fractionation, an introduction to the physical and chemical occurrence of wastewater components is given first (Section 3.2). This chapter further contains an overview of typical characteristics for various types of wastewater (Section 3.11), and touches upon their dynamic behaviour and its implications (Section 3.12). The chapter is concluded with a summary of available wastewater characterization protocols for modelling purposes (Section 3.13).

3.2 PHYSICAL AND CHEMICAL OCCURRENCE OF WASTEWATER COMPONENTS

3.2.1 Soluble *versus* colloidal *versus* particulate constituents

Wastewater components can be classified according to their physical appearance, more specifically the phase in which they occur. This difference is typically reflected in the notation of these components:

- S: soluble
- C: colloidal
- X: particulate

Soluble components are dissolved in water, making up a single liquid phase. Particulate matter is in the form of (very) tiny separate solid-phase particles. In between soluble and particulate components are colloidal components, which form a substance of microscopically dispersed particles in wastewater. A colloidal mixture does not settle or would take a very long time to appreciably do so.

In practical wastewater operation as well as in most wastewater treatment models, only soluble and particulate components are distinguished from each other. The distinction is typically based on filtration over a 0.45 μm filter. Colloids smaller than 0.45 μm are accounted for as soluble components, while colloids larger than 0.45 μm are considered particulate. As an alternative, the distinction between soluble and particulate components is sometimes made by a 0.1 μm filtration or - *quasi* equivalent - flocculation with $Zn(OH)_2$ followed by 0.45 μm filtration (Mamais *et al.*, 1993) . Here colloids smaller than 0.1 μm are considered as soluble components, while colloids larger than 0.1 μm are considered as particulate components.

Table 3.3 summarizes typical values for soluble and particulate components in municipal wastewater. The fractionation between soluble and particulate components influences the wastewater treatment options. A large fraction of the organic matter (expressed in COD or BOD, see Section 3.4) is particulate and can thus be removed through settling. In contrast, the largest share of N- and P-containing components is typically soluble, so these nutrients cannot be removed by settling, filtration, flotation or other means of solid-liquid separation.

Table 3.3 Typical distribution for soluble and particulate components for medium-concentration municipal wastewater (in g/m^3).

Parameter	Soluble	Particulate	Total
COD	200[a)]	550[a)]	750[a)]
BOD	140	210	350
N total	50[a)]	10[a)]	60[a)]
P total	11	4	15[a)]

[a)]Rounded values corresponding with the raw wastewater example in chapters 4-6, Table 3.6 and Table 3.8 (Ekama, 2009).

It should be realized that not all particulate matter is settleable. Particulate matter consists of settleable and non-settleable fractions. The non-settleable fraction comprises colloidal components as well as particulate matter which does not settle for the given settler design and operation.

The simplified view of only considering soluble and particulate components in Activated Sludge Models (ASMs, *e.g.* ASM1 - Henze *et al.*, 1987 and ASM2 – Henze *et al.*, 1995) has some implications which need to be considered. First of all, most models assume soluble organic carbon components (S_S) to be readily biodegradable, while particulate organic carbon components (X_S) are assumed slowly biodegradable. In reality, however, part of the colloidal organic fraction can be measured as soluble (if the colloid size is smaller than the pore size of the filter) and will as such be modelled as readily biodegradable even though in reality colloids are slowly biodegradable. This nuance will not be noticeable while modelling typical wastewater treatment plants in which the sludge retention time (SRT) is sufficiently long for all biodegradable organics (both soluble and particulate, both readily and slowly biodegradable) to be degraded. However, it will play a role in high-rate activated sludge systems, in which the sludge retention time (SRT) is too low for degradation of the soluble but slowly biodegradable colloid matter. In order to accurately describe this phenomenon, one option is to split up the soluble components into readily and slowly degradable fractions and model their degradation separately (*e.g.* Nogaj *et al.*, 2015), while another, simpler approach is to adjust the influent fractionation. In the latter approach, the so-called particulate (X_S) fraction is taken as larger than it would be based on experimental results (filtering), in order to reflect that part of the soluble organics - more specifically the colloidal part - are also slowly biodegradable (Smitshuijzen *et al*, 2016).

Another implication of ignoring the colloid fraction and only distinguishing between soluble and particulate organic matter relies on the adsorption/bioflocculation and settling behaviour. The adsorption of suspended organic matter is a fast process and considered instantaneous in ASM1 (Henze *et al.*, 1987). In activated sludge processes with a long contact time, colloids will typically adsorb in the flocs and subsequently settle, which is satisfactory for colloids which have been modelled as particulate components (namely the ones for which the colloid size is larger than the pore size of the filter). However, in high-rate processes, the short contact time can be too short for colloids to adsorb on larger particles and thus to allow their subsequent settling, even though part of these colloids may have been modelled as particulate components. The efficiency of adsorption is known to be linked to a complex process that drive the sludge characteristics, including the type and fraction of extracellular polymeric substances (EPS) and the presence of storage compounds. One can aim to model all of these phenomena, while explicitly considering colloidal matter as a state variable, as in Nogaj *et al.* (2015). However, even though such a description can be mechanistically more correct, it is definitely more complex and comes at the expense of having to estimate more unknown parameters. A simple alternative is to keep the distinction between soluble and particulate organics only (no colloids), and to lump the adsorption and settling process into a single 'settling efficiency' parameter which can be estimated from routinely measured data in wastewater treatment processes (Smitshuijzen *et al*, 2016). The comparison between both approaches in modelling high-rate activated sludge processes was also touched upon by Jia *et al.* (2020).

The important point is that in most cases, simpler models - in this case only distinguishing between soluble and particulate components, without distinguishing colloids - perform best for overall plant description and optimization. However, as with any model, one needs to be aware of the underlying assumptions.

Modelling colloidal matter explicitly can still be of value if very low effluent concentrations need to be reached, a condition when the behaviour of the colloidal matter becomes significant. Advanced treatment systems including membrane or adsorption processes are increasingly being used for such purposes. The particle size of colloidal matter depends on the purpose of the model used and the method of its determination and can typically be in the range 0.01 to 1 μm. The simulation platform BioWin distinguishes between colloidal (approximately 0.04 to 1.2 μm) and particulate matter (larger than 1.2 μm); the slowly biodegradable COD is considered to be the sum of both the colloidal and particulate fractions.

3.2.2 Organic *versus* inorganic constituents

Wastewater contains organic as well as inorganic constituents. Organic matter can be expressed in terms of COD and can be either biodegradable or unbiodegradable. In addition, a distinction can also be made between soluble and particulate organic matter (see Section 3.4.2). The particulate COD fractions (both particulate biodegradable COD and particulate unbiodegradable COD) can also be expressed in terms of volatile suspended solids (VSS), *e.g.* using the conversion factor from Eq. 3.1, which is in agreement with typical medium wastewater compositions (Table 3.17):

$$1\ \text{gVSS} = 1.48\ \text{gCOD} \tag{3.1}$$

As well as organic matter, wastewater also contains inorganic material. The particulate inorganic material is termed ISS: inert suspended solids. Inorganic (ISS) build-up in the wastewater treatment plan originates directly from inorganics in the influent. Typical influent inorganic concentrations (Ekama, 2009) are approximately 50 mgISS/l for raw wastewater and 10 mg ISS/l for settled wastewater, *i.e.* raw wastewater that has undergone primary settling, which in this case removes approximately 80% of the influent organics. The sum of VSS and ISS is TSS: total suspended solids (see Table 3.4, for typical values).

Table 3.4 ISS, VSS and TSS values (in mg/l), for example raw and settled wastewaters (adapted from Ekama, 2009).

Parameter	Raw wastewater	Settled wastewater
Inorganic suspended solids (ISS)	48	9.5
Volatile (= organic) suspended solids (VSS)	253	69.2
Total suspended solids (TSS)	301	78.7
VSS:TSS ratio (for influent wastewater)	0.84	0.87

It is clear that, as well as carbonaceous matter, other components such as nitrogen and phosphorus can also be present in organic inorganic forms. This is elaborated on in sections 3.5 and 3.6.

3.3 MICROORGANISMS

Protecting human health remains today the prime reason for wastewater treatment. Historically, wastewater handling has been driven by the need to remove the infectious elements to outside the reach of the population in the cities. Back in the 19th century microorganisms were identified as the cause of diseases; it was recognized that a high concentration of microorganisms creates a severe health risk when raw wastewater is discharged to receiving waters.

The microorganisms in wastewater come mainly from human excreta, as well as from the food industry. Table 3.5 gives an idea of the concentration of microorganisms in domestic wastewater. Chapter 8 gives more information on pathogenic microorganisms and their removal from wastewater.

Table 3.5 Concentrations of microorganisms in wastewater (number of microorganisms per 100 ml) (based on Henze *et al.*, 2002).

Microorganisms	Low	High
Escherichia coli	10^6	$5{\cdot}10^8$
Coliforms	10^{11}	10^{13}
Clostridium perfringens	10^3	$5{\cdot}10^4$
Faecal *streptococci*	10^6	10^8
Salmonella	50	300
Campylobacter	$5{\cdot}10^3$	10^5
Listeria	$5{\cdot}10^2$	10^4
Staphylococus aureus	5.10^3	10^5
Coliphages	10^4	$5{\cdot}10^5$
Giardia	10^2	10^3
Roundworms	5	20
Enterovirus	10^3	10^4
Rotavirus	20	100

3.4 ORGANIC MATTER

3.4.1 Characterization: BOD *versus* COD

Organic matter is typically measured in terms of biochemical oxygen demand (BOD) and/or chemical oxygen demand (COD).

3.4.1.1 BOD

BOD analysis measures the oxygen used for oxidation of part of the organic matter. More specifically, it represents the amount of oxygen consumed by microorganisms to biochemically degrade organic matter in a water sample. It relies on microbial metabolisms and can thus be seen as what is termed a 'respiration' test. Standard BOD analysis takes 5 days (BOD_5). During the test, a water sample in a closed bottle is incubated in the dark at a temperature of 20 °C over a 5-day period. The difference between the dissolved oxygen concentration before and after incubation is taken as BOD_5, measured as mgO_2/l. When dealing with wastewater, dilution is necessary to avoid depletion of the oxygen present in the sample for the duration of the incubation period. The BOD test also measures the amount of oxygen used to oxidize reduced forms of nitrogen, *i.e.* the nitrogenous biochemical oxygen demand (NBOD). In order to avoid the latter and only measure the carbonaceous BOD, a nitrification inhibitor (typically ATU: allylthiourea) should be added to the test. Alternative test lengths to determine BOD can be applied as well. A BOD test over only 1 day (BOD_1) can be applied if results need to be obtained quickly, while in Norway and Sweden a BOD test, BOD_7, over 1 week is conventionally used. It is clear that the measured BOD value increases with increased test length since more time is available for organic matter degradation (Figure 3.1). If a measurement of (almost) all the biodegradable material is required, what is called the 'ultimate BOD' (BOD_∞) test is used. The BOD after an incubation period of at least 3 weeks, *e.g.* BOD_{25}, is found to be a good approximation of BOD_∞ (Van Haandel and Van Der Lubbe, 2007). Metcalf and Eddy AECOM (2014) state that 95-99% of the ultimate BOD is reached after 20 days. However, a disadvantage of measuring over a long time period, apart from being time-consuming, is that a larger dilution rate is required to avoid oxygen limitation, which compromises the reliability of the BOD_{25} measurement. It is possible to estimate various BOD values from a single measured value; a typical ratio between BOD_5 and $BOD_{25} \approx BOD_\infty$ is approximately 0.6-0.7 (Table 3.6). The actual ratio between BOD_5 and BOD_{25} depends on the type of wastewater and lies in the range 0.5-0.95 (Roeleveld and Van Loosdrecht, 2002). In this chapter, the term BOD refers to the standard carbonaceous BOD_5 analysis.

Table 3.6 BOD and COD values in urban wastewater.

BOD_1	BOD_5	BOD_7	BOD_{25}	COD
40	100	115	150	210
200	500	575	750	1,100

BOD measurements are also dependent on temperature. For the same incubation period a higher BOD is recorded at higher temperatures (see Figure 3.1). This can be attributed to higher biological activity at increasing temperatures. In view of reproducibility and comparability, it is therefore essential that the BOD test is carried out under standard conditions (20 °C). For more details on BOD analysis, the reader is referred to Spanjers and Vanrolleghem (2016).

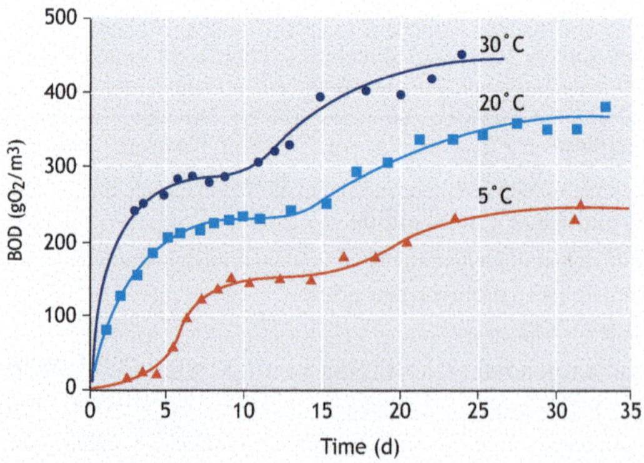

Figure 3.1 BOD analysis results depend on both test length and temperature. The standard is 20 °C and 5 days (Henze *et al.*, 2002).

The typical test length of 5 days makes BOD analysis slow. The method is also cumbersome due to the need for dilution. COD analysis provides a quicker alternative for measuring organic matter in water.

3.4.1.2 COD

COD analysis provides a quick alternative for measuring organic matter. It measures through chemical oxidation by dichromate (processed via hexavalent chromium (Cr(VI)) the majority of the organic matter present in the sample. Mercuric sulphate ($HgSO_4$) can also be added to the sample during COD analysis to prevent interference due to the presence of chloride ions. The excess dichromate is titrated with ferrous ammonium sulphate, the value of which is used to calculate the COD concentration in mgO_2/l. Permanganate can also be used instead of dichromate, however in this case one should beware of a false COD measurement since this method only measures part of the organic matter, and should only be used in relation to planning BOD analysis.

The generally-applied experimental methods for COD determination described above are still based on the use of hazardous chemicals such as mercury (Hg) and hexavalent chromium (Cr(VI)). These chemicals are classified as priority pollutants according to the EU Water Framework Directive and USEPA Clean Water Act and as (possible) human carcinogens. In Japan and the EU there are restrictions on the manufacturing, sale and usage of these substances. Thus, alternative methods for COD determination are currently being considered and examined. The most frequently mentioned alternatives rely on testing for other parameters, such as BOD, TOC or TOD, and using their correlation with COD. However, relationships between these parameters and COD can vary from one sample to another or over time. Hg- and/or Cr(VI)-free COD analysis methods, such as the use of Mn(III), Mn(VII), Ag_2SO_4 and electrochemical methods, are also being studied but have been found to have limitations.

The theoretical COD of a carbonaceous substance is defined as the mass of oxygen needed for its full oxidation to carbon dioxide and water. Its value is calculated from the corresponding oxidation equation. For example, the theoretical COD of ethanol is calculated based on Eq. 3.2:

$$C_2H_6O + 3O_2 \rightarrow 2CO_2 + 3H_2O \tag{3.2}$$

from which it is clear that the full oxidation of 1 mole (or 46 g) of ethanol requires 3 moles (96 g) of oxygen. The theoretical COD of ethanol is thus 96/46 = 2.09.

3.4.2 COD fractionation

Organic matter can be quantified in terms of COD, so components which are expressed in COD are organic components. Given that biological wastewater treatment relies on microbial conversions, a further categorization into biodegradable and unbiodegradable fractions is relevant. An additional distinction between soluble and particulate components is made in view of liquid-solid separation. The particulate COD provides information about the expected sludge production. As a result, four COD fractions are distinguished (Figure 3.2): soluble biodegradable organics (S_S), soluble unbiodegradable organics (S_U), particulate biodegradable organics (X_S) and particulate unbiodegradable organics (X_U).

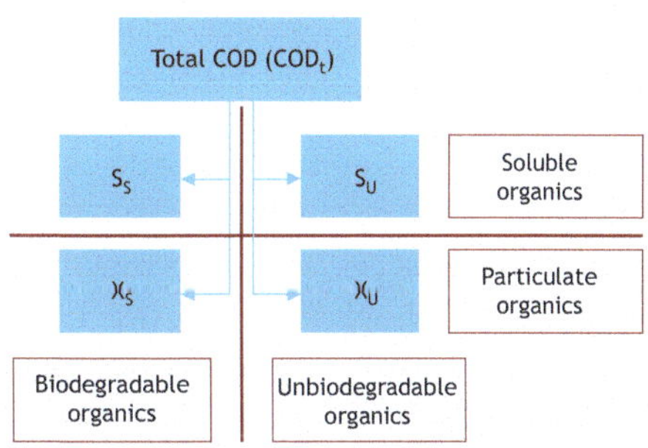

Figure 3.2 Fractionation of COD in wastewater.

In most models, soluble organic carbon components (S_S) are assumed to be readily biodegradable, while particulate organic carbon components (X_S) are assumed to be slowly biodegradable. Some of the implications of these assumptions were discussed in Section 3.2. In most wastewater treatment plants the sludge retention time (SRT) is sufficiently long for all the biodegradable organics (both soluble and particulate) to be degraded, so no distinction needs to be made between readily and slowly degradable organic carbon, at least not as far as the removal of organic material is concerned. In view of nitrogen removal, a distinction between readily biodegradable COD and slowly biodegradable COD is important to accurately calculate the denitrification potential (see Chapter 5). In view of biological phosphorus removal, soluble biodegradable organic carbon is further split into volatile fatty acids (VFA, S_{VFA}) and a fermentable COD fraction (S_F), the uptake of which is divided between phosphorus-accumulating organisms (PAOs) and ordinary heterotrophic organisms (OHOs) (See Chapter 6).

During biological wastewater treatment, all the biodegradable COD can be assumed to be converted for heterotrophic biomass growth while the particulate unbiodegradable COD ends up in the waste sludge. If there is 'perfect' ('ideal') settling, the effluent does not contain particulate material. Thus, only the soluble unbiodegradable COD leaves the system through the effluent.

Typical influent fractionation, *i.e.* to determine the quantities of all the wastewater COD fractions presented in Figure 3.2, takes place according the following procedure:

(1) Quantification of total COD (COD_t) based on standard analytical practices (see Section 3.4.1).
(2) Determination of soluble COD (S_S + S_U) by performing COD analysis on the filtered (typically 0.45 μm) wastewater.
(3) Determination of the particulate (X_S + X_U) COD as the difference between the total COD (COD_t) and the soluble COD (S_S + S_U).
(4) Determination of soluble unbiodegradable influent COD (S_U), based on the measured filtered COD in the effluent of the wastewater treatment plant, all of which is typically assumed unbiodegradable and thus equals the influent soluble unbiodegradable COD (see Chapter 4). Alternatively, S_U can be taken as a fraction (*e.g.*, 0.9) of the measured filtered effluent COD for low-loaded wastewater treatment plants or less for high-loaded ones (Roeleveld and Van Loosdrecht, 2002), which implicitly assumes that the effluent still contains some biodegradable organic matter.
(5) Determination of soluble biodegradable COD (S_S) by subtracting the fraction S_U from the soluble COD.
(6) Determination of the biodegradable COD (COD_b = S_S + X_S. based on standard methods, *e.g.* BOD analysis, with suppressed nitrification (see Section 3.4.1).
(7) Determination of the particulate biodegradable COD (X_S) by subtracting the soluble biodegradable fraction from the biodegradable COD: $X_S = COD_b - S_S$
(8) Determination of particulate unbiodegradable COD (X_U) as $X_U = COD_t - S_U - S_S - X_S$.

It can be noted that the above-mentioned sequence implies that all the errors and inaccuracies are reflected in the fraction X_U, which is thus a very sensitive parameter for modelling (Roeleveld and Van Loosdrecht, 2002).

In studies on biological phosphorus removal, the VFA fraction S_{VFA} can be measured by gas chromatography or by titration; the fermentable COD fraction is then determined as $S_F = S_S - S_{VFA}$.

Table 3.7 gives typical influent characteristics for raw wastewater. A large fraction of COD is particulate and can thus be removed through primary settling. The wastewater after primary settling is termed settled wastewater. Compared to raw wastewater, the COD load of settled wastewater is reduced by approximately 40% or even more, depending on the primary settler efficiency. This constitutes a significant relief for the biological reactors in terms of oxygen demand and secondary sludge production (see Chapter 4), leading to less raw wastewater. Primary settling only affects the particulate COD fraction; it does not influence the soluble concentrations (S_S and S_I). Typical influent characteristics for settled wastewater are given in Table 3.8 (Ekama, 2009). It should be noted that in primary settling a larger proportion of unbiodegradable particulates (over 80%) are removed than biodegradable particulates (removal efficiency of approximately 50%), which is in agreement with observations from practice (Ikumi *et al.*, 2014). A similar observation holds for inorganic components, which are also removed to a larger extent than particulate biodegradable organic matter (approximately 80%, see Section 3.2.2).

Table 3.7 COD fractionation of raw wastewater (Ekama, 2009), corresponding with the design example in Chapter 4. All values are in mg COD/l.

$COD_t = 750$		
$S_S = 146$	$S_U = 53$	Soluble
$X_S = 439$	$X_U = 112$	Particulate
Biodegradable	Unbiodegradable	

Table 3.8 COD fractionation of settled wastewater (40% COD removal efficiency in the primary settler compared to Table 3.7) (Ekama, 2009), corresponding with the design example of Chapter 4. All values are in mgCOD/l.

$COD_t = 450$		
$S_S = 146$	$S_U = 53$	Soluble
$X_S = 233$	$X_U = 18$	Particulate
Biodegradable	Unbiodegradable	

When comparing COD fractionation values for wastewater from different treatment plants, a high variation is noticed in the COD fractions S_S, X_S and X_U, while the fraction S_U does not vary a lot (Roeleveld and Van Loosdrecht, 2002). Table 3.9 summarizes the corresponding COD fractions. Variations in the amounts and ratios of the various COD fractions are related to the contribution of industry to the wastewater, to the sewer type and length, and to the transformation processes in the sewer. The latter are influenced by the sewer redox conditions, hydraulic retention time, and temperature. The amount of readily biodegradable organics typically decreases along long gravity sewers (aerobic) and increases in (anaerobic) pressure mains. Still, despite large variations between wastewater treatment plants, the fractions of COD and nitrogen found in a specific wastewater are quite constant even when concentrations vary (Henze, 1992). It can thus be stated that a specific wastewater exhibits a rather stable 'fingerprint'.

Table 3.9 Minimum, average and maximum COD fractionation results from 21 different wastewater treatment plants (Roeleveld and Van Loosdrecht, 2002).

Influent wastewater fraction	Min.	Avg.	Max.
S_U / COD_t	0.03	0.06	0.10
S_S / COD_t	0.09	0.26	0.42
X_S / COD_t	0.10	0.28	0.48
X_U / COD_t	0.23	0.39	0.50
$COD_b / COD_t = (S_S + X_S)/COD_t$	0.45	0.55	0.68
BOD_5 / COD_t	0.32	0.40	0.51
Particulate COD fraction = $(S_S + X_S)/COD_t$	0.47	0.68	0.88

It is important to note that wastewater also contains a small amount of active biomass, which is not explicitly accounted for in the presented fractionation. Biomass can account for 10-20% of the total organic matter in raw wastewater (Kappeler and Gujer, 1992). Still, this fraction does not need to be explicitly considered for steady state design (see Chapter 4 for more detail). The amount of influent biomass is also neglected in certain calibration protocols for activated sludge models (*e.g.* the STOWA protocol, Hulsbeek *et al.*, 2002), which only assume a small initial concentration of nitrifiers and phosphate-accumulating organisms in the wastewater (Roeleveld and Van Loosdrecht, 2002). With the presented COD fractionation methodology presented here, biomass present in the wastewater will be included in the fractions X_S and X_U. In contrast, some protocols for influent fractionation used for dynamic (ASM-based) modelling (see Section 3.13), do include detailed characterization of heterotrophic and autotrophic biomass. This is the case for the BIOMATH protocol (Vanrolleghem *et al.*, 2003), in which the influent wastewater characterization and parameter estimation of the biological processes heavily rely on respirometric measurements (Spanjers and Vanrolleghem, 2016).

3.5 NITROGEN

The primary source of nitrogen in raw municipal wastewater is human excretion of which some 75% is in the form of urea, which is almost directly hydrolized to ammonia nitrogen, while the rest is organic nitrogen. This ratio is recognized in typical nitrogen fractions (Henze, 1992). The inorganic fraction makes up the largest part of the nitrogen components present in wastewater and is essentially composed of free and saline ammonia (FSA), *i.e.* the sum of free ammonia (NH_3) and the ammonium-ion (NH_4^+). Inorganic nitrogen in the form of nitrate and nitrite is typically negligible for municipal wastewater. In addition, wastewater also contains an organic nitrogen fraction which can be further subdivided into biodegradable and unbiodegradable fractions, each with portions present in soluble and particulate forms (Figure 3.3, Ekama (2009) - see Chapter 5), which directly correspond with the COD fractions in Figure 3.2.

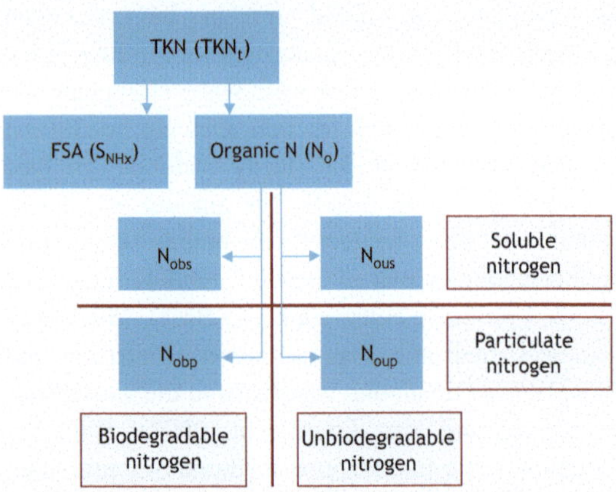

Figure 3.3 Fractionation of nitrogen in wastewater.

For a detailed description of the analytical procedures to determine FSA and total Kjeldahl nitrogen, the reader is referred to standard protocols detailed in Rice *et al.*, 2017.

During biological wastewater treatment, the biodegradable organic nitrogen fractions (both soluble, N_{obs}, and particulate, N_{obp}) are released as ammonia and add up to the FSA pool (in addition to S_{NHx}). Part of the overall FSA pool is incorporated in the biomass, mainly heterotrophic biomass (OHO, with mass MX_{OHOv}), since the mass of autotrophic biomass (MX_{ANOv}) is negligible. During endogenous respiration of the biomass, part of the nitrogen remains fixed, namely in the endogenous residue ($MX_{E,OHOv}$), and the remainder is released again in the form of ammonia. The part of FSA which is not incorporated into biomass is available for conversion to nitrate. The particulate unbiodegradable organic nitrogen fraction (N_{oup}) becomes included in the sludge mass, more specifically in the unbiodegradable organic part of the sludge mass (MX_{Uv}). The soluble unbiodegradable organic nitrogen fraction (N_{ous}) is not removed and thus ends up in the effluent. The relation between the COD and nitrogen fractions in the influent and their contribution to the VSS mass (particulates) in the biological reactor, assumed in Chapter 5, is schematically represented in Figure 3.4. Recall that, apart from the VSS sludge mass, consisting of biomass (MX_{OHOv}), endogenous residue ($MX_{E,OHOv}$) and unbiodegradable organics (MX_{Uv}), the total sludge mass in the reactor also contains an inorganic fraction (ISS) resulting from inorganics in the influent.

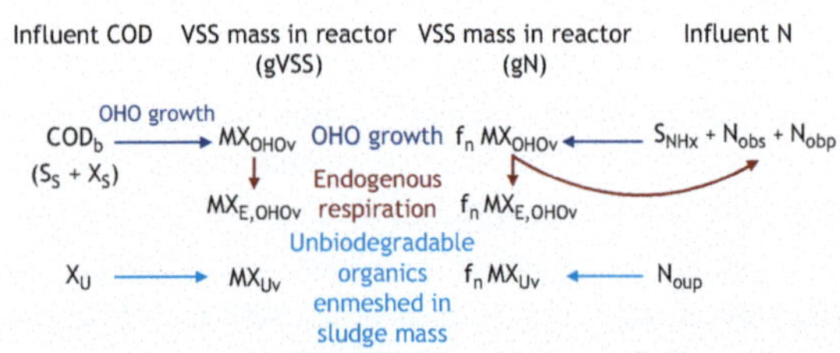

Figure 3.4 Relation between the COD and nitrogen fractions in the influent and their contribution to the VSS mass in the reactor, as presented in Chapter 5.

The organic part of the nitrogen is coupled to the organic carbonaceous matter in the wastewater. The total organic nitrogen fraction is determined as the difference between total Kjeldahl nitrogen and FSA. The total soluble (N_{obs} + N_{ous}) and total particulate

(N_{obp} + N_{oup}) nitrogen fractions are determined from total Kjeldahl nitrogen measurements before and after filtration. The individual soluble and particulate biodegradable and unbiodegradable organic nitrogen fractions cannot be directly measured but can be assumed or deduced from other measurements. A fixed nitrogen fraction for the various COD components (f_n) can be assumed, as long as it is such that the nitrogen content in the sludge of the activated sludge system is predicted with sufficient accuracy. A fixed nitrogen content of f_n = 0.10 mgN/mgVSS is assumed for heterotrophic biomass, endogenous residue and unbiodegradable particulate organics in the design example of Chapter 5. Alternatively, the nitrogen fraction of the soluble and particulate COD compounds can be determined rigorously from FSA and Kjeldahl-nitrogen analysis from filtered and non-filtered influent and effluent samples, as detailed in Roeleveld and Van Loosdrecht (2002). More information on nitrogen fractionation for use in modelling is given by Henze (1992) and Rieger *et al.* (2012). Examples of nitrogen fractionations for raw and settled wastewater are given in Table 3.10 and Table 3.11, respectively (Ekama, 2009). They are applied in Chapter 5 for wastewater treatment plant design for nitrogen removal; the details of the nitrogen fraction are given there as well (sections 5.7.3-5.7.4).

Table 3.10 Nitrogen fractionation of raw wastewater (Ekama, 2009), corresponding with the design example in Chapter 5. All values are in mgN/l.

TKN_t = 60			
S_{NHx} = 45	N_o = 15.0		
	N_{obs} = 1.7	N_{ous} = 1.8	Soluble
	N_{obp} = 3.9	N_{oup} = 7.6	Particulate
	Biodegradable	Unbiodegradable	

Table 3.11 Nitrogen fractionation of settled wastewater (Ekama, 2009), corresponding with the design example in Chapter 5. All values are in mgN/l.

TKN_t = 51			
S_{NHx} = 45	N_o = 6.0		
	N_{obs} = 1.7	N_{ous} = 1.8	Soluble
	N_{obp} = 1.4	N_{oup} = 1.2	Particulate
	Biodegradable	Unbiodegradable	

The TKN/COD ratios of the example raw and settled wastewater amount to ($f_{TKNi/CODi}$ = TKN_i/COD_i) 0.08 and 0.11 gN/gCOD, respectively. The TKN/COD ratio is higher for settled wastewater since the TKN fraction is largely soluble while the COD fraction contains a significant amount of particulates, which are removed during primary settling. The soluble unbiodegradable organic nitrogen fractions ($f_{SU,TKNi}$ = N_{ous}/TKN_t) of the example raw and settled wastewater amount to 0.03 and 0.035, respectively and the soluble unbiodegradable organic nitrogen concentration as such (N_{ous}) are also the same. This nitrogen fraction will end up in the effluent, in addition to the residual ammonia and nitrate (the effluent nitrite concentration is typically negligible).

3.6 PHOSPHORUS

Analogously to the nitrogen components, phosphorus constituents in raw wastewater can also be categorized into organic and inorganic fractions. Raw domestic wastewater contains 6-25 mgP/l, of which approximately 60-70% is inorganic. The inorganic phosphorus fraction is mainly present as orthophosphate. The organic phosphorus fractions are directly coupled to the COD fractions and can be subdivided into biodegradable and unbiodegradable fractions, each with portions present in soluble and particulate forms (Figure 3.5).

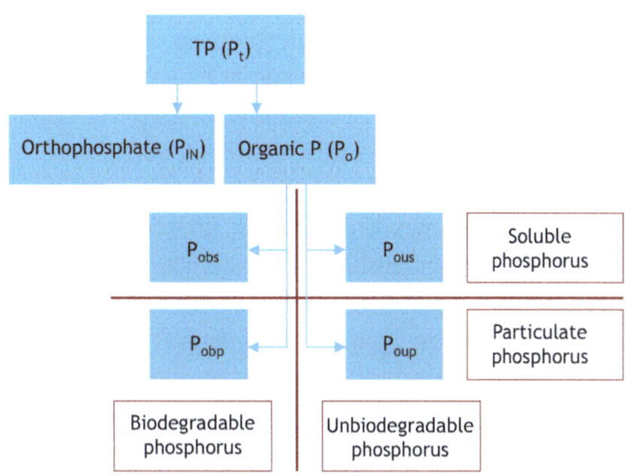

Figure 3.5 Fractionation of phosphorus in wastewater.

Analytical methods based on a combination of filtration and colorimetry are used to determine the different phosphorus fractions shown in Figure 3.5 (Rice *et al.,* 2017). Direct colorimetry is carried out to determine the orthophosphate concentration. For the other fractions, preliminary acid hydrolysis or digestion is performed before colorimetry in order to release phosphorus as orthophosphate (Gu *et al.*, 2011; Rice *et al.,* 2017).

Phosphorus can be removed from wastewater through biological or chemical means, or the combination of the two. Under adequate bioprocess conditions, removal of phosphorus is promoted through incorporation into biomass in quantities beyond that required for cell production. The presence or addition of metal ions, such as iron or aluminium, also leads to phosphorus removal through precipitation. The particulate and soluble unbiodegradable organic phosphorus fractions become included in the waste sludge and effluent, respectively.

The characterization of phosphorus components is analogous to the characterization of nitrogen. The organic part of the phosphorus is coupled to the organic carbonaceous matter in the wastewater. A fixed phosphorus fraction for the various COD components (f_P) can be assumed, matching the phosphorus content in the produced sludge. A fixed phosphorus content of $f_P = 0.025$ mgP/mgVSS can be assumed for the heterotrophic biomass, endogenous residue and unbiodegradable particulate organics (see Chapter 4, Section 4.4.). Alternatively, the phosphorus fraction of the soluble and COD compounds can be determined rigorously from the total phosphate and orthophosphate measurements of filtered and non-filtered influent and effluent samples. The phosphorus fractionation procedure for use in modelling is detailed in Henze *et al.* (1995) and Rieger *et al.* (2012).

3.7 SULPHUR

Sulphur in raw domestic wastewater is mostly present in the form of sulphate, contributing to approximately 97% of the total influent sulphur (Dewil *et al.*, 2008). The typical concentration of sulphate in raw domestic wastewater is 24-72 mg/l (Metcalf and Eddy AECOM, 2014), but can reach as high as 500 mg/l due to significant industrial sources or seawater-based toilets (Lens *et al.*, 1998; Van den Brand *et al.*, 2015). Other inorganic sulphur components that can be present in raw wastewater are in the form of sulphides, thiosulphate and elemental sulphur, dictated by the sewer conditions.

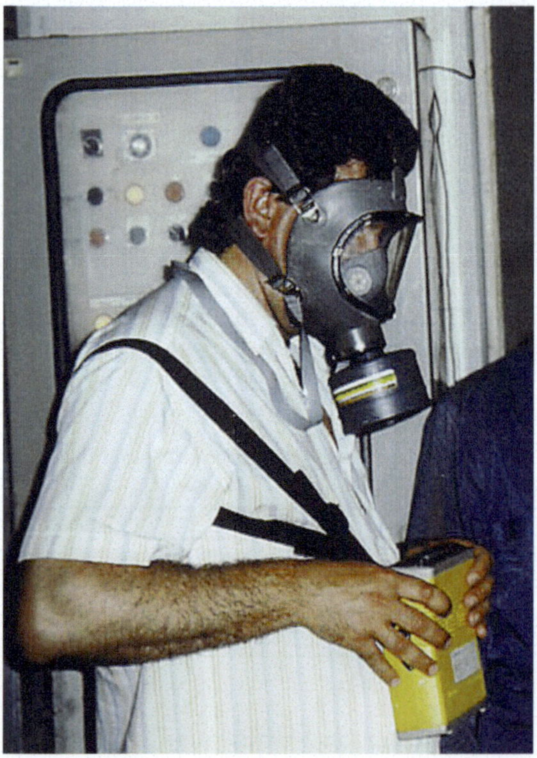

Figure 3.6 Hydrogen sulphide is often present in the influent to treatment plants, especially in the case of pressurized sewers. It is very toxic and precautions must be taken to avoid risk to personnel. The photo shows measurement in a pumping station with high hydrogen sulphide concentration in the air (photo: M. Henze).

Ion chromatography and gravimetric or turbidimetric methods are mostly used to determine the aqueous sulphate concentration, with specified sulphate concentration ranges for which each of the methods is suitable (Rice *et al.,* 2017). Sulphide, on the other hand, should be measured in both the liquid and gas phase. In the liquid phase, sulphide is measured directly using the methylene blue method or estimated through the COD measurement while gas phase measurement is commonly performed through gas chromatography (Lopez-Vazquez *et al.*, 2016).

The distribution of the sulphide species is mainly determined by pH as shown in Figure 3.7, implying that sulphide in municipal wastewater is typically present in H_2S and HS^- forms (pKa = 7.02). During wastewater treatment, sulphur transforms to various forms depending on the operational conditions of the bioprocesses. Aerobic conditions promote the biological conversion of sulphide species to elemental sulphur and sulphate, and the reverse during anaerobic conditions. The presence of metals, such as iron, can lead to precipitation of sulphide and its consequent removal with the waste sludge. Under both aerobic and anaerobic conditions, dissolved H_2S can easily escape to the gas phase due to its high volatility.

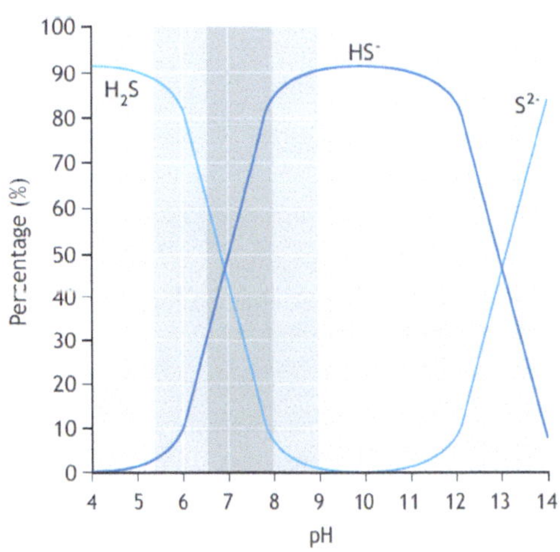

Figure 3.7 The distribution of sulphide species as a function of pH (Lopez-Vasquez *et al.*, 2016).

3.8 CELLULOSE

Cellulose represents a significant fraction of domestic wastewater; 20-30% of the COD load (Reijken *et al.*, 2018) and approximately 35% of suspended solids load (Ruiken *et al.*, 2013) to wastewater treatment plants are attributed to cellulose fibres (Figure 3.8). The main source of cellulose is toilet paper, significant quantities of which are disposed of via the water closet (WC). For an overview of WC-derived sewer solids, the reader is referred to the results of Friedler *et al.* (1996) obtained from a survey over one week of domestic WC usage in the UK.

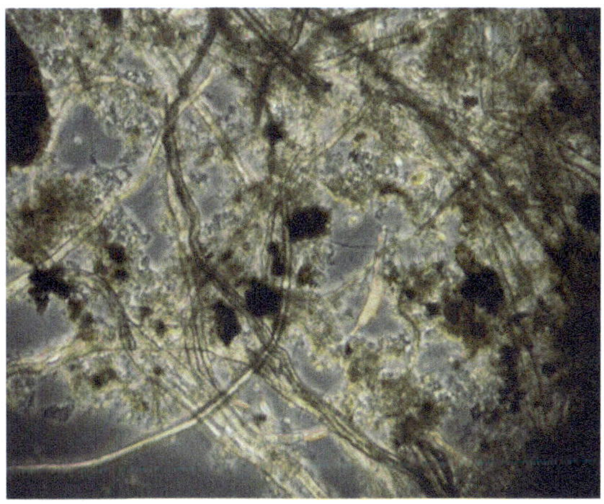

Figure 3.8 Microscopic image of fibres in raw wastewater sediment (Ruiken *et al.*, 2013).

Cellulose can easily be recovered through fine mesh sieves (<0.35 mm) (Ruiken *et al.*, 2013). This results in lower aeration energy requirements and a significant reduction in sludge production, in their turn increasing the treatment capacity of existing plants. The sieved material can be used efficiently in a biomass power plant for energy. Moreover, there is growing interest in recovering cellulose fibres for its potential as raw material for paper products, bioplastics, and road and building materials. Still, the economic viability of cellulose recovery remains to be studied, by integrating its costs, energy and effect with the overall plant efficiency (Solon *et al.*, 2019).

The quantification of cellulose in wastewater and its fate along the wastewater treatment train has been hampered by the lack of a reliable measurement technique which is not influenced by the sludge matrix. Gupta *et al.* (2018) compared four different methods to quantify the cellulose content in municipal wastewater and sludge, namely (*i*) dilute acid hydrolysis, (*ii*) concentrated acid hydrolysis, (*iii*) enzymatic hydrolysis, and (*iv*) the Schweitzer reagent method. Of these, the only reliable method they found was the Schweitzer method, its advantages being its simplicity, reproducibility, accuracy, ideal (100%) recovery, and relative quickness of the test as well as its independence from hydrolysis reactions.

Cellulose shows a specific behaviour in wastewater treatment plants, which is characterized by a very slow hydrolysis rate. This degradation rate is temperature-sensitive, causing a significant difference in cellulose turnover between summer and winter. For wastewater treatment plants with relatively high SRT, it is therefore recommended to include cellulose as a separate state variable (X_{CL}) during activated sludge modelling (Reijken *et al.*, 2018). Considering cellulose as a separate particulate organic carbon component affects the influent fractionation for particulate (slowly) biodegradable (X_S) and particulate unbiodegradable organics (X_U), in a way which is determined by the value of the cellulose hydrolysis constant (Reijken *et al.*, 2018). The latter can, in its turn, only be quantified accurately based on accurate and reliable cellulose measurement methods.

3.9 MICROPOLLUTANTS

Many chemical contaminants are also present in municipal wastewater in trace concentrations measured in the μg/l range and below, making them difficult to detect, hence they are termed 'micropollutants'. Examples of micropollutants are biocides and pesticides, personal care products, solvents and pharmaceuticals (Kümmerer *et al.*, 2013). Although present in quite low concentrations, the effect of many of these micropollutants on the aquatic environment and on human health should not be underestimated. Still, technological limitations and cost aspects to detect these micropollutants were the reason that these components have been ignored in the past. Among the micropollutants, pharmaceuticals have been gaining a lot of attention in the last two decades, especially due to the increasing amount of evidence of their negative biological effects on aquatic animals (Corcoran *et al.*, 2010). Wastewater treatment plants are considered as typical point sources of these micropollutant. Indeed, pharmaceutical components reach wastewater treatment plants through urine and faeces discharged from households and institutions. Whereas wastewater treatment plants are not designed to remove pharmaceutical components, these micropollutants nevertheless undergo biotransformation, sorption, desorption, sequestration and retransformation to the parent compound in the bioprocess units (Plósz *et al.*, 2012) (Figure 3.9). Each pharmaceutical considered can be categorized into four forms. Two forms are in the liquid phase: (*i*) the parent chemical compound and (*ii*) retransformable chemical forms, such as human metabolites, that are biotransformed through the parent chemical form. The other two forms are in the solid phase: (*iii*) sorbed chemical and (*iv*) sequestered (Plósz *et al.*, 2012; Snip *et al.*, 2014).

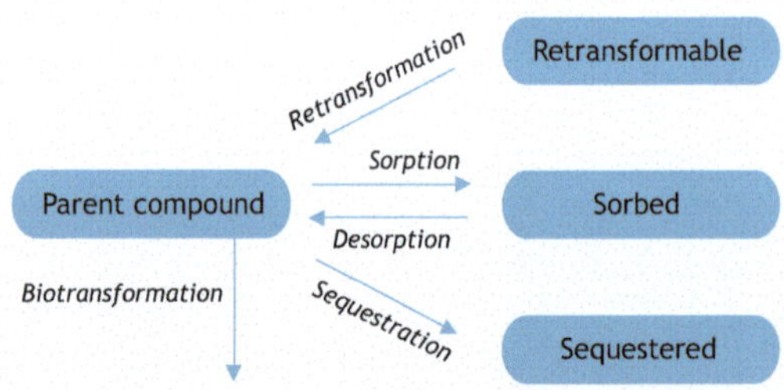

Figure 3.9 Fate of pharmaceuticals during wastewater treatment (adapted from Snip *et al.*, 2014).

The detection and characterization of various pharmaceutical components as well as their metabolites and conjugates are important in order to study their fate and to know how to remove them during wastewater treatment. However, because of their size and low concentrations, measurements of pharmaceutical components in wastewater can be labour-intensive, costly and can require highly trained personnel. Gas chromatography with mass spectrometry (GC-MS), liquid chromatography with tandem mass spectrometry (LC-MS^2) and high performance liquid chromatography (HPLC) are the most frequently employed analytical techniques for determination of various pharmaceutical components in water and wastewater (Fatta *et al.*, 2007).

3.10 OTHER CHARACTERISTICS

Wastewater contains a lot of other components besides the abovementioned, most of which are not the direct target for treatment. However, they can contribute to the toxicity of the wastewater, either in relation to the biological processes in the treatment plant or to the receiving waters. The substances which remain in the effluent of the wastewater treatment plant can end up in a drinking water production facility, if the latter is dependent on surface water extraction.

3.10.1 Metals

The metals in wastewater can influence the possibilities for reuse of the wastewater treatment sludge on farmland. Typical values for metals in municipal wastewater are given in Table 3.12.

Table 3.12 Typical content of metals in municipal wastewater with minor contributions of industrial wastewater (in mg/m³) (Henze, 1982, 1992; Ødegaard, 1992; from Henze *et al.*, 2002).

Metal	Low	Medium	High
Aluminium	350	600	1,000
Cadmium	1	2	4
Chromium	10	25	40
Copper	30	70	100
Lead	25	60	80
Mercury	1	2	3
Nickel	10	25	40
Silver	3	7	10
Zinc	100	200	300

3.10.2 Physical properties of wastewater

Apart from the chemical composition, wastewater is also characterized by physical properties, which can significantly influence the treatment process. Table 3.13 summarizes the ranges of physical parameters for municipal wastewater.

The temperature of the incoming wastewater will significantly influence the temperature of the wastewater treatment basin, and thus the biological activity. The wastewater temperature obviously depends on geographical location and season. For example, a temperature range of 3-27 °C is observed in the United States (Metcalf and Eddy AECOM, 2014). In addition, this temperature is also influenced by greywater sub-streams which can reach elevated temperatures of up to 80 °C (Friedler *et al.*, 2013). The high temperature of wastewater (sub)streams opens up potential for the recovery of heat, at the source (*i.e.* households or buildings), along the sewer line, or at the wastewater treatment plant (Cipolla and Maglionico, 2014).

Municipal wastewater typically has a more or less neutral pH (about 7.0), which is optimal for biological activity. If the influent pH deviates significantly from neutrality, *e.g.* for some industrial wastewater, a neutralization step needs to be added before biological treatment.

Table 3.13 Physical properties of municipal wastewater (from Henze, 1982).

Parameter	Low	Medium	High	Unit
Viscosity	0.001	0.001	0.001	kg/m.s
Surface tension	50	55	60	dyn/cm²
Conductivity	70	100	120	mS/m¹
pH	7.0	7.5	8.0	
Alkalinity	1	4	7	Eq/m³
Specific salt concentration:				
Sulphide	0.1	0.5	10	gS/m³
Cyanide	0.02	0.030	0.05	g/m³
Chloride	200	400	600	gCl/m³

3.10.3 Toxic organic components

Wastewater can also contain specific pollutants such as xenobiotics (Table 3.14), which may cause specific problems (Figure 3.10).

Table 3.14 Special parameters in wastewater, xenobiotics with toxic and other effects (all in mg/l).

Parameter	Low	Medium	High
Phenol	0.02	0.05	0.1
Phthalates, DEHP	0.1	0.2	0.3
Nonylphenols, NPE	0.01	0.05	0.08
PAHs	0.5	1.5	2.5
Methylene chloride	0.01	0.03	0.05
LAS	3,000	6,000	10,000
Chloroform	0.01	0.05	0.01

Figure 3.10 High concentrations of detergents create problems for a wastewater treatment plant operator (photo: M. Henze).

3.11 TYPICAL WASTEWATER CHARACTERISTICS

3.11.1 Population equivalent

The unit Population Equivalent (PE) or 'per capita' is a standard unit referring to the typical contribution of one individual to the wastewater load. PE can be expressed in wastewater flow rate or BOD load, according to eqs. 3.3-3.4.

$$1 \text{ PE} = 0.2 \text{ m}^3/\text{d, Hvitved-Jacobsen (2013)} \quad (3.3)$$

$$1 \text{ PE} = 60 \text{ g BOD/d, EC (2007)} \quad (3.4)$$

These two definitions are based on fixed non-changeable values, independent of the actual wastewater volume produced per person per day or the source of the wastewater. For example, the unit PE can also be used to quantify the load of industrial wastewater streams. However, the actual contribution from a person living in a sewer catchment to the wastewater production, the Person Load (PL), can vary considerably (Table 3.15). The reasons for the variation can be that the working place is outside the catchment, socio-economic factors, lifestyle, type of household installation etc.

Table 3.15 Variations in load per person (Henze *et al.*, 2002).

Parameter	Unit	Range
BOD	g/cap.d	15-80
COD	g/cap.d	25-200
Nitrogen	g/cap.d	2-15
Phosphorus	g/cap.d	1-3
Wastewater flow rate	m^3/cap.d	0.05-0.40

Table 3.16. Load per person in various countries in kg/cap.yr (based on Henze *et al.*, 2002).

Parameter	Brazil	Egypt	India	Turkey	US	Denmark	Germany
BOD	20-25	10-15	10-15	10-15	30-35	20-25	20-25
TSS	20-25	15-25		15-25	30-35	30-35	30-35
N total	3-5	3-5		3-5	5-7	5-7	4-6
P total	0.5-1	0.4-0.6		0.4-06	0.8-1.2	0.8-1.2	0.7-1

The terms Person Equivalent and Person Load are often confused with each other or misunderstood, so one should be careful when using them and be sure to define clearly what they are based upon. PE and PL are both based on average contributions and used to give an impression of the loading of wastewater treatment processes. They should not be calculated from data based on short time intervals (hours or days). The Person Load varies from country to country, as demonstrated by the yearly values given in Table 3.16.

3.11.2 Municipal wastewater composition

The amount and composition of the produced wastewater differs between humans and among industries. The amount and type of waste produced in households is influenced by the behaviour, lifestyle and standard of living of the inhabitants as well as the technical and juridical framework in which people live.

The concentrations found in wastewater are determined by pollutant load and the amount of water with which the pollutant is mixed. The composition of municipal wastewater varies significantly from one location to another. At a given location, the composition will also vary over time. This is partly due to variations in the discharged amounts of substances. Also the retention time in the sewer influences the wastewater composition (Henze, 1992; Huisman and Gujer, 2002). However, the main reasons are variations in water consumption in households and the infiltration and exfiltration during transport in the sewage system.

The composition of typical municipal wastewater with minor contributions of industrial wastewater is shown in Table 3.17. Concentrated wastewater (high) represents cases with low water consumption and/or infiltration, while diluted wastewater (low) represents high water consumption and/or infiltration. Stormwater will further dilute the wastewater as most stormwater components have lower concentrations compared to very diluted wastewater.

Since most wastewater treatment processes are based on biological degradation and conversion of the substances, the degradability of the components is important. Table 3.18 summarizes typical degradability values for municipal wastewater components.

Table 3.17 Typical composition of raw municipal wastewater with minor contributions of industrial wastewater (in g/m^3).

Parameter	Low	Medium	High
COD total	500	750	1,200
COD soluble	200	300	480
COD suspended	300	450	720
BOD	230	350	560
VFAs (as acetate)	10	30	80
N total	30	60	100
Ammonia N	20	45	75
Nitrate + Nitrite N	0.1	0.2	0.5
Organic N	10	15	25
P total	6	15	25
Ortho-P	4	10	15
Organic P	2	5	10
Sulphate	24	36	72
TSS	250	400	600
VSS	200	320	480

Table 3.18 Degradability of medium-concentration municipal wastewater (in mg/l).

Parameter	Biodegradable	Unbiodegradable	Total
COD total	585[a)]	165[a)]	750 [a)]
BOD	350	0	350
N total	50[a)]	10[a)]	60[a)]
P total	14.7	0.3	15[a)]

[a)]The rounded values correspond with the raw wastewater example in chapters 4-6, Table 3.6 and Table 3.8 (Ekama, 2009).

3.11.3 Importance of ratios

The ratio between the various components in wastewater has significant influence on the selection and functioning of the wastewater treatment processes. A wastewater with a low biodegradable organic carbon to nitrogen ratio may need an external carbon source addition for biological denitrification. A variation in the ratio between readily and slowly biodegradable organic carbon (S_S/X_S) for the same total amount of biodegradable organic carbon ($COD_b = S_S + X_S$) can significantly affect the nitrate concentration in the effluent. Wastewater with a relatively high nitrate concentration or a low concentration of volatile fatty acids (VFAs) will not be suitable for biological phosphorus removal. A high COD to BOD ratio in wastewater indicates that a substantial part of the organic matter will be difficult to degrade biologically. When the suspended solids in wastewater have a high organic component (high VSS to TSS ratio) these can be successfully digested under anaerobic conditions.

While most of the pollution load in wastewater originates from households, hospitals, schools and industry, these contribute only partially to the total quantity of sewage. A significant amount of water in sewage can originate from rainwater (in some countries melting snow) or infiltration groundwater. Thus, wastewater components are subject to dilution, which however will not change the ratios between the components. Table 3.19 shows typical component ratios in municipal wastewater.

Given that the ratios in wastewater typically remain relatively constant, the analysis of component ratios over time for a given wastewater stream can be used to detect anomalies that are due to analytical errors or due to special discharges into the sewer system, often from industry. If industrial discharges cause the discrepancy, other components in the wastewater that have not (yet) been analysed can also deviate from their expected values. Since these discrepancies can affect the treatment process the reason for their appearance should be clarified.

3.11.4 Domestic wastewater sub-streams

Domestic wastewater, *i.e.* wastewater originating from households, varies all over the world in terms of amount and composition. These variations are influenced by the climate, socio-economic factors, household technology and other factors. In households most waste will end up as solid and liquid waste, and there are significant possibilities for changing the amounts and composition of the two waste streams generated.

Domestic wastewater consists of different sub-flows, as shown in Table 3.20. These sub-flows each contain specific amounts of organic waste and nutrients. The domestic wastewater composition can be changed by changing the amount and the composition of the wastewater sub-streams.

Table 3.19 Typical ratios in municipal wastewater (adapted from Henze *et al.*, 2002). Ratios for the examples of raw and settled wastewater are taken or calculated from Table 4.2[a)], Table 6.2[b)] and Table 3.3[c)], all in agreement with Ekama (2009). [d)]Typical value, from Dold *et al.* (1980).

Ratio	Low	Medium	High	Example raw	Example settled
COD/BOD	1.5-2.0	2.0-2.5	2.5-3.5		
VFA/COD	0.04-0.02	0.08-0.04	0.12-0.08	0.029[b)]	0.049
COD/TN	6-8	8-12	12-16	12.5[a)]	8.8[a)]
COD/TP	20-35	35-45	45-60	50[a)]	35[a)]
BOD/TN	3-4	4-6	6-8		
BOD/TP	10-15	15-20	20-30		
COD/VSS	1.2-1.4	1.4-1.6	1.6-2.0	1.48[d)]	1.48[d)]
VSS/TSS	0.4-0.6	0.6-0.8	0.8-0.9	0.84[c)]	0.87[c)]
COD/TOC	2-2.5	2.5-3	3-3.5		

Table 3.20 Sources of household wastewater components and their values for a 'non-ecological' lifestyle (from Sundberg, 1995; Henze, 1997).

Parameter	Unit	Toilet		Kitchen	Bath/laundry	Total
		Total[a)]	Urine			
Wastewater	m^3/yr	19	11	18	18	55
COD	kg/yr	27.5	5.5	16	3.7	47.2
BOD	kg/yr	9.1	1.8	11	1.8	21.9
N	kg/yr	4.4	4.0	0.3	0.4	5.1
P	kg/yr	0.7	0.5	0.07	0.1	0.87
K	kg/yr	1.3	0.9	0.15	0.15	1.6

[a)] Including urine.

Options for reducing the physiologically generated amount of waste are not obvious, although diet influences the amount of waste produced by the human organism. However, one could opt to separate the toilet waste (also termed physiological waste or anthropogenic waste) from the waterborne route, leading to a significant reduction in the nitrogen, phosphorus and organic load in domestic wastewater. Waste generated after the separation at source has taken place, however, still has to be transported away from the household, and in many cases, the city.

Urine is the main contributor to nutrients in household waste, thus separating out the urine will significantly reduce the nutrient loads in wastewater (Figure 3.11). Urine separation will reduce nitrogen content in domestic wastewater to a level where nitrogen removal is not needed.

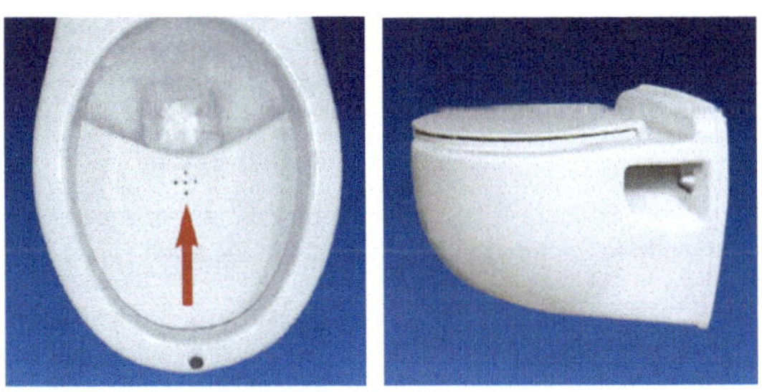

Figure 3.11 Urine-separating toilet.

Kitchen waste contains a significant amount of organic matter which traditionally ends up in wastewater. It is relatively easy to divert some liquid kitchen waste to solid waste by the application of 'cleantech' cooking, thus obtaining a significant reduction in the overall organic load of the wastewater (Danish EPA, 1993). Cleantech cooking means that food waste is discarded into the waste bin and not flushed into the sewer using water from the tap. The diverted part of the solid organic waste from the kitchen can be disposed of together with the other solid waste from the household. The grey wastewater from the kitchen can be used for irrigation or, after treatment, for toilet flushing. Liquid kitchen waste also contains household chemicals, the use of which can affect the composition and load of this type of waste. Wastewater from laundry and bathing or showering carries a minor pollution load only, part of which comes from household chemicals, the use of which can affect the composition and the load of this waste fraction. Waste from laundry and baths or showers can be used together with traditional kitchen wastewater for irrigation. Alternatively, it can be reused for toilet flushing. In both cases considerable treatment is needed.

The compostable fraction of the solid waste from the kitchen can either be kept separate or combined with traditionally waterborne kitchen waste, for later composting or anaerobic treatment at the wastewater treatment plant.

Table 3.21 Examples of the composition of four different industrial wastewaters.

Variable	Units	Pulp and paper industry[a)]	Textile industry[b)]	Winery[c)]	Dairy[d)]
BOD	g/m^3	16-13,300	80-6,000	203-22,418	40-48,000
TSS	g/m^3	-	15-8,000	-	24-7,175
SS	g/m^3	0-23,319	-	66-8,600	-
COD	g/m^3	78-39,800	100-30,000	320-49,105	80-95,000
N total	g/m^3	11-600	70-80	10-415	14-329
P total	g/m^3	0.02-36	<10	2-280	9-132
S total	g/m^3	6-1,270	500-1,000	-	-
pH		2.5-12.3	5.5-11.8	2.5-12.9	4.5-11

[a)] Pokhrel and Viraraghavan (2004), [b)] Yaseen and Scholz (2019), [c)] Ioannou *et al.* (2015), [d)] Kushwaha *et al.* (2011).

Waste disposal units (grinders) are used in many countries for handling the compostable fraction of the solid kitchen waste from households, although sometimes this option is rejected due to the increased waste load to the sewer. However, waste is generated in households, and it must be transported away from households and out of cities by some means. The discharge of solid waste into the sewer does not change the total waste load produced by the household, but it will change the transportation means and the final destination of the waste.

3.11.5 Non-domestic sewage components

Municipal wastewater transported by sewers to wastewater treatment plants originates not only from households but also from industries which do not have their own (decentralized) wastewater treatment facility. In addition, wastewater treatment plants also handle external inputs, such as landfill leachate.

For industries without internal water reuse programmes, 85-95% of the water used will become wastewater; for industries with internal water recycling the wastewater production volume should be estimated on a case-by-case basis (Metcalf and Eddy AECOM, 2014). Wastewater characteristics vary widely among industries, as illustrated in Table 3.21. In addition, these wastewater concentrations fluctuate over time, along with dynamic operation/production schemes.

Another significant external load on a treatment plant can be landfill leachate (Figure 3.12, Table 3.22). Landfill leachate can contain high concentrations of soluble inert COD which passes through the plant without any reduction or change. It can also contain high nitrogen concentrations. In some cases where regulations do not allow discharge of untreated leachate, separate pre-treatment of leachate is required on-site prior to its discharge to a public sewer.

Figure 3.12 Collection and storage of leachate at a sanitary landfill site in Sarajevo in Bosnia and Herzegovina (photo: F. Babić).

Table 3.22 Leachate quality (all in g/m³, except for pH).

Parameter	Low	High
COD total	1,200	16,000
COD soluble	1,150	15,800
BOD total	300	12,000
N total	100	500
Ammonia N	95	475
P total	1	10
TSS	20	500
VSS	15	300
Chloride	200	2,500
H_2S	1	10
pH	6.5	7.2

Table 3.23 Digester supernatant (in g/m³).

Component	Low	High
COD total	700	9,000
COD soluble	200	2,000
BOD total	300	4,000
BOD soluble	100	1,000
N total	120	800
Ammonia N	100	500
P total	15	300
TSS	500	10,000
VSS	250	6,000
H_2S	2	20

3.11.6 Internal loads in wastewater treatment plants

Internal loading at treatment plants is caused by thickening and digester supernatant (Figure 3.13), reject water from sludge dewatering (Figure 3.14), and filter washing water. Digester supernatant often constitutes a significant internal load, especially concerning ammonium (Table 3.23). This can lead to overload of nitrogen in the case of biological nitrogen removal (see also Chapter 5).

Figure 3.13 Digesters produce digester supernatant which often gives rise to problems in wastewater treatment plants due to the high loads of nitrogen and other substances (photo: D. Brdjanovic).

Reject water from sludge dewatering can have relatively high concentrations of soluble material, both organics and nitrogen (Table 3.24).

Table 3.24 Composition of reject water from sludge dewatering (all in g/m³, except for pH) (Gebreeyessus and Jenicek, 2016).

Component	Range
COD total	700-1,400
Organic N	90-187
Ammonia N	600-1,513
P total	trace-130
TSS	<800
pH	7-13

Figure 3.14 Belt filter for sludge dewatering: reject water is collected underneath the machinery (photo: D. Brdjanovic).

Filter wash water can create problems due to high hydraulic overload of the settling tanks in treatment plants. In some cases, this can result in an overload of suspended solids (Table 3.25). Filter wash water in smaller treatment plants should be recycled slowly.

Table 3.25 Filter wash water (in g/m³).

Component	Low	High
COD total	300	1,500
COD soluble	40	200
BOD total	50	400
BOD soluble	10	30
N total	25	100
Ammonia N	1	10
P total	5	50
TSS	300	1,500
VSS	150	900
H_2S	0.01	0.1

3.11.7 Non-sewered (onsite) sanitation flows

This textbook is primarily focused on sewered (urban) sanitation and centralized treatment of sewage/wastewater using conventional or advanced process technology, be it activated sludge, granular sludge or biofilm systems. Situations where practically all the urban areas in a country are serviced by sewerage tend to be the exception rather than the rule and are a privilege of smaller and highly developed countries (*e.g.* Singapore, the Netherlands, Denmark, etc.). Urban and peri-urban areas in low- and middle-income countries (and also in some EU countries) are only partially sewered; the provision of non-sewered (also called onsite) sanitation is a major challenge and as such recognized by the global community and the implementation of the Sustainable Development Goals (SDGs), and in particular Goal 6 - Safe Water and Sanitation.

A non-sewered sanitation service provision chain differs in many aspects from centralized wastewater management but also has several common points where the two sectors meet. The latest developments towards the achievement of SDGs promote a unifying concept of City-wide Inclusive Sanitation (CWIS), an approach to urban sanitation where everyone in the city has equitable access to adequate and affordable improved sanitation services through appropriate systems at all scales (both sewered and non-sewered). This approach urges the consideration of not only sewage but also of materials and resources generated by onsite sanitation facilities dominated by faecal sludge. Faecal sludge management (FSM), which has only started to receive wider scientific attention in the last decade, includes the characterization and quantification of sanitary flows generated in onsite sanitation.

In contrast, activated sludge systems have been studied in detail for more than a century (Jenkins and Wanner, 2014), experimental methods and approaches are well-developed (Van Loosdrecht *et al.*, 2016) and various models have been in use for more than 30 years (Brdjanovic *et al.*, 2015; Van Loosdrecht *et al.*, 2015), while non-sewered sanitation professionals are still struggling to understand and determine the accumulation rates and processes taking place in pit latrines, public ablution blocks, aqua privies and dry toilets (Strande *et al.*, 2014). However, due to the recent attention to non-sewered sanitation, the development and standardization of methods for faecal sludge analysis are rapidly progressing (Velkushanova *et al.*, 2020). Recently, several approaches to obtaining quantitative and qualitative characterization of onsite sanitation flows have been developed and applied in practice (Strande *et al.*, 2018; Velkushanova *et al.*, 2020), and a systematic method, called an excreta flow diagram (or shit flow diagram, SFD), a tool to understand and communicate how excreta 'flow' through a city or town, has been developed and applied in several cities in low and middle income countries (Peal *et al.*, 2020).

In some cases, in urban agglomerations with limited or no sewer systems in place, an effective sanitation practice entails the construction of a designated faecal sludge treatment plant (FSTP). In many countries standard sanitary practices include collection and transport of faecal sludge and its disposal and treatment at a centralized sewage treatment facility. In some cases, these sewage treatment plants are designed to accept such a load; however, the majority are not and even a relatively small contribution of faecal sludge to conventional activated sludge plants like these can lead to deterioration of plant performance (Strande *et al.*, 2014). The main reason for this deterioration is the different composition of faecal sludge compared to sewage. This difference is caused by differences in the physical nature of non-sewered and sewered sanitation facilities, use of water in a particular sanitation system, and retention time of sanitary products in the system. In some cases, a common (but undesirable) practice is that faecal sludge trucks dump sludge into a sewer system: a practice that usually causes a spectrum of operational and environmental issues.

Due to the very high concentrations of organic matter, nitrogen and phosphorus, a sudden full truck load of faecal sludge can cause problems in wastewater treatment plants, and can affect the entire plant including primary treatment, biological treatment and waste sludge treatment. This issue is more likely to occur when the service area contains more zones with decentralised wastewater handling, *e.g.* in peri-urban and non-sewered urban areas. To avoid upsets in small wastewater treatment plants, the septic tank sludge must be unloaded into a storage unit (buffer tank), from which it can be pumped to the plant in periods of low loading (often during the night). As a rule of thumb, for treatment plants operating at over 100,000 person equivalent the unloading of a truck with septic sludge per day will not create direct problems in the plant.

Because large amounts of faecal sludge blended with sewage often reach a sewage treatment plant, it is important to highlight the difference in the composition of domestic sewage and various onsite sanitation flows.

Some readers may not be familiar with the definition of onsite sanitation flows, thus the generally accepted description (adopted from Tilley *et al.*, 2014; Strande *et al.*, 2014; Velkushanova *et al.*, 2020) is provided here:

- *Excreta* consists of urine and faeces that are not mixed with any flush water. Excreta is relatively small in volume but concentrated in both nutrients and pathogens. Depending on the characteristics of the faeces and the urine content, it can have a soft or runny consistency.
- *Faeces* refer to (semi-solid) excrement that is not mixed with urine or water. Depending on diet, each person produces approximately 50-150 l per year of faecal matter of which approximately 80% is water and the remaining solid fraction is mostly composed of organic material. A special form of faeces is diarrhoea, which is loose watery faeces resulting from viral, parasitic protozoan, bacteria or a helminth infection (Type 6 and Type 7 on the Bristol Stool Scale)[1].
- *Urine* is the liquid produced by the body to rid itself of nitrogen in the form of urea and other waste products. In this chapter, the urine product refers to pure urine that is not mixed with faeces or water. Depending on diet, the human urine collected from one person during one year (approx. 300 to 550 l) contains 2-4 kg of nitrogen. Urine can be stored *e.g.* in urine diversion (dry) toilets (UD(D)Ts) and hydrolysed naturally over time, *i.e.* the urea is converted by enzymes into ammonia and bicarbonate.
- *Anal cleansing water* is water used to cleanse the body after defecating and/or urinating; it is generated by those who use water, rather than dry material, for anal cleansing. The volume of water used per cleaning typically ranges from 0.5-3.0 l (but can be more in developed urban areas).
- *Flush water* is the water discharged into the user interface (toilet) to clean it and transport the contents into the conveying system or to the onsite storage. Freshwater, rainwater, recycled

[1] The Bristol Stool Scale is a diagnostic medical tool designed to classify the form of human faeces into seven categories. It is used in both clinical and experimental fields.

greywater, or any combination of these three, can be used as a flush water source. Many sanitation systems do not require flush water (*e.g.* 'dry' sanitation).

- *Blackwater* is the mixture of urine, faeces and flush water with anal cleansing water (if water is used for cleansing) and/or dry cleansing materials.
- *Greywater* is a common name for water generated from washing food, clothes and dishes, as well as from bathing, but not from toilets. It accounts for approximately 65% of the wastewater produced in households with flush toilets.
- *Brown water* is a mixture of faeces and flush water that does not include urine, generated mainly in urine-diverting flush toilets. Its volume depends on the amount of flush water used. It can also include anal cleansing materials.
- *Yellow water* is a mixture of urine and flush water.
- *Faecal sludge* is a mixture of solids and liquids, containing mostly excreta and water, in combination with sand, grit, metals, rubbish and/or various chemical compounds. It can be raw or partially digested, a slurry or semi-solid, and results from the collection and storage/treatment of excreta or blackwater, with or without greywater. Characteristics of fresh faecal sludge (*e.g.* originating from containerized sanitation) and those of old faecal sludge (*e.g.* originating from septic tanks - often called 'septage', an historical term to define sludge removed from septic tanks.) are different for various reasons. Reported faecal sludge production rates vary from 100 to 520 grams of wet sludge per person per day (Rose *et al.*, 2015: Strande *et al.*, 2014; Zakaria *et al.*, 2018).

Many factors affect the composition of faecal sludge such as: (*i*) diet, (*ii*) toilet usage, habits, including social and cultural aspects, (*iii*) storage duration, (*iv*) inflow and infiltration, (*v*) collection method, and (*vi*) climate. It is common knowledge that characteristics of faecal sludge vary both spatially and temporally. There is currently a lack of detailed information on the characteristics of faecal sludge. However, research is actively being conducted in this field. These research results, together with empirical observations, will continue to increase the knowledge on faecal sludge characteristics, and allow more accurate predictions of faecal sludge characteristics using less labour-intensive methods. An overview of the most relevant faecal sludge characteristics from public toilets and septic tanks is presented in Figure 3.26 (adopted from Strande *et al.*, 2014). For more detailed information on faecal sludge characterization, sampling and analysis, the reader is referred to Velkushanova *et al.* (2020). Information presented in Table 3.26 clearly shows that the strength of faecal sludge is much higher than the strength of domestic sewage. In addition, fractionations of different components (such as COD, solids, etc.) of faecal sludge and of sewage usually show even larger differences. Also, there are large variations within various samples of faecal sludge due to reasons already described even if the samples are taken from the same micro-location.

The main conclusions from this section are (*i*) that one should not underestimate the effects of faecal sludge on biological wastewater treatment (and also on primary treatment and waste sludge digestion), and (*ii*) that the characteristics of faecal sludge can vary considerably; therefore, the quantity and quality of faecal sludge needs to be determined with knowledge and skills so that the results are sufficiently accurate to be used with confidence for design, operation and management purposes.

Figure 3.15 Sampling of faecal sludge at FSTP of Faridpur municipality, Bangladesh (photo: IHE Delft).

Table 3.26 Characteristics of faecal sludge from public toilets and septic tanks (adapted from Strande *et al.*, 2014).

	Public toilet	Septic tank	Reference
pH	1.5-12.6		USEPA (1994)
	6.55-9.34		Kengne *et al.* (2011)
Total solids, TS (mg/l)	52,500	12,000-35,000	Koné and Strauss (2004)
	30,000	22,000	NWSC (2008)
		34,106	USEPA (1994)
	≥3.5%	<3%	Heinss *et al.* (1998)
Total volatile solids, TVS (as % of TS)	68	50-73	Koné and Strauss (2004)
	65	45	NWSC (2008)
COD (mg/l)	49,000	1,200-7,800	Koné and Strauss (2004)
	30,000	10,000	NWSC (2008)
	20,000-50,000	<10,000	Heinss *et al.* (1998)
BOD (mg/l)	7,600	840-2,600	Koné and Strauss (2004), NWSC (2008)
	-	190-300	Koné and Strauss (2004): NWSC (2008)
Total nitrogen, TN (mg/l)	3,400	1,000	Katukiza *et al.* (2012)
	3,300	150-1,200	Koné and Strauss (2004)
Total Kjeldahl nitrogen, TKN (mg/l)	2,000	400	NWSC (2008)
NH_4-N (mg/l)	2,000-5,000	<1,000	Heinss *et al.* (1998)
	-	0.2-21	Koottatep *et al.* (2005)
	450	150	NWSC (2008)
Nitrate, NO_3 (mgN/l)	$1 \cdot 10^5$	$1 \cdot 10^5$	NWSC (2008)
Total phosphorus, TP (mgP/l)	2,500	4,000-5,700	Heinss *et al.* (1994)
Faecal coliforms (cfu/100 ml)	20,000-60,000	4,000	Heinss *et al.* (1998)
Helminth eggs (numbers/l)		600-6,000	Ingallinella *et al.* (2002)
		16,000	Yen-Phi *et al.* (2010)

3.12 Dynamics of wastewater characteristics

The wastewater flow rate and its component concentrations vary over time. These dynamics are important for the design, operation, monitoring and control of wastewater treatment plants.

The wastewater flow rate is the volume of wastewater (m^3) per time unit (days, hours or seconds). Various units for wastewater flow rates can be used for the design of different unit processes in a wastewater treatment plant, depending on the hydraulic retention time. For units, for example with screens and a grit chamber, with a short hydraulic retention time, the design flow is represented by m^3/s, while for settling tanks the design flow is usually expressed in m^3/h.

Wastewater flow rates vary over time and place, which complicates their accurate measurement. For design purposes, often the maximum flow rate needs to be known in addition to the average flow rate. The maximum flow rate (Q_{max}) can be calculated by multiplying the average flow rate (Q_{avg}) with a parameter f_{max}, according to Eq. 3.5.

$$Q_{max} = Q_{avg} \cdot f_{max} \tag{3.5}$$

The factor f_{max} depends on the size of the catchment: for cities it will be 1.3-1.7 and for towns 1.7-2.4. The ratio between the maximum and average flow rate is thus higher for smaller catchments, reflecting the smaller retention time in their natural catchment and the associated sewer system.

Variations in wastewater characteristics over time can be significant, as illustrated in Figure 3.16 for the wastewater flow rate, COD and suspended solids concentrations in a specific case. More examples are given in Chapter 15.

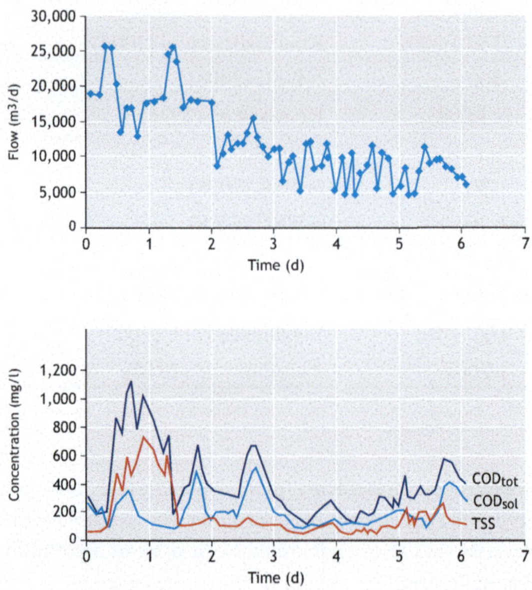

Figure 3.16 Variations in wastewater flow (above) and concentration for COD and suspended solids (below) (Henze *et al.*, 2002).

In many cases daily variations are observed, in some weekly or seasonal variations and others are very likely to be caused by industrial production patterns. If the component load is more or less constant, the component concentration dynamics will largely correlate with the wastewater flow rate. This is quite often the case for organic carbon. Other components such as nutrients can show dynamics independent of the flow rate. For instance, diurnal variations in ammonium concentrations can be caused by urine, which is a very concentrated nutrient stream (Figure 3.17).

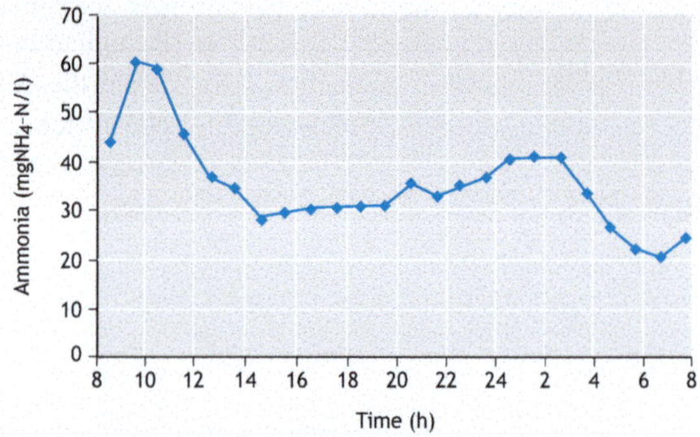

Figure 3.17 Variation by hour in ammonia content in the influent of the Galindo wastewater treatment plant in Spain.

The variations in flow rate and component concentrations make wastewater sampling a challenging task (Figure 3.19). Indeed, the analytical results obtained can vary considerably with the chosen sampling procedure:

- Grab samples, *i.e.* a single sample collected in a recipient at a specific time instant. This type of sampling gives highly variable results.
- Time proportional samples, *i.e.* a number of samples taken at regular time intervals (*e.g.* every hour) which are combined in one final sample. This type of sampling is suitable if the variations in the wastewater characteristics are relatively slow compared to the measurement frequency.
- Flow proportional sampling, *i.e.* a number of samples taken for each specific volume of wastewater flow (typically performed over 24 hours). This gives a more reliable estimate of the

dynamics of the wastewater load components and of its average concentrations over the given time.

The dynamics in wastewater characteristics also have implications for process control. Variations in wastewater flow rate and component concentrations can be seen as disturbances acting on the wastewater treatment plant, which need to be compensated for by feedforward and/or feedback control loops in order to achieve the required effluent quality and also reach other plant objectives (see Chapter 15).

3.13 CALIBRATION PROTOCOLS FOR ACTIVATED SLUDGE MODELLING

Driven by the requirements of mathematical modelling of activated sludge systems, various systematic protocols for activated sludge model calibration have been developed that include different wastewater characterization protocols. The main protocols, developed by four different research groups, range from simplified and relatively practical, to those of increased complexity and more of academic and research interest (Figure 3.18).

- The BIOMATH protocol (Vanrolleghem *et al.*, 2003) uses a combination of respirometry and physical-chemical methods to estimate the influent COD and nitrogen fractions.
- The STOWA protocol (Hulsbeek *et al.*, 2002) is based on a physical-chemical method to determine the soluble COD fractions and BOD measurement in order to characterize the biodegradable fraction of the influent COD. These fractions are then used to find the wastewater characterization required for activated sludge models.
- The Hochschulgruppe (HSG) guidelines (Langergraber *et al.,* 2004) stress the importance of obtaining validated measurement data for use in simulation studies, without mentioning specific influent characterization methods.
- The WERF protocol (Melcer *et al.*, 2003) provides detailed procedures for their respirometric and physical-chemical methods used for influent wastewater characterization and also for determining kinetic and stoichiometric parameters for activated sludge models.

Which protocol to use depends on the purpose of the modelling. Table 3.27 gives a thorough comparison of the four protocols through a SWOT analysis (Sin *et al.*, 2005). For more details on activated sludge treatment modelling the reader is referred to Chapter 14.

Figure 3.18 A course on modelling of activated sludge systems at IHE Delft during which students get familiar with calibration protocols (photo: IHE Delft).

Table 3.27 Comparison of the different characterization protocols (Sin *et al.*, 2005).

Protocol	Strengths	Weaknesses	Opportunities	Threats
BIOMATH	• Detailed settling, hydraulic and biological characterization • Detailed influent characterization • Biomass characterization • Sensitivity analysis/parameter selection • OED for measurement campaign • Structured overview of the protocol • Feedback loops	• Respirometric influent characterization requires model-based interpretation • OED has not been applied yet in practice, but research is ongoing • OED software and specialist skills required • No detailed methodology for data quality check • No practical procedure for parameter calibration	• Generally applicable • Works efficiently once implemented in a simulator • Dynamic measurement campaigns can be designed and compared based on OED	• Not all modelling and simulation software has OED/sensitivity analysis (SA) • High degree of specialization is required for the application
STOWA	• Detailed settling and biological characterization • Process control • Time estimate for different calibration steps • Detailed data quality check • Step-wise calibration of biological process parameters • Structured overview of the protocol • Feedback loops	• No detailed hydraulic characterization • BOD test gives problems with inert particulate biomass fraction • No biomass characterization • No guidance for measurement campaign design • No detailed info on sensitivity analysis • Fixed parameter subsets for calibration of biological processes	• Easy to use • Practical experimental methods • No specialist required • Good for consultants and new modellers •	• No mathematical/statistical approach for parameter selection for calibration • May nor ne applicable for different systems since parameter subset for calibration may change for different WWTPs
HSG	• CFD for hydraulic characterization • Biological characterization • Design of measurement campaign • A standard format for documentation • Data quality check • Structured overview protocol	• No feedback loops in overview diagram • Provides only general guidelines • No detailed settling characterization • No particular methods for influent characterization of parameter estimation • No detailed sensitivity analysis/parameter selection	• Generally applicable • A standard format for thorough documentation/reporting of calibration studies	• Not detailed/practical enough for new practitioners • The free choice of experimental methodologies for influent/kinetic characterization may jeopardize standardization of calibration studies
WERF	• Detailed influent characterization • Detailed $\mu_{ANO,max}$ and b_{ANO} determination • Biomass characterization • Sensitivity analysis/parameter selection • Detailed data quality check • A tiered approach for calibration • Several examples of case studies	• No feedback loops • Settling process less emphasized • Almost no emphasis on other kinetic parameters than nitrification • No structured overview of protocol	• Based on practical experience • A tiered approach for calibration provides different calibration levels for different goals and accuracy of calibration • Good for consultants and new modellers	• Focus on $\mu_{ANO,max}$ determination and influent characterization • Ignoring the significance of the other parts of the full-scale model • Laborious methods

OED: optimal experimental design; CFD: computational fluid dynamics; $\mu_{ANO,max}$: maximum growth rate of autotrophic biomass; b_{ANO}: endogenous decay coefficient of heterotrophs;.

REFERENCES

Brdjanovic D., Meijer S.C.F., Lopez-Vazquez C.M., Hooijmans C.M. and Van Loosdrecht M.C.M. (eds.) (2015). *Applications of Activated Sludge Models*. IWA Publishing, London, UK.

Cipolla S.S. and Maglionico M. (2014). Heat recovery from urban wastewater: analysis of the variability of flow rate and temperature. *Energy Build.* 69, 122-130.

Corcoran J., Winter M.J. and Tyler C.R. (2010). Pharmaceuticals in the aquatic environment: a critical review of the evidence for health effects in fish. *Critical Reviews in Toxicology,* 40(4), 287-304.

Danish Environmental Protection Agency (Danish EPA) (1993). *Domestic Wastewater and Clean Technology*. Miljøprojekt Nr. 2019, Miljøstyrelsen, Copenhagen, Denmark.

Dewil R., Baeyens J., Roels J. and Steene B.V.D. (2008). Distribution of sulphur compounds in sewage sludge treatment. *Environmental Engineering Science,* 25(6), 879-886.

Dold P.L., Ekama G.A. and Marais G.v.R. (1980). A general model for the activated sludge process. *Prog. Wat. Technol.* 12(6), 47–77.

Ekama G.A. (2009). Using bio-process stoichiometry to build a steady state plant wide wastewater treatment plant model. *Water Research,* 43(8) 2101-2120.

European Commission (EC) (2007). *Terms and Definitions of the Urban Waste Water Treatment Directive* (91/271/EEC). European Commission, Brussels, Belgium.

Fatta D., Achilleos A., Nikolaou A. and Meric S. (2007). Analytical methods for tracing pharmaceutical residues in water and wastewater. *Trends in Analytical Chemistry,* 26(6), 515-533.

Friedler E., Butler D. and Alfiya Y. (2013). *Chapter 17: Wastewater Composition*. In: Larsen T.A., Udert K.M. and Lienert J. (eds.). Source Separation and Decentralization for Wastewater Management, pp. 241-254, IWA Publishing, London, UK.

Friedler E., Brown D.M. and Butler D. (1996). A study of WC derived sewer solids. *Water Science and Technology*, 33(6), 17-24.

Gebreeyessus G.D. and Jenicek P. (2016). Thermophilic *versus* mesophilic anaerobic digestion of sewage sludge: a comparative review. *Bioengineering,* 3(2), 15.

Gu A.Z., Liu L., Neethling J.B., Stensel H.D. and Murthy S. (2011). Treatability and fate of various phosphorus fractions in different wastewater treatment processes. *Water Science and Technology*, 63(4), 804-810.

Gupta M., Ho D., Santoro D., Torfs E., Doucet J., Vanrolleghem P.A. and Nakhla G. (2018). Experimental assessment and validation of quantification methods for cellulose content in municipal wastewater and sludge. *Environmental Science and Pollution Research,* 25(17), 16743-16753.

Heinss U., Larmie S.A. and Strauss M. (1998). Solids Separation and Pond Systems for the Treatment of Faecal Sludges in the Tropics. Lessons learnt and recommendations for preliminary design. SANDEC Report No. 5/98. Second Edition. Swiss Federal Institute for Environmental Science and Technology (EAWAG) and Water Research Institute (WRI), Accra/Ghana.

Henze M. (1982). The composition of domestic wastewater. *Stads og Havneingeniøren Journal,* 73, 386-387.

Henze M. (1992). Characterization of wastewater for modelling of activated sludge processes. *Water Science and Technology*, 25(6), 1-15.

Henze M. (1997). Waste design for households with respect to water, organics and nutrients. *Water Science and Technology,* 35(9), 113–120.

Henze M., Grady C.P.L., Gujer W., Marais G.v.R. and Matsuo T. (1987). *Activated Sludge Model No. 1* IAWPRC Scientific and Technical Report No. 1., International Association on Water Pollution Research and Control (IAWPRC), London, UK.

Henze M., Gujer W., Mino T., Matsuo T., Wentzel M.C. and Marais G. (1995). Wastewater and biomass characterization for the activated sludge model no. 2 (ASM2): biological phosphorus removal. *Water Science and Technology*, 31(2), 13-23.

Henze M., Harremoës P., La Cour Jansen J. and Arvin E. (2002). *Wastewater Treatment: Biological and Chemical Processes* (3rd ed.), Springer-Verlag, Berlin, Germany.

Huisman J.L. and Gujer W. (2002). Modelling wastewater transformation in sewers based on ASM3. *Water Science and Technology,* 45(6), 51-60.

Hulsbeek J.J.W., Kruit J., Roeleveld P.J. and Van Loosdrecht M.C.M. (2002). A practical protocol for dynamic modelling of activated sludge systems. *Water Science and Technology,* 45(6), 127-136.

Hvitved-Jacobsen T., Vollertsen J. and Nielsen A.H. (2013). *Sewer Processes: Microbial and Chemical Process Engineering of Sewer Networks* (2nd ed.), CRC Press, Florida, USA.

Ikumi D.S., Harding T.H. and Ekama G.A. (2014). Biodegradability of wastewater and activated sludge organics in anaerobic digestion. *Water Research,* 56(1) 267-279.

Ingallinella A.M., Sanguinetti G., Koottatep T., Montangero A. and Strauss M. (2002). The challenge of faecal sludge management in urban areas – strategies, regulations and treatment options. *Water Science and Technology*, 46(10):285-294.

Ioannou L.A., Puma G.L. and Fatta-Kassinos D. (2015). Treatment of winery wastewater by physicochemical, biological and advanced processes: a review. *Journal of Hazardous Materials,* 286, 343-368.

Jenkins D. and Wanner J. (2014). *Activated Sludge - 100 Years and Counting*. IWA Publishing, London, UK.

Jia M., Solon K., Vandeplassche D., Venugopal H. and Volcke E.I.P. (2020). Model-based evaluation of an integrated high-rate activated sludge and mainstream anammox system. *Chemical Engineering Journal,* 382, 122878.

Kappeler J. and Gujer W. (1992). Estimation of kinetic parameters of heterotrophic biomass under aerobic conditions and characterization of wastewater for activated sludge modelling. *Water Science and Technology,* 25(6), 125-139.

Katukiza A.Y., Ronteltap M., Niwagaba C., Foppen J.W.A., Kansiime F. and Lens P.N.L. (2012). Sustainable sanitation technology options for urban slums. *Biotechnology Advances*, 30(5):964-978.

Kengne I.M., Kengne S.E., Akoa A., Bemmo N., Dodane P.-H. and Kone D. (2011). Vertical-flow constructed wetlands as an emerging solution for faecal sludge dewatering in developing countries. *Journal of Water, Sanitation and Hygiene for Development,* 1(1):13-19.

Kone D. and Strauss M. (2004). Low-cost Options for Treating Faecal Sludges (FS) in Developing Countries – Challenges and Performance. Paper presented to the 9th International IWA Specialist group conference on wetlands systems for water pollution control; and to the 6th International IWA Specialist Group Conference on Waste Stabilisation Ponds, Avignon, France.

Koottatep T., Surinkul N., Polprasert C., Kamal A.S.M., Kone D., Montangero A., Heinss U. and Strauss M. (2005). Treatment of septage in constructed wetlands in tropical climate: lessons learnt from seven years of operation. *Water Science and Technology*, 51(9):119-126.

Kümmerer K. (2013). *Chapter 6: The Issue of Micropollutants in Urban Water*. In: Larsen T.A., Udert K.M. and Lienert J. (eds.). Source Separation and Decentralization for Wastewater Management, pp. 71-83, IWA Publishing, London, UK.

Kushwaha J.P., Srivastava V.C. and Mall I.D. (2011). An overview of various technologies for the treatment of dairy wastewaters. *Critical Reviews in Food Science and Nutrition,* 51(5), 442-452.

Langergraber G., Rieger L., Winkler S., Alex J., Wiese J., Owerdieck C., Ahnert M., Simon J. and Maurer M. (2004). A guideline for simulation studies of wastewater treatment plants. *Water Science and Technology,* 50(7), 131-138.

Lens P.N.L., Visser A., Janssen A.J.H., Pol L.H. and Lettinga G. (1998). Biotechnological treatment of sulfate-rich wastewaters. *Critical Reviews in Environmental Science and Technology,* 28(1), 41-88.

Lopez-Vazquez C.M., Welles L., Lotti T., Ficara E., Rene E.R., Van den Brand T.P.H., Brdjanovic D. and Van Loosdrecht M.C.M. (2016). *Chapter 2: Activated Sludge Activity Tests*. In: Van Loosdrecht M.C.M., Nielsen P.H., Lopez-Vazquez C.M. and Brdjanovic D. (eds.), Experimental Methods in Wastewater Treatment, pp. 7-121, IWA Publishing, London, UK.

Mamais D., Jenkins D. and Pitt P. (1993). A Rapid Physical-Chemical Method for the Determination of Readily Biodegradable Soluble Cod in Municipal Wastewater. *Water Research,* 27(1), 195-197.

Melcer H., Dold P.L., Jones R.M., Bye C.M., Takacs I., Stensel H.D., Wilson A.W., Sun P. and Bury S. (2003). *Methods for Wastewater Characterization in Activated Sludge Modeling.* Water Environment Research Foundation (WERF), Virginia, USA.

Metcalf and Eddy (2014). *Wastewater Engineering: Treatment and Resource Recovery* (5th ed.), McGraw-Hill, Boston, USA.

Nogaj T., Randall A., Jimenez J., Takacs I., Bott C., Miller M., Murthy S. and Wett B. (2015). Modeling of organic substrate transformation in the high-rate activated sludge process, *Water Science and Technology,* 71(7), 971-979.

NWSC (National Water and Sewerage Corporation) (2008). Kampala Sanitation Program (KSP) - Feasibility study for sanitation master in Kampala, Uganda.

Ødegaard H. (1992) Norwegian experiences with chemical treatment of raw wastewater. *Water Science and Technology,* 25(12), 255-264.

Peal A., Evans B., Ahilan S., Ban R., Blackett I., Hawkins P., Schoebitz L., Scott R., Sleigh A., Strande L. and Veses O. (2020). Estimating safely managed sanitation in urban areas - lessons learned from a global implementation of excreta-flow diagrams. *Frontiers in Environmental Science,* 8(1), 1-13.

Plósz B.G., Langford K.H. and Thomas K.V. (2012). An activated sludge modeling framework for xenobiotic trace chemicals (ASM-X): Assessment of diclofenac and carbamazepine. *Biotechnology and Bioengineering,* 109(11), 2757-2769.

Pokhrel D. and Viraraghavan T. (2004). Treatment of pulp and paper mill wastewater - a review. *Science of the Total Environment,* 333(1-3), 37-58.

Reijken C., Giorgi S., Hurkmans C., Pérez J. and Van Loosdrecht M.C.M. (2018). Incorporating the influent cellulose fraction in activated sludge modelling. *Water Research,* 144, 104–111.

Rice E.W., Baird R.B. and Eaton A.D. (eds.) (2017). Standard Methods for the Examination of Water and Wastewater, 23rd edition, *American Public Health Association, American Water Works Association, Water Environment Federation.* ISBN: 9780875532875.

Rieger L., Gillot S., Langergraber G., Ohtsuki T., Shaw A., Takacs I. and Winkler S. (2012). *Guidelines for Using Activated Sludge Models.* IWA Scientific and Technical Report No. 21, IWA Publishing, London, UK.

Roeleveld P.J. and Van Loosdrecht M.C.M. (2002) Experience with guidelines for wastewater characterisation in the Netherlands. *Water Science and Technology,* 45(6), 77-87.

Rose C., Parker A., Jefferson B. and Cartmell E. (2015). The characterization of feces and urine: a review of the literature to inform advanced treatment technology. *Critical Reviews in Environmental Science and Technology,* 45, 1827-1879.

Ruiken C.J., Breuer G., Klaversma E., Santiago T. and Van Loosdrecht M.C.M. (2013). Sieving wastewater - Cellulose recovery, economic and energy evaluation. *Water Research,* 47(1), 43-48.

Sin G., Van Hulle S.W., De Pauw D.J., Van Griensven A. and Vanrolleghem P.A. (2005) A critical comparison of systematic calibration protocols for activated sludge models: A SWOT analysis. *Water Research,* 39(12), 2459-2474.

Smitshuijzen J., Pérez J., Duin O. and Van Loosdrecht M.C.M. (2016). A simple model to describe the performance of highly-loaded aerobic COD removal reactors. *Biochemical Engineering Journal,* 112, 94–102.

Snip L.J., Flores-Alsina X., Plósz B.G., Jeppsson U. and Gernaey K.V. (2014). Modelling the occurrence, transport and fate of pharmaceuticals in wastewater systems. *Environmental Modelling and Software,* 62, 112-127.

Solon K., Volcke E.I.P., Spérandio M. and Van Loosdrecht M.C.M. (2019) Resource recovery and wastewater treatment modelling. *Environmental Science and Water Research,* 5(4), 631-642.

Spanjers H. and Vanrolleghem P.A. (2016). *Chapter 3: Respirometry.* In: Van Loosdrecht M.C.M., Nielsen P.H., Lopez-Vazquez C.M. and Brdjanovic D. (eds.), Experimental Methods in Wastewater Treatment, pp. 133-176, IWA Publishing, London, UK.

Strande L., Ronteltap M. and Brdjanovic D. (2014). *Faecal Sludge Management - Systems Approach for Implementation and Operation.* IWA Publishing, London, UK.

Strande L., Schoebitz L., Bischoff F., Ddiba D. and Niwagaba C. (2018). Methods to reliably estimate faecal sludge quantities and qualities for the design of treatment technologies and management solutions. *Journal of Environmental Management,* 223(1), 898-907.

Sundberg A. (1995) *What is the Content of Household Wastewater?* Report No. 4425, Swedish EPA, Stockholm, Sweden.

Tilley E., Ulrich L., Lüthi C., Reymond P. and Zurbrügg C. (2014). *Compendium of Sanitation Systems and Technologies* (2nd ed.), Swiss Federal Institute of Aquatic Science and Technology (Eawag), Dübendorf, Switzerland.

Van Den Brand T.P.H., Roest K., Chen G.H., Brdjanovic D. and Van Loosdrecht M.C.M. (2015). Potential for beneficial application of sulfate reducing bacteria in sulfate containing domestic wastewater treatment. *World Journal of Microbiology and Biotechnology,* 31(11), 1675-1681.

Van Haandel A. and Van Der Lubbe J. (2007). *Handbook Biological Waste Water Treatment - Design and Optimisation of Activated Sludge Systems*, Quist Publishing, Leidschendam, the Netherlands.

Van Loosdrecht M.C.M., Lopez-Vazquez C.M., Meijer S.C.F., Hooijmans C.M. and Brdjanovic D. (2015). Twenty-five years of ASM1: past, present and future of wastewater treatment modelling. *Journal of Hydroinformatics,* 17(5), 697–718.

Van Loosdrecht M.C.M., Nielsen P.H, Lopez-Vazquez C.M. and Brdjanovic D. (eds.) (2016). *Experimental Methods in Wastewater Treatment*, IWA Publishing, London, UK.

Vanrolleghem P.A., Insel G., Petersen B., Sin G., De Pauw D., Nopens I., Doverman H., Weijers S. and Gernaey K. (2003). A comprehensive model calibration procedure for activated sludge models. In: *Proceedings of the 76th Annual WEF Conference and Exposition,* pp. 210-237, Los Angeles, California, USA, 11-15 October 2003.

Velkushanova K., Strande L., Ronteltap M., Koottatep T., Brdjanovic D. and Buckley C. (2020). *Methods for Faecal Sludge Analysis*, IWA Publishing, London, UK.

Yaseen D.A. and Scholz M. (2019). Textile dye wastewater characteristics and constituents of synthetic effluents: a critical review. *International Journal of Environmental Science and Technology,* 16(2), 1193-1226.

Yen-Phi V.T., Rechenburg A., Vinneras B., Clemens J. and Kistemann T. (2010). Pathogens in septage in Vietnam. *Science of the Total Environment,* 408(9):2050-2053.

Zakaria F., Ćurko J., Muratbegovic A., Garcia H.A., Hooijmans C.M. and Brdjanovic D. (2018). Evaluation of a smart toilet in an emergency camp. *International Journal of Disaster Risk Reduction,* 27, 512-523.

NOMENCLATURE

Symbol	Description	Unit
b_{ANO}	Specific endogenous respiration rate for nitrifiers	d^{-1}
COD_t	Concentration of total COD	mgCOD/l
COD_b	Concentration of biodegradable COD	mgCOD/l
f_n	Nitrogen content of the sludge	gN/gVSS
f_{max}	Maximum flow rate factor	
f_p	Phosphorus content of the sludge	gP/gVSS
$f_{SU,TKNi}$	Influent soluble unbiodegradable organic nitrogen fraction	
$f_{TKNi/CODi}$	TKN/COD ratio in the influent	mgN/mgCOD
MX_{ANO}	Mass of nitrifiers in reactor	mgVSS
MX_{OHOv}	Mass of OHO biomass in reactor	mgVSS
$MX_{E,OHOv}$	Mass of endogenous residue in reactor	mgVSS
MX_{Uv}	Mass of unbiodegradable organics from the influent in the reactor	mgVSS
N_o	Organic nitrogen concentration	mgN/l
N_{obp}	Biodegradable particulate organic nitrogen concentration	mgN/l
N_{obs}	Biodegradable soluble organic nitrogen concentration	mgN/l
N_{oup}	Unbiodegradable particulate organic nitrogen concentration	mgN/l
N_{ous}	Unbiodegradable soluble organic nitrogen concentration	mgN/l
P_t	Total phosphorus concentration	mgP/l
P_{IN}	Inorganic phosphorus concentration	mgP/l
P_o	Organic phosphorus concentration	mgP/l
P_{obp}	Biodegradable particulate organic phosphorus concentration	mgP/l
P_{obs}	Biodegradable soluble organic phosphorus concentration	mgP/l
P_{oup}	Unbiodegradable particulate organic phosphorus concentration	mgP/l
P_{ous}	Unbiodegradable soluble organic phosphorus concentration	mgP/l
Q_{max}	Maximum flow rate	m^3/d
Q_{ave}	Average flow rate	m^3/d
S_F	Fermentable organic matter concentration	mgCOD/l
S_{NHx}	Bulk liquid ammonia concentration	mgN/l
S_S	Soluble readily biodegradable (RB)COD concentration	mgCOD/l
SRT	Sludge retention time	d
S_U	Soluble unbiodegradable COD	mgCOD/l

S_{VFA}	Volatile fatty acids concentration	mgCOD/l
TKN_t	Total Kjeldahl nitrogen concentration	mgN/l
X_{CL}	Cellulose concentration	mgVSS/l
X_U	Unbiodegradable matter in activated sludge	mgVSS/l
X_S	Slowly biodegradable (SB)COD concentration	ngCOD/l

Abbreviation	Description
ATU	Allylthiourea
BOD	Biochemical oxygen demand
BOD_∞	Ultimate BOD
CFD	Computational fluid dynamics
COD	Chemical oxygen demand
CWIS	City-wide Inclusive Sanitation
EPS	Extracellular polymeric substances
FSA	Free and saline ammonia
FSM	Faecal sludge management
FSTP	Faecal sludge treatment plant
GC-MS	Gas chromatography with mass spectrometry
HPLC	High performance liquid chromatography
$LC\text{-}MS^2$	Liquid chromatography with tandem mass spectrometry
NBCOD	Nitrogenous biochemical oxygen demand
ISS	Inert suspended solids
OED	Optimal experimental design
OHOs	Ordinary heterotrophic organisms
PAOs	Phosphorus-accumulating organisms
PE	Population equivalent PE
PL	Person load
SA	Sensitivity analysis
SDGs	Sustainable Development Goals
SFD	Shit flow diagram
TKN	Total Kjeldahl nitrogen
TSS	Total suspended solids
VFAs	Volatile fatty acids
VSS	Volatile suspended solids
WC	Water closet

Greek symbols	Explanation	Unit
$\mu_{ANO,max}$	Maximum specific growth rate of nitrifiers	d^{-1}

Figure 3.19 Sampling is an essential part of wastewater and sludge characterization. The photo shows sample collection from an oil refinery wastewater treatment plant (Photo: D. Brdjanovic).

doi: 10.2166/9781789060362_0111

4

Organic matter removal

George A. Ekama and Mark C. Wentzel†

4.1 INTRODUCTION

4.1.1 Transformations in the biological reactor

For the activated sludge system, it is necessary to characterize wastewater both physically (soluble, non-settleable (colloidal and /or suspended), settleable, organic, inorganic) and biologically (biodegradable, unbiodegradable). The physical, chemical and biological transformations of the organic and inorganic wastewater constituents that take place in the biological reactor are outlined in Figure 4.1. Some of these transformations are important for achieving the required effluent quality while others are not important for the effluent quality but are important for the system design and operation. In Figure 4.1 each of the wastewater organic and inorganic fractions have soluble and particulate fractions, the latter of which subdivides further into suspended (non-settleable) and settleable. Each of the three organic subfractions in turn has biodegradable and unbiodegradable constituents. The inorganic particulate subfraction comprises both settleable and suspended (non-settleable) constituents while the soluble inorganic subfraction comprises both precipitable and non-precipitable and biologically utilizable and non-biologically utilizable constituents.

In the biological reactor the biodegradable organics, whether soluble, non-settleable or settleable, are transformed into ordinary heterotrophic organisms (OHOs, X_{OHO}), which become part of the organic (volatile) suspended solids (VSS) in the reactor. When these organisms die, they leave behind unbiodegradable particulate (but not soluble) organics, called the endogenous residue, comprising mainly unbiodegradable cell wall material ($X_{E,OHO}$). This endogenous residue becomes part of the VSS mass in the reactor. The unbiodegradable suspended and settleable organics from the influent become enmeshed with the OHO and endogenous residue masses. Together these three constituents (X_{OHO} + $X_{E,OHO}$ + X_U) form the organic component of the suspended solids that accumulate in the biological reactor (VSS, X_{VSS}). The inorganic settleable and suspended constituents, together with the precipitable soluble inorganics, form the inorganic component of the suspended solids mass (ISS). The biologically utilizable soluble inorganics are absorbed by the biomass and become part of it or are transformed into the gas phase, in which case they escape to the atmosphere. The non-precipitable and non-biologically utilizable soluble inorganics escape with the effluent. Because of the efficient bioflocculation capability of the organic activated sludge mass, all the solids material, whether biodegradable or unbiodegradable, organic or inorganic, becomes settleable solids. Very little suspended or colloidal (non-settleable) solids mass is formed in the reactor, but when it does it cannot be retained in the system anyway and escapes with the effluent.

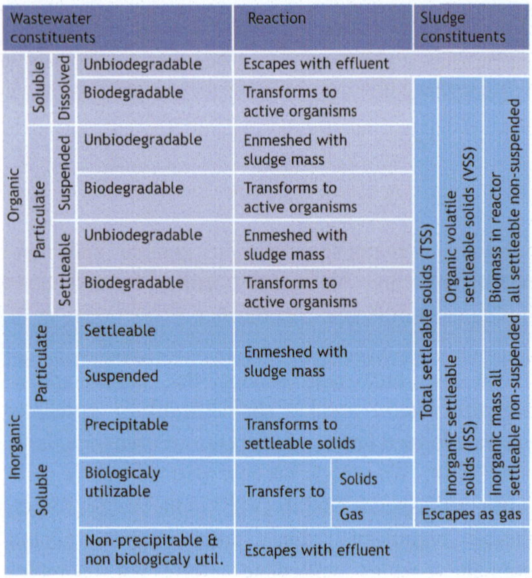

Figure 4.1 Global transformation reactions of organic and inorganic wastewater constituents from the particulate and soluble forms in the solid and liquid phases to the solid phase as sludge constituents, and gas and liquid phases escaping to the atmosphere and with the effluent, respectively.

The degree of wastewater characterization required for the activated sludge system design is not only determined by the physical, chemical and biological processes taking place in the system, but also by the level of sophistication of the design procedures to be applied for design. This is determined largely by the effluent quality required in terms of organic matter (C), nitrogen (N) and phosphorus (P). Generally, the more stringent the effluent quality requirements in terms of C, N and P, the more complex the activated sludge system has to be to achieve the required removals, and the more advanced and realistic the design procedures need to be. The more sophisticated and refined the design procedures are, the more detailed and refined the wastewater characterization needs to be (detailed wastewater fractionation is presented in Section 3.14).

For organic material removal only, with the wastewater strength measured in terms of BOD_5 and suspended solids (SS, settleable and/or non-settleable), little more than a knowledge of the organic load in terms of BOD_5 and SS is adequate. Knowledge of the kind of organics that make up the BOD_5 and SS generally is not required because various empirical relationships have been developed linking the BOD_5 and SS loads to the expected response and performance of the activated sludge system insofar as sludge production and oxygen demand are concerned. Where the organics are assessed in terms of COD, because the COD parameter includes both unbiodegradable and biodegradable organic material, an elementary characterization of the organic material is required, *i.e.* biodegradable and unbiodegradable and soluble and particulate COD concentrations need to be known. The unbiodegradable particulate COD concentration heavily affects sludge accumulation in the reactor and daily sludge production and the unbiodegradable soluble COD concentration fixes the filtered effluent COD concentration from the system. Without nitrification, N removal or P removal, no wastewater N and P characteristics are required. If nitrification is included in the system, knowledge of the components making up the N material in the influent is required (TKN and FSA). With biological nitrogen removal (denitrification), much more information is required: now not only the organic load in terms of COD (not BOD_5) needs to be specified, but also the quality and quantity of some of the organic compounds that make up the total organic (COD) load. Also, the nitrogenous (N) materials need to be characterized and quantified in the same way. With biological P removal, still further specific information characterizing the organic material is required and additionally characterization of the phosphorous (P) materials is required. The quality and quantity of C, N and P compounds entering the nitrogen (N) and nutrient (N and P) removal activated sludge reactor are affected by some unit operations upstream of the reactors, in particular primary sedimentation. It is thus important that the effect of primary sedimentation on the wastewater C, N and P constituents is also determined, to enable the settled sewage characteristics to be estimated.

4.1.2 Steady-state and dynamic-simulation models

For mathematical modelling of wastewater treatment systems, generally two levels of mathematical models have been developed; steady state and dynamic simulation. Steady-state models have constant flows and loads and are relatively very simple. This simplicity makes these models very useful for design or evaluation of existing wastewater treatment plants (WWTPs). In these models complete descriptions of system parameters are not required, but rather the models are oriented to determining the important system design parameters from performance criteria (Ekama, 2009). Dynamic models are much more complex than steady-state ones and have varying flows and loads with the result that time is included as a parameter. Dynamic-simulation models are therefore useful in predicting time-dependent system responses of an existing or proposed system. However, their complexity demands that many more kinetic and stoichiometric constants need to be supplied and all the system design parameters have to be specified. Steady-state models are very useful for calculating the initial conditions required to start dynamic-simulation models such as reactor volumes, recycle and waste flows and values for the various concentrations in the reactor(s) and cross-checking simulation model outputs (Ekama, 2009).

4.2 ACTIVATED SLUDGE SYSTEM CONSTRAINTS

Basically all aerobic biological treatment systems operate on the same principles, *i.e.* trickling filters, aerated lagoons, contact-stabilization, extended aeration, etc. They differ only in the conditions under which the biological reactions are constrained to operate, called system constraints. The activated sludge system comprises the flow regime in the reactor, the sizes and shape, number and configuration of the reactors, recycle flows, influent flow and other features incorporated either deliberately, or present inadvertently or unavoidably. Whereas the response of the organisms is in accordance with their nature *i.e.* biological process behaviour, that of the system is governed by a combination of the organism behaviour and the physical features which define the system, *i.e.* the environmental conditions or system constraints under which the biological processes are constrained to operate.

4.2.1 Mixing regimes

In the activated sludge system, the mixing regime in the reactor and the sludge return are part of the system constraints and therefore influence the response of the system, and hence consideration must be given to reactor mixing regimes. There are two extremes of mixing; completely mixed and plug flow (Figure 4.2).

In the completely mixed regime the influent is instantaneously and thoroughly mixed (theoretically) with the reactor contents. Hence the effluent flow from the reactor has the same compound concentrations as the reactor contents. The reactor effluent flow passes to a settling tank; the overflow from the tank is the treated waste stream, the underflow is concentrated sludge and is recycled back to the reactor. In the completely mixed system the rate of return of the underflow has no effect on the biological reactor except if an undue sludge build-up occurs in the settling tank. The shape of the reactor is approximately square or circular in plan, and mixing is usually by mechanical aerators or diffused air bubble aeration. Examples are extended aeration plants, aerated lagoons, Pasveer ditches and single reactor completely mixed activated sludge plants.

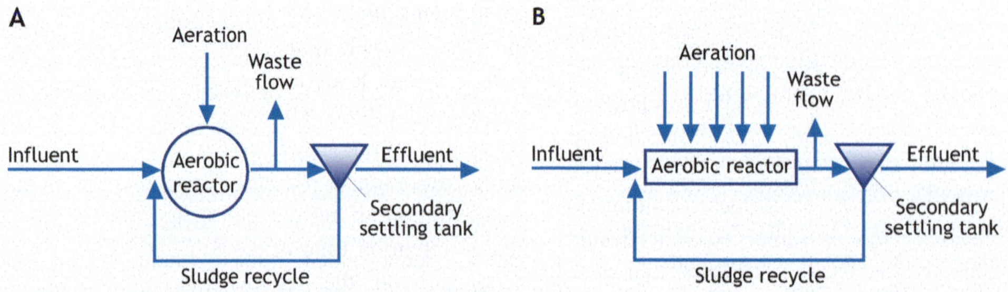

Figure 4.2 Activated sludge systems with (A) a single reactor completely mixed reactor mixing regime, and (B) a plug flow/intermediate reactor mixing regime.

In a plug-flow regime, the reactor is usually a long channel-type basin. The influent is introduced at one end of the channel, flows along the channel axis and is mixed by air spargers set along one side of the channel or horizontal shaft surface aerators. Theoretically each volume element of liquid along the axis is assumed to remain unmixed with the elements leading and following. Discharge to the settling tank takes place at the end of the channel. To inoculate the influent waste flow with organisms, the underflow from the settling tank is returned to the influent end of the channel. This creates an intermediate flow regime deviating from true plug-flow conditions depending on the magnitude of the recycled underflow. Conventional activated sludge plants are of the intermediate flow-regime type with sludge return recycle ratios varying from 0.25 to 3 times the average influent flow rate. If the recycle ratio is very high, the mixing regime approaches that of completely mixed.

Intermediate flow regimes are also achieved by having two or more completely mixed reactors in series, or by step-aeration. In the latter, the influent is fed at a series of points along the axis of the plug-flow-type reactor. Both configurations require, for inoculation purposes, recycling of the settled sludge from the settling tank(s) to the start of the channel reactor.

The mean kinetic response of an activated sludge system, *i.e.* sludge mass, daily sludge production, daily oxygen demand and effluent organics concentration is adequately, indeed accurately, given by assuming the system is completely mixed and the influent flow and load are constant. This allows the reactor volume, the mass of sludge wasted daily, and the average daily oxygen utilization rate to be determined by relatively simple formulations. Peak oxygen utilization rates which arise under cyclic flow and load conditions can be estimated subsequently quite accurately by applying a factor to the average oxygen utilization rate. These factors have been developed from simulation studies with the simulation models on aerobic and anoxic-aerobic systems operated under cyclic and under constant flow and load conditions (Musvoto *et al.,* 2002).

4.2.2 Sludge retention time (SRT)

In the schematic diagrams for the activated sludge system (Figure 4.2), the waste (or surplus) sludge is abstracted directly from the biological reactor. The common practice is that the waste sludge is abstracted from the secondary settling tank underflow. Sludge abstraction directly from the reactor leads to a method of control of the sludge age (sludge retention time or solids retention time: SRT), called the *hydraulic control of sludge age,* which has significant advantages for system control compared to abstracting wastage via the underflow.

The sludge age, SRT in days, is defined by:

$$SRT = \frac{\text{Mass of sludge in reactor}}{\text{Mass of sludge wasted per day}} \quad \text{(d)}$$

By abstracting the sludge directly from the reactor, the sludge concentrations in the waste flow and biological reactor are the same. If sludge age of, say, 10 days is required, one tenth of the volume of the reactor is wasted every day. This can be achieved by a constant waste flow rate, Q_w (l/d), where Q_w is the volume of sludge to be wasted daily. Hence,

$$SRT = \frac{X_{TSS} \cdot V_R}{X_{TSS} \cdot Q_w} = \frac{V_R}{Q_w} \quad \text{(d) (4.1)}$$

where:

V_R volume of the biological reactor (l)
Q_w waste flow rate from reactor (l/d)

Equation 4.1 assumes that the loss of solids with the effluent is negligible and that the mass of sludge in the secondary settling tanks is also negligible relative to that in the biological reactor. This assumption is reasonable when the system is operated at relatively high underflow recycle ratios (~1:1) and the sludge age is longer than approximately 3 days (see Section 4.10).

4.2.3 Nominal hydraulic retention time (HRT_n)

In activated sludge theory, the volume of the reactor per unit of volume of influent flow is known as the nominal hydraulic retention time *i.e.*

$$HRT_n = \frac{V_R}{Q_i} \quad \text{(d) (4.2)}$$

where:

HRT_n average nominal hydraulic retention time (d)
Q_i daily average influent flow rate (l/d)

When the sludge return flow from the secondary settling tank (Q_s) and any other mixed liquor recycle flow entering the reactor (Q_a) are included, the retention time is called the actual hydraulic retention time (HRT_a), *viz.*

$$HRT_a = \frac{V_R}{Q_i + Q_s + Q_a} = \frac{HRT_n}{1 + s + a} \quad \text{(d) (4.3)}$$

where:

HRT_a actual hydraulic retention time (d)
s sludge underflow recycle ratio (Q_s/Q_i)
a mixed liquor recycle ratio (Q_a/Q_i)

4.2.4 Connection between sludge age and hydraulic retention time

From the above definitions, it can be seen that there are two parameters that relate to time in the system: (*i*) the sludge age, which gives the length of time the particulate material remains in the reactor, and (*ii*) the nominal hydraulic retention time, which gives the length of time the liquid and dissolved material remain in the reactor. In activated sludge systems which do not have solid liquid separation with membranes or secondary settling tanks (SSTs), such as aerated lagoons, the sludge age and nominal hydraulic retention time are equal, *i.e.* the liquid/dissolved material and the solids/particulate material remain in the reactor for the same length of time. When solid liquid separation is included, then the liquid and solid retention times are separated and SRT > HRT_n. However, long sludge ages lead to large sludge masses in the reactor, which in turn lead to large reactor volumes (V_R). Therefore, even with solid-liquid separation, as SRT gets longer, so also does HRT_n. This link between SRT and HRT_n is neither proportional nor linear and depends on (*i*) the wastewater organic (COD or BOD_5) concentration and (*ii*) the suspended solids concentration (TSS) in the reactor. For biological nutrient removal activated sludge systems, the sludge age is approximately 10 to 25 days and the nominal hydraulic retention time approximately 10 to 24 h.

4.3 SOME MODEL SIMPLIFICATIONS

4.3.1 Complete utilization of biodegradable organics

The distinction between biodegradable and unbiodegradable is governed by the biomass in the system and the length of time this biomass has to degrade the organics. It has been observed that the difference in the soluble effluent COD concentration from a short (2-3 h) and a very long (18-24 h) hydraulic retention time system is very small, only 10 to 20 mgCOD/l. This indicates that slowly biodegradable soluble organics seem to be very low in concentration in normal municipal wastewater. Therefore, it is reasonable to accept that the soluble organics in municipal wastewater comprise two groups: the biodegradable, which are almost all readily biodegradable and the unbiodegradable. This means that even at very short hydraulic retention times of a few hours, utilization of biodegradable organics is complete leaving only the soluble unbiodegradable organics in the effluent.

The influent particulate biodegradable organics, both settleable and suspended (X_S), are mostly slowly biodegradable. These slowly biodegradable particulate organics (SBCOD), whether settleable or non-settleable, become enmeshed within the activated sludge flocs and become part of the suspended VSS sludge mass in the reactor. As part of the sludge mass, these organics settle out with the sludge mass in the secondary setting tank and are returned to the biological reactor. Undegraded particulate organics therefore do not escape with the effluent but remain part of the sludge VSS mass in the system; the only exit route for the undegraded particulate biodegradable organics is via the waste flow (Q_w) with the waste sludge. The time available for the breakdown of the particulate slowly biodegradable organics by the OHOs is therefore related to the solids retention time or sludge age of the system. Although the biological breakdown of the particulate biodegradable organics is much slower than that of the soluble readily biodegradable organics, this is of little consequence because the solids retention time in the system (SRT) is much longer than the liquid retention time (HRT_n). Once the sludge age is longer than approximately 3 days at 20 °C (4 at 14 °C), the slowly biodegradable organics are virtually completely utilized.

Experimental work has confirmed the above. Short sludge age, and by linked association a short hydraulic retention time, and long sludge age, and by linked association long hydraulic retention time systems yield closely similar influent unbiodegradable soluble and particulate COD fractions ($f_{SU,CODi}$ and $f_{XU,CODi}$). Hence, once the sludge age is longer than approximately 3 to 4 days, the residual biodegradable organic concentration, both soluble (S_S) and particulate (X_S), not broken down that can be accepted is very small. From this an important assumption and simplification can be made for the steady state and simulation models, *i.e.* slowly biodegradable soluble organics and very slowly biodegradable particulate organics can be assumed to be negligibly low in concentration in normal municipal wastewater. However, it must be remembered that, although reasonable, this assumption that all the biodegradable organics are degraded may not be valid for all wastewaters and depends on the type of industries in the catchment of the wastewater treatment plant. When characterizing such wastewaters, any residual biodegradable soluble and particulate organics not degraded in the system are implicitly included with the unbiodegradable soluble and particulate organic fractions, respectively, because this is the way that activated sludge models are structured. For the steady-state model, because all the biodegradable organics are utilized, an additional simplification can be made, *i.e.* it is not necessary to make a distinction between soluble and particulate biodegradable organics; all are transformed into OHO VSS mass. The steady-state activated sludge model equations below are based on this simplification.

4.4 STEADY-STATE SYSTEM EQUATIONS

Once it is recognized that all the organics in the influent, except the soluble unbiodegradable COD, are either utilized by the OHOs to form new OHO mass via growth (X_{OHO}), or remain in the system and accumulate as unbiodegradable (inert) sludge mass ($X_{E,OHO}$ and X_U), it follows that the mass of sludge produced and the carbonaceous oxygen demand in the system are stoichiometric functions of the daily COD mass load; the greater the daily COD mass load, the greater the sludge production and carbonaceous oxygen demand.

The equations below give the masses of sludge generated in the reactor and wasted per day, the average daily oxygen demand and the effluent COD concentration comprising the unbiodegradable soluble organics for organic material removal as a function of the total organic (COD) load per day ($FCOD_i$), the wastewater characteristics, *i.e.* the unbiodegradable soluble and particulate COD fractions ($f_{SU,CODi}$ and $f_{XU,CODi}$) and the sludge age (SRT). The kinetic and stoichiometric constants in the equations, *i.e.* the specific yield coefficient (Y_{OHOv}), the specific endogenous mass loss rate (b_{OHO}), the unbiodegradable fraction of the OHOs ($f_{XE,OHO}$), and the COD/VSS ratio of the sludge ($f_{cv,OHO}$), as well as their temperature dependencies are given in Table 4.1.

4.4.1 For the influent

The mass flows or input fluxes of total organics ($FCOD_i$, mgCOD/d), biodegradable organics ($FCOD_{b,i}$, mgCOD/d), unbiodegradable particulate organics ($FX_{Uv,i}$, mgVSS/d) and inorganic or fixed suspended solids (FSS, $FX_{FSS,i}$, mgISS/d) are:

$$FCOD_i = Q_i COD_i \quad \text{(mgCOD/d)} \quad (4.4)$$

$$FCOD_{b,i} = Q_i COD_{b,i} = Q_i (S_{S,i} + X_{S,i}) \quad \text{(mgCOD/d)} \quad (4.5a)$$

$$FCOD_{b,i} = Q_i COD_i (1 - f_{SU,CODi} - f_{XU,CODi}) \quad \text{(mgCOD/d)} \quad (4.5b)$$

$$FCOD_{b,i} = FCOD_i (1 - f_{SU,CODi} - f_{XU,CODi}) \quad \text{(mgCOD/d)} \quad (4.5c)$$

$$FX_{Uv,i} = Q_i X_{Uv,i} \quad \text{(mgVSS/d)} \quad (4.6a)$$

$$FX_{Uv,i} = Q_i f_{XU,CODi} COD_i / f_{cv,UPO} \quad \text{(mgVSS/d)} \quad (4.6b)$$

$$FX_{Uv,i} = FCOD_i f_{XU,CODi} / f_{cv,UPO} \quad \text{(mgVSS/d)} \quad (4.6c)$$

$$FX_{FSS,i} = Q_i X_{FSS,i} \quad \text{(mgISS/d)} \quad (4.7)$$

Table 4.1 Stoichiometric and kinetic constants and their temperature dependency for the OHOs in the steady-state carbonaceous degradation activated sludge model (Marais and Ekama, 1976).

Constant	Symbol	Temperature dependency	θ	Standard value 20 °C
Yield coefficient (mgCOD/mgCOD)	Y_{OHO}	Remains constant	1	0.67
Yield coefficient (mgVSS/mgCOD)	Y_{OHOv}	Remains constant	1	0.45
Endogenous respiration rate (1/d)	b_{OHO}	$b_{OHO,T} = b_{OHO,20}\, \theta_{b,OHO}^{(T-20)}$	1.029	0.24
Endogenous residue fraction (-)	$f_{XE,OHO}$	Remains constant	1	0.2
ISS content of OHOs	$f_{FSS,OHO}$	Remains constant	1	0.15
[1]COD/VSS ratio of OHO (mgCOD/mgVSS)	$f_{cv,OHO}$	Remains constant	1	1.48
[1]COD/VSS ratio of UPO (mgCOD/mgVSS)	$f_{cv,UPO}$	Remains constant	1	1.48

[1]Different COD/VSS (f_{cv}), N/VSS (f_n) and P/VSS (f_p) mass ratios can be assigned to the OHO biomass (X_{OHO}), OHO endogenous residue ($X_{E,OHO}$) and influent UPO (X_{Uv}). However, in this book these three VSS components are assigned the same f_{cv}, f_n and f_p mass ratios.

4.4.2 For the system

4.4.2.1 Reactor VSS mass

The masses of OHO VSS (MX_{OHOv}, mgVSS), endogenous residue VSS ($MX_{E,OHOv}$, mgVSS), unbiodegradable organics VSS (MX_{Iv}, mgVSS), volatile suspended solids VSS (MX_{VSS}, mgVSS) in the system are given by:

$$MX_{OHOv} = X_{OHOv} V_R \quad \text{(mgVSS) (4.8a)}$$

$$MX_{E,OHOv} = X_{E,OHOv} V_R \quad \text{(mgVSS) (4.8b)}$$

$$MX_{Uv} = X_{Uv} V_R \quad \text{(mgVSS) (4.8c)}$$

$$MX_{VSS} = X_{VSS} V_R \quad \text{(mgVSS) (4.8d)}$$

$$MX_{OHOv} = FCOD_{b,i} \frac{Y_{OHOv} SRT}{(1+b_{OHO,T} SRT)}$$
$$= FCOD_i (1 - f_{SU,CODi} - f_{XU,CODi}) \frac{Y_{OHOv} SRT}{(1+b_{OHO,T} SRT)} \quad \text{(mgVSS) (4.9)}$$

$$MX_{E,OHOv} = f_{XE,OHO} b_{OHO,T} MX_{OHOv} SRT$$
$$= FCOD_{b,i} \frac{Y_{OHOv} SRT}{(1+b_{OHO,T} SRT)} f_{XE,OHO} b_{OHO,T} SRT$$
$$= FCOD_i (1 - f_{SU,CODi} - f_{XU,CODi}) \cdot$$
$$\cdot \frac{Y_{OHOv} SRT}{(1+b_{OHO,T} SRT)} f_{XE,OHO} b_{OHO,T} SRT \quad \text{(mgVSS) (4.10)}$$

$$MX_{Uv} = \frac{FX_{U,i}}{f_{cv,UPO}} SRT = FX_{Uv,i} SRT$$
$$= FCOD_i \frac{f_{XU,CODi}}{f_{cv,UPO}} SRT \quad \text{(mgVSS) (4.11)}$$

$$MX_{VSS} = MX_{OHOv} + MX_{E,OHOv} + MX_{Uv}$$
$$= FCOD_i \frac{Y_{OHOv} SRT}{(1+b_{OHO,T} SRT)} (1 + f_{XE,OHO} b_{OHO,T} SRT)$$
$$+ FX_{Uv,i} SRT$$

$$= FCOD_i \left[\frac{(1 - f_{SU,CODi} - f_{XU,CODi}) Y_{OHOv} SRT}{(1+b_{OHO,T} SRT)} \cdot (1 + f_{XE,OHO} b_{OHO,T} SRT) + \frac{f_{XU,CODi}}{f_{cv,UPO}} SRT \right] \quad \text{(mgVSS) (4.12)}$$

4.4.2.2 Reactor ISS mass

The inorganic suspended solids (ISS) concentration from the influent accumulates in the reactor in an identical way to the unbiodegradable particulate organics (Eq. 4.12), *i.e.* the mass of influent ISS in the reactor is equal to the daily mass flow of ISS into the reactor $FX_{FSS,i}$ times the sludge age (SRT), *viz.*

$$MX_{FSS} = FX_{FSS,i}\ SRT \quad \text{(mgISS) (4.13a)}$$

and:

$X_{FSS,i}$ influent ISS concentration (mg ISS/l)

The influent ISS is only part of the ISS that is measured in the reactor. The OHOs (and PAOs if present) also contribute to the ISS concentration. For fully aerobic and nitrification-denitrification (ND) systems, where only OHOs comprise the active biomass, the OHOs contribute approximately 10% of their OHOCOD mass (15% of their VSS mass) to the ISS (Ekama and Wentzel, 2004). It appears that this ISS mass is intracellular dissolved solids, which, when a sludge sample is dried in the TSS procedure, precipitate as ISS.

Therefore theoretically, this ISS contribution of the OHOs (and PAOs if present) to the TSS strictly should be ignored even though it manifests in the TSS test, because being intracellular dissolved solids, it does not add to the actual ISS flux in the secondary settling tank. However, because this ISS mass has always been implicitly included in the TSS test result in the past, it is retained because SST design procedures have been based on the measured TSS result. In other words, the models are adjusted to accommodate the TSS test 'error'. Including the OHO ISS mass yields for fully aerobic and ND systems,

$$MX_{FSS} = FX_{FSS,i}SRT + f_{FSS,OHO}MX_{OHOv} \quad (4.14a)$$

$$MX_{FSS} = FX_{FSS,i}SRT + f_{FSS,OHO}f_{av,OHO}MX_{VSS} \quad (mgISS/d)\ (4.14b)$$

where:

$f_{av,OHO}$ fraction of the VSS mass that is active OHOs, (see Section 4.4.7)

$f_{FSS,OHO}$ inorganic content of the OHO VSS (0.15 mgISS/mgOHOVSS)

For enhanced biological phosphorus removal (EBPR) systems, the FSS in the phosphorus-accumulating organisms (PAO) also need to be included. For aerobic P uptake EBPR, $f_{FSS,PAO}$ is 1.30 mgFSS/mg PAOVSS *i.e.* 7 times higher than for OHOs when the PAO contain their maximum polyphosphate (PP) content of 0.355 mgPP-P/mgPAOVSS (Wentzel *et al.*, 1990; Ekama and Wentzel, 2004). Therefore, for EBPR systems, the VSS/TSS ratio is significantly lower than for fully aerobic and ND systems.

4.4.2.3 Reactor TSS mass

The total suspended solids (TSS) mass (MX_{TSS}, mgTSS) in the reactor is the sum of the volatile (VSS) and inorganic (ISS) suspended solids masses, *viz.*

$$MX_{TSS} = MX_{VSS} + MX_{FSS} \quad (mgTSS)\ (4.15)$$

The VSS/TSS ratio of the sludge (f_{VT}) is

$$f_{VT} = \frac{MX_{VSS}}{MX_{TSS}} \quad (mgVSS/mgTSS)\ (4.16)$$

If the influent FSS concentration is not known, then the reactor TSS mass (MX_{TSS}) can be calculated from an estimated VSS/TSS ratio (f_{VT}) of the sludge, *i.e.*

$$MX_{TSS} = MX_{VSS} / f_{VT} \quad (mgTSS)\ (4.17)$$

where:

f_{VT} VSS/TSS ratio of the activated sludge

For the same influent FSS concentration ($X_{FSS,i}$), the VSS/TSS ratio of the activated sludge (f_{VT}) does not vary much with SRT even over a wide range from 3 to 30 d (Ekama and Wentzel, 2004).

4.4.2.4 Carbonaceous oxygen demand

The mass of oxygen utilized per day (FO_c, mgO_2/d),

$$FO_c = FCOD_{b,i}\left[\begin{array}{l}(1-f_{cv,OHO}Y_{OHOv})\\ +(1-f_{XE,OHO})b_{OHO,T}\\ \cdot\dfrac{Y_{OHOv}f_{cv,OHO}SRT}{(1+b_{OHO,T}SRT)}\end{array}\right] =$$

$$= FCOD_i(1-f_{SU,CODi}-f_{XU,CODi}) \cdot \left[\begin{array}{l}(1-f_{cv,OHO}Y_{OHOv})\\ +(1-f_{XE,OHO})b_{OHO,T}\\ \cdot\dfrac{Y_{OHOv}f_{cv,OHO}SRT}{(1+b_{OHO,T}SRT)}\end{array}\right] \quad (mgO_2/d)\ (4.18)$$

$$FO_c = V_R O_c \quad (mgO_2/l.d)\ (4.19)$$

where:

O_c carbonaceous oxygen utilization rate (mgO_2/l.d)

V_R volume for the reactor for a fully aerobic AS system (m^3)

From Eq. 4.18, it can be seen that the mass of oxygen utilized by the OHOs per day (FO_c) is the sum of two terms. The first ($1-f_{cv,OHO}Y_{OHOv}$) is the oxygen demand for growth of OHOs. It represents the electrons (COD) that are used in the growth process to generate energy by the OHOs to transform the utilized organics to new biomass (catabolism). The balance of the utilized electrons (COD, $f_{cv,OHO}Y_{OHOv}$) is conserved as new biomass (anabolism). It can be seen that this oxygen demand is proportional to the influent biodegradable organics and does not change with sludge age. This is because all the influent biodegradable organics are utilized and transformed to OHO biomass. The second term is the oxygen demand for endogenous respiration, which increases as sludge age increases. The increase in carbonaceous oxygen demand (FO_c) with sludge age is therefore due to the increasing oxygen demand from endogenous respiration with sludge age. This increases because the longer the OHO VSS mass remains in the reactor, the more of this mass that is degraded via endogenous respiration, and the more of its electrons, carbon and energy that are passed to oxygen, changed to CO_2, and lost as heat, respectively. Therefore, growth is the biological process whereby influent biodegradable organics are transformed to OHO VSS mass (anabolism) with an associated electron transfer to oxygen and an energy loss as heat (catabolism), and endogenous respiration is a process whereby the organism biodegradable organics are degraded via catabolism to CO_2 with a further oxygen demand and energy loss as heat. The electrons transferred to oxygen result in a much lower accumulation of VSS mass in the reactor compared with unbiodegradable particulate organics. All the electrons of these organics are conserved as VSS in the reactor and none are passed to oxygen. Hence the yield of unbiodegradable organics therefore is in effect 1 gCOD/gCOD or $1/f_{cv,UPO}$ gVSS/gCOD.

4.4.3 Reactor volume and retention time

Knowing the mass of total settleable solids (MX_{TSS}) in the reactor, the volume of the reactor is determined from the value specified for the MLSS concentration, X_{TSS} (Section 4.6) *i.e.*

$$V_R = MX_{TSS} / X_{TSS} \qquad (l, m^3 \text{ or Ml}) \quad (4.20)$$

Knowing the volume of the reactor V_R, the nominal hydraulic retention time, HRT_n is found from the design average dry weather flow rate $Q_{i,ADWF}$ from Eq. 4.2.

4.4.4 Irrelevance of HRT

The above design equations lead to the important conclusion that hydraulic retention time (HRT_n) is irrelevant for the design of the activated sludge system. The mass of volatile suspended solids (VSS) in the reactor is a function mainly of the daily COD mass load on it and the sludge age. Similarly, the mass of TSS in the reactor is a function mainly of the daily mass loads of COD and ISS on the reactor and the sludge age. Consequently, insofar as the mass of sludge in the reactor is concerned, it is immaterial whether the mass of COD (and ISS) load per day arises from a low daily flow with a high COD (and ISS) concentration, or a high daily flow with a low COD (and ISS) concentration. Provided $FCOD_i$ (and $FX_{FSS,i}$) is the same in both cases, the mass of VSS and TSS will be virtually identical. However, the hydraulic retention times will differ, being long in the first and short in the second case, respectively. The hydraulic retention time, therefore, is incidental to the COD (and ISS) mass load, the VSS (and TSS) mass and the daily flow: it serves no basic design function for the activated sludge system. Design criteria for the activated sludge reactor volume based on hydraulic retention time should therefore be used with extreme caution because they implicitly incorporate specific wastewater strength and characteristic values typical for the regions for which they were developed.

4.4.5 Effluent COD concentration

Under normal activated sludge system operating conditions, where the sludge ages are in excess of 5 days (to ensure nitrification and biological nutrient removal), the nature of the influent organics in municipal wastewaters is such that the COD concentration in the effluent is inconsequential in the system design. The soluble readily biodegradable organics are completely utilized in a very short time (< 2 h) and the particulate organics, whether biodegradable or unbiodegradable, are enmeshed in the sludge mass and settle out with the sludge in the secondary settling tanks. Consequently, *the effluent COD concentration comprises virtually wholly the soluble unbiodegradable organics (COD)* (from the influent) plus the COD of the sludge particles which escape with the effluent due to imperfect operation of the secondary settling tank. Hence the filtered and unfiltered effluent COD concentration, COD_e, are given by:

$$COD_e = S_{U,e} \quad \text{(filtered mgCOD/l)} \quad (4.21a)$$

$$COD_e = S_{U,e} + f_{cv,OHO}X_{VSS,e} \quad \text{(unfiltered mgCOD/l)} \quad (4.21b)$$

where:

$S_{U,e}$ unbiodegradable COD in the effluent = $S_{U,i}$ = $f_{SU,CODi}$ COD_i (mgCOD/l)

$X_{VSS,e}$ VSS in the effluent (mgVSS/l)

$f_{cv,OHO}$ COD/VSS ratio of the VSS (1.48 mgCOD/mgVSS)

In most cases the effluent VSS and TSS concentrations are too low to measure reliably with the VSS and TSS tests. Alternative methods for measuring low solid concentrations in the effluent have been developed; for the VSS, via the filtered and unfiltered COD concentrations, *i.e.* from Eq. 4.21, and for TSS, via the turbidity once a turbidity *versus* TSS concentration calibration curve for the activated sludge has been prepared (Wahlberg *et al.*, 1994).

$$X_{VSS,e} = \left(COD_{e(unfilt)} - COD_{e(filt)}\right)/f_{cv,OHO} \quad (4.22)$$

4.4.6 The COD (or e⁻) mass balance

In the activated sludge system, COD theoretically must be conserved so that at steady state the COD mass flow out of the system must equal to the COD mass flow into the system over a defined time interval. The COD (e^-) of the influent organics are (*i*) retained in the unbiodegradable particulate and soluble organics, (*ii*) transformed to OHO mass and therefore conserved in a different type of organic material, or (*iii*) passed to oxygen to form water. So in general the COD (or e^-) balance over the activated sludge system at steady state is given by Eq 4.23 where:

COD_e effluent total soluble COD concentration (mgCOD/l)

X_{VSS} VSS concentration in biological reactor (mgVSS/l)

O_c carbonaceous (for organic material degradation) oxygen utilization rate in reactor (mgO_2/l.h)

$$\begin{bmatrix}\text{Flux of COD}(e^-)\\ \text{output}\end{bmatrix} = \begin{bmatrix}\text{Flux of COD}(e^-)\\ \text{input}\end{bmatrix}$$

$$\begin{bmatrix}\text{Flux of soluble}\\ \text{COD in}\\ \text{effluent}\end{bmatrix} + \begin{bmatrix}\text{Flux of soluble}\\ \text{COD in waste}\\ \text{flow}\end{bmatrix} + \begin{bmatrix}\text{Flux of particulate}\\ \text{COD in waste}\\ \text{flow}\end{bmatrix} + \begin{bmatrix}\text{Flux of oxygen utilized}\\ \text{by OHOs for COD}\\ \text{breakdown}\end{bmatrix} = \begin{bmatrix}\text{Flux of}\\ \text{COD input}\end{bmatrix}$$

$$Q_eCOD_e + Q_wCOD_e + Q_wX_{VSS}f_{cv,OHO} + V_RO_c = Q_iCOD_i \quad (4.23)$$

In Eq. 4.23, the first two terms represent the soluble organics that exit the system via effluent and waste flows, the third term the particulate organics that exit the system via the waste flow, and the fourth term the mass of oxygen utilized for biodegradable organic material breakdown by the OHOs. Noting that,

$$(Q_e + Q_w)COD_e = Q_iCOD_e = FCOD_e$$
$$Q_wX_{VSS} = V_RX_{VSS}/SRT = MX_{VSS}/SRT = FX_{VSS}$$
$$V_RO_c = FO_c$$
$$Q_iCOD_i = FCOD_i$$

the general COD mass balance is given by,

$$FCOD_e + f_{cv,OHO} \cdot MX_{VSS}/SRT + FO_c = FCOD_i \quad (4.24)$$

where:

$FCOD_e$ COD flux of soluble organics exiting system via effluent and waste flows (mgCOD/d)

$f_{cv,OHO} \cdot MX_{VSS}/SRT$ COD flux of particulate organics exiting system via waste flow (mgCOD/d)

FO_c flux of oxygen utilized by OHOs for biodegradable organic material degradation, all of those from the influent via the growth process and some of those from the OHO biomass via the endogenous respiration process (carbonaceous) (mgO_2/d)

Substituting Eqs 4.12 and 4.18 for MX_{VSS} and FO_c into Eq 4.24 gives the COD balance *versus* SRT, which is considered in Section 4.5.3 below (Figure 4.3).

The COD mass balance is a very powerful tool for checking (*i*) the data measured in experimental systems (Ekama *et al.,* 1986), (*ii*) the results calculated for design from the steady-state model and (*iii*) the results calculated by dynamic-simulation models. Application of the COD mass balance to nitrification and denitrification activated sludge systems is presented in Chapter 5.

4.4.7 Active fraction of the sludge

The active VSS mass MX_{OHOv} in the reactor is the live OHO mass which performs the biodegradation processes of the organic material. The other two organic solids masses, $MX_{E,OHOv}$ and MX_{Uv} are inactive and unbiodegradable and do not serve any function insofar as the biodegradation processes in the system are concerned. They are given different symbols because of their different origin, the MX_{Uv} is unbiodegradable particulate organics from the influent wastewater and the $MX_{E,OHOv}$ is unbiodegradable particulate organics produced in the reactor via the endogenous respiration process. The active OHO biomass fraction of the volatile solids in the reactor $f_{av,OHO}$ is given by,

$$f_{av,OHO} = \frac{MX_{OHOv}}{MX_{VSS}} \quad \text{(mg OHOVSS/mgVSS)} \tag{4.25}$$

Substituting eqs. 4.9 and 4.12 for MX_{OHOv} and MX_{VSS} and rearranging yields:

$$\frac{1}{f_{av,OHO}} = 1 + f_{XE,OHO} b_{OHO,T} SRT + \frac{f_{XU,CODi}\left(1 + b_{OHO,T} SRT\right)}{f_{cv,OHO} Y_{OHOv}\left(1 - f_{XU,CODi} - f_{SU,CODi}\right)} \tag{4.26}$$

where:

$f_{av,OHO}$ active OHO fraction of the VSS mass

If the total suspended solids mass (TSS) is used as the basis for determining the active fraction, then the active fraction of the sludge mass with respect to the total suspended solids, $f_{at,OHO}$, is given by

$$f_{at,OHO} = f_{VT} f_{av,OHO} \tag{4.27}$$

where:

$f_{at,OHO}$ active OHO fraction of the TSS mass

f_{VT} VSS/TSS ratio of the activated sludge

The active fractions $f_{av,OHO}$ or $f_{at,OHO}$ give an indication of the 'stability' of the waste sludge, which is related to the remaining biodegradable organics in the sludge mass. The only biodegradable organics in the VSS mass are those of the OHOs which, in terms of the steady-state model, is 80% (1 - $f_{XE,OHO}$) of the OHO mass. Hence, the higher the active fraction, the greater the proportion of biodegradable organics remaining in the sludge mass and the greater the utilizable energy content remaining in the sludge mass (sections 4.11 and 4.13). For activated sludge to be stable, the remaining utilizable organics in it should be low so that it will not generate odours through further biological activity. Sludge used as a soil conditioner needs to be stable because its primary purpose is to provide nutrients and unbiodegradable organic content to the soil (Korentajer, 1991); unstable sludges applied to agricentural land lead to an undesirable high oxygen demand in the soil.

4.4.8 Steady-state design

The design equations set out above form the starting point for aerobic and anoxic-aerobic activated sludge system design. They apply to the simple single reactor completely mixed aerobic system and to the more complex multi-reactor anoxic-aerobic systems for biological nitrogen removal. When EBPR is included, the above equations *do not* give an accurate estimate of the VSS and TSS masses in the system. With EBPR, a second group of heterotrophic organisms, the PAO need to be considered, which have different stoichiometric and kinetics constants producing more VSS and TSS mass per mass organics (COD) utilized. Incorporation of the PAOs in the steady-state model is discussed in Chapter 7.

For the more complex anoxic-aerobic systems, the above basic equations apply if the assumptions made in their derivation apply. Provided this is the case, the effects of nitrification (Chapter 5) and nitrification-denitrification (Chapter 5) and the associated oxygen demands can be formulated as additional equations to the basic equations above. The above 'simplified' approach is based on the assumption that the biodegradable organics are completely utilized. This has been established by the close correlation achieved between the mean response of the more complex anoxic aerobic systems predicted by the more complex general kinetic model (and validated experimentally), with that calculated by the above basic equations and the additional equations for nitrification and denitrification. The close correspondence between this simplified steady state model and the more complex general kinetic simulation models such as ASM1 is demonstrated by Sötemann *et al.* (2006) for aerobic and ND systems (including aerobic digestion). Indeed, these simple steady state models can form the basis for 'hand' calculations to (*i*) develop design input information for and (*ii*) check output results from the dynamic-simulation models (Ekama, 2009).

Other assumptions on which the steady state model is based are that (*i*) the mass of active OHOs seeded into the system with the influent is negligible in comparison with that which grows in the reactor, (*ii*) there is no loss of solids in the effluent from the secondary settling tanks, (*iii*) water mass is conserved, (*iv*) a 100% COD balance is achieved, and (*v*) active OHO loss is modelled as endogenous respiration. It is important to understand these assumptions in the model. With regard to the assumption of complete utilization of biodegradable organics, if this is not the case, then the mass of sludge produced per day increases and the carbonaceous oxygen demand decreases below those predicted by the basic equations. The reason for these deviations lies in the kinetics of degradation of the slowly biodegradable particulate material; if, for example, the aerobic fraction of the sludge mass is too small (Chapter 5), the particulate biodegradable organics are only partially utilized and residual particulate biodegradable organics (X_S) accumulate in the system as additional VSS in the same way as the unbiodegradable particulate organics. Concomitantly, the carbonaceous oxygen demand is reduced because less biodegradable organics are utilized. Clearly, for such situations the simulation model results will deviate from steady-state ones. In fact, it would be wise to regard such deviations as a signal for possible incorrect output from the simulation model and start an investigation to find the cause for the deviation.

4.4.9 The steady-state design procedure

The calculation procedure to generate the design results required for a certain sludge age is as follows:

Select the wastewater characteristics $f_{XU,CODi}$ and $f_{SU,CODi}$ which are believed to best represent the unbiodegradable particulate and soluble COD fractions of the wastewater.

Then calculate:

1) $FX_{Uv,i}$ (Eq. 4.6) and $FX_{FSS,i}$ (Eq. 4.7)
2) $FCOD_i$ and/or $FCOD_{b,i}$ (Eq. 4.4 or 4.5)
3) Select the SRT
4) MX_{OHOv} (Eq. 4.9), $MX_{E,OHOv}$ (Eq. 4.10), MX_{Uv} (Eq. 4.11), MX_{VSS} (Eq. 4.12), MX_{FSS} (Eq. 4.14), MX_{TSS} (Eq. 4.15) or select f_{VT}, MX_{TSS} (Eq. 4.17)

5) FO_c (Eq. 4.18) and O_c (Eq. 4.19)
6) V_R (Eq. 4.20)
7) HRT_n, (Eq. 4.2)
8) COD_e (Eq. 4.21)

In this design procedure the input COD and its characteristics will be governed by the specific waste flow. The selection of values for the unbiodegradable soluble and particulate COD fractions is simple but not trivial. Each impacts the design in important areas. The unbiodegradable soluble COD fraction ($f_{SU,CODi}$) has a negligible influence on biological reactor design parameters, such as sludge production and oxygen demand, but has a marked influence on the effluent COD concentration (COD_e). In contrast, the unbiodegradable particulate COD fraction ($f_{XU,CODi}$) has no influence on the effluent COD concentration (COD_e) but has a marked influence on the specific sludge production rate (kgVSS produced/kgCOD load) and specific reactor volume (m^3/kgCOD load per day). The higher the $f_{XU,CODi}$, the larger these values, and as the sludge age increases this influence of $f_{XU,CODi}$ becomes more marked. The system parameter that requires selection is the sludge age. The sludge age selected will depend on the specific requirements from the wastewater treatment plant such as effluent quality, *i.e.* organic COD removal only, nitrification, N removal, biological P removal, and the envisioned sludge treatment facilities *i.e.* whether or not primary settling is included, the stability of the waste activated sludge etc. Specification of the sludge age, therefore, is an important design decision and requires special consideration (Section 4.13).

4.5 DESIGN EXAMPLE

The design procedure developed above for the fully aerobic activated sludge system is demonstrated with a numerical example. Assuming constant flow and load conditions, calculations are shown below how estimates of the system volume requirements, average daily carbonaceous oxygen demand, and daily sludge production are calculated for the treatment of the example raw and settled wastewaters (Table 4.2).

Table 4.2 Examples of raw and settled wastewater characteristics.

Parameter	Symbol	Unit	Raw	Settled
Flow	Q_i	Ml/d	15	14.93
COD concentration	COD_i	mgCOD/l	750	450
Unbiodegradable particulate COD	$f_{XU,CODi}$		0.15	0.04
Unbiodegradable soluble COD	$f_{SU,CODi}$		0.07	0.12
Unbiodegradable soluble OrgN	$f_{SU,TKNi}$		0.03	0.035
TKN concentration	TKN_i	mgN/l	60	51
Total P concentration	P_i	mgP/l	15	12.75
TKN/COD ratio	$f_{TKNi/CODi}$	mgN/mgCOD	0.08	0.117
P/COD ratio	$f_{Pi/CODi}$	mgP/mgCOD	0.02	0.028
Temperature	T_{max}, T_{min}	°C	14-22	14-22
pH	-		7.5	7.5
H_2CO_3 alkalinity	$S_{Alk,i}$	mg/l as $CaCO_3$	250	250
Influent ISS	$X_{FSS,i}$	mgISS/l	47.8	9.5
VSS/TSS of activated sludge	f_{VT}	mgVSS/mgTSS	0.75	0.83

4.5.1 Temperature effects

From Table 4.1, the only kinetic constant in the steady state organic material (COD) degradation model for fully aerobic systems that is affected by temperature is the specific endogenous respiration rate b_{OHO}. This rate decreases by approximately 3% for every 1 °C decrease in temperature (*i.e.* $\theta_{b,OHO}$ = 1.029). From Table 4.1, the rate at 14 °C is 0.202/d and at 22 °C is 0.254/d. The effect of the reduction in the rate with decrease in temperature is that at lower temperatures the daily sludge production is marginally increased and the average carbonaceous oxygen demand is marginally decreased. The differences in sludge production (kgVSS/d) and oxygen demand (kgO_2/d) are less than 5% for an 8 °C change in temperature from 14 to 22 °C. Consequently, the average carbonaceous oxygen demand should be calculated at the maximum temperature and the system volume and sludge production at the minimum temperature in order to find the maximum values of these parameters.

4.5.2 Calculations for organic material degradation

This design example demonstrates the effect of temperature and sludge age on (*i*) the mass of TSS sludge in the system (MX_{TSS}, kgTSS), (*ii*) the average daily carbonaceous oxygen demand (FO_c, kgO_2/d), (*iii*) the active fractions of the sludge with respect to VSS and TSS ($f_{av,OHO}$ and $f_{at,OHO}$) and (*iv*) the mass of TSS sludge wasted per day (FX_{TSS}, kgTSS/d). These four parameters are calculated for the example raw and settled wastewaters at 14 and 22 °C for sludge ages ranging from 3 to 30 days.

Mass of COD treated/d = $FCOD_i$ – Q_i COD_i kgCOD/d

Mass of biodegradable COD treated/d = $FCOD_{b,i}$ = = $(1 - f_{SU,CODi} - f_{XU,CODi})$ COD_i

Mass of unbiodegradable particulate organics flowing into the system as mgVSS /d = $FX_{Uv,i}$ = = $FCOD_i$ $f_{XU,CODi}/f_{cv,UPO}$

Hence for raw wastewater:

$FCOD_i$ = 15 Ml/d · 750 mgCOD/l = 11,250 kgCOD/d

$FCOD_{b,i}$ = (1-0.07-0.15)11,250 = 8,775 kgCOD/d

$FX_{Uv,i}$ = 0.15 · 11,250/1.48 = 1,140 kgVSS/d

$FX_{FSS,i}$ = 15 Ml/d · 47.8 = 717 kgISS/d

and for settled wastewater:

$FCOD_i$ = 15 Ml/d · 450 mgCOD/l = 6,750 kgCOD/d

$FCOD_{b,i}$ = (1-0.117-0.04) · 6,750 = 5,690 kgCOD/d

$FX_{Uv,i}$ = 0.04 · 6,750/1.48 = 182.4 kgVSS/d

$FX_{FSS,i}$ = 15 Ml/d · 9.5 = 142.5 kgISS/d

From eqs. 4.12 and 4.17, the masses of volatile (MX_{VSS}) and total (MX_{TSS}) settleable solids in the system is for the raw wastewater:

$$MX_{VSS} = 8,775\frac{0.45SRT}{\left(1+b_{OHO,T}SRT\right)}\cdot \left(1+0.2b_{OHO,T}SRT\right)+1,140SRT \quad (kgVSS)$$

$$MX_{FSS} = 717\ SRT + 0.15\ f_{av,OHO}\ MX_{VSS} \quad (kgFSS)$$

$$MX_{TSS} = MX_{VSS}/0.75 \text{ or } MX_{FSS} + MX_{VSS} \quad (kgTSS)$$

and for the settled wastewater:

$$MX_{VSS} = 5,690\frac{0.45SRT}{\left(1+b_{OHO,T}SRT\right)}\cdot \left(1+0.2b_{OHO,T}SRT\right)+182.4SRT \quad (kgTSS)$$

$$MX_{FSS} = 142.5\ SRT + 0.15\ f_{av,OHO}\ MX_{VSS} \quad (kgFSS)$$

$$MX_{TSS} = MX_{VSS}/0.83 \text{ or } MX_{FSS} + MX_{VSS} \quad (kgTSS)$$

From Eq. 4.18, the average daily carbonaceous oxygen demand is for the raw wastewater:

$$FO_c = 8,775\cdot\left[\left(0.334\right)+0.533\frac{b_{OHO,T}SRT}{\left(1+b_{OHO,T}SRT\right)}\right] \quad (kgO_2/d)$$

and for the settled wastewater:

$$FO_c = 5{,}690 \cdot \left[(0.334) + 0.533 \frac{b_{OHO,T} SRT}{(1 + b_{OHO,T} SRT)} \right]$$

(kgO_2/d)

From eqs. 4.26 and 4.27, the active fractions with respect to the VSS ($f_{av,OHO}$) and TSS ($f_{at,OHO}$) are for the raw wastewater:

$$f_{av,OHO} = 1 / \left[\begin{matrix} 1 + 0.2 b_{OHO,T} SRT \\ + 0.289 (1 + b_{OHO,T} SRT) \end{matrix} \right]$$

and,

$$f_{at,OHO} = 0.75\ f_{av,OHO}$$

and for the settled wastewater:

$$f_{av,OHO} = 1 / \left[\begin{matrix} 1 + 0.2 b_{OHO,T} SRT \\ + 0.142 (1 + b_{OHO,T} SRT) \end{matrix} \right]$$

and,

$$f_{at,OHO} = 0.83\ f_{av,OHO}$$

From the definition of sludge age (Eq. 4.1), the mass of VSS and TSS secondary sludge produced (or wasted) per day (FX_{TSS}) is for the raw wastewater:

$$FX_{TSS} = Q_w X_{TSS} = MX_{TSS} / SRT$$

$$FX_{TSS} = \frac{8{,}775}{0.75} \frac{0.45}{(1 + b_{OHO,T} SRT)} \cdot (1 + 0.2 b_{OHO,T} SRT) + \frac{1{,}140}{0.75}$$

(kgTSS/d)

and for the settled wastewater:

$$FX_{TSS} = \frac{5{,}690}{0.83} \frac{0.45}{(1 + b_{OHO,T} SRT)} \cdot (1 + 0.2 b_{OHO,T} SRT) + \frac{142.8}{0.83}$$

(kgTSS/d)

The mass of VSS wasted/produced per day FX_{VSS} is simply f_{VT} times the mass of TSS wasted per day,

$$FX_{VSS} = f_{VT} \cdot FX_{TSS} \qquad \text{(kgVSS/d)}$$

Substituting the $b_{OHO,T}$ value for 14 °C (*i.e.* 0.202/d) and 22 °C (*i.e.* 0.254/d) into the above equations allows MX_{VSS}, MX_{TSS}, FO_c, $f_{av,OHO}$, $f_{at,OHO}$, FX_{TSS} and FX_{VSS} to be calculated for sludge ages of 3 to 30 days. The results are shown plotted in Figure 4.3.

From these figures, the mass of sludge in the reactor (TSS or VSS), the average carbonaceous oxygen demand, the mass of TSS sludge produced per day (kgTSS/d) and the active fraction (with respect to MLSS or MLVSS) are only marginally affected by temperature and for these parameters; insofar as design is concerned, temperature effects are not really of much consequence. However, the influent waste type *i.e.* raw of settled sewage, has a significant effect. Raw sewage results in considerably more sludge in the system, a higher oxygen demand and a lower active fraction of the sludge than settled sewage. The difference in the effects of sewage type between raw and settled wastewater depends wholly on the efficiency of the PSTs. The differences apparent in Figure 4.3 arise from a 40% COD removal in the PST(s); the greater the COD removal efficiency, the greater the difference between the parameters for raw and settled sewage.

The results for 20 d sludge age are given in Table 4.3. From Eq. 4.20, for the same reactor TSS concentration, the system volume is proportional to the mass of sludge in it. Hence, for the same TSS concentration, the reactor volume treating settled sewage will be only 33% of that treating raw sewage at 20 days' sludge age. Also, the settled sewage plant requires only 63% of the oxygen the raw sewage plant requires. However, the active fraction with respect to TSS of the sludge in the settled sewage plant is 43%, too high for direct discharge to drying beds, whereas that from the raw wastewater plant is 23%. Clearly the choice of treating settled or raw wastewater requires weighing their advantages and disadvantages, *i.e.* for settled sewage a smaller reactor volume, lower oxygen demand and reduced secondary sludge production, but having to deal with primary and secondary sludge and their stabilization, and for raw sewage, larger reactor volume, higher oxygen demand and increased secondary sludge production, but having no primary sludge to deal with. These aspects are evaluated in further detail in Section 4.11.

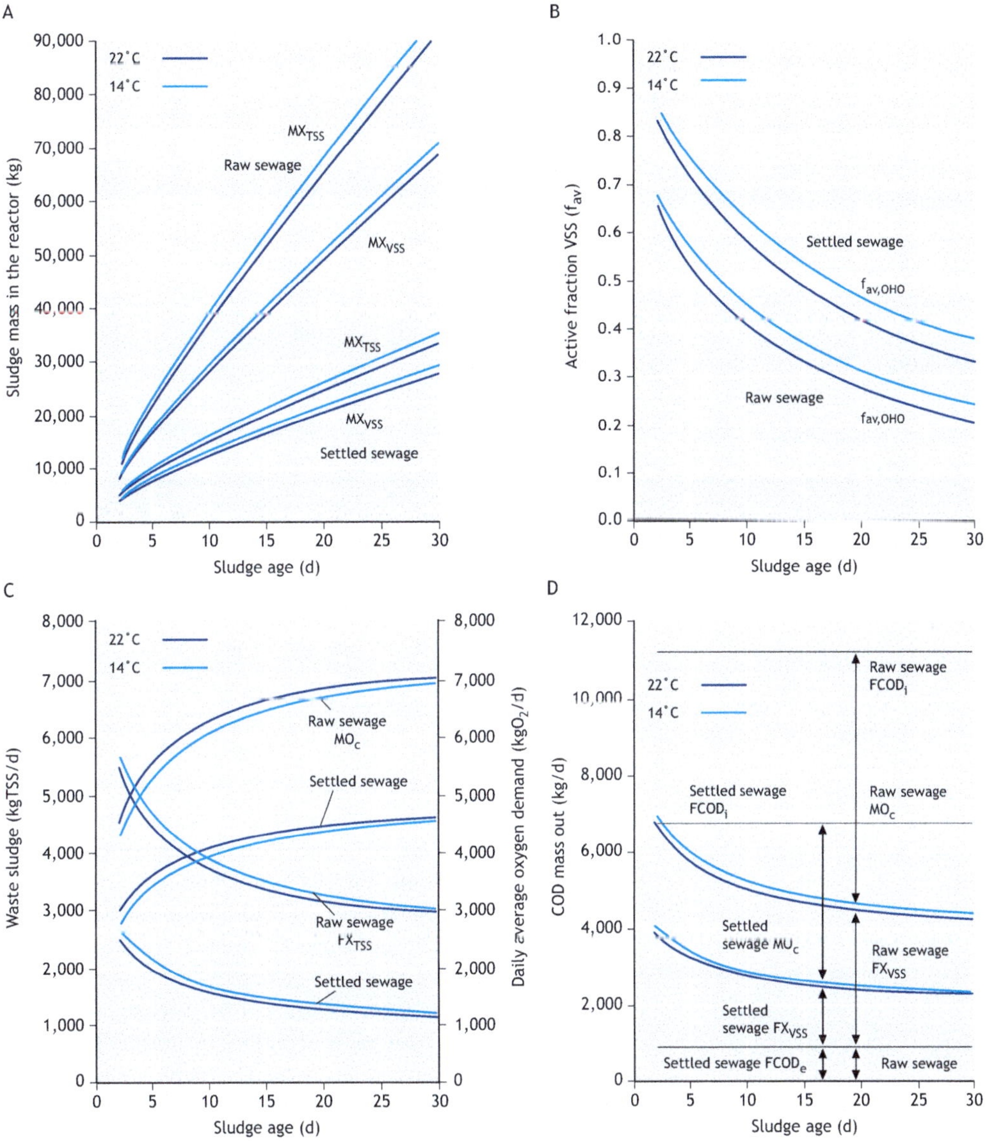

Figure 4.3 Mass of sludge MX_{TSS} (kgTSS) and MX_{VSS} (kgVSS) (A), active fractions of the sludge with respect to VSS ($f_{av,OHO}$) (B), average carbonaceous oxygen demand FO_c (kgO/d) and mass of TSS sludge produced per day (kgTSS/d) (C) and the COD mass balance (D) *versus* sludge age for the example raw and settled sewage at 14 °C and 22 °C. Stoichiometric and kinetic constants and wastewater characteristics are given in tables 4.2 and 4.3.

Table 4.3 Activated system design values for 20 days' sludge age for the example raw and settled wastewaters at 14 and 22°C.

System Parameter	Unit	Raw		Settled[1]	
Wastewater temperature	°C	14	22	14	22
Mass VSS, MX_{VSS}	kgVSS	51,122	48,982	21,918	20,537
Mass TSS, MX_{TSS}	kgTSS	68,162	65,309	26,408	24,743
Mass O_2, FO_c	kgO_2/d	6,679	6,837	4,313	4,415
Active fraction, $f_{av,OHO}$		0.306	0.265	0.461	0.408
Active fraction, $f_{at,OHO}$		0.23	0.199	0.383	0.339
Waste, FX_{VSS}	kgVSS/d	2,556	2,449	1,096	1,027
Waste, FX_{TSS}	kgTSS/d	3,408	3,265	1,320	1,237
Effluent COD, COD_e	mg/l	52.5	52.5	52.5	52.5
COD mass balance					
COD mass out	kgCOD/d				
Soluble in Q_e, $FCOD_e$		743	743	766	766
Oxygen utilized FO_c		6,679	6,837	4,313	4,415
Soluble COD in Q_w		45	17	45	17
COD of VSS in Q_w		3,783	3,625	1,622	1,520
Total COD out	kgCOD/d	11,249	11,249	6,718	6,718
Total COD in[1]	kgCOD/d	11,250	11,250	6,718	6,718
% COD mass balance		100	100	100	100

[1]For settled sewage, based on an influent flow of 14.925 Ml/d to take account of the primary sludge flow of 75 m^3/d (0.5% of influent ADWF)

4.5.3 The COD mass balance

Applying the COD mass balance Eq. 4.24 to the example raw and settled wastewaters yields:

1) soluble COD in effluent and waste flows,

$$(Q_e + Q_w = Q_i): FCOD_e = COD_i Q_i = f_{SU,CODi} COD_i Q_i \quad \text{(kgCOD/d)}$$

2) particulate COD (activated sludge) in waste flow,

$$(Q_w): FCOD_{VSS} = \frac{f_{cv,OHO} MX_{VSS}}{SRT} \quad \text{(kgCOD/d)}$$

where, MX_{VSS} is given by Eq. 4.12

3) carbonaceous oxygen utilized, FO_c, kgO_2/d and given by Eq. 4.18.

4) mass COD entering system,

$$FCOD_i = COD_i Q_i \quad \text{(kgCOD/d)}$$

Activated sludge systems operating at very long sludge ages, called extended aeration (*e.g.* 30 days, Table 4.4), allow the endogenous process to approach completion thereby providing not only sewage treatment in the activated sludge reactor but also a significant measure of aerobic stabilisation of the activated sludge to achieve a low active fraction so that the waste sludge can be discharged directly to drying beds without further treatment. Treating raw wastewater in an extended aeration system therefore obviates the need for sludge treatment, which takes place in the activated sludge reactor also, but this is at the expense of a very large activated sludge reactor and high oxygen utilization and energy consumption rates.

Table 4.4 Comparison of sludge production, stability (biodegradable COD remaining) and oxygen demand treating the example raw and settled wastewaters at long and short sludge ages.

Parameter	Units	Raw	Settled
Temperature	°C	14	14
Sludge age	d	30	8
Activated sludge concentration	mgTSS/l	4,000	4,000
Reactor volume	m^3	23,769	3,544
Oxygen demand	kgO_2/d	6,944	3,758
Primary sludge TSS	kgTSS/d	0	3,335
Primary sludge VSS	kgVSS/d	0	2,468
Primary sludge COD	kgCOD/d	0	4,531
Biodegradable COD remaining	%	0	68.5
Secondary sludge TSS	kgTSS/d	3,169	1,772
Secondary sludge VSS	kgVSS/d	2,377	1,471
Secondary sludge COD	kgCOD/d	3,518	2,177
Active fraction with regard to VSS	kgOHOVSS/kgVSS	0.235	0.662
Biodegradable COD remaining	%	18.8	53
Total sludge TSS	kgTSS/d	3,169	5,107
Total sludge VSS	kgVSS/d	2,377	3,939
Total sludge COD	kgCOD/d	3,518	6,708
Biodegradable COD remaining	%	18.8	63.5

In contrast, treating settled sewage at a very short sludge age (high rate, *e.g.* 8 days), the activated sludge system results in a very small activated sludge reactor and a low oxygen utilization and energy consumption but produces a primary sludge and a very active waste activated sludge, both of which need aerobic or anaerobic treatment for stabilisation to reduce the remaining biodegradable organics.

With anaerobic digestion these sludges can produce a significant amount of methane gas and offset some (or all) of the power consumption for aeration. The remaining biodegradable COD in the WAS, as a proportion of the influent COD flux entering the activated sludge system, for increasing SRT is shown in Figure 4.4 for the example raw and settled wastewater.

It can be seen that the remaining biodegradable COD in the WAS, which is essentially the biodegradable part of the OHO biomass, rapidly decreases as SRT increases, with the result that less and less energy can be extracted from it with anaerobic digestion at the cost of generating a digester dewatering liquor high in ammonia and phosphate (Ekama, 2017). At very short sludge ages, the focus of the activated sludge system therefore is sewage treatment only, with sludge treatment (stabilisation) taking place in separate dedicated aerobic or anaerobic systems for this with possible energy and nutrient recovery. Irrespective of the approach adopted (extended aeration or high rate) for the particular sewage treatment plant, the COD mass must balance over the entire plant, not only over the activated sludge system but over the sludge treatment systems also.

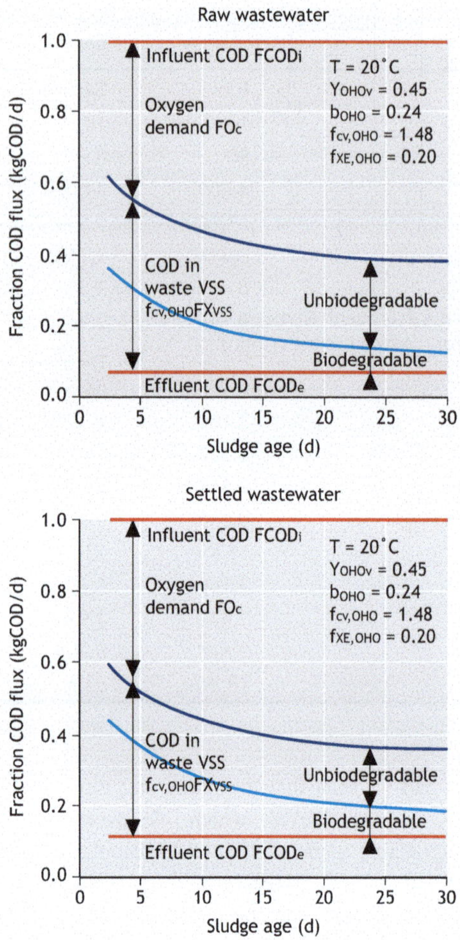

Figure 4.4 Proportion of influent COD flux exiting as COD in effluent, as oxygen demand and COD in waste sludge solids (WAS) and the residual biodegradable COD in the WAS *versus* sludge age for the example raw (Figure 4.4a, top) and settled (Figure 4.4b, bottom) wastewaters at 20 °C.

4.6 REACTOR VOLUME REQUIREMENTS

Once the mass of sludge in the reactor is derived from a specified sludge age and organic COD mass load per day, the reactor volume is determined by 'diluting' this mass of sludge to a specified TSS concentration (X_{TSS}). From the volume, the nominal hydraulic retention time, or aeration time for fully aerobic systems, is fixed (by Eq. 4.2). Hydraulic retention time therefore is immaterial in the design procedure; it is a consequence of the mass of sludge in the reactor and a selected TSS concentration. This point was mentioned earlier but bears repeating because some design procedures lay stress on retention time or aeration time as a basic design parameter, an approach that can result in serious miscalculation of the reactor volume requirements. Compare, for example, two plants operating at the same sludge age, both receiving the same organic load (kgCOD/d) but the first at high influent COD concentration and low flow and the second at a low concentration and high flow. If designed based on a specified hydraulic retention time, the volume of the first will be much smaller than that of the second but the sludge mass in the reactors will be the same. Consequently, the first plant may have an inordinately high TSS concentration which may cause problems in the secondary settling tank. Therefore retention time is a completely inappropriate basis for design and other purposes such as a criterion for comparing the reactor volume requirements of different plants.

Figure 4.5 shows the reactor volume requirements *versus* sludge for the example raw and settled wastewaters obtained from eqs. 4.4 to 4.17 and 4.20. The reactor volume requirements may also be determined from the equivalent COD load per capita or person equivalent (PE), also shown in Figure 4.5 for a raw wastewater COD load of 0.10 kgCOD/PE.d. Hence when treating the example raw wastewater at a sludge age of 20 days and a TSS concentration of 4 kgTSS/m^3, a reactor volume of 145 l per PE is required or 1.45 m^3/kgCOD applied per day to the treatment plant. The comparative reactor volume requirements for settled wastewater per kg COD load per day on the treatment plant also is shown in Figure 4.5 taking due consideration of the COD fraction removed by primary sedimentation (40% for the example of settled wastewater).

From Figure 4.5, treating settled wastewater at a sludge age of 20 days and reactor TSS concentration of 4 kgTSS/m^3 requires a reactor volume of 0.55 m^3/kg raw wastewater COD load per day on the wastewater treatment plant or 55 l per PE. Comparing the reactor volume requirements treating raw or settled sewage, it can be seen that a significant reduction in reactor volume can be obtained by means of primary sedimentation: 62% for the example raw and settled sewage at 20 days' sludge age.

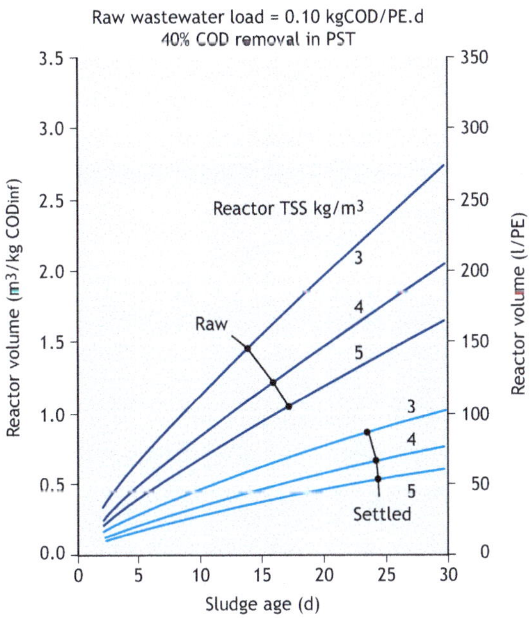

Figure 4.5 Reactor volume requirements in m³/kg COD raw wastewater load per day *versus* sludge age at different average reactor TSS concentrations for raw and settled wastewater (assuming 40% COD removal by primary sedimentation). Reactor volume requirements in l/capita or l/person equivalent (PE) is also given on the right-hand vertical axis based on a raw wastewater COD contribution of 0.10 kg COD/person equivalent.

4.7 DETERMINATION OF REACTOR TSS CONCENTRATION

The choice of the reactor concentration can be done empirically from past experience with similar wastewaters or selected from design guidelines such as those from Metcalf and Eddy (2014), *e.g.* for conventional systems (with primary sedimentation) 1,500 to 3,000 mgTSS/l or extended aeration (without primary sedimentation) 3,000 to 6,000 mgTSS/l. Differences in the reactor TSS concentration for raw and settled wastewaters arise because (*i*) the wastewater flow per kg COD load on the reactor for raw wastewater is significantly greater than that for settled wastewaters and (*ii*) sludge settleability in conventional systems can be poorer than with extended aeration systems: in a survey of 45 full-scale activated sludge plants in the Netherlands, Stofkoper and Trentelman (1982) found significantly higher DSVIs in settled wastewater systems than in raw wastewater systems (Ekama and Marais, 1986).

The effect of sewage strength and sludge settleability, as well as other factors such as the peak wet weather (PWWF) to average dry weather flow (ADWF) ratio (or peak flow factor $f_q = PWWF/ADWF$), wastewater and activated sludge characteristics ($f_{XU,CODi}$, $f_{SU,CODi}$, f_{VT}) and construction costs, can all be taken into account by determining the reactor concentration from a construction cost minimization analysis (Hörler, 1969; Dick, 1976; Riddell *et al.*, 1983; Pincince *et al.*, 1995). In such an analysis, the construction cost of the reactor(s) and the secondary settling tank(s) (SSTs) are determined as functions of the reactor TSS concentration. The reactor concentration at which the combined construction cost of the reactor(s) and the SST(s) is at a minimum is the design reactor concentration.

4.7.1 Reactor cost

For selected wastewater and activated sludge characteristics ($f_{XU,CODi}$, $f_{SU,CODi}$, f_{VT}), sludge age and organic COD load on the reactor ($FCOD_{i,Reactor}$), the mass of TSS in the reactor (MX_{TSS}) can be determined from eqs. 4.15 or 4.17 and remains constant, *e.g.* for the example raw sewage at 20 days sludge age and 14 °C, MX_{TSS} = 68,162 kgTSS. The reactor volume (V_R) in a single module (V_M) as a function of the reactor TSS concentration X_{TSS} is found from Eq. 4.20, *viz.*

$$V_M = MX_{TSS} / X_{TSS} = \frac{68,162}{N_{AS} X_{TSS}} \quad (m^3)$$

where:

X_{TSS} reactor concentration ($kgTSS/m^3$)

N_{AS} number of AS modules with one reactor

per module in order to keep the reactor volume of one module within a maximum reactor volume limit, *i.e.* $V_R=N_{AS}\cdot V_M$.

To estimate the cost of the reactor from the volume, empirical functions relating the construction cost of the reactor to the volume are required. Such functions take the form

$$\text{Reactor costs} = N_{AS}C_{br}(V_R)^{P_{br}} \quad (4.28)$$

where:
C_{br}, P_{br} constants for a particular reactor design.

4.7.2 Secondary settling tank cost

On the basis of the flux theory, Ekama *et al.* (1997) show that provided the underflow recycle ratio s is above the critical minimum value, the surface area of the secondary settling tanks (A_{SST}) is a function of only the reactor (or feed) solids concentration (X_{TSS}) and the sludge settleability. If the reactor concentration increases or the sludge settleability deteriorates, the required surface area for the secondary settling tanks (SSTs) gets larger. Therefore, as the biological reactor gets smaller with increasing X_{TSS}, so A_{SST} gets larger. Hence, the construction cost of the SSTs increases with increases in X_{TSS}.

To determine the surface area of the SSTs, two parameters need to be specified for the design, *viz.* (*i*) the sludge settleability and (*ii*) the peak flow factor f_q (= PWWF/ADWF ratio). The idealized 1D flux theory requires the sludge settleability to be specified in terms of the V_0 and r_{hin} values in the zone settling velocity (V_s, m/h) *versus* solids concentration (X_{TSS}, kgTSS/m^3) relationship *i.e.*

$$V_s = V_0 \exp(-r_{hin}X_{TSS}).$$

Values for V_0 and r_{hin} are not readily available but relationships between different simpler sludge settleability parameters such as the sludge volume index (SVI), stirred specific volume index (SSVI) and diluted sludge volume index (DSVI) have been proposed by various authors (see Ekama *et al.*, 1997). These relationships allow calculation of the flux V_0 and r_{hin} values from the SVI, SSVI or DSVI sludge settleability indices. However, there is considerable variation in these relationships and selection for a particular activated sludge plant needs very careful consideration. For this example, the relationships developed by Ekama and Marais (1986) are accepted:

$$SSVI_{3.5} = 0.67DSVI \quad \text{(ml/g) (4.29a)}$$

$$V_0 / r_{hin} = 67.9\exp(-0.016SSVI_{3.5}) \quad \text{(kgTSS/m}^2\text{.h) (4.29b)}$$

$$r_{hin} = 0.88 - 0.393\log(V_0 / r_{hin}) \quad \text{(m}^3\text{/kgTSS) (4.29c)}$$

$$V_0 = (V_0 / r_{hin})r_{hin} \quad \text{(m/h) (4.29d)}$$

From the 1D idealized flux theory, the maximum permissible overflow rate at PWWF ($q_{i,PWWF}$) is given by,

$$q_{i,PWWF} = 0.8V_s \text{ at } X_{TSS} = V_0 \exp(-r_{hin}X_{TSS}) \quad \text{(m/h) (4.30a)}$$

where:
$q_{i,PWWF}$ overflow rate at PWWF (m/h)
0.8 flux factor

$$q_{i,PWWF} = Q_{i,PWWF} / A_{SST} = f_q Q_{i,ADWF} / A_{SST} \quad \text{(m/h) (4.30b)}$$

From a calibration of the 1D idealized flux theory SST design procedure against full-scale SST performance data, Ekama and Marais (2004) showed that the maximum permissible solids loading rate (SLR, kgTSS/m^2.h) should only be 80% of that estimated by the 1D idealized flux theory and is the reason for the 0.8 in Eq 4.30. This reduction appears to be a consequence of the significant deviation of the hydrodynamics in real SSTs compared with that assumed in the 1D idealized flux theory such as horizontal flows of liquid and solids, turbulence, short circuiting and density currents (Ekama *et al.*, 1997).

Taking the 25% (1/0.80) reduction into account, the surface area of the SST(s), A_{SST}, in terms of X_{TSS} is given by,

$$A_{SST} = \frac{1{,}000 f_q Q_{i,ADWF} / 24}{0.8 V_0 \exp(-r_{hin} X_{TSS}) N_{AS} N_{SST}} \quad (m^2) \quad (4.31)$$

where:

$Q_{i,ADWF}$ average dry weather flow (ml/d)

N_{SST} number of SSTs per AS module and governed by the maximum diameter of one SST.

Functions for the construction cost function in terms of diameter (φ, m) for circular SSTs of given depth may take the form:

$$SSTcost = N_{AS} N_{SST} C_{sst} (\varphi)^{P_{sst}} \quad (4.32)$$

where:

C_{sst}, P_{sst} constants for a particular design.

4.7.3 Total cost

The total cost of the reactor-SST system is the sum of the reactor and SST costs. Qualitative results for the example raw and settled wastewaters are given in Figure 4.6, ignoring that the reactor volume and SST diameter may have upper and lower size restrictions. For full-scale plants, the reactor and/or SST may need to be split into two or more equal-sized modules to bring the volume and diameter within the limit ranges. From cost minimization analyses such as that above, generally it will be found that the range of reactor concentration for minimum construction cost (*i*) is higher for higher influent wastewater strengths (BOD_5, COD), (*ii*) is higher for longer sludge ages, and (*iii*) is higher for raw wastewater than settled wastewater at the same strength, because these three changes all increase the size of the biological reactor relative to that of the settling tank, (*iv*) is lower for higher peak flow factors (f_q), and (*v*) is lower for poorer settling sludges because these two changes all increase the size of the settling tank relative to that of the biological reactor. A universal optimum therefore cannot be specified. In countries with low wastewater strengths and short sludge-age plants (*e.g.* North America), the reactor concentration will tend to be low (2,000-3,000 mgTSS/l) and in countries with high wastewater strengths and long sludge-age plants (*e.g.* South Africa), the reactor concentration will tend to be high (4,000-6,000 mgTSS/l) as the example wastewaters demonstrate.

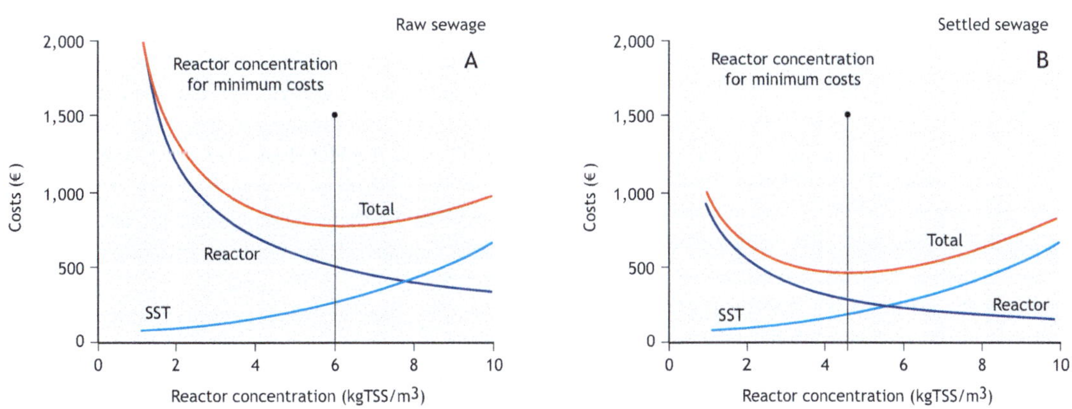

Figure 4.6 Reactor, secondary settling tank (SST) and total construction costs to estimate the reactor concentration for minimum cost for the example raw (A) and settled (B) wastewaters in single-reactor and SST units.

4.8 CARBONACEOUS OXYGEN DEMAND

4.8.1 Steady-state (daily average) conditions

The mean daily carbonaceous oxygen demand per kg COD load on the reactor ($FO_c/FCOD_{i,Reactor}$) is calculated from Eq. 4.18. For sludge ages longer than 15 days the increase in $FO_c/FCOD_{i,Reactor}$ is small with further increase in sludge age for both raw and settled wastewater. The $FO_c/FCOD_{i,Reactor}$ for raw and settled wastewater is usually within 10% of each other, with the demand for settled wastewater being the higher value. This is because compared with raw wastewater, a higher percentage of the total organics (COD) in settled wastewater is biodegradable. For the example wastewaters at 20 days' sludge age, the $FO_c/FCOD_{i,Reactor}$ is 0.604 $kgO_2/kgCOD$ for raw wastewater and 0.653 $kgO_2/kgCOD$ for settled wastewater. Although there is only a small difference in $FO_c/FCOD_{i,Reactor}$ between raw and settled wastewaters, there is a large difference in the oxygen demand per kgCOD load on the plant (Figure 4.3C). For settled wastewater, this is given by $0.653{\cdot}(1-0.40)$ for 40% COD removal in PSTs. This gives 0.38 $kgO_2/kgCOD$ load on the plant. For the raw wastewater it would remain 0.604 $kgO_2/kgCOD$ load on the treatment plant, making the settled wastewater oxygen demand 37% lower than that for the raw wastewater. Clearly primary sedimentation leads to significant aeration energy savings; because primary settling tanks remove about 30 to 50% of the raw influent COD, the carbonaceous oxygen demand for settled wastewater generally will be about 30 to 50% lower than that for raw wastewater.

The carbonaceous oxygen demand is the oxygen demand for the oxidation of the influent organics (COD) and the associated OHO endogenous process only. In N removal systems, oxygen is also required for nitrification, which is the biological oxidation of ammonia to nitrate by autotrophic nitrifiers. However, with denitrification, which is the biological reduction of nitrate to nitrogen gas by facultative heterotrophic organisms, some of the biodegradable organics are utilized with nitrate as electron acceptor, for which oxygen is then not required. Thus denitrification leads to a reduction in the oxygen demand. The total oxygen demand for a N-removal system therefore is the sum of the carbonaceous and nitrification oxygen demands, less that saved by denitrification. The procedures for calculating the oxygen demand for nitrification and the oxygen saved by denitrification are discussed in Chapter 5. The equations given here for calculating the carbonaceous oxygen demand are based on the assumption that all the biodegradable organics are utilized with oxygen as the electron acceptor, *i.e.* for fully aerobic systems.

4.8.2 Daily cyclic (dynamic) conditions

Owing to the daily cyclic nature of the organic (COD) load on the reactor, the carbonaceous oxygen demand will vary concomitantly over the day. The TKN load on the reactor also varies over the day in an approximately similar fashion as the organic load. Generally, the COD and TKN loads on the reactor increase in the morning due to increases in both flow and COD and TKN concentration reaching a peak at around noon. Thereafter, the COD and TKN loads decrease reaching a minimum during the night-time hours of 2:00 to 4:00 due to decreases in both flow and COD and TKN concentration. The peak to average and minimum average load ratios and the time of day these occur depend on the catchment that the particular treatment plant serves, such as size of population, layout of the catchment and industrial activity. Generally, the smaller the catchment, the lower the flow and COD and TKN loads but the greater the peak to average flow and load ratios then the lower the minimum-to-average flow and load ratios. Because the TKN load and its variation over the day and the nitrification process have a profound influence on the daily average and peak total oxygen demands, empirical methods to estimate the peak oxygen demand from the average for fully aerobic nitrifying systems are discussed in Chapter 5. Fully aerobic activated sludge systems with sludge ages longer than 3 days are likely to nitrify at temperatures > 14 °C. Moreover, a sludge age of 3 days is around the limit of validity for the steady state activated sludge model because at sludge ages lower than this the assumption that all the biodegradable organics are utilized is not valid. So there is little merit in developing empirical methods for estimating the peak oxygen demand for fully aerobic systems without nitrification.

4.9 DAILY SLUDGE PRODUCTION

The mass of sludge produced per day by the activated sludge system is equal to the mass of sludge wasted per day from it via the waste flow and is called waste activated sludge (WAS) or secondary sludge. From the definition of sludge age (see Eq. 4.1), the mass of sludge TSS produced per day FX_{TSS} is given by the mass of sludge in the system MX_{TSS} divided by the sludge age *i.e.*:

$$FX_{TSS} = MX_{TSS} / SRT \qquad \text{(mgTSS/d) (4.33)}$$

Substituting eqs. 4.12 and 4.17 for MX_{TSS} and simplifying, yields the sludge produced per day per mg COD load on the biological reactor, *i.e.*

$$\frac{FX_{TSS}}{FCOD_i} = \frac{1}{f_{VT}}\left[\frac{(1-f_{SU,CODi}-f_{XU,CODi})Y_{OHOv}}{(1+b_{OHO,T}SRT)} \cdot (1+f_{XE,OHO}b_{OHO,T}SRT)+\frac{f_{XU,CODi}}{f_{cv,UPO}}\right]$$

$$\text{(mgTSS/d per mgCOD/d) (4.34)}$$

A plot of the daily total sludge mass (TSS) produced per unit COD load on the biological reactor (Eq. 4.34) *versus* sludge age is shown in Figure 4.7 for the example raw and settled wastewaters.

It can be seen that the mass of sludge produced in the activated sludge system (per unit COD load on the biological reactor) decreases as the sludge age increases for both raw and settled wastewater but the rate of decrease is negligible at sludge ages longer than approximately 20 days. Treating settled wastewater results in lower secondary sludge production per unit COD load on the biological reactor than treating raw wastewater. This is because the unbiodegradable particulate COD fraction ($f_{XU,CODi}$) and inorganic content ($X_{FSS,i}/COD_i$) in settled wastewater are significantly lower than that in raw wastewater.

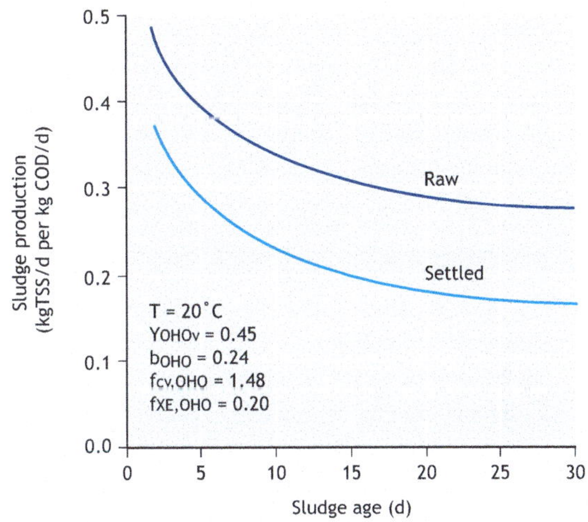

Figure 4.7 Daily sludge production in kgVSS/d and kgTSS/d per kgCOD load per day on the biological reactor for the example raw and settled wastewaters at 14 °C.

Temperature effects on secondary sludge production are small: sludge production at 14 °C is about 5% greater than at 22 °C, a difference which is completely masked by the uncertainty in the estimates of the wastewater characteristic $f_{XU,CODi}$ and the VSS/TSS ratio (f_{VT}) of the sludge if the influent ISS concentration ($X_{FSS,i}$) is not measured.

Although the secondary sludge production treating settled wastewater is lower than that treating raw wastewater, the *total* sludge mass treating settled wastewater is higher because the total sludge production includes both the primary and secondary sludges; at plants treating raw wastewater, only secondary sludge is produced.

In the system treating raw wastewater, the primary sludge is in effect treated in the activated sludge reactor itself. From the COD balance, the more oxygen that is utilized in the system, the lower the sludge production and the lower the active fraction of the sludge (Figure 4.3B,C). Therefore, because the carbonaceous oxygen demand is much higher when treating raw wastewater, the overall sludge production is much lower compared with settled wastewater.

Generalizing the above observations and taking into account the active fraction of the waste sludge as an indication of the remaining biodegradable organics in the waste sludge, there are two extremes in approach to designing wastewater treatment plants with activated sludge for biological treatment:

(*i*) treating settled wastewater at a short sludge age (say 8 days): this results in a very small activated sludge system with low oxygen demand and a high sludge production with high energy content, *i.e.* high remaining biodegradable organics in both the primary and secondary (waste activated) sludges requiring further stabilization treatment before disposal, or,

(*ii*) treating raw wastewater at a long sludge age (say 30 days): this results in a very large activated sludge system with high oxygen demand and a low sludge production with a low energy content *i.e.* no primary sludge and low remaining biodegradable organics (low active fraction) in the secondary sludge, not requiring further stabilization treatment before disposal.

The daily production of secondary and primary sludges is the mass of sludge that needs to be treated and disposed of by downstream sludge handling methods. Sludge treatment and disposal, for biological nutrient removal (BNR) systems in particular, should not be seen as separate from the design of the activated sludge system. In fact, all unit operations of the treatment plant from raw wastewater pumping to ultimate disposal of the sludge, should be viewed as an integrated system where the design of one unit operation depends on the unit operations before it, and decisions on its design may affect the design of unit operations following it. For example, while approach 1 above provides an opportunity for low energy consumption and high energy generation by anaerobic digestion, the impact of the ammonia and phosphate in the digested sludge dewatering liquor on the activated sludge reactor needs to be considered unless N and P removal from the dewatering liquor is included in the sludge treatment scheme (Ekama, 2017). N removal from dewatering liquor is considered in Chapter 5.

4.10 FOOD-TO-MICRO-ORGANISM (F/M) RATIO AND LOAD FACTOR

The food-to-microorganism (F/M) ratio and load factor (LF) have been used for decades to size activated sludge reactors. The F/M seeks to quantify the flux of biodegradable organics (food, COD, BOD_5) on the activated sludge microorganisms (m) or OHO biomass. Because there was no alternative, it was accepted that the VSS, being organic, satisfactorily quantified this OHO biomass. Without expressly referring to biodegradable organics and micro-organisms, the load factor carries the same idea as the F/M by expressing as a ratio the flux organic load (kgCOD/d or kgBOD_5/d) on the sludge mass in the reactor (kgVSS or kgTSS). However, both these ratios are seriously flawed from a modelling perspective. To mention a few (Wentzel *et al.*, 2003): (*i*) while the BOD_5 is an approximation of the biodegradable organics, it measures neither the electron-donating capacity of the biodegradable organics nor the unbiodegradable particulate organics (UPO), which accumulate as VSS in the activated sludge reactor. (*ii*) The BOD_5 is not mass conservative and so cannot be used to make a BOD mass balance around the activated sludge system. (*iii*) Although the COD is not the 'food' because it includes the unbiodegradable organics, most importantly it is mass conservative and so a COD mass balance can be made around the activated sludge system. (*iv*) As the AS model in the previous sections shows, the VSS (and less so the TSS) are not measures of the OHO biomass. The F/M and LF ratios are entirely empirical and for the same activated sludge system, have different numerical values depending on the units (COD, BOD_5, VSS, TSS) selected to express them. Also, the ratios vary with the wastewater characteristics (particularly the UPO fraction) and temperature. In fact, as long ago as 1976, Marais and Ekama show that the F/M or LF add no additional information to the AS system that sludge age (SRT) does not provide because by selecting an F/M or LF, one is implicitly selecting a system SRT. To make matters worse, selecting the same F/M or LF expressed in the same units for AS systems treating different raw or settled wastewater, yields different SRTs depending on the wastewater characteristics. Yet F/M or LF-based empirical design procedures continue to be taught in environmental engineering programs and used in consulting practices as the basis for activated sludge system design. Because SRT forms the basis for modeling and simulation of AS systems, the link between F/M or LF and SRT is given below. This will demonstrate that the F/M and LF are implicitly embedded in and can be completely replaced by the much better SRT as the fundamental basis for activated sludge system sizing and modelling. This will allow the results from the empirical F/M or LF-based design procedures to be compared with the ASM1-aligned steady-state activated sludge model presented in the above sections. Using the TSS and COD for the reactor suspended solids and influent organics concentrations, and considering fully aerobic or nitrification denitrification (ND) systems only, then from eqs. 4.12 and 4.14,

$$A_{ND,TSS} = \frac{MX_{TSS}}{FCOD_i} = \left(1 - f_{SU,CODi} - f_{XU,CODi}\right) \cdot \frac{Y_{OHO}SRT}{\left(1 + b_{OHO,T}SRT\right)}\left(1 + f_{XE,OHO}b_{OHO,T}SRT + f_{FSS,OHO}\right) + \left[\frac{f_{XU,CODi}}{f_{cv,UPO}} + \frac{X_{FSS,i}}{COD_i}\right]SRT \quad (4.35)$$

where:

$A_{ND,TSS}$ or $MX_{TSS}/FCOD_i$ is the inverse of the F/M or LF in units kgTSS/(kgCOD.d).

Note that $A_{ND,TSS}$ is units-specific and AS system type-specific. It is different for ND and ND-enhanced biological P removal (EBPR) systems and varies with the magnitude of biological P removal (Ramphao *et al.*, 2006, see Chapter 7, Section 7.8). Hence the A value as expressed in Eq 4.35 is valid for ND systems only and has units kgTSS/(kgCOD/d). Dividing Eq 4.35 by the SRT gives the specific sludge production ($SP_{TSS,COD}$) as kgTSS produced/d per kgCOD load/d, which is the same as Eq. 4.34, *i.e.*,

$$\frac{1}{\left(F/M\right)_{COD,TSS}} = \frac{1}{LF_{COD,TSS}} = SP_{TSS,COD}R_s = \frac{MX_{TSS}}{FCOD_i} = \frac{V_R X_{TSS}}{Q_i COD_i} = A_{ND,TSS} \quad (4.36a)$$

$$\left(F/M\right)_{COD,TSS} = LF_{COD,TSS} = \frac{1}{A_{ND,TSS}} \quad \text{(kgCOD/kgTSS.d)} \quad (4.36b)$$

where:

MX_{TSS}	mass of TSS in reactor (kgTSS) $= V_R \cdot X_{TSS,ave}$
V_R	volume of reactor (m^3)
$X_{TSS,ave}$	average TSS concentration in reactor (kgTSS/m^3)
$SP_{COD,TSS}$	specific sludge production factor in kgTSS/d/kgCOD.d, Metcalf & Eddy (2014).

Equation 4.36 shows that choosing an F/M or LF (as some of the obsolete design procedures do) is the same as choosing in a single lumped parameter the SRT and wastewater characteristics UPO and USO COD fractions ($f_{XU,CODi}$, $f_{SU,CODi}$) and influent ISS concentration ($X_{FSS,i}$). Therefore, if the wastewater characteristics $f_{XU,CODi}$, $f_{SU,CODi}$ and $X_{FSS,i}$ are known, then the SRT implicit in the selected F/M can be calculated from eqs 4.35 and 4.36. This is shown in figures 4.8a and 4.8b, in which the F/M ratio in kgBOD$_5$/(kgTSS.d) is plotted *versus* SRT for the raw and settled wastewater characteristics in Table 4.2 and COD/BOD$_5$ ratios for raw and settled wastewater of 2.2 and 1.8, respectively (Marais and Ekama, 1976). The right-hand side of figures 4.8A and 4.8B show the ranges of F/M or LF ratios for AS systems for carbon (COD) removal only, C removal with ND and extended aeration, which usually means raw wastewater treatment at long SRT to obviate waste-activated sludge (WAS) stabilization.

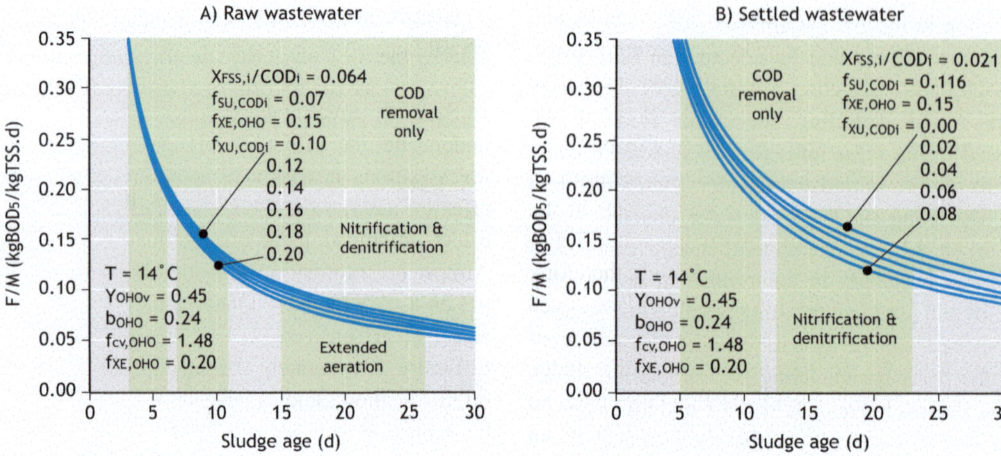

Figure 4.8 F/M ratio *versus* sludge age for different unbiodegradable particulate COD fractions (UPO, $f_{XU,CODi}$) for raw wastewater with a COD/BOD$_5$ ratio of 2.2 (A) and settled wastewater with a COD/BOD$_5$ ratio of 1.8 (B).

The F/M in terms of kgVSS in reactor per kgCOD load/d is obtained from Eq 4.35 by setting to zero the terms relating to ISS, *i.e.* $X_{FSS,i}$=0 and $f_{FSS,OHO}$=0, *viz.*

$$\frac{1}{\left(F/M\right)_{COD,VSS}} = SP_{VSS,COD}SRT = A_{ND,VSS} = \frac{MX_{VSS}}{FCOD_i}$$

$$= \left(1 - f_{SU,CODi} - f_{XU,CODi}\right)\frac{Y_{OHO}SRT}{\left(1 + b_{OHO,T}SRT\right)} \cdot \left(1 + f_{XE,OHO}b_{OHO,T}SRT\right) + \left[\frac{f_{XU,CODi}}{f_{cv,UPO}}\right]SRT \quad (4.37)$$

where:

$SP_{COD,VSS}$ the specific sludge production factor in kgVSS/d/kgCOD/d,

$A_{ND,VSS}$ mass of VSS in the reactor per kgCOD load per day for fully aerobic and ND systems, which is a function of SRT and the wastewater characteristics.

4.11 CAPACITY ESTIMATION OF AS SYSTEMS

For a given biological reactor volume (V_R, m^3) and SST surface area (A_{SST}, m^2) the average dry weather flow ($Q_{i,ADWF}$) capacity in Ml/d can be calculated for a particular wastewater with known characteristics ($f_{XU,CODi}$, $f_{SU,CODi}$ and $X_{FSS,i}$). With the required system SRT known, *e.g.* the balanced SRT of a Modified Ludzack-Ettinger (MLE) ND system (Chapter 5, Section 5.9.3.2), the mass of TSS in the reactor (MX_{TSS}, kgTSS) per kgCOD/d load applied to the reactor [$A_{ND,TSS}$ kgTSS/(kgCOD/d)] can be determined from Eq. 4.35,

$$MX_{TSS} = V_R X_{TSS,ave} = A_{ND,TSS} Q_{i,ADWF} COD_i \quad (kgTSS) \quad (4.38)$$

where $X_{TSS,ave}$ is the volume-weighted average reactor TSS concentration (kgTSS/m^3).

In some BNR system reactors the TSS concentration is not the same in each zone of the reactor, *e.g.* in a UCT EBPR system, the TSS concentration in the anaerobic compartment ($X_{TSS,AN}$) is lower than that in the aerobic compartment

($X_{TSS,OX}$). This makes the volume-weighted average reactor TSS concentration ($X_{TSS,ave}$=MX_{TSS}/V_R) lower than the aerobic compartment TSS concentration ($X_{TSS,OX}$) fed to the SST (Parco *et al.,* 2018).

Rearranging Eq 4.38 yields,

$$Q_{i,ADWF} = \frac{V_R X_{TSS,ave}}{A_{ND,TSS} COD_i} \quad (Ml/d) \ (4.39)$$

Equation 4.39 has two unknowns, $Q_{i,ADWF}$ and $X_{TSS,ave}$, so one more fact is required to determine $Q_{i,ADWF}$. This is obtained from the overflow rate on the SST. The overflow rate at PWWF must not be greater than the flux rating (0.8) times settling velocity (V_S, m/h) of the sludge at the SST feed or aerobic zone TSS concentration as stated by Eq 4.30. Rearranging Eq 4.31 to make $Q_{i,ADWF}$ the subject yields,

$$Q_{i,ADWF} = \frac{24 A_{SST} 0.8 V_0}{1{,}000\ f_q e^{-r_{hin} X_{TSS,OX}}} \quad (Ml/d) \ (4.40)$$

where V_0 and r_{hin} are in units m/h and m^3/kgTSS and are given by Eq 4.29.

Setting Eq 4.40 equal to Eq 4.39 requires an equation linking $X_{TSS,ave}$ and $X_{TSS,OX}$. This is obtained from the BNR system configuration. For fully aerobic, MLE and 4 stage Bardenpho ND systems without step feeding, the TSS concentrations in all compartments of the reactor are the same. Hence $X_{TSS,ave}$=$X_{TSS,OX}$ and setting Eq 4.39 = Eq 4.40 yields,

$$X_{TSS,ave} = B_{ND} e^{-r_{hin} X_{TSS,ave}} \quad (kgTSS/m^3) \ (4.41a)$$

where

$$B_{ND} = \frac{A_{ND,TSS} COD_i A_{SST} 0.8 V_0 24}{f_q V_R 1{,}000} \quad (kgTSS/m^3) \ (4.41b)$$

where $A_{ND,TSS}$, COD_i, A_{SST}, V_0, 24 and V_R have units (kgTSS.d)/kgCOD, mgCOD/l. m^2, m/h and h/d and 0.8, f_q and 1,000 are dimensionless.

With $X_{TSS,ave}$ known, $Q_{i,ADWF}$ is given in Ml/d either by Eq 4.39 or Eq 4.40.

The equation linking $X_{TSS,ave}$ and $X_{TSS,OX}$ for a particular BNR system configuration can be derived from TSS mass balances around each compartment of the reactor and SST, for which the waste flow rate (Q_w) can be set to zero to simplify the derivation. Equations for several different ND and NDEBPR systems with SST (and membranes) are given (Parco *et al.,* 2018) in chapters 5 and 6, respectively.

For a sludge settleability DSVI of 120 ml/gTSS, a PWWF/ADWF ratio (f_q) of 2.5, the mass of TSS (MX_{TSS}) in the reactor treating raw and settled WW at 20d SRT and 14 °C from Table 4.3 and a reactor TSS concentration that minimizes the total cost of the reactor and SST of 5.22 and 4.02 kgTSS/m^3 from Section 4.7 above, yields reactor volumes and SST areas for the raw and settled WW systems listed in Table 4.5.

Table 4.5 Optimum AS reactor volume and SST area for minimum total cost giving maximum ADWF capacity.

Wastewater type	Raw WW	Settled WW
Mass TSS in reactor (MX_{TSS}, Table 4.3)	68,162	26,408
Reactor TSS content for minimum total cost (kgTSS/m^3)	5.22	4.02
Sludge settleability (DSVI, ml/gTSS)	120	120
Reactor volume (V_R, m^3)	13,049	6,577
SST area (A_{SST}, m^2)	1,972 (50.1)	1,248 (39.9)
SST volume at 3.5 m avg. depth (m^3)	6,903	4,367
Total AS-SST volume (m^3)	19,952	10,944

If the SST are 3.5 m deep, then the reactor is 65% and 60%, respectively of the total AS-SST volume for the raw and settled WW AS-SST systems. Accepting these AS systems are each in a single module with one SST, then from eqs. 4.38 to 4.41, the ADWF capacity of the raw and settled WW systems for deteriorating sludge settleability (increasing DSVI from 60 to 180 ml/gTSS) are shown in Figure 4.9. As expected, the ADWF capacity decreases qualitatively as sludge settleability deteriorates, but eqs 4.38 to 4.41 quantify this decrease.

If the SST are assigned an average depth of say 3.5 m then the total volume of the raw and settled WW AS-SST systems are 19,952 and 10,944 m^3, respectively, making the reactor volume 65% and 60% of the total system volume (Table 4.5). Figure 4.10 shows the ADWF capacity of the total AS-SST system volume for increasing the % volume assigned to the reactor and therefore decreasing the % volume assigned to the SST.

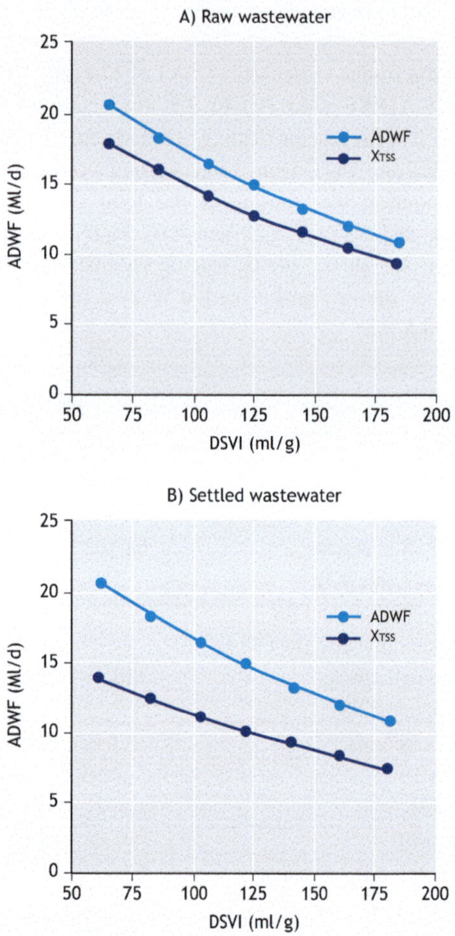

Figure 4.9 ADWF treatment capacity versus DSVI for the raw wastewater (WW) (A) and settled WW (B) AS systems given in Table 4.5.

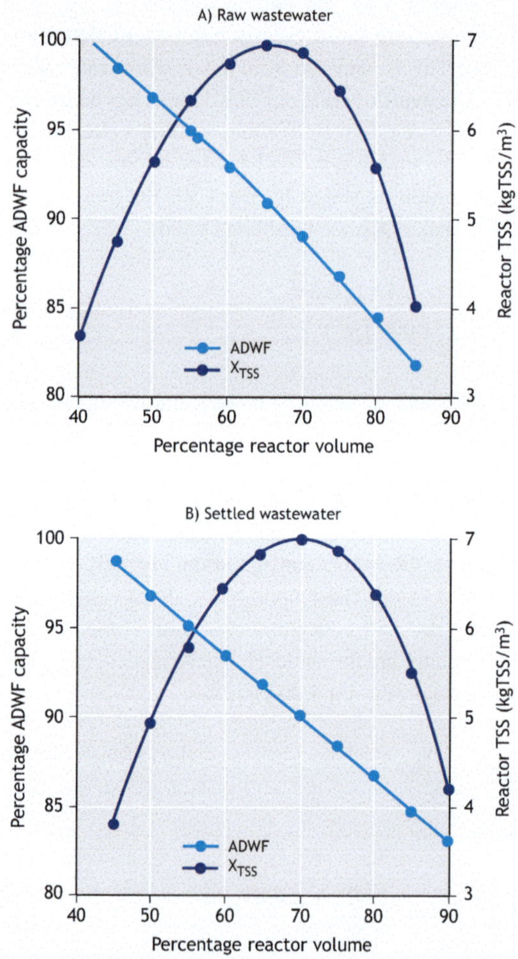

Figure 4.10 ADWF capacity, as a % of the maximum, and reactor TSS concentration for increasing % volume of the total AS-SST system volume assigned to the reactor for raw WW (A) and settled WW (B).

As the reactor occupies a larger and larger % of the total system volume, the reactor TSS concentration decreases and the SST area gets smaller. Figure 4.10 shows that for both raw and settled WW, within ±13%, *i.e.* over a range of 26% of the 65% and 60% for the total volume for the reactor volume, the ADWF capacity is above 95% of the maximum capacity. The reason for this is because the total cost of the AS-SST system is reasonably flat over quite a wide TSS concentration range (see Figure 4.6). This indicates that obtaining a precise reactor TSS concentration to determine the reactor volume and SST area is not so critically important; a TSS concentration reasonably close (within approximately ±0.5 $kgTSS/m^3$) to the optimum that minimizes the total system cost is sufficient. From Figure 4.6, it can be seen that as the cost of the SST increases relative to that of the reactor, this tolerance range for the reactor TSS concentration narrows.

For the reactor volume and SST areas in Table 4.5, the ADWF capacity for raw and settled WW for increasing SRT is shown in figures 4.11a and b. At very short SRT (< 5 d), the ADWF capacity is almost double that at long SRT (> 20). Hence selecting the shortest possible SRT ensures the highest treatment capacity for a given AS reactor volume and SST area. While for organics removal only, there are not well-defined quantified criteria to select the shortest possible system SRT, for ND systems there are. It will be shown in Chapter 5 that the SRT of ND systems is based on sustaining the nitrifiers in the AS system. The autotrophic nitrifiers are the slowest growing functional organisms in the NDEBPR system and therefore the SRT, and hence the reactor volume is governed by their maximum specific growth rate. Calculation of the shortest possible (minimum) SRT for ND (and NDEBPR) systems is therefore possible and is presented in chapters 5 and 6.

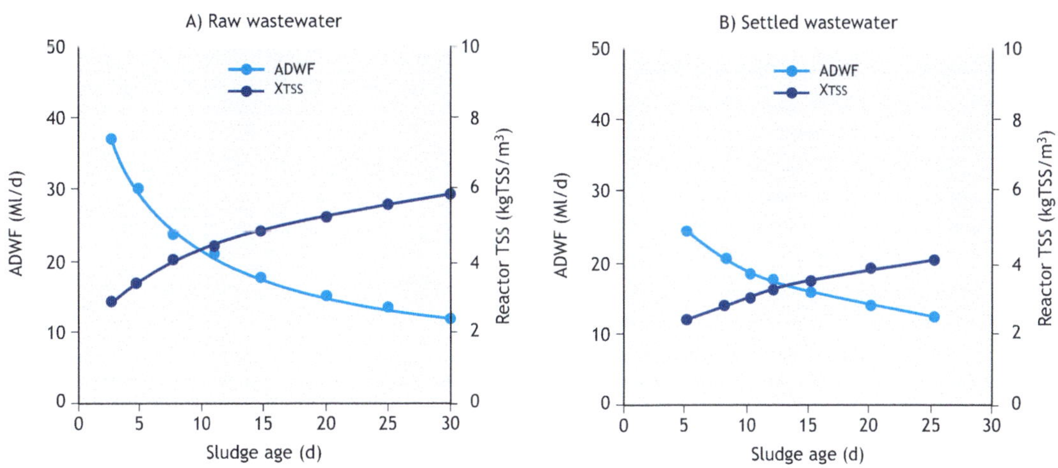

Figure 4.11 ADWF capacity in Ml/d and reactor TSS concentration for increasing SRT for the reactor volume and SST area given in Table 4.5 for raw WW (A) and settled WW (B).

4.12 SYSTEM DESIGN AND CONTROL

The parameter of fundamental importance in the design and control of the activated sludge system is the sludge age, which governs the mass of sludge to be wasted daily from the system. The sludge age can and should completely replace the food-to-microorganism ratio (F/M, kg BOD or COD load per day per kg MLSS or MLVSS in the reactor) or equivalently the load factor (LF) as a reference and control parameter, *in particular if nitrification is required.* The sludge age can be fixed by a simple control procedure if the system is appropriately designed. This control procedure is simpler and operationally more practical and reliable than procedures based on the F/M or LF, which basically seek to control the mass of sludge in the system by controlling the reactor MLSS concentration at some specified value.

4.12.1 System sludge mass control

By far the most common activated-sludge-system control procedure involves keeping the sludge MLSS concentration in the reactor at some specified value. At best this sludge concentration is specified by design or at worst, established from operational experience of the plant behavior, which is usually the concentration that can be contained in the system by the secondary settling tanks (SSTs). This approach does not control sludge age, only the mass of sludge in the system. In fact, in some instances, it may not even be the sludge mass that is controlled via the reactor concentration, but the settled volume at 30 min (SV_{30}) in the 1 l measuring cylinder. If the SV_{30} is greater than say 450 ml/l, then sludge is wasted until it reaches this value again. This approach was developed to obviate the need for measuring the reactor sludge concentration and with it, the sludge concentration in the reactor varied with the sludge settleability (SVI). This approach was acceptable before nitrification became obligatory and at least ensured that the sludge could be contained in the system while maintaining a low effluent suspended solids (ESS) concentration. However, with this method there is no control of the F/M, LF, sludge mass, reactor concentration or sludge age, a situation which is completely untenable when nitrification is required. While nitrification is a simple process to cater for in the design (just make the sludge age long enough and provide sufficient oxygen), it imposes a completely different control regime on the operation of the system. It requires the sludge age to be controlled at a fixed value.

If the F/M or LF is controlled, then in order to keep these parameters within the desired limits, not only does the reactor concentration need to be measured regularly, but also the daily BOD_5 (or COD) load. This requires extensive sampling and testing of the influent BOD_5 (or COD) concentration and flow pattern over the day to determine the daily COD (or BOD) *mass* load. Controlling the sludge age requires measuring the reactor MLSS concentration and the mass of sludge wasted per day. Usually the waste sludge is abstracted from the SST underflow to benefit from its thickening function. However, the sludge concentration of the underflow varies considerably over the day with the daily cyclic flow through the plant (figures 4.12 and 4.13).

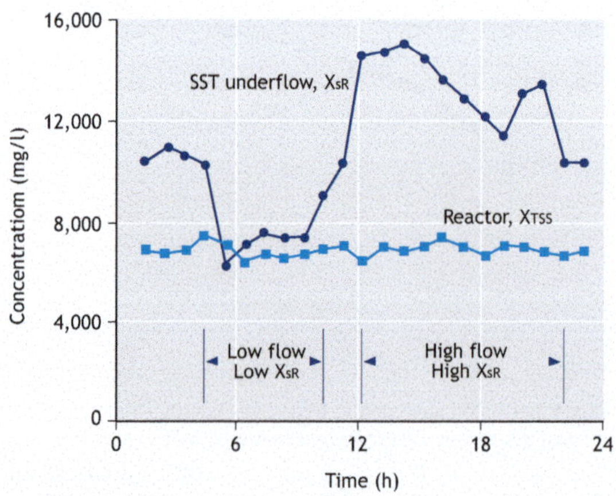

Figure 4.12 Experimental data from a full-scale activated sludge plant illustrating the virtually constant reactor concentration compared with the widely varying SST underflow concentration over the day (data: Nicholls, 1975).

Therefore, in order to know the sludge mass wasted via the underflow, it is necessary to measure the underflow concentration, waste flow rate and duration each time sludge is wasted. Thus to know the LF or sludge age, intensive testing of the influent and/or reactor and underflow concentrations are required. This may be manageable at large plants where the technical capacity is adequate, but at small plants, both the LF and sludge age are usually not known. As a consequence, nitrification is sporadic, partial or stops altogether during periods of poor sludge settleability, which results in high sludge wastage and hence short sludge age.

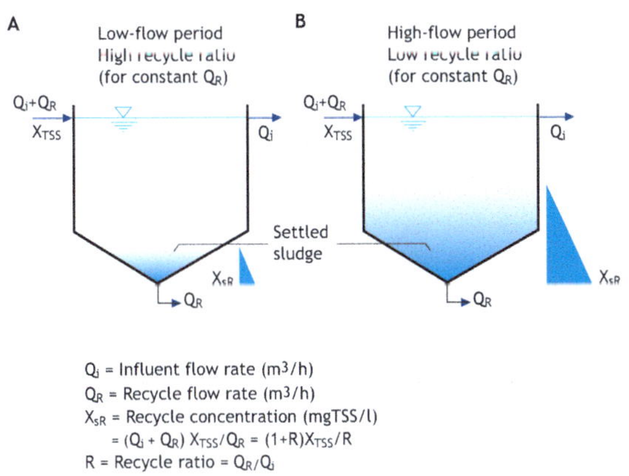

Figure 4.13 Increased sludge accumulation and higher underflow sludge concentration at high influent flow periods (B) than at low influent flow periods (A) at constant recycle flow rate.

Even if the reactor concentration were accurately controlled with modern control equipment such as automated wasting and on-line reactor concentration measurement, this still does not control the sludge age. With reactor concentration control at the same value throughout the year and a stable organic load on the plant (zero urban development), the sludge age decreases during winter because sludge production per kg COD load increases with decrease in temperature due to the lower endogenous respiration rate. While the decrease is relatively small, decreasing the sludge age is nevertheless the opposite of what should be done to the system in winter to keep the ammonia concentration low. This is particularly relevant to plants operated at sludge ages close to the minimum for nitrification (Chapter 5), which is common practice in developed countries to squeeze as much capacity as possible out of the plant because space for extensions is limited. If the reactor concentration is controlled and the organic load on the plant progressively increases, which is usually the case in developing countries where urban growth is often constrained by plant capacity, the sludge age decreases progressively with time (Figure 4.14). Inevitably, on one cold winter's day, nitrification will have stopped. The operators' response is that an industrial discharge has inhibited the nitrifiers and so they do not change their operating procedures and wait for nitrification to return, which it does when the wastewater warms again in the summer.

When nitrification is required, not only should sludge age be controlled, but also the SST can no longer serve the dual purpose of clarifier and waste activated sludge (WAS) thickener. To obtain high WAS concentrations, the underflow recycle ratio must be low (< 0.25:1), which results in long sludge residence times in the SST (Figure 4.8).

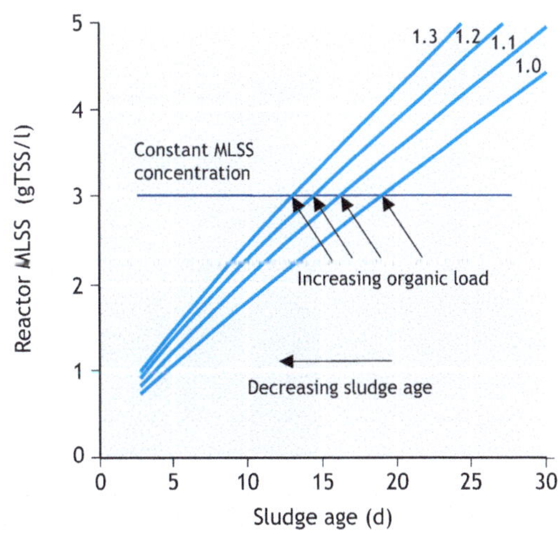

Figure 4.14 Decreasing SRT with increasing organics load when controlling reactor MLSS concentration.

The long sludge residence time stimulates denitrification in the SSTs causing floating (or rising) sludge on the SST surface, in particular in summer when wastewater temperature is high (> 20°C). In fact, in the tropics, the climatic region of most developing countries, it may not be possible to operate an activated sludge system that does not nitrify even at very short sludge ages. So the problem of rising sludge due to denitrification can take place in plants even where nitrification is not a requirement. This happened at the Brasilia wastewater treatment plant (Figure 4.15A and B), which had a low return sludge ratio (0.25:1): it nitrified even at 3 days sludge age and suffered from floating sludge all the time. If the sludge age was reduced to stop nitrification, the COD removal deteriorated below an acceptable level. So once nitrification takes place, whether intentionally by design or unavoidably, one must cater for denitrification in appropriate reactor (anoxic) zones and increase the underflow recycle ratio (~1:1) to minimize rising sludge in the SSTs due to denitrification.

Figure 4.15A One of two wastewater treatment plants in Brasilia, Brasil (photo: R. Brummer).

Figure 4.15B Rising sludge due to denitrification in the SST at one of the wastewater treatment plants in Brasilia, Brasil (photo: M. Henze).

Clearly, once nitrification takes place, whether as a requirement for N removal or unavoidably due to system conditions, one is forced to abandon using the SSTs as WAS thickeners. If one has to thicken WAS in a separate unit, whether from the underflow or reactor, one may as well waste sludge directly from the reactor and derive the significant operational benefit of hydraulic control of sludge age. It is simple, requires very little testing and establishes the sludge age almost exactly. It results in stable all-year-round nitrification and is strongly recommended for activated sludge systems where nitrification is required, even where sophisticated reactor concentration control measures can be applied.

4.12.2 Hydraulic control of sludge age

Hydraulic control of sludge age was first proposed and implemented in a generalized form by Garrett in 1958, and is based on a method of 'modified wastewater aeration' implemented by Setter *et al.* (1945). If a sludge age of 10 days is specified, 1/10th of the reactor volume is wasted daily, and if 20 d, 1/20th is wasted daily, *i.e.* $Q_w = V_R/SRT$ (Eq. 4.1). For plants with low levels of technical support, a satellite settling tank or a dewatering drying bed completely independent of the SSTs can be provided to which the daily WAS flow from the reactor is discharged; for plants with a higher level of technical support a dissolved air flotation unit would be best (Bratby, 1978), which also minimizes P release from EBPR sludges (Pitman, 1999). The supernatant is returned to the reactor and the thickened sludge is pumped to the sludge treatment/disposal part of the plant. This procedure establishes very closely the desired sludge age because the mixed liquor concentration does not change significantly over the day (Figure 4.12).

An important aspect about hydraulic control of the sludge age is that *irrespective of the flow through the plant,* if a fixed fraction of the volume of the reactor is wasted every day, the sludge age is fixed. If the COD mass load per day on the plant remains constant, the sludge concentration will automatically remain constant. If the COD mass load increases, the sludge concentration will increase automatically, to maintain the same sludge age. Thus, by monitoring the reactor concentration and its changes at a fixed sludge age, an indirect measure is obtained of the long-term changes in COD load on the plant. With time the reactor concentration may increase, indicating that the organic load on the plant is increasing (Figure 4.16). Hydraulic control of sludge age is very easy for the operator: (s)he just needs to check that the flume/pipe is not blocked and running at the correct flow rate, and the reactor MLSS concentration does not even have to be measured very often.

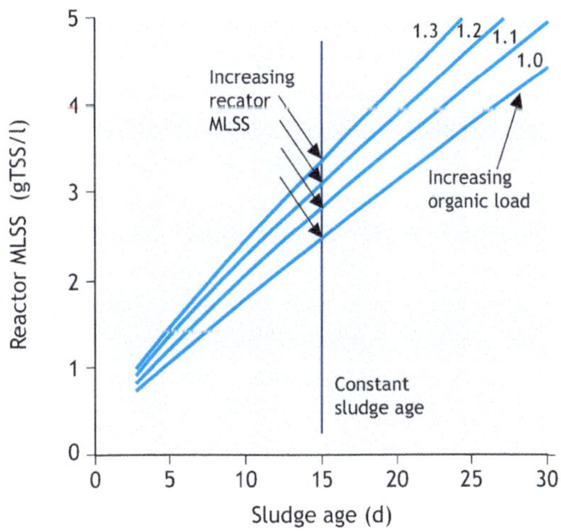

Figure 4.16 Constant SRT but increasing reactor MLSS concentration with hydraulic control of SRT.

By means of the hydraulic control procedure, the sludge age may be changed by simply changing the volume wasted per day. If say, the sludge age is reduced from 25 days to 20 days by hydraulic control, the full effect of the change will become apparent only after about half a sludge age. Thus, the biomass has an opportunity to adapt gradually to the change in F/M and LF. Hydraulic control of sludge age is particularly relevant to plants with sludge ages longer than about 5 days because for these plants the mass of sludge contained in the SSTs is a relatively small fraction of the total mass of sludge in the system. At sludge ages shorter than 5 days the mass of sludge in the SSTs can become appreciable with respect to the total mass of sludge in the system, particularly when the sludge settleability becomes poor (DSVI>150 ml/g). When the mass of sludge in the SSTs is significant, hydraulic control will have to take this into account and accuracy of the control will require additional testing.

Hydraulic control of sludge age devolves a greater responsibility onto the designer and removes responsibility from the plant operator; often operator ingenuity had to work around design inadequacies by force fitting the biological processes into the designed constraints to achieve the best effluent quality. It becomes essential that the designer calculates the sludge mass more exactly, to provide sufficient reactor volume under the design organic load to allow for the required reactor concentration at the specified sludge age. Also, the settling tank surface area, underflow recycle ratio and aeration capacity must be accurately sized for the particular wastewater and sludge age of the system. If these aspects are catered for adequately, then with hydraulic control of the sludge age, plant control is simplified and, for small-scale plants, may even do away with the requirements for solids and SVI tests except at long intervals. Hydraulic control of sludge age makes parameters such as LF and F/M redundant and introduces an entirely different attitude to system control. It is eminently practical and establishes the desired sludge age to ensure all-year-round nitrification. When nitrification is a requirement, sludge age control becomes a requirement, and then hydraulic control of sludge age is the easiest and most practical way to do this. Moreover, with hydraulic control of sludge age the mode of failure of the plant is completely different than with solids mass control. With the solids mass control the plant fails due to nitrification stopping and a high effluent ammonia concentration, a non-visible dissolved constituent which is also difficult to remove by other means. With sludge age control, the plant fails more obviously: sludge over the secondary settling tank effluent weirs. At plants managed with low levels of technical capacity, this is more likely to prompt remedial action.

4.13 SELECTION OF SLUDGE AGE

Selection of the sludge age is the most fundamental and important decision in the design of an activated sludge system. The sludge age selected for a plant depends on many factors, some of which are listed in Table 4.6 such as stability of the system, sludge settleability, whether or not the waste sludge needs to be suitable for direct discharge to drying beds, and most important of all, the quality of effluent required *i.e.* is COD removal only acceptable, must the effluent be nitrified, is nitrogen and phosphorus removal required. Several of the factors have already been discussed earlier and will not be repeated here. Only a few clarifying and additional comments on Table 4.6 will be made below.

4.13.1 Short sludge ages (1 to 5 days)

4.13.1.1 Conventional plants

These plants are operated in the conventional configuration *i.e.* a semi plug-flow configuration, but modified systems such as contact stabilization, step aeration, step feed and others are also implemented. Short sludge-age plants were extensively used in Europe and North America before N (and P) removal became requirements. Their main objective is COD removal only, for which sludge ages of 1 to 3 days are sufficient. BOD_5 or COD reductions range from 75 to 90%. The removal achieved depends on the wastewater characteristics, the operation of the plant in particular the management of the transfer of the sludge between the reactor and SSTs, and the efficiency of the SSTs. Because predatory activity of protozoan organisms on the free swimming bacteria is limited at short sludge ages, the non-settling component (or dispersion) of the activated sludge flocs is high which causes turbidity and high effluent COD (Chao and Keinath, 1979; Parker *et al.*, 1971).

Table 4.6 Some important considerations in the selection of sludge age for the activated sludge system.

Sludge age	Short (1to 5 days)	Intermediate (10 to15 days)	Long (>20 days)
Types	High rate, step feed, aerated lagoons, contact stabilization, pure oxygen	Similar to high rate but with nitrification and sometimes denitrification. BNR systems	Extended aeration Orbal Carousel BNR systems
Objectives	COD removal only	COD removal Nitrification Biological N removal and/or biological P removal	COD removal Biological N removal Biological P removal
Effluent quality	Low COD High ammonia High phosphate Variable	Low COD Low ammonia Low nitrate High/low phosphate Relatively stable	Low COD Low ammonia Low nitrate Low phosphate Usually stable
Primary settling	Generally included	Usually included	Usually excluded
Activated sludge quality	High sludge production Very active Stabilization required	Medium sludge production Quite active Stabilization required	Low sludge production Inactive No stabilization required
Oxygen demand	Very low	High due to nitrification	Very high due to nitrification and long sludge age
Reactor volume	Very small	Medium to large	Very large
Sludge settleability	Generally good, but bulking by non-low F/M filaments such as *S. natans*, 1701, *Thiothrix* possible.	Good at low sludge age and high aerobic mass fractions; but generally poor due to low F/M filament growth such as *M. parvicella*.	Can be good with high aerobic mass fractions, but generally poor due to low F/M filament growth particularly *M parvicella*
Operation	Very complex due to AS system variability and primary and secondary sludge treatment.	Very complex with BNR and 1^{ry} and 2^{ry} sludge treatment	Simple if without primary and secondary sludge treatment, but BNR system is complex.
Advantages	Low capital costs Energy self-sufficient with anaerobic digestion	Good biological N (and P) removal at relatively low capital cost.	Good biological N (and P) removal No primary and stable secondary sludge Low sludge handling costs
Disadvantages	High operation costs effluent quality variation	Complex and expensive sludge handling costs	Large reactor, high oxygen demand, high capital cost

It is accepted in Table 4.6 that short-sludge-age plants would not normally nitrify. For temperate and high latitude regions, where wastewater temperatures are generally below 20 °C, this would be the case. However, in tropical and low-latitude regions, where wastewater temperatures can exceed 25 to 30 °C, short-sludge systems would normally nitrify; in fact, it would be difficult to stop them doing so. For these situations, it is best to accept nitrification as inevitable and design the system accordingly. Furthermore, it would be advantageous to include a small primary anoxic zone (~15-20% anoxic mass fraction, see Chapter 5) in the system to denitrify a considerable proportion of the nitrate generated even if N removal is not required; this increases the minimum sludge age for nitrification, reduces oxygen demand, recovers alkalinity and reduces the risk of sludge flotation and high effluent COD due to denitrification on the SST bottom.

Bio-P removal is possible at short sludge ages of 3 to 5 days; the phosphate-accumulating organisms (PAOs) are relatively fast-growing heterotrophs. In the absence of nitrification, an unaerated zone would be anaerobic (*i.e.* no nitrate or oxygen present or entering it) and provided the readily-biodegradable (RB) COD and short-chain fatty acids (SCFAs) are available from the influent, biological excess P removal will take place. The original Phoredox system developed by Barnard (1976) is based on such a two-reactor anaerobic-aerobic system. The minimum sludge age for EBPR is temperature-dependent, increasing as temperature decreases and is approximately 3 to 5 days at 14 to 20 °C (Mamais *et al.*, 1992). At these temperatures, the minimum sludge age for nitrification is significantly longer than that for EBPR, so that nitrification would generally not take place with the result that the adverse effect of nitrate on the EBPR would be absent. However, in warmer climates the minimum sludge ages for nitrification and EBPR are similar, and ensuring a low nitrate recycle to the anaerobic reactor by including also anoxic zones is essential if EBPR is required (Burke *et al.* 1986). If EBPR is not required, the nitrification changes the two-reactor unaerated-aerated system from a P-removal one to an N-removal one.

Currently there is renewed interest in the 1st or A stage of the 2-stage or AB process, where the A stage is a high-rate activated-sludge (HRAS) system for organics removal (SRT<1.5 d) and the B stage a long sludge-age system for N removal (Haider *et al.*, 2006). This renewed interest has driven low energy wastewater treatment by minimizing energy consumption by aeration and maximizing organics capture as primary and/or secondary sludge for anaerobic digestion and energy recovery (Jimenez *et al.*, 2015; De Graaff *et al.*, 2016) and replacing the B stage by mainstream anammox for N removal (Zhang *et al.*, 2019).

4.13.1.2 Aerated lagoons

Aerated lagoons, different from aerated oxidation ponds where oxygenation is supplemented by algae, are essentially high-rate activated sludge systems because the oxygen demand is totally supplied by aerators. There are essentially two types of aerated lagoons, suspension mixed and facultative. Suspension-mixed aerated lagoons have sufficient energy input per unit volume by the aeration equipment to keep the sludge in suspension. In facultative lagoons this energy input is insufficient and settlement of solids onto the lagoon floor takes place. The biodegradable solids in the sludge layer so formed degrade anaerobically, as in an oxidation pond.

Kinetically, suspension-mixed lagoons are flow-through activated sludge systems, and can be modelled as such (Marais *et al.*, 2017). Their nominal hydraulic retention time equals their sludge age and the waste (Q_w) and effluent (Q_e) flows are one and the same, and equal to the influent flow (Q_i). Hence the volume of the aerated lagoon per unit COD load is very large compared with the conventional short sludge-age systems, which have hydraulic retention times approximately 1/20th of the sludge age.

The effluent from a suspension mixed aerated lagoon has the same constituents as the mixed liquor in the lagoon. The COD removed from the system via the oxygen demand is relatively small so that the COD in the effluent is generally unacceptable for discharge into receiving waters. In fact, the principal objective of all short-age plants is to act as biologically-assisted flocculators. This biologically transforms the influent soluble biodegradable organics to settleable organism mass, and enmeshes with this the influent biodegradable and unbiodegradable particulate organics to form a settleable sludge which allows effective liquid-solid separation. In conventional short-sludge-age plants, the waste sludge is transferred to the sludge treatment facility; in the aerated lagoon systems, the effluent (with the waste sludge) usually flows to a second pond, *i.e.* an oxidation pond or a facultative aerated lagoon, to allow the now readily settleable particulate material to settle to the lagoon floor to produce a relatively solids-free and low COD effluent. The sludge that accumulates on the tank floor undergoes anaerobic stabilization. Aerated lagoons find application principally as low-technology industrial waste treatment systems where organic strengths are high, the load varies seasonally and nitrification is not required. However, treating these waste waters in different types of anaerobic digestion systems is becoming more important to benefit from their better effluent quality, water re-use, energy recovery and reduced greenhouse-gas emissions.

4.13.2 Intermediate sludge ages (10 to 15 days)

Where nitrification is obligatory because of a strict effluent FSA concentration standard, this will govern the minimum sludge age of the activated sludge system. For nitrification, the sludge ages required are 5 to 8 times longer than those for COD removal only, depending on the temperature. In temperate regions where water temperatures can fall below 14 °C, the sludge age is not likely to be less than 10 to 15 days, taking due consideration of some unaerated zones in the reactor for denitrification (and biological P removal). In this range of sludge age, the effluent COD concentration no longer plays a role in the design. For sludge ages longer than approximately 4 days, protozoan organism predation of free swimming bacteria is high and flocculation good so particle dispersion is low. Also, virtually all soluble biodegradable organics are broken down, with the result that the effluent COD (or BOD) concentration remains approximately constant at its lowest achievable value, *i.e.* the unbiodegradable soluble COD concentration. The effluent ammonia concentration also plays a minor role in design because the nitrification kinetics are such that once nitrification is achieved, it is virtually complete provided sufficient oxygen is supplied. Even though the effluent standards may require an effluent ammonia concentration, say <10 mgFSA-N/l, once nitrification takes place the concentration is not likely to be greater than 2 to 4 mgN/l. Consequently, for nitrification the sludge age of the system is fixed principally by the requirement for nitrification. The method for calculating the minimum sludge age for nitrification is given in Chapter 5, Section 5.1.7. Once a sludge age of say 25% longer than the minimum is selected, the effluent FSA concentration is affected more by the system operating conditions than by the nitrification process itself, *i.e.* oxygen supply limitations, variation in ammonia load, uncontrolled loss of sludge, and pH of the mixed liquor. With low alkalinity wastewaters, nitrification can cause a significant reduction in effluent pH, often as low as 5. This not only causes problems with the nitrification process itself, *i.e.* non-compliance with the effluent ammonia standard, but also produces aggressive effluents which can do considerable damage to concrete surfaces. To reduce these problems and derive the other advantages of oxygen and alkalinity recovery (see below), the policy of deliberate biological denitrification is advocated whenever nitrification is likely, even if N removal is not required. However, once nitrification is required and biological denitrification is incorporated in the system, sludge ages longer than 10 to 15 days may be required and the system falls into the long-sludge-age category.

In nitrifying aerobic activated sludge plants, there is always the possibility of denitrification in the SST. This problem is exacerbated by the system control procedure of abstracting the waste sludge from the settling tank underflow (see Section 4.12.1). At low underflow recycle ratios, sludge retention in the SST is long which leads to denitrification (Figure 4.13). Henze *et al.* (1993) estimated that between 6-8 and 8-10 mgN/l nitrate needs to be denitrified to cause sludge flotation at 10 and 20°C, respectively. The concentration of nitrate denitrified increases as (*i*) sludge retention time in the SST increases, which is dependent on the recycle ratio and peak flow conditions, (*ii*) the active fraction of the sludge increases, *i.e.* is larger at shorter sludge ages (Figure 4.3B), (*iii*) temperature increases, and (*iv*) the mass of unutilized enmeshed biodegradable organics increases which is higher at shorter sludge ages and largest at the peak load condition (Ekama *et al.*, 1997).

The above demonstrates that for plants where nitrification takes place, the SST should not serve the dual purpose of solid-liquid separation and thickening of waste sludge, that hydraulic control of sludge age should be employed and that deliberate denitrification should be included in the system (see Section 4.12). These modifications will ameliorate the problem of sludge flotation by denitrification in the SST, but may not completely eliminate the root cause, *i.e.* high nitrate concentrations in the mixed liquor.

In order to reduce the construction cost of the activated sludge system, reductions in sludge age need to be made. Moreover, a reduction in sludge age also increases both biological N and P removal per mass organic load (WRC, 1984; Wentzel *et al.*, 1990) and this would be particularly beneficial for low-temperature wastewaters (10-15 °C) where nitrification is required. Reduction in SRT has been achieved in various ways such as adding nitrifier biofilm carriers to the aerobic reactor or selectively wasting the light fraction of the activated sludge with hydro-cyclones which enriches the activated sludge with the heavier fraction containing aerobic ammonia oxidizing bacteria (AOB), PAO and anaerobic ammonia oxidizing bacteria (AMX) (Ekama, 2015).

To try to reduce the sludge age required for nitrification, and hence the biological reactor volume per Ml of wastewater treated, internal fixed media have been placed in the aerobic reactor (Wanner *et al.*, 1988; Sen *et al.*, 1994; Ødegaard *et al.*, 2014). The nitrifiers that grow on these fixed media are not subject to the mixed liquor sludge age and aerobic mass fraction with the result that both can be reduced. However, the effectiveness of the internal fixed media has not been as good as expected, and they yield a rather low benefit/cost ratio.

Successful reduction of sludge age down to 8 to 10 days has been achieved with external nitrification (Bortone *et al.*, 1996; Sorm *et al.*, 1997; Hu *et al.*, 2000) and this system is starting to find application at full-scale (Vestner and Günthert, 2001; Muller *et al.*, 2006). With external nitrification, the nitrification process is removed completely from the suspended activated sludge and transferred to an external fixed medium system such as a trickling filter. With nitrification independent of the BNR activated-sludge mixed liquor, the sludge age can be reduced to around 8 to 10 d. Such a reduction reduces the biological reactor volume requirement per Ml of wastewater treated by approximately a 1/3rd without negatively impacting either biological N or P removal. Moreover, the sludge settleability improves significantly (DSVI~60-80 ml/g) compared with conventional BNR systems, which further increases the capacity of the system (Hu *et al.*, 2000).

Comparing intermediate sludge age plants with high rate plants, the oxygen demand per kg COD (including nitrification) is doubled (except with external nitrification, for which it is halved), the system volume is 3 to 4 times larger, the daily sludge mass wasted is reduced by 40%, and the active fraction is much lower. Intermediate-sludge-age plants are much more stable than high-rate plants, requiring less sophisticated control techniques or operator intervention (excepting external nitrification) thereby making these plants more suitable for general application.

At intermediate sludge ages, the active fraction of the waste sludge is still too high for direct discharge to drying beds. Consequently some form of waste sludge stabilization would need to be incorporated in the wastewater treatment plant, *e.g.* aerobic or anaerobic digestion. The former has the advantage of ease of operation and if operated at high MLSS concentration (> 2%) and with intermittent aeration, produces low N and P concentrations in the dewatering liquor (Mebrahtu *et al.*, 2008), but has the disadvantage of high energy costs for oxygen supply. The latter has the advantage of energy generation from the biogas but the disadvantage of complexity of operation and high N and P concentrations in the dewatering liquor. Even with energy recovery by anaerobic digestion of waste sludge, because of the low mass of sludge wasted from the activated sludge plant and high oxygen demand per kg COD load, energy self-sufficiency at intermediate sludge ages is not possible. However, in large plants (approximately

500,000 PE) where technical supervision and operator expertise are of a high level, energy costs can be reduced by gas production from anaerobic digesters and can probably be justified economically, particularly if energy costs continue to increase as they have over the past decade. Ekama (2009) found that the carbon dioxide emission originating from the influent organics from two widely-differing treatment plants treating the same wastewater is virtually the same if the residual biodegradable organics (COD) in the final disposed sludge is the same, *viz* (*i*) a long sludge-age (30 d, Table 4.6) extended-aeration activated sludge system treating raw wastewater and (*ii*) a short sludge-age (8 d, Table 4.6) activated sludge system treating settled wastewater with anaerobic digestion of primary sludge and aerobic or anaerobic digestion of wastewater activated sludge with beneficial combustion/flaring of methane gas.

4.13.3 Long sludge ages (20 days or more)

4.13.3.1 Aerobic plants

Long-sludge-age aerobic plants are usually called extended aeration plants. The principal objective of long sludge systems is to obviate primary (1^{ry}) and secondary (2^{ry}) sludge treatment. These plants therefore treat raw wastewater and the sludge age is chosen so that the active fraction (or residual biodegradable organics) of the waste sludge is sufficiently low to allow its direct discharge to sludge drying beds. The sludge age required to produce sludge sufficiently stable so as not to generate odour problems is uncertain and will depend on the temperature and climatic conditions, *i.e.* whether or not the sludge can be dried sufficiently quickly before it starts smelling, probably exceeding 30 days.

Interestingly, from a survey of the residual biodegradable organics in wastewater sludges treated by different sludge stabilization systems, Samson and Ekama (2000) found that aerobically-digested waste activated sludge contained the lowest residual biodegradable organics (10%) compared with wet-air oxidized (Zimpro) and anaerobically-digested primary sludges (25-60 %, Figure 4.17).

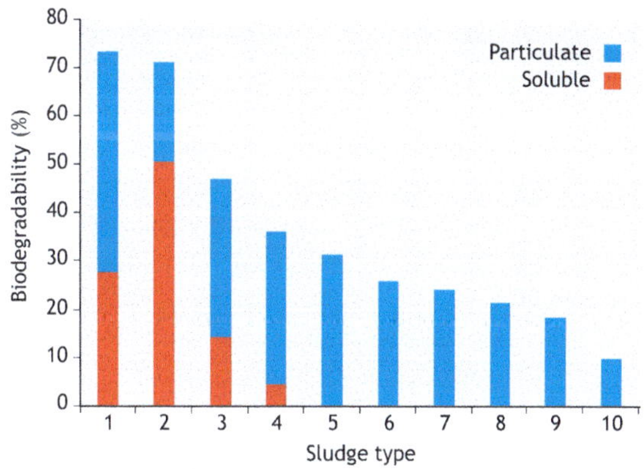

Figure 4.17 Percentage of residual biodegradable organics remaining in stabilised wastewater sludges treated with different stabilization system types. X-axis markers: (1) raw unsettled wastewater, (2) Zimpro humus + 1o - high soluble COD, (3) anaerobically-digested 1o + WAS - high VFA, (4) anaerobically-digested 1o only - high VFA, (5) anaerobically-digested 1o, 1st stage - low VFA, (6) Zimpro humus + 1o - low soluble COD, (7) anaerobically-digested 1o, 2nd stage - low VFA, (8) DAF thickened WAS, (9) anaerobically-digested 1o + WAS, single stage - low VFA and, (10) aerobically-digested WAS.

4.13.3.2 Anoxic-aerobic plants

Once the sludge age exceeds 20 to 25 days, nitrification is inevitable and it is advisable for the reasons cited above to incorporate denitrification in the system, which at these long sludge ages would not affect the stability of nitrification. Furthermore, if required, EBPR can also be included for little extra cost. In fact, biological N and P removal rates are significantly higher with raw wastewater than the settled wastewater due to the higher organic load. To include N (and P) removal, the reactor is subdivided into unaerated (anoxic and anaerobic) and aerated zones in a variety of configurations. Denitrification takes place in the unaerated but mixed zones receiving nitrified mixed liquor via recycles from the aerated zones to give what is referred to as nitrification denitrification (ND) systems. The ND systems include 4 stage Bardenpho, which incorporates primary and secondary anoxic reactors, Modified Ludzack Ettinger (MLE), which incorporates only a primary anoxic reactor, Orbal®, Carousel® and oxidation ditch systems in which the anoxic zones created are along different lengths of the same long channel reactor, or in intermittently decanted extended aeration (IDEA) systems. While incorporation of denitrification imposes some additional constraints on the design at long sludge age, these are minor provided the aeration capacity of the plant is sufficient to ensure efficient nitrification under all expected conditions (see Chapter 5).

4.13.3.3 Anaerobic-anoxic-aerobic plants

When the EBPR is required, an initial anaerobic reactor is included in the configuration that receives the influent wastewater but minimal oxygen and nitrate via the sludge recycles. For EBPR, assurance of a zero-nitrate discharge to the anaerobic zone is critical for achieving good P removal and is an additional constraint on the design when including EBPR in extended-aeration systems. The extent of EBPR achieved will depend on a number of factors, mainly the influent readily-biodegradable (RB) COD concentration, the TP/COD ratio and the degree to which nitrate can be excluded from the anaerobic reactor, which depends on the influent TKN/COD ratio.

The waste sludge from extended aeration systems including EBPR has the potential to release high P concentrations. This can be dealt with in specially designed dewatering/drying beds with sand filters under drains and weir overflows, which allow the drying bed also to operate as a dewatering system. While discharging waste sludge directly to the drying bed, the under drain and overflow are monitored for P concentration and when this gets to say 5 mgP/l, sludge wastage to the drying bed and the return of supernatant to the head of the works is stopped. The relatively small volume of high P liquor that drains from the drying bed thereafter is either chemically treated or irrigated at the plant site. The dewatering capability of the drying bed allows significantly more sludge to be discharged to it than drying beds without these dewatering features.

4.13.4 Dominant drivers for activated sludge system size

In the above section, some of the considerations for selection of the activated sludge system sludge age were set out because this is the most fundamental and important decision in its design. Sludge age is the main driver that governs effluent quality and size of the activated sludge system. Generally, the higher the effluent (and waste sludge) quality required from the system, the longer the sludge age, the larger the biological reactor and the more wastewater characteristics that need to be known (Figure 4.18).

For organic-matter removal only, the sludge age of the system is short and hence the reactor volume small. Essentially only the organic (COD) load and unbiodegradable particulate and soluble COD fractions need to be known. The organic load and unbiodegradable particulate COD concentration heavily affect sludge mass in the reactor and daily sludge production and the unbiodegradable soluble COD concentration fixes the filtered effluent COD concentration from the system. Also the organic load fixes the daily oxygen demand and the peak hydraulic load fixes the secondary settling tank surface area. If nitrification is required from the system, more wastewater characteristics are required to be known. The most important of these are the maximum specific growth rate of the nitrifiers at the standard wastewater temperature of 20 °C ($\mu_{ANO,20}$) and the minimum wastewater temperature (T_{min}), both of which fix the minimum sludge age for nitrification ($SRT_{min,NIT}$). The system sludge age (SRT) must be selected longer than the minimum for nitrification and the higher the system - to - minimum - sludge - age ratio ($SRT/SRT_{min,NIT}$), the lower the effluent ammonia concentration and the more damped its variation in response to nitrogen-load variation. Also required for nitrifying systems is the daily nitrogen load (both TKN and FSA) so that the components making up the N material in the influent can be determined. Note that for nitrification the maximum specific growth rate of the nitrifiers is regarded as a wastewater characteristic and not a model kinetic constant because it is different in different wastewaters.

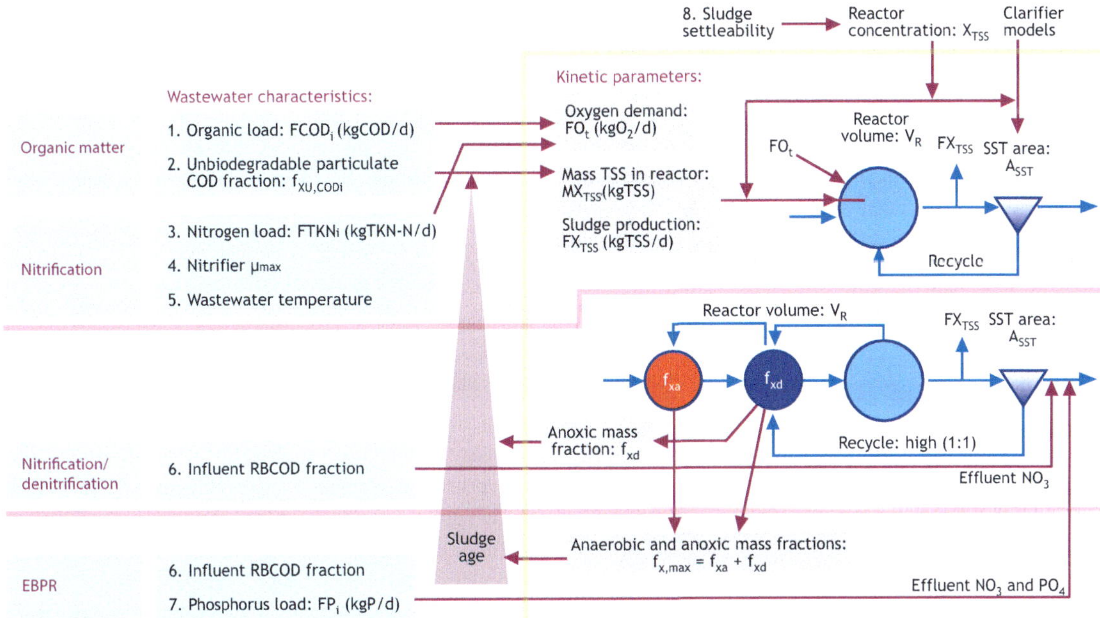

Figure 4.18 Important wastewater characteristics required to be known for different activated sludge systems: fully aerobic, nitrification, nitrification-denitrification and EBPR, and also the inter relationships that affect sludge age and effluent quality.

With biological nitrogen removal (nitrification and denitrification, ND), a part or the biological reactor (the anoxic mass fraction, f_{xd}) is intentionally not aerated. The larger the anoxic mass fraction, the more nitrate can be denitrified but the longer the minimum sludge age for nitrification becomes over that for fully aerobic conditions. Thus for ND systems, the biological reactor becomes larger because the required sludge ages become longer. Also an additional wastewater characteristic needs to be known, *i.e.* the influent RBCOD concentration because a high proportion (up to half) of the nitrate denitrified in the primary anoxic reactor is due to this wastewater constituent: if the influent RBCOD concentration is not known, the effluent nitrate concentration cannot be calculated accurately.

With EBPR, the daily wastewater phosphorus load (both total P and ortho-P) needs to be known so that the components making up the P material in the influent can be determined. With EBPR the influent RBCOD concentration is very important and establishes the extent of biological P removal that can be achieved. If the influent RBCOD concentration is not known, the biological P removal that can be achieved cannot be calculated accurately. The influent RBCOD is indirectly the food source for the phosphate-accumulating organisms (PAOs) that mediate the EBPR process. The purpose of the anaerobic zone, which receives the influent wastewater, is to allow the PAOs to take up the volatile fatty acid (VFA) fermentation products generated from the influent RBCOD. Nitrate (or dissolved oxygen, DO) which enters the anaerobic zone results in utilization of some of the influent RBCOD by ordinary heterotrophic organisms (OHOs), which reduces the VFA products available to the PAOs and hence the biological P removal. The difference between the influent P concentration and the EBPR that can be achieved establishes the effluent P concentration.

Very low nitrate (and DO) concentrations in the recycles entering the anaerobic zone are essential for maximum EBPR. This imposes important requirements on the denitrification required in the anoxic zones. If the influent TKN/COD concentration ratio is too high, then low nitrate concentrations cannot be achieved in the anoxic zone(s) and therefore methanol dosing may be required. High N removals in the anoxic zones requires large anoxic reactor(s), which together with the anaerobic zone, results in large unaerated mass fractions, which in turns requires long sludge ages to ensure nitrification. Unless specific strategies are applied to keep the sludge age low, such as external nitrification or adding fixed media into the aerobic zone to reduce the system sensitivity to the minimum sludge age for nitrification, NDEBPR systems will have long sludge ages, especially where wastewater minimum temperatures are low.

The above overview demonstrates that wastewater characteristic determination is the most important aspect of modelling wastewater treatment plants, whether using steady-state or dynamic-simulation models. Uncertainty in wastewater characteristics (and sludge settleability) results in a commensurate uncertainty in oxygen demand, sludge production, reactor volume and effluent quality. So uncertainty/sensitivity analyses should be applied to the wastewater characteristics rather than to the kinetic and stoichiometric parameters of the model(s). In fact, only rarely should the kinetic and stoichiometric parameters of the model be changed (except the maximum specific growth rate of nitrifiers which is regarded as a wastewater characteristic.) Fitting all the effluent quality concentrations, sludge production and oxygen demand to laboratory, pilot and full-scale plant data can be achieved by changing the wastewater characteristics only, provided the data conform to mass balances (water, COD, N and P). However, more often than not, model predictions cannot be made to conform to measured data because the measured data do not conform to mass balance and continuity principles. Only when the data conform to mass balance and continuity principles and changing the wastewater characteristics cannot yield a close correlation between the model predictions and the measured data, should kinetic and stoichiometric parameters of the model be changed, but such change(s) should be based on bioprocess basics and not simply because 'it makes the model fit'.

4.13.5 Some general comments

In BNR systems of any sludge age, aeration control is a particularly vexing problem under cyclic flow and load conditions, because the system is affected by either too high or too low DO concentrations in the aerobic zone. Too high DO concentrations are unnecessarily expensive and result in oxygen recycle to the anoxic (and the anaerobic zone if EBPR is included), thereby reducing the potential for N and P removal; too low DO concentrations cause nitrification efficiency to decline and possibly poor settling sludges to develop.

While some effective DO control systems have been developed over the years, the cost of providing aeration capacity and SST surface area for the peak flow has prompted research into alternative control solutions such as flow and load equalization. Furthermore, most of the diurnal variation in system variables such as ammonia, nitrate and phosphate concentration is not induced by the biological processes but by the hydraulic flow variation. To minimize hydraulic flow variation, an equalization tank is provided upstream of the activated sludge system and outflow from this tank is controlled in such a manner that the cyclic fluctuations in flow and load are damped to very small values. The tank is controlled by a microcomputer which calculates the tank outflow rate which best damps the projected inflow of the next 24 hours. This flow-equalization approach has been tested at the Goudkoppies BNR plant (Johannesburg, RSA) and showed great potential for reducing aeration and other control problems in nutrient removal plants (Dold *et al.*, 1982, 1984).

REFERENCES

Barnard J.L. (1976). A review of biological phosphorus removal in the activated sludge process. *Water SA,* 2(3), 136-144.

Bratby J. (1978). Aspects of sludge thickening by dissolved-air flotation. *Water Pollution and Control,* 77(3), 421-432.

Brink I.C., Wentzel M.C. and Ekama G.A. (2007). New developments in modelling waste water treatment processes - Using stoichiometry to build a plant wide mass balance based steady state WWTP model. *Procs. 10th IWA conference on Design, Operation and Economics of Large Wastewater Treatment Plants*. Vienna, 9-13 Sept, 243-250.

Bortone G., Saltarelli R., Alonso V., Sorm R., Wanner J. and Tilche A. (1996). Biological anoxic phosphorus removal - The DEPHANOX process. *Water Science and Technology,* 34(1/2), 119-128.

Burke R.A., Wentzel M.C., Dold P.L., Ekama G.A. and Marais, G.v.R. (1989). Biological excess phosphorus removal in short sludge age activated sludge systems. *Procs. 1st biennial WISA Conference*, Cape Town, March.

Chao A.C. and Keinath T.M. (1979). Influence of process loading intensity on sludge clarification and settling characteristics. *Water Research,* 13(12), 1213-1224.

De Graaff M.S., Van den Brand T.P.H., Roest K., Zandvoort M.H., Duin O. and Van Loosdrecht M.C.M. (2016). Full-Scale Highly-Loaded Wastewater Treatment Processes (A-Stage) to Increase Energy Production from Wastewater: Performance and Design Guidelines. *Environmental Engineering Sci*ence, 33(8), doi.org/10.1089/ees.2016.0022.

Dold P.L., Buhr H.O. and Marais G.v.R. (1982). Design and control of equalization tanks. *Research Report No. W 42*, Dept. of Civil Eng., Univ of Cape Town, Rondebosch, 7701, RSA.

Dold P.L., Buhr H.O. and Marais G.v.R. (1984). An equalization control strategy for activated sludge process control. *Water Science and Technology,* 17 (Amsterdam), 221-234.

Dick R.I. (1976). Folklore in the design of final settling tanks. *Journal WPCF,* 48(4), 633-644.

Ekama G.A., Dold P.L. and Marais G.v.R. (1986). Procedures for determining influent COD fractions and the maximum specific growth rate of heterotrophs in activated sludge systems. *Water Science and Technology,* 18(6), 91-114.

Ekama G.A. and Marais G.v.R. (1986). Sludge settleability and secondary settling tank design. *Water Pollution and Control,* 85(2), 100-113.

Ekama G.A., Barnard J.L., Günthert F.W., Krebs P., McCorquodale J.A., Parker D.S. and Wahlberg E.J. (1997). Secondary settling tanks: Theory, design, modelling and operation. IAWQ STR No 6, pp.216, International Association on Water Quality, London.

Ekama G.A. and Marais P. (2004). Assessing the applicability of the 1D flux theory to full scale secondary settling tank design with a 2D hydrodynamic model. *Water Research,* 38(3), 495-506.

Ekama G.A. and Wentzel M.C. (2004). A predictive model for the reactor inorganic suspended solids concentration in activated sludge systems. *Water Research,* 38(19), 4093-4106.

Ekama G.A. (2009). Using bio-process stoichiometry to build a steady state plant wide wastewater treatment plant model. *Water Research,* 43(8) 2101-2120.

Ekama G.A. (2017). Optimizing water and resource recovery facilities (WRRF) for energy recovery without compromising effluent quality. IWA NRR-LWWTP2017 conference - Sustainable wastewater treatment and resource recovery: Research, planning, design and operation. Chongqing, China, 6-9 Nov.

Ekama G.A. (2015). Developments in biological nutrient removal. *Water SA*, 41(4) 515-524.

Garrett M.T. (1958) Hydraulic control of activated sludge growth rate. *Sew. and Ind. Waste,* 30, 253.

Haider S., Svardal K., Vanrolleghem P.A. and Kroiss H. (2003). The effect of low sludge age on wastewater fractionation (SS, SI). *Water Science and Technology*, 47(11) 203–209.

Henze M., Dupont R., Grau P. and De La Sola A. (1993). Rising sludge in secondary settlers due to denitrification. *Water Research,* 27(2), 231-236.

Hörler A. (1969). Discussion of "Performance of (activated sludge) secondary settling tanks" by Pflanz P (1969) *Procs. 4th IAWPR Conference*, Prague, Ed. Jenkins SH, Pergamon Press, Oxford, 569-593.

Hu Z.R, Wentzel M.C. and Ekama G.A. (2000). External nitrification in biological nutrient removal activated sludge systems. *Water SA,* 26(2), 225-238.

Jimenez J., Miller M., Bott C., Murthy S., De Clippeleir H. and Wett B. (2015). High-rate activated sludge system for carbon management - Evaluation of crucial process mechanisms and design parameters. *Water Research*, 87, 476-482.

Korentajer L. (1991). A review of the agricultural use of sewage sludge: Benefits and potential hazards. *Water SA,* 17(3), 189-196.

Mamais D. and Jenkins D. (1992). The effects of MCRT and temperature on enhanced biological phosphorus removal. *Water Science and Technology,* 26(5/6),

955-965.

Marais G.v.R. and Ekama G.A. (1976). The activated sludge process part 1 - Steady state behaviour. *Water SA,* 2(4), 163-200.

Marais G.v.R., Ekama G.A. and Wentzel M.C. (2017). Application of the general activated sludge model to aerated lagoons. *Water SA*, 43(2), 238-257.

Mebrahtu M.K. and Ekama G.A. (2008). Aerobic digestion of waste activated sludge from biological nutrient removal activated sludge systems. *Procs 10th biennial Water Institute of Southern Africa conference*, Sun City, 18-21 May.

Metcalf and Eddy Inc. (2014). Wastewater Engineering: Treatment and resource recovery, 5th edn., McGraw-Hill, New York. ISBN 978-0-07-340118-8.

Muller A.W., Wentzel M.C. and Ekama G.A. (2006). Estimation of nitrification capacity of rock media trickling filters in external nitrification BNR. *Water SA,* 32(5), 611-618.

Musvoto E.V., Samson K., Taljard M., Fawcett K. and Alexander W.V. (2002). Calculation of peak oxygen demand in the design of full scale nutrient removal activated sludge plants. *Water SA,* (Special Edition: WISA Proceedings 2002) 56-60.

Nicholls H.A. (1975). Internal Thesis Report, Dept. of Civil Eng., Univ. of Cape Town, Rondebosch, 7701, RSA.

Ødegaard H., Christensson M. and Sørensen K. (2014). Chapter 15 - Hybrid systems. In: Activated Sludge 100 years – and counting. Eds: Jenkins D. and Wanner J., IWA publishing, London. ISBN13 9781780404936.

Parco V., Du Toit G.J.G. and Ekama G.A. (2018). Biological nutrient removal activated sludge with membranes. Chapter 2 in *Advances in wastewater treatment*, Eds. Mannina G., Ekama G.A., Ødegaard H. and Olson G. IWA Publishing. ISBN 9781780409702.

Parker D.S., Kaufman W.J. and Jenkins D. (1971). Physical conditioning of activated sludge floc. *Journal WPCF,* 43(9), 1817-1833.

Pincince A.B., Braley B.G., Sangrey K.H. and Reardon R.D. (1995). Minimizing costs of activated sludge systems: Application to the Deer Island treatment plant, *Procs. 68th Annual WEF Conference and Exposition*, Miami, 1, 693-699.

Pitman A.R. (1999). Management of biological nutrient removal plant sludges - Change the paradigms? *Water Research,* 33(5), 1141-1146.

Ramphao M.C., Wentzel M.C., Merritt R., Ekama G.A., Young T. and Buckley C.A. (2005). The impact of membrane solid-liquid separation on the design of biological nutrient removal activated sludge systems. *Biotechnology & Bioengineering*, 89(6), 630-646.

Riddel M.D.R., Lee J.S. and Wilson T.E. (1983). Method for estimating the capacity of an activated sludge plant. *Journal WPCF,* 55(4), 360-368.

Samson K.A. and Ekama G.A. (2000). An assessment of sewage sludge stability with a specific oxygen utilization rate (SOUR) test method. *Water Science and Technology,* 42(9), 37-40.

Setter L.R., Carpenter W.T. and Winslow G.C. (1945). Practical application of modified sewage aeration. *Sewage Works Journal,* 17(4), 669-691.

Sen D., Mitta P. and Randall C.W. (1994). Performance of fixed film media integrated in activated sludge reactors to enhanced nitrogen removal. *Water Science and Technology,* 30(11), 13-24.

Sorm R., Bortone G., Saltarelli R., Jenicek P., Wanner J. and Tilche A. (1996). Phosphate uptake under anoxic conditions and fixed film nitrification in nutrient removal activated sludge system. *Water Research,* 30(7), 1573-1584.

Sötemann S.W., Wentzel M.C. and Ekama G.A. (2006). Mass balances based plant wide wastewater treatment plant models - Part 4: Aerobic digestion of primary and waste activated sludges. *Water SA,* 32(3), 297-306.

Stofkoper J.A. and Trentelman C.C.M. (1982). Richtlijnen voor het dimensioneren van ronde nabezinktanks voor actiefslibinstallaties. *H_2O,* 15(14), 344-354.

Vestner R.J. and Gunthert F.W. (2001). Upgrading of trickling filters for biological nutrient removal with an activated sludge stage. *GWF - Water• Wastewater,* 142(15), 39-46.

Wahlberg E.J., Keinath T.M. and Parker D.S. (1994). The influence of flocculation time on secondary clarification. *Water Environmental Research,* 66(6), 779-786.

Wanner J., Kucman K. and Grau P. (1988). Activated sludge process with biofilm cultivation. *Water Research,* 22(2), 207-215.

Wentzel M.C., Dold P.L., Ekama G.A. and Marais G.v.R. (1990). Biological excess phosphorus removal - Steady state process design. *Water SA,* 16(1), 29-48.

Wentzel M.C., Ekama G.A. and Loewenthal R.E. (2003). Fundamentals of biological behaviour and wastewater strength measurement; Chapter 9 in *Handbook of water and wastewater microbiology*; pp. 131-157. Eds Mara D. and Horan N. Academic Press, Elsevier Science Ltd, London.

WRC (1984). Theory, *Design and Operation of Nutrient Removal Activated Sludge Processes.* Water Research Commission, Private Bag X03, Gezina, 0031, RSA.

Zhang M., Wang S., Ji B. and Liu Y. (2019). Towards mainstream deammonification of municipal wastewater: Partial nitrification-anammox *versus* partial denitrification-anammox. *Science of the Total Environment*, 692, 393-401.

NOMENCLATURE

Symbol	Description	Unit
a	Mixed liquor recycle ratio (Q_a /Q_i)	-
A_{SST}	Surface area of the secondary settling tank	m^2
b_{OHO}	Specific rate of endogenous mass loss of OHOs	d^{-1}
C_{br}	Bioreactor cost constant	-
$COD_{b,I}$	Influent biodegradable COD	mgCOD/l
COD_i	Influent total COD	mgCOD/l
COD_e	Effluent total COD	mgCOD/l
$COD_{e(filt)}$	Effluent soluble COD	mgCOD/l
$COD_{e(unfilt)}$	Effluent total COD	mgCOD/l
C_{sst}	Secondary settling tank cost constant	-
DSVI	Diluted sludge volume index	ml/gTSS
f_a	Fraction of OHOs in the activated sludge	mgVSS/mgVSS
$f_{at,OHO}$	Fraction of OHOs in the sludge as TSS	mgVSS/mgTSS
$f_{av,OHO}$	Fraction of OHOs in the sludge	mgVSS/mgVSS
$f_{cv,OHO}$	COD to VSS ratio of the sludge	mgCOD/mgVSS
$f_{FSS,OHO}$	Inorganic content of OHOs	mgISS/mgCOD
f_q	Peak flow factor (PWWF/ADWF)	l/l
$f_{SU,CODi}$	Soluble unbiodegradable fraction of total influent COD	-
f_{VT}	Ratio of VSS over TSS of the sludge	mgVSS/mgTSS
$f_{XU,CODi}$	Particulate unbiodegradable fraction of total influent COD	-
$f_{XE,OHO}$	Unbiodegradable fraction of the OHOs	mgCOD/mgCOD
$FCOD_{b,i}$	Daily flux of influent biodegradable COD	mgCOD/d
$FCOD_e$	Flux of effluent COD	mgCOD/d
$FCOD_i$	Daily flux of influent total COD	mgCOD/d
$FCOD_{VSS}$	Daily flux of particulate organic matter produced	mgCOD/d
FO_c	Daily flux of oxygen utilised	mgO_2/d
$FX_{FSS,i}$	Daily flux of influent particulate inorganic matter	mgISS/d
FX_{TSS}	Daily flux of total solids produced	mgTSS/d
$FX_{U,i}$	Daily flux of influent particulate unbiodegradable COD	mgCOD/d
$FX_{Uv,i}$	Daily flux of influent particulate unbiodegradable matter	mgVSS/d
FX_{VSS}	Daily flux of volatile solids produced	mgVSS/d
ISS	Inorganic suspended solids matter of activated sludge	mgISS/l
HRT_a	Actual hydraulic retention time	d

HRT_n	Nominal hydraulic retention time	d
MX_{OHOv}	Mass of OHO in the bioreactor	mgVSS
$MX_{E,OHOv}$	Mass of endogenous residue in the bioreactor	mgVSS
MX_{FSS}	Mass of influent particulate inorganic matter in the bioreactor	mgISS
MX_{TSS}	Mass of solids in the bioreactor	mgTSS
MX_{Uv}	Mass of influent unbiodegradable matter in the bioreactor	mgVSS
MX_{VSS}	Mass of volatile suspended solids in the bioreactor	mgVSS
O_c	Carbonaceous oxygen utilisation rate	mgO_2/l.d
P_{br}	Bioreactor power cost constant	-
P_{sst}	Secondary settling tank power cost constant	-
Q_a	Mixed liquor recycle flowrate	l/d
Q_e	Effluent flowrate	l/d
Q_i	Influent flowrate	l/d
$Q_{i,ADWF}$	Influent flowrate (for average dry weather)	l/d
$q_{i,PWWF}$	SST overflow rate at peak wet weather flow	m/h
$Q_{i,PWWF}$	SST flowrate at peak wet weather flow	m^3/h
Q_s	Sludge recycle flowrate	l/d
Q_w	Wastage flowrate from bioreactor	l/d
r_{hin}	Sludge settling constant	l/g
SRT	Sludge retention time	d
s	Sludge underflow recycle ratio (Q_s /Q_i)	-
S_S	Soluble (readily) biodegradable (RB) COD	mgCOD/l
$S_{U,e}$	Effluent (unbiodegradable) soluble COD	mgCOD/l
$S_{U,i}$	Influent soluble unbiodegradable COD	mgCOD/l
$SSVI_{3.5}$	Stirred specific sludge volume index at 3.5 g TSS/l	mL/gTSS
V_0	Initial settling velocity	m/h
V_R	Volume of bioreactor	l
V_s	Zone settling velocity	m/h
X	Particulate matter of activated sludge	mgTSS/l
X_{OHOv}	OHO biomass	mgVSS/l
$X_{E,OHOv}$	Endogenous residue from OHOs in activated sludge	mgVSS/l
$X_{FSS,i}$	Influent inorganics concentration	mgISS/l
X_{sR}	Suspended solids concentration in the sludge recycle from the SST	mgTSS/l
X_S	Slowly biodegradable (SB) particulate influent COD	mgCOD/l
$X_{S,i}$	Influent particulate unbiodegradable COD	mgCOD/l
X_{TSS}	Particulate matter of activated sludge	mgTSS/l
X_U	Influent unbiodegradable matter in activated sludge	mgCOD/l
X_{Uv}	Influent unbiodegradable matter in activated sludge	mgVSS/l
X_{VSS}	Organic matter of activated sludge	mgVSS/l
$X_{VSS,e}$	Effluent particulate volatile matter	mgVSS/l
Y_{OHO}	COD yield of OHOs	mgCOD/mgCOD
Y_{OHOv}	VSS yield of OHOs	mgVSS/mgCOD

Abbreviation	Description
ADWF	Average dry weather flow
AS	Activated sludge
BOD	Biological oxygen demand
BNR	Biological nutrient removal
COD	Chemical oxygen demand
DSVI	Diluted sludge volume index
DO	Dissolved oxygen
EBPR	Enhanced biological phosphorus removal
ESS	Effluent suspended solids
F/M	Food to microorganisms ratio
HRT	Hydraulic retention time
FSA	Free and saline ammonia
IDEA	Intermittently decanted extended aeration
ISS	Inorganic component of the settleable solids mass
LF	Load factor
MLSS	Mixed liquor suspended solids
MLVSS	Mixed liquor volatile suspended solids
OHOs	Ordinary heterotrophic organisms
ND	Nitrification-denitrification
PAOs	Phosphorus accumulating organisms
PE	Person equivalent
PST	Primary settling tank
PWWF	Peak wet weather flow
RBCOD	Readily biodegradable COD
SRT	Sludge retention time (sludge age)
SS	Suspended solids
SST	Secondary settling tank
SVI	Sludge volume index
SV	Settled volume
SSVI	Stirred sludge volume index
TKN	Total Kjeldahl nitrogen
TSS	Total suspended solids
VFAs	Volatile fatty acids
VSS	Volatile suspended solids
WAS	Waste activated sludge

Greek symbols	Explanation	Unit
$\theta_{b,OHO}$	Arrhenius temperature coefficient for the endogenous respiration rate of OHOs	-
φ	Secondary settling tank diameter	m

Figure 4.19 'The night shift' - microorganisms 'at work', WWTP Blue Plains, Washington DC (Photo: D. Rosso)

doi: 10.2166/9781789060362_0161

5
Nitrogen removal

George A. Ekama, Mark C. Wentzel† and Mark C.M. van Loosdrecht

5.1 INTRODUCTION TO NITRIFICATION

The term nitrification describes the biological process whereby free and saline ammonia (FSA) is oxidized to nitrite and nitrate. Nitrification is mediated by specific chemical autotrophic organisms with behavioural characteristics that differ significantly from heterotrophic (OHO) organisms. Whereas OHOs obtain their carbon (anabolism) and energy (catabolism) requirements for biomass synthesis from the same organic compound(s), autotrophic nitrifying organisms obtain their carbon requirement (anabolism) from dissolved CO_2 and their energy requirement (catabolism) for biomass synthesis from oxidizing ammonia to nitrite and nitrite to nitrate. This difference results in the autotrophic nitrifiers having much lower biomass growth coefficients (1/5th) than the OHOs. The objectives in this chapter are to review briefly the kinetics of nitrification, to highlight the factors that influence this biological process and to set out the procedure for designing a nitrifying aerobic activated sludge system. It has been well established that nitrification is due to two specific genera of autotrophic bacteria, ammonia-oxidizing organisms (AOOs) and nitrite-oxidizing organisms (NOOs). Originally it was thought that only Nitrosomonas and Nitrobacter mediate nitrification but recent molecular techniques have shown that there are several genera of nitrifying organisms.

Nitrification takes place in two sequential oxidation steps: (*i*) AOOs convert free and saline ammonia to nitrite, and (*ii*) NOOs convert nitrite to nitrate. The nitrifiers utilize ammonia and nitrite principally for synthesis energy requirements (catabolism) but some ammonia is also used anabolically for synthesis of cell mass nitrogen requirements. The ammonia requirement for synthesis, however, is a negligible fraction of the total ammonia nitrified to nitrate by the nitrifiers, at the most 1%. Consequently, in steady-state models it is usual to neglect the synthesis nitrogen requirements of the nitrifiers and to consider that the nitrifiers simply act as biological catalysts in the nitrification process. This stoichiometric approach greatly simplifies the description of the kinetics of the process. The two basic stoichiometric redox reactions in nitrification are:

$$NH_4^+ + \frac{3}{2}O_2(\text{AOOs}) \rightarrow NO_2^- + H_2O + 2H^+ \quad (5.1a)$$

$$NO_2^- + \frac{1}{2}O_2(NOOs) \rightarrow NO_3^- \quad (5.1b)$$

Stoichiometrically the oxygen requirements for the first and second reactions are 3/2 · 32/14 = 3.43 and 1/2 · 32/14 = 1.14 mgO_2/mgN (also written as $mgO_2/mgFSA\text{-}N$). Hence the stoichiometric conversion of ammonia to nitrate, both expressed as N, requires 2 · 32/14 = 4.57 mgO_2/mgN utilized. Taking into account the ammonia utilized for synthesis of nitrifier cell mass, the oxygen requirement per mg FSA-N nitrified is slightly less, with reported values down to 4.3 $mgO_2/mgFSA$. This approach is adopted in the simulation models such as ASM1 (Henze *et al.,* 1987) and is one reason for the small difference in the predicted results between steady-state stoichiometric models and more complex simulation models.

5.2 BIOLOGICAL KINETICS

5.2.1 Growth

In order to formulate nitrification behaviour it is necessary to understand the basic biological growth kinetics of AOOs. The rate of conversion of ammonia to nitrite by AOOs is generally much slower than that of nitrite to nitrate by NOOs. Therefore, under most circumstances in municipal wastewater treatment plants, any nitrite that is formed is converted virtually immediately into nitrate. As a consequence generally very little nitrite (<1 mgN/l) is observed in the effluent from a plant operating on an influent that does not contain substances that inhibit the NOOs. The limiting rate in the two-step nitrification sequence is therefore the ammonia conversion to nitrite by the AOOs. So from a steady-state modelling point of view, one only needs to consider the kinetics of this organism group. Because the nitrite produced is virtually immediately further nitrified to nitrate, it is assumed that the AOOs nitrify ammonia to nitrate directly and the kinetics of nitrification reduce to the kinetic behaviour of the AOOs.

Experimental investigations by Downing *et al.* (1964) showed that the nitrification rate can be formulated in terms of the Monod equation. In fact, Monod kinetics was applied to nitrification before it was applied to model the kinetics of organic material breakdown by heterotrophic organisms. The successful application to nitrification prompted Lawrence and McCarty (1972) to apply it to activated sludge. Monod established that (*i*) the mass of organisms generated is a fixed fraction of the mass of substrate (in this case ammonia) utilized and (*ii*) the specific rate of growth, i.e. the rate of growth per unit mass of organisms, is related to the concentration of substrate surrounding the organisms.

From (*i*):

$$M\Delta X_{ANO} = Y_{ANO} M\Delta S_{NHx} \quad (5.2)$$

where:

$M\Delta X_{ANO}$ mass of nitrifiers generated (mgVSS)
$M\Delta S_{NHx}$ mass of ammonia as utilized N (mgFSA-N)
Y_{ANO} nitrifier yield coefficient (mgVSS/mgN)

Taking the changes over a time interval t and assuming the changes are very small, one can write:

$$\frac{dX_{ANO}}{dt} = Y_{ANO}\left[-\frac{dS_{NHx}}{dt}\right] \quad \text{(mgANOVSS/l.d)} \quad (5.3)$$

From (*ii*) Monod developed the following relationship, known as the Monod equation,

$$\mu_{ANO} = \frac{\mu_{ANO,max} S_{NHx}}{K_{ANO,T} + S_{NHx}} \quad \text{(mgVSS/mgVSS.d)} \quad (5.4)$$

where:

μ_{ANO} specific growth rate at ammonia concentration (1/d)
$\mu_{ANO,max}$ maximum specific growth rate (mgANOVSS/mgANOVSS.d)
K_{ANO} half-saturation constant, i.e. the concentration at which μ_{ANO} = ½ $\mu_{ANO,max}$ (mgN/l)

S_{NHx} bulk liquid ammonia concentration (mgN/l)

The Monod constants maximum specific growth rate $\mu_{ANO,max}$ and half-saturation coefficient (also known as the affinity coefficient) K_{ANO} for the AOOs are sensitive to temperature, generally decreasing as temperature decreases. An additional subscript T on the symbols refers to temperature in °C.

The growth rate is given by the product of the specific growth rate and the ANO concentration (X_{ANO}):

$$\frac{dX_{ANO}}{dt} = \mu_{ANO,T} X_{ANO} = \frac{\mu_{ANO,max,T} S_{NHx}}{K_{ANO,T} + S_{NHx}} X_{ANO} \quad \text{(mgANOVSS/l.d) (5.5)}$$

The rate of ammonia conversion is found by combining eqs. 5.3 and 5.5, viz,

$$\frac{dS_{NHx}}{dt} = -\frac{1}{Y_{ANO}} \frac{\mu_{ANO,max,T} S_{NHx}}{K_{ANO,T} + S_{NHx}} X_{ANO} \quad \text{(mgFSA-N/l.d) (5.6)}$$

Because in the steady-state model the nitrification process is accepted to be stoichiometric, i.e. the nitrifying organisms act only as a catalyst to the process, the rate of nitrate formation is equal to the rate of FSA conversion, i.e.:

$$\frac{dS_{NO3}}{dt} = -\frac{dS_{NHx}}{dt} = \frac{1}{Y_{ANO}} \frac{\mu_{ANO,max,T} S_{NHx}}{K_{ANO,T} + S_{NHx}} X_{ANO} \quad \text{(mgNO}_3\text{-N/l.d) (5.7)}$$

where:

S_{NO3} nitrate concentration ($mgNO_3$-N/l)

The oxygen utilization rate associated with nitrification is based on the stoichiometric oxygen requirement of 4.57 mgO_2/mgFSA-N nitrified to nitrate calculated above, viz.

$$O_{NIT} = 4.57 \frac{dS_{NHx}}{dt} = 4.57 \frac{dS_{NO3}}{dt} \quad \text{(mgO}_2\text{/l.d) (5.8)}$$

However, assuming stoichiometric conversion of FSA to nitrate as in eqs. 5.7 and 5.8 above slightly overestimates the nitrate generation and oxygen consumption because a small proportion (1%) of the FSA taken up by the nitrifiers is used for cell synthesis. Based on the empirical organism cell mass formula $C_5H_7O_2N$, Brink *et al.* (2007) show that for 1 mgFSA-N taken up, 0.99 mgN nitrate and 0.076 mgANOVSS are generated and 4.42 mgO_2 are utilized.

Application of the Monod growth kinetics to nitrification by Downing *et al.* (1964) is probably one of the most successful applications of microbiological kinetic research in wastewater treatment; so much so that today Monod kinetics are commonly used to express the rates of many biological processes in terms of the growth-limiting nutrient concentrations. Monod growth kinetics require three constants to be known: the yield coefficient (Y_{ANO}), the maximum specific growth rate ($\mu_{ANO,max}$) and the half-saturation coefficient (K_{ANO}).

The yield coefficient for nitrifying organisms represents the net organism mass produced per unit mass of substrate nitrogen utilized. Evidence that this coefficient is not constant but can vary with the conditions of growth was presented in the 1960s when the nitrification model was developed. However, Downing *et al.* (1964) stated that the different VSS concentrations obtained from different Y_{ANO} values are inconsequential to the experimentally determined maximum specific growth rate, $\mu_{ANO,max}$, provided a consistent pair of $\mu_{ANO,max}$ and Y_{ANO} is used. This is because the $\mu_{ANO,max}$ is obtained from an observed maximum specific nitrification rate, $K_{NIT,max}$ mg FSA-N nitrified per mg ANOVSS per day, which is equal to $\mu_{ANO,max}/Y_{ANO}$. If Y_{ANO} is selected low, the $\mu_{ANO,max}$ will be low and vice versa. To avoid confusion about the experimentally determined $\mu_{ANO,max}$ rates, a standard Y_{ANO} = 0.10 mgVSS/mgFSA-N or 0.15

mgCOD/mgFSA-N has been adopted in steady-state and dynamic simulation activated sludge models for municipal wastewater treatment plants.

5.2.2 Growth behaviour

In Figure 5.1 the relationship between the specific growth rate, μ_{ANO}, the specific substrate (FSA) utilization or nitrification rate, K_{NIT}, and the bulk liquid FSA concentration, S_{NHx} is shown, as described by the Monod equation Eq. 5.4.

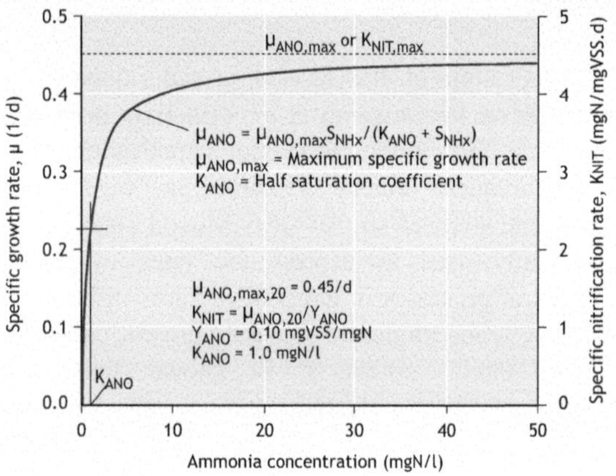

Figure 5.1 The Monod specific growth rate equation for nitrification at 20 °C.

The rate constants selected are $\mu_{ANO,max,20}$ = 0.45/d, Y_{ANO} = 0.10 mg ANOVSS formed per mg FSA-N nitrified, making $K_{ANO,max}$ = 4.5 mgFSA-N/mgANOVSS.d and $K_{ANO,20}$ = 1.0 mgN/l. The interesting feature of this nitrifier growth behaviour is that, because K_{ANO} is so low at ~1 mgFSA-N/l, the nitrification rate is virtually at maximum for concentrations >2 mgFSA-N/l. However, at concentrations < 2 mgN/l, the rate rapidly declines to zero. The implication of this is that when nitrification takes place, it will be nearly complete (provided all other requirements are met; see below) but the ammonia concentration is not readily reduced to zero.

5.2.3 Endogenous respiration

It is generally accepted that all organisms undergo some form of biomass loss due to maintenance or endogenous energy requirements. This behaviour manifests when a biomass has completely utilized its external substrate; its VSS decreases and it continues to utilize oxygen over time. This process is called endogenous respiration. Different organisms have different endogenous respiration rates. For OHOs, it is quite high ($b_{OHO,20}$ = 0.24 /d), whereas for AOOs, it is low ($b_{ANO,20}$ = 0.04 /d). The endogenous respiration process for AOOs is modelled in exactly the same way as that for OHOs, i.e.

$$\frac{dX_{ANO}}{dt} = -b_{ANO,T} X_{ANO} \qquad \text{(mgANOVSS/l.d)} \quad (5.9)$$

where:

$b_{ANO,T}$ specific endogenous mass loss rate for nitrifiers at T °C, (mgANOVSS/mgANOVSS.d).

5.3 PROCESS KINETICS

The basic activated sludge system modelled for nitrification is the single completely-mixed reactor system with hydraulic control of sludge age (see Figure 4.2). This system under steady-state conditions provides the information necessary for design of nitrification. The principal steady-state solution required for this is the effluent ammonia concentration ($S_{NHx,e}$). This solution forms the basis for the analysis of the nitrification process behaviour and provides the information for the design of an activated sludge system including this process. This information is also sufficient to understand the modelling of the nitrification process in activated sludge simulation models such as ASM1.

5.3.1 Effluent ammonia concentration

A mass balance on the change in nitrifier mass $M\Delta X_{ANO}$ over the completely mixed system at steady state (Figure 4.5) is given by:

$$M\Delta X_{ANO} = V_R \Delta X_{ANO} = \frac{\mu_{ANO,max,T} S_{NHx}}{K_{ANO,T} + S_{NHx}} \cdot$$
$$\cdot X_{ANO} V_R \Delta t - b_{ANO,T} X_{ANO} V_R \Delta t - X_{ANO} Q_w \Delta t$$
$$\text{(mgANOVSS)} \quad (5.10a)$$

where:

V_R reactor volume (l)

Q_w waste sludge flow rate from the reactor (l/d)

Dividing by $V_R \Delta t$ yields,

$$\frac{\Delta X_{ANO}}{\Delta t} = \frac{\mu_{ANO,max,T} S_{NHx}}{K_{ANO,T} + S_{NHx}} \cdot \quad (5.10b)$$
$$\cdot X_{ANO,T} - b_{ANO,T} X_{ANO} - \frac{Q_w}{V_R} X_{ANO}$$

Under steady-state (constant flow and load) conditions $\Delta X_{ANO} / \Delta t$ is zero and from Eq. 4.1, $Q_w / V_R = SRT$.

Substituting these and solving for the reactor ammonia concentration (S_{NHx}), and therefore also from the definition of completely mixed conditions, the effluent ammonia concentration ($S_{NHx,e}$), yields,

$$S_{NHx} = S_{NHx,e} = \frac{K_{ANO,T}(b_{ANO,T} + 1/SRT)}{\mu_{ANO,max,T} - (b_{ANO,T} + 1/SRT)}$$
$$\text{(mgN/l)} \quad (5.11)$$

From Eq. 5.11, the ammonia concentration (S_{NHx}) in the reactor and effluent ($S_{NHx,e}$) are independent of the specific yield coefficient (Y_{ANO}) and the influent ammonia concentration ($S_{NHx,i}$). Using $\mu_{ANO,max,20}$ = 0.33 /d and $K_{ANO,20}$=1.0 mgN/l at 20 °C, and taking $b_{ANO,T}$ = 0.04/d (Figure 5.1), a plot of Eq. 5.11 with $S_{NHx,e}$ versus sludge age SRT is given in Figure 5.2. At long sludge ages $S_{NHx,e}$ is very low and remains so until the sludge age is lowered to approximately 4 d. Below 4 d, $S_{NHx,e}$ increases rapidly and in terms of Eq. 5.11 can exceed the influent FSA concentration, $S_{NHx,i}$. This clearly is not possible so the limit of validity of Eq. 5.11 is S_{NHx} = $S_{NHx,i}$. Substituting $S_{NHx,i}$ for S_{NHx} in Eq. 5.11 and solving for SRT gives the minimum sludge age for nitrification, SRT_{min} below which, theoretically, nitrification cannot be achieved, i.e.:

$$SRT_{min} = \frac{1}{(1 + \frac{K_{ANO,T}}{S_{NHx,i}})\mu_{ANO,max,T} - b_{ANO,T}} \quad \text{(d)} \quad (5.12)$$

This minimum sludge age varies slightly with the magnitude of $S_{NHx,i}$ (Figure 5.2); higher $S_{NHx,i}$ gives a slightly lower SRT_{min}. The effect of $S_{NHx,i}$ on SRT_{min} is very small because the magnitude of $K_{ANO,T}$ is very small relative to $S_{NHx,i}$ (<5%). So for $S_{NHx,i}$ > 20 mgN/l (rarely will it be lower than this), and noting that $K_{ANO,20}$ ~1 mgN/l, then $K_{ANO,T}/S_{NHx,i}$ is negligibly small with respect to 1 (<5%). So substituting zero for $K_{ANO,T}/S_{NHx,i}$ in Eq. 5.12 yields,

$$SRT_{min} = \frac{1}{\mu_{ANO,max,T} - b_{ANO,T}} \quad \text{(d)} \quad (5.13)$$

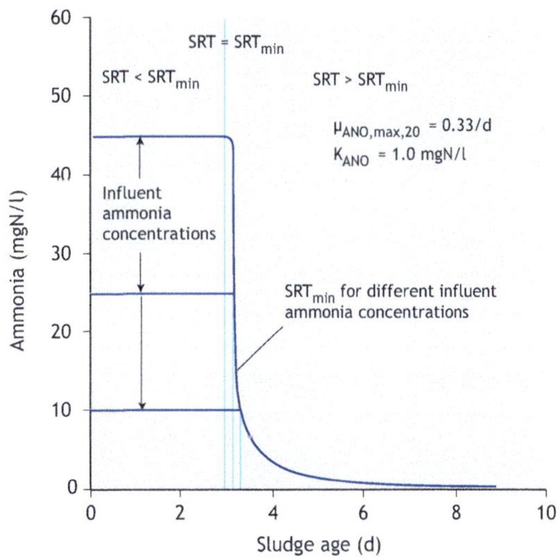

Figure 5.2 Effluent ammonia concentration versus sludge age for the steady-state nitrification model.

For all practical purposes, taking into account the uncertainty in $\mu_{ANO,max}$, Eq. 5.13 adequately defines the minimum sludge age for nitrification. Conceptually, Eq. 5.13 states that if the net nitrifier multiplication rate (inverse of the net maximum specific growth rate, $\mu_{ANO,max} - b_{ANO}$) is slower than the harvesting rate of the nitrifiers via the sludge waste flow rate, then the nitrifiers cannot be sustained in the system and nitrification cannot take place. At sludge ages lower than the minimum for nitrification, nitrifiers are 'washed out' of the system and so are called 'washout' sludge ages. This concept of 'washout' can be applied to any group of organisms in a bioreactor, and defines the sludge age below which the bio-process will not take place because the organisms mediating this process are not sustained in the system.

The virtually constant value for SRT_{min} insofar as the influent FSA concentration is concerned (for the fixed values of $\mu_{ANO,max,T}$ and $b_{ANO,T}$) and the rapid decrease in effluent FSA concentration at sludge ages slightly longer than SRT_{min} is due to the very low Monod half-saturation concentration for the nitrifiers ($K_{ANO,20}$). This feature means that in a particular plant, as the sludge age is increased, once SRT > SRT_{min}, a high efficiency of nitrification will be observed, provided the FSA is the growth-limiting nutrient for the AOOs, i.e. all other requirements such as oxygen are met. Consequently, under steady-state conditions with increasing sludge age, kinetically, one would expect an activated sludge system either not to nitrify at all, or, if it nitrifies, to nitrify virtually completely depending on whether the sludge age is longer or shorter than the minimum (SRT_{min}), respectively. Conversely, as sludge age decreases, one would expect an activated sludge system to nitrify virtually completely and then quite suddenly cease to nitrify, depending on whether the sludge age is shorter or longer than the minimum (SRT_{min}), respectively. This behaviour sometimes happens at full-scale activated sludge systems, where for many years the system has nitrified virtually completely, and suddenly one winter it stops nitrifying and produces very high effluent FSA concentrations. Provided the oxygen supply is not limiting, what happens in these situations is that over the years, the organic (COD) load on the system has increased and in order to maintain the reactor VSS concentration at the required level, the sludge wastage rate (Q_w) has been increased, which has reduced the sludge age. Then, coupled with a low winter temperature, the system sludge age falls below the minimum and nitrification ceases. This cannot happen with hydraulic control of sludge age, where a fixed proportion of the reactor volume is wasted daily to establish a constant sludge age. However, the secondary settling tank may become overloaded as the reactor TSS concentration increases with time, depending on the settleability with the activated sludge (see Chapter 4, Section 4.10). An operator therefore can choose the way an activated sludge system fails with increasing organic loading; it does not have to be with nitrification, and so also with N removal.

5.4 FACTORS INFLUENCING NITRIFICATION

From the discussion above, it can be seen that there are a number of factors that affect the nitrification process, the minimum sludge age required to achieve it, and the effluent FSA concentration from the activated sludge system, i.e.:

1. The magnitude of the kinetic 'constant' $\mu_{ANO,max,20}$ because this rate can vary considerably in different wastewaters.
2. Temperature because temperature decreases the $\mu_{ANO,max,20}$ rate and increases the $K_{ANO,20}$ coefficient.
3. Unaerated zones in the reactor because AOOs are obligate aerobes and can only grow under aerobic conditions.
4. Dissolved oxygen (DO) concentration because Monod kinetics presumes FSA is the growth-limiting nutrient implying that the oxygen supply must be adequate.
5. Cyclic flow and load conditions because FSA is dissolved and therefore the reactor (and effluent) FSA concentration is affected by the instantaneous actual hydraulic retention time;

most FSA not nitrified during the actual hydraulic retention time escapes with the effluent.

6. pH in the reactor because the $\mu_{ANO,max,20}$ is strongly suppressed by pH outside the 7 to 8 range.

These six factors are discussed further below.

5.4.1 Influent source

The maximum specific growth rate constant $\mu_{ANO,max,T}$ has been observed to be quite specific for wastewater and also to vary between different batches of the same wastewater source. This specificity is so marked that $\mu_{ANO,max,T}$ should not be classified as a kinetic constant but rather as a wastewater characteristic. The effect appears to be of an inhibitory nature due to some substance(s) in the influent wastewater. It does not appear to be a toxicity problem because a high efficiency of nitrification can be achieved even with a low $\mu_{ANO,max,T}$ value if the sludge age is increased sufficiently. These inhibitory substances are more likely to be present in municipal wastewater flows that have some industrial contribution. In general, the higher the industrial contribution, the lower $\mu_{ANO,max,T}$ tends to be, but the specific chemical compounds that cause the reduction of $\mu_{ANO,max,T}$ have not been clearly defined.

A standard temperature of 20 °C has been adopted for reporting $\mu_{ANO,max}$ rates to take into account the effect of temperature. A range of $\mu_{ANO,max,20}$ values have been reported from 0.30 to 0.75 /d for municipal wastewaters. These two limits will have a significant effect on the magnitude of the minimum sludge age for nitrification. Two systems, having these respective $\mu_{ANO,max,20}$ values, will have SRT_{min} values that differ by 250%. Clearly, due to the link between the sludge age and $\mu_{ANO,max,T}$, the latter's value should always be estimated experimentally for optimal design. In the absence of such a measurement, a low value for $\mu_{ANO,max,T}$ will necessarily need to be selected to ensure that nitrification takes place. If the actual $\mu_{ANO,max}$ is higher, the sludge age of the system will be longer and the reactor volume larger than necessary. However, the investment in the large reactor is not lost because in the future the plant will be able to treat a higher organic load at a shorter sludge age. Experimental procedures to determine $\mu_{ANO,max,20}$ are given in the literature, e.g. WRC (1984).

The $b_{ANO,20}$ rate is taken as constant for all municipal wastewater flows at $b_{ANO,20}$ = 0.04 /d. Its effect is small so that there is no need to enquire closely into all the factors affecting it. Little information on the effects of inhibitory agents on $K_{ANO,T}$ is available; it is very likely that $K_{ANO,T}$ will increase with inhibition.

5.4.2 Temperature

The $\mu_{ANO,max,T}$, $K_{ANO,T}$ and $b_{ANO,T}$ 'constants' are sensitive to temperature with high temperature sensitivity for the first two while the endogenous rate is accepted to have the same low temperature sensitivity as that for OHOs, viz.

$$\mu_{ANO,max,T} = \mu_{ANO,max,20}\theta_{NIT}^{(T-20)} \qquad (1/d)\ (5.14a)$$

$$K_{ANO,T} = K_{ANO,20}\theta_{NIT}^{(T-20)} \qquad (mgN/l)\ (5.14b)$$

$$b_{ANO,T} = b_{ANO,20}\theta_{b,ANO}^{(T-20)} \qquad (1/d)\ (5.14c)$$

where:

θ_{NIT} temperature sensitivity for nitrification = 1.123

$\theta_{b,ANO}$ temperature sensitivity for endogenous respiration for AOOs = 1.029

The effect of temperature on $\mu_{ANO,max,T}$ is particularly strong. For every 6 °C drop in temperature, the $\mu_{ANO,max,T}$ value halves which means that the minimum sludge age for nitrification doubles. Design of systems for nitrification, therefore, should be based on the minimum expected system temperature. The temperature sensitivity of $K_{ANO,T}$ is also strong, doubling for every 6 °C increase in temperature. This does not affect the minimum sludge age for nitrification, but it does affect the effluent FSA concentration; the higher the K_{ANO} value, the higher the effluent FSA at SRT >> SRT_{min}. However, the faster $\mu_{ANO,max,T}$ rate at the higher temperature compensates for the higher $K_{ANO,T}$ value so that the effluent FSA decreases with increase in temperature.

5.4.3 Unaerated zones

The effect of unaerated zones on nitrification can be formulated based on the following assumptions:

1. Nitrifiers, being obligate aerobes, grow only in the aerobic zones of a system.
2. Endogenous mass loss of the nitrifiers occurs under both aerobic and unaerated conditions.
3. The proportion of the AOOs in the VSS in the unaerated and aerated zones is the same so that the sludge mass fractions of the different zones of the system also reflect the distribution of the nitrifier mass.

From assumptions 1 to 3 above, it can be shown that if a fraction f_{xt} of the total sludge mass is unaerated, i.e. $(1 - f_{xt})$ is aerated; the effluent ammonia is given by

$$S_{NHx,e} = \frac{K_{ANO,T}(b_{ANO,T} + 1/SRT)}{\mu_{ANO,max,T}(1 - f_{xt}) - (b_{ANO,T} + 1/SRT)} \quad (5.15)$$

Equation 5.15 is identical in structure to Eq. 5.11, if one views the effect of the unaerated mass (f_{xt}) as reducing the value of $\mu_{ANO,max,T}$ to $\mu_{ANO,max,T}(1 - f_{xt})$, which conforms to assumptions 1 to 3 above. This sludge mass fraction approach is compatible with the nitrification kinetics in the activated sludge simulation models such as ASM1 and ASM2 (Henze *et al.,* 1987, 1995). In these models nitrifier growth takes place only in the aerobic zone and endogenous respiration in all the zones. This sludge mass fraction approach is not compatible with the aerobic sludge-age approach, which is used in some nitrification-denitrification activated sludge system (NDAS) design procedures (WEF, 1998; Metcalf and Eddy, 1991). In the aerobic sludge-age approach, it is assumed that the growth and endogenous processes of the nitrifiers are active only in the aerobic zone, with neither process active in the unaerated zone(s). This aerobic sludge age approach is not compatible with ASM1 and ASM2 simulation models and so significantly different predictions can be expected for the nitrification behaviour from the aerobic sludge-age-based design procedures and ASM models.

Following the same reasoning as that preceding Eq. 5.13, it can be shown that the minimum sludge age for nitrification SRT_{min} in a ND system having an unaerated mass fraction, f_{xt}, is

$$SRT_{min} = \frac{1}{\mu_{ANO,max,T}(1 - f_{xt}) - b_{ANO,T}} \quad (5.16)$$

Alternatively, if SRT is specified, then the minimum aerobic sludge mass fraction $(1 - f_{x,max})$ that must be present for nitrification to take place is found by substituting SRT for SRT_{min} and $f_{x,max}$ for f_{xt} in Eq. 5.16 and solving for $(1 - f_{x,max})$, i.e.

$$(1 - f_{x,max}) = (b_{ANO,T} + 1/SRT)/\mu_{ANO,max,T} \quad (5.17)$$

or equivalently, from Eq. 5.17, the maximum allowable unaerated sludge mass fraction at a sludge age of SRT is

$$f_{x,max} = 1 - (b_{ANO,T} + 1/SRT)/\mu_{ANO,max,T} \quad (5.18)$$

For a fixed sludge age, SRT, the design value for the minimum aerobic sludge mass fraction $(1-f_{x,max})$ should always be significantly higher than that given by Eq. 5.18, because nitrification becomes unstable and the effluent ammonia concentration increases when the aerated sludge mass fraction decreases to near the minimum value as given by Eq. 5.18. This situation is exacerbated by cyclic flow and ammonia load conditions (see below). Consequently, to ensure low effluent ammonia concentrations, the maximum specific growth rate of nitrifiers must be decreased by a factor of safety, S_f to give the minimum design aerobic sludge mass fraction; from Eq. 5.18,

$$(1-f_{x,max}) = (b_{ANO,T} + 1/SRT)/(\mu_{ANO,max,T}/S_f) \quad (5.19a)$$

The corresponding maximum design unaerated sludge mass fraction from Eq. 5.19a is

$$f_{x,max} = 1 - S_f(b_{ANO,T} + 1/SRT)/\mu_{ANO,max,T} \quad (5.19b)$$

With the aid of the temperature dependency equations for nitrification (Eq. 5.14), the maximum unaerated sludge mass fraction ($f_{x,max}$) from Eq. 5.19 is shown in Figure 5.3 for S_f = 1.25 and $\mu_{ANO,max,20}$ rates from 0.25 to 0.50 at 14 °C.

This shows that $f_{x,max}$ is very sensitive to $\mu_{ANO,max,T}$. Unless a sufficiently large aerobic sludge mass fraction ($1 - f_{x,max}$) is provided, nitrification will not take place and consequently nitrogen removal by denitrification is not possible. In fact, the selection of the maximum unaerated sludge mass fraction to achieve near-complete nitrification and the required degree of N removal is the single most important decision that is made in the design of the BNR activated sludge system because it defines the system sludge age and, for a selected reactor MLSS concentration, also the reactor volume.

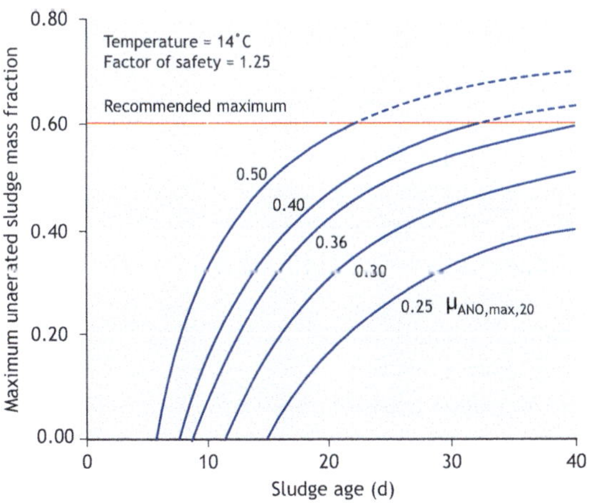

Figure 5.3 Maximum unaerated sludge mass fraction required to ensure nitrification versus sludge age for maximum specific growth rates of nitrifiers: $\mu_{ANO,max,20}$ of 0.25 to 0.50/d at 14 °C for S_f = 1.25.

From eqs. 5.15 and 5.19, it can be shown that for constant flow and ammonia load (i.e. steady state conditions)

$$S_{NHx,e} = K_{ANO,T}/(S_f - 1) \qquad (mgN/l)\ (5.20)$$

From Eq. 5.20, if S_f is selected at say 1.25 or greater at the minimum wastewater temperature, the effluent ammonia concentration ($S_{NHx,e}$) will be lower than 2 mgFSA-N/l at 14 °C for $K_{ANO,20}$ = 1.0 mgN/l. Although K_{ANO} is higher at higher temperature, $S_{NHx,e}$ will decrease with the increase in temperature because at constant sludge age, S_f increases with an increase in $\mu_{ANO,max,T}$. Consequently, for design the lower expected temperature should be selected to determine the sludge age and the aerobic mass fraction. If this is done, using say S_f = 1.25, then it can be accepted from Eq. 5.20 that the effluent ammonia concentration is below 2 mgN/l at the lowest temperature and around 1 mgN/l at 20 °C. In this way explicitly calculating $S_{NHx,e}$ with Eq. 5.15 is not necessary because provision for near-complete nitrification has been made by selection of S_f. Clearly selection of the $\mu_{ANO,max,20}$ and S_f values has major consequences for the effluent FSA concentration and economics of the ND-Nitrification and Denitrification activated sludge system.

5.4.3.1 Maximum allowable unaerated mass fraction

The equations above allow the two most important decisions in the design of an NDAS system to be made, the maximum unaerated sludge mass fraction and sludge age to ensure near-complete nitrification. Evidently from Figure 5.3, for $\mu_{ANO,max,20}$ > 0.50 the unaerated mass fraction at 14 °C can be as high as 0.7 at a sludge age of 40 days. Such a high unaerated mass fraction is apparently also acceptable at SRT = 10 days or longer at 20 °C.

However, there are additional considerations that constrain the unaerated mass fraction-sludge age selection:

1) Experience with laboratory-scale ND[a] (and NDEBPR[b]) systems have shown that at unaerated

[a] Nitrification – Denitrification

[b] Nitrification – Denitrification – Enhanced Biological Phosphorus Removal

mass fractions greater than 0.40, the filamentous bulking can become a problem, in particular at low temperatures (<16 °C). Systems with low unaerated mass fractions of <0.30 show greater tendency to have good settling sludges (Musvoto *et al.*, 1994; Ekama *et al.*, 1999; Tsai *et al.*, 2003).

2) In design of BNR plants for high N and P removal, the unaerated sludge mass fraction $f_{x,max}$ usually needs to be high (> 40%). If the $\mu_{ANO,max,20}$ value is low (< 0.40 /d, which will be the usual case in design where insufficient information on the $\mu_{ANO,max,20}$ is available), the necessary high $f_{x,max}$ magnitudes will be obtained only at long sludge ages (Figure 5.3). For example, if $\mu_{ANO,max,20}$ = 0.35 /d, then with S_f = 1.3 at T_{min} = 14 °C, an $f_{x,max}$ = 0.45 (Eq. 5.19b) gives a sludge age of 25 days and for $f_{x,max}$ = 0.55 a sludge age of 37 days. Long sludge ages require large reactor volumes; increasing SRT from 25 to 37 days increases the reactor volume by 40% whereas $f_{x,max}$ increases only by 22%. Also, for the same P content in the sludge mass, the P removal is reduced as the sludge age increases because the mass of sludge wasted daily decreases as the sludge age increases. Consequently, for low $\mu_{ANO,max,20}$ values, the increase in N and P removal that can be obtained by increasing the unaerated sludge mass fraction above 0.50 to 0.60 might not be economical due to the large reactor volumes this will require, and might even be counterproductive insofar as it affects P removal. A sludge age of 30 days is probably near the limit of economic practicality which for low $\mu_{ANO,max,14}$ = 0.16 values will limit the unaerated mass fraction to approximately 0.5. At higher $\mu_{ANO,max,14}$ values, the sludge ages allowing 50% unaerated mass fractions decrease significantly, again indicating the advantages of determining experimentally the value of $\mu_{ANO,max,20}$ to check whether a higher value is acceptable.
3) An upper limit to the unaerated mass fraction is evident also from experimental and theoretical modelling of the BNR system. Experimentally at 20 °C with SRT = 20 d, if $f_{x,max}$ > 0.70, the mass of sludge generated is found to increase sharply. Theoretically, this happens for $f_{x,max}$ > 0.60 at T = 14 °C and SRT = 20 days. The reason is that for such high $f_{x,max}$, the exposure of the sludge to aerobic conditions becomes insufficient to utilize the adsorbed and enmeshed particulate biodegradable organics. This leads to a decrease in active mass and oxygen demand and a build-up of enmeshed non-degraded organics. When this happens, the system still functions in that the COD is removed from the wastewater, but the degradation of the COD is reduced; the system begins to behave as a contact reactor of a contact-stabilization system, i.e. a bio-flocculation with minimal degradation. This critical state occurs at lower $f_{x,max}$ as the temperature is decreased and the sludge age is reduced.

From the discussion above it would appear that the unaerated mass fraction should not be increased above an upper limit of approximately 60%, as indicated in Figure 5.3, unless there is a very specific reason.

5.4.4 Dissolved oxygen concentration

High dissolved oxygen concentrations, up to 33 mgO_2/l, do not appear to affect nitrification rates significantly. However, low oxygen concentrations reduce the nitrification rate. Stenstrom and Poduska (1980) suggested formulating this effect as follows:

$$\mu_{ANO,O} = \mu_{ANO,max,O} \frac{S_{O2}}{K_O + S_{O2}} \qquad (1/d) \quad (5.21)$$

where:

S_{O2}	oxygen concentration in liquid (mgO_2/l)
K_O	half-saturation constant (mgO_2/l)
$\mu_{ANO,max,O}$	maximum specific growth rate at DO of O (1/d)
$\mu_{ANO,O}$	specific growth rate at DO of O (1/d)

The value of K_O ranges from 0.3 to 2 mgO_2/l, i.e. at DO values below K_O, the growth rate will decline to less than half the rate where oxygen is present in adequate concentrations. The wide range of K_O probably has arisen because the concentration of DO in the bulk liquid is not necessarily the same as inside the biological floc where the oxygen consumption takes place. Consequently, the value will depend on the floc size, mixing intensity and oxygen diffusion rate into the floc. Furthermore, in a full-scale reactor the DO will vary over the reactor volume due to the discrete points of oxygen input (with mechanical aeration) and the impossibility of achieving instantaneous and complete mixing. For these reasons it is not really possible to establish a generally applicable minimum oxygen value; each reactor will have a value specific to the conditions prevailing in it. In nitrifying reactors with bubble aeration, a popular DO lower limit, to ensure unimpeded nitrification, is 2 mgO_2/l at the surface of the mixed liquor.

Under cyclic flow and load conditions the difficulties of ensuring an oxygen supply matching the oxygen demand and a lower limit for the DO are compounded (Chapter 4, Section 4.8.2). Where storm flows are not of long duration, flow equalization is a practical way to facilitate control of the DO concentration in the reactor. In fact, most of the diurnal variation in the dissolved concentrations in a reactor is a direct consequence of diurnal flow variation; negligibly little is due to the kinetic rates of the biological processes, especially at long sludge ages. In the absence of flow equalization, amelioration of the adverse effects of low DO concentration during peak oxygen-demand periods is by increasing the sludge age to significantly longer than the minimum necessary for nitrification, i.e. in effect increasing S_f.

5.4.5 Cyclic flow and load

It is well known both experimentally and theoretically with simulation models, that under cyclic flow and load conditions the nitrification efficiency of the AS system decreases compared with that under steady-state conditions. From simulation studies, during the high flow and/or load period, even though the nitrifiers are operating at their maximum rate, it is not possible to oxidize all the ammonia available, and an increased ammonia concentration is discharged in the effluent. This in turn reduces the mass of nitrifiers formed in the system. Equivalently, the effect of diurnal variation in flow and load is to reduce the system sludge age. The average effluent ammonia concentration from a system under cyclic flow and load conditions is therefore higher than that from the same system under constant flow and load (steady-state conditions). The adverse effect of the diurnal flow variation becomes more marked as the fractional amplitudes of the flow and load variation increase, and is ameliorated as the safety factor S_f increases. Simulation studies of the diurnal flow effect show a relatively consistent trend between the maximum and average effluent FSA concentrations under diurnal conditions as a ratio of the steady-state effluent FSA concentration and the system sludge age as a ratio of the minimum sludge age for nitrification (SRT/SRT_{min}). For $\mu_{ANO,max,20}$ = 0.45 /d (other constants in Figure 5.1), figures 5.4 and 5.5 show the maximum (average not shown) effluent FSA concentration as a ratio of the steady-state effluent FSA concentration versus the system sludge age as a ratio of the minimum sludge age for nitrification (SRT/SRT_{min}) for a single reactor fully aerobic system receiving cyclic influent flow and FSA load as in-phase sinusoidally varying flow and ammonia concentration, both with amplitudes of 0.25, 0.50, 0.75, 1.00 and 0.0 (steady state) at 14 °C (Figure 5.4) and 22 °C (Figure 5.5). For example, at 14 °C (Figure 5.4) if the system sludge age is twice the minimum for nitrification, the maximum effluent FSA concentration is eight times the steady-state value. From Figure 5.4, the latter is 0.8 mgN/l so the maximum is $8 \cdot 0.8 = 6.4$ mgN/l.

From figures 5.4 and 5.5, clearly the greater the diurnal flow variation and the lower the temperature, the higher the maximum (and average) effluent ammonia concentrations. This can be compensated by increasing S_f, which has the effect of increasing the sludge age or decreasing the unaerated mass fraction of the system. This obviously has an impact on the effluent quality and/or economics of the system.

The importance of the selection of $\mu_{ANO,max}$ cannot be over-emphasized. If the value of $\mu_{ANO,max}$ is selected higher than the actual value, even with a safety factor S_f of 1.25 to 1.35, the plant is likely to produce a fluctuating effluent ammonia concentration, with reduced mean efficiency in nitrification. Hence conservative estimates of $\mu_{ANO,max}$ (low) and S_f (high) are essential for ensuring nitrification and low effluent ammonia concentration.

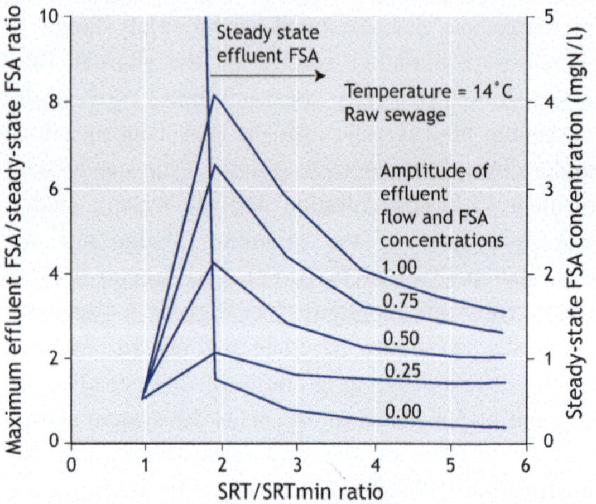

Figure 5.4 Maximum to steady-state effluent FSA concentration ratio versus sludge age to minimum sludge age for nitrification ratio for influent flow and ammonia concentration amplitude (in phase) of 0.0 (steady state) 0.25, 0.50, 0.75 and 1.0 at 14 °C.

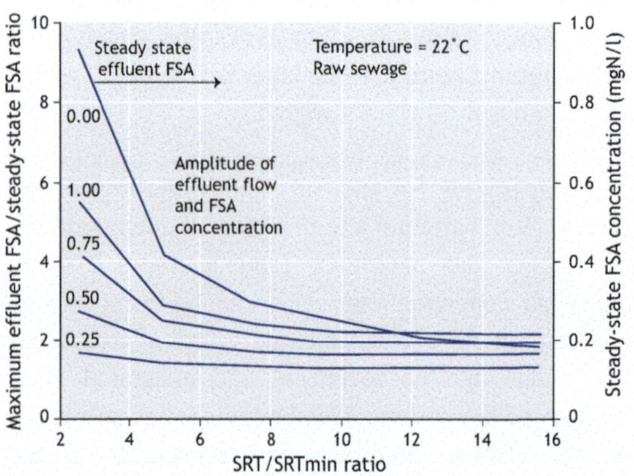

Figure 5.5 Maximum to steady-state effluent FSA concentration ratio versus sludge age to minimum sludge age for nitrification ratio for influent flow and ammonia concentration amplitude (in phase) of 0.0 (steady-state) 0.25, 0.50, 0.75 and 1.0 at 22 °C.

5.4.6 pH and alkalinity

The $\mu_{ANO,max}$ rate is extremely sensitive to the pH of the mixed liquor when outside the 7-8 range. It seems that the activities of both the hydrogen (H^+) and hydroxyl (OH^-) ions act as inhibitors when their respective concentrations increase too much. This happens when the pH increases above 8.5 (increasing (OH^-)) or decreases below 7 (increasing (H^+)); optimal nitrification rates are expected for $7 < pH < 8.5$ with sharp declines outside this range.

From the overall stoichiometric equations for nitrification (Eq. 5.1a), nitrification releases hydrogen ions which in turn decreases alkalinity of the mixed liquor. For every 1 mg FSA-N that is nitrified, $2 \cdot 50/14 = 7.14$ mg alkalinity (as $CaCO_3$) is consumed. Based on the equilibrium chemistry of the carbonate system (Loewenthal and Marais, 1977), equations linking the pH with alkalinity for any dissolved carbon dioxide concentration can be developed. These relationships are shown plotted in Figure 5.6.

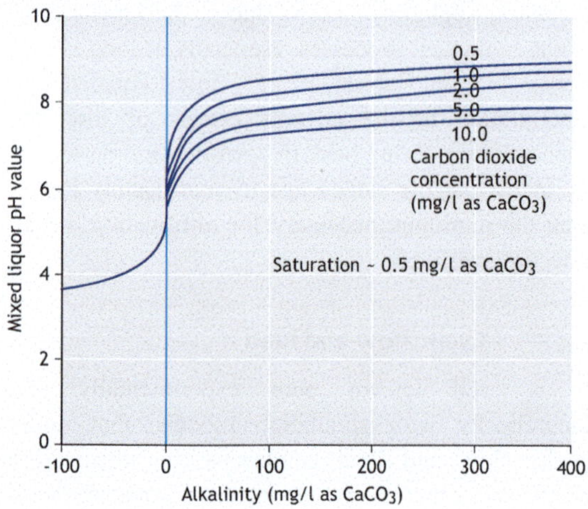

Figure 5.6 Mixed liquor pH versus alkalinity for different concentrations of carbon dioxide.

When the alkalinity falls below approximately 40 mg/l as $CaCO_3$ then, irrespective of the carbon dioxide concentration, the pH becomes unstable and decreases to low values. Generally, if nitrification causes the alkalinity to drop below approximately 40 mg/l (as $CaCO_3$), problems associated with low pH will arise at a plant, such as poor nitrification efficiency, effluents aggressive to concrete, and the possibility of development of bulking (poor settling) sludges (Jenkins *et al.,* 1993).

For any particular wastewater, the effect of nitrification on pH can be readily assessed, as follows. For example, if a wastewater has an alkalinity of 200 mg/l as $CaCO_3$ and the expected production of nitrate is 24 mgN/l, then the expected alkalinity in the effluent will be $(200 - 7.14 \cdot 24) = 29$ mg/l as $CaCO_3$. From Figure 5.6, such an effluent will have a pH < 7.0.

Wastewaters with low alkalinity are often encountered where the municipal supply is drawn from areas underlain with sandstone. A practical approach to treating such wastewaters is to (*i*) dose lime, or better (*ii*) create an anoxic zone(s) to denitrify some or the entire nitrate generated. In contrast to nitrification, denitrification takes up hydrogen ions, which is equivalent to generating alkalinity. By considering nitrate as an electron acceptor, it can be shown that for every mg nitrate denitrified, there is an increase of $1 \cdot 50/14 = 3.57$ mg alkalinity as $CaCO_3$. Hence incorporating denitrification in a nitrification system causes the net loss of alkalinity to be reduced usually sufficiently to maintain the alkalinity above 40 mg/l and consequently the pH above 7. In the example above, where the alkalinity in the system is expected to decline to 29 mg/l as $CaCO_3$, if 50% of the nitrate were denitrified, the gain in alkalinity will be $(0.5 \cdot 24 \cdot 3.57) = 43$ mg/l as $CaCO_3$ and will result in an alkalinity of $(29 + 43) = 72$ mg/l as $CaCO_3$ in the system. In this event the pH will remain above 7. For low alkalinity wastewaters it is imperative, therefore, that denitrification is built into nitrifying plants, even if N removal is not required. Incorporation of unaerated zones in the system influences the sludge age of the system at which nitrification takes place so that cognizance must be taken of the effect of an anoxic or unaerated zone in establishing the sludge age of a nitrifying-denitrifying plant (Section 5.4.3 above).

In the activated sludge systems treating reasonably well buffered wastewaters, quantifying the effect of pH on nitrification is not critical because pH reduction can be limited or completely obviated by including anoxic zones, thereby ensuring alkalinity recovery via denitrification. However, in poorly buffered wastewaters, or wastewaters with high influent N (such as anaerobic digester liquors), the interaction between the biological processes, pH and nitrification is the single most important factor for the N removal activated sludge system. Hence, it is essential to include the effect of pH on the nitrification rate for such wastewaters to quantify this important interaction.

From Eq. 5.4, the specific growth rate of the AOOs (μ_{ANO}) is a function of both $\mu_{ANO,max}$ and K_{ANO}. It was shown above that the minimum sludge age is dominated by the magnitude of $\mu_{ANO,max,T}$; it is only very slightly influenced by $K_{ANO,T}$. At SRT >> SRT_{min}, the effluent ammonia concentration ($S_{NHx,e}$), although low, is relatively speaking significantly higher for larger $K_{ANO,T}$ values. For example, if $K_{ANO,T}$ increases by a factor of two, the effluent ammonia concentration will increase correspondingly by the same factor (Eq. 5.15). Consequently the value of $K_{ANO,T}$ is significant insofar as it governs the effluent ammonia concentration once nitrification takes place at SRT >> SRT_{min}.

Several investigations have been done to understand the effect of pH on $\mu_{ANO,max,T}$. These investigations generally have not separated out the effect of $\mu_{ANO,max,T}$ and $K_{ANO,T}$ so that most data are in effect combined parameter estimates of $\mu_{ANO,max,T}$. Almost no information is available on the effect of pH on $K_{ANO,T}$ by itself. Quantitative modelling of the effect of pH on $\mu_{ANO,max}$ has been hampered by the difficulty of accurately measuring the effects of pH on nitrification. Studies have shown that $\mu_{ANO,max}$ can be expressed as a percentage of the highest value at optimum pH. Accepting this approach and that

$\mu_{ANO,max}$ is highest and remains approximately constant in the pH range for 7.2 < pH < 8.0 but decreases as the pH decreases below 7.2 (Downing *et al.,* 1964; Loveless and Painter 1968; Sötemann *et al.,* 2005) modelled the μ_{ANO} - pH dependency as (for 5<pH<7.2):

$$\mu_{ANO,max,pH} = \mu_{ANO,max,7.2}\theta_{ns}^{(pH-7.2)} \quad (5.22a)$$

where:

θ_{ns} pH sensitivity coefficient 2.35

Declining $\mu_{ANO,max}$ values at pH > 8.0 have been observed and it has been noted that nitrification effectively ceases at a pH of approximately 9.5 (Malan and Gouws, 1966; Wild *et al.,* 1971; Antoniou *et al.,* 1990). Accordingly, for pH > 7.2, Sötemann *et al.* (2005) proposed Eq. 5.22b to model the decline in the $\mu_{ANO,max}$ from pH > 7.2 to 9.5 as a function of $\mu_{ANO,max,7.2}$ using inhibition kinetics as follows:

$$\mu_{ANO,max,pH} = \mu_{ANO,max,7.2}K_I\frac{K_{max} - pH}{K_{max} + K_{II} - pH} \quad (5.22b)$$

where:

K_I 1.13

K_{max} 9.5

K_{II} ≈ 0.3

The overall effect of pH on $\mu_{ANO,max}$ is modelled by combining eqs. 5.22a and 5.22b, which is given by Eq. 5.22c and shown in Figure 5.7. It can be seen that in the range pH = 7.2 to 8.3, the change in $\mu_{ANO,max,pH}$ is small, with $\mu_{ANO,max,pH}/\mu_{ANO,max,7.2} > 0.9$.

$$\mu_{ANO,max,pH} = \mu_{ANO,max,7.2}2.35^{(pH-7.2)}K_I\frac{K_{max} - pH}{K_{max} + K_{II} - pH} \quad (5.22c)$$

where:

$2.35^{(pH-7.2)}$ is set = 1 for pH > 7.2,

$$K_I\frac{K_{max} - pH}{K_{max} + K_{II} - pH} = 1 \text{ for pH} < 7.2 \text{ and,}$$

$\mu_{ANO,max,pH}$ 0 for pH > 9.5

Experimental data from the literature are also shown in Figure 5.7 to provide some quantitative support for Eq. 5.22c. At low pH (<7.2) data from Wild *et al.* (1971) and Antoniou *et al.* (1990) fit the equation reasonably well. Very few data are available for pH>8.5, but the few points from Antoniou *et al.* (1990) show reasonable agreement with Eq. 5.22c.

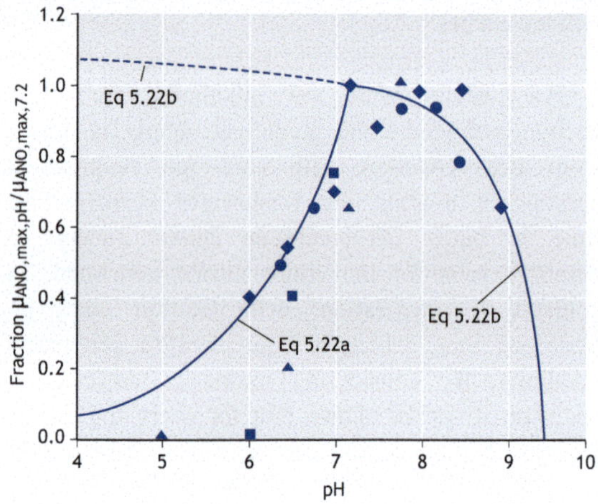

Figure 5.7 Maximum specific growth rate of nitrifiers, as a fraction of the rate at pH 7.2, versus pH of the mixed liquor. The model prediction is given by the solid line. Data from Malan and Gouws (1966); Downing et *al.* (1964); Wild et *al.* (1971); Antoniou et *al.* (1990).

Accordingly, Eq. 5.22c was accepted to calculate $\mu_{ANO,max,pH}$ in the pH range 5.5 to 9.5. From Eq. 5.22c, the minimum sludge age for nitrification (SRT_{min}) at different pH and temperature (T) and unaerated mass fraction ($f_{x,max}$) is given by

$$SRT_{min} = 1/[\mu_{ANO,pH,T}(1 - f_{x,max}) - b_{ANO,T}] \text{ (d)} \quad (5.23)$$

The problem with nitrification in low alkalinity wastewater is that the pH obtained is not known, because it is interactively established between the degree of nitrification, loss of alkalinity, pH and $\mu_{ANO,pH,T}$. To investigate this interaction, the biological kinetic ASM1 model for carbon (C) and nitrogen (N) removal was integrated by Sötemann *et al.* (2005) with a two-phase (aqueous-gas) mixed weak acid/base chemistry kinetic model to extend the application of ASM1 to situations where an estimate for pH in the biological reactor is important. This integration, which included CO_2 (and N_2) gas generation by the biological processes and their stripping by aeration, made a number of additions to ASM1, inter alia the above effect of pH on the autotrophic nitrifiers (AOOs). From simulation of a long sludge age NDAS system with incrementally decreasing influent H_2CO_3 alkalinity, when the effluent H_2CO_3 alkalinity fell below approximately 50 mg/l as $CaCO_3$, the aerobic reactor pH dropped below 6.3, which severely retarded nitrification and caused the minimum sludge age for nitrification (SRT_{min}) to increase up to the operating sludge age of the system. The simulation confirmed the earlier conclusion that when treating low H_2CO_3 alkalinity wastewater, (*i*) the minimum sludge age for nitrification (SRT_{min}) varies with temperature and reactor pH, and (*ii*) for low effluent H_2CO_3 alkalinity (< 50 mg/l as $CaCO_3$), nitrification becomes unstable and sensitive to dynamic loading conditions resulting in increases in effluent ammonia concentration, reduced nitrification efficiency, and as a result lower N removal. For effluent H_2CO_3 alkalinity < 50 mg/l, lime should be dosed to the influent to raise the aerobic reactor pH and stabilize nitrification and N removal.

5.5 NUTRIENT REQUIREMENTS FOR SLUDGE PRODUCTION

All live biological material and some unbiodegradable organic compounds contain nitrogen (N) and phosphorus (P). The organic sludge mass (VSS) that accumulates in the biological reactor comprises active organisms (X_{OHO}), endogenous residue ($X_{E,OHO}$) and unbiodegradable particulate organics (X_I), each of which contains N and P. From TKN and VSS tests conducted on activated sludge, it has been found that the N content (as N with respect to VSS, f_n, mgN/mgVSS) ranges between 0.09 and 0.12 with an average of approximately 0.10 mgN/mgVSS. Similarly, from total P and VSS tests, the P content (as P with respect to VSS, f_p, mgP/mgVSS) of activated sludge in purely aerobic and anoxic aerobic systems ranges between 0.01 and 0.03 with an average of approximately 0.025 mgP/mgVSS. From the steady-state model, the relative proportions of the active organisms (X_{OHO}), endogenous residue ($X_{E,OHO}$) and unbiodegradable particulate organics (X_I) change with sludge age. Yet it has been found that the f_n value of the VSS is relatively constant at 0.10 mgN/mgVSS. This indicates that the N content of the active organisms (X_{OHO}), endogenous residue ($X_{E,OHO}$) and unbiodegradable particulate organics (X_I) are closely the same; if they were significantly different, it would be observed that f_n changes in a consistent fashion with sludge age. Likewise, for fully aerobic systems, the P content of the three constituents of activated sludge is approximately similar at 0.025 mgP/mgVSS.

5.5.1 Nitrogen requirements

The mass of N (or P) incorporated into the sludge mass is calculated from an N balance over the completely mixed activated sludge system (Figure 4.2) under steady-state conditions over 1 day.

TKN mass out = TKN mass in
TKN mass in $= Q_i\,TKN_i = FTKN_i$ (mgN/d)
TKN mass out = TKN mass Q_e and Q_w
$= TKN_e\,Q_e + TKN_e\,Q_w + f_n\,X_{VSS}\,Q_w$

Noting that $Q_w + Q_e = Q_i$ and $Q_w = V_R/SRT$ yields:

$$Q_i TKN_e = Q_i TKN_i - f_n X_{VSS} V_R / SRT$$

from which:

$$TKN_e = TKN_i - f_n M X_{VSS} / (Q_i SRT) \quad \text{(mgN/l)} \tag{5.24}$$

where:

TKN_e effluent TKN concentration (mgN/l)

The last term in Eq. 5.24 is denoted N_s and is the concentration of influent TKN in mgN/l that is incorporated into sludge mass and removed from the system bound in the particulate sludge mass in the waste flow (Q_w),

$$N_s = f_n MX_{VSS} / (Q_i SRT) \qquad (mgN/l_{influent}) \quad (5.25)$$

From the N mass balance, this N_s concentration does not include the N in dissolved form in the waste flow. The soluble TKN concentration in the waste flow is the same as the effluent TKN concentration, TKN_e, which is soluble N in the form of ammonia ($S_{NHx,e}$) and unbiodegradable soluble organic N ($N_{ous,e}$). So from Eq. 5.24, provided nitrifiers are not supported in the activated sludge reactor so that nitrification of ammonia to nitrate does not take place, the effluent TKN concentration TKN_e is given by:

$$TKN_e = TKN_i - N_s \qquad (mgN/l) \quad (5.26)$$

From Eq. 5.24, under daily average conditions, the concentration of N per influent required for incorporation into sludge mass is equal to the N content of the mass of sludge (VSS) wasted per day divided by the influent flow. Substituting Eq. 4.12 relating the mass of sludge (VSS) in the reactor (MX_{VSS}) to the daily average organic load on the reactor ($FCOD_i$), cancelling Q_i and dividing by NIT_c yields the concentration of N required per influent for sludge production per mgCOD/l organic load on the reactor, viz.

$$\frac{N_s}{COD_i} = f_n \left[\frac{(1 - f_{SU,CODi} - f_{XU,CODi}) Y_{OHOv}}{(1 + b_{OHO} SRT)} \cdot (1 + f_{XE,OHO} b_{OHO} SRT) + \frac{f_{XU,CODi}}{f_{cv}} \right] \qquad (mgN/mgCOD) \quad (5.27)$$

The influent TKN comprises ammonia and N bound in organic compounds of a soluble and particulate and biodegradable and unbiodegradable nature. The unbiodegradable organics, some of which contain organic N, are not degraded in the AS system. The influent unbiodegradable soluble organic N ($N_{ous,i}$) exits the system with the effluent (and waste flow) streams. The unbiodegradable particulate organics are enmeshed with the sludge mass in the reactor and so the organic N associated with these organics exit the system via the daily waste sludge (VSS) harvested from the system. The N bound in the biodegradable organics ($N_{obs,i}$ and $N_{obp,i}$) are released as FSA when these organics are broken down. This FSA adds to the FSA in the reactor from the influent. Some of the FSA in the reactor is taken up by the OHOs to form new OHO biomass. Some of the OHO biomass in the reactor is lost via the endogenous respiration process. The N associated with the biodegradable part of the OHO biomass is released back to the FSA pool in the reactor but the N in the unbiodegradable endogenous residue part remains as organic N bound in the endogenous residue VSS. Due to these interactions it is possible that the effluent FSA concentration from a non-nitrifying AS system is higher than the influent FSA concentration; this happens when the influent TKN comprises a high biodegradable organic N fraction. If the conditions are favourable for nitrification, the net FSA concentration in the reactor is available for the AOOs for growth with the associated generation of nitrate.

Unless taken up for OHO growth or nitrified, the FSA remains as such and exits the system with the effluent. So in the absence of nitrification, the effluent ammonia concentration $S_{NHx,e}$ is given by:

$$S_{NHx,e} = S_{NHx,i} + N_{obs,i} + N_{obp,i} - (N_s - N_{oup,i}) \qquad (mgN/l) \quad (5.28)$$

and the effluent TKN (TKN_e) concentration by:

$$TKN_e = N_{ous,e} + S_{NXx,e} \qquad (mgN/l) \quad (5.29)$$

The same approach is applied for the phosphorus (P) requirement for sludge production. Accepting that the P content of the activated sludge in the fully aerobic system without biological excess P removal is 0.025 mgP/mgVSS, the effluent total P (TP) concentration P_e is given by:

$$P_e = P_i - P_s \qquad \text{(mgP/l) (5.30)}$$

where:

$$\frac{P_s}{COD_i} = f_p \frac{MX_{VSS}}{Q_i SRT COD_i} = \frac{f_p N_s}{f_n COD_i} \qquad \text{(mgP/mgCOD}_i\text{) (5.31)}$$

5.5.2 N (and P) removal by sludge production

A plot of eqs. 5.27 and 5.31 versus sludge age is given in Figure 5.8 for f_n = 0.10 mgN/mgVSS, f_p = 0.025 mgP/mgVSS for the example raw and settled wastewaters. It is evident that higher concentrations of TKN and TP are required for sludge production of raw than of settled wastewaters. This is because greater quantities of sludge are produced per mg COD organic load on the reactor at the same sludge age when treating raw wastewaters (see Chapter 4, Section 4.9). Also, the N and P requirements decrease as the sludge age increases because net sludge production decreases as sludge age increases. Generally for sludge ages greater than approximately 10 days, the N removal from the wastewater attributable to net sludge production is less than 0.025 mgN/mgCOD load on the reactor. As influent TKN/COD ratios for domestic wastewater are in the approximate range 0.07 to 0.13 (Figure 5.8), it is clear only a minor fraction of the influent TKN (A in Figure 5.8) is removed by incorporation into sludge mass. Additional N removal (B in Figure 5.8) is obtained by transferring the N from the dissolved form in the liquid phase to the gas phase by autotrophic nitrification and heterotrophic denitrification, which transforms the nitrate to nitrogen gas in anoxic (non-aerated) reactor(s). The details of heterotrophic denitrification are presented below. From Figure 5.8, normal P removal by incorporation into biological sludge mass is limited at approximately 0.006 and 0.004 mgP/mgCOD for raw and settled wastewaters, respectively, effecting a TP removal of approximately 20 to 25% from average municipal wastewaters. As transformation of dissolved ortho-P to a gaseous form is not possible, to increase the P removal from the liquid phase, additional ortho-P needs to be incorporated into the sludge mass. This can be achieved in two ways, (*i*) chemically and/or (*ii*) biologically. With chemical P removal, iron or aluminium chlorides or sulphates are dosed to the influent (pre-precipitation), to the activated sludge reactor (simultaneous precipitation) or to the final effluent (post-precipitation). The disadvantage of chemical P removal is that it significantly increases (*i*) the salinity of treated wastewater, (*ii*) the sludge production due to the inorganic solids formed, and (*iii*) the complexity and cost of the wastewater treatment plant.

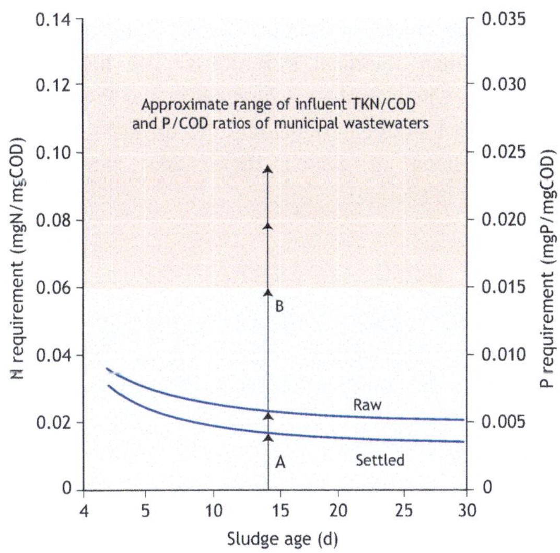

Figure 5.8 Approximate minimum nutrient N and P requirements as mgN/$l_{influent}$ TKN and mgP/$l_{influent}$ total P per mgCOD/l organic load on the activated sludge reactor versus sludge age for the example raw and settled wastewaters at 20 °C together with influent TKN and TP to COD concentration ratio ranges for municipal wastewater.

With biological P removal, the environmental conditions in the biological reactor are designed in such a way that a specific group of heterotrophic organisms (called phosphate-accumulating organisms, PAOs) grow in the activated sludge reactor. With the accumulated polyphosphates these organisms have a much higher P content than ordinary heterotrophic organisms (OHOs), as high as 0.38 mgP/mgPAOVSS (Wentzel *et al.,* 1990). The more PAOs that grow in the reactor, the higher will be the mean P content of the VSS sludge mass in the reactor and therefore the higher the P removal via the wasted sludge. With a significant mass of PAOs present, the mean P content of the VSS sludge mass can increase from 0.025 mgP/mgVSS in aerobic systems to 0.10 to 0.15 mgP/mgVSS in biological N and P removal systems. The advantage of biological P removal over chemical P removal is that (*i*) the salinity of the treated wastewater is not increased, (*ii*) sludge production is increased only between 10 and 15% and (*iii*) the system is less complex and costly to operate. A disadvantage of biological P removal is that, being biological, it is less dependable and more variable than chemical P removal. The biological processes which mediate biological N and P removal in activated sludge systems and the different reactor configurations in which these take place are described in Chapter 7.

5.6 DESIGN CONSIDERATIONS

The kinetic equations describing the interactions between the FSA and the organic N are complex and have been developed in terms of the growth-death-regeneration approach in activated sludge simulation models such as ASM1 and ASM2. However, for steady state conditions assuming (*i*) all the biodegradable organics are utilized in the reactor, and (*ii*) a TKN mass balance over the AS system, a simple steady-state nitrification model can be developed from the nitrification kinetics and the N requirements for sludge production considered above. This model is adequate for steady-state design and from it some general response graphs are developed below for the example raw and settled wastewaters. Detailed system responses can be determined with the simulation models once (*i*) the AS system has been designed and sludge age, zone and reactor volumes and recycle flows are known, and (*ii*) the steady-state concentrations have been calculated to serve as initial conditions for the simulation.

In the nitrifying AS system design, the (*i*) effluent FSA, TKN and nitrate concentrations, and (*ii*) the nitrification oxygen demand need to be calculated.

5.6.1 Effluent TKN

The filtered effluent TKN (TKN_e) comprises the FSA ($S_{NHx,e}$) and the unbiodegradable soluble organic N ($N_{ous,e}$). Once $\mu_{ANO,max,20}$, f_{xt}, SRT and S_f have been selected, the equations for these concentrations are:

1) Effluent FSA ($S_{NHx,e}$): $S_{NHx,e}$ is given by Eq. 5.15, which applies only if SRT > SRT_{min}, which will be the case for S_f >1.0.
2) Effluent soluble biodegradable organic nitrogen concentration ($N_{obs,e}$): the biodegradable organics (both soluble and particulate) are broken down by the OHOs releasing the organically bound N as FSA. In the steady-state model it is assumed that all the biodegradable organics are utilized. Hence, the effluent soluble biodegradable organic N concentration ($N_{obs,e}$) is zero.
3) Effluent soluble unbiodegradable organic nitrogen concentration ($N_{ous,e}$): being unbiodegradable, this concentration of organic N flows through the AS system with the result that the effluent concentration ($N_{ous,e}$) is equal to the influent concentration ($N_{ous,i}$), i.e.,

$$N_{ous,e} = N_{ous,i} \tag{5.32}$$

where:

$N_{ous,i}$ influent soluble unbiodegradable organic nitrogen, mgOrgN-N/l = $f_{SU,TKNi}$ TKN_i where $f_{SU,TKNi}$ is the soluble unbiodegradable organic N fraction of the influent TKN (TKN_i).

The two non-zero effluent TKN concentrations (FSA, $S_{NHx,e}$ and OrgN, $N_{ous,e}$) are soluble and so escape with the effluent (and waste flow). The soluble (filtered) TKN in the effluent (TKN_e) is given by their sum, i.e.

$$TKN_e = S_{NHx,e} + N_{ous,e} \quad \text{(filtered TKN)} \quad (5.33)$$

If the effluent sample is not filtered, the effluent TKN will be higher by the concentration of TKN in the effluent VSS, i.e.

$$TKN_e = S_{NHx,e} + N_{ous,e} + f_n X_{VSS,e} \quad \text{(unfiltered TKN)} \quad (5.34)$$

where:

$X_{VSS,e}$	effluent VSS concentration (mgVSS/l)
f_n	N content of VSS ~ 0.1 (mgOrgN-N/mgVSS)

5.6.2 Nitrification capacity

From a TKN mass balance over the AS system and SRT > SRT_{min}, the concentration of nitrate generated in the system ($S_{NO3,e}$) with respect to the influent flow is given by the influent TKN (TKN_i) minus the soluble effluent TKN (TKN_e) and the concentration of influent TKN incorporated in the sludge wasted daily from the AS system (N_s), i.e.

$$S_{NO3,e} = NIT_c = TKN_i - TKN_e - N_s \quad (5.35)$$

The N_s concentration is determined from the mass of N incorporated in the VSS mass harvested from the reactor per day (Eq. 5.27). The mass of VSS in the reactor (MX_{VSS}) does not have to include the VSS mass of nitrifiers because this mass, as mentioned earlier, is negligible (< 2-4%).

In Eq. 5.35, NIT_c defines the nitrification capacity of the AS system. The nitrification capacity (NIT_c) is the mass of nitrate produced by nitrification per unit average influent flow, i.e. $mgNO_3$-N/l. In Eq. 5.27, the effluent TKN concentration (TKN_e) depends on the efficiency of nitrification. In the calculation for the maximum unaerated sludge mass fraction ($f_{x,max}$) at a selected sludge age, if the factor of safety (S_f) has been selected >1.25 to 1.35 at the lowest expected temperature (T_{min}), the efficiency of nitrification will be high (> 95%) and $S_{NHx,e}$ generally will be less than 1 to 2 mgN/l. Also, with S_f >1.25 at T_{min}, $S_{NHx,e}$ will be virtually independent of both the system configuration and the subdivision of the sludge mass into aerated and unaerated mass fractions. Consequently, for design, with S_f > 1.25, TKN_e will be approximately 3 to 4 mgN/l provided that there is reasonable assurance that the actual $\mu_{ANO,max,20}$ value will not be less than the value accepted for design and that there is sufficient aeration capacity so that nitrification is not inhibited by an insufficient oxygen supply. Accepting the calculated $f_{x,max}$ and selected sludge age (SRT) at the lower temperature, then at higher temperatures the nitrification efficiency and the factor of safety (S_f) will both increase so that at summer temperatures (T_{max}), TKN_e will be lower, approximately 2-3 mgN/l.

Dividing Eq. 5.35 by the total influent COD concentration (COD_i) yields the nitrification capacity per mgCOD applied to the biological reactor, NIT_c/COD_i, viz.

$$NIT_c / COD_i = TKN_i / COD_i - TKN_e / COD_i - N_s / COD_i \quad (5.36)$$

where:

NIT_c/COD_i	nitrification capacity per mgCOD applied to the AS system (mgN/mgCOD)
TKN_i/COD_i	influent TKN/COD concentration ratio of the wastewater
N_s/COD_i	nitrogen required for sludge production per mgCOD applied (from Eq. 5.27)

The nitrification capacity to influent COD concentration ratio (NIT_c/COD_i) of a system can be estimated approximately by evaluating each of the terms in Eq. 5.36 as follows:

TKN_i/COD_i	This ratio is a wastewater characteristic and obtained from the measured influent TKN and COD concentrations; it can range from 0.07 to 0.10 for raw municipal wastewater and 0.10 to 0.14 for settled wastewater

TKN_e/COD_i Provided the constraint for efficient nitrification is satisfied at the lowest temperature (T_{min}), the effluent TKN at T_{min} (TKN_e) will be low at ~2-3 mgN/l, i.e. for influent COD concentrations (COD_i) ranging from 1,000 to 500, TKN_e/COD_i will range from 0.005 to 0.010. At T_{max}, TKN_e 1-2 mgN/l making the TKN_e/COD_i ratio lower

N_s/COD_i Given by Eq. 5.27

A graphical representation of the relative importance of these three ratios to the nitrification capacity, NIT_c/COD_i, is shown in figures 5.9A (for 14 °C) and 5.9B (for 22 °C). These were generated by plotting NIT_c/COD_i versus sludge age for selected influent TKN/COD (TKN_i/COD_i) ratios of 0.07, 0.08 and 0.09 for the example raw wastewater and settled wastewater for 40% COD and 15% TKN removal in primary settling, viz. 0.113, 0.127 and 0.141. Also shown are the minimum sludge ages for nitrification at unaerated sludge mass fractions of 0.0, 0.2, 0.4 and 0.6 for the example $\mu_{ANO,max,20}$ value of 0.45 /d. For a particular unaerated sludge mass fraction, the plotted values of NIT_c/COD_i are valid only at sludge ages longer than the corresponding minimum sludge age. These figures show the relative magnitudes of the three terms that affect the nitrification capacity versus sludge age and temperature.

1) Temperature: to obtain complete nitrification at 14 °C (for a selected $f_{x,max}$), the sludge age required is more than double that at 22 °C. The corresponding nitrification capacities per influent COD at 14 °C show a marginal reduction from those at 22 °C, because sludge production at 14 °C is slightly higher than at 22 °C due to the reduction in endogenous respiration rate of the OHOs.
2) Sludge age: for a selected influent TKN/COD ratio (TKN_i/COD_i), the nitrification capacity (NIT_c/COD_i) increases as the sludge age increases because the N required for sludge production decreases with sludge age, making more FSA available for nitrification. However, the increase is marginal for SRT > 10 days.
3) Influent TKN/COD ratio (TKN_i/COD_i): clearly, for both raw and settled wastewater, at any selected sludge age, the nitrification capacity (NIT_c/COD_i) is very sensitive to the influent TKN /COD ratio (TKN_i/COD_i). An increase of 0.01 in TKN_i/COD_i causes an equal increase of 0.01 in NIT_c/COD_i. For the same TKN_i/COD_i ratio for

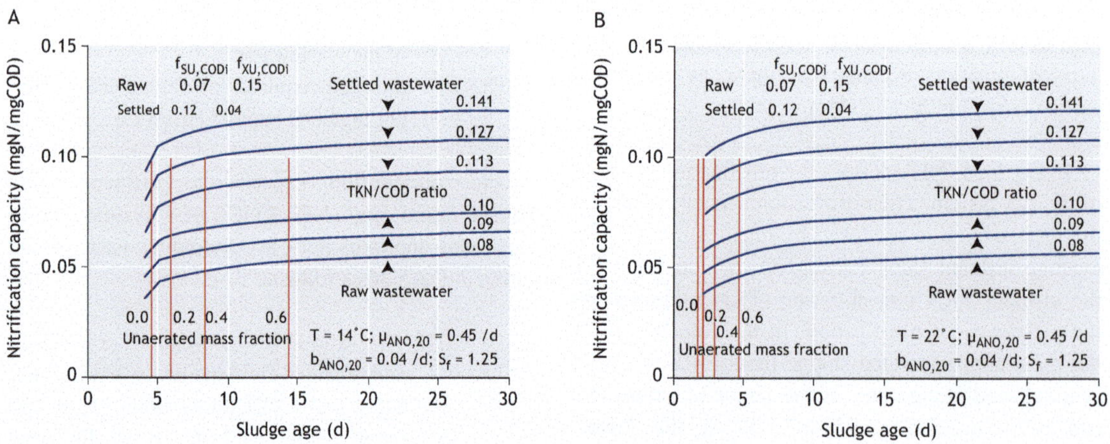

Figure 5.9 Nitrification capacity per mgCOD applied to the biological reactor versus sludge age for different influent TKN/COD concentration ratios in the example raw and settled wastewaters at 14 °C (A) and 22 °C (B). Also shown as vertical lines are the minimum sludge ages required to achieve nitrification for S_f = 1.25 for unaerated sludge mass fractions of 0.0, 0.2, 0.4 and 0.6.

raw or settled wastewater, the nitrification capacity (NIT_c/COD_i) for raw wastewater is lower than for settled wastewater because more sludge (VSS) is produced per unit COD load from raw wastewater than from settled wastewater because the unbiodegradable particulate COD fraction ($f_{XU,}$ COD_i) in raw water is higher than in settled wastewater. Apart from this difference, an increase in influent TKN/COD ratio will result in an equal increase in nitrate concentration (nitrification capacity) per influent COD. This decreases the likelihood, or makes it impossible, to obtain complete denitrification using the wastewater organics as electron donor. This will become clear when denitrification is considered below. Because primary settling increases the influent TKN/COD ratio, N removal via nitrification denitrification is always lower with settled wastewater than with raw wastewater. However, this lower N removal comes with the advantage of a smaller biological reactor and lower oxygen demand resulting in significant savings in reactor and oxygenation costs.

5.7 NITRIFICATION DESIGN EXAMPLE

Design of a nitrification AS system without denitrification is considered below. For the purpose of comparison, the nitrifying AS system is designed for the same wastewater flow and characteristics accepted for the design of the AS system for organics (COD) removal (see Chapter 4, Section 4.5). The wastewater characteristics for the raw and settled wastewaters are listed in Table 4.3 and the additional characteristics needed for nitrification are listed in Table 5.1.

5.7.1 Effect of nitrification on mixed liquor pH

An important initial consideration is the possible effect of mixed liquor pH on the $\mu_{ANO,max,20}$ value. In Section 5.4.6 above it was stated that nitrification consumes alkalinity - 7.14 mg/l as $CaCO_3$ for every mgN/lFSA nitrified to nitrate, and if there is insufficient alkalinity in the influent, the mixed liquor pH decreases below 7 causing a reduction in the $\mu_{ANO,max,20}$ value (Eq. 5.22).

The influent TKN/COD ratio of the raw wastewater is 0.08 mgN/mgCOD (Table 5.1). With a relatively low $\mu_{ANO,max,20}$ rate of 0.45/d, the sludge age needs to be 7 days or longer to ensure nitrification (S_f= 1.3) at a minimum temperature of 14 °C in a purely aerobic process ($f_{x,max}$ = 0) (Eq. 5.19). At this sludge age the nitrification capacity is approximately 0.037 mgN/mgCOD for a TKN/COD ratio of 0.08 mgN/mgCOD (Figure 5.9A). Hence the

Table 5.1 Raw and settled wastewater characteristics required for calculating effluent N concentrations from nitrification AS systems.

Influent wastewater characteristic	Symbol	Unit	Raw sewage	Settled sewage[a)]
Influent TKN	TKN_i	mgN/l	60	51
Influent TKN/COD ratio	$f_{TKNi/CODi}$		0.08	0.113
Influent FSA fraction	f_{aN}		0.75	0.88
Unbiodegradable soluble orgN fraction	$f_{SU,}TKN_i$		0.03	0.034
Unbiodegradable VSS N content	f_n		0.1	0.1
Influent pH			7.5	7.5
Influent alkalinity	S_{Alk}	mg/l as $CaCO_3$	200	200
ANO maximum specific growth rate at 20 °C	$\mu_{ANO,max}$		0.45	0.45
Influent flow rate	Q_i	Ml/d	15	15

[a)]Settled wastewater characteristics must be selected/calculated to be consistent with the raw wastewater characteristics and mass balances over the primary settling tanks, e.g. soluble concentrations must be the same in settled wastewater as in raw wastewater.

nitrate concentration produced (per litre influent) will be approximately = 0.037 · 750 = 28 mgN/l. This will cause an alkalinity reduction of 7.14 · 28 = 200 mg/l as $CaCO_3$. Because the influent alkalinity is only 200 mg/l as $CaCO_3$, the mixed liquor alkalinity will drop below 40 mg/l as $CaCO_3$, causing the mixed liquor pH value to drop below 7 (Figure 5.6). The low mixed liquor pH value will cause, inter alia, unstable and incomplete nitrification and produce an aggressive effluent, which over a few years can cause considerable damage to the concrete surfaces of the treatment plant (see Section 5.4.6 above). This simple approximate calculation can make the designer aware at an early stage of possible adverse consequences in a proposed design. Continuing this example, consideration should be given to designing a nitrification-denitrification (ND) system to recover some of the alkalinity and maintain a near-neutral pH. If as little as 12 mgN/l (~half) of the nitrate were denitrified in an anoxic reactor, the effluent alkalinity will remain above 50 mg/l as $CaCO_3$. In general a high TKN/COD ratio with low alkalinity in the influent are reliable indicators warning of potential problems in fully aerobic nitrifying systems.

5.7.2 Minimum sludge age for nitrification

For the purposes of demonstrating nitrification under purely aerobic conditions, it will be accepted that the influent alkalinity is sufficiently high to maintain an effluent alkalinity above 50 mg/l as $CaCO_3$. No adjustment to $\mu_{ANO,max,20}$ for pH will be made. The adjustment of the ANO kinetic constants for temperature is given in Table 5.2. For a completely aerobic system ($f_{x,max}$ = 0) with $\mu_{ANO,max,20}$ = 0.45 /d and S_f= 1.3, the minimum sludge age for nitrification (SRT_{min}) is found from Eq. 5.19;

$$SRT_{min} = S_f / (\mu_{ANO,max,T} - b_{ANO,T}) =$$

$$= 2.5 \text{ d at } 22\ °C \ (1.9 \text{ d with } S_f = 1.0)$$

$$= 6.9 \text{ d at } 14\ °C \ (5.3 \text{ d with } S_f = 1.0)$$

Clearly, to ensure nitrification throughout the year for the relatively low $\mu_{ANO,max,20}$ rate of 0.45/d, the sludge age of a purely aerobic process should be approximately 8 to 10 days.

5.7.3 Raw wastewater N concentrations

The influent TKN concentration of the raw wastewater is 60 mgN/l (Table 5.1). Accepting an FSA fraction of the influent TKN (f_{aN}) of 0.75 and an unbiodegradable soluble organic nitrogen fraction ($f_{SU,TKNi}$) of 0.03 for the raw wastewater, gives the influent ammonia concentration ($S_{NHx,i}$) as

$$S_{NHx,i} = f_{aN}\, TKN_i = 0.75 \cdot 60 = 45 \text{ mgN/l}$$

and the unbiodegradable soluble organic nitrogen concentration ($N_{ous,i}$) as

$$N_{ous,i} = f_{SU,TKNi}\, TKN_i = 0.03 \cdot 60 = 1.80 \text{ mgN/l}$$

Accepting the N content of the unbiodegradable particulate organics in the influent (f_n) as 0.10 mgN/mgVSS, then the OrgN concentration associated with the unbiodegradable particulate organics ($N_{oup,i}$) is

$$N_{oup,i} = f_n\, f_{XU,NITc}\, NIT_c / f_{cv} =$$
$$= 0.10\ (0.15 \cdot 750)/1.48 = 7.6 \text{ mgN/l}$$

Hence the influent biodegradable organic N concentration ($N_{ob,i}$), both soluble and particulate ($N_{ob,i} = N_{obs,i} + N_{obp,i}$) which is converted to ammonia:

$$N_{ob,i} = 60\ (1 - 0.75 - 0.03) - 7.6 = 5.6 \text{ mgN/l.}$$

Table 5.2 Temperature adjustment of nitrification kinetic constants.

Constant	20 °C	θ	22 °C	14 °C
$\mu_{ANO,max,20}$	0.45	1.123	0.568	0.224
$K_{ANO,20}$	1	1.123	1.26	0.5
$b_{ANO,20}$	0.04	1.029	0.0425	0.034

5.7.4 Settled wastewater

Following the above procedure for settled wastewater, i.e. f_{aN} = 0.83, $f_{SU,TKNi}$ = 0.034 (see Table 5.1) yields:

TKN_i = 51.0 mgN/l
$S_{NHx,i}$ = 0.88 · 51.0 = 45.0 mgN/l
$N_{ous,i}$ = 0.034 · 51.0 = 1.80 mgN/l
$N_{oup,i}$ = 0.10 (0.04 · 450)/1.48 = 1.2 mgN/l
$N_{ob,i}$ = 51.0 – 45.0 – 1.8 – 1.2 = 3.0 mgN/l

Because the settled wastewater is produced from the raw wastewater, the soluble concentrations must be the same as in raw wastewater. Because the COD and TKN concentrations change with primary settling, the soluble constituent fractions increase with primary settling.

5.7.5 Nitrification process behaviour

In the steady-state model it is accepted that all the biodegradable organics are degraded and their N content released as ammonia. The effluent soluble biodegradable organic N concentration ($N_{obs,e}$) therefore is zero.

From Eq. 5.32, the unbiodegradable soluble organic nitrogen in the effluent is (for raw and settled water)

$$N_{ous,e} = N_{ous,i} = 1.8 \qquad \text{(mgN/l) (5.37)}$$

The ammonia concentration available for nitrification ($S_{NHx,NIT}$) is the influent TKN concentration (TKN_i) minus the N concentration required for sludge production (N_s) (Eq. 5.27) and the soluble organic N concentration in the effluent ($N_{ous,e}$), viz.

$$S_{NHx,NIT} = TKN_i - N_s - N_{ous,e} \qquad \text{(mgN/l) (5.38)}$$

If the sludge age of the system is shorter than the minimum required for nitrification (SRT < SRT_{min}), no nitrification takes place and the effluent nitrate concentration ($S_{NO3,e}$) is zero. The effluent ammonia concentration ($S_{NHx,e}$) is equal to the nitrogen available for nitrification ($S_{NHx,NIT}$, Eq. 5.38). If SRT > SRT_{min} for S_f= 1.0, most of the FSA available for nitrification is nitrified to nitrate and the effluent nitrate concentration ($S_{NO3,e}$) is the difference between $S_{NHx,NIT}$ (Eq. 5.38) and the effluent FSA concentration given by Eq. 5.15. For both SRT < SRT_{min} and SRT > SRT_{min}, the effluent TKN concentration (TKN_e) is the sum of effluent ammonia and unbiodegradable soluble organic nitrogen concentrations ($TKN_e = S_{NHx,e} + N_{ous,e}$).

For SRT< SRT_{min}, no nitrification takes place so the effluent nitrate concentration ($S_{NO3,e}$) is zero i.e.

$$S_{NO3,e} = 0 \qquad \text{(mgN/l) (5.39a)}$$

and the effluent ammonia concentration ($S_{NHx,e}$) is:

$$S_{NHx,e} = S_{NHx,NIT} = TKN_i - N_s - N_{ous,e} \qquad \text{(mgN/l) (5.40a)}$$

The effluent TKN concentration (TKN_e) is:

$$TKN_e = S_{NHx,e} + N_{ous,e} \qquad \text{(mgN/l) (5.41a)}$$

The nitrifier sludge mass (MX_{ANO}) and the nitrification oxygen demand (FO_{NIT}) are both zero, i.e.

$$MX_{ANO} = 0 \qquad \text{(mgVSS) (5.42a)}$$

$$FO_{NIT} = 0 \qquad \text{(mgO}_2\text{/d) (5.43a)}$$

With increasing sludge age starting from SRT=0, $S_{NHx,e}$ from Eq. 5.15 is first negative (which is of course impossible) and then > $S_{NHx,NIT}$ (which is also not possible). For a sludge age slightly longer than SRT_{min}, the $S_{NHx,e}$ falls below $S_{NHx,NIT}$. From this sludge age, nitrification takes place and for further (even small) increases in sludge age, the $S_{NHx,e}$ rapidly decreases to low values (< 4 mgN/l).

Hence for SRT > SRT_{min}:

The effluent ammonia concentration ($S_{NHx,e}$) is:

$$S_{NHx,e} = \frac{K_{ANO,T}(b_{ANO,T} + 1/SRT)}{\mu_{ANO,max,T}(1 - f_{xt}) - (b_{ANO,T} + 1/SRT)} \quad (mgN/l) \ (5.40b)$$

the effluent TKN concentration (TKN_e) is:

$$TKN_e = S_{NHx,e} + N_{ous,e} \quad (mgN/l) \ (5.41b)$$

and the effluent nitrate concentration ($S_{NO3,e}$) is:

$$S_{NO3,e} = S_{NHx,NIT} - S_{NHx,e} = TKN_i - N_s - TKN_e \quad (mgN/l) \ (5.39b)$$

Analogous to the concentration of active heterotrophic organisms (see Eq. 4.9), the nitrifier organism mass is given by:

$$MX_{ANO} = FS_{NO3,e}Y_{ANO}SRT/(1 + b_{ANO,T}SRT) \quad (mgVSS) \ (5.42b)$$

where:

$FS_{NO3,e}$ mass of nitrate generated per day
$= (Q_e + Q_w)\, S_{NO3,e} = Q_i\, S_{NO3,e}$ (mgN/d)

The oxygen demand for nitrification is simply 4.57 mgO_2/mgN times the mass of nitrate produced per day, i.e.

$$FO_{NIT} = 4.57 FS_{NO3,e} \quad (mgO_2/d) \ (5.43b)$$

Substituting the influent N concentrations for raw and settled wastewaters and the values of the kinetic constants at 14 °C into eqs. 5.38 to 5.43, the results at different sludge ages were calculated. Figure 5.10A shows the different effluent concentrations of N from the system versus sludge age for raw and settled wastewater at 14 °C. Figure 5.10C shows the nitrifier sludge mass (as a % of the reactor VSS mass) and nitrification oxygen demand for raw and settled wastewater at 14 °C. Also shown in Figure 5.10C are the carbonaceous and total oxygen demands for raw and settled wastewater at 14 °C. The calculations were repeated for 22 °C and shown in figures 5.10B and 5.10D. Figures 5.10A and 5.10B show that once the sludge age is approximately 25% longer than the minimum required for nitrification, nitrification is virtually complete (for steady-state conditions) and comparing the results for raw and settled wastewater, there is little difference between the nitrification oxygen demand and the concentrations of ammonia, nitrate and TKN in the effluent. The reasons for this similar behaviour are: (*i*) the primary settling tank removes only a small fraction of the influent TKN, and (*ii*) settled wastewater results in lower sludge production, so that the FSA available for nitrification in raw and settled wastewater are nearly the same. Once nitrification takes place, temperature has relatively little effect on the different effluent N concentrations. However, a change in temperature causes a significant change in the minimum sludge age for nitrification.

Considering figures 5.10A and 5.10B, for $SRT < SRT_{min}$, the effluent ammonia concentration ($S_{NHx,e}$) and hence the effluent TKN concentration (TKN_e), increase with increasing sludge age up to SRT_{min} because N_s decreases for increases in SRT. For $SRT > SRT_{min}$, $S_{NHx,e}$ decreases rapidly to < 2 mgN/l so that for $SRT > 1.3 \cdot SRT_{min}$, the effluent TKN concentration is < 4 mgN/l. The increase in nitrate concentration ($S_{NO3,e}$) with increase in sludge age for $SRT > 1.3 \cdot SRT_{min}$ is mainly due to the reduction in N required for sludge production (N_s). This is important for BNR systems; increasing the sludge age increases the nitrification capacity (see Section 5.5.2 above) so more nitrate has to be denitrified to achieve the same N removal.

Figures 5.10C and 5.10D show that the nitrification oxygen demand increases rapidly once $SRT > SRT_{min}$ but for $SRT > 1.3 \cdot SRT_{min}$, further increases are marginal irrespective of the temperature or wastewater type, i.e. between sludge ages of 10 and 30 days approximately 2,600 to 2,900 kgO_2/d are required for nitrification. This nitrification oxygen demand represents an increase of 42% and 65% above the carbonaceous (COD) oxygen demand for the raw and settled wastewater. However, the total oxygen demand for treating settled wastewater is only 75% of that for treating raw wastewater.

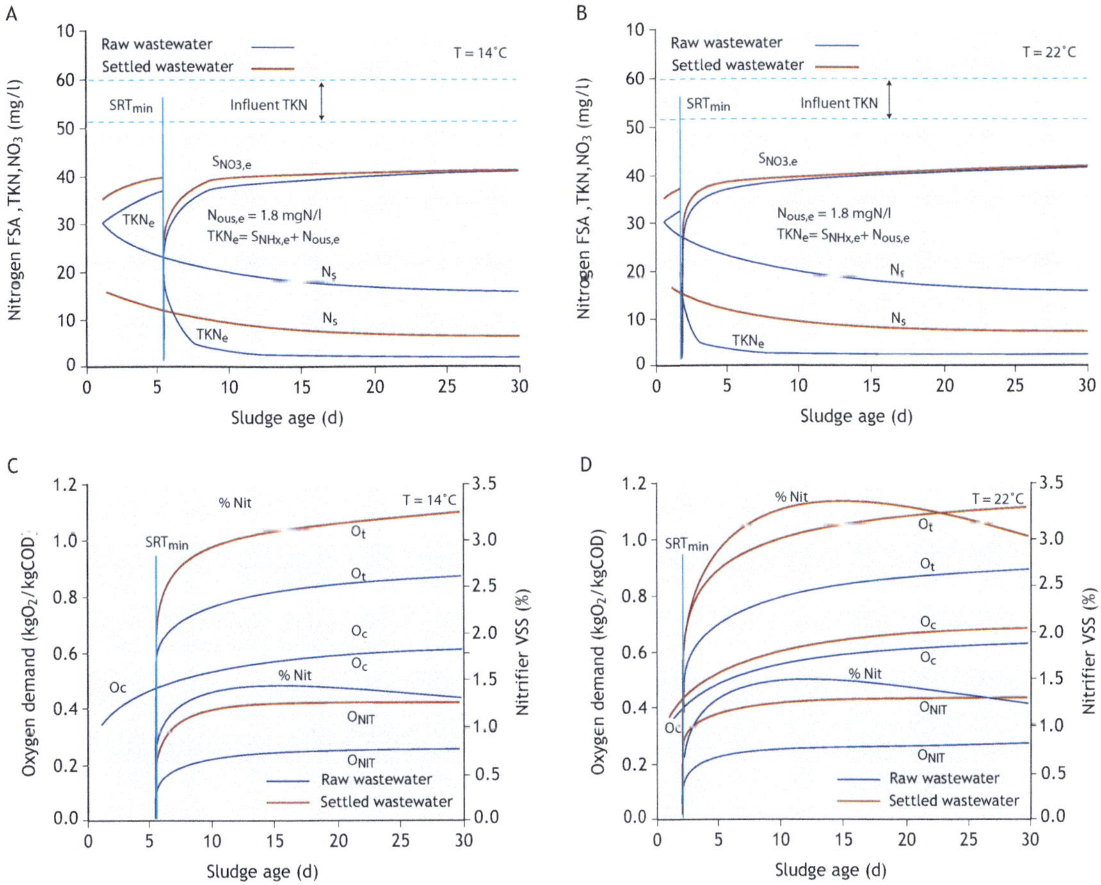

Figure 5.10 Effluent ammonia ($S_{NHx,e}$), TKN (TKN_e) and nitrate ($S_{NO3,e}$) concentrations and N required for sludge production (N_s) versus sludge age at 14 °C (A) and 22 °C (B) and nitrification (O_{NIT}), carbonaceous (O_c) and total (O_t) oxygen demand in $kgO_2/kgCOD$ load and % nitrifier VSS mass versus sludge age at 14 °C (C) and 22 °C (D) for the example raw and settled wastewater.

In order that nitrification can proceed without inhibition by oxygen limitation, it is important that the aeration equipment is adequately designed to supply the total oxygen demand; generally heterotrophic organism growth takes precedence over nitrifier growth when oxygen supply becomes insufficient. This is because heterotrophic organisms can grow adequately with dissolved oxygen concentrations of 0.5 to 1.0 mgO_2/l whereas nitrifiers require a minimum concentration of 1 to 2 mgO_2/l. In the same way that the effluent FSA concentration rapidly decreases for SRT > SRT_{min}, so the nitrifier sludge mass rapidly increases once SRT > SRT_{min}, is slightly higher at 14 °C than at 22 °C due to the lower endogenous respiration rate and it is approximately the same for raw or settled wastewater (420 and 940 kgVSS at 10 and 30 days sludge age, respectively). Comparing the nitrifier sludge mass to the heterotrophic sludge mass as in figures 5.10C and 5.10D, even at high TKN/COD ratios for settled wastewater, the nitrifier sludge mass comprises < 4% of VSS mass and so is ignored in the determination of the VSS concentration in the AS reactor treating domestic wastewater.

It is worth repeating that primary sedimentation removes only a minor fraction of the TKN but a significant fraction of COD (15% and 40%, respectively, in this example). However, even though settled wastewater has a lower TKN concentration than raw wastewater, the effluent nitrate concentration does not reflect this difference. This is because the N removal for sludge production is lower for settled than for raw wastewater. Consequently, the nitrate concentration for settled wastewater is nearly the same as for raw wastewater; for different wastewater characteristics it can be higher than raw wastewater. In contrast, the maximum N removal by denitrification using the wastewater organics as electron donor, called the denitrification potential, is mainly dependent on the influent COD concentration and this concentration is significantly reduced by primary sedimentation. This may result in a situation where it may be possible to obtain near-complete nitrate removal when treating raw wastewater but not when treating settled wastewater. The difference in COD and TKN removal in PSTs therefore has a significant effect on the design of BNR systems.

5.8 BIOLOGICAL NITROGEN REMOVAL

5.8.1 Interaction between nitrification and nitrogen removal

Nitrification is a prerequisite for denitrification; without it biological N removal is not possible. Once nitrification takes place, N removal by denitrification becomes possible and should be included even when N removal is not required, (see Chapter 4, Section 4.11) by incorporating zones in the reactor that are intentionally unaerated. Because the nitrifiers are obligate aerobes, nitrification does not take place in the unaerated zone(s), so to compensate for this, the system sludge age needs to be increased for situations where nitrification is required. For fully aerobic systems and a wastewater temperature of 14 °C, a sludge age of 5-7 days may be sufficient for complete nitrification, taking due consideration of the requirement that the effluent FSA concentration should be low even under cyclic flow and load conditions ($S_f > 1.3$). For anoxic-aerobic systems, a sludge age of 15 to 20 days may be required when a 50% unaerated mass fraction is added (Figure 5.3). Therefore, for plants where N removal is required, invariably the sludge ages are long due to (*i*) the uncertainty in the $\mu_{ANO,max,20}$ value, (*ii*) the need for unaerated zones, and (*iii*) the guarantee of nitrification at the minimum average winter temperature (T_{min}). For plants where nitrification is a possibility and not obligatory, uncertainty in the $\mu_{ANO,max,20}$ value is not important and unaerated zones can be smaller, with the result that sludge ages can be in the usual fully aerobic short sludge age range of 3 to 6 days. Unaerated zones should still be incorporated to derive the benefits of denitrification in the event that nitrification does take place. When it does not, the unaerated zone will be anaerobic (no input of DO or nitrate) instead of anoxic, and some excess biological phosphorus removal (EBPR) may take place. Because EBPR is not required and therefore not exploited fully, whether or not it takes place is not important because it does not affect the system behaviour very much. With some EBPR, the sludge production will be slightly higher (<5%) per COD load, the VSS/TSS ratio and oxygen demand both somewhat lower (by approximately 5%). However, EBPR may result in mineral precipitation problems in the sludge treatment facilities if the WAS is anaerobically digested.

5.8.2 Benefits of denitrification

In the design of fully aerobic systems discussed above, it was suggested that when nitrification is not obligatory but a possibility, unaerated zones should still be incorporated in the system to derive the benefits of denitrification.

These benefits include (*i*) reduction in nitrate concentration which ameliorates the problem of rising sludge from denitrification in the secondary settling tank (Chapter 4, Section 4.11), (*ii*) recovery of alkalinity (Section 5.4.6), and (*iii*) reduction in oxygen demand. With regard to (*iii*), under anoxic conditions, nitrate serves as an electron acceptor instead of dissolved oxygen in the degradation of organics (COD) by facultative heterotrophic organisms. The oxygen equivalent of nitrate is 2.86 $mgO_2/mgNO_3$-N which means that 1 mg NO_3-N denitrified to N_2 gas has the same electron-accepting capacity as 2.86 mg of oxygen. In nitrification to nitrate, the FSA donates 8 electrons (e^-)/mol, the N changing from an e^- state of -3 to +5. In denitrification to N_2, the nitrate accepts 5 e^-/mol, the N changing from an e^- state of +5 to 0. Because 4.57 mgO_2/mgFSA-N are required for nitrification, the oxygen equivalent of nitrate in denitrification to N_2 is $5/8 \cdot 4.57 = 2.86$ $mgO_2/mgNO_3$-N (Table 5.3). Therefore, for every 1 mg NO_3-N denitrified to N_2 gas in the anoxic zone, during which approximately $2.86/(1\text{-}Y_{OHO}) = 8.6$ mgCOD is utilized, 2.86 mg less oxygen needs to be supplied to the aerobic zone. Because the oxygen requirement to form the nitrate from ammonia is 4.57 $mgO_2/mgNO_3$-N, and 2.86 $mgO_2/mgNO_3$-N is 'recovered' in denitrification to N_2 gas, a maximum of 2.86/4.57 or 5/8ths = 0.63 of the nitrification oxygen demand can be recovered. A comparison of the nitrification and denitrification reactions is given in Table 5.3. Under operating conditions it is not always possible to denitrify the entire nitrate formed with the result that the nitrification oxygen recovery by denitrification is approximately 50% (see Figure 5.11).

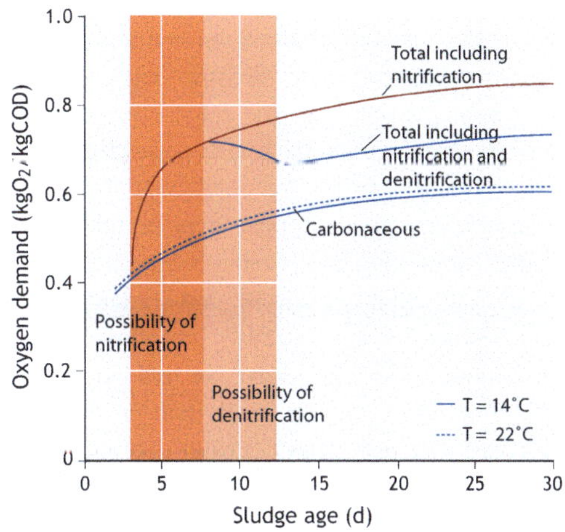

Figure 5.11 Carbonaceous, total including nitrification and total including nitrification and denitrification oxygen demand per unit COD load on the biological reactor versus sludge age for the example raw wastewater.

Table 5.3 Comparison of nitrification and denitrification processes in single-sludge activated sludge systems.

	Nitrification	Denitrification
Form:	Ammonia (NH_4^+)	Nitrate (NO_3^-)
Function:	Electron donor	Electron acceptor
Half reaction:	Oxidation	Reduction
Organisms:	Autotrophs	Heterotrophs
Environment:	Aerobic	Anoxic

Compound:	NH_4^+	N_2	NO_2^-	NO_3^-
Oxidation state:	-3	0	+3	+5

Nitrification (oxidation) →

8 e^- atom N = 4.57 mgO_2/mgN

Net loss ← Denitrification (reduction)

5 e^- atom N = 2.86 mgO_2/mgN

Nitrification: 4.57 $mgO_2/mgNH_4$-N nitrified to NO_3–N

Denitrification: 2.86 mgO_2 recovered/mg NO_3-N denitrified to N_2 gas

Therefore, denitrification allows at best 62.5% (5/8 or 2.86/4.57) recovery of the nitrification oxygen demand

Therefore, once the possibility of nitrification exists, it is always worthwhile to consider including intentional denitrification because of the recovery of alkalinity and oxygen. With regard to oxygen, if the oxygen supply is insufficient to meet the combined carbonaceous and nitrification requirement, areas in the aerobic reactor will become anoxic. Under oxygen limited conditions, the aerobic mass fraction in the 'aerobic' reactor will vary depending on the COD and TKN load on the plant over the day. At minimum load, oxygen supply may be adequate so that nitrification may be complete whereas as at peak load, oxygen supply may be insufficient so that nitrification may cease (partially or completely) and denitrification will take place on the accumulated nitrate. This behaviour is exploited in the single reactor nitrification denitrification configurations such as the ditch or Carousel-type systems.

5.8.3 Nitrogen removal by denitrification

In biological N removal systems, the N is removed by transfer from the liquid phase to the solid and gas phases. About 20% of the influent N is incorporated in the sludge mass (Figure 5.8) but the bulk of the N, i.e. approximately 75% when complete denitrification is possible, is removed by transfer to the gas phase via nitrification and denitrification (Figure 5.12).

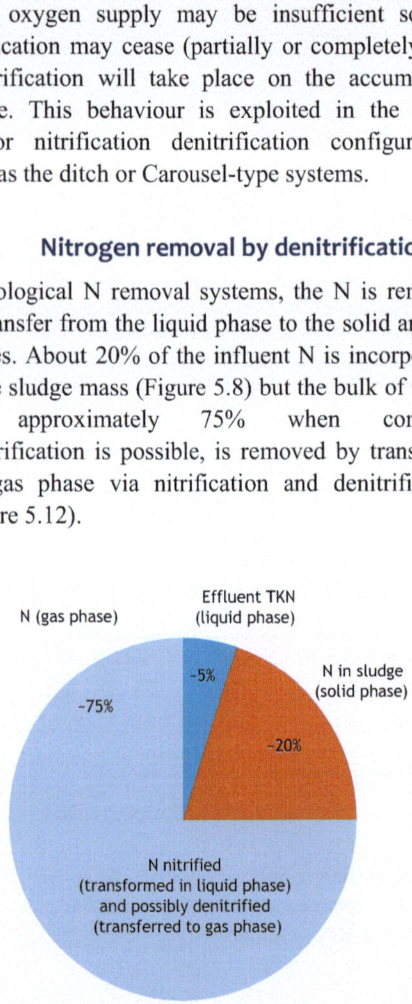

Figure 5.12 Exit routes for nitrogen in single-sludge nitrification-denitrification activated sludge systems.

In the nitrification step the N remains in the liquid phase because it is transformed from ammonia to nitrate. In the denitrification step it is transferred from the liquid to the gas phase and escapes to the atmosphere. When complete denitrification is achieved a relatively small fraction of the influent TKN (~5%) remains in the liquid phase and escapes as total N (TKN + nitrate) with the effluent.

For aerobic conditions, the designer's problem is to calculate the mass of oxygen electron acceptor required by the OHOs (and AOOs) for the utilization of the known mass of electron donors (organics and ammonia) available. However, for anoxic conditions, the problem is the opposite. Here the problem is to calculate the mass of electron donors (COD) that are required to denitrify a known mass of electron acceptors (nitrate). If sufficient electron donors (COD) are not available then complete denitrification cannot be achieved. The calculation of the nitrogen removal is essentially a reconciliation of electron acceptors (nitrate) and donors (COD) taking due account of (*i*) the biological kinetics of denitrification, and (*ii*) the system operating parameters (such as recycle ratios, anoxic reactor sizes) under which the denitrification is constrained to take place.

The electron donors (or COD or energy) for denitrification can come from two sources, (*i*) internal or (*ii*) external to the activated sludge system. The former are those present in the system itself, i.e. those in the incoming wastewater or generated within the biological reactor by the activated sludge itself; the latter are organics imported to the activated sludge system and specifically dosed into the anoxic zone(s) to promote denitrification, e.g. methanol, acetate, molasses, etc. (Monteith *et al.*, 1980). Here the focus is on internal COD sources for denitrification, but the principles and procedures are sufficiently general to be adaptable to also include external COD (energy) sources.

5.8.4 Denitrification kinetics

There are three internal organics sources, two from the wastewater and one from the activated sludge mass itself. The two in the wastewater are the two main forms of organics, i.e. readily biodegradable organics (RBCOD) and slowly biodegradable organics (SBCOD). The third is slowly biodegradable organics generated by the biomass itself through death and lysis of the organism mass (also known as endogenous mass loss/respiration). This self-generated SBCOD is utilized in the same way as the wastewater SBCOD, but is recognized separately because of its different source and rate of supply to that of the influent. The RBCOD and SBCOD (influent or self-generated) are degraded via different mechanisms by the OHOs.

The different RBCOD and SBCOD degradation mechanisms lead to different COD utilization rates. The RBCOD comprises small simple dissolved organic compounds that can pass directly through the cell wall into the organism, e.g. sugars, short chain fatty acids. Accordingly, the RBCOD can be used at a high rate which does not change significantly whether nitrate or oxygen serve as the terminal electron acceptor (Ekama *et al.,* 1996a,b). Simulation models use the Monod equation to model the utilization of RBCOD by OHOs under both aerobic and anoxic conditions. The SBCOD comprises large particulate or colloidal organic compounds, too large to pass into the organism directly. These organics must be broken down (hydrolysed) in the slime layer surrounding the organism into smaller components, which then can be transferred into the organism and utilized. The extracellular SBCOD hydrolysis rate is slow and forms the limiting rate in the utilization of SBCOD. This hydrolysis rate is much slower under anoxic conditions than under aerobic conditions: only about 1/3rd (Stern and Marais, 1974; Van Haandel *et al.,* 1981). This introduces a reduction factor (η) into the SBCOD hydrolysis rate equation for anoxic conditions (Eq. 5.45 below). Research has indicated that the utilization of RBCOD is simultaneous with the hydrolysis of SBCOD. Also the rate of RBCOD utilization is considerably faster (7 to 10 times) than the rate of SBCOD hydrolysis so the denitrification rate with influent RBCOD is much faster than with SBCOD. Therefore, the influent RBCOD is the preferred organic for denitrification and the higher this concentration in the influent with respect to the total COD, the greater the N removal.

5.8.5 Denitrification systems

As a result of the different degradation mechanisms and rates of RBCOD and SBCOD utilization, the position of the anoxic zone in the biological reactor significantly affects the denitrification that can be achieved. There are many different configurations of single-sludge nitrification-denitrification (ND) systems but from the point of view of the source of the organics (electron donors), these can be simplified into two basic types of denitrification or combinations of these. The two basic types utilizing internal organics are (*i*) post-denitrification, which utilizes self-generated endogenous organics and (*ii*) pre-denitrification, which utilizes influent wastewater organics.

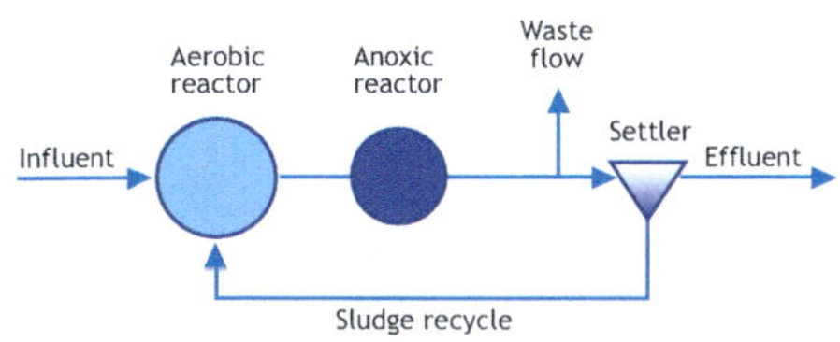

Figure 5.13A The post-denitrification single-sludge biological nitrogen removal system.

With post-denitrification (Figure 5.13A), the first reactor is aerobic and the second is unaerated. The influent is discharged to the aerobic reactor where aerobic growth of both the heterotrophic and nitrifying organisms takes place. Provided the sludge age is sufficiently long and the aerobic fraction of the system is adequately large, nitrification will be complete in the first reactor. The mixed liquor from the aerobic reactor passes to the anoxic reactor, also called the secondary anoxic reactor, where it is mixed by stirring. The outflow from the anoxic reactor passes through a secondary settling tank

(SST) and the underflow is recycled back to the aerobic reactor.

The SBCOD organics released by the sludge mass via the death of organisms provides the energy source for denitrification in the anoxic reactor. However, the rate of release of energy is low, so that the rate of denitrification is also low. So to obtain a meaningful reduction of nitrate, the anoxic mass fraction of the system, i.e. the fraction of the mass of sludge in the system in the anoxic reactor, must be large and this may cause, depending on the sludge age, cessation of nitrification. Thus, although theoretically the system has the potential to remove all the nitrate, from a practical point this is not possible because the anoxic mass fraction will need to be so large that the conditions for nitrification cannot be satisfied particularly if the temperatures are low (<15 °C). Furthermore, in the anoxic reactor, ammonia is released through organism death and lysis, some of which passes out with the effluent thereby reducing the total nitrogen removal of the system. To minimize the ammonia content of the effluent, a flash or re-aeration reactor is sometimes placed between the anoxic reactor and the SST. In this reactor, N_2 gas is stripped from the mixed liquor to avoid possible sludge buoyancy problems in the SST and the ammonia is nitrified to nitrate to assist with compliance of ammonia standards but this reduces the overall efficiency of the nitrate reduction of the system. For these reasons post-denitrification has not been widely applied in practice, except where it is used in combination with chemical dosing.

5.8.5.1 The Ludzack-Ettinger system

Ludzack and Ettinger (1962) were the first to propose a single-sludge nitrification-denitrification system utilizing the biodegradable organics in the influent as organics for denitrification. It consisted of two reactors in series, partially separated from each other. The influent was discharged to the first or primary anoxic reactor which was maintained in an anoxic state by mixing without aeration. The second reactor was aerated and nitrification took place in it. The outflow from the aerobic reactor passed to the SST and the SST underflow was returned to the aerobic (second) reactor. Due to the mixing action in both reactors, an interchange of the nitrified and anoxic liquors was induced. The nitrate which entered the primary anoxic reactor was denitrified to nitrogen gas. Ludzack and Ettinger reported that their system gave variable denitrification results, probably due to the lack of control of the interchange of the contents between the two reactors. In 1973, Barnard proposed an improvement to the Ludzack-Ettinger system by completely separating the anoxic and aerobic reactors, recycling the underflow from the SST to the primary (first) anoxic reactor, and providing a mixed liquor recycle from the aerobic to the primary anoxic reactor (Figure 5.13B).

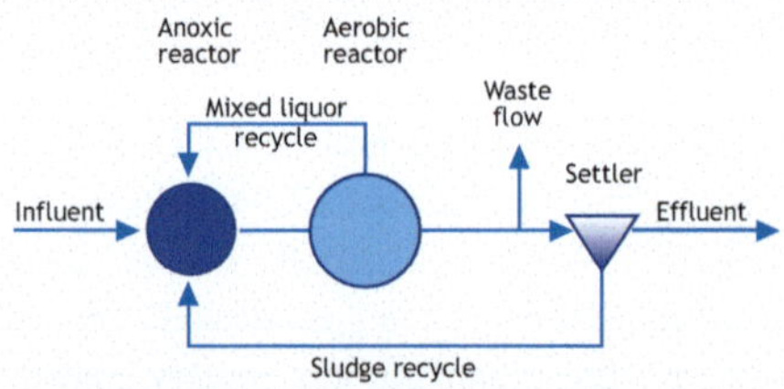

Figure 5.13B The modified Ludzack-Ettinger single-sludge biological nitrogen removal system proposed by Barnard (1973), including the primary anoxic reactor only.

These modifications allowed a significant improvement in control over the system's N removal performance of the system with the mixed liquor recycle flow. The RBCOD and SBCOD organics from the influent stimulated high rates of denitrification in the primary anoxic reactor and much higher reductions of nitrate could be achieved than with post-denitrification, even when the pre-denitrification reactor of this system was substantially smaller than the post-denitrification reactor. In this system, called the Modified Ludzack-Ettinger (MLE) system, complete nitrate removal cannot be achieved because a part of the total flow from the aerobic reactor is not recycled to the anoxic reactor but exits the system with the effluent. To reduce the possibility of flotation of sludge in the SST due to denitrification of residual nitrate, the sludge accumulation in the SST needed to be kept to a minimum. This was achieved by having a high underflow recycle ratio from the SST, equal to the mean influent flow (1:1).

5.8.5.2 The 4-stage Bardenpho system

In order to overcome the deficiency of incomplete nitrate removal in the MLE system, Barnard (1973) proposed including a secondary anoxic reactor in the system and called it the 4-stage Bardenpho system (Figure 5.13C).

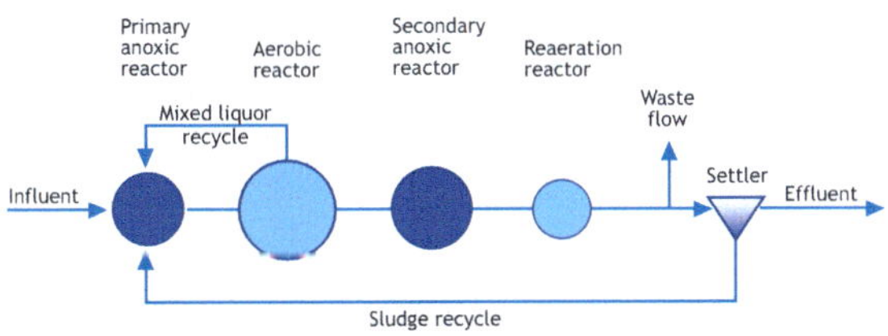

Figure 5.13C The 4-stage Bardenpho single-sludge biological nitrogen removal system, including primary and secondary anoxic reactors.

Barnard considered that the low concentration of nitrate discharged from the aerobic reactor to the secondary anoxic reactor will be denitrified to produce a relatively nitrate-free effluent. He included a flash or re-aeration reactor to strip the nitrogen gas and to nitrify the ammonia released during the denitrification.

Although in concept the Bardenpho system has the potential for complete removal of nitrate, in practice this is not possible except when the influent TKN/COD concentration ratio is quite low, < 0.09 mgN/mgCOD for normal municipal wastewater at 14 °C. The low denitrification rate and ammonia release (approximately 20% of the nitrate denitrified) results in an inefficient use of the secondary anoxic sludge mass fraction. As a result of the competition between the aerated and unaerated sludge mass fractions from the requirement to nitrify (see Section 5.4.3), usually it is better to exclude the secondary anoxic (and re-aeration) reactor, and enlarge the primary anoxic reactor and increase the mixed liquor recycle ratio.

5.8.6 Denitrification rates

The denitrification behaviour in the primary and secondary anoxic zones is best explained by considering these reactors as plug-flow reactors. However, this explanation is equally valid for completely mixed reactors because the denitrification kinetics are essentially zero order with respect to nitrate concentration (Van Haandel *et al.,* 1981; Ekama and Wentzel, 1999). Owing to the two different kinds of biodegradable COD (RBCOD and SBCOD) in the influent wastewater, the denitrification in the primary anoxic reactor follows two phases (Figure 5.14A): an initial rapid phase where the rate is defined by the simultaneous utilization of RBCOD and SBCOD ($K_1 + K_2$) and a second slower phase where the specific denitrification rate (K_2) is defined by the utilization of only SBCOD originating from the influent and self-generated by the sludge through organism death and lysis. In the secondary anoxic reactor only a single slow phase of denitrification takes place (Figure 5.14B, right), the specific rate (K_3) being approximately two-thirds of the slow rate (K_2) in the primary anoxic reactor (Stern and Marais 1974; Van Haandel *et al.,* 1981). In the preceding aerobic reactor all the RBCOD and most of the SBCOD of the influent has been utilized with the result that in the secondary anoxic reactor the only biodegradable COD available is SBCOD from organism death and lysis; the slow rate of supply of this SBCOD governs the K_3 rate and causes this rate to be slower than the K_2 rate. The values of the K rates are given in Table 5.4.

A further specific K rate (K_4) has been defined for denitrification in intermittently-aerated anoxic aerobic digestion of waste activated sludge (WAS) (Warner *et al.,* 1986). This rate is only two-thirds of the K_3 rate in the secondary anoxic reactor (Table 5.4), but sufficiently high to denitrify the entire nitrate generated in aerobic digestion of WAS if the 4 to 6 hour aeration cycle is 50% anoxic and 50% aerobic. Denitrification in anoxic-aerobic digestion adds the benefits of denitrification to this system, i.e. zero alkalinity consumption, oxygen recovery, improved pH control, reduced chemical dosing (Dold *et al.,* 1985) and additionally a nitrogen-free dewatering liquor. This last advantage is significant considering the high N content of WAS compared with primary sludge.

The constancy of K_1, K_2, K_3 (and K_4) specific denitrification rates under constant flow and load conditions can be explained in terms of the kinetics of RB and SB organics utilization included in the activated sludge simulation models such as ASM1 developed later (Chapter 14, Section 14.4). The utilization of RB organics is modelled with the Monod equation and expressing the K_1 rate in terms of this yields,

$$K_1 = \frac{(1 - Y_{OHO}) f_{cv} \mu_{OHO}}{2.86 Y_{OHO}} \frac{S_S}{K_S + S_S} \quad (mgNO_3\text{-}N/mgOHOVSS.d) \quad (5.44)$$

where:

$$\frac{S_S}{K_S + S_S} \approx 1$$

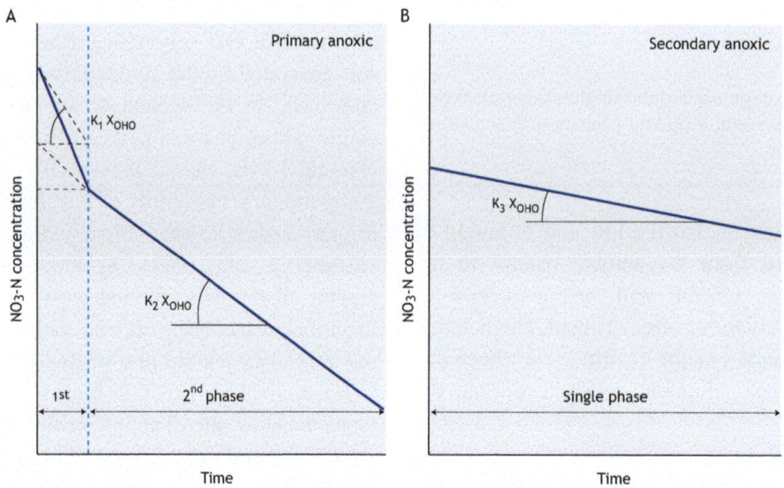

Figure 5.14 Nitrate concentration versus time profiles in primary anoxic (A) and secondary anoxic (B) plug-flow reactors, showing the three phases of denitrification associated with the K_1, K_2 and K_3 rates. In the primary anoxic the initial rapid rate K_1 is attributable to the utilization of the influent RBCOD and the second slower rate K_2 to the utilization of SBCOD from the influent wastewater and self-generated by organism death and lysis. In the secondary anoxic reactor the rate K_3 is attributable to the utilization of the self-generated SBCOD only.

Table 5.4 K denitrification rates and their temperature sensitivity.

Symbol	20 °C	θ	14 °C	22 °C
[a]$K_{1,20}$	0.720	1.200	0.241	1.036
[a]$K_{2,20}$	0.101	1.080	0.064	0.118
[a]$K_{3,20}$	0.072	1.029	0.061	0.076
[a]$K_{4,20}$	0.048	1.029	0.040	0.051

[a] Units: $mgNO_3$-N/mgOHOVSS.d

In plug-flow and completely mixed primary anoxic reactor, the Monod term $S_S/(K_S+S_S)$ remains close to 1 down to low RBCOD concentrations because the half-saturation concentration (K_S) is low. Accepting Y_{OHO} = 0.67 mgCOD/mgCOD and f_{cv} = 1.48 mgCOD/mgVSS yields, K_1 = 0.26 μ_{OHO} mgNO3-N/mgOHOVSS.d. So for the measured K_1 = 0.72 mgNO3-N/mgOHOVSS.d (Table 5.4), the μ_{OHO} must have been approximately 2.8/d. This μ_{OHO} rate is in the range of μ_{OHO} rates measured in activated sludge systems. In investigating the kinetics of RBCOD utilization in aerobic and anoxic selectors, Still *et al.* (1996) and Ekama *et al.* (1996a,b) found μ_{OHO} values ranged between 1.0/d in completely mixed reactor systems and 4.5/d in selector reactor systems, which yields K_1 denitrification rates of approximately 0.26 mgNO3-N/mgOHOVSS.d for completely mixed type systems and 1.17 mgNO3-N/mgOHOVSS.d for systems in which a selector effect (high μ_{OHO}) has been stimulated in the OHO biomass.

The utilization of SBCOD is expressed in terms of the active-site surface hydrolysis kinetic formulation, which has the form of a Monod equation, except the variable is the adsorbed SBCOD to active OHO ratio (X_S/X_{OHO}), not the bulk liquid SBCOD concentration.

Hence the K_2, K_3 (and K_4) rates are given by,

$$K_2 = K_3 = K_4 = \frac{(1-Y_{OHO})f_{cv}}{2.86Y_{OHO}}\frac{\eta K_h(X_s/X_{OHO})}{[K_x+(X_s/X_{OHO})]}$$

(mgNO3-N/mgOHOVSS.d) (5.45)

where:

X_S/X_{OHO} is progressively lower in primary (K_2), secondary (K_3), and anoxic-aerobic digestion (K_4)

In constant flow and load primary and secondary anoxic plug-flow reactors, the (X_S/X_{OHO}) ratio changes very little due to the reduced anoxic hydrolysis rate. The reason for the K_2 being higher than K_3 arises from different concentrations of adsorbed SB organics relative to the active OHO concentration (X_S/X_{OHO}) (Figure 5.15). In the primary anoxic reactor the ratio is high because adsorbed SBCOD originates from the influent and OHO death. In the secondary anoxic, the ratio is lower because SBCOD originates only from OHO death. For the K_2 and K_3 denitrification rates, there is no simple relationship between the K rates and the ηK_h because the adsorbed SBCOD to OHO ratio (X_S/X_{OHO}) is different in the primary and secondary anoxic reactors (and aerobic digester) and varies somewhat with sludge age and unaerated sludge mass fraction.

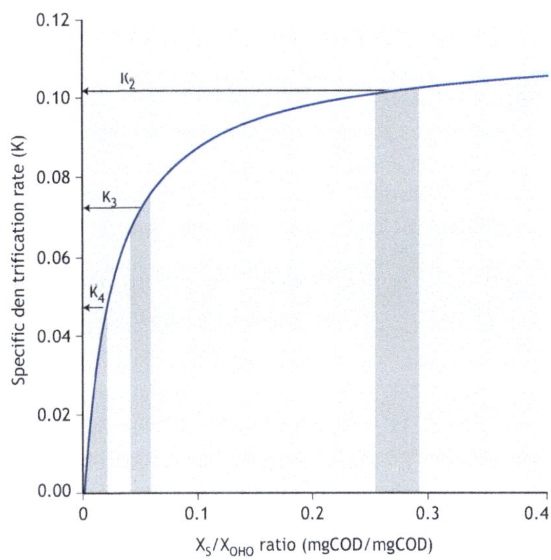

Figure 5.15 Specific denitrification rate (K) versus adsorbed SB organics to OHO biomass concentration ratio (X_S/X_{OHO}), showing the specific primary anoxic (K_2), secondary anoxic (K_3) and anoxic-aerobic digestion (K_4) denitrification rates.

It was concluded that the K_1, K_2, K_3 and K_4 denitrification 'constants' have no direct kinetic significance; their constancy is the result of a combination of kinetic reactions which show little variation with sludge age in the range 10 to 30 days (Figure 5.15).

Although temperature affects the K rates, once these have been adjusted for temperature, again the K rates show little variation at different sludge ages (Van Haandel *et al.,* 1981). It can be concluded from both experimental observation and theoretical kinetic points of view, that acceptance of constant K_2 and K_3 rates is acceptable for steady-state design. This is in fact done to estimate the denitrification potential (D_p) of an anoxic reactor under constant flow and load conditions.

With regard to K_1, this rate can change significantly because the RBCOD utilization rate can change appreciably depending on the mixing regime in the anoxic (or aerobic) reactor (Ekama *et al.,* 1996, a, b and Still *et al.,* 1996). However, its variation does not affect ND design significantly because primary anoxic reactors are normally sufficiently large to allow complete utilization of RBCOD even when the utilization rate (μ_{OHO}) is low. In fact, the denitrification design procedure requires that all the RBCOD is utilized in the primary anoxic reactor which introduces a minimum primary anoxic sludge mass fraction ($f_{x1,min}$) and a minimum a-recycle ratio (a_{min}) to ensure this. These concepts can also be used for anoxic selector reactor design (Ekama *et al.,* 1996a,b). The simulation model was also applied to anoxic-aerobic digestion of waste activated sludge (WAS). It was found that the model accurately predicted both aerobic and anoxic-aerobic digester behaviour under constant and cyclic flow and load conditions and validated the K_4 specific denitrification rate (Warner *et al.,* 1986); no significant adjustment to the kinetic constant values was necessary.

5.8.7 Denitrification potential

The concentration of nitrate (per litre influent flow Q_i) that an anoxic reactor can denitrify biologically is called that reactor's denitrification potential. It is called a potential because whether or not it is achieved depends on the nitrate load on the anoxic reactor(s). If too little nitrate is recycled to the anoxic reactor, the entire recycled nitrate will be denitrified and the actual removal of nitrate, i.e. denitrification performance, will be lower than the potential. In this case the denitrification is system-limited (or more accurately, recycle-limited). An increase in the system recycle ratios will increase nitrate load on the anoxic reactor and hence also the denitrification. Once the recycle rates are such that the nitrate load on the anoxic reactor(s) equals the denitrification potential of the reactor, then the system denitrification performance is optimal and the recycle ratios are at their optimum values. At this point the anoxic reactor and its outflow nitrate concentrations are just zero and the lowest possible, respectively. Increasing the recycle rates to above the optimum increases the nitrate concentration in the anoxic reactor outflow above zero but this does not improve the denitrification performance because the system has now become biological or kinetic-limited. The denitrification potential of the anoxic reactor(s) has been achieved and more nitrate cannot be denitrified by the particular anoxic reactor(s) and wastewater. Indeed, increases in the recycle ratios above the optimum values are uneconomical due to unnecessary pumping costs and they also introduce unnecessary additional dissolved oxygen into the anoxic reactors which causes an undesirable reduction in denitrification performance and increase in effluent nitrate concentration. The principle of denitrification design therefore hinges around (*i*) calculating the denitrification potential of the anoxic reactor(s), (*ii*) setting the nitrate load imposed on the anoxic reactor(s) equal to the denitrification potential, and (*iii*) calculating the recycle ratios associated with this condition. The recycle ratios calculated in this way are the optimum values.

From the discussion above it is clear that calculation of the nitrate load and denitrification potential is critical in the design for denitrification. The nitrate load is calculated from the nitrification capacity, which is the concentration of nitrate per influent flow (Q_i) generated by nitrification (Section 5.6.2, Eq. 5.35). The nitrification capacity (NIT_c, $mgN/l_{influent}$) was shown above to be approximately proportional to the influent TKN concentration (TKN_i). The denitrification potential is calculated separately for the utilization of the RBCOD and

SBCOD. The RBCOD gives rise to a rapid denitrification rate so it can be assumed that it is all utilized in the primary anoxic reactor. This is in fact an objective in the design. Accordingly, the contribution of the RBCOD to the denitrification potential is simply the catabolic component of its electron-donating capacity in terms of nitrate as N. Therefore, in the complete utilization of the influent RBCOD, a fixed proportion ($1-Y_{OHO}$) of the RBCOD electrons (catabolic component) will be donated to NO_3^- reducing it to N_2. Thus, knowing the influent RBCOD concentration and assuming it is all utilized, the denitrification potential of this RBCOD is:

$$D_{p1RBCOD} = f_{SS}COD_{b,i}(1 - f_{cv}Y_{OHOv}) / 2.86 \quad (mgNO_3\text{-}N/l_{influent}) \quad (5.46)$$

where:

$D_{p1\ RBCOD}$	denitrification potential of the influent RBCOD in the primary anoxic reactor
$COD_{b,i}$	influent biodegradable COD (mgCOD/l)
f_{SS}	RBCOD fraction of $COD_{b,i}$
Y_{OHOv}	OHO yield (0.45 mgVSS/mgCOD)
2.86	oxygen equivalent of nitrate

For the SBCOD, this substrate contributes to denitrification in the primary anoxic reactor and the secondary anoxic reactor. The denitrification potentials for the SBCOD are formulated on the basis of the specific denitrification rates K_2 and K_3, respectively. These K rates are a simplification of the kinetic equations describing the utilization of SBCOD from the influent and/or from organism death and lysis and have a basis in the fundamental biological kinetics incorporated in the activated sludge simulation models such as ASM1 (Van Haandel *et al.*, 1981; Henze *et al.*, 1987). The K rates define the denitrification rate as $mgNO_3$-N denitrified per day per mgOHOVSS mass in the anoxic reactor. So to determine the denitrification potential contributed by the SBCOD, the mass of OHO VSS produced per influent flow and the proportion of this mass in the primary and/or secondary anoxic reactors needs to be calculated and multiplied by the K_2 or K_3 rates.

From the steady-state activated sludge model for organics removal (Chapter 4, Section 4.4.2.1), the OHO mass in the system (MX_{OHOv}) is calculated from the biodegradable COD load (Eq. 4.9). Of this MX_{OHOv} mass, a fraction f_{x1} and/or f_{x3} is continuously present in the primary and/or secondary anoxic reactors, respectively, i.e. f_{x1} and f_{x3} are the primary and secondary anoxic sludge mass fractions, respectively. The OHOVSS mass in the primary and/or secondary anoxic reactors per influent flow is therefore given by:

$$f_{x1}MX_{OHOv} / Q_i = f_{x1}COD_{b,i}Y_{OHOv}SRT / (1 + b_{OHO}SRT) \quad (mgOHOVSS/l) \quad (5.46a)$$

for the influent in primary anoxic reactors and

$$f_{x3}MX_{OHOv} / Q_i = f_{x3}COD_{b,i}Y_{OHOv}SRT / (1 + b_{OHO}SRT) \quad (mgOHOVSS/l) \quad (5.46b)$$

for the influent in secondary anoxic reactors.

Multiplying these masses by the respective K rates gives the primary and secondary anoxic reactor denitrification potentials attributable to SBCOD ($D_{p1\ SBCOD}$, $D_{p3\ SBCOD}$), viz.

$$\begin{aligned} D_{p1SBCOD} &= K_2 f_{x1} MX_{OHOv} / Q_i \\ &= K_2 f_{x1} COD_{b,i} Y_{OHOv} SRT / (1 + b_{OHO}SRT) \end{aligned} \quad (5.47)$$

$$D_{p3SBCOD} = K_3 f_{x3} COD_{b,i} Y_{OHOv} SRT / (1 + b_{OHO}SRT) \quad (5.48)$$

This approach is valid because the K_2 and K_3 rates are continuous for the entire sludge residence time in the anoxic reactor(s), provided the nitrate concentration does not decrease to zero (Figure 5.14). Combining the denitrification potential components of the RBCOD and SBCOD yields the total denitrification potential of the primary and secondary anoxic reactors, i.e.

$$D_{p1} = D_{p1RBCOD} + D_{p1SBCOD}$$
$$= f_{SS}COD_{b,i}(1-f_{cv})/2.86 +$$
$$COD_{b,i}K_2f_{x1}f_{cv}SRT/(1+b_{OHO}SRT)$$
$$= COD_{b,i}\{f_{SS}(1-f_{cv})/2.86 +$$
$$K_2f_{x1}f_{cv}SRT/(1+b_{OHO}SRT)\}$$
$$(mgN/l_{influent}) \quad (5.49)$$

$$D_{p3} = D_{p3RBCOD} + D_{p3SBCOD}$$
$$= 0 + COD_{b,i}K_3f_{x3}Y_{OHOv}SRT/(1+b_{OHO}SRT)$$
$$(mgN/l_{influent}) \quad (5.50)$$

In eqs. 5.49 and 5.50, the K_2, K_3 and b_{OHO} rates are temperature-sensitive and decrease as temperature decreases. The temperature sensitivity of these rates has been measured and are defined in Tables 5.4 and 4.2. From eqs. 5.49 and 5.50, it can be seen that the denitrification potentials are directly proportional to the biodegradable COD concentration of the wastewater ($COD_{b,i}$). This is expected because in the same way that the oxygen demand is directly related to the COD load, so also is the nitrate demand (which is called the denitrification potential) since both oxygen and nitrate act as an electron acceptor for the same organic degradation reactions. For the same size anoxic reactor, the primary anoxic reactor has a much higher denitrification potential (by approximately 2 to 3 times) than the secondary anoxic reactor because (*i*) K_2 is larger than K_3 and (*ii*) more importantly, the RBCOD makes a significant contribution to the denitrification potential in the primary anoxic reactor. For this reason the RBCOD needs to be accurately specified to ensure reliable estimates of the N removal that can be achieved. For a normal municipal wastewater with an RBCOD fraction (f_{SS}) of approximately 25% (with respect to biodegradable COD), the RBCOD contributes approximately 1/3 to 1/2 of D_{p1} depending on the size of the primary anoxic reactor and temperature. In a system where a high degree of N removal is required, between 1/4 and 1/3 of the carbonaceous oxygen demand is met by denitrification, which reduces the carbonaceous oxygen demand in the aerobic reactor by the same amount. As mentioned earlier, this reduction represents approximately half of the oxygen that was required to produce the nitrate by nitrification (see Figure 5.11).

From Eq. 5.50, the influent RBCOD contribution to the denitrification potential of the secondary anoxic reactor is zero. This is because all the RBCOD is utilized either in the preceding primary anoxic and/or aerobic reactors. However, the $D_{p3RBCOD}$ term has been included in Eq. 5.50 in the event that an external carbon source such as methanol, acetic acid or high strength organic wastewater is dosed into the secondary anoxic reactor to improve the denitrification. $D_{p3RBCOD}$ is identical to Eq. 5.46 where $f_{SS}COD_{b,i}$ is the COD concentration of the dosed organic in mgCOD/l influent. With methanol the Y_{OHOv} value is significantly lower than 0.56 mgVSS/mgCOD.

The sludge mass fraction approach above is valid because the fraction of the VSS (MX_{VSS}) or TSS (MX_{TSS}) masses that is OHO mass (MX_{OHOv}) is constant for specified wastewater characteristics and sludge age and equal to the active fraction ($f_{at,OHO}$ or $f_{av,OHO}$ - eqs. 4.26 and 4.27) and almost the same in the anoxic and aerobic reactors of the system. Therefore the anoxic and aerobic sludge mass fraction are the same whether calculated from the VSS, TSS or OHO masses; e.g. in a MLE system with anoxic and aerobic reactor volumes of 3,000 and 6,000 m^3, respectively, very close to 1/3 of the OHO, VSS and TSS masses in the system are in the anoxic reactor, and hence the anoxic sludge mass fraction is 0.33.

5.8.8 Minimum primary anoxic sludge mass fraction

In Eq. 5.49 it is assumed that the initial rapid rate of denitrification is always complete, i.e. the actual retention time in the primary anoxic reactor is always longer than the time required to utilize all the influent RBCOD. This is because in Eq. 5.49, the denitrification attributable to the influent RBCOD is stoichiometrically expressed, not kinetically; it gives the concentration of nitrate that the K_1 rate removes when allowed sufficient time to reach completion. By considering the actual retention time (say t_1) required to complete the 1st phase of denitrification (Figure 5.14A), and noting that $t_1(a+s+1)$ is then the minimum nominal hydraulic retention time to achieve this, it can be shown that the minimum primary anoxic sludge mass fraction $f_{x1,min}$ to remove all the influent RBCOD at a rate of K_1 mgNO3-N/mgOHOVSS.d is:

$$f_{x1,min} = \frac{f_{SS}(1-f_{cv}Y_{OHOv})(1+b_{OHO,T}SRT)}{2.86K_{1,T}Y_{OHOv}SRT} \quad (5.51)$$

Substituting the values of the kinetic constants into Eq. 5.51 yields $f_{x1,min} < 0.08$ for SRT > 10 days at 14 °C. This value is much lower than most practical primary anoxic reactors so that Eq. 5.51 will be valid in most cases. Equation 5.51 also applies to sizing anoxic selectors provided K_1 (or μ_{OHO}) is appropriately selected (see Section 5.8.6, Eq. 5.44).

5.8.9 Denitrification - influence on reactor volume and oxygen demand

From the design approach to nitrification (Eq. 5.19) and denitrification (eqs. 5.49 and 5.50), it can be seen that the design for N removal is done entirely using sludge mass fractions and does not require the volume of the reactor to be known. The volume of the reactor is obtained in the identical fashion as for the fully aerobic system and follows from the selection of the TSS concentration (X_{TSS}) for the reactor (Chapter 4, Section 4.7). The volume of the reactor so obtained is then subdivided in proportion to the calculated primary and/or secondary anoxic and aerobic sludge mass fractions. Consequently N removal design is grafted directly into the aerobic system design and for the same design reactor TSS concentration and sludge age, a fully aerobic system and an anoxic-aerobic system for N removal will have the same reactor volume. Research has indicated that there are many factors that influence the mass of sludge generated for a given sludge age and daily average COD load, and alternating anoxic-aerobic conditions is one of them. However, relative to the uncertainty in the organic (COD) load and unbiodegradable particulate COD fraction and their daily and seasonal variation, these influences are not large enough from a design point of view to be given specific consideration in the design procedure. From a design point of view, the only significant difference between aerobic and anoxic-aerobic conditions is the oxygen demand and this difference needs to be taken into account for an economical design (Figure 5.11).

5.9 DEVELOPMENT AND DEMONSTRATION OF DESIGN PROCEDURE

It was concluded above that the influent wastewater characteristics that need to be accurately known are the influent TKN/COD ratio and RBCOD fraction. These have a major influence on the nitrification capacity and denitrification potential, respectively, and hence on the N removal performance and minimum effluent nitrate concentration that can be achieved by biological denitrification. The effect of these two wastewater characteristics on design will be demonstrated below with numerical examples generated from the example raw and settled wastewaters with different influent TKN concentrations and RBCOD fractions. The design of biological N removal is developed and demonstrated below by continuing the calculations with the example raw and settled wastewaters. The wastewater characteristics are listed in tables 4.3 and 5.1. The only additional characteristic required for the denitrification design is the influent RBCOD fraction (f_{SS}), which is 0.25 and 0.385 of the biodegradable COD ($COD_{b,i}$) for the raw and settled wastewaters, respectively. The results obtained so far for the COD removal and nitrification calculations are listed in Table 5.5.

Table 5.5 Summary of the COD removal and nitrification design calculations for N removal at 20 days' sludge age and 14 °C and 22 °C for the example raw and settled wastewaters (see tables 4.3 and 5.1 for wastewater characteristics).

Parameter	Symbol	Unit	Raw sewage		Settled sewage	
Wastewater characteristics						
Influent flow	Q_i	Ml/d	15.00		14.93	
Influent COD concentration	COD_i	mgCOD/l	750		450	
Influent TKN concentration	TKN_i	mgN/l	60		51	
TKN/COD ratio	f_{ns}	mgTKN/mgCOD	0.080		0.113	
RBCOD fraction	f_{SS}	mgCOD/mgCOD	0.25		0.385	
Wastewater temperature	T	°C	14	22	14	22
Carbonaceous material removal (Chapter 4)						
Influent biodegradable COD mass	$FCOD_{b,i}$	kgCOD/d	8,775	8,775	5,664	5,664
Residual biodegradable COD mass	$FCOD_b$	kgCOD/d	0	0	0	0
Active organism mass	MX_{OHOv}	kgVSS	15,659	12,984	10,107	8,381
Endogenous residue mass	$MX_{E,OHOv}$	kgVSS	12,663	13,198	8,174	8,519
Unbiodegradable organic mass	MX_{Iv}	kgVSS	22,804	22,804	3,649	3,649
Volatile suspended solids mass	MX_{VSS}	kgVSS	51,126	48,986	21,930	20,549
Total suspended solids mass	MX_{TSS}	kgTSS	68,168	65,315	26,421	24,757
Active fraction - VSS	$f_{av,OHO}$		0.306	0.265	0.461	0.408
Active fraction - TSS	$f_{at,OHO}$		0.230	0.199	0.383	0.339
Mass carbonaceous oxygen demand	FO_c	kgO_2/d	6,679	6,838	4,311	4,413
Mass nitrogen into sludge production	FN_s	kgN/d	255.6	244.9	109.7	102.7
Mass TSS wasted	FX_{TSS}	kgTSS/d	3,408	3,266	1,321	1,238
Nitrification (Section 5.6)						
Permissible unaerated sludge mass fraction	$f_{x,max}$		0.534	0.80	0.534	0.80
Design unaerated sludge mass fraction	f_{xt}		0.534	0.534	0.534	0.534
Factor of safety	S_f		1.25	2.88	1.25	2.88
Effluent biodegradable organic N	$N_{ob,e}$	mgN/l	0.0	0.0	0.0	0.0
Effluent unbiodegradable soluble organic N	$N_{ous,e}$	mgN/l	1.80	1.80	1.80	1.80
Effluent ammonia	$S_{NHx,e}$	mgN/l	2.0	0.7	2.0	0.7
Effluent TKN	TKN_e	mgN/l	3.8	2.5	3.8	2.5
N concentration into sludge production	N_s	mgN/l	17.0	16.3	7.4	6.9
Nitrification capacity	NIT_c	mgN/l	39.2	41.2	39.9	41.6
Mass nitrifiers	MX_{ANO}	kgVSS	702	669	711	673
Nitrification oxygen demand	FO_{NIT}	kgO_2/d	2,685	2,824	2,719	2,840
Total oxygen demand	FO_t	kgO_2/d	9,364	9,661	7,030	7,254

5.9.1 Review of calculations

For the raw wastewater characteristics, (i.e. $f_{XU,CODi}$ = 0.15 mgCOD/mgCOD, $f_{SU,CODi}$ = 0.07 mgCOD/mgCOD, T_{min} = 14 °C, COD_i = 750 mgCOD/l; see Table 4.3) and 20 days' sludge age, and accepting the nitrogen content of the volatile solids (f_n) to be 0.10 mgN/mgVSS, the nitrogen required for sludge production N_s = 17.0 mgN/l (Eq. 5.27).

From Section 5.7.5, the effluent biodegradable and unbiodegradable soluble organic nitrogen concentrations ($N_{obs,e}$ and $N_{ous,e}$, Eq. 5.37) are 0.0 and

1.80 mgN/l, respectively. From Eq. 5.15 the effluent ammonia concentration $S_{NHx,e}$ is 2.0 mgN/l. The effluent TKN concentration (TKN_e) is the sum of $N_{ous,e}$ and $S_{NHx,e}$ (Eq. 5.33) and hence TKN_e = 3.8 mgN/l (Table 5.5).

The nitrification capacity (NIT_c) is found from Eq. 5.35 and for the example raw wastewater (TKN_i = 48.0 mgN/l; TKN/COD = 0.08 mgN/mgCOD) at 14 °C is NIT_c = 48.0 – 17.0 – 3.8 = 39.2 mgN/l.

The nitrification oxygen demand FO_{NIT} is found from Eq. 5.43, i.e.

$FO_{NIT} = 4.57\, NIT_c\, Q_i = 4.57 \cdot 39.2 \cdot 15 \cdot 10^6$ mgO_2/d
$= 2{,}687$ kgO_2/d

and the mass of nitrifier VSS in the reactor is given by Eq. 5.42, i.e.

$MX_{ANO} = 0.1 \cdot 20 / (1 + 0.034 \cdot 20) \cdot 39.2 \cdot 15 \cdot 10^6$
= 702 kgVSS.

The above calculations for N_s, $S_{NHx,e}$, TKN_e, NIT_c and FO_{NIT} and MX_{ANO} for the raw and settled wastewater at 14 °C and 22 °C are listed in Table 5.5.

In the design, because it is intended to reduce the nitrate concentration as much as possible, the alkalinity change in the wastewater will be minimized; assuming that 80% of the nitrate formed is denitrified, the H_2CO_3 alkalinity change = $-7.14NIT_c$ + 3.57 (nitrate denitrified) = – 7.14 · 39.2 + 3.57 · 0.80 · 39.2 = – 168 mg/l as $CaCO_3$. With an influent H_2CO_3 alkalinity of 250 mg/l as $CaCO_3$ the effluent H_2CO_3 alkalinity = 250 – 168 = 82 mg/l as $CaCO_3$, which, from Figure 5.6, will maintain a pH above 7 (see Section 5.4.6).

5.9.2 Allocation of unaerated sludge mass fraction

In nitrogen removal systems, the maximum anoxic sludge mass fraction available for denitrification ($f_{xd,max}$) can be set equal to the maximum unaerated sludge mass fraction $f_{x,max}$ at the minimum temperature, i.e.

$$f_{xd,max} - f_{x,max} \tag{5.52}$$

where $f_{x,max}$ is given by Eq. 5.19 for selected SRT, $\mu_{ANO,max,T}$ and T_{min}.

This is because for N removal systems, unaerated sludge mass need not be set aside for the anaerobic reactor. In N and P removal systems some of the unaerated sludge mass (0.12-0.25) needs to be set aside for the anaerobic reactor to stimulate EBPR. This sludge mass fraction, called the anaerobic sludge mass fraction and denoted as f_{xa}, cannot be used for denitrification. For EBPR to be as high as possible, no nitrate should be recycled to the anaerobic reactor so that zero denitrification takes place in this reactor. So, for the purposes of this development and demonstration of denitrification behaviour, it will be accepted that the maximum unaerated sludge mass fraction available at 20 days' sludge age ($f_{x,max}$) is all allocated to anoxic conditions, i.e. $f_{xd,max} = f_{x,max} = 0.534$.

5.9.3 Denitrification performance of the MLE system

5.9.3.1 Optimum a-recycle ratio

In the MLE system, the anoxic sludge mass fraction is all in the form of a primary anoxic reactor, i.e. f_{x1} = $f_{xd,max}$ = $f_{x,max}$. The denitrification potential of the primary anoxic reactor D_{p1} is found from Eq. 5.49, i.e. for the example raw wastewater at 14 °C and $f_{x,max} = f_{xd,max} = f_{x1} = 0.534$, D_{p1} = 52.5 mgN/l. The D_{p1} values for the example raw and settled wastewaters at 14 °C and 22 °C are listed in Table 5.6.

Table 5.6 Summary of the denitrification design calculations for the Modified Ludzack-Ettinger (MLE) N removal system at 20 days' sludge age and 14 oC and 22 oC for the example raw and settled wastewaters (see tables 4.3 and 5.1 for other characteristics).

Parameter	Symbol	Unit	Raw sewage		Settled sewage	
Wastewater characteristics						
Influent flow	Q_i	Ml/d	15.00		14.93	
Influent COD concentration	NIT_c	mgCOD/l	750		450	
Influent TKN concentration	TKN_i	mgN/l	60		51	
TKN/COD ratio	$F_{TKNi/CODi}$	mgTKN/mgCOD	0.080		0.113	
RBCOD fraction	f_{SS}	mgCOD/mgCOD	0.25		0.385	
Wastewater temperature		°C	14	22	14	22
MLE system design features						
Primary anoxic mass fraction	f_{x1}		0.534	0.534	0.534	0.534
Denitrification potential	D_{p1}	mgN/l	52.5	71.5	40.1	52.4
Minimum primary anoxic mass fraction	$f_{x1,min}$		0.068	0.019	0.105	0.029
DO in a recycle	$SO_{2,a}$	mgO_2/l	2.0	2.0	2.0	2.0
DO in s recycle	$SO_{2,s}$	mgO_2/l	1.0	1.0	1.0	1.0
Underflow recycle ratio	s		1.0	1.0	1.0	1.0
Performance at example TKN/COD ratio			0.080	0.080	0.113	0.113
Optimum a-recycle ratio	a_{opt}		21.6	44.1	6.5	17.9
Effluent nitrate at a_{opt}	$S_{NO3,e,aopt}$	mgN/l	1.7	0.9	4.7	2.1
Practical a-recycle ratio	a_{prac}		5.0	5.0	5.0	5.0
Effluent nitrate at a_{prac}	$S_{NO3,e,aprac}$	mgN/l	5.6	5.9	5.7	5.9
Oxygen recovered by denitrification	FO_{DENIT}	kgO_2/d	1,440	1,515	1,458	1,524
Net total oxygen demand	FO_{tDENIT}	kgO_2/d	7,924	8,147	5,572	5,730
At TKN/COD ratio where $a_{opt}=a_{prac}=5:1$(balanced)			0.104	0.132	0.119	0.148
Effluent nitrate at a_{opt}	$S_{NO3,e,aopt}$	mgN/l	8.1	11.3	6.0	8.1
Effluent TKN	TKN_e	mgN/l	4.3	3.6	3.9	3.0
Total effluent N		mgN/l	12.4	14.9	9.9	11.1
% nitrogen removal			84.1	84.9	81.5	83.3
Nitrification oxygen demand	FO_{NIT}	kgO_2/d	3,894	5,411	2,884	3,862
Oxygen recovered by denitrification	FO_{DENIT}	kgO_2/d	2,089	2,902	1,547	2,072
Net total oxygen demand	FO_{tDENIT}	kgO_2/d	8,485	9,346	5,648	6,204

In the MLE system, if the nitrate concentration in the outflow of the anoxic reactor is zero, then the nitrate concentration in the aerobic reactor ($S_{NO3,ar}$) is equal to $NIT_c/(a + s + 1)$ i.e. the nitrification capacity in mgN/l influent flow diluted by the total (containing no nitrate) flow entering the aerobic reactor which is (a + s + 1) times the influent flow where a and s are the mixed liquor and underflow recycle ratios (with respect to the influent average dry weather flow Q_i), respectively. Accepting that there is no denitrification in the secondary settling tank (which needs to be minimized anyway due to the problem of rising sludges), the aerobic reactor and system effluent nitrate concentrations ($S_{NO3,ar}$ and $S_{NO3,e}$, respectively) are equal and given by:

$$S_{NO3,e} = S_{NO3,ar} = NIT_c / (a + s + 1) \tag{5.53}$$

Knowing $S_{NO3,e}$ and $S_{NO3,ar}$ and taking into account DO concentrations in the a and s recycles, i.e. $S_{O2,a}$

and $S_{O2,s}$ mgO_2/l, respectively, the equivalent nitrate load on the primary anoxic reactor ($S_{NO3,p1}$) by the a and s recycles is:

$$S_{NO3,p1} = [S_{NO3,ar} + \frac{S_{O2,a}}{2.86}]a + [S_{NO3,e} + \frac{S_{O2,s}}{2.86}]s \qquad (5.54)$$

The optimum denitrification, i.e. lowest effluent nitrate concentration, is obtained when the equivalent nitrate load on the anoxic reactor is equal to the denitrification potential of the anoxic reactor, i.e. $D_{p1} = S_{NO3,p1}$, viz.

$$D_{p1} = [\frac{NIT_c}{(a+s+1)} + \frac{S_{O2,a}}{2.86}] \cdot a + [\frac{NIT_c}{(a+s+1)} + \frac{S_{O2,s}}{2.86}] \cdot s \qquad (5.55)$$

Solving Eq. 5.55 for a yields the a-recycle ratio which exactly loads the primary anoxic reactor to its denitrification potential with nitrate and DO.

This a value is the optimum because it results in the lowest $S_{NO3,e}$ i.e..

$$a_{opt} = [-B + \sqrt{B^2 + 4AC}]/(2A) \qquad (5.56)$$

where:

A	$S_{O2,a}/2.86$
B	$NIT_c - D_{p1} + \{(s+1)S_{O2,a} + s\ S_{O2,s}\}/2.86$
C	$(s+1)(D_{p1} - s\ S_{O2,s}/2.86) - s\ NIT_c$

and

$$S_{NO3,e,min} = S_{NO3,e,aopt} = NIT_c / (a_{opt} + s + 1) \qquad (mgN/l)\ (5.57)$$

For $a = a_{opt}$, Eq. 5.57 for $S_{NO3,e}$ is valid and will give the minimum $S_{NO3,e}$ attainable. When $a \leq a_{opt}$ then Eq. 5.57 also is valid because the assumption on which Eq. 5.56 is based is valid, i.e. $S_{NO3,p1} \leq D_{p1}$ or equivalently, zero nitrate concentration in the outflow of the anoxic reactor. However, for $a > a_{opt}$ this assumption is no longer valid and $S_{NO3,e}$ increases as the a-recycle ratio increases due to increasing DO mass flow rates entering the anoxic reactor. For $a > a_{opt}$, $S_{NO3,e}$ is given by the difference between the equivalent nitrate load on the anoxic reactor (which is the sum of the nitrification capacity NIT_c and the nitrate equivalent of the oxygen concentration with respect to the influent flow) and the denitrification potential D_{p1}, viz.

$$S_{NO3,e} = NIT_c + \frac{aS_{O2,a}}{2.86} + \frac{sS_{O2,s}}{2.86} - D_{p1} \qquad (mgN/l)\ (5.58)$$

Because NIT_c, D_{p1}, $S_{O2,s}$ and $S_{O2,a}$ are constant, the increase in $S_{NO3,e}$ with increasing a above a_{opt} is linear with slope $S_{O2,a}/2.86$ mgN/l. At $a = a_{opt}$, eqs. 5.57 and 5.58 give the same $S_{NO3,e}$ concentrations.

Accepting the design sludge age of 20 days, which allows a maximum unaerated sludge mass fraction $f_{x,max}$ of 0.534, the denitrification behaviour of the MLE system is demonstrated below for the example raw and settled wastewaters at 14 °C and 22 °C. In the calculations the DO concentrations in the a and s recycles $S_{O2,a}$ and $S_{O2,s}$ are 2 and 1 mgO_2/l, respectively, and the underflow recycle ratio s is 1:1. This s-recycle ratio is usually fixed at a value such that satisfactory settling tank operation is obtained. Details of secondary settling tank theory, design, modelling and operation are discussed by Ekama *et al.* (1997) and presented in Chapter 12.

Substituting the values for the nitrification capacity NIT_c and denitrification potential D_{p1} (tables 5.5 and 5.6) into eqs. 5.56 and 5.57, the optimum mixed liquor recycle ratio a_{opt} and minimum effluent nitrate concentration $S_{NO3,e,aopt}$ are obtained, e.g. for the settled wastewater at 14 °C.

A	= 2/2.86 = 0.70
B	= 39.6 – 40.1 + {(1 + 1)2 + 1·1}/2.86
	= + 1.52
C	= (1 + 1)·(40.1 – 1·1/2.86) – 1·39.6
	= + 39.61

Hence a_{opt} = 6.5 and $S_{NO3,e,min}$ = 4.7 mgN/l. The above results, as well as those for the example raw and settled wastewater at 14 °C and 22 °C, are listed in Table 5.6.

The results in Table 5.6 show that for all 4 cases a_{opt} exceeds 5. Although the calculations include the discharge of DO to the anoxic reactor, recycle ratios above 5 to 6 are not cost-effective. The small decreases in $S_{NO3,e}$ which are obtained for even large increases in the a-recycle ratio above approximately 5:1 do not warrant the additional pumping costs.

This is illustrated in Figure 5.16 which shows $S_{NO3,e}$ versus the a-recycle ratio for the example raw (Figure 5.16A) and settled (Figure 5.16B) wastewater at 14 °C and 22 °C plotted from eqs. 5.57 and 5.58. For the settled wastewater (Figure 5.16B) at 14 °C and s = 1:1, for a < a_{opt}, the anoxic reactor is underloaded with nitrate and DO and as the a recycle increases up to a_{opt}, so the equivalent nitrate load increases towards the anoxic reactor's denitrification potential. Initially, $S_{NO3,e}$ decreases sharply for increases in a, but as a increases so the decrease in $S_{NO3,e}$ becomes smaller. At 14 °C with a = a_{opt} = 6.5, the anoxic reactor is loaded to its denitrification potential by the a and s recycles and a $S_{NO3,e,min}$ = $S_{NO3,e,aopt}$ = 4.7 mgN/l is achieved. At a = a_{opt} = 6.5, the greatest proportion of the anoxic reactor's denitrification potential is used for denitrification and therefore yields the minimum effluent nitrate concentration ($S_{NO3,e,aopt}$). This is shown in figures 5.17A and 5.17B for the raw and settled wastewaters at 14 °C. For the settled wastewater at 14 °C (Figure 5.17B) at a_{opt} = 6.5, 88% of the equivalent nitrate load (i.e. (a + s) $S_{NO3,e,min}$ = 35.2 mgN/l out of a D_{p1} = 40.1 mgN/l) is nitrate and therefore 88% of the denitrification potential of the anoxic reactor is utilized for denitrification and 12% for DO removal. The higher the a-recycle ratio, the greater the proportion of the denitrification potential is utilized for DO removal. At 14 °C, for a > a_{opt}, the equivalent nitrate load exceeds the denitrification potential and as the a recycle increases so $S_{NO3,e}$ increases due to the increased DO mass flow to the anoxic reactor. From Eq. 5.58, at a = 15, $S_{NO3,e}$ = 10.6 mgN/l and 27% of the denitrification potential is required to remove DO, leaving only 73% for denitrification (figures 5.16B and 5.17B).

For 14 °C, the plots of $S_{NO3,e}$ versus a at underflow s-recycle ratios of 0.5:1 and 2.0:1 are also given in Figure 5.16 and show that a_{opt} is not significantly different at different s-recycle ratios. Also, at low a-recycle ratios, changes in s have a significant influence on $S_{NO3,e}$, but at high a-recycle ratios, even significant changes in s do not

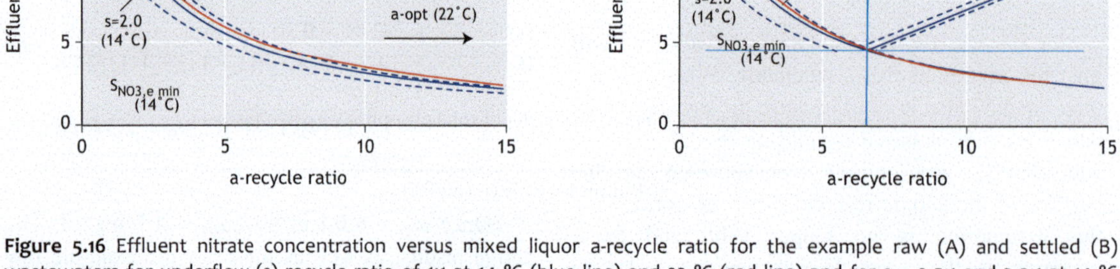

Figure 5.16 Effluent nitrate concentration versus mixed liquor a-recycle ratio for the example raw (A) and settled (B) wastewaters for underflow (s) recycle ratio of 1:1 at 14 °C (blue line) and 22 °C (red line) and for s = 0.5:1 and 2.0:1 at 14 °C (dashed lines).

significantly change $S_{NO3,e}$. This is because at high a, most of the nitrate is recycled to the anoxic reactor by the a recycle, so that changes in s do not significantly change the nitrate load on the anoxic reactor. Hence, for the MLE system, decreases in s can be compensated for by increases in a; it makes little difference which recycle brings the nitrate to the anoxic reactor as long as the anoxic reactor is loaded as closely as practically possible to its denitrification potential in order to minimize $S_{NO3,e}$.

For the settled wastewater at 22 °C and s = 1:1 (Figure 5.16B), $S_{NO3,e}$ versus a is similar to that at 14 °C up to a = 6.5. This is because NIT_c at 14 °C and 22 °C for the example raw and settled wastewater are almost the same (i.e. 39.9 and 41.6 mgN/l at 14 °C and 22 °C, respectively, see Table 5.5). However, at 22 °C, the denitrification potential is significantly higher than at 14 °C (40.1 mgN/l at 14 °C and 52.4 mgN/l at 22 °C, see Table 5.6) so that a higher a_{opt} is required (viz. 17.9) at 22 °C to load the anoxic reactor to its denitrification potential than at 14 °C. Therefore at 22 °C, as the a recycle increases above 6.5, $S_{NO3,e}$ continues to decrease until a_{opt} = 17.9 is reached. The increase in a from 6.5 to 17.9 reduces $S_{NO3,e}$ from 4.9 to 2.1, i.e. only 2.8 mgN/l. This small decrease in $S_{NO3,e}$ is not worth the large increase in pumping costs required to produce it. Consequently for economic reasons, the a-recycle ratio is limited at a practical maximum (a_{prac}) of say 5:1, which fixes the lowest practical effluent nitrate concentration ($S_{NO3,e,prac}$) from the MLE system between 5 and 10 mgN/l, depending on the influent TKN/COD ratio.

From the design procedure demonstrated so far, it is clear that the procedure hinges round balancing the equivalent nitrate load with the denitrification potential by appropriate choice of the a-recycle ratio. For selected system design parameters (sludge age, anoxic mass fraction, underflow recycle ratio, etc.) and wastewater characteristics (temperature, readily biodegradable COD fraction, TKN/COD ratio etc.), the denitrification potential of the MLE system is fixed. With all the above fixed, the system denitrification performance is controlled by the a-recycle ratio, and this performance is optimum when the a-recycle ratio is set at the optimum a_{opt}. For $a < a_{opt}$, the performance will be below optimum because the equivalent nitrate load is less than the denitrification potential (Figure 5.17); for $a = a_{opt}$, the performance is optimal because the equivalent nitrate load equals the denitrification potential; and for $a > a_{opt}$, the performance is again suboptimal because now the equivalent nitrate load is greater than the

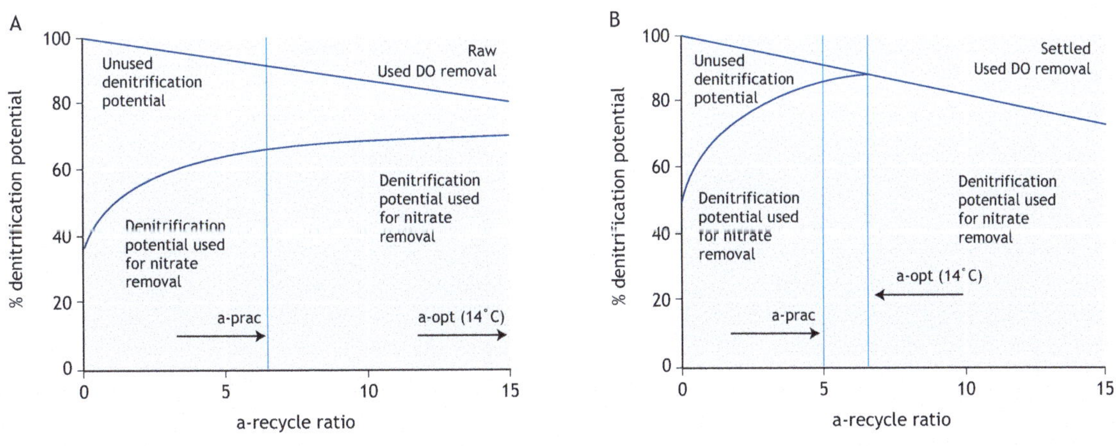

Figure 5.17 Percentage of unused denitrification potential , used by dissolved oxygen in the recycles and for denitrification versus the a-recycle ratio for the example raw (A) and settled (B) wastewaters for the underflow (s) recycle ratio of 1:1 at 14 °C.

denitrification potential and more than necessary DO is recycled to the anoxic reactor which reduces the denitrification (see figures 5.16 and 5.17).

If a practical limit on a is set at say a_{prac} = 5:1 and a_{opt} is significantly higher, then a significant proportion of the anoxic reactor's denitrification potential is not used (Figure 5.17). There are two options to deal with this unused denitrification potential: (1) change the design, i.e. decrease the sludge age (SRT) and/or unaerated sludge mass fraction ($f_{x,max}$), or (2) leave the system as designed (i.e. SRT = 20 d and $f_{x,max}$ = 0.534) and keep the unused denitrification potential in reserve as a factor of safety against changes in wastewater characteristics, such as (*i*) increased organic load, which will require a reduction in sludge age, (*ii*) increased TKN/COD ratio, which will load the anoxic reactor with nitrate at lower a-recycle ratios or (*iii*) decreased RBCOD fraction, which decreases the anoxic reactors' denitrification potential.

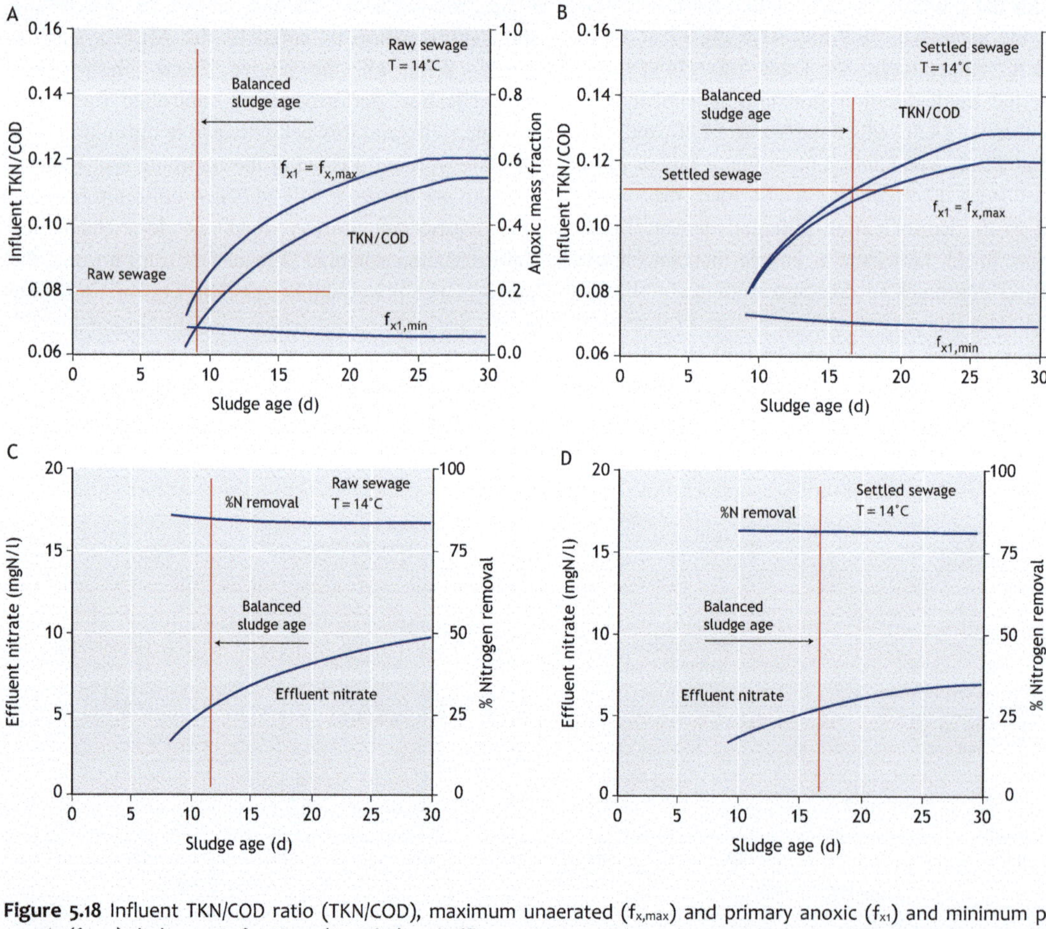

Figure 5.18 Influent TKN/COD ratio (TKN/COD), maximum unaerated ($f_{x,max}$) and primary anoxic (f_{x1}) and minimum primary anoxic ($f_{x1,min}$) sludge mass fractions (A and B) and effluent nitrate concentration and %N removal (C and D) for balanced MLE systems with a 5:1 practical upper limit to the a-recycle ratio for the example raw (A and C) and settled (B and D) wastewaters at 14 °C.

5.9.3.2 The balanced MLE system

With option (1) the anoxic sludge mass fraction f_{x1} is decreased to eliminate the unused denitrification potential. The decrease in f_{x1} increases the aerobic mass fraction and therefore the factor of safety (S_f) on nitrification. To maintain the same S_f, the sludge age of the system can be reduced to that value at which the lower f_{x1} is equal to the maximum unaerated sludge mass fraction $f_{x,max}$ allowed (i.e. $f_{x1} = f_{x,max}$) for the selected $\mu_{ANO,max,20}$ and T_{min}. An MLE system with a sludge age (SRT) and influent TKN concentration (TKN_i) in which $f_{x1} = f_{x,max}$ and $a_{opt} = a_{prac}$ (say 5:1) so that this a_{prac} loads the anoxic reactor exactly to its denitrification potential is called a balanced MLE system. This approach to design of the MLE system was proposed by Van Haandel *et al.* (1982) and gives the most economical activated sludge reactor design, i.e. the lowest sludge age, and therefore the smallest reactor volume, and the highest denitrification with the a-recycle ratio fixed at some maximum practical limit. The influent TKN/COD ratio, $f_{x,max} = f_{x1}$, $f_{x1,min}$, $S_{NO3,e}$ and %N removal ($\%N_{rem}$) versus sludge age for balanced MLE systems for the example raw and settled wastewaters at 14 °C and 22 °C are shown in figures 5.18 and 5.19, respectively.

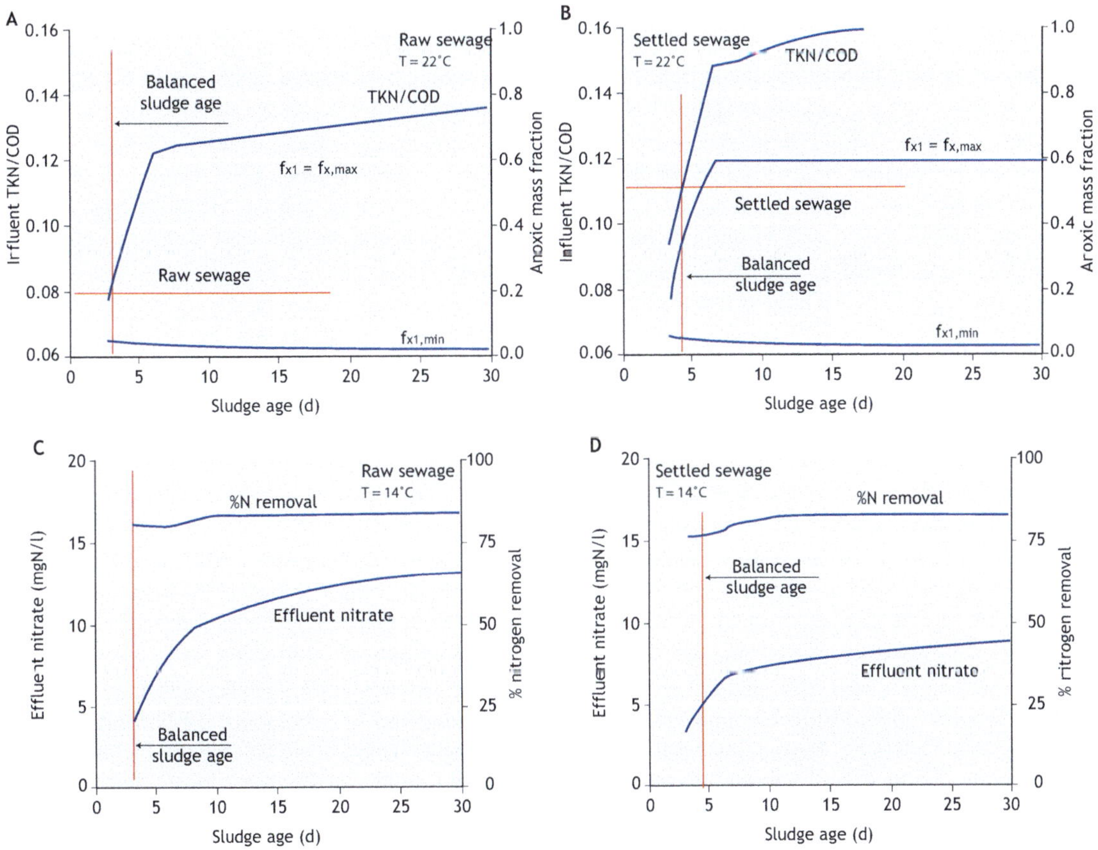

Figure 5.19 Influent TKN/COD ratio (TKN/COD), maximum unaerated ($f_{x,max}$) and primary anoxic (f_{x1}) and minimum primary anoxic ($f_{x1,min}$) sludge mass fractions (A and B) and effluent nitrate concentration and %N removal (C and D) for balanced MLE systems with a 5:1 practical upper limit to the a-recycle ratio for the example raw (A and C) and settled (B and D) sewage at 22 °C.

The sludge age which balances the MLE system for given wastewater characteristics and a_{prac} cannot be calculated directly. It is easiest to calculate the influent TKN concentration for a range of sludge ages and choose the sludge age which matches the wastewater TKN concentration (TKN_i). The procedure for calculating TKN_i for a balanced MLE system is as follows: from the design $\mu_{ANO,max,20}$, T_{min} and S_f and a selected sludge age, $f_{x,max}$ is calculated from Eq. 5.19. Provided $f_{x,max} > f_{x1,min}$ (Eq. 5.51), f_{x1} is set equal to $f_{x,max}$. Knowing f_{x1} and the wastewater characteristics, D_{p1} is calculated from Eq. 5.49. This D_{p1} and a selected value for a_{prac} is then substituted into Eq. 5.55, which sets the equivalent nitrate load on the anoxic reactor equal to the denitrification potential and hence a_{opt} equals the selected a_{prac}. With D_{p1} and a known, NIT_c is calculated from Eq. 5.55. Once NIT_c is known, TKN_i is calculated from $TKN_i = TKN_e + N_s + NIT_c$ (Eq. 5.35) where $TKN_e = N_{ous,e} + S_{NHx,e}$ (Eq. 5.33) and $S_{NHx,e}$ is given by Eq. 5.21 because with S_f fixed the SRT - $f_{x,max}$ relationship is fixed. With NIT_c and TKN_i known, the effluent nitrate concentration $S_{NO3,e}$ and % nitrogen removal ($\%N_{rem}$) are found from Eq. 5.57 and $\%N_{rem} = 100\ (TKN_i - (S_{NO3,e} + TKN_e))/TKN_i$, respectively. This calculation is repeated for different sludge ages. The shortest sludge age allowed is the one which gives $f_{x1} = f_{x,max} = f_{x1,min}$.

In Figure 5.18 for 14 °C, for the raw wastewater (Figure 5.18A and C), it can be seen that $f_{x1} (= f_{x,max})$ increases from approximately 0.09 at 8 d sludge age, at which $f_{x,max}$ is slightly greater than $f_{x1,min}$, to 0.60 at 26 d sludge age, at which $f_{x,max}$ is equal to the upper limit set for it. As f_{x1} increases so the influent TKN/COD ratio increases from 0.061 at 8 d sludge age to 0.115 at 26 d sludge age. With the increase in TKN/COD ratio, the nitrification capacity NIT_c increases and hence $S_{NO3,e}$ increases from approximately 3.2 mgN/l at 8 d sludge age to 9.3 mgN/l at 26 d sludge age because the a and s recycle ratios remain at 5:1 and 1:1, respectively (see Eq. 5.58). The %N removal, which includes the N removed via sludge wastage N_s, decreases marginally from 85 to 82% as the influent TKN/COD ratio and sludge age increase for the balanced MLE system.

For the settled wastewater at 14 °C (Figure 5.18B and D), the influent TKN/COD ratio, f_{x1} and $f_{x1,min}$ results are similar to those for the raw wastewater, i.e. for the same sludge age approximately the same TKN/COD ratio is found for the balanced MLE system. For the settled wastewater, the $S_{NO3,e}$ is slightly lower, increasing from approximately 3.2 to 6.7 mgN/l from 8 d to 26 d sludge age; also the %N removal is somewhat lower, around 78% mainly due to the lower N removal via sludge wastage N_s. However, it must be remembered that the TKN/COD ratio and RBCOD fraction of a settled wastewater are higher than those of the raw wastewater from which it is produced, viz. TKN/COD ratio 0.113 and 0.080 mgN/mgCOD and RBCOD fraction (f_{SS}) 0.385 and 0.25 for the example settled and raw wastewaters, respectively. Therefore at 14 °C, while the raw wastewater can be treated in a balanced MLE system at approximately 11 d sludge age (Figure 5.18A), the sludge age for the settled wastewater balanced MLE system is approximately 17 d (Figure 5.18B). A comparison of the balanced MLE systems for the example raw and settled wastewaters is given in Table 5.7.

From Table 5.7 it can be seen that $S_{NO3,e}$ is less than 1 mgN/l higher for the settled wastewater but the reactor volume and total oxygen demand are significantly lower compared with the raw wastewater. Therefore, from an activated sludge system point of view, treating settled wastewater would be more economical than treating raw wastewater for a comparable effluent quality. Also, both systems require sludge treatment; for the raw wastewater because 11 d sludge-age waste sludge is not stable (high active fraction, $f_{at,OHO}$) and for the settled wastewater, the primary sludge needs to be stabilized. The 11 d sludge-age waste sludge can be stabilised with anoxic aerobic digestion which allows the N released in digestion to be nitrified and denitrified (Warner *et al.,* 1986; Brink *et al.,* 2007) and primary sludge can be anaerobically digested to benefit from gas generation. The decision to treat either raw or settled wastewater therefore does not depend so much on the effluent quality or the economics of the activated sludge system itself, but

Table 5.7 Comparison of balanced MLE systems treating the example raw and settled wastewaters at 14 °C.

Parameter	Symbol	Unit	Raw sewage	Settled sewage
Influent TKN/COD ratio	$f_{TKNi/CODi}$		0.080	0.113
Influent RBCOD fraction	f_{SS}		0.25	0.385
Unaerated mass fraction	$f_{x,max}$		0.306	0.485
Anoxic mass fraction	f_{x1}		0.306	0.485
Minimum anoxic fraction	$f_{x1,min}$		0.079	0.108
a recycle ratio ($a_{prac} = a_{opt}$)	a		5:1	5:1
Sludge age	SRT	d	11	17
Effluent nitrate	$S_{NO3,e}$	mgN/l	5.1	5.7
Effluent TKN	TKN_e	mgN/l	4.3	4.1
Effluent total N ($S_{NO3,e}$+TKN_e)		mgN/l	9.4	9.8
Reactor volume at 4.5 gTSS/l		m^3	9,484	5,264
Carbonaceous O_2 demand	FO_c	kgO_2/d	6,156	4,251
Nitrification O_2 demand	FO_{NIT}	kgO_2/d	2,492	2,685
O_2 recovered	FO_{DENIT}	kgO_2/d	1,327	1,437
Total O_2 demand	FO_{tDENIT}	kgO_2/d	7,321	5,499
%N removal			84.3	80.9
Mass TSS wasted	FX_{TSS}		3,880	1,394
Active fraction with regard to TSS	$f_{at,OHO}$		0.316	0.414

on the economics of the whole wastewater treatment plant including sludge treatment. Because the minimum wastewater temperature (T_{min}) governs the activated sludge system (and sludge treatment) design, the balanced MLE system results for 22 °C are not particularly relevant to temperate climate regions.

However, in equatorial and tropical regions, where wastewater treatment is becoming a matter of increasing concern, high wastewater temperatures are encountered. For this reason and for illustrative purposes also, the balanced MLE results for the raw and settled wastewaters are shown in Figure 5.19.

Compared with 14 °C, the upper limit to $f_{x,max}$ = 0.60 is reached already at 7 d sludge-age at 22 °C and significantly higher influent TKN/COD ratios can be treated at equal sludge ages. These higher TKN/COD ratios result in higher $S_{NO3,e}$, which for the raw wastewater increases from 3 to 13 mgN/l and for the settled wastewater from 3 to 9 mgN/l for increases in sludge age from 4 to 30 days. If T_{min} were 22 °C, the example raw and settled wastewaters could be treated at 3 and 4 d sludge age, respectively, yielding $S_{NO3,e}$ of 5 and 6.5 mgN/l, respectively. This reinforces the conclusion in Section 5.8.1 that in equatorial and tropical climates it is highly likely that activated sludge plants will nitrify even at very short sludge ages (1 to 2 d) and therefore it is important to design for denitrification for operational reasons if not for effluent quality reasons.

5.9.3.3 Effect of influent TKN/COD ratio

When the unused denitrification potential in the anoxic reactor is kept in reserve as a safety factor (Option 2), the sludge age and unaerated (anoxic) mass fraction are not changed. For this situation it is useful to conduct a sensitivity analysis to see the influence of changing the influent TKN/COD ratio and RBCOD fraction on the a-recycle ratio and effluent nitrate concentration. Continuing with the design for the example raw and settled wastewaters for fixed sludge age at 20 d and unaerated (anoxic) mass fraction at 0.534, a plot of the optimum a-recycle ratio a_{opt} and minimum effluent nitrate concentration $S_{NO3,e,aopt}$ for underflow recycle ratios s of 0.5, 1.0 and 2.0 versus influent TKN/COD ratio from 0.06 to 0.16 is given in Figure 5.20 for the raw (A and C) and settled (B and D) wastewaters at 14 °C (A and B) and 22 °C (C and D).

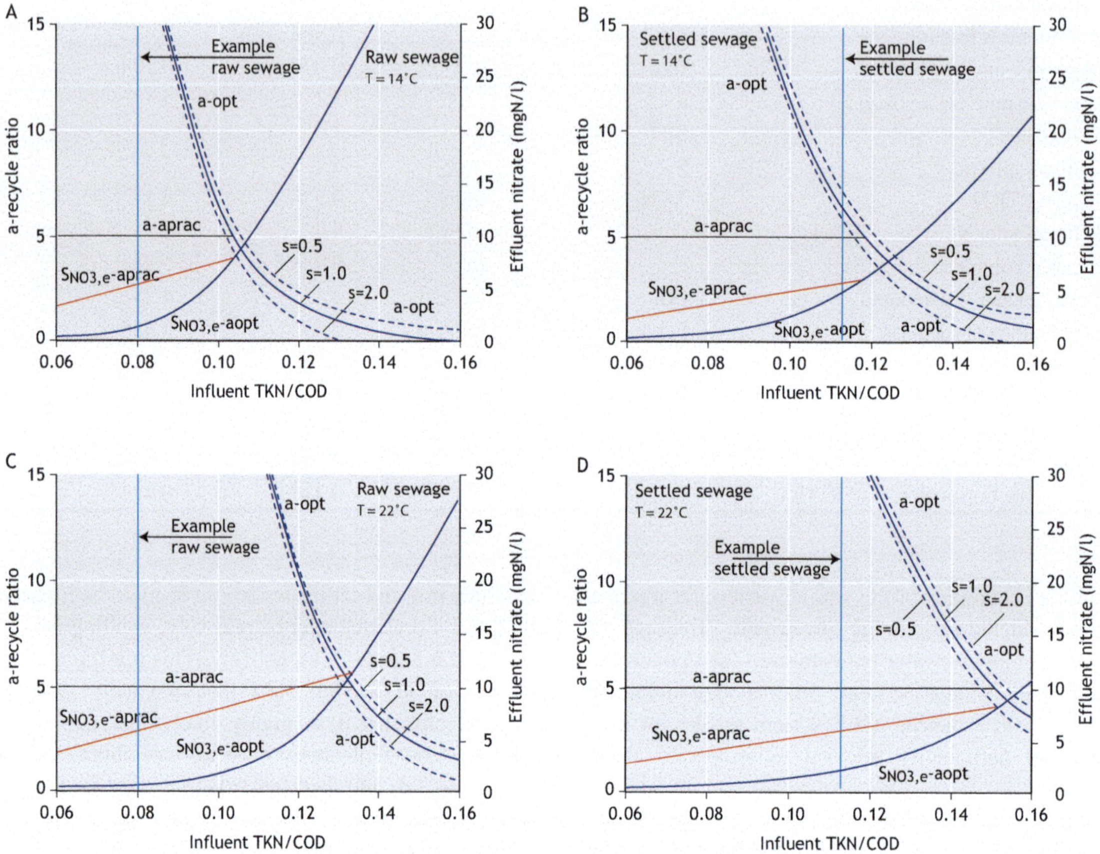

Figure 5.20 Optimum (a_{opt}) and practical upper limit (a_{prac} =5:1) a-recycle ratios (bold lines) and effluent nitrate concentration at a_{opt} ($S_{NO3,e,aopt}$, blue line) and a_{prac} ($S_{NO3,e,aprac}$, red line) versus influent TKN/COD ratio at underflow (s) recycle ratio of 1:1 for the example raw (A and C) and settled (B and D) wastewaters at 14 oC (A and B) and 22 oC (C and D). The optimum a-recycle ratio (a_{opt}) at underflow recycle ratios of 0.5:1 and 2:1 are also shown (dashed lines).

From Figure 5.20, it can be seen that as the influent TKN/COD increases so a_{opt} decreases and $S_{NO3,e,aopt}$ increases. The a_{opt}-$S_{NO3,e,aopt}$ lines in Figure 5.20 give the system denitrification performance when the denitrification potential of the anoxic reactor is fully used, i.e. the system denitrification performance is equal to its denitrification potential and the nitrate concentration is the lowest possible. Also large increases in the underflow recycle ratio s (i.e. from 0.50:1 to 1.0:1 or 1.0:1 to 2.0:1) decrease a_{opt} but do not change $S_{NO3,e,aopt}$ because the DO in the a and s recycles do not differ much in their influence on the anoxic reactor. Therefore it matters little which recycle flow brings the nitrate load to the anoxic reactor. As long as the anoxic reactor is loaded close to its denitrification potential, the same minimum effluent nitrate concentration ($S_{NO3,e,aopt}$) will be obtained at a_{opt}. The a_{opt} – $S_{NO3,e,aopt}$ lines therefore give the system denitrification performance when the denitrification potential of the anoxic reactor is fully used (Figure 5.17B), i.e. the system's denitrification performance is equal to its potential. A better denitrification performance is not possible; the denitrification is kinetic-limited and the biomass (and so also the system) is doing the best it can (for the given K_2 denitrification rate).

From Eq. 5.57, the system denitrification performance with increasing influent TKN/COD ratio at a fixed practical operating a-recycle ratio (a_{prac}) of 5:1 is shown also in Figure 5.20 as the a_{prac} and $S_{NO3,e,aprac}$ lines. It can be seen that $S_{NO3,e,aprac}$ increases linearly with the increase in the influent TKN/COD ratio. For low influent TKN/COD ratios, a_{prac} is considerably lower than a_{opt} and the system denitrification performance is lower than its denitrification potential. This is evident from $S_{NO3,e,aprac}$ being greater than $S_{NO3,e,aopt}$. As the TKN/COD ratio increases, a_{opt} decreases until a_{opt} = a_{prac} = 5.0:1. For the raw wastewater at 14 °C (Figure 5.20A), this happens at an influent TKN/COD ratio of 0.104. This is the influent TKN/COD ratio which balances the MLE system for the selected design conditions viz. 20 d sludge age, $f_{x,max}$ = 0.534 and a_{prac} = 5:1 for the example raw wastewater at 14 °C. For influent TKN/COD ratios > 0.104, the a-recycle ratio should be set at a_{opt}, which fully uses the anoxic reactor's denitrification potential and is now lower than a_{prac} = 5:1. Therefore for a_{prac} set at 5:1, only when the influent TKN/COD ratio is > 0.104, is the denitrification potential of the anoxic reactor fully used.

This same conclusion can be made from Figure 5.18A at 20 d sludge age, i.e. $f_{x,max}$ = 0.534, TKN/COD ratio = 0.104. Therefore for influent TKN/COD ratios < 0.104, while a_{prac} < a_{opt}, the system denitrification performance is lower than its denitrification potential because not all the denitrification potential of the anoxic reactor is used. Once the TKN/COD ratio increases above that value which balances the MLE system, a_{opt} is < a_{prac} and a should be set at a_{opt} to achieve the lowest effluent nitrate concentration ($S_{NO3,e,aopt}$). For these influent TKN/COD ratios, the denitrification potential of the anoxic reactor is fully used and the system denitrification performance is defined by the a_{opt} - $S_{NO3,e,aopt}$ lines.

Figure 5.20 is useful because it combines the system denitrification performance (a_{prac}-$S_{NO3,e,aprac}$ lines) and the denitrification potential (a_{opt}-$S_{NO3,e,aopt}$ lines) in the same diagram as influent TKN/COD ratio increases for a particular wastewater and system design (SRT = 20 d and $f_{x,max}$ = 0.534). The intersection point of the straight $S_{NO3,e,aprac}$ line and the curved $S_{NO3,e,aopt}$ line, i.e. at a_{opt} = a_{prac} = 5:1, gives the influent TKN/COD ratio for the balanced MLE system for the selected a_{prac} = 5:1. The system performance at the TKN/COD ratios which balance the system at 20 d sludge age and $f_{x,max}$ = 0.534 at 14 °C and 22 °C for the example raw and settled wastewaters are given in Table 5.6.

From Table 5.6 and Figure 5.20A, for the raw wastewater at 14 °C, the MLE system (at 20 d sludge age and $f_{x,max}$=0.534) with a recycle ratio 5:1 can maintain effluent nitrate concentrations below 8.1 (total N 12.4) mgN/l for influent TKN/COD ratios below 0.104 (TKN_i = 78.0 mgN/l). With settled wastewater at 14 °C (Figure 5.20B), the MLE system with a 5:1 ratio can maintain effluent nitrate concentrations below 11.3 (total N 14.9) mgN/l for influent TKN/COD ratios up to 0.132 (TKN_i = 59.4 mgN/l). Similarly, from figures 5.20C and D, with raw and settled wastewater at 22 °C, the MLE system with a 5:1 ratio can maintain effluent nitrate concentrations below 6.0 and 8.1 mgN/l (total N 9.9 and 11.1 mgN/l) for influent TKN/COD ratios up to 0.119 (TKN_i = 89.3 mgN/l) and 0.148 (TKN_i = 66.6 mgN/l). These results show that the MLE system treating settled wastewater delivers lower $S_{NO3,e}$ (by 2-3 mgN/l) than when treating raw wastewater and at influent TKN/COD ratios significantly higher. However, it should be noted that (*i*) the influent TKN concentrations (given above) for the raw wastewater are considerably higher than for the settled wastewater and (*ii*) a settled wastewater with a TKN/COD ratio of 0.119 (14 °C) or 0.148 (22 °C) would be produced from a raw wastewater with considerably lower influent TKN/COD ratio than 0.104 (14 °C) and 0.132 (22 °C).

5.9.3.4 MLE sensitivity diagram

In Figure 5.20, the system denitrification performance at a selected $a_{prac} = 5$ is combined with the system's denitrification potential at $a = a_{opt}$ for varying influent TKN/COD ratio and a single influent RBCOD fraction value. This influent TKN/COD ratio sensitivity diagram can be extended by adding the $S_{NO3,e,aopt}$ lines for other influent RBCOD fractions. A sensitivity analysis of the system at the design stage is useful for evaluating the denitrification performance under varying influent TKN/COD ratio and RBCOD fractions. These two wastewater characteristics can vary considerably during the life of the plant and have a major impact on the N removal performance of the system.

The denitrification potential and system performance are combined for the varying influent TKN/COD ratio and RBCOD fraction in Figure 5.21. For the fixed system design parameters (i.e. SRT = 20 d, $f_{xd,max} = f_{x,max} = 0.534$, s = 1.0), the curved (dark blue) lines give $S_{NO3,e,aopt}$ when the anoxic reactor is loaded to its denitrification potential, i.e. $S_{NO3,e}$ for $a = a_{opt}$ for varying TKN/COD ratio from 0.06 to 0.16 and RBCOD fractions from 0.10 to 0.35 for the example raw and settled wastewaters at 14 °C (Figure 5.21A and B) and 22 °C (Figure 5.21C and D). The same $S_{NO3,e,aopt}$ lines are given in Figure 5.20A and C for the raw wastewater RBCOD fraction (f_{SS}) = 0.25. These $S_{NO3,e,aopt}$ lines are calculated from eqs. 5.56 and 5.57. The straight lines in Figure 5.21 give $S_{NO3,e,aprac}$ for fixed a-recycle ratios at the indicated values ranging from 0.0:1 to 10:1. These straight $S_{NO3,e,aprac}$ lines give the system performance for selected a-recycle ratios and are calculated with the aid of Eq. 5.57 from the nitrification capacity value at the given TKN/COD ratio, fixed s-recycle ratio at 1.0:1 and the selected a-recycle ratio.

The $S_{NO3,e,aprac}$ lines for $a = a_{prac} = 5:1$ are the same as the dotted lines in Figure 5.20. At the intersection points of the straight $S_{NO3,e,aprac}$ and curved $S_{NO3,e,aopt}$ lines, the system performance equals the denitrification potential and represents balanced MLE designs, i.e. $a_{opt} = a_{prac}$. For example, for the raw wastewater at 14 °C, at a = 5.0:1 and $f_{SS} = 0.25$, the TKN/COD ratio needs to be 0.104 to give an optimal design, i.e. $a_{opt} = 5:1$ and at this TKN/COD ratio, $S_{NO3,e} = 8.1$ mgN/l. This is the TKN/COD ratio that balances the MLE system at SRT = 20 d and $f_{x,max} = 0.534$ (see Figure 5.18A and C).

For TKN/COD ratios < 0.104, a_{opt} increases above 5:1, but if a is maintained at 5:1 (i.e. $a = a_{prac} = 5:1$), then $S_{NO3,e}$ versus the TKN/COD ratio is given by the a = 5:1 straight $S_{NO3,e,aprac}$ line. For TKN/COD ratios > 0.104, a_{opt} decreases below 5:1, and $S_{NO3,e}$ versus the TKN/COD ratio is given by the curved $S_{NO3,e,aopt}$ (dark blue) line. The a_{opt} value at a particular TKN/COD ratio is given by the a-recycle ratio value of the intersection point between the vertical influent TKN/COD ratio line and the curved $S_{NO3,e,aopt}$ line, e.g. for the example raw wastewater ($f_{SS} = 0.25$) at 14 °C (Figure 5.21A) at a TKN/COD ratio of 0.12, a_{opt} = 2:1 and $S_{NO3,e}$ is 16.0 mgN/l.

The usefulness of Figure 5.21 is that it gives a performance evaluation of a MLE system at a specified sludge age and anoxic mass fraction for a varying influent TKN/COD ratio and RBCOD fraction taking due account of an upper a-recycle ratio limit of a_{prac}. For the example raw wastewater at 22 °C with a RBCOD fraction (f_{SS}) of 0.10 (Figure 5.21C), the influent TKN/COD ratio needs to be greater than 0.113 for a to be < 6.0:1. If a is fixed at $a_{prac} = 6.0:1$ and the TKN/COD is < 0.113, then the anoxic reactor is underloaded with nitrate and the denitrification potential is not achieved. The system performance for influent TKN/COD < 0.113 is given by the straight $S_{NO3,e}$ line for a = 6:1. At influent TKN/COD = 0.113, the straight $S_{NO3,,e}$ line for a = 6:1 cuts the curved $S_{NO3,e,aopt}$ line, $a = a_{opt} = 6:1$ and the system performance equals the denitrification potential. If a is maintained at 6:1 for TKN/COD > 0.113, then the anoxic reactor is overloaded with nitrate and optimal denitrification is not achieved due to the unnecessarily high DO load on the anoxic reactor (similar to that shown in Figure 5.16B for a > 6.7). The a-recycle ratio therefore should be reduced to a_{opt} for influent TKN/COD ratios >0.113, where a_{opt} is given by the a value along the curved $S_{NO3,e}$ line, which represents system performance equal to denitrification potential. For example, if the TKN/COD ratio = 0.120, a = a_{opt} = 4:1 and this a-recycle ratio loads the anoxic reactor to its denitrification potential giving $S_{NO3,e}$ of 12.0 mgN/l. Therefore, for TKN/COD ratios > 0.113, the system performance and $S_{NO3,e}$ is given by the curved $S_{NO3,e}$ line provided the a-recycle ratio is set to a_{opt}, which is given by the a-recycle ratio line which passes through the intersection point of the vertical TKN/COD ratio line and the curved $S_{NO3,e}$ line.

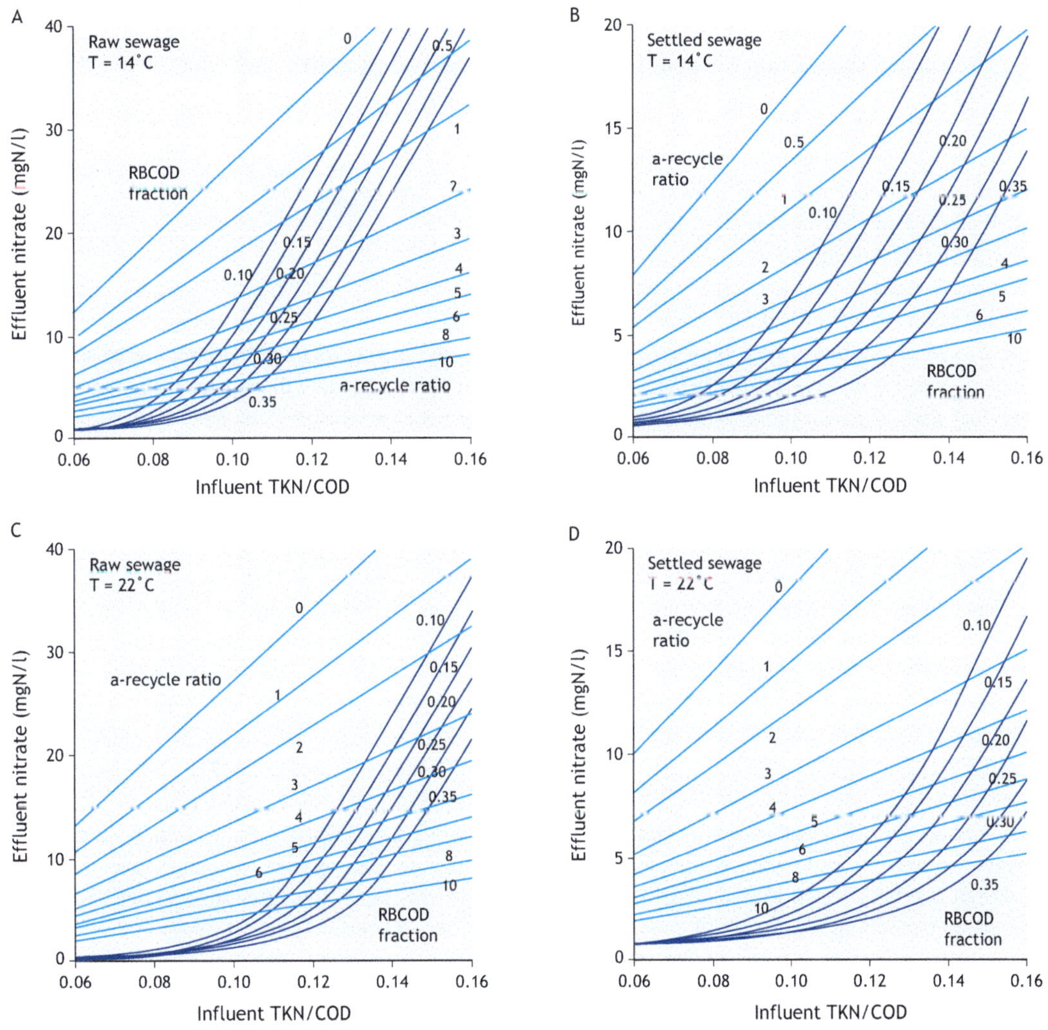

Figure 5.21 Effluent nitrate concentration versus influent TKN/COD ratio for influent readily biodegradable (RBCOD) fractions of 0.10, 0.15, 0.20, 0.25, 0.30 and 0.35 and mixed liquor a-recycle ratios from 0 to 10 for the example raw (A and C) and settled (B and D) wastewaters at 14 oC (A and B) and 22 oC (C and D).

From the above it can be seen that the system performance is only equal to the denitrification potential on the curved $S_{NO3,e}$ line for the particular RBCOD fraction and the a_{opt} that produces this is given by the a-recycle ratio line which passes through the intersection point of the vertical TKN/COD ratio line and the curved $S_{NO3,e}$ line. This curved $S_{NO3,e}$ line (for which $a = a_{opt}$) marks the boundary between underloaded and overloaded conditions in the anoxic reactor. In the domain above the curved $S_{NO3,e}$ line the anoxic reactor is underloaded (left of a_{opt} in figures 5.16 and 5.17) and the system performance ($S_{NO3,e}$) for a particular TKN/COD ratio is given by the intersection point of the vertical TKN/COD ratio line and the straight a-recycle ratio line. In the domain below the curved $S_{NO3,e}$ line, the anoxic reactor is overloaded (right of a_{opt} in figures 5.16 and 5.17). The $S_{NO3,e}$ values obtained from this domain are not valid, but if the a-recycle ratio is reduced to a_{opt} (i.e. the a value of the intersection point of the vertical TKN/COD ratio line and the curved $S_{NO3,e}$ line), then the $S_{NO3,e}$ value is valid again. Valid $S_{NO3,e}$ system performance values are therefore given in Figure 5.21 only on or above the curved $S_{NO3,e}$ boundary line. From Figure 5.21, it can be seen that with an MLE system at the design SRT = 20 d, $f_{x,max}$ = 0.534 and the a-recycle ratio limited at say 5:1 for economic reasons, then the system is best suited to treating high TKN/COD ratios, depending on the RBCOD fraction; > 0.091 for f_{SS} = 0.10 and > 0.117 for f_{SS} = 0.35. This is because with only a primary anoxic reactor, the MLE system cannot produce a low effluent nitrate concentration (< 4 to 6 mgN/l) at a recycle ratio limited at 5.0:1.

If obtaining low effluent nitrate concentrations is not required at low TKN/COD ratios, then a balanced MLE design can be selected by reducing the sludge age as demonstrated in figures 5.18 and 5.19. If obtaining low effluent nitrate concentrations is important at low (< 0.10) TKN/COD ratios, then this can be achieved at high a-recycle ratios ($a_{opt} > a_{prac}$) in MLE systems or at low a-recycle ratios by including a secondary anoxic reactor. Incorporation of a secondary anoxic reactor (and a re-aeration reactor for practical reasons; see Section 5.8.5) produces the 4-stage Bardenpho system (Figure 5.13C). However, because the K_3 denitrification rate is so low and needs to be reduced by at least 20% to account for the ammonia released during endogenous denitrification (which is re-nitrified in the re-aeration reactor), the net additional nitrate removal achieved in a secondary anoxic reactor is very low, too low for secondary anoxic reactors to be included in N-removal systems, unless the influent TKN/COD ratio is unusually low. Secondary anoxic reactors are usually only included where methanol dosing is required to achieve very low total effluent N concentrations (< 5 mgN/l).

5.10 SYSTEM VOLUME AND OXYGEN DEMAND

5.10.1 System volume

Having determined the subdivision of the sludge mass into anoxic and aerobic fractions to achieve the required N removal, the actual sludge mass in the system needs to be calculated to determine the volumes of the different reactors. The mass of sludge, total (MLSS) or volatile (MLVSS) in the system for selected sludge age and wastewater characteristics for N-removal systems is the same as for fully aerobic (COD-removal) systems. The equations given in Chapter 4, Section 4.4.2 therefore also apply to N-removal systems. For the example raw and settled wastewaters, the design parameters for the MLE system are listed in Table 5.8. The MLSS mass in the system at 20 d sludge age and 14 °C is 68,168 and 26,422 kgTSS, respectively. Selecting an MLSS concentration of 4,500 mg/l (4 kg/m^3) (see Chapter 4, Section 4.7) makes the volume of the system treating raw wastewater 15,148 m^3 and that treating settled wastewater 5,871 m^3. Because the sludge mass in the N-removal systems is usually uniformly distributed in the system, i.e. each reactor in the system has the same MLSS concentration, the volume fraction of each reactor is equal to its sludge mass fraction. For the example raw and settled wastewaters at 14 °C, the volume of the anoxic reactors are 0.534 · 15,148 =

8,089 m^3 and 0.534 · 5,871 = 3,135 m^3, respectively. The nominal and actual hydraulic retention times of the anoxic and aerobic reactors are calculated from the reactor volumes divided by the nominal (influent) and total flows passing through them (eqs. 4.2 and 4.3 and Table 5.8). Note that the reactor nominal retention time is a consequence of the mass of sludge generated from the influent COD load, the selected MLSS concentration and the sludge mass fraction; the retention time per se has no significance in the kinetics of, and design for, nitrification and denitrification (see Chapter 4, Section 4.4.4).

5.10.2 Daily average total oxygen demand

The total oxygen demand in a nitrogen removal system is the sum of that required for organic material (COD) degradation and nitrification, less that recovered by denitrification. The daily average oxygen demand for (*i*) organic material removal (FO_c) is given by Eq. 4.18 (see Chapter 4, Section 4.4.2.4) and (*ii*) for nitrification (FO_{NIT}) is given by Eq. 5.43 (see Chapter 5, Section 5.7.5). These oxygen demands in the MLE system at 20 d sludge age for the example raw and settled wastewaters at 14 °C and 22 °C are 9,364 and 7,030 kgO_2/d, respectively (Table 5.8).

The oxygen recovered by denitrification (FO_{DENIT}) is given by 2.86 times the nitrate mass denitrified (Section 5.8.2) where nitrate flux denitrified is the product of the daily average influent flow Q_i and the

Table 5.8 Design details of MLE systems treating the example raw and settled wastewaters at 14 °C at 20 d sludge and 0.534 unaerated sludge mass fraction.

Parameter	Symbol	Unit	Raw	Settled
Influent TKN/COD ratio	$f_{TKNi/CODi}$		0.080	0.113
Influent RBCOD fraction	f_{SS}		0.25	0.385
Unaerated mass fraction	$f_{x,max}$		0.534	0.534
Anoxic mass fraction	f_{x1}		0.534	0.534
Minimum anoxic fraction	$f_{x1,min}$		0.068	0.105
a-recycle ratio ($a_{prac} = a_{opt}$)			5:1	5:1
Sludge age	SRT	d	20	20
Effluent nitrate	$S_{NO3,e}$	mgN/l	5.16	5.7
Effluent TKN	TKN_e	mgN/l	3.8	3.8
Effluent total N ($S_{NO3,e} + TKN_e$)		mgN/l	9.4	9.5
System volume at 4.5 gTSS/l		m^3	15,148	5,871
Anoxic volume		m^3	8,089	3,135
System nominal hydraulic retention time	HRT_n	h	24.2	9.4
Aerobic nominal hydraulic retention time	HRT_n^{AER}	h	11.2	4.4
Aerobic actual hydraulic retention time	HRT_a^{AER}	h	1.60	0.63
Anoxic nominal hydraulic retention time	HRT_n^{AX}	h	12.9	5.0
Anoxic actual hydraulic retention time	HRT_a^{AX}	h	1.85	0.72
Carbonaceous O_2 demand	FO_c	kgO_2/d	6,679	4,311
Nitrification O_2 demand	FO_{NIT}	kgO_2/d	2,685	2,719
O_2 recovered	FO_{DENIT}	kgO_2/d	1,440	1,458
Total O_2 demand	FO_{tDENIT}	kgO_2/d	7,924	5,572
%N removal			84.4	81.4
Mass TSS wasted	$M\Delta X_{TSS}$		3,408	1,321
Active fraction with respect to TSS	$f_{at,OHO}$		0.230	0.383

nitrate concentration denitrified. The nitrate concentration denitrified is given by the difference in the nitrification capacity NIT_c and the effluent nitrate concentration. Hence,

$$FO_{DENIT} = 2.86(NIT_c - S_{NO3,e})Q_i \qquad (mgO_2/d) \ (5.62)$$

From the denitrification performance of the MLE system in Table 5.8, the oxygen recovered by denitrification for the example raw and settled wastewaters at 14 °C are 1,440 and 1,458 kgO_2/d, respectively (Table 5.8).

For the raw wastewater, Table 5.8 shows that (*i*) the nitrification oxygen demand (FO_{NIT}) is approximately 40% that required for COD removal (FO_c), (*ii*) approximately 55% of FO_{NIT} can be recovered by incorporating denitrification, (*iii*) the additional oxygen demand by incorporating nitrification and denitrification is only 20% of that required for COD removal only and (*iv*) the effect of temperature on the total oxygen demand is marginal: less than 3% (see also Figure 5.11). For the settled wastewater, Table 5.8 shows that (*i*) the nitrification oxygen demand is approximately 63% of that required for COD removal, (*ii*) approximately 54% of the nitrification oxygen demand can be recovered by denitrification, (*iii*) the additional oxygen demand by incorporating nitrification and denitrification is approximately 30% of that required for COD removal only, and (*iv*) the effect of temperature on the total oxygen demand is marginal: less than 3% more at the lower temperature.

Comparing the oxygen demand for the raw and settled wastewaters, the total oxygen demand for the latter is approximately 30% less than that of the former. This saving is possible because primary sedimentation removes 35 to 45% of the raw wastewater COD. Furthermore, for the settled wastewater, the nitrification oxygen demand is a greater proportion of the total, and also, less of the nitrification oxygen demand can be recovered by denitrification compared with the raw wastewater. These effects are due to the higher TKN/COD ratio of the settled wastewater. Knowing the average daily total oxygen demand, the peak total oxygen demand can be roughly estimated by means of a simple design rule (Musvoto *et al.,* 2002). From a large number of simulations with ASM1, it was found that provided the factor of safety on nitrification (S_f) is greater than 1.25 to 1.35, the relative amplitude (i.e. (peak-average)/average) of the total oxygen demand variation is the fraction 0.33 of the relative amplitude of the total oxygen demand (TOD) of the influent COD and TKN load (i.e. $Q(COD_i + 4.57\ TKN_i)$). For example, with the raw wastewater case, if the peak influent total oxygen demand potential is obtained at a time of day when the influent flow rate, COD and TKN concentrations are 25 Ml/d, 1,250 mgCOD/l and 90 mgN/l, respectively, i.e. $25 \cdot (1{,}250 + 4.57 \cdot 90) = 41{,}532$ kgO_2/d, and the average total oxygen demand potential is $15 \cdot (750 + 4.57 \cdot 60) = 15{,}363$ kgO_2/d, the relative amplitude of the total influent oxygen demand potential is $(41{,}532 - 15{,}363)\ /\ 15{,}363 = 1.70$; hence the relative amplitude of the total oxygen demand is approximately $0.33 \cdot 1.70 = 0.56$; from Table 5.8 the average daily total oxygen demand is 7,924 kgO_2/d and hence the peak oxygen demand is $(1 + 0.56) \cdot 7{,}924 = 12{,}378$ kgO_2/d. As with all simplified design rules, the above rule should be used with discretion and caution, and where possible, the peak total oxygen demand is best estimated by means of the activated sludge simulations models. It is strongly recommended that the simulation model outputs be compared with the steady-state model results and when there are significant differences, that sources and reasons for the differences be established. This is a good way of finding errors in simulation and steady-state model results.

5.11 SYSTEM DESIGN, OPERATION AND CONTROL

Because the steady-state modes for nitrification and denitrification conform to, and are fully integrated with, the organics removal model, all the design, operation and control issues discussed in Chapter 4, sections 4.9 to 4.11 for the fully aerobic system apply equally to nitrification-denitrification systems and should be referred to there.

5.12 NOVEL NITROGEN REMOVAL PROCESSES

The previous sections describe the background of conventional nitrification and denitrification processes. Activated sludge systems have been successfully used for carbon and nitrogen removal for almost a century. The growing public concern for environmental protection has led to the implementation of continuously increasing effluent standards, which has in turn impelled efforts to achieve better effluent quality especially with regard to nitrogen removal. Existing wastewater treatment plants will require upgrading and thus large expansions to meet new standards. Research into new techniques for upgrading treatment plants without the need for expansion of the existing volumes has therefore increased in the last decade. Treatment of internal process flows with high ammonia concentration provides good potential for upgrading treatment plants. Internal flows at the wastewater treatment plant (especially the return water from digested sludge treatment) account for 10-30% of the total N load at the treatment plant. At larger plants with centralised sludge treatment facilities this contribution to the nitrogen load is especially high. These flows originate from *e.g.* digester effluents or sludge drying facilities. They are highly concentrated with ammonium; consequently relatively small tank volumes will be required for treatment. In addition these internal flows generally have a high temperature (20-35 °C) compared to the main treatment process. Due to the higher maximum growth rate of bacteria at higher temperatures, operation at short solid retention times (SRT) will be possible. Removing the ammonium from these internal flows by physical or biological processes can lead to a significant improvement in the final effluent quality. Finally, if side streams are treated separately the process can be designed with respect to removed load instead of a good effluent quality, since the effluent will be discharged to the main treatment plant. All these factors allow for the design of different and efficient processes treating sludge treatment effluents.

Although this chapter mainly focuses on treatment of sludge digester effluents at municipal wastewater treatment plants, the discussed methods can also be applied to treatment of other concentrated nitrogen-containing wastewaters *e.g.* industrial effluents, leachate or effluents from anaerobic digesters in general. Several processes and technologies have been studied and brought to the activated sludge market over the last years: SHARON®, a simple system for N removal over nitrite (Hellinga *et al.,* 1998); ANAMMOX® fully autotrophic N removal (Mulder *et al.,* 1995); CANON®, a combination of nitritation and anaerobic ammonia oxidation (Third *et al.,* 2001); BABE®, bio-augmentation with endogenous nitrifiers (Salem *et al.,* 2002), etc. Especially Anaerobic ammonium oxidation (anammox) based processes are currently under study for application in the main water line of wastewater treatment plants. This is mainly driven by energy optimisation; when ammonium is removed fully autotrophically, the influent COD can be converted into biogas (Kartal *et al.,* 2010).

Several examples of physical or chemical processes have been evaluated for the treatment of side-stream processes, but are generally less favourable than biological processes. Often a pre-treatment is needed in order to remove the carbonates which might otherwise accumulate in the process equipment. This requires chemicals and additional process steps, and leads to the production of more chemical sludge at the plant. Also, in general the costs of physical-chemical techniques are significantly higher than those of biological treatment. The costs of the SHARON® process with methanol for pH correction were estimated for wastewater treatment plants in the Netherlands to be 0.9-1.4 Euro per kgN removed (STOWA, 1996), while physical-chemical processes were 5-9 times more costly. This is mainly related to differences in investment costs and manpower requirements. Finally, a limitation of physical process steps at a municipal wastewater treatment plant is the requirement of different types of operation and maintenance as compared to a biological step, which would have consequences for the training of operators at the treatment plant.

The most widely used physical technique for the treatment of ammonium-containing wastewater is stripping of the ammonium. After an increase in pH, ammonium can be stripped and recovered. This process can be made more efficient by increasing the stripping temperature by using steam instead of regular air. When there is waste heat available near the site, this might be a feasible option if there is no other use for the steam. The pH increase will lead to precipitation of carbonates which can be prevented by acidifying the water and stripping CO_2 as pre-treatment. However, this approach will further increase the use of chemicals.

Magnesium ammonium phosphate (MAP) precipitation is another option to remove nitrogen or phosphate. Generally, the P/N ratio in the water is such that phosphate can be removed without any addition of chemicals. However, in order to remove ammonium, additional magnesium phosphate has to be added. Also, in this case the carbonates will have to be removed first; otherwise they end up in the MAP precipitates. Addition of magnesium phosphate can be minimized by removing the ammonium from the MAP sludge by heat treatment. The ammonium will volatise and can be recovered, while the magnesium phosphate can be reused. MAP or the ammonium can in principle be reused.

Ammonium recovery is often used as an argument for the application of physical methods. The problem is that a relatively small amount of ammonium is recovered as compared to the general use of ammonium *e.g.* fertiliser. Moreover, these techniques in general require more energy than for nitrification/denitrification and ammonium production in the industry. Because of all these factors, biological treatment of the nitrogen load of side streams has become the most popular process choice.

5.12.1 IMPACT OF SIDE-STREAM PROCESSES

The nitrogen load in a side stream is typically around 10-15% of the total influent nitrogen load. At plants where external sludge is treated, this fraction increases significantly. This flow is typically only 1% or less of the influent flow rate. The impact of treatment for nitrogen on the effluent depends on the limitation for nitrogen removal in the main treatment process. If ammonium is not fully converted in the main process then each kilogram of ammonium removed in the side stream will lead to a kilogram less of ammonium in the effluent of the treatment plant. The decrease in nitrate is however not equivalent to the removed load in the side-stream process. This depends heavily on local conditions, but will generally be in the range of a decreased load equivalent to 40-70% of the removed nitrogen in the side-stream process.

Side-stream processes are especially useful when existing plants require upgrading due to increased effluent requirements or increased load. With a relatively small addition of reactor volume, the effluent concentration can be decreased. An extra advantage is gained by the possibility of building the reactor independently from the main treatment process; in this case the construction work is simpler than when the existing tanks need expansion.

Application of side-stream processes also change the attitude of the operators towards sludge treatment and handling. For example, a SHARON® reactor was installed at the Rotterdam wastewater treatment plant in 1998. At that time the nitrogen load treated was 500 kg/d. Because it was profitable to remove nitrogen in the side-stream reactor, the operators started to optimise the sludge production and handling. Eight years after the start-up of the system the nitrogen load in the side-stream process was 700 kg/d, while the total load to the plant had not changed. Not only extra nitrogen was removed in the side stream, but also extra sludge was digested and the methane production increased. In short, addition of a side-stream process can be even more beneficial when the process operation pays attention to increasing the load to the digesters (*e.g.* by applying more efficient thickening of primary and secondary sludge).

5.12.2 THE NITROGEN CYCLE

The discovery of new organisms is making the N cycle increasingly complicated (Figure 5.22). Traditionally the N cycle is described by nitrification (ammonia is oxidised to nitrate via nitrite), denitrification (conversion of nitrate or nitrite to nitrogen gas), and N fixation. From a process-engineering point of view it would be better to keep nitrate outside this cycle. Recently it was found that conventional 'aerobic' ammonia oxidizers could perform other processes as well.

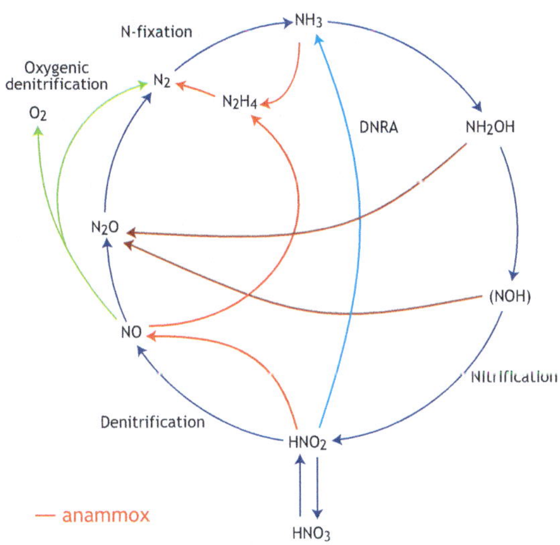

Figure 5.22 The nitrogen web (the red line depicts anammox conversions; DNRA stands for Dissimilatory Nitrate Reduction to Ammonium).

When oxygen is limiting (micro-aerophilic conditions), Nitrosomonas can combine hydroxylamine with nitrite to give dinitrogen-oxide gas (Bock, 1995). This process was referred to as aerobic deammonification (Hippen *et al.,* 1997). Several new organisms have been discovered in the past decades. Of these the anaerobic ammonium oxidisers or anammox bacteria (Mulder *et al.,* 1995) are the most interesting. They oxidize ammonia to dinitrogen gas with nitrite as the electron acceptor in the absence of oxygen.

Nitrification has always been considered to be a process performed by two bacteria, but recently bacteria have been discovered that perform complete ammonium oxidation (comammox) to nitrate (Costa *et al.*, 2006; Koch *et al.,* 2019). These comammox organisms are especially observed in drinking water filtration and aquaculture systems. There are no specific advantages over conventional nitrifiers.

Another intriguing discovery is the anaerobic oxidation of methane (Scheller *et al.*, 2020). This can be based on the conversion of methane into CO_2 and H_2 and subsequent use of H_2 by sulphate reducers or denitrifying bacteria. Alternatively there are denitrifying bacteria that can convert the denitrification intermediate NO into oxygen and nitrogen gas. The oxygen is used to convert methane into methanol, and methanol is then used in a conventional denitrification reaction. The application of this conversion is still under investigation (Lee *at al.*, 2018).

The higher temperature and concentration of *e.g.* digester effluents allow several alternatives for biological nitrogen removal. Firstly, there are better options to stop the nitrification at the nitrite level, saving oxygen and carbon sources. It is even possible to have fully autotrophic N removal by using anammox bacteria. N-rich sludge water could also be used to cultivate nitrifiers in the side stream, which are inoculated in the main stream process. This allows the main stream process to run at sub-optimal aerobic SRT.

The large number of processes for the treatment of sludge treatment effluents sometimes makes it complicated to make a proper choice. The choice of the treatment process for the sludge water treatment will always be highly site-specific. It depends mainly on the limitations in the treatment plant. For example, if the aeration capacity in the treatment plant is limited, then ammonium removal will be the correct process to be chosen. On the other hand, if increasing denitrification capacity is the goal to be achieved, then either N removal or bio-augmentation would be suitable methods. When the treatment plant has limited sludge age, then only augmentation would be the proper process to be selected. Also the choice for treatment over nitrate or nitrite depends on local conditions and sludge water composition. Besides pure process engineering aspects, other aspects also have to be taken into account, such as start-up time, risk of failure, flexibility, etc. The prime decision is whether nitrification or denitrification is limiting in the main treatment process (Figure 5.23).

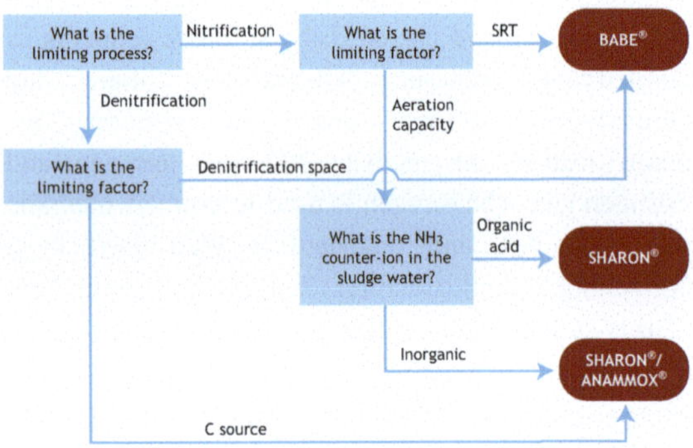

Figure 5.23 Chart for the selection of sludge water treatment process.

If nitrification is limited by the SRT, or if the denitrification is limited by anoxic retention time, then bio-augmentation is the best option. In the case of limited denitrification space, bio-augmentation allows the aerated volume to decrease and the anoxic volume to increase in the main activated sludge process. If, however, aeration (nitrification) or COD (denitrification) is limiting, then nitritation-based process variants could be used. Besides the difference in potential augmentation, the main difference is that for augmentation one should have complete nitrification because otherwise only ammonium oxidisers would be augmented, leading to potential build-up of nitrite in the effluent of the wastewater treatment plant. Special attention should be given to the counter-ion for ammonium.

By nature, the counter-ion after normal anaerobic digestion will be bicarbonate. Already this bicarbonate can supply 50% of the required alkalinity for N removal. The remainder has to come from denitrification (either by adding methanol or using the COD in the return sludge in bio-augmentation processes). pH control by adding alkalinity directly is less cost-effective than using methanol/denitrification.

Of course, the anammox denitrification process can also be used to control the pH. In certain cases, however, used sludge treatment methods lead to another counter-ion for ammonium. If in sludge processing, for example, iron salts are used for increasing dewatering, chloride will become the counter-ion, leading to greater need for alkalinity control. It could be wise in such cases to change the sludge treatment process. If sludge drying is employed, usually fatty acids will form the dominant counter-ion. In such cases there is no need to add an anammox step since there is sufficient COD in the sludge water.

Denitrification is not really needed when the side-stream process is implemented to reduce the ammonium load in the plant effluent. Sludge water treated with nitrate is likely to be fed back to the influent of the wastewater treatment plant where it will be readily denitrified. However, denitrification is a cost-effective way to control the pH, and is often implemented as such. Since the side-stream processes operate at high concentrations and temperature, there is no direct need for sludge retention when conventional nitrification/denitrification is applied. Sludge retention might lead to a somewhat smaller reactor (by a factor of 2-4 depending on the concentration of ammonium), but will also lead to a more complicated operation and a higher investment in the mechanical works. Nitrification and denitrification can be performed in a system with a two-tank set-up

Table 5.9 Decision matrix for selection of a sludge water treatment process at WWTP Beverwijk.

Criterion	Weight (%)	BABE®	SHARON®	SHARON®/ANAMMOX®
Exploitation	27	+	++	+
N-removal efficiency	5	++	++	+
Energy use	15	–	0	++
Odour emission	1	+	+	+
Realisation time	20	+	+	—
Management	1	+	++	++
Construction	1	0	0	0
Flexibility	10	–	+	++
Innovativeness	10	+	+	++
Risk of failure	10	+	+	–
Total	100	3.5	4.1	3.5

Note: ++ =5; + =4; 0 =3; – =2; — =1 Maximum 5 points

(aerated and anoxic tanks with a large recycle) or a one-tank system (with sequential aerated and non-aerated periods). A one-tank system is cheaper to build, but the investment in aeration equipment will be larger since for part of the time it is not used, while the total oxygen input should be the same as in a two-tank system. Also, the process control is different. Set-point control for *e.g.* dissolved oxygen in a two-tank system is different because although in both tanks a steady-state situation will be created, the one-tank lay-out gives more flexibility for process control due to the intrinsic dynamic conditions.

The selection process is further illustrated using an example from the wastewater treatment plant (WWTP) in Beverwijk in the Netherlands. This plant has a capacity of 320,000 P.E. with an N load from sludge treatment (methane digestion and thermal sludge drying) of 1,200 kgN/d. According to new legalisation, effluent N total is not allowed to exceed 10 mgN/l or 75% total N removal and therefore the treatment plant needed to be upgraded to reach this new standard. Removal of the N content in the sludge water would be enough to meet the required standards. The question was which side-stream process is, in this case, the best choice. A comprehensive comparison was made between bio-augmentation (BABE®), nitritation-denitritation (SHARON®) and a combined nitritation-ANAMMOX® process. This sludge water has acetate as the counter-ion of NH_4^+ (which is normal for thermal sludge drying). Laboratory experiments showed that the SHARON® reactor at 1 day aerobic SRT without pH correction performed nitrification and denitrification with efficiency > 90%, via the nitrite route (Schemen *et al.*, 2003).

Besides pure engineering aspects, many other aspects were, in this case and most other cases, considered important. These aspects, which include energy use, management and construction, are shown in the decision matrix in Table 5.9 (Schemen *et al.*, 2003). Clearly a large variety of decision criteria exist, which might change for each specific plant. Moreover each decision aspect has a different weight, again depending on local considerations. The total score per system was calculated. Based on this matrix the final decision was to select the SHARON® technology to be applied for the treatment of sludge water in Beverwijk (Figure 5.25). The aspects to be considered in the decision matrix differ from one treatment plant to another and also the weights given to each aspect will probably differ from one case to another. This could easily lead to different choices for a similar plant at a different location or managed by another organisation.

5.12.3 NITRITE-BASED N-REMOVAL

Nitritation-denitritation techniques have been considered very promising for quite a while. During nitritation or partial nitrification, ammonia is converted to nitrite while further oxidation to nitrate is prevented. The stoichiometric equations for nitrification and denitrification have been given in sections 5.1 and 5.8. The reduced need for oxygen means that partial nitrification needs 25% less aeration. Denitrification of nitrite requires 40% less carbon source than for nitrate. This definitely substantially reduces costs when a low C/N ratio of the wastewater requires the addition of an external electron donor, such as methanol. Finally, the amount of sludge produced in a nitritation-based process is also approximately 40% lower.

Forcing the biological conversion to follow the nitrite route can be obtained by two main approaches: (*i*) by using selective pressures and (*ii*) by keeping oxygen concentration low or applying sub-optimal pH, nitrite or ammonium conditions. Ammonia oxidation is more sensitive to temperature changes than nitrite oxidation (Hellinga *et al.*, 1998). At temperatures above approximately 20 °C ammonium-oxidising bacteria have a higher growth rate than nitrite-oxidising bacteria. The SHARON® process makes use of this different growth rate at warmer temperatures, by selecting a SRT in between the requirements for ammonium and nitrite oxidation. A second major selection factor is also high ammonium, nitrite or salt concentrations in the reactor. Nitrite oxidisers have a lower tolerance and cannot grow in these conditions. This is especially relevant for some industrial wastewaters. As mentioned above, the second possibility to prevent nitrite oxidation is to keep oxygen concentration low or apply sub-optimal pH, nitrite or ammonium concentrations. In this case, nitrite oxidation will in general be only partly inhibited. By combining such a weak factor for nitrite oxidation inhibition with denitrification, a full conversion over nitrite can be obtained. Due to the denitrification of nitrite, nitrite-oxidising bacteria become deprived of their substrate and are washed out of the system.

At low oxygen concentrations nitrite and nitrate are produced due to oxygen limitation. The actual optimal concentration for nitrite accumulation in such processes will be lower for thinner biofilms and smaller flocs or lower loading rates (Hao *et al.*, 2002). However, a low oxygen concentration alone is not sufficient since nitrate will also be formed (Picioreanu *et al.*, 1997). Figure 5.24 illustrates the nitrite/nitrate formation behaviour at different dissolved oxygen concentrations in a biofilm system as experimentally observed by Garrido *et al.* (1997) and theoretically explained by Picioreanu *et al.* (1997).

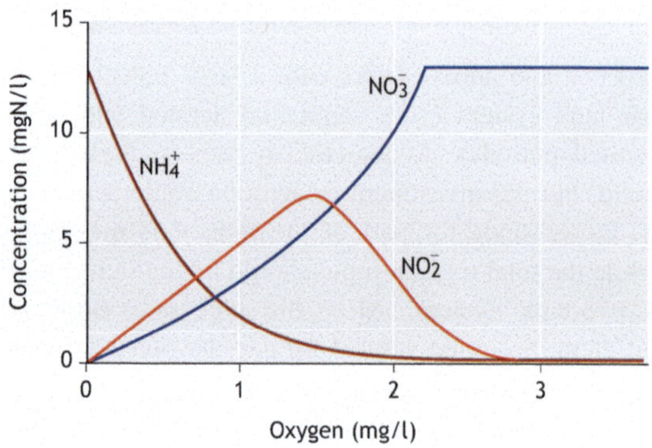

Figure 5.24 The effect of oxygen on accumulation of NO2-N in a biofilm system.

The only way the conversion can be completely directed towards nitrite is by out-competing the nitrite oxidation by competition for oxygen and nitrite simultaneously. This means a direct coupling of nitrification and denitrification. Examples are using a strong recirculation between aerobic and anoxic reactors (*e.g.* Van Benthum *et al.*, 1998) or including denitrification in the deeper layers of a biofilm system (*e.g.* Kuai *et al.*, 1998; Hao *et al.*, 2002).

The SHARON® process (Hellinga *et al.*, 1998) was the first method which was developed and scaled-up to full scale for dedicated treatment of ammonium-rich water with nitrite as an intermediate. SHARON stands for Single-reactor High Activity ammonia Removal Over Nitrite. The process takes advantage of the high temperature of the sludge water, enabling high specific growth rates, making operation without sludge retention possible (Hellinga *et al.,* 1998; Mulder *et al.*, 2001). The process is based on the higher maximum growth rate of ammonia oxidisers relative to nitrite oxidisers at higher temperatures. The SRT is relatively short (around 1 day) to select for ammonia oxidation only without allowing nitrite oxidation.

Figure 5.25 WWTP Beverwijk; the red circle shows the area used for the SHARON® process that treats 1,200 kg nitrogen per day.

The relatively short aerobic SRT (1 day) needed means it is possible to construct the process without biomass retention. The produced sludge ends up in the effluent, which is no problem since the effluent will be sent to the influent of the main wastewater treatment plant. If biomass retention is applied, the aeration time will become the limiting factor for the reactor design due to the large amount of oxygen which needs to be added. It is the final economic balance between reactor volume and retention equipment that sets the proper choice for the application of a sludge retention system. In practice, it appears that at concentrations above 0.4-0.5 gN/l a system without biomass retention is less expensive. Moreover, a system without biomass retention also requires less maintenance.

An important difference between reactors operating with or without biomass retention is the response to changes in concentration in the influent. In retention-based systems the sludge loading is important for the conversion. The system will have to be designed for peak loading. In a system without biomass retention the growth rate is the design factor. Variable loading rates will lead to variable sludge amounts in the reactor and thereby keep the effluent concentration constant and independent of the influent concentrations.

In the SHARON® reactor the denitrification process is usually used as a pH control (Hellinga *et al.,* 1998). Using methanol or waste organics to produce alkalinity through denitrification is cheaper than buying alkalinity as bicarbonate or hydroxide. There is no direct need for full denitrification since the side-stream process is not subjected to effluent restrictions. The main aim is to remove a large amount of nitrogen. For denitrification of regular effluents from sludge digesters, denitrification for pH control already implies a removal percentage of around 95%.

As mentioned above, pH control is the important aspect of the design of nitritation-denitritation processes for digester effluents. The pH is maintained by stripping CO_2 from the liquid. This means that high tanks (above 4-5 meters of water depth) would have to be designed on CO_2 stripping instead of O_2 supply. Also it is preferable to maintain a pH where the bicarbonate equilibrium is shifted towards CO_2, *i.e.* below or around 7.0, despite that for nitrifiers the optimal growth pH might be 7.5-8.0.

Effective process control can also be preferentially based on pH measurements. Both the excessive nitrification and the limited denitrification will rapidly lead to a pH decrease, while too much denitrification leads to a pH increase. There will be a big pH response since in general the system will be run under conditions where hardly any buffer capacity remains. A conversion of 7 mg ammonium-N will generate 1 mM of protons, or a pH decrease to pH 3 in the absence of a buffer; *i.e.* small changes in conversion have a large impact on the system pH.

Nitrification and denitrification are exothermic reactions; this means that the conversion process has a significant impact on the temperature in the reactor. Figure 5.26 gives an overview of the factors which contribute to the temperature in the reactor. Accurate temperature control is not needed; however, it is preferable to operate the process at temperatures above 25 ºC and below 40 ºC.

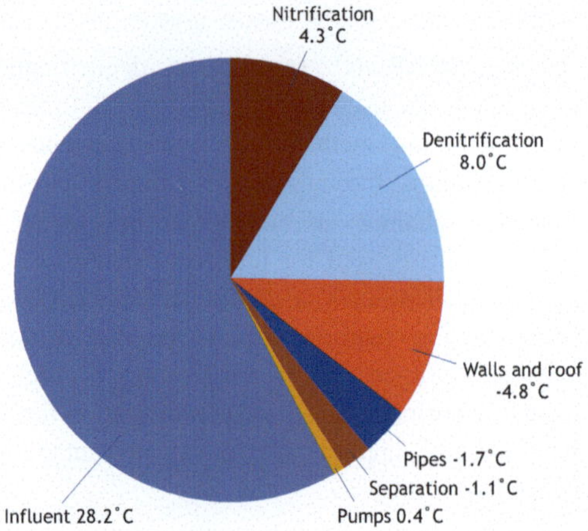

Figure 5.26 Contributions to the heat balance of a full-scale SHARON® reactor at the WWTP Dokhaven in Rotterdam.

The cost of the SHARON® process is mainly influenced by operational factors, energy, and carbon source, as depicted in Figure 5.27. Investment costs are relatively low due to a simple reactor design and operation.

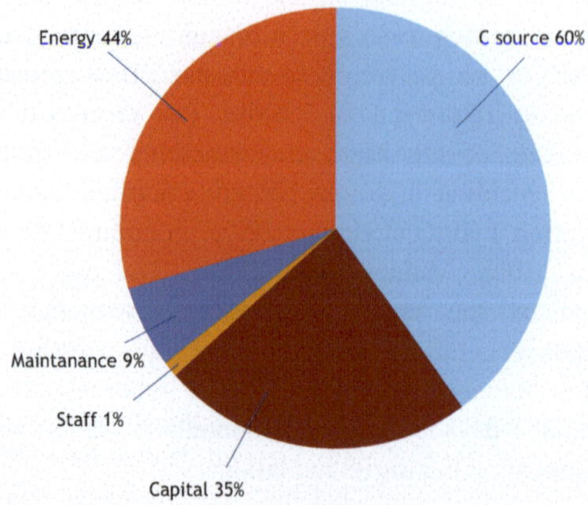

Figure 5.27 Cost build-up for side-stream N-removal in a SHARON® process.

5.12.4 ANAEROBIC AMMONIA OXIDATION

Anammox is an acronym of ANaerobic AMMonium Oxidation. It is a fully autotrophic method for N removal. The microbial process was discovered in the 1980s (Mulder *et al.,* 1989), although studies on its use in wastewater treatment started fully in the 1990s. In fact anammox conversion is a kind of short-cut in the nitrogen cycle. The process converts ammonium directly into dinitrogen gas under anoxic conditions with nitrite as an electron acceptor. The bacteria use CO_2 as a carbon source in the same way as regular ammonium-oxidising bacteria. For more details on the anammox discovery and its applications, the reader is referred to the review paper of Kuenen (2020).

The advantages of using anammox in nitrogen removal are evident. There is no need for external organic carbon sources, only 50% of the ammonium has to be oxidized to nitrite, and there is a low biomass yield (or sludge production). When the anammox process is coupled with a partial nitritation process, the overall conversion is in principle a direct oxidation of ammonium to dinitrogen gas.

The reduced need for energy and no need for an organic electron donor means that use of an anammox process heavily contributes to increasing the sustainability of wastewater treatment operations. Table 5.10 gives an indicative comparison between nitrogen removal by conventional denitrification and the anammox process. In the conventional process approximately 4.7 tons of CO_2 are released per ton of nitrogen removed, whereas for the SHARON®/ANAMMOX® process this is only 0.7 ton of CO_2 per ton of nitrogen removed. This reduction in CO_2 emission might be an additional incentive for the application of anammox in industry. Implications of anammox for large-scale municipal and industrial wastewater treatment plants are summarized in tables 5.10 and 5.11, respectively.

Anammox is currently only applied at higher temperatures (*e.g.* digester effluents), however since in nature it is present everywhere there is no real limit to its application at normal wastewater treatment plants for nitrogen removal. Applying autotrophic nitrogen removal in a 'normal' wastewater treatment plant makes it possible to maximize the sludge production (*e.g.* by primary sedimentation and flocculation) and thereby increase the methane production in a sludge digester. This leads to the substantially improved energy efficiency of the system (Kartal *et al.*, 2010).

The anammox bacteria form a separate and distinct group in the microbial world (Figure 5.28). Their catabolic reactions take place on a cell's internal membrane. In all other bacteria the energy is generated on the outer membrane. Anammox bacteria have a unique intermediate in their catabolism, hydrazine. The exact role of hydroxylamine is still under debate; nowadays NO rather than hydroxylamine is the suggested intermediate. Whereas normal denitrifying bacteria have N_2O as the intermediate, this compound is absent in the anammox physiology. This means that this greenhouse gas will not be produced by anammox bacteria. The anammox growth yield is similar to that of nitrifying bacteria. In the anammox growth process nitrate is produced. This is due to oxidation of nitrite to nitrate, which compensates for the reduction of CO_2 to cellular organic matter. Therefore, in a fully autotrophic process this anoxic nitrate generation is a measure of the biomass growth of anammox and an indication of anammox activity.

Table 5.10 Comparison between a conventional N-removal system and the SHARON®/ANAMMOX® N-removal process.

Item	Unit	Conventional treatment	SHARON®/ANAMMOX®
Power	kWh/kgN	2.8	1.0
Methanol	kg/kgN	3.0	0
Sludge production	kgVSS/kgN	0.5-1.0	0.1
CO_2 emission	kg/kgN	>4.7	0.7
Total costs [1]	€/kgN	3.0-5.0	1.0-2.0

[1] Total costs include both operational costs and capital charge

Table 5.11 Implication of anammox: comparison of nitrogen removal between municipal wastewater treatment plants with conventional denitrification and with an anammox process in the Netherlands (total treatment capacity of 25 million P.E.).

Item	Unit	Conventional	Pre-treatment, anaerobic treatment, anammox	Difference
Energy production (CH_4)	MW	0	40	40
CO_2 emission	kton /year	400	6	394
Power consumption	MW	80	41	39
Sludge production	ktonVSS /year	370	270	100

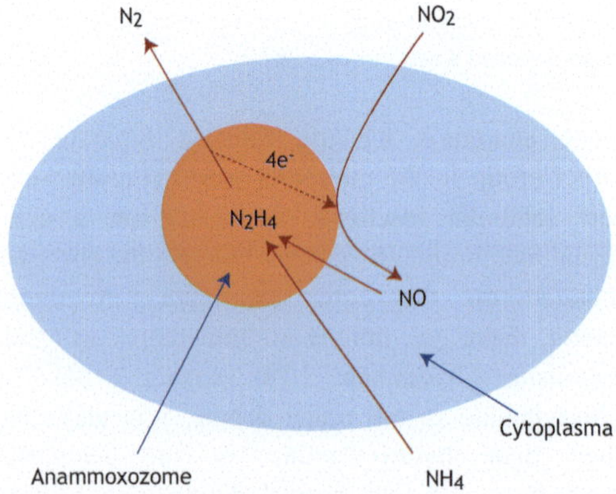

Figure 5.28 Anammox metabolism.

Anammox organisms have always been described as having a very low growth rate (0.069 d^{-1}, Van de Graaf *et al.,* 1996). However, this slow growth rate is no limitation towards high reactor capacities; 5-10 $kgN/m^3.d$ is easily achieved due to the fact that the organisms easily immobilise themselves in compact biofilms or granules allowing very high biomass contents in the reactor. Anammox sludge has a characteristic reddish colour (Figure 5.29). Recent research by Lotti *et al.* (2015) has indicated that the growth can be as high as 1 d^{-1}. This will make anammox more feasible at the lower temperatures of mainstream wastewater treatment.

Anammox conversion needs nitrite for effective ammonium removal; nitrate cannot be used by the anammox organisms. The process needs therefore a first partial nitritation step. A straightforward strategy is to combine the anammox process with a process similar to the SHARON® process (Van Dongen *et al.,* 2001). The ANAMMOX® process is preceded by a SHARON® reactor where only partial nitrification takes place.

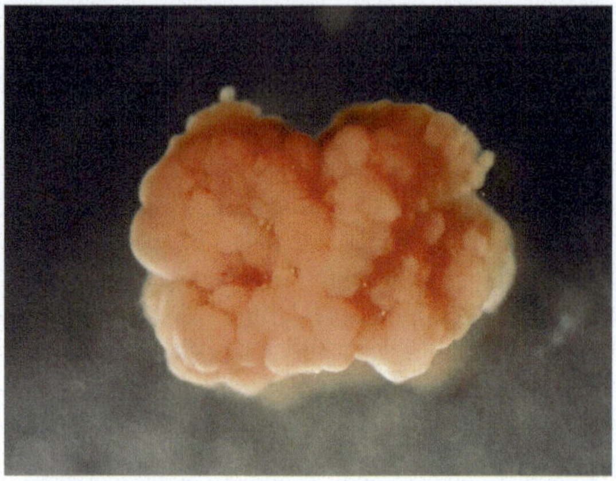

Figure 5.29 Typical anammox sludge granule (photo: Water Board Hollandse Delta).

The SHARON® reactor works under conditions of 1 day HRT, temperature 25-40 °C and pH = 6.6-7.0. Without controlling the pH in the reactor, a mixture of 50% ammonium and 50% nitrite will be obtained in the effluent of the SHARON® reactor. This is due to the fact that in normal digester effluents the counter-ion of ammonium is bicarbonate. When 50% of ammonium is oxidised, all the carbonate buffer is used, and the process inhibits itself due to a pH decrease. The first full-scale application of

Table 5.12 Implication of anammox: comparison of nitrogen removal between industrial wastewater treatment plants with conventional denitrification and with anammox process in the Netherlands (total industrial treatment capacity 21 ktonCOD/yr and 2 ktonN/yr).

	Unit	Conventional	Pre-treatment, anaerobic treatment, anammox	Difference
Energy production (CH_4)	MW	0	2	2
CO_2 emission	kton /year	30	6	24
Power consumption	MW	2.3	0.3	2
Sludge production	ktonVSS /year	30	4	26

SHARON®/ANAMMOX® technology took place at WWTP Dokhaven in Rotterdam, the Netherlands (470,000 P.E., N load 830 kg/d, Van der Star *et al.*, 2007, Figure 5.30).

Figure 5.30 (A) Aerial view of sludge treatment facilities at the WWTP Dokhaven and view of the ANAMMOX® (B) and SHARON® (C) reactors taken from the top of the sludge digester (photo: Water Board Hollandse Delta).

The internal circulation-type reactor used for the anammox process in Rotterdam is especially suited for use of granular sludge. The bottom compartment is a sludge blanket compartment which is well mixed by the functioning of an airlift pump that is driven by the nitrogen gas produced and collected on top of this bottom compartment. In this way, the nitrite concentrations can be kept at a low level in the full reactor compartment despite the high influent concentrations. The upper compartment also contains sludge and is mainly used for achieving a low effluent concentration due to its plug-flow characteristics.

The nitritation and the anammox conversion can also be combined in one reactor. In this case biomass immobilisation is needed because of the very low growth rate of anammox bacteria. The wash-out criterion for preventing nitrite oxidation cannot be used. It is however possible, when oxygen and nitrite are both made limiting, that anammox bacteria effectively out-compete nitrite oxidises from an N-removal process (Hao *et al.,* 2002). The easiest method is operating biofilm or a granular sludge-based autotrophic ammonium oxidation process under oxygen limitation such that approximately 50% of the ammonium can be oxidised. If the biofilm is stable, an anammox population will automatically develop in the deeper biofilm layers as illustrated in Figure 5.31.

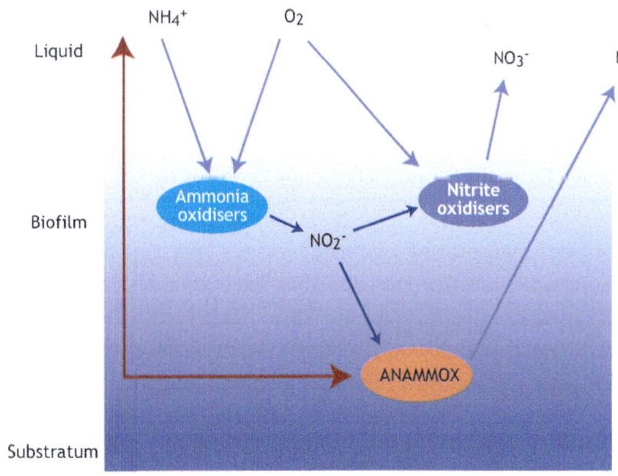

Figure 5.31 Schematic representation of autotrophic N removal in a biofilm process.

This process has occurred spontaneously in several cases (*e.g.* Siegrist *et al.,* 1998). It has been given several acronyms in the literature: Oxygen Limited Autotrophic Nitrification-Denitrification (OLAND®, Kuai *et al.,* 1998); aerobic deammonification (Hippen *et al.,* 1997) and Completely Autotrophic Nitrogen removal Over Nitrite (CANON®, Strous, 2000). The first two names suggest denitrification under aerobic conditions, which is actually not correct; therefore CANON® can be suggested as a general description for the process. Effectively the first two names relate to the original assumption that the denitrification is executed mainly by Nitrosomonas bacteria, although later it was clearly shown that effectively the anammox bacteria are responsible for the conversion (Peynaert *et al.,* 2003). In the meantime, a whole suite of anammox-based processes has been proposed in the literature (Table 5.13, Van der Star *et al.* (2007); Lackner *et al.* (2014). Anammox processes are well developed for higher temperature (25-40 °C) conditions, while development is ongoing to apply anammox under mainstream conditions (Kumwimba *et al.*, 2020). The advantage of this would be that application would make WWTPs self-sufficient for energy (Kartal *et al.* 2010). There have been reports of spontaneous anammox observations in full-scale wastewater treatment plants (*e.g.* in Changi, Singapore and Xi'an, China). Together with several pilot tests this shows that the anammox conversion can potentially function; however, the main challenge currently lies in obtaining a stable nitritation and robust process design.

Table 5.13 Process options and names for nitrogen removal systems involving the anammox process.

Proposed process name	Source of nitrite		Alternative process names	First reference
Two-reactor nitritation-anammox process	NH_4^+	Nitritation	SHARON[a),b)]-ANAMMOX®	Van Dongen *et al.* (2001)
			Two-stage OLAND	Wyffels *et al.* (2004)
			Two-stage deammonification	Treta *et al.* (2004)
One-reactor nitritation-anammox	NH_4^+	Nitritation	Aerobic deammonification	Hippen *et al.* (1997)
			OLAND[c)]	Kuai and Verstraete (1998)
			CANON[d)]	Third *et al.* (2001)
			Aerobic/anoxic deammonification	Hippen *et al.* (2001)
			Deammonification	Seyfried *et al.* (2001)
			SNAP[e)]	Lieu *et al.* (2005)
			DEMON[f)]	Wett (2006)
			DIB[f,g)]	Ladiges *et al.* (2006)
One-reactor denitrification-anammox process	NO_3^-	Denitrification	Anammox[h)]	Mulder *et al.* (1995)
			DEAMOX[i)]	Kalyuzhnyi *et al.* (2006)
			DENAMMOX[j)]	Pathak and Kazama (2007)

a) Single-reactor High Activity Ammonia Removal Over Nitrite; the name only refers to nitritation where nitrite oxidation is avoided by choice of residence time and operation at elevated temperature.
b) Sometimes nitrification-denitrification over nitrite is addressed by this term.
c) Oxygen-Limited Autotrophic Nitrification Denitrification.
d) Completely Autotrophic Nitrogen removal Over Nitrite.
e) Single-stage Nitrogen removal using Anammox and Partial nitritation.
f) Name refers to the deammonification process in an SBR under pH control.
g) Deammonification in Interval-aerated Biofilm systems.
h) System where anammox was originally found. The whole process was originally designated as anammox.
i) DEnitrifying AMmonium OXidation; this name only refers to denitrification with sulphide as the electron donor.
j) DEnitrification-Anammox process; this name only refers to denitrification with organic matter as the electron donor.

5.12.5 BIO-AUGMENTATION

The main design criterion for a nitrifying treatment plant is the required aerobic solid retention time (SRT) for the nitrifying bacteria. The required SRT substantially increases in colder climates. By adding nitrifying bacteria to the activated sludge system it is possible to decrease the required SRT. This can be used as an upgrading option. It can also be used to make highly-loaded systems denitrify, or to free up space in well-nitrifying plants for denitrification. Bio-augmentation can be carried out by externally cultivated nitrifying sludge. However, this has two potential disadvantages. Firstly, the bacteria added might not be the optimal type of nitrifiers for the specific treatment plant. Secondly, if suspended cells are added, these are removed effectively by the protozoa in the sludge. It is therefore of interest to produce the nitrifying bacteria in a reactor continuously inoculated with sludge from the aeration basin and fed with the digester effluent. In this way the nitrifiers which grow in the sludge are those that belong to the system, they grow inside the sludge flocs, and are therefore not grazed by protozoa. Moreover, the N load to the treatment plant decreases.

There have been three different proposals to integrate such a bio-augmentation process. The InNITRI® process (Kos, 1998) is a process where the nitrifiers are produced on the digester effluent (Figure 5.32A). Floc formation is obtained by applying biomass retention. This process carries the risk that an adequate nitrifying microbial population is not produced for bio-augmentation, and application of a longer SRT is not desired since it will minimise the production of sludge/nitrifiers. The two other process options use the return sludge as inoculum for the process. In this case there is no sludge retention needed because the bacteria are already growing in the sludge, and moreover the required COD for denitrification can be derived from the return sludge COD. However, denitrification is needed in order to maintain a good pH in the side-stream reactor. The BAR® process (Bio Augmentation Regeneration; Novak *et al.,* 2003) is derived from a process where the return sludge is aerated in order to obtain sludge mineralization; by adding the digester effluent to this reactor nitrifiers will be produced in this reactor compartment (Figure 5.32B).

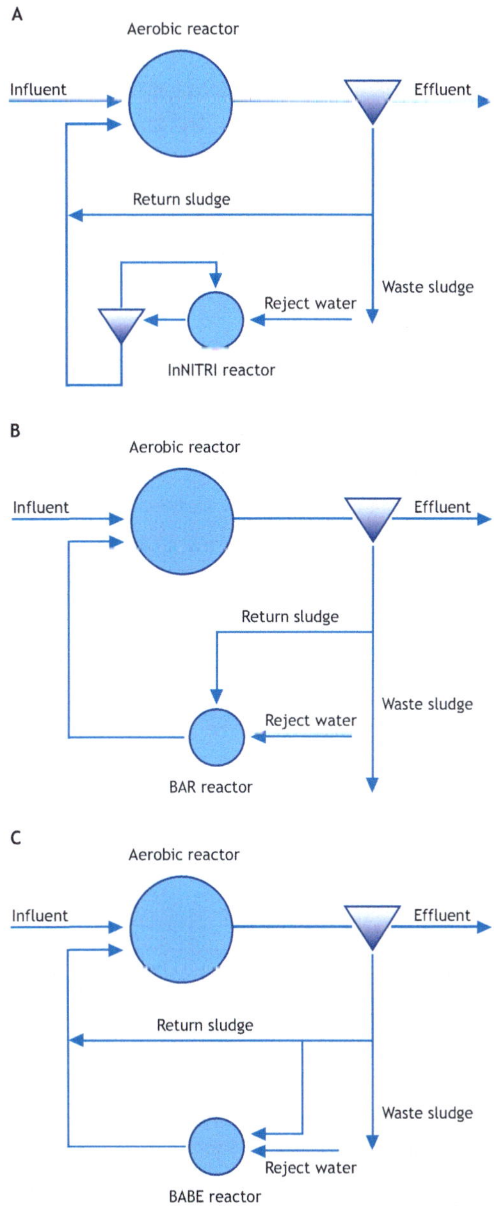

Figure 5.32 Schemes of different bio-augmentation processes: (A) InNITRI®, (B) BAR®, and (C) BABE®.

A disadvantage is that the reactor will have the same low temperature as the activated sludge process itself. Therefore, it is more advantageous to take only a fraction of the return sludge and mix it *e.g.* in a 1:1 ratio with the warm digester effluent (Figure 5.32C). Due to the increased temperature and lower sludge load, the tank can be designed more compactly. The latter process has been dubbed the BABE® process (Bio Augmentation Batch Enhanced; Zilverentant, 1999) because the original process is a repeated batch type of reactor.

The BABE® process has been developed and designed completely based on model simulations. This was feasible since the process did not rely on unknown bacteria but on conventional nitrogen-removing processes. Therefore existing and well-tested activated sludge modelling could be applied (Salem *et al.,* 2003). Model-based design was also needed because of the complexity of the process and because of the cost savings in the process development. The process has many design variables which can be optimised. The side-stream reactor influences nitrification in the main stream reactor and vice-versa. The effect of nitrification in the side-stream reactor on the effluent quality of the main plant has to be evaluated. However, it is not possible to make a proper laboratory or even pilot system since the volume ratio between the main reactor and side-stream reactor is too large.

The minimal SRT is normally determined by the difference between maximum growth rate of the nitrifiers minus their decay rate. In the case of bio-augmentation the maximum growth rate can be summed with the specific addition rate (amount of nitrifiers augmented per unit of nitrifiers per unit of time). A different approach for the BAR® and BABE® processes is to consider the sludge in the side-stream reactor as an integral part of the total sludge inventory; effectively this prolongs the total aerated retention time. However, for evaluating the minimal SRT, it is necessary to take the different sludge concentrations and temperatures into account.

Simulations of the BABE® process have indicated several characteristics of the process (Salem, 2003). A higher fraction of ammonium treated in the side-stream reactor leads to an improved effluent ammonium concentration. The effect is of course only notable when the system SRT is approximately the same as or shorter than the minimum SRT. The biggest impact is observed at 50% of the minimum SRT for nitrification in the activated sludge plant. The temperature effect on the nitrification in the system is less noticeable and even at very low temperatures ammonium concentrations in the effluent decrease. Overall, the process has therefore most impact on highly-loaded treatment plants. The extra ammonium removal from the effluent is due to removal of nitrogen from the digester effluent and from the augmentation of nitrifiers in the main aeration reactor. The augmentation effect contributes 50-70% of the extra nitrogen removal. In low loaded systems the augmentation process allows for an increase in the denitrification space of 10% of the total activated sludge volume.

The simulations also revealed that aiming at maximum N removal (*i.e.* also maximum augmentation) in the side-stream process is not needed. In almost any case there is an optimum conversion in the side-stream reactor above which the effluent of the main wastewater treatment plant no longer changes, while the cost of the side-stream process continues to increase. Also the biomass retention time in the side-stream reactor should be optimised. An increasing retention time leads to a decreased ammonium level in the effluent of the main plant. However, if the retention in the BABE® reactor is increased above a certain optimal time, the effluent in the main plant will deteriorate again. This is due to the fact that increasing the biomass retention time will lead to a lower sludge loading rate and, therefore, sludge production. While this is generally advantageous for a normal wastewater treatment plant, it is not desired for a side-stream process which is supposed to produce nitrifying sludge.

A model-based evaluation of the concept for upgrading WWTP Walcheren (140,000 P.E.) in the Netherlands was carried out and results showed a 50% reduction in area requirement if the BABE® technology was used as compared to conventional upgrading (extending aerated and anoxic volumes) (Salem *et al.,* 2002). A cost analysis showed that using the BABE® technology for upgrading would save around 115,000 Euro per year. There are large savings in construction costs and some savings in energy demand. In this case it was assumed that full denitrification in the BABE® reactor had to be achieved by methanol addition. This leads to extra costs for buying methanol and for extra sludge produced. Building the BABE® reactor a little larger allows significant use of endogenous substrate in the return sludge. For each plant a full cost optimisation will have to define the exact design of the system.

A full-scale evaluation of the BABE® technology took place at WWTP Garmerwolde in the Netherlands (300,000 P.E.). This high-loaded system was run with three parallel lines: one which received the normal N load, one which was operated with only N removal in the digester effluent, and one line which was operated with bio-augmentation according to the BABE® technology. The augmentation effect of the BABE® process improved the nitrification rate of the sludge by almost 60% (Salem *et al.,* 2004), which was in line with model predictions. However, because the SRT at the plant was too short the BABE® process could, in this case, not lead to full nitrification in winter.

A second full-scale application of the process was at the WWTP's-Hertogenbosch, again in the Netherlands (Figure 5.33). This plant was operating at a SRT giving good effluent in summer and no nitrification in winter. The return water from the sludge treatment contained approximately 15% of the N load to the plant. The constructed BABE® process was less than 1% of the existing volume for the activated sludge process. With this implementation, the nitrification could be maintained during the winter months and, moreover, a large extension of the activated sludge basins could be prevented.

Figure 5.33 Full-scale application of the BABE® technology: WWTP 's-Hertogenbosch in the Netherlands (at the front is the bio-augmentation reactor; photo: Royal HaskoningDHV).

REFERENCES

Antoniou P., Hamilton J., Koopman B., Jain R., Holloway B., Lyberatos G. and Svoronos S.A. (1990). Effect of temperature and pH on the effective maximum specific growth rate of nitrifying bacteria. *Water Res.* 24(1), 97-101.

Barnard J.L. (1973). Biological denitrification. Water Pollution Control, 72(6), 705-720.

Brink I.C., Wentzel M.C. and Ekama G.A. (2007). A plant wide stoichiometric steady state WWTP model. Procs. 10th IWA Specialized conference Design, Operation and Economics of large WWTP, Vienna, 9-13 Sept, 243-250.

Bock E., Stüven R., Schmidt I., and Zart D. (1995). Nitrogen loss caused by denitrifying Nitrosomonas cells using ammonium or hydroxylamine as electron donors and nitrite as electron acceptor. *Arch. Microbiol.* 163, 16-20.

Costa E., Pérez J. and Kreft J.U. (2006). Why is metabolic labour divided in nitrification?. *Trends in microbiology*, 14(5), 213-219.

Dold P.L., Ekama G.A. and Marais G.v.R. (1985). pH control and cost savings in aerobic digestion. Procs. IAWPRC specialized seminar on wastewater systems control, Houston, April. In *Advances in Water Pollution Control: Instrumentation and Control of Water and Wastewater and Transport Systems*. (ed. R.A.R. Drake), pp. 375-382. Pergamon Press, Oxford.

Downing A.L., Painter H.A. and Knowles G. (1964). Nitrification in the activated sludge process. *Journal of the Proceedings of the Institute of Sewage Purification*, 130-153.

Ekama G.A., Dold P.L. and Marais G.v.R. (1986). Procedures for determining influent COD fractions and the maximum specific growth rate of heterotrophs in activated sludge systems. *Wat. Sci. Techn*, 18(6), 91-114.

Ekama G.A., Wentzel M.C., Casey T.G. and Marais G.v.R. (1996a). Filamentous organism bulking in nutrient removal activated sludge systems. Paper 3: Stimulation of the selector effect under anoxic conditions. *Water SA*, 22(2), 119-126.

Ekama G.A., Casey T.G., Wentzel M.C. and Marais G.v.R. (1996b). Filamentous organism bulking in nutrient removal activated sludge systems. Paper 6: Review, evaluation and consolidation of results. *Water SA*, 22(2), 147-152.

Ekama G.A., Barnard J.L., Guenthert F.W., Krebs P., McCorquodale J.A., Parker D.S. and Wahlberg E.J. (1997). Secondary settling tanks: Theory, design, modelling and operation. IAWQ STR No. 6, p. 216, International Association on Water Quality, London.

Ekama G.A. and Wentzel M.C. (1999). Denitrification kinetics in biological N & P removal activated sludge systems treating municipal wastewaters. *Wat. Sci. Techn*, 39(6), 69-77.

Henze M., Grady C.P.L., Gujer W., Marais G.v.R. and Matsuo T. (1987). Activated sludge model No.1. IAWPRC Scientific and Technical Reports No 1, International Association on Water Pollution Research and Control, IAWPRC, London, ISSN 1010-707X, p. 33.

Garrido J.M., Van Benthum W.A.J., Van Loosdrecht M.C.M and Heijnen J.J. (1997). Influence of dissolved oxygen concentration on nitrite accumulation in a biofilm airlift suspension reactor. *Biotech. Bioeng.* 53, 168-178.

Hao X., Heijnen J.J. and Van Loosdrecht M.C.M. (2002). Sensitivity analysis of a biofilm model describing a one-stage completely autotrophic nitrogen removal (CANON) process. *Biotech. Bioeng.* 77(3), 266-277.

Hellinga C., Schellen A.A.J.C., Mulder J.W., Van Loosdrecht M.C.M. and Heijnen J.J. (1998). The SHARON process: An innovative method for nitrogen removal from ammonium-rich wastewater. *Wat. Sci. Tech.* 37(9), 135-142.

Henze M., Gujer W., Mino T., Matsuo T., Wentzel M.C and Marais G.v.R. (1995). Activated sludge model No. 2, IAWQ Scientific and Technical Report No. 3. IAWQ, London. ISSN 1025-0913, p. 32.

Hippen A., Rosenwinkel K.-H., Baumgarten G. and Seyfried C.F. (1997). Aerobic deammonification: a new experience in the treatment of wastewater. *Wat. Sci. Tech.* 35(10), 111-120.

Hippen A., Helmer C., Kunst S., Rosenwinkel K.H. and Seyfried C.F., (2001) Six years' practical experience with aerobic/anoxic deammonification in biofilm systems. *Wat. Sci. Techn.* 44(2-3), 39-48.

Jenkins D., Daigger G.T. and Richard M.G. (1993). Manual on the causes and control of activated sludge bulking and foaming. 2nd edition, Lewis Publishers, Chelsea, MI, USA.

Kalyuzhnyi S., Gladchenko M., Mulder A., Versprille B., (2006) DEAMOX—new biological nitrogen removal process based on anaerobic ammonia oxidation coupled to sulphide-driven conversion of nitrate into nitrite.*Water Res.* 40(19), 3637–3645.

Kartal B., Kuenen J.G. and Van Loosdrecht, M.C.M. (2010). Sewage treatment with anammox. *Science*, 328(5979), 702-703.

Koch H., Van Kessel M.A. and Lücker S. (2019) Complete nitrification: insights into the ecophysiology of comammox Nitrospira. *Applied microbiology and biotechnology*, 103(1), 177-189.

Kos P. (1998) Short SRT (solids retention time) nitrification process/flowsheet. *Wat. Sci. Tech.* 38(1), 23–29.

Kuai L.P. and Verstraete W. (1998) Ammonium removal by the oxygen-limited autotrophic nitrification-denitrification system. *Appl. Env. Micr.* 64, 4500-4506.

Kuenen J.G. (2020). Anammox and beyond. *Environmental Microbiology* 22(2) 525–536.

Kumwimba M.N., Lotti T., Şenel E., Li X. and Suanon F. (2020). Anammox-based processes: How far have we come and what work remains? A review by bibliometric analysis. *Chemosphere*, 238, 124627.

Lackner S., Gilbert E.M., Vlaeminck S.E., Joss A., Horn H. and van Loosdrecht M.C. (2014) Full-scale partial nitritation/anammox experiences–an application survey. *Wat. Res.* 55, 292-303.

Ladiges G., Thierbach R.D., Beier M., Focken (2006). Versuche zur zweistufigen Deammonifikation im Hamburger Kla¨rwerksverbund. 6. Aachener Tagung mit Informationsforum. Aachen (Ger), ATEMIS GmbH.p. Fachbeitrag 13.

Lawrence and McCarty (1972). Unified basis for biological treatment design and operation. Journal of the Sanitary Engineering Division, *ASCE*, 96(SA3), 757-778.

Lee H.S., Tang Y., Rittmann B.E. and Zhao H.P. (2018). Anaerobic oxidation of methane coupled to denitrification: fundamentals, challenges, and potential. *Critical reviews in environmental science and technology*, 48(19-21), 1067-1093.

Lieu P.K., Hatozaki R., Homan H. and Furukawa K. (2005). Singlestage nitrogen removal using Anammox and partial nitritation (SNAP) for treatment of synthetic landfill leachate. *Jpn. J. Wat. Treat. Biol.* 41(2), 103.

Loewenthal R.E. and Marais G.v.R. (1977). Carbonate chemistry of aquatic systems: Theory and application. Ann Arbor Science, Michigan, USA.

Lotti T., Kleerebezem R., Abelleira-Pereira J.M., Abbas B. and Van Loosdrecht M.C.M. (2015). Faster through training: the anammox case. *Water Res.* 81, 261-268.

Loveless J.E. and Painter H.A. (1968). The influence of metal ion concentration and pH value in the growth of Nitrosomonas strain isolated from activated sludge. *Journal of General Microbiology*, 52, 1-14.

Ludzack F.J. and Ettinger M.B. (1962). Controlling operation to minimize activated sludge effluent nitrogen. *JWPCF*, 34, 920-931.

Malan W.N. and Gouws E.P. (1966). Geaktiveerde slyk vir rioolwatersuiwering op Bellville. Research report, Council for Scientific and Industrial Research, Nov. 1966.

Metcalf & Eddy (1991). Wastewater engineering: Treatment, disposal, reuse. McGraw-Hill International Series, New York.

Monteith H.D., Bridle T.R. and Sutton P.M. (1980). Industrial waste carbon sources for biological denitrification. *Progress in Water Technology*, 12(Tor), 127- 141.

Mulder, A., Gist Brocades NV, (1989) Anoxic ammonia oxidation. Patent WO88200204.1

Mulder A., Van De Graaf A.A., Robertson L.A., Kuenen J.G. (1995). Anaerobic ammonium oxidation discovered in a denitrifying fluidized bed reactor *FEMS Microb. Ecology*, 16 (3), 177-184.

Mulder J.W., Van Loosdrecht M.C.M., Hellinga C. and Van Kempen R. (2001). Full-scale application of the SHARON process for treatment of reject water of digested sludge dewatering. *Wat. Sci. Tech.* 43(11), 127-134.

Musvoto E.V., Samson K., Taljard M., Fawcett K. and Alexander W.V. (2002). Calculation of peak oxygen demand in the design of full-scale nitrogen removal activated sludge plants. Water SA, Special Edn Wisa 2002, 56-60.

Musvoto E.V., Casey T.G., Ekama G.A., Wentzel M.C. and Marais G.v.R. (1994). The effect of incomplete denitrification on anoxic-aerobic (low F/M) filament bulking in nutrient removal activated sludge systems. *Wat. Sci. Techn*, 29(7), 295-299.

Novák L., Wanner J. and Kos M. (2003). A method for nitrification capacity improvement in an activated sludge process for biological wastewater treatment. CZ patent No. 291 489.

Pathak B.K., Kazama F., (2007) Influence of temperature and organic carbon on denammox process. Proceedings CD, Nutrient Removal 2007, Baltimore, March 4, 2007, pp. 402–413

Pynaert K., Smets B.F., Wyffels S., Beheydt D., Siciliano S.D., and Verstraete W. (2003). Characterization of an Autotrophic Nitrogen-Removing Biofilm from a Highly Loaded Lab-Scale Rotating Biological Contactor. *Appl. Env. Micr.* 69, 3626-3635.

Picioreanu C., Van Loosdrecht M.C.M. and Heijnen J.J. (1997). Modelling the effect of oxygen concentration on nitrite accumulation in a biofilm airlift suspension reactor. *Wat. Sci. Tech.* 36 (1), 147-156.

Salem S., D.H.J.G. and Van Loosdrecht M.C.M. and Heijnen J.J. (2002). Model-based evaluation of a new upgrading concept for N-removal. *Wat. Sci. Tech.* 45(6), 169-176.

Salem S., Berends D.H.J.G. and Van Loosdrecht M.C.M. and Heijnen J.J. (2003). Bio-augmentation by nitrification with return sludge. *Wat. Res.* 37(8), 1794-1804.

Salem S., Berends D.H.J.G.,Van der Roest H.F., Van der Kuijl R.J. and Van Loosdrecht M.C.M. (2004). Full-scale application of the BABE process. *Water Sci. Technol,* 50(7), 87-96.

Scheller S., Ermler U. and Shima S. (2020). Catabolic pathways and enzymes involved in anaerobic methane oxidation. In: *Anaerobic Utilization of Hydrocarbons, Oils, and Lipids*, pp.31-59.

Schemen R., Van der Spoel H., Salem S. and Van Kempen R. (2003). Unieke combinatie op rwzi Beverwijk. *H2O,* 10,17-19.

Seyfried C.F., Hippen A., Helmer C., Kunst S. and Rosenwinkel K.H., (2001) One-stage deammonification:nitrogen elimination at low costs. *Wat. Supply,* 1(1), 71–80

Siegrist H., Reithaar S., Koch G. and Lais P. (1998). Nitrogen loss in a nitrifying rotating contactor treating ammonia-rich wastewater without organic carbon. *Wat. Sci. Tech.* 38(8-9), 241-248.

Sötemann S.W., Musvoto E.V., Wentzel M.C. and Ekama G.A. (2005). Integrated chemical, physical and biological processes modelling Part 1 - Anoxic aerobic C and N removal in the activated sludge system. *Water SA*, (4) 529-544.

Stenstrom M.K. and Poduska R.A. (1980). The effect of dissolved oxygen concentration on nitrification. *Water Research*, 14(6), 643-649.

Stern L.B. and Marais G.v.R (1974). Sewage as the electron donor in biological denitrification. Research Report W7, Dept. of Civil Eng., Univ. of Cape Town, Rondebosch, 7701, Cape, RSA.

Still D.A., Ekama G.A., Wentzel M.C., Casey T.G. and Marais G.v.R. (1996). Filamentous organism bulking in nutrient removal activated sludge systems. Paper 2: Stimulation of the selector effect under aerobic conditions. *Water SA*, 22(2), 97-118.

STOWA (1996). One reactor system for ammonia removal via nitrite, Report no 96-01, Utrecht.

Strous M. (2000). Microbiology of anaerobic ammonium oxidation. PhD thesis, Department of Biotechnology, the Technical university of Delft, The Netherlands.

Third K.A., Sliekers A.O., Kuenen J.G. and Jetten M.S.M. (2001). The CANON system (completely autotrophic nitrogen-removal over nitrite) under ammonium limitation: Interaction and competition between three groups of bacteria. *System. Appl. Microbio.* 24, 588-596.

Treła J., Plaza E., Szatkowska B., Hultman B., Bosander J. and Dahlberg A.-G., (2004) Deammonifakation som en ny process for behandling av avloppsstrommar med hog kvavehalt *Vatten* 60 (2), 119–127.

Tsai M-W., Wentzel M.C. and Ekama G.A. (2003). The effect of residual ammonia concentration under aerobic conditions on the growth of Microthrix parvicella in biological nutrient removal plants. *Water Research*, 37(12), 3009-3015.

Van Benthum W.A.J., Derissen B.P., Van Loosdrecht M.C.M. and Heijnen J.J. (1998). Nitrogen removal using nitrifying biofilm growth and denitrifying suspended growth in a biofilm airlift suspension reactor coupled to a chemostat. *Wat. Res.* 32, 2009-2018.

Van de Graaf A.A., Mulder A., de Bruijn P., Jetten M.S.M, Robertson L.A. and Kuenen J.G. (1996). Autotrophic growth of anaerobic ammonium oxidizing micro-organisms in a fluidised bed reactor. *Microbiology* 142, 2187-2196.

Van der Star W.R.L., Abma W.R., Blommers D., Mulder J.-W., Tokutomi T., Strous M., Picioreanu C. and Van Loosdrecht M.C.M. (2007). Startup of reactors for anoxic ammonium oxidation: Experiences from the first full-scale Anammox reactor in Rotterdam. *Wat. Res.* 41, 4149-4163.

Van Dongen L.G.J.M., Jetten M.S.M. and Van Loosdrecht M.C.M (2001). The combined SHARON/Anammox process: A sustainable method for N-removal from sludge water. STOWA, Utrecht.

Van Haandel A.C., Dold P.L. and Marais G.v.R (1982). Optimization of nitrogen removal in the single sludge activated sludge process. *Water Science and Technology*, 14, 443.

Van Haandel A.C., Ekama G.A. and Marais G.v.R. (1981). The activated sludge process part 3 - Single sludge denitrification. *Water Research*, 15, 1135-1152.

Warner A.P.C., Ekama G.A. and Marais G.v.R. (1986). The activated sludge process Part 4 - Application of the general kinetic model to anoxic-aerobic digestion of waste activated sludge. *Water Research*, 20(8), 943-958.

WEF (1998). Design of Municipal WWTP, 4th edn., Water Environment Federation (WEF) Manual of Practice MOP 8, WEF, Alexandria, VA, American Society of Civil Engineers (ASCE) Manual and Report on Engineering Practice No. 76, *ASCE*, Vol. 3, ASCE, Reston.

Wentzel M.C., Ekama G.A., Dold P.L. and Marais G.v.R. (1990). Biological excess phosphorus removal - steady state process design. *Water SA*, 16(1), 29-48

Wett B. (2006). Solved upscaling problems for implementing deammonification reject water. *Wat. Sci. Techn.*, 53(12), 121-128.

Wild H.E., Sawyer C.N. and McMahon T.C. (1971). Factors affecting nitrification kinetics. *Journal of Water Pollution Control Federation*, September.

WRC (1984). Theory, design and operation of nutrient removal activated sludge processes. Published by the Water Research Commission, Private Bag X03, Gezina, 0031, South Africa.

Wyffels S., Van Hulle S.W.H., Boeckx P., Volcke E., van Cleemput O., Vanrolleghem P.A. and Verstraete W. (2004). Modeling and simulation of oxygen-limited partial nitritation in a membrane-assisted bioreactor (MBR). *Biotechnol. Bioeng.* 86, 531e542.

Zilverentant A. (1999). Process for the treatment of wastewater containing specific components e.g. ammonia. Patent PCT/NL99/00462,WO0005177.

NOMENCLATURE

Symbol	Description	Unit
a	Mixed liquor recycle ratio from the aerobic to the primary anoxic reactor	-
S_{Alk}	Alkalinity with respect to the H_2CO_3 reference solution	mg/l as $CaCO_3$
a_{min}	Minimum a-recycle ratio to utilize all RBCOD in primary anoxic	-
a_{opt}	Optimum a-recycle ratio from the aerobic to the primary anoxic reactor	-
a_{prac}	Maximum practical a-recycle ratio	-
$b_{ANO,20}$	Specific endogenous respiration rate for nitrifiers at 20 °C	/d
$b_{ANO,T}$	Specific endogenous respiration rate for nitrifiers at T °C	/d
b_{OHO}	Specific endogenous respiration rate of ordinary heterotrophic organisms	/d
$b_{OHO,20}$	Specific endogenous respiration rate for OHOs at 20 °C	/d
$b_{OHO,T}$	Specific endogenous respiration rate for OHOs at T °C	/d
$COD_{b,i}$	Influent biodegradable COD concentration	mgCOD/l
COD_i	Influent total COD concentration	mgCOD/l
D_p	Denitrification potential	mgN/l
D_{p1}	Primary anoxic denitrification potential	mgN/l
$D_{p1RBCOD}$	Primary anoxic denitrification potential due to RBCOD	mgN/l
$D_{p1SBCOD}$	Primary anoxic denitrification potential due to SBCOD	mgN/l
D_{p3}	Secondary anoxic denitrification potential	mgN/l
$D_{p3RBCOD}$	Secondary anoxic denitrification potential due to RBCOD dosed	mgN/l
$D_{p3SBCOD}$	Secondary anoxic denitrification potential due to SBCOD	mgN/l
f_{aN}	Influent ammonia to TKN concentration ratio	mgN/mgN
$f_{at,OHO}$	Fraction of OHOs in the sludge as TSS	mgVSS/mgTSS
$f_{av,OHO}$	Fraction of OHOs in the sludge as VSS	mgVSS/mgVSS
$FCOD_b$	Flux of biodegradable COD exiting reactor	kgCOD/d
$FCOD_{b,i}$	Flux of influent biodegradable COD entering reactor	kgCOD/d
f_{cv}	COD to VSS ratio of the sludge	mgCOD/mgVSS
f_n	Nitrogen content of VSS	mgN/mgVSS

$f_{SU,TKNi}$	Fraction of influent TKN that is unbiodegradable soluble organic N	-
FS_{NHx}	Mass per day (flux) of free and saline ammonia (FSA) utilized as N	mgN/d
$FS_{NO3,e}$	Mass per day (flux) of nitrate as N produced by nitrification	$kgNO_3$-N/d
$f_{TKNi/CODi}$	Influent wastewater TKN/COD concentration ratio	mgN/mgCOD
FN_S	Mass per day (flux) of nitrogen required for sludge production	kgN/d
FO_c	Mass per day (flux) of oxygen required for organics (COD) removal	kgO_2/d
FO_{DENIT}	Mass per day (flux) of oxygen recovered by denitrification	kgO_2/d
FO_{NIT}	Mass per day (flux) of oxygen required for nitrification	kgO_2/d
FO_{tDENIT}	Total mass per day (flux) of O_2 required less that recovered by denitrification	kgO_2/d
f_p	P content of VSS	mgP/mgVSS
f_{SS}	RBCOD fraction with respect to influent biodegradable COD	-
$f_{SU,CODi}$	Soluble unbiodegradable fraction of total influent COD	-
$f_{XE,OHO}$	Unbiodegradable fraction of OHOs	mgCOD/mgCOD
$f_{XU,CODi}$	Particulate unbiodegradable fraction of total influent COD	-
f_{x1}	Primary anoxic sludge mass fraction	-
$f_{x1,min}$	Minimum primary anoxic sludge mass fraction	-
f_{x3}	Secondary anoxic sludge mass fraction	-
FX_{ANO}	Mass per day (flux) of nitrifiers generated	mgVSS/d
$f_{xd,max}$	Maximum anoxic sludge mass fraction	-
$f_{x,max}$	Maximum unaerated sludge mass fraction	-
f_{xt}	Fraction of total sludge mass in reactor not aerated	-
FX_{TSS}	Mass per day (flux) of TSS wasted from reactor	kgTSS/d
K_1	Initial rapid specific rate of denitrification in primary anoxic reactor	$mgNO_3$-N/mgOHOVSS.d
K_2	Second specific rate of denitrification in primary anoxic reactor	$mgNO_3$-N/mgOHOVSS.d
K_3	Specific rate of denitrification in secondary anoxic reactor	$mgNO_3$-N/mgOHOVSS.d
K_4	Specific rate of denitrification in anoxic-aerobic digester	$mgNO_3$-N/mgOHOVSS.d
K_{NIT}	Specific nitrification rate	mgN/mgVSS.d
$K_{NIT,max}$	Maximum specific nitrification rate	mgN/mgVSS.d
K_h	Maximum specific uptake rate of SBCOD by OHOs under aerobic conditions	mgCOD/mgCOD.d
K_I	Nitrification pH sensitivity inhibition coefficient	-
K_{II}	Nitrification pH sensitivity inhibition coefficient	-
K_{max}	Nitrification pH sensitivity inhibition coefficient	-
K_{ANO}	Half-saturation constant for nitrifiers	mgN/l
$K_{ANO,20}$	Half-saturation constant for nitrifiers at 20 °C	mgN/l
$K_{ANO,T}$	Half-saturation constant for nitrifiers at T °C	mgN/l
K_O	Half-saturation constant for dissolved oxygen	mgO_2/l
K_S	Half-saturation concentration for RBCOD utilization	mgCOD/l
K_X	Half-saturation concentration for utilization SBCOD by OHOs	mgCOD/l
MX_{ANO}	Mass of nitrifiers in reactor	mgVSS
MX_{OHOv}	Mass of OHO biomass in reactor	kgVSS
$MX_{E,OHOv}$	Mass of endogenous residue in reactor	kgVSS
MX_{Iv}	Mass of unbiodegradable organics from the influent in the reactor	kgVSS

MX_{TSS}	Mass of TSS in reactor	mgTSS/l
MX_{VSS}	Mass of organic matter of activated sludge in reactor	mgVSS/l
NIT_c	Nitrification capacity	mgN/l
NIT_c/COD_i	Nitrification capacity per mg COD applied to reactor	mgN/mgCOD
$N_{ob,e}$	Effluent residual biodegradable organic nitrogen	mgN/l
$N_{ob,i}$	Influent biodegradable organic nitrogen	mgN/l
$N_{obp,i}$	Influent biodegradable particulate organic nitrogen	mgN/l
$N_{obs,e}$	Effluent biodegradable soluble organic nitrogen	mgN/l
$N_{obs,i}$	Influent biodegradable soluble organic nitrogen	mgN/l
$N_{oup,i}$	Influent unbiodegradable particulate organic nitrogen	mgN/l
$N_{ous,e}$	Effluent unbiodegradable soluble organic nitrogen (= $N_{ous,i}$)	mgN/l
$N_{ous,i}$	Influent unbiodegradable soluble organic nitrogen	mgN/l
N_s	Concentration of N in influent required for sludge production	mgN/l
N_s/COD_i	N required for sludge production to influent COD concentration ratio	mgN/mgCOD
O_c	Carbonaceous oxygen utilization rate	mgO_2/l.h
O_{NIT}	Nitrification oxygen utilization rate	mgO_2/l.h
O_t	Total oxygen utilization rate	mgO_2/l.h
P_s	Concentration of influent P required for sludge production	mgP/l
P_e	Effluent total P concentration	mgP/l
P_i	Influent total P concentration	mgP/l
Q_e	Effluent flow rate	l/d
Q_i	Influent flow rate	l/d
Q_w	Waste flow rate from the reactor	l/d
s	Sludge underflow recycle ratio from the SST to the primary anoxic reactor	-
S_f	Factor of safety on maximum specific growth rate of nitrifiers	-
S_{NHx}	Bulk liquid ammonia concentration	mgN/l
$S_{NHx,e}$	Effluent ammonia concentration	mgN/l
$S_{NHx,i}$	Influent ammonia concentration	mgN/l
$S_{NHx,NIT}$	Ammonia concentration per litre influent available for nitrification	mgN/l
S_{NO3}	Nitrate concentration	mgN/l
$S_{NO3,ar}$	Equivalent nitrate concentration loaded onto the primary anoxic reactor	mgN/l
$S_{NO3,e}$	Effluent nitrate concentration	mgN/l
$S_{NO3,e,aopt}$	Effluent nitrate concentration at a_{opt}	mgN/l
$S_{NO3,e,aprac}$	Effluent nitrate concentration at a_{prac}	mgN/l
$S_{NO3,p1}$	Nitrate concentration in the aerobic reactor	mgN/l
S_{O2}	Dissolved oxygen concentration in bulk liquid	mgO_2/l
$S_{O2,a}$	Dissolved oxygen concentration in the a recycle	mgO_2/l
$S_{O2,s}$	Dissolved oxygen concentration in the s recycle	mgO_2/l
S_S	Soluble readily biodegradable (RB)COD concentration	mgCOD/l
SRT	Sludge age	D
SRT_{min}	Minimum sludge age for nitrification	D
T	Temperature	^oC
t_1	Duration (actual retention time) of 1st phase of denitrification	D

$t_1(a+s+1)$ d	Duration (nominal retention time) of 1st phase of denitrification	D
TKN_e/COD_i	Effluent TKN to influent COD concentration ratio	mgN/mgCOD
TKN_e	Effluent TKN concentration	gN/l
TKN_i	Influent TKN concentration	gN/l
TKN_i/COD_i	Influent wastewater TKN/COD concentration ratio	mgN/mgCOD
T_{max}	Maximum wastewater temperature	°C
T_{min}	Minimum wastewater temperature	°C
V_R	Reactor volume	L
$X_{OHO,i}$	OHOVSS concentration in reactor per litre influent flow	mgOHOVSS/l
X_{OHOv}	OHO biomass concentration	mgVSS/l
$X_{E,OHOv}$	Endogenous residue from OHOs in activated sludge	mgVSS/l
X_I	Influent unbiodegradable matter in activated sludge	mgVSS/l
X_S/X_{OHO}	SBCOD/OHO concentration ratio	mgCOD/mgCOD
X_S	Slowly biodegradable (SB)COD concentration	mgCOD/l
X_{TSS}	TSS concentration in reactor	mgTSS/l
X_{VSS}	Organic matter concentration of activated sludge in reactor	mg VSS/l
$X_{VSS,e}$	Effluent particulate volatile matter	mg VSS/l
Y_{ANO}	Yield coefficient for nitrifiers	mgVSS/mgFSA-N
Y_{OHO}	Yield of OHOs in terms of COD (= $f_{cv}Y_{OHOv}$)	mgCOD/mgCOD
Y_{OHOv}	Yield of OHOs in terms of VSS	mgVSS/mg COD

Greek symbol	Explanation	Unit
η	Reduction factor for utilization of SBCOD under anoxic conditions	-
θ_b	Temperature sensitivity coefficient for endogenous respiration	-
θ_{NIT}	Temperature sensitivity coefficient for nitrification	-
θ_{ns}	pH sensitivity coefficient for nitrification	-
μ_{ANO}	Specific growth rate of nitrifier	/d
$\mu_{ANO,20}$	Specific growth rate of nitrifier at 20 °C	/d
$\mu_{ANO,max}$	Maximum specific growth rate of nitrifier	/d
$\mu_{ANO,max,20}$	Maximum specific growth rate of nitrifier at 20 °C	/d
$\mu_{ANO,max,7.2}$	Maximum specific growth rate of nitrifier at pH=7.2	/d
$\mu_{ANO,max,pH}$	Maximum specific growth rate of nitrifier at pH	/d
$\mu_{ANO,max,pH,T}$	Maximum specific growth rate of nitrifier at pH and temperature T °C	/d
$\mu_{ANO,max,T}$	Maximum specific growth rate of nitrifier at T °C	/d
$\mu_{ANO,O}$	Specific growth rate of nitrifier at O mgO_2/l	/d
$\mu_{ANO,T}$	Specific growth rate of nitrifier at T °C	/d
μ_{OHO}	Maximum specific growth rate of OHO	/d

Abbreviation	Description
AOO	Ammonia-oxidizing organism
ANO	Autotrophic nitrifying organism
BABE	Bio-augmentation batch enhanced
BAR	Bio-augmentation regeneration
BNR	Biological nutrient removal
CANON	Completely autotrophic nitrogen removal over nitrate
COD	Chemical oxygen demand
DO	Dissolved oxygen
DIB	Deammonification in interval-aerated biofilm systems
DENAMMOX	Denitrification-Anammox process
DEAMOX	Denitrifying ammonium oxidation
DEMON	Deammonification system
EBPR	Enhanced biological phosphorus removal
HRT	Hydraulic retention time
FSA	Free and saline ammonia
ISS	Inorganic component of the settleable solids mass
MAP	Magnesium ammonium phosphate
MLSS	Mixed liquor suspended solids
MLVSS	Mixed liquor volatile suspended solids
ND	Nitrification and Denitrification
NDEBPR	Nitrification and Denitrification Enhanced Biological Phosphorus Removal
NOO	Nitrite-oxidizing organism
OHO	Ordinary heterotrophic organism
OLAND	Oxygen-limited autotrophic nitrification denitrification
NDAS	Nitrification-denitrification activated sludge
PAO	Phosphorus-accumulating organism
PST	Primary settling tank
RBCOD	Readily biodegradable COD
SBCOD	Slowly biodegradable COD
SRT	Sludge retention time (sludge age)
SHARON	Single-reactor High Activity ammonia removal over nitrite
SS	Suspended solids
SST	Secondary settling tank
SNAP	Single-stage nitrogen removal using the Anammox and partial nitritation
TKN	Total Kjeldahl nitrogen
TSS	Total suspended solids
VFA	Volatile fatty acid
VSS	Volatile suspended solids
WAS	Waste activated sludge
WWTP	Wastewater treatment plant

Figure 5.34 Award-winning (IWA's Europe 2008 Project Innovation Award) wastewater treatment plant Garmerwolde in The Netherlands applying SHARON® process (two white reactors in the front; photo: Grontmij Nederland N.V.).

doi: 10.2166/9781789060362_0239

6

Enhanced biological phosphorus removal

Carlos M. Lopez-Vazquez, Mark C. Wentzel†, Yves Comeau, George A. Ekama, Mark C.M. van Loosdrecht, Damir Brdjanovic and Adrian Oehmen

6.1 INTRODUCTION

Phosphorus is the key element that needs to be removed from aquatic environments in order to limit the growth of aquatic plants and algae, and thus, to control eutrophication. Unlike nitrogen, which can be fixed from the atmosphere and which contains approximately 80% nitrogen gas, phosphorus can only come from upstream of aquatic systems (neglecting atmospheric deposition). Diffuse sources of phosphorus, *e.g.* from agricultural fields, are best controlled by fertilisation plans, while point sources of phosphorus, *e.g.* from wastewater treatment plants, can be removed by chemical or biological processes. Considering the benefit to aquatic environments, stricter regulations are now being applied for phosphorus removal from wastewater.

The enhanced biological phosphorus removal (EBPR) phenomenon, insofar as it pertains to P removal in activated sludge systems, was noted first in the late 1950s. In the decades since, the understanding, conceptualization and application of the phenomenon have grown from initial incidental observations into well-structured biochemical and mathematical descriptions that are applied in the design and control of major full-scale works. However, the impetus for these developments did not stem from a pure scientific interest, but almost wholly from the recognition, albeit slowly, in the 1960s of the pivotal role that P plays in eutrophication of aquatic environments. This recognition, together with the massive increase in phosphorus loads on the aquatic environment since 1950, gave rise to an urgent need to develop effective countermeasures to limit the discharge of P. One such countermeasure is EBPR.

The expression enhanced biological phosphorus removal (EBPR) is used in this chapter to describe what is also referred to in the literature as biological enhanced phosphorus removal or biological excess phosphorus removal (BEPR), or sometimes biological phosphorus removal (BPR), where a wastewater treatment biomass removes phosphorus beyond its anabolic requirements by accumulating intracellular polyphosphate reserves. In addition to P removal for cell synthesis, further phosphorus (P) removal may also take place by chemical precipitation either with chemicals present in the wastewater or added to the treatment system.

Achieving low concentrations of total phosphorus in effluents can be achieved by combining various processes as indicated in Table 6.1. For example, two combinations of processes can be used to reach 0.5 mgP/l, EBPR with sand filtration, without (combination D) or with chemical coagulation (combination E). Biological phosphorus removal combined with a limited supply of chemicals can achieve effluent values below 0.1 mgP/l, with coagulation and filtration being mainly used to remove the phosphate bound in the effluent suspended solids.

In this chapter, the intention is to present the mechanisms of biological P removal, to trace the development of practical systems for biological P removal, and to set out guidelines for the design of biological P removal systems. To facilitate the development of design guidelines for this textbook, the concepts are presented for strictly aerobic phosphorus-accumulating organisms (aerobic PAOs) which can use only oxygen as the electron acceptor for energy production. Considering that some denitrifying PAOs (DPAOs) exist and may have a significant impact on the performance of the process, their influence is discussed where appropriate.

Considering the potential benefits of removing phosphorus biologically rather than chemically, along with organic matter and nitrogen, EBPR has stimulated much interest in the study of biochemical mechanisms, the microbiology of the systems, the process engineering and optimization of plants, and in mathematical modelling. Reviews of the development of EBPR have been regularly published over the years (Marais *et al.,* 1983; Arvin, 1985; Wentzel *et al.,* 1991; Jenkins and Tandoi, 1991; Van Loosdrecht *et al.,* 1997; Mino *et al.,* 1998; Blackall *et al.,* 2002; Seviour *et al.,* 2003; Oehmen *et al.,* 2007; Gebremariam *et al.,* 2011; Yuan *et al.,* 2012; Zheng *et al.,* 2014; Guo *et al.,* 2019; Liu *et al.,* 2019; Nielsen *et al.,* 2019).

6.2 PRINCIPLES OF ENHANCED BIOLOGICAL PHOSPHORUS REMOVAL (EBPR)

Enhanced biological phosphorus removal (EBPR) is the biological uptake and removal by activated sludge systems in excess of the amount that is removed by conventional, completely aerobic activated sludge systems. This is in excess of the 'normal' P requirements for growth of activated sludge. In the completely aerobic activated sludge system, the amount of P typically incorporated in the sludge mass is approximately 0.02 mgP/mgVSS (0.015 mgP/mgTSS). By the daily wastage of surplus sludge, phosphorus is thus effectively removed (Figure 6.1).

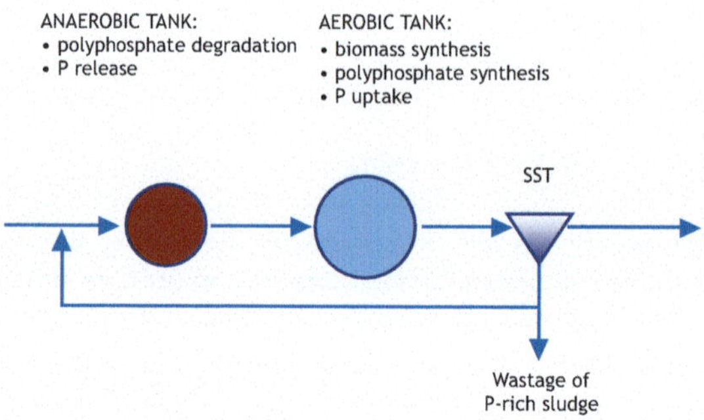

Figure 6.1 Observations of the behaviour of PAOs in an EBPR system (adapted from Tchobanoglous *et al.*, 2003).

This can give a P removal rate of approximately 15-25% in many municipal wastewaters. In an EBPR activated sludge system, the amount of P incorporated in the sludge mass is increased from the normal value of 0.02 mgP/mgVSS to values around 0.06-0.15 mgP/mgVSS (0.05-0.10 mgP/mgTSS). This is achieved by system design or operational modifications that stimulate, in addition to the 'ordinary' heterotrophic organisms present in activated sludge, the growth of organisms that can take up large quantities of P and store them internally in long chains called polyphosphates (sometimes known as poly P); generically these organisms are called phosphorus-accumulating organisms (PAOs).

Table 6.1 Combinations of processes required to achieve given effluent total phosphorus concentration for municipal effluents (adapted from Barnard and Steichen, 2007).

Treatment processes	P limit to achieve (mgP/l)						
		< 1		< 0.5	< 0.1	< 0.05	< 0.01
Combination	A	B	C	D	E	F	G
Chemical coagulation	•		•		•		
EBPR[1)]		•	•	•	•	•	•
Post-coagulation						•	•
Sand filtration				•	•	•	
Adsorption							•
Membrane filtration							•

[1)] with efficient final settling >99.9%.

PAOs (sometimes also called polyphosphate-accumulating organisms) can incorporate up to 0.38 mgP/mgVSS (0.17 mgP/mgTSS). In the biological P removal system both the ordinary heterotrophic organisms (OHOs, which do not remove P in excess) and the PAOs coexist; the larger the proportion of PAOs that can be stimulated to grow in the system, the greater the percentage phosphorus content of the activated sludge and, accordingly, the larger the amount of P that can be removed from the influent. Thus, the challenge in design is to increase the amount of the PAOs relative to the OHOs present in the activated sludge as this will increase the capacity for P accumulation and thereby the phosphorus removal efficiency. The relative proportion of the two organism groups depends, to a large degree, on the fraction of the influent wastewater biodegradable COD that each organism group obtains. The greater the proportion of influent biodegradable COD the PAOs obtain, the greater will be the fraction of PAO in the mixed liquor, the greater the %P content of the activated sludge and the greater the EBPR. This is shown graphically in Figure 6.2.

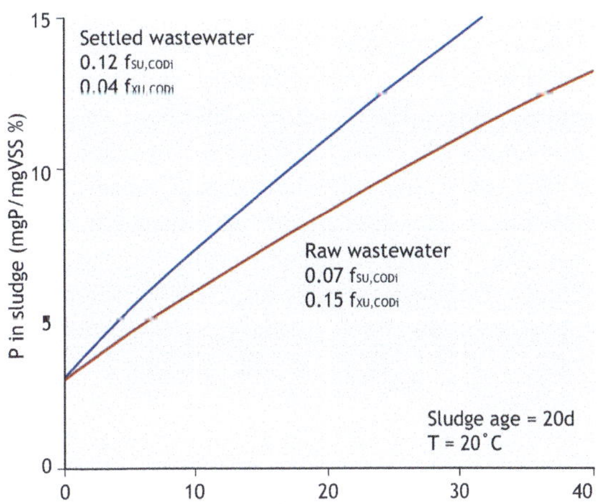

Figure 6.2 Percentage P in VSS mass *versus* the proportion of biodegradable COD mass (as %) obtained by PAOs.

Design and operational procedures are oriented towards maximizing the growth of PAOs. In an appropriately designed EBPR system, the PAOs can make up approximately 11% of the TSS (or 15% of the VSS), and this system can usually remove approximately 10-12 mgP/l per 500 mg influent COD/l.

However, from the first publications reporting enhanced P removal in some activated sludge systems, there has been some controversy as to whether the mechanism is a precipitation of inorganic compounds, albeit perhaps biologically mediated, or biological through formation and accumulation of P compounds in the organisms. The objective of this chapter is not to discuss the evidence that supports the biological nature of enhanced P removal, but to briefly describe the theory of biological P removal as understood by the authors and to demonstrate how this theory can be used as an aid for the design of biological P removal activated sludge systems. This does not imply that precipitation of P due to chemical changes resulting from biological action, *e.g.* alkalinity, pH, does not take place. Such inorganic precipitation can certainly take place, but it would appear that in the treatment of municipal wastewaters by an appropriately designed activated sludge system, within the normal ranges of pH, alkalinity and calcium concentrations in the influent, enhanced P removal is principally caused by a biological mechanism (Maurer *et al.*, 1999; De Haas *et al.*, 2000).

6.3 EBPR MICROBIOLOGY

Polyphosphates can be accumulated by a wide range of bacteria. In general, they are accumulated as a phosphate reserve in relatively low amounts intracellularly. The general definition of what constitutes a PAO has changed in recent years, but is currently most frequently defined as an organism that metabolises and stores organic carbon sources under anaerobic conditions, usually with polyphosphate hydrolysis serving as the primary energy source for this process, and then taking up phosphorus and storing it as polyphosphate under aerobic or anoxic conditions.

The classical PAO definition that existed for many years principally involved the hydrolysis of polyphosphates to accumulate volatile fatty acids (VFAs) as the main carbon source that can be used by PAOs, which sequester VFAs as poly-ß-hydroxyalkanoates (PHAs) under anaerobic conditions (in the absence of an external electron acceptor such as oxygen or nitrate). The PHAs are then oxidised in the presence of an electron acceptor (*i.e.* oxygen, nitrate or nitrite) for cell growth and polyphosphate storage. This definition has since been expanded to include other PAO phenotypes, particularly those involving the anaerobic/aerobic (and/or anoxic) metabolism of polyphosphate-storing organisms based on other carbon sources, such as sugars and amino acids. These mechanisms do not necessarily imply the PHA cycling associated with VFA-driven EBPR.

Linking the microbiological identity of bacteria with their function in EBPR systems has been a challenge for many years, but in recent times it has been facilitated by the development and application of advanced microbiological methods. In the original research on EBPR microbiology conducted with cultivation studies, it was incorrectly considered that PAOs were of the genus *Acinetobacter* (Fuhs and Chen, 1975; Buchan, 1983; Wentzel *et al.*, 1986), *Microlunatus phosphovorus* (Nakamura *et al.*, 1995) or *Lampropedia* (Stante *et al.*, 1997). However, since then culture-independent methods have shown the importance of other organisms, including *Candidatus* Accumulibacter phosphatis (or *Accumulibacter*). A member of the genus *Rhodocyclus* within the *Betaproteobacteria*, *Accumulibacter* is widely regarded to be one of the most important PAOs corresponding to the classical PAO definition described previously. Techniques such as fluorescence *in-situ* hybridisation probes (FISH) combined with chemical staining to detect polyphosphate cycling have shown this organism to correspond to the PAO phenotype and to be an important PAO in full-scale EBPR processes worldwide, as well as lab-scale enrichments fed with VFA (Wagner *et al.*, 1994; Hesselmann *et al.*, 1999; Crocetti *et al.*, 2000; Martin *et al.*, 2006; Oehmen *et al.*, 2007). *Accumulibacter* is the most understood PAO in terms of its microbiological and biochemical characteristics, based on many years of study. *Accumulibacter* clades or sub-groups have also been identified (He *et al.*, 2007; Camejo *et al.*, 2016) and correlations have been attempted to relate the specific identity with observed metabolic behaviour (Flowers *et al.*, 2009; Oehmen *et al.*, 2010; Acevedo *et al.*, 2012; Camejo *et al.*, 2016, 2019; Rubio-Rincon *et al.*, 2017, 2019). It has also been suggested that other organisms, such as *Accumulimonas* and *Dechloromonas*, behave similarly to the classical PAO phenotype and are relevant in at least some EBPR plants (Stokholm-Bjerregaard *et al.*, 2017; Wu *et al.*, 2019), though further investigation is needed to determine the importance of these and other PAOs.

In recent years, another group of PAOs has been found to be of importance in EBPR systems, which belong to the *Tetrasphaera* (Stokholm-Bjerregaard *et al.*, 2017; Liu *et al.*, 2019; Nielsen *et al.*, 2019). *Tetrasphaera* PAOs have been found to be of high abundance in numerous full-scale EBPR systems, particularly under certain configurations, such as sidestream EBPR processes (see Section 6.6.11). *Tetrasphaera* are believed to metabolise mainly sugars and amino acids within EBPR anaerobic zones, though some VFA uptake has also been observed (Nguyen *et al.*, 2011). Thus far, studies have suggested that most *Tetrasphaera* are unlikely to be capable of PHA storage and degradation, quite unlike *Accumulibacter* (Kristiansen *et al.*, 2013). *Tetrasphaera* probably obtain most of their energy for anaerobic organic carbon uptake from fermentation of these carbon sources, with storage of sugars (as *e.g.* glycogen) and certain amino acids suggested to contribute to aerobic P uptake, instead of PHA oxidation (Kristiansen *et al.*, 2013; Nguyen *et al.*, 2015). While anaerobic P release is typically also observed by *Tetrasphaera*, suggesting that polyphosphate is an additional source of energy under anaerobic conditions, results have also suggested that fermentation of certain substrates can yield sufficient energy to contribute towards anaerobic P uptake (Marques *et al.*, 2017). Denitrifying P removal by *Tetrasphaera* is currently believed to be of lesser significance for most members of this group of organisms as compared to aerobic P uptake, which is also unlike previous findings for *Accumulibacter* (Marques *et al.*, 2018). Due to the uncertainty surrounding numerous mechanistic aspects related to *Tetrasphaera* PAOs, the classical PAO phenotype and characteristics, and how they impact EBPR design and operation, will encompass the main focus throughout the remainder of this chapter.

6.4 EBPR MECHANISMS

6.4.1 Background

The basic requirement for EBPR is the presence in the activated sludge system of microorganisms which can accumulate P in excess of normal metabolic requirements, in the form of polyphosphate stored in granules called volutins. In the design procedures presented in this chapter, all the organisms in the activated sludge system accumulating polyphosphate in this fashion and exhibiting the classical observed EBPR behaviour - anaerobic P release, aerobic P uptake and associated PHA production and consumption processes - are grouped together and represented by a generic PAO group. It remains unclear if or how the activity of other PAOs (such as *Tetrasphaera*) can or should be incorporated into EBPR design, thus their implications require further study.

Historically, several research groups have made a number of important contributions towards elucidating the mechanisms of enhanced biological phosphorus removal (EBPR), including Fuhs and Chen (1975), Nicholls and Osborn (1979), Rensink (1981), Marais *et al.* (1983), Comeau *et al.* (1986), Wentzel *et al.* (1986, 1991), Van Loosdrecht *et al.* (1997), Mino *et al.* (1987, 1994, 1998), Kuba *et al.* (1993), Smolders *et al.* (1994a, b, 1995), Maurer *et al.* (1997), Seviour *et al.* (2003), Martin *et al.* (2006), Oehmen *et al.* (2007), Lopez-Vazquez *et al.* (2009b), Oyserman *et al.* (2016), Fernando *et al.* (2019), Rubio-Rincon *et al.* (2019), and Nielsen *et al.* (2019). In this section, an explanation of the basic concepts underlying the more sophisticated mechanistic models for the biological P removal phenomenon is presented. For detailed description of the mechanisms, the reader is referred to the references above.

6.4.2 Prerequisites

As noted earlier, to achieve EBPR in activated sludge systems, the growth of organisms that accumulate polyphosphate (PAOs) has to be stimulated. To accomplish this, two conditions are essential, namely: (*i*) an anaerobic then aerobic (or anoxic) sequence of reactors/conditions, and (*ii*) the addition or formation of VFAs in the anaerobic reactor.

6.4.3 Observations

With the prerequisites for EBPR present, the following observations have been made at full-, pilot- and laboratory-scale (Figure 6.3).

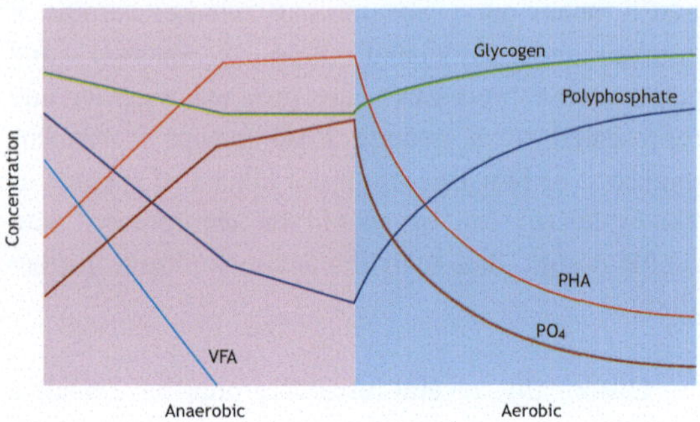

Figure 6.3 Schematic diagram showing the changes as a function of time in concentrations of volatile fatty acids (VFA), phosphate (PO_4), polyphosphate, poly-ß-hydroxyalkanoate (PHA) and glycogen through the anaerobic aerobic sequence of reactors in an EBPR system.

Under anaerobic conditions, bulk solution VFAs and intracellular polyphosphate decrease, soluble phosphate, Mg^{2+}, K^+ and intracellular PHA increase (Rensink, 1981; Hart and Melmed, 1982; Fukase *et al.,* 1982; Watanabe *et al.,* 1984; Arvin, 1985; Hascoet *et al.,* 1985; Wentzel *et al.,* 1985, 1988; Comeau *et al.,* 1986, 1987; Murphy and Lötter, 1986; Gerber *et al.,* 1987; Satoh *et al.,* 1992; Smolders *et al.,* 1994a; Maurer *et al.,* 1997). In many cases, glycogen also decreases, though this is not a strict requirement (see Section 6.4.4).

Under aerobic conditions; intracellular polyphosphate increases; soluble phosphate, Mg^{2+}, K^+ and intracellular PHA decrease (Fukase *et al.,* 1982; Arvin, 1985; Hascoet *et al.,* 1985; Comeau *et al.,* 1986; Murphy and Lötter, 1986; Gerber *et al.,* 1987; Wentzel *et al.,* 1988a; Satoh *et al.,* 1992; Smolders *et al.,* 1994b; Maurer *et al.,* 1997). Glycogen is also produced in most cases.

6.4.4 Biological P-removal mechanism

In describing the mechanisms of EBPR, a clear distinction is made between the PAOs and the organisms not able to accumulate polyphosphate, termed ordinary heterotrophic organisms (OHOs). In the anaerobic/aerobic sequence of reactors, it is considered that VFAs are present in the influent waste stream entering the anaerobic reactor or produced in the anaerobic reactor by fermenting bacteria.

6.4.4.1 In the anaerobic reactor

The reactions taking place in PAOs under anaerobic conditions are illustrated in a simplified biochemical model (Figure 6.4), in a biochemical model showing more explicitly the sources and uses of energy and carbon (Figure 6.5) and in a quantitative model obtained from an enriched culture grown on acetate as sole carbon source at an SRT of 8 days and grown at 20 °C (Figure 6.6).

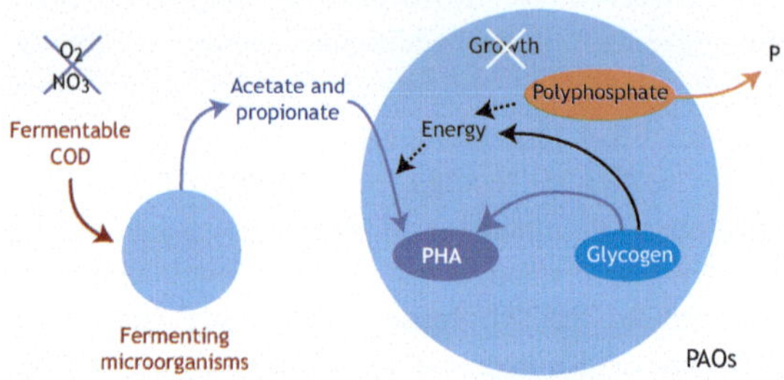

Figure 6.4 Simplified biochemical model for PAOs under anaerobic conditions. Anaerobic uptake of volatile fatty acids (VFAs), originating from the influent or from fermentation in the anaerobic reactor, and storage of polyhydroxyalkanoates (PHAs) by the PAOs take place with associated P release.

The OHOs cannot utilize the VFAs due to the absence of an external electron acceptor, oxygen or nitrate. The PAOs, however, can take up the VFAs from the bulk liquid and store them internally by linking the VFAs together to form complex long-chain carbon molecules of poly-ß-hydroxyalkanoates (PHAs). The two common PHAs are poly-ß-hydroxybutyrate (PHB: a 4-carbon compound synthesized from two acetate molecules) and polyhydroxyvalerate (PHV: a 5-C from Ac + Prop) but poly-ß-hydromethylbutyrate (PH_2MB: 5-C from Ac + Prop) and poly-ß-hydroxymethylvalerate (PH_2MV: 6-C from two Prop) can also be present as minor constituents.

Forming PHAs from the VFAs requires energy for three functions: active transport of VFAs across the cell membrane, energisation of VFAs into coenzyme A compounds (*e.g.* acetylCoA), and reducing power (NADH) for PHA formation. Polyphosphate degradation is associated with formation of ADP from AMP, and with the phosphokinase enzyme 2 ADP are converted into ATP and AMP (Van Groenestijn *et al.*, 1987). When ATP is used orthophosphates are released and accumulate in the cell interior together with the counter-ions of polyphosphate (potassium and magnesium). The efflux of these compounds might be related to building a proton motive force, which either can help in the uptake of acetate or in the generation of a small amount of extra ATP. It is observed (Smolders *et al.* 1994a) that the energy requirements for acetate uptake increase with increasing pH. This can be associated with the fact that the energy needed for acetate transport increases with pH. ATP is used, notably, for the energisation of acetate and propionate into acetyl-CoA and propionyl-CoA. When glycogen degradation occurs, this also results in ATP formation, NADH production and intermediates that are transformed into acetyl-CoA (or propionyl-CoA). However, the ATP and NADH generated through glycogen degradation can be replaced by additional polyphosphate degradation and the activity of the TCA cycle anaerobically, respectively, as observed in certain situations (Zhou *et al.*, 2010; Majed *et al.*, 2012; Lanham *et al.*, 2013). Finally, acetyl-CoA and propionyl-CoA are stored as PHA (Comeau *et al.*, 1986; Wentzel *et al.*, 1986; Mino *et al.*, 1998; Smolders *et al.*, 1994a; Martin *et al.*, 2006; Oehmen *et al.*, 2007; Saunders, 2007).

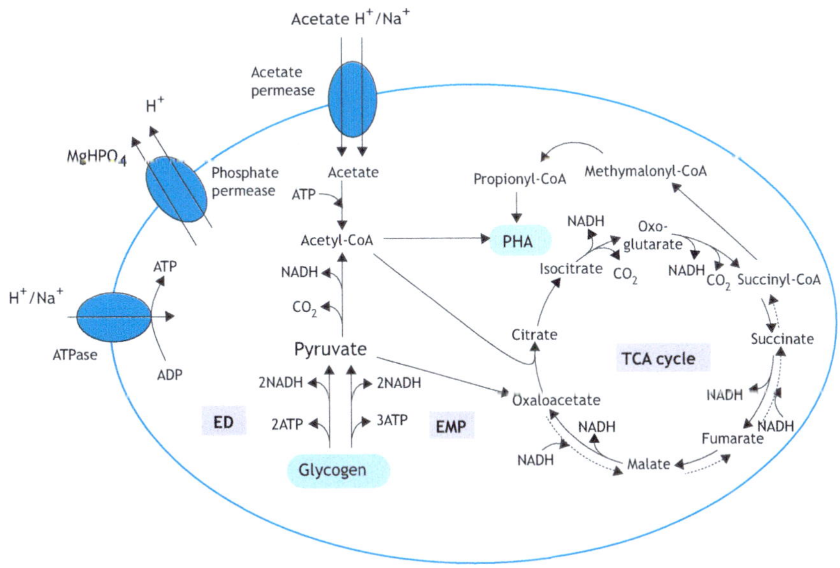

Figure 6.5 Anaerobic metabolic pathways and substrate uptake mechanisms in *Accumulibacter*. Acetate is taken up across the cell membrane, where the energy necessary for transport of acetate is generated by proton (H^+) or sodium (Na^+) efflux, which is promoted by different enzymes depending on the organism. In addition to acetate transport, the occurring ATP and NADH production (and consumption) processes are shown, including P release, glycolysis, and the oxidative and reductive TCA cycle pathways performed by certain PAOs (adapted from Oehman *et al.*, 2010).

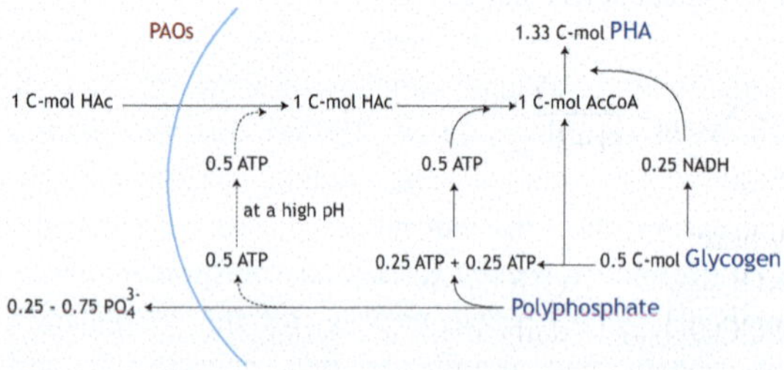

Figure 6.6 Quantitative biochemical model for PAOs under anaerobic conditions (adapted from Smolders *et al.*, 1994a). Values were obtained from an enriched culture grown on acetate as the sole carbon source at 20 °C at an SRT of 8 days.

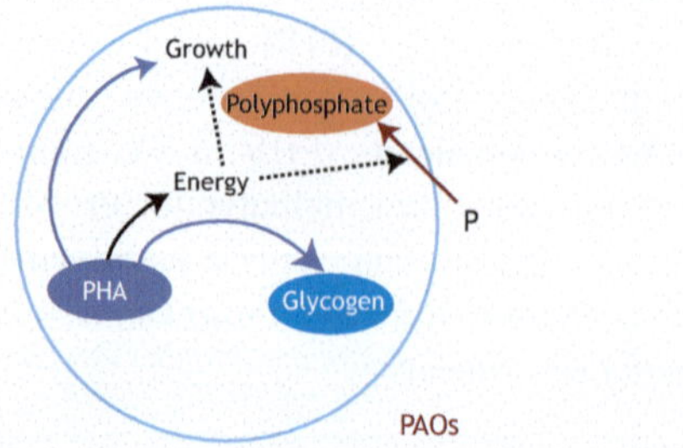

Figure 6.7 Simplified biochemical model for PAOs under aerobic (or anoxic) conditions.

Thus, the PAOs in the anaerobic reactor have taken up the VFAs for their exclusive use under conditions (anaerobic) where ordinary heterotrophic organisms are unable to use this COD. To accomplish this, some of the stored polyphosphate has been consumed and P released to the bulk solution. To stabilize the negative charges on the polyphosphate, the cations Mg^{2+}, K^+ and sometimes Ca^{2+} are complexed. When polyphosphates are consumed and P is released, mainly Mg^{2+} and K^+ cations are released in the approximate molar ratio P:Mg^{2+}:K^+ of 1:0.33:0.33 (Comeau *et al.*, 1987; Brdjanovic *et al.*, 1996; Pattarkine and Randall, 1999).

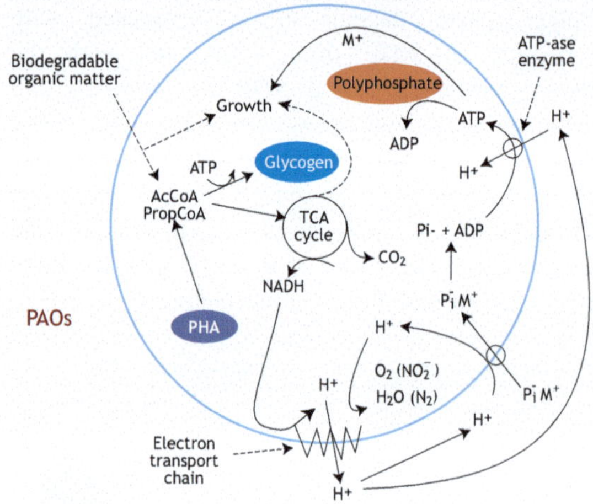

Figure 6.8 Biochemical model for PAOs under aerobic (or anoxic) conditions (adapted from Comeau *et al.* 1986). For design purposes, it is considered that only PHA reserves are used, via acetylCoa and propionylCoaA for PAO growth and not external biodegradable organic matter. The TCA is used to produce carbon intermediates for growth, energy (such as ATP) and reducing power (such as NADH). NADH, in the presence of an electron acceptor such as oxygen or nitrate, is then used to expel protons through the electron transport chain, creating a proton motive force that is used for phosphate (Pi^-) transport with metallic cations (M^+) and ATP synthesis that serves for PAO growth and polyphosphate storage.

6.4.4.2 In the subsequent aerobic reactor

In the presence of oxygen (or of nitrate under anoxic conditions) as an external electron acceptor, the PAOs utilize the stored PHA as a carbon and energy source for energy generation, growth of new cells and typically for regenerating the glycogen consumed in the anaerobic period (figures 6.7 and 6.8).

The stored PHA is also used as an energy source to take up P from the bulk solution to regenerate the polyphosphate used in the anaerobic reactor, and to synthesize polyphosphate in the new cells that are generated.

The uptake of P to synthesize polyphosphate in the new cells generated means that more P is taken up than is released in the anaerobic reactor, giving a net removal of P from the liquid phase in the activated sludge system. Accompanying the P uptake, the cations Mg^{2+} and K^+ are also taken as countercharge for the negatively charged polyphosphate polymer, in the approximate molar ratio $P:Mg^{2+}:K^+$ of 1:0.33:0.33. The PAOs, with stored polyphosphate, are removed from the aerobic reactor of the system (where the internally stored polyphosphate concentration in the PAOs is the highest in the system) via the waste sludge stream (wastage from the underflow recycle stream is possible, but not desirable for hydraulic control of sludge age, see Chapter 4). At steady state the mass of PAOs wasted per day (with stored polyphosphate) equals the mass of new PAOs generated per day (with stored polyphosphate). Thus, for a fixed sludge age, loading and system operation, the mass of PAOs in the biological reactors remains constant, so that in the activated sludge system at steady state there is neither a build-up nor a loss of PAOs, and the P/VSS ratio stays approximately constant. The mass of new PAOs formed depends on the mass of stored substrate (PHA) available to the PAOs. Accordingly, the enhanced P removal attained will depend on the mass of PHA stored in the anaerobic reactor.

6.4.4.3 Quantitative anaerobic-aerobic PAO model

A quantitative model for PAOs subjected to anaerobic and aerobic conditions is shown in Figure 6.9. This model was determined using an enriched PAO culture system operated at an SRT of 8 days, pH of 7.0 and with acetate as the sole carbon source (Smolders *et al.,* 1994a, b). Under anaerobic conditions, influent acetate is taken up by PAOs with energy coming from polyphosphate and glycogen degradation that result in PHB (or PHA) formation and some CO_2 production. Under aerobic conditions, oxygen is consumed for the synthesis of polyphosphate, glycogen and biomass, and for cell maintenance. These aerobic processes result in PHB formation and CO_2 production. With biomass wastage to maintain the SRT, each 1 C-mol of acetate results in 0.04 mol of P removed in the form of polyphosphate.

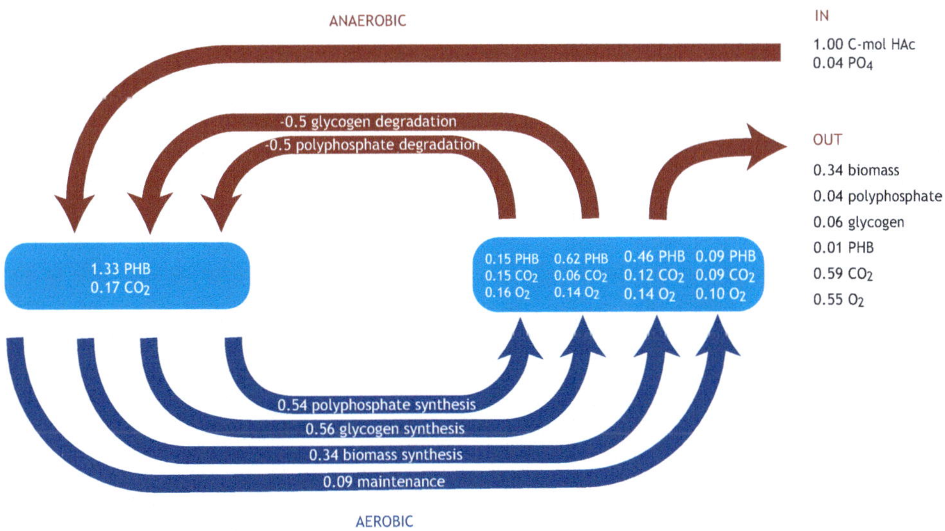

Figure 6.9 Quantitative model for PAOs subjected to anaerobic and aerobic conditions (adapted from Smolders *et al.*, 1994b). Each C-mol acetate corresponds to 0.5 mol of acetate. For acetate, 1 C-mol thus corresponds to 32 gCOD. All other carbon compound concentrations are expressed in C-mol units.

6.4.5 Fermentable COD and slowly biodegradable COD

As indicated above, under anaerobic conditions, PAOs can only store VFAs (S_{VFA}). Some wastewater which contains very few VFAs, however, exhibits significant EBPR which is related to the rapidly biodegradable COD (S_S) which is composed of both S_{VFA} and fermentable COD (S_F) (Siebritz *et al.,* 1982; Wentzel *et al.,* 1985; Nicholls *et al.,* 1985; Pitman *et al.,* 1988; Wentzel *et al.,* 1990; Randall *et al.,* 1994). Thus, it is considered that VFAs coming from the influent and those fermented from S_F are available for anaerobic storage by PAOs.

Slowly biodegradable COD (X_S), even though it can be hydrolyzed into rapidly biodegradable COD under anaerobic conditions, has been shown not to be linked to anaerobic phosphate release. This aspect is of crucial importance as it will influence both the design and operation of BNR systems, such as sizing and determining the number of anaerobic reactors, inclusion of primary sedimentation and maximum EBPR achievable. For the purpose of this design chapter, the experimental evidence linking EBPR to S_S is accepted, and hence a significant conversion of X_S to VFAs is considered unlikely. Accordingly, where VFA production does occur, this will be essentially from the rapidly biodegradable COD. One exception to this consideration is when primary sludge fermentation is provided upstream of the anaerobic reactor, which favours the hydrolysis of some X_S into S_S and VFAs.

6.4.6 Functions of the anaerobic zone

From the description of the mechanisms above, with typical domestic wastewater as influent, the anaerobic zone/reactor serves two functions:

(i) It stimulates conversion of fermentable COD into VFAs by heterotrophic organisms, *i.e.* facultative acidogenic fermentation.

(ii) It enables the PAOs to sequester the VFAs by taking them up and storing them as PHA. In effect this process enables the PAOs to take up and store some of the substrate under conditions (no external electron acceptor, anaerobic) where it is not available to the OHOs. The PAOs then do not have to compete for substrate when an external electron acceptor becomes available (anoxic/aerobic).

Of the above two processes, the former is the slower and determines the size of the anaerobic reactor. Should primary sludge fermentation be implemented at the treatment plant, the first process would not be needed as much and the size of the anaerobic reactor could be decreased.

6.4.7 Influence of recycling oxygen and nitrate on the anaerobic reactor

As noted by numerous investigators (*e.g.* Barnard, 1976a, 1976b; Venter *et al.,* 1978; Rabinowitz and Marais, 1980; Hascoet and Florentz, 1985), recycling of oxygen and/or nitrate to the anaerobic reactor causes a corresponding decrease in EBPR. In terms of the mechanisms described above, if oxygen and/or nitrate is recycled to the anaerobic reactor, the OHOs are able to utilize the fermentable COD for energy and growth using the oxygen or nitrate as external electron acceptor. For every 1 mg O_2 recycled to the anaerobic reactor, 3 mg COD of fermentable COD are consumed and for every 1 mg N of nitrate recycled, 8.6 mg COD of fermentable COD are consumed by the OHOs. The ratio of 3 mg S_F per mg O_2 consumed comes from considering a net yield of 0.67 mg VSS-COD produced per mg COD consumed with the rest, 0.33 mg per mg, serving for energy production using oxygen. Thus, for every mg of O_2 consumed, 3 times as much S_F is consumed. Similarly, considering that 1 mg of nitrate is the equivalent of 2.86 mg of oxygen, a ratio of 8.6 mg COD consumed by mg NO_3-N reduced is obtained.

The fermentable COD consumed is no longer available for conversion by OHOs to VFAs and, therefore, the amount of VFAs generated and released to the solution is reduced, by the amount of RBCOD consumed by the OHOs. Consequently, the mass of VFAs available to the PAOs for storage is reduced, and correspondingly so is the P release, P uptake and net P removal.

Should the influent RBCOD already consist of VFAs and oxygen and/or nitrate be recycled, then the PAOs and OHOs will compete for the VFAs, the PAOs to sequester the VFAs and the OHOs to metabolize it. Accordingly, even in this situation, recycling of oxygen and/or nitrate will reduce the EBPR.

Thus, preventing the recycling of oxygen and nitrate to the anaerobic reactor is one of the primary considerations in the design and operation strategy for EBPR systems (Siebritz *et al.*, 1980).

6.4.8 Denitrification by PAOs

The extent of denitrification with associated anoxic P uptake by the PAOs appears to be very variable (Ekama and Wentzel, 1999), from near-zero anoxic P uptake (*e.g.* Clayton *et al.*, 1991) to anoxic P uptake dominant over aerobic P uptake (*e.g.* Sorm *et al.*, 1996). Experimental evidence tends to suggest that the magnitude of anoxic P uptake is influenced by the anoxic mass fraction and the mass of nitrate loaded on the anoxic reactor relative to its denitrification potential (Hu *et al.*, 2001, 2002). Only some PAOs are actually capable of complete denitrification, reducing nitrate all the way to nitrogen gas (Camejo *et al.*, 2016, 2019). Most appear capable of reducing nitrite to nitrogen gas, where the conversion of nitrate to nitrite can be carried out by flanking organisms (Rubio-Rincon *et al.*, 2017, 2019).

For the purpose of design it will be considered that anoxic P uptake is not significant. Anoxic P uptake decreases the magnitude of P removal in the system (Ekama and Wentzel, 1999), and from a design point of view in which maximising P removal is a priority, anoxic P uptake should be avoided in the system. Hence, in this design chapter anoxic P uptake will not be considered. It must be emphasized, however, that due to the anaerobic storage of RBCOD by PAOs, the kinetics of denitrification do change when an anaerobic reactor is included in the system.

6.4.9 Relationship between influent COD components and sludge components

The relationship described above between influent COD components and the various sludge organic masses (active, endogenous and inert) is shown in Figure 6.10.

6.5 FACTORS IMPACTING EBPR PROCESS PERFORMANCE

6.5.1 Total influent COD (COD_i)

As discussed previously, the total influent COD has a direct influence on the EBPR process since it contains the main substrates required by PAO to perform such a process. To illustrate the effect of the total influent COD on the EBPR process, the stroichiometric steady-state model presented in Section 6.8 is used. Using this model, figures 6.11 and 6.12 plot the effect of a total influent COD concentration of 500 mgCOD/l (in Figure 6.11) and 1,000 mgCOD/l (in Figure 6.12) on the P-removal process at different sludge ages.

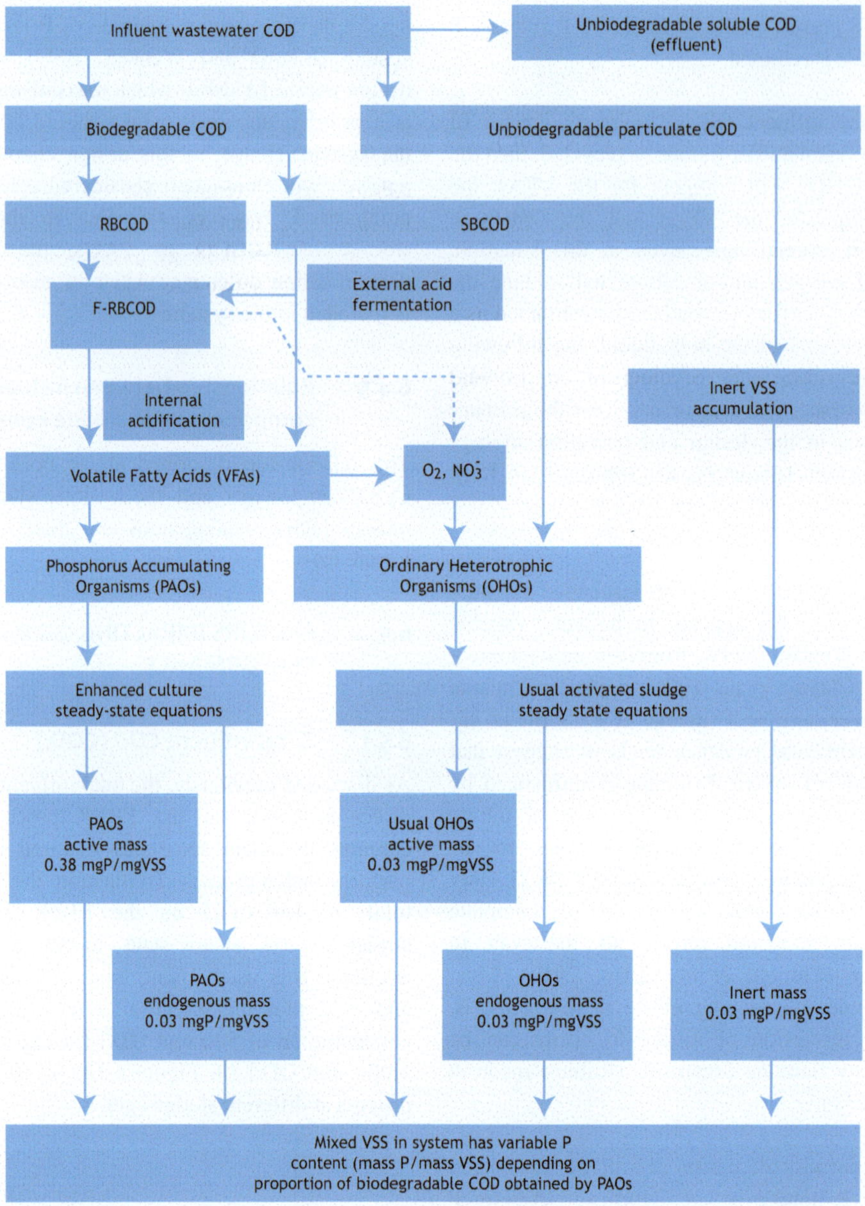

Figure 6.10 Schematic diagram showing the fate of various influent COD fractions in relation to the various active, endogenous and inert masses of the sludge.

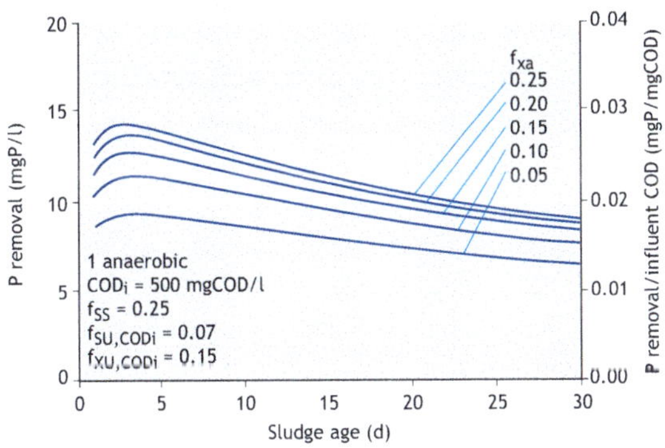

Figure 6.11 Predicted P removal *versus* sludge age for various anaerobic mass fractions (f_{xa}), for a single anaerobic reactor system treating unsettled wastewater with a total COD of 500 mgCOD/l, with characteristics as shown.

To assist a comparison of the effects of the different influent COD concentrations, the right axis is given as P removal/COD_i. It appears that with an increase in COD_i, the P removal efficiency (*i.e.* P removal/COD_i) increases. This is due to the increased magnitude of fermentable COD concentration (influent RBCOD fraction constant at f_{SS} = 0.25), and conversion with increased COD_i as a result of the higher OHO biomass.

6.5.2 Raw or settled sewage

The effect that settling the wastewater has on P removal is illustrated in Figure 6.13 where P removal is plotted against the sludge age for various anaerobic mass fractions f_{xa}, given an influent COD_i of 1,000 mgCOD/l and subject to primary sedimentation (resulting in a settled wastewater with a strength of 600 mgCOD/l). Comparing the P removal for the original unsettled wastewater (Figure 6.12) with that for the settled wastewater (Figure 6.13), it is evident that settling will reduce the P removal of the system. This reduction is due to the decrease in the mass of biodegradable COD entering the activated sludge system which causes a reduction in the fermentable COD converted and in the mass of OHOs generated. However, P removal per influent COD entering the biological reactor is higher for the settled than for the unsettled wastewater. This is apparent from figures 6.12 and 6.13, by comparing the P removal/COD_i on the right-hand axes. This arises because the ratio $S_{S,i}/COD_i$ is higher for settled (f_{SS} = 0.38) than for unsettled wastewater (f_{SS} = 0.25, it should be noted that it is assumed no $S_{S,i}$ is removed in settling. Although this will not be strictly correct the $S_{S,i}$ removal in settling appears to be minimal).

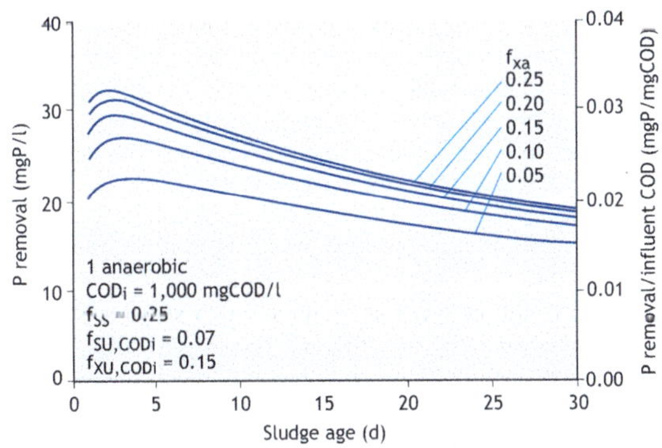

Figure 6.12 Predicted P removal *versus* sludge age for various anaerobic mass fractions (f_{xa}), for a single anaerobic reactor system treating unsettled wastewater with a total COD of 1,000 mgCOD/l, with characteristics as shown.

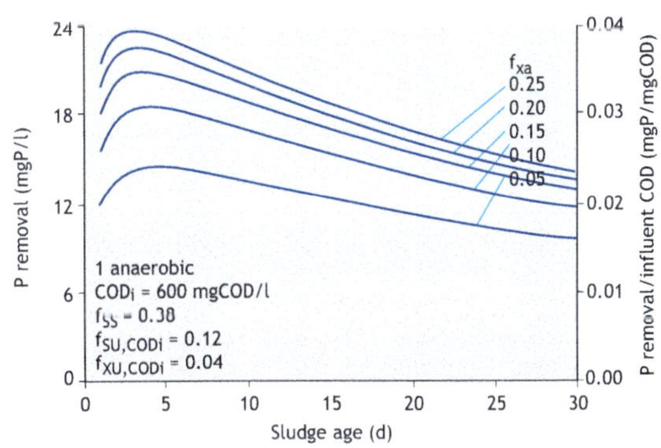

Figure 6.13 Predicted P removal *versus* sludge age for various anaerobic mass fractions (f_{xa}), for a single anaerobic reactor system treating settled wastewater with a total COD of 600 mgCOD/l, with characteristics as shown (c.f. Figure 6.12 for the original unsettled wastewater).

6.5.3 Influence of influent RBCOD fraction

Assuming a zero discharge of nitrate into the anaerobic reactor and considering that VFA ($S_{VFA,i}$) and other fermentable organics ($S_{F,i}$) compose the RBCOD ($S_{S,i}$) present in an influent wastewater, the effect of the influent RBCOD fraction with respect to biodegradable COD ($f_{SS} = S_{S,i}/COD_{b,i}$) is illustrated in Figure 6.14. In this figure, the theoretical P removals are plotted *versus* f_{SS} for a system with two-in-series anaerobic reactors, an SRT of 20 days and anaerobic mass fraction f_{xa} (with regard to the total mass; $f_{xa} = X_A V_A / X_{TSS} V_R$) and wastewater characteristics as shown. It appears that for a selected f_{xa}, as the RBCOD fraction (f_{SS}) increases, the P removal also increases. If needed and as previously discussed, one option to improve the P removal efficiency can be to increase the availability of RBCOD through, for example, acid fermentation of the primary sludge or the addition of external carbon sources (Pitman *et al.* 1983; Barnard 1984; Osborn *et al.* 1989; Vollertsen *et al.*, 2006; Barnard *et al.*, 2017).

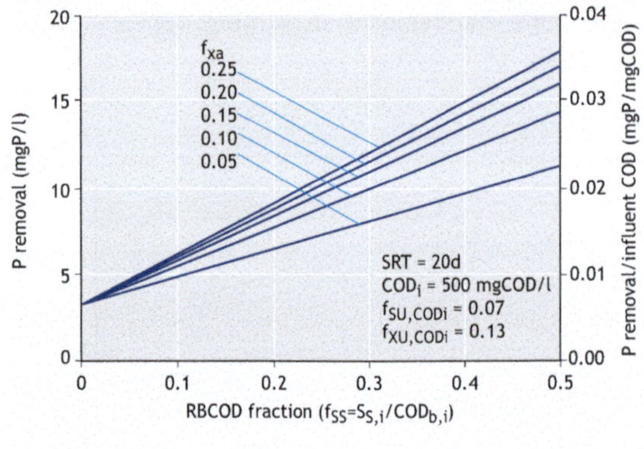

Figure 6.14 Predicted P removal *versus* readily biodegradable COD (RBCOD, $S_{S,i}$) as a fraction of the biodegradable ($COD_{b,i}$) COD ($f_{SS} = S_{S,i}/COD_{b,i}$) for various anaerobic mass fractions (f_{xa}) for a two-in-series anaerobic reactor system at 20 d sludge age, treating unsettled wastewater of 500 mgCOD/l, with characteristics as shown.

6.5.4 Influence of recycling nitrate and oxygen on the anaerobic reactor

The influence of nitrate recycled to the anaerobic reactor is illustrated in Figure 6.15 using the stoichiometric EBPR model where theoretical P removals are plotted *versus* nitrate concentration recycled for a system with two-in-series anaerobic reactors, recycle ratio 1:1, SRT of 20 days and f_{xa} and wastewater characteristics as shown. It appears that recycling of nitrate has a markedly deleterious influence on the magnitude of P removal (in agreement with numerous experimental observations). As the nitrate concentration recycled to the anaerobic reactor increases, the P removal decreases, as explained below.

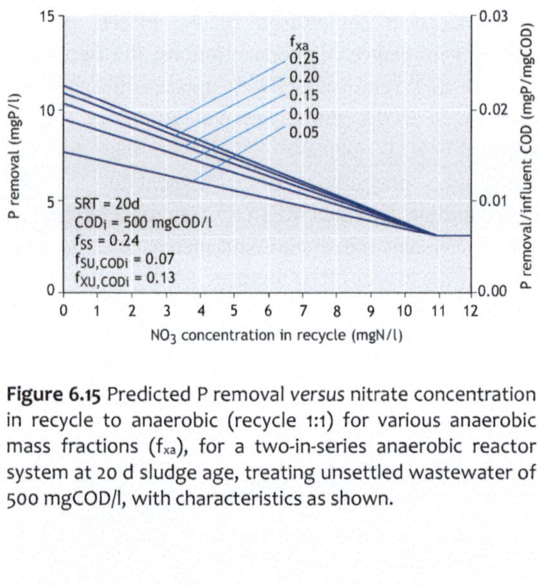

Figure 6.15 Predicted P removal *versus* nitrate concentration in recycle to anaerobic (recycle 1:1) for various anaerobic mass fractions (f_{xa}), for a two-in-series anaerobic reactor system at 20 d sludge age, treating unsettled wastewater of 500 mgCOD/l, with characteristics as shown.

If oxygen and/or nitrate is recycled to the anaerobic reactor, the OHOs no longer transform fermentable COD into VFAs but are able to utilize it for energy and growth using the oxygen or nitrate as an external electron acceptor. For every 1 mgO_2 and 1 $mgNO_3$-N recycled to the anaerobic reactor, 3.0 and 8.6 mgCOD, respectively, are utilized. Consequently, allowing oxygen and/or nitrate to enter the anaerobic reactor reduces the mass of VFAs available to the PAOs for storage, and correspondingly reduces the P release, P uptake and P removal.

From Figure 6.15, when the nitrate concentration in the recycle exceeds approximately 11 mgN/l, the P removal remains constant at approximately 3 mgP/l; for this condition all the influent RBCOD for this wastewater is denitrified by OHOs with the result that no VFAs are released, no COD is available to the PAOs and EBPR no longer takes place; the P removal obtained is due to wastage of sludge with normal metabolic P content (0.03 mgP/mgVSS). If the influent RBCOD concentration increases or decreases, the mass of recycled nitrate that completely consumes the RBCOD will increase or decrease correspondingly below 11 mgN/l (provided the recycle ratio remains unchanged).

From the above it is clear that one of the principal orientations in any design for EBPR is to minimize oxygen entrainment and recycling of nitrate to the anaerobic reactor. In conditions where nitrification is obligatory, a number of different system configurations have been developed specifically to prevent this by incorporating complete denitrification, or passing the underflow recycle through anoxic zones before discharge to the anaerobic reactor.

6.5.5 The effects of the SRT

Using the characteristics of a typical unsettled municipal wastewater with a total influent COD of 250 mgCOD/l, and assuming that no nitrate enters the anaerobic reactor and that a recycle ratio to the anaerobic of 1:1 is present, P removal *versus* sludge age is shown in Figure 6.16 for a single anaerobic reactor with f_{xa} of 0.05; 0.10; 0.15; 0.20 and 0.25. On the same plots P removal/COD_i is also shown.

The plots indicate that the effect of SRT on P removal is complex. For SRT < 3 days, the P removal increases with an increase in the SRT. However, for SRT > 3 days, P removal decreases with an increase in SRT. The reason for this is that an increase in SRT causes an increase in the system OHO mass, which in turn causes an increase in fermentable COD conversion and, therefore, an increase in P release and P uptake. However, the increased SRT also causes a decrease in P uptake due to the lower PAO active mass (and its associated P content) wasted per day. At SRT < 3d, the former effect dominates the P removal, while at SRT > 3d the latter dominates, giving rise to the shape of the curve. The latter effect, that is the decrease in both PAO and OHO active masses with increase in SRT, would be crucially affected by the specific endogenous mass loss rate of the PAOs. However, should the endogenous mass loss rate of the PAOs (0.04 d^{-1}) have been the same as that of the OHOs (0.24 d^{-1}), virtually no EBPR would have been obtained.

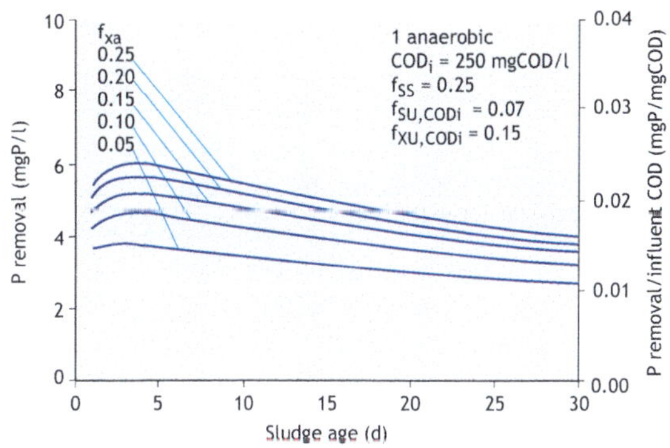

Figure 6.16 Predicted P removal *versus* sludge age for various anaerobic mass fractions (f_{xa}), for a single anaerobic reactor system treating unsettled wastewater with a total COD of 250 mgCOD/l, with characteristics as shown.

In EBPR systems, the aerobic SRT plays an important role due to the aerobic conversion processes that help to replenish the intracellular polymers. In particular, this is because the behaviour of the three storage pools in the cells (PHA, polyphosphate and glycogen) is highly dynamic and is determined by their conversion during the anaerobic and aerobic (or anoxic) phase. The PHA content of the biomass depends on the biomass concentration in the reactor. The biomass concentration can be easily controlled by the manipulation of the substrate loading and SRT. While the anaerobic PHA production depends on the substrate loading to the system, the aerobic PHA consumption depends on the PHA level inside the biomass and on the kinetics of four PHA utilizing processes. The PHA formed under anaerobic conditions must be consumed during the aerobic phase. Otherwise, the PHA level in the cells will increase until a maximum level is reached. From that moment on, no substrate uptake will occur under anaerobic conditions, leading to the deterioration of the EBPR process.

In activated sludge systems designed for the removal of organic matter and nitrogen the SRT is directly linked to the growth rate of the microorganisms; the minimum required SRT corresponds to the maximum growth rate ($SRT_{min}= 1/\mu_{max}$). However, in the EBPR systems where storage materials play an important role in bacterial metabolism, the determination of the total SRT_{min} (defined as a sum of the minimum required anaerobic and aerobic SRT: $SRT_{min,total} = SRT_{min,AN} + SRT_{min,OX}$) depends on the process kinetic rates and on a number of process conditions, notably the time needed for anaerobic RBCOD conversion to PHA, the time required for PHA consumption under aerobic or anoxic conditions, the biomass specific substrate loading rate, the temperature, the operation of the system, and the cell maximum PHA content. Since growth only occurs under aerobic conditions, only the aerobic EBPR process (and therefore only the $SRT_{min,OX}$) is considered here. Clearly there does exist a minimum aerobic oxidation time below which the anaerobically produced PHA cannot be further oxidised. The model for the prediction of the minimum required aerobic SRT as a function of process parameters was developed and compared with experimental data used to evaluate several operational aspects of EBPR in a SBR system (Brdjanovic *et al.*, 1998b). The model was proved as capable of describing them satisfactorily (Figure 6.17).

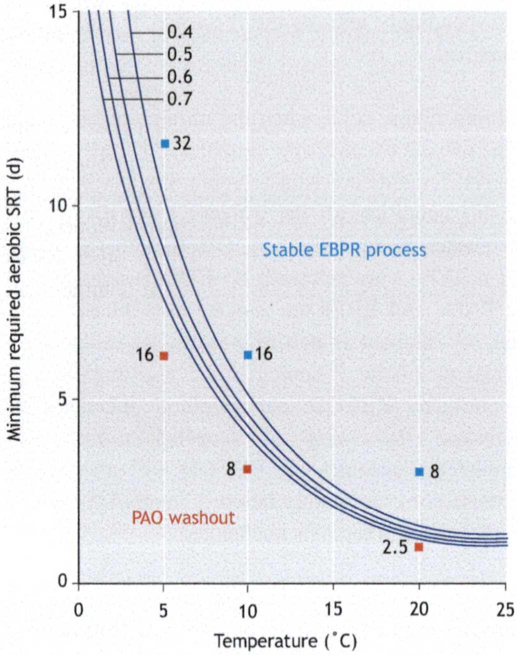

Figure 6.17 Minimum required aerobic SRT as a function of maximum PHA storage capacity of enriched culture biomass (0.4 - 0.7 gCOD-PHA/gCOD-active biomass) and temperature. The symbols indicate the aerobic SRT of several laboratory-scale SBR systems (Smolders *et al.*, 1994c, Brdjanovic *et al.*, 1998b). The numbers next to square symbols indicate the total SRT of the systems. A good EBPR was achieved at the SRTs marked as blue, while the EBPR failed at SRTs marked as red due to too short SRT.

6.5.6 Influence of the anaerobic stage

6.5.6.1 Effect of the anaerobic mass fraction (f_{xa})

The effect of f_{xa} on P removal is also shown in Figure 6.16. For a selected SRT, an increase in f_{xa} gives rise to an increase in P removal. This is due to the increased conversion of fermentable COD with larger anaerobic mass fractions. The improvement in P removal, however, diminishes with each step increase in f_{xa}, due to the first-order nature of the conversion kinetics. From the plot, with a single anaerobic reactor one should select $f_{xa} > 0.15$ because the modest increase in P removal for $f_{xa} > 0.20$ does not seem warranted.

6.5.6.2 Effect of the number of anaerobic reactors (n)

The effect of subdividing the anaerobic reactor is shown in Figure 6.18. The plot is similar to Figure 6.11, but with the anaerobic zone subdivided into two equal reactors. Comparing the P removal behaviour in Figures 6.11 and 6.18, the series operation of the anaerobic zone significantly improves the P removal. This improvement is due to the increased fermentable COD conversion with the in-series anaerobic reactor operation as a result of the first-order nature of the conversion kinetics. A comparison (not shown) between single, two-in-series and four-in-series anaerobic reactors indicates that the main improvement is from single to two-in-series reactors. For design, at least two equally sized in-series anaerobic reactors should be used.

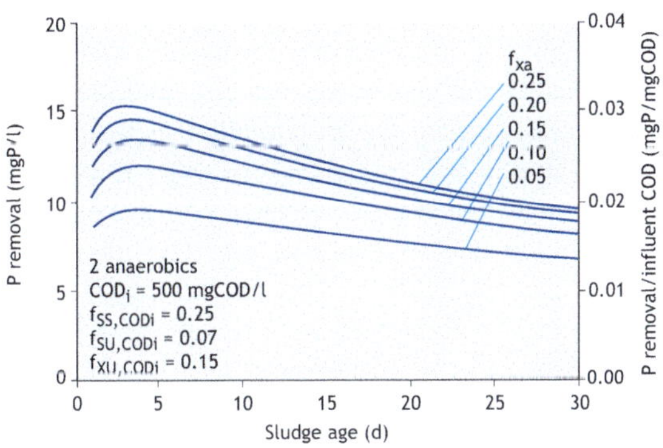

Figure 6.18 Predicted P removal *versus* sludge age for various anaerobic mass fractions (f_{xa}), for a two-in-series anaerobic reactor system treating unsettled wastewater with a total COD of 500 mgCOD/l, with characteristics as shown (c.f. Figure 6.11 for the single reactor system).

6.5.7 Presence of GAOs

Glycogen Accumulating Organisms (GAOs) have a metabolism similar to that of PAOs. However, unlike PAOs, GAOs only rely on the glycolysis of their intracellular glycogen pool as their energy and carbon source for the anaerobic storage of VFAs as PHA (Filipe *et al.*, 2001a; Zeng *et al.*, 2003). Thus, GAOs do not exhibit the typical anaerobic P release and subsequent aerobic P uptake. Therefore, from an EBPR process perspective, GAOs are undesirable microorganisms since they are able to take up VFAs under anaerobic conditions, competing with PAOs for the same carbon source, without contributing to phosphorus removal. So far, two of the best known GAO are *Competibacter* and *Defluviicoccus* (Oehmen *et al.*, 2006a, 2006b; Burow *et al.*, 2007; Lanham *et al.*, 2008). These two GAO are able to compete against PAO under certain environmental and operating conditions as discussed in this and the following subsections.

Different operating and environmental conditions have been identified as important factors to understand the PAO-GAO competition: type of carbon source (acetate and/or propionate), pH, temperature, and influent P/COD ratio. The type of carbon source plays an important role in the competition between PAOs and GAOs because the different PAO and GAO strains identified so far have different preferences, affinities and substrate uptake rates. However, overall a mix of different VFAs (*e.g.* 75 % HAc and 25 % HPr) tend to favour PAO over GAO (Lopez-Vazquez *et al.*, 2009b; Carvalheira *et al.*, 2014). Such observations have been confirmed in full-scale systems with much broader mixtures of carbon sources and amino acids being able to support a satisfactory EBPR process even at high temperatures (Qiu *et al.*, 2019).

pH has a major influence on the anaerobic metabolism of PAOs and GAOs. At higher pH levels (> 7.0), the energy required (ATP) for the transportation of substrate through the cell membrane increases (Smolders *et al.*, 1995; Filipe *et al.*, 2001a). This results in a higher degree of utilization of the intracellular storage fractions of polyphosphate and glycogen. Different reports have described the dominance of PAOs and, thus, stability of the EBPR process performance at high pH levels (pH > 7.25) and the dominance of GAOs at lower pH (pH < 7.25) (Filipe *et al.*, 2001a, 2001b; Schuler and Jenkins, 2002; Oehmen *et al.*, 2005a). These observations suggest that at higher pH either the hydrolysis of glycogen is the limiting metabolic process in GAO metabolism or that PAOs have metabolic advantages over GAOs because they not only rely on the glycolytic pathway but also on the hydrolysis of polyphosphate (Filipe *et al.*, 2001a).

Temperature has a major impact on the competition and occurrence of PAOs and GAOs in activated sludge systems. At moderate and lower temperature (<20 °C) PAOs tend to be the dominant microorganisms and have considerable metabolic advantages over GAOs whereas the opposite occurs at higher temperature (>20 °C). This can be explained by considering that at temperatures lower than 20 °C, PAOs have higher biomass growth rates than GAOs (Lopez-Vazquez *et al.*, 2008b, 2009a) and lower anaerobic maintenance requirements (Lopez-Vazquez *et al.*, 2007) potentially limiting the occurrence of GAOs in wastewater treatment systems operated at lower temperature (Lopez-Vazquez *et al.*, 2008a). At temperatures higher than 20 °C, however, GAOs have higher substrate uptake rates than PAOs (Whang and Park, 2006; Lopez-Vazquez *et al.*, 2007; 2009a) favouring their occurrence when warm wastewaters (>20 °C) are treated. Nevertheless, the applicability of a high pH level and a broader mixture of carbon sources and amino acids appears to give competitive advantages to PAO despite high activated-sludge temperatures (Whang *et al.*, 2007; Lopez-Vazquez *et al.*, 2009b; Qiu *et al.*, 2019).

6.5.9 Carbon sources

Aiming at obtaining highly enriched PAO cultures, several carbon sources have been used for the operation of lab-scale bioreactors (Jeon and Park, 2000; Jeon *et al.*, 2001; Oehmen *et al.*, 2004, 2007; Puig *et al.*, 2008), with a strong focus on the use of acetate and propionate as sole carbon sources due to their abundance in influent wastewaters (Oehmen *et al.*, 2004, 2007). However, the sole use of either acetate or propionate as a single carbon source has proven to lead to EBPR process instability because of the growth and proliferation of GAO (such as *Competibacter* or *Defluviicoccus*) (Oehmen *et al.*, 2006a, 2006b; Burow *et al.*, 2007; Lanham *et al.*, 2008). In view of the capabilities of PAO to take up either acetate or propionate at a relatively similar kinetic rate, Oehmen *et al.* (2006a) and Lu *et al.* (2006) proposed to alternate these two carbon sources periodically to outcompete GAO and obtain a highly enriched PAO culture. By applying a metabolic model, Lopez-Vazquez *et al.* (2009) forecasted that an influent containing 75% acetate and 25% propionate could be sufficient to obtain an enriched PAO culture (*Accumulibacter*). Following the outcome of Lopez-Vazquez *et al.* (2009), Carvalheira *et al.* (2014) conducted a long-term cultivation study that confirmed that a ratio of 75% acetate to 25% propionate can indeed lead to a highly enriched PAO culture (>80% based on FISH analysis).

In addition to the use of acetate and propionate, other carbon sources such as glucose, lactate and alcohols have also been tested (Jeon and Park, 2000; Jeon *et al.*, 2001; Puig *et al.*, 2008). Regarding the sole supply of glucose, the results are somewhat contradictory. While some studies indicate that glucose can lead to the deterioration of the EBPR process due to the proliferation of other organisms that can outcompete PAO by taking up glucose and storing it intracellularly as glycogen (Griffiths *et al.*, 2002; Xie *et al.*, 2017), other studies suggest that glucose can be fermented and produce VFA in the anaerobic stage of EBPR process configurations, resulting in a stable EBPR process performance. Similarly, the supply of lactate or alcohols seems to be associated with a stable EBPR performance whenever the anaerobic stage is long enough to favour their fermentation (Jeon and Park, 2000; Jeon *et al.*, 2001; Puig *et al.*, 2008). In addition, Tu and Schuler (2013) documented that PAO have a higher affinity for C uptake than GAO, outcompeting GAO at low C-source concentrations. Then, in an anaerobic stage, if the fermentation products are formed within the flocs, the higher affinity of PAO can help them to take the C sources as soon as they are generated, to the detriment of GAO. These observations underline the need to ensure that the anaerobic stage volume must be large enough to facilitate the fermentation of such carbon sources in favour of PAO.

Besides the presence and availability of VFA and other fermentable compounds, amino acids are also present in raw wastewaters. Their fermentation and consumption has been associated with the growth and proliferation of *Tetrasphera* (Nguyen *et al.*, 2011; Marques *et al.*, 2017; Liu *et al.*, 2019), which seems to be able to contribute to the P-removal process, but performing a different metabolism than *Accumulibacter* (Liu *et al.*, 2019). While their actual direct contribution to the EBPR process (as a fermentative organism and potential PAO) is still under study (Liu *et al.*, 2019; Rubio-Rincon *et al.*, 2019), their presence has been associated with the stable operation of EBPR plants.

In full-scale WWTP, to ensure that PAO have enough carbon to perform the EBPR process, the addition of molasses to prefermenters has been evaluated (Zeng *et al.*, 2006), resulting in a higher availability of propionate and to an improved and stable EBPR process. Also, as presented in Section 6.6, certain process configurations have been developed to favour the self-production of VFA by fermenting (a fraction of) the RAS (Vollertsen *et al.*, 2006; Barnard *et al.*, 2017). This practice has led to a higher availability of different VFA species (Vollertsen *et al.*, 2006) and also to an increased EBPR performance. In another full-scale WWTP operated at high wastewater temperature (Ong *et al.*, 2014; Qiu *et al.*, 2019), the broad availability of different carbon sources and amino acids led to a stable EBPR operation, even though the temperature tended to be unfavourable for the EBPR process (Lopez-Vazquez *et al.*, 2009) (as discussed in Section 6.5.11). Overall, the availability of different VFA species and carbon sources contributes to an improved EBPR process stability, in contrast to the use of a single carbon source.

6.5.10 Influent COD/P ratio

Another factor affecting the metabolism of PAO is the influent P/COD ratio. Several studies have been conducted indicating that high P/COD ratios favour the metabolism of PAO while low P/COD ratios favour the metabolism of GAO. Schuler and Jenkins (2003a, 2003b) cultivated a PAO culture at different P/COD ratios. They observed a change in the metabolism of the enriched EBPR culture from a Phosphorus-Accumulating Metabolism (PAM) to a Glycogen-Accumulating Metabolism (GAM) as the P/COD ratio decreased. This implied a shift in the metabolism from the utilization of the TCA cycle at high P/COD ratios to a higher involvement of the glycolytic pathway as the P/COD ratio decreased. It is important to notice that in spite of the higher involvement of the TCA cycle at high P/COD ratios, the enriched EBPR culture continued to make use of the glycolytic pathway. Such observations were corroborated by similar research studies (Barat *et al.*, 2006; Zhou *et al.*, 2008; Acevedo *et al.*, 2012; Welles *et al.*, 2015, 2016). In studies performed in full-scale EBPR plants, Pijuan *et al.* (2008) and Lanham *et al.* (2014) also observed similar behaviour in particular at high P/COD ratios. Moreover, Welles *et al.* (2015, 2016) were able to cultivate an enriched PAO culture at a low influent P/COD ratio. They noticed that the PAO culture, due to the low intracellular polyphosphate content, made use of glycogen as the main intracellular compound for the anaerobic uptake of carbon, whereas after being exposed to a higher influent P/COD ratio, the PAO culture could shift their metabolism and take up orthophosphate aerobically.

The influence of the influent P/COD ratio is directly reflected in the anaerobic P-release-to-COD ratio. At higher P/COD ratios, the intracellular polyphosphate contents are higher which results in a higher anaerobic P-release-to-COD-consumption ratio. This seems to be caused by a higher involvement of the TCA cycle (Schuler and Jenkins, 2003a, 2003b; Pijuan *et al.*, 2008; Welles *et al.*, 2015, 2016). Meanwhile, at lower P/COD ratios, the intracellular polyphosphate contents decrease which increases the role of the glycolytic pathway and the anaerobic P-release-to-COD-consumption ratio decreases. In addition, Welles *et al.* (2015) observed that together with the phosphate release rates, acetate uptake rates also increased up to an optimal poly-phosphate/glycogen ratio of 0.3 P-mol/C-mol. At higher poly-phosphate/glycogen ratios (obtained at influent P/C ratios above 0.051 P-mol/C-mol), the acetate uptake rates started to decrease. However, Welles *et al.* (2015) noticed that such metabolism was dependent on the specific type of microorganism present, with *Accumulibacter Type II* being more flexible than *Accumulibacter Type I*, and therefore more flexible and able to perform a GAM metabolism at low P/COD ratios with higher acetate uptake rates.

Overall, the influent P/COD ratio plays a major role in the EBPR process, affecting both the stoichiometry and the kinetics of known EBPR cultures. As such, these effects need to be considered when assessing or evaluating the EBPR process. For practical purposes and to comply with the required P-effluent discharge limits, in principle, EBPR systems tend to be designed considering a low influent P/COD ratio (a rather common practice since most of the EBPR designs and configurations aim at maximizing the influent COD provided to PAO).

6.5.11 pH effects

Smolders *et al.* (1994a) assessed the effects of pH finding a direct correlation between the anaerobic P-release/HAc-uptake ratio and the pH measured in the bulk liquid ($f_{PO4,rel}$ = 0.18 pH – 0.81, in gP/gCOD units). Such a relationship is a reflection of the increased energy required at higher pH levels for HAc uptake. Filipe et al. (2001a) amended such an equation ($f_{PO4,rel}$ = 0.15 pH – 0.53, in gP/gCOD units) suggesting that their research probably had a higher PAO abundance.

6.5.12 Temperature effects

With the application of mathematical modelling on biological processes, the role of temperature coefficients becomes more important. Mathematical models that include the presence of EBPR in activated sludge systems, such as Activated Sludge Model No. 2 (ASM2) (Henze *et al.,* 1994), University of Cape Town Activated Sludge Model (UCTPHO; Wentzel *et al.,* 1992) or the metabolic model of EBPR (TUDP model; Smolders *et al.,* 1994c, 1995) rely on stoichiometric and kinetic coefficients valid in a narrow temperature range or a single temperature value only. In ASM2, process coefficients are defined for two different temperatures (10 and 20 ºC). In this model the stoichiometric coefficients are temperature-independent, while the kinetic coefficients are affected by temperature changes. The processes in ASM2 are classified into four groups based on their temperature dependency (zero, low, medium and high dependency). Identical values at 10 and 20 °C were assigned to many of the coefficients. This was justified by the scarcity of data available or by the low sensitivity of the particular parameters to variations in temperature. In this classification, the EBPR processes are considered to have a low degree of temperature dependency in comparison with other processes incorporated in ASM2. ASM2 is in general recommended for application on wastewater treatment by activated sludge at temperatures between 10 and 25 °C, and the authors of ASM2 are cautious about its applicability outside this range. Similarly, the UCTPHO model has process parameters based on 20 °C. For other operational temperatures the adjustment of the values is computed from the input temperature and Arrhenius temperature constants. In the metabolic model all parameters are determined at 20 °C, but no information is available on their temperature dependency. It has been suggested that the temperature impact on the PAOs can be modelled with the same coefficients as for heterotrophic organisms. However, due to large differences in metabolism and the involvement of storage products, this can be erroneous. As municipal wastewater treatment plants, including those operating with EBPR, may experience mixed-liquor temperatures as low as 5 °C or as high as 35 °C, there was a significant need for a systematic study of the impact of temperature on EBPR systems which should also take into account the specific requirements of mathematical models and their application in different climates.

It was thought some years ago that *Acinetobacter* sp. are the microorganisms most responsible for

EBPR, and consequently, most of the scientific research concerning the impact of temperature was related to these particular bacteria. However, it was then shown that *Acinetobacter* plays a limited role in EBPR (Wagner *et al.,* 1994) and, therefore, information on the P metabolism of *Acinetobacter* alone is to be considered less relevant.

There are several publications reporting the effect of temperature on the efficiency (the difference in the influent and the effluent quality) of EBPR using activated sludge but with inconsistent results. Improved EBPR efficiency at higher temperatures (in the range 20-37 °C) was observed by Jones *et al.* (1987), Yeoman *et al.* (1988), McClintock *et al.* (1993) and Converti *et al.* (1995). In contrast, high or even only comparatively higher levels of P-removal efficiency at lower temperatures (in the range 5-15 °C) was reported by Sell *et al.* (1981), Kang *et al.* (1985b), Krichten *et al.* (1985), Barnard *et al.* (1985), Viconneau *et al.* (1985) and Florentz *et al.* (1987). However, when the kinetics of EBPR processes were studied, such inconsistencies did not exist. Increased P-release and/or P-uptake rates with increased temperature was reported by Shapiro and Levin (1967), Boughton *et al.* (1971), Spatzierer *et al.* (1985), and Mamais and Jenkins (1992). In addition to P-release and P-uptake rates, Mamais and Jenkins (1992) also reported increased growth and substrate consumption rates with an increase in temperature (10-33 °C). The different results on the temperature effect on EBPR with activated sludge can be explained by the use of different substrates, activated sludge and/or measurement methods.

Temperature influences a variety of processes in activated sludge systems (lysis, fermentation, nitrification, etc.) which may influence EBPR processes. These effects complicate the determination of the effect of temperature on EBPR. In addition, most of the findings presented in the paragraph above are based on a black-box approach, comparing the influent and effluent phosphorus concentrations of wastewater treatment plants at different wastewater temperatures. At that time no structured study of the effect of temperature on stoichiometry and kinetics of the EBPR processes under defined laboratory conditions was available. All these factors explain conflicting results in the past, which were difficult to interpret correctly.

The effect of temperature on the stoichiometry of EBPR processes had not been studied in great detail until Brdjanovic *et al.* (1997, 1998c) carried out a systematic study on the effects of temperature changes on both the anaerobic and the aerobic stoichiometry and kinetics. This study comprised experiments to investigate the effects of short-term (hours) temperature changes on the physiology of the EBPR system, and long-term (weeks) temperature changes on the ecology of the EBPR system. Enriched PAO cultures fed with synthetic wastewater were used in an anaerobic-aerobic-settling sequencing batch reactor (SBR) under controlled (laboratory) conditions. The main results from this work are highlighted below.

6.5.12.1 Short-term temperature effects on the physiology of EBPR

P-removing sludge was enriched in an anaerobic-aerobic, acetate-fed, sequencing batch reactor (SBR) at 20 °C. Conversion of relevant compounds for biological phosphorus removal was studied at 5, 10, 20 and 30 °C in separate batch tests during the period of a few hours. The stoichiometry of the anaerobic processes was found to be insensitive to temperature changes while some effects on aerobic stoichiometry were observed. In contrast, temperature had a major influence on the kinetics of the processes under anaerobic as well as aerobic conditions. The anaerobic P release (or acetate uptake) rate showed a maximum at 20 °C. However, a continuous increase was observed in the interval 5-30 °C for the conversion rates under aerobic conditions. Based on these experiments, temperature coefficients for the different reactions were calculated. An overall anaerobic and aerobic temperature coefficient θ was found to be 1.078 and 1.057 (valid in the ranges 5 °C≤T≤20 °C and 5 °C≤T≤30 °C), respectively.

6.5.12.2 Long-term temperature effects on the EBPR process

Steady-state conversion of relevant compounds for EBPR was studied in one reactor subsequently at 20, 30, 20, 10 and 5 °C. The temperature coefficient for metabolic conversions obtained from long-term temperature tests was similar to the temperature coefficient observed in short-term (hours) tests (θ = 1.085 and θ = 1.078, respectively). Temperature had a moderate impact on the aerobic P-uptake process rate (θ = 1.031) during the long-term tests. However, a major temperature effect on other metabolic processes of the aerobic phase, such as PHA consumption (θ = 1.163), oxygen uptake (θ = 1.090) and growth ($\theta > 1.110$), was observed. Different temperature coefficients were obtained for the aerobic phase from long-term and short-term tests, probably due to a change in population structure. This change was also visible from molecular ecological studies. The different temperature coefficient found for P uptake compared to the other metabolic processes of the aerobic phase underlines that in complex processes such as EBPR, it is dangerous to draw conclusions only from easily observable parameters (such as phosphate). Such consideration can easily lead to underestimation of the temperature dependency of other metabolic processes of the aerobic phase of EBPR. Meijer (2004) incorporated temperature coefficients obtained from studies of Brdjanovic *et al.* (1997, 1998c) into the TUDP model. Lopez-Vazquez *et al.* (2008b, c) repeated the original experiments of Brdjanovic *et al.* (1997, 1998c) and extended the TUDP model with coefficients for two additional temperatures, namely 15 °C and 35 °C. In general, this study confirmed the results of Brdjanovic *et al.* (1997, 1998c) for the temperature range 5-30 °C. In addition, Lopez-Vazquez *et al.* (2008c) also determined the temperature dependencies in the short- and long-term of a GAO culture enriched with *Competibacter* and then introduced them into the TUDelft model to assess the PAO-GAO competition at different temperatures (Lopez-Vazquez *et al.*, 2009).

6.5.13 Dissolved oxygen and aeration

The DO concentration plays an important role in the EBPR process. Carvalheira *et al.* (2014) defined that DO lower than 1.5 mg/l are preferable to favour the growth of PAO over GAO due to their higher DO affinity, whereas higher DO concentrations tend to favour GAO which can compete for C sources anaerobically with PAO. With more interest in reducing the energy costs of WWTP through the optimization of the aeration process and thus, operating the WWTP at low DO, such a practice is favourable for PAO.

On the other hand, while the system needs to comply with a minimum aerobic SRT to ensure the growth of PAO (Brdjanovic *et al.*, 1998b), excessive aeration can have a detrimental impact on the EBPR process due to the depletion of the intracellular PHA compounds as an energy source resulting in the aerobic hydrolysis of polyphosphate, and consequently on the aerobic release of P (Lopez *et al.*, 2006). This affects the effluent quality as the P effluent concentrations will rise. Extreme aeration events can take place after heavy rainfalls or during weekends (where the influent wastewater dilutes and the carbon source concentrations decrease leading to less PHA accumulation). To avoid such deleterious effects, the aeration needs to be adjusted and decreased during such conditions.

6.5.14 Inhibitory compounds

The absence or presence of certain compounds has proven to be deleterious for the EBPR process. Saito *et al.* (2004) observed that nitrite was inhibitory for the anoxic and aerobic P-uptake process. Rather than nitrite, Zhou *et al.* (2007) observed that the anoxic phosphorus uptake rates decreased when the free nitrous acid (FNA) concentrations increased from 0.002 to 0.02 mg HNO_2-N/l, being completely inhibited at 0.02 mg HNO_2-N/l. Concerning the aerobic P uptake, Pijuan *et al.* (2010) observed a 50 % inhibition on all anabolic processes at FNA concentrations of approximately $0.5 \cdot 10^{-3}$ mg HNO_2-N/l (equivalent to 2.0 mg NO_2^--N/l at pH 7.0), while full inhibition occurred at FNA concentrations of approximately $6.0 \cdot 10^{-3}$ mg HNO_2-N/l.

Potassium limitation has also been shown to have a deleterious effect on the EBPR process (Brdjanovic *et al.*, 1996) while the excess of calcium (and other salts) for chemical P removal also leads to deterioration of the EBPR process (Barat *et al.*, 2006; Jobagy *et al.*, 2006) since P precipitates chemically and is no longer available to form the intracellular polyphosphate pools required by PAO. Thus, the wastewater composition has a direct impact on the efficacy of the EBPR process.

6.6 EBPR PROCESS CONFIGURATIONS

In this section, the EBPR optimisation concepts are first discussed and the development of the main EBPR activated sludge systems is then reviewed in a historical context.

6.6.1 Phosphorus removal optimization principles

An overview of EBPR and chemical phosphorus removal optimisation principles are presented in Figure 6.19. The principles of EBPR and P-removal process optimisation can be grouped into six categories. A number of configurations or processes that are based on these principles are identified by their specific names.

(i) Oxygen entrainment in the anaerobic reactor should be minimised. For this purpose, mixing vortexes, upstream cascades and screw pumps or air lift pumps should be avoided.

(ii) Nitrate (and nitrite) entrainment in the anaerobic reactor should be minimised. As explained in the next section, a number of named configurations were developed precisely for this purpose. To this end, an anaerobic–aerobic configuration (*e.g.* A/O configuration) can be improved by inserting an anoxic reactor in which aerobic sludge is recirculated for denitrification (*e.g.* A^2/O, modified Phoredox configurations). Also, the return activated sludge from the secondary settling tank can be denitrified either via an anoxic reactor on the sludge recycle line (*e.g.* JHB configuration) or via an anoxic reactor located downstream of the anaerobic zone from where another internal recirculation to the anaerobic reactor is located (*e.g.* UCT configurations). This anoxic reactor can be divided into two zones to provide return sludge denitrification in the first zone and aerobic sludge denitrification in the downstream anoxic reactor (*e.g.* MUCT configuration). Adding a second anoxic zone, downstream of the aerobic zone, is another way of reducing the nitrate concentration in both the effluent and return sludge (*e.g.* Modified Bardenpho configuration).

(iii) VFA uptake by PAOs in the anaerobic reactor should be maximized. Primary sludge fermentation is an efficient way to increase the VFA content of the influent even though it also contributes to an increased loading of ammonia in the activated sludge system. Sodium acetate or fermentable industrial wastes can be added directly to the anaerobic reactor or prefermentation tank. The hydraulic retention time of the anaerobic reactor or prefermenter can be increased to favour *in-situ* fermentation of the influent or added fermentable organic matter. RAS fermentation can also be employed to augment the organic matter to be consumed by PAOs (see Section 6.6.11).

(iv) Effluent particulate phosphorus should be minimized by removing total suspended solids efficiently. The particulate phosphorus content can reach as high as 18% gP/gTSS for enriched cultures. With a more typical 5% content, every 10 mgTSS/l in the effluent will contribute 0.5 mgP/l. Thus, efficient secondary clarification, avoiding floating sludge from denitrification in the settling tank, sand filtration, and even ultrafiltration (in a membrane bioreactor) are all means of reducing the effluent TSS concentration.

(v) Effluent soluble phosphorus should be minimized. Besides optimising the EBPR process, chemical coagulants such as iron (*e.g.*

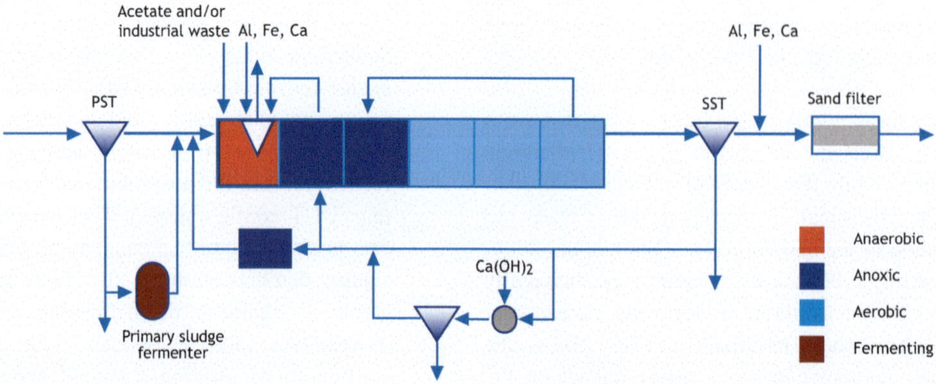

Figure 6.19 Overview of EBPR and P-removal process optimisation. Note: PST: primary settling tank; SST: secondary settling tank.

$FeCl_3$), aluminium (*e.g.* alum) or calcium (*e.g.* lime) salts can be added either in the mainstream for pre-, co- or post-precipitation (in the primary settling tank, in the activated sludge process, downstream of the secondary settling tank, respectively). Extracting the supernatant from the anaerobic tank or taking some sludge from the return activated sludge and coagulating them can also lead to lower effluent soluble phosphorus (*e.g.* the BCFS® process; Van Loosdrecht *et al.,* 1998). Sidestream lime precipitation of phosphate released anaerobically from the return sludge can also be carried out. More efficient phosphate release can be achieved in this sidestream tank by diverting some influent containing readily biodegradable COD (*e.g.* the PhoStrip® process). Should anaerobic or aerobic digestion be performed with the wasted secondary sludge, essentially all of the polyphosphates will be degraded and the phosphate released in solution. Phosphorus recovery in the form of struvite ($MgNH_4PO_4$) or hydroxyapatite [$Ca_{10}(PO_4)_6OH_2$], which can be used as fertilizers, are also means of reducing the loading of soluble phosphate back to the activated sludge process and, eventually, to the effluent.

(*vi*) Phosphorus uptake for cell synthesis can be maximized. Although more limited than the other optimisation concepts in its potential efficiency, maintaining the sludge retention time (SRT) as short as possible will result in an increase in phosphorus removal by cell synthesis. Another benefit of reducing the SRT will be that PAOs will degrade less polyphosphate for cell maintenance. The constraint on manipulating the SRT is maintaining a minimum sludge age to maintain nitrification, while stimulating PAOs.

6.6.2 EBPR process discovery

The removal of P by activated sludge systems in excess of that required for normal organism metabolism was first noted, independently, by two research groups, Srinath *et al.* (1959) in India and Alarcon (1961) in America. Although both groups demonstrated P uptake in aerobic batch tests, they proposed no explanation as to why sludges from certain plants exhibited enhanced P-uptake behaviour and others not, or whether the removal was a biological or physical/chemical phenomenon. Srinath *et al.* (1959) noted that an oxygen deficiency in the upstream zone of plug flow aeration tanks was associated with the occurrence of high phosphate concentration and that this ‘problem’ could be solved by increasing aeration. These observations initiated research into EBPR which ultimately led to the full-scale implementation of EBPR technology.

Since then, a number of processes and configurations have been developed and different modifications and engineering plant-layouts have been proposed for their implementation all around the world. Here, the most fundamental EBPR plant configurations and their principles are presented in a chronological order (Figures 6.20A to 6.20I).

6.6.3 PhoStrip® system

The first structured investigation into the P uptake phenomenon was by Levin and Shapiro (1965). In extensive batch studies on the effect of oxygen tension, pH and inhibitors, they demonstrated the biological nature of EBPR removal. Further, in batch tests on two underflow recycle mixed-liquor samples, in which one was aerated and the other not, the aerated sample took up P while the unaerated sample released P. Shapiro and Levin (1965) focussed attention on the P release in anaerobic batch tests and found that the process of P release under anaerobic conditions could be reversed to a process of P uptake if the batch was subsequently aerated. Levin and Shapiro (1965) utilized the phenomena of P release under anaerobic conditions and P uptake under aerobic conditions to patent the first commercial system for P removal, the PhoStrip process (Figure 6.20G, marketed by Biospherics, USA).

Levin *et al.* (1972) report details of this system: 'The process is based on findings that the aeration of mixed liquor can induce activated sludge microorganisms to take up dissolved phosphorus in excess of the amount required for growth. If the air supply is turned off and the sludge organisms are permitted to consume all of the dissolved oxygen the phosphorus previously taken up is released to the liquid phase. However, when aeration is resumed, the microorganisms again take up the dissolved phosphorus'. The PhoStrip process consists of a single aeration tank with clarifier; a sidestream (typically 10-30 per cent of the influent flow rate) from the underflow of the clarifier passes to an anaerobic 'stripping tank' where the sludge settles and P is released. The 'stripped' sludge is returned to the activated sludge system, while the supernatant is dosed chemically (usually with lime) in a precipitator tank, to precipitate released P which is settled and wasted. The supernatant from the precipitator tank is returned to either the influent or the effluent flow.

The PhoStrip combines chemical and biological P removal, applies to non-nitrifying systems, and is a sidestream process. Later modifications proposed to the PhoStrip include the addition of a part of the influent flow to the stripper tank to promote P release (PhoStrip II; Levin and Della Sala, 1987), elutriation of the released P from the 'stripped' sludge by recycling around the stripper tank (Levin and Elster, 1985) and inclusion of an anoxic tank upstream of the aeration tank with recycle of mixed liquor from the aerobic to the anoxic tanks, to apply the PhoStrip principle to nitrifying activated sludge plants. Since the PhoStrip systems include chemical P removal, design procedures for this system will not be considered in this chapter. In the BCFS process the stripper function has been integrated in the anaerobic tank of activated sludge tanks (Van Loosdrecht *et al.*, 1998).

6.6.4 Modified Bardenpho

Although by the early 1970s the phenomenon of EBPR had been observed at a number of full-scale works (*e.g.* Vacker *et al.*, 1967; Scalf *et al.*, 1969; Witherow, 1970; Milbury *et al.*, 1971) and the first commercial EBPR system (the PhoStrip system) had been developed, there was little confidence in EBPR as a potential practical technology. Mulbarger (1970) went so far as to state 'specialized activated sludge plant design for high level P removal should be avoided and treated as a bonus when and if it occurs'. However, from research in the mid-1970s (Fuhs and Chen, 1975; Barnard, 1974a, 1974b, 1975a, 1975b, 1976a, 1976b) one conclusion emerged that made widespread exploitation of the EBPR phenomenon possible: biological P removal is stimulated by subjecting the activated sludge organisms to a sequence of anaerobic and aerobic conditions. Quantification of the anaerobic state for design and operation, however, presented major problems.

Fuhs and Chen (1975), in a microbiological investigation of EBPR, concluded that the phenomenon is mediated by either a single organism group or several closely related groups. They implicated *Acinetobacter* as the principal organism genus. They concluded that 'anaerobic conditions preceding aerobiosis in sewage treatment could well be related to the appearance of *Acinetobacter* spp. During anaerobiosis, this flora would tend to produce compounds such as ethanol, acetate and succinate, which serve as a carbon source for *Acinetobacter* spp.'. However, Fuhs and Chen did not quantify the 'anaerobic conditions' and developed no practical method for implementation of EBPR.

The first practical mainstream system for EBPR was developed from the work of Barnard (1974a, b, 1975a, b, 1976a, 1976b) and Nicholls (1975). Barnard (1975a, 1975b) while investigating the nitrification/denitrification response of a system he developed for this purpose, the 4-stage Bardenpho system, noted that the system removed more P then expected. Barnard (1974a, 1974b) postulated that 'the essential requirement for phosphorus removal in biological systems is that during some stage before the final stage of the process, the sludge or mixed liquor must pass through an anaerobic stage, during which phosphates may or may not be released, followed by a well aerated aerobic stage, during which the phosphates will either be taken up by the organisms or be precipitated as a result of the change in redox potential'.

Noting the observations of Barnard (1974a), Nicholls (1975) experimented at full-scale with the Alexandra and Olifantsvlei activated sludge systems (Johannesburg, South Africa). He created anaerobic zones in different parts of the two activated sludge systems and concluded that 'good phosphate removal could be expected in the modified Bardenpho system (actually a 5-stage Bardenpho) when an anaerobic basin is placed prior to the activated sludge system'.

Barnard (1976a), in enlarging on the postulations he developed in 1974, concluded that the organism mass 'must pass through an anaerobic phase where the oxygen demand exceeds the supply of both oxygen or nitrates ...' He proposed to produce the anaerobic phase by including an anaerobic reactor prior to the inlet to the plant to 'allow the mixed liquor to become anaerobic through the action of the incoming sewage'. Barnard termed this principle the 'Phoredox' method, and applied it (amongst others) to the 4-stage Bardenpho system; he included an anaerobic reactor before the primary anoxic reactor in the 4-stage Bardenpho, the anaerobic reactor receiving the influent flow and underflow recycle from the secondary settling tanks; this configuration has become known as the 5-stage Modified Bardenpho (Figure 6.20C). Barnard also proposed that when less nitrogen removal is required, the second anoxic and reaeration reactors can be excluded, to give the 3-stage Modified Bardenpho (Figure 6.20A); this configuration has also been called the anaerobic/anoxic/aerobic (A^2O) [1] configuration. To explain the enhanced P removal phenomenon, Barnard (1976a) hypothesized that it is not the P release *per se* that stimulates the P removal, but that the release indicates that a certain low redox potential has been established in the anaerobic zone, *i.e.* that the low redox potential stimulates the enhanced P removal. Barnard (1976a) recognized the difficulties associated with redox potential measurement, and proposed that measurement of P release in the anaerobic zone could serve as a substitute to indicate that conditions necessary for enhanced P removal prevailed.

[1] The original nomenclature of Barnard for anaerobic and anoxic is adopted for use in this chapter; *i.e.* anoxic: a state in which nitrate is present but no oxygen; anaerobic: a state in which neither nitrate nor oxygen is present. The inadequacies of these definitions are apparent when attempting to compare the state of two reactors of the same size, one completely mixed and the other plug flow. A completely mixed anaerobic reactor, for example, will have no nitrate in the reactor; the equivalent plug flow reactor however may contain nitrate for a considerable portion of the reactor length, *i.e.* be partly 'anoxic', partly 'anaerobic' - the inadequacy arises in that no indication is given as to the intensity of the state.

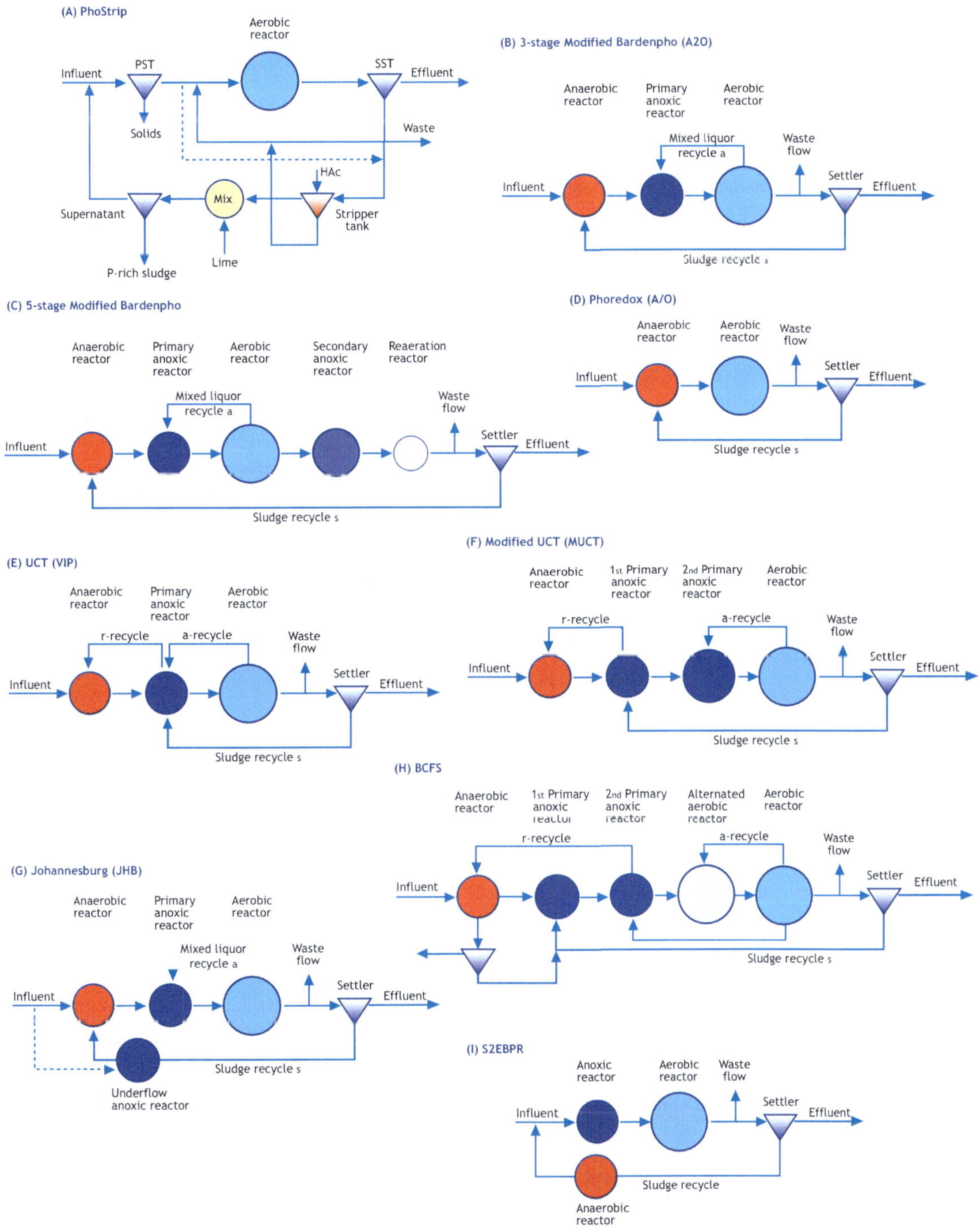

Figure 6.20 System configurations for EBPR: (A) PhoStrip, (B) 3-stage Modified Bardenpho (A2O), (C) 5-stage Modified Bardenpho, (D) Phoredox (A/O), (E) UCT (VIP), (F) MUCT (Modified UCT), (G), Johannesburg (JHB), (H) BCFS, and (I) S2EBPR.

In terms of Barnard's hypothesis, nitrate recycled via the underflow to the anaerobic reactor in the 5-stage Modified Bardenpho will restrain, in some degree, the level to which the redox potential can be lowered and, consequently, nitrate recycled can be expected to influence enhanced P removal adversely, as noted earlier by Barnard (1975b).

Barnard (1976a) apparently accepted that the Modified Bardenpho plant should reduce the nitrate sufficiently that any nitrate in the underflow would not prevent the attainment of the low redox potential necessary for P release in the anaerobic reactor. In any event he considered that nitrate entering the anaerobic reactor could be countered by increasing the retention time of this reactor. For design of the anaerobic reactor, Barnard (1976a) suggested a nominal retention time of one hour. At this stage no rational method for predicting nitrogen and phosphorus removal was available; for design, removals were estimated largely from experience gained in operating experimental systems.

The legal requirement for nitrification in South Africa focussed attention on nutrient removal systems (*i.e.* nitrogen and phosphorus) rather than phosphorus removal only. Consequently, considerable research effort was directed at investigating the Modified Bardenpho system. Early investigations into the 3- and 5-stage Modified Bardenpho systems (McLaren and Wood, 1976; Nicholls, 1978; Simpkins and McLaren, 1978; Davelaar *et al.,* 1978; Osborn and Nicholls, 1978) concurred that nitrate recycled to the anaerobic reactor had a deleterious effect on EBPR and identified that evaluation of the nitrate in the recycle to the anaerobic reactor could be crucial in assessing the success of a nitrifying system in stimulating P release and in determining the magnitude of the P removal. However, none of these investigations provided a reliable model to predict the magnitude of denitrification in order to quantify the nitrate recycled.

Marais and his group (Stern and Marais, 1974; Martin and Marais, 1975; Wilson and Marais, 1976; Marsden and Marais, 1977) recognized the importance of quantifying the nitrate removal. To obtain information on the magnitude and kinetics of denitrification, they replaced the completely mixed reactors in the 5-stage Bardenpho system by plug flow reactors and measured the nitrate along the reactor axes under constant flow and load conditions. Their findings on denitrification kinetics are reviewed in Chapter 5. With regard to relevance to P removal, they found that it was not possible to increase the anoxic zones in this system in order to ensure low nitrate in the effluent and underflow recycles; if, for a selected sludge age and temperature the unaerated mass fraction of the sludge was increased beyond a certain magnitude, the system stopped nitrifying. They showed that the maximum anoxic mass fraction allowable was determined by the maximum specific growth rate of the nitrifiers at the lowest temperature the system would be required to operate, and the sludge age. Limiting the anoxic mass fraction (to ensure nitrification) necessarily limits the magnitude of denitrification achievable. With the systems of the Marais group, when operated at unaerated mass fractions that allowed nitrification, the denitrification was incomplete and the effluent nitrate was high.

Taking note of the findings on the denitrification kinetics, Rabinowitz and Marais (1980) commenced a study on P removal using the Modified Bardenpho system treating unsettled municipal wastewater from the city of Cape Town. They selected the 3-stage Modified Bardenpho (Figure 6.20A) as the basic configuration in preference to the 5-stage (Figure 6.20C) because first, the wastewater source did not allow an unaerated mass fraction of greater than 40 per cent at 14 °C for a sludge age of 20 days if efficient nitrification was to be maintained, and second, taking account that the anaerobic reactor cannot contribute to the system denitrification potential, the 5-stage system could not reduce the nitrate to zero for the measured TKN/COD ratio of the wastewater. Consequently (as discussed earlier in

Chapter 5) the secondary anoxic reactor volume was added to the primary anoxic to obtain the maximum nitrate removal and hence the minimum nitrate concentration in the underflow recycle. The findings of this investigation (Rabinowitz and Marais, 1980) can be summarized as follows:

(i) When the nitrate concentration in the effluent (and underflow recycle) was low, usually P release and enhanced removal were observed. The enhanced P removal decreased quite disproportionately as the nitrate in the underflow recycle increased, a behaviour noted by previous workers (*e.g.* Barnard, 1975b; Simpkins and McLaren, 1978).
(ii) With different batches of wastewater having the same influent COD, with the same concentration of nitrate in the underflow recycle, one wastewater batch gave relatively high rates of P release and enhanced P removal whereas the next gave no (or little) P release and no (or little) enhanced P removal. The reason for this behaviour was not apparent.

In the 3-stage Modified Bardenpho configuration, P removal was disappointing; the system did not give enhanced P removal over lengthy periods of time, and when enhanced P removal was obtained, generally it tended to be low and erratic due to the effects of (*i*) and/or (*ii*) above. Increasing the anaerobic mass fraction during periods of low P removal was found to be counterproductive as this could only be done at the expense of anoxic mass fraction which in turn gave rise to increased nitrate in the recycle. It was finally concluded that for the wastewaters used in the experimental investigation, the Modified Bardenpho type system did not seem suitable for EBPR. This did not imply that the system might not be suitable for other wastewaters, but the investigation did bring to light that there were constraints, not adequately recognized before, that may prevent high P removal. For example, for any selected sludge age and minimum temperature, the requirement for complete nitrification imposes an upper limit on the unaerated mass fraction. This limitation on the unaerated mass fraction correspondingly limits the concentration of nitrate that can be removed. For the 5-stage Modified Bardenpho system, if the nitrate generated is higher than the denitrification achievable, nitrate will appear in the effluent and will be recycled to the anaerobic reactor. For the 3-stage Modified Bardenpho system, complete denitrification is not possible, and nitrate will always be present in the underflow recycle to the anaerobic tank. Recycling of nitrate for both systems will adversely influence the P removal.

Overall, the 3-stage and 5-stage Modified Bardenpho systems showed limitations to achieve stable and reliable P-removal efficiencies. Nevertheless, the work by Barnard led to the development of systems that appeared to incorporate the essential requirements for EBPR even though these requirements were not explicitly understood. This stimulated extensive research into this phenomenon, to gain experience on its behaviour, to delineate more precisely the factors influencing EBPR, and to develop design criteria.

6.6.5 Phoredox or anaerobic/oxic (A/O) system

In the Modified Bardenpho system, the configuration developed by Barnard for EBPR was heavily influenced by the legal requirement for nitrification in South Africa. Should nitrification not be required, the need for anoxic zones (to denitrify) and for long sludge ages (to ensure nitrification) falls away. These aspects were recognized by Barnard (1976a) who also applied the Phoredox method to a non-nitrifying activated sludge system. The configuration for this system was simplified to an anaerobic reactor, receiving the influent and underflow recycle, followed by an aerobic reactor (Figure 6.20B). The sludge age and aerobic tank are designed and controlled to prevent nitrification, *i.e.* short sludge age, high rate plant. This system has become known in South Africa as the Phoredox.

Timmerman (1979) proposed a system, essentially the same as the Phoredox system, which was designated the anaerobic/oxic (A/O) system. The basic A/O configuration is identical to that of the Phoredox, but with the A/O it is specifically proposed that the anaerobic and aerobic zones are partitioned to give a configuration of a series of reactors that approaches plug flow conditions.

Although proposed conceptually in 1976, the requirement for nitrification in South Africa has prevented implementation of the Phoredox or A/O system. The performance of the system under South African conditions was investigated at laboratory-scale by Burke *et al.* (1986), who found difficulty in preventing nitrification at a temperature of 20 °C even at sludge ages as low as 3 days with an unaerated mass fraction of 50 %.

The A/O system has found wider application in the USA and has been investigated by several researchers (*e.g.* Hong *et al.*, 1983; Kang *et al.*, 1985a, etc.) with mixed results.

6.6.6 University of Cape Town (UCT; VIP) system

Rabinowitz and Marais (1980), in reviewing their unsuccessful endeavours to obtain EBPR consistently in the Modified Bardenpho system, came to the conclusion that, irrespective of other factors that may affect the P removal, the recycling of nitrate to the anaerobic reactor via the underflow recycle appeared to be of great significance (the deleterious effect of the presence of nitrate in the anaerobic reactor subsequently was demonstrated directly by Hascoet and Florentz, 1985, among others). If the nitrate in the underflow to the anaerobic reactor could be kept at a low concentration, then there was a high expectation that consistent EBPR could be obtained. The principal obstacle to attaining this desirable end in the Modified Bardenpho system appeared to be that the nitrate discharged to the anaerobic reactor is linked directly to the concentration of nitrate in the effluent. If, for any reason, the effluent nitrate concentration increased while the COD remained constant, *i.e.* if the influent TKN/COD ratio increased, the system appeared to offer little option to reduce this by operational means. The only operational means available was to reduce the magnitude of the underflow recycle, but this was a risky option as the settleability of the mixed liquor in the plants tended to be poorer than in purely aerobic systems. Accordingly, Rabinowitz and Marais (1980) investigated different system configurations that would shield the anaerobic reactor of any input of nitrate, that is, to make the anaerobic reactor independent of the effluent nitrate concentration. This led to the development of the University of Cape Town (UCT) system (Figure 6.20E).

In the UCT system, the underflow sludge recycle (s) is discharged into the primary anoxic reactor. A further recycle (the r-recycle) draws mixed liquor from the primary anoxic reactor and discharges it into the anaerobic reactor. Mixed liquor is also recycled from the aerobic to the primary anoxic reactor (the a-recycle). By manipulation of the a-recycle ratio, the nitrate in the anoxic reactor can be maintained at zero, and thus no nitrate will be recycled to the anaerobic reactor. Consequently, the anaerobic state in the reactor can be maintained irrespective of the effluent nitrate concentration, even if the influent TKN/COD ratio to the plant varies. This desirable condition is achieved in the UCT system at the expense of (*i*) the anaerobic reactor volume; in the UCT system to maintain the same fraction of sludge in the anaerobic reactor as in the Modified Bardenpho system, the volume of the anaerobic reactor in the UCT system would have to be increased by the proportion (1+r)/r, and (*ii*) the inability to achieve complete denitrification.

Laboratory-scale tests on the UCT system using waste flow from Cape Town showed improved EBPR in both magnitude and consistency over that obtained in the Modified Bardenpho systems. However perhaps the most important achievement from a research point of view was that with the UCT system it was possible to eliminate the confounding effect on P removal of nitrate in the recycle to the anaerobic reactor, so that other factors influencing EBPR could be studied with greater ease (Siebritz *et al.*, 1982). From experimental response data the effects of these other factors became clearly evident:

(*i*) for the same influent COD, one batch of sewage gave high levels of P removal, another gave low, an observation previously presumed to be due to the nitrate effect but not explicitly proven as such and (*ii*) the magnitude of the EBPR appeared to be linked to some characteristic of the wastewater, as yet not identified.

6.6.7 Modified UCT system

Experience with the UCT system (Siebritz *et al.,* 1980, 1982) indicated some problems in system control.

The mixed liquor a-recycle ratio needs to be carefully controlled so that the primary anoxic reactor is slightly underloaded with nitrate to avoid a nitrate discharge to the anaerobic reactor. However, under full-scale operation such careful control of the a-recycle ratio is not possible due to uncertainty in the TKN/COD ratio, particularly under cyclic flow and load conditions.

To simplify the operation of the UCT system, a modification was sought whereby careful control of the a-recycle would not be necessary. This led to a modification of the UCT system called the Modified UCT system (Figure 6.20F). In the Modified UCT system, the anoxic reactor is subdivided into two reactors, the first having a sludge mass fraction of approximately 0.10 and the second having the balance of anoxic mass fraction available. The first anoxic reactor receives the underflow s-recycle and the r-recycle to the anaerobic reactor is taken from it. The second anoxic reactor receives the a-recycle. The minimum a-recycle is that which introduces the minimum nitrate to the second anoxic reactor that is sufficient to load it to its denitrification potential. Any recycle higher than the minimum will not remove additional nitrate so that at higher recycles more nitrate is introduced than removed in the second anoxic reactor and nitrate will appear in the effluent from this reactor. This, however, is immaterial insofar as it affects the nitrate in the aerobic reactor which remains constant once $a > a_{min}$. Consequently, one can raise the a-recycle to any value greater than a_{min}, to give the required actual retention time, without affecting the nitrate recycled to the first anoxic reactor; careful control of the a-recycle is no longer necessary[2]. This improvement however is obtained at a cost (WRC, 1984): the maximum TKN/COD ratio to give zero nitrate to the anaerobic reactor is reduced from ± 0.14 in the UCT system to ± 0.11 in the Modified UCT system. However, a TKN/COD of 0.11 mgN/mgCOD includes most raw and settled municipal wastewaters. Furthermore, by making provision that the r-recycle can be taken from either the first or second anoxic reactor, the system can be operated either in a Modified UCT or a UCT configuration, as may be required.

Another variation of the UCT system has been proposed, namely the Virginia Initiative Plant (VIP; Daigger *et al.,* 1987). The basic configuration for this system is identical to that of the UCT, but two specific proposals are made, (*i*) multiple series of mixed reactors are used, and (*ii*) the system is operated at short sludge ages of 5 to 10 days.

6.6.8 Johannesburg (JHB) system

Taking note of the adverse influence of nitrate recycled to the anaerobic reactor in the 5-stage Modified Bardenpho system as reported by Barnard (1976a), Osborn and Nicholls (1978) in a pilot-scale study at Johannesburg Northern Works proposed to alter the configuration of the 5-stage Modified Bardenpho by moving the secondary anoxic zone from the mainstream flow and repositioning it in the underflow recycle stream. The resulting 4-stage system (anoxic, anaerobic, anoxic, and aerobic) has become known as the Johannesburg (JHB) system

[2] Although from a N and phosphate removal point of view, careful control of the a-recycle is not necessary, the appearance of nitrate and/or nitrite in the effluent from the second anoxic reactor has been linked to the problem of low F/M bulking in nutrient removal systems. Thus, to control low F/M bulking careful control of the a-recycle would be necessary which effectively eliminates the advantage of the MUCT over the UCT system. In BCFS systems the SVI is 100-120 *i.e.* this problem does not exist or the redox control is indeed effective.

(Burke *et al.,* 1986; Nicholls, 1987). In the JHB system (Figure 6.20D), by repositioning the secondary anoxic reactor in the underflow stream, the mass of nitrate that needs to be removed in the secondary anoxic zone to give zero nitrate discharge to the anaerobic reactor is reduced to s/(1+s) times that which needs to be removed in the secondary anoxic zone of the 5-stage Modified Bardenpho system (where s is the sludge recycle ratio with regard to the influent flow rate). That is, to protect the anaerobic reactor from recycling of nitrate, in the JHB system only the nitrate in the s-recycle (underflow) stream has to be removed whereas in the 5-stage Modified Bardenpho system the nitrate in the s-recycle plus effluent streams has to be removed. Also, by positioning the anoxic reactor in the underflow s-recycle the sludge concentration in the secondary anoxic reactor of the JHB system is increased by a factor (1+s)/s compared to the secondary anoxic of the 5-stage Modified Bardenpho system, enabling reduction in reactor size to achieve the same anoxic mass fraction. However, unlike the 5-stage Modified Bardenpho, the JHB system (as for the UCT) cannot achieve complete denitrification. Although the JHB system does overcome the problem in the UCT system of increased anaerobic volume for the same mass fraction, the denitrification is at a lower rate than the UCT primary anoxic reactor. Therefore protection of the anaerobic reactor from nitrate can only be achieved at lower influent TKN/COD ratios than the UCT system, although most wastewaters will fall into the range of operation of the JHB system. Extensive full-scale investigation on the performance of the JHB system has been reported (*e.g.* Nicholls, 1987; Pitman *et al.,* 1988; Pitman, 1991).

6.6.9 Biological-chemical phosphorus removal (BCFS® system)

A further adaptation of the MUCT system was developed in the late 20th century in the Netherlands. This system, named BCFS® (Figure 6.20H), was developed to support the biological process by phosphate stripping and potential recovery in the main line, stabilising the sludge-settling properties and optimising the control of nitrogen removal. In this system a third recycle is added from the aerated reactor to the first anoxic reactor in order to maximise denitrification and to be able to aerate the second anoxic reactor during peak flows. In this way both ammonium and nitrate can be better maintained at low effluent values (ammonium typically below 0.5 gN/l and nitrate around 5-8 mgN/l). The recycle flows are controlled by a simple redox electrode-based controller (Van Loosdrecht *et al.,* 1998). The compartmentation contributes to a stable low SVI (approximately 120 ml/g) (Kruit *et al.,* 2002).

Biological phosphorus removal can be supplemented by addition of precipitants to the anaerobic tank. Since phosphate concentrations are high in this tank the precipitants are used effectively. Dosing chemicals, however, should be done carefully. Too much precipitation will make the phosphate unavailable for PAOs and deteriorate the EBPR efficiency. A complicating factor is that the wastewater treatment plant will respond rapidly to changes in chemical addition whereas the biological phosphorus removal process might have a response time of several days if not weeks. In the BCFS process a small baffle is placed at the end of the plug flow anaerobic tank. The sludge will locally settle back into the anaerobic tank and a clear supernatant can be withdrawn for phosphate precipitation. The phosphorus can then be recovered (Barat and Van Loosdrecht, 2006) or the chemical sludge produced can be prevented from accumulating in the activated sludge which would limit the overall capacity of the plant by reducing the sludge age.

In order to efficiently construct all the tanks in these complex biological nutrient-removal systems, it is possible to change the design from rectangular tanks to one round tank divided into rings for the different aerobic/anoxic/anaerobic zones. In this way the amount of concrete needed is minimised since the inner walls require much less strength then the outer walls of the construction (see Figure 10.1).

6.6.10 Sidestream EBPR (S2EBPR) systems

Due to the relatively low concentrations of RBCOD and VFA present in raw wastewater to perform EBPR and N removal, the fermentation of the primary settled sludge (PST) and/or (a fraction) of the RAS has showed an increase in the EBPR efficiency in certain plants in North America since the 1970s. This has resulted in a robust operation of the EBPR plants leading to effluent P concentrations lower than 1 mg/l, and even below the detection limit (Barnard *et al.,* 2017). The increased and robust EBPR performance is associated with the hydrolysis and fermentation of the PST and/or RAS streams. Moreover, in Denmark in the 1990s, several sidestream EBPR (S2EBPR) configurations were implemented (Figure 6.20I) (Vollertsen *et al.,* 2006) as a means to increase the low availability of RBCOD and VFA in raw wastewater (accounting for only around 150 mg RBCOD/l and 1 mgVFA/l, respectively) (Vollertsen, 2002) to support the EBPR process. In the Danish S2EBPR configurations, a fraction of the RAS is directed (typically 4 to 7 %) to an anaerobic sidestream tank with a hydraulic retention time of 30 to 40 h. The anaerobic conditions and relatively long HRT and high solids concentration (of around 10,000 to 11,000 mgTSS/l) can lead to considerably high hydrolysis and fermentation activity. This can result in the production of sufficient VFA (*e.g.* 136-149 mgRBCOD/l) to promote the EBPR and even support the N-removal process (Vollertsen *et al.,* 2006). Furthermore, if needed, the RBCOD production can be increased through the addition of primary sludge, molasses or other carbon sources. Smolders *et al.* (1996) carried out a steady-state model analysis. They found out that the acetate requirements for a S2EBPR process (in a P- stripping tank) are lower than those requirements needed in a mainstream process and that the active PAO biomass required to support the EBPR process can be up to 10 times lower.

The increasingly stricter N and P discharge limits and the limited availability of RBCOD in raw wastewater have promoted the implementation of S2EBPR worldwide since 2000, contributing to a broad development and assessment of different configurations, and to deeper and more thorough physiological and molecular analysis. Whereas the increased EBPR performance appears to be linked to a higher abundance of *Tetrasphera* in several S2EBPR plants (due to its fermentative capabilities) (Stokholm-Bjerregaard *et al.,* 2017), *Accumulibacter* has still also been seen in relatively abundant quantities (Onnis-Hayden *et al.,* 2019; Wang *et al.,* 2019), although the role, interaction and contribution of these microorganisms is still not fully elucidated. Nevertheless, less GAO have been observed in S2EBPR systems, which could be reflected in a more stable EBPR performance (Stokholm-Bjerregaard *et al.,* 2017; Onnis-Hayden *et al.,* 2019; Wang *et al.,* 2019).

However, although the actual mechanisms that favour the EBPR process in S2EBPR processes are still unknown, overall the higher RBCOD availability, combined with the lower VFA and PAO biomass requirements to achieve EBPR, and the lower occurrence of GAO seem to contribute to the increased robustness of the S2EBPR systems.

6.7 MODEL DEVELOPMENT FOR EBPR

6.7.1 Early developments

When the first mainstream nitrification-denitrification EBPR (NDEBPR) system was proposed, the 5-stage Modified Bardenpho system (Barnard, 1976b), initial conceptualization extended little beyond recognition of (*i*) the necessity of an anaerobic/aerobic sequence of reactors, and (*ii*) the adverse influence of nitrate recycled to the anaerobic zone. Design procedures were based on empirically-based estimates for sizing denitrification and anaerobic reactors in terms of nominal hydraulic retention time and sizing of the anaerobic reactor appeared to be linked to reduction of the redox potential below some critical value. No rational method for predicting N and P removal was available and for design, removals were estimated largely from experience gained in operating experimental systems similar to the proposed systems.

6.7.2 Readily biodegradable COD

In seeking an explanation for the different P release and enhanced P removal behavioural patterns in lab-scale Modified UCT and MLE systems, Siebritz *et al.* (1980, 1982) applied the concept of readily biodegradable COD (RBCOD) (see Section 6.4.5) developed in denitrification and aerobic studies (Dold *et al.*, 1980) to EBPR systems. They noted that the only evident difference between the Modified UCT and MLE systems lay in the concentration of RBCOD surrounding the organisms in the anaerobic reactor (S_{VFA}). In the Modified UCT system the RBCOD concentration in the anaerobic reactor (S_{VFA}) is the maximum possible as no nitrate is recycled to the anaerobic reactor. In contrast, in the MLE system sufficient nitrate is recycled to the anoxic reactor to utilize all the RBCOD, *i.e.* S_{VFA} = 0. Therefore, the different behavioural patterns of the processes would be consistently described if it is assumed that the concentration of RBCOD in the anaerobic reactor (S_S) surrounding the organisms is the key parameter determining whether or not P release and enhanced P removal takes place. In terms of our present understanding of EBPR, the parameter S_{VFA} is theoretical and cannot be measured; from the mechanisms of EBPR, the concentration of RBCOD surrounding the organisms in the anaerobic reactor does not equal S_{VFA} due to the conversion of the fermentable COD to VFAs by OHOs and the storage of VFAs by PAOs in the anaerobic reactor (see Section 6.4.5).

6.7.3 Parametric model

Extensive research over a year into the validity of the readily biodegradable COD (RBCOD) hypothesis by Siebritz *et al.* (1983) established that P release appears to be induced if the RBCOD in the anaerobic reactor, S_S, exceeds approximately 25 mg/l, and the P release and enhanced removal increases as (S_{VFA} - 25) increases. That is, the P removal is linearly related to the RBCOD concentration in the anaerobic reactor. This opened the way for enquiry into other factors affecting the P release and enhanced removal, and quantification of the enhanced P removal. They came to the conclusion that enhanced P removal depends on three factors (*i*) (S_{VFA} - 25), (*ii*) the fractional mass of sludge in the system passing through the anaerobic reactor, and (*iii*) the actual time a unit of sludge is retained in the anaerobic reactor.

They hypothesized that if any one of these is zero, no EBPR is obtained. Empirically these three factors are combined in a phosphorus-removal propensity factor. It was found that the mass of phosphorus in the sludge relative to the active mass could be functionally related to the P-removal propensity factor. Further investigation showed that in the modified Bardenpho and UCT (and although not considered, JHB) systems, with their respective recycles and their interactive effects on the anaerobic retention time, the factors (*ii*) and (*iii*) above could be combined by a single parameter, that is, the three parameters could be reduced to two key parameters: (*i*) (S_{VFA} - 25) and (*ii*) the anaerobic mass fraction, defined by (mass of sludge in the anaerobic reactor)/(total mass of sludge in the system).

Based on this observation, EBPR was formulated empirically in terms of the two key parameters and the mass of sludge (active, endogenous and inert) wasted per day, to give the parametric model.

Extensive testing of the concepts embodied in the parametric model did, in general, verify the utility of the model. At laboratory-scale, employing the UCT system, the concepts were tested at different sludge ages, temperatures, anaerobic mass fractions and influent COD concentrations in which the RBCOD fraction of the influent (unsettled municipal sewage) was augmented by addition of glucose or acetate. All these tests gave results consistent with the predictions based on the RBCOD concept embodied in the parametric model.

At full-scale, in a joint research project with Johannesburg City Council on the Goudkoppies and

Northern Works, analysis of the systems in terms of these concepts provided a consistent explanation when efficient or inefficient P removal was obtained (Nicholls, 1982). Thus, for the first time the parametric model allowed a quantitative approach to the design of N and P removal plants, and a basis for evaluating the performance of existing plants (Ekama *et al.,* 1983). For a detailed treatise on the parametric model the reader is referred to WRC (1984).

6.7.4 Comments on the parametric model

The parametric model described above was developed from observed data on experimental systems operated under a range of conditions, as follows:

- Influent COD concentrations: 250-800 mgCOD/l
- Readily biodegradable COD: 70-220 mgCOD/l, *i.e.* representing a f_{SS} fraction of between 0.12-0.27 mgRBCOD/mgCOD_{total}
- TKN/COD ratio: 0.09-0.14 mgN/mgCOD
- Sludge age: 13 and 25 days
- Temperature: 14 °C and 20 °C
- Anaerobic mass fraction: 0.10-0.20 gVSS_{AN}/gVSS_{sys}

Observations under these conditions formed the basis for structuring the formulations for estimating the enhanced P removal, and the equations thus derived were calibrated against the observed data. A comparison of the theoretically predicted and experimentally measured P removal data for the conditions set out above show a good correlation. However, despite the evident utility of the parametric model, it is still an empirical one; it links observable parameters but does not provide any explanation as to why these parameters are important to the phenomenon and it is independent of any formal hypothesis on the biological mechanisms driving the process. Accordingly, application of the parametric model had to be limited strictly to within the ranges of system parameters and wastewater characteristics listed above. What was required was a model with a more fundamental basis.

Essentially, up to this time the description of nitrification denitrification biological P removal (NDEBPR) system behaviour did not recognize the presence of any specific organism implicated in EBPR. The parametric model in fact considered the active sludge as a whole, to constitute a surrogate sludge with a propensity for P removal; variation in EBPR between different systems was hypothesized to be due to changes in the propensity for P removal of this surrogate sludge, caused by changes in influent RBCOD concentration, anaerobic mass fraction and/or nitrate discharge to the anaerobic reactor. However, parallel research in the natural sciences had identified specific organism groups that have the propensity to store large quantities of P in the form of polyphosphates. This led to a shift in the approach to modelling EBPR in NDEBPR systems, from a surrogate sludge to specific organism groups that are responsible for EBPR process, generically termed polyphosphate organisms (Wentzel *et al.,* 1986), bio-P organisms (Comeau *et al.,* 1986) and phosphorus-accumulating organisms (PAOs; Henze *et al.*, 1999).

6.7.5 NDEBPR system kinetics

Wentzel *et al.* (1988) set out to develop a general model that describes NDEBPR system behaviour. They assumed that in an NDEBPR system treating municipal wastewater, a mixed culture would develop which could be categorized into three groups of organisms: (*i*) heterotrophic organisms able to accumulate polyphosphate, termed phosphorus-accumulating organisms (PAOs), (*ii*) heterotrophic organisms unable to accumulate polyphosphate, termed ordinary heterotrophic organisms (OHOs), and (*iii*) autotrophic organisms mediating nitrification, termed nitrifying organisms (NIT). Wentzel *et al.* (1985, 1988) recognized that development of an activated sludge model to describe the behaviour of NDEBPR systems would require inclusion of all three organism groups, and their interactions. With regard to OHOs and NIT, they accepted the nitrification denitrification (ND) steady-state model described earlier and the general ND kinetic model (Dold *et al.,* 1991) (chapters 4 and

5). These models now needed to be extended to incorporate PAO behaviour in order to develop models that would include all three organism groups: general NDEBPR, steady state, and kinetic models. To achieve this objective, the kinetic and stoichiometric characteristics of the PAOs in the activated sludge environment needed to be established. From attempts to obtain information on the characteristics of the PAOs using mixed liquor from NDEBPR systems treating municipal wastewaters, Wentzel *et al.* (1988) concluded that, in these mixed culture systems, the OHO behaviour tends to dominate and mask the PAO behaviour. Accordingly, they endeavoured to isolate the PAO characteristics by enhancing the growth of the PAOs in the mixed-culture activated sludge systems. By enhanced culture is meant a culture in which (*i*) the growth of PAOs is favoured to the extent that they become the dominant primary organism and their behaviour dominates the system response, and (*ii*) the growth of competing organisms is curtailed but not positively excluded; neither are predation or other interaction effects. Wentzel *et al.* (1988) proposed to achieve a PAO-enhanced culture by taking mixed liquor from a mixed-culture NDEBPR system and selecting a substrate and a set of environmental conditions in the activated sludge system that would greatly favour PAO growth and enrichment.

6.7.6 Enhanced PAO cultures

6.7.6.1 Enhanced culture development

From the biochemical models, Wentzel *et al.* (1988) were able to identify conditions to be imposed in an NDEBPR activated sludge system to produce an enhanced PAO culture: an anaerobic/aerobic sequence with an adequate anaerobic mass fraction; influent fed to the anaerobic reactor with acetate as substrate and with adequate macro- and micronutrients, in particular Mg^{2+}, K^{+} and to a lesser degree Ca^{2+}, and pH control in the aerobic reactor. Using the UCT and 3-stage Modified Bardenpho systems, with system sludge ages ranging from 7.5 to 20 days, they developed enhanced cultures of PAOs with greater than 90 % of the organisms cultured aerobically being identified as *Acinetobacter* spp. using the Analytical Profile Index (API) 20NE procedure[3]. The response of the enhanced culture systems indicated that significant concentrations of PAOs developed. For example, the UCT system (anaerobic mass fraction 15%, sludge age 10 days and influent of acetate at 500 mgCOD/l) gave phosphate release of ≈ 253 mgP/l, phosphate uptake of ≈ 314 mgP/l and phosphate removal of ≈ 61 mg/l, all as mgP/l influent flow. This EBPR behaviour was much higher than observed in mixed culture NDEBPR systems with municipal wastewater influent of 500 mgCOD/l giving a phosphate release of ≈ 45 mg/l, phosphate uptake of ≈ 57 mg/l and phosphate removal of ≈ 12 mgP/l. The enhanced culture of mixed liquor in the aerobic zone contained ≈ 0.38 mgP/mgMLVSS and had a VSS/TSS ratio of ≈ 0.46 mgVSS/mgTSS, much higher than for mixed culture systems at a P/MLVSS ratio of ≈ 0.1 and a VSS/TSS fraction of ≈ 0.75. The low VSS/TSS ratio for the enhanced culture systems was due to the large amounts of polyphosphate with associated counter ions stored by the PAOs (Ekama and Wentzel, 2004).

6.7.6.2 Enhanced culture kinetic model

From experimental observations on the enhanced culture steady-state systems, and on batch tests in which mixed liquors drawn from the steady-state systems were subjected to a wide variety of conditions, Wentzel *et al.* (1989a) elucidated the characteristics and kinetic response of the active PAO biomass. Two characteristics of the PAOs in the enhanced cultures were of particular interest:

(*i*) Uncertain capacity to denitrify so that no provision for this process needed to be made in modelling PAO behaviour (the uncertain denitrification activity of PAOs has implications

[3]The API 20NE procedure has subsequently been shown to overestimate *Acinetobacter* spp. numbers due to the testing technique (Venter *et al.*, 1989) and selection in culturing (*e.g.* Wagner *et al.*, 1994). However, for the development of the design and simulation models, exact identification of the PAOs in the enhanced cultures has been of minor consequence as the models are based on quantitative experimental observations.

for modelling denitrification in mixed-culture NDEBPR systems, see Section 6.11).

(ii) An extremely low endogenous mass loss rate, 0.04 mgAVSS/mgAVSS.d which is much lower than that of OHOs in an aerobic activated sludge system at 0.24 mgAVSS/mgAVSS.d (McKinney and Ooten, 1969; Marais and Ekama, 1976). A similar observation had been made by Wentzel *et al.* (1985) in studies on mixed-culture NDEBPR systems treating municipal wastewaters; they noted from plots of phosphate uptake *versus* phosphate release for various sludge ages that, for a given phosphate release, the phosphate uptake was relatively insensitive to sludge age. To explain this observation, Wentzel *et al.* (1985) proposed that the PAOs 'suffer little or no endogenous mass loss'. The high specific endogenous mass loss rate with OHO systems had been attributed to a high rate of predation and regrowth, formulated as death regeneration in the ND kinetic model by Dold *et al.,* (1980). However, the low specific endogenous mass loss rate with PAOs in enhanced culture systems and the observations of Wentzel *et al.* (1985) led Wentzel *et al.* (1989a) to conclude that PAOs are not predated to the same degree as OHOs. Accordingly, in modelling PAO endogenous mass loss, Wentzel *et al.* (1989a) used the classical endogenous respiration approach, except that provision was made for situations where no external electron acceptor is available.

Taking note of the above, Wentzel *et al.* (1989a) developed a conceptual model for PAO behaviour in the enhanced cultures incorporating the characteristics, processes and compounds identified as important from the experimental investigation. Using the conceptual model as a basis, Wentzel *et al.* (1989a) formulated mathematically the process rates and their stoichiometric interactions with the compounds, to develop a kinetic model for the enhanced cultures of PAO. As recommended by the IAWPRC Task Group (Henze *et al.,* 1987), this model was presented in a matrix format with the kinetic and stoichiometric constants of the enhanced cultures being quantified by a variety of experimental procedures (Wentzel *et al.,* 1989b). Thus the PAO model, when integrated with the OHO and NIT simulation model, became known as UCTPHO (Wentzel *et al.,* 1992).

With these constants, application of the kinetic model to the various test responses observed with the enhanced cultures gave a good correlation between observations and simulations (figures 6.21 to 6.23). The model was then applied to simulate the steady-state behaviour of the enhanced culture UCT and 3-stage Modified Bardenpho systems, for which good correlation was obtained again (Wentzel *et al.,* 1989b).

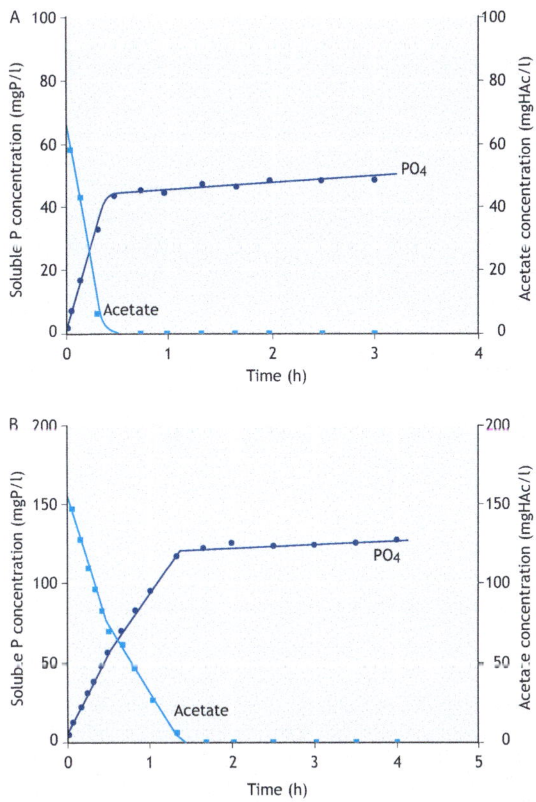

Figure 6.21 Experimentally observed and simulated total soluble phosphorus (PO_4) and acetate concentration-time profiles with the anaerobic addition of (A) 0.11 mgCOD$_{acetate}$/mgVSS and (B) 0.265 mgCOD/mgVSS to a mixed liquor drawn from an enhanced Bardenpho culture system (after Wentzel *et al.*, 1989b).

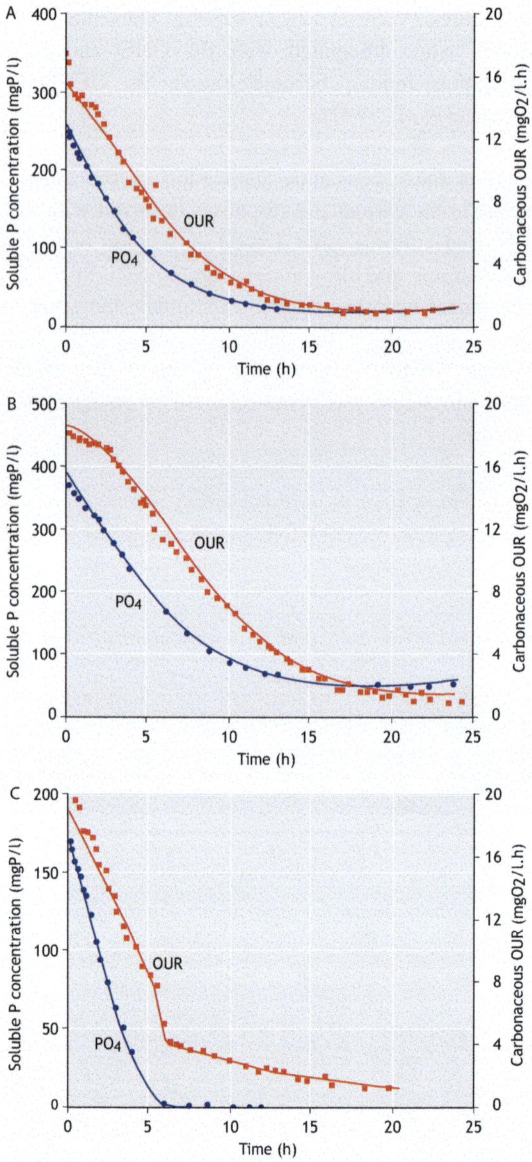

Figure 6.22 Experimentally observed and simulated total soluble phosphorus (PO_4) concentration and carbonaceous oxygen utilisation rate (OUR)-time profiles on aeration following the anaerobic acetate addition of (A) 0.207 mgCOD/mgVSS, (B) 0.363 mgCOD/mgVSS and (C) 0.22 mgCOD/mgVSS to mixed liquor drawn from an enhanced Bardenpho culture system. The PO4 concentration fell to zero during the course of the (C) test (after Wentzel *et al.*, 1989b).

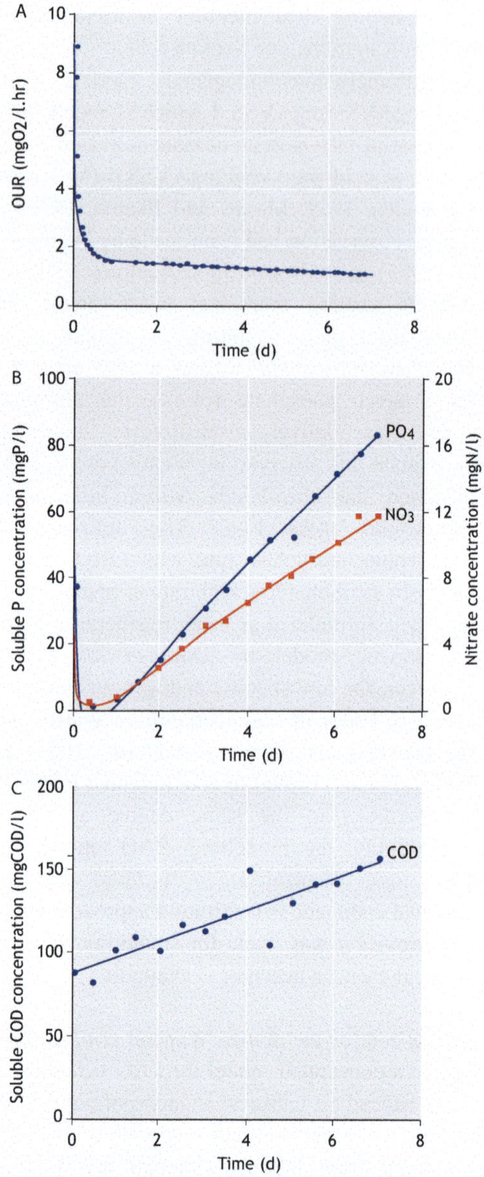

Figure 6.23 Experimentally observed and simulated (A) oxygen utilization rate (OUR), (B) total soluble phosphorus (PO_4) and nitrate (NO_3) concentrations and, (C) filtered COD concentrations with time in a batch digestion of mixed liquor from an enhanced culture system (after Wentzel *et al.*, 1989a).

6.7.6.3 Simplified enhanced culture steady state model

Wentzel *et al.* (1990) simplified the enhanced culture kinetic model to develop a steady-state model for the enhanced culture systems under constant flow and load conditions. From an examination of the kinetics of the processes under steady-state conditions, they found many of the processes to be virtually complete, these kinetic relationships no longer served any function and could be replaced by stoichiometric relationships. For example:

- The anaerobic mass fractions provided in the enhanced culture systems were sufficient to ensure that all the acetate substrate was sequestered in the anaerobic zone, *i.e.* the kinetics of acetate storage need not be incorporated.
- Virtually all the substrate sequestered in the anaerobic zone was utilized in the subsequent aerobic zone, *i.e.* the kinetics of PHA substrate utilization (and polyphosphate storage) did not need to be incorporated.

They noted that these simplifications implied that for a given sludge age, a constant relationship exists between the mass of acetate fed to the system and the mass of PAOs formed with stored polyphosphate. Further, they made an assumption which simplified the development of the steady state model:

- P release for anaerobic maintenance energy requirements is always small compared to phosphate release for VFA storage energy requirements, *i.e.* the kinetics of phosphate release for anaerobic maintenance energy did not need to be incorporated.

They further rationalized that the simplifications and assumptions imply that under steady state the polyphosphate content of the PAOs in the activated sludge is constant at 0.38 $gP/gVSS_{PAO}$, and independent of sludge age. What does vary is the relative proportion of PAOs (with stored polyphosphate) in the activated sludge. Taking due account of the simplifications and assumptions, Wentzel *et al.* (1990) developed a number of steady-state equations for the enhanced cultures, for PAO active and endogenous masses and for phosphate release, uptake and removal due to these masses. These equations provided the means for quantifying the PAO population in mixed-culture NDEBPR systems receiving municipal wastewaters as influent.

6.7.7 Steady-state mixed-culture NDEBPR systems

6.7.7.1 Mixed-culture steady-state model

Having developed the steady-state model for enhanced culture systems, Wentzel *et al.* (1990) extended this model to incorporate mixed cultures of PAOs and OHOs present in NDEBPR systems receiving domestic wastewater as influent, to give a steady-state mixed-culture model. This extension proved possible because (*i*) enhanced cultures rather than pure cultures were used to establish the kinetic and stoichiometric characteristics of the PAOs. In the enhanced cultures, PAOs present in mixed-culture activated sludge were enriched and no single species were artificially selected (as in pure cultures), (*ii*) competing organisms and predators were not artificially excluded (as in pure cultures) so that the PAOs were subjected to the same selective pressures in enhanced as in mixed cultures, (*iii*) the PAOs were also subjected to the same conditions present in mixed-culture activated sludge systems (*e.g.* anaerobic/aerobic sequencing, long SRT > 5 days, etc.), and (*iv*) the PAOs exhibited the same behavioural patterns in the enhanced cultures as they did in mixed-culture activated sludge systems (*i.e.* P release/uptake, PHA/polyphosphate accumulation, etc.). In fact, the similar, though 'magnified' behaviour of the enhanced culture compared to the mixed-culture systems was one criterion used to establish that the correct enhanced cultures had been established.

In extending the model one aspect that emerged was the difference in the endogenous mass loss rate between PAO-enhanced culture sludges and the 'normal' aerobic-OHO activated sludge. As noted earlier, the high rate of specific endogenous mass loss with OHO systems had been attributed to a high rate of predation and regrowth, formulated as death regeneration in the ND kinetic model by Dold *et al.* (1980). However, the low rate of specific endogenous mass loss with PAOs in the enhanced-culture systems led Wentzel *et al.* (1989a) to conclude that the PAOs were not predated to the same degree as OHOs, and to adopt an endogenous respiration approach in modelling PAO endogenous mass loss[4]. The low predation rate on the PAOs, and the fact that the PAOs and OHOs essentially do not compete for the same substrate, implied that PAO and OHO populations act virtually independently of each other in 'normal' mixed-culture NDEBPR systems. In developing the steady-state model for mixed-culture NDEBPR systems, Wentzel *et al.* (1990) noted that this implied that analysis of the two population groups could be largely separated. However, two significant interactions were identified for inclusion in the mixed-culture NDEBPR steady-state model, both in the anaerobic reactor, as follows:

(i) In many 'normal' municipal wastewaters the acetate or other volatile fatty acid (VFA) content is small or not present (Wentzel et al., 1988). Wentzel et al. (1985) showed that in the anaerobic reactor the RBCOD component of the influent is converted into VFAs by acid fermentation by the OHOs, thereby making VFAs available to the PAO mass for storage. The rate of conversion is much slower than the rate of storage for PHA, so that the rate of conversion controls the rate of storage. Hence, the mass of VFA substrate that becomes available in the anaerobic reactor to the PAOs is governed by the kinetics of conversion mediated by the OHOs. The work of Meganck *et al.* (1985) and of Brodisch (1985) supported this conversion hypothesis as they showed that anaerobic/aerobic systems develop organisms which convert sugars and similar compounds into VFAs in the anaerobic reactor.

(ii) If nitrate (or oxygen) is recycled to the anaerobic reactor, RBCOD is utilized preferentially by the OHOs with nitrate (or oxygen) as the external electron acceptor, thereby reducing the mass of RBCOD converted to VFAs.

Wentzel *et al.* (1985) recognized the above points and formulated a kinetic model for conversion of RBCOD to VFAs, and hence for storage of these VFAs. Wentzel *et al.* (1990) accepted this model, but made provision to include situations where VFAs are present in the influent by noting that:

- The RBCOD needs to be subdivided into two fractions, VFAs/RBCOD (*e.g.* acetate) and fermentable S_F/RBCOD (*e.g.* glucose). Both these fractions will be measured as RBCOD in conventional bioassay (*e.g.* Ekama *et al.*, 1986; Wentzel *et al.*, 1995) and filtration (*e.g.* Dold *et al.*, 1986; Mamais *et al.*, 1993; Wentzel *et al.*, 1995) tests, *i.e.*:

$$\text{RBCOD} = \text{VFAs} + \text{fermentable COD} \qquad (6.1a)$$

or, in symbols

$$S_S = S_{VFA} + S_F \qquad (6.1b)$$

- The rate of VFA storage is so rapid that all influent VFAs will be sequestered by the PAOs in the anaerobic reactor for anaerobic mass fractions greater than 10 % and sludge ages greater than 10 days (this can be verified from the kinetics of storage).
- The fermentable COD is converted into VFAs by the OHOs in the anaerobic reactor, and the resultant VFAs are available for storage by the PAOs. The model for conversion is given by Wentzel *et al.* (1985).

This theory provided Wentzel *et al.* (1990) with the means of calculating the mass of VFA substrate (from the influent and from conversion of fermentable COD) sequestered by the PAOs in the anaerobic reactor. Knowing the mass of substrate

[4] From subsequent simulations with the steady-state mixed culture model, it was found that if the PAOs were subjected to a high predation rate, then significant EBPR in the mixed-culture NDEBPR system would not be possible; the rate of death of the PAOs would be so high that no significant mass of these organisms could accumulate in the system, and EBPR would be near zero.

sequestered by the PAOs, the mass of substrate remaining that is available to the OHOs could be calculated. In effect, Wentzel *et al.* (1990) split the biodegradable influent COD into two fractions, one eventually to be utilized by the PAOs and the other to be utilized by the OHOs. Because of the independent action of these two groups of organisms, they could use:

(i) The simplified PAO-enhanced culture steady-state model for calculating the PAO active and endogenous masses formed from the sequestered substrate, and the P release, uptake and removal performed by these masses.

(ii) The steady-state activated sludge model (Marais and Ekama, 1976; WRC, 1984, Chapter 4) to calculate the OHO active and endogenous masses formed from the remaining substrate, the rate of conversion of fermentable COD to VFAs in the anaerobic reactor, the inert VSS accumulated from the influent, and the P requirement of, and hence P removal associated with, the active, endogenous and inert masses. Note that in this steady-state activated sludge model, the endogenous mass loss is modelled using the classical endogenous respiration approach; this approach is simpler and under steady-state conditions gives results very close to the death regeneration approach.

The total P removal for the system was calculated by summation of the individual P removals.

Wentzel *et al.* (1990) evaluated the predictive power of the steady-state mixed-culture EBPR model against observations made of 30 laboratory scale NDEBPR systems over a six-year period. The system configurations were Phoredox, 3-stage Modified Bardenpho, UCT, MUCT and Johannesburg with system sludge ages ranging from 3 to 28 days. For the evaluation, the measured nitrate in the recycle to the anaerobic zone was used to estimate the fermentable COD removal in the anaerobic zone by the OHOs with nitrate as the external electron acceptor. The fermentable COD remaining was available for conversion in the anaerobic reactor to VFAs, for storage as PHA by the PAOs. Plots of the predicted and measured P releases, P removals and VSS concentrations (figures 6.24 to 6.26) show good correlation.

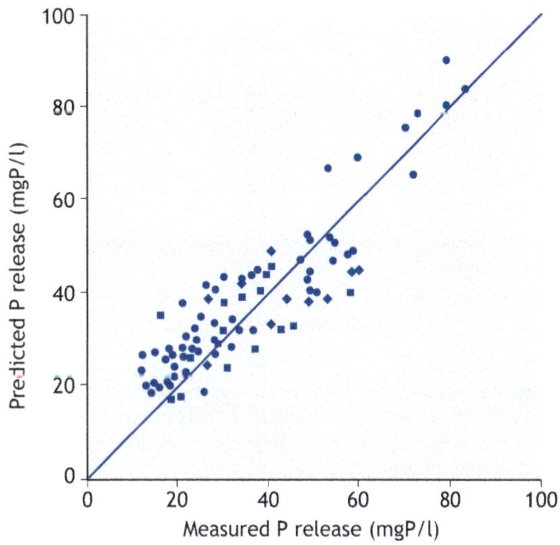

Figure 6.24 Predicted *versus* measured P release in a variety of EBPR systems with various configurations for SRTs from 3 to 28 d (after Wentzel *et al.*, 1990).

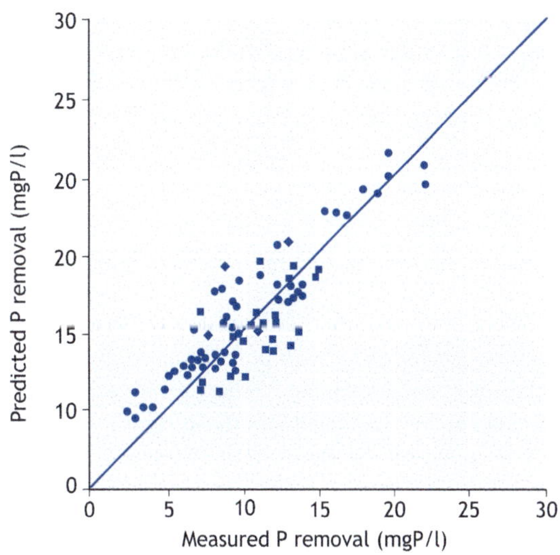

Figure 6.25 Predicted *versus* measured P removal in a variety of EBPR systems with various configurations for SRTs from 3 to 28 d (after Wentzel *et al.*, 1990).

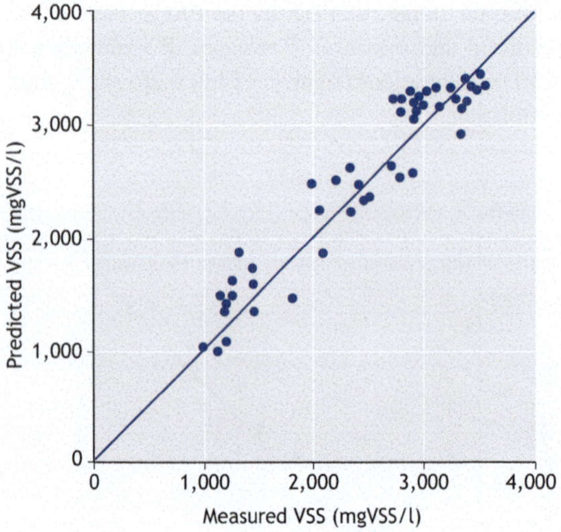

Figure 6.26 Predicted *versus* measured VSS concentration in a variety of biological enhanced P-removal systems with various configurations for SRTs from 3 to 28 d (after Wentzel *et al.*, 1990).

6.8 MIXED-CULTURE STEADY-STATE MODEL

6.8.1 Principles of the model

The fundamental principle underlying the mixed-culture steady-state model is to divide the activated sludge between three population groups:

1. NIT, the nitrifiers;
2. OHOs, the ordinary heterotrophic organisms; and
3. PAOs, the phosphorus-accumulating organisms.

Then, knowing the P content of the sludge fractions generated by each population group (active, endogenous and inert), the P removal for each sludge fraction can be calculated and the system P removal will be given by the summation of the individual P removals.

Procedures for quantification of the NIT are presented in Chapter 5; these procedures can be retained unmodified for nitrifying and denitrifying EBPR systems provided the unaerated mass fraction (f_{xa}) is extended to include both the anoxic and anaerobic reactors. The relatively small contribution that the NIT make to the sludge mass (< 3 per cent) means that the P removal due to this population group can be neglected.

With regard to the OHOs and the PAOs, the principle is to split the biodegradable COD between the two population groups and to calculate the masses that result from the two COD fractions (Figures 6.10 and 6.27); knowing the P content of each mass, then the P removal can be calculated. Procedures for quantification of the OHOs (including inert mass) are presented in Chapter 4; these can be applied to nitrifying and denitrifying EBPR systems, but need to be modified to take account of the biodegradable COD reduction due to COD storage by the PAOs, see below. In this section, procedures will be presented for quantification of the PAOs and OHOs and for dividing the biodegradable COD between the PAOs and OHOs.

The relationship between the flux of the influent biodegradable COD components, their fate in the treatment system and the active biomass produced is illustrated in Figure 6.28 and explained in the following sections.

The sludge biomass is composed of active and inactive particulate fractions. The active fractions include the biomass components of PAOs, OHOs and other biomasses such as nitrifiers that do not need to be calculated in this design example. Inactive components include particulate inert organics and particulate inorganics from the influent, and particulate endogenous residues generated by cell decay.

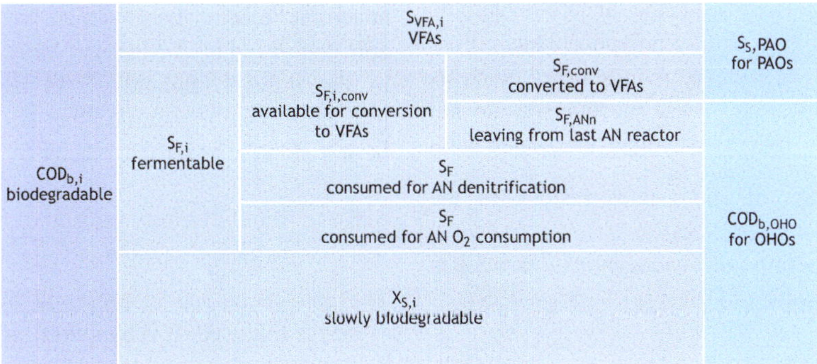

Figure 6.27 Division of influent biodegradable COD between PAOs and OHOs.

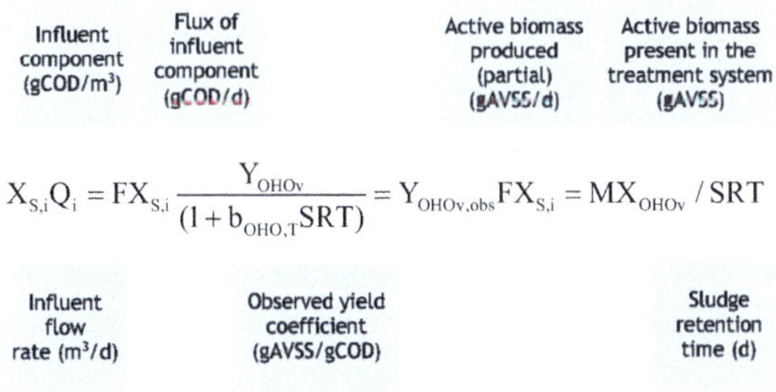

$$X_{S,i}Q_i = FX_{S,i}\frac{Y_{OHOv}}{(1+b_{OHO,T}SRT)} = Y_{OHOv,obs}FX_{S,i} = MX_{OHOv}/SRT \qquad (6.1)$$

Figure 6.28 Relationships between the influent components, flux and biomass produced and present in the system.

6.8.2 MASS EQUATIONS

6.8.2.1 PAOs

Biological active mass:

$$MX_{PAOv} = \frac{Y_{PAOv}}{(1+b_{PAO,T}SRT)}FS_{S,PAO}SRT \qquad (6.2)$$

where:

MX_{PAOv}	biological active mass of PAO (gAVSS)
Y_{PAOv}	PAO biomass yield (gAVSS/gCOD)
$FS_{S,PAO}$	daily mass of substrate stored by PAOs in the anaerobic reactor (gCOD/d)
$b_{PAO,T}$	specific endogenous mass loss rate constant of PAO at temperature T (gEVSS/gVSS.d)
SRT	sludge age (d)

Endogenous mass:

$$MX_{E,PAOv} = f_{XE,PAO}b_{PAO,T}MX_{PAOv}SRT \qquad (6.3)$$

where:

$MX_{E,PAOv}$	PAO endogenous mass (gEVSS)
$f_{XE,PAO}$	fraction of endogenous particulate residue of PAOs (gEVSS/gAVSS)

6.6.2.2 OHOs

Biological active mass:

$$MX_{OHOv} = \frac{Y_{OHOv}}{(1+b_{OHO,T}SRT)}FCOD_{b,OHO}SRT \qquad (6.4)$$

where:

MX_{OHOv}	OHO active biomass (gAVSS)
$FCOD_{b,OHO}$	daily mass of biodegradable substrate available to OHOs (gCOD/d) $= FCOD_{b,i} - FS_{S,PAO}$
$FCOD_{b,i}$	daily mass of influent biodegradable COD (gCOD/d) $= FCOD_i\,(1 - f_{SU} - f_{XU})$
Y_{OHOv}	OHO yield (gAVSS/gCOD)
$b_{OHO,T}$	specific endogenous mass loss rate constant of OHO at temperature T (/d)

Endogenous mass:

$$MX_{E,OHOv} = f_{XE,OHO} b_{OHO,T} MX_{OHOv} SRT \quad (6.5)$$

where:

$MX_{E,OHOv}$ mass of endogenous residue in the system (gEVSS)

$f_{XE,OHO}$ fraction of endogenous particulate residue of OHOs (gEVSS/gAVSS)

6.8.2.3 Inert mass

Inert organic matter from the influent accumulates in the system:

$$MX_{Uv} = \frac{f_{XU,CODi} FCOD_i SRT}{f_{cv}} \quad (6.6)$$

where:

MX_{Uv} mass of inert organic matter in the system coming from the influent (gIVSS)

$f_{XU,COD,i}$ fraction of influent COD that is particulate and unbiodegradable

$FCOD_i$ daily mass of influent total COD

6.8.3 Division of biodegradable COD between PAOs and OHOs

- From the mechanisms for EBPR (Section 6.3 above), only VFA substrate can be stored by the PAOs in the anaerobic reactor. Accordingly, the influent RBCOD ($S_{S,i}$) needs to be subdivided into two fractions, namely VFAs ($S_{VFA,i}$) and fermentable COD ($S_{F,i}$). Thus, $S_{S,i} = S_{VFA,i} + S_{F,i}$.

The VFAs in the influent ($S_{VFA,i}$) are directly available to the PAOs for storage in the anaerobic reactor. Wentzel *et al.* (1985) have shown that the fermentable component ($S_{F,i}$) is converted to VFAs in the anaerobic reactor by the OHOs, thereby making additional VFAs available to the PAOs for storage. The rate of conversion is much slower than the rate of storage, so that the rate of conversion controls the rate of storage of generated VFAs. Hence, the mass of VFA substrate that becomes available in the anaerobic reactor is governed by the kinetics of conversion and by the mass of VFA substrate present in the influent. Should VFAs be present in the influent, it can be assumed that all these VFAs will be stored in the anaerobic reactor by the PAOs.

6.8.3.1 Kinetics of conversion of fermentable organics to VFAs

The conversion model proposed by Wentzel *et al.* (1985) is followed. It is hypothesized that:

- only fermentable COD (S_F) can be converted to a form suitable for storage by the PAOs (*i.e.* VFAs); within the timescale of residence of the mixed liquor in the anaerobic reactor the conversion of slowly biodegradable COD (X_S) to VFAs is assumed to be negligible (see Section 6.3.6.1).
- the conversion is performed by the OHO mass in the anaerobic reactor.
- all VFAs generated from conversion of fermentable COD are immediately stored by the PAOs.
- all fermentable COD not converted to VFAs in the anaerobic reactor is utilized subsequently for OHO metabolism.
- the rate of conversion of fermentable COD is given by:

$$\frac{dS_{F,AN}}{dt} = -k_{F,T} S_{F,AN} X_{OHOv,AN} \quad (6.7)$$

where:

$dS_{F,AN}/dt$ rate of conversion of fermentable organics (gCOD m^3/d)

$k_{F,T}$ first-order fermentation rate constant at temperature T (0.06 m^3/gVSS.d at 20 °C)

$S_{F,AN}$ fermentable COD concentration in the anaerobic reactor (gCOD/m^3)

$X_{OHOv,AN}$ concentration of OHOs in the anaerobic reactor (gAVSS/m^3)

- all VFAs present in the influent to the anaerobic reactor will be immediately stored by the PAOs.

6.8.3.2 Effect of recycling nitrate or oxygen

Should nitrate or oxygen enter the anaerobic reactor via recycle or with the influent, the conversion of fermentable COD to VFAs is further complicated. It is hypothesized that any oxygen or nitrate entering the anaerobic reactor is utilized as the electron acceptor by the OHOs with RBCOD (S_S) as the electron donor (substrate). It is not clear whether the fermentable COD or the influent VFAs will be used preferentially as the electron donor. For the purpose of the steady-state mixed-culture model it is assumed that the influent fermentable COD will serve as the electron donor. The implication is that the VFAs generated by conversion are no longer released, but are metabolized directly by the OHOs, until the oxygen or nitrate is depleted. In the conversion model this can be accommodated by reducing the amount of fermentable COD available for conversion as follows:

$$S_{F,i,conv} = S_{F,i} - 8.6\,(s\,S_{NO3,s} + S_{NO3,i}) - 3.0\,(s\,S_{O2,s} + S_{O2,i}) \tag{6.8}$$

where:

$S_{F,i,conv}$	fermentable COD available for conversion per volume of influent ($gCOD/m^3$)
$S_{F,i}$	fermentable COD influent concentration ($gCOD/m^3$)
s	sludge recycle ratio to anaerobic reactor based on influent flow
$S_{NO3,s}$	nitrate concentration in the sludge recycle to the anaerobic reactor ($gNO_3\text{-}N/m^3$)
$S_{O2,s}$	oxygen concentration in the sludge recycle to the anaerobic reactor (gO_2/m^3)
$S_{NO3,i}$	nitrate concentration in the influent to anaerobic reactor ($gNO_3\text{-}N/m^3$)
$S_{O2,i}$	oxygen concentration in the influent to anaerobic reactor (O_2/m^3)
8.6	mass of COD removed per unit of nitrate denitrified ($gCOD/gNO_3\text{-}N$); $2.86 / (1 - f_{cv} \cdot Y_{OHOv}) = 2.86/(1 - 1.48 \cdot 0.45) = 8.6$
3.0	mass of COD removed per unit of oxygen utilized ($gCOD/gO_2$); $1 / (1 - f_{cv} \cdot Y_{OHOv}) = 1/(1 - 1.48 \cdot 0.45) = 3.0$

6.8.3.3 Steady-state conversion equations

Steady-state equations for the conversion of fermentable COD to VFAs can be developed by applying eqs. 6.7 and 6.8 in mass balances for the nth anaerobic reactor in a series of N anaerobic reactors of equal volume. This yields an equation to calculate the concentration of fermentable COD in the effluent from the nth anaerobic reactor:

$$S_{F,ANn} = \frac{S_{F,i,conv} / (1+s)}{\left[1 + k_{F,T} \dfrac{f_{xa}}{N} \dfrac{MX_{OHOv}}{Q_i\,(1+s)}\right]^n} \tag{6.9}$$

where:

$S_{F,ANn}$	concentration of fermentable COD in the effluent of the nth anaerobic reactor ($gCOD/m^3$)
f_{xa}	anaerobic mass fraction (gVSS/gVSS)
N	total number of anaerobic reactors of equal volume in the series n = 1,2.....N
MX_{OHOv}	mass of OHOs in the whole NDEBPR system (gAVSS)
Q_i	influent flow rate (m^3/d)

Eq. 6.9 provides the means to calculate the fermentable COD converted to VFAs in a series of N anaerobic reactors, *i.e.*:

$$FS_{F,CONV} = Q_i[S_{F,i,conv} - (1+s)S_{F,ANn}] \tag{6.10}$$

where:

$FS_{F,CONV}$	daily mass of fermentable COD converted to VFAs in the anaerobic reactors (gCOD/d)

However, to calculate $S_{F,ANn}$ the term MX_{OHOv}/Q_i needs to be determined.

Now, MX_{OHOv} is synthesized from the total mass of biodegradable influent COD less the mass of COD stored by the PAOs. From the mechanisms of EBPR and the hypothesis for conversion, all the VFAs generated by conversion and all the VFAs in the influent are stored by the PAOs., *i.e.* the mass of COD stored by the PAO, $FS_{S,PAO}$, is given by:

$$FS_{S,PAO} = FS_{F,CONV} + Q_i S_{VFA,i} \quad (6.11)$$

$$FS_{S,PAO} = Q_i[S_{F,i,conv} - (1+s)\,S_{F,ANn}] + Q_i\,S_{VFA,i} \quad (6.12)$$

where:

$FS_{S,PAO}$ daily mass of S_S stored by the PAOs (gCOD/d)

The COD available to the OHOs, is the biodegradable COD not stored by the PAOs:

$$FCOD_{b,OHO} = FCOD_{b,i} + FS_{S,PAO} \quad (6.13)$$

where:

$FCOD_{b,OHO}$ daily mass of biodegradable COD available to the OHOs (gCOD/d)

Accordingly, and as presented earlier, the equation to estimate the mass of ordinary heterotrophic organisms takes account of the reduced COD available:

$$MX_{OHOv} = \frac{Y_{OHOv}}{(1+b_{OHO,T}\,SRT)} FCOD_{b,OHO}\,SRT \quad (6.14a)$$

The production of OHOs can also be expressed as mass synthesized per volume of influent by substituting eqs. 6.12 and 6.13 into eq. 6.14a and dividing by the influent flow rate:

$$\frac{MX_{OHO}}{Q_i} = \frac{Y_H}{(1+b_{OHO,T}\,SRT)} \cdot (COD_{b,i} - (1+s)\,S_{F,ANn} + S_{VFA,i})\,SRT \quad (6.14b)$$

where:

MX_{OHOv}/Q_i equivalent concentration of OHOs produced per volume of influent ($gAVSS/m^3$)

Eqs. 6.9 and 6.14a need to be solved simultaneously to calculate the concentration of fermentable COD ($S_{F,ANn}$) leaving the last anaerobic reactor (ANn); the following iterative procedure can be used:

- Assume $S_{F,ANn} = 0$ mgCOD/l.
- Calculate MX_{OHOv} using Eq. 6.14a.
- Using the calculated value for MX_{OHOv}, calculate $S_{F,ANn}$ from Eq. 6.9.
- Recalculate MX_{OHOv} using the calculated value for $S_{F,ANn}$.
- Repeat the last two steps until $S_{F,ANn}$ and MX_{OHOv} are constant.

Similar equations could be derived for the behaviour of denitrifying PAOs (DPAOs) for anoxic conditions. However, the interaction with strictly aerobic PAOs and ordinary denitrifiers would require that the kinetics of substrate consumption and storage by each group of microorganisms be considered, a task that can best be managed through dynamic modelling.

6.8.3.4 Implications of conversion theory

The conversion theory set out above provides the means for calculating the mass of VFAs generated per day by the OHOs. Accepting that all the VFAs from conversion and from the influent are stored by the PAOs, the mass of substrate available to the OHOs is the remaining biodegradable COD. In effect the influent biodegradable COD is split into two fractions, one to be utilized by the PAOs and the other to be utilized by the OHOs. Because of the independent action of the two groups of organisms, the equations set out earlier (eq. 6.1 to 6.3) can be applied to calculate the active and endogenous PAO masses, and the equations in Chapter 4 can be used to calculate the OHO active, endogenous and inert masses appropriately modified as in eqs. 6.4 to 6.6. Then, knowing the P content of each of these mass fractions, the P removal due to each can be calculated (see below).

6.8.4 P release

The phosphorus release by PAOs as a result of VFA storage does not need to be quantified for the steady-state design of EBPR systems but nevertheless can be useful information to obtain. From the mechanisms for P removal (Section 6.3), for every mole of VFAs stored by PAOs, it is considered that one mole of P is released (recognising that this ratio is pH-dependent, Smolders *et al.,* 1994a; Filipe *et al.,* 2001c) to provide energy to polymerise and store the VFAs as PHA. Accordingly, the P release will be given by:

$$FS_{PO4,rel} = f_{PO4,rel}\, FS_{S,PAO} \quad (6.15a)$$

where:

$FS_{PO4,rel}$ daily mass of P release by PAOs (gP/d)

$f_{PO4,rel}$ ratio P release/VFA uptake = 1.0 molP/molCOD = 0.5 gP/gCOD at pH 7.0

or, in concentration units:

$$S_{PO4,rel} = f_{PO4,rel} \frac{FS_{S,PAO}}{Q_i} \quad (6.15b)$$

where:

$S_{PO4,rel}$ P released (gP/m^3 of influent)

$S_{S,PAO}$ concentration of readily biodegradable COD stored by PAOs ($gCOD/m^3$)

$$S_{PO4,rel} = f_{PO4,rel} \frac{FS_{S,PAO}}{Q_i} \quad (6.15c)$$

If the $f_{PO4,rel}$ coefficient needs to be estimated as a function of the pH in the bulk liquid, the following expression developed by Smolders *et al.* (1994a) could be applied:

$$f_{PO4,rel} = 0.18\ pH - 0.81 \quad (6.15d)$$

where:

$f_{PO4,rel}$ ratio P release/VFA uptake in gP/gCOD

pH pH measured in the bulk liquid of the anaerobic stage of an EBPR system.

6.8.5 P removal and effluent total phosphorus concentration

The P removal is calculated for the individual sludge fractions, with the total P removal being given by the summation of the individual P removal of each biomass. Overall, only the PAO biomass, MX_{PAOv}, is able to store higher P concentrations (of up to 0.38 g P/gVSS of active PAO biomass) than those needed for biomass synthesis. On the other hand, the rest of the biomasses (namely MX_{OHOv}, $MX_{E,OHOv}$, $MX_{E,XUv}$, and even the endogenous PAO biomass $MX_{E,PAOv}$) accumulate the typical P content observed in VSS, f_p, of 0.03 gP/gVSS. It is important to notice that the endogeneous PAO biomass, $MX_{E,PAOv}$, being an endogenous residue, contains only the P present in the residue (presumably used to make the cell tissues and other organic compounds) and therefore is far less than the $f_{P,PAO}$ fraction that the active PAO biomass can actually store.

Thus, the different biomasses contribute to the overall potential P removal as follows:

PAOs:

$$\Delta P_{PAO} = f_{P,PAO} \frac{MX_{PAOv}}{Q_i \cdot SRT} \quad (6.16)$$

where:

ΔP_{PAO} P removal due to PAOs (gP/m^3)

$f_{P,PAO}$ fraction of PAO active mass that is P = 0.38 gP/gVSS with regard to the active PAO biomass, MX_{PAOv}

OHOs:

$$\Delta P_{OHO} = f_p \frac{MX_{OHOv}}{Q_i \cdot SRT} \qquad (6.17)$$

where:

ΔP_{OHO} P removal due to OHOs (gP/m^3)

f_p fraction of OHO active mass that is P = 0.03 gP/gVSS with regard to the active OHO biomass, MX_{OHOv}, which corresponds to the typical P content present in VSS

Endogenous residue mass (from any biomass, including PAOs and OHOs) is defined by:

$$\Delta P_{XE} = f_p \frac{MX_{E,OHOv} + MX_{E,PAOv}}{Q_i \cdot SRT} \qquad (6.18)$$

where:

ΔP_{XE} P removal due to endogenous residue mass (gP/m^3)

f_p fraction of endogenous mass that is P (gP/gVSS) = 0.03 gP/gVSS of endogenous residues ($MX_{E,OHOv}$ and $MX_{E,PAOv}$), which is similar to the typical P content present in VSS

The influent inert mass is:

$$\Delta P_{XU} = f_p \frac{MX_U}{Q_i \cdot SRT} \qquad (6.19)$$

where:

ΔP_{XU} P removal due to influent unbiodegradable particulate organics, $X_{U,i}$ (gP/m^3)

f_p fraction of influent unbiodegradable particulate organics that is P (gP/gVSS) = 0.03 gP/gVSS of $X_{U,i}$, corresponding to the typical P content present in VSS

The total P removal potential by the system, neglecting chemical phosphorus precipitation (typically due to aluminium, calcium or iron salts present in the influent or added to the system), is:

$$\Delta P_{SYS,pot} = \Delta P_{PAO} + \Delta P_{OHO} + \Delta P_{XE} + \Delta P_{XU} \qquad (6.20)$$

where:

$\Delta P_{SYS,pot}$ potential total P removal by the system (gP/m^3)

The actual P removal by the system is the lowest of the total P removal potential and the influent total phosphorus:

$$\Delta P_{SYS,actual} = \min(\Delta P_{SYS,pot}; P_i) \qquad (6.21)$$

where:

$\Delta P_{SYS,actual}$ actual total P removal for the system (gP/m^3)

Any suspended solids in the effluent contributes to increasing the particulate phosphorus concentration in the effluent:

$$X_{P,e} = f_{P,TSS}\, TSS_e \qquad (6.22)$$

where:

$f_{P,TSS}$ average P content of the activated sludge (gP/m^3)

TSS_e total suspended solids concentration of the effluent ($gTSS/m^3$)

The effluent total P concentration is calculated by subtracting the actual total P removal for the system and adding any particulate P contributed by the suspended solids in the effluent:

$$P_e = P_i - \Delta P_{SYS,actual} + X_{P,e} \qquad (6.23)$$

where:

P_i influent total P concentration (gP/m^3)

P_e effluent total P concentration (gP/m^3)

6.8.6 VSS and TSS sludge masses and P content of TSS

6.8.6.1 Actual P content in active PAO biomass

Although the active PAO biomass, MX_{PAOv}, can have the potential to remove a high P concentration, logically the P concentration that MX_{PAOv} can remove (store intracellularly) cannot be higher than the concentration present in the wastewater influent. This principle is similar to the one that defines the total P removal potential ($\Delta P_{SYS,pot}$) and the actual P removal ($\Delta P_{SYS,actual}$) of the system. As such, the actual P content of the active PAO biomass, $f_{P,PAO,act}$, can be estimated as the difference between the actual P mass removed by the system minus the P mass accumulated by the VSS biomass excluding the contribution of MX_{PAOv}, as follows:

$$f_{P,PAO,act} = \frac{\left[Q_i \cdot SRT \cdot \Delta P_{SYS,actual}\right] - \left[f_p (MX_{VSS} - MX_{PAOv}\right]}{MX_{PAOv}} \tag{6.24a}$$

where:

$f_{P,PAO,act}$ Actual P content stored by the active PAO biomass (gP/gVSS of active PAO biomass, MX_{PAOv})

For design purposes, it is important to estimate the actual P content stored by PAO $f_{P,PAO,act}$ since it can affect the estimation of different parameters of utmost importance: (*i*) the determination of MX_{FSS}; (*ii*) the total mass MX_{TSS} accumulated in the system; (*iii*) as a consequence of the influence on MX_{FSS} and MX_{TSS}, it can affect the calculation of the total volume of the plant V_R, and (*iv*) also the estimation of the $f_{P,TSS}$ content which can affect the correct estimation of the actual total P concentration present in the effluent.

6.8.6.2 VSS sludge mass

The VSS sludge mass in the system is calculated in the same fashion used for aerobic and anoxic/aerobic systems, by summing the contributions of the individual VSS fractions, *i.e.*:

$$MX_{VSS} = MX_{PAOv} + MX_{OHOv} + MX_{E,PAOv} + MX_{E,OHOv} + MX_{Uv} \tag{6.24b}$$

$$MX_{VSS} = V_R \; VSS \tag{6.24c}$$

where:

MX_{VSS}	VSS mass in system (gVSS)
VSS	VSS concentration in system ($gVSS/m^3$)
V_R	system process volume (m^3)

As for aerobic and anoxic/aerobic systems, the TSS sludge mass in the system is calculated from the VSS via the VSS/TSS ratio. However, for the PAO mixed-liquor fractions the VSS/TSS ratio will differ substantially from the value for the OHO fractions. This is due to the large amount of inorganic polyphosphate stored internally in the PAOs, with associated counterions. The counterions are required to neutralise the negative charges on the polyphosphate, thereby stabilising it. These counterions are principally Mg^{2+} and K^+, and to a lesser extent Ca^{2+} (Fukase *et al.,* 1982; Arvin *et al.,* 1985; Comeau *et al.,* 1986; Wentzel *et al.,* 1989a).

6.8.6.3 FSS sludge mass

The fixed (inorganic) suspended solids (FSS) sludge mass in the system comes from various sources (Ekama and Wentzel, 2004):

- Intracellular components of active biomass contain salts that are left as inorganic residue by combustion at 550 °C. A fraction of 0.15 gFSS/gVSS is considered for OHOs. Nitrifiers have a similar FSS fraction but they can often be neglected as they normally compose less than 2 % of the biomass;
- PAOs contain both the standard 0.15 gFSS/gVSS fraction plus their polyphosphates and cationic counterions that contribute considerably to the FSS content of the PAOs. For aerobic PAOs containing 38% gP/gVSS, an FSS content of 1.30

gFSS/gVSS is reported by Ekama and Wentzel (2004). However, if the PAO cells have not reached their maximum intracellular P storage of 38% gP/gVSS, it can be assumed that the FSS content can be reduced proportionally. This aspect is taken into account by considering the actual P content stored by the active PAO biomass ($f_{P,PAO,act}$) with regard to their maximum P content ($f_{P,PAO}$) in the determination of MX_{FSS} (see Eq. 6.24d).

- Endogenous and inert organic residues are considered not to contain inorganics as the salt content of these components should have been dissolved upon cell lysis.
- Slowly biodegradable particulate organic matter is also considered not to contain inorganics.
- Influent FSS which is accumulated onto the activated sludge.
- Precipitation of minerals and dissolution of FSS are neglected. Should chemical precipitation take place, however, mineral accumulation into the sludge should be considered.

Thus, the FSS sludge mass in the system is given by:

$$MX_{FSS} = f_{FSS,OHO} MX_{OHOv} + f_{FSS,PAO} \frac{f_{P,PAO,act}}{f_{P,PAO}} MX_{PAOv} + Q_i \cdot SRT \cdot X_{FSS,i} \quad (6.24d)$$

where:

MX_{FSS}	mass of fixed suspended solids in the system (gFSS)
$f_{FSS,OHO}$	fraction of FSS in the OHO active biomass = 0.15 gFSS/gVSS (giving a $f_{VT,OHO}$ of 0.87 gVSS/gTSS)
$f_{FSS,PAO}$	fraction of FSS in the PAO active biomass = 1.30 g FSS/gAVSS for aerobic PAOs (giving a $f_{VT,PAO}$ of 0.44 gVSS/gTSS)
$FX_{FSS,i}$	daily mass of influent FSS (gFSS/d)

6.8.6.2 TSS sludge mass and sludge VSS/TSS ratio

The TSS sludge mass in the system is given by the sum of VSS and FSS:

$$MX_{TSS} = MX_{VSS} + MX_{FSS} \quad (6.25a)$$

$$MX_{TSS} = V_R\ X_{TSS} \quad (6.25b)$$

where:

MX_{TSS}	mass of total suspended solids in the system (gTSS)

and the sludge VSS to TSS ratio:

$$f_{VT} = \frac{MX_{VSS}}{MX_{TSS}} \quad (6.25c)$$

where:

f_{VT}	VSS/TSS ratio for the sludge.

6.8.6.4 P content of TSS

The average phosphorus content of the biomass is calculated by considering each mass contributing to the TSS, and in particular, the actual P content stored by the active PAO biomass since it may contribute with the highest P content. The fraction of phosphorus in the fixed suspended solids can vary significantly depending on the presence of aluminium, iron and calcium salts either present in the influent or added to the system for phosphorus precipitation.

$$f_{P,TSS} = \frac{f_P \cdot (MX_{OHOv} + MX_{E,OHOv} + MX_{E,PAOv} + MX_{Uv})}{MX_{TSS}} + \frac{f_{P,PAO,act} \cdot MX_{PAOv}}{MX_{TSS}} + \frac{f_{P,FSS} \cdot MX_{FSS}}{MX_{TSS}} \quad (6.26)$$

where:

$f_{P,TSS}$	P fraction of total suspended solids mass (gP/gTSS)
$f_{P,FSS}$	P fraction of fixed (inorganic) suspended solids mass (gP/gFSS)

= 0.02 gP/gFSS (proposed value; it would need to be corrected should there be a significant presence of salts that coagulate P such as Al, Fe or Ca salts).

6.8.7 Process volume requirements

As set out in Chapter 4, process volume requirements are determined from the mass of sludge in the system and the selected required sludge concentration either as TSS or as VSS:

$$V_R = MX_{TSS} / X_{TSS,OX} \tag{6.27a}$$

where:

V_R	process volume (m^3)
$X_{TSS,OX}$	selected desired TSS concentration in the aerobic reactor ($gTSS/m^3$)

or, alternatively:

$$V_R = MX_{VSS} / X_{VSS,OX} \tag{6.27b}$$

where:

$X_{VSS,OX}$	the selected desired VSS concentration in the aerobic reactor ($gVSS/m^3$)

The process volume requirements (V_R) is the effective volume, *i.e.* the volume that would be required if the sludge was at uniform concentration throughout the system. However, with some nitrifying and denitrifying EBPR system configurations, this is not true and the sludge concentrations differ between the different zones. For example, the sludge concentration in the anaerobic zone of the UCT/MUCT configuration is reduced by the factor s/(1 + s) compared to the other zones (anoxic and aerobic). In these cases the volume must be adjusted to take this into account.

6.8.8 Nitrogen requirements for sludge production

The form of the equation for calculating the nitrogen requirement for sludge production is:

$$FN_s = f_n \, MX_{VSS} / SRT \tag{6.28a}$$

where:

FN_s	daily mass of nitrogen required for sludge production (gN/d)
f_n	nitrogen content of the sludge = 0.10 gN/gVSS

However, for the EBPR system the term MX_{VSS} needs to take account of the changes in VSS constituents, that is, it must be calculated using Eq. 6.24a.

Expressed on the basis of influent concentration, the nitrogen requirement for sludge production is:

$$TKN_{i,s} = FN_s / Q_i \tag{6.28b}$$

6.8.9 Oxygen demand

6.8.9.1 Carbonaceous oxygen demand

The carbonaceous oxygen demand (FO_c) is given by the sum of oxygen demands due to the PAOs and OHOs. From a COD mass balance point of view, any removed COD not converted into biomass or endogenous residue is consumed for energy production. For example, 1 unit of biodegradable COD (COD_b; such as S_{VFA}) removed will produce ($f_{cv} \cdot Y_{PAOv}$) units of X_{PAO} with energy provided by respiring ($1 - f_{cv} \cdot Y_{PAOv}$) of COD_b. The factor f_{cv} (gCOD-active biomass/gVSS active biomass) is used to convert the units of Y_{PAOv} from gVSS-active biomass/gCOD-substrate into gCOD-active biomass/gCOD-substrate. Thus, 1 unit of COD_b equals ($f_{cv} \cdot Y_{PAOv} + 1 - f_{cv} \cdot Y_{PAOv}$) and the COD mass balance is maintained.

Oxygen demand for PAOs

The oxygen demand for PAOs comes from respiration to provide energy for biomass synthesis and for endogenous respiration.

$$FO_{PAO} = FO_{PAOsynthesis} + FO_{PAOendogenous\ respiration} \quad (6.29a)$$

$$FO_{PAO} = (1 - f_{cv}\, Y_{PAOv})\, FS_{S,PAO} + f_{cv}\,(1 - f_{E,PAO})\, b_{PAO,T}\, MX_{PAOv} \quad (6.29b)$$

or, more explicitly as a function of the daily mass of substrate stored by the PAOs:

$$FO_{PAO} = FS_{S,PAO}\left[(1 - f_{cv}\, Y_{PAOv})\right] + \left[f_{cv}(1 - f_{XE,PAO}) b_{PAO,T} \cdot \frac{Y_{PAOv}}{(1 + b_{PAO,T}\, SRT)} SRT \right] \quad (6.29c)$$

where:

FO_{PAO} daily mass of oxygen consumed by PAOs (gO_2/d)

f_{cv} COD/VSS ratio of the sludge (gCOD/gVSS)

Oxygen demand for OHOs

Similarly, for OHOs:

$$FO_{OHO} = FO_{OHOsynthesis} + FO_{OHOendogenous\ respiration} \quad (6.30a)$$

$$FO_{OHO} = (1 - f_{cv}\, Y_{OHOv})\, FCOD_{b,OHO} + f_{cv}\,(1 - f_{E,OHO})\, b_{OHO,T}\, MX_{OHOv} \quad (6.30b)$$

or, more explicitly as a function of the daily mass of substrate stored by the OHOs:

$$FO_{OHO} = FCOD_{b,OHO}\left[(1 - f_{cv}\, Y_{OHOv})\right] + \left[f_{cv}\,(1 - f_{XE,OHO}) \cdot b_{OHO,T} \frac{Y_{OHOv}}{(1 + b_{OHO,T}\, SRT)} SRT \right] \quad (6.30c)$$

where:

FO_{OHO} daily mass of oxygen consumed by OHOs(gO_2/d)

Total oxygen demand

The total carbonaceous oxygen demand (gO_2/d) is:

$$FO_c = FO_{PAO} + FO_{OHO} \quad (6.31a)$$

where:

FO_c daily mass of carbonaceous oxygen demand (gO_2/d)

Now, assuming that $Y_{PAOv} \approx Y_{OHOv}$, that ($FS_{F,PAO}$ + $FCOD_{b,OHO}$) $\approx$ $FCOD_{b,i}$ and that $f_{XE,PAO}$ (0.20) $\approx$ $f_{XE,OHO}$ (0.25), Eq. 6.31b may be simplified (gO_2/d):

$$FO_c = (1 - f_{cv}\, Y_{OHOv})\, FCOD_{b,i} + f_{cv}\,(1 - f_{XE,OHO}) \cdot (b_{PAO,T}\, MX_{PAOv} + b_{OHO,T}\, MX_{OHOv}) \quad (6.31b)$$

6.8.9.2 Nitrification oxygen demand

Taking due account of the change in nitrogen requirements for sludge production (FN_S) and nitrification capacity (NIT_c), the nitrification oxygen demand FO_{NIT} is given in Chapter 5.

6.8.9.3 Total oxygen demand

For a non-nitrifying EBPR system the total oxygen demand FO_t is given by FO_c, while for a nitrifying EBPR system, FO_t is given by the sum of FO_c and FO_{NIT}. Including nitrification in the EBPR system necessarily means that denitrification must also be included; the effect of nitrification and denitrification on the total oxygen demand will be considered later.

$$FO_t = FO_c + FO_{NIT} \quad (6.31c)$$

where:

FO_t daily mass of total oxygen demand (gO_2/d).

6.9 DESIGN EXAMPLE

6.9.1 Steady-state design procedure

The procedure for the steady-state design of an EPBR process is shown in Figure 6.29. First, the wastewater needs to be characterized in terms of its flow rate and daily fluxes of COD, nitrogen, phosphorus, inorganic solids and oxygen concentration. A treatment configuration is selected which is operated at a given SRT, temperature and with appropriate kinetic and stoichiometric constants. Then, the influent RBCOD is divided between PAOs and OHOs which allows the calculation of their biomass (and endogenous residue) production as VSS and the phosphorus-removal capacity of the system. From total VSS and TSS estimation, the bioreactor process volume can be calculated as well as the nitrogen and oxygen requirements. Finally, a calculation check can be made with the COD mass balance.

6.9.2 Information provided

The raw wastewater (without primary settling) to be treated has a similar composition to that presented in chapters 4 and 5 on organic matter and nitrogen removal, respectively. The influent composition COD fractions are summarized in Tables 6.2 and 6.3. A flow rate of 15 MLD is selected for ease of transposition. The total influent COD is 750 g/m^3 and the influent total phosphorus is 17 g/m^3. The fractionation of the influent COD is illustrated in Figure 6.30. The kinetic and stoichiometric parameters are presented in Table 6.4.

The EBPR process selected (Table 6.5) is a Johannesburg configuration which is operated at 14 °C, with 2 anaerobic zones, an SRT of 20 days, an anaerobic mass fraction of 0.10, a sludge recycle ratio of 0.75 with respect to the influent flow, an aerobic to anoxic recycle ratio of 1.5, a sludge recycle entering the anaerobic zone containing no dissolved oxygen but 0.5 gNO$_3$-N/m^3, a total suspended solids in the effluent of 5 mg/l, and a design aerobic mixed-liquor solids concentration of 4,000 gTSS/m^3.

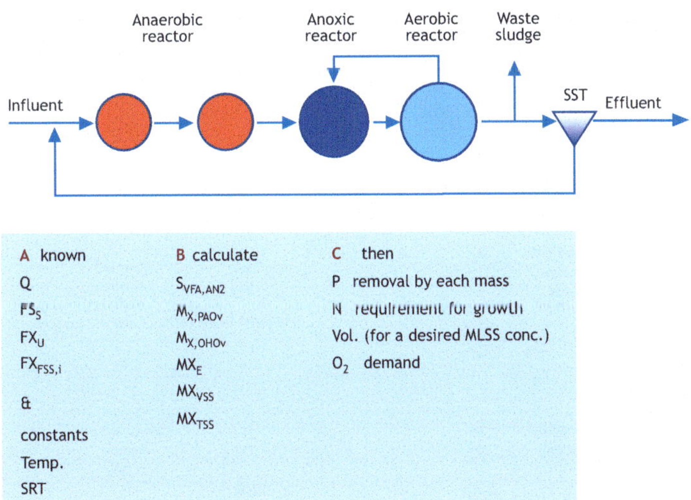

Figure 6.29 Design procedure overview for the EBPR system. A Johannesburg configuration is illustrated. The anaerobic reactor is divided into two cells (not illustrated).

Table 6.2 Influent characteristics for the EBPR design example (raw wastewater).

Description	Symbol	Value	Units	Calculations
Flow rate	Q_i	15	MLD	
Total COD	COD_i	750	$gCOD/m^3$	
COD concentrations				
- readily biodegradable COD	$S_{S,i}$	146	$gCOD/m^3$	$= 750 \cdot 0.195$
- volatile fatty acids	$S_{VFA,i}$	22	$gCOD/m^3$	$= 146 \cdot 0.15$
- fermentable COD	$S_{F,i}$	124	$gCOD/m^3$	$= 146 - 22$
- slowly biodegradable COD	$X_{S,i}$	439	$gCOD/m^3$	$= 750 \cdot (1 - 0.195 - 0.07 - 0.15)$
- inert soluble COD	$S_{U,i}$	53	$gCOD/m^3$	$= 750 \cdot 0.07$
- inert particulate COD	$X_{U,i}$	113	$gCOD/m^3$	$= 750 \cdot 0.15$
Nitrate	$S_{NO3,i}$	0	gN/m^3	
Dissolved O_2	$S_{O2,i}$	0	gO_2/m^3	
Total P	P_i	17.0	gP/m^3	
Fixed (inorganic) SS	$X_{FSS,i}$	49	$gFSS/m^3$	
P fraction of influent FSS	$f_{P,FSS,i}$	0.02	gP/gFSS	
Alkalinity	S_{Alk}	250	$gCaCO_3/m^3$	

Table 6.3 COD fractions of raw wastewater for the EBPR design example.

Description	Symbol	COD fractions	Units
Type of WW		Raw	
COD fractions			
Fraction of RBCOD	$f_{SS,CODi}$	0.195	g/gTCOD
S_{VFA} fraction of RBCOD	$f_{SVFA,SSi}$	0.15	$g/gCOD_{SS}$
Fraction of soluble unbiodegradable COD	$f_{SU,CODi}$	0.07	g/gTCOD
Fraction of particulate unbiodegradable COD	$f_{XU,CODi}$	0.15	g/gTCOD

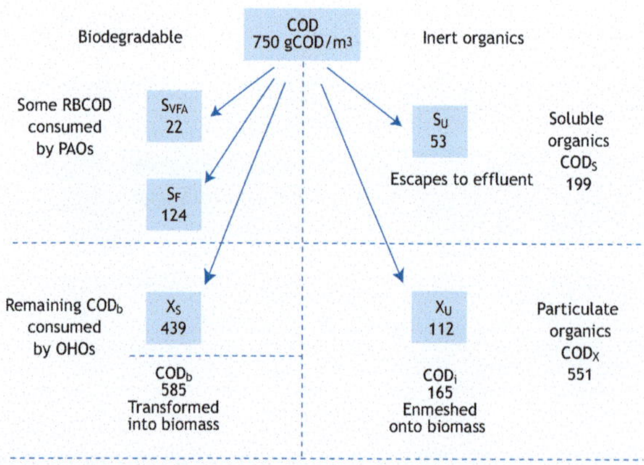

Figure 6.30 Influent COD fractionation for the EBPR design example.

Table 6.4 Kinetic and stoichiometric parameters for the EBPR design example.

	Parameter	Symbol	Value	Units
Kinetic	*OHO*			
	First-order fermentation rate constant at T = 20 °C	$k_{F,20}$	0.06	m^3/gVSS.d
	Temperature coefficient for $k_{F,T}$	$\theta_{k,F}$	1.029	
	First-order fermentation rate constant at temperature T[(a)]	$k_{F,T}$	0.051	m^3/gVSS.d
	Specific endogenous mass loss rate of the OHOs at 20 °C	$b_{OHO,20}$	0.24	gVSS/gVSS.d
	Temperature coefficient for $b_{OHO,T}$	$\theta_{b,OHO}$	1.029	
	Specific endogenous mass loss rate of the OHOs at temperature T	$b_{OHO,T}$	0.202	gVSS/gVSS.d
	PAO			
	PAO-specific endogenous mass loss rate constant at T = 20 °C	$b_{PAO,20}$	0.04	gVSS/gVSS.d
	Temperature coefficient for $b_{PAO,T}$	$\theta_{b,PAO}$	1.029	
	PAO-specific endogenous mass loss rate constant at temperature T	$b_{PAO,T}$	0.034	gVSS/gVSS.d
Stoichiometric	*OHO*			
	Biomass yield of OHOs	Y_{PAOv}	0.45	gVSS/gCOD
	Fraction of endogenous residue of the OHOs	$f_{XE,OHO}$	0.20	gVSS/gVSS
	Fraction of P in the active OHO mass	$f_{P,OHO}$	0.03	gP/gVSS
	Fraction of P in the endogenous mass (OHO and PAO)	f_p	0.03	gP/gVSS
	Fraction of fixed (inorganic) suspended solids of OHOs	$f_{FSS,OHO}$	0.15	gFSS/gVSS
	PAO			
	Biomass yield of PAOs	Y_{PAOv}	0.45	gVSS/gCOD
	Fraction of endogenous residue of the PAOs	$f_{XE,PAO}$	0.25	gVSS/gVSS
	Fraction of P in the active PAO mass	$f_{P,PAO}$	0.38	gP/gVSS
	Fraction of P in the endogenous mass (OHO and PAO)	f_p	0.03	gP/gVSS
	VSS/TSS ratio for PAO active mass	$f_{VT,PAO}$	0.46[(b)]	gVSS/gTSS
	Ratio of P release/VFA uptake	$f_{PO4,REL}$	0.50	gP/gCOD
	Fraction of fixed (inorganic) suspended solids of PAOs	$f_{FSS,PAO}$	1.30	gFSS/gVSS
	Inerts			
	Fraction of P in the inert mass	f_p	0.03	gP/gIVSS
	General			
	COD/VSS ratio of the sludge	f_{cv}	1.48	gCOD/gVSS
	VSS/TSS ratio for OHO active and endogenous masses PAO endogenous mass, and inert mass	f_{VT}	0.80[(b)]	gVSS/gTSS
	Nitrogen content of active biomass	f_n	0.10	gN/gVSS

[(a)] $k_T = k_{20} \cdot \theta^{(T-20)}$; example: $k_{F,14} = 0.060 \cdot 1.029^{(14-20)} = 0.051$
[(b)] These values are not required if the FSS is calculated from Eq. 6.24c

Table 6.5 Biological system characteristics for the EBPR design example (Johannesburg configuration).

Description	Symbol	Value	Units
Temperature	T	14	°C
Number of anaerobic zones	n	2	reactors
Sludge retention time	SRT	20	d
Anaerobic mass fraction	f_{xa}	0.10	gVSS/gVSS
Sludge recycle ratio based on influent flow	s	0.75	$m^3.d/m^3.d$
Aerobic to anoxic recycle ratio	a	1.5	$m^3.d/m^3.d$
Dissolved O_2 in the sludge recycle	$S_{O2,s}$	0	gO_2/m^3
Nitrate concentration in the sludge recycle	$S_{NO3,s}$	0.5	$gNO_3\text{-}N/m^3$
Total suspended solids in the effluent	TSS_e	5	$gTSS/m^3$
Design aerobic TSS concentration	$X_{TSS,OX}$	4,000	$gTSS/m^3$

6.9.3 Calculations

Following the same procedure as presented in Section 6.6, the detailed calculations are shown in Table 6.6 over the following pages. Each step is presented with symbols, values, units, symbol definition, equations used to calculate a given parameter, and the detailed calculation with the numerical values for each parameter. At the end, a COD mass balance is made as a validation check of the calculations.

Note that in step 3.2, the fermentable COD leaving in the effluent of the last anaerobic reactor is calculated by iteration.

Table 6.6 Detailed calculations for the EBPR design example.

1. System configuration			
Johannesburg configuration operated at 14 °C			
2. Influent and sludge recycle composition (from previous tables)			
Q_i	15	MLD	influent flow rate
2.1 Influent concentrations			
Influent and bioreactor data			
COD_i	750	$gCOD/m^3$	influent concentration of total COD
$S_{S,i}$	146	$gCOD/m^3$	influent concentration of RBCOD
$S_{VFA,i}$	22	$gCOD/m^3$	influent concentration of VFAs
$S_{F,i}$	124	$gCOD/m^3$	influent concentration of fermentable COD
$X_{S,i}$	439	$gCOD/m^3$	influent concentration of slowly biodegradable COD
$COD_{b,i}$	585	$gCOD/m^3$	influent concentration of biodegradable COD ($S_{S,i}$ + $X_{S,i}$)
$S_{U,i}$	53	$gCOD/m^3$	influent concentration of soluble inert COD
$X_{U,i}$	113	$gCOD/m^3$	influent concentration of particulate inert COD
$S_{NO3,i}$	0	$gNO_3\text{-}N/m^3$	influent concentration of nitrate
$S_{O2,i}$	0	gO_2/m^3	influent concentration of dissolved oxygen

$X_{FSS,i}$	49	$gFSS/m^3$	influent concentration of fixed (inorganic) suspended solids
P_i	17	gP/m^3	influent concentration of total P

2.2 Influent fluxes used for calculations (= Qi · influent concentration of component)

$FCOD_i$	11,250	kgCOD/d	influent daily flux of total COD
$FS_{S,i}$	2,190	kgCOD/d	influent daily flux of RBCOD
$FS_{VFA,i}$	330	kgCOD/d	influent daily flux of VFAs
$FS_{F,i}$	1,860	kgCOD/d	influent daily flux of fermentable COD
$FCOD_{b,i}$	8,770	kgCOD/d	influent daily flux of biodegradable COD ($S_{S,i} + X_{S,i}$)
$FX_{U,i}$	1,688	kgCOD/d	influent daily flux of particulate unbiodegradable COD
$FS_{U,i}$	795	kgCOD/d	influent daily flux of soluble unbiodegradable COD
$FX_{FSS,i}$	735	kgFSS/d	influent daily flux of fixed (inorganic) suspended solids

2.3 Sludge recycle characteristics

s	0.75	$m^3.d/\ m^3.d$	sludge recycle ratio based on influent flow
$S_{O2,s}$	0	gO_2/m^3	dissolved O_2 in the sludge recycle
$S_{NO3,s}$	0.5	$gNO_3\text{-}N/m^3$	nitrate concentration in the sludge recycle

3. Division of $S_{S,i}$ between PAOs and OHOs

3.1 Fermentable COD available for conversion into VFAs after denitrification reactor (and O_2 consumption) in AN reactor (in units of $gCOD/m^3$ of influent)

$S_{F,i,conv}$ $= S_{F,i} - 8.6 \cdot (s \cdot S_{NO3,s} + S_{NO3,i}) - 3 \cdot (s \cdot S_{O2,s} + S_{O2,i})$

$= S_{F,i} -$ COD for denitrification - COD for D.O.

$= 124 - 8.6 \cdot (0.75 \cdot 0.5 + 0) - 3 \cdot (0.75 \cdot 0 + 0)$

COD for denit.	3.2	$gCOD/m^3$
COD for D.O.	0.0	$gCOD/m^3$
$S_{F,i,conv}$	121	$gCOD/m^3$

3.2 Fermentable COD lost in the effluent of the last anaerobic reactor

N 2 the 2nd AN reactor

calculation done by iterations

a- suppose a seed1 $S_{F,ANn}$ value of 0. This value is used to calculate MX_{OHOv}

b- type the calculated MX_{OHOv} calculated value as seed2 value

c- repeat steps a and b until the seed2 $S_{F,ANn}$ equals the calculated $S_{F,ANn}$

$S_{F,ANn}$ $= S_{F,i,conv} / (1+s) / (1 + (k_{F,T} \cdot (f_{xa} \cdot MX_{OHOv} / (N \cdot Q_i \cdot (1 + s)))))^n$

$= 121 / (1 + 0.75) / (1 + (0.051 \cdot (0.10 \cdot 12{,}500 / (2 \cdot 15 \cdot (1 + 0.75)))))^2$

seed1:

$S_{F,ANn}$	14.2	14.2	$gCOD/m^3$
	↓	↑	
		seed2:	
MX_{OHOv}	12,490	12,490	kgCOD

$= Y_{OHOv} / (1 + b_{OHO,T} \cdot SRT) \cdot FCOD_{b,OHO} \cdot SRT$ (note that $FCOD_{b,OHO}$ is calculated in step 3.4)

$= 0.45 / (1 + 0.202 \cdot 20) \cdot 7{,}000 \cdot 20$

3.3 VFAs stored by PAOs

$FS_{S,PAO}$	$= Q_i \cdot (S_{F,i,conv} - (1 + s) \cdot S_{F,ANn}) + Q_i \cdot S_{VFA,i}$	
	$= 1 \cdot (121 - (1 + 0.75) \cdot 14.3) + 1 \cdot 22$	
$FS_{S,PAO}$	1,770	kgCOD/d

3.4 Remaining biodegradable COD available to OHOs

$FCOD_{b,OHO}$	$= FCOD_{b,i} - FS_{S,PAO}$	
	$= 8{,}770 - 1{,}770$	
$FCOD_{b,OHO}$	7,000	kgCOD/d

4. Biomass (VSS) equations

Corresponds to the biological mass present in the system as synthesized from the influent COD (in g/d) taking into account the cumulative effect of SRT [(g/d) · d = g in the system]

4.1 PAOs

Active mass

Y_{PAOv}	0.45	gVSS/gCOD	
$Y_{PAOv,obs}$	$= Y_{PAOv} / (1 + b_{PAO,T} \cdot SRT)$		
	$= 0.45 / (1 + 0.034 \cdot 20)$		
$Y_{PAO,vobs}$	0.269 gVSS / gCOD		
MX_{PAOv}	$= Y_{PAOv,obs} \cdot FS_{S,PAO} \cdot SRT$		
	$= 0.269 \cdot 1{,}770 \cdot 20$		
MX_{PAOv}	9,511	kgVSS in the system	

Endogenous mass

$MX_{E,PAOv}$	$= f_{XE,PAO} \cdot b_{PAO,T} \cdot MX_{PAOv} \cdot SRT$	
	$= 0.25 \cdot 0.0337 \cdot 9511 \cdot 20$	
$MX_{E,PAOv}$	1,603	kgVSS

4.2 OHOs

Active mass

Y_{OHOv}	0.45	gVSS/gCOD	
$Y_{OHOv,obs}$	$= Y_{OHOv} / (1 + b_{OHO,T} \cdot SRT)$		
	$= 0.45 / (1 + 0.202 \cdot 20)$		
$Y_{OHOv,obs}$	0.089	gVSS/gCOD	
MX_{OHOv}	$= Y_{OHOv,obs} \cdot FCOD_{b,OHO} \cdot SRT$		
	$= 0.089 \cdot 7{,}000 \cdot 20$		
MX_{OHOv}	12,490	kgVSS	(this value is the calculated MX_{OHOv} value of step 3.2)

Endogenous mass

$MX_{E,OHOv}$	$= f_{XE,OHO} \cdot b_{OHO,T} \cdot MX_{OHOv} \cdot SRT$	
	$= 0.20 \cdot 0.202 \cdot 12{,}490 \cdot 20$	
$MX_{E,OHOv}$	10,100	kgVSS

4.3 Inert mass

MX_{Uv}	$= f_{XU,COD,i} \cdot FCOD_i \cdot SRT / f_{cv}$		
	$= 0.15 \cdot 11{,}250 \cdot 20 / 1.48$		
MX_{Uv}	22,804	kgVSS	

5. P removal

5.0 P release

$S_{PO4,rel}$	$= f_{PO4,rel} \cdot FS_{S,PAO} / Q_i$		
	$= 0.5 \cdot 1{,}770 / 15$		
$S_{PO4,rel}$	59	gP/m^3	gP/m^3 of influent, not gP/m^3 of AN reactor

5.1 ΔP by PAOs

ΔP_{PAO}	$= f_{P,PAO} \cdot MX_{PAOv} / (SRT \cdot Q_i)$	
	$= 0.38 \cdot 9{,}511 / (20 \cdot 15)$	
ΔP_{PAO}	12.05	gP/m^3

5.2 ΔP by OHOs

ΔP_{OHO}	$= f_p \cdot MX_{OHOv} / (SRT \cdot Q_i)$	
	$= 0.03 \cdot 12{,}500 / (20 \cdot 15)$	
ΔP_{OHO}	1.25	gP/m^3

5.3 ΔP by endogenous mass

ΔP_{XE}	$= \Delta P_{XE,PAO} + \Delta P_{XE,OHO}$	
$\Delta P_{XE,PAO}$	$= f_p \cdot MX_{E,PAOv} / (SRT \cdot Q_i)$	
	$= 0.03 \cdot 1{,}603 / (20 \cdot 15)$	
$\Delta P_{XE,PAO}$	0.16	gP/m^3
$\Delta P_{XE,OHO}$	$= f_p \cdot MX_{E,OHOv} / (SRT \cdot Q_i)$	
	$= 0.03 \cdot 10{,}100 / (20 \cdot 15)$	
$\Delta P_{XE,OHO}$	1.01	gP/m^3
ΔP_{XE}	1.17	gP/m^3

5.4 ΔP by influent inert mass

ΔP_{XU}	$= f_p \cdot MX_{Uv} / (SRT \cdot Q_i)$	
	$= 0.03 \cdot 22{,}804 / (20 \cdot 15)$	
ΔP_{XU}	2.28	gP/m^3

5.5 ΔP by chemical P precipitation due to salts present in the influent or added to the system

Not considered

5.6 Potential total P removal

$\Delta P_{SYS,pot}$	$= \Delta P_{PAO} + \Delta P_{OHO} + \Delta P_{XE} + \Delta P_{XU}$	
	$= 12.05 + 1.25 + 1.17 + 2.28$	
$\Delta P_{SYS,pot}$	16.8	gP/m^3

5.7 Actual total P removal

P_i	17.0	gP/m^3
$\Delta P_{SYS,actual}$	$= \min(\Delta P_{SYS,pot}; P_i) = \min(16.8; 17.0)$	
$\Delta P_{SYS,actual}$	16.8	gP/m^3

5.8 Particulate P in the effluent

To calculate after step 6.5 where the P content of TSS is calculated

$X_{P,e}$	$= f_{P,TSS} \cdot TSS_e$	
	$= 0.124 \cdot 5$	
$X_{P,e}$	0.6	gP/m^3

5.9 Effluent total P

P_e	$= P_i - \Delta P_{SYS,actual} + X_{P,e}$	
	$= 17.0 - 16.8 + 0.6$	
P_e	0.8	gP/m^3

6. VSS and TSS

6.1 VSS and active fraction

MX_{bio}	$= MX_{PAOv} + MX_{OHOv}$	
	$= 9{,}511 + 12{,}490$	
MX_{bio}	22,000	kgVSS
MX_{VSS}	$= MX_{PAOv} + MX_{OHOv} + MX_{E,PAOv} + MX_{E,OHOv} + MX_{Uv}$	
	$= 9{,}511 + 12{,}490 + 1{,}603 + 10{,}100 + 22{,}804$	
MX_{VSS}	56,506	kgVSS
$f_{bio,VSS}$	$= MX_{bio} / MX_{VSS}$	
	$= 22{,}000 / 56{,}506$	
$f_{bio,VSS}$	39%	

6.2 FSS

$f_{PAO,act}$	$= [(Q_i \cdot SRT \cdot \Delta P_{SYS,actual}) - (f_p \cdot (MX_{VSS} - MX_{PAOv})] / MX_{PAOv}$	
	$= [(15 \cdot 20 \cdot 16.8) - (0.03 \cdot (56{,}506 - 9{,}511)] / 9{,}511$	
$f_{PAO,act}$	0.38	gP/gVSS
MX_{FSS}	$= f_{FSS,OHO} \cdot MX_{OHOv} + f_{FSS,PAO} \cdot (f_{P,PAO,act}/f_{P,PAO}) \cdot MX_{PAOv} + FX_{FSS,i} \cdot SRT$	
	$= 0.15 \cdot 12{,}490 + 1.3 \cdot (0.38/0.38) \cdot 9{,}511 + 735 \cdot 20$	
MX_{FSS}	28,938	kgFSS

6.3 TSS

MX_{TSS}	$= MX_{VSS} + MX_{FSS}$	
	$= 56{,}506 + 28{,}938$	
MX_{TSS}	85,443	kgTSS

6.4 f_{VT}

f_{VT}	$= MX_{VSS} / MX_{TSS}$	
	$= 56{,}506 / 85{,}443$	
f_{VT}	0.66	gVSS/gTSS

6.5 P content of TSS

$f_{P,TSS}$ $= [(f_{P,OHO} \cdot MX_{OHOv} + f_p \cdot (MX_{E,OHOv} + MX_{E,PAOv}) + f_p \cdot MX_{Uv}) / f_{VT}$
$+ (f_{P,PAO} \cdot MX_{PAOv}) / f_{VT,PAO} + f_{P,FSS,i} \cdot MX_{FSS})] / MX_{TSS}$
$= [(0.03 \cdot 12{,}489 + 0.03 \cdot (10{,}109 + 1{,}603) + 0.03 \cdot 22{,}804) / 0.66 + (0.38 \cdot 9{,}517) / 0.46 + 0.02 \cdot 28{,}947)]/ 85{,}443$

$f_{P,TSS}$ 0.124 gP/gTSS

7. Process volume (based on TSS; may also be based on VSS)

Note that the influent flow rate needs to be appropriate

$X_{TSS,OX}$ 4,000 gTSS / m^3

V_R $= MX_{TSS} / X_{TSS,OX}$
$= 85{,}443/ 4{,}000$

V_R 21,361 m^3

The volume of the anaerobic zone (divided in two sections) depends on the anaerobic mass fraction.

$V_{R,AN}$ $= f_{xa}\ V_R$
$= 0.10 \cdot 21{,}361$

$V_{R,AN}$ 2,136 m^3

The anoxic and aerobic mass fractions, and thus the volume of these zones, should be estimated according to the procedure presented in Chapter 5 on nitrogen removal and in Ramphao *et al.* (2005) where the equations that relate the volume fractions to the mass fractions according to recycle ratios are given for various types of reactor configurations, including the JHB. Using an estimate of an aerobic and a total anoxic mass fraction of 0.45 each, the volume (m^3) for each zone would be approximately: AN1: 1,060, AN2: 1,060, AX: 7,000, OX: 10,500, AX-RAS: 1,750, for a total volume of 21,370 m^3. Note that this preliminary approximation does not take into consideration that sludge concentration in the RAS-anoxic zone is 2.3 times more concentrated than in the mainstream zones ((1+r)/r) which results in approximately one third of the anoxic mass being in the RAS-anoxic zone and a lower total process volume requirement.

8. Nitrogen requirement

FN_s $= f_n \cdot MX_{VSS} / SRT$
$= 0.10 \cdot 56{,}506 / 20$

FN_s 283 kgN/d

$TKN_{i,s}$ $= FN_s / Q_i$
$= 283 / 15$

$TKN_{i,s}$ 18.8 gN/m^3

9. Oxygen demand (OD)

OD by PAOs: for synthesis and endogenous respiration

FO_{PAO} $= FO_{PAO,s} + FO_{PAO,endo}$

$FO_{PAO,s}$ $= FS_{S,PAO} \cdot (1 - f_{cv} \cdot Y_{PAOv})$
$= 1{,}770 \cdot (1 - 1.48 \cdot 0.45)$

$FO_{PAO,s}$ 591 kgO_2/d

$FO_{PAO,endo}$ $= FS_{S,PAO} \cdot f_{cv} \cdot (1 - f_{XE,PAO}) \cdot b_{PAO,T} \cdot Y_{PAOv,obs} \cdot SRT)$
$= 1{,}770 \cdot 1.48 \cdot (1 - 0.25) \cdot 0.0337 \cdot 0.268 \cdot 20$

$FO_{PAO,endo}$ 356 kgO_2/d

FO_{PAO} 947 kgO_2/d

OD by OHOs: for synthesis and endogenous respiration

FO_{OHO}	$= FO_{OHO,s} + FO_{OHO,endo}$	
$FO_{OHO,s}$	$= FCOD_{b,OHO} \cdot (1 - f_{cv} \cdot Y_{OHOv})$	
	$= 7{,}000 \cdot (1 - 1.48 \cdot 0.45)$	
$FO_{OHO,s}$	2,338	
$FO_{OHO,endo}$	$= FCOD_{b,OHO} \cdot f_{cv} \cdot (1 - f_{XE,OHO}) \cdot b_{OHO,T} \cdot Y_{OHOv,obs} \cdot SRT$	
	$= 7{,}000 \cdot 1.48 \cdot (1 - 0.20) \cdot 0.202 \cdot 0.0892 \cdot 20$	
$FO_{OHO,endo}$	2,990	kgO_2/d
FO_{OHO}	5,327	kgO_2/d

OD total (carbonaceous)

FO_c	$= FO_{PAO} + FO_{OHO}$	
	$= 947 + 5{,}327$	
FO_c	6,274	kgO_2/d

or in a simplified form:

FO_c	$= (1 - f_{cv} \cdot Y_{OHOv}) \cdot FCOD_{b,i} + f_{cv} \cdot (1 - f_{XE,OHO}) \cdot (b_{PAO,T} \cdot MX_{PAOv} + b_{OHO,T} \cdot MX_{OHOv})$	
	$= (1 - 1.48 \cdot 0.45) \cdot 8{,}770 + 1.48 \cdot (1 - 0.20) \cdot (0.0337 \cdot 9{,}511 + 0.202 \cdot 12{,}490)$	
FO_c	6,288	kgO_2/d

COD mass balance verification					
Input					
$FCOD_i$		11,250	kgCOD/d	100%	IN
Output					
O_2 demand for synthesis and endogenous respiration					
FO_c		6,274	kgCOD/d	55.8%	
Soluble inerts leaving by the effluent					
$FS_{U,i}$		795	kgCOD/d	7.1%	
Sludge	gVSS	gCOD/d			
		(= gVSS · f_{cv} / SRT = gVSS · 1.48 / 20 = gVSS · 0.0740)			
MX_{PAOv}	9,511	704	kgCOD/d	6.3%	
MX_{OHOv}	12,490	925	kgCOD/d	8.2%	
MX_{bio}	22,000	1,628		14.5%	
$MX_{E,PAOv}$	1,603	119	kgCOD/d	1.1%	
$MX_{E,OHOv}$	10,100	747	kgCOD/d	6.6%	
MX_{Uv}	22,804	1,688	kgCOD/d	15.0%	
$MX_{endo+inert}$	34,506	2,553		22.7%	
MX_{VSS}	56,506	4,181	kgCOD/d	37.2%	
	Sum:	11,250	kgCOD/d	100%	OUT
	Delta (OUT-IN):	0	kgCOD/d	0%	

The 100% mass balance for COD indicates that all the influent COD is accounted for in the calculated values of oxygen demand and sludge production. From the COD mass balance, and for the conditions of the design example, the fate of the influent COD is as follows: 56% is oxidized with oxygen, 7% escapes in the effluent as soluble unbiodegradable organics and 37% becomes activated sludge. The sludge is composed of 39% (1,628/4,181) active biomass and 61% (2,553/4,181) inactive particulate matter of which 40% (1,688/4,181) are influent inerts and 21% ((119+747)/4,181) endogenous residue) on a COD basis. A summary of the EBPR system design results is presented in Table 6.7.

Table 6.7 Summary of EBPR system design results (Johannesburg configuration).

Description	Parameter	Units	Value
1. Influent and bioreactor			
Type of wastewater	raw/settled		raw
Temperature	T	ºC	14
Influent flow rate	Q_i	MLD	15
Influent total COD	COD_i	$gCOD/m^3$	750
Influent rapidly biodegradable COD	$S_{S,i}$	$gCOD/m^3$	146
Influent biodegradable COD	$COD_{b,i}$	$gCOD/m^3$	585
Influent total P	P_i	gP/m^3	17
Sludge retention time	SRT	d	20
Sludge recycle ratio	s	$m^3.d /m^3.d$	0.75
Aerobic recycle ratio	a	$m^3.d /m^3.d$	1.5
Nitrate concentration in sludge recycle	$S_{NO3,s}$	gN/m^3	0.5
2. Portion of $S_{S,i}$ for PAOs and of $COD_{b,i}$ for OHOs			
Concentration of fermentable COD in the last AN reactor	$S_{F,ANn}$	$gCOD/m^3$	14.3
Flux of $S_{S,i}$ for PAOs	$FS_{S,PAO}$	kgCOD/d	1,770
Flux of $COD_{b,i}$ for OHOs	$FCOD_{b,OHO}$	kgCOD/d	7,000
3. System biomass (VSS) equations			
Mass of PAOs	MX_{PAOv}	kgVSS	9,511
Mass of endogenous residue from PAOs	$MX_{E,PAOv}$	kgVSS	1,603
Mass of OHOs	MX_{OHOv}	kgVSS	12,490
Mass of endogenous residue from OHOs	$MX_{E,OHOv}$	kgVSS	10,110
Mass of unbiodegradable organics from influent	MX_{Uv}	kgVSS	22,804
4. P removal			
PO_4 release	S_{PO4_rel}	gP/m^3	59.0
P removal by PAOs	ΔP_{PAO}	gP/m^3	12.1
P removal by OHOs	ΔP_{OHO}	gP/m^3	1.3
P removal by endogenous residue	ΔP_{XE}	gP/m^3	1.2
P removal by X_U	ΔP_{XU}	gP/m^3	2.3

Potential P removal by system	$\Delta P_{SYS,pot}$	gP/m^3	16.8
Actual P removal by system	$\Delta P_{SYS,actual}$	gP/m^3	16.8
Effluent particulate P (from TSS_e)	$X_{P,e}$	gP/m^3	0.6
Influent total P	P_i	gP/m^3	17.0
Effluent total P	P_e	gP/m^3	0.9
5. Volatile and total suspended solids (VSS and TSS) in system			
Mass of active biomass	MX_{bio}	kgVSS	22,000
Mass of VSS	MX_{VSS}	kgVSS	56,506
Ratio of AVSS/VSS	$f_{bio,VSS}$	gAVSS/gVSS	0
Mass of fixed SS	MX_{FSS}	kgFSS	28,938
Mass of TSS	MX_{TSS}	kgTSS	85,443
Ratio of VSS/TSS	f_{VT}	gVSS/gTSS	0.66
Fraction of P in TSS	$f_{P,TSS}$	gP/gTSS	0.12
6. Bioreactor total volume			
Bioreactor volume	V_R	m^3	21,361
7. N requirement			
N requirement for synthesis	$TKN_{i,s}$	kgN/d	18.8
8. Oxygen demand			
Flux of O_2 demand by PAOs	FO_{PAO}	kgO_2/d	947
Flux of O2 demand by PAOs	FO_{OHO}	kgO_2/d	5,327
Flux of carbonaceous O_2 demand	FO_c	kgO_2/d	6,274
COD output/COD input	COD mass balance	gCOD/gCOD	100 %

Flow rate is in m^3/d and mass fluxes in g/d
For a flow rate 1,000 or greater mass fluxes can be read in kg/d

6.10 INFLUENCE OF OPERATIONAL FACTORS ON FULL-SCALE EBPR WWTP

6.10.1 Influence on volatile and total suspended solids and oxygen demand

The model for EBPR systems presented above enables the volatile suspended solids (VSS) and total suspended solids (TSS) of the mixed liquor (eqs. 6.23 and 6.24, respectively) and the carbonaceous oxygen demand (Eq. 6.31) to be calculated. A comparison of the mass of VSS and TSS generated and carbonaceous oxygen demand with and without EBPR per kg COD load on the bioreactor *versus* sludge age are shown in Figure 6.31 and Figure 6.32 for raw and settled wastewaters, respectively, with characteristics as shown. These characteristics were an EBPR system with two anaerobic reactors in-series with a total anaerobic mass fraction (f_{xa}) of 15% and no nitrate recycled to the anaerobic reactor operated at 20°C. From this comparison it appears that including EBPR in the activated sludge system increases the VSS only slightly, by about 5 to 12% and 15-25% for raw and settled wastewaters, respectively (depending on sludge age). This increase in VSS is due to the lower endogenous mass loss/death rate of the PAOs (0.04 d^{-1} at 20 °C) compared to the OHOs (0.24 d^{-1} at 20 °C). However, the TSS is increased substantially, by about 20 to 25% and 45 to 55% for raw and settled wastewaters, respectively (depending on sludge age). This higher TSS production is due to the large quantities of stored inorganic polyphosphate and the associated inorganic cations necessary to stabilize the polyphosphate chains - principally Mg^{2+} and K^{+} (Fukase *et al.,* 1982; Arvin *et al.,* 1985; Comeau *et al.,* 1986; Wentzel *et al.,* 1989a; Ekama and Wentzel, 2004). The high inorganic content of the PAO biomass causes the VSS/TSS to be much lower than that of the OHOs, 0.46 mgVSS/mgTSS compared to 0.75 to 0.85 mgVSS/mgTSS. Thus, the higher the PAO fraction of the mixed liquor, the higher the EBPR and the lower the VSS/TSS ratio of the mixed liquor.

The increase in TSS with the inclusion of EBPR needs to be taken into account in the design of the bioreactor volume (Eq. 6.27) and daily sludge production. Also, since the inorganic cations that stabilize the polyphosphate are derived from the influent wastewater, there must be sufficient concentrations of these cations in the influent; otherwise the EBPR may be adversely affected (Wentzel *et al.,* 1988; Lindrea *et al.,* 1994). Further, because the VSS mass generated per kg COD load is greater with EBPR than without, the oxygen demand with EBPR is correspondingly reduced, by approximately 5-6% and 8-9% for raw and settled wastewaters, respectively (depending on sludge age, Figure 6.32).

Although there is only a small difference in VSS production between an EBPR and a non-EBPR system, the constituent sludge fractions for the two systems differ markedly. This can be readily demonstrated by comparing the percentage composition of the VSS mass generated in systems exhibiting EBPR to those that are not: to illustrate, percentage composition of the VSS mass are shown in Figure 6.33 for systems at 20 °C with no EBPR and with EBPR, respectively treating wastewater with characteristics as shown. Note that the EBPR system has a smaller OHO active mass than the non-EBPR system, but that the EBPR system has a significant concentration of PAO biological active mass.

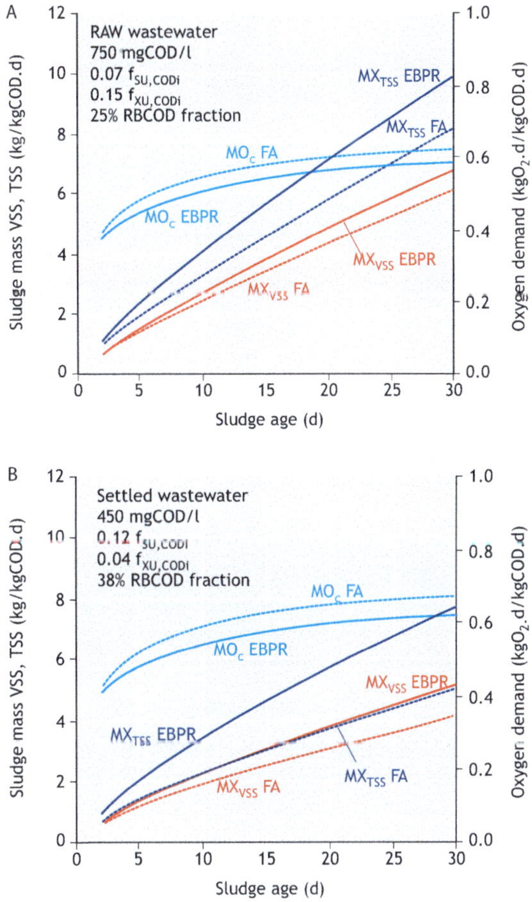

Figure 6.31 and 6.32 Approximate masses of volatile solids (MX_{VSS}) and total solids (MX_{TSS}) and daily carbonaceous oxygen demand (MO_c) per kg COD load on the biological reactor in fully aerobic (FA) and enhanced biological P-removal activated sludge systems treating (A, Fig 6.31) raw and (B, Fig 6.32) settled wastewater.

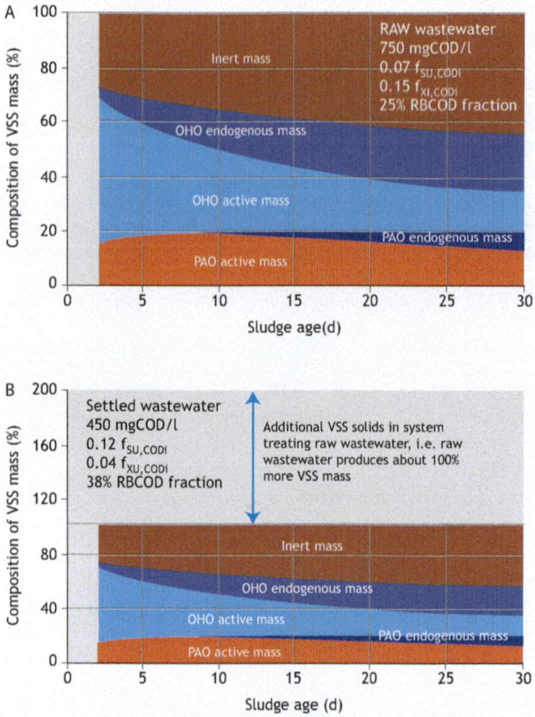

Figure 6.33 Percentage composition of VSS mass for EBPR systems treating (A) raw and (B) settled wastewater.

6.10.2 P/VSS ratio

A parameter often used to evaluate the EBPR performance of an activated sludge system is the P/VSS (or P/TSS) ratio of the mixed liquor. In Figure 6.34, calculated P/VSS ratios for a system with two-in-series anaerobic reactors and wastewater characteristics as shown are plotted *versus* sludge age. A zero discharge of nitrate to the anaerobic reactor is assumed.

From Figure 6.34, as the system sludge age increases, the P/VSS ratio increases up to a sludge age of approximately 10 days. Further increase in sludge age causes a decrease in P/VSS ratio. The initial increase in P/VSS with sludge age can be ascribed to increasing OHO active mass with sludge age. This produces an increased fermentable COD to VFA conversion efficiency in the anaerobic reactor and accordingly an increased PAO active mass (with associated P content of 0.38 mgP/mgVSS). The decrease in P/VSS can be ascribed to the endogenous respiration effect on PAOs.

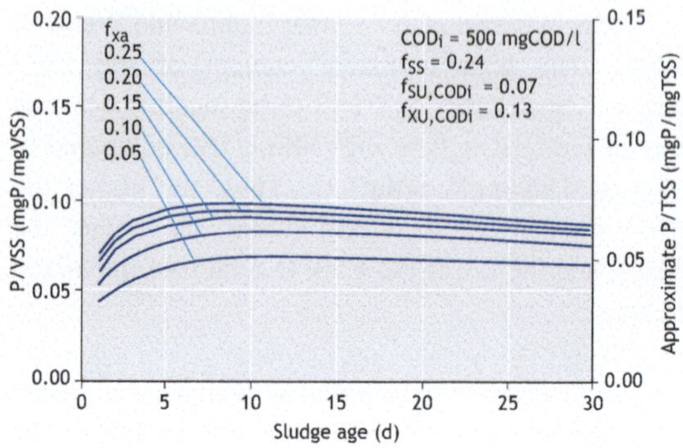

Figure 6.34 Predicted phosphorus to volatile (P/VSS) and total (P/TSS) suspended solids ratios *versus* sludge age for mixed liquor in a biological enhanced P-removal system with various anaerobic mass fractions (f_{xa}) treating wastewater with the characteristics shown.

It appears that the P/VSS ratio is a consequence of the selection of the fundamental design parameters that are sludge age and anaerobic mass fraction. Also, the P/VSS ratio is a function of the wastewater characteristics (*e.g.* RBCOD fraction). Accordingly, the parameter P/VSS ratio can fulfil a function in design only if a prior experimental relationship between the ratio and the design parameters has been established for the wastewater to be treated. It cannot be used reliably as a basic design parameter.

6.11 INTEGRATED DESIGN OF NDEBPR SYSTEMS

6.11.1 Background

In some countries legislation on permissible effluent ammonia concentrations necessitates that nitrification be incorporated in an EBPR removal activated sludge system. In the steady mixed-culture EBPR model, the nitrate recycled to the anaerobic reactor needs to be known due to the adverse influence of recycling nitrate to the anaerobic reactor on P removal. Indeed, one of the principal orientations in any design procedure for P removal is to prevent nitrate recycling. This can be achieved by preventing nitrification in a simple configuration such as the Phoredox or the A/O systems but this option is not available in some countries. Accordingly, reliable and accurate quantification of denitrification in NDEBPR systems is essential for P removal design, in addition to N removal design. One approach that has been used to quantify denitrification in NDEBPR systems was to estimate the denitrification using the theory and procedures for nitrification-denitrification (ND) systems, as set out in Chapter 5 (WRC, 1984). Experimental data indicated that this approach appeared to predict the observed denitrification quite closely (Nicholls, 1982). However, from the mechanisms for EBPR and the development of EBPR kinetic theory, an inconsistency in this approach became evident: the RBCOD appeared to be used twice; in the anaerobic reactor where it is converted to VFAs which are sequestered and stored as PHA by the PAO, and in the primary anoxic reactor for denitrification. This situation would be possible only if the PAOs denitrified significantly using most of the VFAs internally stored as PHA in the upstream anaerobic reactor as electron donor in the downstream anoxic reactor, which implies that the principal P uptake should be in the primary anoxic reactor, not the aerobic reactor. Although this behaviour was not observed in some earlier lab-scale NDEBPR systems and enhanced culture work conducted at the University of Cape Town (Wentzel *et al.,* 1989a), it was clearly shown by Vlekke *et al.* (1988, Kuba *et al.* (1996), Hu *et al.* (2007) and integrated in the Activated Sludge Model 2d: ASM2d (Henze *et al.,* 1999). While ASM2d models PAO PHA utilization under anoxic conditions, it does not address the changes in EBPR behaviour with anoxic P uptake; EBPR P removal declines by as much as a third (Ekama and Wentzel, 1999). ASM2d allows P uptake to commence in the anoxic reactor, but the predicted P removal is the same as if the uptake had taken place only in the aerobic reactor. Subsequent model modifications have sought to address this, *e.g.* Hu *et al.* (2007).

Clearly for completeness, denitrification had to be incorporated into the steady-state mixed-culture model, an aspect omitted up to this stage. Using plug flow anoxic reactors and batch tests, Clayton *et al.* (1991) undertook an experimental investigation into the kinetics of denitrification in NDEBPR systems. They found that in NDEBPR systems:

- In the primary anoxic reactor, (*i*) the rapid rate of denitrification associated with RBCOD was much reduced or absent, and (*ii*) the slower rate of denitrification associated with SBCOD was approximately 2.5 times the rate measured in primary anoxic reactors of ND systems.
- In the secondary anoxic reactor, the denitrification rate was approximately 1.5 times the rate measured in secondary anoxic reactors of ND systems.

From an extensive enquiry into the causes, Clayton *et al.* (1991) concluded that the increased denitrification rates were not due to:

- Denitrification by PAOs, for the systems investigated, PHA and P measurements indicated that the PAOs did not denitrify, which remains controversial.
- Modification of the sewage in the anaerobic zone; wastewater that had not passed through an anaerobic zone induced the same denitrification response as sewage that had passed through the anaerobic zone.

The above observations led Clayton *et al.* (1991) to conclude that the increased rate was due to a stimulation in the active sludge mass of an increased rate of hydrolysis of SBCOD in the anoxic reactors of the NDEBPR systems, apparently induced by the presence of the anaerobic reactor in these systems.

6.11.2 Denitrification potential in NDEBPR systems

The denitrification potential is the maximum amount of nitrate that can be removed by biological means in the anoxic reactors. Since the experimental investigation into denitrification kinetics in NDEBPR systems indicated that the formulation developed for ND systems can be applied to NDEBPR systems, the techniques set out in Chapter 5 to develop equations for denitrification potentials in ND systems can also be followed for NDEBPR systems. Note that a prime (') symbol is added to specific denitrification rate constants to indicate that the parameter value is different between a ND system (without prime) and a NDEPBR system (with a prime) (Clayton *et al.*, 1991; Ekama and Wentzel, 1999).

$$dS_{NO3} / dt = K_T \cdot X_{OHO} \quad (mgNO_3^-\text{-N/l.d}) \quad (6.32)$$

dS_{NO3}/dt	rate of denitrification ($mgNO_3^-$-N/l.d)
K_T	specific denitrification rate at temperature T for a NDEBPR system ($mgNO_3^-$-N/mg AVSS.d)

6.11.2.1 Denitrification potential of the primary anoxic reactor

Denitrification in the primary anoxic reactor is via utilization of any RBCOD leaking through the anaerobic reactor, and SBCOD. Procedures to determine the amount of RBCOD leaking through the anaerobic reactor to the primary anoxic reactor are set out in Section 6.8.3.3, where $S_{F,ANn}$ is the concentration of fermentable COD in the anaerobic reactor outflow, and $S_{F,Ann}$ (1+recycle ratio) the mass per litre influent flow. These procedures take into account the utilization of RBCOD in the anaerobic reactor due to storage by PAOs (either directly or following conversion) or to denitrification / aerobic respiration by OHOs. Accordingly, the denitrification potential in the primary anoxic reactor (D_{p1}) can be expressed as:

$$D_{p1} = S_{F,ANn}(1+r)(1-f_{cv}Y_{OHOv})/2.86 + K_{2,T}X_{OHO}HRT_{np} \quad (mgN/l_{inf}) \quad (6.33)$$

where:

D_{p1}	denitrification potential in the primary anoxic reactor (mgN/l_{inf})
$K_{2,T}$	specific denitrification rate in the primary anoxic reactor of a NDEBPR system on SBCOD at temperature T and ~ 0.23 $mgNO_3^-$-N/mgAVSS.d (Clayton *et al.,* 1991; Ekama and Wentzel, 1999) *i.e.* ~ 2.5 times higher than in ND systems (K_{2T})
HRT_{np}	nominal hydraulic retention time of the process (d)

Following the procedures set out in Chapter 5, Eq. 6.33 can be modified and simplified to give:

$$D_{p1} = S_{F,ANn}(1+r)(1-f_{cv}Y_{OHOv})/2.86 + \frac{f_{x1}K_{2,T}(COD_{b,i} - S_{s,PAO})Y_{OHO}\,SRT}{(1+b_{OHO,T}\,SRT)} \quad (6.34a)$$

or,

$$D_{p1} = \alpha + f_{x1}K_{2,T}\beta \quad (6.34b)$$

where:

f_{x1} primary anoxic reactor mass fraction

$$\alpha = S_{F,ANn}(1+r)(1-f_{cv}Y_{OHOv})/2.86 \quad (6.35a)$$

$$\beta = \frac{(COD_{b,i} - S_{S,PAO})\,Y_{OHO}\,SRT}{(1+b_{OHO,T}\,SRT)} \quad (6.35b)$$

In Eq. 6.34 it is assumed that the initial rapid rate of denitrification ($K_{2,T}$) on RBCOD leaking through the anaerobic reactor $S_{F,Ann}(1+r)$ is always complete, *i.e.* the actual retention time in the primary anoxic reactor is longer than the time required to utilize this RBCOD. As with ND systems, an equation can be developed to determine the minimum primary anoxic mass fraction $f_{x1,min}$ to deplete this RBCOD:

$$f_{x1,min} = \frac{S_{F,ANn}(1+r)(1-f_{cv}\,Y_{OHOv})\,(1+b_{OHO,T}\,SRT)}{(COD_{b,i} - S_{S,PAO})\,2.86\,K_{1,T}\,Y_{OHO}\,SRT} \quad (6.36a)$$

$$f_{x1,min} = \alpha / (\beta \cdot K_{1,T}) \quad (6.36b)$$

Where $K_{1,T}$ is the initial rapid rate of denitrification in the primary anoxic reactor of a NDEBPR system on RBCOD at T °C and equal to that in a ND system K_{1T}.

Substituting the values for the constants into Eq. 6.36 and assuming 80 per cent of the influent RBCOD is sequestered by the PAOs in the anaerobic reactor, $f_{x1,min} < 0.02$ for SRT > 10 days at 14 °C with $COD_{b,i}$ = 800 mgCOD/l and f_{SS} = 0.24. This value of 2% of anoxic mass fraction is much lower than actual primary anoxic reactors mass fraction so that for nearly all cases eqs. 6.34 and 6.35 will be valid.

However, eqs. 6.34a and 6.34b are not without complication. To calculate the primary anoxic denitrification potential (D_{p1}), the concentration of RBCOD in the outflow from the anaerobic reactor ($S_{F,ANn}$) is required. To calculate $S_{F,ANn}$, the concentration of nitrate recycled to the anaerobic reactor is required which in turn requires D_{p1} to be known. This aspect will be dealt with in more detail in Section 6.11.3.2 below.

6.11.3.2 Denitrification potential of the secondary anoxic reactor

The denitrification potential of the secondary anoxic reactor (D_{p3}) is found by following the principles set out in Chapter 5, and is given by:

$$D_{p3} = \frac{f_{x3} K_{3,T} (COD_{b,i} - S_{S,PAO})\, Y_{OHO}\, SRT}{(1 + b_{OHO,T}\, SRT)} \tag{6.37a}$$

$$D_{p3} = f_{x3} K_{3,T} \beta \tag{6.37b}$$

where:

f_{x3}	secondary anoxic reactor mass fraction
$K_{3,T}$	specific denitrification rate in the secondary anoxic reactor at temperature T and ~ 0.10 $mgNO_3^-$-N/mgAVSS.d (Clayton *et al.,* 1991; Ekama and Wentzel, 1999) *i.e.* ~ 1.5 times higher than in ND systems ($K_{3,T}$).

Eq. 6.37 applies to secondary anoxic reactors situated both in the mainstream (*e.g.* 5-stage Modified Bardenpho) and in the underflow recycle (*e.g.* JHB system). However, in applying Eq. 6.37 to secondary anoxic reactors situated in the underflow recycle, care must be taken in evaluating f_{x3}, because the mixed-liquor concentration is increased by a factor (1+s)/s in the underflow anoxic reactor compared to the mainstream reactors.

The higher $K_{2,T}$ and $K_{3,T}$ denitrification rates in a NDEBPR system compared with ND systems require the case of higher η values on the OHO hydrolysis/growth processes of SBCOD under anoxic conditions in ASM2 and ASM2d.

6.11.3 Principles of denitrification design procedures for NDEBPR systems

In NDEBPR systems, design is oriented to achieve the following in a single sludge system:

1. COD removal,
2. N removal (nitrification/denitrification), and
3. P removal (EBPR).

Conflict between these objectives may arise, in particular between N and P removal, *e.g.* unaerated mass required for anoxic reactors (N removal) and anaerobic reactors (P removal). For each design, the priorities for treatment need to be assessed and a compromise reached to optimize the system.

In some countries, the design of NDEBPR systems usually focuses on EBPR with denitrification as a secondary design priority, because legislation limits effluent P concentrations, and only in selected cases are effluent nitrate concentrations limited. Accordingly, in such situations the fundamental principle in denitrification design for NDEBPR systems is to ensure that the anaerobic reactor is protected from recycling of nitrate. This fundamental principle will determine the selection of the system configuration (5-stage Modified Bardenpho, JHB and UCT/MUCT considered in this chapter) and provides procedures for sizing the anoxic reactors.

When selecting a system configuration for EBPR, it is necessary to establish whether complete denitrification can be achieved. For the wastewater characteristics, *i.e.* influent TKN and COD concentrations (TKN_i and COD_i), maximum specific growth rate of the nitrifiers at 20 °C ($\mu_{ANO,max,20}$) and the average minimum water temperature, the maximum unaerated sludge mass fraction ($f_{x,max}$) and the nitrification capacity (NIT_c) can be calculated for a selected sludge age (SRT), see Chapter 5. This $f_{x,max}$ needs to be divided between anaerobic (for EBPR) and anoxic (for denitrification) mass fractions. Consequently, the maximum anoxic sludge mass fraction ($f_{xd,max}$) is the difference between the maximum unaerated mass fraction ($f_{x,max}$) and the selected anaerobic sludge mass fraction (f_{xa}), *i.e.*

$$f_{xd,max} = f_{x,max} - f_{xa} \tag{6.38}$$

where:

$f_{xd,max}$ maximum anoxic mass fraction
$f_{x,max}$ maximum unaerated mass fraction

The value of $f_{x,max}$ is given by Eq (5.19), Chapter 5 for a selected SRT, $\mu_{ANO,max,20}$, S_f and T_{min}.

The value of $f_{xd,max}$ can then be subdivided between primary and secondary anoxic sludge mass fractions (f_{x1} and f_{x3}) and this division fixes the denitrification potential of these two reactors (D_{p1} and D_{p3}) and hence also of the system. If the denitrification potential of the system exceeds the nitrification capacity (*i.e.* $D_{p1} + D_{p3} > NIT_c$) then complete denitrification is possible and the secondary anoxic reactor is situated in the mainstream, the 5-stage Modified Bardenpho. If complete denitrification is not possible, with the 5-stage Modified Bardenpho nitrate will appear in the effluent and be recycled via the s-recycle to the anaerobic reactor. Accordingly, the secondary anoxic reactor is moved to the underflow recycle, the JHB system, in which event the denitrification potential of the secondary anoxic reactor (D_{p3}) must exceed the nitrate and oxygen loads via the underflow s-recycle. If this requirement is not met, nitrate will 'leak' through the underflow secondary anoxic reactor into the anaerobic reactor. In this event, since the denitrification potential of the primary anoxic reactor (D_{p1}) is greater than that of the secondary anoxic reactor (D_{p3}) for equal anoxic mass fractions, incorporation of a secondary anoxic reactor becomes an inefficient utilization of anoxic mass fraction and the secondary anoxic mass fraction is added to the primary anoxic reactor, the UCT/MUCT system. Alternatively, if very low effluent nitrate concentrations are required, the secondary anoxic reactor can be retained and methanol can be added to it.

6.11.4 Analysis of denitrification in NDEBPR systems

Analysis of the denitrification behaviour in the NDEBPR system is essentially the same as for the ND system (Chapter 5) except that:

- The mass fraction for denitrification ($f_{xd,max}$) for the NDEBPR system is given by Eq. 6.38, whereas $f_{xd,max}$ for the ND system is given by Eq. 5.56. Hence, for the same maximum unaerated sludge mass fraction ($f_{x,max}$), the NDEBPR system has a lower mass fraction than the ND system, by an amount equal to f_{xa}.
- The specific denitrification rates for ND systems (K_2 and K_3, Chapter 5) are substituted with the rates measured for NDEBPR systems ($K_{2,T}$ and $K_{3,T}$, Section 6.11.2).
- The denitrification potentials for the primary and secondary anoxic reactors are modified from Chapter 5 for the ND system to those given by eqs. 6.34 and 6.37 for the NDEBPR system to take account of the storage of COD by the PAOs in the anaerobic reactor, and the non-participation of the PAOs in denitrification.

The objective of the simplified steady state model presented below is to obtain an estimate of the a-recycle ratio to load the anoxic reactor to its denitrification potential. A detailed analysis of EBPR systems can be realized with simulation programs. Taking account of the above, denitrification equations are developed below for the UCT system.

6.11.4.1 UCT System

In the UCT system the denitrification behaviour is very similar to that in the MLE system, so that, taking due account of the effect of incorporating the anaerobic reactor, the design equations and procedures developed for the MLE system can be readily adapted for application to the UCT system.

In this application, the following principles are of importance:

- Since complete denitrification is not possible, the entire anoxic mass fraction available is used, in the form of a primary anoxic reactor.
- The a-recycle ratio determines the split of nitrate between the primary anoxic reactor and the effluent. The a-recycle ratio is selected so that the equivalent nitrate loads to the primary anoxic reactor via the a and s recycles just load the reactor to its denitrification potential.

Taking account of the above, design equations are developed below for the UCT system.

- Denitrification potential (D_{p1}): The denitrification potential of the primary anoxic reactor (D_{p1}) is found from Eq. 6.34 with $f_{x1} = f_{xd,max}$, *i.e.*:

$$D_{p1} = \alpha + f_{xd,max} K_{2,T} \beta \tag{6.39}$$

- Effluent nitrate concentration ($S_{NO3,e}$): If the nitrate concentration in the outflow of the primary anoxic reactor is zero, then:

$$S_{NO3,e} = NIT_c / (a + s + 1) \tag{6.40}$$

- Optimum a-recycle ratio (a_{opt}): Due to the similarities between the MLE and UCT systems, an equation for a_{opt} for the UCT system can be developed by following the procedure for the MLE system: *i.e.* a_{opt} is the a-recycle that just loads the primary anoxic to its denitrification potential (D_{p1}). From a mass balance around the primary anoxic reactor, the equivalent nitrate load on this reactor ($FS_{NO3,x1}/Q_i$) is given by:

$$\frac{FS_{NO3,x1}}{Q_i} = s\left[S_{NO3,e} + \frac{S_{O2,s}}{2.86}\right] + a\left[S_{NO3,e} + \frac{S_{O2,a}}{2.86}\right] \tag{6.41}$$

where:
$S_{O2,s}$ and $S_{O2,a}$ are the dissolved O_2 concentration in the s and the a recycles, respectively.

Equating Eq. 6.41 to the denitrification potential given by Eq. 6.39, recognising the a = a_{opt} and solving for a_{opt} gives:

$$a_{opt} = [\,-B + \sqrt{B^2 - 4AC}\,] / (2A) \tag{6.42}$$

where:

A	$S_{O2,a} / 2.86$
B	$NIT_c - D_{p1} + \{(s+1)\, S_{O2,a} + s\, S_{O2,s}\} / 2.86$
C	$s\, NIT_c - (s + 1)\, (D_{p1} - s\, S_{O2,s}/2.86)$

At a = a_{opt}, Eq. 6.42 will give the minimum $S_{NO3,e}$ achievable. Eq. 6.42 is valid for all a ≤ a_{opt} because for all a ≤ a_{opt} the assumption on which Eq. 6.42 is based is valid, *i.e.* zero nitrate concentration in the outflow from the primary anoxic reactor. If the system is operated with a > a_{opt}, the equivalent nitrate load on the primary anoxic reactor via the a- and s-recycles exceeds the denitrification potential, and nitrate will also be recycled via the r-recycle to the anaerobic reactor, to the detriment of EBPR. Furthermore, if nitrate does 'leak' through the primary anoxic reactor then the nitrate concentration in the outflow from the primary anoxic reactor is no longer zero, and consequently, Eq. 6.40 for the effluent nitrate concentration ($S_{NO3,e}$) is not valid.

6.11.5 Maximum nitrate recycled to anaerobic reactor

The design procedures for denitrification in the previous section have been developed assuming that the increased denitrification rates ($K_{2,T}$ and $K_{3,T}$) apply, *i.e.* that the system is exhibiting EBPR. However, recycling nitrate or oxygen to the anaerobic reactor has a detrimental effect on EBPR.

In a case where so much nitrate or oxygen is recycled that all the fermentable COD is consumed for denitrification, none would remain available for conversion to VFAs. In this case, in Eq. 6.8 setting $S_{F,i,conv} = 0$ and solving for $S_{NO3,s}$ gives:

$$S_{NO3,s} = \left[\left\{\frac{S_{F,i}}{8.6} - \frac{(sS_{O2,s} + S_{O2,i})}{2.86}\right\} - S_{NO3,i}\right]/s \qquad (6.43)$$

This nitrate concentration effectively sets the maximum amount of nitrate that can be recycled to the anaerobic reactor with the equations in this chapter remaining valid. At this $S_{NO3,s}$ concentration if there is any VFAs present in the influent, EBPR will still be obtained.

Should $S_{NO3,s}$ be exceeded, a competition between the PAOs and the OHOs for the VFAs develops (for storage and denitrification, respectively) and a kinetic model will be required to determine system performance, and the equations developed in this chapter are not valid for this situation.

6.12 CONCLUSIONS

Enhanced biological phosphorus removal (EBPR) has been developed to assist in the control of eutrophication by removing phosphorus from wastewaters without the use of chemicals. The high phosphorus content of the biomass wasted from EBPR processes makes it amenable to phosphorus recovery by struvite formation (magnesium ammonium phosphate: $MgNH_4PO_4$) especially when an anaerobic digester is used, or as hydroxyapatite [$Ca_{10}(PO4)_5(OH)_2$] when little ammonia is available.

In some sensitive water bodies, very low phosphorus (and nitrogen) discharge limits have been promulgated, sometimes below 0.1 g total P per m3. To consistently achieve such low levels, coagulants and filtration or ultrafiltration systems need to be used.

Phosphorus-accumulating organisms (PAOs) have been studied in order to understand the biochemical mechanisms of their anaerobic, anoxic and aerobic metabolism. From these studies, process optimisation principles have been derived and mathematical models have been developed for steady-state design analysis and incorporated into software programs to study various scenarios and facilitate the design, optimisation and development of EBPR systems. The effect of nitrate in sludge or internal recycles, and the effect of dynamic changes in loadings (*e.g.* organic surcharges after a weekend or the addition of industrial wastes) can best be quantified with such software programs.

Future developments in the field should come from improved understanding of the biochemical mechanisms of different groups of PAOs, GAOs and filamentous organisms to propose practical control strategies to favour the dominance of PAOs. Some PAO groups, such as *Tetrasphaera*, appear to be of high importance in many EBPR processes, where much remains to be learned about how they function and how to best exploit their activity to benefit P removal. With a better fundamental understanding of biochemical processes, improved parametric and metabolic models could then be developed that would lead to more accurate models and more robust EBPR processes.

REFERENCES

Acevedo B., Oehmen A., Carvalho G., Seco A., Borrás L. and Barat R. (2012). Metabolic shift of polyphosphate-accumulating organisms with different levels of polyphosphate storage. *Water Research*, 46(6), 1889-1900.

Alarcon G.O. (1961). Removal of phosphorus from sewage. Unpublished Master, Johns Hopkins University, Baltimore.

Arvin E. (1985). Biological removal of phosphorus from wastewater. *CRC Crit. Rev. Env. Control* 15, 25-64.

Barat R. and Van Loosdrecht M.C.M. (2006). Potential phosphorus recovery in a WWTP with the BCFS® process: Interactions with the biological process. *Water Research,* 40(19), 3507-3516.

Barat R., Montoya T., Borras L., Seco A. and Ferrer J. (2006). Calcium effect on enhanced biological phosphorus removal. *Water Science and Technology*, 53(12), 29-37.

Barnard J.L. (1974a). Cut P and N without chemicals. *Water and Wastes Engineering,*11, 33-36.

Barnard J.L. (1974b). Cut P and N without chemicals. *Water and Wastes Engineering,* 11, 41-44.

Barnard J.L. (1975a). Biological nutrient removal without the addition of chemicals. *Water Research,* 9, 485-490.

Barnard J.L. (1975b). Nutrient removal in biological systems. *Water Pollut. Control,* 74(2), 143-154.

Barnard J.L. (1976a). A review of biological phosphorus removal in the activated sludge process. *Water SA,* 2, 136-144.

Barnard J.L. (1976b). Nutrient removal in biological systems. *Water Pollution Control,* 74, 143-154.

Barnard J.L. and Steichen M.T. (2007). Optimizing BPR plant operations for achieving sustainable low effluent phosphorus. *Proceedings of the Nutrient Removal - The State of the Art, Specialty Conference,* Baltimore, MD, pp. 903-926, Eds Water Environment Federation and International Water Association.

Barnard J.L., Stevens G.M. and Leslie P.J. (1985). Design strategies for nutrient removal plant. *Water Science and Technology*, 17(11/12), 233-242.

Barnard J. L., Dunlap P. and Steichen M. (2017). Rethinking the Mechanisms of Biological Phosphorus Removal: Barnard *et al. Water Environment Research*, 89(11), 2043-2054.

Blackall L.L., Crocetti G., Saunders A.M. and Bond P.L. (2002). A review and update of the microbiology of enhanced biological phosphorus removal in wastewater treatment plants. *Antonie Van Leeuwenhoek Int. J. General Molec. Microbiol.* 81(1-4), 681-691.

Bortone G., Saltarelli R., Alonso V., Sorm R., Wanner J., and Tilche A. (1996). Biological anoxic phosphorus removal - The Dephanox process. *Water Science and Technology,* 34(1-2), 119-128.

Boughton W.H., Gottfried R.J., Sinclair N.A. and Yall I. (1971). Metabolic factors affecting enhanced phosphorus uptake by activated sludge. *Applied Microbiology,* 22, 571-577.

Brdjanovic D., Hooijmans C.M., Van Loosdrecht M.C.M., Alaerts G.J. and Heijnen J.J. (1996). The dynamic effects of potassium limitation on biological phosphorus removal. *Water Research,* 30(10), 2323–2328.

Brdjanovic D., Logemann S., Van Loosdrecht M.C.M., Hooijmans C.M., Alaerts G.J. and Heijnen J.J. (1998c). Influence of temperature on biological phosphorus removal: process and molecular ecology study. *Water Research,* 32(4), 1035-1048.

Brdjanovic D., Moussa M.S., Mithaiwala M., Amy G. and Van Loosdrecht M.C.M. (2007). Use of Modeling for Optimization and Upgrade of a Tropical Wastewater Treatment Plant in a Developing Country. *Water Science and Technology,* 56(7), 21-31.

Brdjanovic D., Slamet A., Van Loosdrecht M.C.M., Hooijmans C.M., Alaerts G.J. and Heijnen J.J. (1998a). Impact of excessive aeration on biological phosphorus removal from wastewaters. *Water Research,* 32(1), 200-208.

Brdjanovic D., Van Loosdrecht M.C.M., Hooijmans C.M., Alaerts G.J. and Heijnen J.J. (1997). Temperature Effects on Physiology of Biological Phosphorus Removal Systems. *Journal of Environmental Engineering,* ASCE, 123(2), 144-154.

Brdjanovic D., Van Loosdrecht M.C.M., Hooijmans C.M., Alaerts G.J. and Heijnen J.J. (1998b). Minimal aerobic sludge retention time in biological phosphorus removal systems. *Biotechnology and Bioengineering,* 60(3), 326-332.

Brdjanovic D., Hooijmans C.M., Van Loosdrecht M.C. M., Alaerts G.J. and Heijnen J.J. (1996). The dynamic effects of potassium limitation on biological phosphorus removal. *Water Research*, 30(10), 2,323-2,328.

Brodisch K.E.U. (1985). Interaction of different groups of microorganisms in biological phosphate removal. *Water Science and Technology,* 17(11/12), 89-97.

Buchan L. (1983). Possible biological mechanism of phosphorus removal. *Water Science and Technology,* 15(3/4), 87-103.

Burke R., Dold P.L. and Marais G. (1986). *Biological*

excess phosphorus removal in short sludge age activated sludge systems. Research Report No. W58, University of Cape Town, South Africa.

Burow L. C., Kong Y., Nielsen J. L., Blackall L.L. and Nielsen P. H. (2007). Abundance and ecophysiology of Defluviicoccus spp., glycogen-accumulating organisms in full-scale wastewater treatment processes. *Microbiology*, 153(1), 178-185.

Camejo P.Y., Owen B.R., Martirano J., Ma J., Kapoor V., Domingo J.S., McMahon K.D. and Noguera D.R. (2016). Candidatus Accumulibacter phosphatis clades enriched under cyclic anaerobic and microaerobic conditions simultaneously use different electron acceptors. *Water Research,* 102, 125-137.

Camejo P.Y., Oyserman B.O., McMahon K.D. and Noguera D.R. (2019). Integrated Omic Analyses Provide Evidence that a "Candidatus Accumulibacter phosphatis" Strain Performs Denitrification under Microaerobic Conditions. *Msystems,* 4(1), 1-23.

Carvalheira M., Oehmen A., Carvalho G. and Reis M. A. (2014). The effect of substrate competition on the metabolism of polyphosphate accumulating organisms (PAOs). *Water Research*, 64, 149-159.

Clayton J.A., Ekama G.A., Wentzel M.C. and Marais G.V.R. (1991). Denitrification kinetics in biological nitrogen and phosphorus removal activated-sludge systems treating municipal wastewaters. *Water Science and Technology,* 23(4-6), 1025-1035.

Comeau Y., Hall K.J., Hancock R.E.W. and Oldham W.K. (1986). Biochemical model for biological enhanced phosphorus removal. *Water Research,* 20, 1511-1521.

Comeau Y., Rabinowitz B., Hall K.J. and Oldham W.K. (1987). Phosphorus release and uptake in enhanced biological phosphorus removal from wastewater. *J. Water Pollut. Control Fed.,* 59, 707-715.

Converti A., Rovatti M. and Borghi del M. (1995). Biological removal of phosphorus from wastewaters by alternating aerobic and anaerobic conditions. *Water Research,* 29(1), 263-269.

Crocetti G.R., Banfield J.F., Keller J., Bond P.L. and Blackall L.L. (2002). Glycogen accumulating organisms in laboratory-scale and full-scale wastewater treatment processes. *Microbiology*, 148, 3353-3364.

Crocetti G.R., Hugenholtz P., Bond P.L., Schuler A., Keller J., Jenkins D. and Blackall L.L. (2000). Identification of polyphosphate-accumulating organisms and design of 16S rRNA-directed probes for their detection and quantitation. *Applied and Environmental Microbiology,* 66(3), 1175-1182.

Daigger G.T., Randall C.W., Waltrip G.D., Romm E.D. and Morales L.M. (1987). Factors affecting biological phosphorus removal for the VIP process, a high-rate University of Cape Town process. *Paper presented at the IAWPRC Biological Phosphate Removal from Wastewater*, Rome, Italy.

Davelaar D. (1978). Biological removal of phosphorus from wastewater in a nitrifying/denitrifying activated sludge system. MSc thesis, Agricultural University, Wageningen, the Netherlands.

De Haas D.W., Wentzel M.C. and Ekama G.A. (2000). The use of simultaneous chemical precipitation in modified activated sludge systems exhibiting biological excess phosphate removal Part 1: Literature review. *Water SA,* 26(4), 439-452.

Dold P.L., Bagg W.K. and Marais G.V.R. (1986). *Measurement of the readily biodegradable COD fraction (Sbs) in municipal wastewater by ultrafiltration*. Research report no. W57 University of Cape Town, South Africa.

Dold P.L., Ekama G.A. and Marais G.v.R. (1980). A general model for the activated sludge process. *Progr. Wat. Technol.,* 12, 47-77.

Dold P.L., Wentzel M.C., Billing A.E. and Marais G.v.R. (1991). Activated sludge system simulation programs: Nitrification and nitrification / denitrification systems, Version 1 (UCTOLD, IAWPRC). *WRC Report no. TT 52/91,* Pretoria, South Africa.

Ekama G.A. and Wentzel M.C. (1999). Denitrification kinetics in biological N and P removal activated sludge systems treating municipal wastewaters. *Water Science and Technology,* 39(6), 69-77.

Ekama G.A. and Wentzel M.C. (2004). A predictive model for the reactor inorganic suspended solids concentration in activated sludge systems. *Water Research,* 38(19), 4093-4106.

Ekama G.A., Dold P.L. and Marais G.v.R. (1986). Procedures for determining influent COD fractions and the maximum specific growth rate of heterotrophs in activated sludge systems. *Water Science and Technology,* 18, 91-114.

Ekama G.A., Siebritz I.P. and Marais G.v.R. (1983). Considerations in the process design of nutrient removal activated sludge processes. *Water Science and Technology,* 15(3/4), 283-318.

Fernando E.Y., McIlroy S.J., Nierychlo M., Herbst F.A., Petriglieri F., Schmid M.C., Wagner M., Nielsen J.L. and Nielsen P.H. (2019). Resolving the individual contribution of key microbial populations to enhanced biological phosphorus removal with Raman-FISH. *Isme Journal,* 13(8), 1933-1946.

Filipe C.D.M., Daigger G.T. and Grady Jr C.P.L. (2001a). A metabolic model for acetate uptake under anaerobic conditions by glycogen-accumulating organisms: Stoichiometry, kinetics and effect of pH. *Biotechnology and Bioengineering,* 76(1), 17-31.

Filipe C.D.M., Daigger G.T. and Grady C.P.L. (2001b). pH as a key factor in the competition between glycogen-accumulating organisms and phosphorus-accumulating. *Water Environment Research*, 73(2), 223-232.

Filipe C.D., Daigger G.T. and Grady Jr C.L. (2001). Stoichiometry and kinetics of acetate uptake under anaerobic conditions by an enriched culture of phosphorus-accumulating organisms at different pHs. *Biotechnology and Bioengineering*, 76(1), 32-43.

Florentz M, Caille D., Bourdon F. and Sibony J. (1987). Biological phosphorus removal in France. *Water Science and Technology*, 19(4), 1171-1173.

Flowers J.J., He S.M., Yilmaz S., Noguera D.R. and McMahon, K.D. (2009). Denitrification capabilities of two biological phosphorus removal sludges dominated by different 'Candidatus Accumulibacter' clades. *Environmental Microbiology Reports,* 1(6), 583-588.

Fuhs G.W. and Chen M. (1975). Microbiological basis of phosphate removal in the activated sludge process for the treatment of wastewater. *Microbial Ecology,* 2, 119-138.

Fukase T., Shibata M. and Miyaji Y. (1982). Studies on the mechanism of biological removal of phosphorus. *Japan J. Water Pollut. Res.,* 5, 309-317.

Gebremariam S.Y., Beutel M.W., Christian D. and Hess T.F. (2011). Research Advances and Challenges in the Microbiology of Enhanced Biological Phosphorus Removal-A Critical Review. *Water Environment Research,* 83(3), 195-219.

Gerber A., Devilliers R.H., Mostert E.S. and Winter C.T. (1987). Interactions between phosphate, nitrate and organic substrate in biological nutrient removal processes. *Water Science and Technology,* 19(1-2), 183-194.

Griffiths P. C., Stratton H. M. and Seviour R. J. (2002). Environmental factors contributing to the "G bacteria" population in full-scale EBPR plants. *Water Science and Technology*, 46(4-5), 185-192.

Groenestijn J.W.v., Deinema M.H. and Zehnder A.J.B. (1987). ATP production from polyphosphate in *Acinetobacter* strain 210A. *Arch. Microbiol.,* 148, 14-19.

Gujer W. and Henze M. (1991). Activated sludge modelling and simulation. *Water Science and Technology*, 23(4-6), 1011-1023.

Guo G., Ekama G.A., Wang Y.Y., Dai J., Biswal B.K., Chen G.H. and Wu D. (2019). Advances in sulfur conversion-associated enhanced biological phosphorus removal in sulfate-rich wastewater treatment: A review. *Bioresource Technology,* 285, 121303.

Hart M.A. and Melmed. L.N. (1982). Microbiology of nutrient removing activated sludge. *Poster pres. Special in Seminar of the IAWPRC*, Pretoria, South Africa.

Hascoët M.C. and Florentz M. (1985). Influence of nitrates on biological phosphorus removal from wastewater. *Water SA,* 11, 1-8.

Hascoët M.C., Florentz M. and Granger P. (1985). Biochemical aspects of enhanced biological phosphorus removal from wastewater. *Water Science and Technology*, 17(11/12), 23-41.

He S., Gall D.L. and McMahon K.D. (2007). "Candidatus accumulibacter" population structure in enhanced biological phosphorus removal Sludges as revealed by polyphosphate kinase genes. *Applied and Environmental Microbiology,* 73(18), 5865-5874.

Henze M., Gujer W., Mino T., Matsuo T., Wentzel M.C. and Marais G.v.R. (1994) Activated sludge model no.2. *IAWQ Scientific and Technical Report No.3.*

Henze M., Grady C.P.L.J., Gujer W., Marais G.v.R., and Matsuo T. (1987) *Activated sludge model no.1. IAWPRC, Sci. and Technical Report No.1.* International Association on Water Pollution Research and Control, London, U.K.

Henze M., Gujer W., Mino T., Matsuo T., Wentzel M.C., Marais G.V.R. and Van Loosdrecht M.C.M. (1999). Activated sludge Model No.2d, ASM2d. *Water Science and Tech*nology, 39(1), 165-182.

Hesselmann R.P.X., Werlen C., Hahn D., van der Meer J.R. and Zehnder A.J.B. (1999). Enrichment, phylogenetic analysis and detection of a bacterium that performs enhanced biological phosphate removal in activated sludge. *Systematic Appl. Microbiol.,* 22(3), 454-465.

Hong S.N., M.L. Spector R.P.S. and Galdieri J.V. (1983). Recent advances on biological nutrient control by the A/O process. Water Pollution Control Federation Research Symposium, Oct., Atlanta., Georgia, USA.

Hu Z.R., Wentzel M.C. and Ekama G.A. (2001). External nitrification in biological nutrient removal activated sludge systems. *Water Science and Technology*, 43(1), 251-260.

Hu Z.R., Wentzel M.C. and Ekama G.A. (2002). The significance of denitrifying polyphosphate accumulating organisms in biological nutrient removal activated sludge systems. *Water Science and Technology,* 46(1/2), 129-138.

Hu Z.R., Wentzel M.C. and Ekama G.A. (2007). A general kinetic model for biological nutrient removal activated sludge systems: Model development. *Biotechnology and Bioengineering,* 98(6), 1242-1258.

Jenkins D. and Tandoi V. (1991) The applied microbiology of enhanced biological phosphate removal - Accomplishments and needs. *Water Research,* 25(12), 1471-1478.

Jeon C.O. and Park J.M. (2000). Enhanced biological phosphorus removal in a sequencing batch reactor supplied with glucose as a sole carbon source. *Water Research*, 34(7), 2160-2170.

Jeon C.O., Lee D.S. and Park J.M. (2001). Enhanced biological phosphorus removal in an anaerobic–aerobic sequencing batch reactor: characteristics of carbon metabolism. *Water Environment Research*, 73(3), 295-300.

Jobbagy A., Literathy B., Wong M. T., Tardy G. and Liu W. T. (2006). Proliferation of glycogen accumulating organisms induced by Fe (III) dosing in a domestic wastewater treatment plant. *Water Science and Technology*, 54(1), 101-109.

Jones P.H., Tadwalkar A.D. and Hsu C.L. (1987). Enhanced uptake of phosphorus by activated sludge - effect of substrate addition. *Water Research*, 21(5), 301-308.

Kang S.J., Horvatin P.J. and Briscoe L. (1985b). Full-scale biological phosphorus removal using A/O process in a cold climate. In Proc. Int. Conf. Management strategies for Phosphorus in the Environment. Selper Ltd., UK.

Kang S.J. and Horvatin P.J. (1985a). Retrofit of a full scale municipal treatment plant at Pontiac, Michigan for biological phosphorus removal. *Pres. Technol. Transfer Sem. BPR in Municipal Wastewater Treatment*, Penticton, BC, Canada.

Krichten D.J., Hong S.N., and Tracy K.D. (1985). Applied biological phosphorus removal technology for municipal wastewater treatment by the A/O process. In Proc. Int. Conf. Management strategies for Phosphorus in the Environment. Selper Ltd., UK.

Kristiansen R., Nguyen H.T.T., Saunders A.M., Nielsen J.L., Wimmer R., Le V.Q., McIlroy S.J., Petrovski S., Seviour R.J., Calteau A., Nielsen, K.L. and Nielsen, P.H. (2013). A metabolic model for members of the genus Tetrasphaera involved in enhanced biological phosphorus removal. *Isme Journal,* 7(3), 543-554.

Kruit J., Hulsbeek J. and Visser A. (2002). Bulking sludge solved?! *Water Science and Technology,* 46 (1-2), 457-464.

Kuba T. and Van Loosdrecht M.C.M. (1996). Phosphorus and nitrogen removal with minimal COD requirement by integration of denitrifying dephosphatation and nitrification in a two-sludge system. *Water Research,* 30(7), 1702-1710.

Kuba T., Murnleitner E., Van Loosdrecht M.C.M., and Heijnen J.J. (1996) A metabolic model for biological phosphorus removal by denitrifying organisms. *Biotechnology and Bioengineering,* 52, 685-695.

Kuba T., Smolders G., Van Loosdrecht M.C.M. and Heijnen J.J. (1993). Biological phosphorus removal from wastewater by anaerobic-anoxic sequencing batch reactor. *Water Science and Technology,* 27(5-6), 241-252.

Lanham A.B., Oehmen A., Saunders A.M., Carvalho G., Nielsen P.H. and Reis M.A.M. (2013). Metabolic versatility in full-scale wastewater treatment plants performing enhanced biological phosphorus removal. *Water Research,* 47(19), 7032-7041.

Lanham A. B., Oehmen A., Saunders A. M., Carvalho G., Nielsen P. H. and Reis M. A. (2014). Metabolic modelling of full-scale enhanced biological phosphorus removal sludge. *Water Research*, 66, 283-295.

Lanham A.B., Reis M.M. and Lemos P.C. (2008). Kinetic and metabolic aspects of Defluviicoccus vanus-related organisms as competitors in EBPR systems. *Water Science and Technology*, 58(8), 1693-1697.

Levin G.V. and Elster B. (1985). Recent Phostrip® Process Results in U.S.A. and Europe. *Water Science and Technology*, 17(11-12), 283–284.

Levin G.V. and Sala U.D. (1987). PhoStrip process - a viable answer to eutrophication of lakes and coastal sea waters in Italy. *Advances in Water Pollution Control, Proc. Rome Specialist Conf. on Biological Phosphate Removal from Wastewater*, Sept. 28-30.

Levin G.V. and Shapiro J. (1965). Metabolic uptake of phosphorus by wastewater organisms. *Journal of the Water Pollution Control Federation,* 37, 800-821.

Levin G.V., Topol G.J., Tarnay A.C. and Samworth R.B. (1972). Pilot plant tests of a phosphorus removal process. *Journal of the Water Pollution Control Federation,* 44(10), 1940-1954.

Lindrea K.C., Pigdon S.P., Boyd B. and Lockwood G.A. (1994). Biomass characterization in a nitrification-denitrification biological enhanced phosphorus removal (NDEBPR) plant during start-up and subsequent periods of good and poor phosphorus removal. *Water Science and Technology,* 29(7), 91-100.

Liu W.T., Nakamura K., Matsuo T. and Mino T. (1997). Internal energy-based competition between polyphosphate- and glycogen-accumulating bacteria in biological phosphorus removal reactors-effect of P/C feeding ratio. *Water Research,* 31(6), 1430-1438.

Liu R., Hao X., Chen Q. and Li J. (2019). Research advances of Tetrasphaera in enhanced biological phosphorus removal: A review. *Water Research*, 166, 115003.

Lopez-Vazquez C.M., Hooijmans C.M., Brdjanovic D., Gijzen H.J. and van Loosdrecht M.C.M. (2008a). Factors affecting the microbial populations at full-scale enhanced biological phosphorus removal (EBPR) wastewater treatment plants in The Netherlands. *Water Research,* 42(10-11):2349-2360.

Lopez-Vazquez C.M., Hooijmans C.M., Brdjanovic D., Gijzen H.J. and van Loosdrecht M.C.M. (2009a). Temperature effects on the metabolism of glycogen accumulating organisms. *Water Research,* 43(11):2852-2864.

Lopez-Vazquez C.M., Oehmen, A., Hooijmans C.M., Brdjanovic D., Gijzen H.J., Yuan Z. and Van Loosdrecht M.C.M. (2009b). Modeling the PAO–GAO competition: effects of carbon source, pH and temperature. *Water Research*, 43(2), 450-462.

Lopez-Vazquez C.M., Song Y.I., Hooijmans C.M., Brdjanovic D., Moussa M.S., Gijzen H.J. and van Van Loosdrecht M.C.M. (2007). Short-term temperature effects on the anaerobic metabolism of glycogen accumulating organisms. *Biotechnology and Bioengineering,* 97(3), 483-495.

Lopez-Vazquez C.M., Song Y.I., Hooijmans C.M., Brdjanovic D., Moussa M.S., Gijzen H.J. and van Van Loosdrecht M.C.M. (2008b). Temperature effects on the aerobic metabolism of glycogen accumulating organisms. *Biotechnology and Bioengineering,* 101(2):295-306.

Lu H., Oehmen A., Virdis B., Keller J. and Yuan Z. (2006). Obtaining highly enriched cultures of *Candidatus Accumulibacter phosphatis* through alternating carbon sources. *Water Research,* 40(20), 3838-3848.

Majed N., Chernenko T., Diem M. and Gu A.Z. (2012). Identification of Functionally Relevant Populations in Enhanced Biological Phosphorus Removal Processes Based On Intracellular Polymers Profiles and Insights into the Metabolic Diversity and Heterogeneity. *Environmental Science & Technology,* 46(9), 5010-5017.

Mamais D. and Jenkins D. (1992). The effects of MCRT and temperature on enhanced biological phosphorus removal. *Water Science and Technology,* 26(5-6), 955-965.

Mamais D., Jenkins D. and Pitt P. (1993). A rapid physical-chemical method for the determination of readily biodegradable soluble COD in municipal wastewater. *Water Research,* 27(1), 195-197.

Marais G.v.R. and Ekama G.A. (1976). The activated sludge process Part I - steady state behaviour. *Water SA,* 2, 164-200.

Marais G.v.R., Loewenthal R.E., and Siebritz I.P. (1983). Review: Observations supporting phosphate removal by biological excess uptake. *Water Science and Tech*nology, 15(3/4), 15-41.

Marques R., Santos J., Hien N., Carvalho G., Noronha J.P., Nielsen P.H., Reis M.A.M. and Oehmen A. (2017). Metabolism and ecological niche of Tetrasphaera and Ca. Accumulibacter in enhanced biological phosphorus removal. *Water Research,* 122, 159-171.

Marques R., Ribera-Guardia A., Santos J., Carvalho G., Reis M.A.M., Pijuan M. and Oehmen A. (2018). Denitrifying capabilities of Tetrasphaera and their contribution towards nitrous oxide production in enhanced biological phosphorus removal processes. *Water Research,* 137, 262-272.

Marsden M.G. and Marais G.v.R. (1977). *The role of the primary anoxic reactor in denitrification and biological phosphorus removal.* Research Report No.W19, Dept. of Civil. Eng., University of Cape Town, South Africa.

Martin H.G., Ivanova N., Kunin V., Warnecke F., Barry K.W., McHardy A.C., Yeates C., He S.M., Salamov A. A., Szeto E., Dalin E., Putnam N.H., Shapiro H.J., Pangilinan J.L., Rigoutsos I., Kyrpides N.C., Blackall L.L., McMahon K.D. and Hugenholtz P. (2006). Metagenomic analysis of two enhanced biological phosphorus removal (EBPR) sludge communities. *Nature Biotech.*, 24(10), 1263-1269.

Martin K.A.C. and Marais G.v.R. (1975). *Kinetics of enhanced phosphorus removal in the activated sludge process.* No. Research Report W14, Dept. of Civil Eng., University of Cape Town, South Africa.

Maszenan A.M., Seviour R.J., Patel B.K.C., Schumann P., Burghardt J., Tokiwa Y. and Stratton H.M. (2000). Three isolates of novel polyphosphate-accumulating Gram-positive cocci, obtained from activated sludge, belong to a new genus, *Tetrasphaera gen. nov.*, and description of two new species, *Tetrasphaera japonica sp, nov* and

Tetrasphaera australiensis sp nov. Internat. J. Systematic Evolutionary Microbiol., 50, 593-603.

Maurer M., Abramovich D., Siegrist H., and Gujer W. (1999). Kinetics of biologically induced phosphorus precipitation in waste-water treatment. *Water Research,* 33(2), 484-493.

Maurer M., Gujer W., Hany R. and Bachmann S. (1997). Intracellular carbon flow in phosphorus accumulating organisms from activated sludge systems. *Water Research,* 31(4), 907-917.

McClintock S.A., Randall C.W. and Pattarkine V.M. (1993). Effects of temperature and mean cell residence time on biological nutrient removal processes. *Water Environment Research*, 65(5), 110-118.

McKinney R.E. and Ooten R.J. (1969). Concepts of complete mixing activated sludge. *Proceedings of the 19th Sanitary Engineering Conference*, University of Kansas, USA, 32-59.

McLaren A.R. and Wood R.J. (1976). Effective phosphorus removal from sewage by biological means. *Water SA, 2*(1), 47-50.

Meganck M., Malnou D.P.L.F., Faup G.M. and Rovel J.M. (1985). The importance of the acidogenic microflora in biological phosphorus removal. *Water Science and Technology,* 17(11/12), 199-212.

Meijer S.C.F. (2004). Theoretical and practical aspects of modelling activated sludge processes. PhD thesis Delft University of Technology, the Netherlands. ISBN 90-9018027-3.

Meyer R.L., Saunders A.M. and Blackall L.L. (2006). Putative glycogen accumulating organisms belonging to alphaproteobacteria identified through rRNA-based stable isotope probing. *Microbiology-SGM,* 152, 419–429.

Milbury W.F., McCauley D. and Hawthorne C.H. (1971). Operation of conventional activated sludge for maximum phosphorus removal. *Journal of the Water Pollution Control Federation,* 43, 1890-1901.

Mino T., Arun V., Tsuzuki Y. and Matsuo T. (1987). Effect of phosphorus accumulation on acetate metabolism in the biological phosphorus removal process. Paper presented at the IAWPRC Biological Phosphate Removal from Wastewater, Rome, Italy.

Mino T., Van Loosdrecht M.C.M. and Heijnen J.J. (1998). Microbiology and biochemistry of the enhanced biological phosphate removal process. *Water Research,* 32(11), 3193-3207.

Mino T., Wen-Tso L., Kurisu F. and Matsuo T. (1994). Modelling glycogen storage and denitrification capability of microorganisms in enhanced biological phosphate removal processes. Spec. conference on modelling of wastewater treatment, Kollekolle, Denmark.

Mulbarger M.C. (1970). The Three Sludge System for Nitrogen and Phosphorus Removal. Proc. 44th Annual Conference of the Water Pollution Control Federation, San Francisco, California, U.S.A.

Murphy M. and Lötter L.H. (1986). The effect of acetate on polyphosphate formation and degradation in activated sludge with particular reference to *Acinetobacter calcoaceticus*: A microscopic study. *Water SA,* 12, 63-66.

Nakamura K., Ishikawa S. and Kawaharasaki M. (1995). Phosphate-uptake and release activity in immobilized polyphosphate-accumulating bacterium *Microlunatus phosphovorus* strain NM-1. *Journal of Fermentation and Bioengineering,* 80(4), 377-382.

Nguyen H.T.T., Le V.Q., Hansen A.A., Nielsen J.L. and Nielsen P.H. (2011). High diversity and abundance of putative polyphosphate-accumulating Tetrasphaera-related bacteria in activated sludge systems. *FEMS Microbiology Ecology*, 76(2), 256-267.

Nguyen H.T.T., Kristiansen R., Vestergaard M., Wimmer R. and Nielsen P.H. (2015). Intracellular Accumulation of Glycine in Polyphosphate-Accumulating Organisms in Activated Sludge, a Novel Storage Mechanism under Dynamic Anaerobic-Aerobic Conditions. *Applied and Environmental Microbiology,* 81(14), 4809-4818.

Nicholls H.A. (1975). Full scale experimentation on the new Johannesburg extended aeration plant. *Water SA,* 1(3), 121-132.

Nicholls H.A. (1978). Kinetics of phosphorus transformations in aerobic and anaerobic environments. *Prog. Wat. Tech., 10.*

Nicholls H.A. (1982). Application of the Marais-Ekama activated sludge model to large plants. *Water Science and Technology,* 14, 581-598.

Nicholls H.A. (1987). Improvement to the stability of the biological phosphate removal process at the Johannesburg Northern Works. *Proc. BPR from Wastewaters,* Rome, Italy. 261-272.

Nicholls H.A. and Osborn D.W. (1979). Bacterial stress: Prerequisite for biological removal of phosphorus. *Journal of the Water Pollution Control Federation,* 51, 557-569.

Nicholls H.A., Pitman A.R. and Osborn D.W. (1985). The readily biodegradable fraction of sewage: Its influence on phosphorus removal and measurement. *Water Science and Technology,* 17(11/12), 73-87.

Nielsen P.H., McIlroy S.J. Albertsen M. and Nierychlo M. (2019). Re-evaluating the microbiology of the enhanced biological phosphorus removal process.

Current Opinion in Biotechnology, 57, 111-118.

Oehmen A., Lemos P.C., Carvalho G., Yuan Z., Keller J., Blackall L.L. and Reis M.A.M. (2007). Advances in enhanced biological phosphorus removal: From micro to macro scale. *Water Research,* 41(11), 2271-2300.

Oehmen A., Saunders A.M., Vives M.T., Yuan Z. and Keller J. (2006a). Competition between polyphosphate and glycogen accumulating organisms in enhanced biological phosphorus removal systems with acetate and propionate as carbon sources. *Journal of Biotechnology,* 123(1), 22-32.

Oehmen A., Vives M.T., Lu H., Yuan Z. and Keller J. (2005a). The effect of pH on the competition between polyphosphate-accumulating organisms and glycogen-accumulating organisms. *Water Research,* 39(15), 3727-3737.

Oehmen A., Yuan Z., Blackall L.L. and Keller J. (2005b). Comparison of acetate and propionate uptake by polyphosphate accumulating organisms and glycogen accumulating organisms. *Biotechnology and Bioengineering,* 91(2), 162-168.

Oehmen A., Zeng R.J., Saunders A.M., Blackall L.L, Keller J. and Yuan Z. (2006a). Anaerobic and aerobic metabolism of glycogen accumulating organisms selected with propionate as the sole carbon source. *Microbiology,* 152(9), 2767-2778.

Oehmen A., Saunders A.M., Vives M.T., Yuan Z. and Keller J. (2006b). Competition between polyphosphate and glycogen accumulating organisms in enhanced biological phosphorus removal systems with acetate and propionate as carbon sources. *Journal of Biotechnology*, 123(1), 22-32.

Oehmen A., Yuan Z., Blackall L.L. and Keller J. (2004). Short-term effects of carbon source on the competition of polyphosphate accumulating organisms and glycogen accumulating organisms. *Water Science and Technology*, 50(10), 139-144.

Oehmen A., Carvalho G., Lopez-Vazquez C.M., Van Loosdrecht M.C.M. and Reis M.A.M. (2010). Incorporating microbial ecology into the metabolic modelling of polyphosphate accumulating organisms and glycogen accumulating organisms. *Water Research,* 44(17), 4992-5004.

Ong Y.H., Chua A.S.M., Fukushima T., Ngoh G.C., Shoji T. and Michinaka A. (2014). High-temperature EBPR process: The performance, analysis of PAOs and GAOs and the fine-scale population study of Candidatus "Accumulibacter phosphatis". *Water Research*, 64, 102-112.

Onnis-Hayden A., Srinivasan V., Tooker N.B., Li G., Wang D., Barnard J.L., Bott C., Dombrowski P., Schauer P., Menniti A., Shaw A., Stinson B., Stevens G., Dunlap P., Takács I., McQuarrie J., Phillips H., Lambrecht A., Analla H., Russell A. and Gu A.Z. (2019). Survey of full scale sidestream enhanced biological phosphorus removal (S2EBPR) systems and comparison with conventional EBPRs in North America: Process stability, kinetics, and microbial populations. *Water Environment Research*, 92(3).403-417.

Osborn D.W. and Nicholls H.A. (1978). Optimisation of the activated sludge process for the biological removal of phosphorus. *Prog. Wat. Tech.* 10(1/2), 261-277.

Oyserman B.O., Noguera D.R., Del Rio T.G., Tringe S.G. and McMahon K.D. (2016). Metatranscriptomic insights on gene expression and regulatory controls in Candidatus Accumulibacter phosphatis. *Isme Journal* 10(4), 810-822.

Pattarkine V.M. and Randall C.W. (1999). The requirement of metal cations for enhanced biological phosphorus removal by activated sludge. *Water Science and Technology,* 40(2), 159–165.

Pijuan M., Oehmen A., Baeza J.A., Casas C. and Yuan Z. (2008). Characterizing the biochemical activity of full scale enhanced biological phosphorus removal systems: a comparison with metabolic models. *Biotechnology and Bioengineering*, 99(1), 170-179.

Pijuan M., Ye L. and Yuan Z. (2010). Free nitrous acid inhibition on the aerobic metabolism of poly-phosphate accumulating organisms. *Water Research*, 44(20), 6063-6072.

Pinzon A., Brdjanovic D, Moussa M., Lopez-Vazquez C.M., Meijer S., Van Straaten H., Janssen A., Van Loosdrecht MCM and Amy G. (2007). Modelling Oil Refinery Wastewater Treatment Plant. *Environ. Technol.,* 29(11).

Pitman A.R. (1991). Design considerations for nutrient removal activated sludge plants. *Water Science and Technology,* 23(4-6), 781-790.

Pitman A.R., Vandalsen L. and Trim B.C. (1988). Operating experience with biological nutrient removal at the Johannesburg Bushkoppie works. *Water Science and Technology,* 20(4-5), 51-62.

Puig S., Coma M., Monclús H., Van Loosdrecht M.C.M., Colprim J. and Balaguer M.D. (2008). Selection between alcohols and volatile fatty acids as external carbon sources for EBPR. *Water Research*, 42(3), 557-566.

Qiu G., Zuniga-Montanez R., Law Y., Thi S.S., Nguyen T.Q.N., Eganathan K., Liu X., Nielsen P.H.,

Williams R.B.H. and Wuertz S. (2019). Polyphosphate-accumulating organisms in full-scale tropical wastewater treatment plants use diverse carbon sources. *Water Research*, 149, 496-510.

Rabinowitz B. and Marais G.v.R. (1980). Chemical and biological phosphorus removal in the activated sludge process. MASc thesis, Univ. Cape Town, South Africa, Res. Rep. No. W32.

Ramphao M.C., Wentzel M.C., Merritt R., Ekama G.A., Young and Buckley C.A. (2005). The impact of solid-liquid separation on design of biological nutrient removal activated sludge systems. *Biotechnology and Bioengineering*, 89(6), 630-646.

Randall A.A., Benefield L.D. and Hill W.E. (1994). The effect of fermentation products on enhanced biological phosphorus removal, polyphosphate storage, and microbial population dynamics. *Water Science and Technology,* 30(6), 213-219.

Rensink J.H., Donker H.J.G.W. and Vries H.P.D. (1981). Biological P-removal in domestic wastewater by the activated sludge process. Proc. 5th Eur. Sewage and Refuse Symp. Munich, Germany.

Rubio-Rincon F.J., Lopez-Vazquez C.M., Welles L., Van Loosdrecht M.C.M. and Brdjanovic D. (2017). Cooperation between Candidatus Competibacter and Candidatus Accumulibacter clade I, in denitrification and phosphate removal processes. *Water Research,* 120, 156-164.

Rubio-Rincon F.J., Weissbrodt D.G., Lopez-Vazquez C.M., Welles L., Abbas B., Albertsen M., Nielsen P.H., Van Loosdrecht M.C.M. and Brdjanovic D. (2019). "Candidatus Accumulibacter delftensis": A clade IC novel polyphosphate-accumulating organism without denitrifying activity on nitrate. *Water Research,* 161, 136-151.

Rubio Rincon F.J., Welles L., Lopez Vazquez C.M., Abbas B., Van Loosdrecht M.C.M. and Brdjanovic D. (2019). Effect of lactate on the microbial community and process performance of an EBPR system. *Frontiers in Microbiology*, 10, 125.

Saito T., Brdjanovic D. and Van Loosdrecht M.C.M. (2004). Effect of nitrite on phosphate uptake by phosphate accumulating organisms. *Water Research*, 38(17), 3760-3768.

Satoh H., Mino T. and Matsuo T. (1992). Uptake of organic substrates and accumulation of polyhydroxyalkanoates linked with glycolysis of intracellular carbohydrates under anaerobic conditions in the biological excess phosphate removal processes. *Water Science and Technology,* 26(5-6), 933-942.

Saunders A.M., Mabbett A.N., McEwan A.G. and Blackall L.L. (2007). Proton motive force generation from stored polymers for the uptake of acetate under anaerobic conditions. *FEMS Microbiol. Lett.* 274(2), 245-251.

Scalf M.R., Pfeffer F.M., Lively L.D., Witherow J.O. and Priesing C.P. (1969). Phosphate removal in Baltimore, Maryland. *J. Sanitary Eng. Div. Am. Soc. Civil Eng.,* 95(SA5) 817-827.

Schuler A.J. and Jenkins D. (2003). Enhanced biological phosphorus removal from wastewater by biomass with different phosphorus contents, Part 1: Experimental results and comparison with metabolic models. *Water Environment Research,* 75(6), 485-498.

Schuler A.J. and Jenkins D. (2002). Effects of pH on enhanced biological phosphorus removal metabolisms. *Water Science and Technology*, 46(4-5), 171-178.

Schuler A.J. and Jenkins, D. (2003a). Enhanced biological phosphorus removal from wastewater by biomass with different phosphorus contents, part I: experimental results and comparison with metabolic models. *Water Environment Research*, 75(6), 485-498.

Schuler A.J. and Jenkins D. (2003b). Enhanced biological phosphorus removal from wastewater by biomass with different phosphorus contents, part III: anaerobic sources of reducing equivalents. *Water Environment Research*, 75(6), 512-522.

Sell R.L., Krichten D.J., Noichl O.J. and Hartzog D.G. (1981). Low temperature biological phosphorus removal. In 54th WPCF Conference, Detroit, USA.

Seviour R.J., Mino T. and Onuki M. (2003). The microbiology of biological phosphorus removal in activated sludge systems. *FEMS Microbiol. Rev.* 27(1), 99-127.

Shapiro J. and Levin G.V. (1967). Anoxically induced release of phosphate in wastewater treatment. *Journal of the Water Pollution Control Federation,* 39, 1810-1818.

Siebritz I.P., Ekama G.A. and Marais G.v.R. (1980). Excess biological phosphorus removal in the activated sludge process at warm temperature climate. *Proc. WasteTreatment Utilization. 2, 233-251*, Pergamon Press, Toronto, Canada.

Siebritz I.P., Ekama G.A. and Marais G.v.R. (1982) A parametric model for biological excess phosphorus removal. *Water Science and Technology,* 15(3/4), 127-152.

Simpkins M.J. and McLaren A.R. (1978). Consistent biological phosphate and nitrate removal in an activated sludge plant. *Prog. Wat. Tech.* 10(5/6), 433-442.

Smolders G.J.F., Van den Meij J., Van Loosdrecht M.C.M. and Heijnen J.J. (1995). A structured metabolic model for anaerobic and aerobic stoichiometry and kinetics of the biological phosphorus removal process. *Biotechnology and Bioengineering,* 47, 277-287.

Smolders G.J.F., Van der Meij J., Van Loosdrecht M.C.M., and Heijnen J.J. (1994a). Model of the anaerobic metabolism of the biological phosphorus removal process: Stoichiometry and pH influence. *Biotechnology and Bioengineering,* 43, 461-470.

Smolders G.J.F., Van der Meij J., Van Loosdrecht M.C.M. and Heijnen J.J. (1994b). Stoichiometric model of the aerobic metabolism of the biological phosphorus removal process. *Biotechnology and Bioengineering,* 44(7), 837-848.

Smolders G.J.F., Van Loosdrecht M.C.M., Heijnen J.J., Henze M. and Gujer W. (1994c). A metabolic model for the biological phosphorus removal process. *Water Science and Technology,* 31(2), 79-93.

Smolders G.J.F., Van der Meij J., Van Loosdrecht M.C.M. and Heijnen J.J. (1994d). Model of the anaerobic metabolism of the biological phosphorus removal process: stoichiometry and pH influence. *Biotechnology and Bioengineering*, 43(6), 461-470.

Smolders G.J.F., Van Loosdrecht M.C.M. and Heijnen J.J. (1996). Steady-state analysis to evaluate the phosphate removal capacity and acetate requirement of biological phosphorus removing mainstream and sidestream process configurations. *Water Research*, 30(11), 2748-2760.

Sorm R., Bortone G., Saltarelli R., Jenicek P., Wanner J. and Tilche A. (1996). Phosphate uptake under anoxic conditions and fixed-film nitrification in nutrient removal activated sludge system. *Water Research,* 30(7), 1573-1584.

Spatzierer G., Ludwig C. and Matsche N. (1985). Biological; phosphorus removal in combination with simultaneous precipitation. *Water Science and Technology,* 17(11/12), 163-176.

Srinath *e.g.*, Sastry C.A. and Pillai S.C. (1959). Rapid removal of phosphorus from sewage by activated sludge. *Experientia,* 15, 339-340.

Stante L., Cellamare C.M., Malaspina F., Bortone G. and Tilche A. (1997) Biological phosphorus removal by pure culture of *Lampropedia* spp. *Water Research,* 31(6), 1317-1324.

Stern L.B. and Marais G.v.R. (1974). *Sewage as the electron donor in biological denitrification.* Research Report No.W7, Dept. of Civil Eng., University of Cape Town, South Africa.

Stokholm-Bjerregaard M., McIlroy S.J., Nierychlo M., Karst S.M., Albertsen M. and Nielsen P.H. (2017). A critical assessment of the microorganisms proposed to be important to enhanced biological phosphorus removal in full-scale wastewater treatment systems. *Frontiers in Microbiology*, 8, 718.

Timmerman M.W. (1979). Biological phosphate removal from domestic wastewater using anaerobic/aerobic treatment. *Development Ind. Microbiology,* 20.

Tu Y. and Schuler A.J. (2013). Low acetate concentrations favor polyphosphate-accumulating organisms over glycogen-accumulating organisms in enhanced biological phosphorus removal from wastewater. *Environmental Science and Technology*, 47(8), 3816-3824.

Vacker D., Connell C.H. and Wells W.N. (1967). Phosphate removal through municipal wastewater treatment at San Antonio, Texas. *Journal of the Water Pollution Control Federation*, 39, 750-771.

Van Loosdrecht M.C.M., Brandse F.A. and De Vries A.C. (1998). Upgrading of waste water treatment processes for integrated nutrient removal - The BCFS process. *Water Science and Technology,* 37(9), 209-217.

Van Loosdrecht M.C.M., Smolders G.J., Kuba T. and Heijnen J.J. (1997). Metabolism of micro-organisms responsible for enhanced biological phosphorus removal from wastewater - Use of dynamic enrichment cultures. *Antonie Van Leeuwenhoek Int. J. Gen. Molec. Microbiology,* 71(1-2), 109-116.

Van Loosdrecht M.C.M., Nielsen P.H, Lopez-Vazquez C.M., and Brdjanovic D. Ed. (2016) Experimental Methods in Wastewater Treatment. *IWA Publising,* pg. 350, ISBN 9781780404745 (Hardback) 9781780404752 (eBook).

Venter S.L., Halliday J. and Pitman A.R. (1978). Optimisation of the Johannesburg Olifantsvlei extended aeration plant for phosphorus removal. *Prog. Wat. Tech.* 10(1/2), 279-292.

Viconneau J.C., Hascoet M.C. and Florentz M. (1985). The first application of biological phosphorus removal in France. In Proc. Int. Conf. Management Strategies for Phosphorus in the Environment, Selper Ltd. UK.

Vlekke G.J.F.M., Comeau Y. and Oldham W.K. (1988) Biological phosphate removal from wastewater with oxygen or nitrate in sequencing batch reactors. *Environ. Tech. Lett.,* 9, 791-796.

Vollertsen, J. (2002) Biologiske processer i den mekaniske rensedel – OUR ma ˚linger pa ˚

Renseanlæg Vest, January–April 2002. In Report to Aalborg Municipality, pp. 14 (in Danish).

Vollertsen J., Petersen G. and Borregaard V. R. (2006). Hydrolysis and fermentation of activated sludge to enhance biological phosphorus removal. *Water Science and Technology*, 53(12), 55-64.

Wagner M., Erhart R., Manz W., Amann R., Lemmer H., Wedi D. and Schleifer K.H. (1994). Development of an rRNA-targeted oligonucleotide probe specific for the genus *Acinetobacter* and its application for *in situ* monitoring in activated sludge. *Applied and Environmental Microbiology,* 60(3), 792-800.

Wang D., Tooker N.B., Srinivasan V., Li G., Fernandez L.A., Schauer P., Menniti A, Maher C., Bott C.B., Dombrowski P., Barnard J.L., Onnis-Hayden A. and Gu A.Z. (2019). Side-stream enhanced biological phosphorus removal (S2EBPR) process improves system performance - A full-scale comparative study. *Water Research*, 167, 115109.

Watanabe A., Miya A. and Matsuo Y. (1984). Laboratory scale study on biological phosphate removal using synthetic waste water: Removal performance and the investigation of enhanced phosphorus accumulating organisms. IAWPRC, *Newsletter of the Study Group on Phosphate Removal in Biological Sewage Treatment Processes,* 2(1), 40-43.

Welles L., Lopez-Vazquez C.M., Hooijmans C.M., Van Loosdrecht M.C.M. and Brdjanovic D. (2016). Prevalence of 'Candidatus Accumulibacter phosphatis' type II under phosphate limiting conditions. *Amb Express*, 6(1), 44.

Welles L., Tian W.D., Saad S., Abbas B., Lopez-Vazquez C.M., Hooijmans C.M., van Loosdrecht M.C.M. and Brdjanovic D. (2015). Accumulibacter clades Type I and II performing kinetically different glycogen-accumulating organisms metabolisms for anaerobic substrate uptake. *Water Research*, 83, 354-366.

Wentzel M.C., Dold P.L., Ekama G.A. and Marais G.v.R. (1985). Kinetics of biological phosphorus release. *Water Science and Technology,* 17(11/12), 57-71.

Wentzel M.C., Dold P.L., Ekama G.A. and Marais G.v.R. (1989a). Enhanced polyphosphate organism cultures in activated sludge systems 3. Kinetic model. *Water SA,* 15(2), 89-102.

Wentzel M.C., Ekama G.A. and Marais G.v.R. (1992). Processes and modelling of nitrification-denitrification biological excess phosphorus removal systems – a review. *Water Science and Tech*nology, 25(6), 59-82.

Wentzel M.C., Ekama G.A., Dold P.L. and Marais G.v.R. (1990). Biological excess phosphorus removal - Steady-state process design. *Water SA,* 16(1), 29-48.

Wentzel M.C., Ekama G.A., Loewenthal R.E., Dold P.L. and Marais G.v.R. (1989b). Enhanced polyphosphate organism cultures in activated-sludge systems .2. Experimental behavior. *Water SA,* 15(2), 71-88.

Wentzel M.C., L.H. Lötter Loewenthal R.E. and Marais G.v.R. (1986). Metabolic behaviour of *Acinetobacter* spp. in enhanced biological phosphorus removal - A biochemical model. *Water SA,* 12, 209-224.

Wentzel M.C., Lotter L.H., Ekama G.A., Loewenthal R.E. and Marais G.v.R. (1991). Evaluation of biochemical-models for biological excess phosphorus removal. *Water Science and Technology,* 23(4-6), 567-576.

Wentzel M.C., Marais G.v.R., Loewenthal R.E. and Ekama G.A. (1988). Enhanced polyphosphate organism cultures in activated-sludge systems - Part 1. Enhanced culture development. *Water SA,* 14(2), 81-92.

Wentzel M.C., Mbewe A. and Ekama G.A. (1995). Batch test for measurement of readily biodegradable COD and active organism concentrations in municipal wastewaters. *Water SA,* 21(2), 117-124.

Whang L.M. and Park J.K. (2006). Competition between polyphosphate- and glycogen-accumulating organisms in enhanced biological phosphorus removal systems: Effect of temperature and sludge age. *Water Environmntal Research,* 78(1), 4-11.

Whang L.M. and Park J.K. (2002). Competition between polyphosphate- and glycogen-accumulating organisms in biological phosphorus removal systems - effect of temperature. *Water Science and Technology,* 46(1-2), 191-194.

Whang L.M., Filipe C.D.M. and Park J.K. (2007). Model-based evaluation of competition between polyphosphate- and glycogen-accumulating organisms. *Water Research,* 41(6), 1312-1324.

Wilson D.E. and Marais G.v.R. (1976). *Adsorption phase in biological denitrification.* Research Report No.W11, Dept. of Civil Eng., University of Cape Town, South Africa.

Witherow J.L. (1970). Phosphorus removal in activated sludge. Proc. 24th Industrial Waste Conference, Purdue University, U.S.A.1169.

Wong M.T., Tan F.M., Ng W.J. and Liu W.T. (2004). Identification and occurrence of tetrad-forming alphaproteobacteria in anaerobic–aerobic activated sludge processes. *Microbiology-SGM,* 150, 3741–3748.

WRC (1984) *Theory, design and operation of nutrient removal activated sludge processes.* Water Research Commission, Pretoria, South Africa.

Wu L., Ning D., Zhang B., Li Y., Zhang P., Shan X., Zhang Q., Brown M.R., Li Z., Van Nostrand Y.D., Ling F., Xiao N., Zhang Y., Vierheilig J., Wells G.F,, Yang Y., Deng Y., Tu Q., Wang A., Global Water Microbiome Consortium., Zhang T., He Z., Keller J., Nielsen P.H., Alvarez P.J.J., Criddle C.S., Wagner M., Tiedje J.M., He Q., Curtis T.P., Stahl D.A., Alvarez-Cohen L., Rittmann B.E., Wen X. and Zhou J. (2019). Global diversity and biogeography of bacterial communities in wastewater treatment plants. *Nature Microbiology,* 4(7), 1183-1195.

Xie T., Mo C., Li X., Zhang J., An H., Yang Q., Wang D., Zhao J., Zhong Y. and Zeng, G. (2017). Effects of different ratios of glucose to acetate on phosphorus removal and microbial community of enhanced biological phosphorus removal (EBPR) system. *Environmental Science and Pollution Research*, 24(5), 4494-4505.

Yeoman S., Hunter M., Stephenson T., Lester J.N. and Perry R. (1988). An assessment of excess biological phosphorus removal during activated sludge treatment. *Env. Tech. Lett.*, 9, 637-646.

Yuan Z.G., Pratt S. and Batstone D.J. (2012). Phosphorus recovery from wastewater through microbial processes. *Current Opinion in Biotechnology,* 23(6), 878-883.

Zeng R.J., Van Loosdrecht M.C.M., Yuan Z. and Keller J. (2003). Metabolic model for glycogen-accumulating organisms in anaerobic/aerobic activated sludge systems. *Biotechnology and Bioengineering,* 81(1), 92-105.

Zeng R.J., Yuan Z. and Keller J. (2006). Effects of solids concentration, pH and carbon addition on the production rate and composition of volatile fatty acids in prefermenters using primary sewage sludge. *Water Science and Technology*, 53(8), 263-269.

Zheng X.L., Su P.D., Han J.Y., Song Y.Q., Hu Z.R., Fan H.Q. and Lu S.Y. (2014). Inhibitory factors affecting the process of enhanced biological phosphorus removal (EBPR) - A mini-review. *Process Biochemistry,* 49(12), 2207-2213.

Zhou Y., Pijuan M. and Yuan Z. (2007). Free nitrous acid inhibition on anoxic phosphorus uptake and denitrification by polyphosphate accumulating organisms. *Biotechnology and Bioengineering*, 98(4), 903-912.

Zhou Y., Pijuan M., Zeng R.J., Lu H. and Yuan Z. (2008). Could polyphosphate-accumulating organisms (PAOs) be glycogen-accumulating organisms (GAOs)? *Water Research*, 42(10-11), 2361-2368.

Zhou Y., Pijuan M., Oehmen A. and Yuan Z. (2010). The source of reducing power in the anaerobic metabolism of polyphosphate accumulating organisms (PAOs) - a mini-review. *Water Science and Technology,* 61(7), 1653-1662.

NOMENCLATURE

Symbol	Description	Unit
a	Mixed liquor recycle ratio based on influent flow	$m^3.d/m^3.d$
a_{opt}	A-recycle ratio that gives a minimum N_{ne}	$m^3.d/m^3.d$
b_{OHO}	Specific endogenous mass loss rate of the OHOs	gEVSS/gVSS.d
$b_{OHO,T}$	OHO specific endogenous mass loss rate at temperature T	gEVSS/gVSS.d
b_{PAO}	Specific endogenous mass loss rate of the PAOs	gEVSS/gVSS.d
$b_{PAO,T}$	PAO specific endogenous mass loss rate at temperature T	gEVSS/gVSS.d
COD_b	Concentration of biodegradable COD	$gCOD/m^3$
$COD_{b,i}$	Concentration of biodegradable COD in the influent	$gCOD/m^3$
$COD_{b,OHO}$	Concentration of biodegradable COD available to the OHOs	$gCOD/m^3$
D_{P1}	Denitrification potential of the primary anoxic reactor	$gNO_3\text{-}N/m^3$ influent
D_{P3}	Denitrification potential of the secondary anoxic reactor	$gNO_3\text{-}N/m^3$ influent

f_{xa}	Anaerobic mass fraction	gVSS/gVSS
f_{x1}	Primary anoxic reactor mass fraction	gVSS/gVSS
$f_{x1,min}$	Minimum primary anoxic mass fraction	gVSS/gVSS
f_{x3}	Secondary anoxic reactor mass fraction	gVSS/gVSS
$FCOD_{b,i}$	Daily mass of influent biodegradable organics	gCOD/d
$FCOD_{b,OHO}$	Daily mass of biodegradable substrate available to OHOs	gCOD/gCOD
$FCOD_{i}$	Daily mass of influent COD	gCOD/d
f_{CV}	COD/VSS ratio of the sludge	gCOD/gVSS
$f_{FSS,OHO}$	Inorganic content of OHOs	gFSS/gTSS
$f_{FSS,PAO}$	Inorganic content of PAOs	gFSS/gTSS
f_{n}	N content of the sludge	gN/gVSS
FN_{s}	Daily mass of nitrogen required for sludge production	gN/d
FO_{c}	Daily mass of carbonaceous oxygen demand	gO_2/d
FO_{OHO}	Daily mass of oxygen consumed by OHOs	gO_2/d
FO_{PAO}	Daily mass of oxygen consumed by PAOs	gO_2/d
FO_{t}	Daily mass of total oxygen demand	gO_2/d
$f_{P,FSS}$	Fraction of P in the fixed (inorganic) suspended solids	gP/gFSS
$f_{P,FSS,i}$	Fraction of P in the influent FSS	gP/gFSS
$f_{P,OHO}$	Fraction of P in the active OHO mass	gP/gAVSS
$f_{P,PAO}$	Fraction of P in the active PAO mass	gP/gAVSS
$f_{P,TSS}$	P content with respect to TSS	gP/gTSS
f_{p}	P content with respect to VSS	gP/gVSS
f_{p}	Fraction of P in the OHO endogenous mass	gP/gEVSS
f_{p}	Fraction of P in the PAO endogenous mass	gP/gEVSS
f_{p}	Fraction of P in the inert mass	gP/gIVSS
$f_{PO4,rel}$	Ratio of P release/VFA uptake	gP/gCOD
$FS_{F,CONV}$	Daily mass of fermentable COD converted into VFAs in the anaerobic reactors	gCOD/d
$f_{SU,CODi}$	Influent unbiodegradable soluble COD fraction	gCOD/gCOD
$FS_{PO4,rel}$	Daily mass of P released by PAOs	gP/d
$f_{SS,CODi}$	Influent readily biodegradable fraction of influent total COD	gCOD/gCOD
f_{SS}	Influent readily biodegradable fraction of the influent biodegradable COD	gCOD/gCOD
$FS_{S,PAO}$	Daily mass of S_S stored by PAOs in the anaerobic reactor	gCOD/d
$FS_{VFA,i}$	Daily mass of influent VFAs	gCOD/d
$f_{SVFA,SSi}$	Fraction of VFAs of the readidly biodegradable COD	gCOD/g COD
f_{VT}	VSS/TSS ratio for OHO active and endogenous masses, PAO endogenous mass and inert mass	gVSS/gTSS
$f_{VT,PAO}$	VSS/TSS ratio for PAO active mass	gVSS/gTSS
$f_{xd,max}$	Maximum anoxic mass fraction	gVSS/gVSS
$f_{XE,OHO}$	Fraction of endogenous residue of the OHOs	gEVSS/gAVSS
$f_{XE,PAO}$	Fraction of endogenous residue of the PAOs	gEVSS/gAVSS
$FX_{FSS,i}$	Daily mass of influent inorganics	gFSS/d

$f_{XU,CODi}$	Fraction of influent unbiodegradable particulate COD	g COD/gCOD
$f_{x,max}$	Maximum unaerated mass fraction	g VSS/gVSS
$FX_{S,i}$	Daily mass of influent slowly-biodegradable COD	g COD/d
HRT_{np}	Average nominal hydraulic retention time of the process	d
$K_{1,T}$	Specific denitrification rate in primary anoxic reactor of NDEBPR system on RBCOD at temperature T	gNO_3^--N/gOHOVSS.d
$K_{2,T}$	Specific denitrification rate in primary anoxic reactor of NDEBPR system on SBCOD at temperature T	gNO_3^--N/gOHOVSS.d
$K_{3,T}$	Specific denitrification rate in secondary anoxic reactor of NDEBPR system on SBCOD at temperature T	gNO_3^--N/gOHOVSS.d
$k_{F,T}$	First-order fermentation rate constant at temperature T	m^3/gOHOVSS.d
K_T	Specific denitrification rate of OHOs for an NDEBPR system (') at temperature T	gNO_3^--N/gOHOVSS.d
$MX_{E,OHOv}$	Mass of OHO endogenous residue in the system	gEVSS
$MX_{E,PAOv}$	Mass of PAO endogenous residue in the system	gEVSS
MX_{FSS}	Mass of fixed (inorganic) suspended solids in the system	gFSS
MX_{OHOv}	Mass of OHOs in the system	gAVSS
MX_{PAOv}	Mass of PAO in the system	gAVSS
MX_{TSS}	TSS mass in the system	gTSS
MX_{Uv}	Mass of inert organic matter in the system, coming from the influent	gVSS (or gIVSS)
MX_{VSS}	Mass of volatile suspended solids in the system	gTSS
n	Number of the anaerobic reactor from a series	-
N	Total number of anaerobic reactors of equal volume in the series n = 1,2.....N	-
NIT_c	Nitrification capacity of the bioreactor	gNO_3-N/m^3
Q_i	Daily average influent flow rate	m^3/d
$Q_{i,ADWF}$	Average dry weather flow	Ml/d
r	Mixed-liquor recycle ratio from the aerobic to anoxic (or anaerobic) reactor based on influent flow	m^3.d/m^3.d
s	Return activated sludge recycle ratio based on influent flow	m^3.d/m^3.d
S_{Alk}	Alkalinity concentration	$mgCaCO_3$/l
S_F	Fermentable organic matter concentration	gCOD/m^3
$S_{F,ANn}$	Fermentable organic matter conc. in the nth AN reactor	gCOD/m^3
$S_{F,conv}$	Fermentable organic matter converted into VFAs per volume of influent	gCOD/m^3
$S_{F,DENIT}$	Fermentable substrate consumed by denitrification in the anaerobic reactor	gCOD/m^3
$S_{F,i}$	Fermentable organic matter concentration in the influent	gCOD/m^3
$S_{F,i,conv}$	$S_{F,i}$ available for conversion into VFAs per volume of influent	gCOD/m^3
$S_{F,OXID}$	Fermentable substrate consumed by aerobic oxidation in the anaerobic reactor	gCOD/m^3
$S_{NO3,e}$	Effluent nitrate concentration	gNO_3-N/m^3
$S_{NO3,i}$	Influent nitrate concentration (to the AN reactor)	gNO_3-N/m^3
$S_{NO3,s}$	Nitrate conc. in the sludge recycle to the AN reactor	gNO_3-N/m^3

S_{O2}	Dissolved oxygen concentration	gO_2/m^3
$S_{O2,a}$	Oxygen conc. in the anoxic recycle to the AN reactor	gO_2/m^3
$S_{O2,i}$	Influent oxygen concentration	gO_2/m^3
$S_{O2,s}$	Oxygen concentration in the sludge recycle to the AN reactor	gO_2/m^3
$S_{PO4,rel}$	Concentration of P released	gP/m^3
SRT	Sludge age	d
$S_{S,i}$	Influent readily biodegradable COD concentration	$gCOD/m^3$
$S_{S,PAO}$	Concentration of S_S stored by PAOs	$gCOD/m^3$
$S_{U,i}$	Influent inert soluble organic matter concentration	$gCOD/m^3$
S_{VFA}	Volatile fatty acids concentration	$gCOD/m^3$
$S_{VFA,i}$	VFA concentration in the influent	$gCOD/m^3$
t	Time	h
T	Temperature	°C
TKN	Total Kjeldahl nitrogen concentration	gN/m^3
$TKN_{i,s}$	Influent TKN required for biomass synthesis	gN/m^3
T_{min}	Minimum temperature	°C
P_e	Effluent total phosphorus concentration	gP/m^3
P_i	Influent total phosphorus concentration	gP/m^3
TSS	Total suspended solids	$gTSS/m^3$
V_R	Volume of biological process (bioreactor)	l
VSS	VSS concentration	$gVSS/m^3$
$X_{FSS,i}$	Influent fixed suspended solids (FSS) concentration	$g\ FSS/m^3$
X_{OHO}	Ordinary heterotrophic organism concentration	$gCOD/m^3$
$X_{OHO,AN}$	Concentration of OHOs in the anaerobic reactor	$gCOD/m^3$
X_{PAO}	Phosphorus-accumulating organisms	$gCOD/m^3$
X_S	Slowly biodegradable organics concentration	$g\ COD/m^3$
$X_{S,i}$	Influent slowly biodegradable organics concentration	$g\ COD/m^3$
X_{TSS}	Reactor total suspended-solids concentration	$g\ TSS/m^3$
$X_{TSS.OX}$	Selected required TSS concentration in the aerobic reactor	$gTSS/m^3$
$X_{U,i}$	Influent inert particulate matter concentration	$gCOD/m^3$
X_{VSS}	Reactor volatile suspended solids concentration	$gVSS/m^3$
$X_{VSS.OX}$	Selected desired TSS concentration in the aerobic reactor	$gVSS/m^3$
Y_{OHOv}	OHO biomass yield	gAVSS/gCOD
ΔP_{OHO}	P removal due to OHOs	gP/m^3 influent
ΔP_{PAO}	P removal due to PAOs	gP/m^3 influent
ΔP_{SYS}	Total P removal by the system	gP/m^3 influent
$\Delta P_{SYS.actual}$	Total P actual removal by the system	gP/m^3 influent
$\Delta P_{SYS,pot}$	Total P potential removal by the system	gP/m^3 influent
ΔP_{XE}	P removal due to endogenous residue mass	gP/m^3 influent
ΔP_{XU}	P removal due to inert mass	gP/m^3 influent

Abbreviation	Description
A/O	Anaerobic/oxic process
A^2O	Anaerobic, anoxic, aerobic process
AN	Anaerobic
AX	Anoxic
AVSS	Active volatile suspended solids
BNR	Biological nitrogen removal
DDGGE	Dry denaturing gradient gel electrophoresis
e	Effluent
EBPR	Enhanced biological phosphorus removal
EM	Electron microscopy
EVSS	Endogenous residue as volatile suspended solids
FISH	Fluorescence *in situ* hybridisation
FSS	Fixed (inorganic) suspended solids
HRT	Hydraulic retention time
IVSS	Inert volatile suspended solids
i	Influent
JHB	Johannesburg process
MLE	Modified Ludzack-Ettinger process
MLSS	Mixed-liquor suspended solids
MLVSS	Mixed-liquor volatile suspended solids
MUCT	Modified UCT process
NIT	Nitrifying organisms
ND	Nitrification-denitrification
NDEBPR	Nitrification-denitrification EBPR
OHO	Ordinary heterotrophic organism
OUR	Oxygen uptake rate
OX	Aerobic
PAO	Phosphate-accumulating organism
PHA	Poly-β-hydroxyalkanoates
PHB	Poly-β-hydroxybutyrate
PHV	Poly-β-hydroxyvalerate
PO_4	Phosphate
RAS	Return activated sludge
RBCOD	Readily biodegradable COD
SBCOD	Slowly biodegradable particulate organic matter
SBR	Sequencing batch reactor
SRT	Sludge retention time
SST	Secondary settling tank
TCA	Tricarboxylic acid cycle
TKN	Total Kjeldahl nitrogen
TN	Total nitrogen
TP	Total phosphorus

TSS	Total suspended solids
UCT	University of Cape Town process
VFA	Volatile fatty acid
VSS	Volatile suspended solids
VIP	Virginia initiative plant process
w	Sludge wastage from the aerobic reactor
ws	Sludge wastage from the sludge recycle line

Greek symbols	Description	Units
α	Constant alpha	
β	Constant beta	
$\mu_{ANO,max,20}$	Maximum specific growth rate of nitrifiers at 20 °C	d^{-1}
$\theta_{k,F}$	Arrhenius temperature coefficient for k_F	-
η	Reduction factor for aerobic hydrolysis/growth process rates on SBCOD for anoxic conditions	
$\theta_{b,OHO}$	Arrhenius temperature coefficient for b_{OHO}	-
$\theta_{b,PAO}$	Arrhenius temperature coefficient for b_{PAO}	-

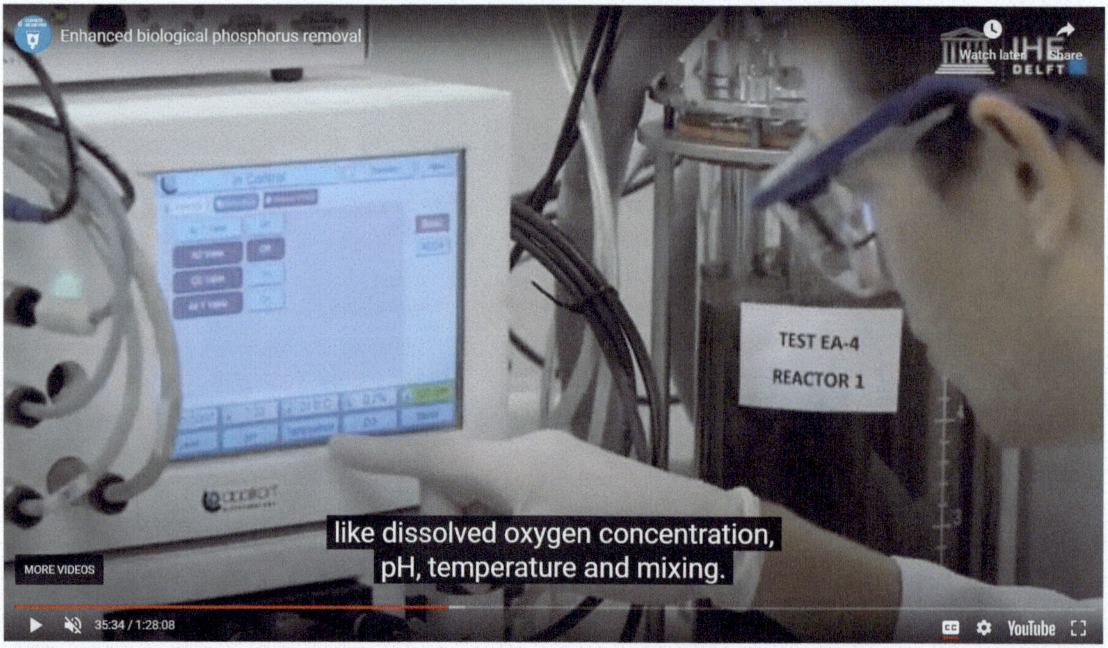

Figure 6.35 Fundamental research using PAO cultures enriched in laboratory-scale sequencing batch reactors (SBRs) have contributed significantly to development of metabolic models. The photo is a screenshot of the online course on EBPR (https://experimentalmethods.org/courses/activated-sludge-activity-tests/) based on Van Loosdrecht *et al.*, 2016.

doi: 10.2166/9781789060362_0327

7

Innovative sulphur-based wastewater treatment

Hui Lu, Di Wu, Tianwei Hao, Ho Kwong Chui, George A. Ekama, Mark C.M. van Loosdrecht and Guanghao Chen

7.1 INTRODUCTION

Sulphur, the least common of the five macro-elements in the biosphere, is widely transformed and translocated via biological and chemical reactions in the biosphere. It presents mainly in the form of mercapto groups in sulphur-containing amino acids (1% wt./dry wt.) in cells. Gypsum ($CaSO_4$), metal sulphides (FeS_2) in rocks and sediments ($7{,}800 \cdot 10^{18}$ g sulphur), and sulphate in seawater ($1{,}280 \cdot 10^{18}$ g sulphur) are the major sources of sulphur in nature (Muyzer and Stams, 2008). Both natural (seawater intrusion and volcanic eruption) and anthropogenic processes (industrial production) emit a large quantity of sulphur into the environment. Across Europe, groundwater in more than 100 regions in 10 countries is affected by seawater intrusion (EEA, 2006); additionally, sulphate-laden wastewater is produced by various industries,

such as farming, flue-gas desulphurization (FGD) at power stations, pulp and paper-making, brewing, pharmaceutical, food production, fish farming, tanning, and petrochemical and mining industries (Lens *et al.*, 2003). Out of these, acid mine drainage (AMD) from the mining industry is currently recognized as one of the most serious water and soil pollution sources in the world (Huisman *et al.*, 2006). It is estimated that the total costs of sulphur-induced treatment range from 2% to 3% of the gross domestic product in developed countries (Kruger, 2011); the annual costs of rehabilitating and preventing sulphur-associated sewer corrosion are estimated at billions of dollars (Pikaar *et al.*, 2014). Control of sulphate-reducing bacteria (SRB) plays a pivotal role in mitigating these problems. Nevertheless, despite half a century of research and development efforts, a widely accepted conclusion has been drawn that no practical methods exist to prevent sulphate reduction caused by anaerobic treatment (Lens *et al.*, 1998). Selective inhibition of SRB by molybdate, transition elements, or antibiotics have been unsuccessful in full-scale implementations. However, alternative SRB-based bioprocesses and biotechnologies have been in development in the last three decades, *e.g.* biogenic sulphide processes for heavy-metal removal from metal-laden wastewater and AMD (THIOTEQ and SULPHATEQ), as well as biological desulphurization of natural gas, flue gas, and liquefied petroleum gas (Hao, 2014).

In treatment of domestic wastewater, elemental sulphur-dosed post-denitrification methods through induced autotrophic denitrification have recently been developed to improve total nitrogen (TN) removal (*e.g.* sulphur: the limestone autotrophic denitrification (SLAD) process) (Zhang and Lamphe, 1999). Such organic independent autotrophic denitrification is particularly attractive when retrofitting and upgrading existing wastewater treatment plants (WWTPs) treating carbon-deficient domestic wastewater that is typical in developing nations. In the case of China, the COD/TN ratio of municipal wastewater ranges between 5.4-10.9, which is lower than the 8-12 required for satisfactory removal of nutrients (nitrogen and phosphorus) (Sun *et al.*, 2016).

In Hong Kong, China (hereafter referred to as Hong Kong), an area that is well-known for being water-scarce, 20% of the total water supply comes from seawater for toilet flushing, covering 80% of the total population (7.5 million) in 2019, saving over 760,000 m^3 of freshwater every day. This system generates a large amount of saline wastewater that provides an opportunity to develop a unique wastewater treatment technology different from the centuries-old freshwater wastewater treatment technology employed elsewhere. Since the introduction of the nitrogen removal process in the 1960s (see Chapter 1), the fundamental biological wastewater treatment theory has been built on electron transfer from organic carbon to free oxygen (dissolved oxygen) and/or combined oxygen (NO_3^- and NO_2^-) for nitrogen and organic removal, as seen in Figure 7.1.

In general, 50-60% of organic carbon in wastewater is converted into CO_2 and the rest becomes sludge (Tchobanoglous *et al.*, 2003). Sludge treatment (thickening, stabilization, dewatering and disinfection) and disposal demand the most complex and costly processes in WWTPs, accounting for over half of the operational cost (Saby *et al.*, 2003; Ekama *et al.*, 2011). Various techniques, therefore, have been investigated to reduce this cost by minimizing excess sludge production at source via a variety of chemical, mechanical/physical, chemical and thermal treatment methods; however, all of them result in extra cost and/or space in application (Saby *et al.*, 2003; Foladori *et al.*, 2010).

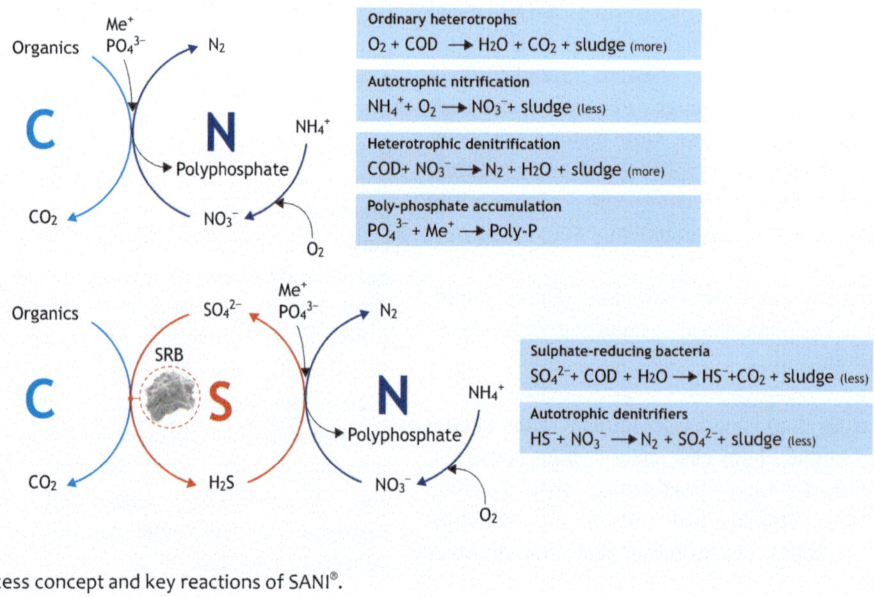

Figure 7.1 Process concept and key reactions of SANI®.

With the above in mind, since 2003 the authors have been developing a new biological wastewater treatment process, namely the Sulphate reduction, Autotrophic denitrification, Nitrification Integrated (SANI®) process. The novelty of this sulphur-based biological treatment system lies in the use of sulphidogenic bacteria, *i.e.* sulphate-reducing bacteria (SRB) for carrying out sulphidogenesis (sulphate reduction). In this anaerobic reaction, sludge production becomes minimal because 96% of organics can be theoretically converted into CO_2 according to the process reaction chemistry and electron balance analysis (refer to Eq. 7.21 in Section 7.4.4). The dissolved sulphide produced offers adequate electron donors driving the autotrophic denitrification reaction without organic carbon, producing minimal sludge as compared with a heterotrophic denitrification reaction which receives electrons from organic carbon. Integration of these two sulphur-based bioprocesses, the sulphate-reducing bioprocess and the sulphide-oxidizing (autotrophic denitrification) bioprocess, forms the SANI® process. The concept of this novel biological treatment process builds on sulphate→sulphide→sulphate conversion-mediated organic carbon (C) and nitrogen (N) removal, which moves C and N removal integration into C, N, and S removal integration (for more detail see Section 7.4).

7.2 SULPHATE-REDUCING BIOPROCESS

7.2.1 Fundamental of this bioprocess

Sulphate present in the ocean and other water bodies can undergo rapid biological conversions mediated by SRB that are ubiquitous in natural and engineered environments. SRB use oxidized forms of sulphurs (*e.g.* mainly elemental sulphur (S^0), thiosulphate ($S_2O_3^{2-}$), and sulphate (SO_4^{2-})) as electron acceptors and sulphide as the end product (Muyzer and Stams, 2008). SRB at one time were believed to be capable of growing only on organic matter, *e.g.* acetate (CH_3COO^-), pyruvate, lactate and propionate, while energy was conserved mainly or exclusively via substrate-level phosphorylation (see Eq. 7.1). However, Badziong and Thauer (1978) reported that *Desulfovibrio vulgaris* proliferates with hydrogen (H_2) as the electron donor and SO_4^{2-} as the electron acceptor in demonstration of the electron transport phosphorylation in the metabolism of SRB for the first time (see Eq. 7.2). As well as oxidized sulphur species, other competent electron acceptors for SRB are alkylbenzene sulphonates, dimethylsulphoxide, NO_3^-, NO_2^-, iron (III), oxygen, fumarate, acrylate, arsenate, chromate, and uranium (Rabus *et al.*, 2013).

$$CH_3COO^- + SO_4^{2-} \rightarrow 2HCO_3^- + HS^-$$
$$-47.6\ (\Delta G^{0'}\ \text{kJ/reaction}) \qquad (7.1)$$

$$4H_2 + SO_4^{2-} + H^+ \rightarrow HS^- + 4H_2O$$
$$-151.9\ (\Delta G^{0'}\ \text{kJ/reaction}) \qquad (7.2)$$

7.2.1.1 Sulphate-reducing pathways

Considering the significance of biological sulphur conversion in environmental engineering, it is advisable to define the sulphate reduction in assimilatory and dissimilatory categories. Before sulphur compounds can be assimilated into biosynthetic pathways, they need to be reduced to hydrogen sulphide (H_2S). The pathways catalysing the reduction of sulphate to sulphide for the purpose of incorporation into synthesized molecules are named the 'pathways of assimilatory sulphate reduction', whereas the pathways for the purpose of energy production are referred to as 'pathways of dissimilatory sulphate reduction'.

Assimilatory sulphate reduction

Sulphate uptake and its incorporation in many biological systems has been well-documented (Bothe and Trebst, 1981). Sulphate uptake is accomplished through active transport mediated by a carrier enzyme system. The cellular metabolism of sulphate begins with a series of activation reactions; the reduction of sulphate to sulphite requires two electrons at a standard redox potential ($E^{0'}$) of -516 mV, which is higher than the physiological electron carriers (Thauer *et al.*, 1977). In this case, sulphate pro-actively participates in the formation of *adenosine* 5'-*phosphosulphate* (APS) or 3'-*phosphoadenylyl-sulphate* (PAPS). In many bacteria and lower eukaryotes, assimilatory sulphate reduction proceeds via activation of sulphate to APS by the enzyme of sulphate adenylyltransferase (*i.e.* ATP sulphurylase).

Subsequently, three pathways proceed with the sulphite reduction (see Figure 7.2). The first pathway is in enteric bacteria, *e.g.* cyanobacteria and yeast; the produced APS is phosphorylated to generate PAPS, which is catalysed by the *adenylyl-sulphate kinase* (APS kinase), and the generated PAPS is then reduced to sulphite with the assistance of thioredoxin (or glutaredoxin)-dependent *phosphoadenylyl-sulphate* reductase (thioredoxin) (PAPS reductase), and finally a reduction of sulphite to sulphide, catalysed by the assimilatory sulphite reductase (NADPH). The second pathway is that enzymes of adenylyl-sulphate reductase (glutathione) and adenylyl-sulphate reductase (thioredoxin) catalyse the second step, *i.e.* direct reduction of APS to sulphite, and subsequently NADPH is used as an electron donor (for most bacteria) with the catalysis of plant sulphite reductases and the enzyme from *Allochromatium vinosum* to reduce sulphite to sulphide; in bacteria and plants sulphide is finally incorporated into O-acetyl-L-serine by cysteine synthase yielding L-cysteine and later converting L-methionine, which are the two major sulphur-containing precursor compounds. The third pathway is that a thioredoxin-dependent enzyme catalyses the direct reduction of APS to sulphite. Pathways 2 and 3 share similar procedures except for the APS reduction enzyme. In both of these pathways, sulphite is reduced to sulphide by the action of assimilatory sulphite reductase (ferredoxin) viz. a ferredoxin-dependent enzyme and a NADPH-dependent enzyme (Lillig *et al.*, 2001).

Dissimilatory sulphate reduction

As one of the most common electron acceptors for anaerobic respiration, dissimilatory sulphate reduction is performed by prokaryotes, a

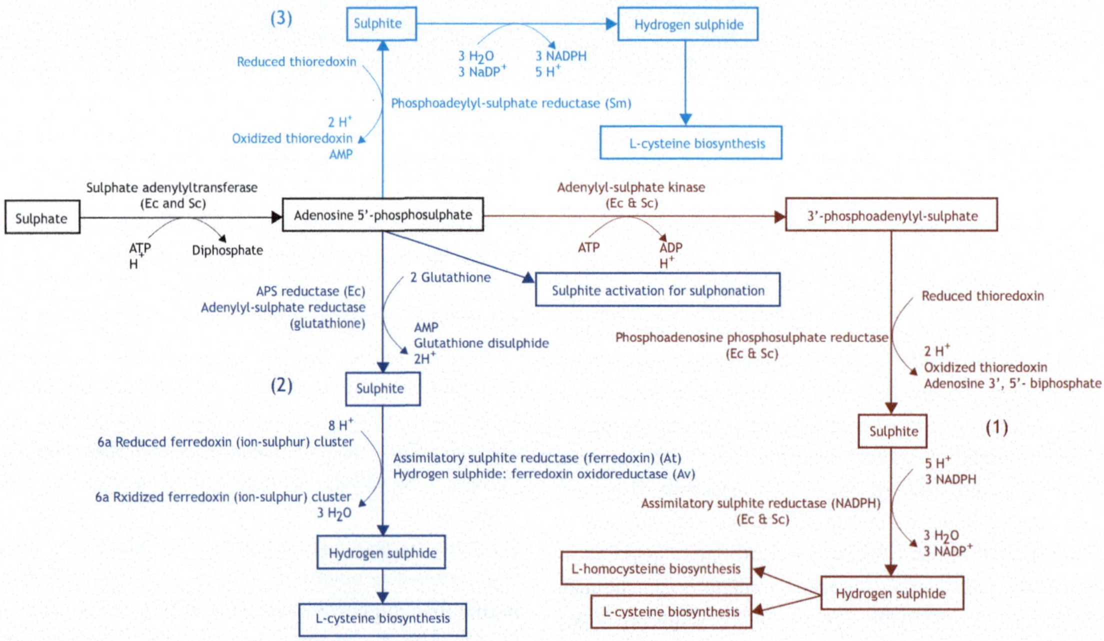

Figure 7.2 Pathways of assimilatory sulphate reduction.

heterogeneous group of bacteria and archaea consisting of diverse phyla, such as Archaeoglobi, Proteobacteria, Firmicutes, Nitrospirae and others, that can use sulphate as a terminal electron acceptor. As all the steps in the sulphate reduction occur in the cytoplasm, similar to the assimilatory conversion, sulphate must be transported across the cytoplasmic membrane into the cells. The overall process of dissimilatory sulphate reduction can be considered as two phases: firstly, the reduction of sulphate to bisulphite through the intermediate compound APS, and secondly, the reduction of bisulphite to sulphide. The biochemistry of the first phase, with respect to the pathway by which sulphate is reduced to sulphite, has been firmly established. However, in contrast, the pathway of bisulphite reduction to sulphide is still uncertain (see Figure 7.3).

Reduction of SO_3^{2-} to S^{2-} is presently regarded as a 'black box' and two most commonly accepted mechanisms are proposed: (*i*) SO_3^{2-} is reduced to HS^- in a single step by a single enzyme, bisulphite reductase (Peck and LeGall, 1982), which was found present in *Desulfovibrio*, and (*ii*) an alternative pathway proposed by Fitz and Cypionka (1989) involves several enzymes (*i.e.* sulphite reductase, trithionate reductase and thiosulphate reductase) and intermediates in the final step of sulphate reduction in which trithionate and thiosulphate are produced (named the trithionate pathway). The trithionate pathway was further confirmed by using in vitro assays of *Desulfovibrio vulgaris* in the investigation of how dissimilatory sulphite reductase (DsrAB) reduce SO_3^{2-} (Santos *et al.*, 2015). In this pathway, the reduction of SO_3^{2-} to DsrC trisulphide is first performed by DsrAB, which is a common enzyme involved in dissimilatory sulphur metabolism. The produced DsrC trisulphide is then reduced to sulphide and DsrCr by the DsrMKJOP complex, as shown in Figure 7.3. DsrC acts as a physiological partner of DsrAB in the reduction of SO_3^{2-}. When DsrC is absent (*i.e.* only DsrAB is present) or in limited supply (*i.e.* there is much more SO_3^{2-} than DsrC), $S_2O_3^{2-}$ may become dominant due to the reaction between SO_3^{2-} and S^{2-} (in vivo) or the partial reduction of sulphate with DsrAB (in vitro) (Leavitt *et al.*, 2015). Despite the remaining uncertainties of this pathway, current findings offer overwhelming support for the argument that DsrC and DsrAB are the central and key proteins in the metabolism dissimilating SO_3^{2-} to S^{2-} by SRB. Recent advances in analytical techniques have enabled intracellular sulphur metabolites and their isotopic compositions such as ^{34}S and ^{33}S to be examined to reveal the sulphur transfer pathway. The detection accuracy, however, is still limited due to the interference by the abundant phosphate.

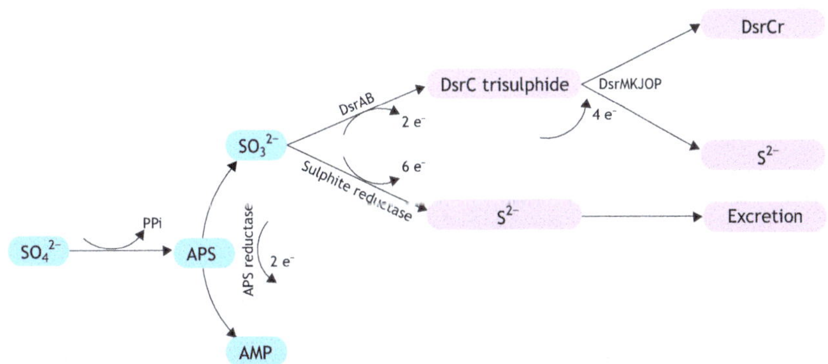

Figure 7.3 Pathways of dissimilatory sulphate reduction.

7.2.1.2 Biochemical reactions involved in sulphate-reducing bioprocesses

All of the main reactions associated with sulphur reactions involving living organisms are closely related to the sulphur and carbon cycles. On the other hand, the amount of carbon involved in the fluxes of the sulphur cycle through biogenic processes varies depending on the characteristics of the organisms undertaking the metabolism. Four main types of reductions are involved in heterotrophic growth of SRB which include: (*i*) complete oxidation of acidified intermediates to CO_2, (*ii*) incomplete oxidation of acidified intermediates to CH_3COO^-, (*iii*) syntrophic degradation of intermediates by acetogenic SRB that are subjected to the efficiency of hydrogen-utilizing bacteria; and (*iv*) fermentative growth of SRB in the presence of propionate and ethanol. Generally, 24 g of organic carbon is mineralized for each 32 g of reduced sulphate through acetyl CoA or a modified tricarboxylic acid cycle (TCA) pathway in heterotrophic sulphate reduction. Many intermediate products originating from anaerobic fermentation/hydrolysis can be metabolized by SRB, such as amino acids, sugars, long-chain fatty acids, aromatic compounds, lactate, butyrate, propionate and CH_3COO^- (see Figure 7.4). Typical reactions involved in sulphidogenesis with electron donors from macromolecule (sugar) to micromolecule (CH_3COO^-) are summarized in Table 7.1. The organic substrate is degraded to either CO_2 or CH_3COO^- depending upon the microbial strains and reaction completion degree. In freshwater or a low sulphur-containing water environment, SRB can play a role in fermentation and oxidation of organic compounds. Many *Desulfovibrio* and *Desulfomicrobium* species can grow on degrading pyruvate to form CH_3COO^-, CO_2 and H_2 as products. With the assistance of hydrogen-consuming methanogens that efficiently remove H_2 and keep a low H_2 pressure in the system, SRB are able to oxidize lactate and ethanol to CH_3COO^-, thereby resulting in a syntrophic growth between sulphate reducers and methanogens under anaerobic conditions. Furthermore, by manipulating the sulphate level in an engineered anaerobic environment (*e.g.* anaerobic bioreactors), SRB can perform like acetogens, which can be applied to improve CH_3COO^- production during the methanogenesis.

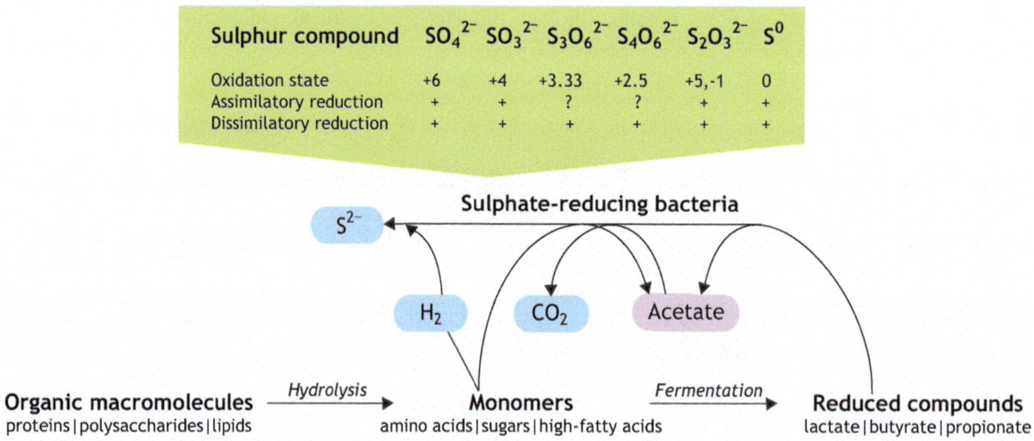

Figure 7.4 Bioreaction of microbial degradation of organic matter through sulphate reduction (+: confirmed; ?: uncertain).

Table 7.1 Stoichiometric reactions involved in the sulphidogenesis.

Sulphur reduction	$\Delta G^{0'}$ (kJ/reaction)
Glucose + SO_4^{2-} → 2 CH_3COO^- + HS^- + 2 HCO_3^- + 3 H^+	-385.2
2 $CH_3CHOHCOO^-$ + SO_4^{2-} → 2 CH_3COO^- + HS^- + 2 HCO_3^- + H^+	-160.1
2 $CH_3CHOHCOO^-$ + 3 SO_4^{2-} → HS^- + 6 HCO_3^- + H^+	-255.3
$CH_3(CH_2)_2COO^-$ + 0.5 SO_4^{2-} → 2 CH_3COO^- + 0.5 HS^- + 0.5 H^+	-27.8
$CH_3(CH_2)_2COO^-$ + 3 SO_4^{2-} + 2 H_2 → CH_3COO^- + HS^- + HCO_3^- + 2 H_2O	-198.4
$CH_3CH_2COO^-$ + 0.75 SO_4^{2-} → CH_3COO^- + 0.75 HS^- + HCO_3^- + 0.25 H^+	-37.7
$CH_3CH_2COO^-$ + 1.75 SO_4^{2-} → 1.75 HS^- + 3 HCO_3^- + 0.25 H^+	-85.4
$CH_3CH_2COO^-$ + SO_4^{2-} + H_2 → CH_3COO^- + HS^- + HCO_3^- + 2 H_2O	-75.8
2 CH_3CH_2OH + SO_4^{2-} → 2 CH_3COO^- + HS^- + H_2O + H^+	-22
2 CH_3OH + SO_4^{2-} → 2 $HCOO^-$ + HS^- + 2 H_2O + H^+	-108.3
CH_3COO^- + 4 S^0 + 4 H_2O → 4 H_2S + 2 HCO_3^- + 2 H^+	-6.9
3 CH_3COO^- + 4 SO_3^{2-} → 1 CO_3^{2-} + 5 HCO_3^- + 4 HS^-	-80

7.2.2 Key microorganisms driving sulphate reduction

SRB are a group of microbes composed of sulphate-reducing bacteria and sulphate-reducing archaea that can perform respiration with oxidized sulphur compounds in their energy metabolism. With over a century's investigation, more than 120 species and 40 genera belonging to four bacterial classes and two archaeal classes have been recorded, namely (*i*) mesophilic δ -proteobacteria with the genera *Desulfovibrio*, *Desulfobacterium*, *Desulfobacter* and *Desulfobulbus*; (*ii*) thermophilic Nitrospirae with the genus *Thermodesulfovibrio*; (*iii*) Clostridia with the genera *Desulfotomaculum* and *Ammonifex*; (*iv*) thermophilic Thermodesulfobacteria with the genera *Caldimicrobium* and *Thermodesulfobacterium*; (*v*) archaeal Thermoprotei with the genus *Caldivirga*; and (*vi*) archaeal Archaeoglobi with the genus *Archaeoglobus* (Barton and Hamilton, 2007). Among the 40 genera, 16 genera are incomplete organic oxidizers, 22 genera are complete oxidizers, and the remaining 2 genera, *i.e.* *Desulfotomaculum* and *Desulfomonile*, cannot be simply aligned in either of the two divisions as they appear to have both complete and incomplete oxidizing species (see Table 7.2). Besides the common oxidized sulphur compounds, NO_3^-, NO_2^-, ferric, manganese oxide and other compounds (*e.g.* fumarate, DMSO) can also serve as an electron acceptor for some SRB species.

Regarding their physiological property, most SRB genera belong to mesophiles with growth temperature ranging from 25 to 40 °C. Growth temperature is a fundamental variable governing the possible application of SRB-based technologies for environmental remediation or industrial production. For instance, *Desulfovibrio*, *Desulfobulbus*, *Desulfomicrobium* and *Desulfobacter* are the most commonly found SRB genera for sulphur-laden wastewater treatment. *Desulfovibrio*, *Desulfobulbus* and *Desulfomicrobium* belong to the group of incomplete organics oxidizers, while *Desulfobacter* is the only dominant complete organics oxidizer identified. Thus, CH_3COO^- residual of sulphate-reducing bioreactors is an inherent feature of SRB.

Table 7.2 Main SRB genera, electron acceptor and optimum temperature (updated from Hao *et al.*, 2014).

	Genus	Cell form	Electron acceptor for growth (other than SO_4^{2-})	Optimum temperature (°C)
Incomplete organic oxidizers	*Desulfovibrio*	Vibrio	$SO_3^{2-}/S_2O_3^{2-}$/fumarate/ Fe(III)/$MnO_2/NO_2^-/NO_3^-/O_2$	30-38
	Desulfomicrobium	Oval to rod	$SO_3^{2-}/S_2O_3^{2-}/NO_2^-$/fumarate/ DMSO	28-37
	Desulfohalobium	Rod	$SO_3^{2-}/S_2O_3^{2-}/S^0$	37-40
	Desulfonatronum	Vibro	$SO_3^{2-}/S_2O_3^{2-}$	37-40
	Desulfobotulus	Vibrio	SO_3^{2-}	34
	Desulfocella	Vibrio	-	34
	Desulfofaba	Vibrio	$SO_3^{2-}/S_2O_3^{2-}$	7
	Desulforegula	Rod	Desulfoviridin	25-30
	Desulfobulbus	Lemon/onion	SO_3^{2-}/ $S_2O_3^{2-}$ /$NO_2^-/NO_3^-/O_2$/ Fe(III)/ graphite	28-39
	Desulfocapsa	Rod	$SO_3^{2-}/S_2O_3^{2-}$/ S^0	20-30
	Desulfofustis	Rod	SO_3^{2-}/S^0	28
	Desulforhopalus	Rod	$SO_3^{2-}S_2O_3^{2-}/NO_3^-$	18-19
	Desulfotalea	Rod	$SO_3^{2-}/S_2O_3^{2-}/S^0$/Fe(III)-citrate	10
	Thermodesulfobacterium	Rod	$SO_3^{2-}/S_2O_3^{2-}$	65-70
	Thermodesulfovibrio	Curved rod	$SO_3^{2-}/S_2O_3^{2-}$/Fe(III)/arsenate	65
	Desulfosporosinus	Straight/curved rod	$SO_3^{2-}/S_2O_3^{2-}/S^0$/Fe(III)	30-37
	Desulfotomaculum[1)]	Vibrio	$SO_3^{2-}/S_2O_3^{2-}/S^0$	30-38; 50-65[2)]
	Desulfomonile[1)]	Rod	3-chlorobenzoatefumarate SO_3^{2-}/ $S_2O_3^{2-}$/ S^0/ NO_3^-	37
Complete organics oxidizers	*Desulfothermus*	Rod to curved	$SO_3^{2-}/S_2O_3^{2-}$	60-65
	Desulfobacter	Rod to ellipsoidal	$SO_3^{2-}/S_2O_3^{2-}$	28-32
	Desulfobacterium	Oval to rod	$SO_3^{2-}/S_2O_3^{2-}$/fumarate	20-35
	Desulfobacula	Oval to curved	$SO_3^{2-}/S_2O_3^{2-}$	28
	Desulfococcus	Sphere	$SO_3^{2-}/S_2O_3^{2-}$	28-36
	Desulfofrigus	Rod	$SO_3^{2-}/S_2O_3^{2-}$/Fe(III)-citrate	10
	Desulfonema	Filaments	$SO_3^{2-}/S_2O_3^{2-}/NO_3^-$	30-32
	Desulfosarcina	Irregular shape/aggregate	$SO_3^{2-}/S_2O_3^{2-}/S^0$	33
	Desulfospira	Curved	$SO_3^{2-}/S_2O_3^{2-}/S^0$	26-30
	Desulfotignum	Rod to curved	$SO_3^{2-}/S_2O_3^{2-}/CO_2$	28-32
	Desulfatibacillum	Rod	$SO_3^{2-}/S_2O_3^{2-}$	28-30
	Desulfarculus	Vibrio	$SO_3^{2-}/S_2O_3^{2-}$	35-39
	Desulforhabdus	Rod to ellipsoid	$SO_3^{2-}/S_2O_3^{2-}$	37
	Desulfovirga	Rod	$SO_3^{2-}/S_2O_3^{2-}/S^0$	35
	Desulfobacca	Oval to rod	$SO_3^{2-}/S_2O_3^{2-}$	37
	Desulfospira	Curved	$SO_3^{2-}/S_2O_3^{2-}/S^0$	26-30
	Desulfacinum	Oval	$SO_3^{2-}/S_2O_3^{2-}/S^0$	60
	Desulfonauticus	Curved rod	$SO_3^{2-}/S_2O_3^{2-}/S^0$	45
	Desulfonatronovibrio	Vibrio	$SO_3^{2-}/S_2O_3^{2-}/S^0/O_2$	37
	Thermodesulforhabdus	Rod	SO_3^{2-}	60
	Thermodesulfobium	Rod	$S_2O_3^{2-}/NO_2^-/NO_3^-$	55
	Archaeoglobus	Irregular coccoid	$SO_3^{2-}/S_2O_3^{2-}$	82-83

[1)] Partial species in the genus metabolizing organics completely; DMSO: dimethyl sulfoxide; [2)] thermophilic species

7.2.3 Electron donors for sulphate-reducing bioprocess

Diverse carbon sources and electron donors drive the metabolism of SRB according to the type of the organisms' growth (autotrophic or heterotrophic). Most electron donors are fermentation products, monomers, or cell components from other sources. Compounds, which can be used as electron donors for SRB, are inorganic substrates (H_2 and CO) and organic substrates (Hansen, 1993).

Hydrogen (an electron donor) serves as an efficient energy source for autotrophic SRB growth and reproduction using sulphate as the electron acceptor. When H_2 and CO_2 are co-utilized as the substrates, high sulphate-reducing rates can be achieved in both mesophilic and thermophilic conditions within 10 days, in which CO_2 serves as a carbon source for cell synthesis. CO is another inorganic electron donor for SRB, while the toxicity against SRB by CO deserves consideration at a concentration range of 2-70% vol., when a synthetic gas mixture of H_2, CO_2 and CO is used. Heterotrophic SRB utilize organic compounds as the electron donors and carbon sources, through a series of enzymatic catalyses. The concentration of organic substrate is normally quantitatively expressed in COD in a bio-reaction; stoichiometrically 0.67 mol of COD are needed for complete reduction of 1 mol sulphate. In a system with deficient COD, extra electron donors or carbon sources must be added to drive the complete reduction of sulphate when sulphate is the target for removal.

Sulphate reduction rates, related reactions and the benefits/drawbacks of using different electron donors in the relevant bioprocesses are summarized in Table 7.3. Electron donor type has a substantial impact on the rate of sulphate reduction as the hydraulic retention times (HRTs) of the bioreactors vary in a wide range of 1-480 h. High heterotrophic sulphate removal rates can be achieved with manure, formate and CH_3COO^- as the carbon sources. In comparison with other fermentive compounds, lactate and propionate have been proven to be more favourable organic substrates than H_2, methanol, ethanol, CH_3COO^- or methane in terms of biomass yield, energy release and production of alkalinity. However, complete lactate oxidation is not achieved by most *Desulfobacter* species or some *Desulfobacterium* species. *Desulfonema magnum* does not grow on lactate. Moreover, direct use of sugars and long-chain fatty acids by SRB has been documented as well (Muyzer and Stams, 2008). *Desulfotomaculum antarcticum* has been reported to use glucose, whereas *Desulfovibrio* and *Desulfotomaculum nigrificans* are able to grow on fructose (Alvarado, 2016). However, molasses cannot be directly degraded by SRB. It has to be fermented by microorganisms such as *Lactobacilli*, prior to being consumed by SRB. Meanwhile, the accumulation of unbiodegradable products through caramelization reduces the biomass activity, thus resulting in a high concentration of COD residual in the effluent (Liamleam and Annachhatre, 2007). Special care should be paid to molasses driving sulphate reduction. The benefits and drawbacks of these electron donor sources are variable, and their selections are dependent on particular reaction requirements.

Due to the diversity of SRB species, a mixture of electron donors is recommended for SRB growth. The mixture of different types of waste containing both relatively readily biodegradable (animal manure, compost, sludge) and recalcitrant cellulosic materials (sawdust or wood chips) gives a better removal efficiency of organics and sulphate by SRB. This implies that locally available organic waste, *i.e.* food/seafood processing industries, animal manure, sewage sludge, molasses, and compost providing complex organic matters, can serve as effective electron donor sources, while agricultural waste, reed canary grass (*Phalaris arundinacea*), sawdust, wood chips, rice straw and leaf compost provide suitable high cellulosic organic matters (Hao *et al.*, 2014).

Table 7.3 Sulphate reduction rates, reactions and benefits/drawbacks of various electron donors for biological sulphate reduction (updated from Hao *et al.*, 2014) (Continued).

Electron donor	Reaction	Sulphate reduction rate (gSO_4^{2-}/l.d)	Pro (+), con (−)
H_2/CO	• $4H_2 + SO_4^{2-} \rightarrow S^{2-} + 4H_2O$ • $4CO + SO_4^{2-} \rightarrow S^{2-} + 4CO_2$	0.4-1.9	+ Low cost + Most SRB can use H_2 as energy source + No organic residual in the effluent + Sufficient supply – Toxicity of CO towards SRB – H_2 mass transfer limits the reaction rate – Induce competition with other organisms (methanogens)
H_2/CO_2	• $4H_2 + SO_4^{2-} + CO_2 \rightarrow HS^- + HCO_3^- + 3H_2O$ If homoacetogenic bacteria present • $4H_2 + 2CO_2 \rightarrow CH_3COO^- + 2H_2O + H^+$ • $CH_3COO^- + SO_4^{2-} \rightarrow HS^- + 2HCO_3^-$	4.5-30.0	+ SRB can outcompete methanogens for H_2 + No organic residual in the effluent – Induce competition with other organisms (methanogens, homoacetogenic bacteria) – Formation of methane lower H_2 utilization efficiency – H_2 safety requirements
Synthetic gas ($H_2 + CO_2 + CO$)	• $4H_2 + SO_4^{2-} \rightarrow S^{2-} + 4H_2O$ • $4CO + SO_4^{2-} \rightarrow S^{2-} + 4CO_2$ • $4H_2 + SO_4^{2-} + CO_2 \rightarrow HS^- + HCO_3^- + 3H_2O$	9.6-14.0	+ Low cost + Some SRB have a much higher tolerance for CO than those documented byprevious studies – Availability may be limited
Methane	• $CH_4 + SO_4^{2-} \rightarrow HCO_3^- + HS^- + H_2O$	$0.4 \cdot 10^{-3}$-0.24	+ Sufficient reserve – Low biomass growth rate
Methanol	• $4CH_3OH + 3SO_4^{2-} \rightarrow 4HCO_3^- + 3HS^- + 4H_2O + H^+$	0.4-20.5	+ Relatively low cost + Require a simple design reactor + SRB can outcompete methanogens at high temperatures (55-70 °C) – Methanogens dominate the community under mesophilic conditions – Only a few SRB strains can utilize methanol
Ethanol	• $C_2H_5OH + 0.5SO_4^{2-} \rightarrow CH_3COO^- + 0.5HS^- + 0.5H^+ + H_2O$ • $CH_3COO^- + SO_4^{2-} \rightarrow 2HCO_3^- + HS^-$ • $4H_2 + SO_4^{2-} + H^+ \rightarrow HS^- + 4H_2O$	0.45-21.00	+ Relatively cheap reagent + Easily converted by SRB – Low biomass yield – Incomplete oxidation to CH_3COO^- leading to high effluent COD concentration
Formate	• $SO_4^{2-} + 4HCOO^- + H^+ \rightarrow HS^- + 4HCO^-$	≤29	+ Produce less CH_3COO^- during formate utilization + Many SRB capable of growing on H_2 can also grow on formate as a sole energy source + A safe alternative for H_2 – Methanogens outcompete SRB at 65-75 °C
Acetate	• $CH_3COO^- + SO_4^{2-} \rightarrow HS^- + 2HCO_3^-$	≤65	+ Methanogens can outcompete SRB for + CH_3COO^- – Only a few SRB can oxidize CH_3COO^- – CH_3COO^- inhibit sulphate reduction at concentrations above 15 mmol/l Low biomass yield
Glucose/acetate	• Glucose + $SO_4^{2-} \rightarrow HS^-$ + • $2\ CH_3COO^- + 2HCO_3^- + 3H^+$	0.9-2.2	– Low pH of system due to fermentation

Table 7.3 Sulphate reduction rates, reactions and benefits/drawbacks of various electron donors for biological sulphate reduction (updated from Hao *et al.*, 2014) (Continued).

Electron donor	Reaction	Sulphate reduction rate (gSO_4^{2-}/l.d)	Pro (+), con (−)
Lactate	• $CH_3CHOHCOOH + 0.5H_2SO_4 \rightarrow CH_3COOH + CO_2 + 0.5H_2S + H_2O$	0.36-5.76	+ Wide spectrum of SRB can grow on lactate + Generate a large amount of alkalinity + Relieve the sulphide toxicity + Preferred carbon source for SRB − High cost
Sucrose/peptone	• $C_{12}H_{22}O_{11} + 5H_2O + 4SO_4^{2-} \rightarrow 4CO_2 + 8H_2 + 4HS^- + 8HCO_3^- + 4H^+$ • $8H_2 + 2SO_4^{2-} + 2H^+ \rightarrow 2HS^- + 8H_2O$	0.6-12.4	+ Sucrose acidification is not inhibited by sulphide + Suitable carbon and energy sources for SRB − CH_3COO^- accumulated in the effluent
Molasses	• $C_{12}H_{22}O_{11} + H_2O \rightarrow 4CH_3CHOHCOOH$ • $CH_3CHOHCOOH + 0.5H_2SO_4 \rightarrow CH_3COOH + CO_2 + 0.5H_2S + H_2O$	1.20-7.22	+ Inexpensive and abundantly available + The acidification products can easily be used by SRB − Partial complex compounds in molasses are barely decomposed resulting in high COD in the effluent − The accumulation of un-biodegradable compounds − Not suitable for SRB growth − Accumulation of volatile fatty acids
Fructose	• $C_6H_{12}O_6 + SO_4^{2-} \rightarrow 2CH_3COO^- + HS^- + 2HCO_3^- + 3H^+$	-	− Only few SRB use fructose − The SRB growth on fructose is slow
Benzene/benzoate	• $C_6H_5COO^- + 0.75SO_4^{2-} + 4H_2O \rightarrow 3CH_3COO^- + 0.75HS^- + HCO_3^- + 2.25H^+$ • $C_6H_5COO^- + 3.75SO_4^{2-} + 4H_2O \rightarrow 7HCO_3^- + 3.75HS^- + 2.25H^+$	0.038	+ Can be completely oxidized to CO_2 without extracellular intermediates − Long degrading time − Cannot be used by some SRB species
Algal extracellular products/algal biomass	-	0.0030-0.0058	+ Low-cost carbon source + Easily utilized by SRB + No available limitation − Cannot be used directly, need fermentative bacteria to collaborate together − May cause high COD in the effluent
Cheese whey	-	0.34	+ Low-cost carbon source + No negative impact on the bacteria − Cannot be used directly, need fermentative bacteria to collaborate together − May cause high COD in the effluent
Watermelon rind	-	0.15-0.24	+ Low cost − Availability may be limited − May cause high COD in the effluent

Table 7.3 Sulphate reduction rates, reactions and benefits/drawbacks of various electron donors for biological sulphate reduction (updated from Hao *et al.*, 2014) (Continued).

Electron donor	Reaction	Sulphate reduction rate (gSO_4^{2-}/l.d)	Pro (+), con (−)
Plant materials:			
– Phalaris-arundinacea	-	2.2-3.3	+ Low cost + Suitable for bioremediation application – May cause high COD in the effluent
– Mixture of wood chips, leaf compost and poultry manure	-	0.01	
– Mushroom compost, wood chips, sawdust, and rice straw	-	0.33-0.57	
Other waste products:			
– Primary sewage sludge	-	2.4	+ Low cost – Some organic matters cannot be used directly – May cause high COD in the effluent
– Animal manure	-	40.3	+ Low cost + Efficient biodegradable substrate – Availability may be limited

7.2.4 Application domain and model parameter

Owing to their physiological properties and wide environmental distribution, SRB-based technologies have been developed and implemented in wastewater and off-gas treatments, metal/radionuclide reduction, hydrocarbon degradation, oil field remediation, concrete corrosion and so on (Rabus *et al.*, 2015; Qian *et al.*, 2015). The major sulphate-reducing biotechnologies developed over the past decades focus on pollutant (organics and sulphate) removal and metal remediation.

7.2.4.1 Sulphur-laden wastewater treatment

Sulphate-reducing processes can be applied for removal and reuse of sulphur compounds from sulphur-laden wastewaters (*e.g.* leather tanning, kraft pulping, food processing, etc.) (Pol *et al.*, 1998), and from FGD wastewater (Roest *et al.*, 2005). For high organic strength sulphur-laden wastewaters (*e.g.* starch, organic acid, ignin, etc.), SRB are primarily intended for removing sulphate from wastewater and concurrently enhancing alkalinity via the carbonization of organic matters (see Figure 7.5A). FGD wastewater lacks sufficient organics for sulphate reduction, demanding external electron donors, such as H_2 and citrate, etc., to be supplied to drive sulphate reduction (see Figure 7.5B). The FGD and its wastewater treatment can be described by the following equations:

$$SO_2 + OH^- \rightarrow HSO_3^- \tag{7.3}$$

$$HSO_3^- + 0.5O_2 \rightarrow SO_4^{2-} + H^+ \tag{7.4}$$

$$HSO_3^- + 6[H] \rightarrow HS^- + 3H_2O \tag{7.5}$$

$$SO_4^{2-} + 8[H] \rightarrow HS^- + 3H_2O + OH^- \tag{7.6}$$

Under a micro-aeration or bio-sulphide oxidation condition, the produced sulphide can be recovered as elemental sulphur. Full-scale applications in treatment of FGD wastewater and other sulphur-laden industrial wastewaters have been applied in China, Brazil and the Netherlands (Roest *et al.*, 2005; Sarti *et al.*, 2011). In Brazil, a sulphidogenic sequencing batch biofilm reactor was adopted for treating sulphonation process wastewater at a sulphate-loading rate of up to 1.3 kgSO_4^{2-}/h, with domestic sewage and ethanol introduced as the external carbon source (Sarti *et al.*, 2011). Figure 7.5 summarizes different sulphate-reducing treatment processes in various applications.

7.2.4.2 Bioremediation of toxic metals

SRB are applied for metal removal/remediation from surface water and industrial-mining wastewater (see Figure 7.5C). SRB-based processes for treatment of wastewater containing dissolved metals involve biological and chemical treatment stages. In the biological treatment stage, biological sulphate reduction is utilized to generate sulphide via dosing with H_2 or organic or sulphur depending on the types of metal-laden wastewater; at the subsequent chemical treatment stage dissolved H_2S forms metal sulphide to be separated via gravity settling and dewatering for refining (see Figure 7.5D). There are other related processes for treatment of metal-containing wastewater reported by Van Houten *et al.* (2009) and DiLoreto *et al.* (2016).

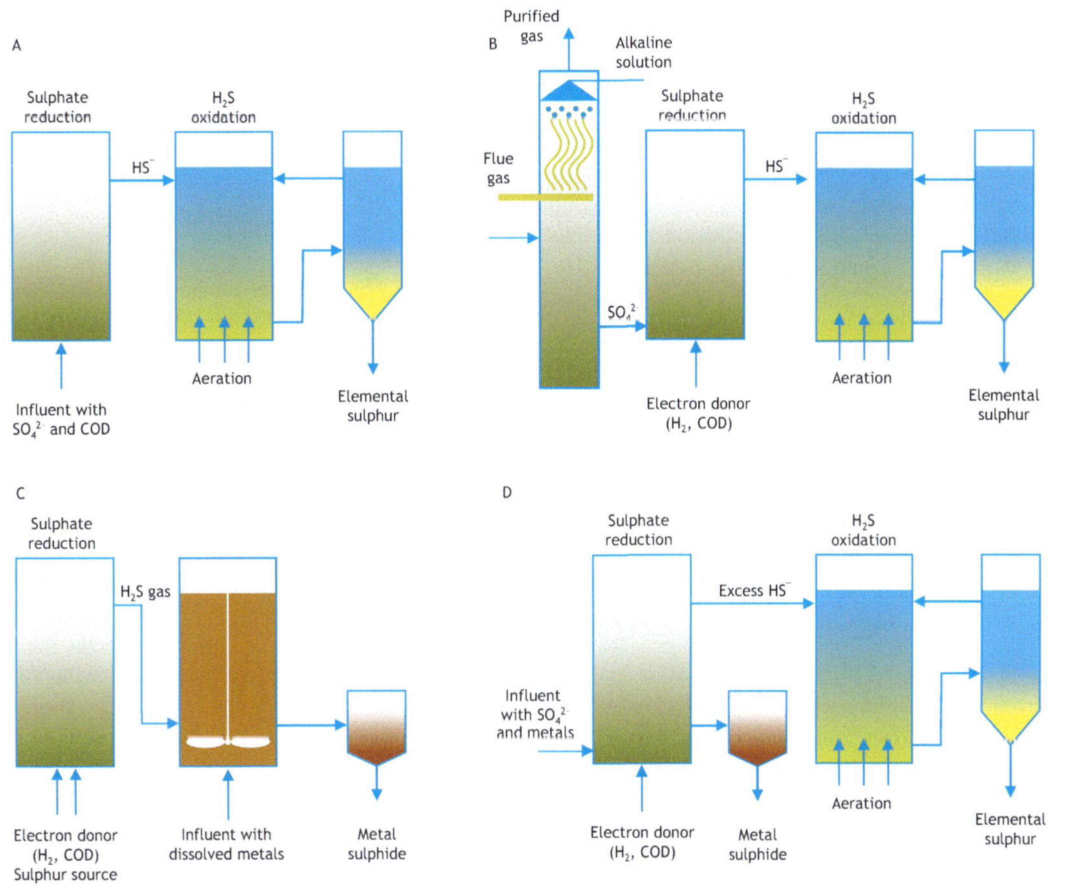

Figure 7.5 Sulphate-reducing processes for treatment of A) wastewater containing sulphate and organics; B) FGD wastewater containing sulphate and insufficient electron donors; C) wastewater containing metal without a sulphur source but with electron donors; and D) wastewater containing sulphate and metal.

Table 7.4 Kinetic parameters of sulphate reduction for different electron donors.

Electron donor	SRB type	μ_{max} (1/d)	Yield (kgVSS/kgCOD)	Ks for eD (kgCOD/m^{3})	Ks for eA (kgSO_4^{2-}/m^{3})
C_4VFA	c4SRB	0.22-0.45	0.023-0.030	0.009-0.100	0.01-0.02
Propionate	pSRB	0.20-0.92	0.022-0.035	0.11-0.30	0.0074-0.0190
CH_3COO^-	aSRB	0.13-0.87	0.024-0.082	0.024-0.220	0.0096-0.0190
H_2	hSRB	0.39-2.80	0.023-0.077	2.50E^{-05}-1.00E^{-04}	0.0009-0.0190

eD: electron donor; eA: electron acceptor

7.2.4.3 Process kinetic parameters

Modelling is generally adopted in reactor design and operation. A dual-term Monod-type kinetic equation is often applied when dealing with simulation of a rate-limiting electron acceptor (sulphate) and/or a rate-limiting electron donor (organic substrate or H_2) in the competition between SRB and methanogenesis. However, in an SRB-dominated system there is no competition between SRB and methanogenesis, thus the dual-term kinetics become unnecessary. Moreover, SRB can use many of the intermediates formed in anaerobic digestion; hence, sulphate reduction is often modelled as general reaction kinetics, for instance, in the extended ADM1 (Fedorovich *et al.*, 2003; Barrera *et al.*, 2015) that includes up to four additional groups of SRB utilizing butyrate/valerate, propionate, CH_3COO^- and H_2 as electron donors, respectively. The kinetic parameters for these SRB are listed in Table 7.4.

7.2.5 Factors that affect sulphate reduction

Despite the ability of SRB to grow in a wide range of environmental conditions (*e.g.* 0-100 °C, low to high salinity, and 1.0-9.8 pH), SRB-based treatment technologies are mostly applied under physiologically suboptimal conditions (Mackenzie, 2005). From the engineering design and operation perspective, improving and optimizing the performance of SRB-based processes have been the key target since the 1990s. However, due to the diverse physiological properties of SRB (*e.g.* growth, temperature, salinity, DO and toxicant tolerance, etc.) and complexity of organic donors, the resultant balance among different variables towards better performance should be the critical norm. Apart from substrate types (carbon and sulphur) and microbial communities, Table 7.5 summarizes other factors that could affect the sulphate reduction efficiency, with the following key factors discussed below.

Temperature

In terms of physiological temperature, SRB can be classified into mesophiles (growth temperature <40 °C), moderate thermophiles (growth temperature 40-60 °C) and extreme thermophiles (>60 °C) (Brahmacharimayum *et al.*, 2019). Mesophilic species predominating in engineered environments are typically *Desulfobacter hydrogenophilus*, *Desulfobacter curvatus*, *Desulfovibrio latus*, *Desulfovibrio vibrioformis* and *Desulfovibrio halotolerans*. Under a natural ambient temperature, the sulphate-reducing rate decreases by approximately 60% when it drops to 10-15 °C. However, the rate can fully recover and improve when the temperature increases to 20-35 °C, although it declines again when the temperature exceeds 40 °C (Alvarado, 2016). In a low temperature application, increasing diversity and abundance of SRB is crucial for maintaining treatment efficiency. On the other hand, in thermophilic processes, the produced H_2S can be easily stripped off, thereby possibly relieving its inhibition while enriching thermophilic SRB (*e.g.* *Thermodesulfobacterium commune* and *Thermodesulfobacterium hveragerdense*).

Table 7.5 Physiological and operational factors affecting the performance of sulphate-reducing reactors (updated from Hao et al., 2014).

Factor	Effect(s)	Condition(s) preferred
Influent components		
Sulphate concentration	• Affects SRB growth and activity • Can be outcompeted by other organisms at a low concentration • High concentrations inhibit SRB activity	• Typical COD/SO_4^{2-} values range between 0.7 and 1.5
Trace element	• Fe, Cu, Zn, Co, Mo, Ni are needed in electron transport • Synthesis of redox-active metalloenzymes • High Mo level inhibits SRB metabolism	• High levels of Fe in culture media are necessary due to sulphide precipitation • Mo above 2 mM completely inhibits SRB[1)]
Metal concentration	• Elevated heavy-metal concentration can reduce or terminate SRB activity	• Desired concentration and the order of decreasing toxicity (mg/l) Cu <4, Cd <11, Ni <13, Zn <16.5, Cr <35, Pb <80
NO_3^- concentration	• NO_3^- is a major inhibitor in the growth and activity of SRB	The impact level: • 70 mM NO_3^- significantly inhibits growth • Long-term injection of 0.25-0.33 mM inhibits the number and activity of SRB
pH	• Affects the growth and activity • Influences the SRB species diversity and competition with methanogens • Affects quantity of dissolved sulphide	• pH range for SRB: 5.5-10
Salinity	• Influences the species of SRB present • Sulphate-reducing rate is inversely correlated with salinity	• Optimum salinity range 6-12%
Operational conditions		
Substrate/sulphate	• Affects growth, activity and microbial diversity • Proper C/S ratio enables SRB to out-compete other organisms	• Optimal COD/SO_4^{2-} ratio for COD removal is 0.6-1.2 and for sulphate removal is 2.4-4.8
Oxidation-reduction potential (ORP)	• Affects the competition between SRB and other organisms *e.g.* methanogens • Affects the activity of SRB	• Suitable ORP for SRB is -50 to -300 mV; Optimal ORP readings of -270 mV using a standard hydrogen probe
Temperature	• Controls the activity and growth • Initial cultivation temperature affects SRB diversity • Lower H_2S solubility at high temperature	• SRB tolerate temperatures between -5 and 75 °C • Optimum temperature for most SRB ranges between 28-40 °C
Sludge retention time (SRT)	• Affects the reactor's performance and sludge production • Affects the competition between SRB and methanogens/homoacetogenic bacteria	• Elevated SRTs delay the methanogens being out-competed, and methanogens can be rapidly removed by applying a short SRT
Hydraulic retention time (HRT)	• Influences SRB activity • Biomass concentration • Competition with other organisms	• Overall optimum HRT of 20-30 h for SRB activity
H_2S concentration	• High H_2S directly and reversibly affects SRB, and inhibits the activity	• Nitrogen purging • Decrease the activity when H_2S is higher than 60-70 mg/l[1)]
Mixing condition	• Mixing frequency significantly impacts SRB activity • Affects the SRB distribution and detachment causing hydraulic loss of biomass	

[1)] Negative impact limitation.

pH and sulphide concentration

Most SRB show a high specific activity in the pH range of 5.0-8.0 and prefer a pH of 7.5-8.0. However, various acid-tolerant SRB species have been found at a working pH as low as 3 (Zeng *et al.*, 2019). Some species have even been reported to be alkali-tolerant, *e.g. Desulfovibrio alkalitolerans* sp. nov. (Abildgaard *et al.*, 2006). On the other hand, the highest pH for SRB can be up to 10.5 (Zeng *et al.*, 2019). The activity of SRB may be subjected to two threshold inhibition levels: one for the un-dissociated sulphide (H_2S) and the other for the total sulphide, although the inhibiting concentrations of sulphide and hydrogen sulphide, respectively, are not conclusive. In general, at pH levels below 7.2 un-dissociated H_2S will be dominant and reach the inhibition threshold, while at pH levels above 7.2, total sulphide will result in an inhibitory effect (Perry, 1984); SRB are less sensitive to total sulphide when the pH is between 6.8 and 8.0 but they will be more sensitive to un-dissociated sulphide. Hydrogen sulphide can combine with iron in ferredoxin, cytochrome, and other essential iron-containing compounds in cells, thereby ceasing the electron transport systems (Koschorreck, 2008). Therefore, in applications the impact of pH should be carefully considered.

COD/SO_4^{2-} mass ratio

In most SRB-driving heterotrophic reductions, the electron donor and the carbon source often share the same compounds except that H_2 and/or CO are also involved. The mass ratio of COD/SO_4^{2-} is recognized as a key factor in these reactions because this ratio regulates the competition between SRB and other organisms for substrate. For instance, 0.6-1.2 is suitable for COD removal while 2.4-4.8 is favourable for sulphate removal; in between 1.7 and 2.7 there is active competition for organic substrate utilization between SRB and methanogens; at <1.5 SRB activity is dominated (Choi and Rim, 1991). Competition between SRB and methanogens can also be influenced by a specific organic substrate or a type of electron donor, owing to the different kinetic properties of SRB and other elector donors; in the case of trimethylamine and methionine as substrate, methanogens will out-compete SRB (Oremland and Polcin, 1982).

7.3 SULPHUR-DRIVEN AUTOTROPHIC DENITRIFICATION

7.3.1 Introduction

Depletion of groundwater resources, concerns with nutrient enrichment in natural water bodies, decreases in carbon footprint, and stringent discharge limits for nutrients are all encouraging the development of new bioprocesses for nitrogen removal from water and wastewater streams (Kostrytsia *et al.*, 2018b). Since heterotrophic denitrification typically requires the supply of organic matters as the electron donor, this usually results in higher sludge production, thus higher operational costs, and increases the residual organics in the treated effluents. Therefore, the inorganic-based autotrophic denitrification bioprocess has attracted increasing attention in the last decade, especially sulphur-driven autotrophic denitrification (SdAD). This is carried out by chemolithoautotrophic denitrifying microorganisms, *i.e.* sulphur-oxidizing bacteria (SOB) such as *Thiobacillus denitrificans*. These are capable of utilizing various inorganic sulphur matters in reduced forms (*e.g.* S^0, TDSd, $S_2O_3^{2-}$, and SO_3^{2-}) as the electron donors for denitrification of NO_3^- and NO_2^- as well as CO_2 sequestration (Cardoso *et al.*, 2006).

The SdAD process possesses several inherent advantages such as: (*i*) no requirement for an exogenous organic carbon source; (*ii*) it eliminates problems associated with residual organics; (*iii*) a lower biomass yield coefficient relative to a conventional heterotrophic denitrification process (*i.e.* 0.15-0.57 g autotrophic biomass production vs. 0.71-1.2 g heterotrophic biomass production per 1 g of NO_3^--N denitrified (Yang *et al.*, 2016a)), thus reducing sludge production for costly treatment and disposal; (*iv*) a significant reduction in nitrous oxide (N_2O) accumulation and emission during denitrification when sufficient sulphur is available (Yang *et al.*, 2016a); and (*v*) it facilitates simultaneous removal of NO_3^-/NO_2^- and sulphur compounds (Moraes *et al.*, 2012). SdAD can be a cost-effective alternative to conventional heterotrophic denitrification in treatment of carbon-deficient sulphur-laden water and wastewaters, such as landfill leachate, aquaculture wastewater, wastewater from the petrochemical industry, effluent polishing of anaerobic reactors, biogas desulphurization, anti-corrosion treatment of sewer systems, and anti-souring treatment of oil reservoirs and petroleum refinery facilities.

7.3.2 Biochemical reactions in the SdAD process

Considering the significance of biological sulphur oxidation in the SdAD process, it is essential to understand the biochemical reactions involved in this process. The stoichiometric biochemical reactions related to simultaneous sulphur oxidation and denitrification from NO_3^- and NO_2^- under anoxic conditions are listed below (see eqs. 7.7-7.12), in consideration of complete denitrification and neglecting the biomass growth (Cardoso *et al.*, 2006; Sierra-Alvarez *et al.*, 2007).

$$S^{2-} + 1.6NO_3^- + 1.6H^+ \rightarrow SO_4^{2-} + 0.8N_2 + 0.8H_2O$$
$$\Delta G^{0\prime} = -743.9 \text{ kJ/mol} \quad (7.7)$$

$$S^{2-} + 2.67NO_2^- + 2.67H^+ \rightarrow SO_4^{2-} + 1.33N_2 + 1.33H_2O$$
$$\Delta G^{0\prime} = -920.3 \text{ kJ/mol} \quad (7.8)$$

$$S^0_{chem} + 1.2NO_3^- + 0.4H_2O \rightarrow SO_4^{2-} + 0.6N_2 + 0.8H^+$$
$$\Delta G^{0\prime} = -547.6 \text{ kJ/mol} \quad (7.9)$$

$$S_2O_3^{2-} + 1.6NO_3^- + 0.2H_2O \rightarrow 2SO_4^{2-} + 0.8N_2 + 0.4H^+$$
$$\Delta G^{0\prime} = -776 \text{ kJ/mol} \quad (7.10)$$

$$S^{2-} + 0.4NO_3^- + 2.4H^+ \rightarrow S^0_{bio} + 0.2N_2 + 1.2H_2O$$
$$\Delta G^{0\prime} - \quad 191.0 \text{ kJ/mol} \quad (7.11)$$

$$S^{2-} + 0.67NO_2^- + 2.66H^+ \rightarrow S^0_{bio} + 0.33N_2 + 1.33H_2O$$
$$\Delta G^{0\prime} = -240.3 \text{ kJ/mol} \quad (7.12)$$

In the sulphide-oxidizing process, sulphide, as an electron donor, can be oxidized to biogenic elemental sulphur (S^0_{bio}) or to sulphate (SO_4^{2-}), as indicated by eqs. 7.7 and 7.11, respectively. In this chapter, we define the sum of H_2S (aq), HS^- and S^{2-} as the total dissolved sulphide (TDSd), whereas elemental sulphur indicates either chemical (S^0_{chem}) (*i.e.* exogenously added) or biogenic (S^0_{bio}) (*i.e.* endogenously produced) elemental sulphur. The complete oxidation of sulphide to SO_4^{2-} (see Eq. 7.7), where eight electrons per atom of sulphur oxidize, are donated from sulphide to oxygen. This reaction is among the most energetic reactions for chemoautotrophs (Di Capua *et al.*, 2019). The complete sulphide oxidation with nitrogen oxides (NO_x^-) as electron acceptor is thermodynamically favourable as indicated by the standard Gibbs free energy changes (see eqs. 7.7 and 7.8). However, elemental sulphur formation as an intermediate of autotrophic denitrification depends on sulphide concentration, especially when NO_3^- is the electron acceptor (Cardoso *et al.*, 2006). This could be attributable to the conversion of sulphide to biogenic elemental sulphur (S^0_{bio}) consuming four times less NO_x^- than its complete oxidation to SO_4^{2-} (see eqs. 7.7, 7.8, 7.11 and 7.12), resulting in a preferential process under electron acceptor limitation. In these conditions, the oxidation of excess sulphide is favoured due to its higher bioavailability to denitrifiers compared with S^0_{bio} that accumulates over time. Based on the biochemical reactions with the intermediate formation of sulphur compounds, NO_2^- reduction is evidently thermodynamically more favourable than NO_3^- reduction from the higher negative standard Gibbs free energy changes (see eqs. 7.11 and 7.12). It is worth noting that the fraction of electrons contributed to growth is neglected in the above eqs. 7.7, 7.9 and 7.10. Generally, the actual amount of sulphur compounds consumed is 30-38% higher than the stoichiometric values, considering the cell growth of SOB. Consequently, eqs. 7.7, 7.9 and 7.10 are extended to eqs. 7.13, 7.14 and 7.15 to represent the stoichiometric biochemical reactions for complete denitrification from NO_3^- in the use of inorganic sources as well as consideration of the associated biomass growth in anabolism. The biomass growth yield (Y, gcells/gNO_3^--N) of sulphur-oxidizing denitrifiers depends on the types of sulphur compounds consumed in the biochemical reactions, as shown in the following equations (Mora *et al.*, 2014).

$$HS^- + 1.23NO_3^- + 0.573H^+ + 0.438HCO_3^- + 0.027CO_2 + 0.093NH_4^+ \rightarrow 0.093C_5H_7O_2N + SO_4^{2-} + 0.614N_2 + 0.866H_2O$$
$$Y = 0.61 \text{ gcells/gNO}_3^-\text{-N} \quad (7.13)$$

$$S^0 + 0.876NO_3^- + 0.343H_2O + 0.379HCO_3^- + 0.023CO_2 + 0.080NH_4^+ \rightarrow 0.080C_5H_7O_2N + SO_4^{2-} + 0.44N_2 + 0.824H^+$$
$$Y = 0.74 \text{ gcells/gNO}_3^-\text{-N} \quad (7.14)$$

$$S_2O_3^{2-} + 1.24NO_3^- + 0.11H_2O + 0.45HCO_3^- + 0.09NH_4^+ \rightarrow 0.09C_5H_7O_2N + 2SO_4^{2-} + 0.62N_2 + 0.40H^+$$
$$Y = 0.59 \text{ gcells/gNO}_3^-\text{-N} \quad (7.15)$$

Based on the stoichiometries of sulphide-, S^0_{chem}- and $S_2O_3^{2-}$-based denitrification (see eqs. 7.7, 7.8 and 7.9), reduction of 1 mgNO_3^--N generates 5.58, 7.83 and 11.07 mg SO_4^{2-}, respectively. This may limit the application of SdAD technology in restoring ground and surface waters contaminated by NO_3^- and NO_2^-.

7.3.3 Microorganisms in the SdAD process

Microbial consortia involved in the SdAD process are phylogenetically diverse and mainly consist of sulphur-oxidizing and nitrate-reducing bacteria, predominantly belonging to phyla Proteobacteria, Firmicutes, Chloroflexi and Chlorobi (Zhang *et al.*, 2015). The microorganisms belonging to the genera *Thiobacillus*, *Thiomonas* and *Sulphurimonas* are the most well-known SOB for autotrophic denitrification which obtains energy for denitrification and biomass synthesis mainly from sulphur oxidation. The microbial composition and diversity of SOB vary significantly depending upon the type of electron donor. Several predominant genera and their optimum growth conditions driven by different electron donors are summarized in Table 7.6, which are *Thiobacillus* and *Sulphurimonas* as obligate and/or facultative chemolithotrophs. Facultative chemolithotrophic SOB can grow autotrophically or mixotrophically on complex organic compounds as their concurrent carbon and energy sources (Muyzer *et al.*, 2013). To date, many SOB species have been isolated from sediments, hydrothermal vents, oceans, soda lakes and hypersaline lakes, and several of them are found in the SdAD process. Table 7.7 summarizes these SOB and their optimum growth conditions. SOB species can grow at different pH values, *e.g. Thiomicrospira* grow from pH 5.5 to 8.5, and *Thiobacillus denitrificans* prefer neutral conditions. Some SOB are acidophiles and can grow at a pH lower than 5.0 and some (*e.g. Thialkalivibrio nitratireducens* and *Thiobacillus versutus*) only grow in an alkaline environment. For most SOB their optimal growth temperatures are around 30 °C, while a few SOB have a wide range of optimal growth temperature, *e.g. Paracoccus* grow under either psychrophilic or thermophilic conditions.

An analysis of the predominant genera gives a deeper comprehension of the microbial community composition and function in the SdAD process under different operational conditions. Microbial community composition can be shaped significantly by changing operational parameters, such as the initial concentrations of sulphur and NO_3^-, reactor temperature, pH and HRT. Electron-donating substrates also play an important role in determining the microbial community structure. For instance, *Chlorobaculum, Thiobacillus* and *Sulphurimonas* are the predominant genera in a $S_2O_3^{2-}$-driven SdAD system, while *Acinetobacter, Thiobacillus* and *Thiothrix* are predominant in a sulphide-driven SdAD system. Some genera in low relative abundance can change to a higher relative abundance depending on the type of electron donor, for instance, *Thauera* has a much higher relative abundance (2.2%) in a S^0_{chem}-based SdAD system than that in a $S_2O_3^{2-}$-based SdAD system (0.4%) (see Table 7.6). Moreover, different functional microorganisms can be enriched respectively at a low, medium or high sulphide-loading rate, which is an important factor affecting the microbial communities in a simultaneous desulphurization and denitrification SdAD process (Yu *et al.*, 2014). The pH value in the biological reactor also affects the enrichment of microorganisms, thereby NO_3^- removal efficiency depends on different operational pH values with different types of sulphur substrate (*e.g.* electron donor) in a SdAD system. SOB-enriched activated sludge can also be granulated in a continuous-flow SdAD bioreactor, with predominant microbial populations (Betaproteobacteria and Epsilonproteobacteria) that are quite different from flocculent sludge (Actinobacteria and Alphaproteobacteria) (Yang *et al.*, 2016a).

Table 7.6 Predominant genera of the bacterial consortium involved in the SdAD process driven by different electron donors.

Electron donor	Predominant genus	Relative abundance (%)	Type of reactor	pH	Temperature (°C)	Electron acceptor
S^0_{chem}	*Thiobacillus*	45.1	AnFB-MBR	-	25-31	NO_3^-
	Ignavibacteriales	25.4				
	Sulphurimonas	7.0				
	Longilinea	4.1				
TDSd	*Thiobacillus*	32.6	GSAD	7.3-7.7	28-32	NO_3^-
	Sulphurimonas	31.3				
	Paracoccus	0.1				
TDSd	*Acinetobacter*	12.8	Up-flow biofilter	7.2-7.3	5-35	NO_3^-
	Thiobacillus	3.6				
	Thiothrix	1.6				
$S_2O_3^{2-}$	*Chlorobaculum*	9.7	Up-flow biofilter	6.7-7.0	5-35	NO_3^-
	Thiobacillus	8.7				
	Sulphurimonas	1.2				
S^0_{chem}	*Dechloromonas*	18.0	Up-flow biofilter	6.5-7.0	5-35	NO_3^-
	Thiobacillus	8.7				
	Thauera	2.2				
S^0_{chem}	*Thiobacillus*	22.6	SBR	7.0-8.5	35	NO_3^-
	Sulphurimonas	11.8				
	Thioahalobacter	11.0				
S^0_{chem}	*Sulphurimonas*	82.2	-	7.5	25	NO_3^-
	Thauera	2.7				
	Simplicispira	2.3				
	Pseudomonas	1.6				
TDSd, $S_2O_3^{2-}$	*Maribius*	11.3	MBBR	7.2-7.8	21-23	NO_3^-
	Paracoccus	10.0				
	Thiobacillus	9.3				
	Sedimenticola	8.3				
	Thioalbus	7.6				
	Sulphurimonas	1.0				
S^0_{chem}	*Thiobacillus*	11.0	-	8.0	31-35	NO_3^-
	Sulphurimonas	5.4				

TDSd: total dissolved sulphide (H_2S, HS^-, S^{2-}); S^0_{chem}: chemically-added elemental sulphur; AnFB-MBR: anaerobic fluidized-bed membrane bioreactor; GSAD: granular-sludge autotrophic denitrification; SBR: sequencing batch reactor; MBBR: moving-bed biofilm reactor.

Table 7.7 Physiology, taxonomy and optimum growth conditions for different isolated chemolithotrophic denitrifying species.

Species	Isolation site	pH	Temperature (°C)	Electron donor	Electron acceptor
Thialkalivibrio nitratireducens	Soda lake	10.0	-	$S_2O_3^{2-}$, S^{2-}, Sn^{2-}	NO_3^-
Thiobacillus versutus	Soda lake	10.1-10.2	35	$S_2O_3^{2-}$, S^{2-}, Sn^{2-}, S^0, $S_4O_6^{2-}$	NO_3^-
Thiobacillus denitrificans	Anaerobic sludge	7.0	27-29	S^{2-}	NO_3^-
Sulphurimonas denitrificans	Deep-sea hydrothermal vent, polychaete nest	6.1	30	S^0, S^{2-}, H_2	NO_3^-, NO_2^-, O_2
Halothiobacillus neapolitanus	Cattle manure	6.5-6.9	28-32	S^{2-}, S^0, $S_2O_3^{2-}$	NO_3^-
Thiohalorhabdus denitrificans	Sediment	6.5-8.2	33-35	$S_2O_3^{2-}$	NO_3^-, O_2
Thiobacillus thiophilus	Oil-contaminated sediment	7.5-8.3	25-30	$S_2O_3^{2-}$	NO_3^-, O_2
Pseudomonas stutzeri	Anoxic sludge	6.5-8.0	30	S^{2-}	NO_3^-, O_2
Thiomicrospira sp.	Oil field	5.5-8.5	5-35	S^{2-}, CH_3COO^-	NO_3^-, NO_2^-
Sulphurimonasgotlandica	Pelagic redox zone of sea	6.7-8.0	15	S^{2-}, S^0, $S_2O_3^{2-}$, H_2, organics	NO_3^-, NO_2^-
Paracoccus sp.	Hot spring	7.0-9.0	20-50	$S_2O_3^{2-}$	NO_3^-
Thiobacillus thioparus	Sandy soil	7.0	28	$S_2O_3^{2-}$	O_2

Sn^{2-}: polysulphide.

7.3.4 Biochemistry of the SdAD process

SOB work independently or in co-culture mode in the SdAD microbial ecological system. The key biochemical reactions involved in sulphur-nitrogen transformation are catalyzed by specific functional enzymes: notably, both nitrogen oxide reductases (nitrogen-reducing enzymes) and sulphur-oxidizing enzymes. SOB oxidize the reduced-sulphur inorganic compounds (the electron donors) for the microbial growth in the anabolism and for the production of energy through the dissimilatory autotrophic denitrification in the catabolism.

7.3.4.1 Sulphur-oxidizing enzymes

TDSd, S^0_{chem} and $S_2O_3^{2-}$ are the most common electron donors for SOB. The metabolic pathways and functional enzymes involved in oxidation of sulphur compounds and energy conservation by SOB have been biochemically characterized, as shown in Figure 7.6. The three main pathways that have been identified so far are: (*i*) the sulphur-oxidizing (Sox)-independent pathway involving distinctive oxidases and hydrolases for TDSd, S^0_{chem} and $S_2O_3^{2-}$ oxidation; (*ii*) the tetrathionate (SI_4) pathway of $S_2O_3^{2-}$ oxidation; and (*iii*) the Sox-dependent multiple enzymatic pathway. The initial bio-oxidation of sulphide can be mediated by the sulphide: quinone oxidoreductase (SQR), a flavoprotein responsible for obligatory biological sulphide oxidation in the photo- and chemolithotrophic bacteria. Moreover, flavocytochrome c sulphide dehydrogenase (FCSD), which shares an ancestor with SQR that has a similar functional pathway, is an alternative sulphur-oxidizing enzyme located in the periplasm. The primary product of the SQR reaction is soluble

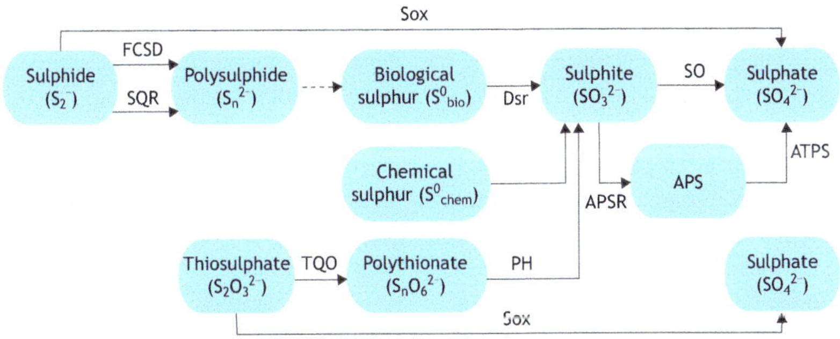

Figure 7.6 Schematic overview of microbial sulphur oxidation pathways and the key functional enzymes involved. FCSD = flavocytochrome c sulphide dehydrogenase; SQR = sulphide: quinone oxidoreductase; Dsr = dissimilatory sulphite reductase; SO = sulphite oxidase; TQO = thiosulphate: quinone oxidoreductase; PH = polythionate hydrolase; APS = adenosine phosphosulphate; APSR = adenosine-5′-phosphosulphate reductase; ATPS = ATP sulphurylase; Sox = sulphur-oxidizing enzyme complex governed by a conserved Sox operon.

polysulphide which is finally converted into sulphur globules (*i.e.* S^0_{bio}) by a purely chemical, spontaneous process. The periplasmically-formed sulphur globules can be further transported into the cytoplasm by a sulphur-carrier enzymatic system (SCS) and oxidized into SO_3^{2-} by Dsr. The subsequent oxidation from SO_3^{2-} to SO_4^{2-} primarily uses one of the two confirmed sub-pathways performed by two types of sulphite dehydrogenases. The first sub-pathway employs the sulphite oxidase that oxidizes SO_3^{2-} and transfers the electrons to cytochrome c; in the second sub-pathway, sulphur chemolithotrophs oxidize SO_3^{2-} to SO_4^{2-} through the reverse activity of the enzyme APS (adenosine phosphosulphate) reductase.

$S_2O_3^{2-}$ oxidation can also be carried out by the Sox-independent enzyme system via the intermediate formation of tetrathionate catalyzed by thiosulphate: quinone oxidoreductase, and then followed by tetrathionate degradation into SO_3^{2-} via hydrolysis mediated by cytoplasmic polythionate hydrolase. This $S_2O_3^{2-}$ oxidation pathway via polythionate formation is commonly found in several β and γ proteobacteria, especially in obligatory chemolithotrophs (*e.g.* *Acidithiobacillus*).

The Sox-dependent enzyme complex, namely the Sox α-proteobacterial multiple enzyme complex, is governed by a conserved Sox operon. The cluster of soxXYZABCD genes encodes the structural components of the sulphur-oxidizing microbial system. These seven genes make the four proteins (SoxXA, SoxYZ, SoxB, and SoxCD) required for sulphur-dependent cytochrome c reduction. This multiple enzyme complex can oxidize various reductive sulphur substrates (*e.g.* TDSd, $S_2O_3^{2-}$, and S^0) into SO_4^{2-}. For detailed information of the biochemical sulphur oxidation pathways and enzymes, please refer to the book authored by Hell *et al.* (2008).

7.3.4.2 Nitrogen-reducing enzymes

Biological autotrophic denitrification is a combination of four microbial nitrogen-reductive reactions: dissimilatory NO_3^- reduction, NO_2^- reduction, NO reduction, and N_2O reduction. All the steps within this metabolic pathway are catalyzed by complex multisite metalloenzymes, which requires 10 mol electrons and 12 mol H^+ for complete reduction of 1 mol NO_3^- to N_2. In the first step, the two electron reduction of NO_3^- to NO_2^- is catalyzed by molybdenum-containing nitrate reductase (Nar). The enzyme is either located in the cytoplasm (Nar GH) or in the periplasm (Nap AB). NO_2^- is further reduced to NO in the second step catalyzed by nitrite reductase (Nir) located in the periplasm (see Figure 7.7). In the third step, two molecules of nitric oxide are conjugated to form N_2O and water with the addition of two electrons. This reaction is catalyzed by a membrane-bound nitric oxide reductase (Nor) that comprises heme c, heme b and non-heme iron cofactors. Finally, a copper-containing metalloenzyme, *i.e.* nitrous oxide reductase (Nos), is responsible for the last step of denitrification, *i.e.* N_2O reduction into dinitrogen (N_2) (see Figure 7.7). For more detailed information on the structure and function of nitrogen oxide-reducing enzymes, readers can refer to the books by Van Spanning *et al.* (2007) and Moura *et al.* (2016).

7.3.4.3 Electron distribution and competition in the SdAD process

This section provides fundamental insights into electron distribution in the SdAD process by considering both the electron supply (*i.e.* sulphur substrate oxidation) and electron consumption and competition among nitrogen oxide reductases (Nar, Nir, Nor and Nos). The electron distribution pathway in the SdAD process is shown in Figure 7.7. Dissolved sulphide can be converted into S^0_{bio} by the SQR or into SO_4^{2-} by the Sox enzyme complex. The electrons generated from the oxidation of inorganic sulphur electron donors facilitate the membrane-bound Nar to perform NO_3^- reduction to NO_2^-. The electrons released from the oxidation can be fed into the electron transport chain, which is directly connected with the quinone pool (UQ/UQH$_2$) at the cytoplasmic membrane. Electrons are delivered to periplasmic Nir by a mono-heme c-type cytochrome, *i.e.* cytochrome c_{550}, or a cupredoxin protein called pseudoazurin. These two periplasmic and water-soluble proteins are reduced by an integral membrane complex called the cytochrome bc_1 complex, which in turn is reduced by ubiquinol. The NO generated must be restricted to a low concentration because of its potential toxicity to cells. The Nor is an integral membrane protein which receives electrons from either pseudoazurin or cytochrome c_{550} in common with the Nir. The final step (from N_2O to N_2) is catalyzed by Nos, another periplasmic enzyme which receives electrons from the intermediate electron donor proteins, *e.g.* pseudoazurin and cytochrome c_{550} (Van Spanning *et al.*, 2007).

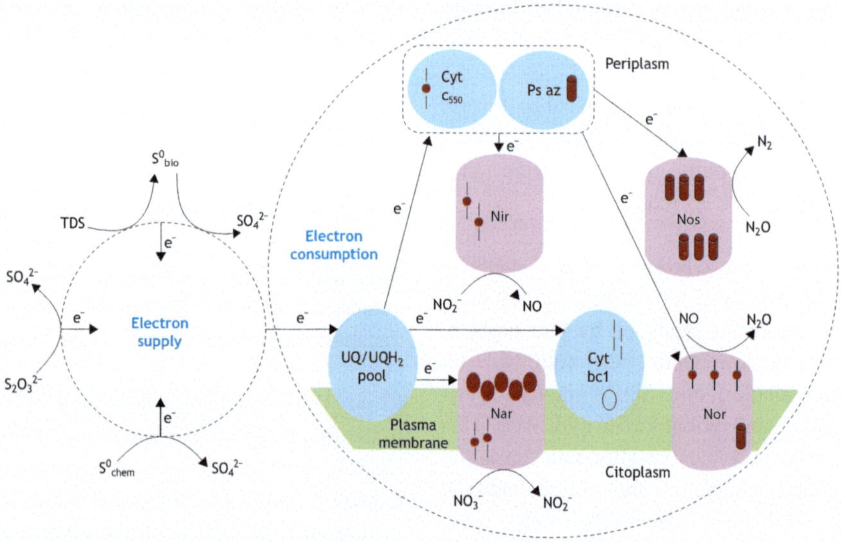

Figure 7.7 Schematics of the electron distribution in the SdAD process. The electrons for each denitrification step mediated by different nitrogen oxides reductases (Nar, Nir, Nor and Nos) originate from the common electron supply source, *i.e.* the ubiquinol pool (UQ/UQH$_2$) which is replenished by the electron input from the oxidation of sulphur substrates (*i.e.* TDSd and S^0_{chem}). TDSd: total dissolved sulphide; S^0_{bio}: biogenic sulphur (endogenously produced); S^0_{chem}: chemical elemental sulphur (externally added); $S_2O_3^{2-}$: thiosulphate, SO_3^{2-}: sulphite; SO_4^{2-}: sulphate; Nar: nitrate reductase; Nir: nitrite reductase; Nor: nitric oxide reductase; Nos: nitrous oxide reductase; Ps az: pseudoazurin; Cytbc1: cytochrome bc1; Cytc550: cytochrome c550.

The competition between different nitrogen oxide reductases (*i.e.* Nar, Nir, Nor, and Nos) commonly exists in the respiratory chain of denitrification because the electrons for nitrogen compound reduction originate from the common electron supply source, *i.e.* the ubiquinol pool (UQ/UQH_2) as discussed above. The UQ/UQH_2 can be replenished by electrons from the oxidation of inorganic electron donors (*e.g.* TDSd and S^0_{chem}). The relative ability to compete for the electrons among Nar, Nir, Nor, and Nos regulates the electron distribution among the denitrification steps and determines the possible accumulation of nitrogen compounds (denitrifying intermediates) during the SdAD reactions. Additionally, the sulphur substrates can possess different oxidation rates and subsequent electron flows in the electron transport chain, imposing electron competition in the SdAD reactions. Electron competition among the nitrogen oxide reductases has been found to affect the reduction rates of the nitrogen oxides and to be responsible for accumulation of the intermediates (*i.e.* NO_2^- and N_2O) during the long-term operation of an SdAD reactor (Pan *et al.*, 2013; Yang *et al.*, 2016b). N_2O emission from the denitrification process during wastewater treatment has recently caused considerable environmental concern, because of its 265-fold higher global warming potential relative to CO_2 and its ability to deplete the stratospheric ozone. Interestingly, the N_2O accumulation and emission in the SdAD process are significantly lower than in conventional heterotrophic denitrification (Zhang *et al.*, 2015; Yang *et al.*, 2016b). This is due to the fact that the SdAD process differs from the reported heterotrophic and hydrogenotrophic denitrification processes due to its unique electron partitioning pathways, as shown in Figure 7.7.

7.3.5 Operational conditions governing the SdAD process

The SdAD process involves different sulphur substrates (*i.e.* S^0_{chem}, S^0_{bio}, TDSd and $S_2O_3^{2-}$) which provide the electron donors for SOB to remove nitrogen oxides (*i.e.* NO_3^- and NO_2^-) from water, wastewater and gaseous streams when sources of organic substrates are scarce. However, sulphur substrates can have different oxidation rates which influences the nitrogen removal efficacy of the SdAD process. For specific sulphur substrates (*e.g.* S^0_{chem}, S^0_{bio}, TDSd and $S_2O_3^{2-}$), the sulphur-to-nitrogen (S/N) mass ratio can influence the accumulation of intermediates (*i.e.* N_2O, NO_2^- and S^0_{bio}). The selection of different bioreactor configurations and biomass types (*i.e.* activated sludge, biofilms and granules) is required to facilitate biomass retention in the reactor so as to maintain an effective denitrification rate of SOB. Meanwhile, optimum alkalinity in the bioreactor is deemed necessary for supplementing the alkali consumed by autotrophic denitrification with S^0_{chem}, S^0_{bio} or $S_2O_3^{2-}$ except using TDSd as the electron donor. Besides the above parameters, other operational parameters, such as dissolved oxygen (DO) and temperature, also influence the rate and efficiency of the SdAD process.

Sulphur substrates and nitrogen oxides

S^0_{chem} is a commonly used electron donor in the SdAD process since it is cost-effective, easily available and less toxic compared with other sulphur substrates. Sulphur-limestone typically offers a dual source of S^0_{chem} and alkalinity for SdAD application and enables SOB-enriched biofilm to develop on the S^0_{chem} particles packed in a bioreactor. However, the main drawback of S^0_{chem} is its low aqueous solubility and its low mass transfer rate in the reactor, thus impeding the electron transportation and autotrophic denitrification. Hence a biofilm reactor with S^0_{chem} as carrier is often recommended for improving the mass transfer efficiency of the reactor. Compared with S^0_{chem}, S^0_{bio} is more reactive and bioavailable for autotrophic microorganisms, owing to the orthorhombic structure of sulphur crystal covered by a hydrophilic layer of naturally-charged polymers (Kostrytsia *et al.*, 2018a).

TDSd has been employed as the electron donor in connecting FGD with the SANI® process for wastewater treatment (Jiang *et al.*, 2013), because its SdAD rate is greater than that with S^0_{chem}, as shown in Table 7.8. It should be noted that TDSd is toxic to SOB when its concentration exceeds 200 mgS/l, affecting Nar activity (Fajardo *et al.* 2014; Lu *et al.*,

2018). Therefore, in TDSd-based SdAD reactors a NO_3^- volumetric loading rate is recommended to be lower than that with S^0_{chem} or S^0_{bio}.

$S_2O_3^{2-}$, present in some industrial effluents (*e.g.* the mining industry) acts as the most effective sulphur substrate for SdAD (Cardoso *et al.*, 2006; Di Capua *et al.*, 2019) and the least inhibitory to SOB whose tolerance can be up to 2,200 mgS/l (Di Capua *et al.*, 2016; Cui *et al.*, 2019), 10 times greater than that of TDSd. However, the instability of $S_2O_3^{2-}$ (*i.e.* it is readily decomposed) may limit its application as an electron donor source. For large-scale engineering applications, selection of the most suitable substrate should be based on kinetics, cost, availability to microorganisms, applicability and potential toxicity. Apart from NO_3^-, NO_2^- can also be used as an electron acceptor. During the reduction of NO_3^-, NO_2^- is an inhibitory intermediate and its accumulation depends on the S/N mass ratio and the concentration of NO_3^- (An *et al.*, 2010). The accumulation of NO_2^- greatly inhibits the denitrification process even at a low level of 200 mgNO_2^--N/l (Claus and Kutzner, 1985).

Sulphur-to-nitrogen (S/N) mass ratio

The S/N mass ratio is closely correlated with the accumulation of intermediates (*i.e.* N_2O, NO_2^- and S^0_{bio}). When the S/N mass ratio is close to 2.1 gS/gN, it starts significantly reducing the emission of N_2O, a potent greenhouse gas, in the SdAD process with sulphide as the electron donor (Yang *et al.*, 2016b). At a low S/N mass ratio, the lack of an electron source can lead to the accumulation of NO_2^-, which inhibits denitrification (Cui *et al.*, 2019). A higher S/N mass ratio favours S^0_{bio} as the main final product, while SO_4^{2-} is the preferential product of TDSd oxidation accompanied by complete denitrification when the S/N mass ratio is below the stoichiometric value (1.4 gS/gN). A minimum S/N mass ratio of 1.6 gS/gN is necessary for achieving >90% NO_3^- removal, which is higher than the theoretical stoichiometric S/N mass ratio of 1.4 gS/gN (Wang *et al.* 2009; Lu *et al.*, 2012b).

Different sludge types

Biomass types (*i.e.* activated sludge, biofilms and granules) in different bioreactor configurations, such as UASB (up-flow anaerobic sludge bed) and biofilm reactor, *i.e.* PBR (packed-bed reactor) and MBBR (moving-bed biofilm reactor), significantly influence the biomass concentration and the denitrification rate. It is important to maintain high biomass to improve the denitrification rate. Biofilm reactor is proven to be robust and to perform well in SdAD. PBR is particularly suitable for the NO_3^- removal at a lowconcentration, since a long retention time (>3 h) is required for complete denitrification. MBBR is suitable for maintaining high biomass content of slow-growing SOB (Cui *et al.*, 2018). Even under severe NO_3^- limitation, MBBR can still achieve complete NO_3^- removal and it has high resiliency (Khanongnuch *et al.*, 2019). Another method to improve the performance of the SdAD process is to granulate the SOB, which can facilitate both high biomass concentration and robustness for efficient and stable autotrophic denitrification (Yang *et al.*, 2016a). Details of different bioreactor configurations and their applications in the SdAD process are summarized in Table 7.8.

Alkalinity, oxygen and temperature

The ratio of alkalinity-to-removed nitrogen should be kept at approximately 4.00-4.46 mg$CaCO_3$/mgN (Zhang and Shan, 1999; Oh *et al.*, 2003) to correspond to the SOB optimal pH range of 6 to 8 (Oh *et al.*, 2001). When S^0_{chem} is used as the electron donor, 4.6 g of alkalinity (as $CaCO_3$) is consumed per gNO_3^--N removed (Kim and Bae, 2000). Some easily available alkaline sources such as limestone, calcite, dolomite, and oyster shell can be considered for addition. In cases of sulphur-limestone systems, the optimal volumetric ratio of the limestone to S^0_{chem} is 1:1-1:3 m^3/m^3. Oxygen affects the growth of autotrophic denitrifiers. DO concentration between 0.1-0.3 mgO_2/l can inhibit the Nos activity, while >1.6 mgO_2/l completely inhibits the SdAD process (Cui *et al.*, 2019). For most autotrophic denitrifiers, the suitable temperature range is generally 25-35 °C (Shao *et al.*, 2010; Fajardo *et al.*, 2014). The SdAD process with sulphide as the electron donor achieves the highest denitrification efficiency (>97%) at 35 °C, which, however, decreases to 36-59% when temperature decreases to 15 °C (Fajardo *et al.*, 2014).

Table 7.8 Different bioreactor configurations containing floc, biofilm and granules for the SdAD process.

Sludge type	Bioreactor configuration	Electron donor	Electron acceptor	Influent pH	Temperature (°C)	Denitrification rate ($gN/m^3.d$)	Scale
Floc	Batch	$S_2O_3^{2-}$	NO_3^-	8.0	30	52.5	Lab
		S^0_{bio}				12.0	
		S^0_{chem}				7.0	
	UASB	$S_2O_3^{2-}$	NO_3^-	7.0	28-32	105.8	Lab
		TDSd				23.1	
		S^0_{chem}				11.2	
	UASB	TDSd	NO_3^-	-	30	4,964.2	Lab
Biofilm	PBR	S^0_{chem}	NO_3^-	7.2-7.5	15-25	360.0	Lab
	PBR	S^0_{chem}	NO_3^-	7.1-7.4	20-25	94.5-232.0	Lab
	FMFAR	TDSd	NO_3^-	-	25	100.8	Pilot
	MBBR	TDSd	NO_3^-	-	20	350.0	Full
	UFCR	S^0_{chem}	NO_3^-	7.1-7.5	25-35	720.4	Lab
	MBBR	TDSd	NO_3^-	7.2-7.8	21-23	180.0	Lab
	MBBR	TDSd	NO_3^-	-	-	370.0	Lab
Granular	GSAD	TDSd	NO_3^-	7.3-7.7	28-32	310.0	Lab

UASB: up-flow anaerobic sludge bed; PBR: packed-bed reactor; FMFAR: fixed-media-filled anoxic reactor; MBBR: moving-bed biofilm reactor; UFCR: up-flow column reactor; GSAD: granular sludge autotrophic denitrification reactor; TDSd: total dissolved sulphide (TDSd=H_2S+HS^-+S^{2-}).

7.3.6 Implications of the SdAD process

The key to successful application of this process is to attain sufficient slow-growing SOB communities in the reactor. Biofilm (such as an MBBR) or a granular sludge reactor are preferred. The SdAD process can used to upgrade or retrofit existing WWTPs treating sulphur-laden wastewater or normal wastewater by dosing the appropriate amount of sulphur substrates into an existing or additional anoxic tank for enhancing TN removal. It could also be applicable in the treatment of pharmaceuticals in wastewater as SOB possess a high tolerance against various pharmaceutical compounds during biodegradation by SOB (Jia *et al.*, 2017). In addition to municipal wastewater, the SdAD process probably contributes to TN removal in the restoration of groundwater and surface water, the treatment of landfill leachate and brackish water, and biogas desulphurization wastewater.

7.4 SANI® PROCESS DEVELOPMENT, MODELLING AND APPLICATION

7.4.1 Introduction

7.4.1.1 The Hong Kong water tale

Today, 40% of the world's population resides in coastal urban areas. Being a highly populous coastal city, Hong Kong's population density is as high as 6,659 people/km^2, which imposes a high water stress of only 143 m^3/cap/year (Lu *et al.*, 2012b) far below the world's absolute water scarcity level of 1,000 m^3/cap/year (UN, 2018). To tackle this survival issue, Hong Kong started using seawater for toilet flushing (SWTF) in the late 1950s and now this covers 85% of the population, saving approximately 280 million m^3 of precious freshwater per annum, equivalent to approximately 20% of the total freshwater demand (WSD, 2019a). This amount is almost equal to the annual reclaimed water used for agricultural irrigation in California (USEPA, 2012). A typical SWTF system in Hong Kong starts with abstraction of seawater from a suitable seafront site, passing through a coarse

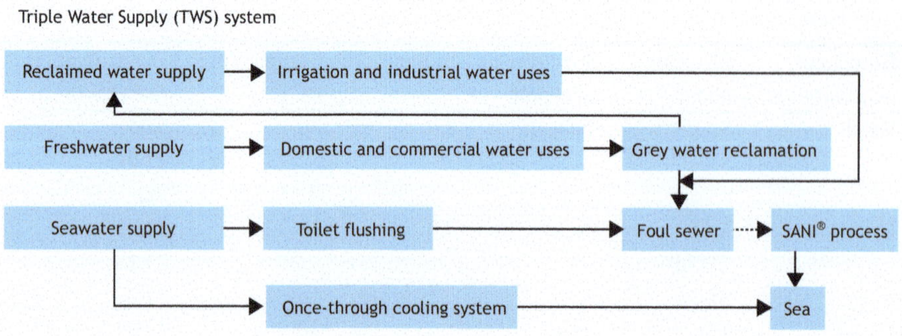

Figure 7.8 The TWS schematic at HKIA (modified from Van Loosdrecht *et al.*, 2012).

screen with mesh size 5-10 mm to protect the pumps, followed by subsequent *in-situ* electro-chlorination (3-6 mg/l chlorine) to disinfect and control microbial growth in the supply networks. The energy consumption of SWTF including abstraction, pre-treatment and distribution is estimated at 0.32 kWh/m^3 (Tang *et al.*, 2007), significantly lower than that of water reclamation from biologically-treated domestic sewage using membrane filtration or a membrane bioreactor coupled with the reverse osmosis (RO) process (0.8-1.2 kWh/m^3 and 1.2-1.5 kWh/m^3, respectively) (Pearce, 2007), as well as seawater desalination (2.5-4.0 kWh/m^3) (World Bank, 2004). Mild steel and ductile iron-piping lines coated with internal cement-mortar lining are applied in the mains pipes, whereas in buildings and housings polyethylene pipes are mainly employed. For details, please refer to the design information of the Hong Kong Water Supplies Department (WSD, 2019b). To maximize energy efficiency, Hong Kong International Airport (HKIA, 70 million total passengers/year) developed a triple-water-supply (TWS) system by extending the dual-water system to grey water reuse in combination with seawater building cooling (see Figure 7.8). This TWS system can save up to 52% of the freshwater demand at HKIA (Leung *et al.*, 2012). This system is rated to be one of the most energy-efficient water systems in the world (Grant *et al.*, 2012). Hong Kong's 60-year successful practice of SWTF offers adequate experience in the direct use of seawater as an economic and sustainable alternative water resource for populous coastal areas and islands. However, there are some limitations for applying this cost- and energy-effective dual-water-supply system in coastal areas or on islands, which is that the seawater intake distance should be under 10 km and population density be at least 3,000 people/km^2 (Liu *et al.*, 2016). Hong Kong's dual-water-supply experience might extend to potential application in water-scarce inland cities that have access to salty or brackish water as an alternative water source for toilet flushing.

7.4.1.2 Principle of the SANI® process

The SWTF practice results in, on average, 167 $mgSO_4^{2-}$-S/l and 5,000 $mgCl^-$/l in Hong Kong saline sewage (Van Loosdrecht *et al.*, 2012), inevitably causing odour issues in the sludge treatment. An ideal solution is to minimize sludge production. Based on this rationale, a novel sulphur-based process for saline wastewater treatment, namely the SANI® process, was developed by the authors from lab-, pilot- and full-scale studies between 2004 and 2017 (Lau *et al.*, 2006; Wang *et al.*, 2009; Lu *et al.*, 2012b; Wu *et al*, 2016). Comparison between the conventional biological nitrogen removal (BNR) process and the SANI® process in terms of process concept and reactions are shown in Figure 7.9 and Table 7.9, respectively.

Since the introduction of the nitrogen removal process in the 1960s, biological wastewater treatment has relied on the electron flow from organic carbon to oxygen through an integrated carbon and nitrogen cycle, as shown in Figure 7.9a.

Table 7.9 Comparison of the key reactions in conventional biological nitrogen removal process and the SANI® process.

Conventional BNR process	Reaction	ΔG°	Theoretical yield
Aerobic organic oxidation	$\frac{1}{24}C_6H_{12}O_6+\frac{1}{4}O_2\rightarrow\frac{1}{4}CO_2+\frac{1}{4}H_2O$ (7.16)	–120.10	0.49 gVSS/gCOD
Heterotrophic denitrification	$\frac{1}{24}C_6H_{12}O_6+\frac{1}{5}NO_3^-+\frac{1}{5}H^+\rightarrow\frac{1}{4}CO_2+\frac{1}{10}N_2+\frac{7}{20}H_2O$ (7.17)	113.63	0.8-1.2 gVSS/gN
Nitrification	$\frac{1}{8}NH_4^++\frac{1}{4}O_2\rightarrow\frac{1}{8}NO_3^-+\frac{1}{4}H^++\frac{1}{8}H_2O$ (7.18)	–43.64	0.17 gVSS/gN
	SANI® process		
Anaerobic sulphate reduction	$\frac{1}{24}C_6H_{12}O_6+\frac{1}{8}SO_4^{2-}\rightarrow\frac{1}{4}HCO_3^-+\frac{1}{8}HS^-$ (7.19)	–20.69	0.07 gVSS/gCOD
Autotrophic denitrification	$\frac{1}{16}H_2S+\frac{1}{16}HS^-+\frac{1}{5}NO_3^-+\frac{1}{80}H^+\rightarrow\frac{1}{8}SO_4^{2-}+\frac{1}{10}N_2+\frac{1}{10}H_2O$ (7.20)	–92.94	0.4-0.6 gVSS/gN
Nitrification	$\frac{1}{8}NH_4^++\frac{1}{4}O_2\rightarrow\frac{1}{8}NO_3^-+\frac{1}{4}H^++\frac{1}{8}H_2O$ (7.18)	–43.64	0.17 gVSS/gN

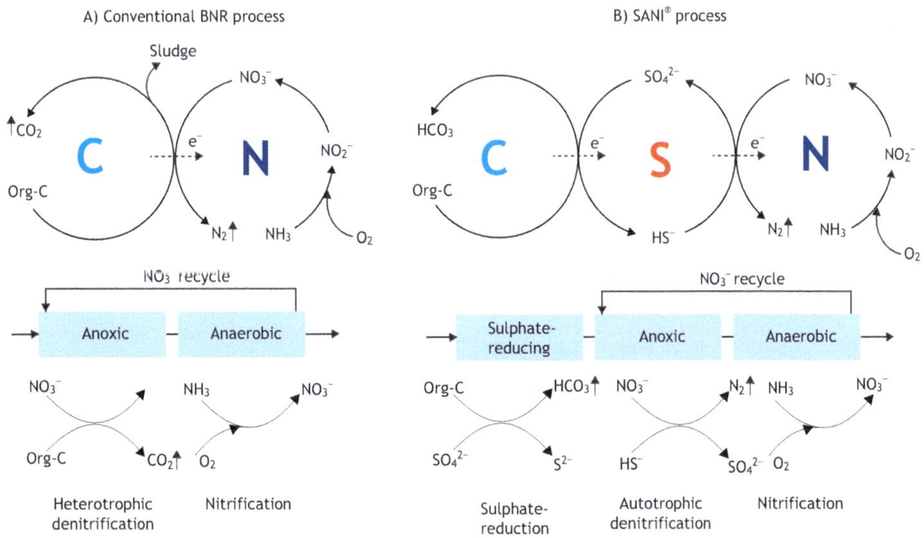

Figure 7.9 Comparison of the conventional biological nitrogen removal process and the SANI® Process (Wu *et al.*, 2016).

However, the conventional BNR process generates a large quantity of excess sludge due to high theoretical sludge yields resulting from organics removal by aerobic and anoxic oxidation (more details in Table 7.9). In contrast, the SANI® process integrates anaerobic organic removal by sulphate reduction and anoxic nitrogen removal by autotrophic denitrification for NO_3^- removal, minimizing biological sludge production as much as possible (see the theoretical sludge yields of the relevant reactions in Table 7.9). Stoichiometric study has showed that biological sludge production of this new BNR can potentially be 1/10 of conventional BNR (Lu *et al.*, 2012b). In this novel process, organic carbon is oxidized to CO_2 by SRB which increases the reactor pH to up to 8.0 at which the vast majority of H_2S generated from the anaerobic reaction dissolves in the aqueous phase of the sulphate-reducing bioreactor (Poinapen and Ekama, 2010a). This thereby provides an adequate amount of dissolved sulphide (the electron donor) for subsequent autotrophic denitrification, during which sulphide is converted back to SO_4^{2-} by SOB. The TDSd residue, if any, could be completely oxidized to sulphate during aerobic nitrification. From the electron transfer point of view, the SANI® process extends a carbon-nitrogen-cycle-based BNR process into a sulphur-carbon-nitrogen-cycle-based BNR process (see Figure 7.9b), which offers an opportunity for sludge-minimized treatment of saline and brackish or sulphur-laden wastewater.

7.4.2 SANI® process development

7.4.2.1 Laboratory study

To prove the SANI® process concept, a laboratory study was conducted at the Hong Kong University of Science and Technology from 2004 to 2009 (see Figure 7.10). The lab-scale system was operated for more than 500 days (Wang *et al.*, 2009; Hao *et al.*, 2014). The influent contained 265 mgCOD/l (mainly glucose, CH_3COO^- and yeast extract), 30 mgNH_4^+-N/l, 400-600 mg/l of SO_4^{2-} and 3,000-4,000 mg/l of salinity, similar to the effluent quality of a primary sedimentation tank at a typical saline sewage biological treatment plant in Hong Kong. In this system, organic carbon is first oxidized to CO_2 by SRB in the sulphate-reducing up-flow sludge blanket (SRUSB) (Reactor 1). This process subsequently increases the pH to an alkaline level (>7.5-8) due to sulphate (strong acid) being reduced to sulphide (weak acid), dissolving the product hydrogen sulphide in water mainly in the form of TDSd for autotrophic denitrification in the subsequent anoxic bio-filter (Rector 2), where NO_3^- is reduced to N_2. Through this reaction dissolved sulphide will be converted back to SO_4^{2-} by SOB. The remaining NH_4^+ and residual sulphide are oxidized to NO_3^- and SO_4^{2-} during nitrification in the aerobic bio-filter (Reactor 3), of which effluent is recirculated to Reactor 2 for nitrogen removal at a recirculation rate of 3Q (Q = influent flow rate), similar to a conventional biological nitrogen removal process.

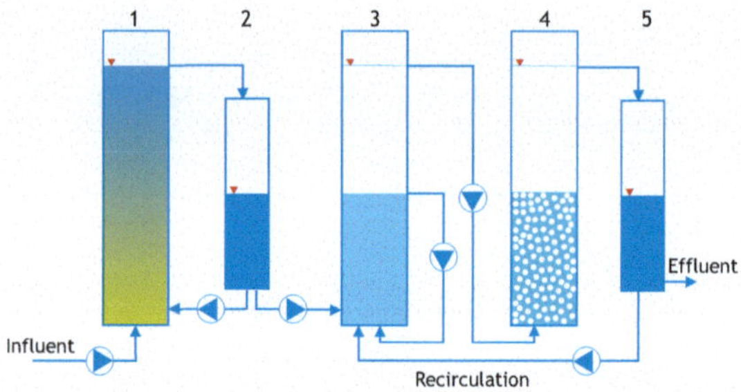

Figure 7.10 A schematic diagram of the experimental setup of the SANI® system. (1) An SRUSB (Reactor 1), (2) a flow-adjusting tank, (3) an anoxic filter (Reactor 2), (4) an aerobic filter (Reactor 3), and (5) an effluent tank.

To enable the three bio-reactions (see eqs. 7.18, 7.19 and 7.20) to carry out individually and effectively, Reactor 1 adopted a typical up-flow anaerobic sludge blanket, and Reactors 2 and 3 were packed with polypropylene plastic media (specific surface area: 215 m^2/m^3) for biofilm development. The system achieved 95% COD, 99% NO_3^-, and 74% TN removal without excess sludge withdrawal throughout the entire operation period (Wang *et al.*,

Table 7.10 The performance of the lab-scale SANI® system under various recirculation ratios (Wang et al., 2009).

Parameters	I	II	III	IV
Anoxic filter/aerobic filter recirculation rate	1Q	2Q	3Q	4Q
Influent COD of anoxic filter (mg/l)	31.8±1.5	15.5±0.8	25.9±1.3	21.4±1.0
Effluent COD of aerobic filter or effluent of the system (mg/l)	14.1±0.7	8.1±0.4	14.7±0.6	11.7±0.6
TN influent of SRUSB (mgN/l)	30±1.5	30±1.3	30±1.5	29±1.5
TN effluent of aerobic filter (mgN/l)	16±0.8	10±0.5	8±0.4	19±0.9
NO_3^- removal in anoxic filter (%)	99±4.1	99±4.5	97±4.6	8±0.4
Nitrification efficiency in aerobic filter (%)	98±4.1	99±4.5	93±4.5	17±0.8
TN removal efficiency (%)	49±2.4	65±3.2	74±3.7	34±1.7
COD removal efficiency (%)	94.4±4.7	96.9±4.8	94.3±4.7	94.2±4.7

Q = influent flow rate; NO_3^- removal in the anoxic filter (%) =NO_3^--N removed in the anoxic filter (gN)/influent NO_3^--N of anoxic filter (gN)·100; Nitrification efficiency in aerobic filter (%) = NH_4^+-N consumed in aerobic filter (gN)/influent NH_4^+-N of the aerobic filter (gN)·100; TN removal efficiency (%) = (influent TN of the SRUSB - effluent TN of the aerobic filter)/influent TN of the SRUSB·100.

2009). The HRT of the aerobic and anoxic bio-filters and the recirculation ratio (R) between these two reactors played an important role in achieving the efficient performance of pollutant removal and sludge minimization. To further understand this novel process, a steady-state model was developed from COD, nitrogen and sulphur mass and charge (electron) balances as well as stoichiometries of sulphate reduction, autotrophic denitrification and autotrophic nitrification (Lu *et al.*, 2009). The modelling approach is detailed in Section 7.4.4. In principal the model prediction agreed well with the measured data in terms of the removal of COD, NO_3 and SO_4^{2-}, sulphide production, and effluent total suspended solids (TSS). The mass balance analyses of COD, sulphur and nitrogen in the three reactors provided a good cross-check for the experimental findings and modelling results. The main performance of the lab-scale SANI® system is summarized in Table 7.10.

7.4.2.2 Pilot-scale study

In order to further demonstrate the feasibility of the SANI® process, a 225-day pilot-scale study was further conducted with 10 m^3/d of real saline sewage (screened by a 6 mm strainer) mainly from Hong Kong International Airport, which contained fractions of complex organics and unbiodegradable COD (Lu *et al.*, 2011; 2012b). The SANI® pilot plant was set up at the Tung Chung Sewage Pump Station (TCSPS) near the airport (see Figure 7.11). Table 7.11 shows the main operating conditions of the SANI® pilot plant during the steady-state period. The pilot plant was fed with saline sewage that contained on average 280 mgTSS/l, 431 mgCOD/l, 588 mgSO_4^{2-}/l and 87 mgTN/l. The COD and nitrogen loading rates were 0.63 kgCOD/m^3/d and 0.12 kgN/m^3/d, respectively, during the entire pilot trial. The SANI® pilot plant was continuously operated for 225 days at steady state and achieved average 87% COD, 87% TSS and 57% TN removal (Lu *et al.*, 2012b). The low nitrogen removal efficiency (57%) was mainly attributed to the high fraction of non-biodegradable soluble organic nitrogen (NSON) in the incoming sewage that was mainly from Hong Kong International Airport (Lu *et al.*, 2011).

Figure 7.11 The SANI® pilot plant at the Tung Chung Sewage Pumping Station (photo: HKUST).

Table 7.11 Main operating conditions of the SANI® pilot plant under steady-state (Lu *et al.*, 2012b).

Reactor	Nominal HRT (NHRT) (h)	Actual HRT (AHRT) (h)	SRT (d)	Influent flow rate	Recirculation ratio	Average temperature (°C)
SRUSB	16.3	4.1	90	Q	-	25
Anoxic bioreactor (BAR1)	9.4	2.7	110	3.5Q	2.5Q	25
Aerobic bioreactor (BAR2)	9.4	2.7	42	3.5Q	-	25

Q = influent flow rate (10 m^3/d)

The ratio of mixed liquor volatile suspended solids (MLVSS) to mixed liquor suspended solids (MLSS) in the SRUSB was stable at 0.7 and the average sludge volume index (SVI) was constantly below 110 ml/g. About 90% of biological sludge reduction was achieved and no sludge was purposely withdrawn from the pilot plant during the 225-day plant operation, which can be attributed to an extremely low observed sludge yield (0.02 kgVSS/kgCOD$_{removed}$) as well as a very long sludge retention time (SRT >100 days) in the SRUSB reactor. This significant biological sludge reduction was verified by the steady-state modelling method stated in Section 7.4.4. H_2S and CH_4 generation were not detected throughout the operation, which was further confirmed by the steady-state model and microbial analysis (Wang *et al.*, 2011; Lu *et al.*, 2012a).

7.4.3 SANI® process demonstration

Based on the experience of the pilot-scale study, the SANI® process was scaled up to 1,000 m^3/d to demonstrate its application potential in Hong Kong. It was built in two vacated rectangular primary sedimentation tanks (SD1 and SD2) at the Sha Tin Wastewater Treatment Works (STWTW) in collaboration with the local wastewater board: the Hong Kong Drainage Services Department (DSD) (see Figure 7.12 and Table 7.12). The demonstration plant consisted of four SRUSB reactors (green) for organic removal in SD1 and one MBBR for nitrogen removal in SD2. In the MBBR, autotrophic denitrification accounted for 54% reactor space and nitrification occupied 46% reactor space. MBBR is a compact technology for wastewater treatment. In this plant, degritted raw saline wastewater passed through 6 mm coarse screen bars and then a fine-mesh sieve (FMS, 350 μm) to replace the primary treatment. Part of the MBBR effluent was further treated by chemical coagulation and settled as a post-treatment option for demonstration purpose (for example, dosed with 12 mgAl/l alum and 1 mg/l anion polymer and operated at 1 h HRT). Later, the full effluent was treated by dissolved air flotation. It should be noted that TSS in the effluent of MBBR was 40-80 mgTSS/l on average, significantly lower than a conventional MBBR system, *e.g.* around 200 mgTSS/l (Ødegaard, 2006), attributable to the significant sludge minimization of the SANI® process.

Table 7.12 Operating conditions of the reactors in the SANI® demonstration plant (Wu *et al.*, 2016).

	SRUSB	Anoxic MBBR	Aerobic MBBR	Total
Phase II - three-month steady-state operation (with an average flow of 800 m^3/d):				
Reactor volume (m^3)	192 (4 × 48)	112	96	400
HRT (h)	5.76	3.36	2.88	12.0
Phase III - one-month dynamic operation (with an average flow of 720 m^3/d, maximum 1,000 m^3/d):				
Reactor volume (m^3)	144 (3 × 48)	112	96	352
HRT (h)	4.8	3.7	3.2	11.7

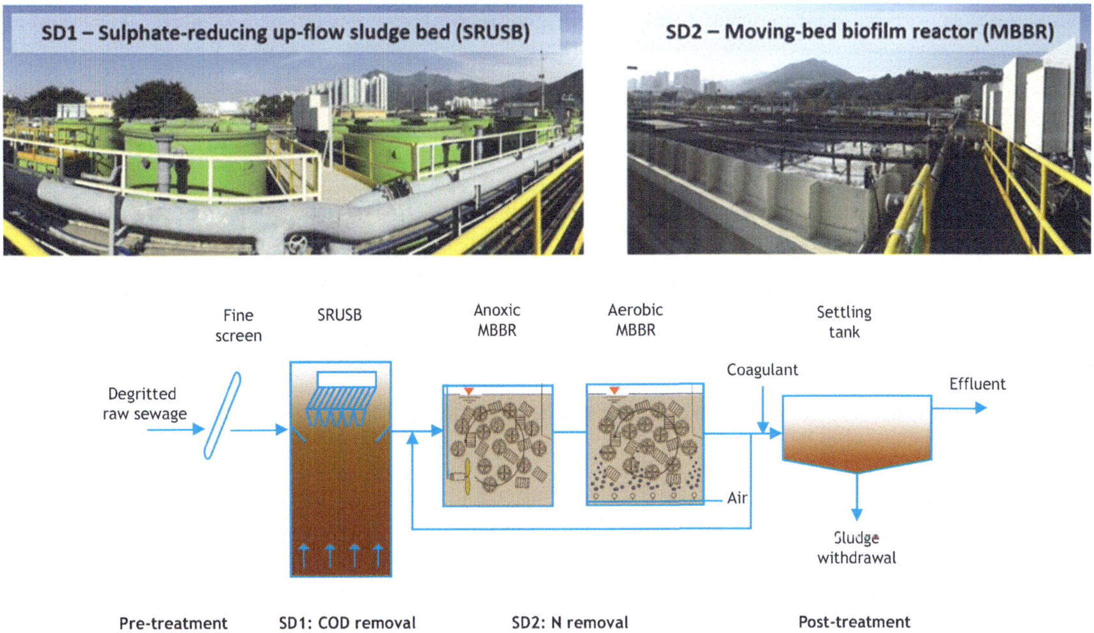

Figure 7.12 The SANI® 1,000 m^3/d demonstration plant (above) and a schematic diagram of the process (below).

The demonstration plant was continuously operated between 2013-2017, in the following phases: (*i*) the plant start-up period (Phase I: days 1-125); (*ii*) a steady-state period with an average flow rate of 800 m^3/d (Phase II: days 126-213); and (*iii*) a dynamic operation period with a diurnal peaking factor of 1.3 at an average flow of 720 m^3/d, *i.e.* maximum flow approaching 1,000 m^3/d (Phase III: days 214-250), as detailed in Table 7.12. Later on, the demonstration plant was operated in a continuously flowrate of 1,000 m^3/d approximately. The demonstration plant confirmed that a total HRT (nominal) of 12-13 h is feasible, which would reduce 30-40% of the total HRT of conventional BNR systems at the local WWTPs.

The average effluent quality (including post-treatment) is: 15±5 mgTSS/l, 5.6±1.8 $mgBOD_5$/l, 8.4±1.6 mgTN/l, and 0.9±0.3 mgTP/l (see Table 7.13). The (biological) sludge production rate was determined to be 0.35±0.08 $gTSS/gBOD_5$ (or 0.19±0.05 gTSS/gCOD). The demonstration plant reduced 60-70% of the sludge production in comparison with the local WWTPs (0.9-1.3 $gTSS/gBOD_5$).

HRT, loading rate, and bioconversion rates affect the design and operation of the SANI® process. The HRT of the SRUSB reactors in this demonstration plant was reduced from 16 h (start) to 4.8 h (stable), concurrently increasing the operational organic loading rate from 0.8 to 2.4 kgCOD/m^3/d, during which around 80% of COD removal was maintained. The minimal HRT of the SRUSB reached 2 h under the diurnal flow pattern, without sludge wash-out owing to little methane gas generation and the unique design of the SRUSB reactor. The HRT of the MBBR was maintained at 4-6 h, corresponding to the nitrogen loading rate of 0.2-0.4 kgN/m^3/d. Along with the maturation of autotrophic nitrifying biofilm and denitrifying biofilm, the specific surface rates of nitrification and SdAD (autotrophic denitrification) could attain up to 1.5 gNH_4^+-N/m^2/d and 2 gNO_x^--N/m^2/d, respectively. However, in consideration of the dynamic changes in wastewater flow and operational loading rate, as well as environmental factors (*e.g.* temperature), conservative specific design values of SO_4^{2-} reduction rate, the SdAD rate and nitrification rate at 20 °C are recommended as 1.5 kgCOD/m^3/d, 1 gNO_x^--N/m^2/d, and 1 gNH_4^+-N/m^2/d, respectively.

The mass balance of SO_4^{2-} to sulphide and back to SO_4^{2-} conversion in this plant was confirmed to be nearly 99%. The elemental sulphur accumulated in the biomass and/or in the reactors might have been involved in the autotrophic denitrification as the sulphur source; hence, the loss of sulphur via elemental sulphur can be trivial in the SANI® process. It should be noted that TDSd will be oxidized to SO_4^{2-} in the conventional COD measurement method. To obtain a reasonable COD value, sample dilution is suggested. A correction method may also be applicable (Poinapen and Ekama, 2010a; Daigger *et al*., 2015).

7.4.4 Steady-state modelling of the SANI® plant

McCarty (1975) proposed the stoichiometric mass balance from the COD (electron) balance; this was later advanced by Gujer and Larsen (1995) and then generalized for use in wastewater treatment by Grau *et al*. (2007). Ekama (2009) introduced plant-wide mass balance-based steady-state modelling for design of WWTPs as well as tracking different products in the solid, liquid and gas streams in WWTPs. In steady-state modelling, many of the bioprocesses can be assumed to reach completion, while other bioprocesses which do not reach completion are either simplified (such as biomass death regeneration to endogenous respiration), or retained because they are the slowest reactions, so to govern the sizing of the system, such as nitrification for the aerobic activated sludge system and hydrolysis for the anaerobic digestion system. Steady-state modelling, therefore, has the merit of demanding fewer input parameters.

Steady-state modelling of conventional BNR processes is based on mass balances of COD, N and P and then C, H, O elements. Steady-state modelling of biological sulphate reduction in a SRUSB reactor fed with primary wastewater sludge was further developed by Pionapen and Ekama (2010a) who validated their steady-state model in co-treatment of primary sludge with acidic mining wastewater. They further demonstrated that most H_2S and CO_2 produced stayed dissolved as HS^- and HCO_3^-, respectively. Without CO_2 gas production and with low ortho-phosphate ($H_2PO_4^-$) issue, the sulphide weak acid/base equilibrium controls the pH of SRUSB reactor. Lu *et al*. (2009) developed a steady-state model for the autotrophic denitrification process by using TDSd as the electron donor and subsequently integrated it with C-N-S bioprocesses; this was verified by a lab-scale SANI® system (see Figure 7.13). This steady-state model was further calibrated by the pilot-scale SANI® study and finally verified by the SANI® demonstration plant.

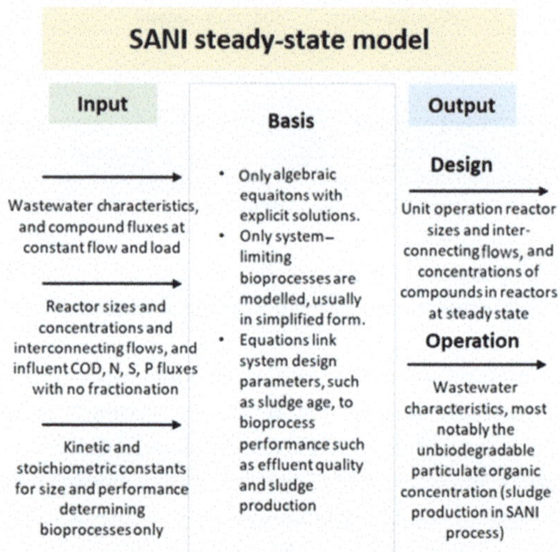

Figure 7.13 Steady-state modelling: input, basis and output.

This steady-state modelling for SANI® plant design consists of two essential parts: (*i*) development of COD, C, H, N, O, P, S mass and charge balances, and stoichiometries to determine the production of biomass and sulphide, ammonia and ortho-phosphate, and total alkalinity from the removal of biodegradable COD (see Chapter 4); and (*ii*) determination of the governing kinetic constants for sulphate reduction, autotrophic denitrification and nitrification. Four common bioprocesses take place under anaerobic, anoxic and aerobic conditions: (*i*) sulphidogenesis in biological sulphate reduction, (*ii*) autotrophic denitrification, (*iii*) nitrification, and (*iv*) aerobic sulphide oxidation. The electron donors and acceptors for these four bioprocesses are listed in Table 7.14.

Table 7.14 Key bioprocesses that take place in the SANI® process.

No.	Bioprocess	Electron donor reactant	Electron donor product	Electron acceptor reactant	Electron acceptor product
1	Sulphidogenesis	$C_xH_yO_zN_aP_b$	$CO_2(g)$; H_2CO_3; HCO_3^-; CO_3^{2-}	SO_4^{2-}; SO_3^{2-}; $S_2O_3^{2-}$; S^0	H_2S/HS^-
2	Autotrophic denitrification	H_2S/HS^-; SO_3^{2-}; $S_2O_3^{2-}$; S^0	SO_4^{2-}	NO_3^-; NO_2^-	N_2
3	Nitrification	NH_4^+	NO_3^-; NO_2^-	O_2	H_2O
4	Aerobic sulphide oxidation	H_2S/HS^-; SO_3^{2-}; $S_2O_3^{2-}$; S^0	SO_4^{2-}	O_2	H_2O

The influent stoichiometric composition, with x, y, z, a and b values in $C_xH_yO_zN_aP_b$ are assigned to each of the main influent wastewater organic fractions: *i.e.* (*i*) biodegradable soluble organics (BSO), (*ii*) biodegradable particulate organics (BPO), (*iii*) unbiodegradable soluble organics (USO), (*iv*) unbiodegradable particulate organics (UPO), and (v) free and saline ammonia (FSA). The defined biomass composition for sulphate reduction by SRB, sulphide oxidation by SOB and ammonia nitrification by autotrophic nitrifier organisms (ANO) are assigned with $C_kH_lO_mN_nP_p$ (representing SRB, SOB or ANO). These elemental compositions are determined based on the equations of Ekama (2009). For example, the organics composition ($C_xH_yO_zN_aP_b$) can be expressed as $C_{f_c/12}H_{f_h/1}O_{f_o/16}N_{f_n/14}P_{f_p/31}$, where f_c, f_h, f_o, f_n, f_p are the mass ratios (g element/g compound) of each element in the organics, and $f_c + f_h + f_o + f_n + f_p = 1$

7.4.4.1 Stoichiometry equations

Stoichiometry of biological sulphate reduction

In biological sulphate reduction, SRB utilize influent COD as their carbon source for biomass growth while SO_4^{2-} is reduced to TDSd, since the COD-to-TOC mass ratio in saline wastewater is basically between 2.5 and 3 (Poinapen and Ekama, 2010a). The general stoichiometry of SO_4^{2-} reduction with biodegradable organics assigned to be $C_xH_yO_zN_aP_b$ and generated biomass composition to be $C_kH_lO_mN_nP_p$ can be expressed by Eq. 7.21 (Lu *et al.*, 2009, 2012a):

$$\begin{aligned}
&C_xH_yO_zN_aP_b + [(1-E)\frac{\gamma_S}{8}]SO_4^{2-} \\
&+[3x - z + 4b - \frac{E\gamma_S}{\gamma_B}(3k - m + n + 4p) - 4(1-E)\frac{\gamma_S}{8}]H_2O \rightarrow \\
&[x - a + b + E\frac{\gamma_S}{\gamma_B}(n - k + p) - (1-E)\frac{\gamma_S}{8}]H_2S \\
&+[2(1-E)\frac{\gamma_S}{8} - E\frac{\gamma_S}{\gamma_B}(n - k - p) - (x - a - b)]HS^- \\
&+[E\frac{\gamma_S}{\gamma_B}]C_kH_lO_mN_nP_p + [a - nE\frac{\gamma_S}{\gamma_B}]NH_4^+ \\
&+[x - kE\frac{\gamma_S}{\gamma_B}]HCO_3^- + f\cdot[b - pE\frac{\gamma_S}{\gamma_B}]H_2PO_4^- \\
&+(1-f)\cdot[b - pE\frac{\gamma_S}{\gamma_B}]HPO_4^{2-}
\end{aligned} \tag{7.21}$$

where *f* is the fraction of $H_2PO_4^-$ in the orthophosphate (OP) species formed (OP = $H_2PO_4^- + HPO_4^{2-}$), γ_S and γ_B are defined as the electrons available for redox reactions per mol of the biodegradable organics, $C_xH_yO_zN_aP_p$, and the electrons per mol of the SRB biomass, $C_kH_lO_mN_nP_p$, respectively:

$$\gamma_S = 4x + y - 2z - 3a + 5b \ \ (e^- \text{ eq/mol}) \tag{7.22a}$$

$$\gamma_B = 4k + 1 - 2m - 3n + 5p \ \ (e^- \text{ eq/mol}) \tag{7.22b}$$

E is defined as the mass of COD produced (as active SRB biomass and endogenous sludge) per day as a fraction of the mass of biodegradable organics that are utilized per day in the SRUSB reactor at a steady state. As adopted from the COD-based anaerobic digester kinetic model of Sötemann *et al.* (2005), E is given by $E = Y_{SRB}/[1 + b_{SRB}SRT(1 - Y_{SRB})]$, where SRT is the sludge age of the SRUSB reactor (d); Y_{SRB} is the yield coefficient of the anaerobic biomass (0.113 $mgCOD_{biomass}/mgCOD_{degraded}$); and b_{SRB} is the endogenous respiration rate of the anaerobic biomass (0.04 1/d).

The concentration of COD/mol and the molar weight (M_w) of the influent organic substrate in composition of $C_xH_yO_zN_aP_b$, are given by:

$$COD = 8 \cdot [y + 2(2x - z) - 3a + 5b] \quad (gCOD/mol) \tag{7.23a}$$

$$Mw = 12x + y + 16z + 14a + 31b \quad (g\ dry\ mass/mol) \tag{7.23b}$$

Stoichiometry of autotrophic denitrification

In the anoxic reactor, sulphide donates electrons to NO_3^- for autotrophic denitrification and thus the stoichiometry of the autotrophic denitrification can be derived from elemental and charge balances (Ekama, 2009; Lu *et al.*, 2009, 2012a), which can be expressed by Eq. 7.24.

$$\begin{aligned}
&\{\frac{1}{8} - \frac{2}{10}(1-E') - \frac{E'}{\gamma'_B}[-n + p(2-f)]\}HCO_3^- \\
&+ \frac{nE'}{\gamma'_B}NH_4^+ + \frac{2}{10}(1-E')NO_3^- + f \cdot p\frac{E'}{\gamma'_B}H_2PO_4^- \\
&+ (1-f) \cdot p\frac{E'}{\gamma'_B}HPO_4^{2-} + \frac{1}{8}HS^- \\
&+ \{\frac{3}{8} - \frac{4}{10}(1-E') - \frac{E'}{\gamma'_B}[2k - m + n + p(2+f)]\}H_2O \rightarrow \\
&\frac{E'}{\gamma'_B}C_{k'}H_{l'}O_{m'}N_{n'}P_{p'} + \frac{1}{8}SO_4^{2-} + \frac{1}{10}(1-E')N_2 \\
&+ \{\frac{1}{8} - \frac{2}{10}(1-E') - \frac{E'}{\gamma'_B}[k - n + p(2-f)]\}CO_2
\end{aligned} \tag{7.24}$$

where γ'_B is the electron-donating capacity per mol of autotrophic denitrifier biomass (*i.e.* SOB), $C_{k'}H_{l'}O_{m'}N_{n'}P_{p'}$; E' is defined as the mass of COD produced (as the autotrophic denitrification biomass and endogenous sludge) per day as a fraction of the flux of nitrate reduced per day in the anoxic bioreactor at a steady state (the net yield of the autotrophic denitrification biomass and endogenous residue), *i.e.* from the equation of $E' = Y_{SOB} / (1 + b_{SOB}SRT_{AX})$, where SRT_{AX} is the sludge age of the anoxic MBBR zone (d), and Y_{SOB} is the yield coefficient of the autotrophic denitrification biomass (0.8 $mgCOD_{biomass}/mgNO_3^--N$); and b_{SOB} is the endogenous respiration rate of the autotrophic denitrification biomass (0.04 1/d).

Stoichiometry of nitrification

The main reaction in the aerobic MBBR zone is nitrification, and its stoichiometry can be described as:

$$\begin{aligned}
&14[\frac{1}{7} - \frac{E''}{8\gamma''_B}(n+p)]HCO_3^- + NH_4^+ + 14[\frac{1}{7} - \frac{E''}{4}(\frac{n''}{\gamma''_B} + \frac{1}{8})]O_2 \\
&+ f \cdot \frac{14p''E''}{8\gamma''_B}H_2PO_4^- + (1-f) \cdot \frac{14p''E''}{8\gamma''_B}HPO_4^{2-} \rightarrow \\
&\frac{14E''}{8\gamma''_B}C_{k''}H_{l''}O_{m''}N_{n''}P_{p''} + 14(\frac{1}{14} - \frac{n''E''}{8\gamma''_B})NO_3^- \\
&+ 14[\frac{1}{7} - \frac{E''}{8\gamma''_B}(k'' + n'' + p'')]CO_2 \\
&+ 14[\frac{3}{14} + \frac{E''}{8\gamma''_B}(2k'' - m'' - 2n'' + 3p'') - \frac{E''}{16}]H_2O
\end{aligned} \tag{7.25}$$

where γ''_B is the electron-donating capacity per mol of autotrophic nitrifier biomass, $C_{k''}H_{l''}O_{m''}N_{n''}P_{p''}$; E'' is defined as the mass of COD produced (as the autotrophic nitrification biomass and endogenous sludge) per day as a fraction of the flux of ammonia oxidized per day in the aerobic zone at a steady state, *i.e.* from the N-based kinetic equation of $E'' = Y_{ANO} /(1 + b_{ANO}\ SRT_{OX})$ and SRT_{OX} is the sludge age of the aerobic zone (d); Y_{ANO} is the yield coefficient of the autotrophic nitrification biomass (0.15 $mgCOD_{biomass}/mgNH_4^+-N$); and b_{ANO} is the endogenous respiration rate of autotrophic nitrifying microorganisms (0.04 1/d).

Stoichiometry of aerobic sulphide oxidation
In practical operation, part of the sulphide is always oxidized by the dissolved oxygen in an aerobic zone. Hence the stoichiometry of the sulphide oxidation to SO_4^{2-} by the dissolved oxygen needs to be expressed as Eq. 7.26 (similar to Eq. 7.24).

$$\frac{1}{8}HS^- + n''' \frac{E'''}{\gamma'''_B} NH_4^+ + f \cdot p''' \frac{E'''}{\gamma'''_B} H_2PO_4^-$$
$$+(1-f) \cdot p''' \frac{E'''}{\gamma'''_B} HPO_4^{2-} + \{\frac{1}{8} + \frac{E'''}{\gamma'''_B}(n''' - p''')\} HCO_3^-$$
$$+\frac{1-E'''}{4} O_2 \rightarrow \{\frac{1}{8} - \frac{E'''}{\gamma'''_B}(k''' - n''' + p''')\} CO_2 + \frac{1}{8} SO_4^{2-}$$
$$+\frac{E'''}{\gamma'''_B} C_{k'''}H_{l'''}O_{m'''}N_{n'''}P_{p'''}$$
$$+\{\frac{1-E'''}{2} - \frac{3}{8} + \frac{E'''}{\gamma'''_B}(2k''' - m''' + n''' + 3p''')\} H_2O \quad (7.26)$$

where γ'''_B is the electron-donating capacity per mol of the aerobic SOB biomass, $C_{k'''}H_{l'''}O_{m'''}N_{n'''}P_{p'''}$; E''' is defined as the mass of COD produced (as the aerobic sulphide-oxidizing biomass and endogenous sludge) per day as a fraction of the flux of ammonia oxidized per day in the aerobic zone at a steady state, i.e. from the N-based kinetic equation of $E''' = Y_{SOB'}/(1+b_{SOB'}\ SRT_{OX})$, where SRT_{OX} is the sludge age of the aerobic MBBR zone (d) and $Y_{SOB'}$ is the yield coefficient of the autotrophic denitrification biomass (0.8 $mgCOD_{biomass}/mgCOD_{oxidzied}$); and $b_{SOB'}$ is the endogenous respiration rate of the autotrophic denitrification biomass (0.04 1/d).

7.4.4.2 Kinetic equations

In contrast to complex dynamic simulation models, steady-state models are comparatively simple and require only constant flows and loads as the inputs to determine the system design parameters. They are based on the slowest process kinetic rates governing the overall behaviour of the system and can relate this process to the system design and operational parameters (Ekama, 2009). In SRUSB, the anaerobic hydrolysis kinetics are the slowest and thus a limiting kinetic constant, which governs the rate of hydrolysis of biodegradable particulate organics (BPO) and determines the biodegradable COD removal and hydrogen sulphide production (Poinapen and Ekama, 2010b).

The sludge age (SRT in days) is the fundamental design parameter. The rate of the change of particulate biodegradable organic concentration (X_S) with time in the SRUSB reactor during the anaerobic hydrolysis can be described by

$$\frac{dX_{SRB}}{dt} = Y_{SRB} \cdot r_{hyd} - b_{SRB} \cdot X_{SRB} \quad (7.27a)$$

$$\frac{dX_S}{dt} = -r_{hyd} + b_{SRB} \cdot X_{SRB} \quad (7.27b)$$

$$\frac{dS_{TDSd}}{dt} = (1 - Y_{SRB}) \cdot r_{hyd} \quad (7.27c)$$

$$\frac{dS_{SO_4}}{dt} = -\frac{1}{2} \cdot \frac{dS_{TDSd}}{dt} = -\frac{1}{2} \cdot (-r_{hyd} + b_{SRB} \cdot X_{SRB}) \quad (7.27d)$$

where,

r_{hyd}	volumetric hydrolysis rate (mgCOD/l/d)
X_{SRB}	biomass COD in the SRUSB reactor (mgCOD/l)
Y_{SRB}	yield coefficient of the SRUSB biomass (*e.g.* 0.113 mgCOD/mgCOD)
b_{SRB}	endogenous respiration rate of the SRB biomass (*e.g.* 0.04 1/d)
X_S	biodegradable particulate COD in the SRUSB reactor (mgCOD/l)
S_{TDSd}	hydrogen sulphide COD in SRUSB (mgCOD/l)
S_{SO4}	sulphate concentration in SRUSB ($mgSO_4^{2-}$-S/l).

Assuming the endogenous residue (f_{ED}=0) and un-biodegradable particulate COD are constant, the change of particulate COD (X_{COD}) is only caused by the changes in biomass (X_{SRB}) and biodegradable particulate COD (X_S), as shown by Eq. 7.28:

$$\frac{dX_{COD}}{dt} = \frac{dX_{SRB}}{dt} + \frac{dX_S}{dt} = (Y_{SRB} - 1) \cdot r_{hyd} \quad (7.28)$$

where the anaerobic hydrolysis rate (r_{hyd}) can be described with saturation kinetics as:

$$r_{hyd} = \frac{K_{SRB,max} \cdot X_S}{K_s + X_S} \cdot X_{SRB} = \frac{\mu_{hyd,max} \cdot X_S}{Y_{SRB}(K_s + X_S)} \cdot X_{SRB} \quad (7.29)$$

where,

r_{hyd}	volumetric hydrolysis rate (mg COD/l/d)
$K_{SRB,max}$	maximum specific hydrolysis rate constant in Monod kinetics (*e.g.* 3.25 mgCOD/mgCOD.d)
K_s	half saturation constant (*e.g.* 500 mgCOD/l at 20 °C)
$\mu_{hyd,max}$	maximum specific growth rate of anaerobic hydrolysis (*e.g.* 0.4 1/d)
X_S	biodegradable particulate COD (mgCOD/l)
X_{SRB}	concentration of SRUSB biomass (mgCOD/l)
Y_{SRB}	yield coefficient of SRUSB biomass (*e.g.* 0.113 mgCOD/mgCOD).

The nitrification process occurs in the aerobic zone, and its behaviour is the same as the conventional autotrophic nitrification process. Hence, the kinetic constants for nitrification can be determined by the Monod kinetics (see Section 5 in Chapter 5). These include nitrifier growth dX_{ANO}/dt (eqs. 5.3, 5.5), the specific growth rate μ_{ANO} Monod constant equation (Eq. 5.4), the ammonia conversion rate (Eq. 5.6), the nitrate formation rate (Eq. 5.7), and the oxygen utilization rate (Eq. 5.8), as well as the endogenous respiration rate for nitrifiers (Eq. 5.9). In the anoxic reactor, the kinetic behaviour of autotrophic denitrification can be simplified to the following eqs. 7.30 and 7.31, by correlating the Monod equation to the parameters of both nitrate (S_{NO3}) and sulphur compounds including dissolved sulphides (S_{TDSd}), elemental sulphur (S_{S^0}), and thiosulphate ($S_{S_2O_3}$).

$$\frac{dS_{NO_3}}{dt} = \frac{1}{Y_{SOB}} \mu_{SOB,max} \left(\frac{S_{TDSd}}{S_{TDSd} + K_{TDSd}} + \frac{S_{S^o}}{S_{S^o} + K_{S^o}} + \frac{S_{S_2O_3}}{S_{S_2O_3} + K_{S_2O_3}} \right) \frac{S_{NO_3}}{S_{NO_3} + K_{NO_3}^{SOB}} X_{SOB} \quad (7.30)$$

$$\frac{dX_{SOB}}{dt} = Y_{SOB} \left[\frac{dS_{NO3}}{dt} \right] - b_{SOB} X_{SOB} \quad (7.31)$$

where,

X_{SOB}	biomass COD in anoxic reactor (mgCOD/l)
Y_{SOB}	yield coefficient of SRUSB biomass (*e.g.* 0.8 mgCOD/mgNO_3^--N)
b_{SOB}	endogenous respiration rate of SOB biomass (*e.g.* 0.04 1/d)
$\mu_{SOB,max}$	maximum specific growth rate of SOB (*e.g.* 0.75 1/d).

As described in Section 7.3, SOB are easily inhibited by exposure to oxygen. Thus, using an attached growth biofilm reactor, *e.g.* MBBR or an integrated fixed-film activated sludge (IFAS) system for the nitrification and SdAD, is ideal to retain most of the functional ANO (nitrifiers) and SOB in the aerobic and anoxic zones, respectively. Modelling of the complex bioprocess kinetics and mass transfer of biofilm in a biofilm reactor for TN removal under different DO levels in the autotrophic reaction zone of the SANI® process should be developed in the future. At the current stage, the complexity is reduced and the general situation is applied with input of empirical kinetic rate constants (see Chapter 18) in the steady-state modelling for design of the biofilm reactors in the SANI® process. Based on the operation of the lab-scale SANI®-MBBR system for over 700 days, the biofilm thickness in both anoxic MBBR and aerobic MBBR zones is typically below 200 μm and the biofilm in both zones has a large porosity (Cui *et al.*, 2019), making the specific loading rates of ammonia, nitrate, oxygen, and sulphide less than the critical substrate flux of conventional biofilm reactors (g/m^2.d). Thus, the biofilm kinetics for autotrophic nitrification and denitrification can be assumed to be zero order with either partial or full penetration. The specific surface conversion rate of NO_3^- and NH_4^+ could thus be used as an input kinetic constant in the SANI® steady-state modelling as obtained from the SANI® demonstration plant, *i.e.* $B_{A,NOx}$ = 1 gNO_x^--N/m².d and $B_{A,NH4}$ =1 gNH_4^+-N/m².d (at 20 °C).

7.4.5 The SANI plant design approach

7.4.5.1 Steady-state plant-wide model

In this section, we verify the above plant-wide model with data from the SANI® demonstration plant (see Figure 7.12b). In the simulation of the primary treatment unit, the solid capture performance was set at 47% to reflect the actual performance of the fine-mesh sieve in this plant; while in the simulation of the performance of the SRUSB reactors, an anaerobic digester unit plus a point clarifier unit were linked in series to determine the sludge bed zones and the three-phase separators of the SRUSB reactors. A recirculation line between the point clarification zone and the anaerobic digestion zone of the SRUSB reactor was built to simulate the reactor internal recirculation. The recirculation flow rate was set at 60% according to the actual operation data of the SANI® demonstration plant. Two MBBR units were applied in the model to simulate both the anoxic and aerobic zones of the MBBR system. The major design parameters applied in the simulation include the tank dimensions, specific surface areas of the carriers, filling ratios of the carriers, and the recirculation flow rates between the aerobic and anoxic zones. The post-treatment was simulated based on the solid capture performance of the pilot unit of the demonstration plant. The above-developed plant-wide model for the SANI® process shows that its design has a lot of potential (see Table 7.15).

Table 7.15 Comparison of the measured (M) values with the model-predicted (P) data of the demonstration plant.

Flow rate (800 m³/d)	Influent		Micro-screen effluent		SRUSB effluent		Anoxic MBBR effluent		Aerobic MBBR effluent		Effluent	
	M	P	M	P	M	P	M	P	M	P	M	P
COD (mg/l)	488	488	352	369	-	69	-	97	112	108	62	45
BOD (mg/l)	231	239	178	188	-	10	-	25	31	31	5.6	9
TSS (mg/l)	362	243	192	149	-	45	-	71	68	79	15	17
VSS (mg/l)	-	193	-	118	-	24	-	47	-	55	-	11
TN (mgN/l)	53	53	48	49	-	48	-	16	16	15	8.4	10.9
NH_4^+-N (mgN/l)	33	35	37	35	-	45	-	10	1.3	2.3	0.4	2.3
NO_3^--N (mg/l)	-	0	-	0	-	0	-	1	7.9	6.2	7.5	6.2

7.4.5.2 Design calculation of SANI reactors

In this section we introduce a design calculation for the reactors to replace the complicated steady-state model. A calculation by hand can reasonably estimate the principal system design and operating parameters, such as sludge age, reactor volume, biofilm carriers etc. Moreover, the calculation results for the reactor volume and sludge age are the key input parameters for the plant-wide steady-state modelling.

SRSUB design

An ideal SRUSB reactor will convert all the biodegradable COD (both readily and slowly degradable COD) into biomass and HCO_3^-, via anaerobic hydrolysis, sulphate reduction and endogenous respiration. The un-biodegradable particulate COD is enmeshed in the sludge bed (and eventually leaves the reactor through sludge wasting), while un-biodegradable soluble COD flows out of the reactor via the effluent. Based on this scenario, the key design parameters include (i) the mass of TSS sludge in the system (MX_{TSS}, kgTSS), (ii) the average daily SO_4^{2-} reduction, (iii) the active fraction of the sludge with respect to VSS ($f_{av,SRB}$), and (iv) the demand for SO_4^{2-} (FSO_4, kgS/d). They can be determined as follows:

Mass of COD treated/d

$$= FCOD_i = Q_i \cdot COD_i \quad (7.32)$$

Mass of biodegradable COD treated per day

$$= FCOD_{b,i} = \left(1 - f_{SU,CODi} - f_{XU,CODi}\right) \cdot COD_i \quad (7.33)$$

Mass of unbiodegradable particulate organics

$$= FX_{Uv,i} = FCOD_i \cdot f_{SU,COD_i} / f_{cv} \quad (7.34)$$

Mass of inorganic suspended solids (ISS)

$$= FX_{FSS,i} = Q_i \cdot X_{FSS} \quad (7.35)$$

$$MX_{VSS} = FCOD_{b,i} \cdot \left[\frac{Y_{SRB} \cdot SRT}{(1 + b_{SRB} SRT)}\right] \cdot (1 + f_{OHO} \cdot b_{SRB} SRT) + FX_{Uv,i} SRT \quad (7.36)$$

$$MX_{FSS} = FX_{FSS,i} \cdot SRT + f_{XU,CODi} \cdot f_{av,SRB} \cdot MX_{VSS} \quad (7.37)$$

$$f_{av,SRB} = 1 / \left[1 + f_{SRB} b_{SRB} SRT + f_{cv}\left(1 + b_{SRB} SRT\right)\right] \quad (7.38)$$

$$MX_{TSS} = MX_{VSS} + MX_{FSS} \quad (7.39)$$

$$f_{VT} = MX_{VSS} / MX_{TSS} \quad (7.40)$$

$$FSO_4 = FCOD_i \left(1 - f_{SU.CODi} - f_{XU,CODi}\right) \cdot \left[\left(1 - f_{cv} Y_{SRBv}\right) + \left(1 - f_{SRB}\right) \cdot b_{SRB} \cdot \left(\frac{\left(Y_{SRBv} \cdot f_{cv} \cdot SRT\right)}{(1 + b_{SRB} SRT)}\right)\right] \quad (7.41)$$

where,

Q_i	influent flow rate (l/d or m^3/d)
CODi	influent COD concentration (mgCOD/l)
$f_{SU,CODi}$	non-biodegradable soluble COD/influent total COD
$f_{XU,CODi}$	non-biodegradable particulate COD/influent total COD
f_{cv}	ratio of COD-to-VSS or COD-to-mass for soluble organics
X_{FSS}	inorganic suspended solid concentration (mgISS/l)
MX_{VSS}	mass of influent particulate inorganic matter in the bioreactor
Y_{SRB}	yield coefficient of SRB biomass
b_{SRB}	endogenous respiration rate of SRB biomass (1/d)
f_{OHO}	endogenous residual fraction of ordinary heterotrophic organisms
MX_{FSS}	mass of volatile suspended solids in the bioreactor
$f_{av,SRB}$	active fraction of SRB
MX_{TSS}	mass of TSS sludge in the system
f_{VT}	VSS/TSS ratio of the activated sludge
f_{SRBv}	endogenous residual fraction of SRB (mgVSS/mgCOD).

In the SRUSB design, the selection of the sludge concentration and the SRT is critical. Based on the above equations, a design curve reflecting the relationships between the designed sludge age (d) and sludge concentration (kgMLSS/m^3) and their effects on COD loading rate (kgCOD/m^3/d) and reactor size (which corresponds to the hydraulic retention time (HRT = $m^3_{reactor\ size}/m^3_{wastewater}$/d)) are shown in Figure 7.14. The considerable SRT and sludge concentration of SRUSB fall in the range of 15-100 days and 5-15 gMLSS/m^3, respectively. The proper design data can be found from this figure; for example, when the influent loading rate is 1.5 kgCOD/m^3/d, an SRUSB can have an HRT at 4.8h, while the SRT should be controlled at 15, 29 and 43 days corresponding to a sludge concentration of 5, 10, and 15 kgMLSS/m^3, respectively. It shall also be mentioned when COD/SO_4^{2-} mass ratio >1.2, the sulphate becomes limiting in the process for COD removal.

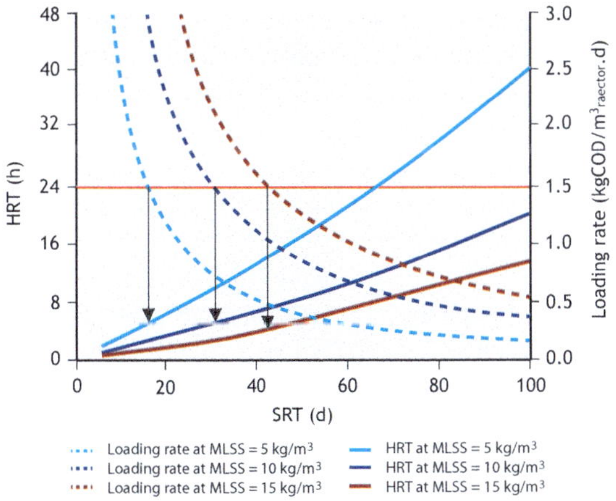

Figure 7.14 SRUSB design curves (for the SANI® full-scale case study; the input data is from Table 7.15).

Design calculation of SdAD and nitrifying reactors

The design calculation of these reactors is limited to the attached growth and based on the data obtained from MBBR in the SANI® demonstration plant. The design principle of these autotrophic denitrifying (anoxic) and nitrifying (aerobic) zones (reactors) is based on the determination of (*i*) the effluent NO_3^- concentration which is correlated with the recirculation flow ratio between these two reactors; (*ii*) the oxygen demand in the aerobic reactor and the aerobic oxidation of TDSd; and (*iii*) the necessary surface area and volume of each reactor, according to the following eqs. 7.42-7.47 (adopted from Chapter 18):

Effluent nitrate concentration

$$= S_{NO_3,e} = NIT_c / (a+1) \quad (7.42)$$

where NIT_c is the nitrification capacity ($=S_{NHx,i}-S_{NHx,e}$), and $S_{NHx,i}$ and $S_{NHx,e}$ are the influent and effluent NH_4^+ concentrations.

The mass of oxygen demand in the aerobic reactor

$$= FO_{NIT} = \left[4.57 \cdot \left(S_{NHx,i} - S_{NHx,e}\right) + S_{TDSd,ane}\right] \cdot Q_i \quad (7.43)$$

The necessary surface area for the nitrifying reactor (also for the aerobic reactor)

$$= A_{F,NH_4} = \left(Q_i \cdot NIT_c\right) / B_{A,NH_4} \quad (7.44)$$

The necessary surface area for SdAD

$$= A_{F,NOx} = \left(Q_i \cdot \left(NIT_c - S_{NO_3,e}\right) B_{A,NOx}\right) \quad (7.45)$$

Volume of aerobic reactor

$$= V_{OX} = A_{F,NH_4} / a_F \quad (7.46)$$

Volume of anoxic reactor

$$= V_{AX} = A_{F,NOx} / a_F \quad (7.47)$$

where a_F is the specific surface area of the biofilm in the respective reactor, which is equal to the carrier-filling percentage multiplied by the raw carrier-specific surface area, *e.g.* 60% filling ratio · 500 m^2/m^3 = 300 m^2/m^3 in the case of the demonstration plant.

7.5 SULPHUR CONVERSION BASED RESOURCE RECOVERY

7.5.1 Introduction

Moving from passive treatment to carbon- and energy-neutral treatment with concomitant resource recovery is the long-held goal of wastewater treatment given that more than 50% of lost/wasted resources in a city end up in wastewater (Van Loosdrecht and Brdjanovic, 2014). Biogas collection through anaerobic digestion has been practiced for over a century, and nutrients are recycled through the land application of bio-solids (USEPA, 1999). Therefore, energy saving and resource recovery from processes involving SO_4^{2-} conversion should be considered in the rapid expansion of biological sulphur conversion-based technologies. The emerging opportunities are outlined as follows (Lin *et al.*, 2018; Zeng *et al.*, 2019).

7.5.2 Metal sulphides

Metal sulphides are extensively applied in industrial processes owing to their unique characteristics, such as hydrocracking in fuel refineries, hydro-processing, and oxygen reduction in fuel cells (Da Costa *et al.*, 2016). At the nanoscale (1 to 100 nm), metal sulphides can be further applied in cancer therapy, photocatalysis, optoelectronics and as anti-microbial agents in textile modification technologies (Zeng *et al.*, 2019). Those metal sulphides that are commercially available are generally synthesized through methods including high gravity, chemical baths, γ radiation, pyrolysis, and thermolysis of metal complexes (Da Costa *et al.*, 2016). High temperature and pressure are essential in those processes which use hazardous chemicals (*e.g.* surface stabilizer trioctylphosphine oxide). In contrast, the biological production of metal sulphides is a safe, cost-effective and environmentally-friendly synthesis method. For instance, cadmium sulphide (CdS) nanoparticles of 8 nm were intracellularly synthesized by photosynthetic bacterium *Rhodopseudomonas palustris*, and *E. coli* are also capable of producing intracellular CdS nanocrystals of 2-5 nm (Bai *et al.*, 2009). Sulphide is the product of sulphate reduction mediated by SRB, which can extracellularly bind to or precipitate with metal ions in solution. Valuable metal sulphides produced via the metabolic pathway of SRB are shown in Table 7.16.

Table 7.16 Potential metallic sulphide products mediated by SRB (Zeng *et al.*, 2019).

Product	Location	Operational pH	Application	Functional agent
CuS	Extracellular	6.0	Photothermal agents for ablation of cancer cells	Sulphidogenesis system
PbS	Extracellular	7.2-7.8	Semiconductors	Sulphidogenesis system
Sb_2S_3	Extracellular	7.0 -9.0	Solar cells	Sulphidogenesis system
FeS	Extracellular	7.5	Groundwater and soil remediation/heavy metal adsorbents	*Desulfovibrio Vulgaris Miyazaki*
CdS	Intracellular/extracellular	7.4	Displays, photodiodes, electronics, solar cells and sensors	*Desulforibrio caledoiensis*
ZnS	Intracellular	7.4	Semiconductors	*Desulfobacteriaceae*

Two strategies are executed when SRB are applied in metal sulphide production: the metal directly reacts with the sulphide in the sulphidogenic system (the in-line system), or sulphide is produced and separated as a metal precipitant agent (the off-line system). Biological-based synthesis demonstrated superiority in products stability; the zinc sulphide (ZnS) precipitates biogenerated by *Desulfovibrio sulfuricans* are more stable than the abiotic ZnS precipitates. The biogenic Ni_3S_2 in a mixed culture of SRB exhibits a better crystallinity than the corresponding abiotic samples (Gramp *et al.*, 2007). The adsorptive extracellular polymeric substances, such as carboxyl, thiol/phosphate, and amino/hydroxyl of SRB and other organisms obstruct the formation of metal sulphides. In addition, aggregation induced by metal-binding polypeptides and proteins of organisms in an in-line system limit the dispersal of metal sulphide particles in the solution, reducing the potential to harvest those particles. In comparison, harvesting biogenic metal sulphides from sulphidogenic off-line system could be a suitable option, particularly in the treatment of metal-containing wastewater. Furthermore, the binding characteristics of metals are also sensitive to pH changes. For example, the nature, surface, and settling characteristics of copper sulphide (CuS) and ZnS are heavily influenced by the bulk pH (Mokone *et al.*, 2010). When precipitation is conducted with excessive sulphide at operational pH below 6, the zeta potential reduces to improve the settling and dewatering properties of ZnS precipitates. A modular bioreactor for selective metal recovery from AMD using an off-line sulphidogenic system was proposed by Hedrich and Johnson (2014), in which pH control is crucial for metal sulphide formation. For instance, with pH below 3, Cu can be selectively precipitated by sulphide while other metals (Fe and Zn) present in high concentrations in leach liquors can be recovered from their sulphide precipitates. The crystallinity of the metal sulphides is of utmost importance particularly when they are used in high-tech industrial processes such as solar panel construction, as the spatial separation of charges is crucial in the design of efficient donor-acceptor heterojunctions (Bansal *et al.*, 2013). The factors determining crystallinity in sulphidogenic systems are not clear yet, although a better crystallinity of biogenic ZnS was obtained by *Desulfovibrio desulfuricans* than in abiogenic tests (Xu *et al.*, 2016). Mineral crystallinity is an attractive approach to achieving metal sulphide biological production in off-line sulphidogenic systems and to realizing the recovery of valuable resources from waste streams.

7.5.3 Elemental sulphur recovery and reuse

As well as metal sulphide, elemental sulphur is also worth recovering and reusing with the assistance of a group of sulphur conversion-based bacteria, namely chemolithotrophic SOB. SOB use O_2 as the electron acceptor to oxidize sulphide to elemental sulphur. The basic principle for selective sulphur oxidation is to regulate the molar ratio of oxygen-to-sulphide (O/S) at approximately 0.7 (Lin *et al.*, 2018). DO control is a common strategy to reach the proper O/S ratio when the reactor is operated under a relatively stable sulphide concentration; however, the dynamic DO concentration resulting from its continuous consumption causes difficulties in maintaining the desired O/S ratio (Janssen *et al.*, 1998). Oxidation-reduction potential (ORP), determined by both sulphide and DO concentrations, is a more feasible, alternative indicator for controlling selective sulphur oxidation.

As one of the most successful sulphur recovery technologies, the THIOPAQ® process has been applied for gas desulphurization for over 25 years (Paques, the Netherlands). This system consists of three parts: the sulphide absorber, the sulphide-oxidizing bioreactor and the sulphur settler. First, the sulphide-containing gas stream contacts an alkaline solution in the absorber. Afterwards, the sulphide-containing liquid is transferred to a sulphide-oxidizing bioreactor where the sulphide is converted biologically to elemental sulphur under weak alkaline conditions (pH 8-9) by SOB, typically dominated by the haloalkaliphilic *Thioalkalivibrio*. Finally, the mixed liquid enters a settler to recover the elemental sulphur. In this process, over 90% of the influent sulphide can be converted to elemental sulphur (Van den Bosch, 2008). The recovered elemental sulphur serves as a raw material for sulphuric acid, fertilizers and pesticide production. In contrast to commercial crystalline sulphur, the sulphur produced from biological sulphide oxidation is characterized by high hydrophilicity and biological activity (Kleinjan *et al.*, 2012). Biogenic elemental sulphur can stimulate sulphuric acid generation and dissolution of sulphide ores relative to chemical elemental sulphur during biological sulphide oxidation in the leaching process (Tichý *et al.*, 1994). Moreover, biogenic sulphur can be used directly as a fertilizer, which is demonstrated with an oil-bearing vegetable, canola. A 50% increase in grain yield was observed after the addition of biogenic sulphur, outperforming the commercial elemental sulphur used in the same study (Cline *et al.*, 2003). In addition, the sulphur-limestone process using elemental sulphur to drive autotrophic denitrification for groundwater remediation has a low reaction rate due to the low solubility of sulphur. Biogenic sulphur may overcome this drawback because its hydrophilicity would accelerate the mass transfer.

7.5.4 Metabolic intermediate recovery

Metabolic intermediates/products by SRB are another opportunity for resource or energy harvesting. However, such attempts can only be based on an understanding of the metabolic mechanism of SRB and syntrophic interaction between SRB and other microorganisms. The feasibility of engineering implementaionwithout jeopardizing the main purpose of sulphur conversion is a research task.

Biohydrogen

A high level of hydrogenase activity is often observed in SRB-enriched systems, especially in environments with low concentrations of sulphur. H_2 production during the breakdown of lactate from the SRB species *Desulfovibrio vulgaris* was first observed by Tsuji and Yagi (1980). With formate as the sole electron donor, the highest H_2 production rate of 0.6 mmol/l_{medium}/h was recorded with the mediation of *Desulfovibrio vulgaris Hildenborough* (Martins and Pereira, 2013). This production rate can significantly increase to 5 mmol/l_{medium}/h by using a sparging reactor to maintain a low H_2 partial pressure (P_{H2}) which facilitates the reaction thermodynamics (Martins *et al.*, 2015). However, whether the energy recovered from the H_2 production is equivalent to the gas supply for sparging is still unclear. The electron transfer pathways of formate-driven H_2 production by *Desulfovibrio* were proposed by Martins *et al.* (2016). The direct electron transfer involving periplasmic enzymes ([NiFeSe] Hase) and the transmembrane electron transfer allow energy metabolism; however, the relative weights of these two pathways are unknown. Conversely, H_2 production by *Desulfovibrio gigas* occurs exclusively via the direct periplasmic enzyme route and transmembrane electron transfer is not involved at all. *Desulfovibrio gigas* and *Desulfovibrio vulgaris* are quite distinct in terms of number, type, and localization of hydrogenase, resulting in variations in their H_2 production performance (0.26 mmol/l/h in *Desulfovibrio vulgaris*; 0.11 mmol/l/h in *Desulfovibrio gigas*). Compared with *Desulfovibrio vulgaris* which contains a large number of hydrogenases, only two [NiFe] Hases are present in *Desulfovibrio gigas*: the energy-conserving hydrogenase (Ech) and the periplasmic HynAB-1 (Morais-Silva *et al.*, 2013).

SRB have also been found to catalyze H_2 production at the cathode of bioelectrochemical systems. A pure culture of *Desulfovibrio vulgaris* Hildenborough immobilized onto a carbon electrode can catalyze H_2 production with methyl viologen as a redox mediator (Tatsumi *et al.*, 1999), while a mediator-less microbial biocathode is more desirable for the stable operation of electrolysis systems. A *Desulfitobacterium*-enriched culture without mediators at the cathode was proposed accordingly, achieving four times higher H_2 production rate than that of the abiotic controls (Villano *et al.*, 2011). Furthermore, the capability of *Desulfovibriocaledoniensis* and *Desulfovibrio paquesii* to receive electrons directly from a graphite cathode for catalysing H_2 production was demonstrated. Additional efforts are required for applying SRB for H_2 production.

Hydrocarbons

For the sulphate-reducing process, incomplete oxidizers usually outcompete complete oxidizers for substrates, such as intermediates of anaerobic degradation of organic matter (*e.g.* H_2 and/or lactate), and incomplete oxidation becomes the dominant metabolic pathway (Muyzer and Stams, 2008). This also results in more intermediate production during their metabolism. Production of aliphatic hydrocarbons of C_{10}-C_{25} by *Desulfovibrio* with organics or fatty acids supplementing seawater containing media was first recorded by Jankowski and Zobell in 1944. n-C_{25}-C_{35}-alkanes could be the major product (up to 80% of the total hydrocarbons) produced by *Desulfovibrio desulfuricans*, and CH_3COO^- and formate are simultaneously formed with fixation of CO_2 (Bagaeva, 1998). It has been hypothesized that this synthesis is initiated by formation of CH_3COO^- and formate through CO_2 fixation. The produced micro-molecules are reduced to aldehydes, which further undergo aldol condensation elongating the carbon chain to produce hydrocarbons. Synthesis of hydrocarbons by *Desulfovibrio desulfuricans* depends largely on the medium composition and the volumetric ratio of H_2 to CO_2 (the optimum is 9:1) in a gaseous phase (Bagaeva, 2000). The supply of H_2 is in favour of bacterial growth and synthesis of reduced products including hydrocarbons. The intracellular hydrocarbon synthesis potential of *Desulfovibrio desulfuricans* ranges from 0.8 to 2.25% dry weight. Apart from *Desulfovibrio desulfuricans*,

Desulfovibrio desulfuricans G20 has also been reported to extracellularly synthesize alkanes (Friedman and Rude, 2012). Overall, many obstacles remain to be conquered before SRB can be used as an economically viable biocatalyst for the industrial manufacture of 'drop-in' biofuels.

Polyhydroxyalkanoates

The storage of polyhydroxyalkanoates (PHAs) in SRB has been investigated for their PHA synthesis ability in different saline water media. Benzoate-grown *Desulfonema magnum* demonstrated the highest PHA content of 88% (wt./wt.) of cell dry matter, which is comparable to some of the best commercially applied organisms (Hai *et al.*, 2004). The PHA contents in *Desulfobacterium autotrophicum*, *Desulfococcus multivorans*, *Desulfosarcina variabilis*, and *Desulfobotulus sapovorans* were 11, 27, 22, and 43% (wt./wt.) of the cell dry matter, respectively. The production cost of PHAs in fed-batch mode using waste materials could reach up to 3-5 USD/kgPHAs (Roland-Holst *et al.*, 2013), which is much higher than that of oil-based plastics of approximately 1 USD/kgPHA (Salehizadeh and Van Loosdrecht, 2004). Most of the cost associated with the biosynthesis of PHAs is related to feedstock, aeration (if required), sterilization (maintaining a pure culture) and extraction of the PHAs. Optimistically, sharing the costs between wastewater treatment and PHA production would be an option when applying SRB-based technologies, for instance.

7.6 CONCLUSION AND PERSPECTIVES

As the first sulphur-based biological treatment process for mainstream wastewater treatment, the SANI® process has been developed to deal with saline wastewater with sludge being significantly minimized. More importantly, the SANI® process offers an opportunity to use seawater for toilet flushing as an economic water resource for water-scarce populous coastal cities and islands. It could also assist in the energy-efficient treatment of sulphur-laden industrial wastewaters for nutrient (including phosphorous in the future) removal.

Our recent work has proven that Anammox bacteria can be enriched in the SANI® process to deal with low-strength saline wastewater or the effluent of chemically-enhanced primary treatment (CEPT) (Deng *et al.*, 2019). This new process, the SMOX® process, will be scaled up in the near future and will optimize the SANI process towards more energy-efficient treatment of saline wastewater. Recovery of scarce biochemical materials (*e.g.* sulphated polysaccharides (SPs)) from the sludge of sulphate-laden wastewater treatment will provide an opportunity for moving wastewater resource recovery into the market effectively (Xue *et al.*, 2019). Technically the sulphur-sulphide-sulphate cycle biological treatment process may be changed to a sulphate-sulphide-elemental sulphur cycle in order to replace the limited provision of oxygen electron acceptors from aeration through innovative biochemical or electrochemical or combined method(s) to overcome the major energy-consuming barrier in WWTPs.

REFERENCES

Abildgaard L., Nielsen M.B., Kjeldsen K.U. and Ingvorsen K. (2006). Desulfovibrio alkalitolerans sp. nov., a novel alkalitolerant, sulphate-reducing bacterium isolated from district heating water. *International Journal of Systemic and Evolutionary Microbiology*, 56(5), 1019-1024.

Alvarado L.R. (2016). Optimization of the electron donor supply to sulphate reducing bioreactors treating inorganic wastewater (Doctoral dissertation).

An S., Tang K. and Nemati M. (2010). Simultaneous biodesulphurization and denitrification using an oil reservoir microbial culture: Effects of sulphide loading rate and sulphide to nitrate loading ratio. *Water Research*, 44(5), 1531-1541.

Badziong W., Thauer R.K. and Zeikus J.G. (1978). Isolation and characterization of Desulfovibrio growing on hydrogen plus sulfate as the sole energy source. *Archives of Microbiology*, 116(1), 41-49.

Bagaeva T.V. (1998). Sulfate-reducing bacteria, hydrocarbon producers (Doctoral dissertation)

Bagaeva T.V. (2000). Effects of composition of the gaseous phase on the formation of hydrocarbons in Desulfovibrio desulfuricans. *Applied Biochemistry and Microbiology,* 36(2), 165-168.

Bai H.J. and Zhang Z.M. (2009). Microbial synthesis of semiconductor lead sulfide nanoparticles using immobilized Rhodobacter sphaeroides. *Materials Letters Journal,* 63(9-10), 764-766.

Bansal N., Reynolds L.X., MacLachlan A., Lutz T., Ashraf R.S., Zhang W. and Hill M.S. (2013). Influence of crystallinity and energetics on charge separation in polymer–inorganic nanocomposite films for solar cells. *International Journal of Scientific Reports*, 3, 1531.

Barrera E.L., Spanjers H., Solon K., Amerlinck Y., Nopens I. and Dewulf J. (2015). Modeling the anaerobic digestion of cane-molasses vinasse: extension of the Anaerobic Digestion Model No. 1 (ADM1) with sulfate reduction for a very high strength and sulfate rich wastewater. *Water Research*, 71, 42-54.

Barton L.L. and Hamilton W.A. (eds.) (2007). Sulphate-reducing bacteria: environmental and engineered systems. Cambridge University Press, UK.

Bothe H. and Trebst A. (1981). Biology of inorganic nitrogen and sulfur. Springer Verlag.

Brahmacharimayum B., Mohanty M.P. and Ghosh P.K. (2019) Theoretical and practical aspects of biological sulfate reduction: a review. *Global NEST Journal,* 21(2), 222-244.

Cardoso R.B., Sierra-Alvarez R., Rowlette P., Flores E.R., Gomez J. and Field J.A. (2006). Sulfide oxidation under chemolithoautotrophic denitrifying conditions. *Biotechnology and Bioengineering,* 95(6), 1148-1157.

Choi E. and Rim J.M. (1991). Competition and inhibition of sulfate reducers and methane producers in anaerobic treatment. *Water Science and Technology,* 23(7-9), 1259-1264.

Claus G. and Kutzner H.J. (1985). Physiology and kinetics of autotrophic denitrification by Thiobacillus denitrificans. *Applied Microbiology and Biotechnology,* 22(4), 283-288.

Cline C., Hoksberg A., Abry R. and Janssen A. (2003). Biological Process for H_2S Removal from Gas Streams: The Shell-Paques/THIOPAQ™ Gas Desulfurization Process. In Proceedings of the Laurance Reid gas conditioning conference.

Cui Y.X., Biswal B.K., Van Loosdrecht M.C.M., Chen G.H. and Wu D. (2019). Long term performance and dynamics of microbial biofilm communities performing sulfur-oxidizing autotrophic denitrification in a moving-bed biofilm reactor. *Water Research,* 166, 115038.

Cui Y.X., Wu D., Mackey H.R., Chui H.K. and Chen G.H. (2018). Application of a moving-bed biofilm reactor for sulfur-oxidizing autotrophic denitrification. *Water Science and Technology,* 77(4), 1027-1034.

Da Costa J.P., Girão A.V., Trindade T., Costa M.C., Duarte A. and Rocha-Santos T. (2016). Biological synthesis of nanosized sulfide semiconductors: current status and future prospects. *Applied Microbiology and Biotechnology,* 100(19), 8283-8302.

Daigger G.T., Hodgkinson A., Aquilina S. and Kim Fries M. (2015). Development and implementation of a novel sulfur removal process from H_2S containing wastewaters. *Water Environment Research,* 87, 618-625.

Deng Y.F., Ekama G.A., Cui Y.X., Tang C.J., Van Loosdrecht M.C.M., Chen G.H. and Wu D. (2019). Coupling of sulfur(thiosulfate)-driven denitratation and anammox process to treat nitrate and ammonium contained wastewater. *Water Research,* 163, Article 114854.

Di Capua F., Ahoranta S.H., Papirio S., Lens P.N.L. and Esposito G. (2016). Impacts of sulfur source and temperature on sulfur-driven denitrification by pure and mixed cultures of Thiobacillus. *Process Biochemistry,* 51(10), 1576-1584.

Di Capua F., Pirozzi F., Lens P.N.L. and Esposito G. (2019). Electron donors for autotrophic denitrification. *Chemical Engineering Journal,* 362,

922-937.

DiLoreto Z.A., Weber P.A., Olds W., Pope J., Trumm D., Chaganti S.R. and Weisener C.G. (2016). Novel cost effective full scale mussel shell bioreactors for metal removal and acid neutralization. *Journal of Environmental Management,* 183, 601-612.

Ekama G.A. (2009). Using bioprocess stoichiometry to build a plant-wide mass balance based steady-state WWTP model. *Water Research,* 43(8), 2101-2120.

Ekama G.A., Wilsenach J.A. and Chen G.H. (2011). Saline sewage treatment and source separation of urine for more sustainable urban water management. *Water Science and Technology*, 64, 1307-1316.

European Environment Agency (EEA) (2006). The changing faces of Europe's coastal areas (No. 6). Office for Official Publications of the European Communities.

Fajardo C., Mora M., Fernández I., Mosquera-Corral A., Campos J.L. and Méndez R. (2014). Cross effect of temperature, pH and free ammonia on autotrophic denitrification process with sulphide as electron donor. *Chemosphere,* 97, 10-15.

Fedorovich V., Lens P. and Kalyuzhnyi S. (2003). Extension of Anaerobic Digestion Model No. 1 with processes of sulfate reduction. *Biotechnology and Applied Biochemistry,* 109(1-3), 33-45.

Fitz R.M. and Cypionka H. (1989). A study on electron transport-driven proton translocation in Desulfovibrio desulfuricans. *Archives of Microbiology,* 152(4), 369-376.

Foladori P., Andreottola G. and Ziglio G. (2010). Sludge Reduction Technologies in Wastewater Treatment Plants. IWA Publishing, London, UK.

Friedman L. and Rude M. (2012). Process for Producing Low Molecular Weight Hydrocarbons from Renewable Resources. WO2008113041.

Gramp J.P., Bigham J.M., Sasaki K. and Tuovinen O.H. (2007). Formation of Ni-and Zn-sulfides in cultures of sulfate-reducing bacteria. *Geomicrobiology Journal,* 24(7-8), 609-614.

Grant S.B., Saphores J.D., Feldman D.L., Hamilton A.J., Fletcher T.D., Cook P.L., Stewardson M., Sanders B., Levin L., Ambrose R., Deletic A., Brown R., Jiang S.C., Rosso D., Cooper W.J. and Marusic I. (2012). Taking the "waste" out of "wastewater" for human water security and ecosystem sustainability. *Science,* 2337(6095), 681-686.

Grau P., Vanrolleghem P.A. and Ayesa E. (2007). BSM2 Plant-wide Model construction and comparative analysis with other methodologies for integrated modelling. *Water Science and Technology,* 56(8), 57-65.

Gujer W. and Larsen T. A. (1995). The implementation of biokinetics and conservation principles in ASIM. *Water Science and Technology,* 31(2), 257-266.

Hai T., Lange D., Rabus R. and Steinbüchel A. (2004). Polyhydroxyalkanoate (PHA) accumulation in sulfate-reducing bacteria and identification of a class III PHA synthase (PhaEC) in Desulfococcus multivorans. *Applied Environmental and Microbiology,* 70(8), 4440-4448.

Hansen T.A. (1993). Carbon metabolism of sulfate-reducing bacteria. In The sulfate-reducing bacteria: contemporary perspectives. Springer, New York, USA. Pp. 21-40.

Hao T.W., Xiang P.Y., Mackey H.R., Chi K., Lu H., Chui H.K., Van Loosdrecht M.C.M. and Chen G.H. (2014). A review of biological sulfate conversions in wastewater treatment. *Water Research,* 65, 1-21.

Hao T. (2014). Granulation of sulfate reducing sludge (Doctoral dissertation)

Hedrich S. and Johnson D.B. (2014). Remediation and selective recovery of metals from acidic mine waters using novel modular bioreactors. *Environmental Science and Technology,* 48(20), 12206-12212.

Hell R., Dahl C., Knaff D.B. and Leustek T. (eds.) (2008). Sulfur Metabolism in Phototrophic Organisms, Springer Netherlands. ISBN 978-1-4020-6863-8.

Huisman J.L., Schouten G. and Schultz C. (2006). Biologically produced sulphide for purification of process streams, effluent treatment and recovery of metals in the metal and mining industry. *Hydrometallurgy,* 83(1-4), 106-113.

Jankowski G.J. and Zobell C.E. (1944). Hydrocarbon production by sulfate-reducing bacteria. *Journal of Bacteriology,* 47, 447.

Janssen A.J.H., Meijer S., Bontsema J. and Lettinga G. (1998). Application of the redox potential for controling a sulfide oxidizing bioreactor. *Biotechnology and Bioengineering,* 60(2), 147-155.

Jia Y, Xuan L, Weiming Y, Qing Z. and Lu H. (2017). Removal ibuprofen in sulfur autotrophic denitrification process. *Chinese Journal of Environmental Engineering,* 11(6), 3461-3467.

Jiang F., Zhang L., Peng G.L., Liang S.Y., Qian J., Wei L. and Chen G.H. (2013). A novel approach to realize SANI process in freshwater sewage treatment–Use of wet flue gas desulfurization waste streams as sulfur source. *Water Research,* 47(15), 5773-5782.

Khanongnuch R., Di Capua F., Lakaniemi A., Rene E.R. and Lens P.N.L. (2019). Long-term performance evaluation of an anoxic sulfur oxidizing moving bed biofilm reactor under nitrate limited conditions. *Water Research and Technology,* 5(6), 1072-1081.

Kim E.W. and Bae J.H. (2000). Alkalinity requirement

and the possibility of simultaneous heterotrophic denitrification during sulfur-utilizing autotrophic denitrification. *Water Science and Technology*, 42(3-4), 233-238.

Kleinjan W.E., De Keizer A. and Janssen A.J. (2012). Biologically produced sulfur. In Elemental sulfur and sulfur-rich compounds I. Springer, Berlin, Heidelberg, Germany.

Koschorreck M. (2008) Microbial sulphate reduction at a low pH. *FEMS Microbiology and Ecology*, 64(3), 329-342.

Kostrytsia A., Papirio S., Frunzo L., Mattei M.R., Porca E., Collins G., Lens P.N.L. and Esposito G. (2018a). Elemental sulfur-based autotrophic denitrification and denitritation: microbially catalyzed sulfur hydrolysis and nitrogen conversions. *Journal of Environmental Management*, 211, 313-322.

Kostrytsia A., Papirio S., Morrison L., Ijaz U.Z., Collins G., Lens P.N.L. and Esposito G. (2018b). Biokinetics of microbial consortia using biogenic sulfur as a novel electron donor for sustainable denitrification. *Bioresource Technology*, 270, 359-367.

Kruger J. and Revie R.W. (2011). Cost of metallic corrosion, pp15-20. Uhlig's corrosion handbook (3rd ed.). Wiley, Hoboken, NJ, USA.

Lau G.N., Sharma K.R., Chen G.H. and Van Loosdrecht M.C.M. (2006). Integration of sulfate reduction, autotrophic denitrification and nitrification to achieve low-cost excess sludge minimization for Hong Kong sewage. *Water Science and Technology*, 53, 227-235.

Leavitt W.D., Bradley A.S., Santos A.A., Pereira I.A. and Johnston D. T. (2015). Sulfur isotope effects of dissimilatory sulfite reductase. *Frontiers in Microbiology*, 6, 1392.

Lens P.N.L., Klijn R., Van Lier J.B. and Lettinga G. (2003). Effect of specific gas loading rate on thermophilic (55 °C) acidifying (pH 6) and sulfate reducing granular sludge reactors. *Water Research*, 37(5), 1033-1047.

Lens P.N.L., Visser A., Janssen A.J.H., Pol L.H. and Lettinga G. (1998). Biotechnological treatment of sulfate-rich wastewaters. *Critical Reviews in Environmental Science and Technology*, 28(1), 41-88.

Leung R.W.K., Li D.C.H., Yu W.K., Chui H.K, Lee T.O., Van Loosdrecht M.C.M. and Chen G.H. (2012). Integration of seawater and grey water reuse to maximize alternative water resource for coastal areas: the case of the Hong Kong International Airport. *Water Science and Technology*, 65, 410-417.

Liamleam W. and Annachhatre A.P. (2007). Electron donors for biological sulfate reduction. *Biotechnology Advances*, 25(5), 452-463.

Lillig C.H., Schiffmann S., Berndt C., Berken A., Tischka R. and Schwenn J.D. (2001). Molecular and catalytic properties of Arabidopsis thaliana adenylyl sulfate (APS)-kinase. *Archives of Biochemistry and Biophysics*, 392(2), 303-310.

Lin S., Mackey H.R., Hao T., Guo G., Van Loosdrecht M.C.M. and Chen G.H. (2018). Biological sulfur oxidation in wastewater treatment: A review of emerging opportunities. *Water Research*, 143, 399-415.

Liu X.M., Dai J, Wu D., Jiang F., Chen G.H., Chui H.K. and Van Loosdrecht M.C.M. (2016). Sustainable Application of a Novel Water Cycle Using Seawater for Toilet Flushing. *Engineering*, 2, 460-469.

Lu H., Ekama G.A., Wu D., Feng J., Van Loosdrecht M.C.M. and Chen G.H. (2012a). SANI® process realizes sustainable saline sewage treatment: steady state model-based evaluation of the pilot-scale trial of the process. *Water Research*, 46, 475–490.

Lu H., Wang J., Li S., Chen G.H., Van Loosdrecht M.C.M. and Ekama G.A. (2009). Steady-state model-based evaluation of sulfate reduction, autotrophic denitrification and nitrification integrated (SANI) process. *Water Research*, 43, 3613–3621.

Lu H., Wu D., Jiang F., Ekama G.A., Van Loosdrecht M.C.M. and Chen G.H. (2012b). The demonstration of a novel sulfur cycle-based wastewater treatment process: sulfate reduction, autotrophic denitrification and nitrification integrated (SANI®) biological nitrogen removal process. *Biotechnology and Bioengineering*, 109, 2778–2789.

Lu H., Wu D., Tang D.T.W., Chen G.H., Van Loosdrecht M.C.M. and Ekama G.A. (2011). Pilot scale evaluation of SANI® process for sludge minimization and greenhouse gas reduction in saline sewage treatment. *Water Science and Technology*, 63, 2149–2154.

Lu H., Huang H., Yang W., Mackey H.R., Khanal S.K., Wu D. and Chen G.H. (2018). Elucidating the stimulatory and inhibitory effects of dissolved sulfide on sulfur-oxidizing bacteria (SOB) driven autotrophic denitrification. *Water Research*, 133, 165-172.

Mackenzie F.T. (2005). Sediments, Diagenesis, and Sedimentary Rocks: Treatise on Geochemistry (7th ed.), Elsevier Science, Amsterdam, the Netherlands.

Martins M. and Pereira I.A. (2013). Sulfate-reducing bacteria as new microorganisms for biological hydrogen production. *International Journal of Hydrogen Energy*, 38(28), 12294-12301.

Martins M., Mourato C. and Pereira I.A. (2015).

Desulfovibrio vulgaris growth coupled to formate-driven H_2 production. *Environmental Science and Technology,* 49(24), 14655-14662.

Martins M., Mourato C., Morais-Silva F.O., Rodrigues-Pousada C., Voordouw G., Wall J.D. and Pereira I.A. (2016). Electron transfer pathways of formate-driven H_2 production in Desulfovibrio. *Applied Microbiology and Biotechnology,* 100(18), 8135-8146.

McCarty P.L. (1975). Stoichiometry of biological reactions. *Prog. Wat. Tech.,* 7(1), 157-172.

Mokone T.P., Van Hille R.P. and Lewis A.E. (2010). Effect of solution chemistry on particle characteristics during metal sulfide precipitation. *Journal of Colloid and Interface Science,* 351(1), 10-18.

Mora M., Guisasola A., Gamisans X. and Gabriel D. (2014). Examining thiosulfate-driven autotrophic denitrification through respirometry. *Chemosphere,* 113, 1-8.

Moraes B.S., Souza T.S.O. and Foresti E. (2012). Effect of sulfide concentration on autotrophic denitrification from nitrate and nitrite in vertical fixed-bed reactors. *Process Biochemistry,* 47(9), 1395-1401.

Morais-Silva F.O., Santos C.I., Rodrigues R., Pereira I.A. and Rodrigues-Pousada C. (2013). Roles of HynAB and Ech, the only two hydrogenases found in the model sulfate reducer Desulfovibrio gigas. *Journal of Bacteriology,* 195(20), 4753-4760.

Moura I., Moura J.J., Pauleta S.R. and Maia L.B. (2016). Metalloenzymes in denitrification: Applications and environmental impacts. *Royal Society of Chemistry,* ISBN 978-1-78262-334-2.

Muyzer G. and Stams A. J. (2008). The ecology and biotechnology of sulphate-reducing bacteria. *Nature Reviews Microbiology,* 6(6), 441-454.

Muyzer G., Kuenen J.G. and Robertson L.A. (2013). Colorless sulfur bacteria. The Prokaryotes. Springer, 555-588. ISBN 978-3-642-30140-7.

Ødegaard H. (2006). Innovations in wastewater treatment: the moving bed biofilm process. *Water Science and Technology*, 53, 17-33.

Oh S.E., Bum M.S., Yoo Y.B., Zubair and Kim I.S. (2003). Nitrate removal by simultaneous sulfur utilizing autotrophic and heterotrophic denitrification under different organics and alkalinity conditions: Batch experiments. *Water Science and Technology,* 47(1), 237-244.

Oh S.E., Yoo Y.B., Young J.C. and Kim I.S. (2001). Effect of organics on sulfur-utilizing autotrophic denitrification under mixotrophic conditions. *Journal of Biotechnology,* 92(1), 1-8.

Oremland R.S. and Polcin S. (1982). Methanogenesis and sulfate reduction: competitive and noncompetitive substrates in estuarine sediments. *Applied Environment and Microbiology*, 44(6), 1270-1276.

Pan Y., Ni B.J., Bond P.L., Ye L. and Yuan Z. (2013). Electron competition among nitrogen oxides reduction during methanol-utilizing denitrification in wastewater treatment. *Water Research,* 47(10), 3273-3281.

Pearce G.K. (2007). UF/MF pre-treatment to RO in seawater and water reuse applications: a comparison of energy costs. *Desalination,* 222(1-3), 66-73.

Peck H.D. and LeGall J. (1982). Biochemistry of dissimilatory sulphate reduction. *Philosophical Transactions of the Royal Society B,* 298(1093), 443-466.

Perry R.H. (1984). Physical and chemical data. Perry's chemical engineers' handbook, 7-374.

Pikaar I., Sharma K.R., Hu S., Gernjak W., Keller J. and Yuan Z. (2014). Reducing sewer corrosion through integrated urban water management. *Science,* 345(6198), 812-814.

Poinapen J. and Ekama G.A. (2010a). Biological sulphate reduction with primary sewage in an upflow anaerobic sludge bed reactor - Part 5: steady state model. *Water SA,* 36(3), 193-202.

Poinapen J. and Ekama G.A. (2010b). Biological sulphate reduction with primary sewage in an upflow anaerobic sludge bed reactor - Part 6: development of a kinetic model for BSR. *Water SA,* 36 (3), 203-213.

Pol L.W.H., Lens P.N., Stams A.J. and Lettinga G. (1998). Anaerobic treatment of sulphate-rich wastewaters. *Biodegradation,* 9(3-4), 213-224.

Qian J., Lu H., Jiang F., Ekama G. A. and Chen G. H. (2015). Beneficial co-treatment of simple wet flue gas desulphurization wastes with freshwater sewage through development of mixed denitrification–SANI process. *Chemical Engineering Journal,* 262, 109-118.

Rabus R., Hansen T.A. and Widdel F. (2013). Dissimilatory Sulfate- and Sulfur-Reducing Prokaryotes. The Prokaryotes. 310-379.

Rabus R., Venceslau S.S., Woehlbrand L., Voordouw G., Wall J.D. and Pereira I.A. (2015). A post-genomic view of the ecophysiology, catabolism and biotechnological relevance of sulphate-reducing prokaryotes. In Advances in microbial physiology. *Academic Press,* 66, 55-321.

Roest K., Heilig H. G., Smidt H., De Vos W.M., Stams A.J. and Akkermans A.D. (2005). Community analysis of a full-scale anaerobic bioreactor treating paper mill wastewater. *Systematic and Applied*

Microbiology, 28(2), 175-185.

Roland-Holst D., Triolo R., Heft-Neal S. and Bayrami B. (2013). Bioplastics in California: Economic Assessment of Market Conditions for PHA/PHB Bioplastics Produced from Waste Methane. University of California, Berkeley. 1-76.

Saby S., Djafer M. and Chen G.H. (2003). Effect of low ORP in anoxic sludge zone on excess sludge production in oxic-settling-anoxic activated sludge process. *Water Research,* 37, 11-20.

Salehizadeh H. and Van Loosdrecht M.C.M. (2004). Production of polyhydroxyalkanoates by mixed culture: recent trends and biotechnological importance. *Biotechnology Advances,* 22(3), 261-279.

Santos A.A., Venceslau S.S., Grein F., Leavitt W.D., Dahl C., Johnston D.T. and Pereira I.A. (2015). A protein trisulfide couples dissimilatory sulfate reduction to energy conservation. *Science,* 350(6267), 1541-1545.

Sarti A., Silva A.J., Zaiat M. and Foresti E. (2011). Full-scale anaerobic sequencing batch biofilm reactor for sulfate-rich wastewater treatment. *Desalination Water Treatment,* 25(1-3), 13-19.

Shao M.F., Zhang T. and Fang H.H.P. (2010). Sulfur-driven autotrophic denitrification: diversity, biochemistry and engineering applications. *Applied Microbiology and Biotechnology,* 88, 1027-1042.

Sierra-Alvarez R., Beristain-Cardoso R., Salazar M., Gómez J., Razo-Flores E. and Field J.A. (2007). Chemolithotrophic denitrification with elemental sulfur for groundwater treatment. *Water Research,* 41(6), 1253-1262.

Sötemann S.W., Ristow N.E., Wentzel M.C. and Ekama G.A. (2005). A steady state model for anaerobic digestion of sewage sludges. *Water SA,* 31(4), 511-527.

Sun Y., Chen Z., Wu G., Wu Q., Zhang F., Niu Z. and Hu H.Y. (2016). Characteristics of water quality of municipal wastewater treatment plants in China: implications for resources utilization and management. *Journal of Cleaner Production,* 131, 1-9.

Tang S.L., Yue D.P.T. and Ku D.C. (2007). Engineering and Costs of Dual Water Supply Systems, International Water Association. IWA Publishing, London, UK.

Tatsumi H., Takagi K., Fujita M., Kano K. and Ikeda T. (1999). Electrochemical study of reversible hydrogenase reaction of Desulfovibrio vulgaris cells with methyl viologen as an electron carrier. *Analytical Chemistry,* 71(9), 1753-1759.

Tchobanoglous G., Burton F.L. and Stensel H.D. (2003). Wastewater Engineering: Treatment and Reuse. Metcalf & Eddy, McGraw-Hill Education.

Thauer R.K., Jungermann K. and Decker K. (1977). Energy conservation in chemotrophic anaerobic bacteria. *Bacteriological Reviews,* 41(1), 100.

Tichý R., Janssen A., Grotenhuis J.T.C., Lettinga G. and Rulkens W. H. (1994). Possibilities for using biologically-produced sulphur for cultivation of Thiobacilli with respect to bioleaching processes. *Bioresource Technology,* 48(3), 221-227.

Tsuji K. and Yagi T. (1980). Significance of hydrogen burst from growing cultures of Desulfovibrio vulgaris, Miyazaki and the role of hydrogenase and cytochrome c3 in energy production system. *Archive of Microbiology,* 125(1-2), 35-42.

UN (2018). Sustainable Development Goal 6: Synthesis Report on Water and Sanitation. The United Nations, New York, USA.

USEPA (1999). Biosolids generation, use and disposal in the United States, EPA530-R-99-009.

USEPA (2012). Guidelines for Water Reuse. EPA/600/R-12/618. Published by the U.S. Environmental Protection Agency, Washington DC, USA.

Van den Bosch P.L. (2008). Biological sulfide oxidation by natron-alkaliphilic bacteria: application in gas desulfurization (Doctoral dissertation)

Van Houten B.H., Van Doesburg W., Dijkman H., Copini C., Smidt H. and Stams A.J. (2009). Long-term performance and microbial community analysis of a full-scale synthesis gas fed reactor treating sulfate-and zinc-rich wastewater. *Applied Microbiology and Biotechnology,* 84, 555-563.

Van Loosdrecht M.C.M. and Brdjanovic D. (2014). Anticipating the next century of wastewater treatment. *Science,* 344(6191), 1452-1453.

Van Loosdrecht M.C.M., Brdjanovic D., Chui S. and Chen G.H. (2012). A source for toilet flushing and for cooling, sewage treatment benefits, and phosphorus recovery: direct use of seawater in an age of rapid urbanisation. *Water,* 17-19.

Van Spanning R.J.M., Richardson D.J. and Ferguson S.J. (2007). Introduction to the Biochemistry and Molecular Biology of Denitrification, Biology of the Nitrogen Cycle. *Elsevier BV,* 3-20.

Villano M., De Bonis L., Rossetti S., Aulenta F. and Majone M. (2011). Bioelectrochemical hydrogen production with hydrogenophilic dechlorinating bacteria as electrocatalytic agents. *Bioresource Technology,* 102(3), 3193-3199.

Wang J., Lu H., Chen G.H., Lau G.N., Tsang W.L. and Van Loosdrecht M.C.M. (2009). A novel sulfate reduction, autotrophic denitrification, nitrification integrated (SANI) process for saline wastewater treatment. *Water Research,* 43, 2363–2372.

Wang J., Shi M., Lu H., Wu D., Shao M.F., Zhang T., Ekama G.A., Van Loosdrecht M.C.M. and Chen G.H. (2011). Microbial community of sulfate-reducing up-flow sludge bed in the SANI® process for saline sewage treatment. *Applied Microbiology and Biotechnology,* 90, 2015-2025.

World Bank (2004). Seawater and Brackish Water Desalination in the Middle East, North Africa and Central Asia: A Review of Key Issues and Experience in Six Countries.

WSD (2019a). Seawater for Flushing. Retrieved from: https://www.wsd.gov.hk/en/core-businesses/water-resources/seawater-for-flushing/index.html.

WSD (2019b). Guide to Application for Water Supply. Retrieved from: https://www.wsd.gov.hk/filemanager/en/content_1805/guide-to-application-for-water-supply-e.pdf.

Wu D., Ekama G.A., Chui H.K., Wang B., Cui Y.X., Hao T.W., Van Loosdrecht M.C.M. and Chen G.H. (2016). Large-scale demonstration of the sulfate reduction autotrophic denitrification nitrification integrated (SANI®) process in saline sewage treatment. *Water Research,* 100, 496-507.

Xu J., Murayama M., Roco C.M., Veeramani H., Michel F.M., Rimstidt J.D. and Hochella Jr. M.F. (2016). Highly-defective nanocrystals of ZnS formed via dissimilatory bacterial sulfate reduction: A comparative study with their abiogenic analogues. *Geochimica et Cosmochimica Acta*, 180, 1-14.

Xue W.Q., Zeng Q., Lin S., Zan F.X., Hao T.W., Lin Y.M., Van Loosdrecht M.C.M. and Chen G.H. (2019). Recovery of high-value and scarce resources from biological wastewater treatment: sulphated polysaccharides. *Water Research,* 163, 114889.

Yang W., Lu H., Khanal S.K., Zhao Q., Meng L. and Chen G.H. (2016a). Granulation of sulfur-oxidizing bacteria for autotrophic denitrification. *Water Research,* 104, 507-519.

Yang W., Zhao Q., Lu H., Ding Z., Meng L. and Chen G.H. (2016b). Sulfide-driven autotrophic denitrification significantly reduces N_2O emissions. *Water Research,* 90, 176-184.

Yu H., Chen C., Ma J., Liu W., Zhou J., Lee D.J., Ren N. and Wang A. (2014). GeoChip-based analysis of the microbial community functional structures in simultaneous desulfurization and denitrification process. *Journal of Environmental Science (China),* 26(7), 1375-1382.

Zeng Q., Hao T.W., Mackey H.R., Van Loosdrecht M.C.M. and Chen G.H. (2019). Recent advances in dissimilatory sulfate reduction: From metabolic study to application. *Water Research,* 150, 162-181.

Zhang L., Zhang C., Hu C., Liu H. and Qu J. (2015). Denitrification of groundwater using a sulfur-oxidizing autotrophic denitrifying anaerobic fluidized-bed MBR: performance and bacterial community structure. *Applied Microbiology and Biotechnology,* 99(6), 2815-2827.

Zhang T.C. and Lampe D.G. (1999). Sulfur: limestone autotrophic denitrification processes for treatment of nitrate-contaminated water: batch experiments. *Water Research,* 33(3), 599-608.

Zhang T.C. and Shan J. (1999). In situ septic tank effluent denitrification using a sulfur-limestone process. *Water Environment Research,* 71(7), 1283-1291.

NOMENCLATURE

Symbol	Description	Unit
a	Recycle ratio	-
a_F	Specific surface area	m^2/m^3
$A_{F,NH4}$	Necessary surface area for the nitrifying/aerobic reactor	m^2/m^3
$A_{F,NOx}$	Necessary surface area for SdAD	m^2/m^3
$B_{A,NH4}$	Surface specific conversion rate of ammonia	gNH_4^+-$N/m^2/d$
$B_{A,NOx}$	Surface specific conversion rate of nitrate/nitrite	gNO_x^--$N/m^2/d$
b_{ADO}	Endogenous respiration rate of autotrophic denitrifying microorganisms	1/d
b_{ANO}	Endogenous respiration rate of autotrophic nitrifying microorganisms	1/d

b_{SOB}	Endogenous respiration rate of SOB biomass	1/d
b_{SRB}	Endogenous respiration rate of SRB biomass	1/d
CH_3COO^-	Acetate	-
CO	Carbon monoxide	-
COD_c	Carbonaceous COD	mg COD/l
COD_i	Influent COD concentration	mgCOD/l
E	Sludge COD produced/COD utilized	-
E	Mass of COD produced (as active SRB biomass and endogenous sludge) per day as a fraction of the mass of biodegradable organics that are utilized per day in the SRUSB reactor at a steady state	-
E'	Mass of COD produced (as the autotrophic denitrification biomass and endogenous sludge) per day as a fraction of the flux of nitrate reduced per day in the anoxic bioreactor at a steady state (the net yield of the autotrophic denitrification biomass and endogenous residue)	-
E''	Mass of COD produced (as the autotrophic nitrification biomass and endogenous sludge) per day as a fraction of the flux of ammonia oxidized per day in the aerobic zone at a steady state	-
E'''	Mass of COD produced (as the aerobic sulphide-oxidizing biomass and endogenous sludge) per day as a fraction of the flux of ammonia oxidized per day in the aerobic zone at a steady state	-
f	Fraction of $H_2PO_4^-$ in the orthophosphate	-
f_{av}	Active biomass fraction	-
$f_{av,SRB}$	Active fraction of SRB	-
f_c	Mass ratio of carbon	-
$F_{CODb,i}$	Mass of biodegradable COD treated	-
F_{CODi}	Mass of COD treated	-
f_{cv}	Ratio of COD-to-VSS or COD-to-mass for soluble organics	-
f_{ED}	Endogenous residue fraction	-
f_h	Mass ratio of hydrogen	-
f_n	Mass ratio of nitrogen	-
f_o	Mass ratio of oxygen	-
f_{OHO}	Endogenous residual fraction of ordinary heterotrophic organisms	-
FO_{NIT}	Oxygen flux for nitrification	-
f_p	Mass ratio of phosphorus	-
FSO_4	Demand for sulphate	-
f_{SRB}	Endogenous residual fraction of SRB	-
$f_{SU,CODi}$	Non-biodegradable soluble COD/influent total COD	-
f_{VT}	VSS/TSS ratio of activated sludge	-
$FX_{FSS,i}$	Mass of inorganic suspended solids	-
$FX_{Uv,i}$	Mass of unbiodegradable particulate organics	-
$f_{XU,CODi}$	Non-biodegradable particulate COD/influent total COD	-
H_2	Hydrogen	-
H_2S	Hydrogen sulphide	-
HRT	Hydraulic retention time	h
K_s	Half saturation constant	-

K_{S^0}	Half saturation constant of elemental sulphur	mgCOD/l
$K_{S_2O_3}$	Half saturation constant of thiosulphate	mgCOD/l
K_{SOB,NO_3}	Half saturation constant of nitrate for SOB	mgN/l
$K_{SRB,max}$	Maximum specific hydrolysis rate constant in Monod kinetics	mgCOD/mgCOD/d
K_{TDSd}	Half saturation constant of sulphide	mgCOD/l
Mw	Molecular weight	-
MX_{FSS}	Mass of influent particulate inorganic matter in the bioreactor	-
MX_{TSS}	The mass of TSS sludge in the system	kgTSS
MX_{VSS}	Mass of volatile suspended solids in the bioreactor	-
N_2	Dinitrogen gas	-
N_2O	Nitrous oxide	-
NH_4^+	Ammonia	-
NIT_c	Nitrification capacity	-
NO	Nitric oxide	-
NO_2^-	Nitrite	-
NO_3^-	Nitrate	-
NO_x^-	Nitrogen oxide	-
Q_i	Influent flow rate	l/d or m^3/d
R	Recirculation ratio	-
r_{hyd}	Volumetric hydrolysis rate	mgCOD/l/d
S^0_{bio}	Biologically-produced sulphur	-
S^0_{chem}	Chemical elemental sulphur	-
$S_2O_3^{2-}$	Thiosulphate	-
$S_4O_6^{2-}$	Tetrathionate	-
$S_{NHx,e}$	Effluent NH_4^+ concentration	-
$S_{NHx,i}$	Influent NH_4^+ concentration	-
S_{NO_3}	Nitrate concentration	-
$S_{NO_3,e}$	Effluent nitrate concentration	-
SO_3^{2-}	Sulphite	-
SO_4^{2-}	Sulphate	-
SRT	Sludge age of the SRUSB reactor	D
SRT_{AX}	Sludge age of the anoxic MBBR	D
SRT_{OX}	Sludge age of the aerobic zone	D
S_{S^0}	Elemental sulphur COD	mgCOD/l
$S_{S_2O_3}$	Thiosulphate COD	mgCOD/l
S_{SO4}	Sulphate concentration in SRUSB	$mgSO_4^{2-}$-S/l
S_{TDSd}	Hydrogen sulphide COD	mgCOD/l
$S_{TDSd,ane}$	Total dissolved sulphide COD in the effluent of the anoxic zone	mgCOD/l
SVI	Sludge volume index	ml/g
V_{AX}	Volume of anoxic bioreactor	m^3
V_{OX}	Volume of aerobic bioreactor	m^3

V_R	Volume of bioreactor	m^3
V_{SRUSB}	Volume of sulphate-reducing upflow sludge bed reactor	m^3
X_{COD}	Particulate COD	mgCOD/l
X_{FSS}	Inorganic suspended solid concentration	mgISS/l
X_S	Biodegradable particulate COD	mgCOD/l
X_{SOB}	Biomass COD in anoxic reactor	mg COD/l
X_{SRB}	Biomass COD in SRUSB	mg COD/l
Y_{ADO}	Yield coefficient of the autotrophic denitrifying microorganisms	mgCOD biomass/mgNO_3^--N reduced
Y_{ANO}	Yield coefficient of the autotrophic nitrification biomass	$mgCOD_{biomass}$/mgNH_4^+-N
Y^i_B	Electrons available for redox reaction per mol autotrophic denitrifying biomass	-
Y_{SOB}	Yield coefficient of SOB biomass	mgCOD/mgNO_3^--N
$Y_{SOB'}$	Yield coefficient of the autotrophic denitrification biomass	$mgCOD_{biomass}/mgCOD_{degraded}$
Y_{SRB}	Yield coefficient of SRB biomass	mgCOD/mgCOD
Y_{SRBv}	Yield coefficient of SRB biomass	mgVSS/mgCOD
γ_B	Electron-donating capacity of SRB biomass	e^- eq/mol
γ'_B	Electron-donating capacity of SOB biomass	e^- eq/mol
γ''_B	Electron-donating capacity of autotrophic nitrifier biomass	e^- eq/mol
γ'''_B	Electron-donating capacity of aerobic SOB biomass	e^- eq/mol
γ_S	Electron-donating capacity of biodegradable organics	e^- eq/mol
μ_{ANO}	Specific growth rate	-
$\mu_{hyd,max}$	Maximum specific growth rate of anaerobic hydrolysis	1/d
$\mu_{SOB,max}$	Maximum specific growth rate of SOB	1/d

Abbreviation	Description
AD	Autotrophic denitrification
ADM	Anaerobic digestion model
ADP	Adenosine diphosphate
AMD	Acid mine drainage
AMP	Adenosine monophosphate
AnFB	Anaerobic fluidized-bed
AnFB-MBR	Anaerobic fluidized-bed membrane bioreactor
ANO	Autotrophic nitrifier organism
APS	Adenosine phosphosulphate
APSR	Adenosine-5′-phosphosulphate reductase
ATP	Adenosine triphosphate
ATPS	Adenosine triphosphatesulphurylase
BNR	Biological nitrogen removal
BOD	Biological oxygen demand
BOD_5	5-day biological oxygen demand

BPO	Biodegradable particulate organics
BSO	Biodegradable soluble organics
CdS	Cadmium sulphide
CEPT	Chemically-enhanced primary treatment
CO_2	Carbon dioxide
COD	Chemical oxygen demand
CuS	Copper sulphide
DMSO	Dimethylsulfoxide
DO	Dissolved oxygen
D_p	Denitrification potential
DSD	Drainage Services Department
Dsr	Dissimilatory sulphite reductase
EQ	Equalization tank
FCSD	Flavocytochrome c sulphide dehydrogenase
FeS	Ferroussulphide
FGD	Flue-gas desulphurization
FMFAR	Fixed-media-filled anoxic reactor
FMS	Fine-mesh sieve
FSA	Free and saline ammonia
GSAD	Granular-sludge autotrophic denitrification
H_2	Hydrogen
Hase(s)	Hydrogenase(s)
HKIA	Hong Kong International Airport
HRT	Hydraulic retention time
IFAS	Integrated fixed-film activated sludge
ISS	Inorganic suspended solids
MBBR	Moving-bed biological reactor
MBR	Membrane bioreactor
MLSS	Mixed liquor suspended solids
MLVSS	Mixed liquor volatile suspended solids
N	Nitrogen
NADPH	Nicotinamide adenine dinucleotide phosphate hydrogen
Nar	Nitrate reductase
Ni_2S_3	Nickel sulphide
Nir	Nitrite reductase
Nor	Nitric oxide reductase
Nos	Nitrous oxide reductase
NSON	Non-biodegradable soluble organic nitrogen
O_2	Oxygen
OP	Orthophosphate
ORP	Oxidation-reduction potential
PAPS	Phosphoadenylyl-sulphate
PBR	Packed-bed reactor

PbS	Lead sulphide
PH	Polythionate hydrolase
PHAs	Polyhydroxyalkanoates
RO	Reverse osmosis
SANI®	Sulfate reduction, autotrophic denitrification and nitrification integrated
Sb_2S_3	Antimony trisulphide
SBR	Sequencing batch reactor
SCS	Sulphur carrier enzymic system
SD	Sedimentation tank
SdAD	Sulphur-driven autotrophic denitrification
SO	Sulphite oxidase
SOB	Sulphur-oxidizing bacteria
Sox	Sulphur-oxidizing enzyme complex
SQR	Sulphide: quinoneoxidoreductase
SRB	Sulphate-reducing bacteria
SRT	Sludge retention time
SRUSB	Sulphate-reducing upflow sludge bed
STWTW	Sha Tin Wastewater Treatment Works
SVI	Sludge volume index
SWTF	Seawater toilet flushing
TCA	Tricarboxylic acid cycle
TCSPS	Tung Chung Sewage Pump Station
TDSd	Total dissolved sulphide
TKN	Total Kjeldahl nitrogen
TN	Total nitrogen
TOC	Total organic carbon
TP	Total phosphorus
TQO	Thiosulphate: quinone oxidoreductase
TSS	Total suspended solids
TWS	Triple Water System
UASB	Up-flow anaerobic sludge bed
UFCR	Up-flow column reactor
UPO	Unbiodegradable particulate organics
UQ	Ubiquinone
USEPA	United States Environmental Protection Agency
USO	Unbiodegradable soluble organics
VFA	Volatile fatty acid
VSS	Volatile suspended solids
WSD	Water Supply Department
WWTP	Wastewater treatment plant
ZnS	Zinc sulphide

doi: 10.2166/9781789060362_0381

8

Wastewater disinfection

Ernest R. Blatchley III

8.1 BACKGROUND

Disinfection of wastewater effluents is applied for purposes of reducing the concentrations of viable or infective microbial pathogens in treated municipal wastewater prior to its release into the environment and more generally for reducing the risks of human exposure to wastewater-associated microbial pathogens. The extent of disinfection required will be influenced by the intended uses of the treated water, as well as the intended uses of the waterway into which the effluent is discharged. Examples of uses span a broad range, from no disinfection requirement at all, to discharge to a waterway that may involve no immediate direct or indirect human contact, then discharge to a waterway that supports recreational water use, to irrigation of edible or non-edible crops, to downstream production of potable water, and ultimately to direct potabilization.

The performance of disinfection systems will be influenced by influent water quality, as well as the characteristics of upstream water treatment processes. Disinfection processes are generally applied at or near the end of a wastewater treatment process train, so as to apply disinfectants to water that contains low concentrations of dissolved and suspended materials that could 'compete' for active disinfection agents.

The inclusion of wastewater disinfection is not universal among developed countries. Even in the United States, where wastewater disinfection is perhaps most frequently applied, there are some facilities that are not required to disinfect, and many others that disinfect seasonally. For example, in temperate parts of the United States, it is common for discharge permit limitations to require effluent disinfection only in 'warm weather' months where human contact with water in receiving streams becomes more likely than in 'cold weather' months.

An interesting case study can be found with the water reclamation facilities that are operated by the Metropolitan Water Reclamation District of Greater Chicago (MWRDGC). In the 1970s, the MWRDGC conducted a landmark study that demonstrated that microbial quality in a receiving stream was essentially identical at locations at least 10 river miles downstream of the point of effluent discharge when chlorine-based disinfection of the MWRDGC effluents was applied and when it was not applied; however, chemical and ecological quality of the receiving stream were actually superior when chlorination was not used. Based on the results of this study, the MWRDGC successfully petitioned the Illinois EPA to allow operation of its facilities without disinfection. Their facilities operated for many years in this manner. More recently, pressure from recreational water users in the Chicago area has resulted in inclusion of disinfection at four of the seven water reclamation facilities currently operated by the MWRDGC.

8.2 INDICATOR ORGANISM CONCEPT

Microbial communities within municipal wastewater effluents tend to be highly complex mixtures of microorganisms from several taxonomic orders. Common microbial pathogens found in effluent samples include a wide range of bacteria (vegetative cells and spores), viruses, and protozoan parasites (including resting stages, such as cysts and oocysts). Because of this, it is impractical to develop a complete inventory of the microbial population, or even of microbial pathogens. Instead, microbial quality is often characterized based on measurements of the concentrations of viable microbes from one or perhaps a few microbial families.

The concept of an indicator organism is applied for this purpose in wastewater samples. Ideally, a microbial indicator will provide a measure of the risk of exposure to microbial pathogens. Desirable characteristics of indicator organisms include: ubiquity in undisinfected effluent samples; they should be non-pathogenic to humans and should be simple, inexpensive, and rapid to quantify using culture-based methods; their presence (or absence) should correlate closely with the presence (or absence) of microbial pathogens; and they should be at least as resistant to disinfectants as microbial pathogens that are likely to be present in an effluent sample. No single microorganism satisfies all of these characteristics, but some microbial groups have many of these characteristics.

Among the most commonly-used indicator organisms used for quantification of effluent microbial quality are coliform bacteria, often measured as *E. coli* or as faecal coliforms. For effluents that are discharged into marine waters, *Enterococcus* is often used as an indicator. Effluent discharge standards for these indicator groups are generally developed to yield microbial water quality that is consistent with the intended uses of the receiving water.

In recent years, recognition of the importance of viruses in disease transmission has motivated examination of the use of viruses or viral surrogates as indicators of microbial quality (EPA, 2015). Common viral indicators that have been proposed for these purposes include F-specific and somatic coliphage, which are viruses that parasitize (coliform) bacterial cells. A number of bacteriophage have been identified that have structural and physiological characteristics that mimic important human viruses. However, since these phage only have the ability to parasitize specific bacterial cells, they are non-pathogenic to humans. Analytical methods are available to quantify the concentrations of ambient, infective phage in water samples; some of these methods are rapid, easy, and inexpensive to apply (EPA, 2015; McMinn *et al.*, 2017).

8.3 DISINFECTION WITH HALOGENS (CHLORINE)

Halogens have chemical characteristics that allow them to function well as disinfectants. Specifically, the elemental forms of halogens have outer electron shells that are one electron short of a completed octet. As such, halogens tend to display chemical behaviour that is similar, in that all of them tend to accept electrons effectively (i.e. they tend to function as effective oxidants). Halogens also are effective substituting agents.

Although halogens tend to display similar chemical behaviour, they do not all function equivalently as disinfectants. Among the halogens, chlorine is by far the most commonly applied, largely because of economics, ease of application, and of roughly a century of accumulated experience working with chlorine as a disinfectant. Because of this, the focus of the discussion on halogens here will be on chlorine. Readers should understand that the behaviour of other halogens that are used as disinfectants (especially bromine and iodine) tends to be similar, though not identical to that of chlorine.

8.3.1 Physical chemistry of chlorine

The term 'free chlorine' refers to forms of oxidized (+1-valent or zero-valent) chlorine that exist in solution; these forms of chlorine tend to have a high affinity for electrons (i.e. they are effective oxidizing agents). Several forms of free chlorine exist and well-defined equilibria govern their distribution. Equations 8.1b-8.4b summarize these equilibria and their respective equilibrium constants, all reported at 25°C. Values of equilibrium constants were taken from Odeh *et al.* (2004) and Deborde and Von Gunten (2008).

$$Cl_2 + H_2O \rightleftharpoons HOCl + H^+ + Cl^- \tag{8.1a}$$

$$K_h = \frac{[HOCl][H^+][Cl^-]}{[Cl_2]} = 1.04 \cdot 10^{-3} M^2 \tag{8.1b}$$

$$HOCl \rightleftharpoons H^+ + OCl^- \tag{8.2a}$$

$$K_a = \frac{[H^+][OCl^-]}{[HOCl]} = 3.39 \cdot 10^{-8} M \tag{8.2b}$$

$$HOCl + HOCl \rightleftharpoons Cl_2O + H_2O \tag{8.3a}$$

$$K_{Cl_2O} = \frac{[Cl_2O]}{[HOCl]^2} = 8.7 \cdot 10^{-3} M^{-1} \tag{8.3b}$$

$$Cl_2 + Cl^- \rightleftharpoons Cl_3^- \tag{8.4a}$$

$$K_{Cl_3^-} = \frac{[Cl_3^-]}{[Cl_2][Cl^-]} = 0.18 M^{-1} \tag{8.4b}$$

These equilibrium expressions indicate that free chlorine speciation will be influenced by fundamental solution chemistry, including pH and chloride ion concentration. Figure 8.1 illustrates the speciation of free chlorine as a function of pH in a sample of water that contains 5.0 mg/l ($7.0 \cdot 10^{-5}$ M) of free chlorine and chloride Cl^- at a concentration of 10^{-4} M (3.6 mg/l), which is representative of tap water or a common municipal wastewater effluent.

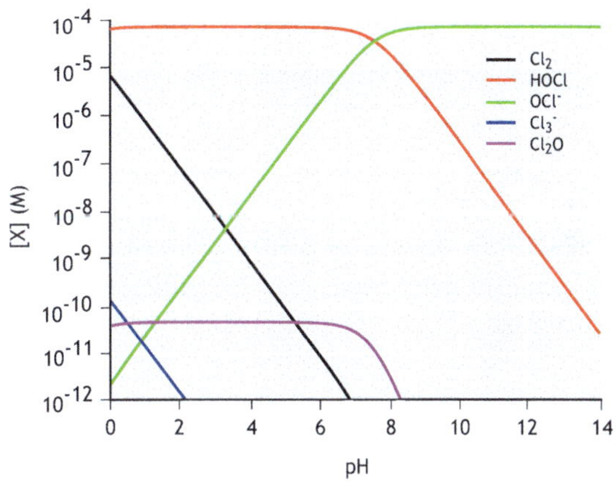

Figure 8.1 Distribution diagram for free chlorine compounds in a sample that contains $7.0 \cdot 10^{-5}$ M free chlorine and chloride ion at a concentration of 10^{-4} M at a temperature of 25 °C. These conditions are representative of free chlorine addition to tap water or a common wastewater effluent.

For the situation illustrated in Figure 8.1, free chlorine is dominated by HOCl and OCl^-, with the distribution between these two forms being governed by pH (relative to $pK_a \approx 7.47$). At near-neutral pH, the molar concentrations of HOCl and OCl^- are at least 5-6 orders of magnitude higher than those of the other forms of free chlorine. For purposes of evaluating the responses of microorganisms to chlorine exposure in these systems, it is likely that HOCl and OCl^- will dominate. In that regard, it is relevant to recognize that HOCl is generally considered to be a much more effective disinfectant than OCl^-. The relatively low concentrations of the other forms of free chlorine will dictate that they are likely to contribute negligibly to disinfection.

On the other hand, the trace forms of free chlorine can be relevant contributors to the chemistry of chlorine-based disinfection systems. One illustration of the importance of these other forms of free chlorine is their roles as substituting agents. In particular, molecular chlorine (Cl_2) and chlorine monoxide (Cl_2O) tend to be better substituting agents than HOCl and OCl^-. Rate constants for reactions involving Cl_2 and Cl_2O can be many orders of magnitude greater than their corresponding rate constants based on HOCl and OCl^- (Blatchley and Cheng, 2010; Sivey *et al.*, 2010). Therefore, Cl_2 and Cl_2O can contribute significantly to chemical dynamics in reactions that involve free chlorine.

Equations 8.5b-8.9b summarize the electrochemical behaviour of the compounds that comprise free chlorine, as well as reported values of the standard reduction potential (SRP) for each compound[1]:

$$Cl_{2,aq} + 2e^- \rightarrow 2Cl^- \tag{8.5a}$$

$$SRP = 1.36V \tag{8.5b}$$

$$HOCl + H^+ + 2e^- \rightarrow H_2O + Cl^- \tag{8.6a}$$

$$SRP = 1.49V \tag{8.6b}$$

$$OCl^- + H_2O + 2e^- \rightarrow Cl^- + 2OH^- \tag{8.7a}$$

$$SRP = 0.90V \tag{8.7b}$$

$$2H^+ + Cl_2O + 4e^- \rightarrow H_2O + 2Cl^- \tag{8.8a}$$

$$SRP = n.a. \tag{8.8b}$$

$$Cl_3^- + 2e^- \rightarrow 3Cl^- \tag{8.9a}$$

$$SRP = n.a. \tag{8.9b}$$

The half-cell reactions that are presented in equations 8.5-8.9 form the basis of electrochemical equivalence for these compounds. Specifically, these reactions indicate that all of the forms of free chlorine indicated in these equations have the ability to accept two electrons per molecule, except chlorine monoxide (Cl2O), which has the ability to accept four electrons per molecule. This information forms the basis of the formal definition of free chlorine:

$$\text{Free chlorine} = [Cl_2] + [HOCl] + [OCl^-] + 2\cdot[Cl_2O] + [Cl_3^-] \tag{8.10}$$

Equation 8.10 is a more detailed expression of free chlorine than is normally presented in the literature, in that it accounts for the contributions of free chlorine forms that are often present at trace concentrations (i.e. Cl_2, Cl_2O, and Cl_3^-). Common definitions of free chlorine usually include only contributions from HOCl, OCl^-, and perhaps Cl_2 because of the dominant roles they play in terms of molar concentration.

The summary of free chlorine behaviour presented above provides information about the distribution of free chlorine and its behaviour when free chlorine dominates the chlorine residual. However, in wastewater effluents, it is likely that the chemistry of chlorine will be influenced by other constituents that are present in solution.

Free chlorine reacts with a wide range of compounds in solution. Among these are reactions between free chlorine and ammonia-N. Collectively, the reactions between free chlorine and ammoniacal-N define the so-called 'chloramination' and 'breakpoint chlorination' mechanisms. These processes are governed by two sets of reactions between free chlorine and ammonia. The first set comprises substitution reactions, in which +1-valent chlorine is substituted for protons (H^+). Collectively, these reactions result in conversion of free chlorine, in the form of HOCl, to inorganic combined chlorine:

---

[1] SRP values are from White (1992).

$$HOCl + NH_3 \rightleftharpoons NH_2Cl + H_2O \quad (8.11)$$

$$HOCl + NH_2Cl \rightleftharpoons NHCl_2 + H_2O \quad (8.12)$$

$$HOCl + NHCl_2 \rightleftharpoons NCl_3 + H_2O \quad (8.13)$$

$$NH_2Cl + NH_2Cl \rightleftharpoons NHCl_2 + NH_3 \quad (8.14)$$

Reactions 8.11-8.13 result in direct formation of the inorganic chloramines: NH_2Cl (monochloramine), $NHCl_2$ (dichloramine), and NCl_3 (trichloramine or nitrogen trichloride). Reaction 8.14, which is sometimes identified as a disproportionation reaction, allows conversion of monochloramine to dichloramine. All four of these substitution reactions are reversible. The forward part of reaction 8.14 is general acid catalysed, which means that it is promoted under any condition in which protons (H^+) are available to be donated, including low pH and solutions that contain high concentrations of proton-donating compounds, such as pH buffers. It is important to recognize that no oxidation/reduction (redox) takes place in reactions 8.11-8.14.

The second group of reactions are redox processes in which +1-valent chlorine is reduced to chloride ion (Cl^-) resulting in oxidation of reduced-N. These redox processes proceed at a wide range of rates. Comprehensive summaries of these reactions and their kinetics are provided in the works of Jafvert and Valentine (1992) and Wahman (2018). An interesting feature of the redox reactions that contribute to this reaction mechanism is that all of them proceed, directly or indirectly, through $NHCl_2$. Therefore, any aspect of process behaviour that promotes formation of $NHCl_2$ will enhance these redox reactions, thereby diminishing the stability of the combined chlorine residual. Specific attributes of process behaviour that will promote $NHCl_2$ formation include reductions in pH and an increase of the Cl:N ratio applied to the system.

Collectively, the substitution reactions and the redox reactions described above will govern the distribution of residual chlorine among the various forms of free and inorganic combined chlorine. With municipal wastewater effluents, there is almost always a small amount of ammonia-N in solution at the influent end of a chlorine contact chamber. Even when biochemical nitrification is applied, ammonia-N will typically be present at a concentration of 0.5-1.0 mg/l (as N) (McCarty, 2018). Therefore, the inorganic chloramines (especially NH_2Cl) are likely to dominate the chlorine residual in wastewater chlorination.

Wahman (2018) developed and made available web-based programs to allow simulations of the reactions between free chlorine and ammoniacal nitrogen. Figure 8.2 illustrates the distribution of residual chlorine in a wastewater effluent that has been chlorinated immediately downstream of a properly-functioning biochemical nitrification system. For this simulation, it was assumed that the NH_3-N concentration was 0.5 mg/l (as N), pH = 7.5; alkalinity = 150 mg/l as $CaCO_3$, with reaction progress being evaluated after 60 minutes of reaction time.

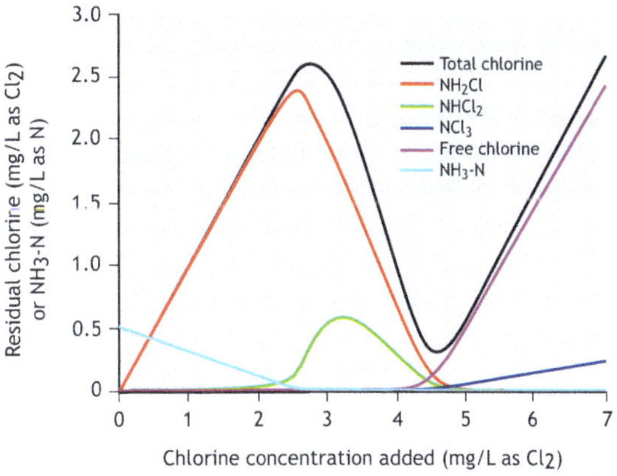

Figure 8.2 Breakpoint chlorination curve after 60 minutes of reaction time for conditions of pH = 7.5, initial NH_3-N concentration = 0.5 mg/l as N, alkalinity = 150 mg/l as $CaCO_3$. Simulation results are from the model of Wahman (2018).

The information presented in Figure 8.2 indicates that the chlorine residual in municipal wastewater effluents is likely to be dominated by NH_2Cl, unless steps are taken to completely remove NH_3-N. For the conditions of the simulation represented in Figure 8.2, the chlorine residual is almost entirely NH_2Cl until the chlorine concentration added to the system exceeds roughly 4.5 mg/l as Cl_2, which corresponds with the breakpoint of this system. This translates to a chlorine:nitrogen molar ratio of approximately 1.8. For free chlorine addition above this value, the residual will exist largely in the form of free chlorine; however, the vast majority of the added chlorine will have been consumed in oxidation of NH_3-N. This represents an inefficient use of chlorine and an unlikely scenario for most practical applications.

The form of residual chlorine will have important implications for disinfection behaviour. NH_2Cl is substantially less effective as a disinfectant than HOCl, although the magnitude of this difference in disinfection efficacy is variable among microorganisms and the conditions of disinfectant application. For example, at 20°C and pH = 7.5, free chlorine is roughly 250 times more effective than NH_2Cl for inactivation of *Giardia* cysts, while under the same conditions free chlorine is roughly 16 times more effective than NH_2Cl for inactivation of viruses (Malcolm Pirnie and HDR Engineering, 1991). Therefore, the presence of even low concentrations of NH_3-N will result in a chlorine residual that is dominated NH_2Cl, not free chlorine. Moreover, the shift to inorganic combined chlorine, essentially all of which will be in the form of NH_2Cl, will diminish disinfection efficacy, relative to the behaviour of free chlorine.

As compared to other disinfectants, +1-valent chlorine compounds tend to be relatively stable. This attribute can be beneficial in applications where it is necessary to prevent re-growth of pathogens downstream of a disinfection process. However, in the context of some wastewater disinfection applications, this characteristic may be detrimental, in that residual chlorine compounds can be toxic to fish and other forms of aquatic life. Because of this, it is sometimes necessary to dechlorinate wastewater.

Dechlorination is accomplished by controlled addition of a chemical that expresses rapid chlorine demand. The most common dechlorinating agents are based on reduced sulphur (S[IV]) compounds, such as sodium bisulphite. The chemistry of dechlorination based on bisulphite is defined by the following stoichiometric expressions:

$$HOCl + HSO_3^- \rightleftharpoons SO_4^{2-} + 2H^+ + Cl^- \tag{8.15}$$

$$NH_2Cl + HSO_3^- + H_2O \rightleftharpoons SO_4^{2-} + H^+ + NH_4^+ \tag{8.16}$$

The equilibria for reactions 8.15 and 8.16 lie far to the right, and forward reactions for both are extremely rapid. These characteristics simplify process control for dechlorination, in that dosing of the dechlorinating agent (bisulphite, HSO_3^-, in this case) can be defined using real-time measurements of residual chlorine concentration and flow rate. In many cases, the dechlorinating agent is added in slight stoichiometric excess of the +1-valent chlorine to ensure complete dechlorination. The bisulphite that remains after dechlorination is fairly benign and does not pose a threat to aquatic life downstream of the point of application.

It should be noted that dechlorination using bisulphite yields protons (H^+) as a product of these reactions. As such, S([IV])-based dechlorination will consume alkalinity. In waters that are poorly buffered, this added acidity can result in a substantial reduction of pH. This can be counteracted by addition of a suitable pH buffer, such as a carbonate-based material.

8.3.2 Disinfection mechanisms: chlorine

Chlorine tends to react broadly and somewhat non-selectively with microorganisms and critical biomolecules (Virto *et al.*, 2005). There are several possible mechanisms by which +1-valent chlorine may inactivate microorganisms; current knowledge of these mechanisms is informed by our understanding of the chemistry of chlorine reactions with these biomolecules.

Free chlorine, especially HOCl, reacts rapidly with amines to yield corresponding N-chlorinated derivatives, often followed more slowly by oxidation/reduction reactions within these compounds. Important biomolecules that contain amine groups include proteins, their constituent amino acids, and nucleobases (purines [adenine and guanine] and pyrimidines [cytosine, uracil, thymine]). HOCl reacts with these biomolecules following mechanisms that mimic breakpoint chlorination (Shang *et al.*, 2000). Similar reaction behaviour has been observed with chlorination of pure cultures of gram-positive and gram-negative bacteria (Shang and Blatchley, 2001). Free chlorine reactions with bacteria may ultimately lead to membrane damage and leakage of cellular constituents into solution. However, membrane damage as a bacterial response appears to require chlorine exposure that greatly exceeds the exposure required to accomplish bacterial inactivation (i.e. loss of the ability to reproduce) (Virto *et al.*, 2005).

Free chlorine has been observed to accomplish rapid inactivation of most viruses, probably because of the structure of virus particles. Specifically, most viruses comprise a nucleic acid, either DNA or RNA, surrounded by a protein shell called a capsid; +1-valent chlorine tends to react rapidly with nucleic acids and proteins.

Although there is a long history of success in this use of chlorine as a disinfectant and oxidant in water and wastewater treatment, there are important drawbacks to these applications as well. Among these are the formation of disinfection by-products (DBPs). More than 700 DBPs have been identified in chlorinated water (Krasner *et al.*, 2006); it is likely that many more DBPs exist but have not yet been identified. Among these compounds, many express one or more forms of toxicity.

A second problem relates to the ineffectiveness of chlorine against some groups of microbial pathogens. Specifically, +1-valent chlorine is minimally effective against many protozoan parasites, particularly when they are present in their resting stages. A prime example is the poor efficacy of chlorine against the oocysts of *Cryptosporidium* species (Murphy *et al.*, 2015).

These important limitations of chlorine-based disinfection have motivated the search for alternative disinfectants. For wastewater disinfection applications, the two most common alternative to chlorination are peracids and ultraviolet (UV) radiation.

8.4 DISINFECTION WITH PERACIDS (PERACETIC ACID)

Peracids are organic disinfectants that include one or more peroxy (–O–OH) groups. These compounds share structural and functional similarities to hydrogen peroxide (H_2O_2) and exist in equilibrium with H_2O_2 in solutions used in practical applications. As described below, synergistic behaviour exists between organic peroxy compounds and H_2O_2 that can be relevant in disinfection applications.

Peracid compounds have been applied in a wide range of applications to accomplish disinfection or oxidation. Examples include: pharmaceutical production, healthcare applications, disinfection of surfaces, bleaching in pulp/paper processing, ballast water treatment, food and beverage production, biofilm control in cooling towers, and more recently in disinfection of wastewater (Block, 2001; Luukkonen and Pehkonen, 2017). These applications are promoted by the broad-spectrum antimicrobial characteristics of the peracids, as well as their low potential to form DBPs. The dominant DBPs that are generated from peracid applications include carboxylic acids that result from their natural degradation, and some aldehydes that form from reactions between peracids and natural organic matter (NOM) (Luukkonen and Pehkonen, 2017; Santoro *et al.*, 2007).

When applied for disinfection of wastewater, peracids can be produced on-site and on-demand by appropriate mixtures of a carboxylic acid and hydrogen peroxide. In the case of peracetic acid

(PAA), the most commonly applied peracid disinfectant, it is also possible to purchase commercial mixtures for these applications. Peracid systems can be retrofitted to existing chlorine contact chambers. As such, their physical appearance and overall footprint are quite similar to those of chlorination systems.

PAA is relatively stable and can be stored for use. Systems based on PAA can be easily turned on or off, making them well-suited to meet variable flow rates and peak flow conditions. A possible mode of application for PAA is in combination with UV disinfection (described below), where a UV disinfection system can be designed for average flow conditions, with PAA being brought online to handle high flow rates or peaking conditions that exceed the design constraints of the UV system. This allows the design and implementation of a UV system to meet most operating conditions, with PAA being used to handle flow rates above the design condition for the UV system. Such an approach will have the benefit of saving on capital costs for the UV system, while having the flexibility to meet variable flow rates that are often observed in municipal wastewater treatment systems, especially those that serve communities with combined sewers.

8.4.1 Physical chemistry of peracids

Peracid disinfectants are prepared by addition of a carboxylic acid to hydrogen peroxide. The process is readily reversible and is represented by the following general reaction:

$$R-COOH + H_2O_2 \rightleftharpoons R-COOOH + H_2O \quad (8.17)$$

The forward reaction in this process is acid-catalysed. The most commonly applied peracid is PAA, CH_3COOH. Its formation is described as follows:

$$CH_3COOH + H_2O_2 \rightleftharpoons CH_3COOOH + H_2O \quad (8.18)$$

Given the reversible nature of these reactions, equilibrium between reactants and products is readily achieved and can be described in quantitative terms for PAA at room temperature as follows (Luukkonen and Pehkonen, 2017):

$$K_{eq} = \frac{[CH_3COOOH]}{[CH_3COOH][H_2O_2]} = 2.10 - 2.91 M^{-1} \quad (8.19)$$

PAA is available as a commercial product, with typical composition (w/w) of 15% PAA, 25% H_2O_2, 35 % acetic acid, and 25% water (Santoro *et al.*, 2007). PAA is a weak acid (Luukkonen and Pehkonen, 2017; Santoro *et al.*, 2007), with a pK_a of 8.0-8.2 at temperatures that are common in disinfection applications. The acid/base behaviour of the peracids is relevant in that the neutral (protonated) forms of these compounds tend to be the most effective as disinfectants. Therefore, peracid application is favoured in disinfection applications when the pH is less than the pK_a of the peracid disinfectant. PAA solutions will undergo slow autodecay; this behaviour can be counteracted by addition of compounds that function as stabilizing agents, such as dipicolinic acid or sodium pyrophosphate (Santoro *et al.*, 2015).

Performic acid (PFA) and perpropionic acid (PPA) have also been applied as disinfectants. However, their applications are less frequent than those involving PAA. In general, the stability of peracids tends to increase as the aliphatic chain length increases, which favours the use of PPA. In fact, PFA is sufficiently unstable that it is usually prepared at the site of application, which complicates its use. On the other hand, as the aliphatic chain length increases, more organic carbon will be present in the product water, which increases the potential for microbial regrowth in the treated water. PAA appears to have the best blend of these characteristics, which is probably why it is the most frequently applied of the peracids for disinfection.

8.4.2 Disinfection mechanisms: peracids

Disinfection and oxidation processes with peracids appear to be strongly linked to their autodecay. Specifically, peracids will undergo homolytic cleavage of the peroxy bond to yield free radicals.

Using PAA as an example, this reaction proceeds as follows:

$$CH_3COOOH \rightarrow CH_3COO^{\cdot} + HO^{\cdot} \quad (8.20)$$

This reaction initiates a series of other, related reactions that result in formation of a spectrum of reactive oxygen species (ROS), including the superoxide anion ($O_2^{\cdot-}$), hydroperoxyl radical ($HO_2^{\cdot}$), as well as acyl and peracyl radicals (Luukkonen and Pehkonen, 2017). The hydroxyl radical ($HO^{\cdot}$) is believed to play a dominant role in disinfection (Block, 2001).

Microbial inactivation by peracids appears to involve attack of several critical biomolecules. Specific reactions that have been identified as contributors to microbial inactivation include oxidation of sulphydryl (–SH) and sulphur (S–S) groups in enzymes, denaturation of proteins, disruption of chemiosmotic functions across membranes, damage to DNA bases, and inactivation of catalase (Block, 2001; Luukkonen and Pehkonen, 2017; Santoro *et al.*, 2007). This last mechanism contributes to synergism that has been reported between peracids and hydrogen peroxide because catalase is effective for deactivation of hydrogen peroxide as a disinfectant. In the absence of catalase, hydrogen peroxide functions well as a disinfectant (Block, 2001).

8.5 DISINFECTION WITH ULTRAVIOLET RADIATION

Ultraviolet (UV) disinfection systems accomplish microbial inactivation through damage to one or more critical biomolecules by photochemical reactions. Therefore, to understand UV disinfection processes, it is necessary to have a basic understanding of UV radiation, as well as the fundamental principles that govern photochemistry and photochemical processes.

8.5.1 Laws of photochemistry

A group of remarkable scientists developed a set of three 'laws' of photochemistry that are helpful in understanding the dynamics of these processes. The Grotthus-Draper law[2] (also known as the first law of photochemistry or the principle of photochemical activation) states that electromagnetic radiation (photons) must be absorbed by a system to bring about a photochemical change. The Stark–Einstein law[3] (also known as the second law of photochemistry, the photochemical equivalence law, or the photo equivalence law) states that for each photon absorbed by a chemical, only one molecule is activated for a photochemical (or photophysical) process. The Bunsen-Roscoe law (sometimes referred to as the third law of photochemistry) indicates that the extent of a photochemical reaction will be governed by the dose (fluence) of radiation absorbed by the target; in turn, the dose of radiation is determined by the rate at which photons are delivered to the photochemical target and the time of exposure. Collectively, the laws of photochemistry establish a one-to-one correspondence between photons and molecules and also indicate that the dose of radiation (defined formally below) is the master variable in governing the extent of a photochemical reaction. Another implication of these laws is that photon energy must meet or exceed bond energy for a photochemical reaction to take place.

Figure 8.3 illustrates the concept of photon-molecule equivalence for wavelengths of radiation in the visible and UV ranges, using a set of common chemical bonds and representative values of their

[2] An important exception to the first law of photochemistry is reactions that involve photosensitization, wherein a molecule in the immediate vicinity of the target molecule will absorb a photon, then transfer the absorbed energy to the target molecule by a subsequent collision.

[3] Some exceptions to the second 'law' have also been observed. An important example involves the use of powerful sources of radiation (*e.g.* some lasers), which can impose photons on a target at a sufficiently fast rate to allow what are referred to as 'multi-photon' processes to occur. In these processes, more than one photon may be absorbed by a target molecule before the energy associated with the first absorbed photon(s) is dissipated through a photochemical or photophysical process. These intense sources of radiation are rarely used in practical applications. As such, the Stark-Einstein law will apply in most practical applications.

bond energy. The data in this figure indicates wavelengths of UV radiation that are sufficiently energetic to promote cleavage of chemical bonds. This correspondence between photon energy and bond energy dictates that most photochemistry involves UV radiation.

Also indicated at the bottom of Figure 8.3 are the nominal wavelength ranges that have been defined for visible radiation (i.e. wavelengths to which the sense of vision will respond in humans), as well as various sub-regions of UV radiation, including UVA, UVB, UVC, and vacuum UV (VUV) radiation. It should be noted that there are variations in the reported ranges of these sub-regions in the literature.

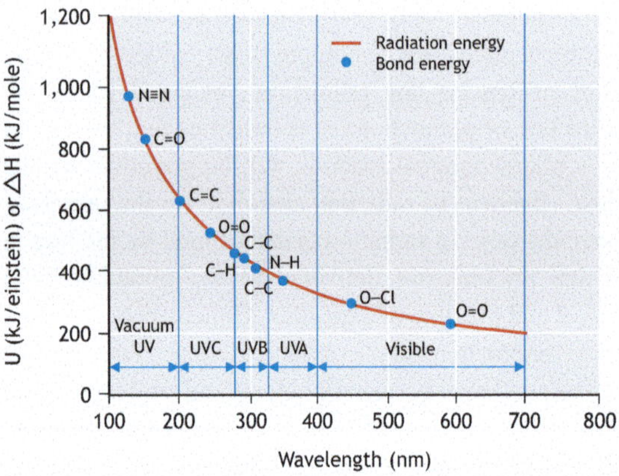

Figure 8.3 Energy of photons (U) as a function of wavelength. Superimposed are reported values of average bond energy (ΔH) for common bonds. Also indicated at the bottom of the graph are wavelength ranges corresponding to definitions of radiation types. Average bond energy values can be obtained from the dedicated website[4].

8.5.2 Principles of photochemical kinetics

The laws of photochemistry indicate that two fundamental conditions must be met for a photochemical reaction to take place. First, a photon must be absorbed by a target molecule. Second, the absorbed photon must have sufficient energy to break an existing bond or form a new one. If both of these conditions are met, then it is possible for a photochemical reaction to take place, though not certain. If either of these conditions is not met, then a photochemical reaction cannot take place.

These qualitative conditions define when it is possible for a photochemical reaction to take place, but they do not provide information to define the rate of such a process. Quantitative descriptions of the rate of a photochemical process are based on logic that is similar to the logic that is used to describe conventional thermal chemical processes, specifically a bimolecular elementary reaction. Table 8.1 provides a summary of rate expressions for elementary bimolecular (thermal) reactions and elementary photochemical reactions.

Considering first the case of elementary bimolecular reactions, at a fundamental level the process involves the collision of two molecules, B and C, to result in the formation of one or more products. This process requires that B and C collide with sufficient energy to cause one or more bonds to be broken or a new bond to be formed. The rate expression can be decomposed to describe these two processes. The concentrations (or more formally activities) of the reactants ([B] and [C]) can be thought of as representations of their availability to participate in collisions with other molecules; the rate of this process increases linearly with [B] and [C]. The product,

$$A \cdot [B] \cdot [C] \tag{8.21}$$

represents the frequency of collisions between B and C. The exponential term,

$$\exp\left(-\frac{E_a}{R \cdot T}\right) \tag{8.22}$$

describes the fraction of these collisions that have sufficient energy to result in a reaction. Often the

[4]www.butane.chem.uiuc.edu/pshapley/environmental/l14/1.html

reaction rate expression is presented in a shortened form:

$$\frac{d[B]}{dt} = -k \cdot [B] \cdot [C] \quad (8.23)$$

where the term k represents a second-order reaction rate constant that can be expanded as follows:

$$k = A \cdot \exp\left(-\frac{E_a}{R \cdot T}\right) \quad (8.24)$$

When applied at a microscopic level, equation 8.24 indicates that the rate of a bimolecular, elementary reaction at a point in space will be governed by the conditions that exist at that point in space, including local temperature and local activities of the reactants. In many chemical and biochemical reactor systems, mixing conditions are such that reactant concentrations are uniform, or nearly so, over large spatial extents.

Photochemical kinetics are described using similar logic, but with terminology and process variables that are relevant to photochemistry. As indicated above, a photochemical process can take place when the energy of a photon is absorbed by a target molecule; that process is described by the basic rate expression for a photochemical reaction indicated in Table 8.1. Note that the term hv is shorthand notation to represent the energy of a photon of frequency v (h = Planck constant). The symbol E_i represents the fluence rate of radiation (at a given wavelength) that is incident on the target molecule; it can be thought of as an expression of the availability of photons at a location. The symbol ε represents the molar absorption coefficient for a molecule, which describes the efficiency by which the target molecule absorbs incident photons of a given wavelength. The symbol Φ represents the quantum yield, a term that describes the fraction of absorbed photons at a given wavelength that result in a reaction. Lastly, the term Q_e describes the energy of radiation at a given wavelength.

As with the case of the elementary bimolecular reaction, the fundamental kinetic expression for a photochemical reaction can be decomposed. The quotient:

$$\frac{2.303 \cdot E_i \cdot \varepsilon \cdot [B]}{Q_e} \quad (8.25)$$

describes the rate at which incident photons are absorbed by the target molecule, while the quantum yield describes the fraction of these photon absorption events that lead to reaction. In many cases, the rate of a photochemical process is defined in a simplified form:

$$\frac{d[B]}{dt} = -k' \cdot E_i \cdot [B] \quad (8.26)$$

where the term k' represents a photochemical reaction rate constant that can be expanded as follows:

$$k' = \frac{2.303 \cdot \varepsilon \cdot \Phi}{Q_e} \quad (8.27)$$

Table 8.1 Summary of fundamental reaction rate expressions for elementary (thermal) chemical reactions and photochemical reactions.

Reaction type	Basic reaction expression	Kinetic expression
Elementary bimolecular (thermal)	$B + C \rightarrow$ products	$\frac{d[B]}{dt} = -A \cdot \exp\left(-\frac{E_a}{R \cdot T}\right) \cdot [B] \cdot [C]$
Elementary photochemical (photolytic)	$B + hv \rightarrow$ products	$\frac{d[B]}{dt} = -\frac{2.303 \cdot E_i \cdot \varepsilon \cdot [B] \cdot \Phi}{Q_e}$

The rate expressions that are used to describe photochemical kinetics apply at a local level. This attribute is critical to understanding the dynamic behaviour of photochemical reactors, including those used for disinfection, because the fluence rate fields within these systems generally demonstrate strong spatial gradients. Since the local rate of a photochemical process is directly proportional to the local fluence rate, this means that the gradients in local fluence rate result in equally strong gradients in local photochemical reaction rates. As will be demonstrated in the discussion of photochemical reactor dynamics, this has profound effects on photoreactor analysis, design, and performance.

8.5.3 Mechanisms of microbial inactivation: ` UV irradiation

The primary mechanism of microbial inactivation resulting from exposure to UVB and UVC radiation is direct, photochemical damage to nucleic acids (DNA and RNA) and proteins. With regard to the former, UVB and UVC radiation (sometimes referred to as 'germicidal UV') are absorbed effectively by nucleic acids.

Figure 8.4 illustrates normalized absorption spectra for DNA extracts from a common marine algal species, a waterborne bacterium, and a protozoan parasite. Despite the differences in these organisms, their DNA absorption spectra show strikingly similar shapes, with peak absorption near 260 nm and generally less efficient absorption at wavelengths immediately above and below this value. Below approximately 240 nm, all three DNA extracts demonstrate strong absorption.

Absorption of UVC and UVB radiation promotes reactions between adjacent bases within a nucleic acid strand; this is especially true among adjacent pyrimidine bases, with thymine being the most efficient. The two most common reaction products to result from UV irradiation of nucleic acids are cyclopyrimidine dimers (CPDs) and 6-4 photoproducts (see Figure 8.5). These and other forms of photochemical damage are sometimes referred to as 'lesions' within a nucleic acid molecule. Accumulation of a sufficient number of lesions can lead to loss of viability or infectivity among microorganisms.

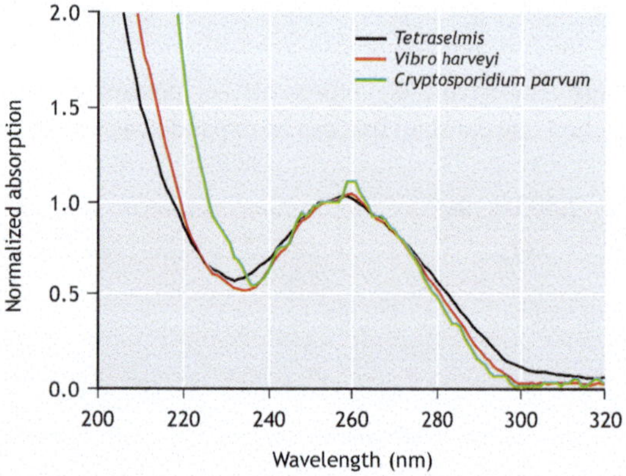

Figure 8.4. Absorption spectra for DNA extracted from three aquatic organisms: Tetraselmis (a marine alga), Vibrio harveyi (a bacterium), and Cryptosporidium parvum (a common protozoan parasite). These absorption spectra were normalized to their values measured at 254 nm.

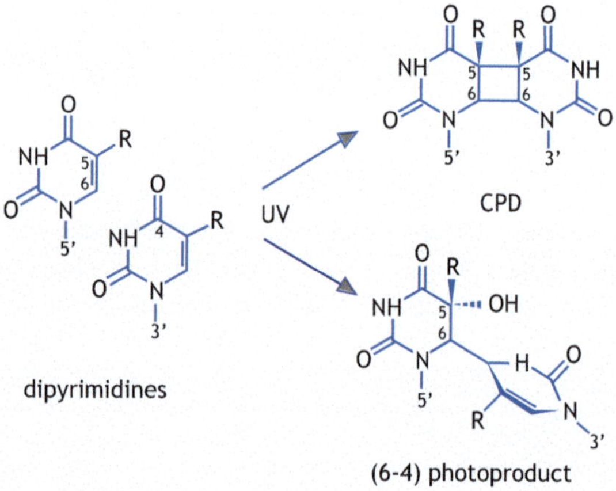

Figure 8.5 Schematic illustration of photoreactions that yield cyclopyrimidine dimer (CPD) and the 6-4 photoproduct. Image from Li *et al.* (2006).

UVC radiation can also cause damage to proteins. This is particularly important with some viruses, whose structure typically involves a proteinaceous shell called a capsid that surrounds a nucleic acid molecule. Damage to capsids tends to become more important at wavelengths below approximately 240 nm, where proteins absorb radiation most effectively. Capsids can play critical roles in the attachment of viruses to their specific host cell. If a virus particle is incapable of attaching to its host, then infection of the host is prevented. Photochemical damage to capsid proteins is believed to be responsible for enhanced sensitivity of some viruses at these relatively short wavelengths (Eischeid and Linden, 2011).

8.5.4 Sources of germicidal UV radiation

The most common source of germicidal UV radiation is a mercury (Hg) discharge lamp. A general schematic of discharge lamp construction is provided in Figure 8.6. Establishment of a voltage between the electrodes in this system causes electrons (e^-) to migrate from the cathode to the anode. Interactions between electrons and the gases inside the lamp lead to emission of radiation.

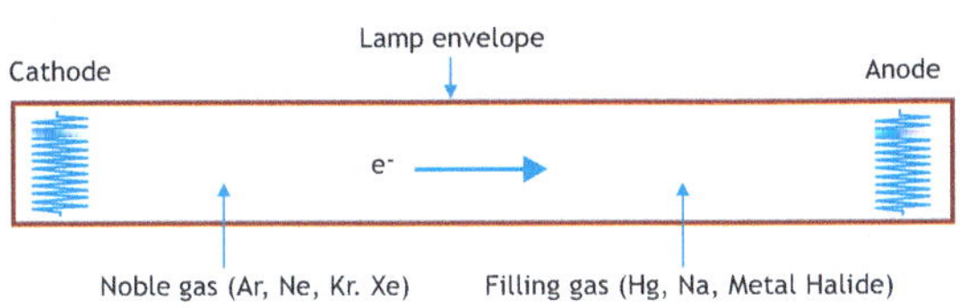

Figure 8.6 Schematic illustration of a gas discharge lamp.

Two common classes of Hg lamps are available. The most frequently applied source of germicidal UV radiation is a low-pressure (LP) Hg lamp, which is characterized by Hg partial pressure on the order of 10^{-3}-10^{-2} Torr. Collisions between electrons, noble gas molecules, and Hg atoms lead to electronic excitation of mercury. Excited Hg atoms can return to lower energy or ground state conditions by releasing a photon. The two dominant emission lines for excited Hg occur at 253.7 nm and 185 nm.

Mercury amalgam lamps have been developed as alternatives to conventional low-pressure Hg lamps. In amalgam lamps, the mercury is blended with other metals to form a solid amalgam. The amalgam serves as a pressure buffer for Hg in the system. Amalgam lamps can operate at higher temperatures (ca. 80°C) than conventional low-pressure Hg lamps. This allows amalgam lamps to generate more power than conventional low-pressure Hg lamps. However, the output spectrum of an amalgam lamp is essentially identical to that of a conventional low-pressure Hg lamp, in terms of the location and relative height of the peaks.

Medium-pressure (MP) Hg lamps operate at Hg partial pressures of roughly 75 Torr (approximately 0.1 atm). Under these conditions, the physics of photon generation will include other processes that are not relevant in low-pressure Hg lamps. As a result, the output spectrum of MP Hg lamps is quite different than that of a LP Hg lamp. In particular, the output spectra of MP Hg lamps are characterized by a broad 'hump' between wavelengths of roughly 200-240 nm. Well-defined lines are evident at wavelengths that are scattered across the entire UV portion of the spectrum and also into the visible portion of the spectrum. Normalized output spectra for LP (amalgam) and MP Hg lamps are presented in Figure 8.7.

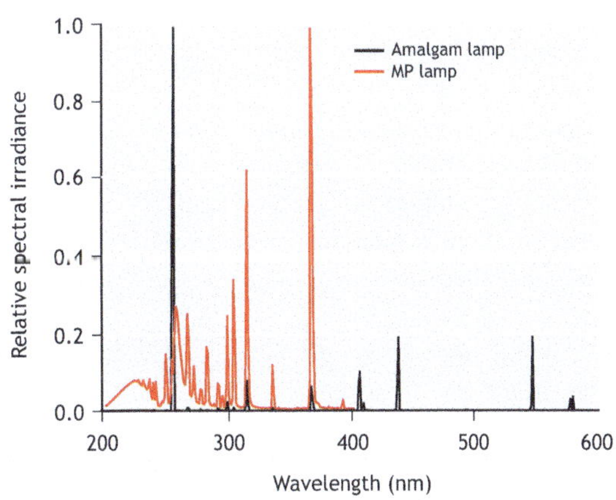

Figure 8.7 Normalized output spectra for a mercury amalgam (low pressure) lamp and a medium-pressure (MP) Hg lamp. The spectra for each lamp are normalized against their respective spectral peaks (data: Trojan Technologies).

LP Hg lamps are characterized by output that is dominated by the line at 254 nm, which is close to the wavelength of optimum absorption by nucleic acids. In addition, LP Hg lamps are characterized by wall plug efficiencies of roughly 35-40%. Despite this relatively high efficiency, conventional LP Hg lamps have modest output power density, resulting in the need for large numbers of lamps in some applications. Amalgam lamps improve that situation, but systems based on amalgam lamps can still involve large numbers of lamps, which leads to system complexity.

MP Hg lamps can have output power that is 1-2 orders of magnitude higher than that of a LP Hg lamp. However, MP lamps have wall plug efficiencies of roughly 10%, so they require much more input electrical power than LP lamps to meet a given output power requirement.

An important drawback of LP and MP lamps is the inclusion of Hg in their construction. Globally, there is increasing pressure to reduce or eliminate uses of Hg. The Minamata Convention on Mercury is an international treaty that calls for reductions or elimination of commercial and industrial uses of Hg (EU, 2011). The Minamata Convention includes exceptions for some applications where alternatives have not emerged or been demonstrated; important among these are so-called 'fluorescent' lamps, which contain Hg. These lighting sources, which are electronically identical to LP Hg lamps, are preferred in many applications because of their high efficiency. To date, there are no provisions within the Minamata Convention or other international treaties to eliminate the use of Hg lamps for generation of UV radiation. However, there is mounting pressure to develop and implement alternative sources of UV radiation.

Among the most promising alternative UV sources are UV light emitting diodes (UV LEDs). Like all LEDs, UV LEDs are semiconducting devices that emit electromagnetic radiation ('light') as a result of recombination of electrons and holes, following the principle of electroluminescence (Held, 2009). UV LEDs contain no mercury. Their output spectra are governed by the elemental composition of the diode material. Figure 8.8 illustrates normalized output spectra for several commercially available UV LEDs.

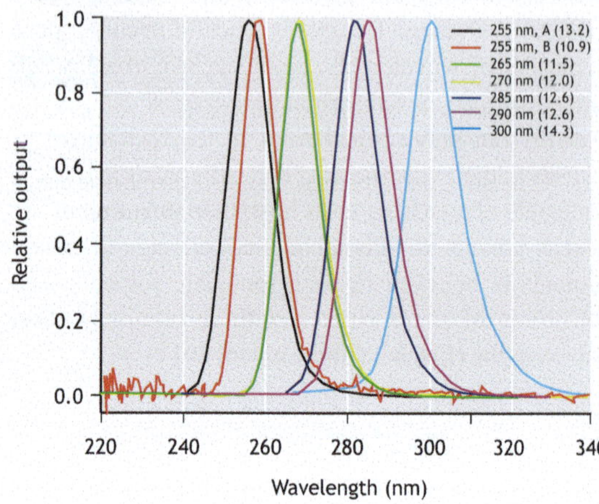

Figure 8.8 Normalized output spectra of some commercially available UV LEDs. Values in the legend indicate the nominal peak wavelength of each LED, as indicated by the manufacturer. Horizontal reference line indicates 50% of maximum relative output. Parenthetical entries in the legend indicate spectrum bandwidth at ½ maximum output. Data provided by AquiSense.

Commercially available UV LEDs offer important advantages of being essentially instantaneous 'on/off' sources, requiring no warm-up time. Also, because their output spectra are governed by the elemental composition of the diode, it is possible to select UV LEDs with output spectra that are tailored to an application. In the case of disinfection applications, one example would be the use of UV LEDs with output centred in the vicinity of 270-280 nm. These UV LEDs tend to have higher wall plug efficiency, last longer, and be less expensive than those that provide output closer to 260 nm where nucleic acid absorption generally peaks. Also, it is common for aqueous solutions to have higher transmittance at 270-280 nm than at 260 nm or below; therefore, the output radiation from these devices tends to penetrate through the fluid being disinfected more effectively than radiation at shorter wavelengths.

Among UV LEDs that are commercially available at present, output power tends to be fairly modest and wall plug efficiency is low, as compared with Hg lamps. However, historical trends suggest that as demand for LEDs increases, their output power and wall plug efficiency increase, while unit costs of the devices tends to decrease. These trends have been followed in recent years as interest in UV LEDs for water treatment applications has increased.

8.6 DISINFECTION KINETICS

The logic used in development of mathematical models to describe the kinetics of disinfection processes is adapted from the fundamental theories that are used to describe the intrinsic kinetics of chemical and photochemical reactions. However, it is important to recognize that disinfection as an endpoint is inherently more complex than an elementary chemical reaction, in that it can involve numerous biochemical reactions. Moreover, the endpoint of the process is a physiological change in the behaviour or response of a microbial population, whereas with a chemical reaction the endpoint of interest is loss of a reactant or gain of a reaction product, which can often be measured more directly and precisely than microbial inactivation.

Another relevant point to consider in these models is that the endpoint that is most relevant to disinfection applications is loss of viability or the ability to infect a host cell. In the context of disinfection, viable bacteria are able to reproduce; evidence of this ability is usually demonstrated through culture-based methods. For parasitic organisms, such as protozoa or viruses, the ability to infect a host cell and to complete a lifecycle is the endpoint of interest. Infectivity assays for viruses and protozoa also involve incubation of samples to confirm infection and completion of the lifecycle. More broadly, viability and infectivity endpoints represent the gold standards for quantification of disinfection efficacy. As such, mathematical models that are used to describe disinfection kinetics should be based on viability or infectivity as their endpoints, unless an alternative analytical method can be demonstrated to yield similar results.

Collectively, these characteristics dictate that mathematical models that are used to simulate the kinetics of disinfection processes will involve a blend of empiricism and theory. Most models of disinfection kinetics involve a theoretical foundation that is based on fundamental principles of chemical or photochemical reaction kinetics. However, these models also need to be adapted to the viability or infectivity endpoint, which almost always involves elements of empiricism.

8.6.1 Disinfection kinetics: chemical disinfectants

As with most disinfectants, the strategy that is used to simulate chemical disinfection kinetics is based on logic that is similar to that used to define the kinetics of elementary chemical reactions. In general terms, the disinfectant molecule (+1-valent chlorine, for example) and the target microbe are viewed as 'reactants' in these models.

The most basic of these models was developed first by Harriet Chick (Chick, 1908), then reinforced and slightly modified by Herbert Watson (Watson, 1908), The Chick-Watson model takes the form:

$$\frac{dN}{dt} = -\Lambda C^{n} N \tag{8.28}$$

where:

N concentration of viable microbes

t time

Λ coefficient of specific lethality (a reaction rate parameter)

C concentration of the chemical disinfectant (*e.g.* chlorine)

n empirical constant (often assumed ≅ 1).

Separation of variables and integration allows definition of an equation to describe the extent of inactivation resulting from disinfectant exposure. If the empirical constant is assigned a value of n = 1, then the integrated form of the Chick-Watson model is:

$$\ln\left(\frac{N}{N_0}\right) = -\Lambda \cdot Ct \tag{8.29}$$

Equations 8.28 and 8.29 are often used as the default expressions to define the kinetics of disinfection with chemical disinfectants, including free chlorine, inorganic chloramines (i.e. +1-valent chlorine), and peracids. Expression 8.29 is often observed to be consistent with inactivation responses of simple microbes, such as viruses and some bacteria, over limited extents of inactivation. It also forms the basis of the 'Ct concept' that is commonly used to define disinfectant exposures required to meet a specific treatment endpoint. The idea behind this concept is that a given extent of inactivation is governed by the product of disinfectant concentration and exposure time. A more formal definition of Ct (which is a representation of a chemical disinfectant dose) is the time-integral of the disinfectant concentration over the period of exposure:

$$Ct = \text{disinfectant dose} = \int_0^{\tau} C(t) \cdot dt \tag{8.30}$$

where:

$C(t)$ time-dependent disinfectant concentration
τ period of exposure.

The Chick-Watson model implies a log-linear (i.e. 1st-order) kinetic response in disinfection. However, important deviations from 1st-order behaviour are often observed in empirical measurements of microbial inactivation kinetics. Two common deviations from Chick-Watson behaviour are the existence of a 'shoulder' and the presence of 'tailing' in the response of a microbial population to disinfectant exposure (see Figure 8.9). The shoulder response has been attributed to the influence of repair mechanisms and the ability of some microbes to withstand a nominal dose of disinfectant exposure without demonstrating a measurable loss of viability or infectivity. Tailing has been attributed to incorporation of microbes into particles, which are then shielded from disinfectant exposure. Tailing may also be caused by phenotypic heterogeneity within a microbial population (Pennell *et al.,* 2008). Regardless of the cause, shouldering and tailing phenomena are evident in measurements of the responses of many microbial populations to disinfectant exposure.

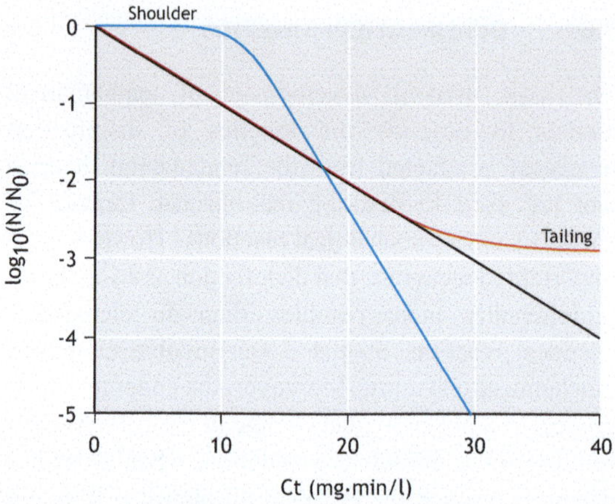

Figure 8.9 Schematic illustration of disinfection kinetics as predicted by the Chick-Watson model (solid black line). Also included are common deviations from Chick-Watson behaviour, including the existence of a 'shoulder' (blue line) and 'tailing' behaviour (red line).

Accurate simulations of disinfection process behaviour require the use of kinetic models that accurately simulate the responses of microbial populations. This need has motivated the development of mathematical models of disinfection kinetics that are consistent with observed responses of microbial populations.

The Collins-Selleck model was developed to account for shoulder and tailing behaviour (Selleck *et al.,* 1978). The model itself has the following mathematical form:

$$\frac{dN}{d(Ct)} = -k \cdot N \tag{8.31}$$

$$k = 0 \text{ for } Ct \leq \tau$$

$$k = \frac{k_0}{b \cdot Ct} \text{ for } Ct > \tau$$

The Collins-Selleck model incorporates two notable modifications of the Chick-Watson model. First, the independent variable is Ct, not time. Second, the rate parameter (k) is allowed to vary. It is assigned a value of zero for disinfectant doses below a threshold value (τ); it is assumed that no microbial inactivation takes place for disinfectant doses below this value. Furthermore, the rate parameter is assumed to decrease as Ct increases. Separation of variables and integration of (8.31) yields an algebraic form of the Collins-Selleck model:

$$\int_{N_0}^{N} \frac{dN}{N} = -\frac{k_0}{b} \int_{\tau}^{Ct} \frac{d(Ct)}{Ct} \tag{8.32}$$

$$\ln\left(\frac{N}{N_0}\right) = -\frac{k_0}{b} \cdot \ln\left(\frac{Ct}{\tau}\right)$$

or

$$\frac{N}{N_0} = \left(\frac{Ct}{\tau}\right)^{-k_0/b}$$

Figure 8.10 provides a graphical illustration of the Collins-Selleck model. For disinfectant doses below the threshold dose ($Ct = \tau$), no change in the concentration of viable or infective microbes is observed. For disinfectant doses beyond the threshold, the concentration of viable or infective microbes decreases monotonically, but the rate of decline reduces as Ct increases.

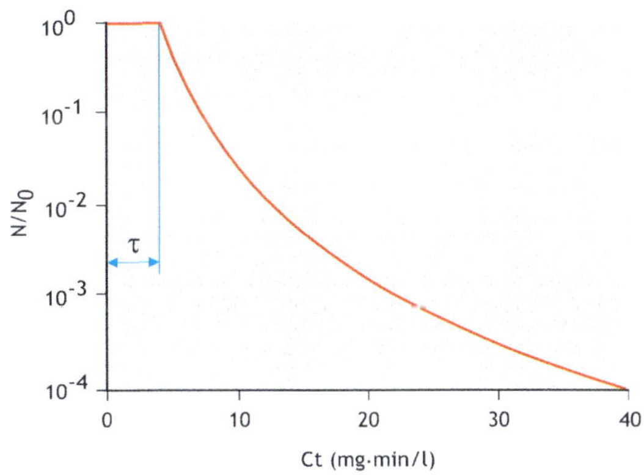

Figure 8.10 Graphical illustration of Collins-Selleck model for chlorine-based disinfection, based on a threshold dose (τ) of 4 mg·min/l and $k_0/b = 4$.

The Collins-Selleck model has been used to fit empirical data for inactivation of indicator bacteria by chemical disinfectants (Hassen *et al.*, 2000; Selleck *et al.*, 1978). Similar logic has allowed application of the Collins-Selleck model to describe disinfection kinetics associated with UV irradiation (Brahmi *et al.*, 2010). However, there are many other models that have been developed to account for deviations from Chick-Watson behaviour. For example, the Hom model has been widely used to simulate disinfection kinetics in wastewater applications. The model takes the basic form:

$$\frac{dN}{dt} = -m \cdot k \cdot C^n \cdot t^{m-1} \cdot N \tag{8.33}$$

where m and n are empirical, dimensionless parameters that are selected (by regression analysis) to fit the model to empirical data. When $m = 1$, the Hom model reduces to Chick-Watson. For $m < 1$, the Hom model demonstrates tailing, while for $m > 1$ the model is consistent with the presence of a shoulder.

In all models of chemical disinfection, the ability to account for decay of the disinfectant or disinfectant 'demand' is helpful. Incorporation of a term to account for 1st-order disinfectant decay yields a closed-form solution to describe disinfection kinetics (i.e. in a batch reactor) (Haas and Joffe, 1994). The existence of such a closed-form solution can be helpful in simulations of disinfection process performance.

8.6.2 Disinfection kinetics: UV irradiation

The third law of photochemistry indicates that the extent of a photochemical reaction is governed by the dose of radiation delivered to the target. By extension of this principle to disinfection, it is evident that the extent of inactivation that is observed within a microbial population is governed by the UV dose to which the population is exposed. As such, UV dose becomes the master variable in governing UV disinfection processes.

The logic used in describing UV disinfection kinetics is adapted from the methods used to describe the kinetics of a pure photochemical process. The simplest of the models that is used to describe UV disinfection kinetics is the single-event model, which assumes that a single photochemical event within a critical biomolecule of a target microbe is sufficient to cause its inactivation. The single-event model, which is analogous to the Chick-Watson model, takes the form:

$$\frac{dN}{dt} = -k \cdot E_i \cdot N \qquad (8.34)$$

where,

N	concentration of viable or infective microbes
t	time (s)
k	inactivation constant (cm^2/mJ)
E_i	incident fluence rate (mW/cm^2).

Separation of variables and integration yields an algebraic form:

$$\ln\left(\frac{N}{N_0}\right) = -k\int_0^{\tau} E_i \cdot dt \qquad (8.35)$$

where:

N_0	concentration of viable or infective microbes prior to UV exposure
τ	period of exposure.

The integral on the right-hand side of equation 8.35 is the dose of UV radiation to which the microbial population has been exposed. Implicit in this derivation is that all microorganisms within the target population receive the same dose, a condition that is closely approximated in laboratory experiments that involve shallow, well-mixed batch reactors placed under a collimated beam of UV radiation. In explicit terms, the formal definition of UV dose is:

$$\text{UV dose} = \int_0^{\tau} E_i \cdot dt \qquad (8.36)$$

The single-event model suggests that a plot of $\ln(N/N_0)$ vs. UV dose should yield a straight line through the origin with a slope of $-k$. This is analogous to behaviour that is predicted by the Chick-Watson expression, except that the independent variable is the UV dose instead of the chemical disinfectant dose (i.e. Ct). As with chemical disinfection, the single-event model is useful for simulation of the inactivation responses of some simple microorganisms, including some viruses and bacteria, over limited extents of inactivation. However, as with chemical disinfection, common deviations from single-event behaviour are observed in empirical data, including the existence of a shoulder or tailing.

The series-event model was developed to account for the presence of a shoulder in the response of a microbial population to UV exposure. The logic behind this model is that inactivation of higher organisms may require multiple units of damage. The series-event model was developed under the assumption that increments of UV-induced damage accumulate as a result of independent, discrete, sequential photochemical events. The model further hypothesizes that a microbe may be able to withstand some level of damage before loss of viability is evident. In other words, microbes are assumed to have a threshold for UV-induced damage, below which they will retain viability. The logic of the series-event model is described below:

$$M_0 \xrightarrow{h\nu} M_1 \xrightarrow{h\nu} \cdots M_{n-1} \xrightarrow{h\nu} M_n \xrightarrow{h\nu} M_{n+1} \cdots \quad (8.37)$$

where M_i represents a microbe that has accumulated i units of damage. Transformation of a microbe from one unit of damage to the next is assumed to take place at a rate that is described by the single-event model (Equation 8.34). A threshold for viability is assumed to exist at n units of damage. Therefore, all microbes with n-1 units of damage or less are assumed to retain viability. By summing the concentrations of all microbes with n-1 units of damage or less as a function of UV dose, it is possible to define the series-event dose-response model (Severin *et al.*, 1983, 1984):

$$\frac{N}{N_0} = \exp(-k \cdot D) \sum_{i=0}^{n-1} \frac{(k \cdot D)^i}{i!} \quad (8.38)$$

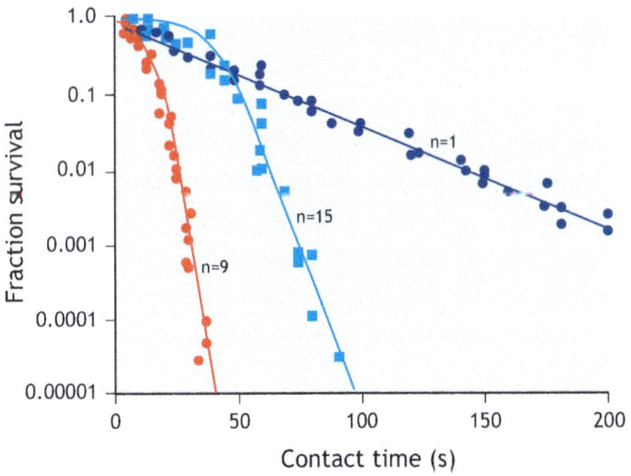

Figure 8.11 Illustration of the series-event model for simulation of UV dose-response behaviour of a bacterium, *E. coli* (•), a fungus, *Candida parapsilosis* (▪), and a bacterial virus, bacteriophage f2 (•) (Severin *et al.*, 1983).

Figure 8.11 illustrates the application of the series-event model for description of the UV dose-response behaviour of three aquatic microbes: a bacterium, a fungus, and a virus (bacteriophage). In the case of the virus, the smallest and simplest of these microbes, a threshold value of n=1 was observed. When n=1, the series-event model reduces identically to the single-event model, yielding 1st-order behaviour. For the more complex organisms, a larger threshold value was observed, as evidenced by a 'shoulder' in the dose-response behaviour of these more complex microbes.

Another common feature of UV dose-response behaviour is tailing, wherein the rate of inactivation slows dramatically. Pennell *et al.* (2008) developed the Phenotypic Persistence and External Shielding (PPES) model to account for this behaviour. The model was developed around the hypothesis that a microbial population could be described as a set of two sub-populations. One of the sub-populations was assumed to be relatively sensitive to microbial inactivation. The behaviour of this sub-population was described using the series-event model, which allowed for a shoulder to be present in the data. The second sub-population was assumed to display greater resistance to UV exposure than the first, resulting in relatively slow inactivation; the dose-response behaviour of this sub-population was also described by a series-event model, but in most cases there are insufficient data to support a threshold value other than n=1 for this subpopulation. Resistance to UV among the second sub-population is assumed to be attributable to the presence of particles (i.e. shielding from UV exposure) or phenotypic variation (i.e. the existence of an inherently resistant fraction of the population). Both factors will yield UV dose-response behaviour that differs from that of the sensitive sub-population, though it is usually difficult to define which of these factors is actually responsible for resistance. Mathematically, the PPES model is a two-population model that can be used to account for the existence of shouldering and/or tailing in a microbial dose-response data set:

$$\frac{N}{N_0} = (1-p) \cdot \exp(-k_1 \cdot D) \sum_{i=0}^{n-1} \frac{(k_1 \cdot D)^i}{i!} + p \cdot \exp(-k_2 \cdot D) \quad (8.39)$$

where:

p — fraction of microbial population that expresses resistance to inactivation

k_1 — inactivation constant for sensitive sub-population

k_2 — inactivation constant for resistant sub-population.

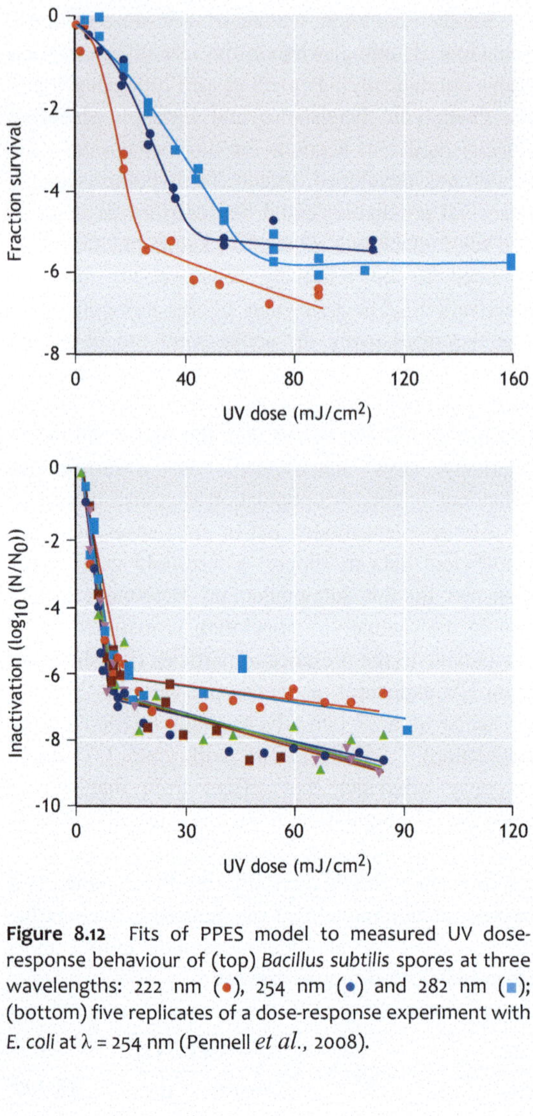

Figure 8.12 Fits of PPES model to measured UV dose-response behaviour of (top) *Bacillus subtilis* spores at three wavelengths: 222 nm (•), 254 nm (•) and 282 nm (▪); (bottom) five replicates of a dose-response experiment with *E. coli* at λ = 254 nm (Pennell *et al.*, 2008).

Figure 8.12 illustrates the application of the PPES model for description of the UV dose-response behaviour among two microbial populations. Populations of spores, such as those that are produced by *Bacillus subtilis*, often display shouldering. At large extents of inactivation, these spore populations often demonstrate tailing. The PPES model is able to account for both of these behaviours (see Figure 8.12, top panel). A somewhat simpler response is seen among vegetative cells of *E. coli*, which often lack an obvious shoulder. The PPES model is able to account for that behaviour as well (see Figure 8.12, bottom panel).

8.6.3 Comparisons of disinfection kinetics among common disinfectants

The ability of disinfectants to inactivate microorganisms depends on the structure of the microorganism, the conditions of application, and the mechanisms by which the disinfectant accomplishes microbial inactivation. Therefore, the effectiveness of disinfectants will be variable among target microbes and the conditions of application. However, broadly speaking, it is possible to assess inactivation efficacy among disinfectants. In quantitative terms, this is often done by measuring the fraction of a microbial population that is inactivated as a result of exposure to a quantifiable dose of a disinfectant. As an illustration, Table 8.2 provides representative values of disinfectant doses that have been reported to accomplish 3.0 $\log_{10}$ units of inactivation of various microorganisms that are often targets of disinfection design.

The four microorganisms that are listed in Table 8.2 are important indictors of microbial quality in municipal wastewater, particularly as related to disinfection processes. *E. coli* is a commonly used indicator organism that often forms the basis for regulation of effluent microbial quality. As such, disinfection systems are often designed to control *E. coli* so as to comply with discharge permit limitations. The oocysts of *C. parvum* are the resting stage of this protozoan parasite. *C. parvum* is responsible for a large fraction of waterborne disease worldwide, so control of this pathogen is relevant for wastewater systems, especially when human contact with treated water is expected, such as when water is discharged to a recreational area or when the effluent is used in irrigation of crops. The human norovirus (HNV) is responsible for a large fraction of waterborne gastrointestinal illness (Wobus *et al.*, 2006). Although a method to culture HNV has been developed (Ettayebi *et al.*, 2016), it has been difficult for others to replicate. In the absence of a practical culture-based method to grow a virus, it is not possible to quantify disinfection kinetics. As a result, the murine norovirus (MNV) has been studied as a surrogate for HNV. MNV shares many biochemical and genetic characteristics of HNV and culture-based methods are available to propagate MNV (Wobus *et al.*, 2006). Consequently, MNV has been used as a surrogate for HNV in experiments to quantify the kinetics of inactivation with various disinfectants. Coliphage MS2 is another viral surrogate that is commonly used in experiments of this type. Like all coliphage, MS2 is parasitic toward bacterial cells (coliforms) and does not express pathogenicity toward human cells, which simplifies their use in experiments. In addition, coliphage are common in wastewater effluents. MS2 is single-stranded RNA virus that is structurally similar to some human viruses. Methods are available to propagate MS2 that are relatively simple, inexpensive, and rapid to apply. Also, MS2 has been shown to be relatively resistant to most common disinfectants. Collectively, these characteristics motivate the use of MS2 as a viral surrogate and also as possible indicator organism.

The data presented in Table 8.2 reveal several trends that are important for these commonly applied disinfectants. HOCl is effective for control of most vegetative bacterial cells and most viruses. However, it is remarkably ineffective for control of some protozoan parasites, especially *C. parvum*. NH_2Cl is also effective as a disinfectant, but often requires 1-2 orders of magnitude larger dose than HOCl to accomplish the same extent of inactivation. PAA generally requires slightly larger doses than NH_2Cl to accomplish a given level of microbial inactivation, and tends to show similar relative effectiveness against the range of target microbes listed in Table 8.2 as HOCl or NH_2Cl. All three of the chemical disinfectants listed in Table 8.2 are effective for control of vegetative bacteria and viruses at doses that can be delivered easily in practical applications; however, control of *C. parvum* is modest at chemical disinfectant doses that are used in most practical applications.

Table 8.2 Summary of disinfectant doses reported to yield 3.0 $\log_{10}$ units of inactivation for common disinfection target microorganisms. Chemical disinfectant doses are reported in units of mg·min/l. UV_{254} doses are presented in units of mJ/cm². References from which values were extracted are indicated in parentheses.

Disinfectant	Microorganism			
	Escherichia coli	*Cryptosporidium parvum* oocysts	Murine norovirus	Coliphage MS2
HOCl	0.10[1)]	5,300[4,A)]	0.179[7)]	0.142[7)]
NH_2Cl	6.9[1)]	14,000[5)]	11[8)]	655[8)]
PAA	80[2)]	n/a[B)]	73[8)]	609[8)]
UV_{254}	5.5[3)]	5.2[6)]	15[9)]	97[9)]

[1)]Chauret *et al.*, 2008; [2)]Rossi *et al.*, 2007; [3)]Sommer *et al.*, 2000; [4)]Murphy *et al.*, 2015; [5)]Rennecker *et al.*, 2000; [6)]Craik *et al.*, 2001; [7)]Lim *et al.*, 2010; [8)]Dunkin *et al.*, 2017; [9)]Weng *et al.*, 2018.

[A)]Inactivation responses for HOCl were estimated by assuming that HOCl was the only form of free chlorine that contributed to inactivation of *C. parvum* oocsysts.

[B)]Reports of the efficacy of peracetic acid for inactivation of *C. parvum* oocysts are widely variable in the literature. Peracetic acid is often reported to be similar to free chlorine for inactivation of *C. parvum* oocysts.

Figure 8.13 includes photomicrographs of several microorganisms that are relevant to wastewater disinfection processes. In addition to the organisms listed in Table 8.2 and described above, images are presented for *Enterococcus* bacteria and human norovirus. *Enterococcus* are often used as an indicator organism in wastewater disinfection processes where treated wastewater is discharged to the ocean or a marine setting. Included in these images are bars to indicate the relative sizes of these microbes.

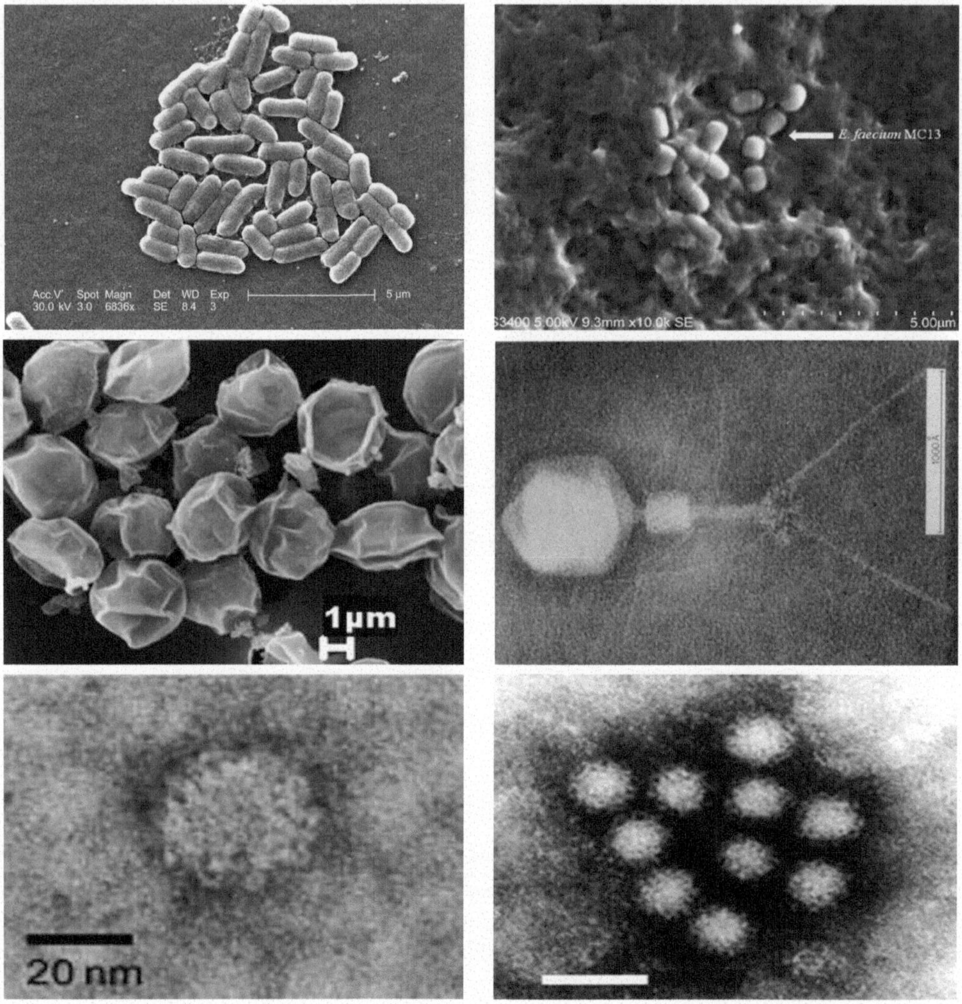

Figure 8.13 Escherichia coli (top left; https://www.ciriscience.org/ph_12-Escherichia-coli-O157H7-BandW; *Enterococcus* (top right, from purification and characterization of enterocin MC13 produced by a potential aquaculture probiont *Enterococcus* faecium MC13 isolated from the gut of Mugil cephalus, Canadian Journal of Microbiology (2011); *Cryptosporidium parvum* oocysts (centre left, from: Solar water disinfection (SODIS): A review from bench-top to roof-top, Article· Literature Review in Journal of Hazardous Materials, 235-236:29-46 · August 2012); Coliphage MS2 (centre right, from Almeida *et al.*, 2018); Murine norovirus particle, (bottom left, from Hsu *et al.*, 2005; human norovirus particle (bottom right, image from Wikipedia; white bar = 50 µm).

For disinfection systems based on UV_{254} irradiation, control of most microbes is accomplished at doses of 40 mJ/cm^2 or less. However, some viruses are recalcitrant to UV_{254} exposure. Inactivation of many viruses is improved through the use of radiation at wavelengths below (approximately) 240 nm, probably because of the ability of this short wavelength radiation to damage capsid proteins in virus particles. It is also notable that the range of UV_{254} doses required to accomplish control of the microbes listed in Table 8.2 is smaller than the corresponding ranges for the chemical disinfectants. This trend holds broadly for most microbes, which implies that design of a disinfection system based on UVC radiation for control of a target microbe is likely to accomplish broader microbial control than would be achieved with a comparable chemical disinfection system.

8.7 PROCESS MODELS

Like all reactor systems, process performance in disinfection reactors is governed by the combined effects of transport behaviour, internal mixing, and the intrinsic kinetics of the reactions of interest. In the case of disinfection systems, the reactions of interest will include disinfection itself, as well as reactions that result in decay or demand on the disinfectant, and production of disinfection by-products. Process models that properly integrate terms to account for each of these effects have been demonstrated to provide accurate predictions of disinfection process performance for essentially all disinfection systems.

8.7.1 Deterministic process models

Among the first models to provide comprehensive, integrated descriptions of disinfection process performance were those originally developed for simulation of disinfection efficacy in UV disinfection systems (Chiu *et al.*, 1999a, 1999b; Lyn *et al.*, 1999). These models provided detailed accounting of process behaviour by integrating the effects of fluid mechanics, including turbulence, as well as the spatial distribution of radiant energy and its effects of local disinfection rates within the reactor. This modelling approach was able to identify aspects of reactor design that limited reactor behaviour. In turn, identification of these limiting features allowed for improvements in reactor design that resulted in improvements in reactor process efficiency.

A similar, parallel effort emerged soon thereafter to provide comprehensive, integrated descriptions of the process efficacy in chemical disinfection systems. This parallel effort led to development of the Integrated Disinfection Design Framework (IDDF) which allows accurate, detailed, deterministic characterization of process performance in chemical disinfection systems (Bellamy *et al.*, 2000; Ducoste *et al.*, 2001).

Contemporary applications of these integrated modelling methods generally involve the use of computational fluid dynamics (CFD) to simulate fluid mechanics (including turbulence) and mixing behaviour. CFD simulations are used to solve the equations of motion and continuity, along with an appropriate turbulence closure model. When applied properly, CFD simulations can provide accurate information about fluid mechanics within reactor systems. With contemporary CFD software and computer hardware, it is possible to develop simulations that capture most of the relevant fluid mechanical behaviour in a disinfection reactor in a relatively short period of time, often as short as a few hours to a day or two of computer simulation time. Together with these simulations of fluid mechanics, many CFD packages allow for inclusion of equations to describe the behaviour of scalar quantities, such as the concentration of viable microorganisms, the concentration of a chemical disinfectant, the spatial distribution of radiant energy (i.e. a fluence rate field within a UV disinfection system), and the rates of reactions that are responsible for the formation of disinfection by-products.

As an aside, considerable improvements in analytical methods and numerical models for measurement and simulation of fluence rate fields in UV photoreactors have been seen in recent years. With regard to measurements, the Micro Fluorescent

Silica Detector (MFSD) allows for accurate, pointwise measurements of fluence rate distributions in UV photoreactors (Li *et al.,* 2011). This device collects these measurements in a manner that is consistent with the fundamental definition of fluence rate and can be applied in a wide range of physical settings. For numerical simulations, the Ray Tracing method has been implemented to provide detailed, accurate simulations of fluence rate fields in a wide range of UV reactor types (Ahmed *et al.*, 2018). The ability of ray tracing to provide accurate simulations of fluence rate fields is governed by the ability of these models to account for the optical phenomena that govern the fluence rate field, including divergence, absorption, reflection, and refraction; these models also account for the geometry of the system.

For chemical and biochemical constituents that are relevant to disinfection applications, accurate simulations of scalar, reactive quantities depend on availability of accurate information to describe the kinetics of reactions that lead to their formation or decay. In the case of a fluence rate field, it is also necessary to incorporate model terms to accurately account for the optical behaviour of the system (i.e. absorbance, dissipation, reflection, refraction) that governs the spatial distribution of radiant energy. When integrated properly, and when accurate information is available to describe other input parameters (*e.g.* flow rate, reactor geometry, reaction kinetics), these integrated models can provide accurate, deterministic predictions of process performance. As such, these numerical tools can be valuable in the design and optimization of disinfection systems.

Figure 8.14 provides an illustration of the ability of these integrated modelling approaches to simulate disinfection process performance in chlorine contact chambers. These model simulations were based on CFD simulations of fluid mechanics in chemical (chlorine) disinfection contact chambers with the same internal volume, but five different internal baffle configurations. Kinetic expressions for decay of chlorine, production of trihalomethanes (as an indicator of DBP formation), and inactivation of *Giardia lamblia* were included in the model to allow direct comparisons of the performance of each reactor geometry.

This approach to simulation of process performance represents a powerful tool for design of disinfection contact chambers and for optimization or modification of reactor design after installation. However, these modelling approaches yield deterministic predictions of process performance. Deterministic approaches can be sufficient for analysis of disinfection process performance in systems where input parameters do not demonstrate strong temporal variability. However, in some applications, including most wastewater applications, influent water quality and flow rate, as well as the kinetics of critical reactions, can demonstrate strong temporal variation. As such, it is appropriate to consider modelling approaches that account for the stochastic nature of these processes.

8.7.2 Probabilistic (stochastic) process models

Process performance in wastewater disinfection systems is affected by many variables, including flow rate and water quality. The effects of these factors on process performance can be accurately accounted for in deterministic models. By extension, if sufficient data are available to define variability in the input parameters that govern model output, then these modelling approaches can be applied in a stochastic sense to yield predictions of process performance that account for variability. An important advantage of stochastic modelling approaches is that the use of safety factors, which are often applied to address sources of uncertainty (variability) in process performance, can be reduced or eliminated. This allows for development of reactor designs that are simultaneously more efficient and more reliable than those that are developed based on deterministic modelling approaches.

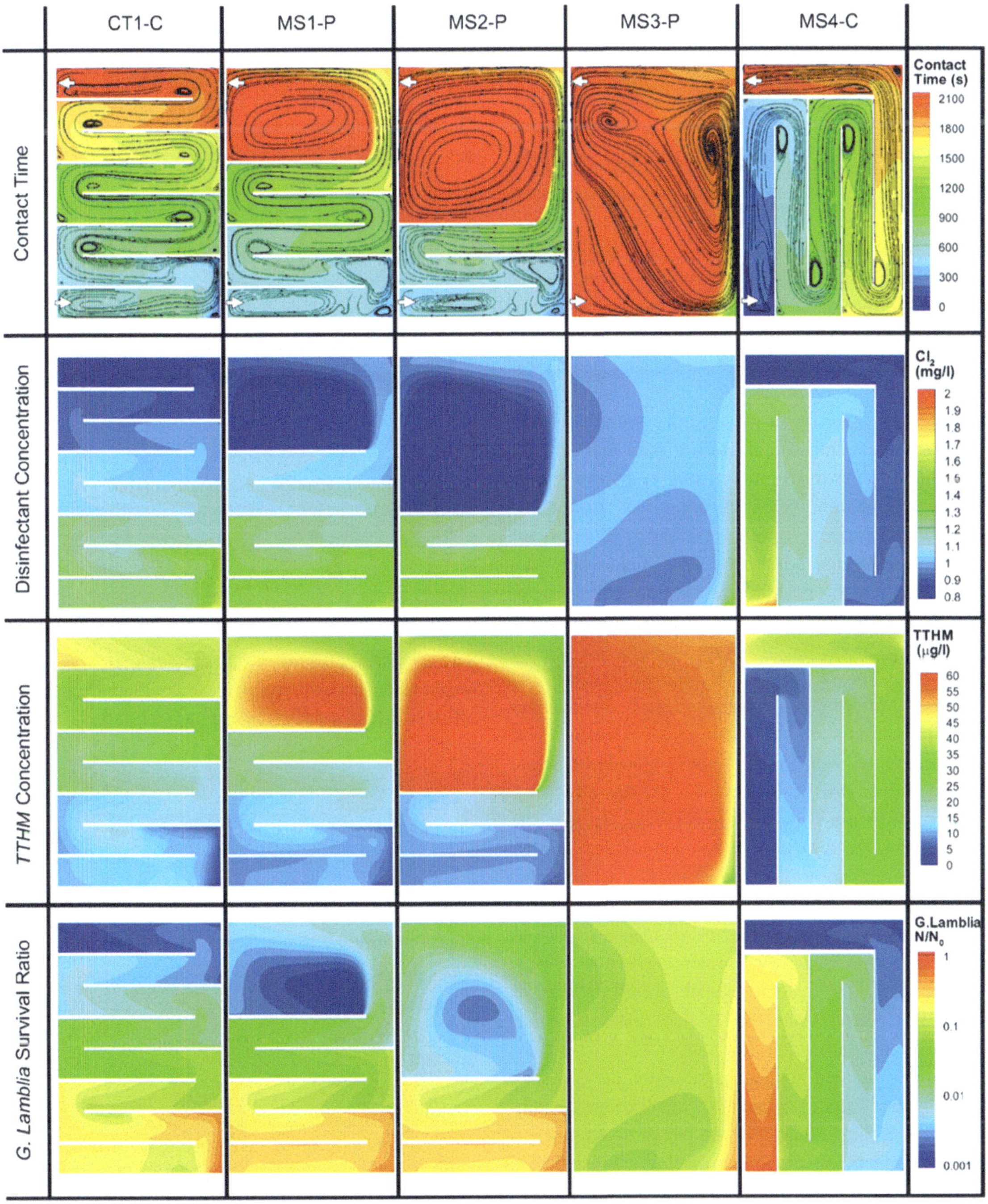

Figure 8.14 Simulations of chlorine-based disinfection process performance for a range of reactor designs based on CFD simulation integrated with terms to describe the kinetics of participating reactions. Colour schemes indicate calculated residence time (top row), disinfectant concentration (2nd row), total THM concentration, as an indicator of DBP formation (3rd row), and fractional survival of G. lamblia (bottom row), based on application of a CFD model (Angeloudis *et al.*, 2014).

Stochastic modelling approaches have been applied recently for analysis and prediction of process performance in disinfection systems that are used in wastewater treatment applications. As an illustration of such an approach, the method of Ahmed *et al.* (2019) for large-scale UV disinfection applications in municipal wastewater treatment is summarized below.

Several variables are known to affect process performance in UV disinfection systems, where process performance is defined by the concentration of viable or infective target organisms that leave in the treated effluent from the disinfection system, N. Factors that directly affect N include the concentration of viable or infective target organisms that enter the reactor, N_0, the intrinsic kinetics of their inactivation, and the UV dose distribution delivered by the reactor to the population of target organisms that move through the system. In turn, the UV dose distribution is directly influenced by the flow field (i.e. fluid mechanics) in the reactor and the fluence rate field. These aspects of the reactor are influenced by the reactor geometry, the flow rate through the system, lamp output power, and the UV transmittance of the water. All of the input variables described above can be measured or simulated accurately and independently. Some of these attributes demonstrate variability or can be controlled. Others are fixed. Lastly, some are dependent on other aspects of the system, but the direction of this dependence and its specific nature are well-defined.

Among the factors that affect UV disinfection process performance, flow rate, lamp power, UV transmittance, and reactor geometry/configuration all play roles in determining the UV dose distribution that is delivered by a reactor. For a given set of operating conditions, defined by these parameters, the UV dose distribution becomes essentially deterministic. On the other hand, the value of N_0 and the intrinsic kinetics of microbial inactivation display considerable variability, as illustrated in Figure 8.15. By extension, for a given operating condition, the vast majority of variability in process performance is attributable to these parameters that are specific to the microbial population and the water being treated.

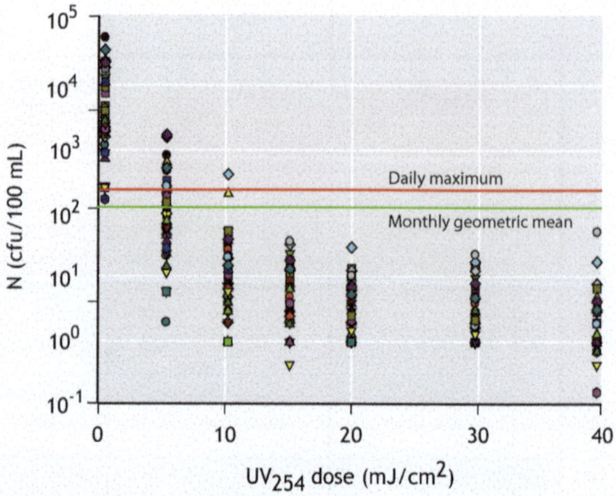

Figure 8.15 Replicate (n=46) measurements of intrinsic kinetics of *E. coli* inactivation by UV_{254} irradiation for effluent samples from a municipal wastewater treatment facility. Horizontal reference lines indicate discharge permit limitations for this facility based on a daily maximum and a monthly geometric mean. From Ortiz (2014).

Considering further the data presented in Figure 8.15, which represent measurements of UV_{254} dose-response behaviour for *E. coli* in 46 undisinfected effluent samples collected from a single facility over roughly a 12-month period, there are several important trends that emerge. First, the value of N_0 varies by roughly 3 $\log_{10}$ units over these samples. The value of N_0 has a direct influence on the concentration of viable *E. coli* that are present in the disinfected water (N). The individual data sets all show evidence of tailing. The PPES model (Pennell *et al.*, 2008) was judged to be consistent with these data sets; therefore, non-linear regression was applied to each of the data sets to develop estimates of the PPES model parameters (see equation 8.39). As such, this analysis yielded 46 independent sets of model parameters.

The data presented in Figure 8.16 also illustrate that a UV_{254} dose of 10-15 mJ/cm^2 will accomplish reliable compliance with discharge permit limitations for the facility from which the samples were collected. This is notable in that conventional design guidelines, such as the Ten States Standards (Managers, 2014), call for a nominal dose of 30 mJ/cm^2 or higher to be used in municipal wastewater applications of UV disinfection. The data presented in Figure 8.15 indicate that this would represent over-design by a factor of 2-3 for this facility.

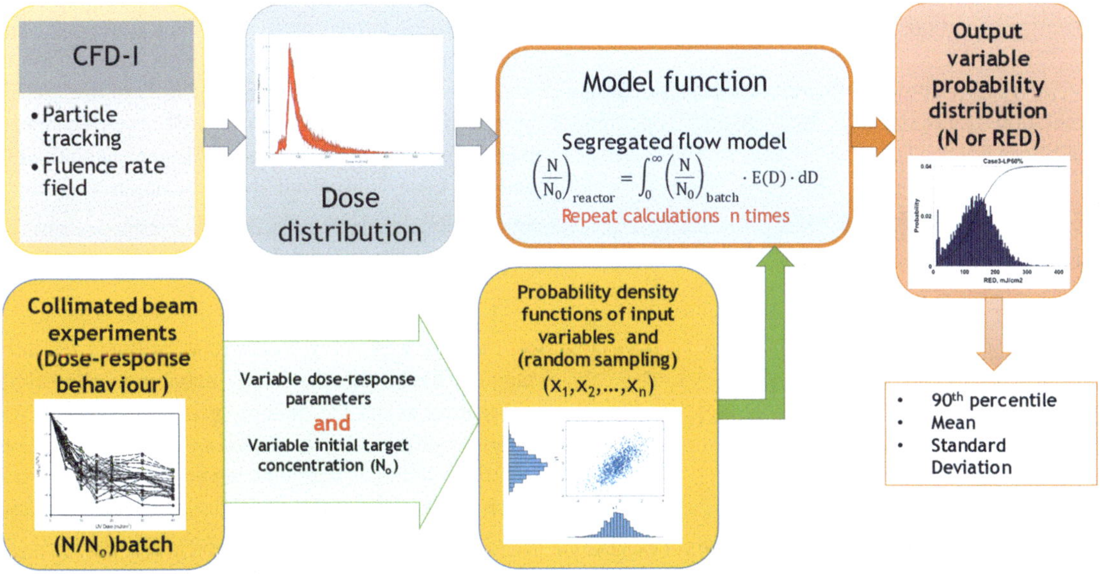

Figure 8.16 Schematic illustration of the Monte Carlo modelling approach used for simulation of UV disinfection at a large-scale municipal wastewater treatment facility (adapted from Ahmed *et al.*, 2019). For a given operating condition, as defined by reactor geometry, flow rate, lamp power, and UV transmittance, the dose distribution delivered by the reactor was calculated based on integrated application of computational fluid dynamics and a fluence rate field simulation (i.e. CFD-I model; upper left-hand side). For each of the dose-response experiments, a kinetic model was applied to develop estimates of dose-response model parameters and N_0 (lower left). Quasi-random sampling of model parameters and N_0 were applied for prediction of reactor performance based on the segregated-flow model. For each simulation, an estimate of N was developed. Collectively, these simulation results yielded estimates of process performance, including variability, for a given operating condition.

A Monte Carlo approach was applied to develop stochastic predictions of process performance of this system. Figure 8.16 provides a schematic illustration of this modelling approach. For any given operating condition, defined by reactor geometry, flow rate, lamp output power, and UV transmittance, the UV dose distribution delivered by the reactor was assumed to be deterministic. The UV dose distribution was calculated by integrated application of a CFD simulation for fluid mechanics and a simulation of the fluence rate field by the ray tracing approach (Ahmed *et al.*, 2018). The PPES model was fit to each of the data sets from the replicate UV dose-response experiments (*e.g.* Figure 8.15); this analysis yielded replicate sets of PPES model parameters and N_0. The segregated-flow model was applied for prediction of reactor performance using the UV dose-distribution and a set of input parameters, including N_0 and the PPES model parameters. Values of N_0 and PPES model parameters were selected by an approach that was quasi-random: for a given value of N_0, remaining model parameters were selected in a manner that accounted for the correlations among model parameters. By repeating this approach for a large number of samples, a statistical process known as 'bootstrapping', it was possible to develop a population of a reactor performance predictions for each operating condition that was evaluated. As such, this process allowed for an assessment of reactor performance, including variability, for a range of operating conditions (Ahmed *et al.,* 2019).

Figure 8.17 illustrates the results of this stochastic simulation approach for two distinct operating conditions.

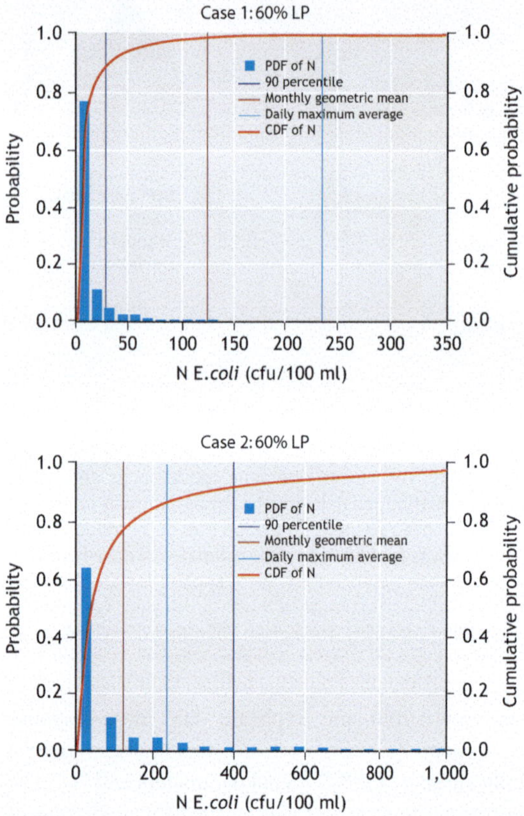

Figure 8.17 Summary of Monte Carlo simulation results for two operating conditions. Both cases are applied to a large-scale UV disinfection system based on Hg amalgam lamps oriented horizontally, parallel to the nominal direction flow. Other details of the operating conditions are given in the text of the original reference (Ahmed *et al.*, 2019).

For both cases, the reactor geometry was a large, open-channel UV disinfection system based on Hg amalgam lamps positioned horizontally, nominally parallel to the direction of flow; the lamps were assumed to be operated at 60 % of their peak output power. The system comprised 7 channels, each with two banks of lamps; each lamp bank comprised an 8x24 array of lamps (i.e. 192 lamps per bank). Case 1 involved a flow rate of 7,880 m^3/hr, resulting in an approach velocity of 0.15 m/s. All seven channels were operating, but only one bank per channel was active. Case 2 involved the same input variables, except that the flow rate was increased to 23,600 m^3/hr, resulting in an approach velocity of 0.47 m/s.

For Case 1, the 90th percentile of the simulation results was well below the regulatory limits for this facility, based on the daily maximum allowable concentration and the monthly geometric mean. For these conditions, the likelihood of a violation of the permit limitation is very low, but non-zero. However, discharge permit limitations for microbial quality often involve a statistical component, such as a monthly geometric mean. As such, the discharge permit limitations are consistent with the stochastic modelling approach. These results suggest that it may be possible to relax the operation of the system by reducing lamp power or reducing the number of channels that accept flow.

For Case 2, the 90th percentile of the simulation results was well above the permit limitations, suggesting that the likelihood of a permit limitation violation was high. For these circumstances, possible courses of action would include increasing lamp output power or increasing the number of operating banks per channel from 1 to 2.

The stochastic modelling approach illustrated in figures 8.15-8.17 allows quantitative analysis of reactor performance, including variability, for UV disinfection systems. This method of analysis can be applied to essentially any UV photoreactor or UV disinfection system, but is probably best suited for application to large-scale systems, where capital and operating costs associated with process uncertainty are higher than with small systems, and where application of such an approach can be justified economically.

This probabilistic method of analysis is not limited to UV disinfection systems. In fact, a conceptually similar Monte Carlo approach has been developed and applied for analysis of process performance in chemical disinfection systems based on peracetic acid (Santoro *et al.,* 2015). In this stochastic analysis, CFD was applied for simulation of fluid mechanics. For each flow rate, the CFD results were interrogated to yield a population of particle trajectories through the reactor. A model was developed to describe the decay of PAA through the reactor; this model was superimposed on the particle trajectories to yield a population of PAA doses for each operating condition. As with the stochastic model of UV disinfection, the authors observed that the dose distribution was essentially deterministic for a given operating condition. The stochastic nature of the process was attributed to variability in disinfection kinetics. Also as with the stochastic model of UV disinfection, the stochastic model of disinfection by PAA yielded estimates of process performance as a function of operating condition, including variability. This modelling approach allows designers and operators of these systems to make informed decisions about the operation of their system to meet treatment objectives.

8.8 DISINFECTION APPLICATIONS IN WASTEWATER TREATMENT

The design and implementation of disinfection systems in a wastewater treatment context are influenced by the quality and flow rate of fluid that enters the system, as well as variability in these parameters and the requirements for microbial quality in the disinfected product water. The quality of water entering a disinfection system will be influenced by the uses of water in the community that is served by the treatment facility (*e.g.* residential, commercial, industrial), as well as the treatment processes that precede the disinfection system. Flow rate will also be influenced by the size of the community and the water use patterns within the community. In communities that have combined sewers, the flow rate and water quality entering the facility can display dramatic changes during runoff events.

The requirements for microbial quality in disinfected water will be influenced by the intended use of the finished water. In cold-weather locations, some facilities have no requirement for microbial quality. In locations where finished water is discharged to a receiving stream that has a low risk of direct human contact, permit limitations are often based on indicator bacteria groups, such as E. *coli,* thermotolerant coliforms, or Enterococcus. These permit limitations may include a daily maximum concentration of viable indicator bacteria, a limitation based on a periodic (*e.g.* monthly) geometric mean concentration of a viable indicator bacterium, or both. In circumstances where disinfected effluent is used for irrigation, the requirements for treatment are likely to be more stringent. In the United States, Federal regulations have not been defined for reuse applications; therefore, regulation of reuse applications is the responsibility of each state. An example of the regulatory framework for wastewater reuse has been defined by the State of California through regulations in Titles 17 and 22 of the California Code of Regulations (CSWRC Board, 2015). Table 8.3 provides a brief summary of these regulatory requirements for various reuse applications in California.

Several treated water categories are listed in Table 8.3. The regulations that govern treatment and use are aimed at minimizing risks of human exposure to microbial pathogens in these applications. The requirements for treatment become more stringent as the level of human contact with the treated water increases. For example, 'Disinfected Secondary-23 Recycled Water' may be used for irrigation of areas where no food crops are grown and minimal human contact is involved, such as freeway landscaping and cemeteries. In contrast, 'Disinfected Tertiary Recycled Water' can be applied in a wide range of applications, including irrigation of food crops, parks, and playgrounds.

Table 8.3 Summary of reuse regulations defined by California State Water Resources Control Board (CSWRC Board, 2015).

Reuse category	Treatment minima		Disinfection		
	Oxidation[1)]	Filtration[2)]	Total coliforms		Other requirements
Disinfected Secondary-23 Recycled Water	✓		$< 23\frac{\text{MPN}}{100\text{ ml}}$	7-d median	
			$< 240\frac{\text{MPN}}{100\text{ ml}}$	No more than one sample in a 30-d period	
Disinfected Secondary-2.2 Recycled Water	✓		$< 2.2\frac{\text{MPN}}{100\text{ ml}}$	7-d median	
			$< 23\frac{\text{MPN}}{100\text{ ml}}$	No more than one sample in a 30-d period	
Disinfected Tertiary Recycled Water	✓	✓	$< 2.2\frac{\text{MPN}}{100\text{ ml}}$	7-d median	Chlorine[3)]
			$< 23\frac{\text{MPN}}{100\text{ ml}}$	No more than one sample in a 30-d period	Other disinfectants[4)]
			$< 240\frac{\text{MPN}}{100\text{ ml}}$	Always	

[1)]'Oxidized wastewater' is defined as wastewater in which organic matter has been stabilized, is nonputrescible, and contains dissolved oxygen.
[2)]'Filtered wastewater' is oxidized wastewater that has also been subjected to one or more physical separation process, which may include granular media filtration and/or membrane-based processes.
[3)]Minimum conditions required for chlorine-based disinfection include disinfectant exposure ≥ 450 mg·min/l and contact time of at least 90 minutes under peak dry-weather flow conditions.
[4)]When alternatives to chlorine disinfection are applied, such as UV irradiation, they must be validated to achieve at least 5 $\log_{10}$ units inactivation of coliphage MS2, poliovirus, or an alternative challenge virus that is at least as resistant as poliovirus.

The performance of disinfection systems is governed by the efficiency by which the disinfecting agent is delivered to the target microbes. However, design and implementation of disinfection systems should involve detailed characterization of the water to be disinfected. This is best accomplished by regular sampling of the water source, with measurements of water quality parameters that will affect disinfection process performance. For chemical disinfection systems, these will include measurements of disinfectant demand, which is often attributable to reduced compounds that are dissolved in the water source. The presence of particles will also affect disinfection processes, as suspended matter will provide shelter for microbes in water; tailing behaviour in disinfection kinetics is often attributed to the effects of suspended particles in water. For UV disinfection, water quality measurements should include UV transmittance at a wavelength corresponding to the UV source that is expected to be used. Often this means measurements of UV transmittance at 254 nm, as this wavelength corresponds to the dominant output line of low-pressure Hg lamps.

8.8.1 Chemical disinfection systems

Chemical disinfection systems generally involve a contact chamber that is designed to approach plug-flow conditions, which corresponds to the theoretical upper limit on reactor efficiency. These conditions are appropriate in disinfection systems because they are often expected to yield overall inactivation performance in the order of 3-4 $\log_{10}$ units or more, which corresponds to much higher reactant conversion than is required in almost any other wastewater treatment process. Conditions that favour the approach to plug flow are long, slender reactor vessels, such as circular pipes or rectangular channels.

Figure 8.18 illustrates plan views of two chemical disinfectant contact chamber types that have been used in wastewater settings. These two examples have the same overall footprint, but different internal baffle structures. Both configurations employ a serpentine structure for internal baffles. This geometry is employed to provide a long, narrow channel while also taking advantage of shared internal walls as a means of reducing construction costs and providing a reactor shape that is generally easier to fit into the overall footprint of a wastewater treatment facility. In both cases, the disinfectant chemical is injected at the upstream end of the contract chamber, usually in an area that is characterized by aggressive mixing, so as to promote uniform distribution of the disinfectant. In chlorination systems, dechlorination is often required; this is accomplished by injection of a dechlorinating agent, usually in the form of an S(IV) compounds, at the effluent end of the reactor. The dechlorinating agent should also be introduced into an aggressively-mixed area so as to promote uniform distribution within the effluent. Dosing of the dechlorinating agent is often conducted based on real-time measurements of flow rate and residual chlorine concentration in the effluent. It is common to apply the dechlorinating agent in slight stoichiometric excess of the requirement (roughly 10% excess) to promote dechlorination. The reactions that lead to dechlorination with S(IV) compounds [see reactions (8.15) and (8.16)] are rapid, so if the S(IV) compound is mixed completely with the effluent, essentially complete dechlorination can be accomplished reliably. Also included in Figure 8.18 are dominant paths of fluid flow through these systems as well as areas that tend to be characterized by poor mixing. These 'dead zones' have a negative effect on overall process performance in disinfection systems. The geometry at the top of Figure 8.18 is generally preferred in chemical disinfection systems in that it tends to be more hydrodynamically efficient. Several factors contribute to this behaviour, including the presence of longer runs and fewer turns within the system. As a rule of thumb, chemical disinfectant contact chambers should be designed such that each pass through the reactor has an aspect ratio (length:width) that approaches 20:1, with two or more passes in the system.

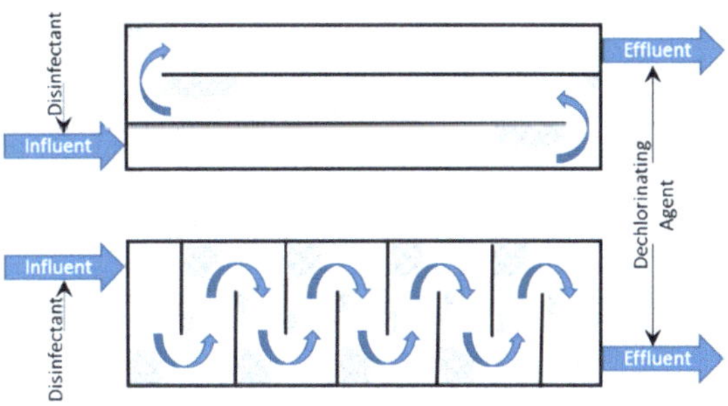

Figure 8.18 Schematic (plan) illustrations of chemical disinfectant contact chamber designs. Arrows are intended to illustrate dominant paths of fluid flow through the system. Stippled areas indicate zones that are generally characterized by poor mixing and circulation, sometimes referred to as 'dead zones', which are parts of these systems that diminish process performance.

The requirements for contact time with chlorine and PAA are similar, so conversion from chlorine-based disinfection to PAA is often accomplished by retrofitting of PAA injection hardware to a system that had been previously used for chlorination. More generally, the geometric characteristics of contact chambers that are used for chlorine- or PAA-based disinfection of wastewater are essentially identical. Typical mean contact times for these systems range from 30-120 minutes, depending on the requirements of the application and local regulatory constraints. Overall contact chamber size should be based on time-series measurements of flow rate to be imposed on the system. Measurements of flow rate should be conducted over a period of time long enough to allow for quantification of flow rate variations that are to be expected in the system.

As indicated in the expressions that are used to describe disinfection kinetics with chlorine or PAA, the disinfectant dose (Ct) represents the master variable that governs disinfection efficacy. As such, a relatively short contact time can be counteracted by a relatively high disinfectant concentration. This relationship can be used to inform a control strategy for the disinfection system. For example, under low-flow conditions, a long residence time in the contact chamber will be observed, which will allow for a relatively low disinfectant concentration to be maintained. Conversely, under high-flow conditions, contact time will be relatively low, so the concentration of disinfectant may need to be increased to maintain adequate disinfection efficacy.

8.8.2 UV disinfection systems

The development of a UV disinfection system design should involve characterization of the history of flow rate and UV transmittance (UVT) for the water to be treated. The flow rate record will provide information to define the range of flow rates to be treated by the system. This information is critical to design as there is generally a requirement for the disinfection system to effectively inactivate target microbes, even under high-flow conditions. Similarly, a historical record of UVT will provide information to indicate the range of conditions that will need to be addressed for system design. Ideally, any correlation between flow rate and UVT should also be identified, as this can play a relevant role in the development of a control algorithm for the system.

Process performance in UV disinfection systems is governed by the UV dose distribution delivered by the reactor to the microbial targets. In turn, the UV dose distribution will be governed by the internal geometry of the reactor, lamp output power, and other parameters related to the reactor itself. Unlike chemical disinfection systems, the actual design of the lamp modules is usually conducted by the manufacturers of UV disinfection systems. As such, the design of an in-place UV disinfection reactor will involve collaboration between the manufacturer, the design engineer, and end-users.

Most UV disinfection systems used for treatment of municipal wastewater involve a modular design and open-channel configurations. Figure 8.19 illustrates two examples of open-channel configurations for UV disinfection systems used in treatment of municipal wastewater: systems with horizontal lamps (parallel to the nominal direction of fluid flow) and systems with diagonally-oriented lamps.

Figure 8.19 Schematic illustrations of two common open-channel UV disinfection system configurations. The top image is an open-channel system with a horizontal lamp configuration, where lamps are oriented parallel to the nominal direction of flow. Also included in this image is an illustration of a liquid-level control device, located downstream of the irradiated zone. The bottom image is a system with lamps oriented diagonally to the nominal flow direction. Liquid-level control will also be required in this system, but is not illustrated in this image (Images: courtesy of Trojan Technologies).

Open-channel systems can be simpler to maintain than closed-vessel systems, in part because they offer easy access to lamp modules. However, a drawback of open-channel devices is that the free surface in the reactor will follow the energy grade line of the system. As such, local energy losses due to head loss will translate to local changes in free surface elevation. This behaviour is complicated somewhat by the variability in flow rates that is often seen in municipal wastewater treatment applications. As a result, open-channel systems will include liquid-level control devices, located downstream of the irradiated zone. These often take the form of a flap gate or an elongated weir.

The modular design that is typical of open-channel UV disinfection systems also facilitates the use of similar reactor geometries over a wide range of operating conditions. For large systems in particular, it is common to build systems with multiple channels, where each channel will contain two or more banks of lamps. Control algorithms for these devices are then implemented to activate channels and/or banks within a channel depending on flow rate, UVT, and perhaps lamp power, based on a given treatment objective. 'Flow pacing' algorithms have been developed to activate channels and banks, and sometimes to call for lamp 'dimming' based on real-time measurements of flow rate and UVT.

8.9 FUTURE DIRECTIONS

Contemporary modelling tools allow for detailed, accurate predictions of disinfection process performance. These tools are generally based on integrated applications of computational fluid dynamics, kinetic expressions to define the rates of disinfectant decay, target microorganism inactivation, and DBP formation, with information regarding water quality and flow rates imposed on the reactor. The recent adaptations of these methods to include stochastic simulations are potentially important advancements that could yield simultaneous improvements in process performance and reliability. These stochastic models have been applied for chemical and UV-based disinfection systems and are particularly relevant in large-scale applications, where capital and energy costs are likely to be highest.

Conventional designs of disinfection systems for wastewater treatment have focused on the use of a single disinfectant. However, the characteristics of disinfectants may allow them to be used more effectively in combination than by themselves. A potentially important example involves the use of UV disinfection for common operating conditions, with inclusion of PAA or another chemical disinfectant to address demanding operating conditions, such as high flow rate or poor water quality.

Historically, the regulatory focus of disinfection systems has been on the concentration of indicator bacteria. However, most vegetative bacteria tend to be easy to inactivate and may not be good indicators of the threats to human health posed by microbial pathogens in wastewater. Interest in the fate of viruses in disinfection systems is increasing, given that they are responsible for the majority of waterborne illness, and some viruses are resistant to disinfectants. It is likely that interest in this area will continue to increase going forward. One area of specific interest is developing assays that allow for assessment of the risks posed by human norovirus, which is responsible for a large fraction of human disease related to water. Possible approaches may include development of a surrogate or indicator virus or development of an assay to measure infective norovirus in water samples. Either approach could result in improvements of our understanding of the behaviour of the human norovirus in treatment systems, which in turn could allow for development of better disinfection system designs.

To date, most practical applications of UV disinfection systems have been based on low-pressure or medium-pressure mercury lamps. However, these UV sources are available only in a narrow range of geometries. Also, political and public pressure to decrease the use of mercury may make it difficult to rely on these UV sources in the future. Among the alternatives to mercury lamps are UV LEDs and excimer lamps. Both of these UV source types offer the potential to select an output wavelength range that is optimized to a given application, while also allowing flexibility in source geometry. As demand for these alternative UV sources increases, their costs of production are likely to decrease, while their output power and wall-plug efficiency are likely to improve. As such, it is likely that UV disinfection systems based on these alternative sources will become more common in the future.

8.10 FINAL REMARKS

Disinfection of municipal wastewater is practiced with the goal of reducing the risk of human exposure to waterborne microbial pathogens. Common process options used today include chlorination (usually coupled with dechlorination), application of peracetic acid, and UV irradiation. All three approaches offer effective control of waterborne bacteria, but may have limited efficacy against other microbial pathogen groups, such as protozoan parasites or viruses. Process modelling can facilitate the design of disinfection systems, allowing for development and implementation of systems that are reliable and efficient. New disinfection strategies are likely to emerge as new microbial pathogens and new disinfectants are identified. Some of these systems may involve combinations of disinfectants that have previously been used alone.

REFERENCES

Ahmed Y.M., Jongewaard M., Li M.K. and Blatchley III E.R. (2018). Ray Tracing for Fluence Rate Simulations in Ultraviolet Photoreactors. *Environ. Sci. Technology*, 52(8), 4738-4745.

Ahmed Y.M., Ortiz A.P. and Blatchley III E.R. (2019). Stochastic Evaluation of Disinfection Performance in Large-Scale Open-Channel UV Photoreactors. *J. Environ. Eng.* 145(10).

Almeida G.M.F., Leppänen M., Maasilta I.J. and Sundberg L.R. (2018). Bacteriophage imaging: past, present and future, *Research in Microbiology*, 169(9), 488-494.

Angeloudis A., Stoesser T. and Falconer R.A. (2014). Predicting the disinfection efficiency range in chlorine contact tanks through a CFD-based approach. *Water Research,* 60, 118-129.

Bellamy W., Carlson K., Pier D., Ducoste J. and Carlson, M. (2000). Determining disinfection needs. *J. Am. Water Works Ass.* 92(5), 44-52.

Blatchley III E.R., and Cheng M. (2000). Reaction Mechanism for Chlorination of Urea. *Environ. Sci. Technol.* 44(22), 8529-8534.

Block S.S. (2001). Peroxygen Compounds. In Disinfection, Sterilization, and Preservation, Fifth ed., Block, S.S., Ed. Lippincott, Williams & Wilkins: Philadelphia, pp. 185-204.

Brahmi M., Belhadi N.H., Hamdi H. and Hassen A. (2010). Modeling of secondary treated wastewater disinfection by UV irradiation: Effects of suspended solids content. *J. Environ. Sci.*, 22(8), 1218-1224.

Chauret C., Smith C. and Baribeau H. (2008). Inactivation of Nitrosomonas europaea and pathogenic Escherichia coli by chlorine and monochloramine. *J. Water Health.*, 6(3), 315-322.

Chick H. (1908). An investigation of the laws of disinfection. *J. Hyg.,* 8(1), 92-158.

Chiu K.P., Lyn D.A., Savoye P. and Blatchley III E.R. (1999b). Effect of UV system modifications on disinfection performance. *J. Environ. Eng.*, 125(5), 459-469.

Chiu K., Lyn D.A., Savoye P. and Blatchley III E.R. (1999a). Integrated UV disinfection model based on particle tracking. *J. Environ. Eng.*, 125(1), 7-16.

Craik S.A., Weldon D., Finch G.R., Bolton J.R. and Belosevic M. (2001). Inactivation of Cryptosporidium parvum oocysts using medium- and low-pressure ultraviolet radiation. *Water Research*, 35(6), 1387-1398.

CSWRC Board (2015). Regulations Related to Recycled Water in Sacramento, CA.

Deborde M. and Von Gunten U. (2008). Reactions of chlorine with inorganic and organic compounds during water treatment - Kinetics and mechanisms: A critical review. *Water Research*, 42(1-2), 13-51.

Ducoste J., Carlson K. and Bellamy W. (2001).The integrated disinfection design framework approach to reactor hydraulics characterization. *J. Water Supply Res T.*, 50(4), 245-261.

Dunkin N., Weng S.C., Schwab K.J., McQuarrie J., Bell K. and Jacangelo J.G. (2017). Comparative Inactivation of Murine Norovirus and MS2 Bacteriophage by Peracetic Acid and Monochloramine in Municipal Secondary Wastewater Effluent. *Environ. Sci. Technol.*, 51(5), 2972-2981.

Eischeid A.C. and Linden, KG. (2011). Molecular Indications of Protein Damage in Adenoviruses after UV Disinfection. *Appl. Environ. Microb.*, 77(3), 1145-1147.

EPA (2015). Review of Coliphages as Possible Indicators of Fecal Contamination for Ambient Water Quality. In *EPA Office of Water, Office of Science and Technology, Health and Ecological Criteria Division:* p 129.

Ettayebi K., Crawford S.E., Murakami K., Broughma J. R., Karandikar U., Tenge V.R., Neill F.H., Blutt S.E., Zeng X.L., Qu L., Kou B., Opekun A.R., Burrin D., Graham D.Y., Ramani S., Atmar R.L. and Estes M.K. (2016). Replication of human noroviruses in stem cell-derived human enteroids. *Science.* 353(6306), 1387-1393.

EU (2011). Directive 2011/65/EU of the European Parliament and of the Council of 8 June 2011 on the restriction of the use of certain hazardous substances in electrical and electronic equipment. EUR-Lex: *Official Journal of the European Union.* L174/88-L174/110.

Haas C.N. and Joffe J. (1994). Disinfection under dynamic conditions: modification of hom model for decay. *Environ. Sci. Technology,* 28(7), 1367-1369.

Hassen A., Heyouni A., Shayeb H., Cherif M. and Boudabous A. (2000). Inactivation of indicator bacteria in wastewater by chlorine - a kinetics study. *Bioresour. Technol.*, 72(1), 85-93.

Held G. (2009). Introduction to Light Emitting Diode Technology and Applications. CRC Press: *Boca Raton*, FL.

Jafvert C.T. and Valentine R.L. (1992). Reaction Scheme for the Chlorination of Ammoniacal Water. *Environ. Sci. Technol.*, 26(3), 577-586.

Krasner S.W., Weinberg H.S., Richardson S.D., Pastor S.J., Chinn R., Sclimenti M.J., Onstad G.D. and Thruston A.D. (2006). Occurrence of a new generation of disinfection byproducts. *Environ. Sci. Technol.*, 40(23), 7175-7185.

Hsu C.C., Wobus C.E., Steffen E.K., Riley L.K. and Livingston R.S. (2005). Development of a Microsphere-Based Serologic Multiplexed Fluorescent Immunoassay and a Reverse Transcriptase PCR Assay To Detect Murine Norovirus 1 Infection in Mice, *Clinical and Diagnostic Laboratory Immunology*, 12(10), 1145-1151.

Li J., Uchida T., Todo T. and Kitagawa, T. (2006). Similarities and differences between cyclobutane pyrimidine dimer photolyase and (6-4) photolyase as revealed by resonance Raman spectroscopy - Electron transfer from the FAD cofactor to ultraviolet-damaged DNA. *J. Biol. Chem.*, 281(35), 25551-25559.

Li M.K., Qiang Z., Li T.G., Bolton J.R. and Liu C.L. (2011). In Situ Measurement of UV Fluence Rate Distribution by Use of a Micro Fluorescent Silica Detector. *Environ. Sci. Technol.,* 45(7), 3034-3039.

Lim M.Y., Kim J.-M. and Ko, G. (2010). Disinfection kinetics of murine norovirus using chlorine and chlorine dioxide. *Water Research,* 44(10), 3243-3251.

Luukkonen T. and Pehkonen S.O. (2017). Peracids in water treatment: A critical review. *Crit. Rev. Env. Sci. Tec.*, 47(1), 1-39.

Lyn D.A., Chiu K. and Blatchley III E.R. (1999). Numerical modeling of flow and disinfection in UV disinfection channels. *J. Environ. Eng.*, 125(1), 17-26.

Malcolm Pirnie I. and HDR Engineering I. (1991). Guidance Manual for Compliance with the Filtration and Disinfection Requirements for Public Water Sources. In Branch, S. a. T., Division, C. a. S., Water, O. o. D., Eds. *US EPA*: Washington DC, p. 580.

Managers, G. L.-U. M. R. B. o. S. a. P. P. H. a. E. (2014). Recommended Standards for Wastewater Facilities: Policies for the Design, Review, and Approval of Plans and Specifications for Wastewater Collection and Treatment Facilities. In *Health Research*, Inc., Health Education Services Division: Albany, NY, p. 175.

McCarty P.L. (2018). What is the Best Biological Process for Nitrogen Removal: When and Why? *Environ. Sci. Technol.*, 52(7), 3835-3841.

McMinn B.R., Huff E.M., Rhodes E.R. and Korajkic A. (2017). Concentration and quantification of somatic and F plus coliphages from recreational waters. *J. Virol. Methods.*, 249, 58-65.

Medema G., Heijnen L., Elsinga G., Italiaander R., and Brouwer A. (2020). Presence of SARS-Coronavirus-2 in sewage. *medRxiv*.
doi: https://doi.org/10.1101/2020.03.29.20045880

Murphy J.L., Arrowood M.J., Lu X., Hlavsa M.C., Beach M.J. and Hill V. R. (2015). Effect of Cyanuric Acid on the Inactivation of Cryptosporidium parvum under Hyperchlorination Conditions. *Environ. Sci. Technol.,* 49(12), 7348-7355.

Odeh I.N., Nicoson J.S., Hartz KE. H. and Margerum D.W. (2004). Kinetics and mechanisms of bromine chloride reactions with bromite and chlorite ions. *Inorg. Chem.,* 43(23), 7412-7420.

Ortiz A.P. (2014) Variability of UV Disinfection of Municipal Wastewater, M.S. Thesis, Purdue University.

Pennell K.G., Aronson A.I. and Blatchley III E.R. (2008). Phenotypic persistence and external shielding ultraviolet radiation inactivation kinetic model. *J. Appl. Microbiol.*, 104(4), 1192-1202.

Ramraj S.K., Paul P., Yuvaraj N., Paari K.A., Pattukumar V. and Arul V. (2011). Purification and characterization of enterocin MC13 produced by a potential aquaculture probiont Enterococcus faecium MC13 isolated from the gut of Mugil cephalus. *Canadian Journal of Microbiology.* 57, 993-1001. 10.1139/w11-092.

Rennecker J.L., Driedger A.M., Rubin S.A. and Mariñas B. J. (2000). Synergy in sequential inactivation of Cryptosporidium parvum with ozone/free chlorine and ozone/monochloramine. *Water Research,* 34(17), 4121-4130.

Rossi S., Antonelli M., Mezzanotte V. and Nurizzo C. (2007). Peracetic acid disinfection: A feasible alternative to wastewater chlorination. *Water Environ. Res.,* 79(4), 341-350.

Santoro D., Crapulli F., Raisee M., Raspa G. and Haas C.N. (2015). Nondeterministic Computational Fluid Dynamics Modeling of Escherichia coli Inactivation by Peracetic Acid in Municipal Wastewater Contact Tanks. *Environ. Sci. Technol.*, 49(12), 7265-7275.

Santoro D., Gehr R., Bartrand T.A., Liberti L., Notarnicol M., Dell'Erba A., Falsanisi D. and Haas C.N. (2007). Wastewater disinfection by peracetic acid: Assessment of models for tracking residual measurements and inactivation. *Water Environ. Res.*, 79(7), 775-787.

Selleck R.E., Saunier B.M. and Collins H.F. (1978). Kinetics of bacterial deactivation with chlorine. *J. Environ. Eng. Div.*, 104(6): 1197-1212.

Severin B.F., Suidan M.T. and Engelbrecht R.S. (1983). Kinetic modeling of uv disinfection of water. *Water Research*, 17(11), 1669-1678.

Severin B.F., Suidan M.T. and Engelbrecht R.S. (1984). Mixing effects in uv disinfection. *J. Water Pollut. Control Fed.,* 56(7), 881-888.

Shang C. and Blatchley III E.R. (2001). Chlorination of pure bacterial cultures in aqueous solution. *Water Research*, 35(1), 244-254.

Shang C., Gong W.L. and Blatchley III E.R. (2000). Breakpoint chemistry and volatile byproduct formation resulting from chlorination of model organic-N compounds. *Environ. Sci. Technol.*, 34(9), 1721-1728.

Sivey J.D., McCullough C.E. and Roberts A.L. (2010). Chlorine Monoxide (Cl2O) and Molecular Chlorine (Cl-2) as Active Chlorinating Agents in Reaction of Dimethenamid with Aqueous Free Chlorine. *Environ. Sci. Technol.,* 44(9), 3357-3362.

McGuigan K.G., Conroy R.M., Mosler H.J., du Preez M., Ubomba-Jaswa E. and Fernandez-Ibanez P. (2012). Solar water disinfection (SODIS): a review from bench-top to roof-top, *Journal of Hazardous Materials*, 235, 29-46.

Sommer R., Lhotsky M., Haider T. and Cabaj A. (2000). UV inactivation, liquid-holding recovery, and photoreactivation of Escherichia coli O157 and other pathogenic Escherichia coli strains in water. *J. Food Protect.*, 63(8), 1015-1020.

Virto R., Manas P., Alvarez I., Condon S. and Raso J. (2005). Membrane damage and microbial inactivation by chlorine in the absence and presence of a chlorine-demanding substrate. *Appl. Environ. Microb.*, 71(9), 5022-5028.

Wahman D.G. (2018). Web-Based Applications to Simulate Drinking Water Inorganic Chloramine Chemistry. *J. Am. Water Works Ass.,* 110(10), E43-E61.

Watson H.E. (1908). A note on the variation of the rate of disinfection with change in the concentration of the disinfectant. *J. Hyg.*, 8(4), 536-542.

Weng S., Dunkin N., Schwab K.J., McQuarrie J., Bell K. and Jacangelo J.G. (2018). Infectivity reduction efficacy of UV irradiation and peracetic acid-UV combined treatment on MS2 bacteriophage and murine norovirus in secondary wastewater effluent. *J. Environ. Manage.,* 221, 1-9.

White G.C. (1992). The handbook of chlorination and alternative disinfectants. 3rd ed., Van Nostrand Reinhold: New York, p xiv, p.1308.

Wobus C.E., Thackray LB. and Virgin H.W. (2006). Murine norovirus: a model system to study norovirus biology and pathogenesis. *J. Virol.*, 80(11), 5104-5112.

NOMENCLATURE

Symbol	Description	Unit
C	Concentration of the chemical disinfectant	mg/L
Ct	Chemical disinfectant dose	mg·min/L
C(t)	Time-dependent disinfectant concentration	mg/L
E_i	The fluence rate of radiation	mW/cm^2
hv	The energy of a photon of frequency "v" (h = Planck constant)	J/photon
k	Inactivation constant	cm^2/mJ
k'	The photochemical reaction rate constant	cm^2/mJ
k_1	Inactivation constant for sensitive sub-population	cm^2/mJ
k_2	Inactivation constant for resistant sub-population	cm^2/mJ
N	Concentration of viable or infective microbes	Cfu/100 mL or MPN/100 mL
n	Empirical constant	-
N_0	Concentration of viable or infective microbes prior to UV exposure	Cfu/100 mL or MPN/100 mL
m	Empirical parameter	-
M_i	The microbe that has accumulated i units of damage	-
p	Fraction of microbial population that expresses resistance to inactivation	-
Q_e	The energy of radiation at a given wavelength	J/einstein
t	Time	Seconds (for UV disinfection) Minutes (for chemical disinfection)
ΔH	Average bond energy	kJ/mole

Greek symbol	Explanation	Unit
Λ	Coefficient of specific lethality	-
ε	The molar absorption coefficient for a molecule	L/mole.cm
Φ	The quantum yield	mole/einstein
τ	Period of exposure	Seconds (for UV disinfection) Minutes (for chemical disinfection)
λ	Wavelength	nm

Abbreviation	Description
CFD	Computational fluid dynamics
CPDs	Cyclopyrimidine dimers
DBPs	Disinfection by-products

HNV	Human norovirus
IDDF	Integrated Disinfection Design Framework
LEDs	Light emitting diodes
LP	Low-pressure
MFSD	Micro Fluorescent Silica Detector
MNV	Murine norovirus
MP	Medium-pressure
MWRDGC	Metropolitan Water Reclamation District of Greater Chicago
NOM	Natural organic matter
PAA	Peracetic acid
PFA	Performic acid
PPA	Perpropionic acid
PPES	Phenotypic Persistence and External Shielding
SODIS	Solar water disinfection
SRP	Standard reduction potential
ROS	Reactive oxygen species
UV	Ultraviolet
UVT	UV transmittance
VUV	Vacuum UV

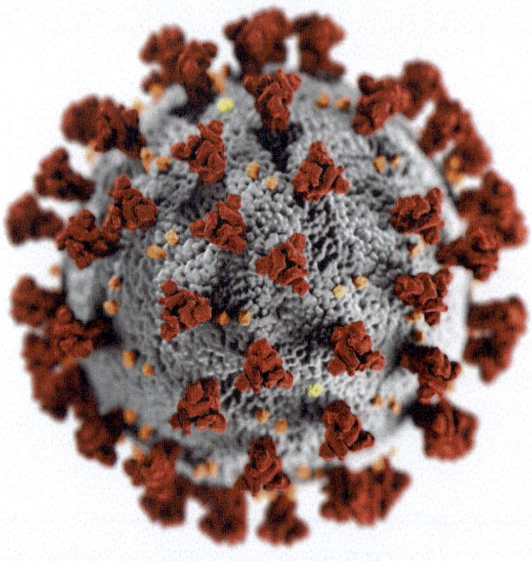

Figure 8.20 This book was completed during the outbreak of the SARS-Coronavirus-2 (SARS-CoV-2) pandemic. The work of Medema *et al.* (2020) demonstrated the presence of SARS-CoV-2 viral RNA in municipal wastewater. According to current knowledge, the novel coronavirus is shed in stools, but it is questionable if it is still infectious after shedding in this manner. SARS-CoV-2 is a member of a group of viruses that do not have a significant waterborne transmission route. Therefore, the consensus based on current knowledge is that standard protocols are sufficiently protective. (Image: A. Eckert and Higgins, caption text adopted from https://www.kwrwater.nl/).

doi: 10.2166/9781789060362_0419

9

Aeration and mixing

Diego Rosso, Michael K. Stenstrom and Manel Garrido-Baserba

9.1 AERATION FUNDAMENTALS AND TECHNOLOGY

9.1.1 Fundamentals and metrics

To meet the microbial oxygen requirements in biological processes, it is necessary to transfer oxygen to the liquid. Aeration systems transfer oxygen into liquid media by releasing air through orifices or pores (*i.e.* bubble diffusers), by shearing the liquid surface with a mechanical device (such as a mixer or turbine), or through direct contact between air and a large water surface (*e.g.* the surface of lagoons). A physical entity central to the classification and understanding of oxygen transfer in water is the interfacial velocity between gas and liquid. Falling droplets and rising coarse bubbles have large interfacial gas-liquid velocity gradients and can be grouped as high-flow regime interfaces, whereas fine bubbles have low interfacial velocity gradients and can be grouped as low-flow regime interfaces (Rosso, 2018; Rosso *et al.*, 2012; Rosso and Stenstrom, 2006b).

When analysing or specifying aeration systems, it is important to define the oxygen transfer efficiency, which is necessary to compare different technologies, as well as to monitor aeration systems in operation over an extended time (ASCE, 1989). While some parameters can be calculated in clean water (in the absence of pollutants) others need to be corrected for the conditions of wastewater, because the effects of pollutants and process conditions exert a strong influence on mass transfer efficiency.

A note to the reader: in this chapter we summarise a simplified report of the equations pertaining oxygen transfer in water. The full treatise is reported elsewhere in journal papers (*inter alia,* Redmon *et al.*, 1983; Philichi and Stenstrom, 1989; Wagner and Pöpel, 1996; Gillot *et al.*, 2005), books (*e.g.* USEPA, 1989; Rosso, 2018), or standard protocols (*e.g.* ASCE, 1997, 2007).

9.1.1.1 Oxygen transfer in clean water

Oxygen is a sparingly soluble gas, implying that the saturation is low and much effort is required to maintain an appropriate level of dissolved oxygen (DO). This also means that at any time we can use Henry's law to describe the equilibrium between atmospheric air and oxygen in the water. The target for aeration design is to provide oxygen to the water in excess to the oxygen requirements, *i.e.* leaving a remainder that corresponds to the DO. The first step is to quantify the oxygen transfer rate in clean water (OTR, kgO_2/h). The OTR quantifies the capacity of the aeration system, *i.e.* the amount of oxygen it can supply to the water per unit of time:

$$OTR = k_L a \cdot (DO_{sat} - DO) \cdot V \tag{9.1}$$

where:

k_La liquid-side mass transfer coefficient (1/h)
DO_{sat} dissolved oxygen in water at saturation (kgO_2/m^3) also represented as C^*_∞
DO dissolved oxygen in water (kgO_2/m^3), also represented as Cr
V water volume (m^3).

Under standard conditions (clean water; new diffusers, *i.e.* no fouling: F=1; dissolved oxygen concentration = 0 mg/L; water temperature = 20°C; and standard atmospheric pressure: 101.3 kPa; relative humidity: 36%), the standard oxygen transfer rate, SOTR, can be derived from equation (9.1) with the appropriate unit conversion:

$$SOTR = k_L a^*_{20} \cdot C^*_{\infty 20} \cdot V_{Tk} \tag{9.2}$$

As can be seen in Eq. 9.1 (and below in Eq. 9.3) the oxygen transfer rate is proportional to the difference between the existing DO concentration and the equilibrium concentration of the oxygen in the water, *i.e.* to the driving force for oxygen transfer through the interfacial film. This can also be expressed by the following equation:

$$\frac{dC}{dt} = k_L a^* \cdot (C^*_\infty - C_r) \cdot V \tag{9.3}$$

The nomenclature followed here conforms to that of the ASCE Standard for Measurement of Oxygen Transfer in Clean Water (ASCE, 2006).

where:

dC/dt the rate of change of oxygen concentration ($mgO_2/L.h$)
k_La^* oxygen transfer coefficient at water temperature T (1/h)
C^*_∞ equilibrium concentration of the oxygen in the water at water temperature T and field atmospheric pressure P_b (mg/L)
C_r dissolved oxygen in water at saturation (kgO_2/m^3).

The OTR defines the capacity of the aeration system regardless of its efficiency, and it is therefore necessary to define additional entities, such as the aeration efficiency (AE, kgO_2/kWh):

$$AE = \frac{OTR}{P} \tag{9.4}$$

where:

P power drawn by the aeration system (kW).

The power drawn used in the calculation of aeration efficiency, its standardised correction, and its definition in field conditions (AE, SAE or αSAE, respectively; Table 9.1) can be defined in different ways. Wire power is most often used because it is the best predictor of the actual power consumption. Using wire power includes the inefficiencies of the blower, motors, and gearboxes even if they are not known or measured. Mechanical (brake) power may be more convenient to use if the motors or gearboxes are being specified separately and because it is related to the adiabatic compression formula that is applied to blowers. To obtain wire power from brake power, losses must be measured and sometimes there is controversy in measurement methods. A significant source of error in aeration system design results from confusing power types. Unless otherwise specified, wire power will be used throughout this chapter.

For subsurface aeration devices (such as fine or coarse bubble diffusers), the oxygen transfer efficiency (OTE, %) can be used, defined as follows:

$$OTE = \frac{OTR}{W_{O2}} \sim \frac{(O_{2,in} - O_{2,out})}{O_{2,in}} \tag{9.5}$$

with OTR from Eq. 9.1, W_{O2} as the mass flow of oxygen fed to the aeration tank (kgO_2/h):

$$W_{O2} = O_2 massflow = Q_{air} \cdot \rho_{air} / air \qquad (9.6)$$

$O_{2,in}$ and $O_{2,out}$ represent the mass fluxes of oxygen in and out of the clean water volume. OTE quantifies the efficiency of operation and is often more convenient since it allows comparison of the aeration system without the complicating issues surrounding the blowers. Blowers are often provided by a different manufacturer or contractor and using OTE and blower efficiency independently simplifies specification and design of the aeration system.

In order to avoid bias due to site-specific environmental and process conditions, standard conditions are used and are defined as zero DO, zero salinity, 20 °C, 1 atm and 36% relative humidity. Therefore, results are typically reported as Standard Oxygen Transfer Efficiency (SOTE, %), Standard Oxygen Transfer Rate (SOTR, kgO_2/h), or Standard Aeration Efficiency (SAE, kgO_2/kWh). It is highly recommended to use these terms, as they are standardized and will avoid confusion. All details on correcting to standard conditions are available in the standard protocols. There are a number of definitions used for describing aeration systems which are listed in Table 9.1 and used throughout this chapter.

9.1.1.2 Oxygen transfer in process water

Translating standard conditions in clean water to process conditions requires the use of several site-specific, empirical parameters. Process water (or wastewater) is characterized by dissolved and suspended contaminants, plus the presence of biomass, which cause a deviation in aerator performance from clean water conditions. The parameter with the greatest impact is the α factor (see Section 9.1.1.3), which is defined as the ratio of the process water mass transfer coefficients to clean water mass transfer coefficients, or

$$\alpha = \frac{(k_L a)_{\text{process water}}}{(k_L a)_{\text{clean water}}} \qquad (9.7)$$

or

$$\alpha = \frac{\alpha SOTE}{SOTE} \qquad (9.8)$$

where:

SOTE oxygen transfer efficiency at standard conditions (%)

αSOTE oxygen transfer efficiency in process water at standard conditions except for the effect of contaminants on the mass transfer coefficient (%).

Note that we called alpha a parameter, while we have now recognised its variable nature. Later in the chapter we will elaborate more on the dynamic nature of alpha.

SOTE has to be measured under clean water conditions and following the guidelines of a standard protocol (e.g. ASCE, 2007)(see Section 9.1.1.1). This value, or an estimation of the real value, should be provided by the manufacturer of the aeration technology. On the other hand, αSOTE can be only measured *a posteriori* and *in-situ* or in a sidestream setup (see Section 9.1.1.2).

Once the alpha factor is known, the OTR under field conditions can be calculated. The k_La needs to be corrected for water temperature (using an Arrhenius-like correction), for process water (using the alpha factor) and for diffuser fouling conditions (*i.e.* the F factor). Similarly, C^*_∞ must be corrected for water temperature, pressure, and process conditions.

The resulting OTR formula for field and standard conditions (αSOTR), with the appropriate unit conversions, is as follows:

$$OTR = \alpha \cdot F \cdot k_L a_{20} \cdot \Theta^{(T-20)} \cdot (\beta \cdot \Omega \cdot \tau \cdot C^*_{\infty 20} - C_r) \cdot V_{Tk} \qquad (9.9)$$

where:

OTR oxygen transfer rate at temperature T, ambient pressure p_b, with process (dirty) wastewater, and residual DO concentration C_r (kg O_2/d)

α ratio of the process water k_La of a new diffuser to the clean water k_La of a new diffuser, also known as the alpha factor (dimensionless)

F ratio of the process water k_La of a diffuser after a given time in service to the process water k_La of a new diffuser (no fouling at t_0: F=1), also known as the fouling factor (dimensionless).

k_La_{20} apparent volumetric oxygen mass transfer coefficient in clean water at temperature T=20°C, also referred to as oxygen mass transfer coefficient (1/h)

Θ Arrhenius temperature coefficient with a typical value of 1.024

T mixed liquor temperature (°C)

β correction factor for dissolved solids and can be assumed to be equal to 0.95 to 1.0 if the plant treats domestic wastewater. For wastewater with a considerable industrial component, its value can be calculated based on the dissolved solids content of the process water. In this case, β should be calculated as the ratio of the DO surface saturation concentration in water with dissolved solids equal to that of the process water, to the DO surface saturation concentration in clean water

Ω correction factor for atmospheric pressure and at tank depths of less than 6 meters it may be approximated by p_b/p_s, where p_b is ambient barometric pressure and p_s is standard pressure at mean sea level

τ correction factor for temperature and can be calculated based on published DO surface saturation values as:

$$\tau = \frac{C^*_S}{C^*_{S,20}} \tag{9.10}$$

where:

C^*_S tabular value for DO surface saturation concentration at water temperature T (*i.e.* that recorded during the test), standard atmospheric pressure p_s, and 100% relative humidity

$C^*_{S,20}$ tabular value for DO surface saturation concentration at water temperature 20 °C, standard atmospheric pressure p_s, and 100% relative humidity ($C^*_{S,20}$ = 9.08 mg/l, Metcalf & Eddy, 2014).

The oxygen transfer efficiency (OTE) under field or process conditions can be calculated as follows:

$$OTE = \frac{OTR}{W_{O2}} \tag{9.11}$$

For as long as the DO remains unchanged (*i.e.* at steady-state conditions), the oxygen uptake rate OUR can be calculated directly as:

$$OUR = \frac{OTR}{V} \tag{9.12}$$

For the purpose of testing full-scale aeration systems using the off-gas method (see Section 2.3.2), the DO remains practically constant during the few minutes of a measurement, which allows for the direct calculation of the real-time *in-situ* OUR.

The alpha factor and the fouling factor (which have a considerable variability), have a tremendous influence on the oxygen transfer rate (OTR) and the oxygen transfer efficiency (OTE) under field conditions. The following sections describe the alpha factor and its influence in treatment plants on the process design, equipment selection, operation and maintenance strategies, and energy costs.

Table 9.1 Summary of all definitions used to specify aeration systems.

Symbol	Definition	Remarks
W_{O2}	Oxygen mass flow fed to the aeration tank	
OTR	Oxygen transfer rate in water	$OTR = k_L a \cdot (DO_{sat} - DO) \cdot V$
SOTR	Oxygen transfer rate in standard conditions in clean water	
OTE	Oxygen transfer efficiency in clean water	$OTE = \frac{OTR}{W_{O2}} \sim \frac{(O_{2,in} - O_{2,out})}{O_{2,in}}$
SOTE	Oxygen transfer efficiency in standard conditions in clean water	
AE	Aeration efficiency in clean water	$AE = \frac{OTR}{P}$
SAE	Aeration efficiency in standard conditions in clean water	
k_La	Liquid-side mass transfer coefficient	Measured in clean water tests
α	Alpha factor, *i.e.* ratio of process- to clean- water mass transfer.	$\alpha = \frac{\alpha SOTE}{SOTE}$ or $\alpha = \frac{(k_L a)_{process\ water}}{(k_L a)_{clean\ water}}$
F	Fouling factor	$F = \frac{\alpha SOTE_{new\_difuser}}{\alpha SOTE_{used\_difuser}}$
αF	Alpha factor for used diffusers	
αSOTE	Oxygen transfer efficiency in standard conditions in process water	
αFSOTE	Oxygen transfer efficiency in standard conditions in process water for used diffusers	
αSAE	Aeration efficiency in standard conditions in process water	
αFSAE	Aeration efficiency in standard conditions in process water for used diffusers	

Standard conditions are defined as 20°C, 1 atm, zero salinity, 36% relative humidity, zero DO in water. Key: P = power drawn; V = water volume.

9.1.1.3 The mysterious alpha factor

The reduction in the oxygen transfer rate caused by contaminants present in wastewater is governed by the alpha factor (α), as can be found in one of the earliest works on the subject by Eckenfelder (1956). This particular label for an efficiency indicator is given by its unknown exact nature. The α factor is, then, reported as the most uncertain aeration process parameter (Karpinska and Bridgeman, 2016) since it is highly variable and is difficult to estimate. The α factor is highly dynamic (Jiang *et al.*, 2017; Leu *et al.*, 2009) and cannot be predicted with certainty under different process conditions, either in a single facility (Amerlinck *et al.,* 2016) or across different facilities (Gillot and Héduit, 2008; Redmond *et al.,* 1992).

Although it is generally assumed that the most significant component affecting the α factor is surface-active agents or surfactants, the alpha factor is actually a composite parameter encompassing a wide range of potential components and conditions, all affecting in different ways and magnitudes. Chemical compounds (*i.e.* surfactants, organic substrates, etc.) as well as physical constraints (*i.e.* bubble coalescence, hydrodynamics, etc.), including the influence of the microbial activity, are some of the main causes leading to the reduction of the aeration efficiency in wastewater. The sludge rheology is particularly important in highly concentrated solid suspensions, such as membrane bio-reactors (MBRs) or aerobic digesters.

The α factor is unknown *a priori* and may only be calculated *a posteriori*, because the wastewater composition and exact reactor conditions cannot be known beforehand. Lower flow regime gas-liquid interfaces (such as the ones produced by fine-pore diffusers) generally have lower α factors than higher flow regime interfaces (such as the ones produced by coarse-bubble diffusers or surface aerators) for similar conditions (Stenstrom and Gilbert, 1981). However, one must remember that high alphas are usually associated with lower SOTE and *vice versa*, due to the geometry and hydrodynamics of bubbles and droplets.

Differences in α factors among aeration systems were noted as early as the 1930s (Kessener and Ribbius, 1934), but were generally forgotten until the energy crisis of the 1970s increased the awareness of energy-efficient technologies. Prior to the 1980s, many plants were designed with an α factor of 0.8, which was considered a 'universal' α factor for all types of aeration systems. However, it has been shown that different aeration methods have different α, and for fine-pore diffusers the initial α factor decreases over time in operation due to fouling or scaling (Boyle and Redmon, 1983). Furthermore, for fine-bubble systems in activated sludge, α is a function of process conditions such as the sludge retention time (SRT, or sludge age, also called mean cell retention time - MCRT) or the airflow rate (Rosso and Stenstrom, 2005). Moreover, the effects of biomass on α has been documented (Cornel *et al.,* 2003; Krampe and Krauth, 2003; Germain *et al.,* 2007; Henkel *et al.,* 2011), providing the experimental evidence that is driving the efforts to study the fluid dynamics of aerated tanks (e.g. Fayolle *et al.,* 2007).

9.1.2 Fine bubbles, coarse bubbles, and droplets

Aeration systems transfer oxygen into a liquid media by either diffusing gas through a gas–liquid interface, or dissolving gas into the liquid solution using a semi-permeable membrane. In other words, environmental technologies can rely on (*i*) interfacial gas transfer, with a gas-liquid interface created by either shearing the liquid surface with a mixer or turbine, or (*ii*) by releasing air through spargers or porous materials. Moreover, a third category of technologies exists: (*iii*) novel devices (referred to as bubbleless, e.g. the MABR) which utilize the intrinsic porosity of 'true' semi-permeable membranes, such as reverse osmosis membranes, for transferring gas into the liquid. These membranes allow the transit of water and air across the membrane, without permitting the passage of solute or suspended matter and without generating an interface. Other aeration equipment also exists, such as submerged turbines or jet diffusers, which can create fine bubbles without using small orifices, and are fine-bubble but not fine-pore aerators. These technologies use mechanical energy to shear large bubbles into fine bubbles. However, due to their inefficiency, they have been largely retired by municipal wastewater treatment facilities.

The oxygen transfer mechanism in the activated sludge process is still far from being fully described or modelled, despite the available knowledge. Nowadays experts acknowledge that the oxygen transfer rate can be evaluated by the overall volumetric oxygen transfer coefficient, k_La, which is composed of the overall oxygen transfer coefficient, k_L, and by the interfacial area available for exchange, a. k_La is generally determined for clean water, following the ASCE standard. It is used as a constant or as a function of airflow rate, based on the assumptions that: (*i*) the aeration tanks are completely mixed; (*ii*) the rheological characteristics of the activated sludge do not heavily impact the oxygen transfer, and (*iii*) aeration is homogeneously provided. These assumptions are known to limit the accuracy of aeration efficiency models and as such hamper any optimization study making use of them (Amaral *et al.,* 2016; Karpinska *et al.,* 2016). Therefore, these aeration models have been extended using empirical relationships considering the aforementioned α factor (see Section 9.2.1.3), to introduce the dependency on operational and environmental conditions of the process.

Moreover, bubble size distribution was found to play a role in the estimation of the oxygen transfer affecting k_L and a. In some studies, a single average bubble size is assumed (Fayolle *et al.*, 2007). These limitations are expected to be slowly overcome, however to date only some studies (at lab scale and not representing real full-scale plants) are trying to go beyond these assumption through photographic quantification (McGinnis and Little, 2002), through the introduction of interaction models which are accounting for bubble coalescence and breakage (Karpinska *et al.*, 2016), or expanding the existing knowledge on the effects of viscosity (Amaral *et al.*, 2016) by using computational fluid dynamics (CFD).

In wastewater treatment there are two regimes of interfacial velocity: high and low, which correspond to coarse-bubble and surface aerators, and fine-pore diffusers, respectively.

High interfacial velocities. Surface aerators shear the wastewater surface producing a spray of fine droplets that land on the wastewater surface within a few seconds and over a radius of a few metres. Falling droplets and rising coarse bubbles have large interfacial gas-liquid velocity gradients and can be grouped as high-flow regime interfaces, whereas fine bubbles have low interfacial velocity gradients and can be grouped as low-flow regime interfaces. Coarse bubbles are larger, customarily produced by orifices 6mm and larger, and may have diameters as large as 50mm. The high turbulence associated with coarse-bubble aerators makes them behave more like surface aerators than fine-bubble aerators. High-turbulence aerators may achieve better transfer rates, but at the expense of greater energy requirement, and lower aeration efficiency.

Low interfacial velocities. Diffusers are nozzles or porous surfaces placed on the tank bottom; they release bubbles that travel towards the tank surface. In general, bubbles are considered to be fine in size when their diameters are less than 5 mm. Polymeric or ceramic porous sintered materials and polymeric punched membranes are usually grouped together and referred to as fine-pore diffusers. While turbines and jet aerators produce bubbles, once these bubbles reach their terminal velocity their interfacial velocity is the same as the same bubble originating from a fine-pore diffuser. Fine-bubble diffusers have greater mass transfer decline in process water than coarse-bubble diffusers or surface aerators.

Turbulence controls gas transfer depression and can be characterized by the interfacial Reynolds or Péclet numbers and the gas transfer by the interfacial Sherwood number. The Sherwood number can be correlated to the Péclet number to account for turbulence and to a dimensionless number, called the interfacial contamination number, to account for interfacial contaminant accumulation (Rosso and Stenstrom, 2006b).

9.1.3 Inside the aeration tank

The next sections present an overview of commercially available aeration systems (excluding the blowers), whose characteristics will be described later in this chapter. The figure below (Figure 9.1) illustrates the main components of the aeration system near or within the tank.

9.1.3.1 Bubble aeration

Bubbles are typically released at full depth and they transfer oxygen as they travel to the surface. The bubbles also induce liquid movement that contributes to mixing. In fact, since 4/5 of air is nitrogen, we can conclude that at least 80% of the energy invested in bubble aeration contributes solely to mixing.

Coarse-bubble diffusers

Coarse-bubble diffusers utilize relatively large orifices or slots to release air bubbles generally of dimensions above 50 mm. Bubbles in that range of diameters released in water cannot exist as spheres, but as spherical caps (which resemble the shape of jellyfish), because water is too thin. Coarse bubbles are turbulent in nature and are characterised by a less severe surfactant interfacial accumulation, in part due to their high rate of surface renewal. Therefore, they have higher α factors when compared to fine-bubble systems (Eckenfelder and Ford, 1968; Kessener and

Ribbius, 1934; Rosso and Stenstrom, 2006a). Coarse-bubble systems are generally installed in full-floor configuration, to optimize air distribution and efficiency. In former times with less expensive energy costs, coarse-bubble diffusers were often installed in a single row on the sides of plug flow tanks (spiral roll) or in two or more rows (cross roll, ridge and furrow). Increasing airflows in these diffuser arrangements typically increases the bulk liquid velocity but has only a limited impact on OTR. These systems require fewer diffusers, reducing capital cost, but are among the least efficiency aeration systems, and are good candidates for upgrading.

Coarse-bubble diffusers are seldom affected by fouling or scaling. This is due to the large dimension and the high turbulence of the discharge orifices, which makes them difficult to clog in practice. On the other hand, these diffusers are always characterised by low SAE (in the range of 0.6-1.5 kgO_2/kWh), because large bubbles travel very rapidly through the water column and have a relatively low surface-to-volume ratio.

Figures 9.2 and 9.3 show two models of coarse-bubble diffusers commercially available. Figure 9.2 shows two views of a sparger. Spargers belong to the first generation of coarse-bubble diffusers, and are essentially capped hollow metal (older) and plastic (newer) bolts with one or more air release holes. The air travels through the air manifold, the drop leg, and inside the sparger, finally to be released through the sparger holes placed under the metal cap that enhances bubble shearing and prevents air channelling (bubbles travelling upwards like pearls of a necklace). A simpler system to produce coarse bubbles with relatively high efficiency is to use a piping grid, usually plastic, with approximately 5 mm diameter holes spaced at less than 0.5 meter intervals along the bottom side of the pipes.

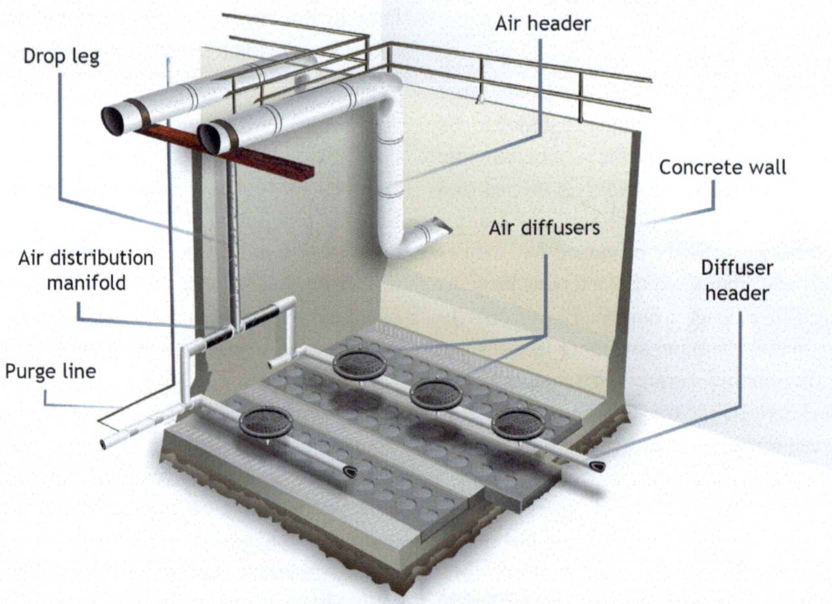

Figure 9.1 Representation of the different components in an aeration system.

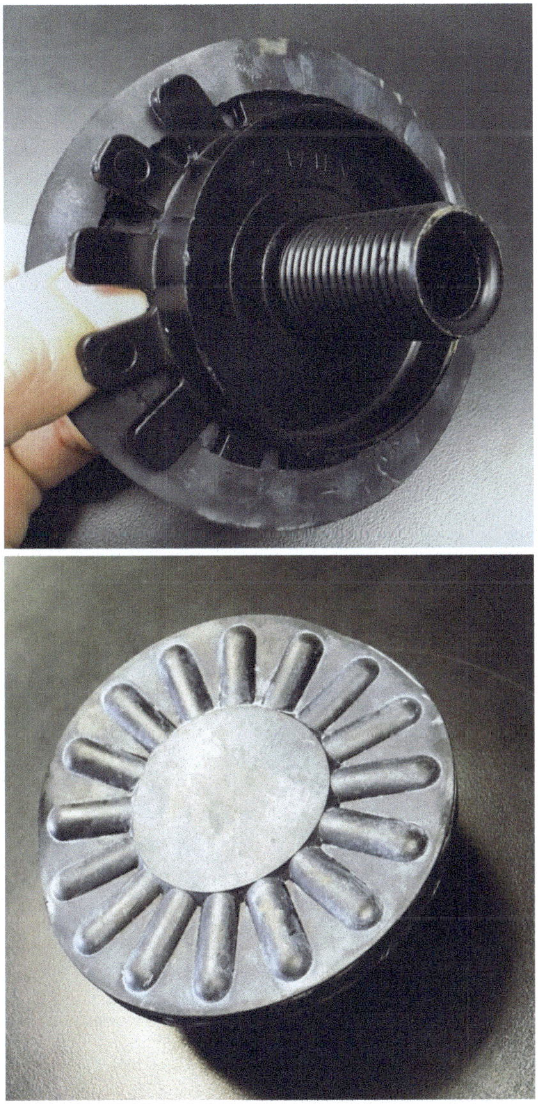

Figure 9.2. Example of an air sparger. On the top is the base, partially disassembled, to show the inner structure. On the bottom is the top of the rubber cap that acts as a check valve.

The 'chicken-feeder' diffuser (Figure 9.3) is a newer technology of coarse-bubble diffusers, characterised by a broader range of gas flowrates. This diffuser has on its side orifices of two sizes that proportionally increase in diameter going from the top to the bottom of the diffuser lateral. When the airflow rate is low, only the top orifices release air, but as the airflow increases the back pressure increases and air flows from the row of larger orifices. At the highest airflow rates, the air is forced through the open slot, which releases large spherical caps (> 100 mm).

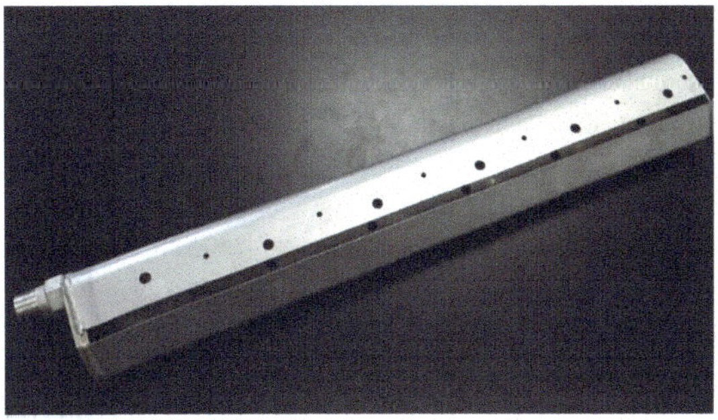

Figure 9.3 A 'chicken-feeder' coarse-bubble diffuser.

Due to their ability to accommodate high air flows, coarse-bubble diffusers have the advantage of being able to provide extremely high OTR within a given tank volume. Coarse-bubble systems generally have low SAE (SAE and OTR in general are inversely proportional). Obtaining large mass transfers (*i.e.* high OTR) requires high airflow rates that result in a short bubble retention time and reduce the area for gas transfer. The maximum OTR of coarse-bubble diffusers can be several times higher than fine-pore or surface aerators and is usually limited by the blower capacity (not the tank bottom surface area as in the case of fine-pore diffusers). Therefore, dense grids of coarse-bubble diffusers are sometimes the technology of choice for high-strength industrial wastewater treatment, where fine-bubble diffusers are a poor choice due to their limited air discharge and their extremely low alpha factors. For treatment systems not requiring high OTR per unit volume, such as municipal plants, coarse-bubble diffusers are a poor choice for energy conservation. Plants with coarse-bubble diffusers began to be replaced after the rapid rise in energy prices in the early 1970s, and nowadays municipal wastewater treatment plants most often use fine-pore diffusers. High coarse-bubble flowrates create aerosols and spray that usually require some type of management.

Fine-bubble diffusers

Fine bubbles can be produced by different technologies, by either releasing air through a porous plate, small orifices, or by mechanically shearing large bubbles into small ones. The latter technology employs submerged turbines or jet diffusers, which create fine bubbles but do so without small orifices, and in both cases mechanical energy is used to shear large bubbles into fine bubbles. Fine bubbles from turbines or jets always have lower SAE (in the range of 1.2-1.8 kgO_2/kWh) than fine bubbles from fine-pore diffusers. Also, their bubbles are released in a few concentrated points, thereby often leading to uneven DO distribution and lower overall efficiency.

An important subset of fine-bubble systems is fine-pore diffusers (figures 9.4, 9.5, 9.6 and 9.7); fine-pore diffusers produce their small bubbles by releasing compressed air through small orifices or pores in either punched membranes or porous material, such as ceramic stones or sintered plastic. Fine-pore diffusers are now the most commonly used diffusers in wastewater treatment in the United States and Europe. They have higher SAE (in the range of 3.6-4.8 kgO_2/kWh), and are routinely used in full-floor configurations, taking maximum advantage of their efficiency. Fine-pore diffuser systems limit stripping of volatile organic compounds by virtue of their increased efficiency, which results in lower airflow rates (Hsieh *et al.*, 1994, 1993). For the same reason, fine-pore diffusers also have reduced heat losses (Lippi *et al.*, 2009; Talati and Stenstrom, 1990).

However, fine-pore diffusers have two important disadvantages: the first is the need for periodic cleaning. Every fine-pore diffuser fouls inevitably and inexorably. Moreover, inorganic scaling and material degradation (*i.e.* ageing) add complexity to the problem of quantifying cleaning and replacement schedules. No diffuser is immune from a combination of these phenomena, the difference is only in the extent of them for a particular installation.

Figure 9.4 The first generation of fine-pore diffusers: ceramic domes (top). Air is fed through a hollow bolt, filling the space enclosed by the dome, and is then released through the pores of the sintered ceramic. On the bottom is a diffuser harvested after a prolonged time in operation, with visible bio-fouling on the outer surface. The discoloured section of the dome interior shows the position of the bolt's air release hole and is due to contaminants in the air supply.

The second important disadvantage of fine bubbles (this one also applies to turbines and jets) is the large negative effect on transfer efficiency from wastewater contaminants, which is most often quantified by the α factor. Fine-pore diffusers generally have lower α factors than coarse-bubble diffusers or surface aerators for similar conditions (Kessener and Ribbius, 1934; Stenstrom and Gilbert, 1981). Prior to the 1980s, many plants were designed with a constant α value of 0.8, which was considered as the 'universal' α factor for all types of aeration systems and for all conditions. In similar cases, nowadays we still witness poor design procedures employing such elevated and outdated values for α. However, it has been shown that different aeration methods have different α, and for fine-pore diffusers the initial α decreases over time in operation due to fouling or scaling (Rosso and Stenstrom, 2006b). Furthermore, for fine-bubble systems α is a function of process conditions such as the mean cell retention time (MCRT) or the airflow rate (Rosso *et al.*, 2005). Finally, diurnal variations in load cause α to behave dynamically (Jiang *et al.*, 2017; Leu *et al.*, 2009).

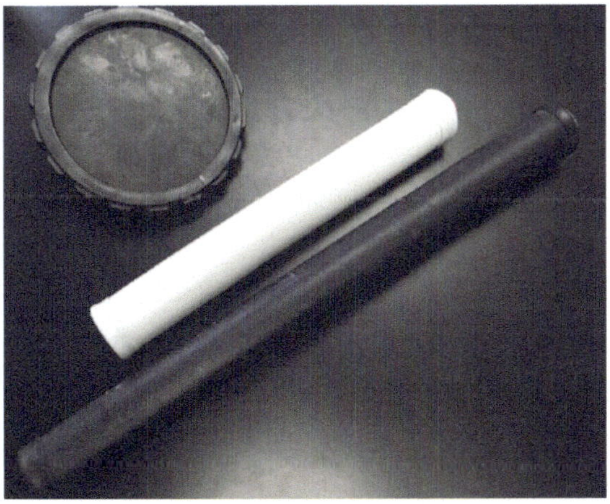

Figure 9.5 Assortment of diffuser models from different manufacturers. From left to right: a EPDM membrane disc, a silicone membrane tube, and an EPDM membrane tube.

Figure 9.6 Examples of full-floor installations of fine-pore diffusers. Top: ceramic discs; bottom left: polyurethane strips; bottom right: polyurethane panels

Figure 9.7 Fine-pore diffuser application in an aerated lagoon. The air pipes (on top of the diffuser unit) distribute the air along the lagoon, and vertical hoses convey the air from the air pipes to the diffuser units (placed close to the lagoon bottom). The image on the right shows a diffuser unit at a trade show. This system can be used in lagoons with uneven depths.

The summary table below compiles the authors' experience spanning the past four decades in many WRRFs (Table 9.2). The numbers are for relative comparison and should not be used for design purposes. Notice that fine-pore diffusers have the potential to achieve the highest AE but give their best performance in processes with longer sludge retention time. This indicates that the efficiency is not only a function of the aeration system, but also of the process where it is installed. Hence our cautionary statement against using these data for design, since the numbers here may correspond to process configurations very different to the case under design.

Table 9.2 Summary of aeration efficiency (AE), standard aeration efficiency (SAE), and airflow weighted α factors for commercially available aeration systems. These ranges are the result of decades of field measurements by the authors in a large number of WRRFs operating under very different configurations. Use at your own peril!

<table>
<tr><th>Aerator type</th><th>SAE kgO_2/kWh</th><th>Low SRT AE at 2 mgO_2/l</th><th>High SRT AE at 2 mgO_2/l</th><th>Alpha factor</th></tr>
<tr><td>High speed</td><td>0.9-1.3</td><td colspan="2">0.4-0.8</td><td>0.47-0.64</td></tr>
<tr><td>Low speed</td><td>1.5-2.1</td><td colspan="2">0.7-1.5</td><td>0.48-0.71</td></tr>
<tr><td>Turbine</td><td>1.2-1.8</td><td>0.4-0.6</td><td>0.6-0.8</td><td>0.30-0.47</td></tr>
<tr><td>Jets</td><td>1.5-1.7</td><td colspan="2">0.8-1.2</td><td>0.37-0.60</td></tr>
<tr><td>Coarse bubble</td><td>0.6-1.5</td><td>0.3-0.7</td><td>0.4-0.9</td><td>0.35-0.64</td></tr>
<tr><td>Fine pore</td><td>3.6-4.8</td><td>0.7-1.0</td><td>2.0-2.6</td><td>0.20-0.55</td></tr>
</table>

9.1.3.2 Mechanical aeration

Surface aerators

Surface aerators belong to the first generation of oxygen transfer technologies. These aerators do not need a blower to perform oxygen transfer. Surface aerators shear the liquid into small droplets, which are spread in a turbulent plume at several metres per second. The travelling droplets are in turbulent contact with the atmospheric air and typically oxygenate to at least half-saturation. As soon as they land onto the free surface of the liquid they mix with the liquid bulk, producing a typical DO pattern as in Figure 8. Since there is no airflow with the surface aerators, the OTE can neither be defined nor measured. The circulation of the liquid drawn into the spray also provides mixing. In some cases, surface aerators are also specified by the liquid pumping rate as well as an OTR. Surface aerators are typically characterised by high OTR and low SAE values (in the range of 0.9-2.1 kgO_2/kWh).

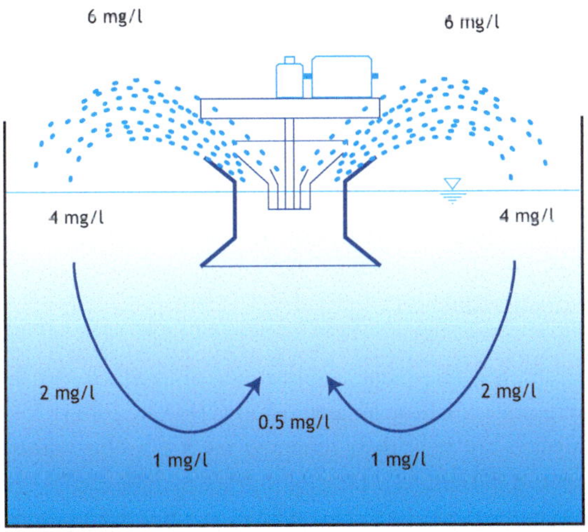

Figure 9.8 Schematic of DO patterns in tanks equipped with surface aerators.

Surface aerators move a liquid spray through air allowing the droplets to cool to the wet bulb temperature of the air (wet bulb temperature is a function of both the temperature and the humidity of air, so that cold, dry air would have the lowest wet bulb temperature) through evaporation. This cooling may be welcome when treating warm industrial effluents, or unwelcome when treating municipal wastewater in cold climates. The heat loss and the formation of aerosol spray are important factors and should be considered when selecting aeration systems. Issues such as spray drift and odour generation are especially important in urban areas.

Surface aerators are provided in two different configurations: high speed (*i.e.* direct drive), and low speed (*i.e.* with a gearbox). High-speed aerators typically rotate at 900-1,200 rpm, and due to the absence of the gearbox they are easy to install and are less expensive. On the other hand, the plume they generate is highly turbulent, which results in higher aerosol formation and potential floc shear. In general, high-speed aerators have lower SAE than low-speed aerators (0.9-1.3 kgO_2/kWh for high speed; 1.5-2.1 kgO_2/kWh for low speed).

The introduction of a gearbox between the electric motor and the impeller allows the aerator to rotate at lower speed (30-60 rpm). This is usually associated with increased capital costs and prolonged time for procurement (typically the gearbox is manufactured only after a purchase order is issued). The higher initial cost and procurement time may be partially compensated for by higher SAE.

Surface aerators cannot be used in deep tanks or lagoons without provisions for mixing the lower sections such as using draft tubes or lower impellers. A draft tube ensures that the upward pumping effect of the surface aerator extends to the lower levels of the tank. A lower impeller is mounted on a long shaft, which positions the lower impeller approximately 1 m above the tank bottom. High-speed and low-speed aerators are seldom used in tanks deeper than 4 m and 5 m, respectively, without draft tubes or lower mixers.

Figure 9.9 A low-speed surface aerator during installation (top) and in operation (bottom). The impeller at the bottom of the tank is used to ensure that solids and DO are distributed through the entire depth. The four structural beams act as support for the aerator as well as baffles to prevent vortex formation.

In the case of lagoons or earthen bottom tanks, the bottom must be protected from surface scour by the liquid. It is not uncommon for surface aerators to create depressions in earthen bottoms that may allow rocks and debris to enter and damage the propellers or create structural instabilities. Additionally, for very shallow and wide lagoons, the aerator zone of influence may not extend to the edge of the lagoon, potentially allowing solids deposition and sludge formation.

A type of surface aerator that provides aeration and mixing, as well as imparting horizontal velocity to the water is the surface brush aerator or rotor, typically found in oxidation ditches (Fig. 9). These low-speed surface aerators generally have elevated specific energy requirements (in other words, low SAE) because they pump the fluid as well as aerate. In oxidation ditches a minimum liquid velocity is required to keep the mixed liquor in suspension. In plug flow tanks there is no need to maintain a liquid velocity. Since water is about three orders of magnitude denser than air, much of the energy of brush aerators is used for pumping liquid and not for aeration. Therefore, ditches with surface brushes are often candidates for retrofitting to fine-bubble diffusers and low-power subsurface mechanical mixers/pumps (such as 'banana-blade' mixers), which may significantly reduce the energy consumption of oxidation ditches.

A less frequent application of surface brush aerators is shallow aerated lagoons (Figure 9.11). The surface brush is then mounted on a floating barge, open at the centre, where the surface aerator carries on mixing and aerating. This type of aeration technology is chosen in lagoons for ease of operation and maintenance, as the barge can be easily tugged ashore for maintenance or repair. Also the aerator can be moved around to prevent the accumulation of sludge on the bottom.

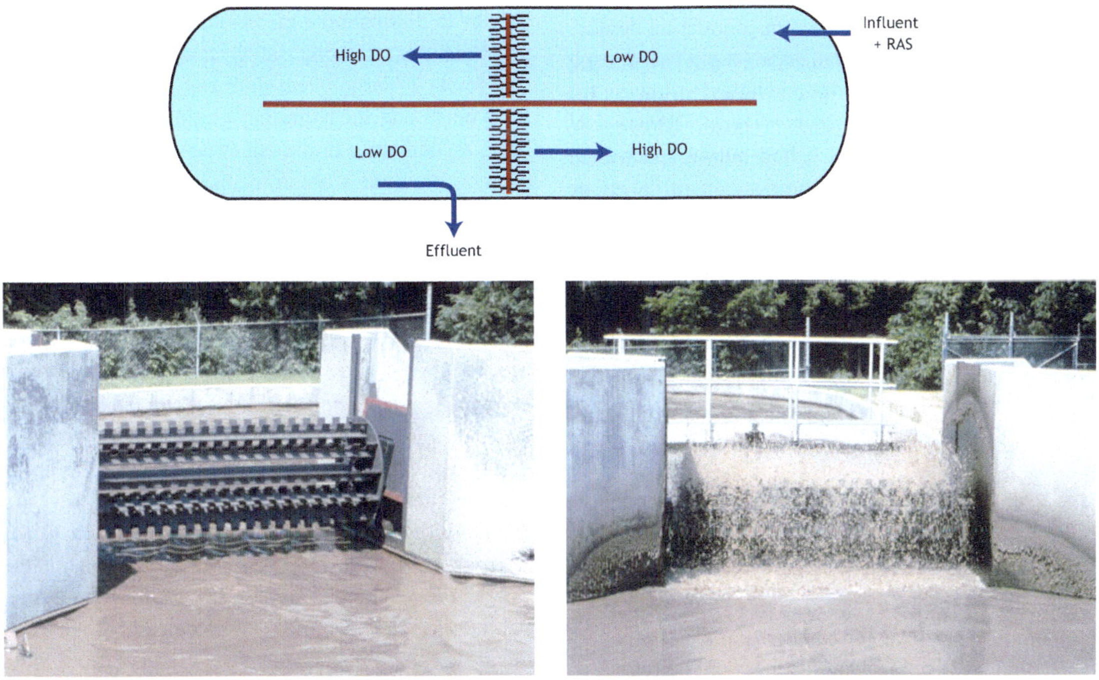

Figure 9.10 An oxidation ditch (shown schematically above) equipped with a surface brush aerator during standby (bottom left) and in operation (bottom right). The aerator's shaft is mounted on rails that can adjust the power drawn and lift the entire aerator above the free water surface, thereby avoiding any obstruction to the liquid flow.

Figure 9.11 A surface brush aerator mounted on a barge in an aerated lagoon.

9.1.4 Air blowers

Blowers are adiabatic fluid machinery used to produce relatively high volumes of low-pressure air or gas for mixing, aeration or combustion purposes. Blowers do not compress air, they blow it. This concept is critical because the limitations of blower discharge pressure can be a crippling constraint in the selection and operation/maintenance of diffusers. The adiabatic compression operated by blowers has a corresponding increase in pressure at the outlet of the blower only by virtue of the fact that we impose a pressure drop onto the aeration system. Without a pressure drop imposed on the air discharge, there is no pressure and only velocity (just like a hair dryer).

All blowers are associated with some degree of inefficiency, with 60-80% being a typical efficiency. The power lost to blower inefficiency elevates the discharge temperature and only this increased temperature can heat the aeration tank (in practice, there is no appreciable water temperature increase). In reality, some heat losses from a hot steel pipe to the surrounding atmosphere should be expected, and often we observe air distribution pipes that are jacketed to conserve heat and prevent contact burns. Since blower ability to overcome pressure drop is very limited, at some point every blower will meet its maximum discharge pressure, which is referred to as surge. The surge zone is always on the upper left area of a blower curve. When a blower surges, it can no longer displace air through the outlet and begins an unstable operation (vibrations begin rapidly, due to the need for the blower to dissipate the mechanical work imparted by its motor) until some safety controllers impose a shutdown. A surge event does not go unnoticed: the loud noise and the vibrating gratings on the blower building announce imminent structural damage, if the blower operation is not immediately halted.

The compressed air is distributed through a piping system to submerged diffusers for release at the floor of the reactor. As the bubbles rise through the bulk liquid, oxygen is transferred into it. A well-designed blower system supplies a wide range of airflows, matching fluctuating process oxygen demand, typically with a relatively narrow pressure range under variable environmental conditions. Blowers, as a key part of aeration systems, need to be carefully evaluated. Particular attention must be paid to their 'turn-up' or 'turn-down' capabilities to ensure that the full range of demands can be met, without any unattainable flows in that range. Remember that 4/5 of the air is composed of a gas that is not contributing to aeration, hence even more attention must be given to the transfer of oxygen; any changes in oxygen requirements multiply the air requirements (and blower power requirements and energy costs) by several fold.

Types of blowers

Blowers are classified into two broad types: dynamic (*i.e.* centrifugal) and positive displacement. There are a number of different configurations of each type of blower as shown in Figure 9.12.

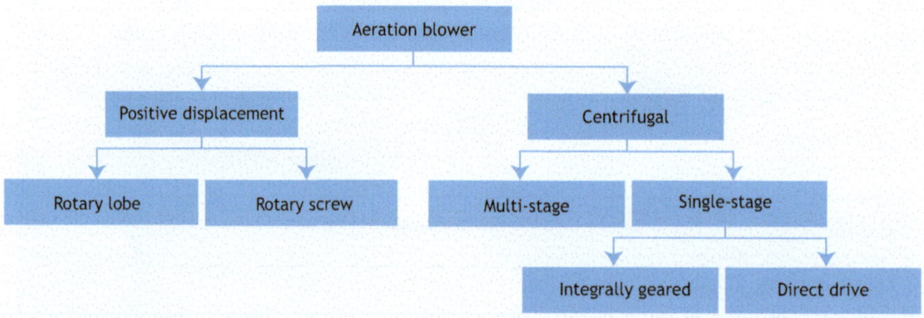

Figure 9.12 Positive displacement and dynamic blower configurations.

Table 9.3 Summary of features characteristic of PD and centrifugal blowers.

Positive displacement blowers	Centrifugal blowers
• More economical at small scale. • Noisy - the rattling associated with the rotary lobes is harder to dampen. Three lobe blowers partially address this shortcoming. • Vibration transmissions to piping and supports is sometimes problematic. • Motor overloads with excessive discharge pressure, requiring motor current protection. • Higher discharge pressures possible.	• Economical at all scales but especially for large installations. • Can also be noisy but the continuous, higher frequency spinning sounds are easier to dampen. • Operation at excessive pressure causes surge, which may result in destruction of the blower. Current protection and vibration detection controls are required for safe operation. • Higher discharge flow possible.

Centrifugal (or dynamic) blowers, like centrifugal pumps, produce variable flow and a narrow range of pressures, using rotating impellers. On the other hand, positive displacement (PD) blowers are generally considered as constant flow and variable pressure devices. PD blowers continuously compress a fixed volume of gas in an enclosed space to increase the pressure and provide some velocity. There are advantages and disadvantages of the two types, which are summarized in Table 9.3.

9.1.4.1 Centrifugal blowers

Centrifugal blowers are more common than positive displacement blowers at larger plants due to larger volumes of low-pressure air available. These blowers rely on the rotational speed of an impeller within an enclosed casing to provide velocity to the gas. The air is continuously discharged radially, and its increased kinetic energy is converted to a pressure increase by imposing the pressure drop on the discharge. Figure 9.13 illustrates the concept. Also, original centrifugal blowers do not have turn-up or turn-down capability and must be operated at constant rotational speed.

Newer technologies include centrifugal blowers with variable intake guide vanes and variable outlet diffusers. By varying the angle of the vanes, the airflow rate can be varied, and the blower acquires tuning capability. Nevertheless, centrifugal blowers have an optimal operating region, and outside that region the efficiency declines.

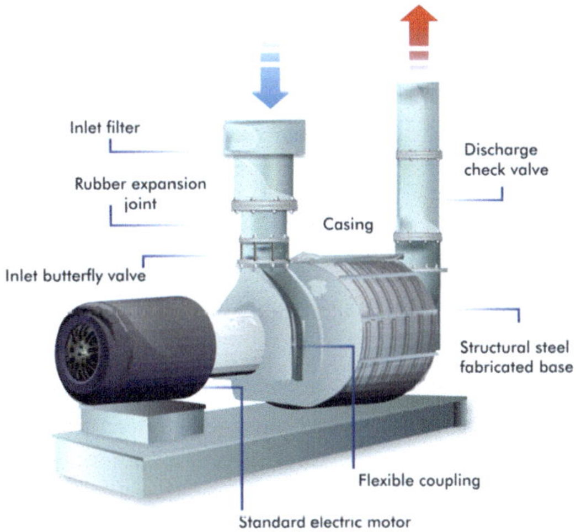

Figure 9.13 One of the most common blowers in WRRFs: the multi-stage centrifugal blower.

As with positive displacement blowers, centrifugal blowers are also subject to losses as gas under pressure escapes between the impellers and casing and around the impellers. Figure 9.14 shows the discharge flow and pressure relationship of a number of commonly used blowers.

The performance of blowers is affected by changes in the temperature of the inlet gas and barometric pressure, which causes a change in density of the compressed gas. The density of the gas or air increases with lower temperature and/or higher pressure, requiring greater power input to compress the same volume of gas. The capacity of centrifugal blowers decreases at high temperatures, and proper design requires that the blower size be evaluated at the maximum expected operating temperature.

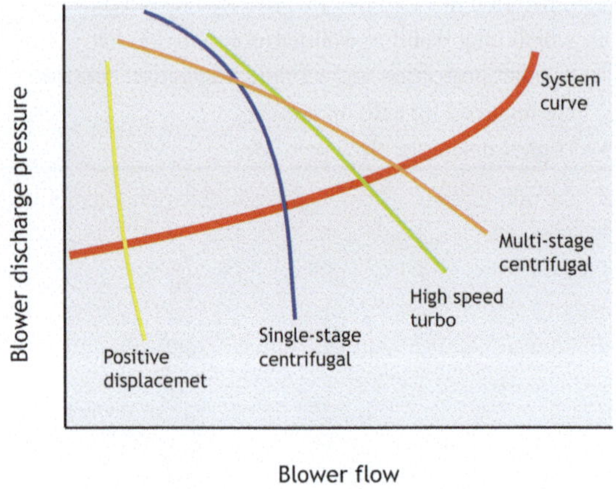

Figure 9.14 Pressure vs flow relationship for commonly used blowers (adapted from Loera, 2013). The surge area is in the upper left corner of the plot.

A typical multi-stage centrifugal blower is shown in Figure 9.15. Under conventional operation, multi-stage blowers use a number of impellers in series to increase the discharge pressure of the gas. In this configuration, variable output can be achieved by installing a control valve on the blower inlet or through the use of a variable speed drive on the blower motor. Turndown of multi-stage centrifugal blowers is limited however because their curve is flat as output approaches the surge point as shown in Figure 9.14. Figure 9.14 shows the typical performance curves for multi-stage centrifugal blowers with a variable speed driver. Note the proximity of the anti-surge control line to the predicted surge limit. The performance curves become quite flat as the capacity is reduced towards the surge line creating a zone of potential instability. It is important to ensure that the anti-surge control line is located sufficiently far from the predicted surge line to maintain stable operation. Location of the anti-surge line identifies the minimum turndown of the blower.

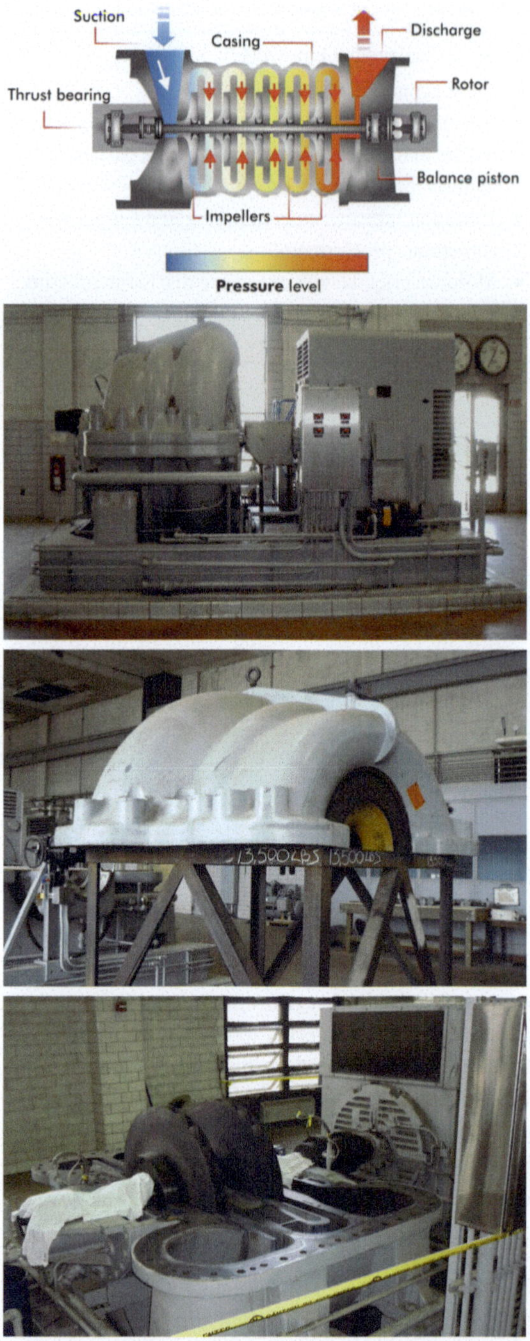

Figure 9.15 Multi-stage centrifugal blower schematic (top) and example of a multi-stage centrifugal blower in operation at a large wastewater treatment plant (second top). Below is a similar unit but during maintenance.

Single-stage centrifugal blowers rely on a single impeller to develop the required discharge pressure and airflow volume. There are a number of different single-stage centrifugal blower. Differences in the various configurations include variable-frequency-driven (VFD) motors *versus* constant-speed motors, or the use of inlet guide vanes and outlet diffuser vanes. Single-stage centrifugal blowers typically operate a much higher rpm than multi-stage blowers and there is usually a gearbox to increase the impeller rotational speed.

The direct-drive, or high-speed turbo single stage centrifugal blower is relatively new and utilizes an impeller very similar to the integrally-geared single-stage blower impeller. They are referred to as direct-drive because the impeller is attached directly to the motor armature shaft. They are also referred to as 'high-speed turbo' because their rotational speed is very high (as high as 70,000 rpm). Because of the high rotational speed, specialized equipment such as a high-speed VFD, an EMC filter, a harmonic filter, and a sinusoidal filter, special bearings, etc., becomes required to complete the full blower package. These blowers were introduced to the market at the turn of the century. Direct-drive single-stage centrifugal blowers are manufactured in two configurations. Due to the way the motor and impeller are closely integrated, direct-drive blowers are ideally suited for installations requiring a single-supplier integrated package.

Figure 9.16 shows the horizontal configuration using airfoil bearings. Since it uses airflow to suspend the rotor axis, the airfoil-bearing type is not a good choice for applications requiring frequent stops and starts. In fact, this is the bearing for jet airplane turbines which is why aircraft in airports do not use turbine power (to avoid wear of the shaft). The other option is to use permanent magnet bearings. The capacity of these blowers is currently limited to approximately 200 kW by the bearing technology and the weight of the armature.

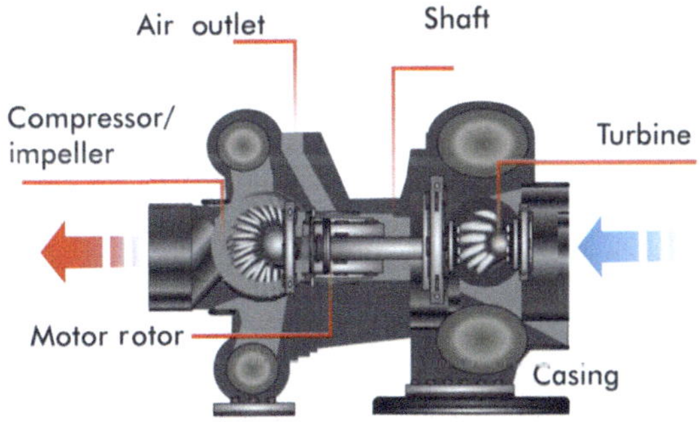

Figure 9.16 Direct-drive or high-speed turbo blower.

An integrally-geared single-stage blower relies on a gearbox located between the drive motor and the blower to increase the speed of the specially designed impeller to provide the required airflow and discharge pressure. The gearbox and impeller are designed based on variable site-specific conditions of airflow, discharge pressure, inlet air temperature, and relative humidity. Figure 9.17 shows a cut-away view of an integrally-geared blower showing the airflow path through the inlet guide vanes, into the impeller, through the outlet diffuser vanes to the discharge pipe. The blower operates at a fixed rotational speed and the complex control system that continuously varies the angle of the inlet and guide vanes allows for variable flow and relatively high efficiency even at turndown.

9.1.4.2 Positive displacement blowers

Positive displacement (PD) blowers produce compressed air at constant capacity and varying pressure. PD blowers compress discrete volumes of air transiting through the cavities of coupled of bi- or tri- lobed gears or rotors. The following drawing illustrates the concept (Figure 9.18). Due to the discreet nature of the process, positive displacement is not as efficient as centrifugal flow, but can achieve higher output pressures for the same airflow rates. Also, the airflow can be varied by varying the speed of the PD blower. The disadvantage of this type of blower is the noise produced by compression, typically recognizable as a loud thumping noise.

Figure 9.17 Integrally-geared single-stage centrifugal blowers. Top left: blower battery; top right: detail of intake while on display with servomotor operating the guide vanes on the left of the intake; bottom left: detail of the inlet vanes; bottom right: detail of the discharge guide vanes.

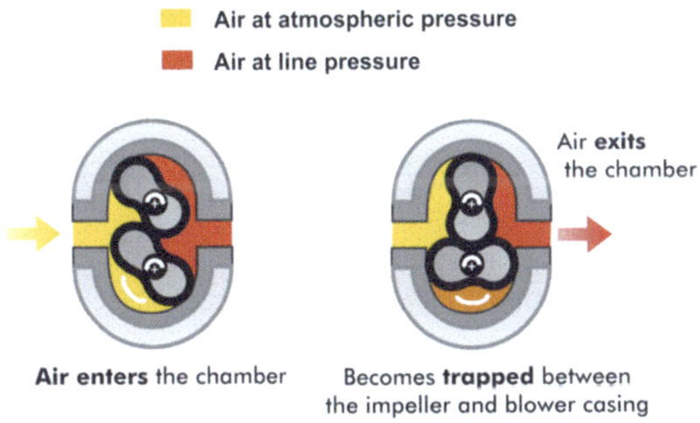

Figure 9.18 Schematic of a positive displacement blower.

As already mentioned, PD blowers are not perfect, and losses reducing the efficiency inevitably occur. Losses in positive displacement blowers occur as air under pressure escapes around the lobes, and between the casing and the lobes.

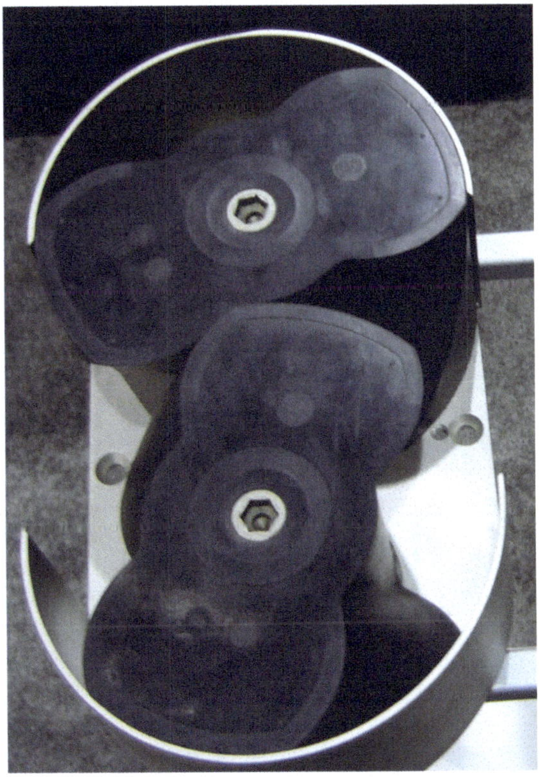

Figure 9.19 Rotary lobe positive displacement blower.

The typical rotary lobe blower, as shown in Figure 9.19, consists of two symmetrical impellers that rotate in opposite directions in an enclosed space, moving gas from the inlet to the outlet. Rotation of the lobes does not compress the gas. Pressure is generated by restricting discharge of the gas from the machine.

Similar to the rotary lobe blower, the rotary screw consists of two helically-shaped shafts rotating in different directions in an enclosed space. The cavity volume between the two shafts progressively decreases, resulting in an increase in pressure and decrease in discharge capacity from the inlet to the outlet of the blower casing. In contrast to the traditional PD blower, the rotary screws operate in continuous mode (not discretely) and are *de facto* a hybrid technology between PD and centrifugal.

9.1.5 The 'elephant in the room': HPO processes

The High Purity Oxygen Activated Sludge process (HPOAS; Figure 9.20) involves the use of pure oxygen in covered stage reactors, instead of air diffusers or mechanical aerators in uncovered basins. The process was developed to combat the need for plant expansion in capacity when limitations were imposed on the physical footprint of the facility. Hence, most of these installations are in densely populated areas and tend to be large processes. In the most common layout, 3 to 5 reactors in series are used to ensure a high utilization of the applied oxygen. Larger plants will typically have several parallel trains of reactors. The process was introduced and developed by the Union Carbide Corporation. For a history of the technology and its development program see the book by the process inventor (McWhirter, 1978).

Figure 9.20 Typical high-purity oxygen activated sludge process. The series of reactors is covered while the secondary clarifiers are open to atmosphere.

The adoption of this technology reached its maturity a generation ago (in the mid-1980s). The age of these facilities, combined with the increased energy costs and interest in nutrient removal/recovery, poses the question of what to do as we approach the end of their useful life. A large portion, if not the majority, of the wastewater flow in the United States is received by a small number of HPOAS facilities. More importantly, approximately one-third of the United States wastewater flow is handled by 47 plants that are larger than 100 MGD (Metcalf & Eddy, 2014), many of which are located in coastal areas (e.g. New York, Boston, Philadelphia, Los Angeles, San Francisco, Miami, Seattle, etc.). In fact, given the small number of HPOAS plants and their relatively large size, the analysis of a small number of plants can impact a large number of people served on a continental scale.

HPOAS processes are almost exclusively performing BOD_5 removal, which is a challenge when nutrient removal requirements are faced by the utility. Traditionally, HPO plants that have required nitrification have been built as two plants in series with inter-stage clarifiers. Moreover, as water reuse becomes an integrated part of the water supply portfolio in various areas, changes to the HPOAS plants are appearing on the horizon and the future of these HPOAS plants needs to be rethought to address the following issues:

- The gas traps between reactors at HPO plants also trap foam. To combat this problem, HPO plants have typically operated at very low SRTs in order to 'wash out' foaming organisms. SRTs in the 1 to 2-day range are common for HPO plants.
- The low SRT prevents nitrification, which is often not needed for large coastal plants. When nitrification is practiced at HPO plants by increasing the SRT, it may be necessary to adjust pH. The closed reactors in HPO plants prevent carbon-dioxide stripping which lowers pH. It is common to observe pHs of 6.0 in HPO reactor effluents, which reduces nitrification rates and may cause excessive corrosion. It will also be necessary to develop new procedures to reduce sludge bulking.
- Operation at low SRT usually produces an effluent that is not suitable for water reuse, due to high effluent TSS and poor removal of emerging contaminants (Leu *et al.,* 2012).
- Operation at low SRT and high shear in the reactors creates other challenges in solids separation, due to poor mixed-liquor thickening and elevated waste sludge volume (*i.e.* low MLSS).
- Energy consumption at HPO plants can be higher than at air plants. Initially when HPO plants were first introduced, they were more energy efficient than air plants because air plants mostly used spiral roll coarse-bubble aeration systems. However, with the advent of fine-pore, full-floor coverage diffusers, the situation is reversed.

- It will be difficult to convert HPO plants to air due to the high OURs that HPO operation supports. The first reactor OUR can be as high as 150 mg/l.h which is well above the practical maximum of 80 mg/l.h provided by efficient fine-pore diffusers. A possible solution is increasing reactor volume, which may not be possible.
- The oxygen generation facilities at many HPO are at the end of their serviceable life and need replacing, which will require capital and a continuing reliance on HPO technology. Recent advances in pressure-vacuum swing technology may moderate this problem, since this new technology is perhaps twice as efficient as earlier oxygen generation technologies.
- Simple conversion of surface aerators to fine-pore diffusers at HPO plants is not feasible with existing equipment. Operation at high oxygen purity is not compatible with an existing diffuser and aeration piping. A new generation of fine-pore diffusers with suitable materials will be required. Also, the low-speed surface aerators used at HPO plants have higher α factors (0.8) at the low SRT. Fine-pore diffusers would have much lower α factors (~0.4) at the low SRT.

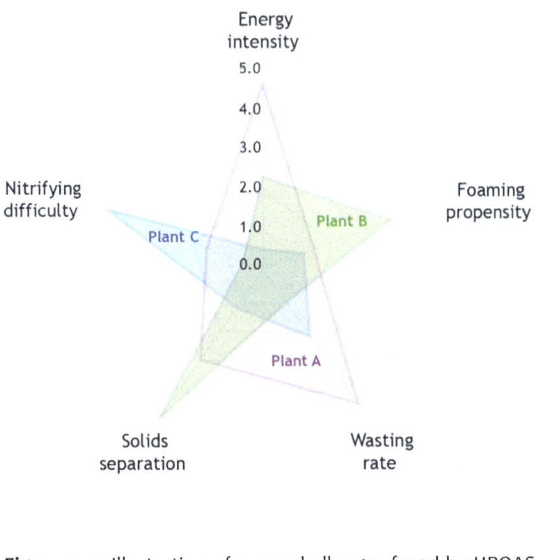

Figure 9. 21 Illustration of some challenges faced by HPOAS processes for three sample plants.

A well-controlled process achieves 70-80% oxygen utilization by the bacteria, which equates to 20-30% of the oxygen generated being wasted. In fact, the original forecasts by Union Carbide of optimal performance reaching ~90% oxygen utilization were rarely achieved. However, the engineers at the San Francisco Oceanside Water Pollution Control Plant (OSP) devised a more efficient control strategy capable of achieving up to ~100% utilization of the oxygen supplied to the process (Miot *et al.,* 2016).

As many of these facilities are reaching the end of their useful life, the open question regarding their future is becoming ever more important. However, each case is being treated on a site-specific basis drawing from this suite of technological options:

- Conversion of an HPOAS facility to air, if land is available.
- Decommission on-site oxygen generation and purchase oxygen from an *ex-situ* supplier.
- Convert partially or completely to atmospheric process. Carbon-dioxide stripping in the final reactor of HPO plants has been successfully used for almost 20 years to raise effluent pH.
- Incorporate state-of-the-art intensification technologies (*i.e.* ballasted processes, biofilm-based processes).
- Nitrify and denitrify in a modified HPO process.
- Add a NDN process downstream from HPO.

At the time this book was written there were no clear guidelines on how to approach the issue systematically. Several large HPOAS plants in North America are evaluating alternatives and developing ways to nitrify and denitrify.

9.2 MIXING IN ACTIVATED SLUDGE

In wastewater treatment plants, aeration provides both oxygen and mixing. However, due to plant hydrodynamics, process conditions in aeration basins can vary significantly throughout the tank volume. The uniformity of the activated sludge suspension, hence of its biological activity, is determined by the mixing conditions (Rosso, 2018). The reduction in conversion efficiency due to short-circuiting can exceed 50% (Ahnert *et al.*, 2010), while the corresponding dead zones can amount to 10-20% of basin volume (De Clercq *et al.*, 1999). Mixing is essential to accomplish the following objectives: (*i*) keep mixed liquor in suspension and ensure that the entire zone volume is utilized; (*ii*) blend different feed streams to produce a homogeneous suspension; (*iii*) encourage flocculation of the mixed liquor; and (*iv*) avoid short-circuiting.

Generally, mixing can be classified according to its duration. The first is rapid (<30 s) and is used where blending is required such as where a chemical is added to the effluent or mixed liquor (*i.e.* addition of coagulants for flocculation, polymers for dewatering, hypochlorite for disinfection, etc.) or different mixed-liquor streams are combined (*i.e.* RAS and PE, internal recycle streams and mainstream flow). Rapid mixing has been extensively studied and modelled (Rosso, 2018). Details, equipment characteristics, and further examples used in wastewater-treatment facilities for rapid mixing can be found in technical manuals and in Metcalf & Eddy (2014). However, the type of mixing process that is less understood and certainly more challenging to characterise and model is continuous (or gentle) mixing. This cannot be decoupled from observed microbial reaction rates and governs two key processes that accelerate biological conversion: flocculation (formation of microbial aggregates or flocs) and maintenance of solids in suspension. Although some researchers have been tempted to distinguish the two, they are similar in most applications, given that solids need to be maintained in suspension in order for flocculation to occur (Pretorius *et al.*, 2015).

Mixing and aerating activated sludge tanks are energy-intensive processes, so therefore it has become important to optimize mixing equipment (Füreder *et al.*, 2018; Sharma *et al.*, 2011). While aeration can make up about 50-75% of the total energy consumption of a WRRF (Reardon, 1995; Rosso and Stenstrom, 2005; WEF, 2009) some authors have shown that mixing consumes 5-20% (Füreder *et al.*, 2018; Krampe, 2011). In addition, up to three quarters of the tanks have an optimization potential in terms of energy consumption (Füreder *et al.*, 2018). Therefore, the study and optimisation of mixing goes hand in hand with any effort to decrease the energy intensity and increase the sustainability of the treatment process.

Aeration is in general not decoupled from mixing, whether surface aeration or diffused air are used. The release of a continuous supply of bubbles satisfies the oxygen demand, but also maintains the mixing conditions within the reactor, which helps to ensure that the corresponding substrate (*i.e.* COD, NO_3^-, VFAs, etc.) is in contact with microorganisms. However, decoupling aeration from mixing may have advantages but in such cases unaerated zone must be provided with minimal mixing. Providing only mixing without aeration can be done in two ways. The most obvious is by mechanical means, usually with impellers. The option of employing air-powered equipment to provide mixing exists, but care must be taken to avoid oxygen transfer (e.g. by mixing anoxic zones with coarse bubble diffusers, à la Neethling). The advent of mechanical devices that can provide both oxygen transfer and mixing (e.g. hyperbolic aerator/mixers) provides the opportunity to fit aerated, swing, and intermittent zones with one equipment unit.

Sometimes limiting or reducing both aeration and mixing conditions can provide improved nutrient removal, such as the *in-situ* fermentation proposed by Barnard *et al.* (2017). Incomplete mixing conditions aim to allow solids blankets to develop on the floor of the unaerated zones, triggering anaerobic conditions in the blanket and promoting fermentation-related processes, which transform complex organic carbon into more easily degradable and available organic carbon. Caution must be used when exploring these configurations, due to the secondary effects of excessive fermentation (Barnard *et al.*, 2017).

9.2.1 Mixing quantification and design

The traditional conservative approach of oversizing mixing systems during the design phase can have major disadvantages (Pretorius *et al.,* 2015). These include excessive capital spent on equipment and too much power used for the entire service life of the mixer, wasting resources and power, and negatively impacting the environment. Further, overly mixing in anaerobic or anoxic zones can lead to entrainment of air, negatively impacting the selector performance and effluent quality, especially in biological nutrient removal (BNR) systems. Moreover, the mixed-liquor floc structure can be damaged, encouraging floc breakup and negatively impacting sludge settleability (Rosso, 2018). Finally, Barnard et al. (2017) stated that overly conservatively sized mixers often inhibit the fermentation of RBCOD in anaerobic zones due to incidental aeration, thereby limiting the potential for biological P removal. All these disadvantages underline the importance of proper design for mixing equipment.

Ensuring proper mixing conditions requires a quantification of the level of mixing. Yet, the question of how to quantify 'homogeneous suspension' in aeration tanks is the subject of an ongoing debate as other authors have defined different benchmarks for this condition (Rosso, 2018). The methods proposed for quantifying mixing efficacy can be divided into two groups: traditional and computational (CFD). In the majority of designs, fictional conditions of complete mixing are usually assumed (Stamou, 2008), but in reality factors can impact this theoretical 'good or complete mixing' in the AS process. Even if some common rules of thumb indicate the minimum air requirements to provide good mixing conditions (Metcalf and Eddy, 2013; Pretorius *et al.,* 2015), none of them can assume the impact of other factors.

It is commonly accepted that the air requirement to ensure effective mixing will vary from 1.2 to 1.8 $m^3/h.m^3$ of tank volume, and that typical power requirements for maintaining a completely mixed flow regime with mechanical aerators varies from 13 to 26 W per m^3 of tank volume (Karpinska *et al.,* 2016). However, as we mentioned, none of these assumptions consider the impact of tank hydraulics (cross-section, depth or presence of baffles), energy input, or any other variable that might affect the mixing.

One of the most common methods used to quantify mixing is the mapping of the Coefficient of Variation (CoV) of the solids concentration, DO or OUR. The CoV method can be used not only for total suspended solids (TSS) but also for other key parameters, ranging from contaminants (*i.e.* BOD_5, nitrite, and nitrate) to process indicators, which can bring useful insights on the existing homogeneity conditions. Moreover, using DO measurement as an indicator of mixing may have also some limitations, especially regarding low DO values. Low DO values (*i.e.* DO<0.5 mg/l) are difficult, if not impossible, to use to define the actual oxygen conditions (aerobic, anoxic, anaerobic) and certainly to discriminate between the anoxic and anaerobic metabolisms, due mainly to the noise-to-signal ratio of DO probes at their low range of operation. Other factors, such as biological conversion rates, substrate concentrations, concentration of various electron acceptors, etc. also affect whether anoxic or anaerobic reactions are taking place.

A promising approach in the monitoring of mixing is tank mapping by using oxygen-uptake rate (OUR) measurements. OUR is an indicator of biomass activity ($mgO_2/l.h$) and it can became an important technique in assessing the distribution of the biological reactions taking place in the aeration tank. The calculation, and corresponding mapping, of the microbial activity distribution by OUR can detect those areas where the suspended biomass is receiving less load due to hydrodynamic/mixing issues. These OUR readings will help to identify those areas where the values are lower than expected in comparison with neighbouring areas or past experience. However, we should clarify the main limitation of this approach, which consists in its bi-dimensionality. The bubbles collected represent the average of the gas transfer phenomena over the water depth, and there is no depth mapping when bubbles are collected at the surface. In theory the off-gas could be collected below the surface, thereby allowing the mapping in 3-D, but the practical limitations of pushing the gas hood in depth, combined with the lower gas transfer at lower depths (while the instrument error remains unaffected by depth), severely limit this approach. OUR mapping via off-gas relies on the collection of off-gas at the surface, which means that only the bubbles reaching the surface at the point of collection are captured by the analyser. These bubbles may have been generated below the off-gas hood or in its vicinity (especially upstream, due to the average forward liquid velocity). Hence, sampling uncertainty is inherently built into this measurement. Figure 9.22 was created to show two typical OUR profiles in aeration tanks (Hodgson *et al.*, 2019). It can be seen how the OUR mapping unveils potential shortcuts in the hydrodynamics which are decreasing the reactive or functional space within the tank.

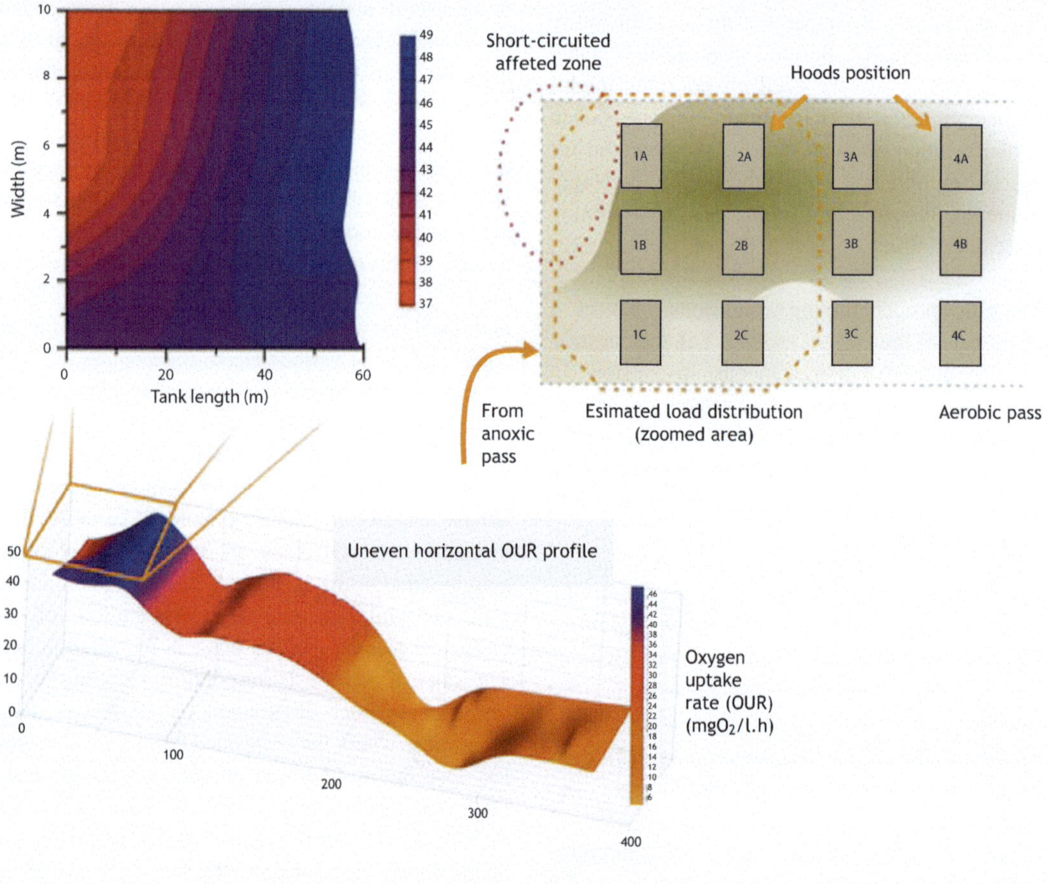

Figure 9.22 OUR mapping. The upper left figure shows the corresponding position of the deployed hood to retrieve the off-gas measurements (OTE, %). The X axis corresponds to the aeration tank length, the Y axis to oxygen uptake rate ($mgO_2/l.h$) and Z to the width of the reactor pass. Note the visible gradient across the reactor width.

Over the last few years the use of Computational Fluid Dynamics (CFD), which combine CFD analysis and biological reaction rates, has been evolving (Fayolle *et al.*, 2007; Karpinska *et al.*, 2016) and attempts are being made to overcome its current limitations (Amaral *et al.*, 2019; Karpinska *et al.*, 2016). Potential possibilities of coupling the CFD data with ASM-based codes have been explored in order to generate a suitable tank-in-series model, where the biokinetic model can be implemented (Karpinska *et al.*, 2016). However, to date, a model capable of a complete three-phase CFD model coupled with an ASM is unavailable. Despite seeking more reliable results than those obtained from the sole ASM, this approach relies on the trade-off between a model which can be implemented and run in a realistic timeframe, and the extent of simplifications and the solution accuracy, which may lead to a series of output errors.

9.2.2 Mixing equipment

Rotating impellers can induce flow in either a radial or an axial direction (Figure 9.23).

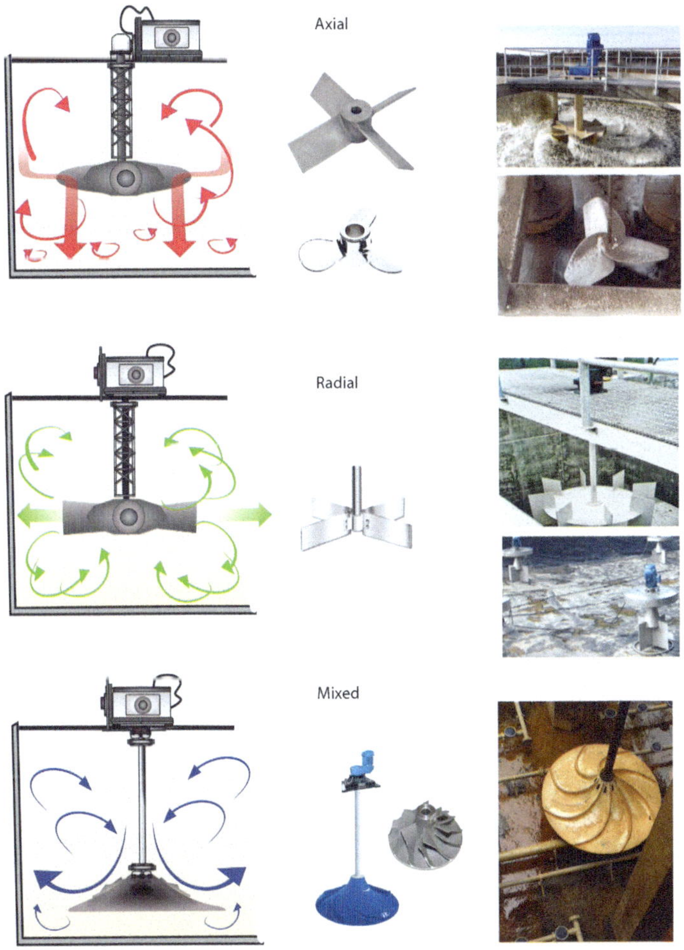

Figure 9.23 Schematics of flow patterns and image of different impellers.

In wastewater applications, axial flow impellers are most commonly used. The axial flow can be forward (away from the shaft through which the impeller is powered) or backward. The chief example of backward flow would be up-pumping vertical impellers. The main exception to axial flow would be the hyperboloid mixer (Figure 9.24) that is designed to induce flow radially from a vertical impeller, specifically along the floor of the zone. Floating vertical impellers may also be used to avoid structural requirements. Horizontal impellers, on the other hand, usually require submerged motors, since wall-penetrating impellers are seldom used in the wastewater industry. Submerged motors have proven to be challenging to keep free of leaks, due to a combination of poor maintenance and manufacturing features.

Figure 9.24 Pictures of the hyperbolic mixer deployed at a WRRF, installed above the existing diffuser grid.

Air-powered mixing is another option, which includes all bubble aeration systems. Many fine-bubble systems are designed to maintain a certain minimum airflow, considered the mixing limit. Many air-powered diffusers are designed specifically to induce bulk flow, in particular large and coarse-bubble diffusers, but also some fine-bubble mixers, for example when used in spiral roll configurations. Air-powered mixers have the advantage of distributing mixing throughout the entire water column without moving parts below the liquid surface. The main benefit of an air mixer is that the effect of the mixing power is distributed vertically through the entire water column from the point of release to the surface. However, the main concern with using air mixers in unaerated zones is incidental oxygen transfer. The concept of intentionally aerating an anoxic zone for the purpose of promoting both nitrification and denitrification was apparently first investigated at the 91st Avenue Plant, in Phoenix, Arizona (Stensel *et al.,* 1994). In subsequent years, simultaneous nitrification-denitrification has been demonstrated as viable. In such systems, the requirement for low-level aeration while maintaining a homogeneous suspension of mixed liquor is obvious.

Air mixing is usually employed in channels to transport mixed liquor from aeration basins to secondary clarifiers. Different researchers have found slightly different optimum power levels, and designing adjustable airflow rates allowing the operators to finetune the exact optimum power level for a given system is recommended (Pretorius *et al.,* 2015). A simple test to verify the effectiveness of using air for mixing unaerated zones is by dropping a sludge judge in the reactor and collecting samples from the reactor, and making a visual observation on the formation of a sludge blanket.

9.3 FACTORS AFFECTING OXYGEN TRANSFER

Different factors impact oxygen transfer in aeration systems, from environmental to process-related. In the following sections, the state-of-the-art regarding the factors affecting oxygen transfer is presented. For example, diffuser-related characteristics such as type, installation depth, distribution, time in operation and the airflow rate will influence the capacity of the system to transfer air in the form of dissolved oxygen. Similarly, aerobic reactor features such as depth, volume, and type of reactor, as well as operating conditions such as mean cell retention time, nutrient removal processes, turbulences, wastewater composition and fouling will influence the oxygen mass transfer (Eckenfelder Jr *et al.*, 1956; Groves *et al.*, 1992; Henkel, 2010). Nevertheless, the efficiency can also be impacted by other factors that can be controlled or predicted such as temperature, barometric pressure, salinity or water pressure.

9.3.1 Sludge retention time

The most important process variable to affect aeration efficiency in activated sludge is the mean cell retention time. SRT is directly related to the biomass concentration, and dictates oxygen requirements. Aeration efficiency and α factors are higher at higher SRTs. Biological nutrient-removal processes, by operating at increased SRTs, have improved aeration efficiency. Furthermore, anoxic and anaerobic selectors in plants with nutrient removal have beneficial effects that go beyond nutrient removal or improved settling characteristics. By utilising the readily available carbon in the wastewater, they more rapidly remove surfactants, which have the most dramatic impact in reducing oxygen transfer.

Literature studies (Rosso and Stenstrom, 2005; USEPA, 1989) showed that oxygen transfer efficiency is directly proportional to SRT, inversely proportional to airflow rate per diffuser, and directly proportional to geometry parameters (diffuser submergence, number and surface area of diffusers). Figure 9.25 illustrates this concept. The SRT determines the net oxygen requirement and relates to the degree of treatment and removal of oxygen-transfer-reducing contaminants. Higher SRT systems remove or sorb the surfactants early in the process, which improves average oxygen transfer efficiency. The net effect of increasing SRT is to increase the oxygen requirements, improve removal of biodegradable organics, and improve the overall oxygen transfer efficiency (Rosso and Stenstrom, 2007). The increase in oxygen requirement is partially or more than offset by the savings produced by the higher transfer efficiency, for processes in warm climates.

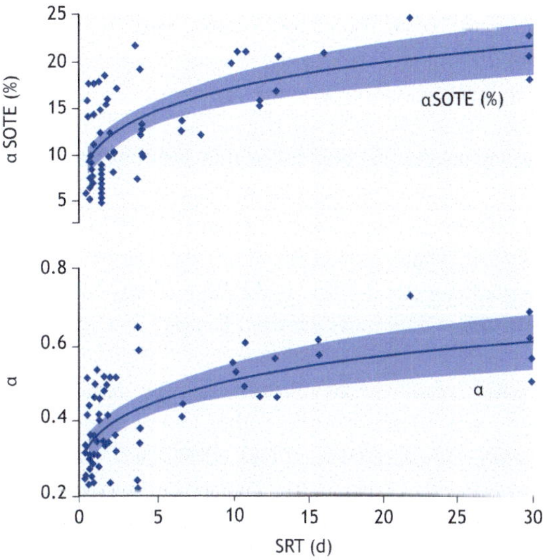

Figure 9.25 Effect of SRT on α and αSOTE. Shaded areas are 95% confidence intervals.

In addition to these advantages, there is growing evidence that processes operating at higher SRT are more efficient in removing anthropogenic compounds, such as pharmaceuticals (Göbel *et al.*, 2007; Soliman *et al.*, 2007). Studies reported removal rates of up to 90% for the endocrine disruptor 17a-ethinylestradiol (EE3) after a wastewater treatment plant was retrofitted to remove nutrients at SRT of 11-13 days (Andersen *et al.*, 2003). Operation at longer SRT in order to enhance the removal of trace organics will become more important as wastewater water reclamation becomes more widely practiced.

9.3.2 Role of selectors

Almost all new activated sludge process designs utilize selectors, either anoxic or anaerobic. The benefit of selectors is the reduction in the number of filamentous organisms (Harper and Jenkins, 2003), which improves the Sludge Volume Index (SVI) and reduces the probability of sludge bulking and rising sludge blankets in secondary clarifiers (Schuler and Jang, 2007). One study surveyed 21 plants with anoxic and anaerobic selectors, and reported that all the plants showed improvement after selector installation (Parker *et al.,* 2003). Among the plants with anoxic selectors, 70% had a sludge volume index (SVI) lower than 200 ml/g. Plants using anaerobic selectors were even better and more than 90% of the plants had SVIs of less than 150 ml/g. Martins *et al.* (2004) reported similar results and concluded that better operation is achieved if the first stage is anaerobic. The activity of phosphorus-accumulating organisms (PAO) was reported to increase even when operating a strictly anoxic selector, with PAO improving the floc structure and biomass density (Tampus *et al.,* 2004).

An advantage of selectors is the removal or sorption of a fraction of the carbonaceous load, *i.e.* the readily biodegradable COD (RBCOD). This is illustrated in Figure 9.26. The RBCOD is partially composed of surfactants, which are typically discharged as fatty acids, oils, soaps, and detergents. The surfactants, because of their amphiphilic nature, accumulate at the air-water interface of rising bubbles, reducing oxygen transfer efficiency. Removal of the RBCOD can improve oxygen transfer efficiency, which reduces operating costs for aeration (Rosso and Stenstrom, 2006a, 2007).

Figure 9.27 shows the results of a survey of many WRRFs tested for oxygen transfer efficiency. The diffusers tested were of all makes and models and had been operating for very variable lengths of time (from brand new to more than 5 years). In this figure we defined the diffusers as NEW when they had been installed within 1 month, USED when they had been in operation between 2 and 24 months, OLD when they had been in operation for more than 24 months, and CLEANED for those that had been cleaned within a month. We clustered the data by process type, *i.e.* BOD removal only (*i.e.* conventional), nitrification only, and NDN. Each point in the plot is the air flow-weighted average of an off-gas test for a tank at a WRRF.

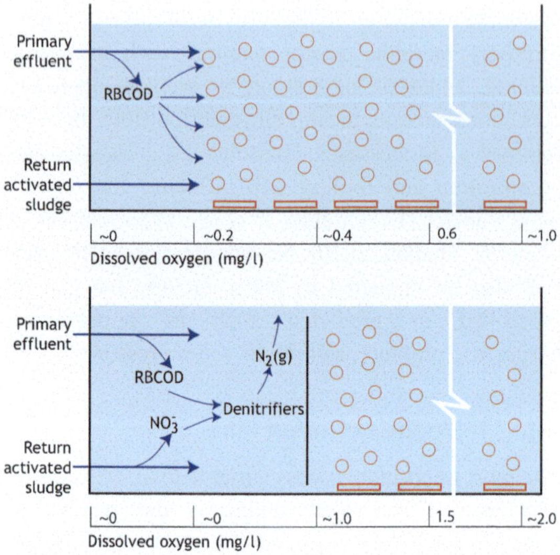

Figure 9.26 Schematic view of the role of selectors on the pathway and fate of readily biodegradable COD (containing surfactants). The graph shows proof of the different process efficiencies. The data points in the graph were measured in a treatment plant with two independent activated sludge trains with independent clarifiers, treating the same wastewater.

The first observation is that for each data cluster the spread is wide: alphas always range ±0.1, indicating a high variability amongst sites and times in operation. However, if we look at the averages and error bars there is a clear upward trend. In fact, conventional processes have to be operated at the lower end of SRT, NDN benefit from a longer sludge age. Also, the SRT range is the same for nitrification-only and NDN and the only difference appears to be the presence of anoxic selectors.

High-rate processes, characterised by very low sludge age (e.g. the A stage in the A/B process), may have a range of alpha factors that is not a downward linear extrapolation of the trend visible in Figure 9.28 (Rosso and Al-Omari, 2019).

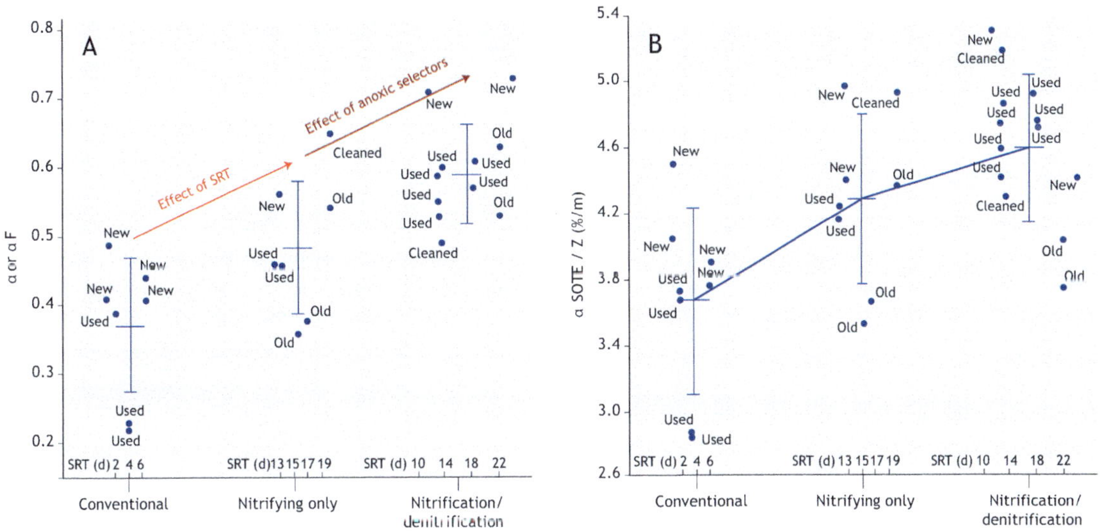

Figure 9.27 A) Normalized Standard Oxygen Transfer Efficiency for selected plants operating with different layouts; B) The evolution of α or αF factors for selected plants operating with different layouts. Labels refer to the diffuser status: NEW (within 1 month of installation), USED (between 1 and 24 months of operation), OLD (over 24 months in operation), and CLEANED (within 1 month from a cleaning event). Note the increase in α or αF with the increase in SRT, and the further increase due to anoxic selectors (the average SRT for N-only and NDN is the same).

For these and other reasons, anoxic selectors for nitrification/denitrification (NDN) should always be evaluated as an alternative to conventional treatment. Our previous analysis (Figure 9.28) showed that NDN operations can have lower total operating cost (aeration + sludge disposal costs - methane credit) than a nitrifying-only or a conventional treatment plant (Rosso and Stenstrom, 2005). If conventional process operating cost was normalized to 1.00, nitrifying-only will have a total cost of 1.13, and NDN operations will have a total cost of 0.88. NDN operation offers an oxygen credit due to its process nature, and higher oxygen transfer efficiencies associated with the higher SRT. The two factors overcome the additional oxygen demand that is created by the longer SRT.

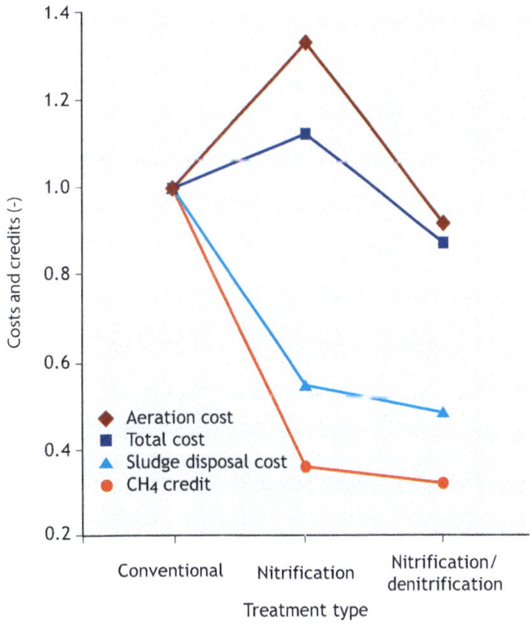

Figure 9.28 Dimensional and dimensionless costs and credits for different process layouts.

9.3.3 Airflow rate

In fine-pore diffusers, the bubble size distribution is linked to the airflow rate, with higher flow per diffuser resulting in larger bubble diameters. Consequently, these larger bubbles result in a smaller interfacial area. Therefore, as the airflow rate per diffuser increases, the clean water performance SOTE decreases (Warriner and Brenner; USEPA, 1989). The cause of more airflow in a plant is an increase in oxygen demand load, which corresponds to higher concentrations of contaminants that depress the alpha factor. Hence, in process water, when higher airflow is required, the lower SOTE compounds with the alpha factor decrease by the more concentrated contaminants, and the overall result is an amplified decrease in αSOTE.

Furthermore, increased aeration rates generally result in increased OTRs, because of the effects of turbulence caused by aeration. Higher airflow rates and mixing turbulence can reduce the depth of the boundary layer, improving the oxygen transfer coefficient k_L and OTR (Ji and Zhou, 2006; Vogelaar *et al.*, 2000).

9.3.4 Diffuser density

Typically, an increase in diffuser density, defined as the area covered by the diffusers in relation to the total area of the tank floor, is expected to result in a higher OTE. However, a maximum value exists for the density of diffusers in which the SOTE increase is minimal (diminishing returns), which is determined by diffuser size, airflow rate, and the space between diffusers (USEPA, 1989).

Care must be applied during design to prevent an overzealous specification of diffusers. If too many diffusers are selected, based on a cost/benefit analysis that indicates that more diffusers would have an overall energy lifecycle cost, there is a risk of excessively low airflow per diffuser. When diffusers are operated at excessively low airflow, the bubble discharge may be uneven and the mixed liquor may no longer be well-mixed.

9.3.5 Reactor depth

The OTE of the aerobic reactor increases with increasing depth because of the longer residence time of the bubbles, and the larger partial pressure of oxygen at the moment of bubble formation. However, Wagner and Pöpel (1998) and Gillot and Héduit (2008) showed how the alpha factor decreases with increasing depth. As the partial pressure increases with diffuser depth, the operating pressure for the blower also increases. Nonetheless, the Standard Aeration Efficiency (SAE) remains constant because an increase in depth reflects as an increase in energy consumption (USEPA, 1989; WEF, 2009). When measuring gas transfer coefficients in deep tank systems, the difference between the true k_La and the apparent k_La (*i.e.* the k_La actually measured) can be large. Researchers discussed these phenomena and the corrections during the period leading to the compilation of the earlier testing standards (Downing and Waste, 1963). The apparent measurement of k_La reflects an analogous departure in the calculation of alpha factors. The magnitude of the effects is usually small, and much less than the experimental error associated with field testing. However, the case of very deep tanks (>10 m) is still unresolved and *in-situ* testing remains the only solution to quantify alpha factors at this time.

9.3.6 Diffuser fouling, scaling, and cleaning

The oxygen transfer efficiency of fine-pore diffusers inevitably decreases over time due to a combination of fouling and scaling (Figure 9.29). Diffusers offer a unique habitat for biofilm development and growth. Both the substrate from the incoming wastewater and the soluble microbiological products (SMP) produced by the planktonic biomass in the aeration tank are readily available for the biofilm, serving as a substrate for the sessile microbial community. The exact microbial composition and its extent will vary from one treatment plant to another, as biofilm formation is affected by operating-process variables (*i.e.* availability of soluble substrate, dissolved oxygen level, iron concentration in the mixed liquor, and organic loading rate) (Kim and Boyle, 1993; Rieth *et al.*, 1990).

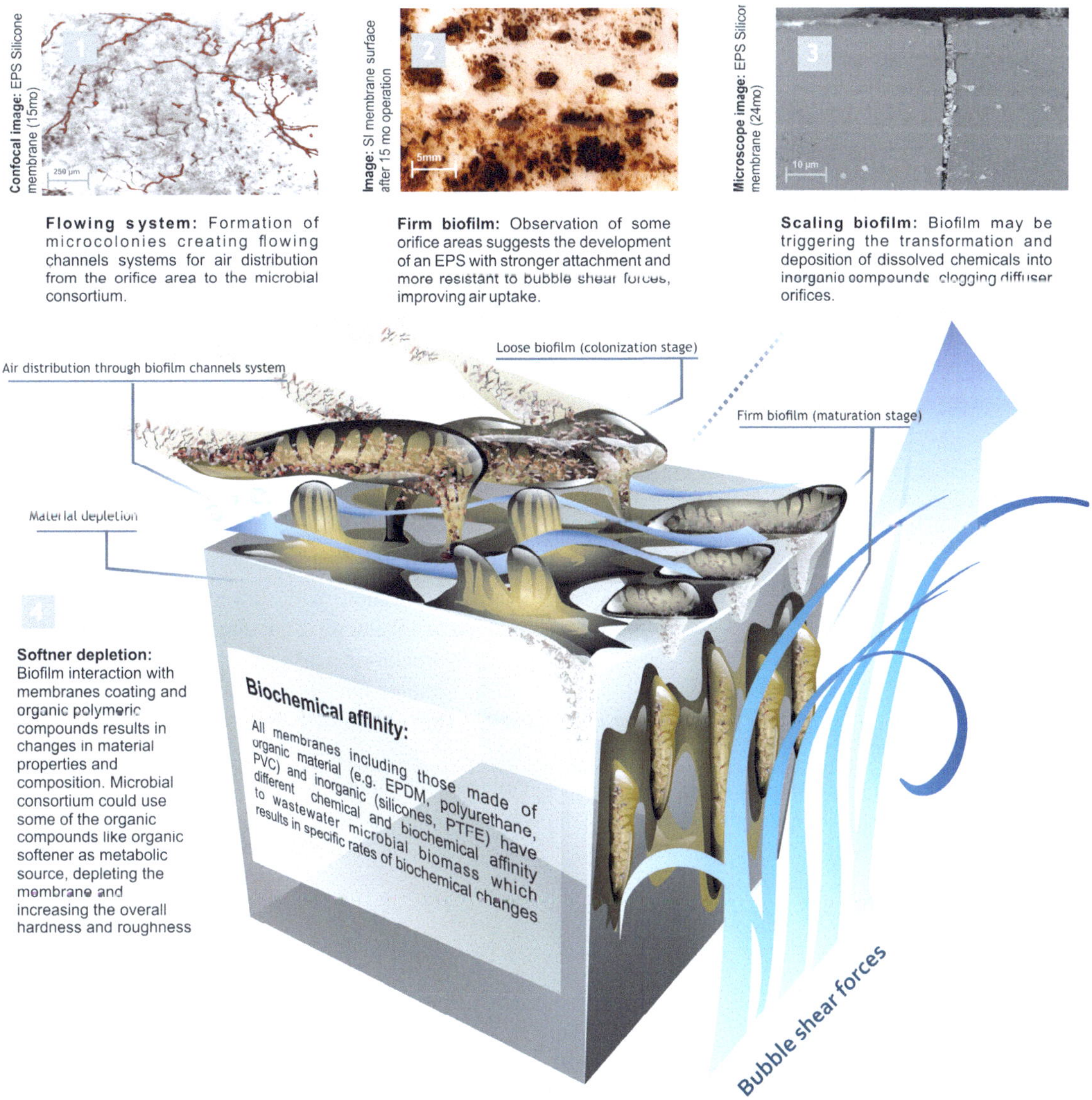

Figure 9.29 Illustration of biofilm interaction with mechanical and biomechanical fine-pore diffuser properties at the orifice surface of fine-pore diffusers, with micrographs showing details of the EPS structure, biofilm attachment and scaling phenomena (adapted from Garrido-Baserba *et al.*, 2018).

At the same time, the back pressure (often called dynamic wet pressure or DWP) usually increases, and in some cases dramatically. This DWP increase is due to clogging of pores in ceramic and membrane diffusers (USEPA, 1989) or is associated with a permanent change in orifice characteristics for polymeric membranes (Kaliman *et al.*, 2008). Both effects account for the decrease in overall process efficiency and power wastage. Cleaning fine-pore diffusers is almost always required and it restores process efficiency and reduces power costs. Observations of 94 field tests show that efficiency decreases with time and the greatest rate of decrease occurs in the first 24 months of operation (Rosso and

Stenstrom, 2006a). Efficiency decline was quantified and included in cost analyses, and the net-present worth was compared to cleaning costs. The cleaning frequency is always higher for higher fouling rates and optimal frequency was as short as 9 months and never more than 24 months.

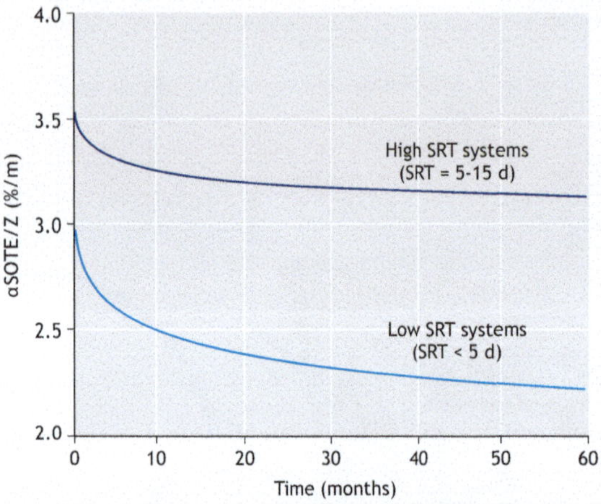

Figure 9.30 Alpha Standard Oxygen Transfer Efficiency per depth (αSOTE/Z) of 103 WWTPs in operation from 0 to 60 months. The WWTPs are divided into those with a SRT of less than 5 days and those with a SRT higher than 5 days.

Due to the chemical nature and the morphology of these materials, they experience fouling and scaling depending on the process conditions, water quality, diffuser type, and time in operation (USEPA, 1985,1989). As a result, fine-pore diffusers need to be routinely cleaned. The choice of cleaning frequency and method determines the long-term efficiency and benefits of using fine-pore aeration. Different methods have been used to clean fine-pore diffusers and these methods vary in complexity and cost. The simplest method is to dewater the aeration tank and wash the diffusers from the tank top, using plant effluent. This form of cleaning, called tank-top hosing, is effective in removing biological slime accumulation, and it generally partially or completely restores efficiency. However, for cases where inorganic precipitates (silica, calcium carbonate, gypsum, etc.) have caused scaling, acid cleaning may be required. In addition, manually washing with low-strength hydrochloric acid (10 to 15% w/w) is popular and acid gas cleaning using hydrochloric acid gas or acetic acid injected to air distribution lines is also possible (Schmit *et al.*, 1989). This technique is limited to ceramic diffusers, due to the incompatibility between currently employed polymeric membranes and strong inorganic acids.

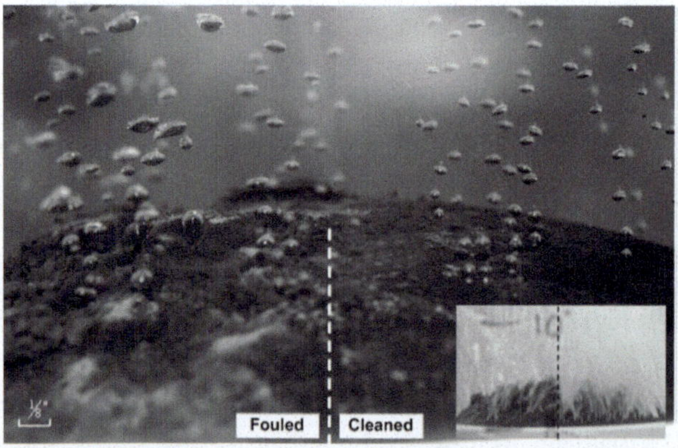

Figure 9.31 A fouled EPDM disc diffuser harvested from a conventionally-operated treatment plant after 2 years in operation (SRT < 2 days), cleaned by surface brushing on its right-hand side. The difference in number and size of bubbles released is due to the bio-fouling layer still present on the left-hand side of the membrane. The air released from the orifices on the left half has to travel through the biological layer, and during this travel time it has the opportunity to coalesce and form fewer and larger bubbles with disadvantaged mass transfer. The detail shows the membrane at operating airflow rate (large photo courtesy of Shao-Yuan Ben Leu).

Figure 9.32 Examples of diffuser cleaning. Top: a dewatered tank equipped with ceramic domes before (left) and during (right) cleaning by means of tank-top hosing; the cleaned diffusers are clearly lighter in colour. Below: cleaning of polyurethane membranes by means of brushing.

Also, pneumatic cleaning is possible for membrane diffusers, labelled flexing and reverse flexing (or relaxing). The practice of flexing consists of periodically increasing the airflow to an excess level for a short period of time, which can be followed by the reverse procedure consisting of releasing the air pressure in the air feed and letting the diffuser membrane collapse onto the diffuser housing. While in theory this practice is effective and relatively simple to implement, it is seldom the case that the blowers supply sufficient discharge pressure to meet the effective flexing requirements.

Specific results will depend upon plant design and provisions for diffuser cleaning (Rieth *et al.*, 1990). For example, it is necessary to have spare capacity or periods of reduced loading or modified operation in order to dewater aeration tanks for cleaning. This is generally possible at large plants but may not be possible at small plants. There are also direct cleaning costs, such as the labour associated with cleaning, chemicals, and replacement parts. Therefore, the choice of cleaning methods and frequencies is important.

9.3.7 Mixed-liquor concentrations

In the absence of respiration, the physical presence of solids has a detrimental effect on the OTR because the layer of solids that accumulates on the bubbles has low permeability and blocks oxygen transfer (Ju and Sundararajan, 1994). The oxygen transfer has, for a long time, been considered inversely related to the MLSS concentrations in the reactor (Krampe and Krauth, 2003; Muller *et al.*, 1995; Ozdemir and Yenigun, 2013). To represent this phenomenon, different authors have proposed relations between MLSS concentration and the α factor. For example, Krampe and Krauth (2003) proposed an inverse exponential relation and Henkel *et al.* (2011) theorized an inverse linear relation.

However, Rosso *et al.* (2005) showed the relationship between the α factor and SRT (hence, MLSS) in activated sludge for fine-pore aeration. To reconcile the different positions, we plotted in Figure 9.33 all the data from Rosso *et al.* (2005) and compared it with other data from fine-pore aeration of highly concentrated MLSS reactors. The data sets plotted in Figure 9.33 are from both pilot and full-scale tests. The fitting line is a double exponential curve with descriptive statistics reported in the plot (see Table 9.4 and Figure 9.33). Note the gap between 4 and 6 g/l, where membrane bioreactors are uneconomical to operate and activated sludge clarifiers are limited by the solid gravity flux. In summary, these relationships allow us to obtain similar results for concentrations less than 10 g/L MLSS, but for higher concentrations, variations merge because of the type of relationship proposed, and the conditions for validity in each case, as shown in Figure 9.33.

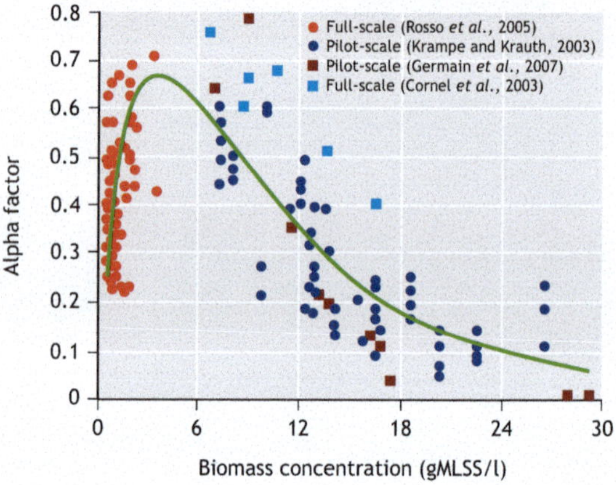

Figure 9.33 Dependence of the αfactor on the MLSS concentration for activated sludge processes (red points) and membrane bioreactors (all other points), equipped with fine-pore diffusers. Note the sharp decrease in alpha due to the heavily non-Newtonian nature of the thick sludge for MLSS > 1% (Baquero-Rodriguez *et al.*, 2018).

Table 9.4 Alpha factor prediction based on MLSS concentration.

Reference	MLSS range	Equation
Krampe and Krauth (2003)	0-30 g/l	$\alpha = e^{-0.08788\ MLSS}$
Henkel *et al.* (2011)	1-12 g/l	$\alpha = 0.062 \cdot MLVSS + 0.972 \pm 0.070$
Baquero-Rodríguez *et al.* (2018)	0-30 g/l	$\alpha = \left(\frac{u}{u-v}\right) \cdot \exp(-v \cdot MLSS) \cdot \exp(-u \cdot MLSS)$ {u=0.507; v=0.104}

As the sludge concentration increases, the bubble coalescence must increase because of the shear-thinning nature of the sludge. In fact, coalesced bubbles (associated with more interfacial shear) can thin the fluid and experience less resistance to rise. Such coalesce bubbles, having a substantially lower 'a' in their k_La, do indeed exhibit lower α factors. Correlations between the aeration performance and the apparent viscosity of the activated sludge have been reported. For stirred tank reactors, an exponential correlation exists between the activated sludge k_La and the Reynolds number (Krampe and Krauth, 2003; Nittami *et al.*, 2013). It can be assumed that a rise in apparent viscosity may lead to the production of larger bubbles at the formation stage. Larger bubbles will lead to a reduction of the specific surface area (Durán *et al.*, 2016).

9.3.8 Temperature and pressure

Changes in water temperature induce changes in the oxygen saturation concentration of water. The oxygen saturation concentration of water decreases with increasing temperature, and the decreased solubility of oxygen at higher temperature is an important factor in aeration system design. Increases in the mixed-liquor temperature generally coincide with increases in ambient air temperature, and decrease the capacities of the blowers in diffused aeration systems (Jenkins, 2013).

Despite remarkable progress in aeration, the effects of barometric pressure on OTE have been largely overlooked (Baquero-Rodríguez *et al.*, 2018). Regarding aeration energy demand, the blower performance is inversely related to the altitude at which the blower is installed, because of the effects of altitude on atmospheric pressure and air density. Thus, atmospheric pressure is considered here.

Three factors decrease the air density: increases in air temperature, decreases in atmospheric pressure, and increases in relative humidity. From a mechanical efficiency standpoint, a decrease in air density results in a greater demand for volumetric airflow production by the blower, to guarantee an equivalent airflow mass with respect to standard conditions. The evaluation of energy requirements depends on the operating conditions, control techniques, and blower type (Water Environment Federation, 2009). A comparison of energy consumption at sea level and at higher altitudes is obtained by dividing the energy consumption under standard conditions by the correction factor of the air density under the desired conditions. For cities with elevations above sea level ranging from 1,600 to 3,600 masl, the corresponding altitude correction factor is in the range between 0.81 and 0.69 (Ludwig, 1997). Such is the case in cities like La Paz, Bolivia (3,640 masl), Quito, Ecuador (2,850 masl), Bogotá, Colombia (2,625 masl), Mexico City, Mexico (2240 masl), and Denver, United States (1,600 masl). Most of the above-mentioned locations are considered developing countries with poor sanitation coverage, and since, at the present moment, the development of their WRRFs is largely underway, we highlight here the shortage of published knowledge on aerated processes in high-elevation regions.

The oxygen saturation concentration in water is important for aeration and represents the maximum amount of oxygen that can be dissolved in water (Jenkins, 2013). The oxygen saturation concentration in water is a function of barometric pressure, water pressure, salinity, and temperature (Jenkins, 2013), and decreases noticeably when the atmospheric pressure is not 1 atm, and/or the salinity (based on the Practical Salinity Scale) or temperature is high. The procedures for estimating the oxygen saturation concentration under these 'standard' conditions can be found in Benson and Krause (1984). As stated before, the driving force for oxygen transfer is the concentration gradient between the oxygen saturation concentration and dissolved oxygen. For all practical purposes, the salinity of potable water can be deemed negligible.

9.3.9 Impact of hydrodynamics

Previous sections of this book summarized the investigations focusing on the impact of environmental and operational conditions of the process on oxygen transfer. There is also an academic interest in exploring the effect on hydrodynamics in gas-liquid transfer driven by both understanding in more depth the transport phenomena within the aeration tank and the desire to model the emission of greenhouse gases.

Although hydraulic design of wastewater treatment systems is a crucial step to assure reliable operation of the process, in the majority of designs flow behaviour is predicted via the ideal reactor model (Stamou, 2008). WRRFs are assumed, then, to have a complete mixed-flow regime in all type of tanks (*i.e.* aerobic, anoxic, and anaerobic). However, the hydrodynamics of a bioreactor can influence how 'well mixed' an AS system actually is. In fact, in an heterogeneous multi-phase real reactor, the fluid flow can be impacted by vessel geometry, the physical properties of its contents (phase, density, viscosity) and operating condition variables (flow rates, distribution of oxygen, temperature, flocs, etc).

In recent years the approach based on dynamic modelling showed the potential to be a powerful tool providing detailed insight into the unit process and system behaviour. Further details on how to measure the impact of hydrodynamics in WRRFs by CFD and their current limitations can be found in the review by Karpinska and Bridgeman (2016).

At the time this book was written, all the work using computational fluid dynamics to determine the influence of hydrodynamics could not yet be considered in its full state of maturity, as different experts and research groups are still addressing a set of fundamental limitations:

- In some models bubbles are considered part of the liquid phase and are accounted for only as a change in density of the liquid, proportional to the gas holdup fraction; when each phase is considered to have its unique velocity vector, an empirical drag model is employed.
- Bubbles are assumed not to change in size during their rise, despite the change in hydrostatic pressure with water depth, with the consequence that bubbles do not accelerate during their rise and do not increase their perturbation of the liquid velocity field, as in fact happens.
- Bubbles are assumed not to interact to each other.
- Viscosity is the only effect of the gas transfer coefficient, not the effect of contaminants (*i.e.* surfactants), which is the equivalent of exclusively considering biomass and not the effect of alpha-depressing agents (surfactants).

The drive for advancing this knowledge is certainly important, as it will elucidate the gas transfer phenomena whose effects are recorded quantitatively by field-testing (e.g. gas transfer measurements). Therefore, the integration of successful biological wastewater treatment modelling combining hydrodynamics, mass transfer, and biokinetics remains one of the major goals in wastewater process modelling (Pereira *et al.,* 2012).

9.3.10 Daily dynamics and alpha

The deleterious effects on oxygen transfer efficiency caused by COD tend to escalate during the daily period of greatest aeration requirements, which also tends to occur when energy and power demand charges are maximum (Aymerich *et al.,* 2015). This amplification frequently exacerbates the aeration costs during peak times, often resulting in a period of 4-6 hours costing as much as the remainder of the day (Emami *et al.,* 2018). This phenomenon occurs consistently in almost all WRRFs, where the influent characteristics fluctuate daily, weekly, and monthly.

To date, the use of constant α factors are the most common practice in process design and modelling, thus promoting inaccuracies in the oxygen provision and the constant need for model recalibration to compensate for a lack in structure (Amaral *et al.,* 2016; Plano *et al.,* 2011). Only recently have process models incorporated the ability to specify time-dependent α factors. The state-of-the-art in aeration design and modelling using commercially available simulators based on the Activated Sludge Model (ASM) family (Henze, 2000; Henze *et al.,* 1987) is to rely on user inputs of SOTE curve-fitting parameters and constant or scheduled α factors. The SOTE curves are typically only corrected for air flux, diffuser density, and submergence (Barker and Dold, 1997; DHI, 2014; Snowling, 2014). Because both OTE and α vary with load, both of them exhibit a dynamic nature. Although numerous attempts at developing descriptive or predictive relationships for α have been published (Gillot *et al.,* 2005; Gillot and Héduit, 2008; Plano *et al.,* 2011; Rosso and Stenstrom, 2005; Wagner and Pöpel, 1998), none of these attempts have been completely successful in describing the true dynamic nature of α. Hence, the use of a dynamic α becomes necessary to address the real process dynamics. The first attempt at modeling α dynamically was presented by the authors Jiang *et al.* (2017). By performing 24-h off-gas tests on the first aerated zone of two WRRFs, a correlation between the alpha factor and the COD applied to the reactor was developed:

$$\alpha = A \cdot e^{-K_\alpha \cdot [COD]} \tag{9.13}$$

where:

A, k_α Fitting parameters (function adjustment by minimization of residuals)

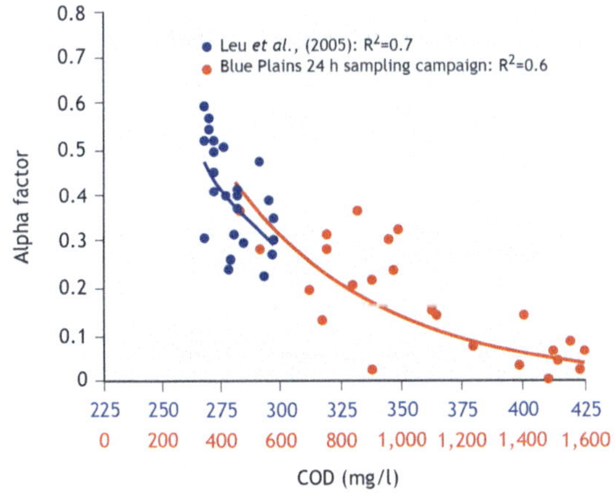

Figure 9.34 Alpha factor (α) and total COD. The negative correlation of the alpha factor (α) as a function of the local COD (*i.e.* measured at the same point where the off-gas is measured) over a diurnal cycle (data from Leu *et al.,* 2009). Aggregate data from both studies is represented by the red dotted line (r2 = 0.7).

The results from these experiments are presented in Figure 9.34 showing the COD and the α factor negatively correlated as a function of the influent COD plant load.

9.4 DESIGN ALGORITHM

To apply what we have illustrated in this chapter to treatment plant design or upgrade, two hypothetical examples are proposed. First, as a design example, an algorithm can be implemented as shown in Figure 9.35.

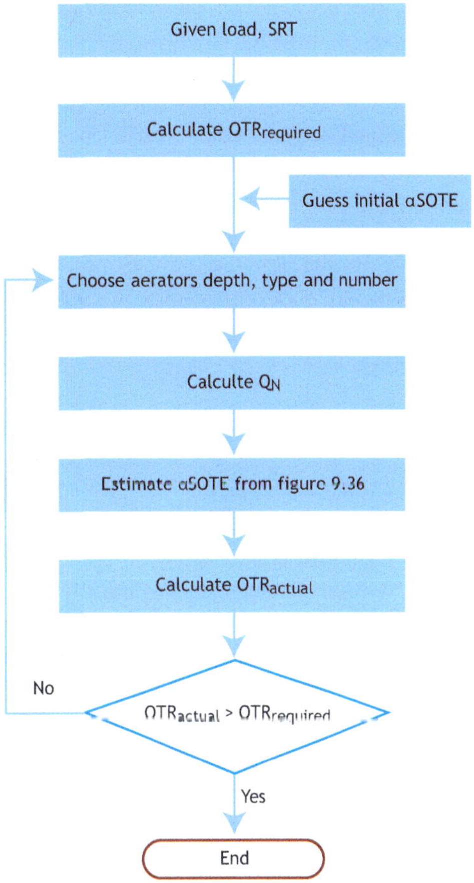

Figure 9.35 Aeration tank design flowchart.

For a fixed wastewater load with a selected SRT, the oxygen demand is first calculated as the required OTR (mass O_2 per unit time):

$$OTR = (\text{plant flow} \cdot \text{plant load}) - \text{wasted sludge} \quad (9.14)$$

Aeration tank size and side water depth are determined based on site-specific geometric and economic restraints. Next, the diffuser type is selected, and an estimated αSOTE is assumed. Based on the diffuser manufacturer's recommended range of airflow per diffuser, the total number of diffusers is estimated. The αSOTE can be selected based on manufacturers' information and literature values of α, or any other information available. The required airflow rate can now be calculated from the oxygen uptake rate and oxygen transfer efficiency, which allows the specific air flux to be calculated. Next, the design point is found in Figure 9.36 by locating the SRT on the horizontal axis, and the contour that corresponds to the specific normalized air flux, defined as:

$$Q_N = \frac{AFR}{a \cdot N_D \cdot Z} \quad (9.15)$$

where:

AFR	airflow rate, (m^3/s)
a	diffuser bubbling area (m^2)
N_D	total diffuser number (-)
Z	diffuser submergence (m)

A new value of αSOTE is determined by reading the ordinate of the design point. If the new αSOTE is different to the assumed αSOTE by more than a small difference (e.g. 0.5%), a new airflow rate and specific air flux must be calculated using the new αSOTE to create a new design point. The new design point is located in Figure 9.36, and a third value of αSOTE is determined and compared to the second αSOTE. The process is repeated until the new αSOTE and previous αSOTE are approximately equal. Should the procedure not converge, a different number of diffusers or even a different diffuser technology should be chosen for the procedure.

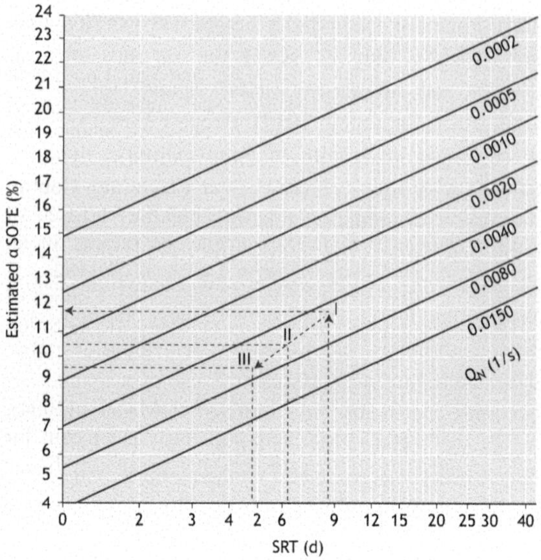

Figure 9.36 Design and verification graph. I: Flow = 0.875 m^3/s (design example), Q_N = 0.0046 1/s, SRT = 8.7 days, $\alpha SOTE_{EST.}$ = 11.9%; II: Flow = 1.094 m^3/s, Q_N = 0.0058 1/s, SRT = 6.3 days, $\alpha SOTE_{EST.}$ = 10.5%; III: Flow = 1.313 m^3/s, Q_N = 0.0069 1/s, SRT = 4.9 days, $\alpha SOTE_{EST.}$ = 9.5%.

A numerical example is provided. Given an influent flowrate of 0.875 m^3/s, with a load of 180 mg/l COD, and assuming a yield of 0.5 and a decay coefficient of 0.06 1/d, the required OTR will be 9,540 kgO_2/d. Considering a hydraulic retention time of 4 h, 3 tanks with dimensions 90 × 9 × 5 m (length × width × depth) each, and an initial αSOTE of 13.5%, the airflow rate will be 0.985 m^3/s. When the design utilises 22.9 cm ceramic discs (a = 0.0373 m^2 per diffuser) operating at $7.87 \cdot 10^{-4}$ m^3/s per diffuser, 1,252 diffusers per tank are required. Given these data, it is possible to calculate Q_N = 0.004152 1/s. This value, together with the SRT value of 8.7 d, is located in Figure 9. 36 (point I). The new αSOTE is 11.9%. A new Q_N is calculated as 0.0051 1/s, and the process converges at 11.7% after one iteration.

The design procedure presented here involves several parties (Figure 9.37). The main role of the manufacturer is to supply the aeration system and to verify its performance using a factory or field test (witnessed by an independent third-party expert) in a repeatable and transparent fashion. This is why standards for testing aeration systems in clean and process water are crucial. In an analogous fashion, blower performance and efficiency should be assessed using the separation of roles illustrated here.

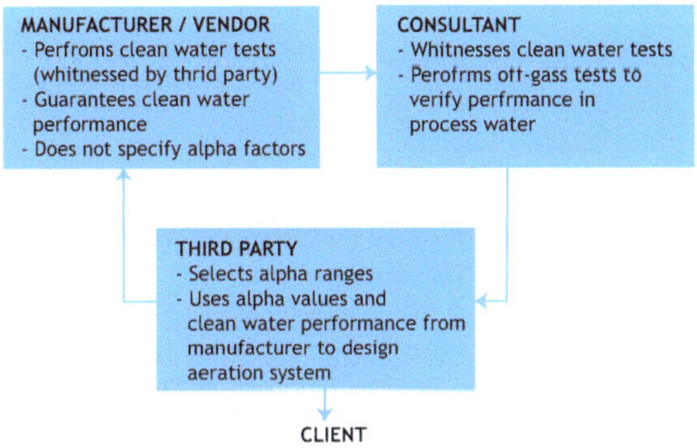

Figure 9.37 Roles and responsibilities during an ideal aeration system design. The arrows broadly describe the flow of information during the project.

The independent witness must bear no conflict of interest with the manufacturer, of course. Also, the consultant who designs the aeration system should request the third-party verification whenever in-house expertise to critically evaluate the equipment warranty is not available (the devil is in the details, as the saying goes). Since the manufacturer best knows the aeration technology provided, they need to guarantee the clean water performance, which is a measure of the equipment's performance and efficiency in a standard condition that is independent of the wastewater. The ratio between oxygen transfer rate in process water and clean water (the mysterious alpha factor) is a function of influent characteristics and operating conditions more than of equipment. Given that understanding the wastewater characteristics and operating conditions is the responsibility of the consulting engineers in charge of the process design, it is their role to select the ranges of alpha factors. The consultant who uses an alpha factor suggestion from the manufacturer without critically addressing its validity is exposing the design team to the potential failure of the system (and probable litigation with the client). In fact, the consulting engineers are hired specifically to help the client decipher the quantitative implications of design assumptions: otherwise, the client could probably carry out the design on their own. If all the roles are separate and the aeration performance is tested according to standard protocols (e.g. the ASCE or European Standards), the design procedure should yield a fully functional aeration system whose performance and efficiency can be verified independently and be within the targeted expectations. The role of the third-party expert is to verify the manufacturer's data and to verify the consultant's assumptions on alpha.

It is of crucial importance to clearly define the efficiency and performance parameters of aeration systems during their analysis or specification. To compare different technologies, as well as to monitor the same aeration system over extended time in operation, a level playing field is necessary. The most basic parameter to define performance is the OTR. This is the design target: should the amount of oxygen transferred to the water per unit time be lower than the target OTR for extended periods of time each day (or especially indefinitely), the biological reactor will experience oxygen deficiency and its performance will be compromised and effluent discharge violations are likely.

The role assumed by the client during design is that of observing the procedure and diligently checking that all parties are fulfilling their roles. When the design procedure is complete, the client must undertake the responsibility of operating the plant in the way it was designed. Departures from the design conditions could be necessary, but an evaluation with the design team is recommended to prevent unforeseen effects and the resulting arguments between the parties.

9.4.1 Verification/upgrade algorithm

A second example is useful to illustrate growth in load on an existing plant and is shown in Figure 9.38. The additional load entering the plant increases oxygen demand, and, to supply a higher oxygen mass, the aerators are operated at a higher airflow rate. This causes an increase in Q_N since the number of diffusers and tank geometry do not change. If the MLSS is not increased, the plant will operate at lower SRT. This will cause two sources of reduced aeration efficiency: lower SRT and higher airflow rate.

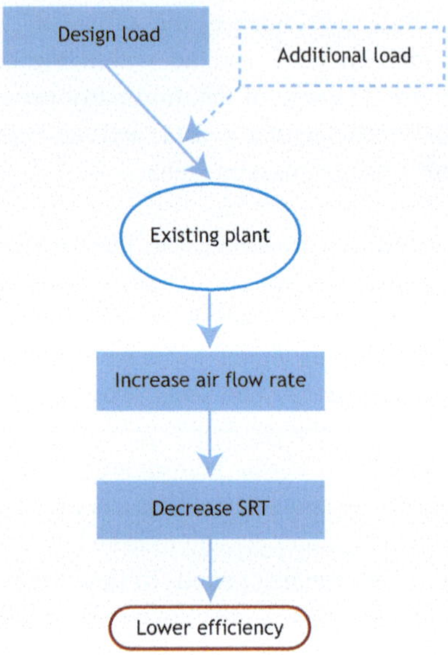

Figure 9.38 Aeration tank verification/upgrade flowchart.

Two scenarios are presented in Figure 9.36: design point I shows the initial design and is the same as the previous example; design point II shows a load increase from 0.875 to 1.094 m^3/s, with a drop in αSOTE from 11.9 to 10.5%. The increase from 1.094 to 1.313 m^3/s to design point III results in an additional drop in efficiency to 9.5%. The practical effect of the load increase will be an increase in electric power consumption per unit of load treated. The importance of the increased load example is to understand that there are two reasons for reduced aeration efficiency: increased airflow rate and reduced SRT.

When plants undergo upgrades or retrofits, there is a need for data and design parameters from manufacturers of candidate diffusers. The candidate manufacturers can provide detailed guidelines to adhere to in order to successfully design the aeration system, but the efficiency parameters provided and warranted by manufacturers should be limited to clean water. The design engineers should perform site-specific tests to quantify the α factor for the existing wastewater, using *in-situ* column evaluations (Figure 9.39).

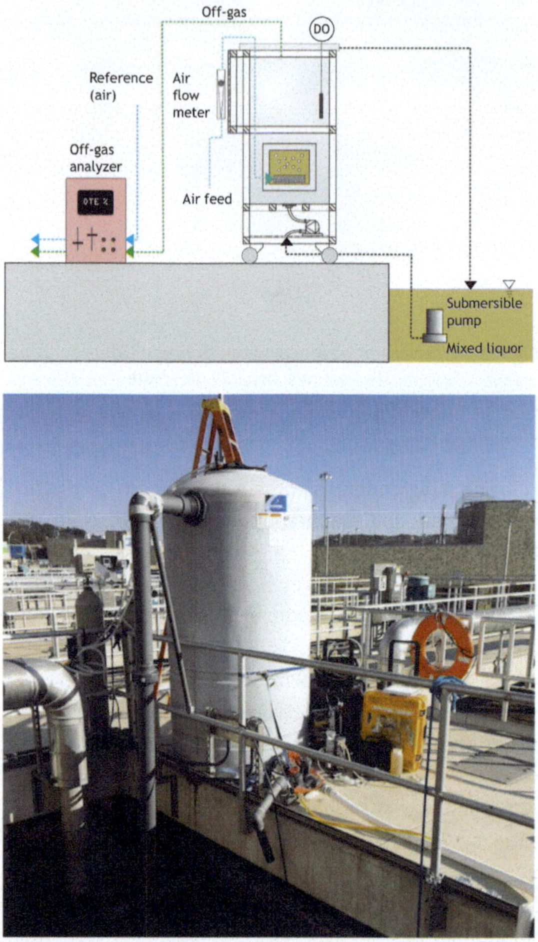

Figure 9.39 Schematic and image of *in-situ* column testing for process water.

The alternative is to estimate α factors on historical experience in other similar installations and on internal experience, but even for a seasoned engineer this choice is risky. The methodology based on an *in-situ* column method is fully documented in

Rosso *et al.* (2013). This testing has to be performed in clean and process water at the treatment facility. The obtained results will help to bypass the uncertainty caused by extrapolating results from different wastewaters and projects. Moreover, an extended fouling test in the plant's wastewater can be carried out within the scheduled term of the design, providing the designers with site-specific aeration performance data. This procedure minimises the risk of underestimating the oxygen transfer rate or the desire to compensate for the lack of certainty on α with a conservative inflated over-design.

9.5 AERATION AND ENERGY

The energy cost of aeration can account for 45-75% of process costs (Reardon, 1995; Rosso and Stenstrom, 2005; Rosso *et al.,* 2008). Energy consumption will increase in the case of enhanced nitrogen removal, thus there is an evident trade-off between the effluent quality reached and the energy investment and environmental footprint. Values consistent with those presented in Figure 9.40 are extensively reported worldwide.

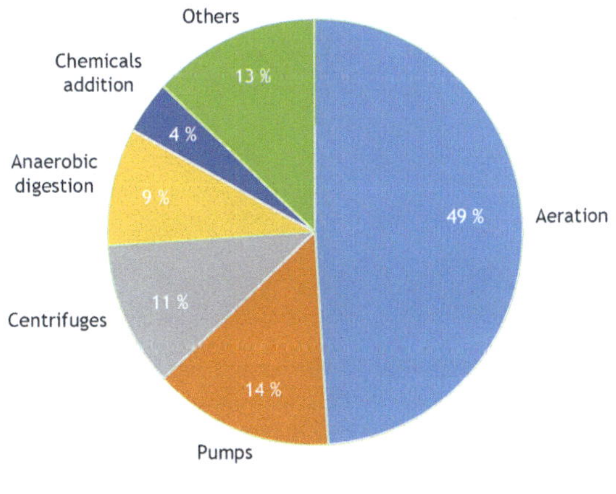

Figure 9.40 Estimated power usage breakdown for a typical 20MGD (74,500m^3/day) (adapted from WEF, 2009).

Hence, aeration has become the main focus of consideration to reduce energy use in the ultimate aspirational goal of achieving net energy neutrality for water resource recovery. As an example, we can see that the energy consumption of WRRFs ranges from 335 MWh per month (WRRFs serving a 100,000 population-equivalent, or PE) up to 6,600 MWh per month (WRRFs serving 3,000,000 PE), while the associated energy costs can range from 45,000 EUR per month to 280,000 EUR per month, respectively.

Due to the elevated energy intensity, it is of the utmost importance to understand aeration efficiency in order to improve design, performance, cost predictions, and process sustainability. Understanding aeration efficiency and lowering operating costs go hand-in-hand (Rosso and Shaw, 2015; Rosso and Stenstrom, 2006a).

An extensive treatise of the relationship between aeration and energy is reported in Rosso (2018). What we summarise here is the compounding effect of diurnal variations in flow, load, air requirements and ultimately greenhouse gas emissions from imported power. The energy requirements during circadian cycles will be further exacerbated due to various factors. Recent studies demonstrated that daily variations in COD, TKN, influent flow, and αSOTE, together with energy tariff structures and current energy production methods have exacerbated air requirements and GHG emissions for WRRFs sited in areas where peak plant load corresponds to peak power rates. The compounding effect of peak flow, peak load, and dynamics of the aeration system performance needs to be considered.

9.6 SUSTAINABLE AERATION PRACTICE

9.6.1 Aeration diagnostics

A very important component of aeration diagnostics is walking the plant looking for symptoms of unhealthy conditions. An example of simple fault detection is the observance of surface boils, similar to small geysers, which indicate damaged diffusers or leaking air headers (Figure 9.41).

Figure 9.41 Surface boil indicating on the surface that at the bottom of the tank a diffuser is either damaged or an air header is leaking.

One may be sufficiently lucky to encounter a dewatered tank at a treatment facility. Certain details, visible only when the tank is dewatered partially or in full, allow for an inspection of the aeration system elements that the full operating tank would not. For example, the selective replacement of a cracked diffuser header (Figure 9.42) produces differential air discharge due to the lower pressure drop associated with the new diffusers.

Another issue that can be identified with visual observation of partially dewatered tanks is the uniform discharge of air throughout a diffuser grid. Figure 9.43 shows a patchy air distribution from a group of diffusers in a new installation, probably due to poor header levelling during on site assembly.

Figure 9.42 Differential air discharge from a new diffuser header installed to replace a damaged header. Note the increased air discharge from the new diffusers, due to their lower pressure drop. The image on the top shows the diffusers and the image on the bottom is the surface of the tank.

Figure 9.43 Patchy air distribution from a group of diffusers in a new installation, probably due to poor header levelling during assembly.

Another relatively easy practice is to track the discharge pressure from the blower house. However, due to the major effect that the hydrostatic head has on the overall discharge pressure, this practice requires good quality pressure gauges and can be applied to diffusers whose pressure drop will increase by an absolute value that can be recorded within the discrimination of the gauge. Therefore, it would be very challenging if not impossible to notice the pressure drop increase of ceramic diffusers.

Diffuser cleaning can be performed as: (*i*) mechanical cleaning: tank top hosing; pressure washing; brushing; pneumatically (flexing/reverse-flexing membranes), and (*ii*) chemical cleaning, using acids (liquid or gas), bases, and/or detergents.

Current knowledge suggests that cleaning once every 24 months is a good practice to curb the decline in efficiency of all diffusers (Rosso and Stenstrom, 2006a); however, cleaning effectiveness and frequency is site-specific and should be evaluated on a case-by-case basis. When using chemicals, care must be applied to prevent damage of the air distribution piping and fittings. Tank-top hosing and pressure washing should be performed with some water above the diffusers, to prevent mechanical damage or destruction of diffuser membranes.

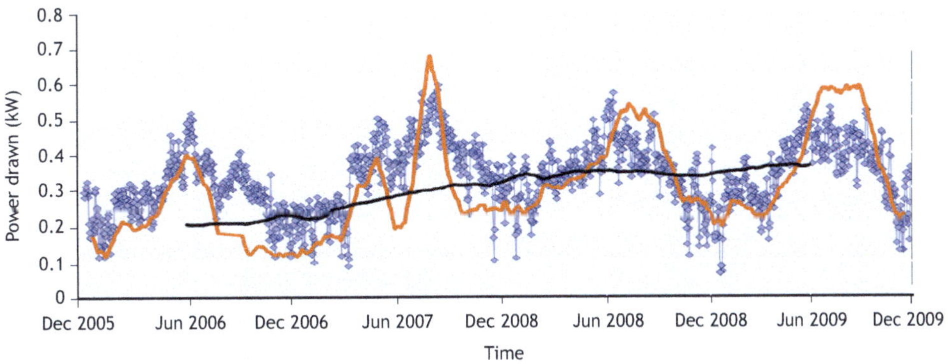

Figure 9.44 Power drawn by the blower house of a full-scale wastewater treatment plant over a period of five years. Note the yearly seasonal cycle, shown by the daily values (blue points) and the 30-day running average (orange line). The 1-year running average (black line) shows the long-term increase in power demand, almost doubling in the period covered by this plot. The calculations to produce this plot were based on process data logs.

Regarding the effectiveness of pneumatic cleaning, a study on diffusers that were flexed and reverse-flexed (*i.e.* the pressure in the air feed was relieved, causing membrane collapse onto the diffuser frame) was performed (Figure 9.45). In this case, pneumatic cleaning proved effective in curbing the increase in DWP, however the decline in αFSOTE showed no statistically significant difference between diffusers that were pneumatically cleaned and those that were not. Nevertheless, this practice is automated and virtually effortless (provided that the blower discharge pressure can actually meet the requirements for effective flexing) and even if only the escalation of DWP were curbed, the operation of the blower battery may spare the events of additional units coming online at peak power demand charge periods.

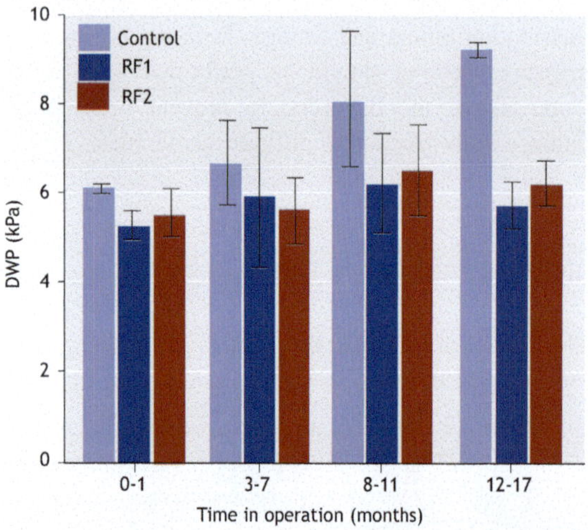

Figure 9.45 Airflow-weighted average of the discharge pressure <DWP> of polyurethane membrane panels over time in operation. The plot shows both the pressure drop of a diffuser not experiencing the prophylaxis of flexing and reverse flexing its membrane (labelled Control), and of diffusers undergoing such a procedure daily (labelled RF1 and RF2). The benefit of flexing the membranes is clear here, despite the lack of effectiveness in curbing the decline in oxygen transfer efficiency for these particular diffusers (data from Odize *et al.*, 2017).

In addition to cleaning diffusers, one important prophylaxis is to ensure good grit removal. Why are we talking about grit removal in a chapter on aeration? Because without good grit removal upstream of the aeration tank, the accumulation of grit on the tank could cripple the aeration diffusers, especially in grids with low air flux (Figure 9.46).

Figure 9.46 Example of the deleterious effects of grit accumulation on the aeration tank floor, with corresponding obstruction and damage to the diffusers.

Finally, one very important element, which is relatively negligible in cost but important in its functionality, is a properly installed purge in order to continuously drain the moisture that may accumulate and the sludge that may enter the air headers. This purge can be opened periodically to drain the collected liquid or can be always left open. Ideally, this purge would be a tube ~10-20 mm in diameter connecting the top of the tank above water with the bottom of the aeration header (*i.e.* the low point where liquid would inevitably accumulate). A good design procedure should always include a purge.

9.6.2 Mechanically-simple aerated treatment systems

Lagoons or stabilization ponds are often inexpensive methods for wastewater treatment with low energy intensity. However, they have an intense land use and therefore their suitability as a sustainable alternative is mainly limited to small communities and/or rural areas. For facultative water resource recovery lagoons, the driver for oxygen transfer and for process energy consumption is mixing. Figure 9.47 illustrates a schematic of an aerated lagoon.

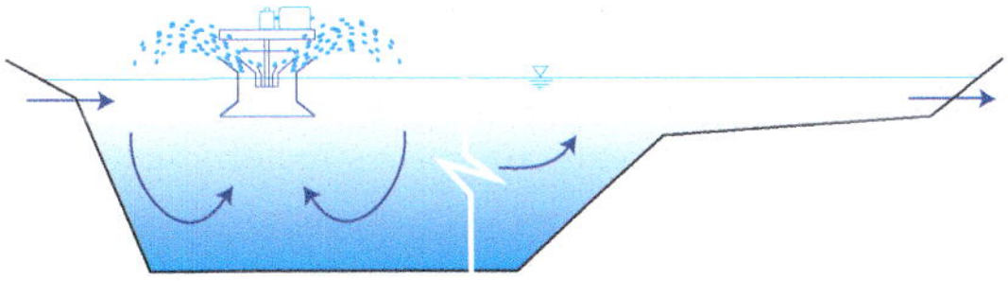

Figure 9.47 Schematic of a section of an aerated lagoon equipped with a surface aerator.

Figure 9.48 Example of aerated lagoons in operation at a large facility. Above is a view of the lagoon from the effluent side; on the left shore is the air manifold with all the floating headers where many diffuser assemblies are attached and floating underwater. Bottom left is an image of the blower house. Bottom centre is a photograph of a damaged assembly being retrieved on the service barge; note the large amount of rags being removed from the surface of the assembly, visible on the left and at the top of the image; one of the diffuser membranes was removed from this assembly for off-site fouling studies. Bottom right is a detail of the said diffuser membrane surface.

Aeration for aerated lagoons is usually performed by mechanical aeration, although diffused aeration is possible. In the case of diffused aeration (Figure 9.48), primary separation is a key element in preventing the destructive accumulation of fibrous material and rags on the diffuser assemblies. For mechanical aeration, high-speed surface aerators are usually employed, typically operating at the low end of the SAE ranges. The additional issue of heat loss in winter, exacerbated by surface aeration systems, contributes to the interest in retrofits to diffused aeration systems. Diffused aeration systems in lagoons tend to employ floating systems rather than diffuser grids anchored to the floor. This facilitates the retrieval of single diffuser assemblies with the service barge and also means that the diffuser system works well in lagoons of uneven depth by levelling the diffusers to the surface. This strategy maintains even the discharge of bubbles across distant assemblies, since the diffuser submergence for all units is pegged to the water surface.

There are trade-offs between the maintenance requirements of the two alternative aeration methods. If surface aerators are used, they must be maintained using boats with personnel trained to work offshore, which may be prohibitive in winter. If fine-pore diffusers are used, the blower maintenance is always performed primarily on-shore, but diffuser maintenance would require offshore methods such as a barge.

9.6.3 Energy-conservation strategies

We present here three case studies performed at municipal wastewater treatment plants to show the implementation of possible energy-conservation strategies. A summary of the data regarding the treatment plants is summarized in Table 9.5.

Plant 1

This treatment plant serves 800,000 people and treats approximately 240,000 m^3/d. This plant uses disk-type ceramic diffusers for aeration. When upgraded to a fine-pore system, this treatment plant used the conventional activated sludge process (CAS) for secondary treatment and then converted to nitrification/denitrification (NDN) process nearly 20 years later to remove ammonia. Off-gas tests were applied continuously in two processes to evaluate diffuser fouling.

Plant 2

The operation and testing scenario of this plant is similar to the first plant: it uses ceramic disks for aeration and the treatment process has been converted from conventional to NDN. The capacity of this plant is relatively smaller than the other two: the population served is approximately 220,000 and the average volumetric flowrate of this plant is 76,000 m^3/d.

Plant 3

Fine-pore diffusers used in this treatment plant are made of EPDM (Ethylene Propylene Diene Monomer rubber), and this plant has not yet been upgraded to a NDN process due to overloading. Similar to Plant 1, the capacity of this plant is 12,500 m^3/d and the population served is 880,000. Until the last test, diffusers in most of the tanks in this plant (8 of 10) had not been cleaned since installation (8 years).

Off-gas tests were performed as in Redmon *et al.* (1983). The mass balance between oxygen-fed (20.95% mole fraction) and off-gas stream returns the OTE, furthermore OTE was standardized to αSOTE. For each test, data were collected from 6 to 8 hood positions evenly distributed on the surface of the aeration tank and then presented as a flow-weighted average. In addition, the OTR, kgO_2/h was calculated by measuring the flowrate of the off-gas and diffuser headloss was measured from plant readings and clean water tests. With OTR and headloss, aeration costs were calculated by the adiabatic function of blowers (Metcalf and Eddy, 2013).

Table 9.5 Background of tested wastewater treatment plants.

Parameter	Plant 1	Plant 2	Plant 3
Average volumetric flow (m^3/h)	10,000	3,150	12,500
BOD_5 in primary effluent (mg/l)	162	132	146
Ammonia in primary effluent (mgN/l)	28	25	28
Tank in operation	18	6	10
Treatment process	NDN[a]	NDN	CAS[b]
Population served	800,000	220,000	880,000

[a] NDN = Nitrification/denitrification process, [b] CAS = Conventional activated sludge process.

Discussion

Figure 9.49 shows the effects of fouling/scaling on aeration efficiency: αSOTE is plotted as a function of time in operation after the installation of fine-pore diffusers. In Figure 9.49A, αSOTE in all of the systems decreases with increased length of time in operation, but the rate of fouling/scaling varies with diffuser types and plant operation. The numeric label on top of each data point is the operating SRT. It can be observed that the fouling is extremely fast when SRT is short. diffusers lose almost 50% of their αSOTE after three months in operation. Figure 9.48B shows the recovery of αSOTE due to diffuser cleaning. In Plant 3, only two out of the ten aeration tanks were dewatered and cleaned during the testing period. When compared with the tanks with no cleaning, the cleaned tanks had 19.5% αSOTE compared to 15.5% for no cleaning, an increase of 4 percentage points or a relative improvement of approximately 30%. In addition, energy savings due to diffuser cleaning can be evaluated by integrating the net increase in aeration efficiency between cleaned and uncleaned diffusers, as shown in the shaded area.

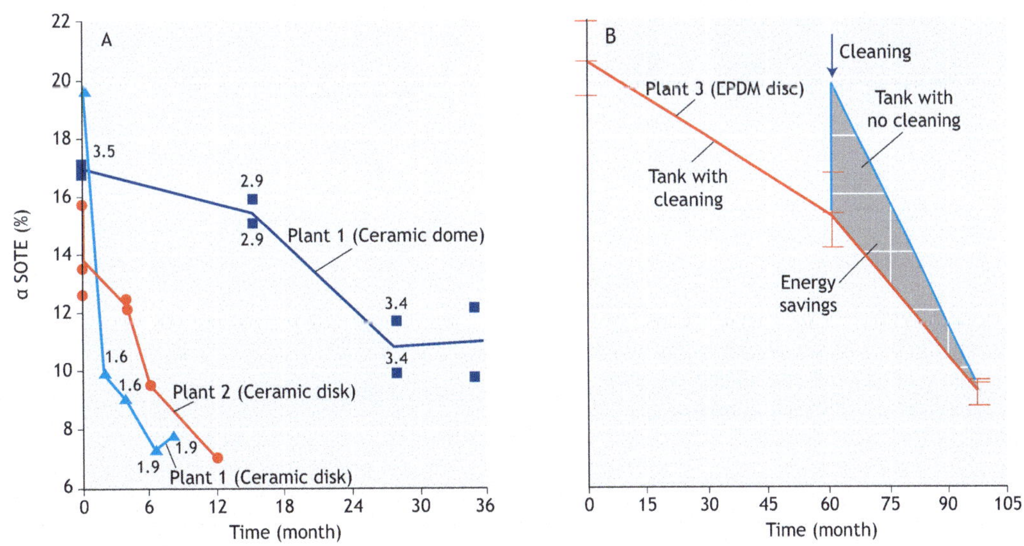

Figure 9.49 (A) Decrease of aeration efficiency of fine-pore diffusers. Depending on the types of diffusers and operation processes the rate of diffuser fouling is site-specific. (B) Diffuser cleaning significantly recovered aeration efficiency, and energy savings were calculated by taking the difference of diffusers with and without cleaning.

Figure 9.50 shows the power costs of aeration in the three plants. Based on cleaning/fouling status, diffusers were grouped into three columns: the light blue column shows newly installed diffusers, the dark blue column shows cleaned diffusers, and the dark red column shows fouled diffusers. The line bar on top of each column shows the standard deviation of the test results, and aeration costs of different processes (conventional or NDN) were plotted separately for plants 1 and 2. Significant improvements on energy usage by diffuser cleaning were observed, which provided an energy saving to approximately 4.5 kWh/PE.yr, or 9,900 kWh/d in Plant 1.

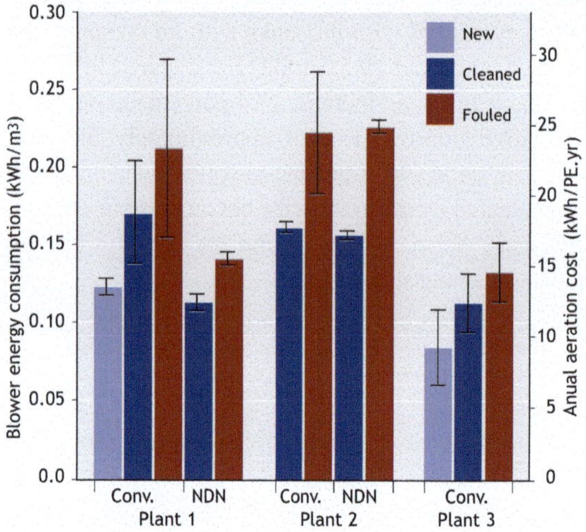

Figure 9.50 Energy consumption of fine-pore diffusers in different treatment plants. Aeration costs were standardized by total oxygen loads and population served.

Figure 9.51 shows the aeration cost under different process conditions. The capital power consumption of Plant 1 was plotted with the sludge retention time (SRT). To compare different processes, treatment loadings were converted to the same bases: the total OTR of the conventional processes was scaled up to the level of NDN. As a result, aeration costs reduced exponentially with SRT. When operating with a short SRT, aeration status varies dramatically and more pumping energy is required due to more aggregate diffuser fouling (in red). Longer SRT provides better OTE and hence the NDN process requires less air to oxidize the same amount of pollutants. This observation confirmed former studies (Rosso and Stenstrom, 2006b) that long SRT and the anoxic environment in NDN process provide a healthier bacteria composition that may reduce the negative impacts of surfactants on oxygen transfer (*i.e.* a higher α factor, Stenstrom *et al.,* 1981). In these cases, off-gas analysis serves as a useful tool for plant operation and diffuser maintenance.

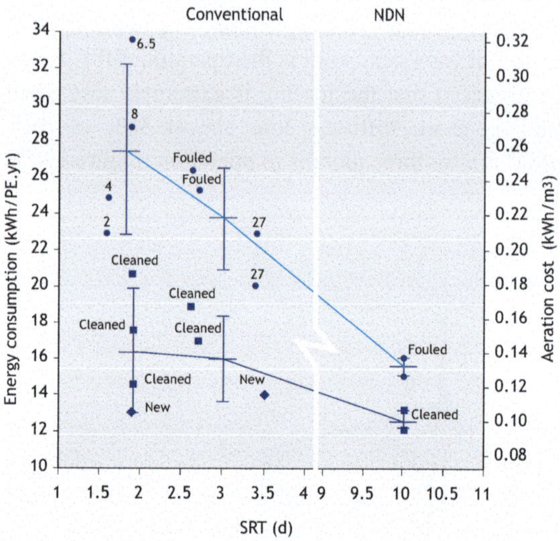

Figure 9.51 Aeration costs of Plant 1, presented as a function of plant operation and diffuser fouling: diffuser fouling was reduced at longer SRT.

REFERENCES

Ahnert M., Kuehn V. and Krebs P. (2010). Temperature as an alternative tracer for the determination of the mixing characteristics in wastewater treatment plants. *Water Research*, 44, 1765–1776.

Amaral A., Gillot S., Garrido-Baserba M., Filali A., Karpinska A.M., Plósz B.G., De Groot C., Bellandi G., Nopens I., Takács I., Lizarralde I., Jimenez J.A., Fiat J., Rieger L., Arnell M., Andersen M., Jeppsson U., Rehman U., Fayolle Y., Amerlinck Y. and Rosso D. (2019). Modelling gas–liquid mass transfer in wastewater treatment: when current knowledge needs to encounter engineering practice and vice versa. *Water Science and Technology,* 80, 607–619.

Amaral A., Schraa O., Rieger L., Gillot S., Fayolle Y., Bellandi G., Amerlinck Y., Mortier S.T.F.C., Gori R., Neves R. and Nopens I. (2016). Towards advanced aeration modelling: from blower to bubbles to bulk. *Water Science and Technology*, 75, 507–517.

Amerlinck Y., Bellandi G., Amaral A., Weijers S. and Nopens I. (2016). Detailed off-gas measurements for improved modelling of the aeration performance at the WWTP of Eindhoven. *Water Science and Technology*, 74.

Andersen H., Siegrist H., Bent H.-S. and Ternes A.T. (2003). Fate of Estrogens in a Municipal Sewage Treatment Plant. *Environmental Science Technology*, 37, 4021–4026.

ASCE (American Society of Civil Engineers). (2007). Measurement of Oxygen Transfer in Clean Water [WWW Document]. URL http://www.asce.org/templates/publications-book-detail.aspx?id=8108 (accessed 25.02.2015).

ASCE (American Society of Civil Engineers). (1997). ASCE Standard: Standard Guidelines for In-Process Oxygen Transfer Testing (18-96). New York, USA.

ASCE (American Society of Civil Engineers). (1989). Proceedings of the ASCE 1989, in: Environmental Engineering Specialty Conference. New York, USA, Texas.

Aymerich I., Rieger L., Sobhani R., Rosso D. and Corominas L. (2015). The Difference Between Energy Consumption and Energy Cost: Modelling Energy Tariff Structures for Water Resource Recovery Facilities. *Water Research*, 81, 113–123.

Baquero-Rodríguez G.A., Lara-Borrero J.A., Nolasco D. and Rosso D. (2018). A Critical Review of the Factors Affecting Modeling Oxygen Transfer by Fine-Pore Diffusers in Activated Sludge. *Water Environmental. Research*, 90, 431–441.

Barker P. and Dold P.L. (1997). General model for biological nutrient removal activated-sludge systems: model presentation. *Water Environmental. Research*, 69, 969–984.

Barnard J.L., Dunlap P. and Steichen M. (2017). Theory to understand all Biological Phosphorus Removal Observations, in: Proceedings of the Water Environment Federation.

Benson B.B. and Krause D. (1984). The concentration and isotopic fractionation of oxygen dissolved in freshwater and seawater in equilibrium with the atmosphere1. *Limnology and Oceanography*, 29, 620–632.

Boyle W.C. and Redmon D.T. (1983). Biological Fouling of Fine Bubble Diffusers: State-of-Art. *Journal of Environmental Engineering*, 109, 991–1005.

Cornel P., Wagner M. and Krause S. (2003). Investigation of oxygen transfer rates in full scale membrane bioreactors. *Water Science and Technology*, 47.

De Clercq B., Coen F., Vanderhaegen B. and Vanrolleghem P.A. (1999). Calibrating simple models for mixing and flow propagation in waste water treatment plants. *Water Science and Technology*, 39, 61–69.

DHI. (2014). WEST Models Guide. Hørsholm, Denmark http://www.mikebydhi.com/ (last accessed in February 2014).

Downing A. and Waste A.B. (1963). Oxygen transfer in the activated sludge process. *Adv. Biol. Wastewater Treatment*, 131–147.

Durán C., Fayolle Y., Pechaud Y., Cockx A. and Gillot S. (2016). Impact of suspended solids on the activated sludge non-Newtonian behaviour and on oxygen transfer in a bubble column. *Chemical Engineering Science*, 141, 154–165.

Eckenfelder Jr. W.W., Raymond W.L. and Lauria D.T. (1956). Effect of various organic substances on oxygen absorption efficiency. *Sewage Ind. Waste*, 28, 1357–1364.

Eckenfelder Jr. W.W. and Ford D. (1968). Engineering aspects of surface aeration design.

Emami N., Sobhani R. and Rosso D. (2018). Diurnal variations of the energy intensity and associated greenhouse gas emissions for activated sludge processes. *Water Science and Technology*, 77, 1838–1850.

Fayolle Y., Cockx A., Gillot S., Roustan M. and Héduit A. (2007). Oxygen transfer prediction in aeration tanks using CFD. *Chemical Engineering. Science*, 62, 7163–7171.

Füreder K., Svardal K., Frey W., Kroiss H. and Krampe J. (2018). Energy consumption of agitators in activated sludge tanks – actual state and optimization potential. *Water Science and Technology*, 77, 800–808.

Garrido-Baserba M., Asvapathanagul P., Park H.D., Kim T.S., Baquero-Rodriguez G.A., Olson B.H. and Rosso D. (2018). Impact of fouling on the decline of aeration efficiency under different operational conditions at WRRFs. *Science of the Total Environment*, 639, 248-257.

Germain E., Nelles F., Drews A., Pearce P., Kraume M., Reid E., Judd S.J. and Stephenson T. (2007). Biomass effects on oxygen transfer in membrane bioreactors. *Water Research*, 41, 1038–1044.

Gillot S., Capela-Marsal S., Roustan M. and Héduit A. (2005). Predicting oxygen transfer of fine bubble diffused aeration systems - Model issued from dimensional analysis. *Water Research*, 39, 1379–1387.

Gillot S. and Héduit A. (2008). Prediction of alpha factor values for fine pore aeration systems. *Water Science and Technology*, 57, 1265–1269.

Göbel A., McArdell C.S., Joss A., Siegrist H. and Giger W. (2007). Fate of sulfonamides, macrolides, and trimethoprim in different wastewater treatment technologies. *Sci. Total Environ.*, 372, 361–371.

Groves K.P., Daigger G.T., Simpkin T.J., Redmon D.T. and Ewing L. (1992). Evaluation of oxygen transfer efficiency and alpha-factor on a variety of diffused aeration systems. *Water Environmental Research*, 64, 691–698.

Harper Jr W.F. and Jenkins D. (2003). The effect of an initial anaerobic zone on the nutrient requirements of activated sludge. *Water Environment Research*, 75(3), 216-224.

Henkel J. (2010). Oxygen transfer phenomena in activated sludge.

Henkel J., Cornel P. and Wagner M. (2011). Oxygen transfer in activated sludge – new insights and potentials for cost saving. *Water Science and Technology*, 63.

Henze M. (2000). Activated sludge models ASM1, ASM2, ASM2d and ASM3. IWA Publishing, London, UK.

Henze M., Grady Jr. C.P.L. and Gujer W. (1987). A general model for single-sludge wastewater treatment systems. *Water Research*, 21, 505–515.

Hodgson B., Subramanian R., Cavanaugh B., Rosso D., Brischke K. and Garrido-Baserba M. (2019). Our blowers are too large! We're wasting too much energy!...hmmm..maybe not! WEFTEC 2019 - 92nd Annu. Water Environ. Fed. Tech. Exhib. Conf. 1415–1435.

Hsieh C.-C., Babcock R.W. and Stenstrom M.K. (1994). Estimating semivolatile organic compound emission rates and oxygen transfer coefficients in diffused aeration. *Water Environmental Research*, 66, 206–210.

Hsieh C.-C., Babcock R.W. and Stenstrom M.K. (1993). Estimating Emissions of 20 VOCs. II: Diffused Aeration. *Journal of Environmental Engineering*, 119, 1099–1118.

Jenkins T.E. (2013). Aeration control system design : a practical guide to energy and process optimization. Wiley.

Ji L. and Zhou J. (2006). Influence of aeration on microbial polymers and membrane fouling in submerged membrane bioreactors. *J. Memb. Sci.*, 276, 168–177.

Jiang L.M., Garrido-Baserba M., Nolasco D., Al-Omari A., DeClippeleir H., Murthy S. and Rosso D. (2017). Modelling oxygen transfer using dynamic alpha factors. *Water Research*, 124, 139–148.

Ju L.-K. and Sundararajan A. (1994). The effects of cells on oxygen transfer in bioreactors: physical presence of cells as solid particles. *Chem. Eng. J. and Biochem. Eng. J.*, 56, B15–B21.

Kaliman A., Rosso D., Leu S.S.Y. and Stenstrom M.M.K. (2008). Fine-pore aeration diffusers: Accelerated membrane ageing studies. *Water Research*, 42, 467–475.

Karpinska A. and Bridgeman J. (2016). CFD-aided modelling of activated sludge systems–A critical review. *Water Research*, 861–879.

Kessener H. and Ribbius F. (1934). Comparison of aeration systems for the activated sludge process. *Sewage Work. J.*

Kim Y. and Boyle W.C. (1993). Mechanisms of Fouling in Fine-Pore Diffuser Aeration. *Journal of Environmental Engineering*, 119, 1119–1138.

Krampe J. (2011). Full scale evaluation of diffuser ageing with clean water oxygen transfer tests. *Water Science and Technology*, 64, 700.

Krampe J. and Krauth K. (2003). Oxygen transfer into activated sludge with high MLSS concentrations. *Water Science and Technology*, 47, 297–303.

Leu S.-Y., Chan L. and Stenstrom M.K. (2012). Toward Long Solids Retention Time of Activated Sludge Processes: Benefits in Energy Saving, Effluent Quality, and Stability. *Water Environmental Research*, 84, 42–53.

Leu S.-Y., Rosso D., Larson L.E. and Stenstrom M.K. (2009). Real-Time Aeration Efficiency Monitoring in the Activated Sludge Process and Methods to Reduce Energy Consumption and Operating Costs. *Water Environmental Research*, 81, 2471–2481.

Lippi S., Rosso D., Lubello C., Canziani R. and Stenstrom M.K. (2009). Temperature modelling and prediction for activated sludge systems. *Water Science and Technology*, 59, 125–131.

Loera J. (2013). High-Speed Turbo Blowers Need … Or Just Want. Wateronline. URL https://www.wateronline.com/doc/high-speed-turbo-blowers-need-or-just-want-0001 (accessed 06.07.2018).

Ludwig E. (1997). Applied process design for chemical and petrochemical plants.

Martins A.M.P., Heijnen J.J. and Van Loosdrecht M.C.M. (2004a). Bulking Sludge in Biological Nutrient Removal Systems (2004) *Biotechnology and Bioengineering*. 86(2), 125-135.

McGinnis D.F. and Little J.C. (2002). Predicting diffused-bubble oxygen transfer rate using the discrete-bubble model. *Water Research*. 36(18):4627-4635.

McWhirter J.R. (1978). The Use of High-Purity Oxygen in the Activated Sludge Process (2 vols.) by J.R. McWhirter, Ed: CRC Press, Inc, West Palm Beach, FL. Appledore Books, ABAA, 2nd ed., CRC Press. Palm Beach, Florida, USA.

Metcalf and Eddy. (2014). Wastewater Engineering: Treatment and Resource Recovery, 5th ed. McGraw-Hill Education.

Metcalf and Eddy. (2013). Wastewater Engineering: Treatment and Resource Recovery.

Miot A., Levitin Y. and Stenstrom M.K. (2016). Novel Control Method to Achieve Higher Oxygen Utilization in High Purity Oxygen Activated Sludge Plants, in: Proceedings of the Water Environment Federation, 2016(9), 3623-3631. pp. 3623-3631.

Muller E.B., Stouthamer A.H., Van Verseveld H.W. and Eikelboom D.H. (1995). Aerobic domestic waste water treatment in a pilot plant with complete sludge retention by cross-flow filtration. *Water Research*, 29, 1179-1189.

Nittami T., Katoh T. and Matsumoto K. (2013). Modification of oxygen transfer rates in activated sludge with its characteristic changes by the addition of organic polyelectrolyte. *Chem. Eng. J.*, 225, 673–678.

Odize V.O., Novak J., De Clippeleir H., Al-Omari A., Smeraldi J.D., Murthy S. and Rosso D. (2017). Reverse flexing as a physical/mechanical treatment to mitigate fouling of fine bubble diffusers. *Water Science and Technology*, 76, 1595–1602.

Ozdemir B. and Yenigun O. (2013). A pilot scale study on high biomass systems: Energy and cost analysis of sludge production. *Journal of Membrane Science*, 428, 589–597.

Parker D., Geary S., Jones G., McIntyre L., Oppenheim S., Pedregon V., Pope R., Richards T., Voigt C., Volpe G., Willis J. and Witzgall R. (2003). Making Classifying Selectors Work for Foam Elimination in the Activated-Sludge Process. *Water Environmental Research*, 75, 83–91.

Pereira J., Karpinska A., Gomes P., Martins A.A., Dias M.M., Lopes J.C.B. and Santos R.J. (2012). Activated sludge models coupled to CFD simulations. Single two-phase flows *Chem. Biomed. Eng.*, 153–173.

Philichi T.L. and Stenstrom M.K. (1989). Effects of dissolved oxygen probe lag on oxygen transfer parameter estimation. *Journal of the Water Pollution Control Federation*, 83–86.

Plano S., Jiang L.-M., Nopens I., Weijers S. and Rosso D. (2011). Increasing the Accuracy of Energy Footprint Predictions for Activated Sludge Processes Using a Rigorous Dynamic Aeration Model. *Proceedings of the Water Environment Federation*, 2011, 6500–6511.

Pretorius C., Wicklein E., Rauch-Williams, T., Samstag R. and Sigmon C. (2015). How oversized mixers became an industry standard, in: *Proceedings of the Water Environment Federation*. pp. 4379–4411.

Reardon D. (1995). Turning down the power. Civ. Eng.

Redmon D., Boyle W.C. and Ewing L. (1983). Oxygen Transfer Efficiency Measurements in Mixed Liquor Using Off-Gas Techniques. *Journal of the Water Pollution Control Federation*, 55, 1338-1347.

Redmond D., Groves K., Daigger G., Simpkin T. and Ewing L. (1992). Evaluation of oxygen transfer efficiency and alpha-factor on a variety of diffused aeration systems. *Water Environmental Research*, 64, 691–698.

Rieth M., Boyle W. and Ewing L. (1990). Effects of selected design parameters on the fouling characteristics of ceramic diffusers. *Journal of the Water Pollution Control Federation*, 877–886.

Rosso D. (2018). Aeration, Mixing, and Energy: Bubbles and Sparks. IWA Publishing, Irvine, California, USA.

Rosso D. and Al-Omari A. (2019). Carbon Capture and Management Strategies for Energy Harvest from Wastewater. The Water Research Foundation. *Water Research Foundation*. URL https://www.waterrf.org/research/projects/carbon-capture-and-management-strategies-energy-harvest-wastewater (accessed 04.11.2019).

Rosso D., Iranpour R. and Stenstrom M.K. (2005). Fifteen years of offgas transfer efficiency measurements on fine-pore aerators: key role of sludge age and normalized air flux. *Water Environmental Research*, 77, 266–273.

Rosso D., Jiang L.-M., Pitt P., Hocking C.S., Stenstrom M.K., Murthy S., Hayden D.M., Zhong J., Coller D.H., Kim A.Y. and Xu H. (2013). Methodology for In Situ Column Testing to Improve Accuracy during Design and Specification of Aeration Systems. *Journal of Environmental. Engineering*, 139, 530–537.

Rosso D., Jiang L.M., Hayden D.M., Pitt P., Hocking C.S., Murthy S. and Stenstrom M.K. (2012). Towards more accurate design and specification of aeration systems using on-site column testing. *Water Science and Technology*, 66, 627–634.

Rosso D. and Shaw A.R. (2015). Framework for Energy Neutral Treatment for the 21st Century through Energy Efficient Aeration. *Water Environ. Res. Found*. (Project INFR2R12).

Rosso D. and Stenstrom M.K. (2006a). Economic implications of fine-pore diffuser aging. *Water Environmental Research*, 78, 810–815.

Rosso D. and Stenstrom M.K. (2005). Economic Implications of Fine Pore Diffuser Aging. *Proc. Water Environ. Fed*.

Rosso D. and Stenstrom M.K. (2007). Energy-saving benefits of denitrification. *Environ. Eng. Appl. Res. Pract*. 3.

Rosso D. and Stenstrom M.K. (2006b). Surfactant effects on α-factors in aeration systems. *Water Research*, 1397-1404.

Rosso D. and Stenstrom M.K. (2005). Comparative economic analysis of the impacts of mean cell retention time and denitrification on aeration systems. *Water Research*, 39, 3773–3780.

Rosso D., Stenstrom M.K. and Larson L.E. (2008). Aeration of large-scale municipal wastewater treatment plants: state of the art. *Water Science and Technology*, 57, 973–8.

Schmit F., Redmon D. and Ewing L. US Patent 4, 889, 620, 1989. In place gas cleaning of diffusion elements. Google Patents 4, 889.

Schuler A.J. and Jang H. (2007). Density effects on activated sludge zone settling velocities. *Water Research*, 41, 1814–1822.

Sharma A.K., Guildal T., Thomsen H.R. and Jacobsen B.N. (2011). Energy savings by reduced mixing in aeration tanks: results from a full scale investigation and long term implementation at Avedoere wastewater treatment plant. *Water Science and Technology,* 64, 1089.

Snowling S. (2014). GPS-X Technical reference.

Soliman M.A., Pedersen J.A., Park H., Castaneda-Jimenez A., Stenstrom M.K. and Suffet I.H. (2007). Human Pharmaceuticals, Antioxidants, and Plasticizers in Wastewater Treatment Plant and Water Reclamation Plant Effluents. *Water Environmental Research*, 79, 156–167.

Stamou A.I. (2008). Improving the hydraulic efficiency of water process tanks using CFD models. *Chem. Eng. Process*, 47, 1179–1189.

Stensel H.D., Albertson O.E. and Hendricks P. (1994). Evaluation of anoxic-aerobic treatment at the phoenix 91st avenue plant, in: *Proceedings of the Water Environment Federation*.

Stenstrom M.K. and Gilbert R.G. (1981). Effects of alpha, beta and theta factor upon the design, specification and operation of aeration systems. *Water Research*, 15, 643–654.

Talati S.N. and Stenstrom M.K. (1990). Aeration-Basin Heat Loss. *Journal of Environmental Engineering*, 116, 70-86.

Tampus M.V., Martins A.M.P. and Van Loosdrecht M.C.M. (2004). The effect of anoxic selectors on sludge bulking. *Water Science and Technology,* 50, 261–268.

USEPA. (1989). Fine Pore (Fine Bubble) Aeration Systems.

USEPA. (1985). Fine Pore (Fine Bubble) Aeration Systems, EPA/625/8-85/ 010.

Vogelaar J.C.T., Klapwijk A., Van Lier J.B. and Rulkens W.H. (2000). Temperature effects on the oxygen transfer rate between 20 and 55°C. *Water Research*, 34, 1037–1041.

Wagner M.R. and Pöpel H.J (1996). Surface active agents and their influence on oxygen transfer. *Water Science and Technology,* 34, 249–256.

Wagner M.R. and Pöpel H.J. (1998). Oxygen transfer and aeration efficiency - Influence of diffuser submergence, diffuser density, and blower type. *Water Science and Technology,* 38, 1–6.

Warriner R. and Brenner R. (1996). Oxygen Transfer Efficiency Surveys at the South Shore Wastewater Treatment Plant 1985-1987. Div. US. EPA, Cincinnati, Ohio, USA.

WEF. (2009). Energy Conservation in Water and Wastewater Treatment Facilities, WEF Manual of Practice. WEF Man. Pract. McGraw-Hill, Inc., New York, NY. URL http://www.amazon.com/Energy-Conservation-Water-Wastewater-Facilities/dp/0071667946 (accessed 25.02.2015).

NOMENCLATURE

Symbol	Description	Unit
a	Diffuser bubbling area	m^2
AE	Aeration efficiency in clean water	kgO_2/kWh
AFR	Airflow rate	m^3/s
DO	Dissolved oxygen in water	kgO_2/m^3
DO_{sat}	Dissolved oxygen in water at saturation	kgO_2/m^3
k_La	Liquid side mass transfer coefficient	1/h
K_La_{cw}	Mass transfer coefficient for clean water	l/h
K_La_{pw}	Mass transfer coefficient for process water	l/h
N_D	Total diffuser number	-
OTE	Oxygen transfer efficiency in clean water	%
OTR	Oxygen transfer rate in clean water	kgO_2/h
P	Power drawn by the aeration system	kW
SAE	Standard aeration efficiency	kgO_2/kWh
SOTE	Oxygen transfer efficiency in standard conditions in clean water	%
SOTR	Oxygen transfer rate in standard conditions in clean water	kgO_2/h
V	Water volume	m^3
Z	Diffuser submergence	m

Abbreviation	Description
AC	Alternating current
ASCE	American Society of Civil Engineers
CAS	Conventional activated sludge process
DWP	Dynamic wet pressure
EE3	17a-ethinylestradiol
EPA	Environmental protection agency
EPDM	Ethylene propylene diene monomer rubber
F	Fouling factor
MLSS	Mixed liquor suspended solids
NDN	Nitrification/denitrification process
PD	Positive displacement
RBCOD	Readily biodegradable COD
Re	Reynolds number
rpm	Revolutions per minute
SRT	Sludge retention time
SVI	Sludge volume index
VFD	Variable frequency drive

Greek symbol	Explanation	Unit
α	Ratio of process- to clean-water mass transfer	-
αF	Alpha factor for used diffusers	-
αFSAE	Aeration efficiency in standard conditions in process water for used diffusers	%
αFSOTE	Oxygen transfer efficiency in standard conditions in process water for used diffusers	%
αSAE	Aeration efficiency in standard conditions in process water	%
αSOTE	Oxygen transfer efficiency in process water at standard conditions in process water	%

Figure 9.52 Aeration system of Donald C. Tillman Water Reclamation Plant of the City of Los Angeles' LA Sanitation (Photo: D. Rosso).

doi: 10.2166/9781789060362_0475

10

Bulking sludge

Mark C.M. van Loosdrecht, Antonio M. Martins and George A. Ekama

10.1 INTRODUCTION

The activated sludge process is the most commonly used technology for biological wastewater treatment. It consists of two stages, a biochemical stage (aeration tank) and a physical stage (secondary clarifier). In the aeration tank, organic carbon, ammonium and phosphate are removed from the wastewater by the activated sludge. The amount of biomass which is produced in wastewater treatment plant is relatively low. An influent carbon content of 500 mg COD generates 200-300 mg suspended solids. Without biomass retention this would also be the actual sludge concentration in the treatment process. Therefore biomass retention is used in order to increase the biomass concentration in the biochemical stage and obtain higher volumetric conversion rates. Since bacteria form flocs which can be separated from the treated wastewater by gravity forces, this energy-friendly and economical option is the standard technology applied for solid-liquid separation. A good separation (settling) and compaction (thickening) of activated sludge in the secondary clarifier is a necessary condition to guarantee a good effluent quality from the activated sludge process. This separation is therefore based on the formation of compact flocs. The relatively low force of gravity means that the required settling area (*i.e.* settling tanks) becomes a large part of the total treatment plant; it easily consists of 30-50% of the total treatment area (Figure 10.1).

Figure 10.1 A modern biological nutrient removal plant in The Netherlands (BCFS® process) showing the importance of sludge separation on the total process layout. The settling tank design was based on a sludge volume index (SVI) of 120 ml/g (photo: Van Loosdrecht *et al.*, 1998).

The relation between sludge settling and settler design is treated in detail in Chapter 12. The sludge volume index (SVI) is used as an empirical measure which links the sludge characteristics and settler design (Ekama and Marais, 1986). This value is obtained by settling a sludge sample in a 1 litre measurement cylinder for 30 minutes. The volume of the sludge layer can be read and divided by the original suspended solids content of the sludge sample. In this way one obtains the volume which is taken by a gram of sludge after settling. The SVI has a major effect on the required settler size (Figure 10.2), and an increase in SVI from 100 to 150 ml/g will result in almost doubling the design area needed for settlers. Details on this measurement are provided in Van Loosdrecht *et al.*, 2016.

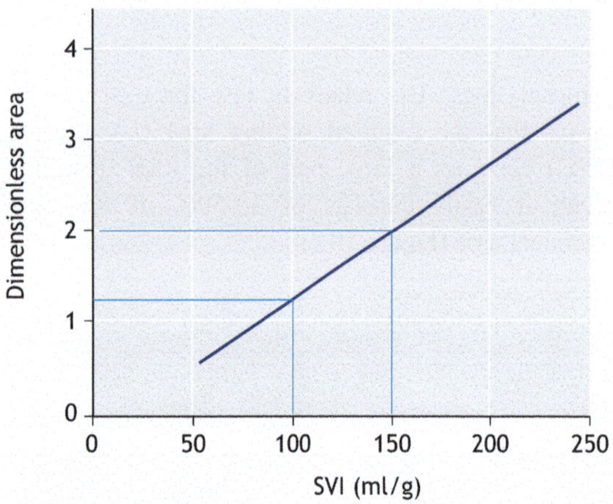

Figure 10.2 Relation between sludge volume index and surface area needed for a settler according to the STOWA (STOWA, 1994) design guidelines for settlers.

Bulking sludge, a term used to describe the excessive growth of filamentous bacteria, is a common and longstanding problem for activated sludge processes (*e.g.* Donaldson, 1932). When the sludge flocs are open and porous the settling is hindered and the settled sludge will contain low amounts of solids. In practice bulking is associated with a high SVI. The critical value for the SVI above which bulking sludge occurs is heavily dependent on the local practise in design and construction for settlers. In effect bulking sludge is defined in general when suspended solids cannot be maintained in the settler, although different regions have different traditions in settler design. For instance, nowadays in the Netherlands an SVI above 120 ml/g is considered bulking sludge since this index value is currently used in the design guidelines for settlers. Bulking sludge is typically an operational or empirical problem, so there is not an exact scientific index value to distinguish bulking sludge from non-bulking sludge. Open and porous sludge flocs settle more slowly so require larger settlers in order to be maintained in the process and/or prevent solids from occurring in the effluent. The growth of filamentous bacteria is especially detrimental and leads to many problems in practice. The volume fraction of filamentous bacteria in the activated sludge community which causes settling problems can be very small; volume fractions of 1-20% are sufficient to cause bulking sludge (Palm *et al.,* 1980; Kappeler and Gujer, 1994b). Although filamentous bacteria are often not the dominant metabolic bacterial group in the treatment plant, they still cause bulking sludge (Figure 10.3).

Despite a vast amount of research on bulking sludge it continues to be a problem in operating wastewater treatment plants. This is likely caused by several conditions which cause filamentous organisms to proliferate. Many filamentous bacteria are not available in pure cultures, preventing a detailed microbiological study of these organisms. The status of the plant operation under which bulking sludge occurs is usually only marginally documented. One reason for not finding a good general solution to bulking sludge might be the absence of a consensus on the exact level at which the problem should be approached. The dominant approach found in the literature is by trying to identify the specific filamentous bacterium in a bulking sludge (Eikelboom, 1975; 2000). By studying and understanding the ecophysiology of the filamentous bacterium (either in pure culture or by applying *in-situ* techniques, such as microautoradiography: MAR), it is hoped that a solution to avoid the occurrence of the specific filament can be found.

Figure 10.3 Nightmares for the wastewater treatment plant operator: (A) filamentous bacteria, (B) bulking sludge leaving the plant with the effluent (photos: D.H. Eikelboom and D. Brdjanovic).

A different approach is by recognising that the general characteristic is the cell morphology. Realising how the microbial cell morphology affects the ecology of the bacteria can lead to a general solution independent of the species involved (Chudoba *et al.,* 1973a; Rensink, 1974). In this approach the occurrence of a specific kind of filamentous bacterium is a second order problem. The problem is therefore that process engineering as well as microbiology knowledge is needed in order to solve the problem and that the solution cannot be obtained from either of these two fields in isolation.

10.2 HISTORICAL ASPECTS

It is not the intention in this chapter to fully describe the history and developments of activated sludge systems. For this the reader is invited to read the reviews provided, for instance, by Allemann and Prakasam (1983) or Albertson (1987). We will just highlight some of the most important historical facts which have contributed to the understanding of the bulking sludge problem.

The activated sludge process was developed in the early 1900s in England (Ardern and Lockett, 1914). Initially fill-and-draw systems were brought into operation but they were quickly converted into continuous flow systems. Despite the increased occurrence of settling problems, continuous flow systems became popular and spread worldwide. Donaldson (1932) suspected that back-mixing in plug-flow aeration basins, which changes the hydraulic behaviour and the substrate regime to a completely mixed mode, was an important factor promoting the development of bulking sludge. As a corrective measure he suggested that the aeration basin should be compartmentalised (*i.e.* a plug flow reactor) to promote the development of satisfactorily settling sludge. Nevertheless, continuously-fed completely mixed activated sludge systems remained the preferred design. Clearly civil engineering advantages in the construction phase prevailed over process engineering advantages during operation. However, the discussion on the effect of the feeding pattern on sludge settleability was reopened in the 1970s. Studies showed the advantage of using compartmentalised tanks with a plug flow pattern over continuously-fed completely mixed systems (Chudoba *et al.,* 1973b; Rensink, 1974; and many others), confirming the earlier recommendations of Donaldson (1932).

Pasveer (1959) went back to the original fill-and-draw technology of Ardern and Lockett, from which he developed the Pasveer or Oxidation Ditch system. This reopened the discussion on the advantages of utilising these systems in the treatment of municipal wastewater. The fill-and-draw oxidation ditch became relatively popular in Europe for a few years, but once more, almost all these systems were soon converted to continuous-flow oxidation ditches by the addition of a secondary settler and solids recycling. Nevertheless, during the 1960s Pasveer showed that intermittently-fed full-scale oxidation ditches produce sludge with better settleability than continuously-fed completely mixed systems (Pasveer, 1969).

In the 1970s Chudoba and his co-workers (1973b) and Rensink (1974) developed the selector reactor, which became the most widespread engineering tool to control bulking sludge. However, although the use of selectors has been successful and has reduced bulking problems in many activated sludge systems, there are still regular reports of their failure.

10.3 RELATIONSHIP BETWEEN MORPHOLOGY AND ECOPHYSIOLOGY

One of the most intriguing and complex questions on bulking sludge is whether microbial morphology, physiology and substrate kinetics are related and how these contribute to the dominance of filamentous bacteria in activated sludge. Is there a general mechanism that could explain the growth of filamentous bacteria or does each filamentous microorganism need to be identified and physiologically, morphologically, kinetically and taxonomically described in order to develop strategies for bulking sludge control? Is it possible to design reactor conditions which prevent all the filaments from proliferating and still achieve the effluent quality required biologically? Even though some plants have never been observed to have bulking sludge, for decades scientists, engineers, and microbiologists have failed to find a definitive answer to these questions. However, some relationships can be inferred and they will be briefly discussed here.

10.3.1 Microbiological approach

The lack of success in finding a general solution to bulking sludge control led many researchers to look at the microbial population and search for the predominant filamentous bacteria responsible for bulking. Identification keys were developed (Eikelboom, 1977, 2000) to identify filamentous bacteria based on microscopic characterisation.

Despite several limitations these identification methods produced a systematic tool which allowed relative confidence in the identification of filaments. The next step was finding relationships between the most predominant filaments and their physiology and the operational conditions (*e.g.* dissolved oxygen concentration - DO, food/microorganism ratio - F/M, etc.) in order to define (specific) strategies for its control (Jenkins *et al.,* 1993) (Table 1.1). The distribution of filamentous microorganisms varies considerably between different geographical areas (Martins *et al.*, 2004a) and seasonally. It can be concluded that *Microthrix parvicella* and types 0092 and 0041/0675 are apparently the major morphotype filaments, mainly responsible for the bulking events observed in biological nutrient removal (BNR) activated sludge systems. These surveys also showed that bulking sludge episodes, supposedly due to the abundance of *Microthrix parvicella*, were more frequent in winter and spring than in summer and autumn (*e.g.* Kruit *et al.,* 2002). It was also confirmed that the morphotypes type 021N, type 0961, *Sphaerotilus natans* and *Thiothrix sp.*, are controlled by anaerobic and anoxic stages, as typical in bio-P and denitrifying systems (Ekama *et al.*, 1996b). However, these conditions seem to be inefficient for the dominant filamentous microorganisms found in biological nutrient removal systems. Curiously the morphotype filamentous bacteria found in biological nutrient removal systems are usually Gram-positive which implies that their likely hydrophobic cell surface could easily adsorb compounds with a low solubility. It is however unclear whether low-loaded systems also can be enriched with Gram-positive floc-forming bacteria.

Table 10.1 Proposed groups of model morphotype filamentous microorganisms (Wanner and Grau, 1989; Jenkins *et al.*, 1993).

Microorganisms	Features	Control
Group I: Low DO aerobic zone growers		
Sphaerotilus natans, type 1701, *Haliscomenobacter hydrossis*	Use readily biodegradable substrates; grow well at low DO concentrations; grow over wide range of SRTs.	Aerobic, anoxic or anaerobic plug flow selectors; increase SRT; increase DO concentration in the aeration basin (> 1.5 mg O_2/l).
Group II: Mixotrophic aerobic zone growers		
Thiothrix sp., type 021N	Use readily biodegradable substrates, especially low molecular weight organic acids; present at moderate to high SRT; capable of sulphide oxidising to stored sulphur granules; rapid nutrient uptake rates under nutrient deficiency.	Aerobic, anoxic or anaerobic plug flow selectors; nutrient addition; eliminate sulphide and/or high organic acid concentrations (eliminate septic conditions).
Group III: Other aerobic zone growers		
Type 1851, *Nostocoida limicola spp.*	Use readily biodegradable substrates; present at moderate to high SRTs.	Aerobic, anoxic or anaerobic plug flow selectors; reduce SRT.
Group IV: Aerobic, anoxic, anaerobic zone growers		
Microthrix parvicella, type 0092, type 0041/0675	Abundant in anaerobic-anoxic-aerobic systems; present at high SRTs; possible growth on hydrolysis of particulate substrates.	Still uncertainty but the most recommended solutions are: install a skimmer to remove particulate substrate; maintain a plug-flow regime in the entire system; several stages (anaerobic/anoxic/aerobic) should be well defined; maintain a relatively high oxygen concentration in the aerobic phase (1.5 mgO_2/l) and a low ammonium concentration (<1mgN /l) (Kruit *et al.*, 2002) and a low nitrate and nitrite in the anoxic reactor before the aerobic reactor (Casey *et al.*, 1999; Musvoto *et al.*, 1994).

During the 1990s molecular methods based on DNA and RNA analyses were introduced into biological wastewater treatment (Chapter 2). These methods enable the correct identification of a filamentous bacteria population. Therefore, it is advisable to apply specific gene probes, whenever they exist, in bulking sludge surveys. Their use, together with filamentous bacteria characterisation and definition of the correct control and operational conditions (*e.g.* selector reactor), are considered major challenges to control bulking sludge.

10.3.2 Morphological-ecological approach

Filamentous bacteria grow preferentially in one or two directions. This morphological feature appears to give competitive advantages to filamentous organisms under substrate limiting concentrations (*e.g.* diffusion-limiting environments). It is thought that these organisms have a higher outward growth velocity and win the competition because they gain easy access to bulk liquid substrate (Martins *et al.*, 2003a). This is in line with some other studies which also connect the excessive growth of filamentous microorganisms with substrate diffusion resistance inside biological flocs (Pipes, 1967; Kappeler and Gujer, 1994a).

Given these views, their morphology as such gives these organisms an ecological advantage. It would also imply that under non-bulking process conditions, filamentous bacteria can still be present inside the floc. If substrate limitation occurs they will then quickly grow out of the floc. The almost ubiquitous presence of filaments in activated sludge has even led to suggestions that actually filamentous organisms form the backbone of activated sludge flocs (Jenkins *et al.*, 1993a). This type of filamentous skeleton structure would promote the attachment of other cells by their extracellular polymeric substances (EPS).

10.4 FILAMENTOUS BACTERIA IDENTIFICATION AND CHARACTERISATION

The basis for understanding and characterising bulking sludge is generally thought to depend on the correct identification of the filamentous bacteria involved. This is briefly discussed below.

10.4.1 Microscopic characterisation *versus* molecular methods

Many types of bacteria have still not been identified and are not recognised taxonomically. Therefore, these bacteria are not documented in standard microbiological identification manuals such as Bergey's Manual of Systematic Bacteriology. Eikelboom (1975, 1977) developed the first identification key to identify filamentous bacteria in activated sludge systems. This identification is mainly based on morphological characteristics and on the response of the filamentous bacteria to a few microscopic staining tests. The procedures, techniques and identification keys were compiled in a microscopic sludge investigation manual (Eikelboom, 2000) that, together with a slightly different manual by Jenkins *et al.* (1993a, 2006), are used as worldwide references on filamentous bacteria identification.

However, although very useful, this type of identification has its limitations. For instance, many filamentous bacteria (*e.g.* the morphotypes *Sphaerotilus natans*, 1701, 0092 and 0961) can change morphology in response to changes in environmental conditions and although some of them can look the same morphologically, they probably vary considerably in their physiology and taxonomy. For instance, the filamentous bacterial morphotype *Nostocoida limicola* encompasses several phylogenetically different bacteria (Seviour *et al.*, 2002) belonging to the following groups: low mol% G+C Gram-positive bacteria, high mol% G+C Gram-positive bacteria, *Planctomycetes*, green non-sulphur bacteria and the alpha subclass of *Proteobacteria* (Martins *et al.*, 2004b). This also applies to the filamentous morphotype Eikelboom type 1863.

Microscopic identification of filamentous bacteria based on morphology requires a well-trained and experienced person; otherwise a wrong judgement can easily be made. Furthermore, about 40 new morphotypes of filamentous bacteria were identified in a survey study in industrial activated sludge systems (Eikelboom and Geurkink, 2002), making the identification of filamentous bacteria even more complex. This misleading and difficult identification by traditional microscopic techniques has directed research towards molecular methods. Molecular methods based on analysing the DNA or RNA of the bacteria have developed rapidly. For activated sludge several methods are presently commonly used. In order to characterise the complexity of a microbial community the 16S rRNA of the bacteria can be used. Details of these methods are out of the scope of this chapter and are briefly treated in Chapter 2. Details on the methodology for microscopic evaluation is described in Van Loosdrecht *et al.* (2016).

10.4.2 Physiology of filamentous bacteria

As stated above, most filamentous organisms are still very poorly characterised and described, mainly due to the problems of cultivation and maintenance of cultures. However, recent developments in combining micro-autoradiography with fluorescent *in-situ* hybridisation (FISH) are promising for elucidating the exact physiology of filamentous bacteria. Nonetheless there exists no obvious intrinsic relation between the morphology of filamentous bacteria and a specific physiology. Therefore, there general ecological behaviour of filamentous bacteria is more linked with the morphology then with a specific metabolism.

A general problem we face is that old physiological data are described for morphotype filamentous bacteria, which are likely to be phylogenetically unrelated bacteria with large physiological differences, and, consequently, old physiological data (*e.g.* the morphotype '*Nostocoida limicola*') might or might not be correct. Therefore, old physiological data should be interpreted with caution and future bacterial physiological studies should unequivocally show the taxonomy of the studied organisms.

The few physiological studies with pure cultures of chemoheterotrophic filamentous bacteria showed that most of them appear to have a strictly aerobic respiratory metabolism, with oxygen as the electron acceptor. To our knowledge, only the morphotypes type 0961, type 1863, type 1851 and *Nostocoida limicola* have the capacity to perform a fermentative metabolism and therefore may have competitive advantages in systems with anaerobic stages. However, these morphotypes are believed to be only minor components of the total microbial population and they are in general not responsible for bulking sludge episodes.

Some filamentous bacteria, such as *Microthrix parvicella*, *Sphaerotilus natans*, *Thiothrix spp.*, type 021N and type 1851, are able to use nitrate as the electron acceptor, reducing it only to nitrite but the substrate uptake rate and denitrification rate for the filamentous bacteria analysed so far (type 021N and *Thiothrix spp.*) are much lower (by more than 80 times) than for floc-forming bacteria (Shao and Jenkins, 1989). Type 0092, a filamentous bacterium dominant in many nutrient removal activated sludge systems, seems to be incapable of using nitrate as an electron acceptor. Furthermore, in the case of *Microthrix parvicella* it is reported that growth is not sustained under anoxic conditions. Anoxic contact zones have been using this physiological information to control bulking sludge particularly due to bulking caused by type 021N and *Sphaerotilus natans* (Ekama *et al.,* 1996a). Of the predominant filamentous bacteria found in biological nutrient removal activated sludge systems only the morphotype type 0092 and *Microthrix parvicella* were grown in a pure culture, and significant difficulties are encountered in the isolation of the latter. *Microthrix parvicella* seems to be the most dominant problematic organism in biological nutrient removal processes (Nielsen *et al.*, 2002). The main difference is that their preferred substrate are long chain fatty acids rather then volatile fatty acids. The organism needs reduced sulphur compounds for protein synthesis and is described as microaerophilic (Slijkhuis and Deinema, 1988; Rossetti *et al.*, 2005). When anoxic or anaerobic–anoxic conditions are introduced to stimulate biological nutrient removal, *M. parvicella* can therefore proliferate. It was observed that selectors indeed could not eliminate *M. parvicella* bulking under anoxic-aerobic conditions. When in a system with a selector the main reactor was made fully aerobic and *M. parvicella* bulking was properly controlled, whereas an anoxic-aerobic main reactor led to the proliferation of *M. parvicella* (Ekama *et al.*, 1996a). This laboratory experience is supported by a full-scale evaluation by Kruit *et al.* (2002). They concluded that the main criterium to prevent *M. parvicella* bulking was to have distinctly separate aerated (DO $>$ 1.5 mg/l) and anoxic stages (no detectable oxygen).

A metagenomics approach (McIllroy *et al.*, 2013) gave further insight into the metabolism of *Microthrix parvicella.* The dominance of long-chain fatty acid accumulation as lipids under anaerobic or microaerophilic conditions was highlighted, including the need for glycerol in the medium. Polyphosphate is likely the main energy supply for this anaerobic/microaerophilic metabolism. The organism produces exocellular lipases, indicating that fats are indeed the main carbon source.

10.5 CURRENT GENERAL THEORIES TO EXPLAIN BULKING SLUDGE

Several hypotheses about bulking sludge have been formulated in the hope of finding a general explanation for this problem. Unfortunately, none of them have led to a definitive solution. Moreover, most of these theories still lack unequivocal experimental verification. Nevertheless, they form the current basic theoretical framework to approach and understand bulking sludge and, therefore, they will be further discussed.

10.5.1 Diffusion-based selection

Several researchers have pointed out that the morphology of filamentous bacteria aids the substrate uptake under low nutrients or oxygen concentrations. Until the early 1970s the competition between filamentous and non-filamentous bacteria was based on the fact that the surface-to-volume (A/V) ratio is higher for filamentous bacteria (Pipes, 1967). Especially at low substrate concentration this high A/V ratio is advantageous to organisms since it apparently facilitates mass transfer to cells with a high A/V ratio. At lower substrate concentrations this would lead to a relatively higher growth rate.

Later theories proposed that filaments could easily penetrate outside the flocs. When the flocs are growing at a low substrate concentration, the filamentous bacteria detect effectively a higher substrate concentration than the floc formers inside the floc (Sezgin *et al.,* 1978; Kappeler and Gujer, 1994a). Micro-gradients of substrate concentration in flocs have been theoretically predicted (*e.g.* Beccari *et al.,* 1992) and experimentally observed in sludge flocs. Later Martins *et al.* (2004c) extended this theory by comparing floc growth with biofilm growth. Van Loosdrecht *et al.* (1995) and Picioreanu *et al.* (1998) indicated that in diffusion-dominated conditions (*i.e.* low substrate concentrations), open, filamentous, biofilm structures arise. At high substrate concentrations compact and smooth biofilms arise. Ben-Jacob *et al.* (1994) showed that the colony morphology of a pure culture also depends on substrate micro-gradients, with low substrate concentrations leading to filamentous colony morphology. Therefore, it can be that the low substrate concentration leads to a floc being more open and filamentous (Martins *et al.,* 2003b). Filamentous bacteria fit excellently in such a structure.

10.5.2 Kinetic selection theory

Similarly to Donaldson (1932), Chudoba *et al.* (1973a) related the settling characteristics with the mixing characteristics of the activated sludge aeration tank. Using mixed cultures with defined substrate under laboratory-controlled conditions, Chudoba *et al.* (1973a) showed that aeration systems with a low degree of axial mixing and higher macro-gradients of substrate concentration along the system suppress the growth of filamentous bacteria and lead to the development of sludge with satisfactory settling properties. The authors concluded that the primary cause of the selection of floc-forming microorganisms in a mixed culture is the macro-gradient of substrate concentration at the inlet part of the system.

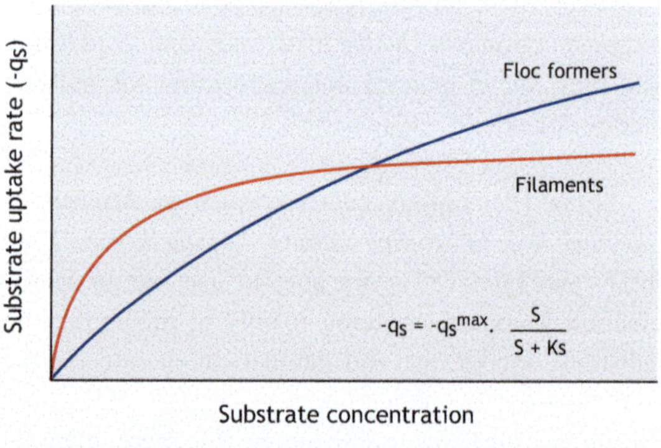

Figure 10.4 Relation between substrate uptake rate (q_S) and substrate concentration (S) for floc-forming and filamentous bacteria according to the kinetic selection theory.

Based on these results, Chudoba *et al.* (1973b) formulated the kinetic selection theory to explain the occurrence or suppression of filamentous bacteria in activated sludge systems. Their explanation was based on a selection criterion for the limiting soluble substrate by filamentous and floc-forming bacteria. Chudoba *et al.* (1973b) hypothesised that filamentous microorganisms (K strategists) are slow-growing organisms which can be characterised as having maximum growth rates (μ_{max}) and an affinity constant (K_S) lower than floc-forming bacteria (r strategists) (Figure 10.4). In systems where the substrate concentration is low (typically $C_S<K_S$), as in continuously-fed completely mixed systems, filamentous bacteria have a higher specific growth rate than floc-forming bacteria, and thereby win the competition for substrate. In systems where the substrate concentration is high, as in plug-flow reactors and sequencing batch reactor (SBR) systems, filamentous bacteria should be suppressed since their growth rate is expected to be lower than that for floc-forming bacteria. Pure culture studies with some filamentous bacteria (*e.g.* Sphaerotilus natans, Haliscomenobacter hydrossis, type 1701, type 021N, Microthrix parvicella) and floc-forming bacteria (Arthrobacter globiformis, Zoogloea ramigera) supported this theory (*e.g.* Van den Eynde *et al.*, 1983). It is however questionable whether these floc-forming bacteria are representative of activated sludge systems. Use of molecular probes has shown that regularly non-dominant bacteria have been enriched from activated sludge. A technique based on quantitative MAR and FISH was applied *in situ* to measure the kinetics of filamentous bacteria (*Candidatus* Meganema perideroedes and *Thiothrix sp.*) (Nielsen *et al.*, 2003). This approach is promising and efforts should be made to extend it to other filamentous and non-filamentous bacteria.

Until now no one has unequivocally shown that filamentous bacteria have in general a lower maximum growth rate than other bacteria present in sludge. Moreover, there is no theoretical explanation why a filamentous morphology would lead to a lower growth rate. The generally lower K_S value for filamentous bacteria as proposed in kinetic selection theory is also not proven yet for the general case. If the K_S is seen as a property of the substrate uptake enzymes there again seems to be no direct relation between K_S and filamentous morphology. If however the K_S is seen as an apparent mass transfer parameter describing mass transfer to the cell, as in the diffusion-based A/V hypothesis of Pipes (1967), then it is fully in agreement with the kinetic selection theory. In flocs the K_S value based on bulk liquid measurements is anyway an apparent coefficient influenced by floc morphology. The more diffusion resistance (the larger and denser the flocs), the higher the measured apparent K_S value (Beccari *et al.,* 1992; Chu *et al.,* 2003). For filaments extending from the floc this would mean a lower apparent K_S value compared to bacteria inside the flocs. Based on this reasoning it might well be that the diffusion-related theories (Pipes, 1967; Sezgin *et al.,* 1978; Kappeler and Gujer, 1994a; Martins *et al.,* 2004c) and the kinetic selection theory (Chudoba *et al.,* 1973b) are two sides of the same coin, and therefore have the same descriptive power.

One experiment which indicates that both theories are potentially correct was performed by Martins *et al.* (2008). When bacteria grow on starch the soluble substrate concentration is always low. The hydrolysis product (maltose) is directly taken up by the actively growing cells. In this case there is substrate uptake at low concentrations, but without forming a substrate gradient because the starch is hydrolysed inside the floc and not in the bulk liquid.

In this case satisfactorily settling sludge was obtained (according to the diffusion-based theory) but the flocs were dominantly formed by *nostocoida* cells (according to the kinetic selection theory). This observation would indicate that the competition between filamentous and non-filamentous bacteria is correctly described by the kinetic selection theory, but the floc morphology is dependent on the same diffusion gradient formation (Figure 10.5) as for biofilm systems (Van Loosdrecht *et al.,* 1995).

Figure 10.5 (A) Left: sludge containing starch stained blue with iodide, right: supernatant, no staining because of the absence of starch, (B) microscopic graph of sludge floc from a starch grown activated sludge culture dominated by *Nostocoida* cells. The blue line represents the substrate concentration for a normal soluble substrate, and the red line the substrate concentration for a hydrolysis product (photos: A.M. Martins).

10.5.3 Storage selection theory

Traditionally non-filamentous microorganisms are supposed to exhibit the ability to store substrate under high substrate concentrations. This ability presumably gives an extra advantage to non-filamentous bacteria in highly dynamic activated sludge systems such as plug-flow reactors, SBR and selector systems (*e.g.* Van den Eynde *et al.,* 1983). However, recent studies have showed that bulking sludge could have a similar or even higher storage capacity than satisfactorily settling sludge (Beccari *et al.,* 1998; Martins *et al.,* 2003b). Studies of pure and mixed cultures also show that some filamentous bacteria, such as *Microthrix parvicella*, can have a high storage capacity under all the possible environmental conditions (*i.e.* aerobic, anoxic and anaerobic) (Nielsen *et al.,* 2002). The filamentous organism *Thiothrix caldifontis* was shown to have the regular metabolism of phosphate-accumulating bacteria, as well as sulphide storage potential (Rubio-Rincon *et al.,* 2017). This stored material can be metabolised for energy generation or protein production during the aerobic famine periods, which would represent a major advantage in selecting these microorganisms instead of other filamentous and non-filamentous bacteria. A lower storage capacity by filamentous bacteria can clearly not be considered as an absolute rule in the selection mechanism for filamentous bacteria but although they may not be prime selection parameters, storage and regeneration (depletion) are intrinsic processes which play a key role in selector-like systems (Van Loosdrecht *et al.,* 1997). Therefore, they should be considered in the description of the metabolic processes which take place in bulking and non-bulking systems.

10.6 REMEDIAL ACTIONS

There are two principal strategies that can be followed to control bulking sludge, *i.e.* non-specific and specific methods. The non-specific methods comprise techniques such as chlorination, ozonation and the application of hydrogen peroxide. The application principle of these methods is quite simple: since filamentous bacteria causing bulking sludge are placed mostly outside the floc, they are more susceptible to oxidants than floc-forming bacteria. Note that this explanation is in line with the diffusion-based hypothesis for competition by filamentous bacteria. Chlorination is widely used in the USA and the procedures for its implementation are well documented (*e.g.* Jenkins *et al.,* 1993b). However, its application in Europe is limited due to environmental concerns about the potential formation of undesirable by-products such as halogenated organic compounds. Another negative aspect is that slow-growing bacteria such as nitrifiers when affected by oxidants take a long time to recover, which could potentially lead to effluent quality deterioration. Furthermore, non-specific methods do not remove the causes of the excessive growth of filamentous microorganisms and their effect is only transient. The same applies to short-term control methods, such as redistribution of biomass from the clarifiers to the aeration tanks and/or increase in the sludge wasting rate. On the other hand, the specific methods are preventive methods which are intended to encourage the growth of floc-forming bacterial structures at the expense of filamentous bacterial structures. The challenge is to find the right environmental conditions in an activated sludge treatment plant to reach this goal. Because these specific methods allow the permanent and sustainable control of bulking in activated sludge systems, they should be developed and adopted in preference to non-specific methods.

Until now preventive action for bulking sludge has not been based on knowledge of the physiology and/or kinetics of a specific type of filamentous bacteria. This is despite the major emphasis on studies to identify the filamentous bacteria present. Generalised preventive actions seem to agree that readily biodegradable substrates need to be consumed at high substrate concentrations. This means that in the entrance part of the activated sludge process a plug-flow type of hydraulics is needed until the readily degradable COD is consumed, and thereafter a completely mixed tank can be used. If oxygen is consumed at low concentrations it leads to bulking sludge in a similar manner as for readily biodegradable COD.

The combined effect of oxygen concentration and readily biodegradable substrate concentration on sludge properties is depicted in Figure 10.6.

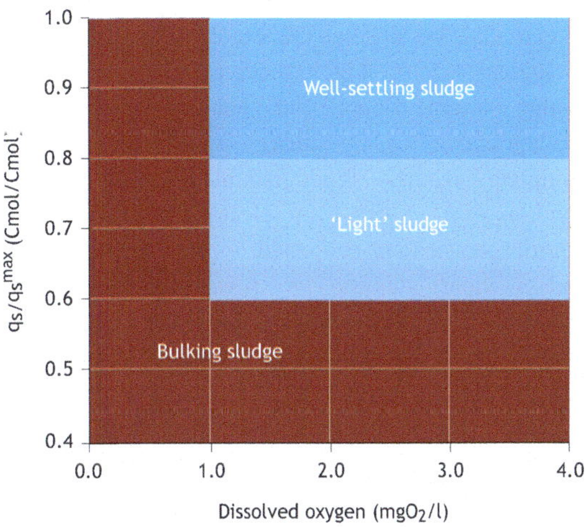

Figure 10.6 Effect of concentration of oxygen and readily available substrate (the latter expressed as the actual rate for substrate uptake relative to its maximum rate) on the type of sludge formed in an activated sludge process (Martins *et al.*, 2003b).

The effective substrate concentration should be seen in relation to the affinity constant for the substrate, and therefore the ratio between the actual and the maximum substrate rate is used here. The dissolved oxygen content seems only to be relevant for the period where readily biodegradable substrate is available. The prerequisite of a plug-flow initial part of the activated sludge process has resulted in the development of selectors to prevent bulking. Both the theories for sludge bulking (A/V or diffusion-based selection as well as kinetic selection theory) support this approach.

10.6.1 Selector

A selector is defined as the initial part of a biological reactor, characterised by a low dispersion number and by an adequate macro-gradient of substrate concentration (Chudoba *et al.*, 1973b; Rensink, 1974). It can also be a small separate initial zone of a biological reactor which receives the influent and sludge return flows and has a high readily biodegradable COD uptake rate, with virtually complete readily biodegradable COD removal (Jenkins *et al.*, 1993a). In selector-like systems, the microorganisms are subjected to periods with (feast) and without (famine or regeneration) external substrate. In essence a pulse-fed SBR or a SBR fed in a static way is the ideal selector system.

It has been shown that indeed in such systems the opposite of bulking sludge, *i.e.* aerobic granular sludge, can be formed (Beun *et al.*, 1999). In the selector the microorganisms are subjected to high growth rate environments and are able to accumulate substrate as internal storage products in their cells (storage). A sufficiently long period without any available external substrate (a low growth rate or famine environment) should then exist (aerobic stage) to re-establish the storage capacity of the cells (Van Loosdrecht *et al.*, 1997; Beun *et al.*, 1999). However, although selectors have been widely installed and are still the most applied engineering tool world-wide in full-scale activated sludge systems for the prevention of the bulking sludge phenomena, nevertheless there are regularly still reports citing selector failure in the control of bulking sludge (*e.g.* in Ekama *et al.*, 1996b). It is unclear if such failures were due to a bad design of the selector tank, to transient conditions in the biological treatment system, or to other factors which have somehow affected the population dynamics, giving competitive advantage to filamentous bacteria. For the control of *M. parvicella*-type bulking in biological nutrient removal processes, selectors seem to fail (Eikelboom, 1994; Ekama *et al.*, 1996b; Kruit *et al.*, 2002) or are not sufficient (see also Section 10.4.2 on physiology). The different selectors and their potential pitfalls will be briefly described here in the following sections. A general overview of selector design guidelines can be found in Table 10.2.

10.6.1.1 Aerobic selectors

Until the end of the 1980s only organic carbon removal was required in most countries, and fully aerobic systems usually with a completely mixed feeding pattern were preferred. In the USA, the systems were mainly a high loading rate with a sludge retention time (SRT) lower than 5 days. Under these conditions the occurrence of bulking sludge was mainly attributed to the excessive growth of filamentous bacteria such as type 021N and type 1701. In Europe and South Africa, low loading rate plants such as oxidation ditch systems and extended aeration systems were constructed. By the 1990s more stringent regulations regarding nutrient emissions, particularly ammonia emissions, were required in Europe and the USA. In order to fulfil these requirements wastewater treatment plants had to be upgraded and improvements for biological nitrification capability were made. The aeration systems were improved and to keep the nitrifying bacteria in the system, the SRT was usually increased to over 10 days. Furthermore, intermittent aeration systems became more common since they allowed a certain degree of denitrification. In these conditions bulking sludge was mainly due to the proliferation of the morphotypes *Microthrix parvicella* and types 021N, 0041/0675 0092, and 0581. These observations led to the definition of what is referred to as the low F/M filamentous bacteria group by Jenkins *et al.* (1993a,b). Aerobic selectors, a small mixing zone (aerobic or anoxic) or contact zone (without aeration), were implemented to control bulking sludge attributed in many cases to the excessive growth of type 021N, *Thiothrix spp.*, *Sphaerotilus natans*, but not always successfully in the case of *Microthrix parvicella*.

The contact time, a typical design parameter for selectors, has a significant and non-linear effect on sludge settleability (Martins *et al.*, 2003a). When the contact time is insufficient, soluble substrate is not fully consumed in the contact zone, and is carried over into the main aeration basin. In this case the growth of filamentous microorganisms will occur due to substrate uptake at a low substrate concentration in the aeration basin. On the other hand, when the contact time is even slightly too long, the concentration of substrate will be low, approaching the typical level of completely mixed tanks, which also favours the growth of filamentous microorganisms. The major effect of a contact tank that is either too large or small on the sludge volume index (SVI) makes a good design difficult (Figure 10.7).

Figure 10.7 Aerobic selector (photo: M. van Loosdrecht).

In systems with highly dynamic feeding patterns, such as temperature and flow, and load variations such as wastewater treatment systems, a good design is not easy and may be a plausible reason for the regular reports on the failing of aerobic selector tanks. Therefore, in practice it is expected that only plug flow systems, as in long channels (length-to-width ratio larger than 10:1), compartmentalised contact tanks, or a SBR fed in a pulse-feed, can guarantee a significant macro-gradient of substrate concentration and will function properly under highly dynamic conditions. Furthermore, proper staging can improve the performance of activated sludge systems which are kinetically limited (Scuras *et al.,* 2001). The necessity to maintain a minimum DO concentration as a function of the soluble organic loading rate or soluble substrate uptake rate in the aeration basin and in the aerobic selector has been recognised and verified in several studies and working diagrams have been proposed (Figure 10.5). Although the recommended contact time in an aerobic selector tank is very short, the amount of oxygen required is approximately 15 to 30% of the soluble COD removed (Jenkins *et al.*, 1993a; Ekama *et al.*, 1996a; Martins *et al.*, 2003b). This underlines the importance of a sufficient oxygen supply in the aerobic selector. If a compartmentalised (plug flow) aerobic selector tank has a too-low aeration rate the negative impacts on the sludge settleability could be worse than with an 'overdesigned' (too large) completely mixed selector tank (Martins *et al.*, 2003b). Furthermore, the aeration control is very important and the sensors should be placed in the first compartment where the oxygen consumption is highest (Table 10.2) and not, as is often the case, at the end of the selector.

10.6.1.2 Non-aerated selectors

As in the aerobic selectors, all the readily biodegradable COD should be removed in anoxic and anaerobic (selector) reactors, preventing any readily biodegradable COD entrance into the aerobic stage, which if it occurs might give advantages to filamentous bacteria (Kruit *et al.,* 2002). Furthermore, oxygen and nitrate should both be absent from the anaerobic reactor, and oxygen from the anoxic reactor. Recycle flows might unintentionally add to the introduction of oxygen in such selectors. In addition to disruption of EBPR and/or denitrifying activity, the presence of microaerophilic conditions in the anaerobic and/or anoxic stages, which for instance can be attributed to the diffusion of oxygen through the liquid surface (Plósz *et al.,* 2003) or to the aeration of the returned sludge/liquid stream in screw pumps or at overflow weirs, can lead to worsening sludge settling characteristics.

Table 10.2 Selector design guidelines recommended for aerobic, anoxic and anaerobic selectors in municipal wastewater treatment systems.

Parameter	Value	Reference
Aerobic selector		
Number of compartments	≥ 3	Jenkins *et al.* (1993a)
Contact time	10-15 min., depending on load, temperature and wastewater composition (*i.e.* fraction of readily biodegradable COD).	Still *et al.* (1996)
Sludge loading rate	12 (1st compartment), 6 (2nd comp.) and 3 (3rd comp.) kgCOD/kgMLSS.d	Jenkins *et al.* (1993a)
Floc loading	50-150 gCOD/kgTSS (1st comp.)	Kruit *et al.* (1994)
DO concentration	≥ 2 mgO_2/l, but it depends on the sludge loading rate, floc loading rate and/or substrate uptake rate. Sensor should be placed in the 1st comp.	Sezgin *et al.* (1978), Albertson (1987), Martins *et al.* (2003b)
Anoxic selector		
Number of compartments	≥ 3	Jenkins *et al.* (1993a)
Sludge loading rate	6 (1st comp.), 3 (2nd comp.) and 1.5 (3rd comp.) kgCOD/kgMLSS.d	Jenkins *et al.* (1993a)
Contact time	45-60 min.	Kruit *et al.* (2002)
$(RBCOD/NO_3\text{-}N)_{consumed}$	Usually around than 7-9 mg readily biodegradable COD per $mgNO_3\text{-}N$ due to substrate storage.	Jenkins *et al.* (1993a), Ekama *et al.* (1996a), Van Loosdrecht *et al.* (1997)
Anaerobic selector		
Number of compartments	≥ 3, long channels (length-to-width ratio larger than 10:1)	Albertson (1987), Kruit *et al.* (2002)
Contact time	1-2 h	Kruit *et al.* (2002)
$(COD_{VFA+fermentable}/PO_4\text{-}P)_i$	9-20 gCOD/gP	Wentzel *et al.* (1990), Smolders *et al.* (1996)

10.6.1.3 Anoxic selectors

The design criterion of anoxic selectors (Table 10.2) is primarily based on the ratio of readily biodegradable COD *versus* nitrate entering the reactor (Ekama *et al.*, 1996b). Since in selectors an important fraction of readily biodegradable COD is expected to be converted into storage products, the ratio is higher than the typical range for direct denitrification (approximately 7-9 mg readily biodegradable COD per mg NO_3-N). The type of mixing has been found to be of less or no influence compared to aerated selectors. Anoxic selector designs are therefore in principle more robust and can therefore cope with variations in flow, as long as nitrate remains in surplus (Martins *et al.,* 2004b). However, in full-scale systems it is difficult to balance the nitrate load to readily biodegradable COD load since there are daily variations and some degree of denitrification takes place in the secondary clarifier.

Periods with lower nitrate concentration or temporarily anaerobic conditions in the anoxic selector are expected. These conditions are not necessarily harmful for the sludge-settling characteristics because in a plug anoxic selector an important fraction of readily biodegradable COD can either be stored by ordinary heterotrophic organisms (Beun *et al.,* 2000), or used by the phosphorus accumulating organisms (PAOs) or by glycogen accumulating non-polyphosphate organisms (GAOs).

However, leakage of readily biodegradable COD into the aeration basin, and subsequently bulking sludge, can occur if the anoxic selector has a reduced storage capacity (*e.g.* in completely mixed systems). More research is needed to uncover the key factors in the competition between these microorganisms. In the meantime, in order to design a reliable full-scale anoxic selector it is advisable to first perform pilot-plant studies and only then to scale up the system. Nevertheless, despite all the above, the struggle for even more reduced effluent nitrate concentrations will anyway lead to the recycling of sludge with low nitrate content, limiting the use of anoxic selectors.

10.6.1.4 Anaerobic selectors

Under strictly anaerobic conditions (*e.g.* in UCT-type processes) the soluble substrate, mainly volatile fatty acids and other simple substrates, are taken up and mostly stored. The design of anaerobic selectors follows the ratio of the readily biodegradable COD uptake rate to the phosphorus release rate, which is needed for phosphorus removal, ensuring that virtually no readily biodegradable COD enters the main aeration basin (Table 10.2). These conditions are created in activated sludge systems to promote the growth of PAOs. However, another group of bacteria, known as glycogen accumulating organisms (GAOs), can grow quite well in similar conditions. Both types of bacteria are capable of taking up simple soluble substrates in the anaerobic stage and store it as polyhydroxyalkanoates (PHA). The energy reserve which allows the uptake and storage mechanisms is however different in both types of bacteria. Polyphosphate is used in the case of PAOs and glycogen in the case of GAOs. This metabolic diversity gives considerable flexibility to the anaerobic selector in removing the organic load, independently of the occurrence of phosphorus removal. However, in spite of the great diversity of PAOs and GAOs, no filamentous bacteria have been unequivocally identified so far as having this metabolism. As a result of the availability and consumption of readily biodegradable COD in the anaerobic stage, PAOs and GAOs accumulate in the sludge and therefore force the aerobic microorganisms, which lack substrate in the aerobic phase, to decrease in number. Thus, the more that substrate is removed from the anaerobic stage, which also means less substrate available in the oxic stage, the better the settling characteristics of the activated sludge should be. Furthermore, sludge rich in polyP bacteria usually settles better because they form dense clusters, and also intracellular polyphosphate, in combination with chemical phosphorus precipitation, increases the sludge density even more. The mixing conditions in anaerobic selectors, as in anoxic selectors, do not seem to be critical. Moreover, the carryover of COD into the aerated stage is much less detrimental than in aerobic conditions; this means that the anaerobic selector design is not critical (Martins *et al.*, 2004a). Recent reports have confirmed the success of anaerobic selectors in controlling sludge bulking, even when *Microthrix parvicella* is the most dominant filamentous bacteria (Kruit *et al.*, 2002). An anaerobic selector, however, cannot always be used. For instance, its application is not recommended for waste streams rich in sulphur compounds. This is because anaerobic conditions can increase the production of reduced sulphur compounds, which can be used in the aerobic stage by filamentous sulphur-oxidising bacteria (Eikelboom, 2000).

Recent studies in the Netherlands have showed that satisfactorily settling sludge (SVI<120 ml/g with common values below 100 ml/g) can be achieved in full-scale biological nutrient removal systems by implementing carefully controlled strictly anaerobic and anoxic plug-flow selectors (Kruit *et al.*, 2002). A potentially important factor which has led to better sludge settleability was the introduction of an aerobic reactor after the anoxic/aerobic stage to create simultaneously a low ammonium concentration (< 1 mgN/l) and a high DO concentration (> 1.5 mgO_2/l) (Kruit *et al.*, 2002; Tsai *et al.*, 2003). An example of a treatment system based on these considerations is the BCFS® concept (Van Loosdrecht *et al.*, 1998) which is currently being successfully applied to twelve full-scale plants in the Netherlands.

10.7 MATHEMATIC MODELLING

To study complex ecosystems such as activated sludge cultures, in which many factors are acting together, mathematical modelling can be a very useful tool. A lot of progress has been achieved in this field in spite of the extreme complexity of activated sludge population dynamics. The Activated Sludge Models (ASM 1, 2, 2d and 3) published by the IWA Task Group on Mathematical Modelling for Design and Operation of Biological Wastewater Treatment are examples of useful models to study population dynamics in activated sludge systems. As the knowledge of bacterial physiology increases these models are continuously being upgraded (Figure 10.8).

Figure 10.8 Modelled floc structure with filamentous and floc-forming bacteria (image: Martins *et al.*, 2004c).

An example is the incorporation of storage processes in ASM 3. This is a first attempt to allow for modelling of storage polymer metabolism and to better describe the conversions occurring in selector-like systems. In addition, recently developed metabolic models provide a better link between the kinetics and the biochemistry of storage (Beun *et al.*, 2000) and will certainly contribute to the description and modelling of the metabolic processes which take place in selectors. However, despite the great detail in these models, the growth of filamentous bacteria and, thus bulking sludge, still cannot be predicted.

Models which can predict the settling characteristics of activated sludge are in an early phase of development. Some models already exist to predict the development of filamentous and non-filamentous bacteria considering either existence of a dual species or a group competition (*e.g.* floc formers, filaments, low-dissolved oxygen filaments, low F/M filaments) for a single substrate or for a group of substrates (readily biodegradable COD or slowly biodegradable COD) (Kappeler and Gujer, 1994a; Takács and Fleit, 1995). These models can be basically categorised into two groups: the first one considering the bacterial physiology and kinetic-biokinetics, and the second considering both the physiology and kinetics, as well as the morphology of the bacteria. Diffusional transport of substrates into the activated sludge flocs is an important mechanism in the competition between floc-forming bacteria and filamentous bacteria. Kappeler and Gujer (1994a) proposed that readily biodegradable COD could favour the growth of filamentous microorganisms due to substrate diffusional resistance in the biological flocs. They suggested the integration of this behaviour in traditional activated sludge models (Chapter 14). Observed readily biodegradable COD half-saturation coefficients for filamentous microorganisms were considered to be lower than those for non-filamentous bacteria to represent the differences in substrate diffusion resistance. This approach gives realistic qualitative results. However, it is still not possible to predict the SVI of the sludge or the sludge settling properties.

Later studies have taken both the micromorphology of the floc and the oriented growth characteristics of the filamentous bacteria (preferential unidirectional growth) into account (Takács and Fleit, 1995). This study was the first attempt to combine the morphological characteristics with the physiology of filamentous and non-filamentous bacteria. Three groups of microorganisms (floc formers, low-dissolved oxygen filaments and low F/M filaments) were considered, with kinetic parameters following the trend indicated by the kinetic selection theory, and different scenarios of soluble substrate and DO were simulated. The simulation of the activated floc structure under diffusion-governed conditions showed, as expected, that filamentous bacteria predominate in soluble substrate and DO-limited environments. The authors did not specifically differentiate between the effect of kinetic parameters and the effect of cell morphology.

More recently Martins *et al.* (2004c) adopted a previous model for predicting biofilm morphology (Picioreanu, 1998) for activated sludge flocs. This approach showed that the diffusion gradient is more important for floc morphology than the differences in affinity constants between different organisms, supporting the diffusion gradient-based theory for selection of filamentous bacteria.

In summary, modelling can be used to better evaluate the role of unidirectional growth of filamentous bacteria together with the expected higher capacity of filamentous bacteria to grow according to the substrate micro-gradient in sludge flocs, under a wide range of kinetic parameters. More research should be carried out on the role of bacterial morphology and diffusion on this competition because the kinetic parameters, namely the intrinsic substrate half-saturation coefficient, storage capacity, and decay rates, are largely unknown. This kind of study may lead to the better understanding of the competition between filamentous and non-filamentous bacteria in the gradient-governed microenvironments that are very typical of activated sludge systems.

10.8 GRANULAR SLUDGE

With the understanding that bulking sludge occurs when readily biodegradable COD is removed under conditions where significant substrate gradients occur over the sludge floc, it was realized that granules should form when these conditions are minimised (Beun *et al.*, 1999). In effect, granular sludge and bulking sludge are at opposite ends of the sludge morphology scale (Figure 10.9).

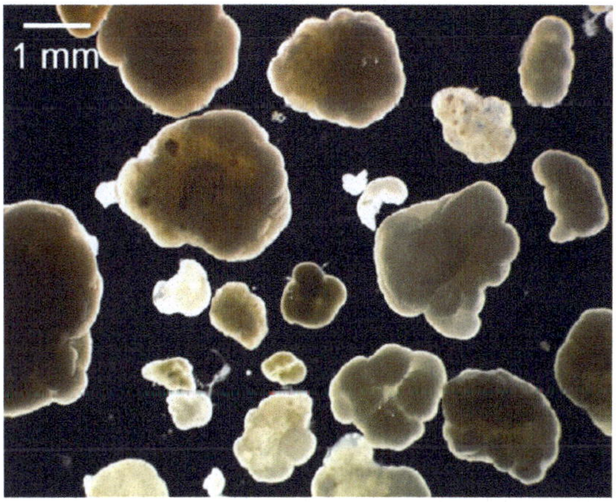

Figure 10.9 Aerobic granular sludge (photo: M.R. de Kreuk).

For biofilms it was already hypothesized that biofilm morphology depends on the ratio between substrate transport rate and biomass growth (Van Loosdrecht *et al.*, 1995, 2002). This not only means that minimising substrate gradients over the sludge floc will improve the SVI, but also selecting slow-growing bacteria will improve the SVI too. Therefore it was *e.g.* always relatively easy to obtain anaerobic granular sludge or nitrifying granular sludge (Figure 10.10), and difficult to achieve this under full aerobic conditions.

The application of anaerobic selectors results in a group of bacteria (phosphate and glycogen accumulating bacteria) with a lower maximum growth rate than ordinary heterotrophic bacteria. They have therefore an additional advantage over aerobic selectors. Selecting for this type of conditions has also been shown to lead to more stable aerobic granular sludge formation (De Kreuk *et al.*, 2004).

A further description of aerobic granular sludge processes is given in Chapter 11.

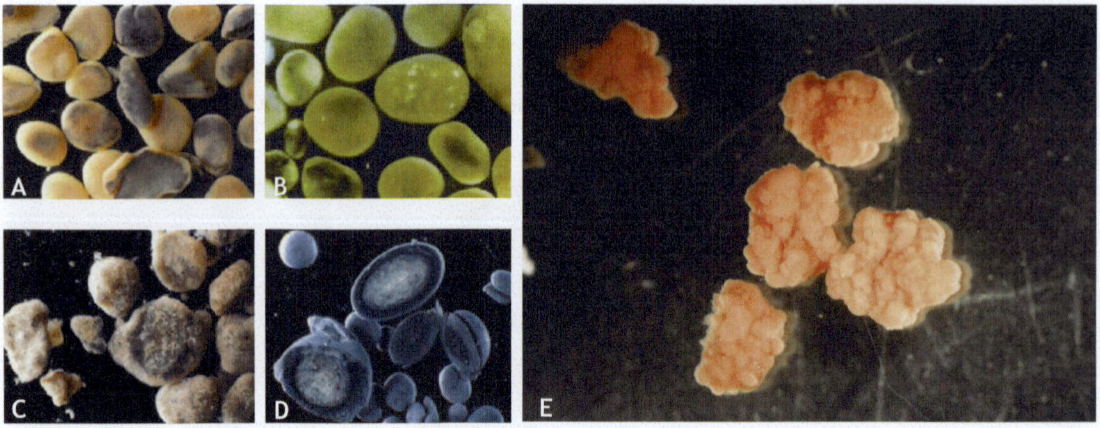

Figure 10.10 Varieties of granular sludge: (A) nitrifying, (B) heterotrophic, (C) denitrifying, (D) methanogenic (photos: Biothane B.V), and (E) anammox (photo: Paques B.V.).

10.9 CONCLUSIONS

Bulking sludge is one of the main problems of activated sludge properties, but in practice there is a sufficient level of understanding at least at the level needed to control the problem. For instance, an activated sludge BNR system designed to minimise bulking sludge problems should have the following general characteristics: (*i*) a pre-treatment step to remove complex substrates (*e.g.* lipids), (*ii*) plug-flow selector reactors to allow a significant macro-gradient of substrate concentration along the system, (*iii*) well-defined anaerobic, anoxic and aerobic plug-flow stages and exclusion of oxygen from the anoxic stage, and nitrate and oxygen from the anaerobic stage, (*iv*) the avoidance of intermittent aeration and microaerophilic conditions, and (*v*) sufficient aeration to maintain high DO concentration (> 1.5 mgO_2/l) and low ammonium concentration (< 1mg N/l) in the final aerobic stage.

The basic ideas outlined in Chapter 11 have even led to processes based on the opposite of bulking sludge: granular sludge. Even in well-designed systems, operational weaknesses can easily lead to cases of bulking sludge. Therefore, as long as the basic processes governing sludge morphology are not fully taken into account, the statement made by Albertson (1987), *'In spite of all we learn and understand, some sludge will still bulk'*, will still be valid (Figure 10.11).

Figure 10.11 Extreme manifestation of sludge bulking (photos: Eikelboom, 2006).

REFERENCES

Albertson O.E. (1987). The control of bulking sludges: from the early innovators to current practice. *J. Water Pollution Control Federation,* 59(4), 172-182.

Alleman J.E. and Prakasam T.B.S. (1983). Reflections on seven decades of activated sludge history. *J. Water Pollution Control Federation,* 55(5), 436-443.

Ardern E. and Lockett W.T. (1914). Experiment on the oxidation of sewage without the aids of filters. *J. of the Society of Chemical Industry,* 33, 523-539.

Beccari M., Di Pinto A.C., Ramadori R. and Tomei M.C. (1992). Effect of dissolved oxygen and diffusion resistances on nitrification kinetics. *Water Research,* 26(8), 1099-1104.

Beccari M., Majone M., Massanisso P. and Ramadori R. (1998). A bulking sludge with high storage response selected under intermittent feeding. *Water Research,* 32(11), 3403-3413.

Ben-Jacob E., Schochet O., Tenenbaum A., Cohen I., Czirók A. and Vicsek T. (1994). Generic modelling of cooperative growth patterns in bacterial colonies. *Nature,* 368(3), 46-49.

Beun J.J., Hendriks A., Van Loosdrecht M.C.M., Morgenroth E., Wilderer P.A. and Heijnen J.J. (1999). Aerobic granulation in a sequencing batch reactor. *Water Research,* 33(10), 2283-2290.

Beun J.J., Paletta F., Van Loosdrecht M.C.M. and Heijnen J.J. (2000). Stoichiometry and kinetics of poly-B hydroxybutyrate metabolism under denitrifying conditions in activated sludge cultures. *Biotechnology and Bioengineering,* 67, 379-389.

Casey T.G., Wentzel M.C., Loewenthal R.E., Ekama G.A. and Marais G.v.R. (1992). A hypothesis for the cause of low F/M filament bulking in nutrient removal activated sludge systems. *Water Research,* 26(6), 867-869.

Chu K.H., Van Veldhuizen H.M. and Van Loosdrecht M.C.M. (2003). Respirometric measurement of kinetic parameters: Effect of activated sludge floc size *Water Science and Technology,* 48(8), 61-68.

Chudoba J., Ottová V. and Madera V. (1973a). Control of activated sludge filamentous bulking - I. Effect of the hydraulic regime or degree of mixing in an aeration tank. *Water Research,* 7(9), 1163-1182.

Chudoba J., Grau P. and Ottová V. (1973b). Control of activated sludge filamentous bulking - II. Selection of microorganisms by means of a selector. *Water Research,* 7(10), 1398-1406.

De Kreuk M.K. and Van Loosdrecht M.C.M. (2004). Selection of slow growing organisms as a means for improving aerobic granular sludge stability *Water Science and Technology*, 49(11-12), 9-17.

Donaldson W. (1932). Use of activated sludge increasing. *Civil Engineering,* 2(3), 167-169.

Eikelboom D.H. (1975). Filamentous organisms observed in activated sludge. *Water Research,* 9, 365-388.

Eikelboom D.H. (1977). Identification of filamentous organisms in bulking activated sludge. *Prog. Wat. Tech*., 8, 153-162.

Eikelboom D.H. (1994). The Microthrix parvicella puzzle, *Water Science and Technology,* 29, 271–279.

Eikelboom D.H. (2000). Process control of activated sludge plants by microscopic investigation, IWA Publishing, London, UK.

Eikelboom D.H. and Geurkink B. (2002). Filamentous microorganisms observed in industrial activated sludge plants. *Water Science and Technology,* 46(1/2), 535-542.

Eikelboom D.H. (2006). Identification and control of filamentous micro-organisms in industrial waste water treatment plants, IWA Publishing, London, UK.

Ekama G.A. and Marais G.v.R. (1986). Sludge settleability and secondary settling tank design procedures. *Water Pollution Control,* 85(1), 101-113.

Ekama G.A., Wentzel M.C., Casey T.G. and Marais G.v.R. (1996a). Filamentous organism bulking in nutrient removal activated sludge. Paper 3: Stimulation of the selector effect under anoxic conditions. *Water SA,* 22(2), 119-126.

Ekama G.A., Wentzel M.C., Casey T.G. and Marais G.v.R. (1996b). Filamentous organism bulking in nutrient removal activated sludge systems. Paper 6: review, evaluation and consolidation of results. *Water SA,* 22(2), 147-152.

Jenkins D., Richard M.G. and Daigger G.T. (1993). *Manual on the causes and control of activated sludge bulking and foaming*, 2nd ed., Michigan: Lewis Publishers, Inc.

Jenkins D., Richard M.G. and Daigger G.T. (2003). *Manual on the Causes and Control of Activated Sludge Bulking, Foaming, and Other Solids Separations Problems*, 3rd ed. Lewis Publishers, Inc.

Kappeler J. and Gujer W. (1994a). Development of a mathematical model for 'aerobic bulking'. *Water Research,* 28(2), 303-310.

Kappeler J. and Gujer W. (1994b). Verification and applications of a mathematical model for 'aerobic bulking'. *Water Research,* 28(2), 311-322.

Kruit J., Boley F., Jacobs L.J.A.M. and Wouda T.W.M. (1994). Prediction of the O2 conditions in the selector. *Water Science and Technology,* 29(7), 229-237.

Kruit J., Hulsbeek J. and Visser A. (2002). Bulking sludge solved?! *Water Science and Technology,* 46(1/2), 457-464.

Martins A.M.P., Van Loosdrecht M.C.M. and Heijnen J.J. (2003a). Effect of feeding pattern and storage on the sludge settleability under aerobic conditions. *Water Research,* 37(11), 2555-2570.

Martins A.M.P., Van Loosdrecht M.C.M. and Heijnen J.J. (2003b). Effect of dissolved oxygen concentration on the sludge settleability. *Applied Microbiology and Biotechnology*, 62(5-6), 586-593.

Martins A.M.P., Heijnen J.J. and Van Loosdrecht M.C.M. (2004a). Bulking Sludge in Biological Nutrient Removal Systems (2004) *Biotechnology and Bioengineering.* 86(2), 125-135.

Martins A.M.P., Pagilla K., Heijnen J.J. and Van Loosdrecht M.C.M. (2004b). Filamentous bulking sludge - A critical review. *Water Research,* 38, 793-817.

Martins A.M.P., Picioreanu C., Heijnen, J.J. and Van Loosdrecht M.C.M. (2004c). Three-dimensional dual-morphotype species modeling of activated sludge floc. *Environmental Science and Technology,* 38, 5632-5641.

Martins A.M.P., Karahan Ö, Van Loosdrecht M.C.M. and Heijnen J.J. (2008). Effect of polymeric substrate on sludge settleability. *Water Research*, 45(1), 263-273.

McIlroy S.J., Kristiansen R., Albertsen M., Karst S.M., Rossetti S., Nielsen J.L., Tandoi V., Seviour R.J. and Nielsen P.H. (2013). Metabolic model for the filamentous 'Candidatus Microthrix parvicella'based on genomic and metagenomic analyses. *The ISME Journal*, 7(6), 1161.

Musvoto E.V., Casey T.G., Ekama G.A., Wentzel M.C. and Marais G.v.R. (1994). The effect of incomplete denitrification on anoxic-aerobic (low F/M) filament bulking in nutrient removal activated sludge systems *Water Science and Technology,* 29 (7), 295-299.

Nielsen P.H., Roslev P., Dueholm T.E. and Nielsen J.L. (2002). Microthrix parvicella, a specialized lipid consumer in anaerobic-aerobic activated sludge plants. *Water Science and Technology,* 46(1-2), 73-80.

Nielsen J.L., Christensen D., Kloppenborg M. and Nielsen P.H. (2003). Quantification of cell-specific substrate uptake by probe-defined bacteria under *in situ* conditions by microautoradiography and fluorescence *in situ* hybridisation. *Environmental Microbiology,* 5(3), 202-211.

Palm J.C., Jenkins D. and Parker D.S. (1980). Relationship between organic loading, dissolved oxygen concentration and sludge settleability in the completely-mixed activated sludge process. *J. Water Pollution Control Federation.,* 52, 2484-2506.

Pasveer A. (1959). A contribution to the development in activated sludge treatment. *J. Proc. Inst. Sewage Purif.,* 4, 436.

Pasveer A. (1969). A case of filamentous activated sludge. *J. Water Pollut. Control Fed.,* 41(7), 1340-1352.

Picioreanu C., Van Loosdrecht M.C.M. and Heijnen J.J. (1998). Mathematical modelling of biofilm structure with a hybrid differential-discrete cellular automaton approach. *Biotechnology and Bioengineering,* 58, 101-116.

Pipes W.O. (1967). Bulking of activated sludge. *Adv. Appl. Micr.,* 9, 185-234.

Plósz B., Jobbágy A. and Grady C.P.L. (2003). Factors influencing deterioration of denitrification by oxygen entering an anoxic reactor through the surface. *Water Research,* 37, 853-863.

Rensink J.H. (1974). New approach to preventing bulking sludge. *J. Water Pollution Control Federation,* 46(8), 1888-1894.

Rossetti S., Tomei M.C., Nielsen P.H. and Tandoi V. (2005). "Microthrix parvicella", a filamentous bacterium causing bulking and foaming in activated sludge systems: A review of current knowledge *FEMS Microbiology Reviews*, 29(1), 49-64.

Rubio-Rincón F.J., Welles L., Lopez-Vazquez C.M., Nierychlo M., Abbas B., Geleijnse M., Van Loosdrecht M.C.M. and Brdjanovic D. (2017). Long-term effects of sulphide on the enhanced biological removal of phosphorus: the symbiotic role of Thiothrix caldifontis. *Water Research*, 116, 53-64.

Scuras S.E, Jobbágy A. and Grady Jr. C.P.L. (2001). Optimization of activated sludge reactor configuration: kinetic considerations. *Water Research,* 35(18), 4277-4284.

Seviour R.J., Liu J.R., Seviour E.M., McKenzie C.A., Blackall L.L. and Saint C.P. (2002). The "*Nostocoida limcola*" story: the phylogeny of this morphotype responsible for bulking in activated sludge. *Water Science and Technology,* 46(1/2), 105-110.

Sezgin M., Jenkins D. and Parker D.S. (1978). A unified theory of filamentous activated sludge bulking. *J. Water Pollution Control Federation,* 50(2), 362-381.

Shao Y.-J. and Jenkins D. (1989). The use of anoxic selectors for the control of low F/M activated sludge bulking. *Water Science and Technology,* 21, 609-619.

Slijkhuis H. and Deinema M.H. (1988). Effect of environmental conditions on the occurrence of

Microthrix parvicella in activated sludge, *Water Research,* 22, 825–828.

Smolders G.J.F., Van Loosdrecht M.C.M. and Heijnen J.J. (1996). Steady-state analysis to evaluate the phosphate removal capacity and acetate requirement of biological phosphorus removing mainstream and sidestream process configurations. *Water Research,* 30(11), 2748-2760.

Still D.A, Ekama G.A., Wentzel M.C., Casey T.G. and Marais G.v.R. (1996). Filamentous organism bulking in nutrient removal activated sludge systems. Paper 2: Stimulation of the selector effect under aerobic conditions. *Water SA,* 22(2), 97-118.

STOWA (1994). *Selector design: the role of the influent characterization*. Report 94-16 (in Dutch).

Takács I. and Fleit E. (1995). Modelling of the micromorphology of the activated sludge floc: low DO, low F/M bulking. *Water Science and Technology,* 31(2), 235-243.

Tsai M.-W., Wentzel M.C. and Ekama G.A. (2003). The effect of residual ammonia concentration under aerobic conditions on the growth of *Microthrix parvicella* in biological nutrient removal plants. *Water Research,* 37(5), 3009-3015.

Van den Eynde E., Geerts J., Maes B. and Verachtert H. (1983). Influence of the feeding pattern on the glucose metabolism of *Arthrobacter sp.* and *Sphaerotilus natans*, growing in chemostat culture, simulating activated sludge bulking. *European. Journal of Applied Microbiology and Biotechnology,* 17, 35-43.

Van Loosdrecht M.C.M., Eikelboom D., Gjaltema A., Mulder A., Tijhuis L. and Heijnen J.J. (1995). Biofilm structures. *Water Science and Technology,* 32(8), 35-43.

Van Loosdrecht M.C.M., Pot M.A. and Heijnen J.J. (1997). Importance of bacterial storage polymers in bioprocesses. *Water Science and Technology, l* 35(1),41-47.

Van Loosdrecht M.C.M., Brandse F.A. and De Vries A.C. (1998). Upgrading of waste water treatment processes for integrated nutrient removal - the BCFS process. *Water Science and Technology,* 37(9), 209-217.

Van Loosdrecht M.C.M., Heijnen J.J., Eberl H., Kreft J. and Picioreanu C. (2002). Mathematical modelling of biofilm structures. *Antonie van Leeuwenhoek, International Journal of General and Molecular Microbiology,* 8, 245-256.

Van Loosdrecht M.C., Nielsen P.H., Lopez-Vazquez C.M. and Brdjanovic D. (eds.). (2016). Experimental methods in wastewater treatment. IWA publishing.

Wanner J. and Grau P. (1989). Identification of filamentous organisms from activated sludge: a compromise between wishes, needs and possibilities. *Water Research,* 23(7), 883-891.

Wentzel M.C., Dold P.L., Ekama G.A. and Marais G.v.R. (1990). Biological excess phosphorus removal – steady state process design. *Water SA,* 16(1), 29-48.

NOMENCLATURE

Symbol	Description	Unit
C_S	Substrate concentration in bulk liquid	mgCOD/l
K_S	Half saturation concentration for substrate utilization	mgCOD/l
q_S	Substrate uptake rate	mgCOD/l.h

Abbreviation	Description
AS	Anaerobic digestion model
ASM	Activated sludge model
A/V	Surface to volume ratio
BOD	Biological oxygen demand
COD	Chemical oxygen demand
EPS	Extracellular polymeric substance
FISH	Fluorescence *in-situ* hybridization
GAO	Glycogen accumulating organism
IWA	International Water Association
MAR	Microautoradiography
MLSS	Mixed liquor suspended solids
PAO	Phosphorus accumulating organism
PHA	Polyhydroxyalkanoate
RBCOD	Readily biodegradable COD
SRT	Sludge retention time
SVI	Sludge volume index
UCT	University Cape Town
VFA	Volatile fatty acid

doi: 10.2166/9781789060362_0497

11

Aerobic granular sludge

Mario Pronk, Edward J.H. van Dijk and Mark C.M. van Loosdrecht

11.1 INTRODUCTION

The aerobic granular sludge process is a wastewater treatment process developed to allow higher sludge concentrations and higher energy efficiency than conventional activated sludge (CAS) processes (Figure 11.1). Conventional wastewater treatment based on the activated sludge process requires large volumes due to the relatively low concentrations of sludge that can be maintained (3-5 g/l). This concentration is limited by the clarification process, which relies on gravity to separate the flocculent sludge from the treated wastewater. Conventional wastewater treatment plants therefore need large secondary clarifiers; the limited sludge loading rates that can be applied result in the large surface area of the clarifiers (Chapter 12).

Several options for designing more compact wastewater treatment plants have emerged over time. These rely mostly on immobilising the biomass on carriers as biofilms (Chapter 18) or by keeping sludge in the reactor by membrane separation in membrane bioreactors (see Chapter 13). Biofilm processes are usually limited by the amount of biofilm mass transfer area and need post-treatment to remove suspended solids, making the reduction in footprint limited. Membrane bioreactors require a relatively high level of investment and extra energy for the membrane separation process. An alternative option is to create sludge with a low sludge volume index (20-70 ml/g) that has a granular morphology, without the need for a carrier. Although this was developed in the early 1970s for anaerobic wastewater treatment (Chapter 16), granular sludge processes have only become available in the early 2000s for aerobic biological nutrient removal processes.

In aerobic granular sludge (AGS), the morphology of the biomass allows for a high settling velocity of the sludge. Flocs are in general small and have a high drag coefficient, whereas granules generally have a larger radius and lower drag coefficient (Figure 11.2). The resulting higher bed settling velocities of aerobic granules (4-10 m/h) compared to activated sludge flocs (0.8-1.4 m/h) allow for the integration of the settler in the treatment reactor and thus a compact reactor design. This is further enhanced by the discrete settling properties of a granular sludge bed. Biomass concentrations of more than 10 g/l are easily achieved, however in practice lower values suffice since the design of wastewater treatment plants is often limited by other constraints of the plant (Chapter 5).

Figure 11.1 Nereda® wastewater treatment plants in the Netherlands in (A) Garmerwolde (140,000 P.E.) and (B) Utrecht (430,000 P.E.). In Utrecht the new 6-tank Nereda® facility (in operation since 2019) is visible on the left with part of the old abandoned plant on the right.

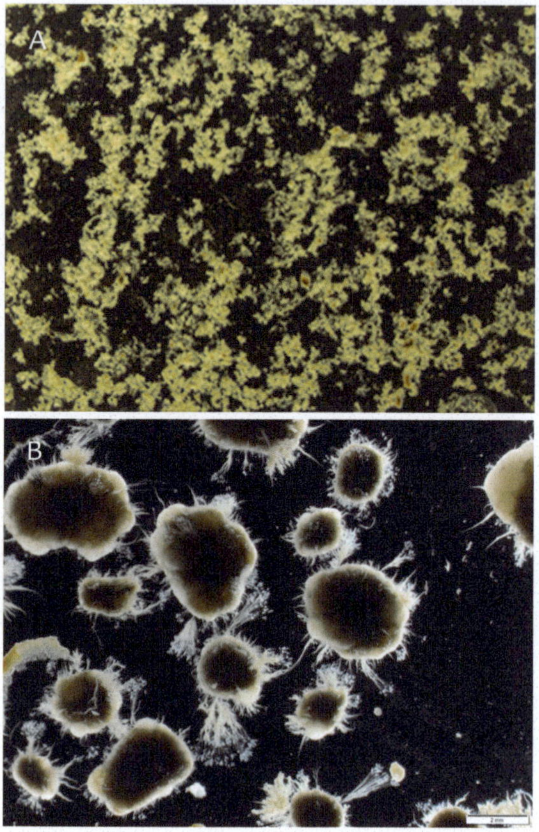

Figure 11.2 (A) Activated sludge from WWTP Harnaschpolder, the Netherlands, and (B) washed aerobic granules with stalked ciliates from the Nereda® plant in Utrecht, the Netherlands.

Since the emergence of anaerobic granular sludge in the 1970s, granulation has mostly been considered to be related to the specific microbiology of methanogenic processes. It was thought that the complex community structures required to convert substrates anaerobically into methane, and the crucial role of interspecies hydrogen transport, form the evolutionary driver for anaerobic organisms to grow in compact aggregates or granules. However, trials to form aerobic granular sludge failed and it was assumed this was because aerobic conversion of substrates in general do not depend on syntrophic interactions.

Due to the introduction of taxes on discharging wastewater in 1970 in the Netherlands, industry started to find compact solutions for treating their wastewater. This led to the Upflow Anaerobic Sludge Bed Reactor (UASB) (Lettinga *et al,* 1975), which was quickly introduced especially in the food industry. During the 1980s the company Gist-Brocades also started to develop a compact wastewater treatment system (Heijnen, 1984; Heijnen *et al.*, 1990). For the design of compact reactors, without the need to rely on granular sludge formation, they developed anaerobic and aerobic treatment technology using biofilms on small suspended carriers (sand grains). This led to well-settling particles with a high specific biofilm surface area. This anaerobic technology rapidly developed into the now widely used expanded granular sludge bed (EGSB) and internal circulation (IC) reactors (Chapter 16). Both rely on granular sludge biofilm processes. For aerobic processes the formation of granules proved more complex and the CIRCOX® reactor technology became the standard compact aerobic technology for industrial wastewater (Heijnen *et al.*, 1993). However, a translation of these technologies for use in municipal wastewater treatment did not occur since the reactor technologies were not suited to handle the large hydraulic variations occurring in municipal wastewater flows.

Research in the 1990s into the morphogenesis of biofilms and flocs (especially for the CIRCOX® reactor development) resulted in the postulation of a general hypothesis. This hypothesis stated that the ratio between biofilm surface loading (or the rate at which new biomass is produced) and shear rate determines the biofilm structure. When shear forces are relatively high, only a patchy biofilm will develop, whereas at low shear rates the biofilm becomes highly heterogeneous with many pores and protuberances. With a correct balance between biofilm surface loading and shear rate, smooth and stable biofilms can be obtained (Van Loosdrecht *et al.*, 1995). This was further evaluated by mathematical models (Van Loosdrecht *et al.*, 2002; Picioreanu *et al.*, 1998). A similar approach could be used in the context of predicting the Sludge Volume Index (SVI) (Martins *et al.*, 2004). From this research it was concluded that slower growing bacteria (methanogens, anammox and nitrifying bacteria) will form granules more naturally than fast-growing bacteria such as aerobic heterotrophs and fermentative bacteria.

The first proof of principle for aerobic granulation was reported by Heijnen and Van Loosdrecht (1998) and a patent was applied for. The granules were grown aerobically on molasses in a sequencing batch reactor (SBR). By applying a critical settling rate of 35 m/h, granulation could be obtained. Fast-settling granules remained in the system while the slow-settling flocs were washed out. However, although granules were formed initially, stable long-term operation was found to be problematic (Morgenroth *et al.*, 1997). The reason for this was not clear at the time. Nevertheless, the follow-up research led to the development of full-scale aerobic granular sludge technology. The specifics are further discussed in this chapter.

The translation from lab-scale observations to commercial full-scale AGS reactors took approximately 12 years. A public-private consortium formed by Delft University of Technology, several Dutch water authorities, the consultancy company Royal HaskoningDHV, supported by national and international innovation programs, was responsible for the development of aerobic granular sludge technology in the Netherlands. This new technology was registered as a trademark under the name Nereda® by Royal HaskoningDHV. The technology began to be applied for industrial wastewater treatment in 2005, and the first demonstration plant for municipal wastewater was designed and constructed in Gansbaai, South Africa, between 2006 and 2008. In 2010 the first commercially implemented full-scale aerobic granular sludge reactor treating domestic (65%) and industrial (35%) (slaughterhouse) wastewater was constructed in Epe, the Netherlands. Another important milestone in the technology development was in 2013 at the Garmerwolde WWTP, the Netherlands, when Nereda® reactors were installed that ranked among the world's largest SBR reactors, establishing it as a fully scaled-up technology. The Garmerwolde Nereda® reactor design and performance is described in detail in Pronk *et al.*, 2015). This technology is currently spreading throughout the world and by 2020 had been adopted by more than 80 plants ranging from 5,000 P.E. to 2,400,000 P.E. (Figure 11.3).

Figure 11.3 Nereda® plants; from Vika, the Netherlands (5,000 P.E.) to Ringsend in Dublin, Ireland (2,400,000 P.E.).

11.2 IMPORTANT CONSIDERATIONS FOR SELECTING AEROBIC GRANULAR SLUDGE

11.2.1 Gradients

Substrate gradients are an important parameter that controls the morphology of flocs and biofilms (Chapter 10). Due to simultaneous substrate consumption and diffusion, different concentrations will occur as a function of the depth in the floc or biofilm. To support large granules, the whole interior needs to receive substrate. If the substrate does not diffuse throughout the interior, the structure will eventually deteriorate due to decay causing unstable granulation. If the substrate uptake rate is lower than the diffusion of the substrate into the granule, it will lead to more uniform bacterial growth throughout. When the substrate uptake rate is higher than the diffusion rate, the substrate only penetrates the outer part of the granule or even only the tips of the protrusions. This relation can be elegantly expressed with the dimensionless factor *G* (growth), which was developed by Picioreanu *et al.* (1998) for planar biofilms. It shows the ratio between maximum biomass growth rate and the maximum substrate transport rate through diffusion.

$$G = L_Y^2 \cdot \frac{\mu_m \cdot C_{x,m}}{D_s \cdot C_{s,o}} \tag{11.1}$$

where:

$C_{s,o}$	concentration of the substrate in the bulk (g/m^3)
D_s	diffusion coefficient of that substrate (m^2/d)
$C_{x,m}$	maximum density of the biomass in the biofilm (g/m^3)
μ_m	maximum specific growth rate (1/d)
L_y	biofilm thickness (m)

High G values will lead to finger-type outgrowths, rougher biofilms and thus slower settling rates. It will also cause unstable biofilm or granule formation. In the case of flocculent sludge it leads to bulking sludge. On the other hand, low G values will result in stable, dense biofilms or granules.

One of the reasons for the difficulty in obtaining aerobic granular sludge is the low solubility of oxygen. Oxygen diffusion into the biofilm is usually rate-limiting for most conversions. For example, at an oxygen concentration of 2 mgO_2/l, oxygen is rate-limiting for nitrification unless the ammonium concentration is less than 0.44 mgN/l (Chapter 17). Typically, at these concentrations the penetration depth of oxygen is not more than 20 to 40 μm.

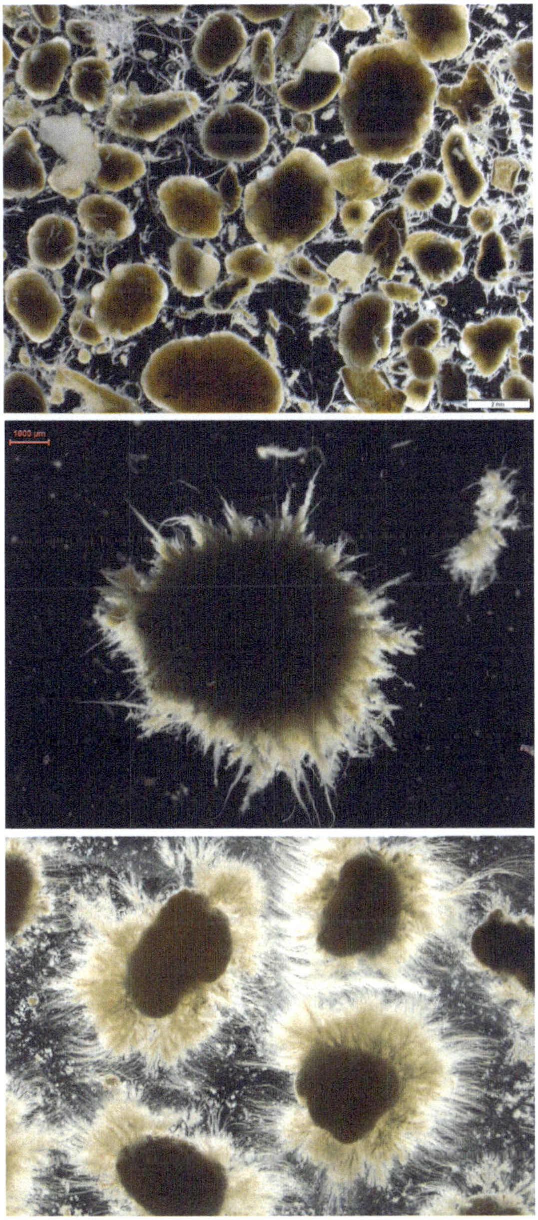

Figure 11.4 Aerobic granules with (top) smooth morphology, (middle) with finger-type outgrowth and (below) filamentous outgrowth.

Low concentrations of readily degradable COD (*i.e.* acetate, glucose etc.), *e.g.* when substrate is supplied to a completely mixed reactor, will lead to finger-type outgrowth on the outside of granules or biofilms. In granular sludge systems such conditions will even lead to filamentous outgrowth when also the oxygen transport rate to the biofilm is limiting (Figure 11.4). This dual oxygen and COD limitation has a negative impact on the settling rate, stability and process performance. This problem has been overcome in Nereda® technology by applying an anaerobic feed of sewage allowing full uptake of readily degradable COD before aeration starts.

Note the G factor (Eq. 11.1) assumes zero-order processes while some conversions will occur at concentrations below the cellular substrate affinity coefficients (usually around 10-100 µg/l). Nevertheless, the G factor gives valuable insight into the granulation process.

11.2.2 Microbial selection

Cultivation and stability of aerobic granular sludge under low oxygen concentrations is vital for the full-scale application of aerobic granular sludge. A low dissolved-oxygen concentration is needed for minimising aeration energy as well as for efficient nitrogen removal by allowing simultaneous nitrification and denitrification. Incorporating an anaerobic feed in the process leads to readily biodegradable COD (RBCOD) to be converted into storage polymers (mainly polyhydroxyalkanoates (PHA)). This is then followed by an aerobic period where growth on the storage polymers occurs, effectively bypassing the mechanism expressed by the G factor via separation of substrate uptake and biomass growth. This makes the granule formation less dependent on the oxygen concentration. In the anaerobic period phosphate-accumulating (PAO) or glycogen-accumulating organisms (GAO) will store the RBCOD as PHA. Since substrate concentrations will be much higher under anaerobic conditions than those in the presence of oxygen the RBCOD is stored over a significant depth. In the aerated phase the inner layers either grow on nitrate (denitrification) or get oxygen later in the process when the storage pools of the organisms in the outer layers are depleted. There is an extra advantage in the conversion of RBCOD into storage polymers (PHA). Bacteria growing on storage polymers have a lower growth rate (μ_{max}) than heterotrophic bacteria that grow directly on RBCOD in the presence of oxygen (Van Loosdrecht *et al.*, 1997). In general, slower-growing bacteria grow in more dense structures than fast-growing bacteria. This growth in dense structures is an extra reason why slow-growing bacteria (methanogens, anammox, nitrifiers, PAO) generate smooth and compact granular sludge. The distribution of substrate combined with slow growth was also found to be the key mechanism for stable, smooth biofilm formation (Picioreanu *et al.*, 1998).

11.2.3 Physical selection

In practice sewage is a complex mixture of substrates. It consists of both dissolved and particulate matter (Chapter 3). The soluble compounds can diffuse into the granular sludge, whereas the particulate matter first has to be converted through hydrolysis. The non-biodegradable particulate matter fraction will accumulate in the reactor. For aerobic granular sludge one could classify the COD into granular, sludge-forming COD, and non-granular, sludge-forming COD. The non-granular sludge-forming COD consists of inert material and COD that is not convertible into PHA anaerobically and eventually leads to floc formation. As a result, granular sludge reactors always contain a fraction of flocculent material consisting of inert particulate COD, biomass grown on the remaining COD after the anaerobic period, and eroded material from the granular sludge. This poor settling fraction needs to be removed to form and maintain the granular fraction in the AGS plant. To remove the slow-settling sludge and at the same time retain granular sludge, selection is applied. To do this the top of the sludge bed is removed after a short settling period. The short settling phase allows fast-settling granules to be retained in the reactor while flocculent material residing at the top of the sludge blanket is spilled. This results in a solid retention time of 0.5-5 days for the flocculent fraction and around 30 days for the granular fraction. More in-depth discussion on settling and its associated parameters can be found in Section 11.3.4.

11.2.4 Shear

Shear can counterbalance the tendency of fast-growing bacteria to form filamentous outgrowth. For instance, for the CIRCOX® reactor mentioned in the introduction, shear is a dominant factor in obtaining smooth biofilms in a reactor which is fed continuously under aerobic conditions (Kwok *et al.*, 1998). When slow-growing bacteria (*i.e.* all the RBCOD is converted anaerobically into PHA) are properly selected, shear plays no significant role in the granule formation process. Also, in practice applying shear would mean extra energy dissipation and costs and is therefore neither desirable nor a generally required parameter for AGS.

11.2.5 Plug-flow feeding

All these aspects described above have led to the operational strategy for granular biomass formation with sewage using the Nereda® process (Figure 11.11). To minimize the substrate diffusion limitation and ensure the best uptake of substrate over the depth of the granules, the influent should not be diluted before contact with the granules. In AGS reactors, this is achieved by feeding the influent through the settled granular sludge bed from the bottom (Figure 11.5) in the absence of aeration.

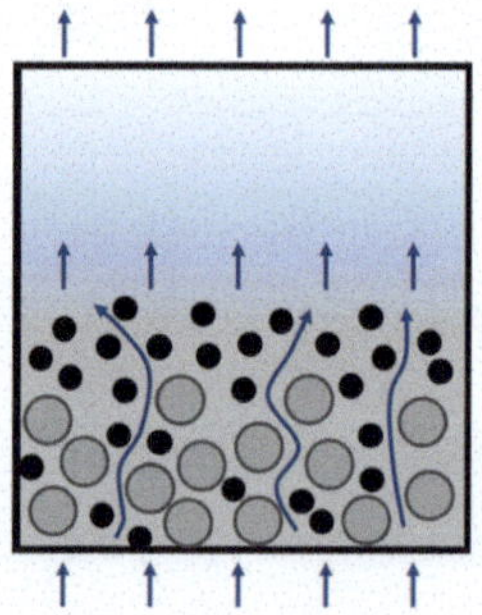

Figure 11.5 Graphical presentation of anaerobic plug-flow feeding through a bed of settled granules in an aerobic granular sludge reactor.

The plug-flow regime ensures a relatively high substrate concentration through the settled sludge bed. Combined with a sufficiently long anaerobic feeding period, the substrate can penetrate deep into the granule. Therefore, growth can take place over the depth of the granular biomass and not primarily at the granule surface. This mode of operation has a self-stabilising effect: larger granules will tend to be at the bottom of the sludge bed and get thereby most of the RBCOD, while none or very limited RBCOD is left for the floc fraction at the top of the settled sludge bed.

11.2.6 Effect of substrate and feeding regime on granule morphology

The applied feeding regime of AGS reactors with different substrates will have a significant impact on the granule morphology and thus on the stability of the system in general. In practice sewage is of course a complex mixture of substrates. Figure 11.6 (A-D) demonstrates this complexity and the impact of various feeding strategies in SBR systems on the formation, morphology and stability of aerobic granular sludge.

(A) Readily biodegradable soluble substrates when fed anaerobically are taken up by PAO or GAO types of bacteria and converted into storage polymers. In a subsequent aerobic period these storage polymers are used for growth at a relatively slow rate. This results in compact granular sludge. Compounds that are not converted anaerobically (*e.g.* butanol, propanol) but are adsorbed in the granular sludge matrix will be converted in the aerobic phase. This also leads to growth throughout the granule and thereby to stable granule formation. This anaerobic feeding strategy not only ensures stable granulation, but also ensures optimal phosphate and nitrogen removal which is important for the treatment of domestic sewage.

(B) Readily biodegradable soluble substrates remaining in the liquid after the anaerobic phase or dosed as a pulse in a fully aerobic granular sludge reactor will lead to substrate uptake under oxygen diffusion limitation. The substrate is used for simultaneous growth and formation of storage polymers and is mainly limited to outside areas of the granule, while the inner regions are deprived of oxygen (Beun *et al.,* 2002). The fast consumption of easy biodegradable substrates in the presence of oxygen on the outside fraction of the granule will lead to formation of finger-type or filamentous outgrowth. This will depend on the oxygen concentration in the bulk and increased shear is needed to ensure smooth and stable granulation (Beun *et al.,* 1999). Granules formed under this regime are more prone to breaking under shear stress since the inner regions are inactive and will eventually decay and weaken the granule (Beun *et al.,* 2002). This leads to unstable granulation and poor settling characteristics combined with higher suspended solids (flocs and loose cells) in the liquid after fast settling. Besides this, the nitrification and phosphorus removal potential are also decreased. The slow-growing nitrifying bacteria will be overgrown and pushed down to oxygen limited layers by the faster-growing heterotrophs (Elenter *et al.,* 2007; Gonenc and Harremoes, 1990). Some substrates also lead to good granulation even if they are converted aerobically. Ammonium and methanol are such substrates. Both these substrates are converted with oxygen by relatively slow-growing bacteria, which leads to a denser biofilm formation (Mosquera-Corral *et al.,* 2003; Villaseñor *et al.,* 2000). In AGS systems, substrates that are converted aerobically by slow-growing bacteria are therefore generally expected to lead to stable granulation.

(C) Particulate substrates (*e.g.* starch, proteins) present another challenge, because of the need for hydrolysis. Particulate substrates are mainly hydrolysed at the surface of the granules (De Kreuk *et al.,* 2010). The hydrolysis products (VFA) will thereafter become available for conversion into storage polymers by PAO and GAO-like organisms. This results in stable granulation. Depending on the anaerobic hydrolysis rate, some aerobic hydrolysis will also occur. Under aerobic conditions the hydrolysis products will be directly used for growth by the organisms present at the granule's surface. This creates steep substrate diffusion limitation gradients, which will induce finger-type outgrowth, less stable granule formation and higher levels of suspended solids in the liquid phase.

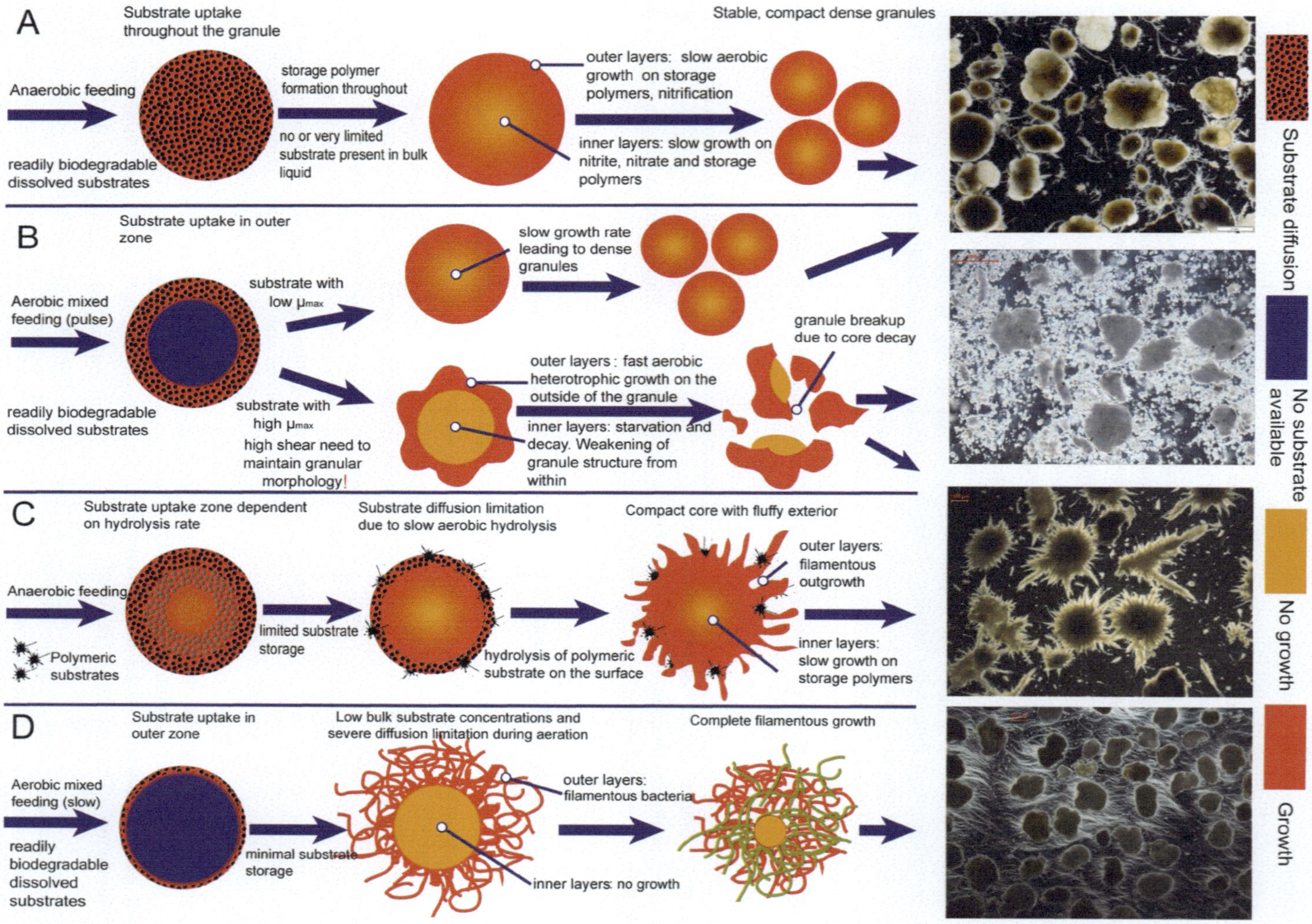

Figure 11.6 Effect on aerobic granule formation with different carbon sources and feeding regimes in sequencing batch reactors selecting for aerobic granular sludge; (A) Readily biodegradable dissolved substrates fed anaerobically, (B) aerobic feeding of readily biodegradable dissolved substrates into a mixed reactor, (C) anaerobic feeding of polymeric substrates through a settled granular bed, (D) slow aerobic feeding of readily biodegradable dissolved substrates in a mixed reactor.

(D) Readily-biodegradable substrates fed slowly in a mixed aerobic environment will lead to severe substrate diffusion limitation gradients. The substrate concentration in the bulk is very low or most likely zero. This provides very good conditions for the proliferation of filamentous organisms as they have the advantage of growth direction and available surface area (Martins *et al.*, 2003; Martins *et al.*, 2011). Aerobic granular sludge fed under these conditions will therefore quickly deteriorate. Breakage of the granules will occur as the inside will not receive any substrate and will die. Filamentous growth will have a detrimental effect on the settling properties of the granules and thus on the effluent quality. Granulation formation is thus unlikely. High shear will help control the outgrowth, but this is not always an option.

11.3 KINETICS OF AEROBIC GRANULAR SLUDGE

11.3.1 Carbon removal

In AGS carbon removal is not much different from normal biological phosphate-removal plants. Overall the same bacteria are active in activated sludge as in granular sludge. A major difference is the distribution of SRT in granular sludge plants. In activated sludge plants the sludge flocs all have the same SRT. Although the average SRT in granular sludge plants is similar to that of activated sludge plants, the SRT of granules is larger than the SRT of the flocculent sludge fraction as shown by Ali *et al.* (2019). Large granules consist of PAO and several other bacterial groups growing on soluble and readily hydrolysable substrates and can have an SRT larger than 30 days. However, the flocculent sludge fraction also contains inert particulate compounds, eroded granular sludge material and slowly-hydrolysable particulates, and has an SRT of 0.5-5.0 days, which is much lower than the SRT of the granular fraction. The composition of the flocculent fraction is comparable to sludge of highly-loaded activated sludge plants. It is not mineralised and therefore gives a relatively high methane yield.

11.3.2 Nitrogen removal

Nitrification is performed by nitrifiers which are present in both the granules and flocs. Due to the difference in SRT between granules and flocs, nitrification can easily be maintained even at cold temperatures. In activated sludge plants the nitrifying activity of sludge decreases due to two factors: the intrinsic effect of temperature on the activity of microorganisms (halving the rate for every 8-10 °C temperature decrease) as well as an accumulation of more suspended COD in the sludge due to decreased hydrolytic rates. However, for granular sludge the effect of temperature is much less as the reduced microbial activity is compensated for by a larger penetration depth of oxygen into the granular sludge. Furthermore due to the preferential removal of suspended COD in an AGS plant (by the selective sludge wasting of the flocculent fraction), the nitrifying fraction in the sludge is not diluted by it. Therefore overall, the sludge-specific nitrification rates for granular sludge are much less affected by temperature (Figure 11.7).

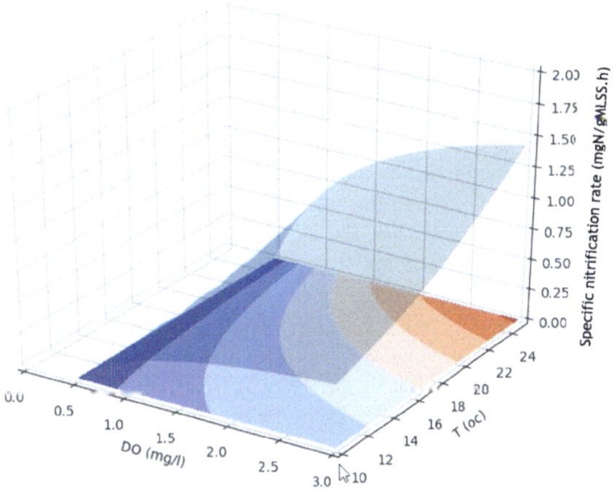

Figure 11.7 General relation of temperature and dissolved oxygen concentration on the volumetric nitrification rate in full-scale aerobic granular sludge reactors.

In sufficiently large granules the oxygen concentration in the inner parts of the granule becomes zero, while the oxygen concentration near the outer layer of the granule is high enough for nitrification. Because of this, nitrification and denitrification can occur simultaneously within the granule. The dissolved oxygen concentration needed for optimal simultaneous nitrification and denitrification is directly related to the size and activity of the granules. Larger granules have more anoxic volume than smaller granules at the same dissolved oxygen concentration in the bulk liquid (Figure 11.8). At lower temperatures, because of lower activity, the penetration depth of oxygen increases. This decreases the volume for denitrification, effectively making it more difficult to maintain simultaneous nitrification-denitrification.

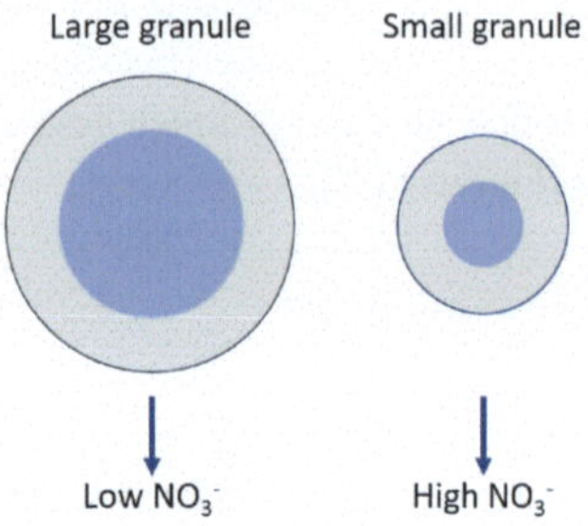

Figure 11.8 Effect of oxygen concentration in the bulk liquid on the available denitrification volume inside granular sludge. The grey zone indicates the oxygen penetration depth. The blue is the anoxic zone.

The efficiency of simultaneous nitrification and denitrification also depends on the amount of storage polymers stored in the deeper parts of the granules. In wastewater with relatively low RBCOD concentrations, denitrifiers will depend more on the slowly-degradable COD or storage polymers present in the outer layer of the granular sludge. In these cases, denitrification can be enhanced by alternating aerobic/anoxic conditions *i.e.* by switching aeration on and off.

11.3.3 Biological phosphorus removal

Biological phosphorus removal in AGS is performed by the same organisms found in conventional activated sludge (CAS) processes. In the AGS plant the growth conditions for PAOs are intrinsically optimal, since stable granulation depends on the effective and full consumption of COD under anaerobic conditions as well.

The phosphorus present in the influent is removed by biomass growth and via storage of polyphosphate in the PAOs. Under anaerobic conditions the PAOs take up VFAs forming PHA while hydrolysing polyphosphate for energy production (Chapter 6). This leads to phosphate release into the bulk liquid. The duration of the feeding phase is defined by the hydraulic need to take in wastewater and typically lasts 0.5-1 hour. In sewer systems where there is significant hydrolysis and fermentation occurring, for example in systems with long pressure pipelines, this time is long enough for complete uptake of the substrates. In sewer systems with less hydrolysis and fermentation in the sewer, for example in a gravity flow system with short pipelines, a prolonged anaerobic period might be needed to convert hydrolysable COD into VFAs to allow growth of PAOs.

In the aerobic phase following the anaerobic feeding phase, the reactor is aerated to allow for growth of the bacteria. Oxygen diffuses from the bulk liquid into the granules, allowing the PAOs and GAOs to start consuming the anaerobically stored PHA for growth. The PAOs will replenish the polyphosphate pool removing the phosphate from the bulk liquid. In the deeper layers of the granules, the PAO will be active using nitrate or nitrite instead of oxygen as the electron acceptor. This makes the phosphate uptake rate less susceptible to changes in the oxygen concentration than nitrification.

11.3.4 Granular sludge properties

The terminal settling velocity of an aerobic sludge granule is mainly dependent on the diameter and can have values up to 100 m/h (Figure 11.9).

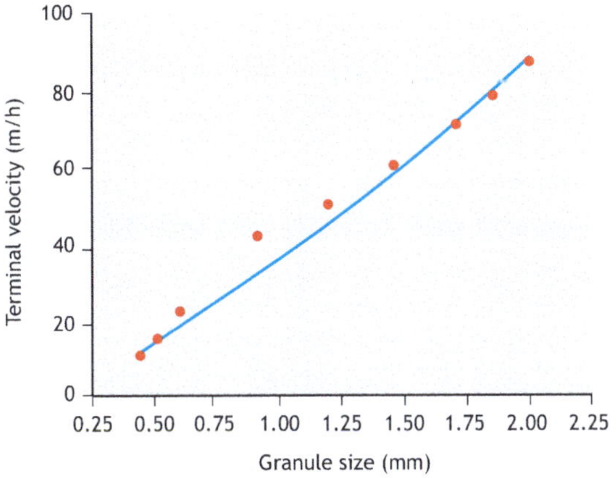

Figure 11.9 Terminal settling velocity of individual granules from the full-scale Nereda® plant in Garmerwolde, the Netherlands (the line with $R^2 = 0.98$ is calculated by a model).

A standard 30-minute measurement of the sludge volume index (SVI) therefore will give limited information about the settling behaviour of granular sludge, because in 30 minutes the granules will already have finished settling for a long time. It is therefore common practice for granular sludge to be measured after 5 minutes (SVI_5) and after 30 minutes (SVI_{30}) of settling. If the SVI_5 approximates the SVI_{30}, this is an indication of a high degree of granulation. The SVI of a mature granular bed typically is in the range between 20 and 40 ml/g. One needs to keep in mind that the SVI measured is a maximum value dictated by the worst-settling fraction present. A small fraction of slower-settling flocs in a traditional SVI test largely determines the SVI. Subsequently the SVI is not a very useful parameter to monitor quality of granular sludge. However, the SVI is important to monitor the amount of biomass that can be maintained without *e.g.* during feeding expanding towards the effluent decant. Another metric for the characterization of aerobic granular sludge is the granule size distribution. A sludge sample is sieved in a series of sieves (between 200 and 2,000 µm). Then the biomass concentration of the sieve fractions is measured. Since a granule is defined as an aggregate larger than 200 µm, all sludge smaller than 200 µm is considered non-granular. Note that this threshold is relatively arbitrary and the 'non-granular' fraction next to flocculent sludge also contains smaller 'baby granules'. A normal value for the flocculent fraction is between 1 and 2 g/l in a full scale AGS reactor. At higher MLSS concentrations the contribution of the fraction larger than 1 mm increases and in some cases this fraction can compose up to 90% of the total biomass concentration in the reactor.

11.3.5 Reactor operation aspects

Current aerobic granular sludge technologies depend on a sequencing batch operation. This has several advantages. The main aspect is the promotion of stable granulation by direct contact of granular sludge and raw wastewater at high concentrations. Also, granules are preferentially fed over flocs and RBCOD is fully converted under the anaerobic conditions.

This batch-wise operation also allows for lower effluent concentrations. The effluent concentrations in continuous processes are equal to the concentrations in the aeration tank. For low effluent requirements this can lead to decreased conversion rates at low operating concentrations. In a sequencing batch operation the aeration phase starts with high concentrations reaching high conversion rates. This means that substrate concentrations during most of the reaction phase are high enough not to be limiting the conversion rate. Thus, the conversion rates only drop at very low substrate concentrations. Because of this, effluent concentrations can reach virtually zero without lowering the overall conversion rates.

Batch operation also enables a continuous monitoring of the kinetics of the granular sludge. Figure 11.10 gives the online monitored concentrations of ammonium, nitrate and phosphate during an operational day in a Nereda® plant. During influent feeding (Q_{inf}) the concentrations (measured at the top of the reactor) are equal to the effluent concentrations. When feeding stops and aeration starts the reactor gets completely mixed leading to a sharp increase in ammonium and phosphate concentrations and decrease in the nitrate. These changes are proportional to the volume exchange rate of the SBR operation. The subsequent decrease in ammonium and phosphate immediately give the rate of conversion at the prevailing oxygen concentration which can be used for process monitoring and control purposes.

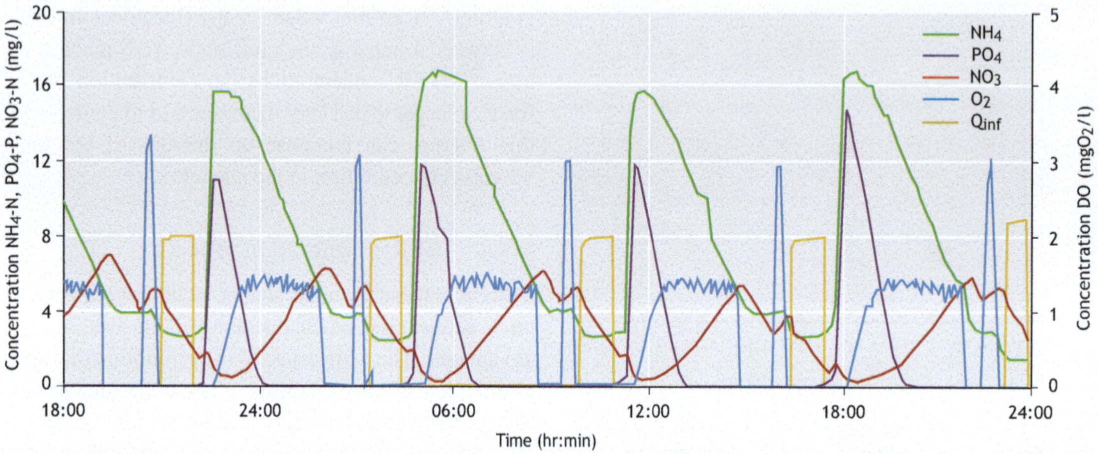

Figure 11.10 Online monitored concentration of ammonium, phosphate, nitrate, oxygen and influent flow rate during an operational day in a Nereda® plant.

In the reaction phase, the NH_4^+ is converted into NO_2^+, NO_3^- and N_2. An advantage of the AGS process is the simultaneous nitrification and denitrification. When oxygen is present in the water phase, NH_4^+ will be converted into NO_2^- by the ammonia-oxidizing bacteria (AOB). Because of the anoxic conditions within the core of the granule, and the presence of PHA stored by the PAO/GAO population, part of the NO_2^- will be converted directly into N_2 gas. The rest of the NO_2^- will be converted into NO_3^- by the nitrite-oxidizing bacteria (NOB). Similar to the NO_2^-, the NO_3^- can also be denitrified in the anoxic core of the granule. The amount of simultaneous denitrification depends on many process conditions, such as temperature, dissolved oxygen concentration in the reactor, and the amount of VFAs in the influent. In practice, this value can vary between 20% and 80%. If enhanced denitrification is required to achieve the treatment targets, simultaneous denitrification can be introduced by including additional denitrification periods (*i.e.* by switching on and off aeration during the reaction phase).

11.4 PROCESS CONTROL

11.4.1 The Nereda® cycle

The aerobic granular sludge process is a sequencing batch process. A reactor goes through a series of steps to clean the wastewater. These consecutive steps (simultaneous feeding from the bottom and drawing effluent from the top, the aeration phase and a short settling period) are called the Nereda® cycle (Figure 11.11).

The steps, or phases, in a Nereda® cycle are not fixed. The length of the different phases can be adapted according to the process circumstances. Also, the total cycle time can be changed by the process control software. More importantly, detailed online monitoring (Figure 11.10) enables one to follow the progress of the biological conversion. This can be used to adjust the phases and process set-points. Therefore the AGS process is highly adaptable if compared to a conventional continuously-fed activated sludge plant. The AGS process is controlled to meet effluent requirements for COD, N and P, to limit the amounts of suspended solids in the effluent, to optimize the SRT and to manage the variations in the influent flow. In the following paragraphs the most common approaches are shown.

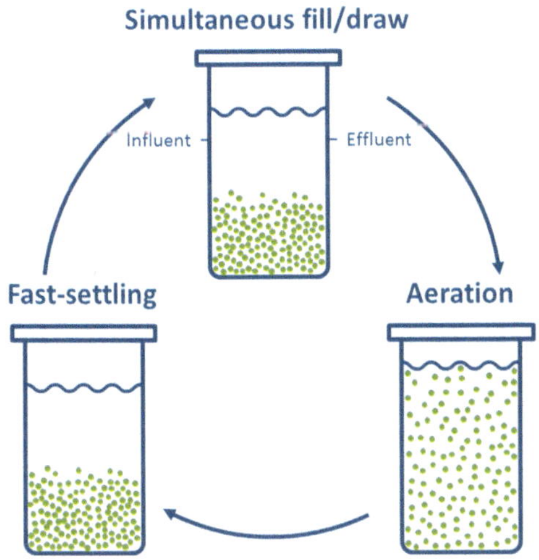

Figure 11.11 The Nereda® cycle

In the first step the influent is fed from the bottom through a settled bed into the reactor, while the cleaned water from the previous batch is pushed out at the top. Due to the good settling properties of aerobic granular sludge this can be done at relatively high up-flow velocities, without the risk of spilling sludge into the effluent. Up-flow velocities in the range of 5 m/h are not uncommon for an AGS reactor.

During the aeration phase the reactor is aerated for conversion of COD and nutrients. Also, anoxic conditions can be applied for the removal of nitrite and nitrate. In practice aeration is controlled based on a series of different strategies, such as oxygen concentration, redox, pH, ammonium, phosphate and NO_x concentrations.

The settling phase has two purposes. Before starting to feed again in the next cycle, the sludge bed must settle to prevent the washout of sludge when the simultaneous feeding and decanting starts. Granules will settle faster than flocs, which form a layer on top of the granular bed. The waste sludge is taken from the top of the sludge blanket at the end of the settling phase. Here the granular sludge is favoured above the flocculent sludge, which is an important step in the granulation process (as discussed in Section 11.2.3).

11.4.2 Batch scheduling

In most wastewater treatment plants the flow towards the plant can vary greatly. It is not uncommon that during a rain event the flow to the treatment plant is 5 times higher than under dry weather conditions, while overall pollution loads often remain the same. To treat this increase in flow rate in an as compact as possible reactor, the duration of each cycle can be adapted. This process is called batch scheduling. An example of a schedule under both dry weather conditions and wet (rainy) weather conditions is given in Figure 11.12. The example shows the scheduling for a system with two reactors and an influent buffer. Under dry weather conditions the reactors run with a total cycle time of 6.5 hours, a feeding phase of 1 hour, a reaction phase of 5 hours and a settling/wasting/idle phase of 0.5 hour. The reactors are fed from the influent buffer that stores the water when one reactor is in the feeding phase.

During wet weather conditions the cycle shortens. The total cycle time is only 3 hours, but the duration of the feeding phase is increased to 90 minutes. Also, there is no longer any time between the feeding phases, so one of the reactors is always feeding. The total time available for aeration is reduced by these changes in the scheduling. This can partly be compensated by increasing the oxygen levels in the reactor. So, depending on the effluent requirements and the local hydrograph, these wet weather conditions can be the determining factor for the reactor sizes in the design.

11.4.3 Nutrient removal

The process control for effective biological phosphorus removal is minimal. The anaerobic time is mainly decided by the hydraulic constraints of the process. In general, the feeding time is long enough to ensure enough COD uptake by PAO to allow good growth and phosphorus uptake under aerobic conditions. Only in the case of wastewater with very low soluble COD could it be necessary to control the length of the anaerobic period. In the aeration stage the phosphorus uptake is in general faster than the nitrification rate and again no specific control is required. As for the stable operation of any enhanced biological phosphorus removal (EBPR) process, it is necessary to minimize aeration after the phosphate is removed from the liquid in order to prevent reduction of PAO-activity by extensive over-aeration. Where necessary the overall biological phosphate removal can be supplemented with some chemical precipitants. This for example might be required for very strict effluent demands or in the case of low COD/P ratio in the wastewater or during periods with excessive hydraulic loading such as during peak wet weather flows. The amount of this supplementary chemical dosing can be easily optimized and controlled effectively. Since phosphorus uptake is linear after the first hour of aeration, and phosphate can be monitored online, the effluent phosphorus concentration achievable by biological removal can be accurately predicted while the treatment batch progresses. This information can be used to determine online if and how much precipitant is required and provide set points for the control of chemical dosing.

Nitrogen removal is controlled in the aeration phase. Nitrification is sensitive to the dissolved oxygen concentration. Online monitoring of ammonium gives the nitrification rate under the applied oxygen concentration. The oxygen concentration can be adjusted by changing the aeration rate. A decrease in aeration rate results in a decreased nitrification rate and increased simultaneous denitrification. With the process control the optimal ratio between both processes can be set. Since PAOs can use both oxygen and nitrate as the electron acceptor there is no big influence of aeration control on the P-removal process. If the simultaneous denitrification is insufficient, extra (post-) denitrification can be obtained by applying on/off aeration in the second phase of the cycle. The batch-wise operation and consequent direct monitoring of the concentrations give useful information for the online control of on/off aeration cycles.

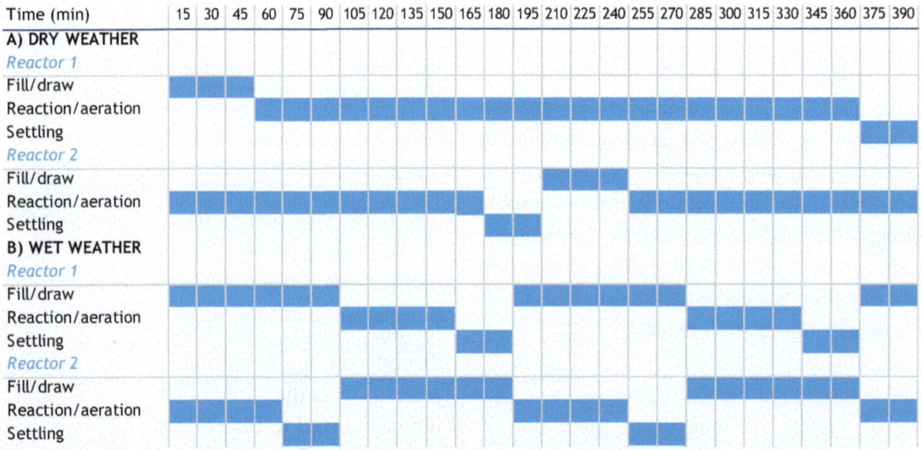

Figure 11.12 Batch scheduling for a plant with two reactors and a buffer under dry weather and wet weather conditions.

11.4.4 Effluent suspended solids

Effluent suspended solids removal performance, and measures to optimize this removal are similar to activated sludge plants. To prevent washout of small floating material with the effluent, as in clarifiers, baffles are placed in front of the fixed effluent weirs. Further, as for activated sludge clarifiers, it is important to prevent undesired denitrification resulting in an increase of effluent suspended solids. Rising sludge due to N_2 gas production is a known problem in clarifiers in conventional activated sludge plants. If no measures are taken, sludge rising due to denitrification can happen in the feeding and decant phase of the Nereda® process. Due to the high conversion rates in an AGS reactor and the plug-flow feeding from the bottom, degasification of N_2 gas can occur more easily. The gas deficit for N_2 gas between a completely stripped-out water phase and saturation is between 5 and 10 mg/l, depending on the water temperature and the ambient pressure. A higher temperature will give a lower solubility of N_2 gas in water and a lower N_2 gas deficit. A higher ambient pressure (atmospheric pressure plus the pressure due to the water column) gives a higher N_2 gas solubility and a higher N_2 gas deficit.

The equilibrium concentration of N_2 gas during aeration can be calculated with Eq. 11.2. With a gas fraction of N_2 gas in air (f) of 0.79, the minimum concentration of N_2 in the water phase can be calculated at every water depth (h). Also the saturation concentration can be calculated at every water depth by setting f to 1. If the N_2 concentration in the water phase becomes higher than the saturation concentration, small gas bubbles will form, leading to rising sludge. The actual N_2 concentration in the water phase is not only influenced by transfer from and to the gas bubbles during aeration and the denitrification process, but also by the mixing of the reactor. This complex process is shown in Figure 11.13.

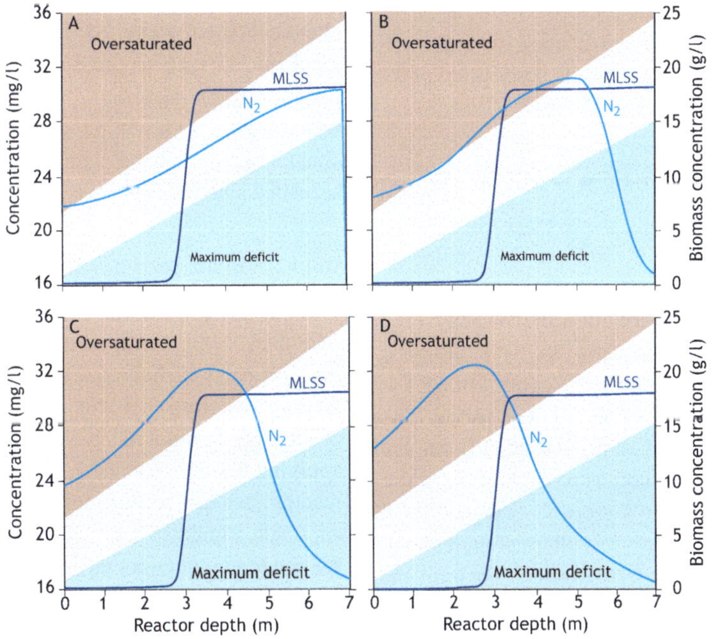

Figure 11.13 Stripping of nitrogen gas from fully saturated situation (A) and after 5 minutes (B), 15 minutes (C) and 60 minutes (D) of stripping at 20 °C. The red area marks an over-saturated concentration and the green area marks a deficit bigger than equilibrium with air.

$$C_{eq} = k_H \cdot f \cdot (P_{atm} + h \cdot \delta \cdot g) \quad (11.2)$$

where:

C_{eq}	equilibrium concentration of N_2 gas (mol/m^3)
k_H	Henry coefficient N_2 gas ($mol/m^3.Pa$)
f	gas fraction of N_2 gas in the gas bubble (-)
P_{atm}	atmospheric pressure (Pa)
h	water depth (m)
ρ	density of water (kg/m^3)
g	gravity acceleration (m/s^2)

During feeding, the potential for degasification is dependent on two processes. Firstly, the anoxic conditions lead to production of N_2 gas due to denitrification and thus a reduction of the N_2 gas deficit. Secondly, the plug-flow feeding from the bottom moves the water upwards in the reactor, lowering the ambient pressure and resulting in a lower saturation concentration. To prevent degasification during the feeding phase, a short stripping phase can be added at the end of the cycle if and when needed. By intensely aerating the reactor for a few minutes, the N_2 gas is stripped out and degasification of N_2 gas will not occur. A more in-depth view and description of the model is given by Van Dijk *et al.*, 2018.

11.4.5 Solids retention time

As mentioned earlier, while the average SRT of the sludge in a typical AGS process is similar to conventional activated sludge, the individual sludge fractions have a wide distribution of various SRTs. Since the granular fraction has an SRT allowing nitrification at all temperatures, the control of SRT is less sensitive than for activated sludge plants. The SRT of the flocculent fraction is not explicitly set, but optimally it should be as short as possible if the waste sludge is to be digested. In a properly designed AGS reactor, preferentially only smaller granules and flocculent sludge are regularly wasted. To limit the size of the granules and maintain optimal biological phosphorus removal efficiencies, it is necessary to control the SRT of the larger granules. This is done by occasionally purging sludge from the bottom of the settled bed. Large granules lower the specific biofilm surface area of the reactor. Thereby the potential aerobic conversion rates are decreased because these are dependent on the oxygen mass transfer rate from liquid to granule. Note that since these large granules are relatively old and well mineralised, the digestibility of this sludge is relatively low. Because the digestibility of the flocculent sludge fraction is higher than that of large granules, the overall digestibility of AGS waste sludge is similar to, or slightly higher than, activated sludge.

11.5 DESIGN CONSIDERATIONS

The AGS process contains similar treatment steps to those required by conventional activated sludge plants, but there are some key differences. The requirements for pre-treatment are similar to activated sludge; they depend on the wastewater characteristics and include screening, grit and if necessary oil, fat and grease removal. For stringent effluent requirements a range of post-treatment steps can be used, both chemical (such as metal dosing, activated carbon) and mechanical (such as sand filter, cloth media filter, ultrafiltration) systems. These pre- and tertiary technologies will not be further addressed in this chapter as they have similar designs to those for activated sludge effluents.

11.5.1 Plant configuration

Since the AGS process is a batch process, the treatment of a continuous influent flow requires multiple reactors. Each reactor goes through a cycle of feeding - reacting - settling. While a reactor is reacting or settling, it cannot receive influent. This is solved by applying multiple reactors, so that there is always one reactor that can receive influent. Subsequently, this approach stipulates that a minimum of three reactors are needed for continuous operation and feeding times will therefore be 1/3 of the total cycle time (Figure 11.14, left).

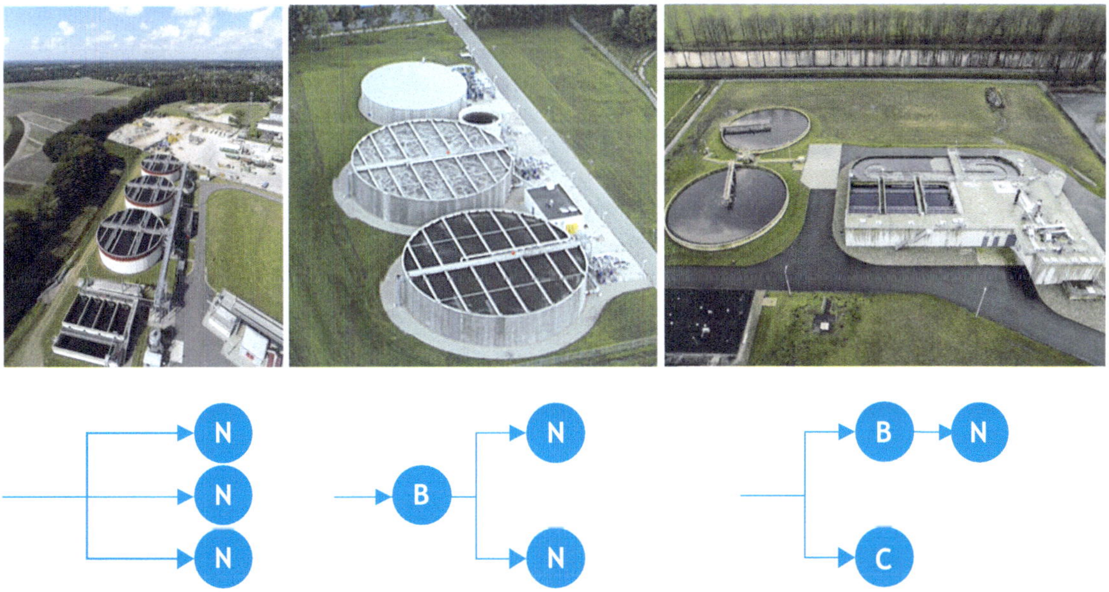

Figure 11.14 Different configurations for Nereda® plants in the Netherlands: (left) in Epe, three reactors (N) without a buffer, (middle) in Garmerwolde, two reactors (N) with a buffer (B), and (right) in Vroomshoop, a hybrid set-up where a Nereda® plant (with a buffer) is operated in parallel with a conventional activated sludge process (C).; here the waste sludge from the granular sludge plant is discharged in the activated sludge plant. Below is a depiction of the configuration.

This could result in feeding times that are longer than required for anaerobic COD uptake and phosphate release and results in larger than strictly necessary reactor volumes. This however makes biological phosphate removal more stable. Such over-design can be limited by applying more reactors (n), reducing the feeding time to $1/n^{th}$ of the total cycle time.

For smaller plants, building three treatment tanks might be uneconomic and an influent buffer can be used to overcome the need to have always one reactor in feed mode (Figure 11.14, middle). Using such an influent buffer only one reactor would suffice for treatment. Also, for larger plants with more than three reactors an influent buffer could be used to minimize reactor tank volume as the feeding phase can be shorter with higher flows. When minimising the anaerobic feeding time the minimal anaerobic time for biological phosphorus removal needs to be characterised which will depend on the local ratios of COD/P, $RBCOD/COD_{total}$ and the fraction of hydrolysable substrate.

Another design option for granular sludge is a hybrid configuration. In this configuration, the AGS reactor is built parallel to a continuously-fed activated sludge reactor (Figure 11.14, right). Certainly when upgrading existing activated sludge treatment plants this is often an interesting option. In the hybrid configuration the AGS reactor treats only part of the total influent and discharges the waste sludge into the activated sludge plant. By doing so, the AGS waste sludge improves the settleability of the activated sludge and the nutrient removal of the activated sludge reactor, making optimal use of both systems.

11.5.2 Design volume

The design volume is mainly dependent on the local influent flow rate and its variation, total COD and nitrogen load, water temperature and effluent requirements. The sum of the times for the three phases in the cycle (feed, aerate and settle) and the amount of wastewater determines the total volume of the AGS reactors.

In the anaerobic feeding phase, wastewater is fed from the bottom through a settled granular bed into the reactor. Simultaneously the clean effluent is pushed out from the top of the reactor. As a result, the volume in the reactor is constant. In most conventional SBR plants this is simply not possible due to the flocculent sludge settling properties. The amount of exchange between fresh influent and clean effluent is expressed in the exchange ratio, which is a number between 0 and 100%. A properly designed AGS reactor can have an exchange ratio up to 65% and sometimes even more. This maximum exchange ratio is influenced by the level of (vertical) plug flow behaviour in the reactor. A sub-optimal plug flow will lower the maximum exchange ratio that can be reached and will lead to a higher required reactor volume. One of the parameters influencing the plug flow is the up-flow velocity (m/h). During the anaerobic feeding period RBCOD is taken up from the influent by the granular bed. To allow for the hydrolysis and uptake of RBCOD, generally feeding times of 0.5-3.0 hours are required, depending on the type of wastewater. In general, the anaerobic time is set by the hydraulic constraints on the specific treatment site.

The processes in the reaction phase depend on the effluent requirements. If the goal is to only remove COD, the reaction phase consists of a relatively short aeration period where only the COD and stored PHA is oxidized. In many cases this will also result in near-complete phosphorus removal, because PAOs play an important role in the granulation process and in COD removal. However, if nitrification is also required, the reaction phase will require a longer aeration period leading to larger volumes. The time needing to be allocated for denitrification is very dependent on the type of wastewater and sewer system used.

The total volume of the plant (Nereda® reactors + influent buffer) depends on the sludge loading rate, the flow variations and cycle configuration. In the following example we have a Nereda® plant with three reactors (on the left in Figure 11.14). To determine the plant volume, first a cycle configuration needs to be established. The total cycle time can be calculated by:

$$t = t_{feed} + t_{react} + t_{settle} \qquad (11.3)$$

where:

t	total cycle time (h)
t_{feed}	feeding phase duration (h)
t_{react}	reaction phase duration (h)
t_{settle}	time for sludge settling and decanting (h)

In a plant configuration with multiple reactors without an influent buffer the total cycle time must be a multiple of the feeding phase duration:

$$t_{feed} = (t_{react} + t_{settle}) / (n_{reactors} - 1) \qquad (11.4)$$

where:

$n_{reactors}$ number of reactors.

Hence in a 3-reactor configuration, the duration of the feeding phase is half of the sum of the duration reaction phase and the settling phase. The feeding phase and the reaction phase scale together, so a longer feeding phase leads to a longer reaction phase. Only the settling/wasting phase has a fixed duration of typically 20-30 minutes. As a result the reactor volume is only marginally influenced by the cycle time chosen. In practice the duration of the feeding phase is determined by parameters such as the exchange ratio, the up-flow velocity, and the minimum anaerobic feeding time.

In the design process the reaction (t_{react}) time can be freely chosen, for example 4 hours. Later, this value can be optimized to minimize the total plant volume, as can be seen in Figure 11.15. Based on equations 11.3 and 11.4 and the chosen value of t_{react}, a total cycle time (t) can be calculated. The average reaction time per day per reactor can be calculated based on the number of cycles per day:

$$n_{cycles} = 24 / t \qquad (11.5)$$

and

$$t_{react,day} = n_{cycles} \cdot t_{react} \tag{11.6}$$

where:

n_{cycles} number of cycles per day per reactor

$t_{react,\,day}$ total reaction time per day (h).

The reaction volume can now be calculated based on the total reaction time per day and the design flow, the COD concentration and the sludge loading rate. The sludge loading rate is typically similar to activated sludge processes and range between 0.1-3.0 kgCOD/kgTSS.d. Nitrification occurs typically until 0.4 kgCOD/kgMLSS.d at moderate temperatures. Higher temperatures can allow nitrification with even higher loading rates.

$$V_{reactor} = \frac{Q \cdot COD}{\frac{t_{react,day}}{24} MLSS \cdot SLR \cdot n_{reactor}} \tag{11.7}$$

where:

$V_{reactor}$ volume of a reactor (m^3)

Q flow (m^3/d)

COD COD concentration (kg/m^3)

MLSS sludge concentration (kg/m^3)

SLR sludge loading rate (kgCOD/kgMLSS.d)

In a configuration with a buffer the duration of the feeding phase can be more freely chosen. Typically, the feeding phase is shorter than in a configuration without a buffer, allowing for a longer reaction phase. The reactor volume will be more efficiently used. The total volume of the buffer and the reactors combined will not be much smaller than in a configuration without a buffer. However, the reduction in reactor volume will give an overall cost reduction. The feeding time t_{feed} is not calculated anymore according to equation 11.4 but it is freely chosen in the design process taking into account upflow velocity limits. Typical values of 0.5-1.0 hour are used. The buffer volume is calculated based on the peak flow during dry weather conditions. During rainy weather conditions the reactor scheduling is changed as described in section 11.4.2 and the buffer will not be emptied during the feeding phase because inflow and outflow are balanced. The buffer volume follows from:

$$V_{buffer} = \frac{Q_{peak}}{n_{reactor} \cdot n_{cycles}} - Q_{peak} \cdot \frac{t_{feed}}{24} \tag{11.8}$$

where:

Q_{peak} DWF peak flow (m^3/d).

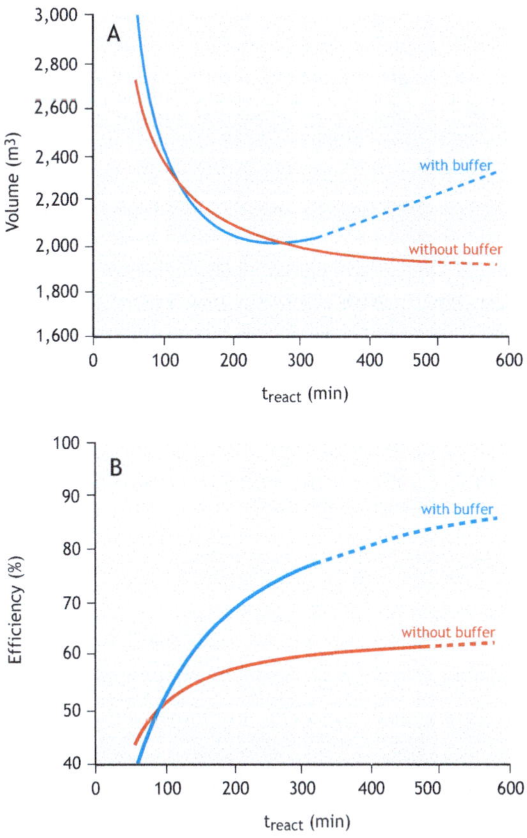

Figure 11.15 The effect of the application of a buffer. The application of a buffer (blue) does not lead immediately to a lower total volume (A), but the efficiency of the reactors (B) expressed as relative time available for reaction increases. The dotted lines show the design is limited by the exchange ratio.

Using all the above calculations, an optimal reactor configuration can be derived. Figure 11.15A shows that the total volume of tanks including the buffer (called the plant volume) does not change significantly when a buffer is applied. In contrast the remaining reactor volume is used more efficiently for aeration as shown in Figure 11.15B. This shows the efficiency, which can be expressed as the number of hours per day that the reactors are used in the reaction phase. Also, there is an optimal cycle configuration, where a minimum amount of reactor volume is needed.

In a hybrid configuration many variations are possible. It can be designed with or without a buffer, with one or many reactors. In the simplest form only one AGS reactor is built; here the conventional plant takes all the load when the AGS reactor is not feeding. Similar to the configuration with a buffer, the duration of the feeding phase and the reactor phase can be chosen based on boundaries from both the AGS process and the conventional process. As an alternative the AGS reactors can be built as a separate lane, continuously being fed a fixed fraction of the total flow coming into the plant. In this case a configuration either with or without a buffer can be applied.

11.5.3 Sludge treatment

Sludge is wasted in an AGS reactor for two reasons: the selection spill contains mainly slow-settling sludge and the SRT spill contains a mixture of granules. At the end of each cycle the worst-settling fraction of the AGS is wasted by removing the top of the sludge bed after a period of settling. This is called the selection spill and it is essential for the granulation process. The selection spill consists of sludge flocs, small granules, sheared off material from big granules and suspended solids from the influent such as cellulose fibres. The selection spill consists of relatively non-mineralised material. The digestibility of this selection spill is generally good. The treatment of the waste sludge is comparable with normal activated sludge, although recent developments show biopolymers can be recovered from the waste sludge (see Section 11.6).

11.5.4 Mixed liquor suspended solids

Aerobic granular sludge plants are generally operated at mixed liquor suspended solids (MLSS) concentrations which are higher than in typical conventional activated sludge plants. This is also an important parameter that drives the compactness of the Nereda® technology. A typical design concentration for an AGS plant is 8 g/l, but values as high as 15 g/l are not uncommon in operational plants. However, at values higher than 8 g/l the biomass concentration is not the limiting factor in the purification process. These high concentrations can be reached because of the dense biomass in the granules. Mature granules can have a MLSS concentration of 50-60 g/l. Therefore, with a typical bed porosity of 50%, a settled granular bed has a MLSS concentration of 25-30 g/l.

11.6 RESOURCE RECOVERY

The primary aim of wastewater treatment is the protection of human health and the environment. Nevertheless, in the last decade resource recovery has also become an important aspect of wastewater treatment process design (Kehrein *et al.*, 2020). The study of granular sludge formation has opened up new options for resource recovery. Especially the polymers that form the matrix of the granular sludge offer many new possibilities. The integration of resource recovery with Nereda® technology is briefly introduced below.

Traditionally energy in the form of biogas is recovered from waste sludge. The granular sludge process has a similar recovery potential for biogas as conventional activated sludge plants. In a conventional activated sludge plant the primary clarification is used to maximize biogas production. For granular sludge processes this has been integrated in the treatment reactor. Particulate COD from wastewater is incorporated in the floc fraction of the sludge. This fraction has a short residence time (0.5-5.0 days) and thereby a very good digestibility.

Granular sludge has a very long SRT and thereby a slightly lower digestibility than secondary flocculent sludge (200 *versus* 230 $mlCH_4/gVSS$). The total excess sludge gives a similar yield of biogas to activated sludge processes with combined primary and secondary sludge digestion (300 $mlCH_4/gVSS$) (Guo *et al.*, 2020).

One of the compounds in the flocculent sludge fraction is cellulose fibres (Figure 11.16) which form a significant fraction of the influent (20-30% of the particulate COD). This fraction ends up in the waste sludge and contributes to the higher biogas production from this fraction. The cellulose can be recovered by a sieving step either in the influent or in the waste sludge. The latter has the advantage of a smaller hydraulic flow (smaller sieve capacity needed). Recovered cellulose can be used to replace new cellulose in industrial processes. It can also be hydrolysed into sugars (and subsequently be fermented in volatile fatty acids) and dosed as substrate in WWTPs where the COD is limiting for phosphate removal or nitrogen removal.

Figure 11.16 Image of sieved (≥ 200μm) excess sludge from a full-scale Nereda® plant showing cellulose fibres meshed with pieces of granular sludge.

Phosphate recovery can be integrated as in conventional activated sludge processes, usually in the form of struvite recovery from the digestate. When iron is dosed or present in the wastewater, vivianite is formed during digestion which can be recovered by magnetic separation (Prot *et al.*, 2019). An extra option in a batch granular sludge process such as in Nereda® plants is given by the large-scale release of phosphate during the influent addition. After the start of the aeration phase, phosphate concentrations are high. If the reactor liquid is retrieved at this stage, it is possible to recover phosphate as calcium mineral or as struvite. However, recovery at this stage should be balanced in order to leave sufficient phosphate available for effective biological phosphorus removal (Barat and Van Loosdrecht, 2006).

The most interesting raw material to recover is formed by the extracellular polymeric matrix of granular sludge. This can be extracted and used as the basis for novel materials. At the time of writing (2020), the extracellular polymeric matrix has in general not been well characterised (Seviour *et al.*, 2019) New analytical tools will unveil the composition in the coming years. Methods that have been used widely for sugar and protein analysis have shown to have large biases (Seviour *et al.*, 2019) and the composition is not a simple polysaccharide and protein mixture. In granular sludge, and probably also other biofilms, bacteria produce glycoproteins, sulphated glucosaminoglycan-like, Hyaluronic acid-like, and sialic acids containing compounds. These polymers are responsible for the very stable polymeric matrix that embeds the microorganisms which produce this matrix (Felz *et al.*, 2020).

Gel-forming polymers are generally extracted from biological sources and not produced from oil-based chemicals. The supply of these polymers is in general limited because separate production of them is too expensive. For instance, alginate supply is limited by the amount of naturally grown seaweed that can be harvested. Cultivation makes the polymer too expensive. The extracellular polymeric matrix of activated sludge or granular sludge can therefore form a very interesting new resource that can be recovered. The polymers from activated sludge have a more flocculent behaviour, while the polymers from granular sludge can form stable gels. The production volume could be much larger than most current biopolymers; 5 kg/P.E.yr of biopolymers could be produced in a Nereda® plant. The harvesting and use of polymers from aerobic granular sludge are currently in the development stage. The first full-scale extraction facility was inaugurated in 2019 in the Netherlands (Figure 11.17) and the polymer is marketed under the brand name Kaumera gum. The current extraction process is similar to the harvesting of alginate from seaweed. First the polymer is solubilised by heating the sludge in an alkaline carbonate solution. After removing the non-solubilised material by centrifuging, the polymer is precipitated by pH neutralisation and calcium addition or by acidifying the liquid to a low pH.

Traditionally biopolymers are used mainly in the food and medical industry. This is due to their limited availability and high price. Wastewater-derived polymers are less suitable for these markets but the potential market volume is stimulating the development of new applications. First applications can be found in agricultural application, mainly also due to observed plant growth stimulation by the polymer. An interesting characteristic is the flame retardancy of the polymer (Kim *et al.,* 2020). This characteristic can be an added value when Kaumera is used for production of materials and coatings. Another interesting characteristic is the use of the polymer for composite material production. Where chemical polymers can be used for composite materials, they can bind up to 10-20% of inorganic filler material. Biopolymers can also be used but can hold up to 80% of filler material. Current investigations have revealed that clay-Kaumera composites resemble nacre materials (structured organic/inorganic material composite materials; *e.g.* mother of pearl). They have a very high tensile strength, are non-flammable and keep their strength up to 180 °C. Composites formed by Kaumera and cellulose produce a material resembling mother of pearl which has a high aesthetic value (Figure 11.18), showing that what is flushed through a toilet can be transformed into attractive products.

Figure 11.17 The Nereda® plant in Zutphen, the Netherlands with a Kaumera-extraction plant. In the background is the existing conventional activated sludge plant.

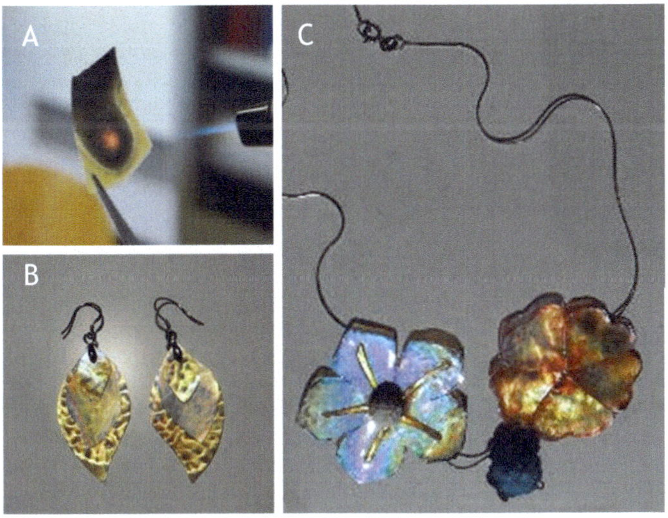

Figure 11.18 A: Flame-retardant composite plastic derived from Kaumera and clay. B and C: wastewater-based earrings and necklace. The main material is Kaumera and cellulose composite. The plastic sphere on the necklace is from PHA. The blue colour is vivianite and the red colour is extracted from anammox sludge (art created by Yuemei Lin).

REFERENCES

Ali M., Wang Z., Salam K.W., Hari A.R., Pronk M., Van Loosdrecht M.C.M., and Saikaly P.E. (2019). Importance of species sorting and immigration on the bacterial assembly of different-sized aggregates in a full-scale Aerobic granular sludge plant. *Environmental Science and Technology,* 53(14), 8291-8301.

Barat R. and Van Loosdrecht M.C.M. (2006). Potential phosphorus recovery in a WWTP with the BCFS® process: Interactions with the biological process. *Water Research,* 40(19), 3507-3516.

Beun J.J., Hendriks A., Van Loosdrecht M.C.M., Morgenroth E., Wilderer P.A. and Heijnen J.J. (1999). Aerobic granulation in a sequencing batch reactor. *Water Research,* 33(10), 2283-2290.

Beun J.J., Van Loosdrecht M.C.M. and Heijnen J.J. (2002). Aerobic granulation in a sequencing batch airlift reactor. *Water Research,* 36(3), 702-712.

De Kreuk M.K., Kishida N., Tsuneda S. and Van Loosdrecht M.C.M. (2010). Behavior of polymeric substrates in an aerobic granular sludge system. *Water Research,* 44(20), 5929-5938.

Elenter D., Milferstedt K., Zhang W., Hausner M. and Morgenroth E. (2007). Influence of detachment on substrate removal and microbial ecology in a heterotrophic/autotrophic biofilm. *Water Research,* 41(20), 4657-4671.

Felz S., Neu T.R., Van Loosdrecht M.C.M. and Lin Y. (2020). Aerobic granular sludge contains Hyaluronic acid-like and sulfated glycosaminoglycans-like polymers. *Water Research,* 169, 115291.

Gonenc E. and Harremoes P. (1990). Nitrification in rotating disc systems - II. Criteria for simultaneous mineralization and nitrification. *Water Research,* 24(4), 499-505.

Guo H., Van Lier J.B. and De Kreuk M. (2020). Digestibility of waste aerobic granular sludge from a full-scale municipal wastewater treatment system. *Water Research,* 173, 115617.

Heijnen J.J. (1984). *Biological industrial wastewater treatment minimizing biomass production and maximizing biomass concentration.* PhD thesis, Delft University of Technology, Delft, the Netherlands.

Heijnen J.J. and Van Loosdrecht, M.C.M. (1998). Netherlands Patent No. WO 98/37027.

Heijnen J.J., Van Loosdrecht M.C.M., Mulder R., Weltevrede R. and Mulder A. (1993). Development and Scale-Up of an Aerobic Biofilm Air-Lift Suspension Reactor. *Water Science and Technology,* 27(5-6), 253-261.

Heijnen J.J., Mulder A., Weltevrede R., Hols P.H. and Van Leeuwen H.L.J.M. (1990). Large-scale anaerobic/aerobic treatment of complex industrial wastewater using immobilized biomass in fluidized bed and air-lift suspension reactors. *Chemical Engineering & Technology - CET,* 13(1), 202-208.

Kehrein P., Van Loosdrecht M.C.M., Osseweijer P., Garfí M., Dewulf J. and Posada J. (2020). A critical review of resource recovery from municipal wastewater treatment plants – market supply potentials, technologies and bottlenecks. *Environ. Sci.: Water Res. Technol.*, 6:877-910.

Kim N.K., Mao N., Lin R., Bhattacharyya D., Van Loosdrecht M.C.M. and Lin Y. (2020). Flame retardant property of flax fabrics coated by extracellular polymeric substances recovered from both activated sludge and aerobic granular sludge. *Water Research,* 170, 115344.

Kwok W.K., Picioreanu C., Ong S.L., Van Loosdrecht M.C.M., Ng W.J. and Heijnen J.J. (1998). Influence of biomass production and detachment forces on biofilm structures in a biofilm airlift suspension reactor. *Biotechnology and Bioengineering,* 58(4), 400-407.

Lettinga G., Jansen A.G.N. and Terpstra P. (1975). Anaerobic treatment of sugar beet wastes (in Dutch). *H2O,* 8(26), 530-536.

Martins A.M., Heijnen J.J. and Van Loosdrecht M.C.M. (2003). Effect of feeding pattern and storage on the sludge settleability under aerobic conditions. *Water Research,* 37(11), 2555-2570.

Martins A.M., Karahan O. and Van Loosdrecht M.C.M. (2011). Effect of polymeric substrate on sludge settleability. *Water Research,* 45(1), 263-273.

Martins A.M.P., Pagilla K., Heijnen J.J. and Van Loosdrecht M.C.M. (2004). Filamentous bulking sludge—a critical review. *Water Research,* 38(4), 793-817.

Morgenroth E., Sherden T., Van Loosdrecht M.C.M., Heijnen J.J. and Wilderer P.A. (1997). Aerobic granular sludge in a sequencing batch reactor. *Water Research,* 31(12), 3191-3194.

Mosquera-Corral A., Montras A., Heijnen J.J. and Van Loosdrecht M.C.M. (2003). Degradation of polymers in a biofilm airlift suspension reactor. *Water Research,* 37(3), 485-492.

Picioreanu C., Van Loosdrecht M.C.M. and Heijnen J.J. (1998). Mathematical modeling of biofilm structure with a hybrid differential- discrete cellular automaton approach. *Biotechnology and Bioengineering,* 58(1), 101-116.

Pronk M., De Kreuk M.K., De Bruin B., Kamminga P., Kleerebezem R. and Van Loosdrecht M.C.M. (2015). Full scale performance of the aerobic granular sludge process for sewage treatment. *Water Research,* 84, 207-217.

Prot T., Nguyen V.H., Wilfert P., Dugulan A.I., Goubitz K., De Ridder D.J., and Van Loosdrecht M.C.M. (2019). Magnetic separation and characterization of vivianite from digested sewage sludge. *Separation and Purification Technology,* 224, 564-579.

Seviour T., Derlon N., Dueholm M.S., Flemming H.-C., Girbal-Neuhauser E., Horn H. and Lin, Y. (2019). Extracellular polymeric substances of biofilms: Suffering from an identity crisis. *Water Research,* 151, 1-7.

Van Dijk E.J.H., Pronk M. and Van Loosdrecht M.C.M. (2018). Controlling effluent suspended solids in the aerobic granular sludge process. *Water Research,* 147, 50-59.

Van Loosdrecht M.C.M., Eikelboom D., Gjaltema A., Mulder A., Tijhuis L. and Heijnen J.J. (1995). Biofilm structures. *Water Science and Technology,* 32, 35-43.

Van Loosdrecht M.C.M., Heijnen J.J., Eberl H. and Picioreanu C. (2002). Mathematical modelling of biofilm structures, *Antonie van Leeuwenhoek*, 81: 245–256.

Van Loosdrecht M.C.M., Pot M.A. and Heijnen J.J. (1997) Importance of bacterial storage polymers in bioprocesses. *Water Science and Technology,* 35(1): 41-47.

Villaseñor J.C., Van Loosdrecht M.C.M., Picioreanu C. and Heijnen J.J. (2000). Influence of different substrates on the formation of biofilms in a biofilm airlift suspension reactor. *Water Science and Technology,* 41, 323-330.

NOMENCLATURE

Symbol	Description	Unit
C_{eq}	Equilibrium concentration of N_2 gas	mol/m^3
$C_{s,o}$	Concentration of the substrate in the bulk	g/m^3
$C_{x,m}$	Maximum density of the biomass in the biofilm	g/m^3
D_s	Diffusion coefficient of the substrate	m^2/s
f	Gas fraction of N_2 gas in the gas bubble	-
g	Gravity acceleration	m/s^2
G	Dimensionless growth factor	-
h	Water depth	m
k_H	Henry coefficient N_2 gas	$mol/m^3.Pa$
L_y	Biofilm thickness	m
$n_{reactors}$	Number of reactors	-
n_{cycles}	Number of cycles per day per reactor	-
t	Total cycle time	h
t_{feed}	Feeding phase duration	h
t_{react}	Reaction phase duration	h
$t_{react,day}$	Total reaction time per day	h
t_{settle}	Time for sludge settling and decanting	h
Q	Flow	m^3/d
Q_{peak}	DWF peak flow	m^3/d
P_{atm}	Atmospheric pressure	Pa
$V_{reactor}$	Volume of a reactor	m^3

Abbreviation	Description
AGS	Aerobic granular sludge
AOB	Ammonia-oxidizing bacteria
CAS	Conventional activated sludge
EBPR	Enhanced biological phosphorus removal
GAO	Glycogen-accumulating organisms
MLSS	Mixed liquor suspended solids
NOB	Nitrite-oxidizing bacteria
PAO	Phosphate-accumulating organisms
PHA	Polyhydroxyalkanoates
RBCOD	Readily biodegradable COD
UASB	Upflow anaerobic sludge bed reactor
VFA	Volatile fatty acids
SLR	Sludge loading rate
SRT	Sludge retention time

Greek symbols	Explanation	Unit
μ_m	Maximum specific growth rate	1/d
ρ	Density of water	kg/m^3

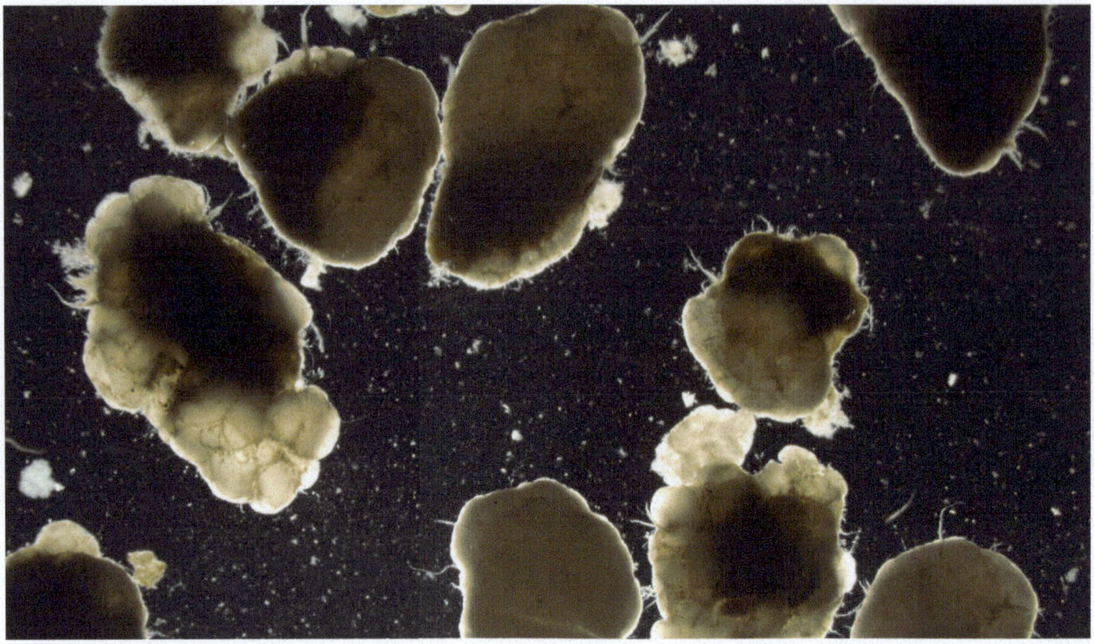

Figure 11.20 Aerobic granular sludge from a Nereda® plant (photo: Royal HaskoningDHV).

Figure 11.21 Data series presented in Figure 11.10 have been transformed into a sequence of notes connected in such a way that they are played or heard separately, the outcome is known as Nereda® Melody produced in cooperation between Royal HaskoningDHV and Foundation Pinta 021 (https://pinta021.org/muzika-vode/) (photo: Foundation Pinta 021).

doi: 10.2166/9781789060362_0523

12

Final settling

Imre Takács and George A. Ekama

12.1 INTRODUCTION

Final settling, at its most fundamental level, separates a heavier solid phase (the sludge mass) and a lighter liquid phase (effluent) from each other using gravity, therefore, it is often termed solid-liquid phase separation. The process is usually implemented in large concrete basins, called final settling tanks or secondary clarifiers. Thousands of books, journal articles and reports contain a wealth of information produced in the last half century related to the theory, design and operation of final settling tanks. The objective of this chapter is to provide an overview of the settling process and its implementation in practice, with special emphasis placed on the practical aspects, the design and operation of these phase separation units.

Gravity induced settling is used in several other situations, for example for producing primary effluent from raw wastewater in primary settlers, or thickening (reducing the water content of) waste sludge in thickeners. The wastewater industry, in addition to gravity settling, uses other methods (*e.g.* membranes or flotation) for phase separation of activated sludge. These applications and their engineering aspects are not discussed in this chapter (see Chapter 13).

12.1.1 Objective of settling

In activated sludge reactors, a concentrated mixture of sludge and wastewater is produced and maintained for the biological treatment of wastewater. Once sufficient biological treatment is achieved, the sludge has to be separated from the treated wastewater, which becomes secondary effluent. The sludge in the reactor consists of micro-organisms (primarily bacteria) and cell debris in the micrometer size range that would normally be difficult to separate from the liquid phase. However, the sludge is flocculent in nature and given the right conditions, readily forms activated sludge flocs which are one to three orders of magnitude larger than individual bacteria. The flocs, with their density slightly higher than water, can be settled out of the wastewater in clarifiers. Since the density difference between the sludge mass and water is small, settling velocities are slow and long hydraulic residence times are required, typically in the order of hours. This, combined with the large sewage flows generated in populated areas, results in large structures visible from space close to large populated areas.

For every 1,000 persons in a city serviced by a proper sewer system approximately 5 to 15 m^2 of secondary clarifier surface area is required, depending on local water usage habits.

12.1.2 Functions of a secondary settling tank

Final settlers play multiple roles in wastewater treatment, providing three distinct functionalities: (*i*) clarification, producing a clear effluent, (*ii*) thickening, providing a concentrated recycle stream, and (*iii*) sludge storage, usually on a temporary basis, during high flow periods.

12.1.2.1 Clarification in secondary settlers

Final effluents leaving a well designed clarifier usually contain only 5 to 15 mg/l of suspended solids. Considering the typical operating MLSS concentration range of 1,500 to 3,500 mg/l, the efficiency of the clarifier is expected to be in the 99% to 99.9% range. There are two key factors to achieve this high efficiency on a consistent basis: (*i*) a clarifier should create conditions which promote flocculation of sludge and capture of small particles within the activated sludge floc; (*ii*) flow conditions in the clarification zone should be uniform, particularly around effluent launders and weirs, to minimize the effect of local currents lifting liquid from deeper layers with higher concentrations and mixing it in the effluent. The engineering aspects of achieving these goals are summarized in the section 12.2. The suspended solids content of an effluent sample contributes to, and sometimes forms the majority of, effluent BOD, COD, TP (and less so with TN). With the low effluent concentration limits which are starting to be required in sensitive areas, it is important to keep effluent solids to the lowest level possible. Even when the secondary effluent is subject to further tertiary treatment (such as various filtration technologies), a well clarified effluent, low in suspended solids, increases the lifetime and decreases the operating costs of the tertiary unit.

12.1.2.2 Thickening in secondary settlers

Sludge settled to the bottom of the clarifier has to be returned to the activated sludge reactor on a continuous basis. The higher the concentration of the return activated solids (RAS), the lower the required return flow rate. A well designed clarifier (provided the biological system is functioning optimally) will generate highly thickened solids (typically 7 to 12 kgTSS/m^3) for the return solids stream. If the thickening performance of the clarifier is not optimal, higher return flows are required for day-to-day operation, which increases energy input into the clarifier through the sludge removal mechanism and the increased MLSS input through the higher recycle flow. If the sludge return flow is too high, there is an increased chance of sludge blanket instability (see next paragraph).

12.1.2.3 Sludge storage in secondary settlers

In typical activated sludge systems, most of the sludge at any one time is in the bioreactors, but there is a continuous sludge mass exchange between the reactors and the clarifier. A sudden increase of influent flow or alternatively decrease of sludge compactibilty will shift some of the sludge in the reactor to the clarifier, producing an elevated sludge blanket. It will take time (and potentially operator intervention) to return the sludge mass stored in the clarifier blanket to the reactor. The sludge storage functionality of the clarifier captures solids and retains it until the sludge return mechanism can cope with the temporary overload.

There are clarifier configurations and operational strategies that explicitly use the sludge storage functionality of a clarifier (see for example blanket filtration in a vertical flow clarifier). This can be beneficial for the effluent quality, but good design and careful operation is required to prevent accidental scouring of the blanket and high effluent solids concentrations.

Figure 12.1 Final settlers are one of the most visible structures on Earth from space (photo: D. Brdjanovic).

12.2 SETTLING TANK CONFIGURATIONS IN PRACTICE

Well-designed final settling tanks should provide quiescent, slow moving conditions for the wastewater they process to achieve the best clarification and compaction possible. At the same time, economic drivers point to opposite objectives; these units are large with significant construction cost and the land they occupy is potentially expensive. The flow pattern inside a clarifier plays an important role in enhancing or hindering the performance of a clarifier. The flow pattern is a consequence of the shape of the clarifier, the position and configuration of the inlet and effluent structures, sludge removal mechanism and internal baffling. Only the general features of the three most popular clarifier types (radial flow, horizontal flow and vertical flow) will be discussed here. Research and operational experience at full scale facilities have shown there is no significant difference in the performance of well designed clarifiers irrespective of their shape. The decision of choosing one or the other is usually not process performance driven. Space, manufacturer or other engineering considerations will determine the best option for a certain location. For example, if space is limited, rectangular clarifiers with common wall construction may be more suitable. Matching existing units or simplifying operations (using the same clarifiers with which operators are already familiar) may also be an important consideration.

12.2.1 Circular clarifiers with radial flow pattern

One of the most popular clarifier shapes, due to its simpler mechanical design, is the circular clarifier

(Figure 12.2). Feed and effluent collection structures can be placed in different locations, but the flow pattern in these units in general is radial, leading to higher linear velocities at the centre of the clarifier, tapering off towards the perimeter. Mixed liquor is usually fed into a flocculation or stilling well in the centre. The design of the well should aid flocculation of the sludge mass. The stilling well ports opening into the clarifier help dissipate the energy of the liquid, providing quiescent conditions in the clarifier. The mixed liquor from the reactor, due to its higher density, will usually flow towards the perimeter just above the bottom on top of the sludge blanket as a 'density current'. Frequently a doughnut shaped recirculation pattern develops with liquid flowing back towards the centre close to the surface. Control of flow patterns in clarifiers is usually achieved by placing flow diversion structures, baffles close to the inlet or outlet points. Effluent is removed through a V-notch weir, and the overflow is collected in a trough. The weir should be constructed level such that the flow is uniformly distributed along the whole length of the weir. Sludge settled onto the tank floor is collected by the sludge collection mechanism and is removed from the sludge hopper.

Figure 12.2 Circular clarifier (photo: EnviroSim).

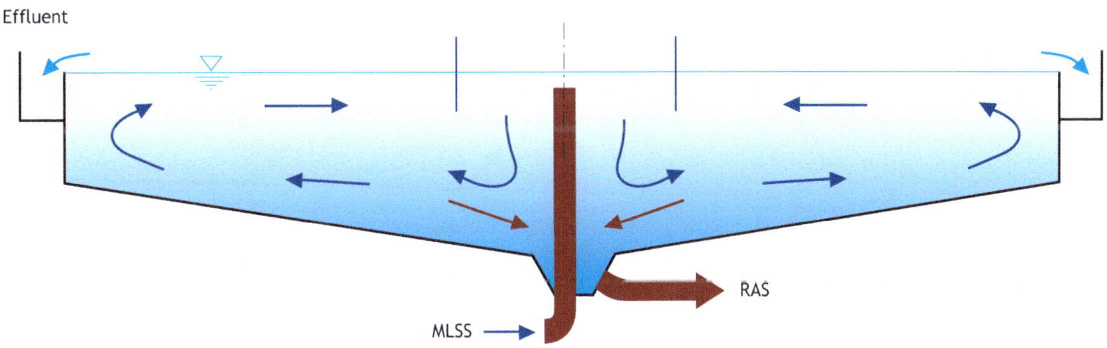

Figure 12.3 Conceptual illustration of a circular settling tank with radial flow pattern.

The two most typical sludge collection mechanisms are scrapers (plows) and suction. Utilizing the advantage provided by the circular shape, both are rotated a few times per hour by a peripheral drive which is rolling on the clarifier outer wall.

A conceptual illustration for a specific, centre-fed, peripheral effluent launder system is shown in Figure 12.3. Black arrows represent general flow pattern in the tank, brown arrows indicate the direction of sludge flow, and blue arrows are used to show clarified effluent flow removal through the launders.

For detailed design of each clarifier component, hydraulic, structural and other types of calculations are necessary, which are beyond the objective of this text.

12.2.2 Rectangular clarifiers with horizontal flow pattern

Rectangular clarifiers can be built sharing common walls which lead to better use of the available area. Therefore large plants are usually designed with rectangular clarifiers (Figure 12.4).

Figure 12.4 Rectangular clarifier (photo: D. Brdjanovic).

Similar to circular clarifiers, feed and effluent collection can be in different locations, but the flow pattern in general is horizontal. A conceptual illustration for a rectangular system is shown in Figure 12.5. In this example, mixed liquor is fed in the inlet and effluent is removed on the opposite side resulting in longitudinal flow. The density current and a recirculation pattern exist, and usually are attenuated using internal baffles. Effluent is collected through overflow weirs in launders. Sludge removal, which in this example is counter-current with respect to the clarified liquid flow, could also be concurrent or cross-current. The sludge removal mechanism is usually a mechanical scraper or flights.

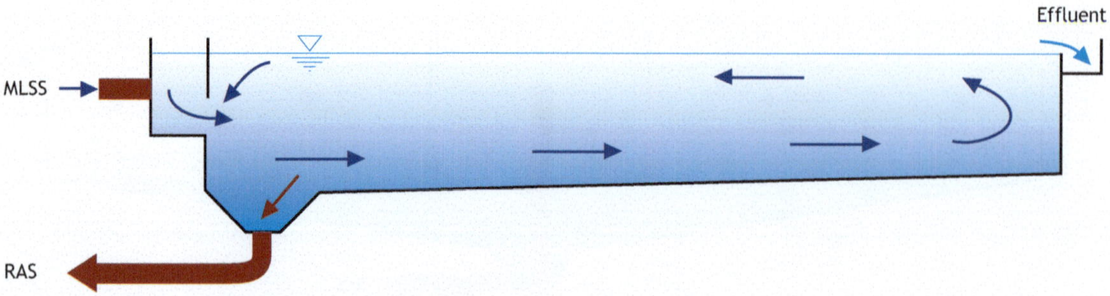

Figure 12.5 Generalized rectangular settling tank with horizontal flow.

12.2.3 Deep clarifiers with vertical flow pattern

If the clarifier is relatively deep compared to its diameter, the flow pattern becomes predominantly vertical. This design is mostly used in Germany (*e.g.* the Dortmund tank). A conceptual drawing shows the main characteristics of these structures (Figure 12.6).

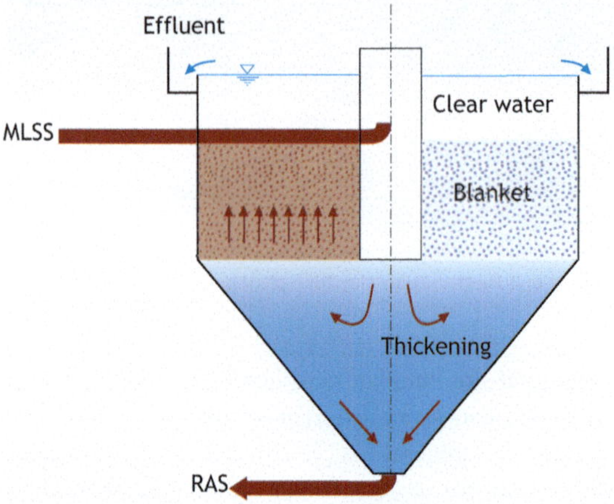

Figure 12.6 Vertical–flow settler (Dortmund tank).

A unique feature of the deep clarifier is the blanket filtration – the MLSS is introduced below the sludge blanket, and while flowing through it in the vertical direction, slightly fluidizes the blanket. In this process fine particles are captured and filtered out. Therefore this clarifier type frequently produces clear effluent with low solids concentration independent of hydraulic loading, as long as the blanket does not expand up to the weirs.

12.2.4 Improvements common to all clarifier types

From the many engineering solutions developed to improve the performance of clarifiers, four items are discussed here with a short explanation: flocculation wells, scum removal, baffles and lamellas placed in settlers.

12.2.4.1 Flocculation well

Not all clarifiers are designed and built with flocculation wells. MLSS can also enter the clarifier through flow stilling openings or gaps directly from the feed pipe.

Figure 12.7 Flocculation well (photo: Brown and Caldwell).

A properly designed flocculation well may significantly reduce the effluent solids concentration by promoting flocculation, since the sludge is under high shear conditions in the aerated reactor and floc break-up is likely. Typical design values for a flocculation well are 20 minutes hydraulic residence time and a mean velocity gradient (G) value of 15 s^{-1}. A flocculation well is shown in Figure 12.7 (the ports on the feed pipe where mixed liquor enters the well are visible).

12.2.4.2 Scum removal

Scum floating on top of the clarifier, if not removed, may deteriorate effluent quality (Figure 12.8).

Figure 12.8 Scum on a clarifier (photo: Black and Veatch).

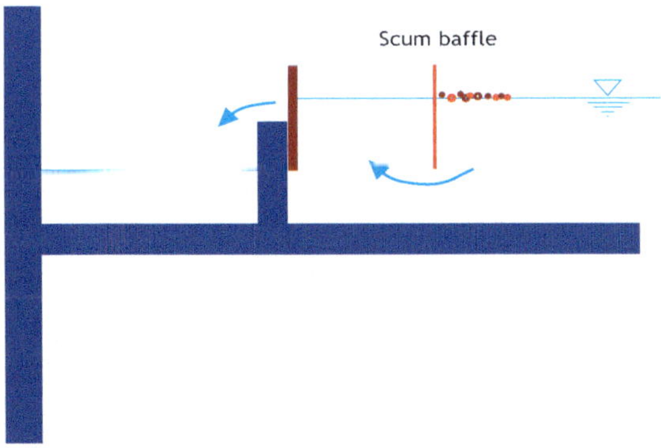

Figure 12.9 Scum baffle (Black and Veatch type).

Most clarifiers are fitted with scum baffles (an example design is shown in Figure 12.9) and various scum removal mechanisms.

Scum can have different origins, mechanical or biological in nature. It may be lighter debris that was not removed earlier in the wastewater treatment plant, lighter biological solids due to gas content (*e.g.* entrapment of nitrogen gas due to denitrification in the sludge blanket), or biologically produced foam (*e.g.* Nocardia). Biological foam usually indicates an operational problem with the activated sludge reactor itself and is best removed before the clarifier. Recycling foam-forming micro-organisms into the biological system may worsen the situation. Scum should be removed from the water stream as soon as possible and destroyed, with the minimum of water usage.

12.2.4.3 Baffles

Baffles are flow diversion and energy dissipation elements, which can be solid or equipped with slits or openings. They are usually placed in the secondary settler close to the inlet or outlet, though baffles have been employed in the middle of tanks also to reduce the density current effect. Baffles are discussed more in detail in the 12.6 section along with CFD modelling.

12.2.4.4 Lamellae

When discrete particles settle slowly through the clarified liquid in a clarifier, their travel distance to reach the sludge blanket, against the flow of water, is significant. Lamellae, slanted tubular or plate structures placed in the clarified zone in the settler, will reduce this travel distance to a few centimetres. Solids will settle to the slanted surface and slide down as one mass, rather than as individual particles. In spite of this, lamellae are not widely used in secondary settlers, due to the potential of sludge accumulation inside slanted surfaces of the tubes or plates, resulting in cleaning requirement and in extreme cases, plugging of the lamella or producing anaerobic sludge clumps with the potential to float.

12.2.5 Operational problems

A wastewater treatment plant is a complex, large-scale industrial environment and properly documented operating procedures need to be followed. This is equally valid for the operation of settlers. A significant part of the maintenance and operation effort is spent on the proper upkeep, cleaning and lubrication of the equipment so that it can perform through its design lifetime with the expected performance. These mechanical issues are not considered in detail in this chapter, rather the focus is on process engineering.

Many operational problems are caused by faulty design, and can only be corrected through a redesign and reconstruction/refurbishment/upgrade project. Other problems in operation require changes that are within the normal operating procedures and can be implemented by plant management, maintenance or operations.

12.2.5.1 Shallow tanks

Due to site conditions or by the designer's choice, settlers are sometimes designed with less than 2.5 – 3.0 m side water depth. These settlers are more susceptible to solids overload, since the sludge blanket can easily get close to the effluent weirs (*e.g.* due to increased loading, deterioration of sludge settling properties or insufficient recycle flow). This can lead to scouring (loss of solids) from the sludge blanket or gross clarifier failure due to the sludge blanket reaching the effluent weirs. Maintaining a low sludge blanket via higher recycle rates is important in shallow clarifiers.

12.2.5.2 Uneven flow distribution

Wastewater treatment plants usually have several final settler units, either a) because the area required is too large and it is mechanically not feasible to construct only one unit, or b) to provide redundancy for maintenance procedures or mechanical failure, such that one clarifier can be taken out of service and emptied while the plant is still operating using the other units. Flow and solids distribution between different clarifiers can sometimes be significantly uneven, due to construction or operational problems. This can lead to one of the settlers processing more flow or producing a more turbid effluent with higher suspended solids concentration than the others. Sometimes visual inspection of flow through weirs can provide a simple clue to hydraulic loading differences. Flow imbalances can frequently be corrected by measuring flows around individual units, correcting uneven wastage (maintaining a more uniform MLSS in different activated sludge units), and adjusting gates in splitting boxes. Splitting flows evenly by simply building the channel with levelled walls does not work in practice.

12.2.5.3 Uneven weir loading

Improperly levelled weirs can cause significantly more flow at a certain segments of the clarifier, creating a localized upwelling and potential to lift or scour the sludge blank et under the affected part. The weir in these cases should be re-levelled which can lower effluent solids concentration at the same loading conditions. This is one of the reasons for using steel V-notch weirs attached to the concrete overflow channel so that the V-notch weir can be fitted precisely with levelling equipment.

12.2.5.4 Effect of wind

Strong, consistent wind can affect the flow circulation pattern in a final settler (Figure 12.10). Particularly large diameter circular tanks are prone to wind exposure. The altered circulation pattern may result in uneven weir loading and elevated suspended solids concentration in the effluent, even if the weirs are perfectly levelled. Wind direction should be taken into account during design of scum removal equipment. If site conditions allow, fences or hedges can reduce the effect of wind.

12.2.5.5 Sudden temperature changes

In certain geographic locations with large daily temperature variation or strong exposure to sunlight temperature induced inversion can occur in the final settler. Settled sludge is normally slightly heavier than wastewater, therefore it occupies the bottom of the settling tank. A significant temperature difference

Figure 12.10 Effect of wind (photo: D. Brdjanovic).

between the water on the surface of the clarifier and the sludge on the bottom can cause an unexpected layer inversion, resulting in flotation of the sludge blanket. This could be due to drastic cooling of the surface on a cold night, warmer sludge entering the settler during daytime operation, or warming of the sludge layer due to direct sunshine. If the problem is persistent, only a cover of the majority of the area or a full enclosure may be able to remediate the situation.

12.2.5.6 Freezing in cold weather

In cold climates, in spite of the natural heat content of wastewater, ice build-up can occur on exposed surfaces. This is particularly prevalent in plants with aeration tanks using surface aeration systems (*e.g.* mechanical rotors) that can decrease the temperature of the wastewater. In these places, covered reactor and settler structures, as well as diffused aeration which transfer some heat to the liquid stream should be considered.

12.2.5.7 Recycle problems

In settlers using suction pipes (organ pipes) for sludge collection and recycle, loss of suction, usually in the longest pipes, can be a problem. In these cases the recycle system has to be run at a high flow which is not always desirable, or the suction pipes can be replaced by scrapers.

12.2.5.8 Algae on weirs

Algae growth will occur on weirs and other exposed places on the settler structure due to natural sunlight and the phosphorus content of the effluent (Figure 12.11). In addition to being unsightly this can have a periodic negative effect on effluent quality or interfere with equipment operation such as scum removal troughs. Algae can be manually cleaned off weirs. Some clarifier designs include brushes which continuously wipe exposed structures. Other solutions include chlorination introduced locally around the weir, or covering the tank.

Figure 12.11 Algae on weirs (photo: D. Brdjanovic).

12.2.5.9 Anaerobic clumps

If dark brown or black, anaerobic floating clumps appear regularly on the surface of the final settler tank, it is possible that the sludge collection mechanism is misaligned or chipped, leaving uncollected sludge masses on the floor. These dense sludge pockets turn anaerobic and due to methane gas generation, float to the surface. Tank cleaning and inspection may be necessary to determine the cause of the floating debris.

12.2.5.10 Birds

In certain areas seagulls and other birds can populate final settlers to the extent that their waste and feathers can noticeably contribute to effluent nutrient loading, as well as hampering the proper operation of equipment or increasing cleaning and maintenance costs. Covered settlers do not have this problem. Another solution can be to stretch visible lines across the clarifier which do not interfere with operation but annoy the birds sufficiently to leave the area.

12.2.5.11 Bulking sludge

Bulking sludge does not compress to the same high concentration levels on the bottom of final settlers as normal sludge. This condition is so frequent and so difficult to overcome that a whole chapter is devoted to it (Chapter 10).

12.2.5.12 Rising sludge

Heterotrophic micro-organisms present in the MLSS continue their metabolism even after the sludge was transferred to the clarifier. If enough substrate and nitrate is available, nitrogen gas, in the form of microscopic or visible bubbles can be generated through denitrification. Gas bubbles attaching to the flocs can drastically reduce their density, resulting in sludge rising (floating) to the surface. This condition can be alleviated by proper treatment, primarily a high level of denitrification in the process, so that only low levels of residual substrate and nitrate are present in the outflow from the reactors.

12.3 MEASURES OF SLUDGE SETTLEABILITY

Biological sludges, depending on their source, history, composition, density and ability to flocculate, do settle and compact differently, and this characteristic should be taken into account during design and operation of clarifiers. Consequently, several measurement methods are used to quantify sludge settleability. The methods can be divided into two categories: (*i*) providing information on settling velocity, and (*ii*) providing information on compactibilty. The most frequently used methods are briefly described below. For performing an actual test, the reader is referred to the relevant standard which contains a detailed description of all experimental circumstances that have to be taken into consideration to produce reliable results that can be compared with other measurements of the same method.

12.3.1 Sludge Volume Index

The most common measurement for plant operations, due to its simplicity, is the Sludge Volume Index, SVI (called Mohlmann Index in several countries). In this test (APHA *et al.,* 2006), a sludge sample is taken in a 1l graduated cylinder, and after initial mixing, left to settle for 30 minutes. The concentration of the sludge is measured from the MLSS test. SVI is calculated taking the volume (in millilitres per litre in the 1 litre cylinder) that the sludge blanket occupies after 30 minutes and dividing by the MLSS concentration of sludge in grams per litre that is in the test cylinder, so essentially the SVI (unit ml/g) describes the volume that 1 gram of sludge occupies after 30 minutes settling under the test conditions.

The SVI test in its original form is the simplest settling test that can be performed but it has several drawbacks. 30 minutes is an arbitrary point on the settling curve, therefore the results are variable. Sludges which settle and compact fast or tests done using lower operating MLSS concentrations are mostly finished settling in 30 minutes and the SVI in these cases is an indicator of the compactibilty of sludges. At higher MLSS concentrations or in the case of slower settling sludges, the settling process is not finished at 30 minutes and in these cases the SVI is an indicator of the settling velocity of the sludge sample. At high MLSS concentrations the test fails – for example, an SVI higher than 150 ml/g cannot be measured if the concentration in a 1l (1,000 ml) cylinder is higher than 1,000/150 or 6.7 g/l. Wall effects can also change the measured SVI since the small cylinders used for the test have a very high wall area to volume ratio compared to full scale clarifiers.

12.3.2 Other test methods

Several methods have been developed to improve and standardize the results of the SVI test. The diluted SVI (DSVI) test requires dilution of the sludge sample with effluent such that the settled volume after 30 minutes falls into the 150-250 ml range. This test avoids the problem of high MLSS concentrations, and can be used as a better indicator of potential filamentous bulking (if DSVI>150 ml/g). The stirred SVI test performed at 3.5 g/l MLSS concentration ($SSVI_{3.5}$) is a further improvement. This test is used as the standard in several countries (*e.g.* in the UK), though it requires a more complex experimental setup. The MLSS is always diluted or concentrated to the same 3.5 g/l, and the settling vessel (which usually has a volume of 5l and a diameter of 120 mm) is slowly stirred at 1-2 rpm. The results of this test are more reproducible than those of the SVI or DSVI tests.

There have been many calls from industry leaders to standardize settling test protocols in the literature over the last 20 years, since this would improve the quality of the data collected and the operation of secondary clarifiers. In spite of all these efforts and all the deficiencies of the SVI test, it is still in widespread use in its original, simplest form.

The Zone Settling Velocity (or Stirred Zone Settling Velocity) measurement is a test designed to provide information on the settling velocity of a sludge sample by recording the actual subsidence rate of the sludge interface at a certain MLSS concentration. Since this test is frequently used in the context of the flux theory, it is described in Section 12.4.1. Other fully automated, continuous test methods (*e.g.* settlometer) are also available.

12.4 FLUX THEORY FOR ESTIMATION OF SETTLING TANK CAPACITY

Solids flux is a special form of mass flow rate - the mass of solids transferred through a unit area per unit time (expressed for example in kg/m^2/h). Flux theory describes the various solids fluxes that affect solids transport in a clarifier, and is used, among other methods, to estimate clarifier area required and operational parameters such as recycle (return or under-flow). This chapter will briefly summarize the sludge settleability measurement method used and the mathematical background of flux theory, as well as other design methods used in practice.

12.4.1 Zone Settling Velocity test

Zone Settling Velocity (ZSV) is the subsidence rate of the solids interface (sludge blanket), as measured in m/h or similar units in a test vessel according to Standard Methods (method 2710 D). The test is called the Stirred Zone Settling Velocity (SZSV) test if the vessel is equipped with a slow stirring mechanism. The SZVS test provides a more accurate measure of sludge settleability, in this case, zone settling velocity, than other settleability measures like the SVI, DSVI and SSVI tests. An illustration of the progression of the test in time is shown on Figure 12.12 (for clarity, an unstirred cylinder is pictured, but the standard ZSV test is required to be stirred at 1-2 rpm).

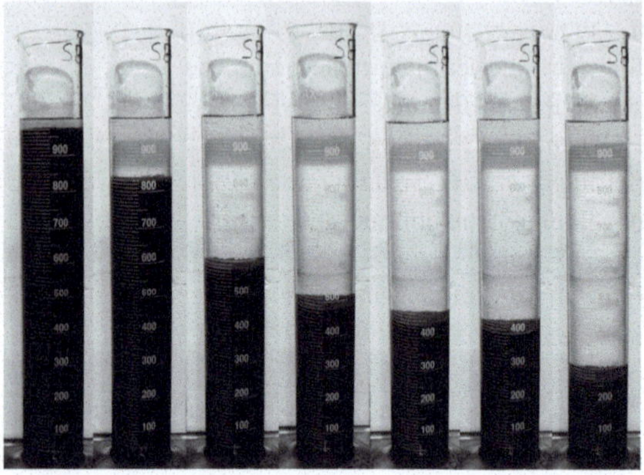

Figure 12.12 Progression of a zone settling velocity test (at 1, 2, 4, 6, 8, 10 and 45 minutes), (photo: Environment Canada).

In this case, an MLSS mixture of 5,400 mg/l (5.4 kg/m^3) was placed in the cylinder, and snapshots of the same cylinder taken at 1, 2, 4, 6, 8, 10 and finally at 45 minutes were pasted together. The interface height is recorded and plotted as a function of time, producing a plot similar to Figure 12.13.

The curve in Figure 12.13 can be divided into three distinct sections:

1) Lag phase lasting a minute or two at the beginning of the test. This is due to acceleration of the sludge particles making up the interface and dissipation of the mixing energy from filling the cylinder.
2) Linear section usually lasting 3-30 minutes, depending on MLSS concentration and height of the settling column. The slope of this linear section gives the Zone Settling Velocity (ZSV) of the sludge at the concentration with which the column was filled. When the column is stirred as is required for the standard settling velocity test, the slope of the linear section is the stirred zone settling velocity (SZSV) at the concentration with which the column was filled.
3) The onset of compression from the bottom of the cylinder results in a gradual decrease in settling velocity following the linear phase.

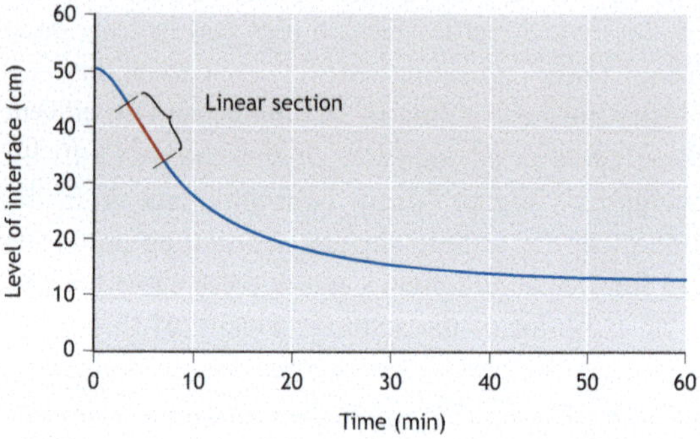

Figure 12.13 Interface height as a function of time in a ZSV test.

A series of SZSV tests at different MLSS concentrations are usually performed to measure sludge settling velocity used in flux theory. The theory and its application are described below. The lowest MLSS concentration where the test can be performed successfully is between 1 and 1.5 g/l. At low concentrations it may be difficult to determine the location of the sludge blanket (also called zone settling) as shown in Figure 12.14.

Also, with dilute concentrations, the subsidence rate is high, and onset of compression from the bottom may start within a few minutes. On the contrary, at high MLSS concentrations, the interface is well defined but depending on the concentration and the compactibilty of the sample, the linear zone settling phase may be overwhelmed by compression very early in the test.

Figure 12.14 Difficult interface (photo: Environment Canada).

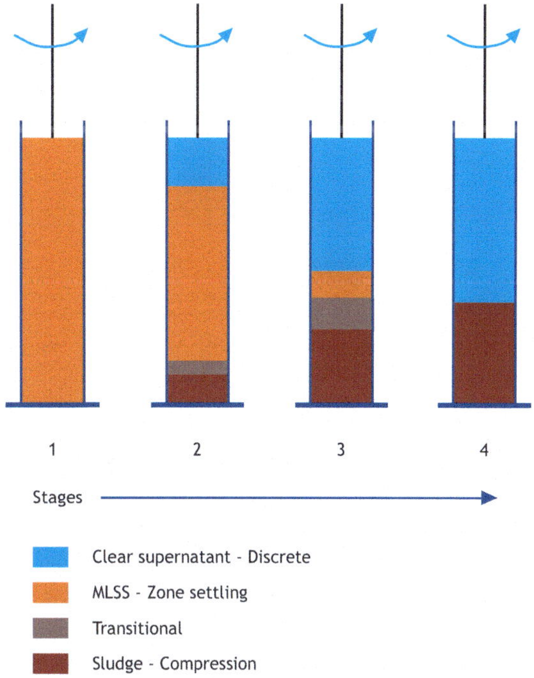

Figure 12.15 Progression of ZSV test.

12.4.2 Discrete, flocculent, hindered (zone) and compression settling

A suspension containing independent discrete non-flocculating particles (of the same type, *e.g.* sand) settles at the same velocity irrespective of its concentration according to Stokes's law. The settling velocity of the individual particles will depend only on their shape, size (diameter) and density. An MLSS mixture from an activated sludge reactor behaves very differently compared to such a sand suspension, primarily due to its flocculent nature. At the beginning of an SVI or Zone Settling Velocity test, the sludge comprising large and small organic flocs and inorganic particles will start to settle as a coagulated matrix (stage one in Figure 12.15) and the settling velocity will be strongly influenced by the MLSS concentration.

This is called the Zone Settling phase and it is maintained through stages 1 to 3 in Figure 12.5 until there is no more blanket left at the original MLSS concentration – it subsides through a transitional region into the compression zone (Stage 4). Compressive settling is distinctly different from zone settling – the particles support each other and compression is achieved by water squeezing out from the sludge matrix. Settling velocity is no longer a function of the sludge concentration, but depends on interstitial pressure, and the compressibility and permeability of the sludge.

12.4.3 The Vesilind settling function

If the interface level is plotted in a SZSV test (Figure 12.13), the zone settling stage can be distinguished by the linear phase where the sludge blanket settles with a constant velocity. The test is repeated at several concentrations as shown in Figure 12.16.

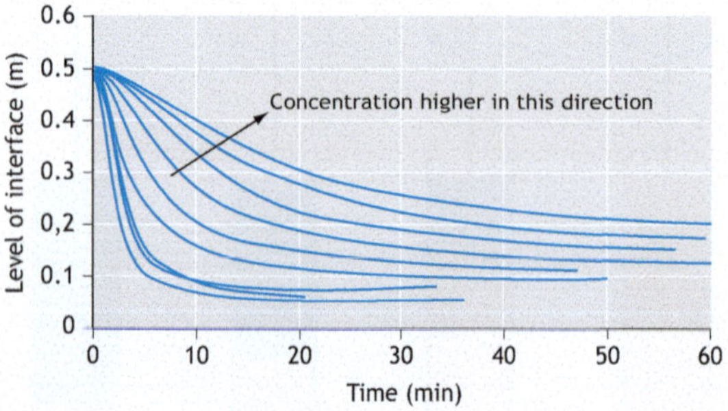

Figure 12.16 Results of SZSV tests at different MLSS concentrations (data: Environment Canada).

The initial lag phase and the linear portion is clearly visible on the plot. Tests performed with low concentration MLSS show high settling velocity, while those with higher concentration samples are settling slower. The extracted zone settling velocities (in m/h) can be plotted as a function of the MLSS concentration (mg/l), resulting in a curve that can be very well approximated by an exponential function (Figure 12.17). This function, called the Vesilind function, is of the form:

$$v_S = v_o \cdot e^{-p_{hin} \cdot X} \qquad (12.1)$$

where:

- v_S settling velocity (m/h)
- v_0 initial settling velocity (m/h), an extension of the curve to the zero concentration intercept
- p_{hin} hindered settling parameter (l/g or m^3/kg)
- X MLSS concentration (g/l or kg/m^3) in the various ZSV tests

It is important to observe that the initial part of the curve in the low concentration region (0 to approximately 1,500 mg/l on Figure 12.17) is purely a mathematical extension of the measured data points and in reality zone settling velocity cannot be extended or measured in that concentration range.

If the same data points from Figure 12.17 are plotted in a semi-log representation, the exponential function is transformed into a linear plot (Figure 12.18) and the slope gives p_{hin} while the intercept is $\ln(v_0)$.

The error distribution of the measurements is different in the original linear and the semi-logarithmic representation of the Vesilind function. From a practical standpoint though, there is not much difference between extracting the v_0 and p_{hin} parameters directly from the exponential curve or its linearized form. In the former the regression is done on the v_s values directly, and in the latter on the $\ln(v_s)$ values. It is assumed that most evaluations these days will be performed using spreadsheet functions (*e.g.* Excel Solver) as opposed to direct graphical methods widely used historically. Slightly different values may be obtained in the two different representations – however the accuracy of the results will mostly depend on the proper reading of the interface during the actual SZSV tests and the selection of the linear region from the interface height – time curve.

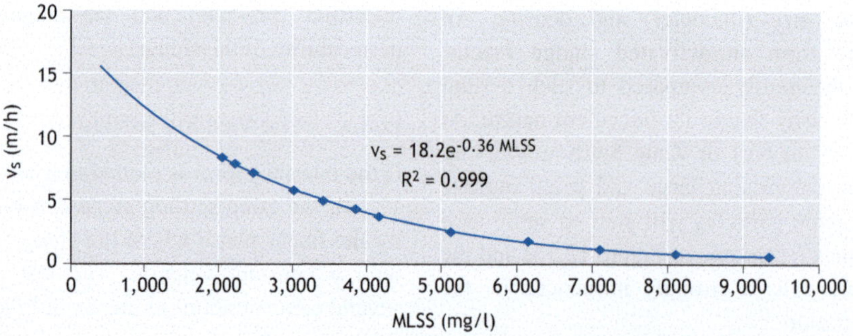

Figure 12.17 The Vesilind relationship between settling velocity and test MLSS concentration (data: Environment Canada).

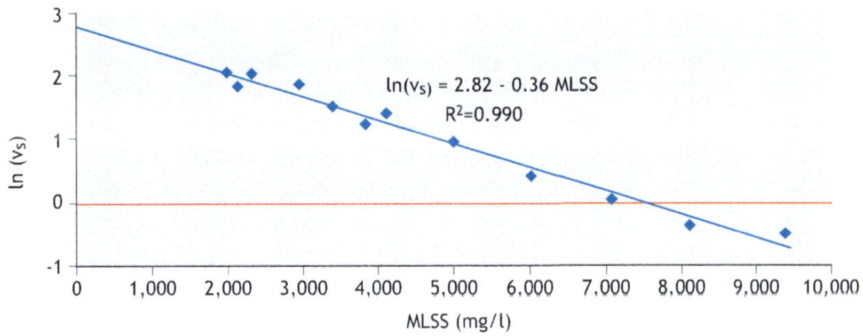

Figure 12.18 The Vesilind function plotted on semi-log scale (data: Environment Canada).

12.4.4 Gravity, bulk and total flux curves

Gravity (settling) flux (JS) is the mass of solids transported under the influence of gravity induced settling, and can be calculated as the product of the settling velocity (v_s) and the solids concentration (X):

$$J_S = v_s \cdot X \tag{12.2}$$

where:

J_S	gravity flux (kg/(m²/h))
v_s	settling velocity (m/h) at X concentration (from Eq. 12.1)
X	solids concentration (kg/m³)

The gravity flux for the previous, well settling sludge with an SVI of about 48 ml/g is plotted in Figure 12.19. The gravity flux curve has a maximum usually at 2 to 3 kg/m³ concentration. Below this concentration the flux decreases due to low solids concentration, while above, it decreases due to reduced settling velocity at higher concentrations.

At a fixed recycle flow the bulk flux is linearly proportional to the solids concentration, X (Figure 12.20), *i.e.* the higher the X, the higher the solids flux to the bottom of the settler generated by the underflow rate.

The total flux transporting solids to the bottom of the clarifier is the sum of the gravity and bulk fluxes (Figure 12.21).

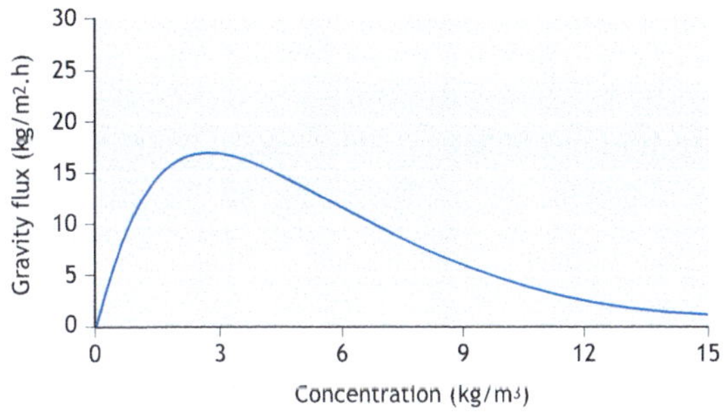

Figure 12.19 Gravity (settling) flux.

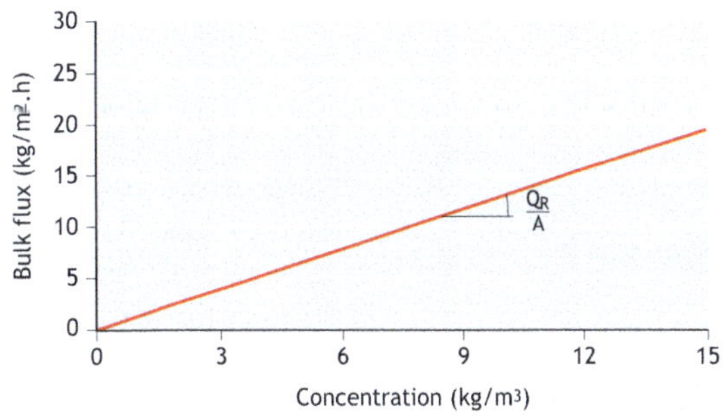

Figure 12.20 Bulk solids flux.

The total flux, for a particular underflow rate, has a minimum at a certain concentration called the "limiting concentration", X_L. If the concentration is lower than the limiting concentration, the increase in settling flux more than compensates for the decrease in bulk flux, while at higher concentrations the increase in bulk solids flux more than compensates for the decrease in settling flux. The layer in the settler that is at the limiting concentration will be the bottleneck because it transports the lowest solids flux, j_L, to the bottom in the whole concentration range. The concentration range spans from the low concentration of the feed concentration to the high concentration of the return sludge.

$$j_L = j_{S(X_L)} + j_{B(X_L)} \tag{12.3}$$

When the applied flux (the mass of solids applied per unit settler area (kg/m²/h)) exactly matches the limiting flux, at that point the clarifier is critically loaded or it is at the point of failure. This condition usually has to be satisfied for peak wet weather flow (PWWF) conditions in settler design. After finding the limiting concentration (X_L kg/m³) from the minimum of the total flux curve, the minimum or limiting flux (j_L kg/m²/h) can be determined.

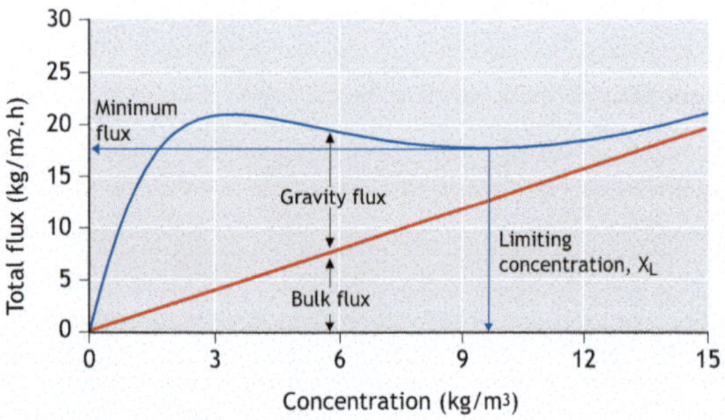

Figure 12.21 Total solids flux.

12.4.5 Solids handling criteria limits of the clarifier

Consider the total flux as depicted in Figure 12.21. If the underflow rate is increased, the bulk flux and the total flux curves will be rotated counter-clockwise around the origin and at a certain underflow rate the local minimum (and therefore the limiting concentration) will disappear (Figure 12.22).

This underflow rate is called the critical underflow ($q_{R,crit} = Q_R/A$ in m/h), and it defines the lowest limiting concentration that can be determined ($X_{L,min}$). The maximum flux that can be transported to the bottom of which the clarifier or maximum solids handling ability of a clarifier are determined by two different criteria, depending on whether the underflow rate is below or above this critical value.

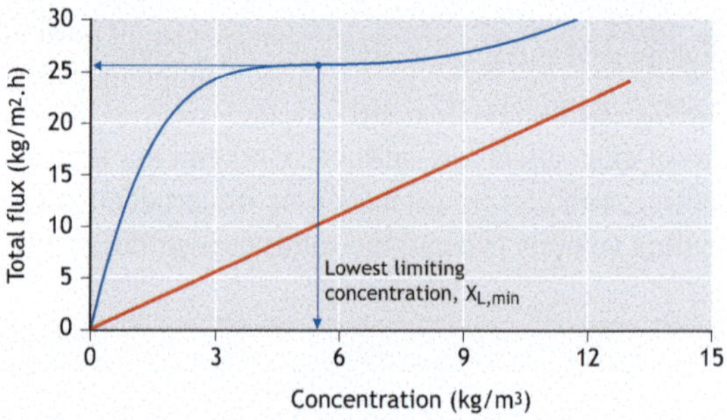

Figure 12.22 Total flux curve at critical recycle flow.

12.4.5.1 Solids Handling Criterion I – minimum solids flux limiting

At underflow rates below the critical underflow rate, a minimum flux at the limiting concentration can always be determined.

This situation is plotted in Figure 12.21. The applied flux to the clarifier has to be less than this minimum flux for the clarifier to be under loaded. The clarifier is critically loaded if the applied flux is equal to the minimum flux. If the flow (PWWF) and MLSS concentration is given, the area and recycle ratio must be selected such that the resulting total flux is equal to or less than the minimum flux.

12.4.5.2 Solids Handling Criterion II – applied flux (overflow rate) limiting

At underflow values above the critical underflow, a limiting concentration cannot be found (Figure 12.23).

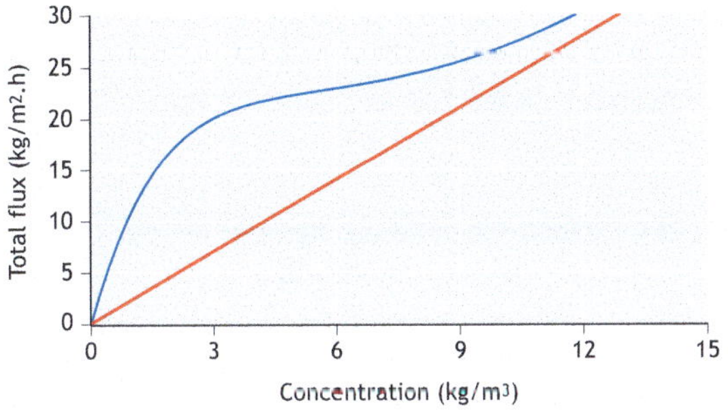

Figure 12.23 Total flux curve without limiting concentration (underflow larger than critical).

In this condition the applied flux (solids loading) to the clarifier must be less than the gravity flux at the feed concentration or equivalently, the applied overflow rate (hydraulic loading, m/h) must be less than the zone settling velocity of the sludge at the feed concentration (MLSS).

$$q_I = \frac{Q_{PWWF}}{A} = v_{s,MLSS} = v_0 \cdot e^{-p_{hin} \cdot MLSS} \tag{12.4}$$

The following sections describe in more detail how the flux theory is applied for clarifier design, specifically establishing required area and minimum recycle ratio.

12.4.6 State Point Analysis

State Point Analysis (SPA) is a convenient visual way to determine the operating condition of the clarifier. SPA is based on solids mass balances around the clarifier expressed graphically. The method inherently contains simplifications such as (*i*) it is based on steady-state conditions, (*ii*) only one (vertical) dimension is considered, no short-circuiting or the details of the sludge withdrawal mechanism is accounted for (*iii*) effects such as compression are not considered (*iv*) effluent solids are neglected. In spite of these simplifying assumptions, SPA is frequently used in pre-design to establish clarifier area and return pump capacity, and in operation, to estimate maximum MLSS and required return flow settings before fine-tuning the plant's operation based on actual performance.

The State Point Diagram (Figure 12.24) is constructed representing various fluxes in a clarifier as a function of solids concentration.

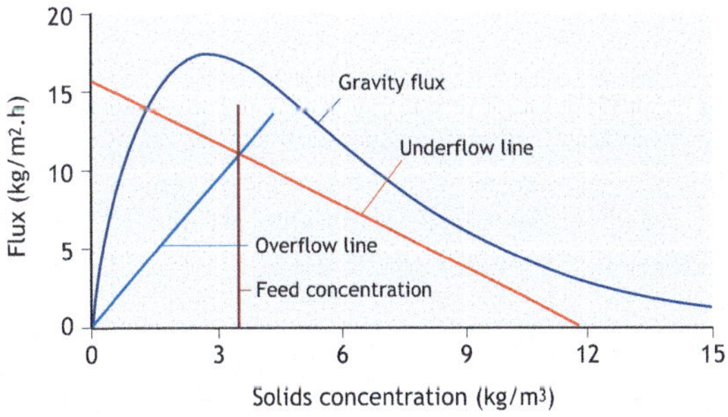

Figure 12.24 State Point Diagram.

The State Point Diagram is based on the gravity (settling) flux curve from Figure 12.19. This curve requires only the two Vesilind constants (v_0 and p_{hin}), to be known. Superimposed on the gravity flux are the 'overflow', 'underflow' and 'feed' lines. The overflow and underflow lines both represent the solids fluxes applied to the clarifier generated by the overflow and underflow rates, in the same units as the gravity flux, but the definition of their vertical (y) axes is different.

The overflow line represents the flux that the overflow applies to the clarifier (in the opposite direction compared to the gravity flux).

$$J_I = \frac{Q_I}{A} \cdot X_F \tag{12.5}$$

The slope of the overflow line is the applied hydraulic loading, that is $q_I = Q_I/A$ (m/h), commonly referred to as the surface overflow rate (SOR).

Note that the overflow flux is not the actual solids loading that the settler receives. That is called the total applied flux (in Eq. 12.8).

The feed concentration line is a vertical line indicating the feed concentration. This line meets the overflow line at the State Point (operating point). The flux on the y axis at this point is the solids loading rate. The *underflow line* is defined similarly to the overflow line:

$$J_R = \frac{Q_R}{A} \cdot X_F \tag{12.6}$$

However, two transformations are performed to increase the method's usefulness.

1) The line is plotted with a negative slope (since it moves in the opposite direction as the overflow).

2) The underflow line (which, originally, according to Eq. 12.7, starts at zero flux at zero concentration) is shifted upwards such that it starts from the total applied flux at the vertical axis (X=0). The total applied flux (also called the solids loading rate) is obtained by adding the overflow flux and the underflow flux at the feed concentration,

$$J_{AP} = \frac{Q_I + Q_R}{A} \cdot X_F = \frac{Q_I}{A} \cdot (1 + R) \cdot X_F \tag{12.7}$$

where:

R recycle ratio (Q_R/Q_I).

Since the total applied flux is removed from the clarifier at the underflow concentration (assuming zero solids in the effluent), the point where the shifted underflow line crosses the X axis (zero "residual" flux) represents the underflow concentration in an under loaded clarifier. When the overflow line, feed concentration line and underflow line all intersect at the state point, then the solids mass balance over the clarifier is satisfied and all the solids entering the clarifier exit the clarifier via the underflow recycle, provided the state point and underflow line are within the envelope of the gravity flux curve. The most important features and concentrations present in a state point diagram are marked on Figure 12.25.

Where:

- X_F feed concentration (kg/m^3)
- X_R recycle concentration (kg/m^3)
- X_L limiting concentration, (kg/m^3)
- q_I hydraulic loading or overflow rate, Q_I/A, (m/h)
- q_R hydraulic underflow rate, Q_R/A, (m/h)
- j_I overflow rate flux, $Q_I/A \cdot X_F$ ($kg/m^2/h$)
- j_R underflow rate flux, $Q_R/A \cdot X_F$ ($kg/m^2/h$)
- j_{AP} total applied flux, $(Q_I + Q_R)/A \cdot X_F$ ($kg/m^2/h$).

State Point is the operating point of the clarifier, overflow flux at feed concentration.

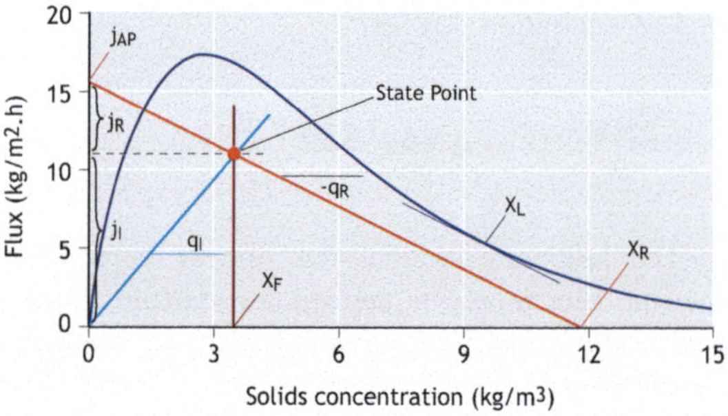

Figure 12.25 Important information about the State Point Diagram.

The position of the state point and the underflow line, relative to the gravity flux curve, determines the operating status of the clarifier, classified by Solids Handling Criterion (SHC) types.

3) If the State Point is above the gravity flux curve, the clarifier is overloaded (SHC II failure). In this condition higher flux is applied than can be processed by the clarifier. This will lead to solids build-up in the clarifier which cannot be maintained on a steady-state basis and will lead to gross solids loss in the effluent.

4) If the State Point lies on the gravity flux curve, the clarifier is at least critically loaded for SHC II, and its status depends on the position of the underflow line relative to the descending limb of the gravity flux curve at higher concentrations:

   - If the underflow line falls below the descending limb of the gravity curve, the clarifier is critically loaded (SHC II critical, SHC I satisfied).
   - If the underflow line crosses the descending limb of the gravity flux curve, the clarifier is overloaded (SHC II critical, SHC I failure).

5) If the State Point lies below the gravity flux curve, the clarifier satisfies SHC II, and its condition will depend on SHC I, the minimum solids flux.

   - If the underflow line falls below the descending limb of the gravity curve, the clarifier is under loaded (both SHC II and SHC I satisfied).
   - If the underflow line is tangential to the descending limb of the gravity curve, the clarifier is critically loaded (SHC II satisfied but SHC I critical).
   - If the underflow line crosses the descending limb of the gravity flux curve, the clarifier is overloaded (SHC II satisfied but SHC I failure).

All potential combinations are presented visually in Figure 12.26, where the blue dot is the state point and the red line is the underflow line. The overflow line (not shown) connects the origin and the state

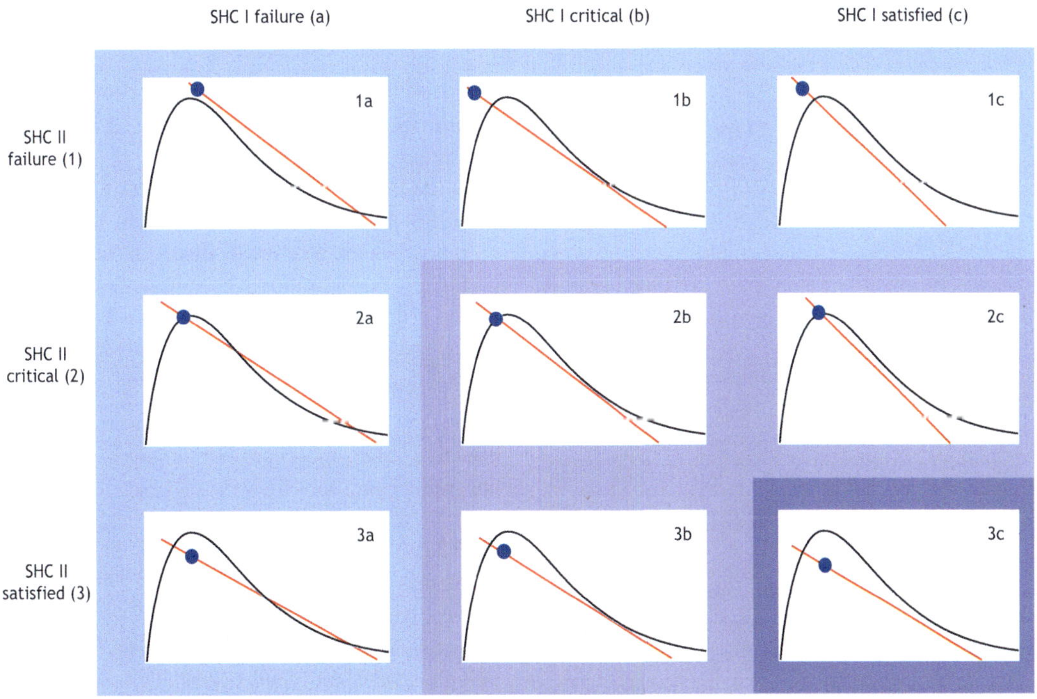

Figure 12.26 State Point Diagrams (overflow and feed line not shown) for different loading conditions.

point. The clarifier has to meet both solids handling criteria in order not to be critically loaded or overloaded. Of the nine cases pictured, 1a, 1b, 1c, 2a and 3a are overloaded, 2b, 2c and 3b are critically loaded, and only one case (3c) is under loaded. The flux theory can be conceptually and graphically expressed in several different ways in addition to the State Point Diagram presented above. These methods are based on the same theory, contain the same Vesilind settling function, and project the same gravity settling flux and overflow, underflow fluxes using different axes. They may be more practical or convenient for a certain design or operational purpose but they will yield the exact same results.

The Ekama Design and Operation (D&O) chart

This chart reorganizes the information available in the flux theory and the State Point Diagram. Overflow rate (Q/A, in m/h) is plotted against the recycle ratio as X axis (Figure 12.27).

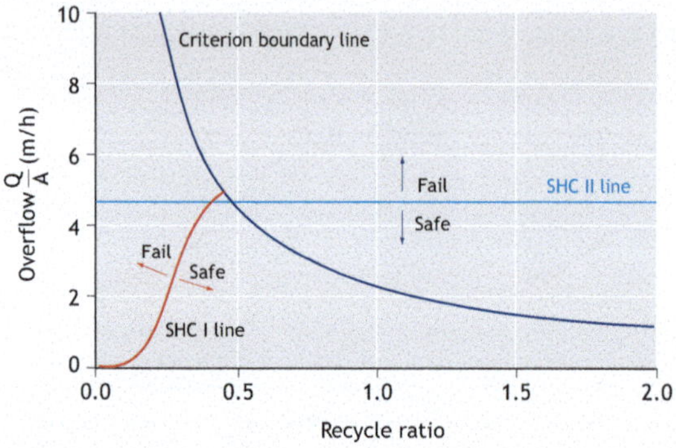

Figure 12.27 The Ekama Design and Operation Chart.

The chart contains three lines that help determine the operating condition of a clarifier based on its hydraulic loading (overflow rate) and recycle ratio.

- *SHC II line.* The straight horizontal line represents the settling velocity at the feed concentration, X_F (based on the two Vesilind parameters, v_0 and p_{hin}), as expressed in Eq. 12.4. In the area above this line Solids Handling Criterion II is not satisfied, the clarifier is overloaded. However, under-loaded conditions are not guaranteed below the horizontal line – that depends on the position of the limiting flux (SHC I) line.

- *SHC I line.* The allowable solids flux according to SHC I increases with increasing recycle ratio. This minimum flux concept was explained earlier on Figure 12.21, and it is equivalent to a series of critical loading conditions, when the recycle line is tangential to the gravity flux of the State Point Diagram. The equation for this line (without detailing the development) is given in Eq. 12.8 and 12.9. If the operating point falls below this line, the SHC I is satisfied.

$$\frac{Q_I}{A} = \frac{v_0}{R} \cdot \frac{1+\alpha}{1-\alpha} e^{\frac{-p_{hin}(1+R)\cdot X_F\cdot(1+\alpha)}{2R}} \tag{12.8}$$

where:

$$\alpha = \sqrt{1 - \frac{4\cdot R}{p_{hin}\cdot(1+R)\cdot X_F}} \tag{12.9}$$

- *Criterion boundary line.* According to the principle illustrated in Figure 12.20, above a certain recycle ratio (R) it is not possible to determine a critical concentration and minimum solids flux. The boundary between lower recycle ratios where a critical flux can be found and higher recycle ratios where it cannot be found is the criterion boundary line. It can be shown that the critical recycle ratio (above which a minimum flux does not exist) is a hyperbolic function:

$$\frac{Q_I}{A} = \frac{v_0}{e^2 \cdot R} \tag{12.10}$$

where:

$q_{R,crit}$ — v_0/e^2 (m/h), and represents the slope of the gravity flux curve at its inflection point, which occurs at double the X values (at $2/p_{hin}$) than the maximum gravity flux (at $1/p_{hin}$).

The nine possible loading cases shown in Figure 12.26 can also be placed on the Ekama D&O chart (Figure 12.28). Only a detail of the whole chart from Figure 12.27 is shown.

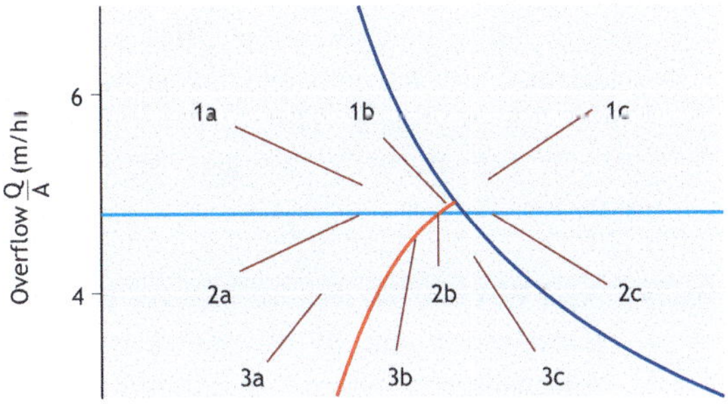

Figure 12.28 Loading examples in the D&O chart.

12.5 OVERVIEW OF THE USE OF FLUX THEORY AND OTHER METHODS FOR DESIGN AND OPERATION

The flux theory is not the only method in use today for settler design. In fact many consulting companies and contractors have their own, experience-based design methods. The use of computational fluid dynamics (see Section 12.6) for detailed settler design is spreading in practice. There are several country-wide design standards as well, for example in the UK (WRC), Germany (ATV, and its new edition, DWA) and the Netherlands (STOWA). These guidelines and design methods are widely used in those and neighbouring countries. The principle of the five most widely used design procedures (including flux theory) is summarized below. The objective is to illustrate a variety of principles that are used in these methods, and not to provide a detailed, step-wise guideline. To implement an actual design procedure the reader must follow the original steps in the referenced guidelines. Since only the most important elements of the design procedures are used in these examples, a clarifier designed with the actual procedure could be significantly different from the results of these examples.

The simplified examples below focus on two specific design parameters: (*i*) Area: how large should the clarifier be to be able to handle peak flows, and (*ii*) Recycle pump capacity: what kind of return pumps should be installed for safe dry weather and wet weather operation.

The methods use relatively simple straightforward algebraic equations, and usually can be calculated by hand or in a simple spreadsheet. To facilitate the use of the book, the user is encouraged to use this spreadsheet[1]. All design methods require the specification of the loading that the clarifier is designed for. In these examples the values as presented in Table 12.1 (Data Tab in the spreadsheet) will be used.

From Table 12.1, the expected diurnal peak in dry weather at this plant will be 1,500 m^3/h and a wet weather peak flow of 2,500 m^3/h has to be accommodated during design. Some design methods account for the mass of solids transferred to the settler during peak flow conditions, resulting in a temporary decrease of the reactor MLSS concentration ($X_{F,PWWF}$).

Various design methods require different measures of the settling properties of the sludge (*i.e.* v_0 and r_{hin}, or SVI, or DSVI, etc.). The following design examples are all based on the same, well settling sludge. The settling properties of this sludge

Table 12.1 Design example – clarifier loading.

Parameter	Symbol	Value	Unit
MLSS at DWF	$X_{F,DWF}$	3.5	kg/m^3
Average dry weather flow	Q_{ADWF}	1,000	m^3/h
Diurnal peaking factor (dry weather)	PF_{DW}	1.5	-
Storm peak factor	PF_{WW}	2.5	-

[1]https://experimentalmethods.org/courses/settling-tests/

sample were measured using all the different required methods, which allow a direct comparison of the methods and the resulting clarifier design. In an actual design case more conservative sludge settling parameters should be taken into account.

12.5.1 Design using flux theory

The Vesilind settling parameters for the sludge were determined in a series of zone settling experiments (Figure 12.16, results in Table 12.2). In the practical application of flux theory, it is recommended that the surface area is increased (the permissible flux is reduced) by a 25% safety factor compared to the theoretical value. This is to account for non-idealities in the real clarifier structure, in contrast to the ideal 1-D approximation as laid out in the flux theory (Ekama and Marais 2004).

The parameters to be used in the design are shown in Table 12.2 (Data Tab in the spreadsheet). No other information (other than Table 12.1 and Table 12.2) is required to use the flux theory for design.

Design steps (in Design Tab in the spreadsheet):

1) Settling velocity at the MLSS concentration (Solids Handling Criterion II) is calculated from Eq. 12.5 ($v_{s,MLSS} = 16.8 \cdot e^{(-0.36 \cdot 3.5)} = 4.8$ m/h).

2) The overflow rate during PWWF must not exceed this velocity, therefore the minimum area required is 523.6 m^2 (= 2,500 / 4.8)

3) The recycle ratio can be selected by two methods (which will provide the same results):

   a) The minimum recycle ratio to satisfy SHC I can be read at the intersection of Lines I and II on the Ekama D&O chart (Ekama D&O Tab in the spreadsheet) (R = 0.45).

   b) Using the State Point Analysis diagram, the recycle ratios (underflow rates) that make the underflow line tangential to the gravity flux curve can be found. This is done both for PWWF (SP PWWF Tab) and PDWF (SP PDWF Tab):

      - At PWWF 0.44 recycle ratio is required (Design tab and SP PWWF Tab in the spreadsheet) resulting in 1,100 m^3/d recycle flow
      - At PDWW 0.31 recycle ratio is required (Design Tab and SP PDWF Tab) resulting in 465 m^3/d recycle flow.

   c) Due to practical consideration, two pumps are chosen, each with 550 m^3/h capacity. Total recycle capacity therefore is 1,100 m^3/h.

4) 25% safety factor is applied (minimum area now is 654.5 m^2).

5) Due to practical considerations (standard design drawings, number of units, etc. the actual area chosen could be somewhat larger than the minimum theoretically required. In this case, 700 m^2 is chosen.

6) The PWWF overflow rate is 2,500/700 = 3.6 m/h.

7) The final chosen area (700 m^2) and recycle flows (550 and 1,100 m^3/h) can be entered in the fields (in Design Tab), and

Table 12.2 Design parameters for flux theory.

Parameter	Symbol	Value	Unit
Initial settling velocity	V_0	16.8	m/h
Hindered settling parameter	p_{hin}	0.36	m^3/kg
Safety factor on area	F_A	25	%

a) The State Point and underflow line position can be verified in the PWWF and PDWF State Point diagrams (comparing with Figure 12.26).
b) The operating point on the Ekama D&O diagram can be checked to assure that it falls in the safe operating zone (below SHC II and to the right of SHC I on Figure 12.27).

12.5.2 Empirical design

Empirical design rules are based on (often local) engineering experience and as such, are quite varied and depend on different countries and areas. The example used here should not be construed as generally valid – it is just an example.

The selection of required settler area can be based on maximum hydraulic loading specification and/or maximum solids loading specification (as presented in this example) or other criteria. In this case, maximum 1 m/h for PDWF and 2.5 m/h for PWWF is envisaged. In addition, the clarifier should not be loaded higher than 6 kg/m^2/h during dry weather and 15 kg/m^2/h temporarily during wet weather. Each of these specifications will lead to a required clarifier area as calculated in the Design Tab in the spreadsheet). The largest area will be selected – in this case 1,108 m^2. Recycle ratios between 0.5 and 1.0 are usually satisfactory. This clarifier design results in a significantly larger clarifier compared to the design based on flux theory (1,100 m^2 vs. 700 m^2). This is in a large part due to the excellent settling properties of the sludge used as example (v_0 = 16.8 m/h, SVI = 60 ml/g, $SSVI_{3.5}$ = 48 ml/g), which are not considered at all in the empirical design criteria. In practice, more conservative sludge settling parameters would be selected, even if there is evidence that an existing biological process consistently produces well-settling sludge.

12.5.3 WRC design

The WRC design procedure is based on the $SSVI_{3.5}$ test, which provides the most reliable measure of settleability as described in Section 12.3. The $SSVI_{3.5}$ of the sludge used in this example was 48 ml/g (very well settling and compacting).The WRC design method is presented and used here with the extension as described in the IWA Scientific and Technical Report No.6 (Ekama *et al.* 1997). The extension consists of an empirical relationship between the critical recycle ratio and the $SSVI_{3.5}$. According to flux theory, which the WRC method is based on, the surface overflow rate can only increase up to the critical recycle rate. The original WRC method does not have that feature - it always gives a higher overflow rate for a higher recycle rate. The modification in step 1 below finds the critical recycle ratio from the SSVI which therefore gives the maximum overflow rate. For recycle flow rates larger than critical value, the overflow rate (hydraulic loading) must not be increased (as per SHC II of the flux theory).

Design steps (in Design Tab in the spreadsheet):

1) The critical recycle ratio is calculated from Eq 12.11a.

$$q_{R,crit} = 1.612 - 0.00793 \cdot SSVI_{3.5} - 0.0015 \cdot \max\left(0, \left(SSVI_{3.5} - 125\right)\right)^{1.115} \quad (12.11a)$$

and, 1.23 m/h is obtained.

2) The required area is calculated from an empirical equation that is derived based on flux settling parameters measured at 30 plants in the UK and correlated to $SSVI_{3.5}$.

$$A = \left(\frac{X_F \cdot Q_{PWWF}}{306.86 \cdot q_{R,crit}^{0.68} \cdot SSVI_{3.5}^{-0.77} - X_F \cdot q_{R,crit}} \right) \quad (12.11b)$$

The area calculated for peak wet weather flow is 642 m^2, which is larger than the area required for PDWF.

3) 25% safety factor is applied (minimum area now is 802 m^2).

4) The PWWF overflow rate is 2,500/802 = 3.1 m/h.

12.5.4 ATV design

The ATV (recently DWA) design guidelines provide detailed, practical guidance for many aspects of final settler design, such as area, depth, recycle ratio, weir loading, compaction time, number of bridges, effluent launder position, scum baffles, etc. The guidelines takes into account dynamic changes, such as solids transferred to the clarifier during a storm event, causing a reduction in MLSS and reduced solids loading. In this chapter only a simplified calculation is provided to illustrate the principle.

The ATV 1976 (and STOWA) design principles are based on the DSVI test. The DSVI test is essentially an SVI test performed under more uniform conditions: diluting the sludge sample with effluent such that the settled volume falls into the 150-250 ml range. Based on the DSVI, two concepts related to settled sludge volume are introduced:

DSV_{30} is the settled volume of the MLSS under the test conditions:

$$DSV_{30} = X_F \cdot DSVI \quad (ml/l) \quad (12.12)$$

Sludge volume loading rate (q_{SV} in $l/m^2/h$) is

$$q_{SV} = Q_i / A \cdot DSV_{30} \quad (l/m^2/h) \quad (12.13)$$

which is the volumetric rate of settled sludge loaded in the clarifier, analogously to solids loading, but expressed in volume instead of mass.

Design steps (in Design Tab in the spreadsheet):

1) The permissible overflow rate depends on DSV_{30}.

$$q_o = 2{,}400 \cdot DSV_{30}^{-1.34} \quad (12.14)$$

and, q_o = 1.86 m/h is calculated

2) q_o must be smaller than 1.6 m/h.

3) The area required for PWWF is 2,500 (m^3/h)/1.6 (m/h) = 1,563 m^2.

4) For practical considerations, 1,500 m^2 is selected.

5) The required recycle flow is based on the compressibility of the sludge, the maximum solids concentration it can reach under the given conditions, which is estimated from the DSVI test.

Under average dry weather conditions:

$$X_{R,ADWF} = \frac{1200}{DSVI} \quad (12.15)$$

Under peak wet weather flow conditions:

$$X_{R,PWWF} = \frac{1200}{DSVI} + 2 \quad (12.16)$$

where:

$X_{R,ADWF}$	20.0 (1,200/60)
$X_{R,PWWF}$	22.0 (1,200/60+2) (g/l) return concentrations are calculated respectively

6) The necessary recycle flow is calculated based on a simple mass balance, presented in Eq. 12.18.

$$(Q_I + Q_R) \cdot X_F = Q_R \cdot X_R \quad (12.17)$$

where:

Q_I	influent flow (m^3/h)
Q_R	recycle flow (m^3/h)
X_F	bioreactor mixed liquor suspended solids concentration (kg/m^3)
X_R	return solids concentration (kg/m^3)

212 and 473 m^3/h is calculated for the two conditions. A 500 m^3/h pump is selected for practical considerations.

12.5.5 STOWA design

The STOWA design procedure is closely based on the ATV design. Design steps (in Design Tab in the spreadsheet)

1) Calculate permissible overflow rate based on Eq. 12.18. The permissible overflow rate depends on DSV_{30}.

$$q_O = \frac{1}{3} + \frac{200}{DSV_{30}} \quad \text{(m/h)} \quad (12.18)$$

q_O = 1.29 m/h is calculated.

2) The sludge volume loading rate is calculated according to Equation (12.13). q_{SV} = 270 l/m²/h is obtained. This rate must be between 300 and 400 l/m²/h so 300 will be used in the surface area calculation.

3) The permissible overflow rate is calculated as $q_O = q_{SV}/DSV_{30} = q_{SV}/(X_F \cdot DSVI) = 300/(60 \cdot 3.5) =$ 1.43 m/h.

4) Therefore, during ADWF conditions, 1,000 (m³/h)/1.43 (m/h) = 700 m² area is required.

5) During wet weather flow, the solids transferred temporarily to the clarifier and the resulting drop in MLSS is taken into account in step 3. The actual calculation is an iterative one until the reduced solids loading balances with the sludge stored in the clarifier blanket, subject to real world conditions. The maximum reduction on X_F allowed is 30%, and this will be used in this simplified example. Therefore, during PWWF conditions, 0.7·2,500 (m³/h)/1.43 (m/h) = 1,225 m² area is required.

6) For practical considerations 1,200 m² area is selected.

7) Recycle flow can be calculated identically to the ATV method, resulting in 500 m³/h pump capacity.

Table 12.3 Design comparison summary table.

Parameter	Unit	Empirical	Flux	WRC	ATV (1976)	STOWA
Settler area	m²	1,108	700	802	1,500	1,200
At ADWF 1,000 m³/h						
Overflow rate	m/h	0.90	1.43	1.25	0.67	0.83
Recycle rate	m/h	0.81	0.79	1.25	0.33	0.42
RAS concentration	kg/m³	7.39	9.86	7.00	10.50	10.50
Solids loading rate	kg/m²/h	6.00	7.75	8.72	3.50	4.38
At PDWF 1,500 m³/h						
Overflow rate	m/h	1.35	2.14	1.87	1.00	1.25
Recycle rate	m/h	1.20	0.64	1.25	0.33	0.42
RAS concentration	kg/m³	9.33	13.05	8.75	14.00	14.00
Solids loading rate	kg/m²/h	7.58	10.25	10.91	4.67	5.83
At PWWF 2,500 m³/h						
Overflow rate	m/h	2.26	3.57	3.12	1.67	2.08
Recycle rate	m/h	0.90	1.57	1.25	0.33	0.42
RAS concentration	kg/m³	12.25	11.45	12.25	21.00	21.00
Solids loading rate	kg/m²/h	11.05	18.00	15.27	7.00	8.75

12.5.6 Comparison of settlers designed using different methods

It is clear from the above examples that significantly different clarifier design principles are used around the world and the simplified demonstration described in this chapter leads to different overflow and underflow rates (Table 12.3). The clarifiers designed based on flux theory and using the WRC principles have a relatively smaller surface area, and larger pumps are used to remove the settled sludge from the bottom, at a lower concentration. The ATV and STOWA guidelines lead to building larger settlers, and count on good sludge compactibilty (as it is the case with the sludge used in this demonstration) which requires lower recycle pumping rates.

12.6 MODELLING OF SECONDARY SETTLERS

Secondary clarifier models, sole or coupled to an activated sludge model, are routinely used in process engineering and design work. Depending on the modelling objectives different levels of conceptualization are available. The two most used in practice are a) flux based one dimensional (1-D) models in conjunction with activated sludge modelling, and b) computational fluid dynamic (CFD) models (2-D or 3-D) that can be used to aid detailed clarifier design. Figure 12.29 illustrates three different types of models that will be briefly introduced in this chapter.

12.6.1 Zero dimensional models

This simple representation is a 'volumeless' clarifier model, without area or depth. The sole purpose of these models is often to retain the MLSS in the system, and the concept is essentially based on an instantaneous mass balance around the clarifier (*e.g.* Eq. 12.17). The return solids concentration, X_R, (if calculated at all) can be expressed from Eq. 12.17, if the flows and the operating MLSS concentration are known. Effluent solids may be ignored. Most early activated sludge simulation models before 1990 used such an approach since their focus was on biological performance and solely on the soluble components in the effluent. It is also possible to calculate effluent solids using simple empirical approaches, usually linked to input MLSS (percent removal) or applied solids flux. In this case solids lost through the effluent have to be included in Eq. 12.17.

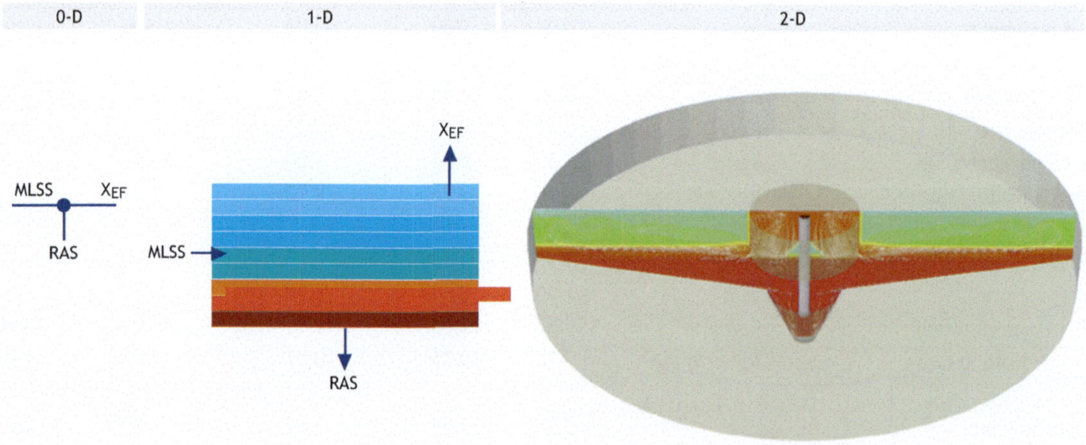

Figure 12.29 Zero-, one- and two-D representation of the same clarifier (image: MMI Engineering).

12.6.2 One-dimensional models

These models take the volume of the settler into consideration. Several variations exist in this category, including simple two-compartment models (only the clarified zone and sludge blanket are considered), or a mixture of mass balance and empirical based models that estimate underflow, effluent and sludge blanket concentrations using various algebraic equations.

However, the most widely used model in this category is the layered 1-D flux model. This model represents the clarifier as a stack of horizontal layers. Horizontal movement is not considered, in agreement with flux theory. Circular and rectangular tanks are not distinguished in 1-D models. Bulk flow and settling flux based dynamic mass balances are implemented in each layer, and the model's output is a vertical solids profile (one concentration for each layer). Although the flux theory as discussed in this chapter forms the basis of these models, since it does not account for discrete and compression settling, models based on it alone cannot predict effluent solids or a stable sludge blanket. There are various additions implemented in 1-D flux models that make their predictions more realistic.

The presence of a sludge blanket is simulated either (*i*) using a small number (8-15) of layers and the "minimum of fluxes" approach between neighbouring layers. In this method, the smaller of two fluxes is used in each layer, one that can be "accepted" based on the current solids concentration present in the layer, or one that can be delivered by the layer above based on its own solids concentration, or (*ii*) by implementing a back mixing or numerical diffusion process acting between layers.

Effluent solids and compressive behaviour are simulated using additions to the Vesilind settling function to account for discrete and compressive settling (*e.g.* the double-exponential model, Takacs *et al.*, 1991). For the discrete settling region (top of the clarifier, *i.e.* effluent solids, further research led to the identifiability of the flocculant settling parameter in this model. The TSS concentration at which discrete settling transitions into flocculant settling is referred to as threshold of flocculation (TOF). TOF is directly linked to the collision efficiency of particles and is thus dependent on the non-settling fraction, particle charge, particle density, particle size, extracellular polymeric substances (EPS) characteristics etc. (Mancell-Egala *et al.*, 2017). In addition, TOF describes the flocculant settling behavior which is the main settling regime in a clarifier. Based on earlier work, a linear correlation between TOF and effluent suspended solids (ESS) was observed under steady state conditions (Figure 12.30) (Ngo *et al.*, 2018). The lower the TOF, the better the collision efficiency leading to better effluent quality. In addition, TOF has been proposed as a measurement method for calculation of the flocculation coefficient of the Takacs et al. (1991) settling model, rather than using the latter parameter as calibration parameter.

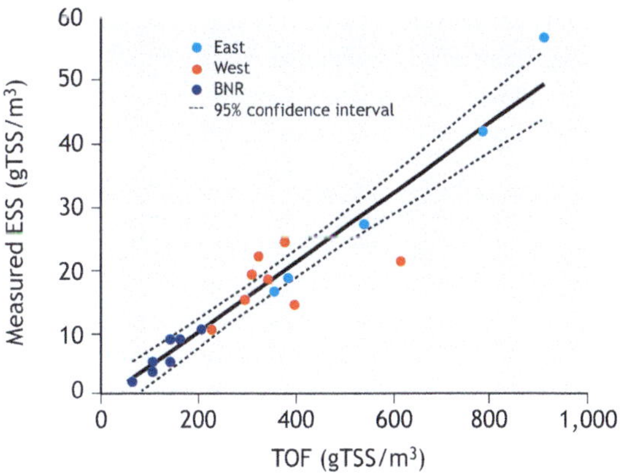

Figure 12.30 The correlation between TOF and effluent solids (ESS) (based on Ngo *et al.*, 2018).

There are research efforts on-going for improving the clarifier capacity estimation methods as well. Flocculent settling (stokesian) is predominant within ideally operating clarifiers, and the shift to 'slower' hindered settling (non-stokesian) causes both failure and poor effluent quality. A new metric for settling characteristics was developed (Mancell-Egala *et al.*,

2017) and classified as Limit of Stokesian Settling (LOSS). This parameter describes the TSS concentration at which a transition of flocculant to hindered settling occurs. The larger the LOSS, the higher the capacity of the clarifier will be as hindered settling happens only at increased concentrations. A wide range of sludge types were characterized with granular sludge, BNR sludge and high-rate activated sludge showing LOSS values of 5,000, 1,200 and 780-1,000 mg TSS/l, respectively. LOSS numbers collected experimentally were validated with the Takacs *et al.* (1991) settling model (Mancell-Egala *et al.*, 2016). When compared to flux curves with small changes in sludge matrix, LOSS was proven to be faster at characterizing hindered settling velocity and was less erratic. A further promising improvement is described in Torfs *et al.* (2017) over traditional settling models. In these models the settling behavior is described as a function of solids concentration only, even though it is well known that the flocculation state (*i.e.* floc density, floc strength, floc size) plays an important role on settling performance as well. Moreover, typically only specific parts of the settling behavior are included in the model structure (*i.e.* hindered and sometimes compression settling for SST models and discrete settling for PST models).

To overcome some of these simplifications, Torfs *et al.* (2017) presented a modelling framework which combines the state-of the art in based SST models (hindered and compression modelling) with state of the art in PST modelling (particle settling velocity distribution concept – Bachis *et al.*, 2015). The resulting framework allows to integrate all different settling regimes (discrete, flocculent, hindered and compression) in a modular way and it allows to account for changes in the flocculation state by considering different floc classes instead of only a single concentration variable. However, the calibration requirements for this modelling framework are high and a robust protocol as well as reliable and innovative measurements are needed.

1-D dynamic models play an important role in conjunction with activated sludge and plant-wide process predictions. Due to their simple structure, they do not add a significant computational load to the process model, and they can reasonably predict the three main functionalities of secondary settlers – clarification, thickening and sludge storage. Effluent solids predicted in these models form an important part of effluent quality. Return solids are used for wastage and affect SRT, thickening and the loading and performance of the solids line. Finally, sludge storage (dynamic sludge blanket prediction) takes into account changes in reactor solids inventory, which may have a significant effect on process performance. In certain conditions, biological or chemical reactions occur in secondary settlers, such as denitrification. 1-D models are, almost exclusively, used to simulate these reactions since 0-D models, in lack of a reactive volume, are not suitable for this purpose, and implementing complex biological models in 2- and 3-D hydrodynamic models presents a prohibitive computational demand. 1-D layered models cannot be used for investigating the details of clarifier structures, *e.g.* tank geometry or baffle placement. 2 or 3-D computational fluid dynamic (CFD) models are required for this purpose.

12.6.3 Computational Fluid Dynamic models

CFD models are based on conservation of fluid mass (continuity), conservation of momentum in the horizontal and vertical directions, conservation of solids mass (transport of suspended solids), conservation of enthalpy (heat balance), and a turbulence model. To achieve a stable numerical solution, the settler has to be discretized into a fine grid, often using tens of thousands of grid elements. In this representation it is possible to account for fine physical details such as baffles, their exact geometry, placement and angle. The above set of equations is subsequently solved at each node at each time step. This represents a significant computational load, but can result in a very detailed picture of solids distribution and flow patterns in the clarifier as shown in Figure 12.31 (in a 2-D CFD model example). Adding the third dimension further increases the complexity and execution time of these models, and should only be included if necessary.

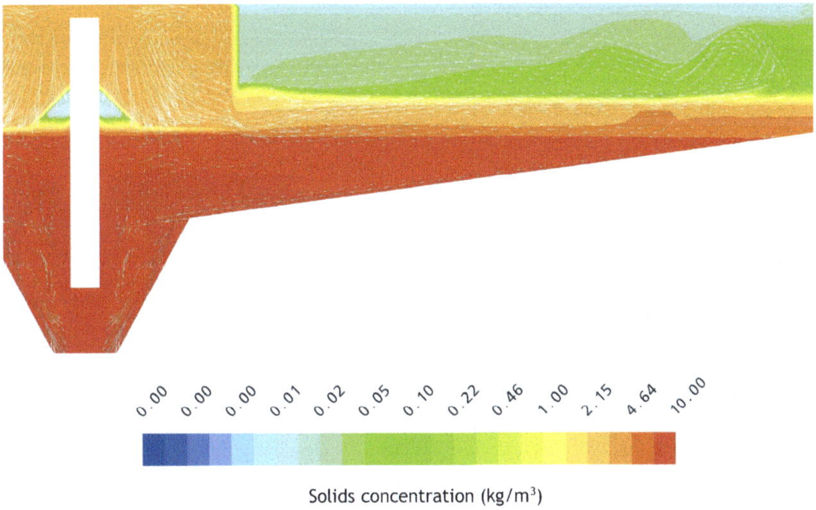

Figure 12.31 2-D CFD results (image: MMI Engineering).

The use of CFD models has accelerated significantly in recent years, due to advances in hydrodynamic modelling and model calibration. Details of clarifier design are frequently subjected to verification or optimization in a CFD model before implementation in the full-scale clarifier. An example is shown in Figure 12.32. In this case, a Stamford baffle, which is designed to deflect flow away from the effluent weir area, is simulated. The effect of the baffle is clearly visible on the flow field; however, in this case the simulation did not predict a significant improvement in effluent suspended solids.

12.7 DESIGN EXAMPLES

Design a clarifier. Establish the required clarifier area and the return pump capacity. Use the simplified methods as described in Section 12.5 to design the clarifier for expected future loading conditions as specified in Table 12.4.

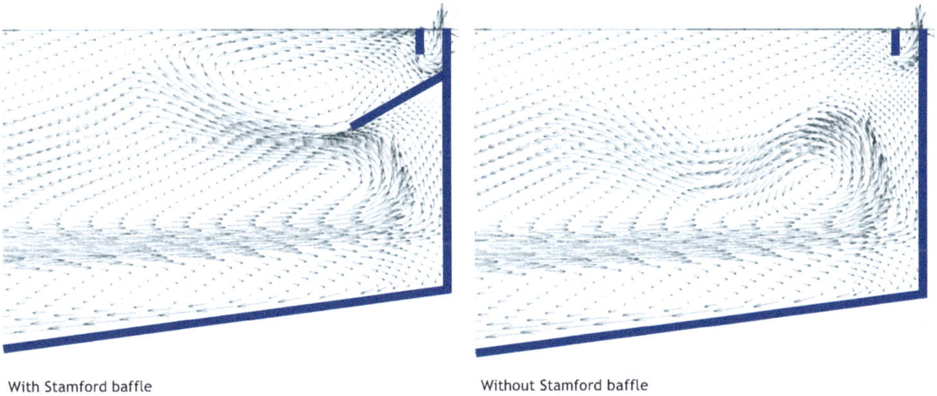

Figure 12.32 Effect of Stamford baffle on flow field around the weirs (image: MMI Engineering).

Table 12.4 Design specifications.

Parameter	Symbol	Value	Unit
MLSS	X_F	3.2	kg/m^3
Average dry weather flow	Q_{ADWF}	240	m^3/h
Diurnal peaking factor (dry weather)	PF_{DW}	1.4	
Storm peaking factor	PF_{WW}	2.8	
Safety factor for flux theory	F_A	1.25	
Safety factor for WRC procedure	F_{WRC}	1.25	

Sludge settling properties are not known at the site, since the new process to be built will implement biological nutrient removal as well as receive 10% industrial contribution in the influent. The current plant at the site is not required to nitrify and has no trade influent input. The average settleability values will therefore be assumed as shown in Table 12.5.

This assignment can easily be calculated by hand or using the earlier mentioned spreadsheet.

Solution steps

Based on Table 12.4, dry weather diurnal peak is 336 m^3/h, and a wet weather peak flow of 672 m^3/h.

1) Design using flux theory
   a) Settling velocity at the MLSS concentration (Solids Handling Criterion II) is calculated from Eq. 12.4 (1.5 m/h).
   b) The minimum area required during PWWF is 672/1.5 = 448 m^2.
   c) The minimum recycle ratio from the Ekama D&O chart is 0.49. This will be a safe recycle ratio since it does not include the safety factor on the area yet.
   d) Using the State Point Analysis diagram, drawing tangents to the gravity curve from the state point, at PWWF 0.49 (329 m^3/h), and at PDWW 0.32 recycle ratio is sufficient. The recycle pump capacity is selected at 330 m^3/h (two pumps).
   e) 25% safety factor is applied (A = 553.4 m^2). This is rounded to 550 m^2.

2) Empirical design
   Recycle pump is selected at 100% recycle for PDWF or 336 m^3/h. Using 1 m/h hydraulic or 5 kg/m^2/h solids loading for PDWF and 2 m/h hydraulic or 10 kg/m^2/h solids loading for PWWF, the 5 kg/m^2/h loading rate results in the largest clarifier area, 3.2·(336+672)/5 = 369 m^2 area is required.

This clarifier design results in a smaller clarifier compared to the design based on flux theory. This is due to the empirical method not taking into account the expected low settling properties of the sludge.

Table 12.5 Assumed sludge settling parameters for the various design methods.

Parameter	Symbol	Value	Unit
SVI	SVI	190	ml/g
DSVI	DSVI	160	ml/g
$SSVI_{3.5}$	$SSVI_{3.5}$	120	ml/g
Initial settling velocity	v_0	5.82	m/h
Hindered settling parameter	p_{hin}	0.42	m^3/kg

3) WRC design

a) Calculate critical recycle rate as 0.66 m/h based on Eq. 12.12a.
b) The required area from Eq. 12.11b for PWWF is 583 m^2 and for PDWF is 292 m^2. The larger one with 25% safety results 729 m^2.
c) PWWF recycle is 481 m^3/h. A 500 m^3/h recycle pump is chosen. The PWWF overflow rate is 672/729 = 0.92 m/h.

4) ATV design

a) DSV_{30} from Eq. 12.12 is 3.2 · 160 = 512 ml/l.
b) The overflow rate from Eq. 12.14 is 0.56 m/h. This is smaller than the maximum, 1.6 m/h.
c) The area required for PWWF is 672 (m^3/h) / 0.56 (m/h) = 1,196 m^2. 1,200 m^2 is selected.
d) The return concentration under PWWF conditions (Eq. 12.15) is 7.5 g/l, and under PWWF conditions (Eq. 12.16) it is 9.5 g/l.
e) Recycle flow, based on the mass balance presented in Eq. 12.18 is 179 m^3/h during ADWF and 341 m3/h during PWWF; 350 m^3/h is selected.

5) STOWA design

a) DSV_{30} from Eq. 12.12 is 3.2 · 160 = 512 ml/l Permissible overflow rate based on Eq. 12.18 is 0.72 m/h
b) The sludge volume loading rate from Eq 12.13 is 371 l/m^2/h (falls between 300 and 400 l/m^2/h). Therefore 0.72 m/h is accepted.
c) Area required is 240/0.72 = 332 m^2 for ADWF and using the 70% maximum MLSS reduction, 0.7 • 672/0.72 = 650 m^2.
d) Recycle flow, based on the mass balance presented in Eq. 12.17, is 179 m^3/h during ADWF and 341 m^3/h during PWWF; 350 m^3/h is selected.

Table 12.6 summarizes the selected settler areas and return pump capacities using the different design methods.

Table 12.6 Summary of solutions.

	Unit	Flux theory	Empirical	WRC	ATV (1976)	STOWA
Settler area	m^2	550	369	729	1,200	650
Recycle pump	m^3/h	330	336	500	350	350

Figure 12.33 Example of properly operated and maintained secondary settling tank which produces effluent of good quality (photo D.H. Eikelboom).

REFERENCES

APHA (2006). Standard Methods for the examination of water and waste water. American Public Health Association. 874pp.

Bachis G., Maruéjould T., Tik S., Amerlinck Y., Melcer H., Nopens I., Lessard P. and Vanrolleghem P.A. (2015). Modelling and characterisation of primary settlers in view of whole plant and resource recovery modelling. *Water Sci. Technol.* 72, 2251–2261.

Ekama G.A., Barnard J.L., Günthert F.W., Krebs P., McCorquodale, J.A., Parker D.S. and Wahlberg E.J. (1997). Secondary Settling Tanks: Theory,

Modeling, Design and Operation. *IAWQ Scientific and Technical Reports* #6, IAWQ London.

Ekama G.A. and Marais P. (2004). Assessing the applicability of the 1D flux theory to full-scale secondary settling tank design with a 2D hydrodynamic model. *Wat. Res.* 38(3), 495-506.

Mancell-Egala W.A., Kinnear D., Jones K.L., De Clippeleir H., Takacs I. and Murthy S.N. (2016). Limit of stokesian settling concentration characterizes settling velocity. *Water Research*. 90, 100-110.

Mancell-Egala W.A., De Clippeleir H., Su C., Takacs I., Novak J.T. and Murthy S.N. (2017). Novel stokesian metrics that quantify collision efficiency, floc strength and discrete settling behavior. *Water Environment Research*. 89(7), 586-597.

Ngo K.N., Van Winckel T., Takács I., Wett B., Massoudieh A., Murthy.S., Al-Omari A. and De Clippeleir H. (2018). Towards more predictive clarification models via experimental determination of flocculent settling coefficient values. Proceedings of WEFTEC, New Orleans, USA.

Takács I., Patry G.G., Nolasco D. (1991). A Dynamic Model of the Clarification/Thickening Process. *Water Research,* 25(10), 1263-1271.

Torfs E., Martí M.C., Locatelli F., Balemans S., Bürger R., Diehl S., Laurent J., Vanrolleghem P.A., François P. and Nopens I. (2016). Concentration-driven models revisited: towards a unified framework to model settling tanks in water resource recovery facilities. *Water Sci. Technol.* 75(3-4):539-551

Water Environment Federation (2005). *Clarifier Design Manual of Practice, No. FD-8*, 2nd Edn., McGraw-Hill, New York.

NOMENCLATURE

Abbreviation	Description
ADWF	Average dry weather flow
CFD	Computational fluid dynamics
DSVI	Diluted sludge volume index
DWF	Dry weather flow
MLSS	Mixed liquor suspended solids
PF_{DW}	Diurnal peaking factor (dry weather)
PF_{WW}	Storm peak factor
PWWF	Peak wet weather flow
RAS	Return activated solids
SBR	Sequencing batch reactor
SOR	Surface overflow rate
SPA	State point analysis
$SSVI_{3.5}$	Stirred sludge volume index test performed at 3.5 g/l MLSS concentration
STOWA	Stichting Toegepast Onderzoek Waterbeheer
SVI	Sludge volume index
SZSV	Stirred zone settling velocity
ZSV	Zone settling velocity

Symbol	Description	Unit
A	Area of the settler	m^2
DSV_{30}	Settled volume of MLSS under test conditions	ml/l
F_A	Safety factor on area	-
G	Velocity gradient	s^{-1}
j_{AP}	Total applied flux	$kg/m^2.h$
J_B	Bulk flux	$kg/m^2/h$
j_I	Overflow rate flux	$kg/m^2.h$
J_L	Limiting flux, corresponding to X_L	$kg/m^2.h$
j_R	Underflow rate flux	$kg/m^2.h$
J_s	Gravity flux	$kg/m^2/h$
p_{hin}	Hindered settling parameter	l/g or m^3/kg
q_I	Hydraulic loading or overflow rate	m/h
Q_I	Influent flow	m^3/h
q_o	Permissible overflow rate	m/h
Q_R	Recycle flow	m^3/h
q_R	Hydraulic underflow rate	m/h
Q_R	Recycle flow	m^3/h
$q_{R,crit}$	Critical underflow	m/h
q_{sv}	Sludge volume loading rate	$l/m^2/h$
R	Recycle ratio (Q_R/Q_I)	-
v_o	Initial settling velocity	m/h
v_s	Settling velocity	m/h
x	MLSS concentration in the various ZSV tests	g/l or kg/m^3
X	Solids concentration	kg/m^3
X_F	Feed concentration	kg/m^3
X_L	Limiting concentration	kg/m^3
X_R	Recycle concentration	kg/m^3

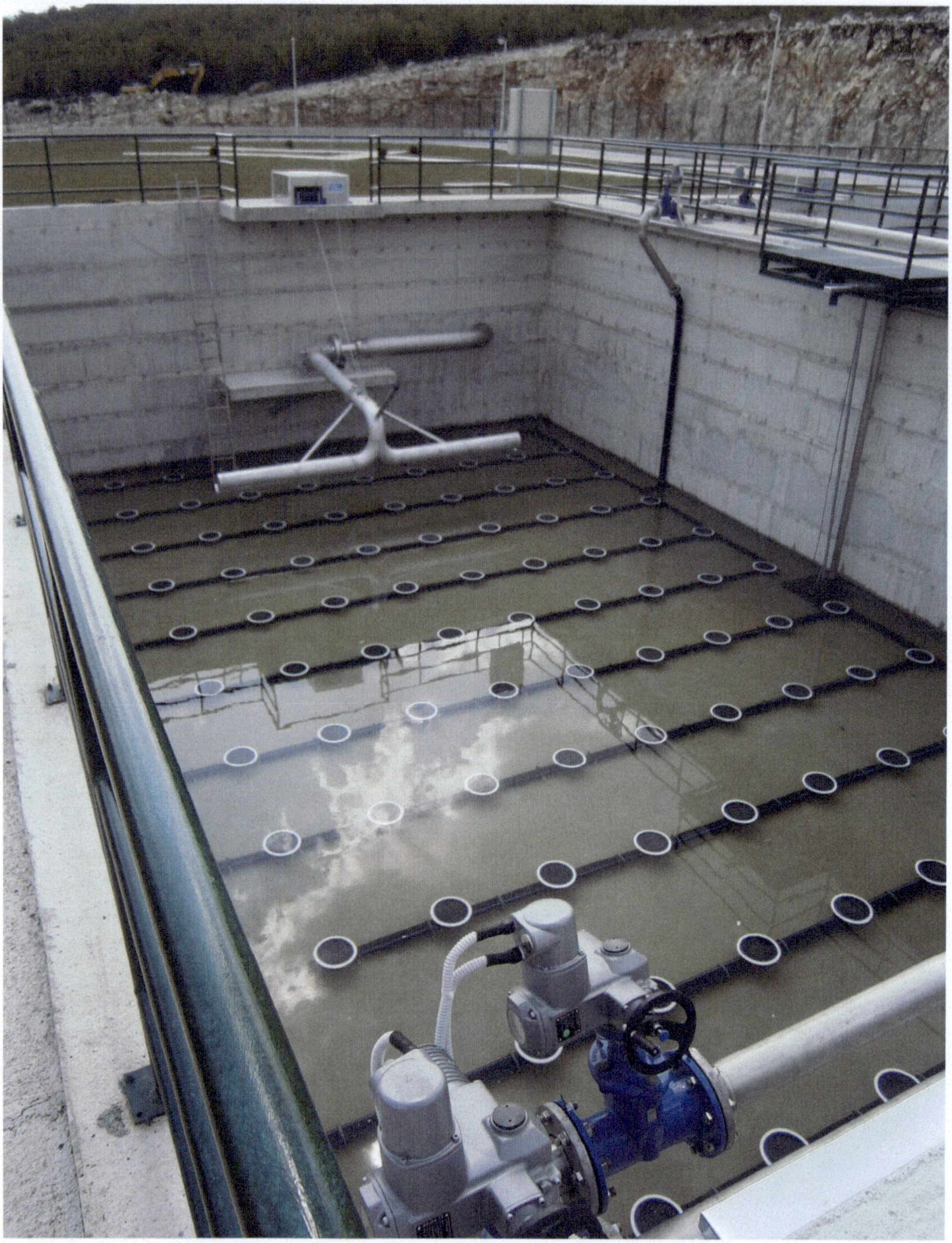

Figure 12.34 A sequencing batch reactor (SBR) serves as a settling tank during the idle phase. This particular design (WWTP Bileća, Bosnia and Herzegovina) entails an adjustable effluent discharge system that determines the volume of the effluent discharged from the reactor after the settling period (photo: D. Brdjanovic).

doi: 10.2166/9781789060362_0557

13

Membrane bioreactors

Xia Huang, Fangang Meng, Kang Xiao, Hector A. Garcia and Jiao Zhang

13.1 MEMBRANE SEPARATION PRINCIPLES

The major function of membrane separation is to separate, purify or concentrate components from a mixture by applying one or more driving forces across a selective permeable membrane so that the designated components can preferentially penetrate the membrane. The selective permeable membrane is the key to guaranteeing effective separation. The mechanisms of membrane separation include size sieving, diffusion/electrodialysis, osmosis, phase transition, etc. Size sieving is a mechanical filtration process using porous membranes, while other mechanisms rely on the physicochemical interaction between the permeating components and the membrane material. Membrane separation processes can be driven by pressure gradient, concentration gradient, temperature gradient, electric potential gradient, etc.

Pressure-driven membrane processes have been successfully applied in water industries including water purification, wastewater treatment, and water reclamation. Microfiltration (MF), ultrafiltration (UF), nanofiltration (NF), and reverse osmosis (RO) are the best examples, rejecting pollutants with a size ranging from several microns (*e.g.* sludge particles) down to several angstroms (*e.g.* salts). The pore sizes of these membranes, in comparison with the particulate/colloidal/soluble pollutant sizes, are schematically shown in Figure 13.1. MF rejects particulate substances with a size larger than 0.1 μm. UF rejects colloids and macromolecules with a size of 0.002-0.100 μm or more; the rejectability of UF membranes is usually measured by molecular weight cut-off. NF rejects organic compounds with a molecular weight of 100-1,000 Da or more as well as some hardness ions. RO rejects salts and small-size solutes in the feed mixture. MF and UF are referred to as low-pressure processes, while NF and RO are referred to as high-pressure processes.

The membrane bioreactor (MBR) is an integrated process of biological treatment and membrane filtration for wastewater treatment and water reclamation. The membrane used in a MBR is a MF or UF membrane, with the major purpose of separating water from the activated sludge solids in order to guarantee the effluent quality. The membrane can retain a full portion of sludge flocs and bacteria cells, and some of the colloidal and soluble biopolymers in the sludge. Aerobic MBR systems are designed to remove organic pollutants, nutrient elements, and most of the pathogens. Anaerobic MBRs (AnMBRs) are designed to recover carbon energy, carbon material, and N/P nutrients, with the assistance of MF or UF membrane rejection.

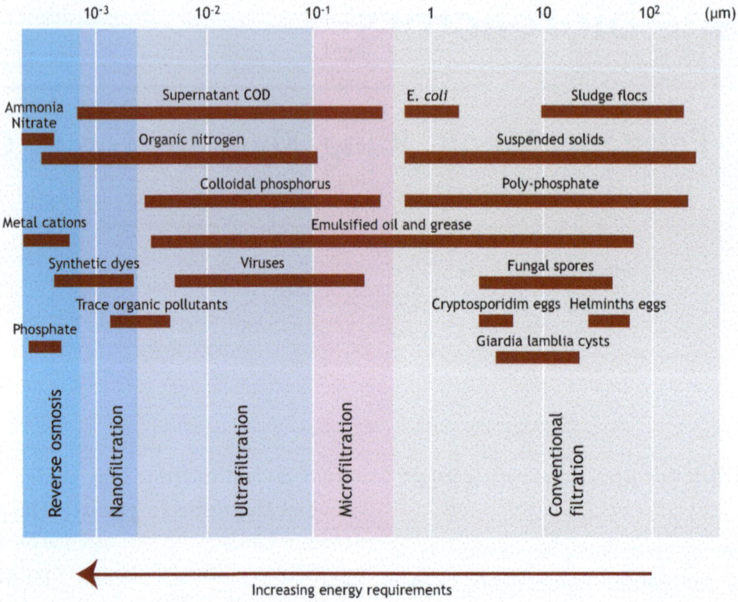

Figure 13.1 Schematic showing pollutant size exclusion by reverse osmosis (RO), nanofiltration (NF), ultrafiltration (UF), microfiltration (MF) membranes, and conventional filtration (CF).

13.2 INTRODUCTION TO MEMBRANE BIOREACTORS

13.2.1 Membrane bioreactor history

Since the discovery of activated sludge in the early 20st century, activated sludge processes have been widely used in wastewater treatment, although they are usually limited by the difficulties in solid/liquid separation. Ultrafiltration membranes were developed in the early 20st century in Europe (Bechhold, 1907) and commercially used (Zsigmondy and Bachmann, 1922) in lab research with the purpose of separation of bacteria or biomolecules. In the late 1960s, MBRs were developed by Dorr-Oliver with a combination of side-stream membranes and activated sludge, aiming to address the issue of solid/liquid separation in conventional activated sludge (CAS) processes (Bemberis *et al.*, 1971). This combined system was initially applied in Japan for small-scale municipal wastewater treatment.

Because high energy consumption was a general limitation of side-stream MBRs, a submerged MBR developed by Yamamoto *et al.* (1989) was a breakthrough in MBR development. Submerged MBRs usually have lower energy consumption due to the lower suction pressure and the absence of a recirculation pump. Since then, MBR-based processes have been widely accepted and applied in wastewater treatment. An active player in MBR application in the 1990s was Kubota; approximately 60 Kubota MBR plants were installed in Japan for domestic and industrial wastewater treatment in the 1990s, resulting in a total treatment capacity of 5.5 MLD (Judd and Judd, 2011). Since 1990, other membrane companies including Mitsubishi Rayon (the SUR MBR membrane module), Norit X-Flow (polymeric multitube membrane modules) and GE-zenon (ZeeWeed® immersed UF membranes) have promoted the commercial application of MBRs. Since then, MBRs have been widely used throughout the world.

Owing to the benefits of membrane separation, MBR-based processes have been retrofitted largely with the use of novel membrane processes such as membrane distillation (MD)-based MBRs and forward osmosis (FO)-based MBRs. MBR systems have also been expanded to other biological processes, such as aerobic granules, anammox, and particularly anaerobic processes. Anaerobic MBRs (AnMBRs) are considered to be a promising process for energy production (*i.e.* methane production) during wastewater treatment.

13.2.2 Membrane bioreactor features

MBRs are a combination of the biological process and the membrane process using MF or UF. MBRs achieve wastewater treatment through the biological transformation of pollutants and the membrane separation of bacteria and treated wastewater. MBRs have the following advantages.

- Reduced footprint because of the elimination of the secondary clarification tank and an increased volumetric loading rate resulting from the high bacteria concentrations (land usage can be reduced by 30-50% depending on the plant capacity).
- Increased water quality of the permeate owing to membrane rejection and the enhanced biological degradation (lower concentrations of pollutants in the effluent).
- Independent control of solids retention time (SRT) and hydraulic retention time (HRT), leading to easier regulation of such parameters and much higher mixed liquor suspended solids (MLSS) concentrations.
- Reduced sludge production as a result of these much higher MLSS concentrations.

A smaller footprint is interesting for wastewater treatment plant (WWTP) design in large cities facing land shortage. Increased permeate quality, which can meet more stringent discharge regulations and the requirement of water reuse, is one of the major reasons for the wide application of MBRs in the world. Reduced sludge production is attractive in regions where waste sludge treatment has not well been conducted in the past. Because of these advantages, MBRs are considered to be an important process for municipal wastewater treatment and industrial wastewater treatment.

13.2.3 Membrane bioreactor configuration

As mentioned above, there are two main MBR configurations with different membrane units and different driven pressure for permeate (Figure 13.2A and 13.2B): the side-stream MBR (sMBR) with positive pressure, and submerged or immersed MBRs (iMBRs) with vacuum pressure.

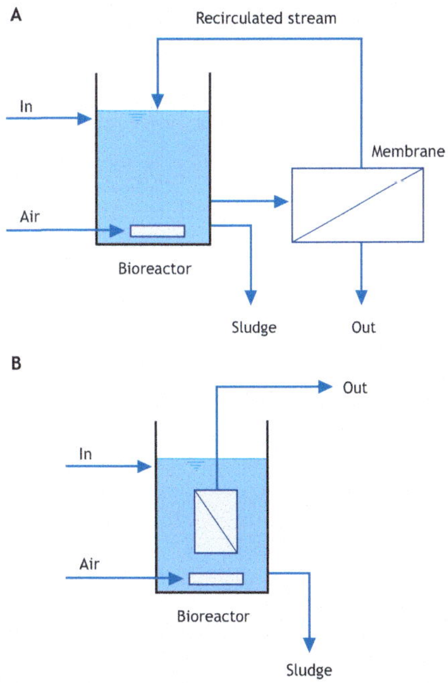

Figure 13.2 MBR process configurations: (A) a side-stream MBR (sMBR) and (B) a submerged or immersed MBR (iMBR).

In the side-stream configuration, the membrane units are installed externally to the bioreactor, which is often located in a plant room (Figure 13.2A). The biomass is either pumped directly through a number of membrane modules in series and back to the bioreactor, or it is pumped to a bank of modules, from which a second pump circulates the biomass through the modules in series. Cleaning and soaking of the membranes can be undertaken *in situ* with the use of an installed cleaning tank, pump and pipework.

With submerged MBRs, the membrane units are installed in either the main bioreactor vessel or in a separate tank (Figure 13.2B). In these MBRs, aeration can provide oxygen for biomass and can control membrane fouling by their scouring effects. The membranes can incorporate an online backflush system which reduces membrane surface fouling by pumping membrane permeate back through the membrane. Similarly, online chemical cleaning can also be conducted by pumping chemical solutions back to the modules without the need of the transfer of either membrane modules or mixed liquor. In systems where the membranes are in a separate tank to the bioreactor, individual trains of membranes can be isolated to undertake cleaning regimes incorporating membrane soaks; however, the biomass must be continuously pumped back to the main bioreactor to limit the increase in MLSS concentrations. Where the membranes are installed in the main bioreactor, the membrane modules are removed from the vessel and transferred to an offline cleaning tank.

iMBRs are generally less energy-intensive than sMBRs, because of the side-effects of aeration for fouling control. However, despite the more favourable energy usage of submerged membranes, there continues to be a market for the side-stream configuration, particularly in smaller flow industrial applications (*i.e.* landfill leachate treatment), where membrane cleaning is more frequent. Note that the design of MBR processes should take full consideration of all the operating factors including membrane flux, transmembrane pressure (TMP), energy usage, and cleaning frequency.

Figure 13.3 shows a number of novel MBRs that have been developed or applied for various purposes, including fouling control, nutrient removal, wastewater reuse, the cultivation of specialized bacteria, flux enhancement, high-strength wastewater treatment, and biological nutrient removal (*i.e.* nitrogen and phosphorous elimination).

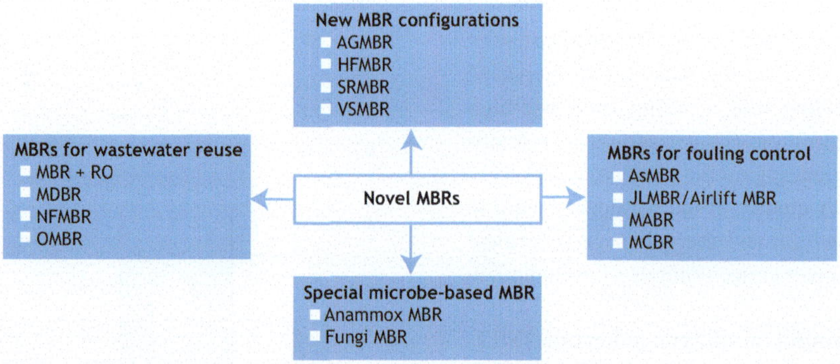

Figure 13.3 Summarized illustration showing novel MBRs developed in recent years: AGMBR (aerobic granular sludge MBR); HFMBR (hybrid biofilm MBR); SRMBR (submerged rotating MBR); VSMBR (vertical submerged MBR); MBR+RO (MBR with reverse osmosis); MDBR (membrane distillation bioreactor); NFMBR (nanofiltration MBR); OMBR (osmotic MBR); AsMBR (air-sparging MBR); JLMBR (jet loop MBR); MABR (membrane adsorption bioreactor); MCBR (membrane coagulation bioreactor); anammox MBR (anaerobic ammonium oxidation MBR) (adopted from Meng *et al.*, 2011a).

13.2.4 Membrane materials and modules

Generally, two different types of materials, namely, polymeric and ceramic, can be used to form membranes. A classic membrane normally has an asymmetric structure, comprising a thin dense layer on the top which provides the perm-selectivity and a thicker porous support layer which provides the mechanical property. A membrane for MBR application should have the following characteristics: (*i*) high porosity, suitable pore size and narrow pore size distribution to provide high selectivity and high water permeability; (*ii*) mechanically robust, *i.e.* it can withstand frequently flushing and air scouring; (*iii*) material that is resistant to chemical attack, *i.e.* resistant to extreme pH and oxidant during chemical cleaning; and (*iv*) ideally fouling-resistant. Several other membrane characteristics such as surface charge, roughness and hydrophilicity/hydrophobicity have also been demonstrated to have an impact on MBR performance, especially on membrane fouling.

The common polymers for commercial MF/UF membranes include cellulose acetate (CA), polyvinylidene difluoride (PVDF), polysulfone (PS), polyethylsulphone (PES), polyacrylonitrile (PAN), polyvinyl chloride (PVC), polyethylene (PE), polytetrafluoroethylene (PTFE) and polypropylene (PP). All the above polymers can be fabricated into membranes with desirable physical properties and reasonable chemical resistance by using specific manufacturing techniques. For the MBR market, PES and PVDF membranes are the dominant products. More recently, PTFE products have also become available. PES membranes present good chemical stability, which can tolerate a pH range of 1.5-13 and show moderate chlorine resistance. PVDF membranes can tolerate strong acids (*e.g.* down to a pH of 1), but they struggle to operate in alkaline conditions (the operation pH should below 11). PVDF membranes present good resistance to chlorine, making them ideal for MBR applications where hypochlorite cleaning is widely used. PTFE membranes show an excellent chemical resistance to either extreme pH conditions or oxidants such as hypochlorite and ozone.

Inorganic materials, such as aluminium, zirconium, silica, and titanium oxide, are used to fabricate ceramic membranes. Ceramic membranes show superior hydraulic, thermal and chemical resistance. Stainless steel membranes are a potential alternative for the treatment of industrial wastewater, because of their tolerance to high temperature and the high permeate flux. However, these inorganic membranes are not the preferred option for large-scale MBR plants due to their high costs. Moreover, inorganic membranes sometimes induce more severe inorganic fouling in anaerobic membrane bioreactors (Trzcinski and Stuckey, 2016). Therefore, inorganic membranes can only be used in some special applications such as high-temperature industrial wastewater treatment.

13.2.5 Commercial membrane module makers

Because of the unique feature of MBRs and particularly the significant reduction in membrane price, MBRs have become increasingly and widely used for wastewater treatment in the last decade (Xiao *et al.*, 2014). More than 40 companies globally provide membrane module products for MBR applications. These products can be categorized based on hollow-fibre (HF), flat-sheet (FS) and multitube (MT) configurations. The basic properties of membrane module products from the main membrane suppliers are listed in Table 13.1.

Table 13.1 Main membrane suppliers for engineering applications of MBRs globally.

Membrane material[a)]	Module configuration	Nominal pore size (μm)	Product name	Manufacturer
PVDF	Hollow fibre	0.035	ZeeWeed 500	Suez (Zenon), USA
PVDF	Hollow fibre	0.4	Sterapore	Mitsubishi Rayon, Japan
PVDF	Hollow fibre	0.1	BSY	Origin Water, China
PVDF	Hollow fibre	<0.1	SMM	Memstar, Singapore
PVDF	Hollow fibre	0.03	PURON	Koch Membrane Systems, USA
PVDF	Hollow fibre	0.1	Microza MUNC	Asahi Kasei, Japan
PVDF	Hollow fibre	0.04	Memcor MemPulse	DuPont, USA
PVDF	Hollow fibre	0.075	Saveyor SVM	Canpure, Canada
		0.05	CPM	
PVDF	Hollow fibre	0.1	SMT-600-BR	Beijing Scinor, China
PVDF	Hollow fibre	<0.1	BF, BT	Tianjin Motimo, China
PVDF	Hollow fibre	0.1-0.2	FMBR	Hangzhou Creflux, China
PVDF	Hollow fibre	0.02	FFM-MBR-20	FFM Inc., USA
HDPE	Hollow fibre	0.4	KSMBR	Econity, South Korea
PVDF	Hollow fibre	0.4	SuperMAK	ENE, South Korea
PVDF	Hollow fibre	0.1	RCM	Philos, South Korea
PVDF & PTFE	Hollow fibre	0.04-0.08	EcoFil, EcoFlon	Ecologix, USA
PTFE	Hollow fibre	0.1, 0.2, 0.45	Poreflon SPMW	Sumitomo, Japan
PVDF	Hollow fibre	0.1	ZENOMEM	Suzhou Vina, China
Chlorinated PE	Flat sheet	0.4	Kubota SMU	Kubota, Japan
PVDF	Flat sheet	0.08	MEMBRAY	Toray, Japan
PVDF	Flat sheet	0.1	SINAP	SINAP, China
PVDF	Flat sheet	0.08-0.30	PEIER	Jiangsu Lantian Peier, China
PES	Flat sheet	0.04	CES SubSnake	Colloide Engineering Systems, Northern Ireland
PES & PVDF	Flat sheet	0.08, 0.1, 0.4	EcoPlate, EcoSepro	Ecologix, USA
PES	Flat sheet	0.01-0.20	Neofil	LG Electronics, South Korea
PES	Flat sheet	0.04	BIO-CEL	Microdyn-Nadir, Germany
PES	Flat sheet	~0.07	U70	A3 Water Solutions GmbH, Germany
PVDF	Flat sheet	0.14	M70	
PVDF	Flat sheet	0.2	MFP2	Alfa Laval, Sweden
PES	Flat sheet	~0.07	MEMBRIGHT	FLI Environmental Group, USA
PVDF	Flat sheet	0.05	VINAP	Suzhou Vina, China
PES	Flat sheet	0.05	MicroClear	Weise, Germany
PVDF	Tubular	0.03	Airlift MBR	Pentair X-Flow (Norit), Netherlands
PVDF	Tubular	0.03, 100 kDa	BioFlow	Berghof, Germany
Ceramic	Flat/multi channels	0.1	CH250	Meidensha, Japan
Ceramic	Flat/multi channels	0.2	CFM	ItN Nanovation, Germany
Ceramic	Tubular	0.02, 0.05, 0.1, 0.2, 0.5	JWCM	Jiangsu Jiuwu Hi-tech, China

[a)]PVDF: polyvinylidene fluoride; PE: polyethylene; HDPE: high-density polyethylene; PVC: polyvinyl chloride; PES: polyethersulfone; PTFE: polytetrafluoroethylene.

13.2.5.1 Immersed HF products

Suez (Zenon)

Suez Water Technologies & Solutions' ZeeWeed® 500 technology was one of the earliest MBR membranes and the first vertical hollow-fibre membrane on the market. Developed by Zenon Environmental Inc. (1980 to 2006) and later acquired by GE Water & Process Technologies (2006-2017), the ZeeWeed® 500 vertical hollow-fibre membrane was first introduced in 1993 (called ZeeWeed® 150) to replace their Permaflow® tubular membrane (introduced in 1983). The current ZeeWeed® 500 membrane (Figure 13.4A) can be used for applications ranging from drinking water to seawater desalination but it is best known for use in MBR applications.

The ZeeWeed® 500 membrane for MBR uses a specifically formulated reinforced PVDF chemistry that results in a UF membrane with a nominal pore size of 0.035 μm. The current product features LEAPmbr® membrane and aeration technology (Figure 13.4B), resulting in total energy usage (aeration + permeation) of less than 0.06 kWh/m^3. Today over 2,300 MBRs globally are using ZeeWeed® 500 membranes, representing more than one million operating modules in the field producing over 26.5 million m^3/day (7 billion gallon/day) of clean water.

Mitsubishi Rayon

Mitsubishi Chemical Aqua Solutions Co., Ltd. (MCAS) is a leading company of MF/UF hollow fiber membrane products in the world (Figure 13.5).

A

B

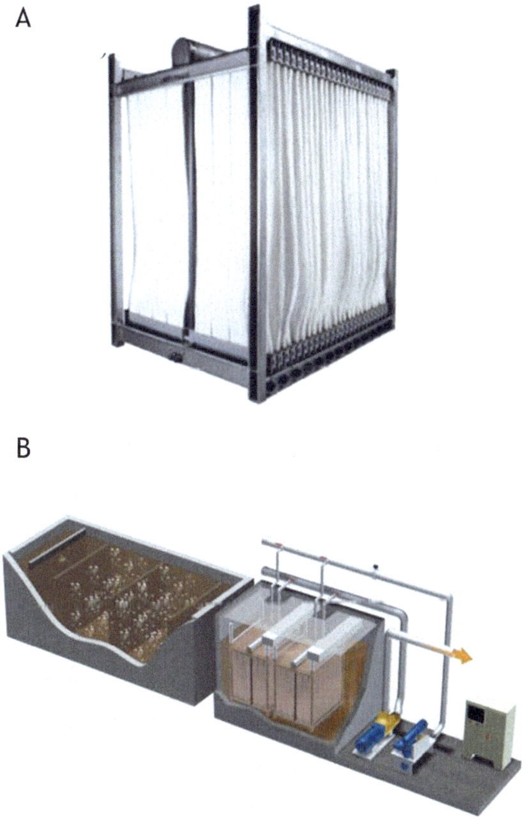

Figure 13.4 ZeeWeed® membrane products: (A) a ZeeWeed® 500D Cassette; (B) a ZeeWeed® 500 membrane train with LEAP aeration (source: Suez).

A

B

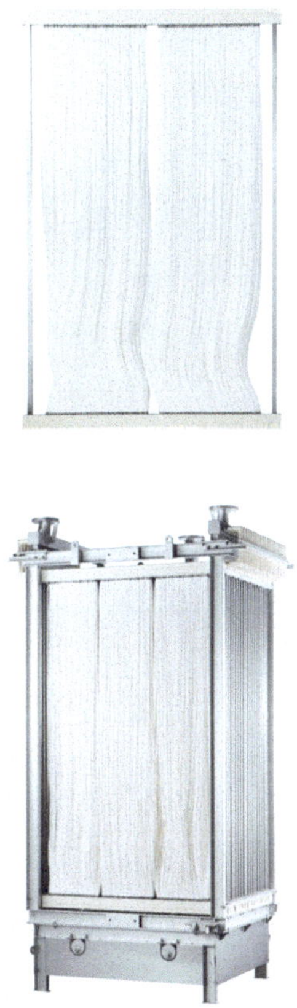

Figure 13.5 Sterapore™ membrane products: (A) a module and (B) a single-deck unit (source: www.thembrsite.org).

MCAS can offer a wide range of MF/UF membrane products for wastewater treatment, drinking and processing water treatment and so on. MCAS main products, STERAPORE™, are submerged MF/UF membranes for the MBR application. STERAPORE™ are reinforced PVDF membranes with the nominal pore sizes of 0.05 and 0.4 μm. Effective membrane areas are from 18 to 2,400 m^2 per module. STERAPORE™ has been installed in more than 5,000 MBR systems for sewage and industrial wastewater treatment plants in worldwide since 1992. As a global leader of MBR membrane supplier, MCAS are now expending membrane business in the world.

Beijing Origin Water

Beijing Origin Water Technology Co. Ltd. (BOW) was founded in 2001. It has become the largest membrane supplier in China, with a market share of about 70% (Figure 13.6).

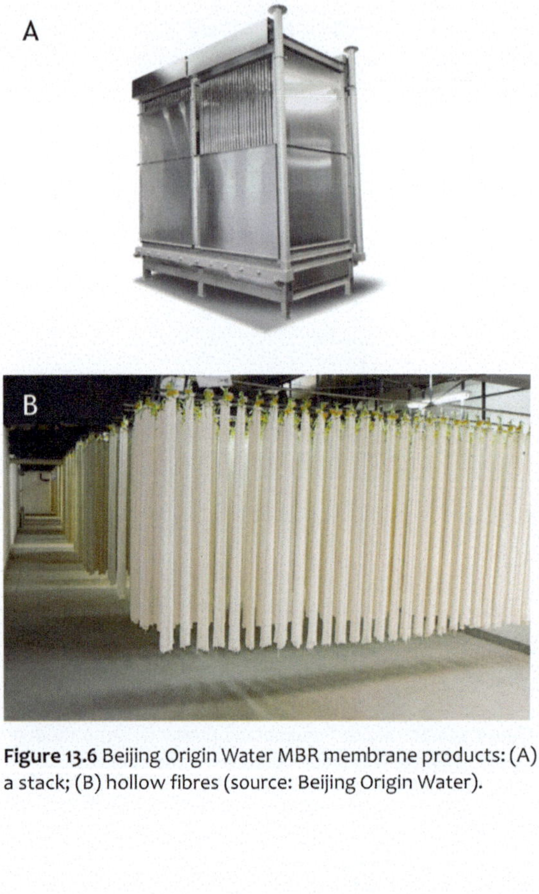

Figure 13.6 Beijing Origin Water MBR membrane products: (A) a stack; (B) hollow fibres (source: Beijing Origin Water).

It is reported that by the end of 2019 BOW had built over 30 MBR projects each with an overall capacity larger than 100,000 m^3/d. The BOW MBR membrane uses reinforced PVDF materials, which provide high chemical stability, and it has a pore size of 0.1 μm.

Memstar

Memstar is the membrane production branch of CITIC Envirotech Limited (CEL) Group, which is a leading membrane-based integrated environmental solutions provider specializing in water and wastewater treatment, water supply and recycling. Memstar specializes in advanced membrane-based water and wastewater treatment and manufactures both non-solvent-induced phase separation (NIPS) and thermally-induced phase separation (TIPS) PVDF hollow fibres; it is the largest manufacturer of TIPS PVDF products globally. Memstar has four manufacturing facilities in Singapore, China and the USA, using state-of-the-art manufacturing processes. The Memstar MBR technology uses PVDF hollow-fibre membranes with a nominal pore size 0.04 μm and 1.3 mm outer diameter. They provide membrane modules with an effective membrane area of 10 (SMM-1010), 12.5 (SMM-1013), 20 (SMM-1520), 25 (SMM-1525) or 30 (SMM-2030) m^2 (Figure 13.7).

Figure 13.7 Memstar membrane products: (A) a module; (B) a stack (or skid) (photo: Judd and Judd, 2011).

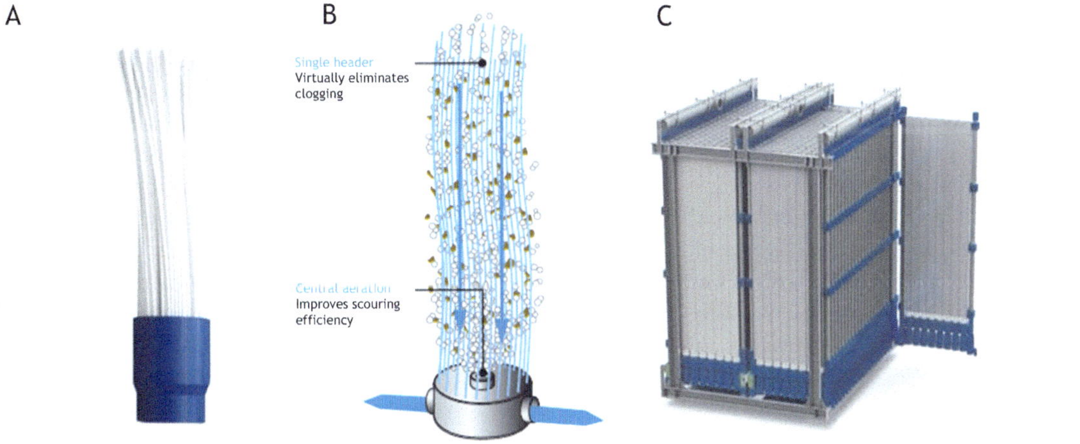

Figure 13.8 PURON® MBR and PULSION MBR membrane products: (A) PURON® hollow fibre membranes, (B) PURON® single header design, (C) PULSION® MBR module (images: Koch Separation Solutions, Inc.).

Koch

Puron GmbH was founded in late 2001 and acquired by Koch Separation Solutions, Inc. (formerly Koch Membrane Systems, Inc.) in 2004. The single-header ultrafiltration PURON® membrane bioreactor has transformed industrial and municipal wastewater treatment technology. This patented module design (Figure 13.8) features reinforced PVDF hollow fibres that are fixed only at the bottom, virtually eliminating the clogging build-up of hair, fibrous materials and sludge solids. Solids and particles, including bacteria, remain on the outside, while permeate is drawn through the membrane to the inside of the fibres. The aeration nozzle is cantered within the fibre bundle to scour the entire fibre length, minimizing power consumption. The next generation PULSION® MBR product offers up to 40% aeration energy reduction and 25% footprint reduction. This innovation pulses a large bubble through a chambered membrane fibre bundle, creating a piston-like pumping action that results in lower air and aeration energy requirements than traditional scouring methods. Improved recirculation of mixed liquor in the membrane module boosts achievable fluxes and overall performance. Optimized design and layout have reduced tank sizing while eliminating the need for air cycle valves. Minimized air flow applied to the membranes has decreased the size of air supply equipment by 50%.

13.2.5.2 Immersed FS products

Kubota

KUBOTA Submerged Membrane Unit® was developed in the late 1980s by the Kubota Corporation, who provides membrane systems for advanced wastewater treatment and regeneration (Figure 13.9).

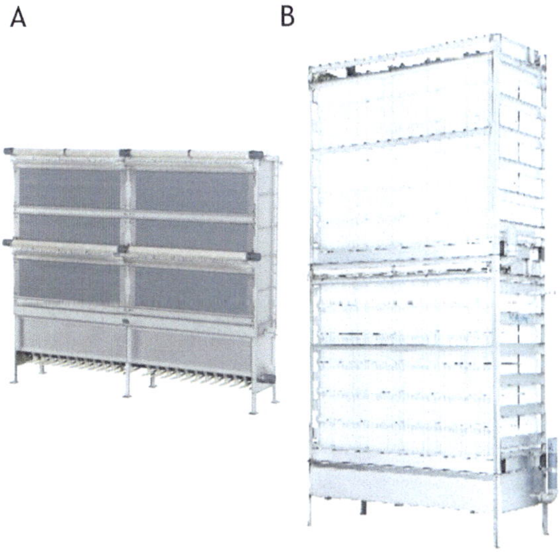

Figure 13.9 Example of Kubota (A) RM and (B) SP series (source: www.kubota.com).

KUBOTA Submerged Membrane Unit® has been employed in plenty of MBR processes for sewage and industrial wastewater treatment all over the world since the 1990s. The Membrane Unit consists of a Membrane Case and a Diffuser Case, and the Membrane Case houses multiple Membrane Cartridges with an average pore size of 0.2 μm. The Membrane Cartridge is Flat Sheet configuration with an effective filtration area of 0.80 m^2/cartridge (FS/FK series) or 1.45 m^2/plate (RM/RW series) (Figure 13.9). As of December 2019, Kubota has more than 6,000 MBR installations all over the world.

Toray

Toray is an established Japanese membrane manufacturer which offers high quality membranes and water treatment products globally (Figure 13.10).

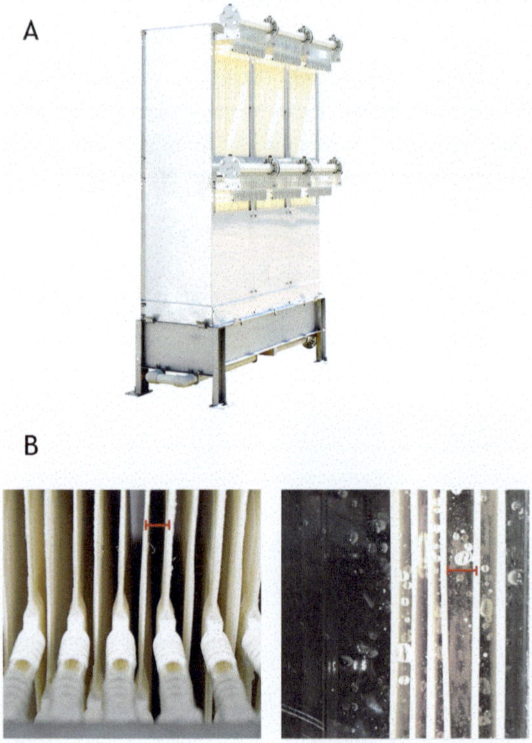

Figure 13.10 The Toray NHP membrane module: (A) the NHP-210-300S module; (B) a lateral view of membrane elements (source: www.toraywater.com).

Toray's MEMBRAY® MBR membranes are PVDF flat-sheet membranes on a polyethylene terephthalate (PET) non-woven support layer with a nominal pore size of 0.08 μm. Each sheet has a membrane area of 0.7-1.4 m^2. Toray has recently developed the new NHP series module. The NHP module has a similar packing density close to a hollow-fibre module, while maintaining the excellent features of a flat-sheet MBR module. This new module design provides higher productivity with a smaller footprint and higher cleaning efficiency at lower energy demand.

Sinap

Shanghai Sinap Membrane Technology Co., Ltd is jointly established by Shanghai Institute of Applied Physics of Chinese Academy of Science (Figure 13.11).

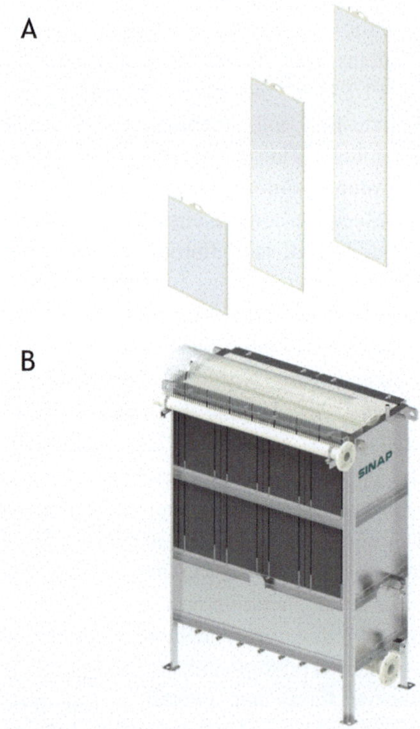

Figure 13.11 The Sinap MBR module: (A) Membrane panels and (B) Sinap module (source: Sinap).

Sinap is the leading flat-sheet membrane manufacturer in China. Sinap flat-sheet membrane uses PVDF as material and pore size is 0.1 μm. Sinap membrane has typical asymmetric structure, providing superior pollution resistance, low membrane intrinsic permeation resistance and better operational stability.

Meidensha

The Meidensha Co. is a Japanese company established in 1897 and engaged in the water processing engineering business. They have provided ceramic flat-sheet membranes for immersed MBRs since 2012 (Figure 13.12).

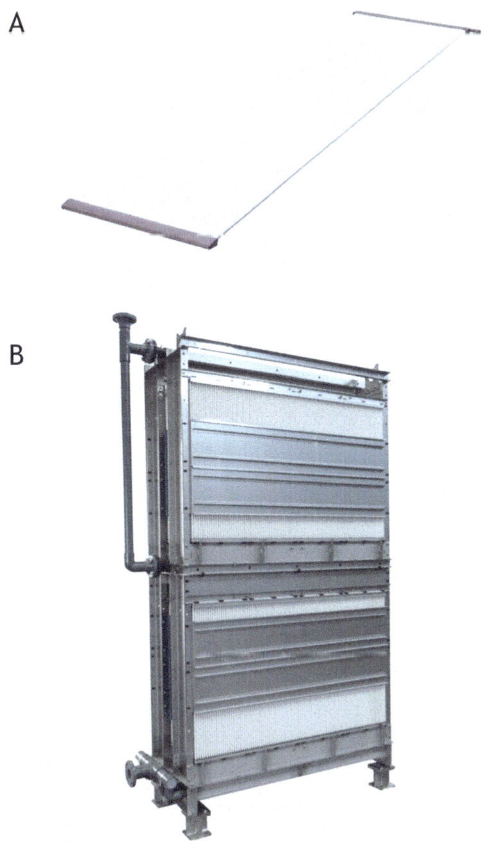

Figure 13.12 The ceramic flat-sheet membrane element and double-stack unit (source: www.meidensha.com).

Compared with conventional MBRs, their system combines the features of ceramic and flat-sheet membranes and offers several obvious advantages: (*i*) it is effective in treating challenged industrial wastewater containing solvents, oil, chemicals and/or suspended solids; (*ii*) the ceramic membrane is robust and exhibits resistance to chemical cleaning and air scrubbing; (*iii*) automatic cleaning can be used to reduce daily maintenance; (*iv*) the hydrophilic and smooth surface can maintain longer filtration performance; and (*v*) a smaller amount of scrubbing air is required to reduce energy consumption. The Meiden CH250 series membrane uses alumina material and the nominal pore size is 0.1 μm.

13.2.5.3 Tubular products

Pentair

X-Flow is the international membrane brand of Pentair and uniquely combines high-quality membranes with outstanding implementation know-how. In 2011, Pentair Inc. acquired X-Flow. The X-Flow AnMBR technology is preferred in a side-stream MBR configuration. The membrane modules are vertically arranged outside of the bioreactor tank in a setup of up to maximum of thirteen modules in parallel and two in series. The membrane material is PVDF and is optimized using the unique Helix technology of enhancing turbulence improving performance. Its mean pore size is 0.03 μm. In the past decade, Pentair X-Flow developed the Crossflow FFR technology especially for AnMBR systems. The Crossflow FFR technology with the aid of absolute barrier of the Compact 33V Helix membrane elements provides an optimal UF performance able to deliver excellent quality effluent that can be reused.

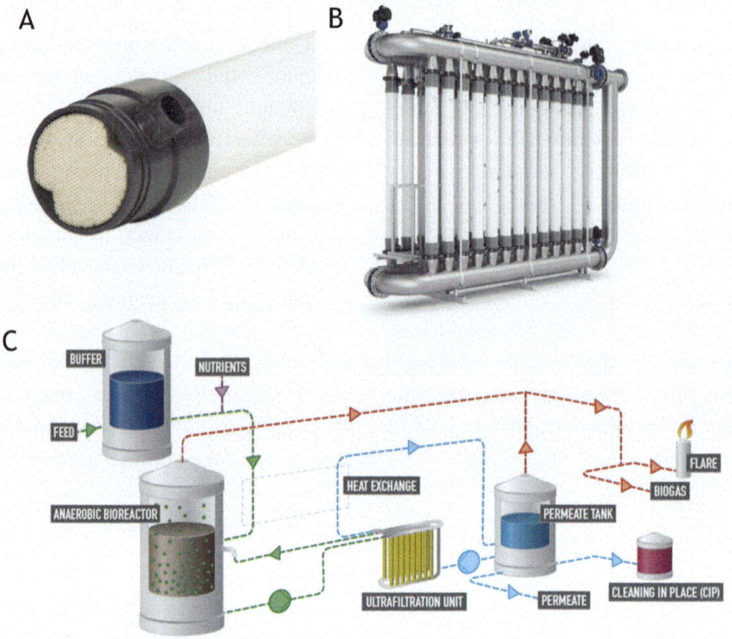

Figure 13.13 Pentair X-Flow module: A) the Compact 33V module; B) Anaerobic MBR; C) schematic of AnMBR process with Crossflow FFR technology (source: www.xflow.pentair.com).

13.3 WASTEWATER TREATMENT PERFORMANCE AND EFFLUENT QUALITY

13.3.1 Ordinary pollutant removal

The ordinary pollutants typically include organic matter (measured by chemical oxygen demand - COD or biological oxygen demand - BOD), ammonium nitrogen (NH_4^+-N), total nitrogen (TN), total phosphorus (TP), suspended solids (SS), and turbidity. The required treatment efficiency differs according to wastewater type (*e.g.* municipal wastewater vs. industrial wastewater), effluent demand (*e.g.* discharge vs. reuse), and specific standards adopted in different countries.

The primary version of MBR, *i.e.* a simple combination of the aerobic unit and membrane (termed O-MBR here), can be used to effectively remove COD, BOD, NH_4^+-N, and SS. The first wave of full-scale MBRs, established in the 1990s or 2000s, were mainly in the form of O-MBR. For TN removal, denitrification is required and, hence, an anoxic zone should be added before or after the aerobic tank, forming an anoxic/aerobic-MBR (AO-MBR) process. For simultaneous removal of the N and P nutrients using a biological method, an anaerobic zone is further added to trigger the microbial P release-and-uptake mechanism and form an anaerobic/anoxic/aerobic-MBR (AAO-MBR) process. The AAO-MBR process is the most prevalent in large-scale MBRs for municipal wastewater treatment. The detailed arrangement of the zones can differ according to the specific design of the process flow (Xiao *et al.*, 2014). In this chapter, the complete biological/membrane treatment process is regarded as MBR as a whole.

A survey of the treatment efficiencies of COD, NH_4^+-N, TN, and TP in full-scale municipal wastewater MBRs, from published literature worldwide, is given in Figure 13.14 (Xiao *et al.*, 2019).

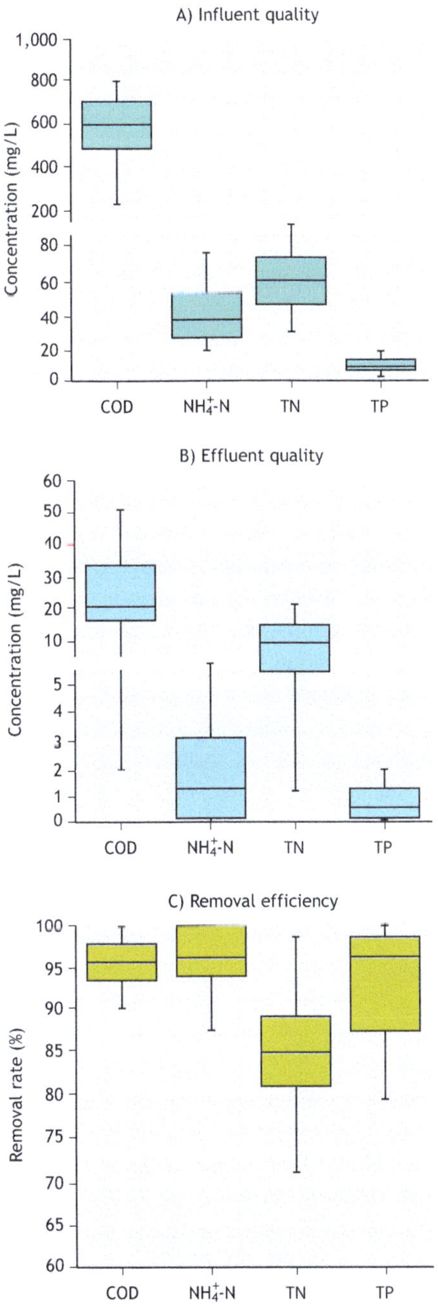

Figure 13.14 (A) influent concentrations, (B) effluent concentrations, and (C) removal rates of COD, NH_4^+-N, TN, and TP in full-scale MBRs treating municipal wastewater (modified from Xiao *et al.*, 2019).

The survey shows that the influent concentrations of COD, NH_4^+-N, TN, and TP were mainly in the ranges of 500-700, 25-50, 40-70, and 6-12 mg/l, respectively. The effluent concentrations of COD, NH_4^+-N, TN, and TP were mainly 10-30, 0-3, 5-15, and 0-1, respectively. An average rate of ~95% was achieved for COD and NH_4^+-N removal. Although with a larger fluctuation, the TN removal rate mostly reached 80-90%. The TP removal rate varied from 80% to 99% (mainly 87-98%), depending on the biological status and chemical dosages into the process for TP removal. In the case of China, the concentrations of COD, NH_4^+-N, TN, and TP in the municipal wastewater were 150-400, 15-30, 25-40, and 2-4 mg/l, respectively (data presented as interquartile ranges) (Sun *et al.*, 2016b). According to statistics for the influent and effluent quality, the average removal rates of BOD, COD, NH_4^+-N, TN, TP, and SS by all full-scale municipal WWTPs (mainly adopting CAS processes) are estimated to be at the level of 92%, 88%, 87%, 65%, 83%, and 93%, respectively (Sun *et al.*, 2016b). Compared with CAS processes, MBR is particularly effective in the removal of SS, COD, and NH_4^+-N, which was also verified by a long-term comparison of a full-scale MBR and CAS operated in parallel (located adjacently and sharing the same stream of wastewater). As for biological nutrient removal (BNR), MBR is also competitive for TN removal. Chemical precipitation of TP is sometimes required to assist in further removal of TP. Overall, MBR is extremely suitable for municipal wastewater treatment.

The application of MBR has also been extended to industrial wastewater treatment. An increasing number of MBRs have been applied to treat a variety of industrial wastewaters (Xiao *et al.*, 2019). The pollutant removal efficiencies vary largely according to the wastewater type. MBR has even been applied for full-scale leachate treatment and a recent survey shows that MBR units can also achieve over 90 % of BOD and COD removal and 95% of NH_4^+-N removal (Zhang *et al.*, 2020). Further research and development are definitely worthwhile.

There are several technical advances of MBR to support its excellence in ordinary pollutant removal (Xiao *et al.*, 2019). Undoubtedly, these membranes can directly reject sludge flocs, microorganism cells, and other suspended solids larger than 0.5 μm. The rejectability of these membranes can be enhanced when the membrane pores are narrowed due to adsorption/blocking of foulants or the membrane surface is covered by a dynamic foulant layer, enabling partial rejection of biopolymer clusters and debris, colloidal COD/N/P, and even soluble substances such as SMP and some less biodegradable solutes.

MF or UF membrane filtration, as a physical process, is not likely to straightforwardly alter the nature of microorganisms. However, it can profoundly enhance the biological process due to complete retention of the microorganisms and partial rejection of the pollutants. The complete separation of HRT and SRT enables MBR to maintain a high sludge concentration and a low food-to-microorganism (F/M) ratio (*i.e.* a lower sludge loading), and to keep the microorganisms in the endogenous respiratory period for more thorough degradation of the pollutants. For municipal wastewater treatment, the sludge loading of MBR is typically 0.03-0.10 $kgBOD_5/kgMLSS.d$, lower than that of CAS which is normally 0.05-0.15 $kgBOD_5/kgMLSS.d$. The co-retention of microorganisms and pollutants is also conducive to microbial co-metabolism for the removal of refractory pollutants. Again, due to membrane rejection, nitrifying bacteria are easier to enrich in the system and more resilient at low temperatures, which is critical to the stable removal of NH_4^+-N and beneficial for overall TN removal. To treat wastewater with low COD/TN ratios, an endogenous carbon source may be employed for denitrification via extension of the anoxic zone, which further reduces sludge production concomitantly.

For biological P removal, MBR acts like a double-edged sword. On the one hand, MBR can possess a high sludge concentration and a high P content per unit mass of sludge (*i.e.* an overall high P concentration in the sludge phase), and the membrane can reject some of the colloidal P as well; on the other hand, the typically longer SRT (15-30 d) of MBR means a slower discharge of the excess sludge from the system. Therefore, it is not conclusive whether MBR can facilitate biological P removal or not, and chemical removal of TP (by adding precipitants or coagulants) is sometimes needed to guarantee the effluent quality. It is noteworthy that adding precipitants or coagulants also has the merit of alleviating membrane fouling.

Finally, due to the complete rejection of biomass by the membrane, the MBR system is more resistant to the sludge bulking problems which usually trouble CAS processes. Compared with CAS, a more stable operation and pollutant removal may be achieved by MBR in cases of high-strength wastewater, a frequent shock of influent quality, fragile microbiota, or unstable ambient conditions. Moreover, MBR is more flexible in process configuration, easier for modularized installation, and more convenient for automatic operation. These provide MBR with additional opportunities for applications in the areas of *e.g.* specific industrial wastewater treatment, decentralized wastewater treatment (such as in rural areas), and anaerobic processes (such as AnMBR).

13.3.2 Hygiene water quality

The presence of pathogens is of primary concern in wastewater reclamation and reuse for the purposes of *e.g.* lavatory flushing, vehicle washing, urban greening, recreational environment reuse, and farmland irrigation. Pathogens cover a wide range of protozoa (such as Cryptosporidium and Giardia), helminths (as well as their eggs), fungal spores, pathogenic bacteria (such as faecal coliform and enterococci), and viruses (such as enterovirus, adenovirus, and norovirus). The health risks due to direct or indirect human exposure to these pathogens should be seriously considered when designing wastewater reclamation processes. Faecal coliform is usually employed as an indicator of pathogenic bacteria in wastewater reclamation and reuse standards. It is noted that pathogenic bacteria are also subject to wastewater discharge standards, due to their profound risks to the ecosystems and public health. F-specific coliphage, somatic coliphage, and phage T4 are often employed as indicators of viruses in MBR research.

It is not surprising that MF or UF membranes can effectively reject protozoa, helminths, fungal spores, and bacteria, because they are larger than the membranes. Figure 13.1 presents a comparison between the sizes of membrane pores and pathogens. Obviously, membrane rejection is superior to gravitational settlement in bacteria removal. A survey of full-scale WWTPs shows that the logarithmic removal of *E. coli* and enterococci by MBR was approximately 3 orders higher than that by CAS, as depicted in Figure 13.15.

Unlike the obvious rejection of bacteria by the membrane via steric exclusion, the rejection of viruses raises uncertainty, due to the comparable or even smaller size of viruses than the membrane pores, as illustrated in Figure 13.1. In practice, full-scale MBRs demonstrate a 3-6 log removal of the tested viruses (allowing some fluctuation), much higher than CAS plants do (1-3 log removal) (Figure 13.15). Unlike those of bacteria, the removal mechanisms of viruses are complicated and have not yet been fully understood. There are several explanations in the literature. A prevailing one is that the membrane and the gel/cake foulant layer together act as a composite filter in tandem to achieve increased rejection of viruses. The membrane itself can partly reject viruses, with the rejection rate depending on the pore size. The pore channels can be narrowed due to adsorption of foulants onto the pore walls. The gel layer is a matrix of biopolymers with high porosity and water content but very low permeability, and it is compressible and becomes denser at higher trans-membrane pressures (TMP). The effective pore size of the calcium alginate gel layer is estimated to be 5-30 nm (Huang *et al.*, 2010). The gel layer can contribute largely to the overall rejection of viruses. It should be noted, however, that overgrowth of the gel layer exacerbates membrane fouling. In addition, the viruses can attach to the suspended solids in the mixed liquor, or be entrapped into the sludge flocs, and then be rejected by the membrane. Microbial inactivation of the viruses in the sludge mixed liquor is also a possible mechanism for the removal.

Compared with chemical disinfection technologies (such as chlorination, UV radiation, and advanced oxidation), membrane-based rejection of the pathogens is a gentle physical process that bypasses the risks of disinfection by-products, germ resistance, and involuntary mutations.

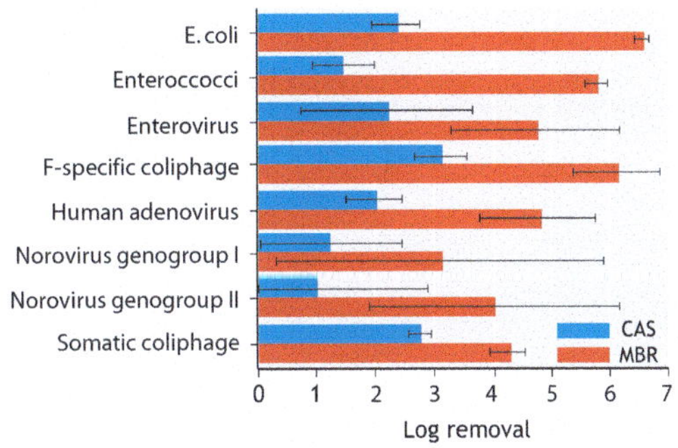

Figure 13.15 Removal efficiencies of bacteria and viruses in full-scale MBR and CAS plants (modified from Xiao *et al.*, 2019).

13.3.3 Emerging pollutant removal

Trace organic pollutants (TrOPs), or micro-organic pollutants, have raised public concerns for water safety. These pollutants can exert acute or chronic toxicity even at trace concentrations, and are usually persistent or pseudo-persistent in the aquatic environment. The occurrence of TrOPs in the effluents of WWTPs should be seriously considered, no matter whether the effluents are to be discharged or reused. The most widely investigated TrOPs include endocrine disrupting compounds (EDCs) and pharmaceutical and personal care products (PPCPs), categorized according to the impacts and usage, respectively.

EDCs include a range of phenolic compounds (such as alkylphenols, and alkylphenol polyethoxylates) and steroidal estrogens. PPCPs can be classified, according to their medical effects, into: anti-inflammatory and analgesic drugs, antibiotics, lipid regulating drugs, musks, among others.

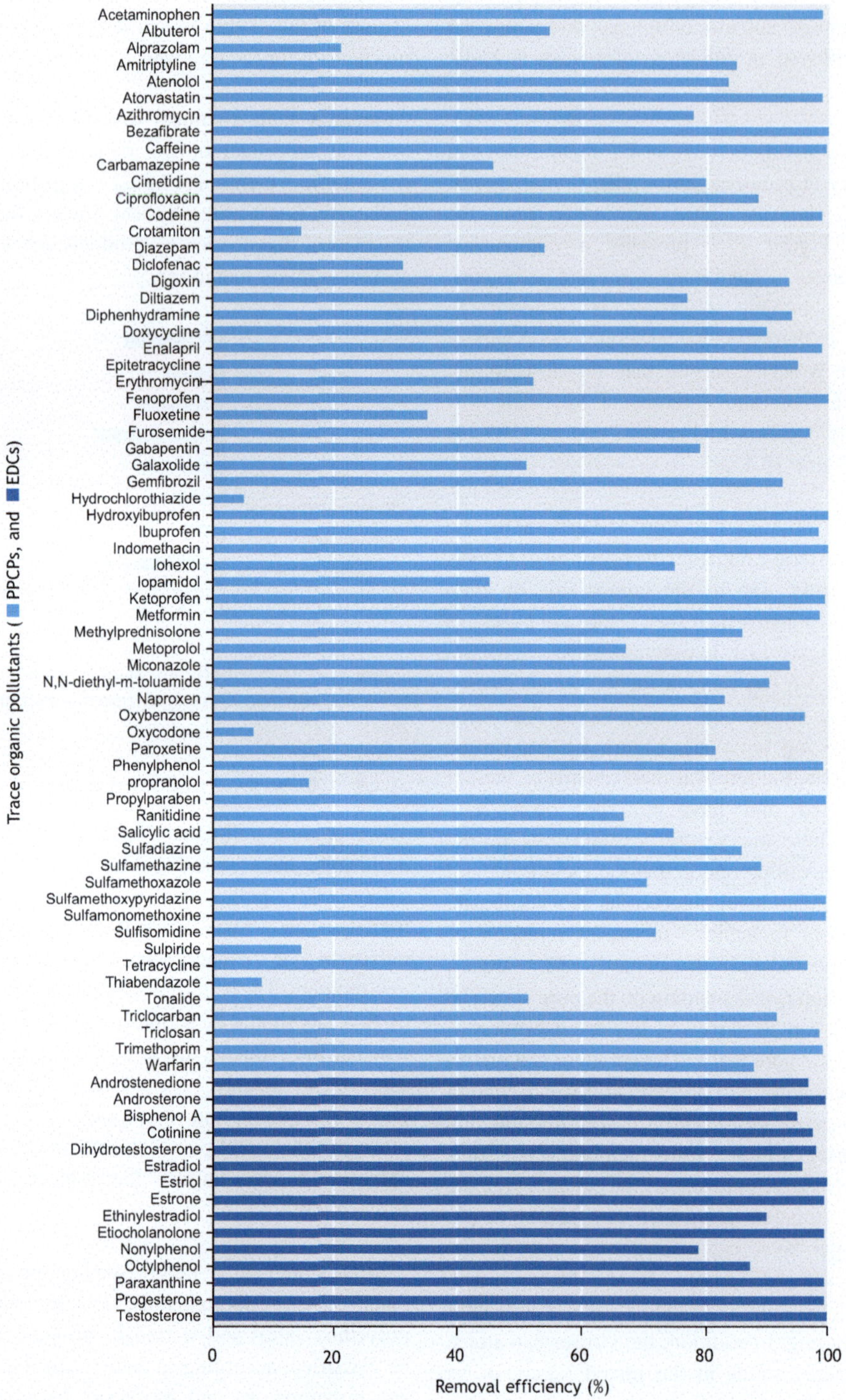

Figure 13.16 Removal efficiencies of EDCs and PPCPs in full-scale MBRs (adopted from Xiao *et al.*, 2019).

The antibiotics can be subdivided into sulfonamides, quinolones, tetracyclines, macrolides, etc. Note that the pollutants listed here are just the tip of the iceberg. There are far more TrOPs, including synthetic chemicals and intermediate metabolites, yet to be properly monitored and unveiled.

The individual concentrations of the known EDCs in the municipal wastewater typically range from 1 ng/l to 1 μg/l (Xue *et al.*, 2010), variable according to the type of EDCs. The individual concentrations of detectable PPCPs are usually 10 ng/l to 10 μg/l in raw wastewater (Sui *et al.*, 2011). Concentrations are quite variable among the different kinds of antibiotics (Gao *et al.*, 2012). The removal efficiencies of EDCs and PPCPs in full-scale MBRs were surveyed, with the results shown in Figure 13.16. About half of the TrOPs exhibited a high removal rate of over 95%, and two thirds of the TrOPs had a removal rate of over 80%, whereas a few PPCPs had a removal efficiency lower than 20%. The large variability in the removal performance was attributable to the widely varied biodegradability and hydrophobicity which ultimately stem from the chemical structure of the compounds.

TrOPs in the MBR system may undergo adsorption, degradation, rejection, and evaporation. They can be adsorbed by the bacterial cells, EPS, suspended solids, and colloids in the mixed liquor. The adsorption is mainly determined by the hydrophobicity of TrOPs which is reflected by the octanol/water partition coefficient (K_{ow}). After being adsorbed into the sludge phase, the pollutants may then undergo biodegradation, and the two-step process can be described by a two-phase fate model where the sludge-water partition coefficient, water-sludge mass transfer rate constant, and biodegradation rate constant are the key parameters (Urase and Kikuta, 2005; Xue *et al.*, 2010). The MF or UF membrane has a marginal effect on the direct retention of TrOPs (via short-term adsorption into the pores); however, the adsorption and size exclusion by the gel/cake layer are non-negligible. The TrOPs can be readily rejected when they are adsorbed in the sludge phase or bound to colloids (Figure 13.1). Additionally, a small portion of TrOPs (particularly the uncharged compounds with large values for Henry's constant) may evaporate under the aeration conditions in the aerobic or membrane tanks.

The removal efficiencies of TrOPs are mainly dependent on their biodegradability and hydrophobicity. For TrOPs which are readily degradable, the removal efficiencies by MBR and CAS are normally comparable, whereas for refractory TrOPs, the MBR is more effective in biological degradation. For TrOPs with strong hydrophobicity that are readily adsorbable by the sludge, the higher sludge concentrations as well as probably higher EPS contents in the MBR are conducive to the TrOPs uptake, but the longer SRT is unfavourable for the discharge of excess sludge. For TrOPs with weaker hydrophobicity but still adsorbable by the particles and colloids in the mixed liquor, the MBR is advantageous in rejecting these chemicals and improving the effluent quality. Overall, the removal efficiencies are jointly influenced by the mixed liquor properties (*e.g.* MLSS and EPS) and the operational conditions (such as SRT).

The adsorption may be strengthened by adding adsorbents such as powdered activated carbon (PAC) into the MBR system. The membrane will guarantee complete rejection of the adsorbent particles. Note that adsorption is merely a physical process that transfers the TrOPs from the aqueous phase to the solid/sludge phase. Ultimate removal of the pollutants still relies on degradation or proper disposal of the excess sludge. For TrOPs that are neither adsorbable nor biodegradable, advanced treatment (such as ozonation and NF) is incorporated into the system to ensure high quality of the effluent.

The biological responses to the pharmaceuticals in wastewater systems are also noteworthy. WWTPs are potential reservoirs of antibiotic residues, antibiotic resistant bacteria (ARB), and antibiotic resistance genes (ARGs), and the release of these contaminants into the environment is considered to be risky to the ecosystems and human health. It has been reported that the MBR outperforms CAS in the removal of most ARB and ARGs, owing to the efficiency of membrane rejection (Le *et al.*, 2018). However, some ARGs still persist in the membrane effluent, and further efforts are required to achieve a thorough elimination of these substances.

Microplastics, commonly defined as plastic particles smaller than 5 mm, have recently raised public concerns for human health and ecological safety due to their potential ecotoxicological impacts and persistence in the environment. These materials are difficult to settle in sedimentation tanks and, as such, have not been satisfactorily removed in conventional WWTPs. In comparison, the MBR can achieve over 99% removal of microplastics (Lares *et al.*, 2018), owing to the evident superiority of membrane rejection. The membrane can serve as an effective barrier to block the discharge of microplastics from WWTPs into the environment.

13.3.4 Energy recovery

Nowadays wastewater is regarded as a stream that contains valuable resources rather than just a contaminant, and energy and resource recovery from wastewater is a vital development aspect of a sustainable wastewater treatment system besides water purification. The anaerobic treatment process is technically promising to realize this goal, and AnMBRs are attracting lots of attention due to their advantages in retention of slow-growing anaerobes, excellent and stable effluent quality, efficient bioenergy production, their small footprint, etc. compared with conventional anaerobic processes. To date, AnMBRs have been successfully applied to treat high-strength wastewater (for example, food-processing wastewater) with COD of 10^2-10^3 mg/l, with pollutant removal efficiencies of mostly >90%. With relatively low organic strength but large water volume, municipal wastewater treatment has recently become a hot application field of AnMBRs, and various configurations and modifications have been proposed to enhance the treatment performance and methane production. In general, AnMBRs produce 0.1-0.5 $m^3CH_4/kgCOD_{removed}$ of methane in municipal wastewater treatment (Wei *et al.*, 2014; Hu *et al.*, 2018).

Methane production in AnMBRs is affected by both operational parameters and the bacterial competition in the reactors. The fermentation temperature, pH, and organic loading rate can significantly influence methane production. Mesophilic (20-42°C) and psychrophilic (0-20°C) conditions are favoured for AnMBRs, and higher methane yield was observed at higher temperatures. Additionally, methane loss, namely dissolved methane in the effluent, is temperature-related. Higher solubility of methane at lower temperature causes more methane loss especially in hydraulic pressure-driven AnMBRs. Besides nearly neutral pH, an appropriate organic loading rate is also a crucial limiting parameter of biogas yield, and a theoretical optimal methane production of 0.382 $m^3CH_4/kgCOD_{removed}$ was proposed at an organic loading rate of 6 $kgCOD/m^3.d$ for AnMBRs treating municipal wastewater (Wei *et al.*, 2014). Internal competition between sulphate-reducing bacteria and methane-producing archaea for organic substances in AnMBRs are affected if insufficient organic load and/or high sulphate content is provided (Chen *et al.*, 2016; Lei *et al.*, 2018).

As mentioned above, methane loss is an existing problem in AnMBRs, and harvesting the dissolved methane in permeate is worth investigating from perspectives of both increasing biogas efficiency and reducing greenhouse gas emission. Besides, membrane fouling in AnMBRs is also unavoidable. To achieve a steady operation of AnMBRs, multiple fouling control strategies can be applied including optimizing operational conditions, dosing additives (such as granular activated carbon), increasing the shear force on the membrane surface by mechanical movement or biogas sparging, regulating chemical cleaning, applying ceramic membranes, etc. It should be noted that, with further research and development, AnMBR has a promising practical application to achieve efficient energy recovery from municipal wastewater treatment.

13.4 MEMBRANE FOULING AND CONTROL

13.4.1 Definition of membrane fouling

Membrane fouling is normally defined as the accumulation of sludge flocs, colloids, solutes as well as inorganic matter in membrane pores and/or the deposition of such compounds on membrane surfaces (Meng *et al.*, 2009). Such behaviour is operationally dependent on various factors such as membrane pore size and the imposed flux as well as the operating time of membrane modules. The consequence of membrane fouling in MBRs is the decline in water production and the requirement of membrane cleaning (*e.g.* physical and chemical cleaning). In addition, the fouling itself and the implemented cleaning both result in accelerated membrane aging, thus shortening the membrane life. Therefore, membrane fouling is one of the major issues limiting the engineering application of MBRs.

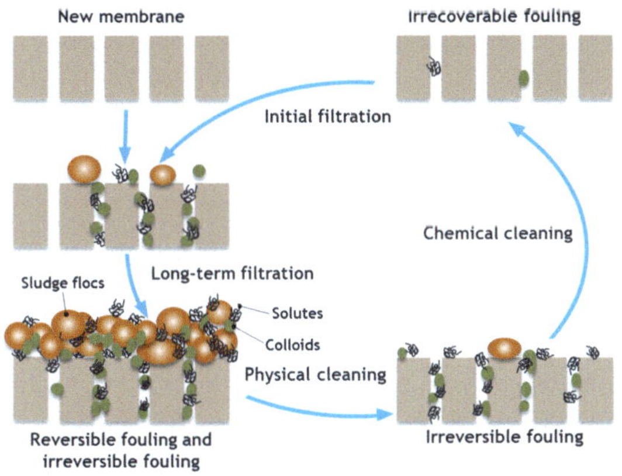

Figure 13.17 Schematic illustration of the formation and removal of membrane foulants in MBRs (adopted from Meng et al., 2009).

According to the removal propensity of foulants by physical and chemical cleaning (Figure 13.17), fouling can be either reversible fouling that can be easily eliminated by physical cleaning, irreversible fouling that cannot be removed by physical cleaning but can be readily removed by chemical cleaning, or irrecoverable fouling that cannot be removed by either. Thus, the development rates of reversible fouling and irreversible fouling determine the implementation of physical and chemical cleaning, respectively. In comparison, the accumulation of irrecoverable fouling can be considered as a potential indicator of the deterioration of membrane performance.

13.4.2 Characterization of membrane fouling

Any compounds occurring in wastewater stream or in the mixed liquor of activated sludge are potential foulants for MBRs. Membrane foulants can be characterized from compound or community level to chemical or species level, depending on the approaches used for fouling characterization (Meng *et al.*, 2017). At compound level, EPS and SMP are widely acknowledged to play dominate roles in membrane fouling development. Polysaccharides are considered the major component of biopolymers (>100 kDa), which are known as the key fouling-causing substances (Rosenberger *et al.*, 2006; Shen *et al.*, 2012). In fact, polysaccharides have higher membrane rejection degrees than proteins and humic substances because of the large size of polysaccharides, leading to the higher accumulation potentials of polysaccharides in both mixed liquor and bio-cake (Meng *et al.*, 2011b). Additionally, polysaccharides have a strong gelling propensity, which is associated with the cross-linked structure of polysaccharide molecules. In particular, the presence of multi-valent cations in wastewater streams can considerably increase the gel layer formation on membranes. Moreover, other biomolecules, such as proteins and humic substances, can also contribute to membrane fouling development via direct deposition in the bio-cake or the inter-molecule interactions (Wang *et al.*, 2015; Zheng *et al.*, 2014).

Bacteria deposition as well as biofilm development on membranes is another key factor leading to membrane fouling development (Blanpain-Avet *et al.*, 2011; Gutman *et al.*, 2013). On the one hand, the deposited bacteria or the developed biofilms on membranes can produce a large amount of EPS or SMP and may contribute to the biopolymer accumulation on membranes. On the other hand, the bacteria can degrade some biodegradable biopolymers from the mixed liquor of activated sludge. Thus, the presence of microbes on membranes makes the fouling behaviour far more complex than we currently know. For instance, sub-visible particles, which are of micro-meter size, are key foulants during the operation of AnMBRs (Zhou *et al.*, 2019). These particles can be considered as both biopolymer clusters and free bacteria.

High-throughput sequencing technology is powerful for characterizing the microbial communities in bio-cake. Nonetheless, identification and characterization of the crucial fouling-causing microbiomes should not be confined simply to their relative quantities but also to their roles in fouling development. Some rare species (with a low relative abundance) have been found to play over-proportional roles in fouling development (Zhang *et al.*, 2018). Bio-cake communities are not randomly assembled, but are regulated by mechanical force, local substrate and species-species associations (Xu *et al.*, 2019). For instance, those bio-cake communities formed at high flux are mainly shaped by deterministic processes, whereas those at low flux are primarily guided by stochastic dispersal from activated sludge. However, the bacterial assemblies of bio-cake communities in anaerobic MBRs are usually formed in a non-selective stochastic manner (Cheng *et al.*, 2019), indicating that aerobic and anaerobic MBRs have distinctly different fouling mechanisms.

In the future, an understanding of the dynamic changes in both biopolymers and bacteria in real MBR plants is needed to more clearly reveal MBR fouling mechanisms. A source tracking method would be of some help in identifying the dynamic process of membrane fouling and the source of membrane foulants during their long-term operation. Comprehensive understanding of the functional and metabolic processes in bio-cake, multi-omics (*e.g.* metatranscriptomics, metaproteomics and metabolomics) is recommended. Such studies would eventually help the development of fouling control strategies.

13.4.3 Comprehensive control strategies for membrane fouling

Normally, the essence of fouling control in most membrane processes is to regulate the interactions between membranes and foulants by various methods (Figure 13.18).

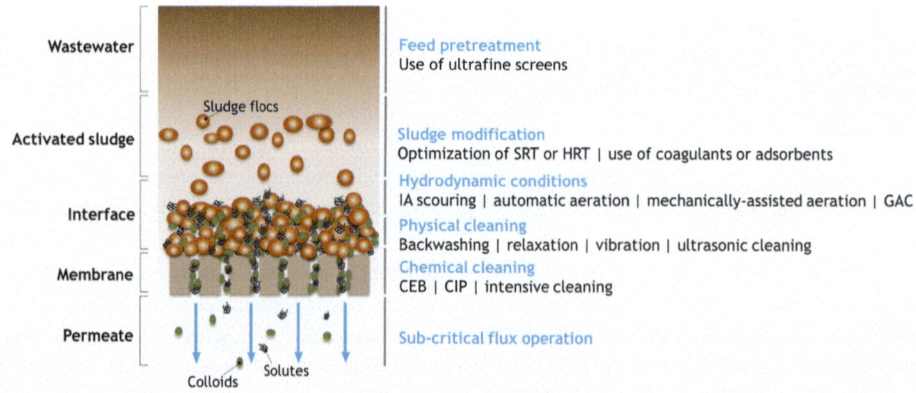

Figure 13.18 Fouling control strategies for MBR operation: SRT, solids retention time; HRT, hydraulic retention time; IA, intermittent aeration; CEB, chemically-enhanced backflush; CIP, cleaning in place; GAC, granular activated carbon.

Overall, fouling control in MBRs can be achieved by the following: (*i*) feed pretreatment to remove potential foulants (*e.g.* fibres, hairs and grits) from wastewater streams; (*ii*) enhancement of the hydrodynamic conditions to regulate the membrane-foulant interactions; (*iii*) determination of a proper operating flux for the membranes; (*iv*) physical and chemical cleaning of the fouled membranes to remove the deposited or accumulated foulants on/in membranes; (*v*) modification of the bulk sludge to improve the filterability of the mixed liquor. Details of these fouling control strategies are presented in the following section (Section 13.4).

13.4.4 Optimization of membrane operation conditions

13.4.4.1 Feed pretreatment

Before being pumped to the bioreactors in WWTPs, wastewater streams should be pre-treated with screens. Compared with conventional WWTPs using coarse screens, MBRs require the use of ultrafine screens (0.2-1.0 mm). This is because coarse screens cannot remove large or long hairs and other debris which tend to clog or accumulate in membrane channels. The clogging not only accelerates fouling development, but also reduces the aeration scouring effects. As the clogging matter cannot be effectively removed by physical and/or chemical cleaning, clogging leads to unscheduled manual maintenance in MBR plants. Details of the pretreatment design in MBRs can be found in Section 13.5.2.

13.4.4.2 Enhancement of hydrodynamic conditions

The mechanical scouring of aeration can put the sludge flocs in a suspended state and mitigate membrane fouling. Thus, air scouring or aeration is a widely used method for membrane cleaning. Improvement of aeration intensity can facilitate physical cleaning due to the increase of cross-flow velocity and shear stress. In order to save energy for MBR operation, the optimization of aeration, including aeration rates, bubble size and aeration modes, is of great importance for fouling control. Intermittent or cyclic aeration has been found to be an efficient fouling control strategy with less energy consumption. The 'eco-aeration' strategy using intermittent aeration with 10/30 intervals saves up to 50% of the energy consumed in aeration (Buer and Cumin, 2010).

Automatic aeration control, including DO-feedback, ammonium feedback and TMP-feedback automatic control, can achieve energy savings (Gabarrón *et al.*, 2015; Sun *et al.*, 2016a). It is of some interest is that the intermittent aeration mode can potentially enhance nutrient removal in MBRs because the functional bacteria responsible for N and P elimination can be enriched as a result of the shift of high/low DO concentrations (Guadie *et al.*, 2014).

The addition of hard particles (granular-activated carbon and plastic beads) into the mixed liquor can considerably intensify the fluid dynamics in the module zone (Rosenberger *et al.*, 2016). Thus, such mechanical cleaning in combination with aeration scouring can achieve efficient fouling control in MBRs. Plastic beads (Microdyn-Nadir GmbH) have been commercially applied for fouling control in flat-sheet membranes (Rosenberger *et al.*, 2016). The plastic beads which are fluidized by aeration scouring can create a cutting effect on bio-cake (Rosenberger *et al.*, 2011). A recent study showed that the addition of plastic granules (polyethylene glycol, 4 mm) with a filling ratio of 5% in the reactor decreases the aeration rate by 50% (Kurita *et al.*, 2015). However, the introduction of plastic particles or beads may cause membrane damage (Kurita *et al.*, 2015; Siembida *et al.*, 2010) and therefore optimization of the shape and hardness of the plastic beads and their dosage is needed during real applications.

13.4.4.3 Optimization of membrane flux

The trade-off between imposed flux and fouling control is very important for MBR operation. A low flux can ensure a low fouling development rate, but with low water production. Operationally, critical flux (defined as 'there exists a flux below which a decline of flux with times does not occur; above it fouling is observed' (Field *et al.*, 1995)) determination is a prerequisite for MBR operation, as a sub-critical flux operation is usually preferred by most membrane suppliers. It should be noted that fouling can inevitably occur even when the membrane is operated below the critical flux. Particularly during long-term operation, membrane foulants accumulate even at very low operating flux. In addition, the critical flux for a given membrane varies with ambient conditions such as temperature, sludge properties, accumulation of irreversible fouling and membrane ageing. Therefore, the critical flux in a given MBR plant should be periodically re-evaluated and the fouling control strategies should be updated accordingly.

13.4.5 Cleaning fouled membranes

Membrane cleaning is usually required during MBR operation to maintain membrane performance. It is well known that membrane cleaning includes physical cleaning and chemical cleaning. Physical cleaning can eliminate loosely attached membrane foulants, often termed 'reversible fouling', while chemical cleaning can remove some stubborn materials on/in the membranes. In MBRs, these two methods can be used separately, but are usually applied in combinations to effectively recover the membrane permeability.

13.4.5.1 Physical cleaning

Maintenance operation of MBRs normally relies on physical cleaning through aeration scouring, backwashing *i.e.* reversing the flow, or relaxation, which is simply ceasing permeation whilst continuing to scour the membrane with aeration. With the increasing research on physical cleaning, a series of new cleaning strategies have been also proposed, for instance, online ultra-sonication, vibration/rotation, etc.

Backwashing

The key cleaning parameters for backwashing using permeate are backwash flux, duration, and frequency. A variety of backwashing frequencies and durations are used in MBRs *i.e.* less frequent and longer backwashing (7-16 min filtration/30-60 s backwashing), or more frequent and shorter backwashing (5-12 min filtration/5-20 s backwashing). An optimal backwashing frequency and duration is dependent on membrane type, operational parameters (flux, temperature, etc.), backwashing flux, the TMP variations, and foulant properties.

Relaxation

Relaxation allows diffusive back transport of foulants away from membrane surfaces driven by a concentration gradient, and is often carried out in conjunction with aeration scouring to enhance the diffusion. It is widely accepted that the intermittent filtration mode (filtration coupled with relaxation) can alleviate membrane fouling, which has been widely incorporated into MBRs as a standard operating strategy for enhancing membrane performance. In some cases, relaxation combined with backwashing can be used to enhance cleaning efficacy.

Ultrasonic cleaning

Ultrasonic cleaning is effective in alleviating the concentration polarization and removing the cake layers on the membrane, and it is used in various membrane filtration processes because of several advantages, *e.g.* no chemical reagents and not interrupting the filtration process. There are several key parameters influencing cleaning efficiencies, *e.g.* ultrasonic frequency, power density, and duration, which can be used for the optimization of ultrasonic cleaning.

Vibration/rotation

Vibration and rotation of membranes can normally generate high shear or turbulence at the membrane surface, which can enable online cleaning of fouled membranes.

13.4.5.2 Chemical cleaning

Chemical cleaning is an essential process for the permeability recovery of severely fouled membranes because of the involvement of chemical reagents. Typically, *in situ* and *ex situ* cleaning can be applied depending on the fouling status. *In situ* cleaning includes online cleaning where both mixed liquor and membrane modules are not transferred, and recovery cleaning that is performed in the membrane tanks after the mixed liquor is transferred out from the membrane tank into other tanks. *Ex situ* cleaning is also a type of recovery cleaning which requires the fouled membranes to be moved to cleaning tanks.

Online chemical cleaning, including cleaning in place (CIP) and chemically enhanced backflush (CEB), is normally aimed to maintain the membrane permeability, thus it is also called maintenance cleaning. Such maintenance cleaning can be intensified by increasing the concentrations of chemical reagents. In comparison, recovery cleaning in either *in situ* or *ex situ* mode aims to remove most foulants on the membrane surface and in pores, which is conducted using high concentrations of chemical reagents and a long exposure time.

CIP

CIP is designed to maintain membrane permeability and reduce the frequency of offline cleaning. CIP is usually carried out at intervals of between 1 and 6 weeks in MBRs. The most common duration for such maintenance cleaning is 1-3 h. NaOCl, the prevalent cleaning reagent in full-scale MBRs, is often combined with citric acid during cleaning. Popular NaOCl and citric acid concentrations for CIP are 300-2,000 mg/l and 0.2-1.5 wt %, respectively. It is evident that CIP requires only moderate chemical concentrations compared to offline cleaning.

CEB

Through adding a low concentration of cleaning agents into the backflush water, the cleaning efficiency of backflushing is enhanced. CEB is also part of maintenance cleaning. It is performed less frequently than normal physical backwashing but more often compared to regular CIP. CEB can be carried out daily or up to every 7-14 d, but usually on a daily basis. The concentration of chemical reagents used in CEB is generally lower than that of CIP and offline cleaning. The typical concentration for CEB using NaOCl is in the range of 100-500 mg/l.

Recovery cleaning

Recovery cleaning is normally conducted once or twice per year. This type of chemical cleaning uses chemical reagents with high concentrations, *e.g.* 0.2-0.3 wt % NaOCl, 0.5-1 wt % oxalic acid, and 0.2-0.3 wt % citric acid. The entire cleaning process can last for a couple of hours depending on the fouling status of membranes. Through recovery chemical cleaning, most foulants on the membrane surface or in membrane pores can be largely removed or damaged due to the exposure to high concentrations of chemical reagents and the long exposure time. Nevertheless, changes or damage to membranes caused by long-term chemical cleaning is unavoidable. These changes will affect the membrane performance such as flux and filtration resistance. With respect to CIP and CEB, microbial activities and bulk sludge can also be impacted by the chemical reagents during membrane cleaning. Moreover, the waste stream should be paid more attention and especially the overdosing of cleaning chemicals should be avoided.

13.4.6 Improving the filterability of mixed liquor

Because of the ultimate roles of EPS/SMP and bacterial species, HRT and SRT are very important in fouling development (Meng *et al.*, 2009). That is, the optimization of HRT and SRT will help to improve the filterability of the mixed liquor in MBRs. However, the design of such reactor parameters is normally targeted on nutrient removal including nitrogen and phosphorous removal.

The addition of adsorbents or coagulants into the mixed liquor can be adopted because of their positive roles in decreasing the level of SMP or EPS. The addition of powdered activated carbon (PAC) is a convenient method for decreasing SMP levels. PAC can not only adsorb biomolecules in the sludge, but also acts as a carrier for biomass growth and it reduces floc breakage (Wang *et al.* 2020). More interestingly, the presence of PAC can enhance the aeration scouring effects on the membrane surface. However, the overuse of such adsorbents may result in the production of more waste sludge.

Because of the charge neutralisation and bridging of coagulants, addition of an optimum coagulant concentration can induce lower SMP concentration, lower hydrophobicity, and better flocculation, which results in decreases in cake resistance and pore blocking. So far, alum, ferric chloride and chitosan have been used as coagulants in MBRs (Wu *et al.*, 2006). Additionally, commercial products, based on cationic polymer-based compounds, have been developed for MBRs, *e.g.* the MPE50 and MPL30 developed by Nalco and KD452 by Adipap.

13.4.7 Other potential fouling control methods

13.4.7.1 Biological methods

In addition to physical cleaning and chemical cleaning, biologically-based methods have also recently been acknowledged to be a sustainable strategy for fouling control in MBRs. The bacteria on membranes can secrete molecules called quorum sensing (QS) signals or autoinducers (AIs) that can regulate their population densities and biofilm formation (Grandclement *et al.*, 2016). Focusing on this phenomenon, the quorum quenching (QQ)-based biological control method was developed (Lee *et al.*, 2016), which can prevent cell-cell or membrane-cell interactions in three possible ways (Grandclement *et al.*, 2016; Lade *et al.*, 2014): (*i*) disturbance of synthesis of QS signals or AIs, (*ii*) degradation of QS signals or AIs, and (*iii*) interdiction of transport or acceptance of QS signals or AIs. A number of lab-scale and pilot-scale studies have shown the strong potential of QQ-based methods in fouling control using QQ bacteria immobilized on magnetic enzyme carriers (MECs) (Yeon *et al.*, 2009), enzyme-immobilized membranes (Kim *et al.*, 2011; Oh *et al.*, 2012) and beads (Kim *et al.*, 2013; Lee *et al.*, 2016).

Moreover, protozoans and metazoans, which are the major consumers of bacteria and ubiquitously occur in wastewater treatment plants, are very significant for fouling control. The movement of such worms leads to the formation of porous and sponge-like structures of bio-cake, as reported in a pilot-scale study (Jabornig and Podmirseg, 2015).

13.4.7.2 Electrically-assisted approaches

Electrically-assisted approaches (*e.g.* electrocoagulation, electrophoresis) have been incorporated into MBRs to improve membrane permeability. The electrophoresis (EP) induced by an electric field yields an electric repulsive force between foulants and membranes (Akamatsu *et al.*, 2010; Hawari *et al.*, 2015). In the EP-MBR a cathode is normally set near the membranes at the permeate side and an anode is submerged in the bulk sludge. The use of conductive membranes, which work as both separation membranes and cathodes, simplifies the system. Conductive membranes can be fabricated by depositing conductive polymers, such as polypyrrole (Liu *et al.*, 2013), on commercial membranes.

Electrocoagulation-based MBRs have also been developed as a novel process for fouling mitigation in MBRs (Bani-Melhem and Elektorowicz, 2010; Wei *et al.*, 2012; Zhang *et al.*, 2015). The electrocoagulation process can generate metal cations as a result of the electro-dissolution of soluble anodes made of aluminium or iron and can contribute to fouling control by two mechanisms: (*i*) charge neutralization and sorption of SMP and EPS, and (*ii*) chemical oxidation of SMP and EPS. Despite no real applications so far, these approaches provide a potential alternative for developing fouling control strategies.

13.4.7.3 Potential fouling mitigation using nanomaterials-based membranes

In general, membrane fouling occurs more readily on hydrophobic membranes than on hydrophilic ones due to the hydrophobic interaction between hydrophobic matter and membranes in the bioreactors. Most commercially used polymeric membranes for MBR applications are hydrophobic in nature; the affinity between foulants and membrane materials might lead to severe membrane fouling. As such, much attention has been paid to reducing membrane fouling by modifying hydrophobic membranes into relatively hydrophilic ones. For example, hydrophilic water-soluble polymers are used as additives to improve the surface hydrophilicity during the membrane fabrication process. Amphiphilic copolymers with a hydrophilic chain segment can also be involved in the fabrication of hydrophilic membranes. An interfacial hydration layer is produced on the hydrophilic membrane surface as a result of the interaction between hydrophilic hydrogen forming functional groups and the surrounding water molecules, which prevents the attachment of the foulants on the membrane surface (Chew *et al.*, 2017).

The incorporation of nanomaterials into/onto membranes is one of the most popular topics in the research and development of novel membranes. Engineered nanomaterials (ENMs) which include graphene oxide (GO), reduced graphene oxide (rGO), carbon nanotubes (CNTs), fullerenes (C_{60}), copper nanoparticles (Cu NPs), silver nanoparticles (Ag NPs), titanium dioxide nanoparticles (TiO_2 NPs) and zinc oxide nanoparticles (ZnO NPs) can be used to develop new membranes or modify the surface properties of commercial membranes so that they have high permeability or strong anti-fouling properties. Most of these ENMs possess distinguishing characteristics such as antimicrobial ability and hydrophilicity, thus resulting in multifunctional effects to alleviate membrane fouling (Qu *et al.*, 2013).

13.5 MBR PLANT DESIGN, OPERATION AND MAINTENANCE

13.5.1 Process composition

The MBR-based process for wastewater treatment consists of pretreatment, biological treatment, membrane filtration, and monitoring and control systems. The membrane filtration system includes membrane modules/cassettes, membrane tanks, and membrane filtration/aeration/cleaning facilities. The design of the whole process should be compatible with the raw wastewater quality and treatment requirements. According to the different required levels of C/N/P pollutant removal, there are several choices of process configuration for the biological treatment, as illustrated in Figure 13.19.

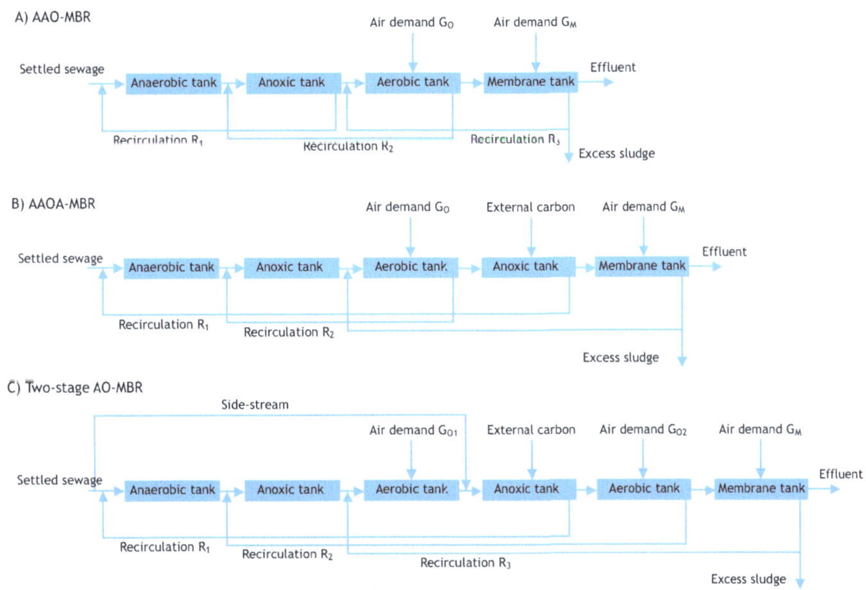

Figure 13.19 Examples of the flow charts of MBR-based processes for municipal wastewater treatment: (A) AAO-MBR for C, N, and P removal; (B) AAOA-MBR for enhanced endogenous denitrification; (C) two-stage AO-MBR for enhanced denitrification. The pretreatment, primary treatment and sludge treatment are not depicted.

When COD and BOD are the main target and there is no requirement for TN or TP removal, a primitive aerobic MBR may be adequate for this purpose. However, when the removal of TN is taken into account, an anoxic zone is required to compose the anoxic/aerobic-MBR (AO-MBR) process. For simultaneous biological removal of N and P, the AAO-MBR process depicted in Figure 13.19A fulfils this purpose.

The AAO-MBR process is most commonly applied for municipal wastewater treatment, and several customized variants of the process have been developed through modifying the spatial locations of the A/O tanks, the temporal alternation of the A/O tanks (*e.g.* an A/O switchable tank), the routes of raw wastewater (*e.g.* part of the wastewater bypasses aerobic zones and goes directly to anoxic zones to save the carbon source for denitrification), and the routes of mixed liquor recirculation (*e.g.* the reverse AAO-MBR and UCT (University of Cape Town)-MBR). For example, the AAOA-MBR process (Figure 13.19B) has an additional anoxic tank positioned after the aerobic zone to enhance endogenous denitrification in the case of insufficient carbon source from the raw wastewater (*i.e.* with a low C/N ratio). The two-stage or multi-stage $(AO)_n$-MBR is also designed to achieve a thorough utilization of the endogenous carbon source, as illustrated in Figure 13.19C.

For the treatment of industrial wastewater (such as mining, petroleum refining, petrochemicals, coal chemicals, fine chemicals, food processing, pharmaceuticals, papermaking, printing and dyeing, and electronic wastewater), the whole process is normally longer than that for municipal wastewater treatment due to the complexity of industrial wastewater and difficulty of the treatment. Unlike municipal wastewater, industrial wastewater may have properties of high strength, poor biodegradability, high toxicity, high salinity, or high oil and grease contents. Prior to MBR-based biological treatment, industrial wastewater needs to be adequately pre-treated by means of *e.g.* oil/water separation, coagulation, floatation, chemical precipitation, pH conditioning, and hydrolysis-acidification. Furthermore, post-treatment of the membrane effluent may be required, such as activated carbon adsorption, advanced oxidation, biological aerated filter treatment, NF, RO, etc. The specific configuration of the whole process depends on the wastewater quality and the treatment goal.

13.5.2 Pretreatment

The pretreatment (and primary treatment) facilities include a screen, a grit chamber, and a primary sedimentation tank. The screen serves to remove large-size suspended and floating objects, fibrous material, and particulate matter, in order to protect the subsequent biological treatment and membrane filtration systems. Conventional coarse and fine screens have a slot width or mesh size of 16-25 mm and 1.5-10.0 mm, respectively. However, they cannot guarantee complete rejection of hair and fibrous material that can get entangled in membrane fibres, clog membrane channels, or block the coarse-bubble aerators in the membrane tank. Therefore, an additional ultrafine screen with a mesh size of 0.2-1 mm (termed an ultrafine screen here) is necessary for the MBR system (especially those with hollow-fibre membrane modules). An ultrafine screen can be installed before or after the grit chamber or primary sedimentation tank, or integrated into the grit chamber to save land.

The grit chamber removes grit larger than 0.2 mm with a relative density of above 2.65 to avoid abrasion of the mechanical equipment and membrane modules and interference in the biological systems (Metcalf & Eddy, 2003). If the grit is not effectively removed, the adverse effects are amplified to a larger degree in the MBR process than the CAS process due to the typically longer SRT in the MBR system. To enhance the grit removal, a longer HRT in the grit chamber is suggested for the MBR process than for the CAS process. Among the various types of grit chamber, the aerated grit chamber has the merit of removing some of the oil and grease, which is favourable for reducing the risk of fouling the membrane caused by oily matter.

The primary sedimentation tank is designed to remove suspended solids, and is required for the MBR process when the influent SS concentration is high (*e.g.* >350 mg/l) (CECS, 2017). The sedimentation time may be 0.5-2.0 h and the hydraulic loading rate may be 1.5-4.5 m^3/m^2.h (CECS, 2017). When the C/N ratio of the raw wastewater is inadequate for biological denitrification, the sedimentation time may be shortened to 0.5-1.0 h (and the hydraulic loading rate is correspondingly 2.5-4.5 m^3/m^2.h) to save the carbon source; another option is to allow a portion of the wastewater to bypass the primary sedimentation tank on its route to the biological treatment units.

The pretreatment (and primary treatment) should ensure sufficient removal of oily substances (for instance, animal and vegetable oils are decreased to <50 mg/l and petroleum-like substances decreased to <5 mg/l). Oil/water separation and floatation facilities may be adopted for the treatment when necessary. When the pH of the wastewater is out of the range of 6-9 and becomes unsuitable for a biological system, a pH conditioning tank is required in the pretreatment section.

For industrial wastewater, the aim of pretreatment is to reduce the concentrations of SS, oily substances, and refractory pollutants, improve the biodegradability, and adjust the pH of the wastewater, which is critical to subsequent biological treatment. In addition to the screens, grit chamber, and primary sedimentation tank, industrial wastewater pretreatment facilities may include oil separation, floatation, flocculation, pH conditioning, chemical precipitation, or hydrolysis-acidification units, according to the wastewater characteristics. For instance, oil/water separation and floatation are normally required for the pretreatment of petroleum refinery and slaughterhouse wastewaters; coagulation can be conducted for highly toxic or refractory wastewaters from coal chemical, petrochemical, and fine chemical industries; pH may need to be adjusted for fine chemical, papermaking, electroplating, and printing and dyeing wastewaters; chemical precipitation can be applied to heavy metal wastewaters from electroplating and electronic industries; and hydrolysis-acidification and anaerobic sludge beds/blankets can be adopted for food processing, papermaking, tobacco, and pharmaceutical wastewaters with high strength and/or low biodegradability.

13.5.3 Biological treatment units and kinetic parameters

13.5.3.1 Overview of the biological treatment units

The design of the biological treatment units of an MBR process mainly includes the volume of the tanks, mixed liquor recirculation routes and ratios, excess sludge production, and aeration demand. The volume of the tanks is related to the wastewater flow rate and quality, sludge concentration and loading rate, sludge yield and retention time, and/or reaction rate of the pollutants. The mixed liquor recirculation routes and ratios can affect the distributions of dissolved oxygen (DO) concentrations, sludge concentrations, and pollutant removal efficiencies in the anaerobic (A_1), anoxic (A_2), aerobic (O), and membrane (M) tanks. For instance, the recirculation from O to A_2 should be high enough for the removal of NO_3^--N, but not too high to disturb the anoxic environment of A_2. The DO in the anoxic zone should be less than 0.2 mg/l to avoid inhibition of the denitrification process. For MBR processes with separate O and M tanks, an adequate ratio of recirculation from M to O (or to A_2) will prevent any excessive accumulation of MLSS in the membrane tank from exacerbating membrane fouling. The DO concentration in the separate O tank is recommended to be 1-2 mg/l. If the biological process is not sufficient to meet the effluent TP standard, chemical precipitation can be conducted (by dosing chemicals into the biological tanks or in an additional reaction tank) as a supplement. For the calculation of the tank volumes, recirculation flow rates, sludge production, and aeration demand, Table 13.2 lists a series of parameters that are mainly adopted from the 'Technical Specification for Design of Membrane Bioreactor Processes for Municipal Wastewater Treatment' (Chinese standard T/CECS 152-2017) (CECS, 2017). This standard was mainly based on the Chinese experience of full-scale MBR engineering. The detailed procedure and equations for the calculation are introduced in the following sections (13.5.3.2-13.5.3.4). There are other approaches to the design of MBRs, for which readers can refer to the books by *e.g.* Judd and Judd (2011), Park *et al.* (2015), and WEF (2012).

Table 13.2 Typical parameters for the design of MBR-based biological treatment of municipal wastewater.

Description	Symbol	Unit	Typical value
Mixed liquor suspended solids (MLSS) concentration in the membrane tank[a)]	X_M	gMLSS/l	6-15 (hollow fibre) 10-20 (flat sheet)
Proportion of volatile MLSS in the total MLSS (MLVSS/MLSS)	y	kgMLVSS/kgMLSS	0.4-0.7
Sludge organic loading rate	L_s	$kgBOD_5/kgMLSS.d$	0.03-0.10
Total sludge age (or SRT)	θ_t	d	15-30
Recirculation ratio (from anoxic to anaerobic tank)	$R_{A2\rightarrow A1}$	-	1-2
Recirculation ratio (from aerobic to anoxic tank)	$R_{O\rightarrow A2}$	-	3-5
Recirculation ratio (from membrane to aerobic tank)	$R_{M\rightarrow O}$	-	4-6
Observed yield of solids[b]	Y_t	$kgMLSS/kgBOD_5$	0.25-0.45 (with primary sedimentation) 0.5-0.9 (without primary sedimentation)
Synthesis yield of biomass[b)]	Y	$kgMLVSS/kgBOD_5$	0.3-0.6
Endogenous decay coefficient[b)]	k_d	d^{-1}	0.05-0.2
Maximum specific growth rate of nitrifying bacteria[b)]	μ_{nm}	d^{-1}	0.66
Maximum specific ammonia utilization rate[b)]	v_{nm}	$kgNH_4^+\text{-}N/kgMLSS.d$	0.02-0.10
Half-velocity constant for ammonia utilization[b)]	K_n	$mgNH_4^+\text{-}N/l$	0.5-1.0
Specific denitrification rate[b)]	K_{dn}	$kgNO_3^-\text{-}N/kgMLSS.d$	0.03-0.06
Phosphorus content in sludge	P_x	kgP/kgMLVSS	0.03-0.07

[a)]The MLSS concentrations in other tanks can be calculated from the mass balance; [b)]at 20°C.

13.5.3.2 Calculation of tank volumes and recirculation flow rates

This section presents the dimensioning of biological treatment units in an MBR process, taking the AAO-MBR process as an example (see Figure 13.19A for the process flow chart).

Aerobic zone

The volume of the aerobic zone can be calculated from the growth rate of nitrifying bacteria and SRT (Option 1) using the equation:

$$V_{OM1} = \frac{Q \cdot (S_0 - S_e) \cdot \theta_{OM1} \cdot Y_t}{1{,}000 \cdot X_{OM1}} \tag{13.1}$$

where V_{OM1} (m^3) denotes the summed volume of the aerobic tank (O) and the membrane tank (M, excluding the membrane cassette volume) where the aerobic biological reactions occur; Q (m^3/d) is the designed flow rate; S_0 and S_e (mg/l) are the BOD_5 concentrations in the feed and effluent of the biological treatment section, respectively (S_e is negligible when the removal rate is higher than 90%); X_{OM1} (g/l) is the weighted average of the O and M tank MLSS concentrations (X_O and X_M, respectively; X_O can be derived from X_M and recirculation ratio via mass balance); and Y_t (kgMLSS/kgBOD$_5$) is the observed yield of solids. θ_{OM1} (d) is the minimum sludge age in the aerobic zone, which can be estimated via:

$$\theta_{OM1} = F \cdot \frac{1}{\mu_n} \tag{13.2}$$

where μ_n (d^{-1}) is the specific growth rate of nitrifying bacteria and F is a safety factor with the value of 1.5-3. μ_n can be derived from the maximum specific growth rate of nitrifying bacteria (μ_{nm}, d^{-1}) according to the Monod relationship, with temperature correction:

$$\mu_n = \frac{\mu_{nm} \cdot N_{OM1} \cdot 1.07^{(T-20)}}{K_n \cdot 1.053^{(T-20)} + N_{OM1}} \tag{13.3}$$

where K_n (mgNH_4^+-N/l) is the half-velocity constant for ammonia utilization; N_{OM1} (mgNH_4^+-N/l) is the ammonia concentration in the aerobic zone; and T (°C) is the temperature of the mixed liquor. See Table 13.2 for typical values of X_M, Y_t, μ_{nm}, and K_n.

The volume of the aerobic zone can also be calculated from the ammonia utilization rate and N mass balance (Option 2) using the equation:

$$V_{OM1} = \frac{Q \cdot (N_{k0} - N_{ke}) - 124 \cdot \Delta X_v}{1{,}000 \cdot X_{OM1} v_n} \tag{13.4}$$

where N_{k0} and N_{ke} (mg/l) are the Kjeldahl N concentrations of the O tank influent and membrane effluent, respectively; ΔX_v (kgMLVSS/d) is the rate of biomass discharge from the system; 124/1,000 is the nitrogen content in the microorganisms with the empirical chemical formula of $C_5H_7NO_2$; and v_n (kgNH_4^+-N/kgMLSS.d) is the specific ammonia utilization rate. ΔX_v can be estimated via:

$$\Delta X_v = y \cdot Y_t \cdot \frac{Q \cdot (S_0 - S_e)}{1{,}000} \tag{13.5}$$

where y is the MLVSS/MLSS ratio. v_n can be derived from the maximum specific ammonia utilization rate (v_n, kgNH_4^+-N/kgMLSS.d), with temperature correction:

$$v_n = \frac{v_{nm} \cdot N_{OM1} \cdot 1.07^{(T-20)}}{K_n \cdot 1.053^{(T-20)} + N_{OM1}} \tag{13.6}$$

See Table 13.2 for typical values of y and v_{nm}.

Anoxic zone

The volume of the anoxic zone (A_2) can be calculated from the denitrification rate and N mass balance using the equation:

$$V_{A2} = \frac{Q \cdot (N_{t0} - N_{te}) - 124 \cdot \Delta X_v}{1{,}000 \cdot X_{A2} \cdot K_{dn} \cdot 1.026^{(T-20)}} \tag{13.7}$$

where V_{A2} (m^3) is the volume of the anoxic zone; N_{t0} and N_{te} (mg/l) are the TN concentrations in the influent and effluent of the biological system, respectively; K_{dn} (kgNO_3^--N/kgMLSS.d) is the specific denitrification rate; and X_{A2} (g/l) is the MLSS concentration in the anoxic tank which can be derived from X_M and recirculation ratios via mass balance. See Table 13.2 for the typical value of K_{dn}.

Anaerobic zone

The volume of the anaerobic zone (A_1) can be calculated from the HRT:

$$V_{A1} = \frac{Qt_{A1}}{24} \tag{13.8}$$

where V_{A1} (m^3) is the volume of the anaerobic zone; and t_{A1} (h) is the HRT of A_1, normally in the range of 1-2 h.

Recirculation flow rates

The recirculation flow rates of the mixed liquor between different tanks can be calculated from the recirculation ratios, using the equations:

$$Q_{M \to O} = Q \cdot R_{M \to O} \tag{13.9}$$

$$Q_{O \to A2} = Q \cdot R_{O \to A2} \tag{13.10}$$

$$Q_{A2 \to A1} = Q \cdot R_{A2 \to A1} \tag{13.11}$$

where $Q_{M \to O}$, $Q_{O \to A2}$, and $Q_{A2 \to A1}$ are the recirculation flow rates from M to O, from O to A_2, and from A_2 to A_1, respectively; and $R_{M \to O}$, $R_{O \to A2}$, and $R_{A2 \to A1}$ are the recirculation ratios from M to O, from O to A_2, and from A_2 to A_1, respectively. See Table 13.2 for typical values of $R_{M \to O}$, $R_{O \to A2}$, and $R_{A2 \to A1}$.

13.5.3.3 Calculation of excess sludge production

The excess sludge production can be calculated from the synthesis yield of biomass, observed yield of solids, or sludge age (solids retention time) (denoted as options 1-3, respectively). For the engineering application of MBRs in China for example, Option 1 is often adopted for process design, Option 2 offers a convenient estimation during practical operation due to its simplicity, and Option 3 adds to a retrospective analysis of the engineering data. The equations for the calculation are given as follows:

$$\Delta X = \begin{cases} \dfrac{YQ(S_0 - S_e)}{1{,}000} - k_d V_t X y + \dfrac{fQ(SS_0 - SS_e)}{1{,}000} & \text{(Opt. 1)} \\ \dfrac{Y_t \cdot Q \cdot (S_0 - S_e)}{1{,}000} & \text{(Opt. 2)} \\ \dfrac{V_t \cdot X}{\theta_t} & \text{(Opt. 3)} \end{cases} \tag{13.12}$$

where ΔX (kgMLSS/d) is the excess sludge production rate; Y ($kgMLVSS/kgBOD_5$) is the synthesis yield of biomass; k_d (d^{-1}) is the endogenous decay coefficient; Y_t ($kgMLSS/kgBOD_5$) is the observed yield of solids; SS_0 and SS_e (mg/l) are the feed and effluent SS concentrations of the biological system, respectively; f is the conversion rate of sludge from SS, typically 0.5-0.7 gMLSS/gSS; X (g/l) is the weighted average of MLSS concentrations; y is the MLVSS/MLSS ratio; V_t (m^3) is the total volume of biological tanks; and θ_t (d) is the total sludge age. See Table 13.2 for typical values of Y, Y_t, k_d, y, and θ_t.

The results of calculations using these different approaches can be compared for mutual verification. When chemical agents such as aluminium or iron salts are added to the biological unit for the purposes of TP removal or membrane fouling control, the excess sludge will contain chemical precipitates or coagulation products; the increased amount of excess sludge can be estimated to be 5 times the mass of aluminium or 3.5 times the mass of iron (CECS, 2017; JSWA, 2009). The excess sludge is usually pumped out from the membrane tank because this is where the MLSS concentration is highest. The average flow rate of excess sludge (Q_w, m^3/d) can be calculated using the following equation for the design of sludge discharge facilities:

$$Q_w = \frac{\Delta X}{X_M} \tag{13.13}$$

13.5.3.4 Calculation of aeration demand for biological reactions

Oxygen demand

The oxygen demand for the aerobic zone includes the oxygen required for organic carbon oxidation and nitrification. For the calculation of the required oxygen amount, the organic carbon consumed for denitrification should be subtracted from the total oxidized organic carbon. It should be noted that the aeration in the membrane tank (usually coarse-bubble aeration for membrane scouring) can contribute considerably to DO in the aerobic zone through mixed liquor recirculation from the membrane tank to the aerobic tank. For the calculation of aeration demand in the aerobic zone, the contribution from membrane aeration should be deducted to avoid overestimation.

The oxygen demand for biological reactions in the aerobic zone (O, kgO_2/d) is expressed in the form of:

$$O = (O_s + O_n - O_{dn})\frac{V_O}{V_O + V_{M1}} - O_m \quad (13.14)$$

where O_s is the total oxygen demand for oxidizing organic carbon, O_n is the oxygen demand for nitrification, O_{dn} is the oxygen demand offset by denitrification, and O_m is the oxygen amount brought by the recirculated liquid from the membrane tank to the aerobic tank. V_O is the volume of the aerobic tank and V_{M1} is the volume of the membrane tank (excluding the membrane cassette volume). O_s, O_n, O_{dn}, and O_m (all in kgO_2/d) can be calculated using the following equations.

$$O_s = \frac{1.47}{1{,}000} Q \cdot (S_0 - S_e) - 1.42 \cdot \Delta X_v \quad (13.15)$$

$$O_n = 4.57 \cdot [\frac{Q \cdot (N_{k0} - N_{ke})}{1{,}000} - 0.124 \cdot \Delta X_v] \quad (13.16)$$

$$O_{dn} = 2.86 \cdot [\frac{Q \cdot (N_{t0} - N_{te})}{1{,}000} - 0.124 \cdot \Delta X_v] \quad (13.17)$$

$$O_m = \frac{1}{1{,}000} Q_{M \to O} \cdot C_{omd} \quad (13.18)$$

where C_{omd} (mg/l) is the DO concentration in the recirculated flow, typically 4-8 mg/l; 1.47 is the ratio of ultimate BOD to 5-day BOD (BOD_5); 1.42 is the oxygen demand per unit MLVSS (expressed as $C_5H_7NO_2$); 4.57 is the oxygen demand per unit Kjeldahl N; and 2.86 is the oxygen demand offset per unit NO_3^--N.

Standard-state oxygen demand

The calculated oxygen demand should then be converted to the standard-state amount (at 20 °C and 1 bar in clean water) based on the local altitude, water depth, temperature, designed DO concentration, and oxygen transfer in the mixed liquor vs. in clean water.

The equation for the standardization of the oxygen demand is given as follows:

$$O_{std} = \frac{OC_{os(20)}}{1.024^{(T-20)} \alpha \cdot [\beta \cdot C_{os(T)} (1 + \frac{\rho \cdot g \cdot h}{2P}) - C_o]} \quad (13.19)$$

where O_{std} (kgO_2/d) is the standard-state oxygen demand; $C_{os(20)}$ and $C_{os(T)}$ (mg/l) are the clean-water saturated DO concentrations at 1 bar at 20°C and T °C, respectively; C_o (mg/l) is the average DO concentration in the aerobic tank, typically 1-2 mg/l; ρ is the density of the mixed liquor, typically 1.002-1.006 g/cm^3; g (m/s^2) is the gravitational acceleration; h (m) is the water depth of the aerobic tank; P (kPa) is the actual atmospheric pressure; α is the ratio of the oxygen transfer coefficient in the sludge to that in clean water, and β is the ratio of saturated DO concentration in the sludge to that in clean water. Particularly, the α factor is an important correction factor for the mass transfer of oxygen in the sludge, which is critical to the design of the aeration amount. It has been reported that the value of the α factor in full-scale MBRs is generally higher than that in CAS processes at the same MLSS concentration (*e.g.* ~10 g/l) (Xu et al., 2017). The relationship between the α factor and MLSS concentration is given as:

$$\alpha = k_1 \cdot \exp(-k_2 \cdot X_O) \quad (13.20)$$

where X_O (g/l) is the MLSS concentration in the aerobic tank. The empirical parameters k_1 and k_2 may be 1.7 and 0.08, respectively, when X_O is in the range of 6-20 g/l, according to an investigation of large full-scale MBRs treating municipal wastewater (Xu et al., 2017).

Air supply

Finally, the operational air supply can be calculated according to the efficiency of the aerator (blower or mechanical aerator). Taking blower aeration as an example, the air supply is expressed as:

$$G_O = \frac{O_{std}}{0.28\eta_A} \cdot \frac{100}{24} \quad (13.21)$$

where G_O (Nm^3/h) is the standard-state air supply; 0.28 (kgO_2/m^3) is the oxygen content in air; and η_A (%) is the oxygen transfer efficiency of the blower.

13.5.4 Membrane filtration system

13.5.4.1 Flux

It is recommended that the membrane filtration be operated in constant-flux mode, that is, the average flux is maintained relatively constant during the filtration. For submerged MBRs, the suction pumps are operated in an on/off intermittent manner to impede membrane fouling development. A filtration cycle consists of a running period (*e.g.* 7-13 min) and a relaxation period (*e.g.* 1-3 min). For side-stream MBRs, the membrane should be back-flushed at appropriate frequency, duration, and intensity according to manufacturer guidelines or operating status.

For the design of the membrane filtration system, there are several fluxes which need to be considered: the average flux, running flux, peak flux, temporarily peak flux, and critical flux. For municipal wastewater treatment, the average flux of the whole filtration cycle is normally 15-25 $l/m^2.h$ for submerged MBRs and 30-45 $l/m^2.h$ for side-stream MBRs. The average flux for industrial wastewater treatment is usually lower and varies according to the specific wastewater being treated (Judd and Judd, 2011). In a filtration cycle, the instantaneous flux is zero in the relaxation period and, as such, the flux in the running period (*i.e.* running flux) is higher than the average flux of the whole cycle. The relationship between the average flux (J_{avg}, $l/m^2.h$) and the running flux (J_1, $l/m^2.h$) in a filtration cycle is expressed as:

$$J_{avg} = J_1 \cdot \frac{\tau_1}{\tau_1 + \tau_0} \quad (13.22)$$

where τ_1 (min) is the running period and τ_0 (min) is the relaxation period. If the membrane is frequently back-flushed at a reverse flux of J_b ($l/m^2.h$), a duration of τ_b (min), and an interval of t_b (min), then the average flux over multiple backflush cycles (during which no chemical cleaning is performed) is calculated as:

$$J_{avg} = J_1 \cdot \frac{\tau_1}{\tau_1 + \tau_0} - J_b \cdot \frac{\tau_b}{t_b + \tau_b} \quad (13.23)$$

Peak flux corresponds to the peak flow rate of the wastewater influent, and temporary peak flux occurs when some of the membrane cassettes are suspended due to membrane cleaning and equipment maintenance. To protect the membrane filtration system, the operation at peak flux or temporary peak flux should not last too long. The designed flow rate of the suction pumps should satisfy the operation at maximum flux. Critical flux refers to the flux below which membrane fouling is not significant and the membrane filtration resistance (or TMP) is relatively stable over time; above it the resistance (or TMP) increases rapidly. It is recommended that the running flux, peak flux, and temporary peak flux should always be less than 80-90% of the critical flux.

13.5.4.2 Membrane area

The total membrane area (A_M, m^2) required for the filtration system can thus be calculated from the total flow rate (Q, m^3/d) and average flux (J_{avg}, $l/m^2.h$), multiplied by a safety factor (F_M) of approximately 1.1-1.2 (or estimated from the ratio of maximum flux to average flux as the case may be), using the following equation:

$$A_M = \frac{Q}{0.024 \cdot J_{avg}} \cdot F_M \quad (13.24)$$

The number of membrane cassettes (n_M) can then be determined via:

$$n_M = \frac{A_M}{A_{M0}} \quad (13.25)$$

where A_{M0} is the membrane area per unit cassette (m^2).

The membrane tank volume (V_M, m^3) can be designed from A_M (m^2) and the apparent packing density (φ_M, m^2/m^3) of the whole membrane tank:

$$V_M = \frac{A_M}{\varphi_M} \tag{13.26}$$

The packing density may depend on the specific configuration of the membrane product and the designed hydraulics in the membrane tank. Sufficient space surrounding the membrane cassettes is required to ensure the circulation of mixed liquor in the tank.

13.5.4.3 Aeration demand

For submerged MBRs, coarse-bubble aeration is normally applied for membrane scouring to mitigate membrane fouling and clogging; the membrane aeration system consists of an aerator, blower, air pipeline, and ancillary equipment. The specific aeration demand with respect to the projection area (i.e. footprint) of the membrane cassette (SAD_f) is usually 60-120 $Nm^3/m^2.h$ in the case of municipal wastewater treatment MBRs (CECS, 2017). The specific aeration demand can also be calculated with respect to membrane area (SAD_m) and permeate flow (SAD_p). For municipal wastewater treatment, the reference SAD_m and SAD_p for hollow-fibre membrane modules are 0.3 $Nm^3/m^2.h$ and 15 Nm^3/m^3-permeate, respectively, while those for flat-sheet membrane modules are double (Judd and Judd, 2011). The flow rate of a blower supplying air for membrane-tank aeration can thus be calculated from SAD_f, SAD_m, or SAD_p using the following equations:

$$G_M = \begin{cases} SAD_f \cdot A_P \\ SAD_m \cdot A_M \\ SAD_p \cdot 24Q \end{cases} \tag{13.27}$$

where G_M (Nm^3/h) is the air supply to the membrane tank, A_P (m^2) is the total footprint of the membrane cassettes, and Q (m^3/d) is the total permeate flow rate. Note that in addition to continuous aeration as calculated above, membrane aeration can also be performed in an intermittent, alternating, or pulsed mode.

13.5.4.4 Chemical cleaning procedure

Chemical cleaning of a fouled membrane includes in situ cleaning and ex situ cleaning according to the location of the cleaning, or maintenance cleaning and recovery cleaning according to the severity of fouling (see Section 13.4.5.2). The online cleaning or in situ maintenance cleaning is usually performed in the CIP or CEB mode. NaOCl and NaOH are typical alkaline cleaning agents to remove organic foulants, and citric acid, oxalic acid, and HCl are typical acid cleaning agents to remove inorganic foulants (*e.g.* metal ions). NaOCl and citric acid are the most commonly applied cleaning agents in MBR engineering.

An online chemical cleaning system should include a back-flushing pump (for the case of CEB) and facilities for chemical storing, dosing, metering, and mixing, and also be under automatic control. Table 13.3 gives examples of procedures for online NaOCl cleaning of hollow-fibre and flat-sheet membrane modules, respectively. For online cleaning using NaOCl, citric acid or oxalic acid, a recommendation for engineering application is that the concentration is smaller than 3-5 g/l, the dosage is less than 4-6 L/m^2, and the contact time is shorter than 1-2 h (CECS, 2017). In cold seasons when fouling is prone to become worse, the online cleaning frequency, concentration, dosage, contact time, and temperature may be increased accordingly.

Table 13.3 Exemplified procedures for *in situ* NaOCl cleaning of hollow-fibre and flat-sheet membrane modules.

Membrane module	Situation	Procedure	Dosage
(a) Hollow fibre	Maintenance cleaning: weekly	Injection (30 min) → immersion (60 min) → aeration (30 min)	Concentration: 0.5-1.0 g/l of chlorine; Dosing volume: 3-5 l/m^2
	Intensified cleaning: monthly, or when TMP > 30 kPa	Injection (30 min) → immersion (60 min) → aeration (30 min)	Concentration: 2-3 g/l of chlorine; Dosing volume: 3-5 l/m^2
(b) Flat sheet	Maintenance cleaning: not mandatory (depending on the manufacturer)		
	Intensified cleaning: once every 3-6 months, or when TMP > 30 kPa	Injection (8-15 min) → immersion (1-2 h)	Concentration: 3-5 g/l of chlorine; Dosing volume: 4-6 l/m^2

The recovery cleaning system should include chemical dosing facilities, lifting devices and cleaning tanks (for *ex situ* recovery cleaning), and sludge transfer facilities (for *in situ* recovery cleaning). Acid and alkaline agents should be transported in separate pipelines. A recommendation for full-scale cases is that recovery cleaning should be carried out regularly once a year, or immediately when the TMP reaches 50 kPa (or when online cleaning fails to reduce the TMP to lower than 30 kPa) (CECS, 2017). For each instance of offline cleaning, for example, the fouled membranes can be immersed in NaOCl (3-5 g/l of chlorine) for 12-24 h, or immersed in NaOCl (3-5 g/l of chlorine) and citric/oxalic acid (10-20 g/l) alternately. The concentration, contact time, and temperature can be adjusted as needed. For more details, Wang *et al.* (2014) provided a survey of practically applied online and offline cleaning protocols in full-scale MBRs.

13.6 PRACTICAL APPLICATION

13.6.1 Overall MBR applications

There have been significant increases in the treatment capacity of full-scale MBRs. According to the MBR site (http://www.thembrsite.com/, accessed on 21 January 2020), the global number of the largest MBRs with a treatment capacity of $\geq 10^5$ m^3/d has reached 62 with a total peak treatment capacity of $147.33 \cdot 10^5$ m^3/d. China, USA, and Singapore have most of the largest MBRs, with 84% of the total global treatment capacity. It is noteworthy that China has the largest share of large-scale MBRs.

MBRs have been extensively applied to treat municipal wastewater, industrial wastewater, and leachate due to increasingly stringent effluent requirements. Most MBRs, especially large-scale applications, are for municipal wastewater treatment, and underground MBR plants are favoured in some areas in order to further reduce their footprint. Besides, although MBRs appear in almost all industrial sectors and in leachate treatment, the high concentration of refractory pollutants in influent requires effective modification of the MBR processes and/or efficient integrations with other technologies. Moreover, energy reduction in MBRs is a permanent development goal. With their increasing capacity and number, the wide-scale application of MBRs is providing useful experience and feedback for their further development. The following examples are provided regarding these issues.

13.6.2 Four full-scale cases of MBR application

This section gives four global cases of MBR application. The influent in these cases includes municipal wastewater, industrial wastewater, and leachate. These cases are typical in advanced technologies and modifications for energy mitigation, footprint reduction, and effluent quality improvement. Treatment process design, actual operating parameters, and performances of these cases are provided below.

Case 1: A large-scale underground MBR for municipal wastewater treatment

The Jingxi underground WWTP in Guangzhou City, the capital of Guangdong Province in South China, was commissioned in 2010, at which point it was the largest underground MBR plant in Asia. It has a treatment capacity of 100,000 m^3/d (peak flow at 120,000 m^3/d), occupying a footprint of 1.82 ha and treating wastewater from a catchment area of 16,500 ha. By placing the WWTP underground (Figure 13.20), the unsightly architecture and unpleasant odours of typical WWTPs are eliminated. The synergy between the smaller footprint due to MBR technology and underground placement of the plant also significantly reduced the capital expenditure on land investment as rapid urbanization has greatly driven up land prices in Guangzhou.

Figure 13.20 Aerial view of the Jingxi WWTP.

At the Jingxi underground WWTP the influent flows into the AAO-MBR system after primary treatment by a fine screen and grit chamber (Figure 13.21).

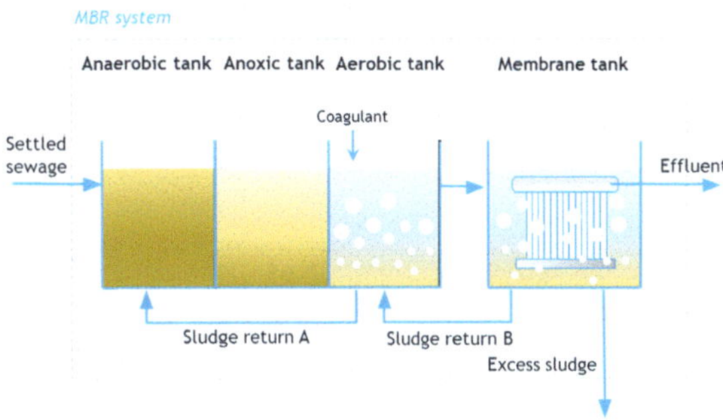

Figure 13.21 Schematic of the MBR-based treatment process in the Jingxi WWTP.

The AAO process guarantees the removal efficiencies of TN and TP (with the aid of chemical coagulants). There is a submerged MBR (Figure 13.22) with PVDF hollow fibre membranes (SMM-1520, Memstar, Singapore). Ten membrane tanks are operated simultaneously, with 20 membrane units for each membrane tank and 88 membrane modules for each unit. The total effective surface area of each membrane unit is 1,760 m^2. The baseline flux is 15 l/m^2.h, and higher flux up to 25 l/m^2.h can be achieved with intensified membrane maintenance cleaning protocols. The specific energy demand of the system can be reduced from 0.22 kWh/m^3 to 0.18 kWh/m^3 by the optimized internal recirculation path and aeration protocol. The plant effluent is used for replenishing the nearby urban river.The HRT of each tank in the AAO-MBR system is set at 1.03 h, 2.09 h, 3 h and 1.24 h, respectively. MLSS in the membrane tank is in the range of 8,000-13,000 mg/l, and SRT is controlled at 12 d. The influent and effluent quality of MBR tank are shown in Table 13.4.

Figure 13.22 Photos of the membrane tank at the Jingxi plant: (A) the installation of membrane modules, and (B) the effluent discharge 'waterfall'.

Table 13.4 Influent and effluent quality at the Jingxi WWTP.

Pollutant	Influent (mg/l)	Effluent (mg/l)
BOD_5	130±30	3±1
COD	210±50	13±5
SS	90±30	2±1
NH_4^+-N	20±5	0.6±0.3
TN	30±5	7.4±0.6
TP	2.5±0.5	0.15±0.04

Case 2: MBRs for municipal wastewater treatment with low energy consumption and no recovery cleaning

There are seven MBRs (total treatment capacity 328,050 m^3/d, total peak flow 522,075 m^3/d) in operation or under commission in Singapore, and one of the largest MBRs worldwide is under construction (Tuas Water Reclamation Plant, average 800,000 m^3/d, peak 1,200,000 m^3/d) as of 2020.

An Integrated Validation and Demonstration Plant was commissioned in 2017, with a treatment capacity of 12,500 m^3/d (peak 18,750 m^3/d). It simulates the treatment process for the Tuas Water Reclamation Plant under construction, demonstrating the concept of low energy consumption and no chemical-recovery cleaning.

The influent (raw used water) of the plant is subject to a biologically-enhanced primary treatment where biosorption is applied before the compact lamella plate primary clarifier. The primary effluent flows through a step-feed AO-MBR system; there are five passes through the system and each pass contains non-aerated and aerated zones. The mixed liquor of the last pass flows into a submerged membrane tank. The MBR effluent is further treated by RO and reused for NEWater (the ultra-clean, high-grade reclaimed water to cushion the water supply in Singapore) production (Figure 13.23). Biologically-enhanced primary treatment maximizes the organic carbon capture and reduces the aeration requirement in the subsequent processes. The step-feed AO-MBR enhances the nutrient removal without the need for internal nitrate/nitrite circulation or external carbon addition. It also promotes denitrifying polyphosphate-accumulating organisms. The operating parameters are shown in Table 13.5. Regarding the membrane tank, PVDF hollow-fibre membranes supplied by Suez Water Technologies and Solutions (nominal pore size of 0.035 μm) are installed, with a total effective surface area of 26,396.2 m^2. The net flux is 20 l/m^2.h and the peak flux can reach 30 l/m^2.h. The MBR effluent quality is shown in Table 13.6.

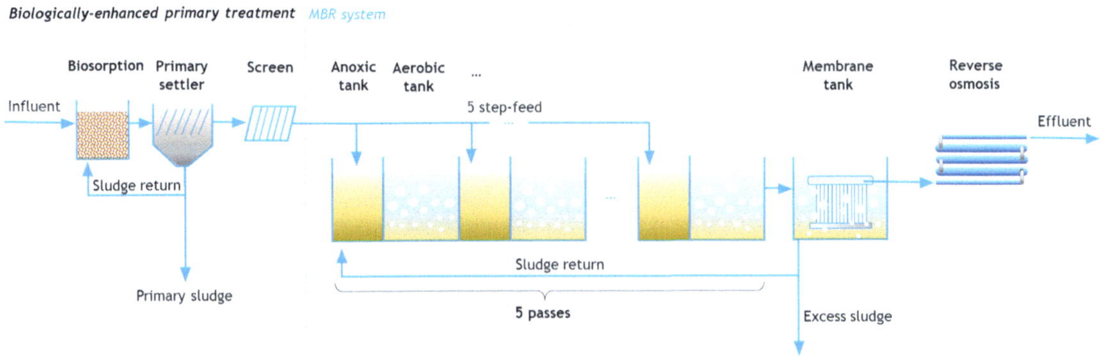

Figure 13.23 Schematic of the MBR-based treatment process in an integrated validation and demonstration plant.

Table 13.5 Key process parameters of an integrated validation and demonstration plant.

Parameter	Unit	Value
Biosorption tank HRT	hour	0.5
Biosorption tank MLSS	mg/l	1,500
Lamella plate setting area/installation floor area	-	10
Lamella plate spacing	mm	100
MBR fine screen	mm	2
Primary sludge recirculation ratio	-	0.1-0.2
Number of passes of the step-feed bioreactor	-	5
MLSS in bioreactor	mg/l	2,000-5,000
Bioreactor HRT	hours	5
Bioreactor SRT	days	5-10
Bioreactor anoxic/aeration volume ratio	-	1:1
MBR sludge recirculation ratio	%	100

Table 13.6 Effluent quality of an MBR tank in an integrated validation and demonstration plant.

Pollutant	Unit	Effluent concentration
Total organic carbon (TOC)	mg/l	6.6
NH_4^+-N	mg/l	2.19
NO_3^--N	mg/l	1.19
NO_2^--N	mg/l	0.51
PO_4^{3-}-P	mg/l	0.17
Conductivity	μS/cm	728

The energy consumption of the whole treatment process is in the range of 0.24-0.28 kWh/m^3, or less than 0.5 kWh/kgCOD, and the air scouring energy consumption in the membrane tank is as low as 0.04 kWh/m^3. Maintenance cleaning using NaOCl (concentration of 200 mg/l) is conducted twice per week, and no recovery cleaning was conducted between the start of operation in August 2017 and February 2020.

Case 3: A carrier-assisted MBR for industrial wastewater treatment

This wastewater treatment plant/facility is owned by Adama Agan Ltd., Ashdod, Israel. The company is an agriculture-related fine chemical producer. Because updated regulations required higher effluent quality, the former CAS processes were retrofitted as a dual aerobic sludge system, where MBR is applied. The project was carried out by Adama Agan and EPT (Environmental Protection Technologies), an Israeli based process engineering firm. The MBR retrofit plant was commissioned in 2012, with a treatment capacity of 7,200 m^3/d (Figure 13.24).

Figure 13.24 Aerial view of the MBR treatment plant belonging to Adama Agan Ltd.

In this dual aerobic sludge system, the first sludge system consists of a micro-aerobic reactor, an aerobic reactor, and a gravity clarifier. The effluent from the first sludge system flows into the second sludge system (*i.e.* the MBR system) which consists of an aerobic tank, post-anoxic tank, and membrane tank (Figure 13.25).

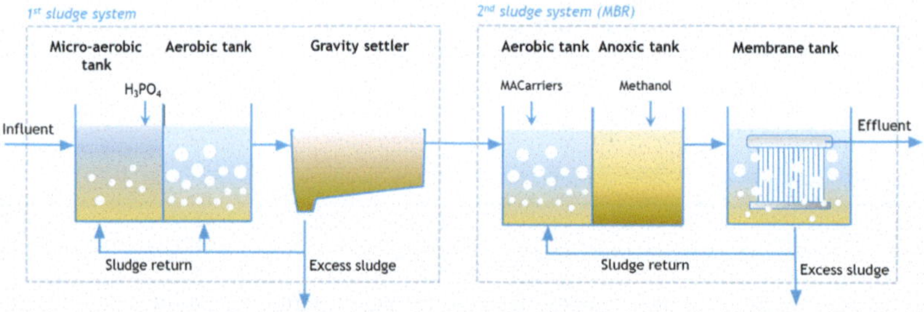

Figure 13.25 Schematic of the wastewater treatment process owned by Adama Agan Ltd.

ZeeWeed membrane modules (SUEZ, France) are installed. In order to enhance the removal efficiencies of recalcitrant COD, MACarriers (membrane-accommodating carriers) are added to the reactor in the second sludge system. Dense and strong biofilm can be quickly formed on the huge bacteria-friendly surface area of MACarriers which can adsorb and pre-concentrate the residual recalcitrant organic compounds in the mixed liquor. This process not only improves the effluent quality but also helps remove toxicity and make the nitrification process more stable.

The influent and effluent quality of the MBR without and with MACarriers are provided in Table 13.7.

Table 13.7 Influent and effluent quality of the wastewater treatment plant owned by Adama Agan Ltd.

Pollutant	Influent (mg/l)	Effluent (mg/l)	
		MBR without MACarriers	MBR with MACarriers
COD	3,500	200-300	Data not available
TOC	Data not available	95-120	80-110
NH_4^+-N	10.5	0.5-10	0.5-2
TN	236	<20	<20
TSS	60	<3	<3
AOX	50.4	5-8	3.5-5

AOX: absorbable organic halogens

Case 4: A side-stream MBR for leachate treatment

The Waste-to-Energy project in the Wuzhong Venous Industrial Park, Jiangsu Province, is one of the largest solid waste incineration plants in China, with a designed treatment capacity of 3,550 t/d of solid waste when it was commissioned. The affiliated leachate treatment plant adopting MBR is owned and operated by Everbright Environmental Energy (Suzhou) Co. Ltd. Commissioned in 2014, it has a designed treatment capacity of 700 m^3/d.

Leachate flows successively into three parallel pretreatment tanks, three parallel internal and outer circulation (IOC) anaerobic reactors (effective volume: 1,900 m^3 each), two parallel anoxic-aerobic tanks (effective volume: 800 m^3 for each anoxic tank; 2,400 m^3 for each aerobic tank), a UF unit, an NF unit and an RO unit (Figure 13.26). The treated water is totally reused, and zero liquid is discharged. This MBR system is designed with a side-stream configuration, consisting of an A/O process and external UF units. The UF units contain 18 tubular PVDF membrane modules (MEMOS, Germany) with a total effective surface area of 490 m^2 (Figure 13.27). Cross-flow filtration is applied, and filtration and online cleaning are automatically controlled. The concentrated liquor is pumped back to the A/O tanks.

Regarding the operational parameters, the MLSS concentration is approximately 15,000 mg/l, and the HRT is set at 7.81 d (1.92 d for the anoxic tank, and 5.89 d for the aerobic tank). The influent and effluent quality in this leachate treatment plant are shown in Table 13.8. As demand for municipal solid waste treatment increases, the waste-to-energy project will expand, and the leachate treatment plant will be expanded to treat 2,500 m^3/d in the future.

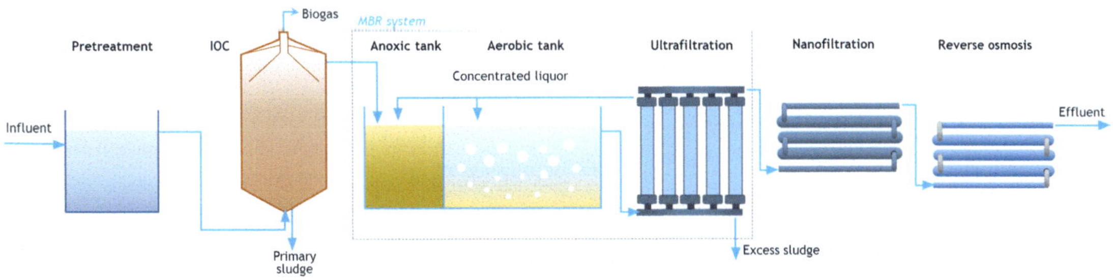

Figure 13.26 Schematic of the treatment process in the leachate treatment plant in Suzhou.

Table 13.8 Influent and effluent qualities for the leachate treatment plant in Suzhou (unit: mg/l).

Water	COD	BOD	NH_4^+-N	TN	SS	Cl^-
Influent	32,000	15,000	1,500	1,800	22,000	6,500
Effluent	13	0	0.22	13	0	200

Note: annual average values from November 2018 to October 2019.

Figure 13.27 Membrane modules in the leachate treatment plant in Suzhou.

13.6.3 Latest developments in MBR systems

13.6.3.1 The high-loaded MBR (HL-MBR) concept

HL-MBR refers to an MBR operated at a high organic-loading rate and high MLSS concentration (15-40 g/l) (Kim *et al.*, 2019). Advantages of HL-MBRs compared to conventional MBRs include: (*i*) increased capacity to treat higher organic-loading rates, (*ii*) increased oxygen uptake rates associated with a higher COD conversion rate, (*iii*) minimized footprint needs and construction costs, and (*iv*) reduced waste solids and handling costs (Barreto *et al.*, 2017; Livingstone, 2010). These advantages may encourage the design of compact and containerized movable/portable MBR systems for the treatment of municipal and industrial wastewater in remote areas without access to sewer systems. In addition, HL-MBR systems may offer on-site alternatives for the provision of urgent sanitation services in the context of either natural or man-made disasters (Barreto *et al.*, 2017; Hai and Yamamoto, 2011; Hai *et al.*, 2014).

Such a HL-MBR system can be achieved by either operating the MBR system at high SRT conditions (*i.e.* not wasting sludge), or by working at high influent organic-loading rates. When operating the system at high SRT conditions, the majority of the MLSS consists of non-biodegradable and/or inorganic suspended solids instead of active biomass without enhancing the treatment capacity of the MBR at a given footprint. On the other hand, operating the MBR system at a low, relatively standard SRT (5 to 15 days) and at a high influent loading rate results in sludge consisting primarily on active biomass and therefore, enhancing the treatment capacity of the MBR at a given footprint. For example, the volumetric organic loading rate can be increased from 4 to 13 kg RBCOD/m^3.d as the MLSS concentration increases in the MBR from 10 to 40 g/l at an SRT of approximately 10 days (Kim *et al.*, 2019). Moreover, the sludge wastage can also be reduced by a factor of 10 when increasing the MLSS concentration from 10 to 40 g/l.

A major challenge for the HL-MBR is the inefficiency of conventional aeration systems in oxygen transfer at higher than usual MLSS concentrations (*e.g.* >20 g/l). Several authors have evaluated the effects of MLSS concentration on oxygen transfer reporting the α factor (the mass transfer ratio of process water to clean water) (Cornel *et al.*, 2003; Durán *et al.*, 2016; Germain *et al.*, 2007; Henkel *et al.*, 2011; Krampe and Krauth, 2003; Muller *et al.*, 1995). These authors agree that the α factor exponentially decreases with the MLSS concentration, even reaching a certain MLSS concentration value at which negligible α factors are reported. Moreover, oxygen transfer efficiency is also proportional to SRT (Rosso *et al.*, 2008); the lower the SRT, the lower the oxygen transfer efficiency and the α factor. Therefore, the design features of the HL-MBR concept (high MLSS concentrations and low SRTs) maximize the oxygen transfer limitations imposed by conventional diffused aeration systems. Because of this, the HL-MBR system cannot be operated at MLSS concentrations of 20 g/l and higher. These limitations imposed by conventional aeration systems cap the design and operation of the HL-MBR system. In this context, there is a need for more efficient oxygen delivery technologies to remove such design limitations.

There are several innovative aeration technologies available to enhance the oxygen transfer process such as deep-shaft (a U-tube contactor or vertical shaft), high purity oxygen (HPO) aeration systems, and supersaturated oxygen aeration systems. Among them, supersaturated oxygen aeration systems exhibit promising advantages for coupling to the HL-MBR concept. They have been developed for working with HPO and in high-pressure conditions. Advantages of such systems include: (*i*) they can be operated with HPO in high-pressure conditions to achieve oxygen transfer efficiencies of above 95% (Ashley *et al.*, 2014), (*ii*) their operational cost is low due to both their simple operation and maintenance needs, (*iii*) they are not affected by the system being submerged (Osborn *et al.*, 2010), (*iv*) they are fully movable/portable which is ideal for coupling with movable/portable wastewater treatment systems, and (*v*) they are very simple to scale up for retrofitting wastewater treatment systems. If properly implemented, supersaturated oxygen aeration systems may be able to eliminate the current limitation imposed by conventional diffused aeration systems on the HL-MBR concept. The two most relevant supersaturated oxygen aeration technologies available on the market are the supersaturated dissolved oxygen (SDOX) system and Speece cone technology. Examples of full-scale devices using such technologies are presented in Figure 13.28.

The SDOX system, consisting of a pressurized chamber operated in high-pressure conditions (approximately 800 kPa and higher) connected to an HPO source, was developed by BlueInGreen LLC in the USA. The influent stream enters at the top of the pressurized chamber where a large gas-liquid interface is created between the liquid and pure oxygen, allowing the dissolved oxygen (DO) concentration to reach up to 350 mg/l in the pressurized chamber or higher, depending on the operational pressure conditions (Osborn *et al.*, 2010). SDOX technology has been mostly evaluated for introducing DO in the context of lake and river restoration; other application includes odour control in municipal sewer systems.

Speece cone technology, consisting of a down-flow conical bubble contact chamber connected to a HPO source, was developed by Dr. Richard Speece in 1971 (McGinnis and Little, 1998). The influent stream and the HPO flow are simultaneously introduced on the top of the Speece cone chamber at a moderate pressure. At a specific downward water velocity, oxygen bubbles are pushed down in the cone and dissolved into the stream. The oxygen-rich supersaturated solution is then returned back into the receiving basin through submerged diffusers. Speece cone technology is mostly applied for introducing DO to restore lakes and rivers (Ashley *et al.*, 2008).

Figure 13.28 (A) Example of the side-stream dissolution process in the SDOX® system; (B) a containerized SDOX® system packaged in a 20-ft ISO shipping container; (C) a ECO_2® Speece cone system for highly effective superoxygenation.

13.6.3.2 Applications of the HL-MBR system

Case 1: An HL-MBR equipped with SDOX

A bench-scale HL-MBR equipped with the SDOX system was established by Kim *et al.* (2020). Figure 13.29 shows the experimental set-up, and Figure 13.30 shows the schematic of the system.

Figure 13.29 Bench-scale experimental set-up of an HL-MBR equipped with the SDOX system. On the left-hand side is the SDOX system provided by BlueInGreen LLC, and at the right-hand side is the MBR system (adopted from Kim *et al.*, 2020).

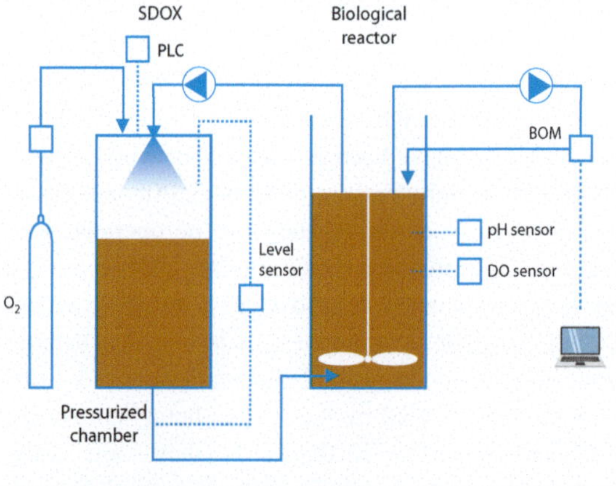

Figure 13.30 Schematic diagram of the SDOX HL-MBR system (adopted from Kim *et al.*, 2020).

The oxygen transfer performances of the SDOX system in a wide range of MLSS concentrations in the context of the HL-MBR were evaluated, including the oxygen mass transfer rate, the oxygen transfer efficiencies and the α factors (Kim *et al.*, 2020). Figure 13.31 shows the α factors as a function of the MLSS concentrations for the SDOX system compared to (*i*) the α factors for fine-bubble diffusers (on the same sludge as in the SDOX evaluation) (Kim *et al.*, 2019) and to (*ii*) the α factors for conventional bubble diffusers (on sludge from full and pilot-scale municipal/industrial MBR systems) (Germain *et al.*, 2007).

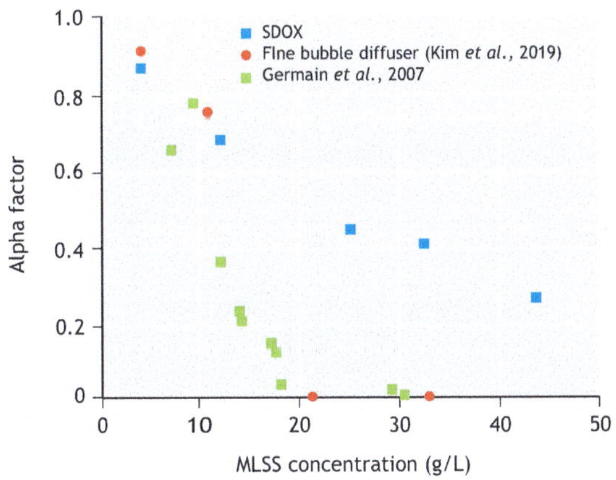

Figure 13.31 α factors as a function of the MLSS concentration for the SDOX system, for fine bubble diffusers (Kim et al., 2019) and for full-scale municipal and industrial MBR plants provided with fine and coarse bubble diffusers (Germain et al., 2007) (adopted from Kim et al., 2020).

Both studies conducted with conventional bubble diffusers showed a dramatic decrease in the α factor as a function of MLSS concentration, and even negligible α factor values at MLSS concentrations above 15 g/l. On the other hand, the SDOX system exhibited much higher α factor values at the high MLSS concentration range of 15-45 g/l. Even at the highest evaluated MLSS concentration of approximately 45 g/l, an α factor of 0.35 was reported. The α factors reported in Figure 13.31 were obtained on sludge with the characteristics expected for a HL-MBR system (*i.e.* high MLSS concentrations and obtained from plants operating at short SRT values). Therefore, it can be concluded from an oxygen transfer performance standpoint that the SDOX system outperforms conventional diffusers at the evaluated conditions and that it removes the limitations imposed by conventional diffused aeration for the HL-MBR concept. It is possible to operate the HL-MBR system at MLSS concentrations as high as 45 g/l.

The SDOX system needs to operate with pure oxygen and to recirculate the sludge through a pressurized vessel, which may introduce additional operational costs. The energy needs for the SDOX system compared to the energy needs for conventional diffusers were evaluated by Kim *et al.* (2020) for a hypothetical MBR system designed to treat an influent municipal wastewater flow rate of 2,000 m^3/d and exhibiting a biological oxygen demand of 2,530 kgO_2/day. The results are presented in Table 13.9. Three different aeration systems were included in the comparison as follows: fine diffusers, coarse diffusers, and the SDOX system. Medium standard aeration efficiency (SAE) values were considered (Henze *et al.*, 2008; www.blueingreen.com). The α factors selected for this evaluation were taken from the literature as described in Table 13.9. When working at standard conventional activated sludge (CAS)-relevant concentrations of approximately 4 g/l, fine bubble diffusers demand less power (28 kW) compared to coarse bubble diffusers (147 kW) and to the SDOX system (80 kW). It should be noted that the α factors stated above for bubble diffusers systems are higher than reported in studies on full-scale operating WWTPs. Rosso *et al.* (2005), based on operating data from approximately 30 WWTPs, reported an average α factor of 0.3 for systems operating at low SRTs. Bubble diffuser systems are prone to fouling when operated in biological systems, and it is common practice to apply a fouling factor when estimating the performance of bubble diffusion systems. When applying a fouling factor for long-term operation, fine bubble diffusers demand similar power (84 kW) to the SDOX system (80 kW) at CAS-relevant MLSS concentrations. However, at a sludge concentration of 10 g/l (*i.e.* a standard MBR concentration range), fine bubble diffusers (33 kW) demand less energy than both coarse bubble diffusers (242 kW) and the SDOX system (86 kW). Applying the fouling factor for long-term operation, fine bubble diffusers demand more power (99 kW) compared to the SDOX system (80 kW). At a concentration of 20 g/l MLSS, the SDOX system (99 kW) significantly outperforms both the fine (672 kW) and coarse bubble (554 kW) aeration systems. This MLSS concentration range applies to the HL-MBR concept. Beyond the 20 g/l MLSS concentration the differences are even more noticeable. Above 20 g/l it is not even feasible to introduce DO by fine bubble diffusers since non-detectable α factors have been reported (Germain *et al.*, 2007; Kim *et al.*, 2019). SDOX technology was only compared to conventional diffused aeration systems since those are the most widely used systems to provide DO into engineered wastewater treatment systems (Mueller *et al.*, 2002). In addition, in this comparison only energy (power) requirements were discussed and additional costs for maintenance activities were not considered. SDOX technology may require less maintenance compared to diffused aeration systems which represents other potential advantages for SDOX technology that may make the system competitive at the lowest range of evaluated MLSS concentrations and be even more advantageous in the high MLSS concentration range.

Table 13.9 Comparison of power needs for aeration systems as a function of the MLSS concentration (adopted from Kim *et al.*, 2020).

MLSS (g/l)	Type of aeration system							
	Fine-buble diffuser				Course-bubble diffuser			
	Alpha factor		Power required (kW) SAE (4.2 Kg/kWh)		Alpha factor	Power required (kW) SAE (1.0 kg/kWh)	Alpha factor	Power required (kW) SAE (4.1 kg/kWh)
	Germain *et al.*, 2007	Kim *et al.*, 2019	Germain *et al.*, 2007	Kim *et al.*, 2019	Gunder, 2000	Gunder, 2000	Kim *et al.*, 2019	Kim *et al.*, 2019
4	1.00	0.90	25	28	0.72	147	0.85	80
10	0.50	0.75	50	33	0.44	242	0.71	86
20	0.04	ND	672		0.19	554	0.53	99
30	ND	ND			0.08	1,271	0.39	116
45	ND	ND			0.02	4,416	0.25	153

Case 2: An HL-MBR equipped with a Speece cone system

The performance of the Speece cone system was also evaluated in the context of HL-MBRs. Barreto *et al.* (2017) evaluated the performance of an MBR provided with a Speece cone system fed with municipal wastewater. Moreover, Barreto *et al.* (2018) evaluated the oxygen transfer performance of the Speece cone at several operational conditions in clean water and in sludge. The pilot set-up with a treatment capacity of ~1 m^3/d was located at the Delft Blue Innovations research hall at the Harnaschpolder wastewater treatment plant in Den Hoorn, The Netherlands. Figure 13.32 depicts the experimental set-up. Figure 13.33 shows the schematic process flow diagram of the Speece cone HL-MBR system.

The operational conditions evaluated by Barreto *et al.* (2018) include the pressure inside the cone, the inlet velocity to the cone, the recirculation flow rate, and the pure oxygen mass flow rate into the cone. The evaluations were performed feeding HPO to the cone at gas flow rates ranging between 5 and 40% of the maximum theoretical oxygen dissolution capacity of the cone. The authors reported that the standard oxygen transfer efficiency (SOTE) is not affected by either the pressure inside the cone, or the inlet velocity, or the recirculation flow rate. However, the standard oxygen transfer rate (SOTR) increases with the pressure inside the cone and with the recirculation rate, not being affected by the inlet velocity. Moreover, the SOTR increases with the HPO mass flow rate into the cone, although the SOTE is negatively affected. Therefore, if a higher SOTR is the desired output, the Speece cone system should be operated at the maximum allowable operational pressure, recirculation flow rate, and HPO gas flow rate. However, the higher the HPO gas flow rate, the lower the SOTE. So the decision at which HPO gas flow rate the cone should be operated should be based both on the oxygen needs of the system and on financial considerations on the capital *versus* the operational costs of the system. SOTE values of approximately 100 % can be achieved, but at the expense of either having a large Speece cone system, or by operating at a high sludge recirculation rate through the Speece cone (*i.e.* high energy needs and operational costs).

Figure 13.32 The pilot-scale Speece cone HL-MBR system located at the Harnaschpolder wastewater treatment plant in Den Hoorn, The Netherlands. On the left-hand side of the photo is the Speece cone aeration device, and on the right-hand side is the MBR (photo: H.A. Garcia).

Compared to fine bubble diffused aeration, the Speece cone is more efficient, since nearly 100% of the supplied oxygen can be effectively delivered into the final receiving solution. Moreover, there is no need for the large amount of energy that is necessary to compress tremendous amounts of air in diffused aeration systems that is released back in the atmosphere carrying several volatile compounds. In addition, the Speece cone is also free of maintenance, since it does not require diffusers to be cleaned or replaced.

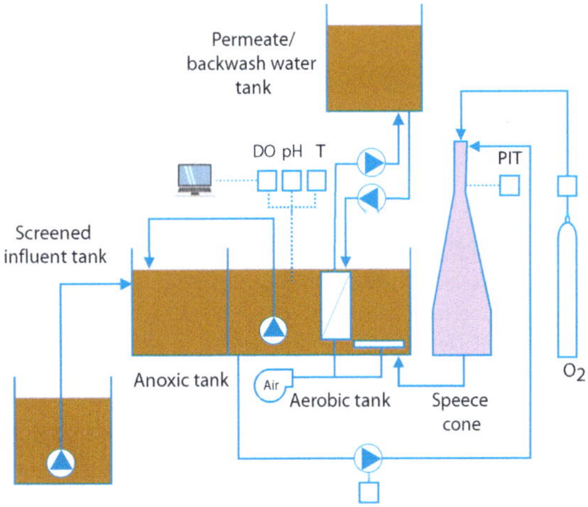

Figure 13.33 Schematic process flow diagram of the Speece cone HL-MBR system (adopted from Barreto *et al.*, 2017).

Barreto *et al.* (2017) also evaluated the performance of the Speece cone HL-MBR at a MLSS concentration of up to approximately 30 g/l. The results are shown in Figure 13.34. The loading rates to the HL-MBR system were continuously increased by mostly increasing the COD influent concentration until reaching the desired MLSS concentration to operate the HL-MBR system. The Speece cone HL-MBR system performed properly, biologically removing all the influent COD until reaching a MLSS concentration of 25 to 30 g/l. At these concentrations the system still exhibited COD removal efficiencies, however, not as well as at the lower MLSS concentration values. The COD removal efficiencies decreased from approximately 100% to 80%. The cause for this reduction was a limiting capacity in the flow of the recirculation sludge pump that limited the amount of DO introduced into the system. Nevertheless, the Speece cone system showed a good overall COD removal performance when pushed at higher MLSS concentrations.

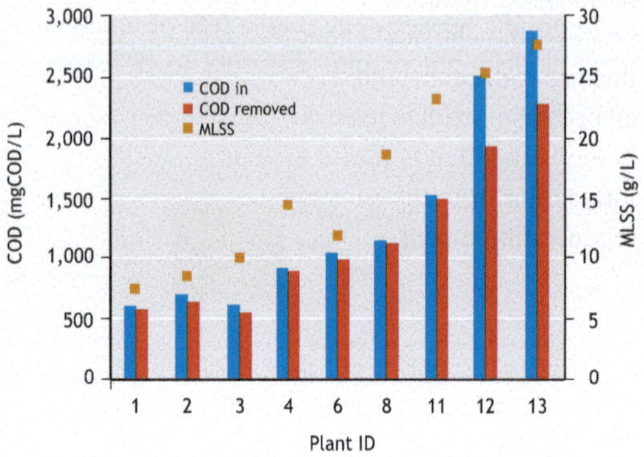

Figure 13.34 COD removal as a function of the MLSS concentration for the Speece cone HL-MBR (adopted from Barreto *et al.*, 2017).

13.7 FUTURE TRENDS IN MBR TECHNOLOGY

After half a century of development, MBRs are acknowledged to be a very promising alternative for municipal and industrial wastewater treatment. As mentioned above, they have several advantages over conventional CAS processes because of the involvement of membrane separation. In all, the application of MBRs can substantially reduce land usage and improve effluent quality, etc. As such, MBRs are an indispensable alternative in the field of wastewater treatment and reclamation. However, challenges remain. Future research should focus on addressing the challenges that include (*i*) controlling membrane fouling, (*ii*) reducing capital and operational costs, (*iii*) improving the synergistic effects between membranes and microbes, and (*iv*) applying MBRs rationally.

Controlling membrane fouling

The occurrence of membrane fouling can ultimately result in unexpected and irregular manual maintenance such as the implementation of intensive cleaning. This leads to additional burdens for an MBR plant. Thus, it is necessary to understand the fouling development patterns in large-scale MBR plants, *e.g.* the seasonal variations of membrane fouling rates. Then, according to the big data of MBR plants, computer-aided diagnosis and control of membrane fouling (*e.g.* with the aid of artificial intelligence) can be developed in the near future. In addition, because chemical cleaning is essential for flux recovery, future work should also focus on looking for more efficient cleaning agents and developing new cleaning modes. Other mechanically-assisted fouling control methods such as mechanical vibration are a potential alternative.

Reducing capital and operational costs

Higher capital and operational costs is another factor limiting the acceptance of MBRs by wastewater sectors. A decrease in capital cost can be achieved through development of low-cost membranes or development of chemical-resistant membranes with a long lifespan. The operational cost can be decreased by optimizing the membrane scouring and bioreactor aeration systems, which are linked with membrane fouling control and pollutant removal, respectively. Regardless, the aeration modes should be updated in order to enhance the fouling control of membranes and to increase oxygen transfer in the mixed liquor, which can lead to a decrease in aeration usage.

Improving the synergistic effects between membranes and microbes

Owing to the independent control of SRT and HRT, there are higher MLSS concentrations in MBRs than in CAS processes. Therefore there is a big opportunity to take advantage of bacterial populations with such a high concentration and high biodiversity. For instance, MBRs can allow the occurrence of endogenous denitrification, the possibility of sludge reduction, and the growth of specialized microorganisms. Enhancing the synergistic effects between membranes and microbes through various methods such as process optimization can potentially decrease the operation cost and improve the pollutant removal of MBRs.

Applying MBRs rationally

Undoubtedly MBRs are preferred in densely populated cities where land usage is relatively expensive or in water-deficient regions. MBRs can also be used for decentralized wastewater treatment (such as in rural areas), because MBRs, compared with CAS processes, are easier for modularized installation, with high flexibility in process configuration and centralized management. Moreover, combined MBRs+RO processes are very appropriate for achieving zero liquid discharge during industrial wastewater treatment. Because of the high selectivity of membranes and the strong biodegradation capacity of biomass, MBRs are also attractive for achieving resource and energy recovery from wastewater streams.

Acknowledgements

The authors express their great appreciation to Dr. Guihe Tao from PUB, Singapore's National Water Agency, Dr. Jinsong Zhang from CITIC Envirotech Ltd., Dr. Min Luo from SUEZ Water Technologies & Solutions, and China Everbright International Ltd. for providing information about the full-scale cases in Section 13.6. The authors would also like to acknowledge Mr. Sang Yeob Kim and Mr. Mauricio Barreto, PhD students at IHE Delft Institute for Water Education, for their technical contribution to developing HL-MBR in Section 13.6.3. Moreover, the authors would like to thank BlueInGreen LLC, OVIVO Water-MBR Systems, and ECO Oxygen Technologies LLC for their financial support to conduct most of the research on HL-MBR presented in Section 13.6.

REFERENCES

Akamatsu K., Lu W., Sugawara T. and Nakao S. (2010) Development of a novel fouling suppression system in membrane bioreactors using an intermittent electric field. *Water Research,* 44(3), 825-830.

Ashley K., Mavinic D. and Hall K. (2008). Oxygenation performance of a laboratory-scale Speece cone hypolimnetic aerator: preliminary assessment. *Canada Journal of Civil Engineering,* 35, 663-675.

Ashley K., Fattah K., Mavinic D. and Kosari S. (2014). Analysis of design factors influencing the oxygen transfer of a pilot-scale Speece cone hypolimnetic aerator. *Journal of Environmental Engineering,* 140, 04013011.

Barreto C.M., Garcia H.A., Hooijmans C.M., Herrera A. and Brdjanovic D. (2017). Assessing the performance of an MBR operated at high biomass concentrations. *Int. biodeter. biodegr.,* 119, 528-537.

Barreto C.M., Ochoa I.M., Garcia H.A., Hooijmans C.M., Livingston D., Herrera A. and Brdjanovic D. (2018). Sidestream superoxygenation for wastewater treatment: Oxygen transfer in clean water and mixed liquor. *Journal of Environmental Management,* 219, 125-137.

Bani-Melhem K. and Elektorowicz M. (2010). Development of a Novel Submerged Membrane Electro-Bioreactor (SMEBR): Performance for Fouling Reduction. *Environmental Science and Technol*ogy, 44(9), 3298-3304.

Bechhold H. (1907). Kolloidstudien mit der Filtrationsmethode. *Zeitschrift für Physikalische Chemie,* 60, 257.

Bemberis I., Hubbard P.J. and Leonard F.B. (1971). Membrane sewage treatment systems potential for complete wastewater treatment, *American Society of Agricultural Engineers Winter Meeting,* 71-878, 1-28.

Blanpain-Avet P., Faille C., Delaplace G. and Benezech T. (2011). Cell adhesion and related fouling mechanism on a tubular ceramic microfiltration membrane using Bacillus cereus spores. *Journal of Membrane Science,* 385(1-2), 200-216.

Buer T. and Cumin J. (2010). MBR module design and operation. *Desalination,* 250(3), 1073-1077.

Chen C., Guo W.S., Ngo H.H., Lee D.J., Tung K.L., Jin P.K., Wang J. and Wu Y. (2016). Challenges in biogas production from anaerobic membrane bioreactors, *Renewable Energy,* 98, 120-134.

Cheng H., Cheng D., Mao J., Lu T. and Hong P.Y. (2019). Identification and characterization of core sludge and biofilm microbiota in anaerobic membrane bioreactors. *Environment International,* 133(Pt A), 105165.

Chew N.G.P., Zhao S., Malde C. and Wang R. (2017). Superoleophobic surface modification for robust membrane distillation performance. *Journal of Membrane Science*, 541, 162-173.

China Association for Engineering Construction Standardization (CECS) (2017). Technical Specification for Design of Membrane Bioreactor Process for Municipal Wastewater Treatment

(T/CECS 152-2017). China Planning Press, Beijing (in Chinese).

Cornel P., Wagner M. and Krause S. (2003). Investigation of oxygen transfer rates in full-scale membrane bioreactors. *Water Science and Technology,* 47(11), 313-319.

Durán C., Fayolle Y., Pechaud Y., Cockx A. and Gillot S. (2016). Impact of suspended solids on the activated sludge non-newtonian behaviour and on oxygen transfer in a bubble column. *Chemical Engineering Science,* 141, 154-165.

Field R.W., Wu D., Howell J.A. and Gupta B.B. (1995). Critical flux concept for microfiltration fouling. *Journal of Membrane Science,* 100(3), 259-272.

Futselaar H., Schonewille H., De Vente D. and Broens L. (2007). NORIT AirLift MBR: side-stream system for municipal wastewater treatment. *Desalination,* 204, 1-7.

Gabarrón S., Dalmau M., Porro J., Rodriguez-Roda I. and Comas J. (2015). Optimization of full-scale membrane bioreactors for wastewater treatment through a model-based approach. *Chemical Engineering Journal,* 267, 34-42.

Gao L., Shi Y., Li W., Niu H., Liu J. and Cai Y. (2012). Occurrence of antibiotics in eight sewage treatment plants in Beijing, China. *Chemosphere,* 86, 665-671.

Germain E., Nelles F., Drews A., Pearce P., Kraume M., Reid E., Judd S.J. and Stephenson T. (2007). Biomass effects on oxygen transfer in membrane bioreactors. *Water Research,* 41, 1038-1044.

Grandclement C., Tannieres M., Morera S., Dessaux Y. and Faure D. (2016). Quorum quenching: role in nature and applied developments. *FEMS Microbiology Reviews,* 40(1), 86-116.

Guadie A., Xia S., Zhang Z., Zeleke J., Guo W., Ngo H.H. and Hermanowicz S.W. (2014). Effect of intermittent aeration cycle on nutrient removal and microbial community in a fluidized bed reactor-membrane bioreactor combo system. *Bioresource Techno*logy, 156, 195-205.

Gutman J., Walker S.L., Freger V. and Herzberg M. (2013). Bacterial attachment and viscoelasticity: Physicochemical and motility effects analyzed using quartz crystal microbalance with dissipation (QCM-D). *Environmental Science and Techno*logy, 47(1), 398-404.

Hai F.I. and Yamamoto K. (2011). Membrane biological reactors. In Wilderer P. (eds.), *Treatise on Water Science,* Elsevier, Oxford, pp. 571-613.

Hai F.I., Yamamoto K. and Lee C.-H. (2014). *Membrane Biological Reactors: Theory, Modeling, Design, Management and Applications to Wastewater Reuse.* IWA Publishing, London.

Hawari A.H., Du F., Baune M. and Thoming J. (2015). A fouling suppression system in submerged membrane bioreactors using dielectrophoretic forces. *Journal of Environmental Science,* 29, 139-145.

Henkel J., Cornel P. and Wagner M. (2011). Oxygen transfer in activated sludge - new insights and potentials for cost saving. *Water Science and Technology,* 63(12), 3034-3038.

Henze M., Van Loosdrecht M.C.M., Ekama G.A., Brdjanovic D. (2008). Biological wastewater treatment: Principles, modelling and design. IWA Publishing, London.

Hu Y.S., Yang Y., Yu S.C., Wang X.C. and Tang J.L. (2018). Psychrophilic anaerobic dynamic membrane bioreactor for domestic wastewater treatment: Effects of organic loading and sludge recycling. *Bioresource Technology*, 270, 62-69.

Huang X., Xiao K. and Shen Y. (2010). Recent advances in membrane bioreactor technology for wastewater treatment in China. *Frontiers of Environmental Science and Engineering China,* 4, 245-271.

Jabornig S. and Podmirseg S.M. (2015). A Novel Fixed Fibre Biofilm Membrane Process for On-Site Greywater Reclamation Requiring No Fouling Control. *Biotechnology and Bioengineering,* 112(3), 484-493.

Japan Sewage Works Association (JSWA) (2009). Sewerage Facility Planning: Design Guidelines and Commentary. JSWA, Tokyo (in Japanese).

Judd S. and Judd C. (2011). The MBR Book: Principles and Applications of Membrane Bioreactors for Water and Wastewater Treatment, 2nd ed. Butterworth-Heinemann, Oxford.

Kim J.H., Choi D.C., Yeon K.M., Kim S.R. and Lee C.H. (2011). Enzyme-immobilized nanofiltration membrane to mitigate biofouling based on quorum quenching. *Environmental Science and Technology,* 45(4), 1601-1607.

Kim S.Y., Garcia H.A., Lopez-Vazquez C.M., Milligan C., Livingston D., Herrera A., Matosic M., Curko J. and Brdjanovic D. (2019). Limitations imposed by conventional fine bubble diffusers on the design of a high-loaded membrane bioreactor (HL-MBR). *Environmental Science and Pollution Research,* 26, 34285-34300.

Kim S.Y., Garcia H.A., Lopez-Vazquez C.M., Milligan C., Herrera A., Matosic M., Curko J. and Brdjanovic D. (2020). Oxygen transfer performance of a supersaturated oxygen aeration system (SDOX) evaluated at high biomass concentrations. Unpublished data.

Kim S.R., Oh H.S., Jo S.J., Yeon K.M., Lee C.H., Lim D.J., Lee C.H. and Lee J.K. (2013). Biofouling Control with Bead-Entrapped Quorum Quenching Bacteria in Membrane Bioreactors: Physical and Biological Effects. *Environmental Science and Technology*, 47(2), 836-842.

Krampe J. and Krauth K. (2003). Oxygen transfer into activated sludge with high MLSS concentrations. *Water Science and Technology,* 47(11): 297-303.

Kurita T., Kimura K. and Watanabe Y. (2015). Energy saving in the operation of submerged MBRs by the insertion of baffles and the introduction of granular materials. *Separation and Purification Technology,* 141, 207-213.

Lade H., Paul D. and Kweon J.H. (2014). Quorum Quenching Mediated Approaches for Control of Membrane Biofouling. *International Journal of Biological Sciences,* 10(5), 547-562.

Lares M., Ncibi M.C., Sillanpaa M. and Sillanpaa M. (2018). Occurrence, identification and removal of microplastic particles and fibers in conventional activated sludge process and advanced MBR technology. *Water Research,* 133, 236-246.

Le T.-H., Ng C., Tran N.H., Chen H. and Gin K.Y.-H. (2018). Removal of antibiotic residues, antibiotic resistant bacteria and antibiotic resistance genes in municipal wastewater by membrane bioreactor systems. *Water Research,* 145, 498-508.

Lee S., Park S.K., Kwon H., Lee S.H., Lee K., Nahm C.H., Jo S.J., Oh H.S., Park P.K., Choo K.H., Lee C.H. and Yi T. (2016). Crossing the Border between Laboratory and Field: Bacterial Quorum Quenching for Anti-Biofouling Strategy in an MBR. *Environmental Science and Technology*, 50(4), 1788-1795.

Lei Z., Yang S.M., Li Y.Y., Wen W., Wang X.C. and Chen R. (2018). Application of anaerobic membrane bioreactors to municipal wastewater treatment at ambient temperature: A review of achievements, challenges, and perspectives. *Bioresource Technology,* 267, 756-768.

Liu L.F., Liu J.D., Bo G., Yang F.L., Crittenden J. and Chen Y.C. (2013). Conductive and hydrophilic polypyrrole modified membrane cathodes and fouling reduction in MBR. *Journal of Membrane Science,* 429, 252-258.

Livingstone D. (2010). Beyond conventional MBRs: Oxygen transfer technology revolutionizing MBR applications. Paper presented at the *Membrane Technology Conference & Exposition.*

McGinnis D.F. and Little J.C. (1998). Bubble dynamics and oxygen transfer in a Speece cone. *Water Science and Technology,* 37, 285-292.

Meng F., Chae S.-R., Drews A., Kraume M., Shin H.-S. and Yang F. (2009). Recent advances in membrane bioreactors (MBRs): Membrane fouling and membrane material. *Water Research,* 43(6), 1489-1512.

Meng F., Chae S.-R., Shin H.-S., Yang F. and Zhou Z. (2011a). Recent advances in membrane bioreactors: configuration development, pollutant elimination, and sludge reduction. *Environmental Engineering Science,* 29, 139-160.

Meng F., Zhou Z., Ni B.-J., Zheng X., Huang G., Jia X., Li S., Xiong Y. and Kraume M. (2011b). Characterization of the size-fractionated biomacromolecules: Tracking their role and fate in a membrane bioreactor. *Water Research,* 45(15), 4661-4671.

Meng F., Zhang S., Oh Y., Zhou Z., Shin H.-S. and Chae S.-R. (2017). Fouling in membrane bioreactors: An updated review. *Water Research*, 114, 151-180.

Metcalf & Eddy (2003). Wastewater Engineering: Treatment and Reuse, 4th ed. McGraw-Hill Higher Education, New York, USA.

Mueller J.A., Boyle W.C. and Popel H.J. (2002). *Aeration: Principles and Practice.* CRC Press, Florida, USA.

Muller E.B., Stouthamer A.H., Van Verseveld H.W. and Eikelboom D.H. (1995). Aerobic domestic wastewater treatment in a pilot plant with complete sludge retention by cross-flow filtration. *Water Research,* 29(4), 1179-1189.

Oh H.S., Yeon K.M., Yang C.S., Kim S.R., Lee C.H., Park S.Y., Han J.Y. and Lee J.K. (2012). Control of membrane biofouling in MBR for wastewater treatment by quorum quenching bacteria encapsulated in microporous membrane. *Environmental Science and Technology*, 46(9), 4877-4884.

Osborn S., McCain A. and Richardson G. (2010). Introduction to SDOX®, in: Jones B. (ed.) *LakeLine 30-2: New Approaches.* North American Lake Management Society.

Park H.-D., Chang I.-S. and Lee K.-J. (2015). Principles of Membrane Bioreactors for Wastewater Treatment. CRC Press, Boca Raton, Florida, USA.

Qu X.L., Alvarez P.J.J. and Li Q.L. (2013). Applications of nanotechnology in water and wastewater treatment. *Water Research*, 47(12), 3931-3946.

Rosenberger S., Laabs C., Lesjean B., Gnirss R., Amy G., Jekel M. and Schrotter J.C. (2006). Impact of colloidal and soluble organic material on membrane performance in membrane bioreactors for municipal wastewater treatment. *Water Research,* 40(4), 710-720.

Rosenberger S., Helmus F.P. and Drews A. (2016). Addition of Particles for Fouling Minimization in Membrane Bioreactors - Permeability Performance, Fluid Dynamics, and Rheology. Chemie Ingenieur Technik 88(1-2), 29-38.

Rosenberger S., Helmus F.P., Krause S., Bareth A. and Meyer-Blumenroth U. (2011). Principles of an enhanced MBR-process with mechanical cleaning. *Water Science and Technology*, 64(10), 1951-1958.

Rosso D., Stenstrom M.K. and Larson L.E. (2008).

Aeration of large-scale municipal wastewater treatment plants: state of the art. *Water Science and Technology,* 57(7), 973-978.

Rosso D., Iranpour R., Stenstrom M.K. (2005). Fifteen years of offgas transfer efficiency measurements on fine-pore aerators: key role of sludge age and normalized air flux. *Water Env. Res.* 77, 266–273.

Shen Y.-X., Xiao K., Liang P., Sun J., Sai S. and Huang X. (2012). Characterization of Soluble Microbial Products in 10 Large-scale Membrane Bioreactors for Municipal Wastewater Treatment in China. *Journal of Membrane Science,* 415-416, 336-345.

Siembida B., Cornel P., Krause S. and Zimmermann B. (2010). Effect of mechanical cleaning with granular material on the permeability of submerged membranes in the MBR process. *Water Research*, 44(14), 4037-4046.

Sui Q., Huang J., Deng S., Chen W. and Yu G. (2011). Seasonal variation in the occurrence and removal of pharmaceuticals and personal care products in different biological wastewater treatment processes. *Environmental Science and Technology,* 45, 3341-3348.

Sun J., Liang P., Yan X., Zuo K., Xiao K., Xia J., Qiu Y., Wu Q., Wu S., Huang X., Qi M. and Wen X. (2016a). Reducing aeration energy consumption in a large-scale membrane bioreactor: Process simulation and engineering application. *Water Research*, 93, 205-213.

Sun Y., Chen Z., Wu G., Wu Q., Zhang F., Niu Z. and Hu H.-Y. (2016b). Characteristics of water quality of municipal wastewater treatment plants in China: Implications for resources utilization and management. *Journal of Cleaner Production,* 131, 1-9.

Trzcinski A.P. and Stuckey D.C. (2016). Inorganic fouling of an anaerobic membrane bioreactor treating leachate from the organic fraction of municipal solid waste (OFMSW) and a polishing aerobic membrane bioreactor. *Bioresource Technology Journal,* 204, 17-25.

Urase T. and Kikuta T. (2005). Separate estimation of adsorption and degradation of pharmaceutical substances and estrogens in the activated sludge process. *Water Research,* 39, 1289-1300.

Wang J., Cahyadi A., Wu B., Pee W., Fane A.G. and Chew J.W. (2020). The roles of particles in enhancing membrane filtration: A review. *Journal of Membrane Science,* 595, 117570

Wang Z., Ma J., Tang C.Y., Kimura K., Wang Q. and Han X. (2014). Membrane cleaning in membrane bioreactors: A review. *Journal of Membrane Science*, 468, 276-307.

Wang Z.G., Cao J. and Meng F.G. (2015). Interactions between protein-like and humic-like components in dissolved organic matter revealed by fluorescence quenching. *Water Research,* 68, 404-413.

Water Environment Federation (WEF) (2012). Membrane Bioreactors. WEF Press, Alexandria, Virginia, USA.

Wei C.H., Harb M., Amy G., Hong P. Y. and Leiknes T. (2014). Sustainable organic loading rate and energy recovery potential of mesophilic anaerobic membrane bioreactor for municipal wastewater treatment, *Bioresource Technology Journal*, 166, 326-334.

Wei V., Elektorowicz M. and Oleszkiewicz J.A. (2012). Electrically enhanced MBR system for total nutrient removal in remote northern applications. *Water Science and Technology*, 65(4), 737-742.

Wu J., Chen F., Huang X., Geng W. and Wen X. (2006). Using inorganic coagulants to control membrane fouling in a submerged membrane bioreactor. *Desalination,* 197, 124-136.

Xiao K., Liang S., Wang X., Chen C. and Huang X. (2019). Current state and challenges of full-scale membrane bioreactor applications: A critical review. *Bioresource Technology Journal,* 271, 473-481.

Xiao K., Xu Y., Liang S., Lei T., Sun J., Wen X., Zhang H., Chen C. and Huang X. (2014). Engineering application of membrane bioreactor for wastewater treatment in China: Current state and future prospect. *Frontiers of Environmental Science and Engineering*, 8, 805-819.

Xu R., Yu Z., Zhang S. and Meng F. (2019). Bacterial assembly in the bio-cake of membrane bioreactors: Stochastic vs. deterministic processes. *Water Research*, 157, 535-545.

Xu Y., Zhu N., Sun J., Liang P., Xiao K. and Huang X. (2017). Evaluating oxygen mass transfer parameters for large-scale engineering application of membrane bioreactors. *Process Biochemistry*, 60, 13-18.

Xue W., Wu C., Xiao K., Huang X., Zhou H., Tsuno H. and Tanaka H. (2010). Elimination and fate of selected micro-organic pollutants in a full-scale anaerobic/anoxic/aerobic process combined with membrane bioreactor for municipal wastewater reclamation. *Water Research,* 44, 5999-6010.

Yamamoto K., Hiasa M., Mahmood T. and Matsuo T. (1989). Direct solid-liquid separation using hollow fiber membrane in an activated sludge aeration tank. *Water Science and Technology*, 21, 43-54.

Yeon K.M., Lee C.H. and Kim J. (2009). Magnetic enzyme carrier for effective biofouling control in the membrane bioreactor based on enzymatic quorum quenching. *Environmental Science and Technology,* 43(19), 7403-7409.

Zhang J., Satti A., Chen X., Xiao K., Sun J., Yan X., Liang P., Zhang X. and Huang X. (2015). Low-voltage electric field applied into MBR for fouling suppression: performance and mechanisms, *Chemical Engineering Journal*, 273, 223-230.

Zhang J., Xiao K. and Huang X. (2020). Full-scale MBR applications for leachate treatment in China: Practical, technical, and economic features. *Journal of Hazardous Materials,* 389, 122138.

Zhang S., Zhou Z., Li Y. and Meng F. (2018). Deciphering the core fouling-causing microbiota in a membrane bioreactor: Low abundance but important roles. *Chemosphere,* 195, 108-118.

Zheng X., Khan M.T. and Croué J.P. (2014). Contribution of effluent organic matter (EfOM) to ultrafiltration (UF) membrane fouling: Isolation, characterization, and fouling effect of EfOM fractions. *Water Research*, 65, 414-424.

Zhou Z., Tao Y., Zhang S., Xiao Y., Meng F. and Stuckey D.C. (2019). Size-dependent microbial diversity of sub-visible particles in a submerged anaerobic membrane bioreactor (SAnMBR): Implications for membrane fouling. *Water Research,* 159, 20-29.

Zsigmondy R. and Bachmann R. (1922). Filter and method of producing the same. *US Patent* 1 421 341.

NOMENCLATURE

Symbol	Description	Unit
A_M	Total membrane area	m^2
A_{M0}	Membrane area per unit membrane cassette	m^2
A_P	Total projection area (footprint) of membrane cassettes	m^2
C_o	Average DO concentration in the aerobic tank	mg/l
C_{omd}	DO concentration in the recirculated flow from membrane to aerobic tank	mg/l
$C_{os(20)}$	Clean-water saturated DO concentration at 1 bar at 20 °C	mg/l
$C_{os(T)}$	Clean-water saturated DO concentration at 1 bar at T °C	mg/l
F	Safety factor for estimation of the minimum sludge age in the aerobic zone	-
F_M	Safety factor for membrane area calculation	-
f	Conversion rate of sludge from SS	gMLSS/gSS
G_M	Standard-state air supply of blower aeration to the membrane tank	Nm^3/h
G_O	Standard-state air supply of blower aeration for biological reactions	Nm^3/h
g	Gravitational acceleration	m/s^2
h	Water depth of the aerobic tank	m
J_{avg}	Average flux of a filtration cycle	$l/m^2.h$
J_b	Backflush flux	$l/m^2.h$
K_{dn}	Specific denitrification rate	$kgNO_3^-$-N/kgMLSS.d
K_n	Half-velocity constant for ammonia utilization	$mgNH_4^+$-N/l
K_{ow}	Octanol/water partition coefficient	-
k_1	The first empirical parameter for estimation of the α factor	-
k_2	The second empirical parameter for estimation of the α factor	-
k_d	Endogenous decay coefficient	d^{-1}
L_s	Sludge organic loading rate	$kgBOD_5$/kgMLSS.d
N_{k0}	Kjeldahl N concentration of the aerobic tank influent	mg/l
N_{ke}	Kjeldahl N concentration of the membrane effluent	mg/l
N_{OM1}	Ammonia concentration in the aerobic zone	mg /l
N_{t0}	TN concentration in the influent of the biological system	mg/l
N_{te}	TN concentration in the effluent of the biological system	mg/l
n_M	Number of membrane cassettes	
O	Oxygen demand for biological reactions in the aerobic zone	kgO_2/d

O_{dn}	Oxygen demand offset by denitrification	kgO_2/d
O_m	Oxygen amount brought by recirculated liquid from the membrane tank to the aerobic tank	kgO_2/d
O_n	Oxygen demand for nitrification	kgO_2/d
O_s	Oxygen demand for oxidizing organic carbon	kgO_2/d
O_{std}	Standard-state oxygen demand at 20 °C and 1 bar in clean water	kgO_2/d
P	Actual atmospheric pressure	kPa
P_x	Phosphorus content in sludge	kgP/kgMLVSS
Q	Designed flow rate	m^3/d
$Q_{A2\rightarrow A1}$	Recirculation flow rate from anoxic to anaerobic tank	m^3/d
$Q_{M\rightarrow O}$	Recirculation flow rate from membrane to aerobic tank	m^3/d
$Q_{O\rightarrow A2}$	Recirculation flow rate from aerobic to anoxic tank	m^3/d
Q_w	Average flow rate of excess sludge discharge	m^3/d
$R_{A2\rightarrow A1}$	Recirculation ratio from anoxic to anaerobic tank	-
$R_{M\rightarrow O}$	Recirculation ratio from membrane to aerobic tank	-
$R_{O\rightarrow A2}$	Recirculation ratio from aerobic to anoxic tank	-
S_0	Feed BOD_5 concentration	mg/l
S_e	Effluent BOD_5 concentration	mg/l
SS_0	Feed SS concentration of the biological system	mg/l
SS_e	Effluent SS concentration of the biological system	mg/l
T	Temperature of the mixed liquor	°C
t_{A1}	HRT of the anaerobic tank	h
t_b	Interval of membrane backflush	minute
V_{A1}	Volume of the anaerobic tank	m^3
V_{A2}	Volume of the anoxic tank	m^3
V_M	Volume of the membrane tank	m^3
V_{M1}	Volume of the membrane tank excluding membrane cassette	m^3
V_O	Volume of the aerobic tank	m^3
V_{OM1}	Volume of the aerobic zone, *i.e.* sum of V_O and V_{M1}	m^3
V_t	Total volume of biological tanks	m^3
v_n	Specific ammonia utilization rate	$kgNH_4^+$-N/kgMLSS.d
v_{nm}	Maximum specific ammonia utilization rate	$kgNH_4^+$-N/kgMLSS.d
X	Weighted average of biological tank MLSS concentrations	g/l
X_{A2}	MLSS concentration in the anoxic tank	g/l
X_M	MLSS concentration in the membrane tank	g/l
X_O	MLSS concentration in the aerobic tank	g/l
X_{OM1}	Weighted average of the aerobic and membrane tank MLSS concentrations	g/l
Y	Synthesis yield of biomass	$kgMLVSS/kgBOD_5$
Y_t	Observed yield of solids	$kgMLSS/kgBOD_5$
y	Proportion of volatile MLSS in the total MLSS	kgMLVSS/kgMLSS
ΔX	Excess sludge production rate	kgMLSS/d
ΔX_v	Biomass discharge rate	kgMLVSS/d

Abbreviation	Description
A_1	Anaerobic tank
A_2	Anoxic tank
AAO-MBR	Anaerobic/anoxic/aerobic-MBR
AAOA-MBR	AAO-MBR with a post-anoxic tank
Ag NP	Silver nanoparticle
AGMBR	Aerobic granular sludge MBR
AI	Autoinducer
anammox	Anaerobic ammonium oxidation
AnMBR	Anaerobic MBR
AO-MBR	Anoxic/aerobic-MBR
AOX	Absorbable organic halogens
ARB	Antibiotic resistant bacterium
ARG	Antibiotic resistance gene
AsMBR	Air-sparging MBR
bCOD	Biodegradable COD
BNR	Biological nutrient removal
BOD	Biochemical oxygen demand
BOW	Beijing Origin Water Technology Co. Ltd.
C_{60}	Fullerenes
CA	Cellulose acetate
CAS	Conventional activated sludge
CEB	Chemically enhanced backflush
CEL	CITIC Envirotech Ltd.
CIP	Cleaning in place
CNT	Carbon nanotube
COD	Chemical oxygen demand
Cu NP	Copper nanoparticle
DO	Dissolved oxygen
EDC	Endocrine disrupting compound
ENM	Engineered nanomaterial
EP	Electrophoresis
EPS	Extracellular polymeric substance
F/M	Food-to-microorganism ratio
FO	Forward osmosis
FS	Flat sheet
GO	Graphene oxide
HF	Hollow fibre
HFMBR	Hybrid biofilm MBR
HL-MBR	High-loaded MBR
HPO	High-purity oxygen
HRT	Hydraulic retention time
IA	Intermittent aeration scouring

iMBR	immersed MBR
IOC	Internal and outer circulation
JLMBR	Jet loop MBR
M	Membrane tank
MBR	Membrane bioreactor
MCBR	Membrane coagulation bioreactor
MD	Membrane distillation
MDBR	Membrane distillation bioreactor
MEC	Magnetic enzyme carrier
MF	Microfiltration
MLSS	Mixed liquor suspended solids
MLVSS	Volatile MLSS
MRC	Mitsubishi Rayon Co. Ltd.
MT	Multitube
NH_4^+-N	Ammonium nitrogen
NM	Nanomaterial
PAC	Powdered activated carbon
PAN	Polyacrylonitrile
PE	Polyethylene
PES	Polyethylsulphone
PET	Polyethylene terephthalate
PP	Polypropylene
PPCP	Pharmaceutical and personal care product
PS	Polysulfone
PTFE	Polytetrafluoroethylene
PVDF	Polyvinylidene difluoride
QQ	Quorum quenching
QS	Quorum sensing
rGO	Reduced graphene oxide
RO	Reverse osmosis
SAD_f	Specific aeration demand with respect to the footprint of the membrane cassette
SAD_m	Specific aeration demand with respect to the membrane area
SAD_p	Specific aeration demand with respect to the permeate flow
SAE	Standard aeration efficiency
SDOX	Supersaturated dissolved oxygen
sMBR	Side-stream MBR
SMP	Soluble microbial product
SRMBR	Submerged rotating MBR
SRT	Solids retention time
SS	Suspended solids
STOE	Standard oxygen transfer efficiency
STOR	Standard oxygen transfer rate
TiO_2 NP	Titanium dioxide nanoparticle

TMP	Trans-membrane pressure
TOC	Total organic carbon
TN	Total nitrogen
TP	Total phosphorus
TrOP	Trace organic pollutant
UCT	University of Cape Town
UF	Ultrafiltration
UV	Ultraviolet
VSMBR	Vertical submerged MBR
WWTP	Wastewater treatment plant
ZnO NP	Zinc oxide nanoparticle

Greek symbols	Explanation	Unit
α	Ratio of the oxygen transfer coefficient in the sludge to that in clean water	-
β	Ratio of saturated DO concentration in the sludge to that in clean water	-
η_A	Oxygen transfer efficiency of the blower	%
θ_{OM1}	Minimum sludge age in the aerobic zone	d
θ_t	Total sludge age (*i.e.* SRT)	d
μ_n	Specific growth rate of nitrifying bacteria	d^{-1}
μ_{nm}	Maximum specific growth rate of nitrifying bacteria	d^{-1}
ρ	Density of the mixed liquor	g/cm^3
τ_0	Relaxation period of a filtration cycle	minute
τ_1	Running period of a filtration cycle	minute
τ_b	Duration of membrane backflush	minute
φ_M	Apparent membrane packing density in the membrane tank	m^2/m^3

Figure 13.35 Hospital buildings in Guiyang, China have an underground, 50,000 m^3/d, MBR plant for wastewater reclamation constructed 30 m beneath them (photo: G. Chen with courtesy of SEI and SUKE).

doi: 10.2166/9781789060362_0613

14

Modelling activated sludge processes

Mark C.M. Van Loosdrecht, George A. Ekama, Carlos M. Lopez Vazquez, Sebastiaan C.F. Meijer, Christine M. Hooijmans and Damir Brdjanovic

14.1 WHAT IS A MODEL?

A model can be defined as a purposeful representation or description (often simplified) of a system of interest (Wentzel and Ekama, 1997). This consequently means that a model never exactly reflects the reality. So, the question 'Does this model describe a wastewater treatment plant?' is senseless, unless what part(s) of the treatment plant the model should describe has already been defined. One never develops a model that describes every single organism, every molecule of water or every detail of the process. Models are used as a simplification of reality in such a way that they describe that part of reality that is relevant to understand and to deal with. It is also important to note that a mathematical model can only be successful if it fulfils the expectations people have of it.

There are two aspects that are very relevant in modelling: the aspects of time and of scale. In general, processes can be separated into three groups from the perspective of time. Processes can be in a frozen state, dynamic state, and steady state. Models are usually made to describe the dynamic state, the state where variations occur as a function of time. When a process is in a frozen state it means that the process will change over time, but not in the time interval that one is interested in. For example, if the daily dynamics at a wastewater treatment plant are of concern, the concentration of ammonia in the effluent will vary over time; the concentration of nitrate will vary in the reactors etc. However, within one day, the ammonia and nitrate concentrations in the sludge digester, which is sometimes part of the total activated sludge model, will not change. Usually the hydraulic retention time is 30 days, resulting in a characteristic time of change in this digester in the order of two to three weeks. Consequently one can consider the processes taking place in the digester being in a kind of frozen state. There are hardly any daily variations in the processes in comparison with those taking place in the wastewater treatment line. On the other hand, there are processes that are so fast that they are in steady state or in equilibrium condition. These processes occur so rapidly that the speed of change by far exceeds the dynamics that one is interested in. The changes that are of usual concern in wastewater treatment are, for example, changes in the ammonia concentration which have a rate value in the order of magnitude of hours. The changes that are relevant to process control have a rate value in the order of magnitude of minutes. However, if one considers chemical precipitation processes, they will occur more or less instantaneously (in a few seconds). These rapid processes do not have to be described in a dynamic way because they proceed so fast that one can assume they are in equilibrium condition or fully performed. Therefore, one of the first considerations in making a model is to consider which processes are of interest, followed by a determination of the relevant timeframe, an assessment of the dynamics of the process, and finally an adequate description of those processes that are time-variable. The other processes, which are in a frozen or steady state, are not of primary importance as they can be introduced in a much quicker, simpler way in the model, or even omitted. This is because they can be considered as continuous processes with stable concentrations under certain conditions (as in digesters). So, the aspect of time is the first major issue in trying to simplify the reality. The recommended approach is to consider the time constants and select those processes that have the dynamics in the order of time constants one is interested in. For wastewater treatment this usually means hourly or daily dynamics, and sometimes yearly dynamics. In the latter case, of course, digestion will become important as over the year the performance of a digester will change because the sludge production will vary during the year.

The second relevant issue for modelling is space resolution. One can theoretically make a model that describes every square inch of an activated sludge plant. However, the question is whether one is interested in such a detailed description in the first place. The answer depends again on the purpose of the model. In general, in wastewater treatment practice, the reactor size is in the order of tens of meters. To describe the concentration gradients of relevant components in the reactor, of which oxygen is the most sensitive one, usually a scale size of a few meters is needed. On a different scale, there is a gradient of concentrations inside the activated sludge floc that can theoretically be described by a model. However, in standard activated sludge modelling it is neglected, as being not relevant enough to be taken into account. This consequently means that activated sludge models are usually not designed to describe a system on the scale of an activated sludge floc but on the scale of a reactor.

The next step in modelling is to look at the relevant level of detail of a microbial model (Figure 14.1).

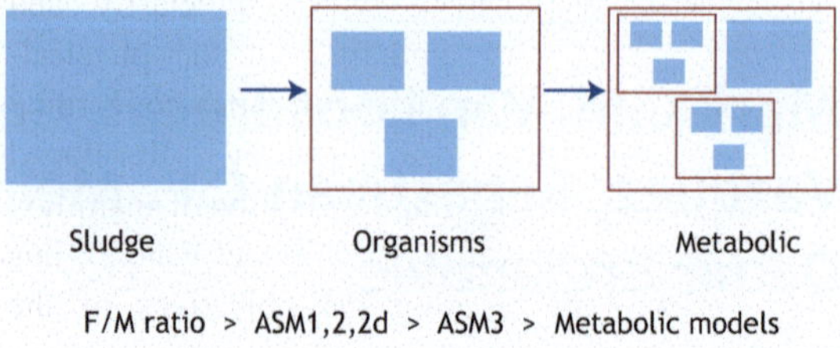

Figure 14.1 Schematic representation of the step-wise refinement of a model (Smolders *et al.*, 1995).

Typical traditional wastewater treatment design methods are based on the 'black-box' approach focusing on plant influent and effluent characteristics, while nothing or very little is known about what is happening inside the wastewater treatment plant. Traditional design parameters, such as the F/M ratio (sludge loading rate), are not based on the understanding of the processes within a wastewater treatment plant. However, one can design a plant reasonably well by applying a proper loading rate, without really knowing what processes are taking place in the plant. So the black-box model can work out well in practice.

The black-box model is not by definition wrong or not scientific, but the application of the black-box approach depends very much on the purpose of the model. If the purpose is to design a wastewater treatment plant, practice has shown that the F/M ratio can be a very good basic approach for the design, despite the fact that it does not give information about the sludge composition. One can refine this approach and move towards grey-box models, as was the case in Activated Sludge Model No.1 (ASM1, Henze *et al.*, 1987a, 1987b), No. 2 (ASM2, Henze *et al.*, 1995), and No. 2d (ASM2d, Henze *et al.*, 1999). Here the sludge was split up into relevant fractions: an inert organic matter fraction, a fraction of nitrifying bacteria, heterotrophic bacteria, denitrifying bacteria, and a fraction of phosphate-removing bacteria. Different functional aspects of the sludge were specified for a population-based model where selected microbial communities were defined inside the activated sludge and as such incorporated in the model.

Furthermore, the metabolism of the organisms and the metabolic routes inside the organisms can also be described. By such an increase in information, the approach becomes close to glass-box modelling (such as Activated Sludge Model No.3: ASM3, Gujer *et al.*, 1999, and the TU Delft EBPR model: TUDP model, Van Veldhuizen *et al.,* 1999). This results in a bigger and more complex model. The challenge here is to figure out for each process what the adequate level of description is. The question is: does the increase in complexity also increase the quality of the model (outputs), in other words, does it provide a better description of the wastewater treatment plant? For example, it has been shown that with an increase in the level of detail in the description of nitrification, a marginal improvement of model performance can be obtained while in the case of phosphate removal, significant improvement can be gained by including a metabolic description. Therefore, the preference for black-, grey-, or glass-box modelling depends very much on the purpose and application of the model. This is the point where modelling often goes wrong as modellers neglect the purpose and make the model itself the purpose of modelling.

Of course, one can even attempt to proceed to the next level of complexity by including fundamentals of microbial genetics and genetic change. Technically and in principle this is possible, but again, must be done depending on the purpose and use of the model. If the model is made too complex and with too many parameters in relation to what one ultimately wants to describe, then such an approach can be generally considered as a waste of time and effort. There is also no absolute need for a model to exactly describe reality. How far reality should be matched depends again on the purpose. For example, if one wants to get an idea about N_2O emission from wastewater treatment plants, maybe three or four theories can be created and incorporated into the model. At this point it is of prime interest to look at the simulation results of different models in terms of trends and to which extent these trends reflect the reality. At this stage one has to focus on trends only and good calibration, exact fitting and accurate knowledge of parameter values are not necessary. On the contrary, as an example, if one needs to predict the performance of a plant to satisfy legislation that requires every sample taken from the effluent to be below 1 $mgNH_4/l$, then the accuracy of the parameters put in the model must be much higher. In this instance one has to guarantee that the model prediction must be below 1 $mgNH_4/l$ exactly. These two examples show again that the model should always be judged relative to the purpose of its use.

Two extremes in the type of mathematical models can be identified: empirical and mechanistic models. An empirical model is based on recognition of the parameters that seem to be essential to describe the behavioural pattern of interest, and linking these by empirical relationships established by observation. The mechanisms and/or processes operating in the system are not known or are ignored: a classical black-box approach. In contrast, a mechanistic model is based on some conceptualization of the biological/physical mechanisms operating in the system, *i.e.* it is based on a conceptual idea (or model). The complexity of this mechanistic model will depend on the degree of understanding of the biological and chemical processes occurring in the system. Because mechanistic models have some conceptual basis, they are often more reliable than empirically based models. Because of their black-box approach, empirical models have an application strictly limited by the boundaries (*e.g.* wastewater characteristics, system parameters) within which the model was developed; only interpolation is possible. Being conceptually based, mechanistic models have greater sureness in application outside the boundaries within which the model was developed; both interpolation and extrapolation are possible. However, ultimately all models are only our rationalization of behavioural patterns of processes we conceive to be of interest.

Owing to this rationalization, any model needs to be rigorously calibrated and adequately verified by appropriate tests. Also, the conditions within which the model is expected to operate successfully need to be firmly delineated. For empirical models these are strictly the conditions within which the model was developed, while for mechanistic models these are the conditions under which the conceptualized behaviour is expected to remain valid. It is evident from the discussion above that mechanistically based models have greater potential for application in wastewater treatment plants, and attention will be focused on these models.

Processes operating in a system and the compounds on which these act must be identified in order to set up a conceptual model on which the mechanistic mathematical model is based. The various interactions between the processes, and between the processes and compounds, are delineated descriptively. To develop the mechanistic model from the conceptual model, the process rates and stoichiometric interactions with the compounds are formulated mathematically. The mathematical equivalent of the mechanistic model very probably will not include all the processes and compounds that are present in the system; only those considered to be of significance for fulfilling the objectives set for the model need be included. The art of constructing conceptual and mechanistic models is in eliminating those processes and compounds that contribute little or nothing to fulfilling the objectives set for the model. It is a waste of time and effort to develop a complicated model where a simpler one is adequate. It is most unlikely that a model can be developed that describes a phenomenon completely. Theoretically a complete description should include aspects down to the most fundamental level. The level of organization is usually set by the objectives for the model. For example, in modelling biological behaviour in wastewater treatment systems, we cannot directly implicate biochemical control mechanisms (such as ADP/ATP and NAD/NADH ratios), or even the behaviour of specific microorganism species. The mixed liquor in the activated sludge system contains a wide diversity of different microorganism species for which identification and enumeration techniques have recently become available. These techniques, however, are time and labour-consuming. Instead, microorganisms that fulfil a particular function in the activated sludge system (*e.g.* aerobic degradation of organics or nitrification) are grouped together as a single entity, which is called a 'surrogate' organism. This surrogate organism is assigned a set of unique characteristics that reflect the behaviour of the group but might not reflect the characteristics of any individual organism or species of organisms in the group. To illustrate, this approach is equivalent to modelling the 'macroscopic' behaviour of a forest of trees as opposed to the 'microscopic' behaviour of each individual tree or species of trees that makes up the forest. In considering the behaviour of the forest, a parameter that could be modelled, for example, is carbon dioxide (CO_2) production. The forest as an entity will have defined CO_2 production and consumption rates. Individual tree species within the forest, or even every single tree, might have specific CO_2 production and consumption rates that deviate significantly from those of the forest entity. However, the effect achieved by modelling the forest as an entity will closely equal the net effect of modelling the cumulative contribution of each individual tree or tree species. The great advantage in modelling the forest as an entity over modelling the individuals is that considerably less information is required to develop the model and to calibrate it. In modelling biological wastewater treatment systems, the utilization of substrate by organisms is a typical example: Monod's Equation (Monod, 1949) is used to relate the specific growth rate of the surrogate organism to the surrounding substrate concentration, whereas the organisms making up the surrogate group might have different specific growth rates or might respond differently to the various substrates present in the influent wastewater. Thus, for modelling wastewater treatment systems the organizational level that is modelled is the mass behaviour of a population or group of selected microorganisms. In the models developed for activated sludge systems, the principle organism groups, their functions and the zones in which these functions are performed are summarized in Chapter 2.

The parameters at that level that need to be included in the mathematical model depend greatly on the objectives for the model taking into account the level of organization described above. For mathematical modelling of wastewater treatment systems two different kinds of mathematical models are generally developed: steady state and dynamic models. Steady state models have constant flows and loads, and tend to be relatively simple. This simplicity makes these models useful for design. In these models complete descriptions of system parameters are not required. They are oriented towards determining the more important system design parameters. On the other hand, dynamic models have varying flows and loads and accordingly include time as a parameter. Dynamic models are more complex than steady state ones and are useful in predicting the time-dependent system response of an existing or proposed system. Their complexity means that for application the system parameters have to be completely defined. For this reason, the use of dynamic models for design is restricted. Often steady state design and dynamic kinetic models evolve interactively. Dynamic kinetic models can provide guidance for the development of steady state design models; they help identify the design parameters that have a major influence on the system response and help eliminate those processes that are not of major importance at steady state. For dynamic models, with their greater complexity, only those parameters that seem to be of importance are considered for inclusion in the model.

For activated sludge systems, selecting the level of organization at the surrogate organism or mass behaviour of populations, until recently the dynamic models have been structured to consider only the net effects as present in the bulk liquid. For example, in using Monod's Equation the kinetic rate was determined by the bulk liquid soluble COD and surrogate organism concentrations. However, with the extensions of the models to include enhanced biological phosphorus removal (EBPR), parameters internal to the surrogate biomass have had to be included, *e.g.* poly-β-hydroxyalkanoate (PHA), glycogen and polyphosphate. With this development, although the model might be at a selected level of the organization, information on processes and behaviour from lower levels of the organization is often essential, particularly to identify the key processes that control the response of the system. Usually information from lower levels of the organization is of a microbiological and/or biochemical nature, and the more complete this information is the more reliable the model. To make use of this information, 'model' organisms that are part of the 'surrogate' are identified and the known microbiological and biochemical characteristics of the organism are used to obtain a greater understanding of the surrogate. More recently the surrogate organism approach to modelling has been found to be inadequate to describe completely some behavioural patterns observed in activated sludge systems; for example, the selector effect (Gabb *et al.*, 1991), substrate utilization inhibition on transfer from anoxic to aerobic zones (Casey *et al.*, 1994), and generation of nitrogen intermediates in denitrification (Casey *et al.*, 1994). To describe these and similar observations, it has been found that a lower level of organization needs to be selected: the synthesis and activity of certain key enzymes and the processes they mediate need to be modelled (Wild *et al.*, 1994). Modelling at this level of the organization has been termed modelling with structured biomass. Detailed microbiological and biochemical information is required for this modelling approach (Wentzel and Ekama, 1997).

It should be noted that there is an essential difference between an activated sludge model and a wastewater treatment (plant) model. The latter term is used to indicate the ensemble of an activated sludge model, hydraulic model, oxygen transfer model and sedimentation tank model, all needed to describe an actual wastewater treatment at a full-scale installation (Gernaey *et al.,* 2004). The wastewater treatment plant model should be furthermore distinguished from a plant-wide model, which combines wastewater treatment models with sludge treatment models.

14.2 WHY MODELLING?

The most important advantages of the use of models in wastewater treatment are: getting insight into plant performance, evaluating possible scenarios for upgrading, evaluating new plant design, supporting management decisions, developing new control schemes, and providing operator training.

Modelling forces the modeller to make their work explicit. Qualitative comparisons are often found in the literature such as 'better', 'larger', 'smaller', 'higher', etc. Such comparisons are not very useful and are of a subjective nature, as, for instance, the perception of 'large' or 'small' by a researcher in the laboratory or by a person operating a wastewater treatment plant is not necessarily the same. When it comes to modelling it is not possible to use descriptive elements, but it is necessary to use quantitative inputs for sizes, rates, and conversions as the model requires numbers as input. This also forces modellers to become quantitative and objective in their approach and thus the process knowledge is better defined. Of course, one can do without modelling, but very often by making a model one makes a framework that takes into account everything which is considered relevant. It furthermore forces structured and more extensive data collection, and encourages the modeller to be organized. It often exposes knowledge and data limitations and/or incorrect data (such as SRT or flows), supports efforts to improve the quality of data, and enhances good plant monitoring practices. Therefore it is not surprising that getting insight into plant performance (quantification of information, mass balances and data reconciliation) and learning about the wastewater treatment plant in question can be even more important than the modelling itself.

The second main reason for using models is the possibility of saving time and money in the process of the technology/process selection. Comparison of the system performances in a quantitative instead of a qualitative approach allows in many cases for easier decision-making and rapid comparison of options. In comparison with a qualitative description such as 'one system is more efficient than the other', model results showing that 'one system is 2% (or 20%) more efficient than the other' is much more informative and useful. If important information or selection criteria are quantified (such as purification efficiency, effluent quality, sludge production, oxygen requirements, etc), application of modelling for evaluating possible scenarios for upgrading will make the comparison more effective and faster than discussion on such issues that are usually empirical, intuitive, long and often cumbersome. For the purpose of evaluating upgrade scenarios, it is not necessary to make a very accurate model by performing an extensive calibration procedure, as the real uncertainty is associated with the model inputs and not with the model parameters. It is considered much more useful to use trends for comparison as small differences are not relevant in the context of the usual design horizon used in wastewater treatment engineering. In the case of evaluating new plant design, again it is not necessary to have a fine-tuned model due to uncertainty in process conditions in the coming 10 or 20 years. For primary plant design usually static (steady state) models are used while dynamic modelling is applied for sensitivity analysis and optimizing the design. An additional challenge is the fact that wastewater has an extremely complex and uncertain composition. Wastewater flow rate and concentrations are of a highly dynamic nature and are very difficult to control, despite certain limited possibilities to influence the wastewater composition (Chapter 3). Many processes take place within the wastewater treatment plant; some of them are relevant to the treatment and many of them are not. However, many of them occur simultaneously, even in a single process unit. In order to handle such a complex situation, there is a need for a model to support the understanding of those relevant processes. Therefore, despite the fact that from the design perspective, modelling as such is usually not strictly necessary, it is becoming increasingly used as a part of the design process. By applying statistical methods to the occurrence of worst-case scenarios, significant savings can be made and the plant can still achieve its effluent quality standards for approximately 95% of the time. In traditional design all the worst-case scenarios are often assumed to occur simultaneously, leading to a highly unlikely scenario.

Another important reason for using models is the possibility of decreasing or minimizing risks. By using models, 'what-if' scenarios can be examined in a quantitative way in respect of what the effects of potential risks are. Such a glass-box type (as opposite to the black-box type) quantification is invaluable in the evaluation and selection of acceptable risk, rejecting risks that cannot be taken, and in the identification of upfront measures that can be taken to mitigate or control such risks. For example, questions like 'What will happen if the flow rate doubles?' and 'What is the effect of such an increase on effluent quality?' can be properly addressed by using models. Furthermore, models allow for minimization of risks that are related to the scaling-up of the systems (lab-scale vs. pilot-scale vs. large-scale). Related risks originate from the fact that, for example, mixing conditions, load variations etc., are different for full-scale and lab-scale installations. From the perspective of process control, pilot-scale gives a much faster response in comparison with full-scale plants with greater inertia.

Furthermore, the application of models improves knowledge transfer and decision-making. Wastewater treatment engineering and environmental engineering in general are multidisciplinary fields requiring knowledge of different disciplines, such as microbiology and biochemistry, as well as physical, biological and mechanical engineering. In addition, each expert group involved, be they operators, engineers or scientists, usually has its own perspective of the same subject. By phrasing the subject into a mathematical context the same communication tool (language) is used. Such a multidisciplinary approach allows for a better description of the reality, each discipline delivering its own input for a better understanding of the reality that can be incorporated in a structured, organized and quantitative way into the model. Model-related communication was greatly improved after the introduction of ASM1 in 1987. Prior to the introduction of ASM1, at least five or six different ways of modelling wastewater treatment plants were described; each model had its different approaches in writing up, in notation, and in the implementation of equations, which made it extremely difficult to understand the models and their results. The uniform context and standardization introduced by ASM1, in terms of notations, symbols and structure, made comparison of results and knowledge transfer much easier and further encouraged modelling applications.

Nowadays models are invaluable tools for training. For example, the plant operator can safely investigate by means of modelling what can happen if one takes certain action at a treatment plant without risking upsetting the operation of the plant. Moreover, models can be used to transfer knowledge from design engineers to operators and, of course, in academia worldwide, where modelling is increasingly becoming a part of the curricula for engineers and scientists. From the perspective of process control, in practice there are no direct model-based controllers functional yet, as their application still remains of scientific interest. In practice simple controllers are tuned based on the model, which allows for much quicker optimization of control strategy at full-scale installations (Chapter 15).

In the framework of integrated urban water system modelling, wastewater treatment modelling is an important component and it is necessary to link up wastewater treatment with the sewer system (to take into account the effects of, for example, combined sewer overflows or processes taking place in the sewerage system) on the one hand, or the receiving water's quality and quantity, on the other hand. Integrated modelling is becoming an increasingly popular tool to support decision-making at the level of urban water system management as it brings objectivity and gives quantitative insight into the relevant differences between options.

14.3 MODELLING BASICS

14.3.1 Model building

Many different types of models exist; these can be broadly categorized into (*i*) physical, (*ii*) verbal or conceptual and (*iii*) mathematical models. The physical model is a spatially scaled representation of the system. For example, the laboratory- and pilot-scale experiments used by scientists and engineers to investigate system response and behaviour are physical models. The verbal or conceptual model provides a qualitative description of the system and is usually developed from detailed observations; these models can be presented as schematic diagrams (*e.g.* flow diagrams) or as a series of narrative statements. Preparation of a verbal model is the most important but also the most complex part of model building. The mathematical model provides a quantitative description of the system. With mathematical models the rates of the processes acting in the system and their stoichiometric interaction with the compounds are formulated mathematically. The mathematical formulations need to be incorporated in a solving procedure that takes account the physical constraints and characteristics imposed by the system in which the processes take place, *e.g.* temperature and mixing conditions. Mathematical models are seldom developed in isolation, but usually evolve interactively from a conceptual model that might be based to some degree on observations made on a physical model, *e.g.* laboratory- or pilot-scale experiments (Wentzel and Ekama, 1997).

Research methodology which combines verbal, mathematical and physical models (Figure 14.2) is very helpful to make rapid progress and to evaluate new systems.

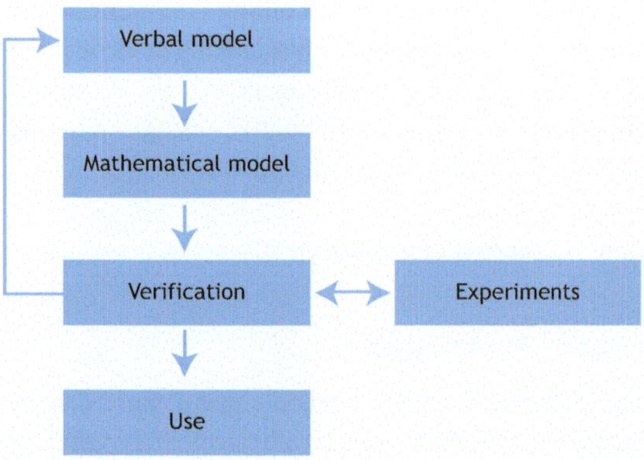

Figure 14.2 The model-building process.

A number of factors are to be considered regarding activated sludge modelling and simulation, and a step-wise approach is needed to progress from the model purpose definition to the point where a wastewater treatment plant model is available for simulations. The following main steps can be distinguished in this process (Coen *et al.*, 1996; Petersen *et al.*, 2002; Hulsbeek *et al.*, 2002):

- definition of the model purpose or the objectives of the simulation study;
- model selection: choice of the models needed to describe the different plant units to be considered in the simulation, *i.e.* selection of the activated sludge model, the sedimentation model, etc.;
- hydraulics, *i.e.* determination of the hydraulic models for the plant or plant tanks;
- wastewater and biomass characterization, including biomass sedimentation characteristics;
- calibration of the activated sludge model parameters;
- model falsification, and
- scenario evaluations.

The methodology is illustrated in detail by Petersen *et al.*, (2002).

14.3.2 General model set-up

Balance equations form the basis of any model description. These equations describe the change in concentration in a reactor over time as the result of chemical and biological conversions and of transport processes. In steady state the change of the concentration as a function of time becomes zero. Transport and conversion processes are two different parts of the model (of a physical and chemical-biological nature, respectively).

The biological processes are only dependent on the concentration in a reactor at the place where the conversion takes place. In essence the conversion processes are therefore independent of the type of reactor or the size of the reactor (microorganisms do not know in which type of rector they are in, concrete or steel, plug-flow or fully-mixed, activated sludge or biofilm reactor, etc.). Therefore the biological and chemical conversions are called micro-kinetics and can be easily studied in the laboratory and will not change at a full-scale installation. This part of a process model is therefore universal and can be formulated as a general activated sludge model, such as the ASM model family. The local concentrations in a reactor depend on the transport of the reacting compounds in the reactor system or treatment plant. When comparing full-scale systems, the essential differences lie in these transport processes.

The advantage of transport processes (such as convective flow, mixing, aeration) is that they are very well studied and described by general rules. They can therefore be relatively well predicted for different types and scales of processes. One can study biology and chemistry in the laboratory (for instance, the effect of temperature, concentration and pressure on microorganisms) and then use physical transport models to predict what is going to happen at full-scale. Acknowledging the fact that microbes will not undergo change between laboratory and full-scale conditions, as opposed to transport processes, helps to understand the processes and their integration in mathematical models. Such integration allows these models to be used in the process design (selection of bio-reactors, types, stability, optimization, automation and control, scale-up, etc.).

The components of a full wastewater treatment model are schematically given in Figure 14.3. First, measurable wastewater parameters have to be transformed into an influent vector with the concentrations of the different model compounds (Chapter 3). The wastewater treatment plant is modelled hydraulically by describing the different zones/reactor compartments of the plant, including the settler. Each reactor compartment is modelled individually for its mixing and mass transfer (*e.g.* aeration) characteristics. Usually a completely mixed tank reactor is used. A mass balance equation is applied for each reactor. Such a mass balance equation includes a bioconversion model. In the overall model all the compartments are coupled by the state vector including the concentrations and flow rates of the links between the compartments. This overall model is usually numerically solved to give the concentrations of all the compounds as a function of time for each compound included in the model. Therefore, effectively we can speak about four models: the process model, the hydraulic model, the

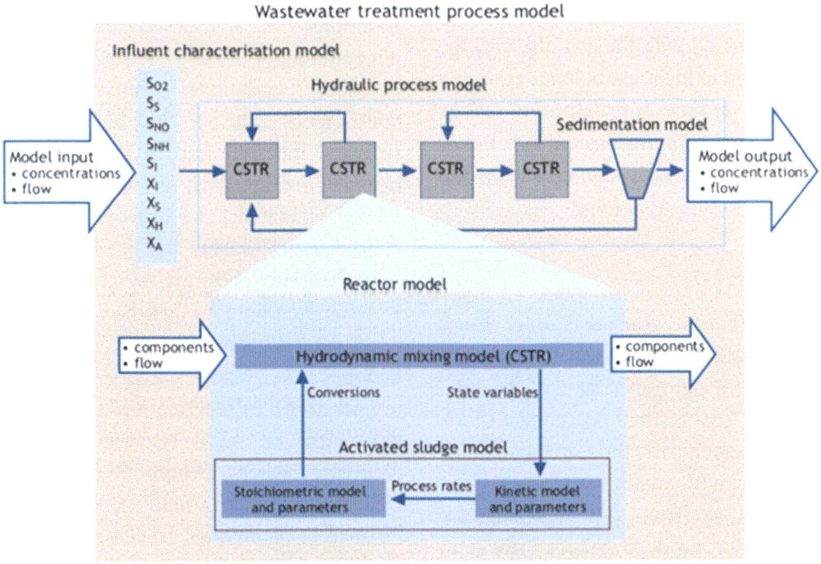

Figure 14.3 Schematic representation of a complete wastewater treatment plant model (Meijer, 2004).

reactor/compartment model and finally the activated sludge model.

The mass balance equation in steady state in mathematical terms reads:

$$\frac{\delta(S_{in} \cdot Q_{in})}{\delta t} = \frac{\delta(S_{out} \cdot Q_{out})}{\delta t} + (\alpha \cdot q \cdot X \cdot V) + (k_l A \cdot (S_{max} - S)) \quad (14.1)$$

where:

α	stoichiometry
A	surface area (m^2)
k_l	external mass transfer coefficient (m/h)
q	specific conversion rate (l/h)
Q_{in}	influent flow (m^3/h)
Q_{out}	effluent flow (m^3/h
S_{max}	saturation concentration (gCOD/m^3)
S	concentration in liquid (gCOD/m^3)
S_{in}	concentration in influent (gCOD/m^3)
S_{out}	concentration in effluent (gCOD/m^3)
t	time (h)
V	volume (m^3)
X	biomass concentration (gCOD/m^3)

Effectively it states that a compound entering a reactor either leaves with the effluent, is converted in the reactor, or is exchanged with the gas phase in the compartment. Each term in the mass balance equation is expressed as mass over time. It is helpful to realize that in order to analyse a complex system it is better to work in these dimensions than with concentration terms.

14.3.3 Stoichiometry

From the system definition one takes only those compounds of the system which are considered important and/or make a significant part of the total system mass (being at least a small percentage of it). For example, in the case of nitrification, in most plants the nitrite concentration will remain very low or close to the detection limit, so from the mass balance perspective there is no need to take nitrite into account. In anaerobic digestion, similarly, there is no need to take hydrogen into account, as the hydrogen content of the gas is very low, as almost everything ends up as methane. Such intermediates will only be specified if considered important, for example, when there is nitrite or hydrogen accumulation. Nitrite is not included in the nitrification process in ASM1, while in the Anaerobic Digestion Model (ADM1, Batstone *et al.*, 2002) hydrogen is included as it plays an important role in the stability of the anaerobic system. ASM models are specifically designed for applications at lower temperatures (5 to 20ºC), in which no significant accumulation of nitrite is expected to take place. Nitrite will only accumulate at higher temperatures or in the case of unusual toxic events. Thus nitrite is left out of the model. Similarly, in denitrification, only a small amount of nitrate turnover is in the form of N_2O, so from the perspective of describing the N removal it is not relevant to include the contribution of N_2O. However, if the plant has to fulfil N_2O gas limits, then it becomes important to take it into account. Again, this depends on the purpose of the application of the model.

Beside the determination of relevant compounds and processes, defining relevant balances is essential. For each conserved balance the number of atoms of a compound entering the plant is equal to those leaving. Examples of conserved balances are nitrogen, phosphorus, COD or alkalinity conversion. Using balance equations, unknown stoichiometry coefficients can be calculated. This substantially reduces the required information for modelling because the approach allows a number of unknowns to be calculated. The use of BOD measurement to characteristic wastewater is declining and, instead, modern approaches rely on COD. BOD-based design is associated with a black-box approach, it cannot be used for balancing as it is not conserved, and it depends on many factors (*e.g.* reaction time, temperature). In practice it is still mostly used to link the output of ASM regarding effluent impact on receiving waters (where BOD is still a relevant indicator of water quality). In contrast, the COD balance is maintained because COD is by definition the amount of electrons which are transferred to oxygen in order to oxidize all the organic matter in the system to CO_2 and water. This is why modelling nowadays is based on COD rather than on BOD.

Stoichiometry can be determined based on relevant compounds involved in the reaction and use of balances to calculate the relevant coefficients. For example, in the reaction for heterotrophic growth the relevant compounds are organic matter, oxygen, ammonia, alkalinity, biomass, carbon dioxide and water. At this stage of equation development it is not necessary to determine which compound is utilized and which is produced as it will simply receive a negative or positive sign; in other words, it is not important on which side of the equation the parameter is. The next step is to choose one coefficient as 1 and to use balances to calculate all the other relevant coefficients. In the example here, five balance equations can be made (for carbon, oxygen, hydrogen, nitrogen and charge) and there are seven unknown coefficients. One of these coefficients can be made equal to 1 and there is only one coefficient that we need to know, for example the amount of oxygen consumed per COD converted or the amount of biomass produced per substrate COD mass utilized (yield coefficients).

This is a general approach for setting up the stoichiometry of the reaction for any biological process *e.g.* if organics (or COD) are utilized aerobically (with O_2).

$$?COD + ?O_2 + ?NH_4^+ + ?HCO_3^- \rightarrow ?Biomass\text{-}COD + ?CO_2 + ?H_2O \quad (14.2)$$

In wastewater treatment systems we are usually not interested in the CO_2 and H_2O conversion, and the COD balance is used to replace one of the elemental balances. If one sets the coefficient for substrate COD as 1, and if the yield coefficient for the biomass is known, Eq.14.2 becomes:

$$1COD + (1-Y_H)O_2 + f_N Y_H NH_4^+ + f_N Y_H HCO_3^- \rightarrow Y_H \text{ Biomass COD} \quad (14.3)$$

where:

Y_H heterotrophic yield (gBiomass COD /gSubstrate COD), and

f_N fraction of nitrogen in biomass (gN/gBiomass COD)

To derive this equation we effectively used the COD, nitrogen and charge balance. The COD balance states that oxygen consumption and biomass production are always coupled; it is impossible to save oxygen and produce less sludge as the substrate (COD) is either oxidized by oxygen or becomes sludge. From the N balance the amount of ammonia needed can be calculated, and from the charge balance the amount of bicarbonate (alkalinity) can be determined, etc. The stoichiometric reaction can be written down as a function of the yield coefficient and, in this particular example, the amount of nitrogen inside the biomass. The stoichiometric coefficients for each compound are included in the model matrix (Table 14.1).

14.3.4 Kinetics

Each reaction has its own rate equation. The rate equation specifies the rate of conversion of the compound with the stoichiometric yield coefficient of 1. The conversion rate of the other compounds follows from multiplying each yield coefficient with the rate equation. The model can be based either on substrate-based kinetics (a substrate stoichiometric coefficient is equal to 1) or growth-based kinetics (a biomass stoichiometric coefficient equal to 1). It is not advisable to use both at the same time in one model. In ASM1 the rates are described based on the growth rate; the biomass coefficient is therefore set as 1. In ASM, a saturation equation is used as the standard rate equation. Saturation (Monod) kinetics includes two main parameters, the maximum rate parameter and affinity or saturation constant (the K value, defined as the concentration at half the maximum rate). The saturation term $S/(K+S)$ can have a value between 0 and 1, and can have a different function in the model. Several affinity terms reflect a real value, *e.g.* the oxygen affinity term is an observed parameter. However, in some cases the saturation term is only a switching term. For example, a switching function is used in the model to stop the growth process when there is no ammonia present (Eq. 14.4). The affinity constant for ammonia is effectively very low and scarcely measurable, so the coefficient placed in the equation has the sole purpose of guaranteeing that there is no further growth when the ammonia is fully consumed. This consequently means that one does not need to calibrate this value. How to distinguish between real measurable parameters and switching functions is a bit vague and inexplicit in activated sludge modelling. Therefore it is important to realize whether the K values are there as real model parameters or as a switching function to stop the process when the relevant compound is no longer present.

$$\mu = \mu^{max} \frac{S}{K_S + S} \cdot \frac{S_O}{K_O + S_O} \cdot \frac{S_N}{K_N + S_N} \cdots \qquad (14.4)$$

To describe inhibition kinetics, a similar approach is applied, but now the affinity constant is called the inhibition constant, and consequently it is possible to define an inhibition term (Eq. 14.5), which again has a value between 0 and 1. The inhibition constant is equal to the substrate concentration at which a 50% decrease in the rate is observed. There are also much more complex inhibition terms, but in ASM this is the term that is usually applied, especially for the substrate inhibition.

$$1 - \frac{S}{K_S + S} = \frac{K_S}{K_S + S} \qquad (14.5)$$

It is important to note that multiplying so many factors causes deviation because these factors are never exactly 1. If one multiplies two factors with a value of 0.9 with the value of a third factor that is 0.5, the result will be 0.4, while the real value should be 0.5 because this is the limiting factor. This consequently means a 20% lower rate value. Therefore it is better to use a logical operator in the model and choose the minimum factor among the terms (Eq. 14.7) instead of multiplying these factors (Eq. 14.6) as it seems that this better approximates the reality.

$$\mu = \mu^{max} \cdot \frac{S}{K_S + S} \cdot \frac{S_O}{K_O + S_O} \cdot \frac{S_{NH}}{K_{NH} + S_{NH}} \cdot \frac{S_{KI}}{K_I + S_{KI}} \qquad (14.6)$$

$$\mu = \mu^{max} \cdot MIN\left(\frac{S}{K_S + S}; \frac{S_O}{K_O + S_O}; \frac{S_{NH}}{K_{NH} + S_{NH}}; \frac{S_{KI}}{K_I + S_{KI}} \right) \qquad (14.7)$$

The reason that Eq.14.6 is used is partly an inherited habit (at the time of early model development in the 1970s, computing logical operators by integral differential equations was difficult and extremely time-consuming and thus was not applied). It does not matter so much which equation is used for activated sludge modelling; the point is to understand the reasons for the choices that were made at different stages of the model development.

14.3.5 Transport

A typical wastewater treatment model has several transport terms, which are often time-dependent (Figure 14.3). The model input is the time-variable flow and composition of the wastewater. The process is described in a hydraulic model, representing the hydraulics of a full-scale plant. An example is given in Figure 14.4.

The main problem is associated with making a hydraulic model of a wastewater treatment plant. A rigorous solution would be to make a full computational fluid dynamics model of the plant, which can exactly describe the flow in the reactors. However in general the details obtained for the flow in this way exceed by far the requirements of most conversion models. Since we are mainly interested in the bioconversion we need to adequately describe changes in concentrations in the treatment plant. Measuring several relevant compounds can help to define the hydraulic model. For activated sludge models these compounds are in general oxygen, ammonium and nitrate and, for phosphate removing systems, phosphate. As a first step a clear division can be made between aerobic and anoxic or anaerobic zones in a treatment plant.

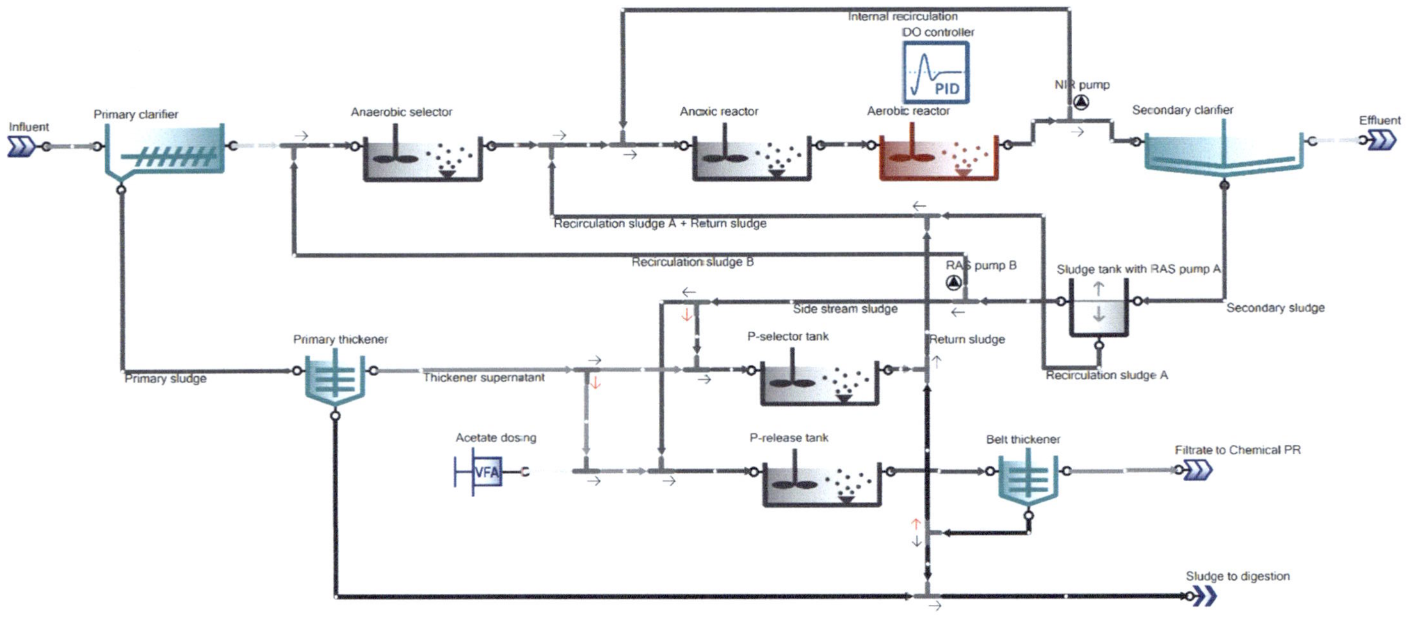

Figure 14.4 Hydraulic process scheme of the PhoStrip® process at WWTP Haarlem Waarderpolder in the Netherlands and its representation in the modelling simulator SUMO (adapted from Brdjanovic *et al.*, 2000 by Dynamita).

Within each zone one then has to observe whether *e.g.* in an aerated tank there exists a gradient in the oxygen concentration. As long as the oxygen concentration is always well above the saturation coefficient for oxygen used in the kinetic equations, there is no direct need to describe the changes in concentration in the aerobic zone, and the tank can be considered fully mixed. If the concentration of the reacting compounds becomes close to or lower than the saturation constants, the hydraulic model should be such that the change in concentrations is clearly described. In general, this means using a plug flow model or describing the system as a series of tanks. If the observed concentration of *e.g.* ammonia in the aeration tank is approximately 4 mg/l across the tank, it can be considered as a fully mixed tank and be represented in the hydraulic scheme of the plant as a single reactor. However, if the observed ammonia concentration changes from 4 mg/l at the inlet to the aeration tank to 0 mg/l at the outlet, this indicates a strong concentration gradient within the tank, and consequently it is much better to model it as a multi-compartmental tank with a number of fully mixed smaller reactors in series. A second aspect taken into consideration is the transfer of compounds between gas and liquid (*e.g.* oxygen) in aerated reactors or between biofilm and liquid. This is described in detail in Chapter 9 and Chapter 17.

14.3.6 Matrix notation

The balance equation (Eq.14.1) can be described for each individual compound. The large number of relevant compounds and conversions make activated sludge modelling complex. A large number of balance equations need to be formulated resulting in a loss of overview. Therefore the IAWQ Task Group (Henze *et al.*, 1987) on 'Mathematical Modelling of Wastewater Treatment' recommends the matrix method for model presentation.

This format facilitates clear and unambiguous presentation of the compounds and processes and their interaction on a single page. In addition, the matrix format allows easy comparison of different models, and facilitates transforming the model into a computer program. The matrix is represented by a number of columns and rows; one column for each compound and one row for each process. A simplified example is given in Table 14.1.

The first step in setting up the matrix is to identify the compounds of relevance in the model. The compounds are presented as symbols listed at the head of the appropriate column including a row with the dimensions.

The second step in setting up a matrix is to identify the biological processes occurring in the system. These are conversions or transformations that affect the compounds considered in the model and are itemized one below the other down the left-hand side of the matrix. The process rates are formulated mathematically and are listed on the right-hand side of the stoichiometry matrix in line with the respective process. Along each process row the stoichiometric coefficient for conversion from one compound to another is inserted so that each compound column lists the stoichiometric coefficients for the processes that influence that compound. If the stoichiometric coefficient equals zero it is for clarity generally not given in the printed matrix. The sign convention used in the matrix for each compound is 'negative for consumption' and 'positive for production'. In this convention the process rates always have a positive sign. Note that oxygen is negative COD because oxygen accepts electrons: electrons are passed from the substrate to oxygen to form water. Care must be taken with the units used in the rate equations for the processes. The stoichiometric coefficients are greatly simplified by working in consistent units.

In the example presented in Table 14.1 the compounds are expressed as COD equivalents. Provided that consistent units are used, continuity can be checked against the stoichiometric parameters by moving across any row of the matrix; the sum of the stoichiometric coefficients must be zero.

This matrix forms a succinct summary of the complex interactions between compounds and processes. It allows alterations in processes,

Table 14.1 Example of a simple stoichiometric matrix for activated sludge modelling (Henze *et al.*, 1987b).

Components i	1: S_O	2: S_S	3: X_H	Process rate equation ρ_j
List of processes j				
Aerobic growth	$-\frac{1}{Y_H}+1$	$-\frac{1}{Y_H}$	+1	$\mu_H^{max} \cdot \frac{S_S}{K_S+S_S} \cdot X_H$
Lysis		+1	−1	$b_H \cdot X_H$
Observed transformation rate r_i	$r_i=\sum_j v_{i,j} \cdot \rho_j$ $[M_iL^{-3}T^{-1}]$			
Definition of stoichiometric parameters: Y_H Heterotrophic yield coefficient $[M_H M_S^{-1}]$	Dissolved oxygen (O_2)	Dissolved organic substrate (COD)	Heterotrophic biomass (COD)	Definition of kinetic parameters: μ_H^{max} Maximum specific growth rate $[T^{-1}]$ K_S Saturation coefficient for substrate $[M_{COD}L^{-3}]$ b_H Rate constant for decay $[T^{-1}]$

compounds, stoichiometry and kinetics to be readily incorporated.

The matrix shows two important process aspects: the reaction equation for each process is represented on the different rows, and in the columns for each compound one can directly observe in which conversions the compound is involved. By multiplying the stoichiometric factors with their respective rate equations one gets the total conversion equation for each compound.

For convenience two extra aspects can be added to the matrix description (Table 14.2). The first aspect is a matrix with the composition in terms of conserved balances, in this case the COD, N and charge balance. Biomass is expressed in the stoichiometric matrix in terms of COD, but it also contains nitrogen. In the composition matrix this is included. Since the composition matrix and the stoichiometry matrix effectively contain all the conserved balances, multiplication of the two matrices therefore leads to zero.

Secondly, we are generally interested not only in the compounds expressed in the dimensions as used in the model, but also in their measured or observed units. For instance, the sludge amount is usually measured as gTSS and not gCOD. The observed matrix contains these conversion numbers between *e.g.* gCOD and gTSS. Other potentially interesting observed quantities are Kjeldahl nitrogen, VSS or BOD.

14.4 STEPWISE DEVELOPMENT OF THE BIOKINETIC MODEL: ASM1

Model development is a step-wise, bottom-up process where only strictly necessary processes for the pre-defined purpose of modelling are included. Starting simple and increasing complexity when needed is the general governing principle in model development. In general, activated sludge models from the ASM family are developed to describe oxygen uptake rate and sludge production (coupled on the COD balance), and N and P conversions at domestic wastewater treatment plants.

Table 14.2 Example of a stoichiometric matrix for activated sludge modelling (adapted from Gujer and Larsen, 1995).

Component	Oxygen	Inert	Substrate	Ammonia	Alkalinity	Biomass	Inert	Substrate	TSS	Rate
Symbol	S_O	S_I	S_S	S_{NH}	S_{HCO}	X_H	X_I	X_S	X_{TSS}	
Unit	gO_2	gCOD	gCOD	gN	mole	gCOD	gCOD	gCOD	gTSS	
Process	STOICHIOMETRY MATRIX									
Hydrolysis			1					–1	–0.75	r_1
Aerobic growth	–0.5		–1.5	-0.08	–0.005714	1			0.9	r_2
Lysis				0.07	0.005	–1	0.2	0.8	–0.12	r_3
Conservatives	COMPOSITION MATRIX									
ThOD-COD	–1	1	1	0		1	1	1		
N		0.02		1		0.08	0.05	0		
Charge				0.071429	–1					
Observables										
TSS						0.9	0.9	0.75		

However, despite the fact that they are designed for practical (and therefore not academic purposes), they are not sanitation models as they do not describe the removal of pathogens. Probably the best way to describe stepwise activated sludge model development is the original approach of Ekama and Marais (1978), later depicted by Dold *et al.*, (1980), and further elaborated in Gujer and Henze (1991). The outcome of this approach is a model which comes close to ASM1 and as such is briefly described here. The experimental system used in this approach comprised a completely mixed activated sludge system using settled domestic wastewater, and the basic influent and sludge characterization and operational conditions are listed in Table 14.3.

The objective of the study was to use the model to correctly describe the biomass content in the system, oxygen uptake by the biomass, and nitrogen conversion. To begin, one can use a very simple model consisting of only three relevant components (dissolved oxygen S_O, dissolved organic substrate S_S, and heterotrophic biomass X_H) and two relevant conversion processes (aerobic biomass growth and

Table 14.3 Experimental system summary data, Ekama and Marais (1978).

Parameter	Value
Feeding regime	12 h/d turned on between 02 and 14 hrs
Flow	18 l/d
Reactor volume	6.73 l
Processes	COD removal and nitrification (fully aerobic)
Biomass content in the reactor	1,375 mgVSS/l or 2,090 mgCOD/l
Sludge retention time (SRT)	2.5 d
Operating temperature	20.4°C
Influent COD concentration	570 mgCOD/l
Influent TKN concentration	46.8 mgN/l

lysis). With an increase in SRT, the biomass (live organisms) as a fraction of the sludge mass (VSS) in the system decreases. To describe this, the lysis process or death regeneration was used *i.e.* disintegration of death cells resulting in the generation of soluble biodegradable substrate available for the generation of new biomass (Chapter 4). Lysis of heterotrophic biomass here summarizes all the processes which lead to a loss of biomass (decay, lysis, endogenous respiration, predation etc). Maintenance or endogenous decay could have also been used here to describe the reduction in biomass. For the aerobic growth process all three components are relevant; dissolved oxygen and organic substrate are utilized by the biomass under aerobic conditions (thus negative coefficient) to produce biomass (positive coefficient). In general, a matrix can be simplified if one can choose to freely assign for each process one stoichiometric coefficient with a value of +1 or –1. The choice of the yield coefficient Y_H (0.67 gCOD/gCOD) together with the COD conservation equation is sufficient to determine all the stoichiometric coefficients for aerobic growth (Figure 14.5). For both processes one can define the rate; for aerobic growth it is a product of the maximum specific growth rate, the affinity for substrate and the biomass concentration (assuming that the oxygen is not limiting the growth).

For lysis, it is a sort of 1st order process where the biomass falls apart in proportion to the biomass concentration present and the constant of proportionality is called the rate constant for decay. Substituting the coefficients in the biokinetic model gives the matrix for Model A (Table 14.4).

If this model is used to compare the experimentally observed oxygen uptake rate (OUR), it can be seen that the experimental observations deviate considerably from the model predictions, except for the period between 0 and 2 hrs and at the very end of the experiment (endogenous respiration). In general, if the model predictions are deviating in terms of the levels of parameter of interest, this can be relatively straightforwardly adjusted by changing the value of the selected model parameter(s). However, if the model predictions in terms of trends and shapes are wrong, very likely the relevant process or processes have been overlooked. In this particular experiment, the difference between the total oxygen consumption observed over 24 hrs and the one predicted by the model seems to be quite close. However, it was the deviation between the model prediction and experimental results that led Dold *et al.* (1980) to suggest splitting the degradation of organic matter in wastewater into two processes (fractions): the relatively rapid process of biodegradation of part of the COD comprising of organics such as VFAs and glucose, and the relatively slow process of COD degradation (cellulose, starch, proteins etc). This fractionation of biodegradable COD into readily and slowly biodegradable COD (RBCOD and SBCOD, respectively) was triggered by the experimental observation of the OUR profile which showed a very sharp drop almost immediately after feeding stopped (14 hrs), followed by a slow decrease observed until the end of the experiment where it reached the steady value observed during the first two hours of the test. Therefore, the lysis process was reasonably well described by the model. It was furthermore concluded that the SBCOD is converted into RBCOD by the relatively slow process of hydrolysis (Figure 14.5).

Table 14.4 Matrix presentation of Model A.

Component	S_O	S_S	X_H	Rate
Growth	–0.5	–1.5	1.0	$\mu_H^{max} \cdot \frac{S_S}{K_S+S_S} \cdot \frac{S_O}{K_{O,H}+S_O} \cdot X_H$
Lysis		+1.0	–1.0	$b_H \cdot X_H$

This implies that there is a need to extend Model A by introducing two types of substrate (RBCOD and SBCOD) and one additional process (hydrolysis). In ASM1 it was also assumed that slowly biodegradable substrate consists entirely of particulate substrate (X_S), which is not necessarily true, but in ASM1 accepted as such. A distinction between soluble (S) and particulate (X) material is necessary in order to determine which compound will settle in the clarifier and which will leave the system with the effluent.

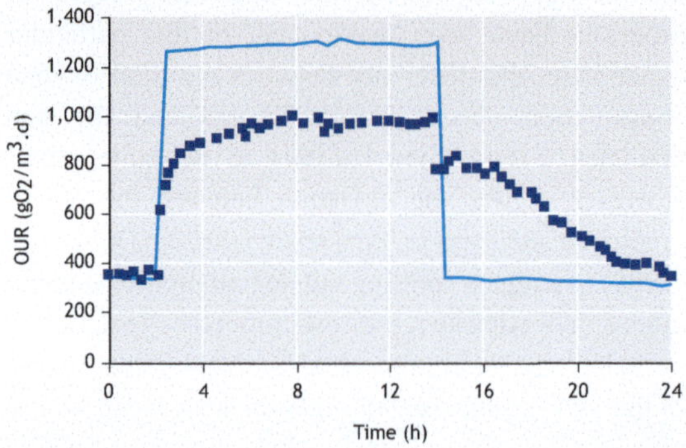

Figure 14.5 Model A: a comparison of experimentally observed values (data points) with a theoretically predicted oxygen uptake rate (continuous line) (adapted from Ekama and Marais, 1978; Gujer and Henze, 1991).

The introduction of a SBCOD fraction (X_S) did not affect the heterotrophic growth process as it is assumed that growth is not directly based on SBCOD. Also, the lysis processes were adjusted by the assumption that the lysis products are slowly biodegradable and, as such, are added to the pool of X_S. These products are made available for aerobic heterotrophic growth by hydrolysis. This means that there are two types of particulate substrate: one derived from the influent wastewater, and the second generated by the biomass decay. In some cases these are lumped together (as in this case), while in some models they are taken into account separately. However, both options result in virtually zero net difference. Furthermore, hydrolysable material X_S is assumed to adsorb onto heterotrophic biomass X_H resulting in a kind of Lagrangian kinetic expression as the rate equation for hydrolysis. Therefore it is the amount of substrate per biomass which is important here (rate limiting) and not the substrate concentration with respect to the bulk liquid as in the case for the RBCOD. By implementing this, Model B was formed (Table 14.5).

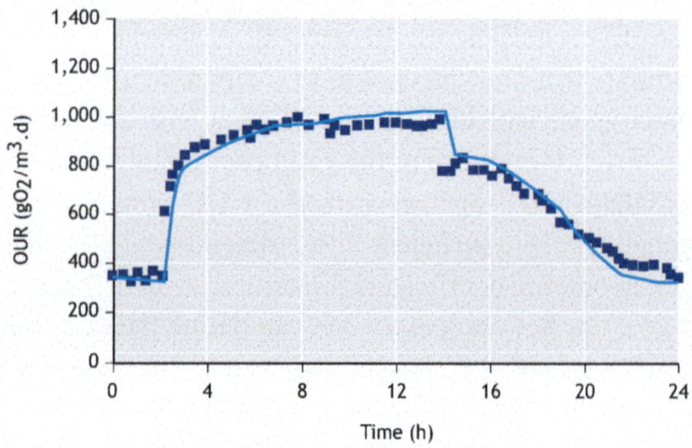

Figure 14.6 Model B: comparison of experimentally observed values (data points) with theoretically predicted oxygen uptake rate (continuous line) (adapted from Ekama and Marais 1978; Gujer and Henze 1991).

Model B describes the OUR very accurately (Figure 14.6). However, the predicted activated sludge concentration was 22% less than the measured concentration. This indicated the need to increase the sludge production by introduction of an influent non-

Table 14.5 Matrix presentation of Model B.

Component	S_O	S_S	X_H	X_S	Rate
Growth	–0.5	–1.5	1.0		$\mu_H^{max} \cdot \frac{S_S}{K_S+S_S} \cdot \frac{S_O}{K_{O,H}+S_O} \cdot X_H$
Lysis			–1.0	+1.0	$b_H \cdot X_H$
Hydrolysis		+1.0		–1.0	$k_H \cdot \frac{(X_S/X_H)}{K_x+(X_S/X_H)} \cdot X_H$

biodegradable fraction of COD (being inert and also particulate organic matter which accumulates in the reactor: X_I). The term 'non-biodegradable' is used in the context of wastewater treatment for a compound which is not degraded by microorganisms during its retention in the treatment system. Materials such as plastics, wood-based and fibrous materials, nails, and hairs are all organic and strictly speaking considered biodegradable, but not in wastewater treatment systems. Even a compound such as cellulose is considered non-biodegradable in high-loaded wastewater treatment plants but biodegradable in low-loaded systems. Besides inert particulate material derived from the influent wastewater, there is also the second component generated by the biomass decay. The latter arises from the fact that cell walls are made of COD which biodegrades very slowly and is considered non-biodegradable, resulting in the experimentally-determined division of lysis process products of 92% being X_s (biodegradable) and 8% X_I (unbiodegradable). Consequently the rates in Model B are not changed given the fact that the OUR profile was described correctly.

Inclusion of X_I resulted in the new Model C shown in Table 14.6. The observed biomass concentration in the reactor was very well predicted by Model C; however, due to higher sludge (COD) production/removal from the system, the oxygen consumption was significantly underestimated by the model (despite the fact that the general OUR profile was well matched, Figure 14.7).

This is to be expected as oxygen consumption and sludge production are coupled via the COD balance, and consequently, an increase in sludge production will cause a decrease in oxygen demand. It was not possible to predict correctly both oxygen consumption and sludge production by Model C.

From the experimental observations (results not shown) it was clear that the effluent from the treatment plant contained nitrate which implied that nitrification should be included.

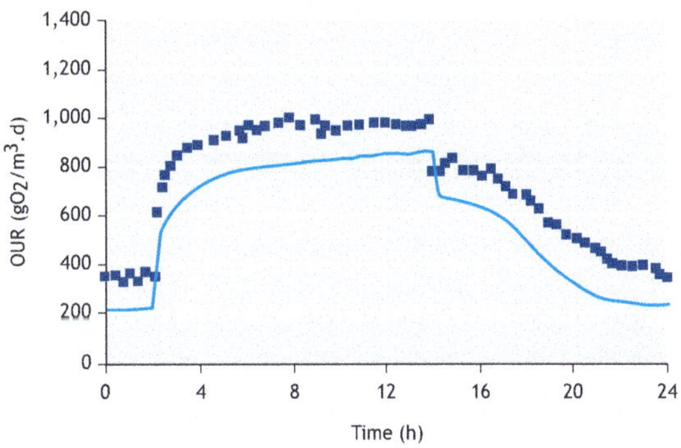

Figure 14.7 Model C: comparison of experimentally observed values (data points) with theoretically predicted oxygen uptake rate (continuous line) (adapted from Ekama and Marais, 1978; Gujer and Henze, 1991).

By inclusion of nitrification the model had to be extended by the addition of three materials and two processes, namely ammonium (NH_4^+, S_{NH}), nitrate (NO_3^-, S_{NO}) and nitrifying autotrophic biomass (X_A), and aerobic (nitrifier) growth and nitrifier lysis. Again, it was necessary to evaluate the influence of each of the additional materials on the existing reactions. Ammonia is not only used in nitrification, but also for cell growth, so therefore it was necessary to add a stoichiometric factor for ammonia in the growth relation. If the biomass contains 8% nitrogen (0.08 mgN/mgCOD), then the factor becomes 0.08.

Table 14.6 Matrix presentation of Model C.

Component	S_O	S_S	X_H	X_S	X_I	Rate
Growth	-0.5	–1.5	1.0			$\mu_H^{max} \cdot \frac{S_S}{K_S+S_S} \cdot \frac{S_O}{K_{O,H}+S_O} \cdot X_H$
Lysis			-1.0	+0.92	+0.08	$b_H \cdot X_H$
Hydrolysis		+1.0		-1.0		$k_H \cdot \frac{(X_S/X_H)}{K_x+(X_S/X_H)} \cdot X_H$

Furthermore, it was assumed that in the lysis process nitrogen remains within the biomass. However, in the hydrolysis process the biomass SBCOD is degraded (*e.g.* proteins into amino acids) resulting in a release of ammonia. As a consequence, besides one unit of substrate, 0.08 units of ammonia is also produced. This also happens with the influent SBCOD: as the protein part of it is hydrolysed, ammonia is released, and the proteins are measured in the influent as organic N, *i.e.* the difference between the TKN and FSA. So for nitrification, a certain amount of oxygen and ammonia are consumed, and nitrate and biomass are produced. The amount of ammonia consumed is not exactly the same as the amount of nitrate produced, due to dual use of the ammonia, namely (*i*) for energy generation in the nitrification process, and (*ii*) as a nitrogen source for the heterotrophic biomass growth. The difference between the ammonia consumed and nitrate formed is 0.08, representing the nitrogen content of the heterotrophic biomass. The overall N balance in this case will fit ($4.25 = 4.17 + 0.08 \cdot 1.0$). Furthermore, the lysis process for the autotrophs was assumed to be the same for the heterotrophs, by which particular substrate and a small amount of inert substrate are produced. The process rate for growth for autotrophs is generated analogously to heterotrophs with saturation terms for ammonia and oxygen. By the inclusion of an additional three materials and two processes, Model D was created which satisfied both the COD and N balance. By having such a model it was possible to split the total oxygen consumption into oxygen consumption for ammonium oxidation and oxygen consumption for COD degradation. This shows the added value of using the model by providing insight into where the oxygen is used and for which processes (known as process analysis).

The results of the simulation of the OUR by Model D are presented in Figure 14.8.

There were a few dilemmas at this stage of the model development such as 'Is the match sufficiently accurate?' or 'Is deviation, for example, in OUR of 5 to 10% at 14.00 h acceptable?'. The answers to these questions depends entirely on the quality of the experimental data. If the COD and N balances of the data are exactly 100%, then it may be worth pursuing refinements in the model to get a better prediction because the data are reliable and accurate. However, if the COD and N balances are not 100% but in the 95-105% range, then there is not much sense in making the model very much more accurate. Making models is relatively easy but getting reliable and accurate experimental data is the most difficult part of developing models for wastewater treatment systems. If the equipment results in an inaccuracy of 5-10% there is no sense in making the model more accurate.

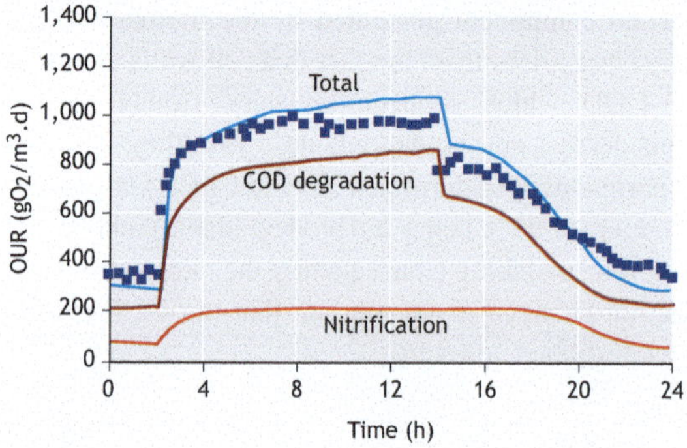

Figure 14.8 Model D: comparison of experimentally observed values (data points) with theoretically predicted total oxygen uptake rate (continuous line). Oxygen uptake for COD degradation and nitrification are separated out (adapted from Ekama and Marais, 1978; Gujer and Henze, 1991).

One should not forget that a model can only be successful if it fulfils the expectations the modeller has of it. If the purpose of the model is to correctly describe general trends, there is no need for further refinements. Of course, fine-tuning and calibration for a more accurate fit are possible, but this increases the model complexity. Here, it was decided not to add extra compounds or processes, as the last part of calibration can be done by straightforward shifting some model parameters. In general, the model simulations showed that the measured line and model line fit rather well: OUR, sludge production and nitrification (data not shown, see Dold *et al.*, 1980; Gujer and Henze, 1991) are predicted correctly. For the pre-defined goal, the model is considered correct,

Table 14.7 Matrix presentation of Model D.

Component	S_O	S_S	S_{NH}	S_{NO}	X_H	X_S	X_I	X_A	Rate
Growth	-0.5	−1.5	−0.08		+1.0				$\mu_H^{max} \cdot \frac{S_S}{K_S+S_S} \cdot \frac{S_O}{K_{O,H}+S_O} \cdot \frac{S_{NH}}{K_{N,H}+S_{NH}} \cdot X_H$
Lysis					−1.0	+0.92	+0.08		$b_H \cdot X_H$
Hydrolysis		+1.0	+0.08			−1.0			$k_H \cdot \frac{(X_S/X_H)}{K_x+(X_S/X_H)} \cdot X_H$
Autotrophic growth	−18.0		−4.25	+4.17				+1.0	$\mu_A^{max} \cdot \frac{S_O}{K_{O,A}+S_O} \cdot \frac{S_{NH}}{K_{N,A}+S_{NH}} \cdot X_A$
Autotrophic lysis						+0.92	+0.08	−1.0	$b_A \cdot X_A$

but this does not necessarily mean that the assumptions used are correct. Indeed, by using those assumptions, a mathematical description sufficiently appropriate for the purpose of its use is obtained. However, Model D omitted some processes which in reality play an important role, such as protozoa activity (Table 14.7). In this case, it was evaluated that inclusion of protozoa in the model would not increase its descriptive power and consequently this process was not included.

The next step in model development was the inclusion of the denitrification process. There are in general two approaches possible that ultimately lead to the same end results. It can be assumed that there is either a special group of bacteria which perform denitrification or that all heterotrophic microorganisms can denitrify, but at a fraction of their rate under aerobic conditions. In other words, there is either a specialized population which can utilize both oxygen and nitrate, and another part of the population which can only use oxygen, or all heterotrophs can denitrify but at a reduced rate, corrected by the η factor (reduction factor for growth rate under anoxic conditions). These are different assumptions conceptually but mathematically they come down to the same equation. Since the last assumption simplifies the model, it has been chosen to use stoichiometry for denitrification and the bacteria that are ordinary heterotrophic organisms. Although the reality is probably much more complex, it was demonstrated that this simplified approach works satisfactorily in practice.

Another important aspect concerned the differentiation between fractions of inert material and nitrogen which is one of the items that differs from one commercial model to another. As mentioned earlier, inert material can originate from influent or from the degradation of biomass, and the end content of the degraded biomass, inert material, might be different in composition to that of inert material in the influent. This can also be taken into account in the model; the inert material can be either separated or lumped together. In principle, it is not strictly necessary to define these fractions separately, but sometimes this is done based on aesthetic reasons or for the specific purpose of the model application. Similar reasoning applies for nitrogen fractionation. The model development as described by Ekama and Marais (1978) is still considered valid and Model D extended with denitrification becomes close to ASM1 (Henze *et al.*, 1987b). For further details on ASM1 the reader is referred to Dold *et al.*, 1980; Van Haandel *et al.*, 1981; Alexander *et al.*, 1983; Warner *et al.*, 1986; Henze *et al.*, 1987a, 1987b, 2000.

One of the most important limitations of ASM is that it does not describe the sludge bulking phenomenon. Therefore, if ASM is used, for example to improve the nitrification process, it is necessary to check if the proposed changes create bulking sludge. Limited aeration, which is beneficial for nitrogen removal, will almost inevitably induce bulking sludge. Sludge bulking itself cannot be modelled in a way that is sufficiently reliable for implementation in commercial software packages despite some attempts described in the literature (Krebs, 1995). This consequently means that the model cannot be accurately applied to predict very low effluent concentrations when highly efficient processes are implemented. Furthermore, this analogous consideration is valid for the denitrification process as well. On the other hand, even if the model is able to predict the concentration of ammonia at 0.5 mg/l, there are always some inaccuracies and imperfections in the analytical procedures for determining ammonia concentration, as well as in the sampling procedure and sample handling.

Another ASM limitation is that it does not take into account the removal of micropollutants such as metals, xenobiotics or oestrogenic endocrine-disrupting compounds. This is partially due to the required increase in the complexity of the model and partially due to lack of knowledge about the microorganisms and biochemical reactions involved in the conversion of these compounds. In some cases, such as in modelling of the wastewater treatment at oil refineries, it is necessary to predict the phenol reduction. To support denitrification, methanol is often added under anoxic conditions and its conversion needs to be included. There are also cases when, for example, one is interested in sulphite reduction. In all these cases a new specialized organism must be included in the model, as the biomass included in ASM1 does not convert these micropollutants. Examples of such extensions can be found in the literature, and nowadays some commercial software packages include for example methanol utilization. In the case of other COD compounds such as volatile fatty acids (VFAs), the ordinary organisms which remove COD from the wastewater will convert them and therefore the model does not have to be extended for this purpose. Beyond the ASM1 level, the model can be extended to take into account oxygen transfer, pH and alkalinity, anaerobic digestion, chemical phosphorus removal and precipitation, additional units (such as settlers etc), side-stream processes, the gas phase etc. Again, whether the model needs to be extended depends on the purpose of the model.

14.5 ACTIVATED SLUDGE MODELS

Over the past twenty years, the knowledge and understanding of wastewater treatment has advanced extensively and moved away from empirically-based approaches to a fundamentally-based 'first principles' approach embracing chemistry, microbiology, physical and bioprocess engineering, and mathematics. Many of these advances have matured to the degree that they have been codified into mathematical models for simulation by computers.

Before the 1980s, several research groups worked independently of each other on developing models for activated sludge. Each group developed and applied their own approach and notation, first in steady-state models, and later on, in dynamic models. Table 14.8 summarizes the essential features of these and several other activated sludge models.

In the early 1980s, Poul Harremoës, President of IAWPRC (the International Association of Water Pollution, Research and Control, which later became IAWQ, the International Association of Water Quality, and which is nowadays the IWA, the International Water Association) initiated the idea of combining the most relevant and applied models and working together at an international level to accelerate development of a common, unified model. As a consequence, in 1982, the 'Task group on mathematical modelling for design and operation of biological wastewater treatment' was established with Gerrit Marais (University of Cape Town), Leslie Grady (Clemson University), Willy Gujer (EAWAG), Tomonori Matsuo (Tokyo University),

Table 14.8 Overview of selected activated sludge models (based on Gernaey *et al.*, 2004).

Model	Nitrification	Denitrification	Heterotrophic autotrophic decay	Hydrolysis	EBPR	Denitrifying PAO	Lysis of PAO / PHA	Fermentation	Chemical P removal	Reactions	State variables	Reference
UCTOLD	●	●	DR, Cst	EA						8	13	Dold *et al.*, 1980, 1991
ASM1	●	●	DR, Cst	EA						8	13	Henze *et al.*, 1987b
ASM3	●	●	ER, EA	Cst						12	13	Gujer *et al.*, 1999
UCTPHO	●	●	DR, Cst	EA	●		Cst	●		19	19	Wentzel *et al.*, 1988, 1989a,b
ASM2	●	●	DR, Cst	EA	●		Cst	●	●	19	19	Henze *et al.*, 1995
ASM2d	●	●	DR, Cst	EA	●	●	Cst	●	●	21	19	Henze *et al.*, 1999
B&D	●	●	DR, Cst	EA	●	●	EA	●		36	19	Barker and Dold, 1997
TUDP	●	●	DR, Cst	EA	●	●	EA	●		21	17	Meijer, 2004
ASM3-bioP	●	●	ER, EA	Cst	●	●	EA			23	17	Rieger *et al.*, 2001

DR, death regeneration concept; EA, electron acceptor dependent ER, endogenous respiration concept; Cst, not electron acceptor dependent

and Mogens Henze (Technical University of Denmark) as chairman. This joint activity resulted in the development of the first dynamic Activated Sludge Model, called in short ASM1 (Henze *et al.,* 1987b). ASM1 can be considered as the reference model, since this model triggered the general acceptance of wastewater treatment modelling, first in the research community and later on also in practice. This evolution was undoubtedly supported by the availability of more powerful computers. ASM1 is in essence a consensus model-compromising result of discussions at the time between different modelling groups. Many of the basic concepts of ASM1 were adapted from the activated sludge model defined by Dold *et al.* (1980). A summary of the research developments that resulted in ASM1 was given by Jeppsson (1996) and in a recent chapter of Ekama and Takács (2014) (in Jenkins and Wanner, 2014). Even today, the ASM1 model is still in many cases the state of the art for modelling activated sludge systems (Roeleveld and Van Loosdrecht, 2002). ASM1 has become a reference for many scientific and practical projects, and has been implemented (in some cases with modifications) in most of the commercial software available for modelling and simulation of plants for N removal. Copp (2002) reports on experiences with ASM1 implementations on different software platforms. In general, activated sludge models from the ASM 'family' are developed to describe the oxygen uptake rate and sludge production (coupled with COD balance), and N and P conversions at domestic wastewater treatment plants. However, despite the fact that they are designed for practical (and therefore not academic) purposes, they are not sanitation models as they do not describe the removal of pathogens. Probably the best way to describe the stepwise activated sludge model development is the original approach of Marais and Ekama (1976) and Ekama and Marais (1978), later depicted by Dold *et al.* (1980), and further elaborated on in Gujer and Henze (1991). The outcome of this approach is a model which comes close to ASM1.

The ASM1 model is a structured model based on Monod kinetics that predicts the processes of biological (bacteriological) reactions. ASM1 models

COD and N removal, oxygen consumption and sludge production. Wastewater is characterized in terms of seven dissolved and six particulate components that are used to describe two biomass groups, seven fractions of COD (organic material) and four fractions of nitrogen (Henze *et al.*, 1987b; Gujer and Henze, 1991). Dissolved oxygen concentration and alkalinity are also included as part of the wastewater characteristics. From the eight processes of the model, three are related to the growth of heterotrophic and autotrophic organisms, two describe the biomass decay (death-regeneration theory, Dold *et al.,* 1980), and three are related to hydrolysis. The model is presented in a matrix format, also known as the Petersen matrix or Gujer matrix (Petersen, 1965; Takács *et al*., 2007). This matrix contains stoichiometric coefficients and a kinetic vector. All the state variables involved in a process are displayed in columns, and all the processes where a state variable is involved are presented in the rows of the matrix. Already in use in chemical modelling (Petersen, 1965), this representation helped to present the model in a condensed form. This facilitated its publication, interpretation and comparison not only between models, but also between processes and compounds. However, certain major limitations of ASM1 are, for example, that it only describes heterotrophic and autotrophic reactions under aerobic and anoxic conditions (in which, for instance, ordinary heterotrophs consume carbonaceous substrates and autotrophic nitrifying organisms oxidize ammonia to nitrate), and it does not include enhanced biological phosphorus removal (EBPR) processes (Gujer and Henze, 1991). Despite the fact that to a great extent knowledge of EBPR processes was already available when ASM1 was developed (Comeau *et al.*, 1986; Van Loosdrecht *et al.*, 1997), EBPR was not included in ASM1 since most of the wastewater treatment plants at that time did not incorporate biologically enhanced (or chemical) phosphorus removal (Fenu *et al.,* 2010).

In recent years several research groups have started to work on the description of EBPR for its incorporation in dynamic activated sludge models, mostly based on directly measurable soluble compounds. From the mid 1980s to the mid 1990s, the EBPR process grew in popularity and the understanding of the underlying bio-chemical mechanisms increased (Henze *et al.*, 2000). In the meantime, in 1990, the composition of the task group changed, when Leslie Grady left and Takashi Mino (Tokyo University) and Mark Wentzel (University of Cape Town) joined. The knowledge acquired on EBPR led to the publication of Activated Sludge Model No. 2 (ASM2) (Henze *et al.,* 1995), which included the EBPR processes. In particular, ASM2 includes Phosphate Accumulating Organisms (PAO), growing only under aerobic conditions, with the correspondingly associated anaerobic, anoxic, and aerobic reactions. ASM2 was a compromise between complexity and simplicity, and between different points of view on how the correct model should look like to be used as a conceptual platform for further model development (Henze *et al.,* 2000). In 1996, Mark Van Loosdrecht (Delft University of Technology) became a member of the task group following the departure of Tomonori Matsuo, Mark Wentzel and Gerrit Marais. Because the occurrence of denitrifying EBPR was well-established (*e.g.* Kuba *et al.,* 1997; Murnleitner *et al.*, 1997) the ASM2 model was expanded in 1999 by the inclusion of denitrifying PAO (DPAO). This version of the model was denoted as ASM2d (Henze *et al.,* 1999). Both ASM2 and ASM2d are similar to ASM1 in that they assume the cell to be a black box, as opposed to using the metabolic approach to modelling the processes that take place inside the cell. However, ASM2d appeared to be overparameterized with respect to the available data, requiring a more systematic approach for calibration (Brun *et al.*, 2002). Although this allowed the model to adapt and describe dynamic changes in the activated sludge community, it still lacked the ability to entirely describe the observed dynamics, particularly with regard to hydrolysis and EBPR processes (Sin and Vanrolleghem, 2006). Parallel to these developments, in 1994 an increasing knowledge of the internal cell biochemistry of PAO resulted in the development of a metabolic model (TUDP model; Smolders *et al.,* 1994a, b; Murnleitner *et al.,* 1997) describing the anaerobic and aerobic phases of EBPR based on intracellular storage compounds. This model was later fully integrated with ASM by Meijer (2004).

At the same time as the ASM2d model was presented, the task group also developed the ASM3 model to correct some of the shortcomings of ASM1. ASM3 was proposed to become the new standard for ASM-based modelling. ASM3 replaced the death-regeneration process for heterotrophic organisms with an endogenous respiration process and also introduced the role of storage of organic substrates (Gujer *et al.*, 1999). In 2000, the task group presented the overview of the ASM models 1 to 3 (Henze *et al.*, 2000). In essence, ASM3 describes the same processes as ASM1, although ASM3 was introduced to correct the deficiencies of ASM1. This is partly based on the observations from the oxygen utilization rate (OUR) tests with activated sludge which revealed the fact that bacteria rapidly take up readily biodegradable COD and store it as internal substrate which will then be converted slowly (conversion of readily biodegradable COD into slowly biodegradable COD). When acetate (defined substrate) is added to the activated sludge, the observed OUR suggests the presence of two substrates; both a rapid and a slow degradation of substrate associated with OUR can be observed (Henze, 1992; Henze *et al.*, 2008).

In ASM1 it appears as if two substrates are present (S_S and X_S) while in the original experiments only acetate (S_S) was dosed. In order to describe the observed OUR by ASM1 in this case, it was necessary to define that the acetate is partly soluble and partly particulate, which is not recommended. This deficiency is solved by the introduction of a storage compound, $X_{STO,S}$, in ASM3. This means that substrate is taken up rapidly and stored, while growth occurs within the stored substrate. Both models will describe the observed OUR, but only ASM3 will accurately describe the uptake. However, there is no problem in using ASM1 for simulation of nitrogen removal systems because nitrification is a slow process, and thus enough time is available for biodegradation of slowly biodegradable COD.

The second reason to introduce ASM3 was that ASM1 proved to be so successful for simulation of wastewater treatment plants that too many started to believe that what was in ASM1 was 100% true and the reality. However, the storage mechanisms exhibited by the biomass show that what is in ASM1 is not all true, but close enough to reality to serve its purpose. Therefore, ASM3 has an added educational value because it demonstrates that there are different (but not necessarily better) ways to model the same treatment plant. However, the most important reason to introduce ASM3 was the recognition of the importance of three rates of oxygen consumption in the process, namely: the rapid rate of oxygen consumption for degradation of readily biodegradable COD (RBCOD), the slow rate associated with degradation of slowly biodegradable COD (SBCOD), and the even slower endogenous OUR. In contrast, in ASM1 there is only one oxygen-consuming process, so it is very difficult to perform calibration as one needs to calibrate other processes that indirectly influence the processes that consume oxygen.

The other problem is the cycling of the COD in the process, as in the decay process particulate COD is produced, hydrolyzed, and used for growth again. This means that if in the process one parameter is changed, it influences all the other processes due to the cycling, and it is difficult to use automated calibration as every parameter has influences on every process. In ASM3 this issue has been solved as the decay process has been replaced by endogenous respiration which eliminates the COD cycle (Figure 14.9). In other words, once the cells are produced, they start to oxidize themselves and by this means the biomass is reduced by the aerobic mineralization process (the classical endogenous respiration). While this has some conceptual controversy, *e.g.* why would an organism oxidize itself (*i.e.* go on a diet) when there is food around, it is useful to eliminate the bioprocess interaction from the substrate recycling of the death–regeneration model.

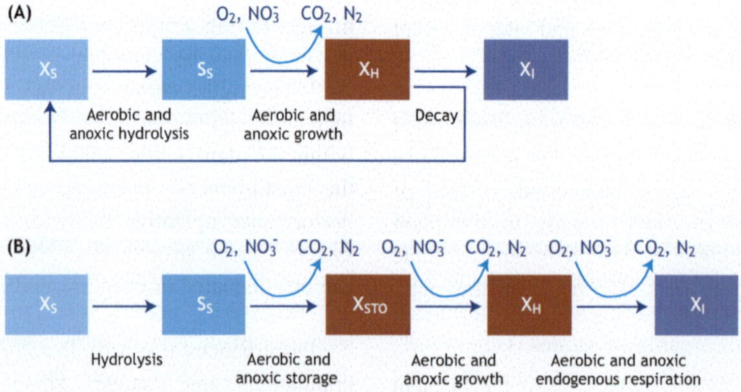

Figure 14.9 Degradation of COD in (A) ASM1 and (B) ASM3.

One of the most important applications of ASM3 is in plug flow reactors, such as selectors (Makinia *et al.*, 2006). If, for example, acetate must be removed in the aerobic selector to prevent sludge bulking, the design of the selector is governed by the time needed to take up the acetate and by the amount of oxygen needed for it. If ASM1 is used instead then the oxygen requirements in the selector will be significantly overestimated. In reality a large proportion of acetate is stored inside the biomass, and once it is stored, there is no longer a problem with bulking sludge. If one wants to design the aerobic selector and include it in the model, then ASM3 is the best model to use.

Another relevant application of ASM3 is for the description of a pre-denitrifying nitrogen removal plant operating at a short SRT (Yuan *et al.*, 2002; Sahlstedt *et al.*, 2004). Here, it makes a substantial difference whether or not readily or slowly biodegradable COD is present or whether COD is stored or not. In systems with a long SRT (10-20 days depending on temperature, which are more common in practice), a large part of the nitrate removal is effectively associated with the slowly biodegradable COD from the influent and death-regeneration in the pre-denitrification reactor and from death–regeneration only in the post-denitrification reactor, so the sensitivity to the exact ratio between readily and slowly biodegradable COD is much less. The same applies for the differentiation between ASM1 and ASM3. In highly loaded systems endogenous respiration is less important and therefore the accumulation of COD in the form of storage polymers and the carry over in the aerated phase of a treatment plant might be significant.

In conclusion, ASM3 is recommended to be used for (*i*) simulating highly loaded nitrification-denitrification systems with short anoxic retention times (volumes), (*ii*) supporting selector modelling, (*iii*) improving aeration demands for tapered systems, during step-feed operations or when high amounts of soluble industrial components are present in the influent, and (*iv*) easing automatic calibration. Otherwise, ASM1 should be equally successful in describing an activated sludge plant.

The consequence of introducing EBPR and Phosphorus Accumulating Organisms (PAO) into ASM is that the model becomes quite complex, as illustrated in Figure 14.10.

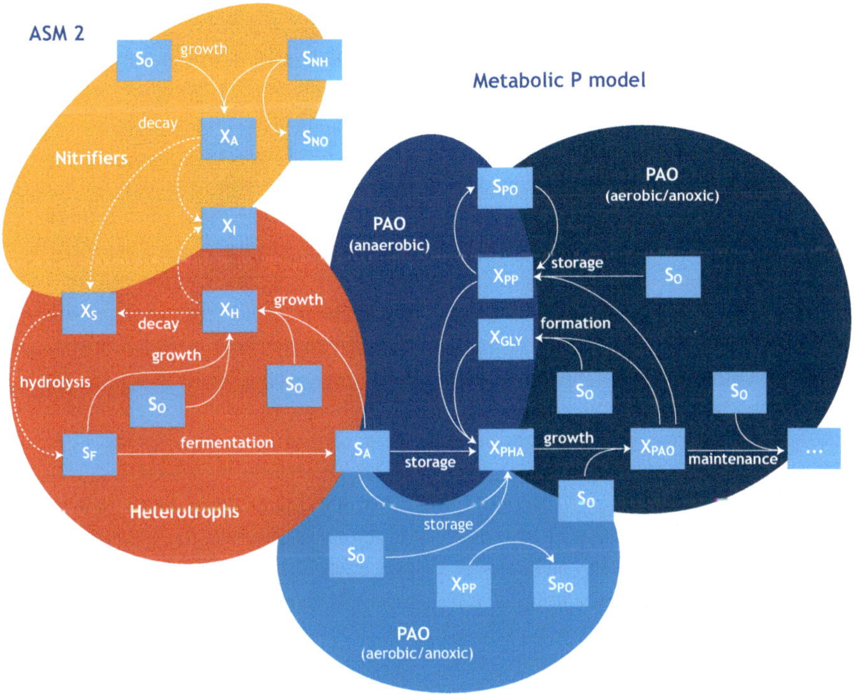

Figure 14.10 Interactions in the integrated ASM2-TUDP model (Meijer, 2004).

The left side of Figure 14.10 depicts the part of conversions carried out by nitrifiers and ordinary heterotrophs, while the right side shows the extension needed for the description of the complex physiology of PAO. The nitrifiers and ordinary heterotrophs use oxygen to oxidize their substrate to form CO_2 or nitrate and biomass. They have a rather simple physiology resulting in simple processes. PAO's physiology includes internal storage polymers (poly-hydroxy-alkanoate: PHA, glycogen and poly-P) and their behaviour under anaerobic, anoxic and aerobic conditions is different. They also behave differently under aerobic conditions depending on whether the substrate is present or not. Obviously, there are lots of possible variations and inclusion of EBPR in the model substantially increases its complexity (the number of processes in ASM increases from 11 to 22). The situation becomes even more complex when glycogen accumulating organisms (GAO) are also included. ASM2 and ASM2d are similar to ASM1 in assuming the cell to be a black box as opposed to using the metabolic approach to modelling which takes into account what is happening inside the cell.

In 1994, increasing knowledge of the internal cell biochemistry of PAO resulted in the development of a metabolic model describing the anaerobic and aerobic phases of EBPR (Smolders *et al.*, 1994a,b; 1995a,b,c). The model was developed and validated using enriched PAO cultures cultivated in lab-scale anaerobic/aerobic (A/O) sequencing batch reactor (SBR) experiments. Why is it useful to use a metabolic model? In the standard model for heterotrophic growth there are seven relevant compounds (substrate, oxygen, charge, carbon dioxide, water, ammonia and biomass), five independent balances (carbon, hydrogen, oxygen, nitrogen and charge), and two degrees of freedom. If one knows one yield and one rate coefficient, it is possible to describe the whole system with one model. If one was to describe COD removal and nitrification at a metabolic level, it would not bring any advantage as the yield and rate coefficients would still be needed. Although the metabolic stoichiometry allows tracking, the C, H, O, N, P and charge flows through a system give more information from a modelling point of view, which makes the model more complex but not more accurate. All the rates are linked through conservation relations (stoichiometry) and, therefore, the choices between the process rate or growth rate, and the substrate uptake rate or oxygen utilization rate, are not important.

Thus the black box approach can be used, as is also the case with ASM1. So for the activated sludge system itself, C, H, O and charge tracking is not required; COD and N is enough, but when the ASMs are integrated with anaerobic digestion (AD) models to form plant-wide models, it becomes important because AD modelling requires C, H, O and charge tracking to predict gas production and composition and alkalinity generation (Brink *et al.*, 2007).

However, if one needs to describe the situation with heterotrophic growth and product (storage polymers) formation as for PAO in EBPR processes, the number of relevant compounds increases; each additional storage polymer brings an extra compound, but the number of balances does not increase, which means that the degrees of freedom (unknown values) increase as a consequence of the increased number of unknown compounds. In this case, one needs to know at least one yield and rate coefficient, and the choice of the process rate becomes important. For example, during aerobic conditions PAO use internally-stored PHA to produce the intermediate compound acetyl-CoA which is used further for biomass growth, glycogen formation and creation of the energy required for these processes, and for poly-P formation.

Obviously, the introduction of storage compounds creates a more complicated network of processes. In the processes with extra storage polymers, extra yield coefficients will also be introduced. The efficiency of the conversion processes would however be the same for all the yields. Within a metabolic model one can link the macroscopic yields to the metabolic yield, which is the efficiency of energy (ATP) generation per unit of substrate oxidized. The substrate oxidation is related to electron transfer to oxygen or nitrate consumption. The yield coefficients are therefore all a function of this basic parameter (ATP produced per pair of electrons transferred) and the number of independent yield parameters is less in a metabolic description for these complex microorganisms.

Initially, the simplest metabolic model kinetics were chosen. Smolders *et al.* (1994b) proposed a kinetic structure in which the oxygen (or nitrate) consumption and PHA degradation are the net result of biomass growth (r_X), poly-P formation (r_{PP}), glycogen formation (r_{GLY}) and maintenance (m_O and m_{PHA}). Their kinetic structure is expressed by linear equations which lead to a set of overall reactions (Meijer, 2004). Soon after, Kuba *et al.* (1996) proposed a metabolic model for denitrifying EBPR. In 1997, Murnleitner *et al.* combined the anaerobic, aerobic and anoxic models, proposing a kinetic structure in which growth was the net result of PHA consumption and poly-P and glycogen formation without changing its original stoichiometry. From an ecological point of view, storage is preferred over growth, suggesting that, in their competition with other micro-organisms, PAO rely on their storage ability. A rapid resupply of storage compounds is a primary condition for long term survival. Thus, the maximum growth rate is no longer an intrinsic property of PAO, but becomes dependent on environmental conditions and the maximum PHA storage capacity (Brdjanovic *et al.*, 1998). With the reformulated kinetic structure, Murnleitner *et al.* (1997) described all the experiments performed by Smolders *et al.* (1994a,b; 1995a,b) and Kuba *et al.* (1996) with one set of model parameters. Nevertheless, one must underline that these reactions cannot be read separately, as they are merely the result of the mathematical reformulation. From the metabolic reactions, an overall anaerobic, aerobic and anoxic stoichiometry was determined. A full description of the TUDP model is given by Meijer (2004) and De Kreuk *et al.* (2007). Overall, the formulation of an overall anaerobic reaction is unambiguous, as there is only one metabolic reaction. As such, by measuring the acetate uptake rate, all the other rates are fixed. Concerning the aerobic and anoxic stoichiometry, five overall reactions (r_X, r_{PP}, r_{GLY}, r_{PHA} and m_{PHA}) are found but the system can be solved if four out of five rates are determined. In 1999, Van Veldhuizen *et al.* integrated the metabolic EBPR model with the heterotrophic, hydrolytic and autotrophic processes from ASM2d (Henze *et al.*, 1999). With this model a full-scale Modified University of Cape Town (MUCT) process for COD, N and P removal was simulated (Veldhuizen *et al.*, 1999). This study showed that the TUDP model was capable of describing full scale conditions without any significant adjustments. To strengthen the full-scale application of the model, a calibration protocol was developed and tested. Using the same model, Brdjanovic *et al.* (2000) simulated a full-scale side-stream P-removing process. After calibrating glycogen formation, the model described the process without the need to further adjust other parameters. Since temperature plays a major role in microbial conversions, Brdjanovic *et al.* (1998) studied the effect of temperature on EBPR. Their results were incorporated in the TUDP model that was used to simulate a full-scale MUCT process optimised for denitrifying EBPR (WWTP Hardenberg, see Meijer *et al.*, 2001). On the basis of all these practical experiments, the updated and validated metabolic TUDP model showed that its stoichiometry is fully reliable and can be used and extrapolated without calibration. To simulate full-scale EBPR, the metabolic model was combined with the heterotrophic, hydrolytic and autotrophic reactions from ASM2d (Henze *et al.*, 1999). Figure 14.11 shows how the different model structures interact.

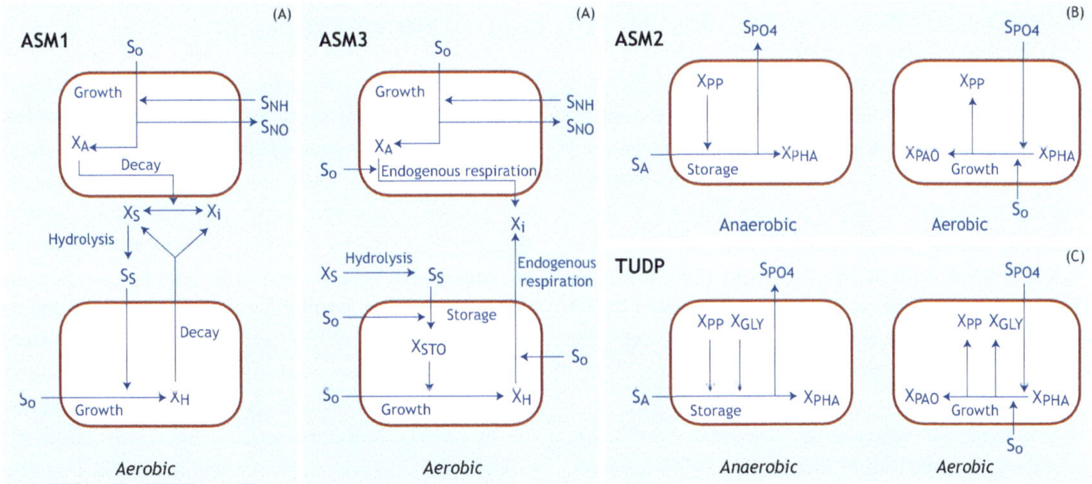

Figure 14.11 Simplified schemes of substrate flows for (A) autotrophic and heterotrophic biomass in the ASM1 and ASM3 models (modified from Gujer *et al.*, 1999), (B) storage and growth of PAO in the ASM2 model (Henze *et al.*, 1995), and (C) storage and aerobic growth of PAO in the TUDP model (Van Veldhuizen *et al.*, 1999; Brdjanovic *et al.*, 2000). Adapted from Gernaey *et al.* (2004).

Although it would be possible to reformulate the auto- and heterotrophic processes of ASM2d in a metabolic form, such a model would have the same number of yields as the original model. Therefore, it would not be smaller and, moreover, it would not improve the model performance. Therefore, in the TUDP model, the ASM2d processes were maintained in their original form and the integration of the two models was relatively simple. This could increase the reliability of the EBPR process description that ASM2d previously appeared to lack (Sin and Vanrolleghem, 2006). Nevertheless, where the two models are merged a new form of substrate competition develops (*e.g.* between Ordinary Heterotrophic Organisms: OHO and PAO). Moreover, with the EBPR also the fermentation and hydrolysis processes in the model become more sensitive and two concepts of endogenous respiration/maintenance are used simultaneously.

In the TUDP metabolic model, the kinetic structure results in a set of atypical model reactions. These reactions are the mathematical result of the kinetic formulation, and cannot be seen independently. For those not aware of this, this could easily lead to misinterpretations of the model matrix, as the individual stoichiometric reactions do not exemplify the actual EBPR process. This should be realised when the model is being used for educational purposes. However, in modelling practice, working with the metabolic concept has important advantages over other model approaches. The main advantage is the solid stoichiometric base of the metabolic model. This solid stoichiometric base is largely owed to the inclusion of glycogen and the simultaneous modelling of the counteracting dynamics of glycogen and PHA.

It is clear that when using metabolic information the degrees of freedom in the model can be reduced. Better understanding of the metabolic processes of the organism will close the gap to a fully white box situation. The increased complexity of processes is consequently reflected in the models. However, improved understanding of the complex interactions within the cell and the introduction of the metabolic approach gives more confidence and consistency in the application of models to describe activated sludge processes. It is in effect gathering information from a lower level of organization to help understand and model the processes at a higher level of organization. For further details on ASM2, ASM2d, ASM3 and metabolic models the reader is referred to Henze *et al.* (2000), Gernaey *et al.* (2004) and Meijer (2004).

Following the work on metabolic modelling by the Delft group, Filipe *et al.* (2001) improved the model for anaerobic acetate uptake. A kinetic poly-P dependency was included, which improved the description of acetate uptake under varying initial poly-P concentrations. Also a different pH dependency for anaerobic acetate uptake was suggested that becomes critical when anaerobic substrate uptake is limiting. In the TUDP model, anaerobic acetate uptake was modelled according to Smolders *et al.* (1994a). Also, Filipe *et al.* (1999) proposed improvements for the anoxic acetate uptake model according to Smolders *et al.* (1994a). However, these improvements were not incorporated in the TUDP model.

14.6 THE ASM TOOLBOX

Currently, activated sludge models are considered reliable and capable of describing complex wastewater treatment plants. From the practical perspective, for most engineering applications, models are considered sufficiently developed. Within the context of model development it is also important to mention the role of hardware. The development of the models and computer capacity (CPU) grew hand-in-hand (Gujer, 2006). From a technical perspective, it became feasible to work with models that contained a large number of process descriptions and variables. In the 1990s, models were increasingly used by researchers, but also mathematical modelling became popular among practitioners. Today, mathematical models are commonly used in North America, Australia and many countries in Europe (Hauduc *et al.*, 2009). To facilitate the application of mathematical models, software has been developed to assist in design, optimisation, operation and training. Modelling simulators provide a better understanding of wastewater treatment plants since they allow users to view the response of the treatment systems to changes in a number of different variables, and are also used to optimize wastewater treatment plants and to train plant operators. Examples of commercial packages are GPS-X, SIMBA, STOAT, SUMO, WEST, BioWin etc. For research and training, SSSP, ASIM, AQUASIM and even Microsoft Excel are regularly used (references to packages are provided at the end of the chapter). Significant benefits are associated with the use of simulators in the analysis, design, and operation of wastewater treatment systems (Meijer and Brdjanovic, 2012).

To make it easier for practitioners to use and facilitate its use in a structured and organized manner, several practical guidelines on how to model a wastewater treatment plant, and protocols on how to characterize the wastewater/sewage and sludge, have been developed around the world over the last few years. In 2004 at the 4th IWA World Water Congress in Marrakech, groups that developed various protocols (Hochschulgruppe, STOWA, BIOMATH and WERF) came together to develop plans to synthesize the best modelling practices available. A new IWA Good Modelling Practice Task Group (GMP-TG) was formed on the use of Activated Sludge Models, parallel to and with the full support of the original task group on mathematical modelling for design and operation of biological wastewater treatment. The website of this and other relevant IWA task groups in the field of mathematical modelling are presented in the annexes. GMP-TG consists of an international team of modellers collecting experience and knowledge on activated sludge modelling to provide guidance to practitioners (Rieger *et al.*, 2012). One of the aims of GMP-TG was to prepare a scientific and technical report to propose simple and effective procedures for the use of ASM-type models (Rieger *et al.*, 2012). In preparation of this report, GMP-TG developed and sent out a questionnaire in 2007 to benchmark and collect relevant information on the practical use of modelling. The objectives were to better define the profile of ASM users, to identify the tools/procedures that are used (models, guidelines, protocols) and to highlight the main limitations encountered while building and using ASM-type models (Hauduc *et al.*, 2009). The outcome of the questionnaire, filled in by 96 respondents, showed that models are used by researchers for optimisation purposes, as well as by modellers employed by private companies to carry out design studies. Modelling is seen as an engineering tool, needing relevant training that is often lacking. The most used biokinetic models were ASM1 (57%) and ASM2d (32%), followed by ASM3, other (non-specified), ASDM (BioWin), Mantis (GPS-X) and the TUDP model. The study also revealed that models are sometimes not properly applied, which might be due to a lack of knowledge and to standardised procedures. The development and improvement of standardised modelling procedures and better knowledge transfer by making available some practical case studies were mentioned as key instruments to addressing certain obstacles such as the complexity of the model theories and procedures, the time consuming steps and finally the reliability of the models.

Besides sending out the questionnaire, several workshops, meetings and courses on activated sludge modelling were also organised, such as wastewater treatment modelling seminars. GMP-TG was involved in the development of a new IWA Model Notation System (Corominas *et al.*, 2010), and interviewed several distinguished modellers (Peter Dold, George Ekama, Willi Gujer, Mogens Henze, Mark Van Loosdrecht, among others). One of the suggestions was to recommend typical values for regularly used ratios, variables and parameters in wastewater modelling. The feedback was compiled in the IWA book 'Guidelines for using activated sludge models', Scientific and Technical Report No. 22 (Rieger *et al.*, 2012). Obviously, IWA played an important role in the evolution of wastewater modelling by facilitating its development, providing a platform for various communities of practice, and promoting modelling research and practice through various publications.

On the same GMP-TG website, adjusted 'Gujer matrices' for seven published models can be downloaded as an MS Excel spreadsheet: (*i*) ASM1 (Henze *et al.*, 1987a,b, 2000); (*ii*) ASM2d (Henze *et al.,* 1999); (*iii*) ASM3 (Gujer *et al.,* 1999); (*iv*) ASM3+BioP (Rieger *et al.,* 2001); (*v*) ASM2d+TUD (Meijer, 2004); (*vi*) the Barker & Dold model (Barker and Dold, 1997); and (*vii*) UCTPHO+ (Hu *et al.,* 2007). On the same website, a comparison of different parameter-naming rules including the new IWA Model Notation System (Corominas *et al.,* 2010) can also be found.

More recently, among other initiatives to improve knowledge transfer by facilitating and making available some practical case studies (Hauduc *et al.,* 2009), IHE Delft published the book 'A practical guide to activated sludge modelling' (Meijer and Brdjanovic, 2012) which is used in the modelling course delivered every year at IHE Delft [1] in cooperation with Delft University of Technology. With a very practical focus, it presents all the steps performed as a part of a modelling project where five WWTPs were subject to upgrade to EU effluent discharge standards in a new EU member state. Besides the general modelling protocols and guidelines for wastewater characterization and fractionation, methods for quantitative influent assessment, addressing different components of the urban wastewater chain and introducing a methodology for quantification of sewage components are presented. Guidelines for plant flows and measurement points necessary for the preparation of an additional sampling program are also shown, as well as an inventory of all the regular day-to-day sampling to be used in a modelling project, a methodology for activated sludge plant assessment and methods to evaluate raw plant data and to filter out possible errors which affect the model reliability and results. Furthermore it includes practical methods for wastewater data evaluation were developed by Meijer *et al.* (2002). New developments on data evaluation are also presented in this book in relation to a well-documented recent case study performed on a plant in the Netherlands. A methodology for the design and assessment of secondary settlers is also described, including the five most commonly used settler design and operation procedures such as the empirical, flux-theory, WRC, ATV (recently DWA) and the STOWA design guidelines (Henze *et al.,* 2008). The last part of the book elaborates on the methodology applied in model calibration and the main steps thereof. In addition, another recent publication describes the selection of fifteen municipal and industrial activated sludge model applications carried out over the last two decades by the Delft modelling group (Brdjanovic *et al.,* 2015). Besides a number of examples from the Netherlands, this book includes some pioneering studies on activated sludge modelling in India (Brdjanovic *et al.,* 2007; Lopez-Vazquez *et al.,* 2013), Bosnia and Herzegovina (Hodzic *et al.,* 2011; Price and Vojinovic, 2010), Mexico (Fall *et al.,* 2012), Croatia (Meijer and Brdjanovic, 2012) and Uruguay (Bentancur, 2014). This demonstrates that efforts are also being undertaken in developing countries to apply the models to existing activated sludge systems (mostly for optimization or upgrade).

Overall, mathematical modelling is becoming a standardized and mature practice, mostly in developed countries, and the efforts made by the IWA GMP-TG and the publication of diverse modelling guidelines (*e.g.* Rieger *et al.,* 2012; Meijer and Brdjanovic, 2012; Brdjanovic *et al.,* 2015) will ultimately contribute to increasing their reliability and application.

14.7 CHALLENGES FOR ASM AND FUTURE TRENDS

Concerning the future development of activated sludge modelling, it is important to take into consideration the current and future needs and developments. The trends regarding existing wastewater process technologies will likely need to further focus on providing a better description of current and novel nutrient removal processes, not only for the sake of the removal of nutrients but also to reduce the associated energy costs and the environmental impact of wastewater treatment plants.

[1] www.un-ihe.org/modelling-wastewater-treatment-processes-and-plants

Such efforts, in combination with the recovery of resources (not only of the nutrients themselves but also of other valuable products) will contribute to moving forward from conventional treatment concepts towards achieving the goals outlined for Water and Resource Recovery Facilities (WRRF), wastewater treatment plants as energy factories, and biorefineries (Kartal *et al.*, 2010; Van Loosdrecht and Brdjanovic, 2014; Energiefabriek, 2015).

With regard to new developments, although the EBPR process can reach relatively high phosphorus removal efficiency rates (with effluent phosphorus concentrations lower than 1 mg/l), it tends to experience process upsets and deterioration due to factors not yet completely understood (Oehmen *et al.*, 2007). In this regard, the appearance of Glycogen Accumulating Organisms (GAO), such as *Competibacter* and *Defluviicoccus*, has been tentatively linked to the suboptimal operation and even failure of the EBPR process performance (Cech *et al.*, 1993; Satoh *et al.*, 1994; Saunders *et al.*, 2003). Thus, GAO are seen as undesirable microorganisms in wastewater treatment since they do not contribute to the EBPR process but compete with PAO in the anaerobic stage for the same carbon source (RBCOD, *e.g.* volatile fatty acids: VFA). In 2009, Lopez-Vazquez *et al.* incorporated the influence of carbon source (such as acetate and propionate), temperature (from 10 to 30°C) and pH dependency of PAO and GAO (from pH 6.0 to 7.5) in the metabolic model amended by Murnleitner *et al.* (1997). Thus, using a mechanistic model, Lopez-Vazquez *et al.* (2009) were able to evaluate the carbon source, pH and temperature influence on the PAO and GAO interaction and their effects on EBPR stability aiming at facilitating improved process efficiency and robustness. They concluded that PAO are favoured by temperatures lower than 20°C and pH levels higher than 7.0. Building on the research carried out by Lopez-Vazquez *et al.* (2009), Oehmen *et al.* (2010) expanded the competition between PAO and GAO to sequential anaerobic-anoxic-aerobic conditions which are typically found in most biological nutrient removal (BNR) systems. This suggested the incorporation of up to six different biomass groups consisting of *Accumulibacter Types I* and *II* and denitrifying and non-denitrifying *Competibacter* and *Defluvicoccus* in accordance to their observed denitrifying capabilities. Their model also included a multistep denitrifying process (from nitrate to di-nitrogen gas). Overall, the model of Oehmen *et al.* (2010) with minimum adjustments was able to successfully describe the EBPR biomass activities observed in lab-scale anaerobic-anoxic-aerobic sequencing batch reactor (SBR) data. However, the application to full-scale conditions of the metabolic model developed by Oehmen *et al.* (2010) is not that straightforward (*e.g.* due to the absence of organic matter and nitrogen oxidation processes as well as practical limitations concerning the elemental balances). Recently, SUMO, a model developed by Dynamita[2] has not only incorporated an extended PAO-GAO module to take into account the interactions between these organisms but also a relatively new type of PAO with fermentative capabilities (arguably *Tetrasphera*, Liu *et al.*, 2019). This 'new type of PAO' appears to be active in side-stream EBPR systems (S2EBPR) and therefore proliferates in these systems, as suggested by practical observations (Dunlap *et al.*, 2016; Barnard *et al.*, 2017; Varga *et al.*, 2018). Following this conceptual model, in addition to the 'conventional PAO' (active in the main-stream line) a second group of PAO, a 'side-stream PAO', has been incorporated into the EBPR models. As such, in the newly added EBPR SUMO module, the 'side-stream PAO' becomes active and is able to take up VFA when the oxido-reduction potential in the side-stream line (calculated in terms of the concentrations of oxygen, nitrate, and sulphate in the liquid phase) decreases to below -200 mV. Despite this, although such a modelling approach would likely require further improvements, mostly because the actual biological mechanisms and the dominant microorganisms involved are not completely known (Onnis-Hayden *et al.*, 2019; Liu *et al.*, 2019), it has so far satisfactorily described the biological P-removal activity observed in S2EBPR systems. Overall, such integrated models

[2] www.dynamita.com

seem to be useful to describe the relevant EBPR microbial activities and the populations of interest with the objective of exploring the environmental and operational conditions beneficial for the EBPR process.

In some cases, such as high pH (>7.5) and high Ca^{++} concentrations, it may be necessary to add biologically induced P precipitation to ASM models (Maurer *et al.*, 1999; Maurer and Boller, 1999). Indeed, under certain conditions the EBPR reactions coincide with a natural precipitation that can account for an important P removal effect that is not related to the EBPR reactions included in the models described thus far. The formation of these precipitates, mostly consisting of calcium phosphate, is promoted by the high P concentration and increased ionic strength during the anaerobic P release by the PAO. Model equations and components necessary to describe this precipitation process were given by Maurer and Boller (1999). Furthermore, Flores-Alsina *et al.* (2015) developed a plant-wide aqueous phase chemistry module able to describe the pH variations in activated sludge systems. Mbamba *et al.* (2016) further studied and applied the model developed by Flores-Alsina *et al.* (2015) to describe the precipitation of minerals with specific emphasis on P precipitates. Furthermore, with an increasing interest in improving the P-removal process in existing activated sludge plants through the implementation of chemical P removal, Mbamba *et al.* (2019) modelled the iron (Fe) addition. The authors successfully described the chemical P-removal process in a pilot plant. Most commercial simulators have already incorporated similar functionalities (like BioWin, SUMO and GPS-X) but future research and developments will likely focus on the implementation of such models in full-scale plant-wide systems with particular interest in P recovery.

Regarding the implementation of innovative wastewater treatment technologies, a stronger focus can be expected on modelling the bioprocesses involved and related to the implementation of the anammox process in the mainstream treatment line and on aerobic granular sludge technology (De Kreuk *et al.*, 2007; Kartal *et al.*, 2010; Wett *et al.*, 2013; Lackner *et al.*, 2014). In the case of the anammox process and aerobic granular sludge technologies, one of the first challenges to overcome is to satisfactorily describe the 2-step nitrification process (the ammonium and nitrite oxidation activities, respectively) as well as their strong interaction with the anammox and EBPR-related organisms (Dapena-Mora *et al.*, 2004; Wyffels *et al.*, 2004). As such, for both mainstream anammox and aerobic granular sludge modelling applications, it becomes rather important to describe the substrate, product reactions and conversions considering the redox gradients within the corresponding flocs or granules and the populations involved, taking into account their symbiosis and/or competition. In this regard, factors such as diffusion (which depends on floc properties and these ones on shear conditions, cohesion, expolymeric substances and turbulence intensity, Arnaldos *et al.*, 2015; Regmi *et al.*, 2019) play an important role in particular in describing the simultaneous nitrification-deamonification process observed in anammox-driven and simultaneous nitrification-denitrification P removal in aerobic granular sludge systems (Kagawa *et al.*, 2015; Baeten *et al.*, 2018a, 2018b). Apparent half-saturation coefficients (that lump the reaction-diffusion processes inside dense flocs or granules) when applying ASM models instead of biofilm models have proven successful in describing the macro-scale activity (*e.g.* predicting effluent concentrations) but mostly under rather stable population dynamics (Baeten *et al.*, 2018a). Therefore, it will be useful when defining the operational ranges to use either a lumped approach (Arnaldos *et al.*, 2015; Baeten *et al.*, 2018a) or a biofilm-based model (Picioreanu *et al.*, 2004; Kagawa *et al.*, 2015). This choice depends considerably on the modelling goal ranging from understanding the micro- and macro-scale phenomena and reactor operation to process optimization and design of microbial activities (Baeten *et al.*, 2018b). However, the choice made will determine whether to use a less or more complex model. So far, existing modelling simulators, such as BioWin and SUMO, have specific modules and extensions to describe the applications of anammox-based and aerobic granular sludge technologies, but still further research needs to focus on the goal definition and consequently on the definition of the level of model complexity.

The application of the sulphur conversion in wastewater treatment will likely continue to be a viable option for the treatment of sulphate-rich waters generated by industry, saline water intrusion, and even the use of seawater for sanitation to alleviate water scarcity (Hvitved-Jacobsen *et al.,* 1998; Huisman and Gujer, 2002; Wang *et al.,* 2009; Hao *et al.*, 2014; Cui *et al.*, 2019).

Undoubtedly, mathematical modelling can be a useful tool to get a better understanding of the factors affecting these processes and facilitate their implementation. However, the application of sulphate or sulphur conversion models in municipal wastewater treatment systems is limited to only a few studies despite the fact that several wastewater treatment plants may be exposed to relatively high influent sulphate concentrations due to saline water intrusion or the use of alternative water sources (Lu *et al.*, 2009), particularly in regions with a warm climate (Sharma *et al.,* 2008; Choubert *et al.,* 2012). Moreover, the potential positive or negative influence of sulphur-driven processes on municipal wastewater treatment systems (Rubio-Rincon *et al.,* 2017; Cui *et al.,* 2019) increases the need to further develop sulphate and/or sulphur models. On the other hand, most of the commercial software simulators (*e.g.* BioWin, GPS-X or SUMO) have suitable additional simulation modules to describe sulphur conversions but only for industrial applications (*e.g.* usually sulphate-rich effluents from petrochemical industries) and are therefore limited to specific applications beyond the industrial sector. The need to describe and assess the corrosion of concrete structures, process upsets or process improvements as well as changes in dominant microbial populations caused by the presence of relatively high influent sulphate concentrations will likely expand the application of sulphur and sulphate-related models to the municipal sector.

In addition to the aforementioned developments, the application of new technologies, such as anammox, sulphur-related processes and aerobic granular sludge, may imply the need to incorporate more complex biokinetic models (Nielsen *et al.,* 2010; Oehmen *et al.*, 2010), elemental balance approaches (Takács *et al.*, 2007; Lu *et al.*, 2009), and is likely also to establish stronger links with genomics, molecular techniques and metabolome analyses (Fiehn, 2001) as well as to develop the required experimental methods in order to determine and understand the microbial activities involved (Van Loosdrecht *et al.,* 2015).

The incorporation of new processes (*e.g.* anammox, aerobic granular sludge and sulphur conversions) will likely follow the IWA GMP-TG concepts (Rieger *et al.*, 2012). The increased complexity of these combined biological processes requires the development of effective process controls in which application of dynamic ASMs could become crucial for successful full-scale application (Regmi *et al.,* 2019).

Modelling of nitrous oxide emissions in nutrient removal plants (Ni *et al.,* 2013; Nopens *et al.*, 2014) is another emerging topic. In this regard, the IWA Task Group on Greenhouse Gas (GHG) is playing a major role in the design and operation of environmentally-friendly wastewater treatment systems. Robust and relevant developments in mathematical modelling have been made to describe N_2O and CH_4 emmissions (Flores-Alsina *et al.,* 2011; Corominas *et al.*, 2012; Mampaey *et al.,* 2019). Such advances have led to the assessment of specific parameters to reduce GHG emissions and also to evaluate the performance of wastewater treatment plants (Flores-Alsina *et al.,* 2014a). Undoubtedly, further modelling efforts will continue to focus on the assessment of the biological mechanisms and operational conditions that lead to GHG emissions and the development of minimization strategies linked to process control and automation (Regmi *et al.,* 2019).

As a consequence of the interaction between the existing but also for the implementation of new technologies, plant-wide modelling will pay special attention to: (*i*) developing an optimal plant control strategy, (*ii*) increasing the efficiency of the removal processes, (*iii*) reducing the operating costs, (*iv*) maximizing energy recovery through biogas production, and (*v*) maximizing the removal and recovery of nutrients in the side-stream processes. In this regard, the IWA Task group on benchmarking of control strategies for wastewater treatment plants (BSM) has played and continues to play a prominent role towards achieving these goals (Copp, 2002; Jeppsson *et al*., 2013; Gernaey *et al*., 2014). As a plant-wide modelling starting point, the mathematical description of the separation processes in the primary settling tanks (PST) affecting the COD fractions needs particular attention (Nopens *et al*., 2014; Vanrolleghem *et al.*, 2014). This is mostly because a PST can contribute to maximizing the recovery of energy via the anaerobic digestion of organics and favouring the conversion of existing plants from 'removal-type systems' towards 'resource-recovery systems' and 'energy factories' (Kartal *et al.*, 2010; Van Loosdrecht and Brdjanovic, 2014; Energiefabriek, 2015). Until a few years ago, most PST separation processes were still modelled as black boxes with lumped and gross removal coefficients assigned to all the particulate organics, whereas unbiodegradable particulate organics have been shown to be subject to higher removal efficiencies than the biodegradable organics (Ikumi *et al*., 2014a,b). Bachis *et al.* (2015) developed a model for a PST based on the settling velocity distributions determined by Maruejouls *et al.* (2012) observing that the removal process that takes place in PST affects the (settled) wastewater characteristics as well as the generation and quality of the settled sludge (which is usually anaerobically digested for biogas generation). In addition to PST modelling, Torfs *et al.* (2016) proposed a unified model framework described with a set of partial differential equations (PDEs) to model the settling processes that occur in PST and secondary settling tanks (SST). This set of equations is able to describe the discrete settling process that takes place in PST as well as the compression settling observed in SST. This proposal was further amended by Bürger *et al.* (2017) who suggested a numerical solution to increase the versatility of the unified settling theory. These advances are contributing to tackling the long-standing limitations of PST and SST which have been neglected up to now but are relatively important in achieving the ultimate goals of the WRRF.

Another important aspect in plant-wide modelling is the coupling of the state variables (Volcke *et al.*, 2006) from the activated sludge process tanks and those from the SST (Bürger *et al.*, 2011, 2012; Torfs *et al.*, 2013). Models for clarifiers use total suspended solids as a state variable, which is not explicitly used in ASM models and needs to be calculated as a composite variable of the activated sludge processes. In addition, due to the different redox conditions created, the bottom of a clarifier needs to be dynamically modelled in a similar way to a bioreactor to take into account the potential redox effects on the active biological processes. Some examples of the need to have a satisfactory modelling description of the operation of SST are rising sludge due to denitrification under anoxic conditions in nitrogen removal plants, secondary P release under anaerobic conditions in EBPR systems, and the description of the sludge settleability. Some of these processes can be mimicked by the addition of an anoxic (denitrifying) tank in addition to the secondary settling tank (Brdjanovic *et al.*, 2000). However, this 'trick' is more an intermediate than the final solution to the problem. Nevertheless, to apply plant-wide modelling the biggest challenges can be found when coupling ASM models with anaerobic digestion models (such as ADM1) (Batstone *et al*., 2002). These challenges are related not only to the description of the sludge digestion processes that take place in sludge thickeners and anaerobic sludge digesters, but also to the physicochemical processes occurring within these systems. One of the first challenges is the different sets of state variables used by ASM models and ADM1. Overall, there are two different ways to deal with this issue: (*i*) the 'super model' approach where a complete set of variables valid for both aerobic and anaerobic environments is defined (Grau *et al.*, 2009), which is also available in *e.g.* BioWin, SUMO or GPS-X simulators, and (*ii*) the use of established interlinked models by applying a set of algebraic transformation equations ('transformers') based on a 'Gujer matrix' description of the two models (Vanrolleghem *et al.*, 2005; Volcke *et al.*, 2006;

Nopens *et al.*, 2009; Solon *et al.*, 2017). Overall, these previous approaches can satisfactorily help to apply a plant-wide model. Recently, plant-wide models have also been developed to describe the P transformations throughout the whole plant including the anaerobic digestion fate of EBPR sludge (highly rich in phosphorus and intracellular compounds) by including the required processes in ADM1 (Ikumi *et al.*, 2014a,b; Ikumi and Ekama, 2019) and the development of a new set of biological (activated sludge, anaerobic digestion), physical-chemical (aqueous phase, precipitation, mass transfer) process models and model interfaces (between water and sludge line) to describe the required tri-phasic (gas, liquid, solid) compound transformations and the close interlinks between the P and the sulphur (S) and iron (Fe) cycles (Flores-Alsina *et al.*, 2016; Mbamba *et al.*, 2016; Solon *et al.*, 2017). These models have contributed not only to expanding the ASM models but also ADM1 (Batstone *et al.*, 2012), including relevant processes to assess the anaerobic digestion of EBPR sludge, chemical precipitation of P compounds, gas composition (involving S-related compounds) and mineral precipitation. In this regard, the IWA Task group on the development of a generalized physical-chemical framework (Batstone *et al.*, 2012) contributed to this achievement. Although the fundamentals of the physical-chemical reactions are well understood, are available from other disciplines, and do not need calibration (since thermodynamics define the end points of the kinetic processes), further research is needed to validate the fate of the organic compounds driving the bioprocesses and the thermodynamics of the precipitation rates in full-scale systems. These plant-wide modelling aspects require additional research involving both experimental and modelling development activities to clarify and achieve a satisfactory modelling of the physical-chemical processes. Together with those concerning the implementation of recently developed technologies (such as the implementation of anammox for the treatment of nitrogen-rich reject waters), this can contribute to reaching the objectives of the WRRF.

Computational fluid dynamics (CFD) can potentially help to improve the description and operation of practically all the process units of wastewater treatment plants (from primary to secondary settling tanks including the aeration and biological processes, Plósz *et al.*, 2012; Laurent *et al.*, 2014; Nopens *et al.*, 2014; Wicklein *et al.*, 2015; Karpinska and Bridgeman, 2016; Samstag *et al.*, 2016). By studying the influence of diffusion limitations and gradients through CFD, the interactions between bacterial morphology, bacterial competition and their physical and hydrodynamic interactions can be better understood and therefore lead to better operation and control practices. However, despite the added complexity, reliable experimental methods and set-ups are being developed to support model development and increase their reliability and application. So far, by coupling CFD with bio-kinetic and chemical models, this has led to a better description of the residence time distribution and cross-flow gas-liquid interactions (Le Moullec *et al.*, 2008), improved mixing and aeration as well as better assessments of the biological reactions in process reactors (Le Moullec *et al.*, 2010; Meister and Rauch, 2016; Rehman *et al.*, 2017) and even to the study of N_2O emissions (Bellandi *et al.*, 2019) and to the improved description of nutrient recovery via struvite formation (Rahaman and Mavinic, 2009). Besides the potential energy savings that hydroinformatics tools (such as CFD) can bring (Rieger *et al.*, 2012), with the increasing need and interest in water reuse and integrated modelling, the biological and physical-chemical removal processes of micropollutants will be another modelling area of future major expansion and development (Gujer *et al.*, 2006; Clouzot *et al.*, 2013) where CFD could also be applied (Radu *et al.*, 2010; Laurent *et al.*, 2014). It is likely that, due to its rather practical approach, further efforts will continue hand-to-hand with stronger collaborative links between practice and research to assess and provide feedback on newly developed models under real-case scenarios. A clear and promising example of such development is the establishment of the AM Team (Advanced Modelling for Process Optimization[3]), a spin-off of Gent University for CFD-based wastewater treatment modelling. This close collaboration between academia and practice has contributed to the application of CFD to improve effluent quality, reduce energy consumption, and increase the capacity and reliability of wastewater treatment and resource recovery. Probably other specialist CFD companies will emerge in the coming years. To support the application of CFD models in a structured and reliable manner, IWA has helped to establish the Working Group on CFD for Unit Processes[4] with the main aim of developing and establishing modelling guidelines for the application of CFD models in wastewater treatment. Towards this goal, Nopens *et al.* (2012), Laurent *et al.* (2014) and Wicklein *et al.* (2015) have proposed structured guidelines and protocols towards the development of good modelling practices when applying CFD in wastewater treatment plants. There is no doubt that CFD modelling will be further developed and expanded, yet its use and application needs to be well-defined by first answering what the goal of modelling is and consequently selecting an appropriate modelling approach, *e.g.* a CFD-based model or a (simpler) ASM model.

A growing interest in integrated (urban) water modelling will continue to motivate the integration of wastewater treatment process models with receiving water quality (RWQM) and sewer models (Gujer, 2006; Vanrolleghem *et al.,* 2014). Until a few years ago, only hydraulics and pollutant transport phenomena in the sewers were taken into account (Hvitved-Jacobsen, 2013). However, recent models and simulators (*e.g.* SUMO) are starting to consider the chemical and biological processes that take place in the sewer system, looking at the sewers as physical, chemical and biological reactors (Rauch *et al.,* 2002). One of the first examples of holistic modelling (a combined sewage network, plants and the recipient/river) using different models (combining Mike Urban, BioWin and HEC-RAS), although carried out in a sequential mode (as opposed to a better and more realistic but much more complex real-time approach), showed great advantages with this type of modelling application (Hodzic *et al.,* 2011, Price and Vojinovic, 2010). At the moment a large holistic modelling study is being carried out in Croatia (Brdjanovic, personal communication) which combines a large sewer system, four wastewater treatment plants and seawater quality modelling along a shoreline of 23 km[5].

Such an integrated approach is of major importance for the design, operation and maintenance of sewer networks, not only from a holistic water management perspective but also from a potential future asset management focus (which needs a satisfactory modelling description of the removal of micropollutants). In collaboration with the University of Cape Town, WEST, hydraulic modelling software developed by Gent University (Vanhooren *et al.,* 2003) and nowadays held by the Danish Hydraulic Institute (DHI), has been upgraded to make one of the prime efforts to link wastewater treatment models with RWQM and sewer models (Ikumi *et al.,* 2014b) in addition to other recent developments (Benedetti *et al.*, 2013a; Langeveld *et al.*, 2013). Together with the plant-wide modelling advances, this could lead to promising opportunities for the development of an integrated urban water model suitable for and capable of describing and optimizing the entire urban water system, including receiving waters (Benedetti *et al.*, 2013b).

However, one should not overlook the fact that despite these significant advances and the development of more (complex and) complete mathematical models, a common issue is still the lack of (quality and reliable) input data to 'feed' the models or the potential influence of regularly dynamic and even extreme scenarios affecting the quality and characteristics of the influent wastewater quality and consequently the reliability of models. Furthermore,

[3] www.am-team.com

[4] www.iwa-connect.org/group/working-group-on-computational-fluid-dynamics-cfd-for-unit-processes/about

[5] http://odvodnjaporec.hr/projekti/projekt-porec/

as underlined by Nopens *et al.* (2014), more methods are needed to assess the probability of compliance, quantify uncertainty and its sources, and evaluate how risks, benefits and costs are or can be distributed among stakeholders (consultants, contractors, operators and owners). One of the objectives of the recently conceived IWA Task Group on Design and Operations Uncertainty (DOUT) is to overcome such limitations with actions such as the development of influent generator models to provide relevant input data and incorporate explicit uncertainty evaluations in model-aid design and operation of wastewater treatment systems (Gernaey *et al.*, 2011; Flores-Alsina *et al.*, 2014b; Nopens *et al.*, 2014).

Bearing in mind that 2.6 billion people still do not have access to sanitation, that most of the population in developing countries is not connected to sewer systems, and that only a small fraction of sewage in developing countries is treated, this raises the issue of holistic modelling where the urban drainage and sewerage models and wastewater treatment models (not only ASM and ADM) will be complemented by and integrated with (de)centralized sanitation models in cities which are not entirely covered by sewerage. There are several examples of this worldwide, especially in developing countries. Recent advances in this direction include the development of an excreta flow diagram (also often described as a shit flow diagram, SFD[6]) which is a tool to help understanding and communicate the management of excreta in a city or town. Another novelty is a decentralized sanitation concept and its model which considers all the major components of the sanitation service chain (Brdjanovic *et al.*, 2015). This non-sewered sanitation decision support tool has been incorporated in a decision support tool called WaMEX by IHE Delft which also includes sewerage (Abbot and Vojinovic, 2009; Sanchez *et al.*, 2013; Vojinovic *et al.*, 2014), and sewage treatment components (Von Sperling and Chernicharo, 2005).

Another interesting modelling aspect is modelling of the influence of faecal and septic sludge discharge (load) to sewage treatment plants, a practice regularly applied in many (especially low and middle income) countries (Strande *et al.*, 2014). More interesting faecal sludge management modelling tools are to be found in the Faecal Sludge Management (FSM) Toolbox[7]; however, FSM modelling falls outside the scope of this book and for more information the reader is referred to Strande *et al.*, 2014 and Velkushanova *et al.*, 2020.

Further, cloud computing has gained in interest lately (Armbrust *et al.*, 2010) by joining efforts and contributing to standardized approaches and notation (Corominas *et al.*, 2010), and therefore sharing wastewater treatment models between researchers, software developers, and practitioners, despite them being in different longitudes and latitudes, may not be far from reality. This could be a powerful tool to facilitate the application of plant-wide and integrated urban water modelling to contribute to optimizing the water quality and quantity transported through the aquatic veins and arteries of an urban settlement.

From a commercial and practical perspective, the incorporation of the processes and approaches described previously will considerably increase the model complexity. However, understandably, practitioners feel uncomfortable working with increasingly complex models. Therefore, vendors with specific modelling skills might appear on the market, since conventional wastewater treatment 'generalists' will not be able to cope with the fast release and development of more complex models for particular applications. Thus, like in other fields, in the near future consultants will outsource their modelling activities to modelling specialists.

It is not impossible to imagine that sooner or later new interfaces and methods of interaction between (very probably less specialized) users and models will be created. These could be in the form of multi-layer serious gaming and use 3D urban water system simulators with simplified 'surface' user interfaces and more complex expert models running 'invisibly'

[6] www.sfd.susana.org

[7] www.fsmtoolbox.com/

in the background. Another expected future development is the use of models built in the data acquisition systems (SCADA) of larger wastewater treatment facilities; thereby the complex knowledge contained in ASMs would be made available to process operators, making more efficient and safe plant operation possible on a daily basis. It is also expected that modelling boundaries will be further extended reaching a trans-disciplinary character as other issues will be included, *e.g.* emergencies, risks, and social aspects (Abbott and Vojinovic, 2010a,b; Vojinovic and Abbott, 2011; Abbott and Vojinovic, 2013; Brdjanovic *et al.*, 2014; Zakaria *et al*., 2015). By doing so, modelling will come closer to the decision makers and increase and facilitate the use of models by different and currently not involved stakeholders. Last but not least, despite all these expected developments (Van Loosdrecht and Brdjanovic, 2014) and the release of more complex models for several wastewater treatment applications, one must keep in mind that a model is still a mere representation of reality, generally applied as a tool for improvement and optimization purposes. A model must by no means be used as a substitute for an educational programme or design criterion, but rather as a complement.

14.8 CONCLUSIONS

Modelling is an important activity in the development of science. Modelling not only requires the explicit and quantitative formulation of theoretical concepts, it also allows transfer of complicated knowledge between scientific disciplines as well as between theoretical and practical applications. For 30 years, activated sludge models have played a crucial role in the development of the activated sludge process. These models are not typically academic; they do not aim to include every potential sub-process involved in the activated sludge process. Instead, they are formulated with the minimum complexity needed to describe the relevant features of the process in practice. They also provide a systemized platform for the description of environmental biotechnological models in general, through the use of standardized notation and a matrix presentation. Over the years, many wastewater research projects have benefitted greatly from the development of activated sludge models. On the one hand, modelling has been expanded through the development of novel theoretical concepts and their application in new fields. On the other hand, models have been used for practical applications. We trust that this chapter will be useful to inspire future engineers to use models as central tools in their work on improving wastewater treatment technology through innovation and optimization.

Table 14.9 A stoichiometric matrix and a component composition matrix (Meijer, 2004).

				1	2	3	4	5	6
			Component →	S_O	S_F	S_A	S_{NH}	S_{NO}	S_{N2}
↓ Process				gO_2/m^3	$gCOD/m^3$	$gCOD/m^3$	gN/m^3	gN/m^3	gN/m^3
1	r_h^O	Aerobic hydrolysis	$gCOD_{XS}/d$		$1-f_{SI}$		$c_{N,1}$		
2	r_h^{NO}	Anoxic hydrolysis	$gCOD_{XS}/d$		$1-f_{SI}$		$c_{N,1}$		
3	r_h^{AO}	Anaerobic hydrolysis	$gCOD_{XS}/d$		$1-f_{SI}$		$c_{N,1}$		
Regular heterotrophic organisms X_H									
4	r_{SF}^O	Aerobic growth on S_F	$gCOD_{XH}/d$	$-(1/Y_H-1)$	$-1/Y_H$		$c_{N,4}$		
5	r_{SA}^O	Aerobic growth on S_A	$gCOD_{XH}/d$	$-(1/Y_H-1)$		$-1/Y_H$	$c_{N,5}$		
6	r_{SF}^{NO}	Anoxic growth on S_F	$gCOD_{XH}/d$		$-1/Y_H$		$c_{N,6}$	$-\frac{(1/Y_H-1)}{2.86}$	$\frac{(1/Y_H-1)}{2.86}$
7	r_{SA}^{NO}	Anoxic growth on S_A	$gCOD_{XH}/d$			$-1/Y_H$	$c_{N,7}$	$-\frac{(1/Y_H-1)}{2.86}$	$\frac{(1/Y_H-1)}{2.86}$
8	r_{fe}^{AN}	Fermentation	$gCOD_{SF}/d$		-1	1	$c_{N,8}$		
9	r_{HL}	Heterotrophic lysis	$gCOD_{XH}/d$				$c_{N,9}$		
Phosphorus-accumulating organisms X_{PAO}									
10	r_{SA}^{AN}	Anaerobic storage of S_A	$gCOD_{SA}/d$			-1			
11	r_M^{AN}	Anaerobic maintenance	gP/d						
12	r_{SA}^{NO}	Anoxic storage of S_A	$gCOD_{SA}/d$			-1		$-\frac{(1-Y_{SA}^{NO})}{2.86}$	$\frac{(1-Y_{SA}^{NO})}{2.86}$
13	r_{PHA}^{NO}	Anoxic PHA consumption	$gCOD_{PHA}/d$				$c_{N,13}$	$-\frac{(1-1/Y_{PHA}^{NO})}{2.86}$	$\frac{(1-1/Y_{PHA}^{NO})}{2.86}$
14	r_{PP}^{NO}	Anoxic storage of polyP	gP/d				$c_{N,14}$	$-\frac{(1/Y_{PP}^{NO})}{2.86}$	$\frac{(1/Y_{PP}^{NO})}{2.86}$
15	r_{GLY}^{NO}	Anoxic glycogen formation	$gCOD_{GLY}/d$				$c_{N,15}$	$-\frac{(1/Y_{GLY}^{NO}-1)}{2.86}$	$\frac{(1/Y_{GLY}^{NO}-1)}{2.86}$
16	r_M^{NO}	Anoxic maintenance	$gCOD_{PAO}/d$				$c_{N,16}$	$-1/2.86$	$1/2.86$
17	r_{PHA}^O	Aerobic PHA consumption	$gCOD_{PHA}/d$	$1/Y_{PHA}^O-1$			$c_{N,17}$		
18	r_{PP}^O	Aerobic storage of polyP	gP/d	$-1/Y_{PP}^O$			$c_{N,18}$		
19	r_{GLY}^O	Aerobic glycogen formation	$gCOD_{GLY}/d$	$1-1/Y_{GLY}^O$			$c_{N,19}$		
20	r_M^O	Aerobic maintenance	$gCOD_{PAO}/d$	-1			$c_{N,20}$		
Autotrophic nitrifying organisms X_A									
21	r_A^O	Autotrophic growth	$gCOD_{XA}/d$	$1-4.57/Y_A$			$c_{N,21}$	$1/Y_A$	
22	r_{AL}	Autotrophic lysis	$gCOD_{XA}/d$				$c_{N,22}$		

			1	2	3	4	5	6
		Component →	S_O	S_F	S_A	S_{NH}	S_{NO}	S_{N2}
↓ Composition			gO_2	gCOD	gCOD	gN	gN	gN
1	COD	gCOD	-1	1	1		-2.86	…
2	TOC/COD	gC/gCOD		…	0.4			
3	Nitrogen	gN		$i_{N,SF}$	$i_{N,SA}$	1	1	1
4	Phosphorus	gP		$i_{P,SF}$	$i_{P,SA}$			
5	Ionic charge	mole			-1/64	+1/14	-1/14	
6	TSS	g						

Table 14.9 ... continued (for definition of the symbols see Meijer (2004).

7	8	9	10	11	12	13	14	15	16	17	18
S_{PO}	S_I	S_{HCO}	X_I	X_S	X_H	X_{PAO}	X_{PP}	X_{PHA}	X_{GLY}	X_A	X_{TSS}
gP/m³	gCOD/m³	mole/m³	gCOD/m³	gCOD/m³	gCOD/m³	gCOD/m³	gP/m³	gCOD/m³	gCOD/m³	gCOD/m³	g/m³
$c_{P,1}$	f_{SI}	$c_{e,1}$		-1							$c_{TSS,1}$
$c_{P,1}$	f_{SI}	$c_{e,1}$		-1							$c_{TSS,1}$
$c_{P,1}$	f_{SI}	$c_{e,1}$		-1							$c_{TSS,1}$
$c_{P,4}$		$c_{e,4}$			1						$c_{TSS,4}$
$c_{P,5}$		$c_{e,5}$			1						$c_{TSS,5}$
$c_{P,6}$		$c_{e,6}$			1						$c_{TSS,6}$
$c_{P,7}$		$c_{e,7}$			1						$c_{TSS,7}$
$c_{P,8}$		$c_{e,8}$									$c_{TSS,8}$
$c_{P,9}$		$c_{e,9}$	$f_{XI,H}$	$1-f_{XI,H}$	-1						$c_{TSS,9}$
Y_{PO}^{AN}		$c_{e,10}$					$-Y_{PO}^{AN}$	Y_{SA}^{AN}	$1-Y_{SA}^{AN}$		$c_{TSS,10}$
1		$c_{e,11}$					-1				$c_{TSS,11}$
Y_{PO}^{NO}		$c_{e,12}$					$-Y_{PO}^{NO}$	Y_{SA}^{NO}			$c_{TSS,12}$
$c_{P,13}$		$c_{e,13}$				$1/Y_{PHA}^{NO}$		-1			$c_{TSS,13}$
$c_{P,14}$		$c_{e,14}$				$-1/Y_{PP}^{NO}$	1				$c_{TSS,14}$
$c_{P,15}$		$c_{e,15}$				$-1/Y_{GLY}^{NO}$			1		$c_{TSS,15}$
$c_{P,16}$		$c_{e,16}$				-1					$c_{TSS,16}$
$c_{P,17}$		$c_{e,17}$				$1/Y_{PHA}^{O}$		-1			$c_{TSS,17}$
$c_{P,18}$		$c_{e,18}$				$-1/Y_{PP}^{O}$	1				$c_{TSS,18}$
$c_{P,19}$		$c_{e,19}$				$-1/Y_{GLY}^{O}$			1		$c_{TSS,19}$
$c_{P,20}$		$c_{e,20}$				-1					$c_{TSS,20}$
$c_{P,21}$		$c_{e,21}$								1	$c_{TSS,21}$
$c_{P,22}$		$c_{e,22}$	$f_{XI,A}$	$1-f_{XI,A}$						-1	$c_{TSS,22}$
7	8	9	10	11	12	13	14	15	16	17	18
S_{PO}	S_I	S_{HCO}	X_I	X_S	X_H	X_{PAO}	X_{PP}	X_{PHA}	X_{GLY}	X_A	X_{TSS}
gP	gCOD	mole	gCOD	gCOD	gCOD	gCOD	gP	gCOD	gCOD	gCOD	g
	1		1	1	1	1		1	1	1	
	…		…	…	…	0.334 (α)		0.334	0.375	…	
	$i_{N,SI}$		$i_{N,XI}$	$i_{N,XS}$	$i_{N,XH}$	$i_{N,BM}$				$i_{N,BM}$	
1	$i_{P,SI}$		$i_{P,XI}$	$i_{P,XS}$	$i_{P,XH}$	$i_{P,BM}$	1			$i_{P,BM}$	
-1.5/31		-1					-1/31				
			$i_{TSS,XI}$	$i_{TSS,XS}$	$i_{TSS,BM}$	$i_{TSS,BM}$	$i_{TSS,PP}$	$i_{TSS,PHA}$	$i_{TSS,GLY}$	$i_{TSS,BM}$	1

REFERENCES

Abbott M.B. and Vojinovic Z. (2009). Applications of numerical modelling in hydroinformatics. *J Hydroinformatics*, 11(3-4), 308-319.

Abbott M.B. and Vojinovic Z. (2010a). Realising Social Justice in the Water Sector: Part 1, *J Hydroinformatics*, 12(1), 110–130.

Abbott M.B. and Vojinovic Z. (2010b). Realising Social Justice in Water Sector: Part 2, *J Hydroinformatics*, 12(2), 225–239.

Abbott M.B. and Vojinovic Z. (2013). Towards a hydroinformatics praxis in the service of social justice, *J Hydroinformatics*, 16(2), 516–530.

Alexander W.V., Ekama G.A. and Marais G.v.R. (1980). The activated sludge process Part 2 - Application of the general kinetic model to the contact stabilization process. *Water Research*, 14, 1737-1747.

Armbrust M., Fox A., Griffith R., Joseph A.D., Katz R., Konwinski A., Lee G., Patterson D., Rabkin A., Stoica I. and Zaharia M. (2010). A view of cloud computing. *Communications ACM*, 53(4), 50-58.

Arnaldos M., Amerlinck Y., Rehman U., Maere T., Van Hoey S., Naessens W. and Nopens I. (2015). From the affinity constant to the half-saturation index: understanding conventional modelling concepts in novel wastewater treatment processes. *Water Research*, 70, 458-470.

Bachis G., Maruéjouls T., Tik S., Amerlinck Y., Melcer H., Nopens I., Lessard P., and Vanrolleghem P.A. (2015). Modelling and characterization of primary settlers in view of whole plant and resource recovery modelling. *Water Science and Technology*, 72(12), 2251-2261.

Baeten J.E., Batstone D.J., Schraa O.J., Van Loosdrecht M.C.M. and Volcke E.I. (2018b). Modelling anaerobic, aerobic and partial nitritation-anammox granular sludge reactors - A review. *Water Research*, 149, 322-341.

Baeten J.E., Van Loosdrecht M.C.M. and Volcke E.I. (2018a). Modelling aerobic granular sludge reactors through apparent half-saturation coefficients. *Water Research*, 146, 134-145.

Barker P.S. and Dold P.L. (1997). General model for biological nutrient removal activated-sludge systems: model presentation. *Water Environmental Research*, 69(5), 969-984.

Barnard J., Dunlap P. and Steichen M. (2017). Rethinking the Mechanisms of Biological Phosphorus Removal. *Water Environmental Research*, 89(11), 2043-2054.

Batstone D.J., Amerlinck Y., Ekama G.A., Goel R., Grau P., Johnson B., Kaya I., Steyer J.-P., Tait S., Takács I., Vanrolleghem P.A., Brouckaert C.J. and Volcke E. (2012). Towards a generalized physicochemical framework. *Water Science and Technology*, 66(6), 1147-1161.

Batstone D.J., Keller J., Angelidaki I., Kalyuzhyi S.V., Pavlostathis S.G., Rozzi A., Sanders W.T.M., Siegrist H. and Vavillin V.A. (2002). The IWA Anaerobic digestion model No. 1 (ADM1). *Water Science and Technology*, 45(10), 65-73.

Bellandi G., De Mulder C., Van Hoey S., Rehman U., Amerlinck Y., Guo L., Vanrolleghem P., Weijers S., Gori R., and Nopens I. (2019). Tanks in series *versus* compartmental model configuration: considering hydrodynamics helps in parameter estimation for an N_2O model. *Water Science and Technology*, 79(1), 73-83.

Benedetti L., Langeveld J., Comeau A., Corominas L., Daigger G., Martin C., Mikkelsen P.S., Vezzaro L., Weijers S. and Vanrolleghem P.A. (2013b). Modelling and monitoring of integrated urban wastewater systems: review on status and perspectives. *Water Science and Technology*, 68(6), 1203-1215.

Benedetti L., Langeveld J., Van Nieuwenhuijzen A.F., De Jonge J., De Klein J., Flameling T., Nopens I., Van Zanten O. and Weijers S. (2013a). Cost-effective solutions for water quality improvement in the Dommel River supported by sewer–WWTP–river integrated modelling. *Water Science and Technology*, 68(5), 965-973.

Bentancur S. (2014). Evaluation and modelling of UPM pulp mill wastewater treatment plant in Fray Bentos, Uruguay. MSc thesis. IHE Delft Institute for Water Education, Delft, the Netherlands.

Brdjanovic D., Logemann S., Van Loosdrecht M.C.M., Hooijmans C.M., Alaerts G.J. and Heijnen J.J. (1998). Influence of temperature on biological phosphorus removal: process and ecological studies. *Water Research*, 32(4), 1035-1048.

Brdjanovic D., Meijer S.C.F, Lopez-Vazquez C.M., Hooijmans C.M. and Van Loosdrecht M.C.M. (2015). Applications of Activated Sludge Models. IWA Publishing, London, UK. ISBN: 9781780404639.

Brdjanovic D., Moussa M.S., Mithaiwala M., Amy G. and Van Loosdrecht M.C.M. (2007). Use of plant-wide modelling for optimization and upgrade of a tropical wastewater treatment plant in a developing country. *Water Science and Technology*, 56(7), 21-31.

Brdjanovic D., Van Loosdrecht M.C.M., Versteeg P., Hooijmans C.M., Alaerts G.J. and Heijnen J.J. (2000). Modelling COD, N and P removal in a full-scale WWTP Haarlem Waarderpolder. *Water Research*, 34(3), 846–858.

Brdjanovic D., Zakaria F., Mawioo P.M., Garcia H.A., Hooijmans C.M., Curko J., Pean Y.P. and Setiadi T. (2014). eSOS® – Emergency Sanitation Operation System. *J Water Sanitation Hygiene Develop*, 5(1), 156-164.

Brink I.C., Wentzel M.C. and Ekama G.A. (2007). New developments in modeling wastewater treatment plants – Using stoichiometry to build a plant-wide mass balance based steady state WWTP model. In: Proceedings 10th IWA Conference on Design, Operation and Economics of WWTP. 9-13 September 2007, Vienna, Austria.

Brun R., Kühni M., Siegrist H., Gujer W. and Reichert P. (2002). Practical identifiability of ASM2d parameters - systematic selection and tuning of parameter subsets. *Water Research*, 36(16), 4113-4127.

Bürger R., Diehl S. and Nopens I. (2011). A consistent modelling methodology for secondary settling tanks in wastewater treatment. *Water Research*, 45, 2247-2260.

Bürger R., Diehl S., Farås S. and Nopens I. (2012). On reliable and unreliable numerical methods for the simulation of secondary settling tanks in wastewater treatment. *Computers & Chemical Engineering*, 41, 93-105.

Bürger R., Diehl S., Martí M.C., Mulet P., Nopens I., Torfs E. and Vanrolleghem P.A. (2017). Numerical solution of a multi-class model for batch settling in water resource recovery facilities. *Applied Mathematical Modelling*, 49, 415-436.

Casey T.G., Ekama G.A., Wentzel M.C. and Marais G.vR (1994). An hypothesis for the causes and control of F/M filamentous organism bulking in nitrogen (N) and nutrient (N & P) removal activated sludge systems. *Water Science and Technology*, 29(7), 203-212.

Cech J.S. and Hartman P. (1993). Competition between phosphate and polysaccharide accumulating bacteria in enhanced biological phosphorus removal systems. *Water Research*, 27, 1219-1225.

Choubert J.M., Rieger L., Shaw A., Copp J., Spérandio M., Sørensen K., Rönner-Holm S., Morgenroth E., Melcer H. and Gillot, S. (2013). Rethinking wastewater characterisation methods for activated sludge systems – a position paper. *Water Science and Technology,* 67(11), 2363-2373.

Clouzot L., Choubert J.M., Cloutier F., Goel R., Love N.G., Melcer H., Ort C., Patureau D., Plósz B.G., Pomiès M. and Vanrolleghem P.A. (2013). Perspectives on modelling micropollutants in wastewater treatment plants. *Water Science and Technology*, 68(2), 448-461.

Coen F., Vanderhaeghen B., Boonen I., Vanrolleghem P.A., Van Eyck L. and Van Meenen P. (1996). Nitrogen removal upgrade of a WWTP within existing reactor volumes: A simulation supported scenario analysis. *Water Science and Technology*, 34(3/4), 339-346.

Comeau Y., Hall K.J., Hancock R.E.W. and Oldham W.K. (1986). Biochemical model for enhanced biological phosphorus removal. *Water Research*, 20(12), 1511–1521.

Copp J.B. (2002). The COST Simulation Benchmark: Description and Simulator Manual. Office for Official Publications of the European Community, Luxemburg, Luxembourg. ISBN: 92-894-1658-0.

Corominas L., Flores-Alsina X., Snip L. and Vanrolleghem P.A. (2012). Comparison of different modelling approaches to better evaluate greenhouse gas emissions from whole wastewater treatment plants. *Biotechnology and Bioengineering*, 109(11), 2854-2863.

Corominas L., Rieger L., Takács I., Ekama G.A., Hauduc H., Vanrolleghem P.A., Oehmen A., Gernaey K.V. and Comeau Y. (2010). New framework for standardized notation in wastewater treatment modelling. *Water Science and Technology*, 61(4), 841–857.

Cui Y.X., Biswal B.K., Guo G., Deng Y.F., Huang H., Chen G.H. and Wu D. (2019). Biological nitrogen removal from wastewater using sulphur-driven autotrophic denitrification. *Applied Microbiology and Biotechnology,* 1-17.

Dapena-Mora A., Van Hulle S. W., Campos J.L., Méndez R., Vanrolleghem P. A. and Jetten M. (2004). Enrichment of Anammox biomass from municipal activated sludge: experimental and modelling results. Journal of Chemical Technology & Biotechnology: *International Research in Process, Environmental & Clean Technology*, 79(12), 1421-1428.

De Kreuk M.K., Picioreanu C., Hosseini M., Xavier J.B. and Van Loosdrecht M.C.M. (2007). Kinetic model of a granular sludge SBR: influences on nutrient removal. *Biotechnology and Bioengineering*, 97(4), 801-815.

Dold P.L., Ekama G.A. and Marais G.v.R. (1980). A General Model for the Activated Sludge Process. *Progress in Water Technology,* 12(6), 47-77.

Dold P.L., Wentzel M.C., Billing A.E., Ekama G.A. and Marais G.v.R. (1991). Activated sludge simulation programs: Nitrification and nitrification/denitrification systems (version 1.0).

Report TT 52/91. Water Research Commission, Pretoria, South Africa.

Dunlap, P., Martin, K. J., Stevens, G., Tooker, N., Barnard, J. L., Gu, A. Z., I Takacs, A Shaw, A Onnis-Hayden, Li, Y. (2016). Rethinking EBPR: what do you do when the model will not fit real-world evidence?. In: *5th IWA/WEF Wastewater Treatment Modeling Seminar*, 39-62.

Ekama G.A. and Marais G.v.R. (1978). The dynamic behaviour of the activated sludge process. Research Report W27, Dept. of Civil Eng, University of Cape Town, Rondebosch, South Africa.

Ekama G.A. and Takács I. (2014). Modelling. In: Activated sludge: 100 years and counting, (eds.) J.Wanner and D. Jenkin. IWA Publishing. London, UK. ISBN: 9781780404936.

Energyfabriek (2015). Energiefabriek: een initiatief van de Nederlandse Waterschappen (in Dutch: *Energy factories: An initiative of the Dutch waterboards*). http://energiefabriek.com.

Fall C., Flores-Alamo N., Esparza-Soto M. and, Hooijmans C.M. (2012). Tracer test and hydraulics modelling of a large WWTP. *Water Practice Technology,* 7(1).

Fenu A., Guglielmi G, Jimenez J., Spèrandio M., Saroj D., Lesjean B., Brepols C., Thoeye C. and Nopens I. (2010). Activated sludge model (ASM) based modelling of membrane bioreactor (MBR) processes: a critical review with special regard to MBR specificities. *Water Research*, 44(15), 4272-4294.

Fiehn O. (2001). Combining genomics, metabolome analysis, and biochemical modelling to understand metabolic networks. *Comparative Functional Genomics*, 2(3), 155-168.

Filipe C.D.M. and Daigger G.T. (1999). Evaluation of the capacity of phosphorus-accumulating organisms to use nitrate and oxygen as final electron acceptors: A theoretical study on population dynamics. *Water Environmental Research*, 71(6), 1140-1150.

Filipe C.D.M., Daigger G.T. and Grady C.P.L. (2001). Stoichiometry and kinetics of acetate uptake under anaerobic conditions by an enriched culture of phosphorus-accumulating organisms at different pHs. *Biotechnology and Bioengineering*, 76(1), 32-43.

Flores-Alsina X., Arnell M., Amerlinck Y., Corominas L., Gernaey K.V., Guo L., Lindblom E., Nopens I., Porro J., Snip L., Vanrolleghem P.A. and Jeppsson U (2014a). Balancing effluent quality, economic cost and greenhouse gas emissions during the evaluation of (plant-wide) control/operational strategies in WWTPs. *Science of the Total Environment,* 466, 616-624.

Flores-Alsina X., Corominas L., Snip L. and Vanrolleghem P.A. (2011). Including greenhouse gas emissions during benchmarking of wastewater treatment plant control strategies. *Water Research*, 45(16), 4700-4710.

Flores-Alsina X., Mbamba C.K., Solon K., Vrecko D., Tait S., Batstone D.J., Jepsson U. and Gernaey K.V. (2015). A plant-wide aqueous phase chemistry module describing pH variations and ion speciation/pairing in wastewater treatment process models. *Water Research*, 85, 255-265.

Flores-Alsina X., Saagi R., Lindblom E., Thirsing C., Thornberg D., Gernaey K.V. and Jeppsson U. (2014b). Calibration and validation of a phenomenological influent pollutant disturbance scenario generator using full-scale data. *Water Research*, 51, 172-185.

Flores-Alsina X., Solon K., Kasadi M., Tait S., Jeppson U., Gernaey K.V. and Batstone D.J. (2016). Modelling phosphorus, sulphur and iron interactions during the dynamic simulation of anaerobic digestion processes. *Water Research* 95, 370–382.

Gabb G.M.D., Still D.A., Ekama G.A., Jenkins D. and Marais G.v.R. (1991). The selector effect on filamentous bulking in long sludge age activated sludge system. *Water Science and Technology*, 23, 867 877.

Gernaey K.V., Flores-Alsina X., Rosen C., Benedetti L. and Jeppsson U. (2011). Dynamic influent pollutant disturbance scenario generation using a phenomenological modelling approach. *Environmental Modeling and Software*, 26(11), 1255-1267.

Gernaey K.V., Jeppsson U., Vanrolleghem P.A. and Copp J.B. (2014). Benchmarking of Control Strategies for Wastewater Treatment Plants. IWA Scientific and Technical Report No. 23. IWA Publishing, London, UK. ISBN: 9781843391463.

Gernaey K.V., Van Loosdrecht M.C.M., Henze M., Lindd M. and Jørgensen S.B. (2004). Activated sludge wastewater treatment plant modelling and simulation: state of the art. *Environmental Modelling and Software*, 19(9), 763-783.

Grau P., Copp J., Vanrolleghem P.A., Takács I. and Ayesa E. (2009). A comparative analysis of different approaches for integrated WWTP modelling. *Water Science and Technology*, 59(1), 141-147.

Gujer W. (2006). Activated sludge modelling: past, present and future. *Water Science and Technology*, 53(3), 111-119.

Gujer W. and Henze M. (1991). Activated sludge modelling and simulation. *Water Science and Technology*, 23, 1011-1023.

Gujer W. and Larsen T.A. (1995). The implementation of biokinetics and conservation principles in ASIM. *Water Science and Technology*, 31(2), 257–266.

Gujer W., Henze M., Mino T. and Van Loosdrecht M.C.M. (1999). Activated Sludge Model No. 3. *Water Science and Technology*, 3, 183-193.

Hao T.W., Xiang P.Y., Mackey H.R., Chi K., Lu H., Chui H.K., Van Loosdrecht M.C.M. and Chen G.H. (2014). A review of biological sulfate conversions in wastewater treatment. *Water Research*, 65, 1-21.

Hauduc H., Gillot S., Rieger L., Ohtsuki T., Shaw A., Takács I. and Winkler S. (2009). Activated sludge modelling in practice – An international survey. *Water Science and Technology*, 60(8), 1943–1951.

Henze M. (1992). Characterization of wastewater for modelling of activated sludge processes. *Water Science and Technology*, 25, 1-15.

Henze M., Grady J.R.C.P.L, Gujer W., Marais G.v.R. and Matsuo T. (1987b). Activated sludge model No 1. IAWPRC Scientific and Technical Report No. 1. IAWQ, London, UK.

Henze M., Grady J.R.C.P.L. and Gujer W. (1987a). A general model for single-sludge wastewater treatment systems. *Water Research*, 21(5), 505-515.

Henze M., Gujer W., Mino T. and Van Loosdrecht M.C.M. (2000). Activated sludge models ASM1, ASM2, ASM2d and ASM3. IWA Scientific and Technical Report No. 9. IWA Publishing, London, UK. ISBN: 1900222248.

Henze M., Gujer W., Mino T., Matsuo T., Wentzel M.C. and Marais G.V.R. (1995). Activated Sludge Model No. 2. IAWQ Scientific and Technical Report No. 3. IAWQ, London, UK.

Henze M., Gujer W., Mino T., Matsuo T., Wentzel M.C., Marais G.V.R. and Van Loosdrecht M.C.M. (1999). Activated Sludge Model No. 2d, ASM2d. *Water Science and Technology*, 39(1), 165–182.

Henze M., Van Loosdrecht M.C.M., Ekama G.A. and Brdjanovic D. (2008). Biological wastewater treatment. Principles, modelling and design. IWA Publishing, London, UK. ISBN: 1843391880.

Hodzic A., Vojinovic Z., Seyoum S.D., Pathirana A., Meijer S.C.F. and Brdjanovic D. (2011). Model-based evaluation of the urban wastewater infrastructure reconstruction options in a developing country: Case study Sarajevo in Bosnia and Herzegovina. In: Proceedings 2nd IWA Development Congress. 21-24 November 2011, Kuala Lumpur, Malaysia.

Hu Z.R., Wentzel M.C. and Ekama G.A. (2007). A general kinetic model for biological nutrient removal activated sludge systems: Model development. *Biotechnology and Bioengineering*, 98(6), 1242-1258.

Huisman, J.L. and Gujer W. (2002). Modelling wastewater transformation in sewers based on ASM3. *Water Science and Technology*, 45(6), 51-60.

Hulsbeek J.J.W., Kruit J., Roeleveld P.J. and Van Loosdrecht M.C.M. (2002). A practical protocol for dynamic modelling of activated sludge systems. *Water Science and Technology*, 45, 127-136.

Hvitved-Jacobsen T., Vollertsen J. and Nielsen A.H. (1998). A process and model concept for microbial wastewater transformations in gravity sewers, *Water Science and Technology*, 37(1), 233–241.

Hvitved-Jacobsen T., Vollertsen J. and Nielsen A.H. (2013). Sewer processes: microbial and chemical process engineering of sewer networks. CRC press. ISBN: 9781439881774.

Ikumi D.S. and Ekama G.A. (2019). Plantwide modelling–anaerobic digestion of waste sludge from parent nutrient (N and P) removal systems. *Water SA*, 45(3), 305-316.

Ikumi D.S., Harding T.H. and Ekama G.A. (2014a). Biodegradability of wastewater and activated sludge organics in anaerobic digestion. *Water Research*, 56, 267-279.

Ikumi D.S., Harding T.H., Vogts M., Lakay M.T., Mafungwa H., Brouckaert C.J. and Ekama G.A. (2014b) Mass balances modelling over wastewater treatment plants III. Final Report. Water Research Commission, South Africa.

Jenkins D. and Wanner J. (2014). Activated sludge – 100 years and counting. IWA Publishing, London, UK. ISBN: 9781780404936.

Jeppsson U. (1996). Modelling aspects of wastewater treatment processes. PhD thesis. Lund University of Technology, Lund, Sweden. ISBN: 91-88934-00-4.

Jeppsson U., Alex J., Batstone D., Benedetti L., Comas J., Copp J.B., Corominas L., Flores-Alsina X., Gernaey K.V., Nopens I., Pons M.-N., Rodriguez-Roda I., Rosen C., Steyer J.-P., Vanrolleghem P.A., Volcke E.I.P. and Vrecko D. (2013). Benchmark simulation models, quo vadis?. *Water Science and Technology*, 68(1), 1-15.

Kagawa Y., Tahata J., Kishida N., Matsumoto S., Picioreanu C., Van Loosdrecht M.C.M. and Tsuneda, S. (2015). Modeling the nutrient removal process in aerobic granular sludge system by coupling the reactor-and granule-scale models. *Biotechnology and bioengineering*, 112(1), 53-64.

Karpinska A. M. and Bridgeman J. (2016). CFD-aided modelling of activated sludge systems – A critical review. *Water Research*, 88, 861-879.

Kartal B., Kuenen J.G. and Van Loosdrecht M.C.M. (2010). Sewage treatment with anammox. *Science*, 328(5979), 702-703.

Krebs P. (1995). Succes and shortcomings of clarifier modelling. *Water Science and Technology*, 32(2), 181-191.

Kuba T., Murnleitner E., Van Loosdrecht M.C.M. and Heijnen J.J. (1996). A metabolic model for biological phosphorus removal by denitrifying organisms. *Biotechnology and Bioengineering*, 52(6), 685-695.

Kuba T., Van Loosdrecht M.C.M., Brandse F.A. and Heijnen J.J. (1997). Occurrence of denitrifying phosphorus removing bacteria in modified UCT-type wastewater treatment plants. *Water Research*, 31, 777–786.

Lackner S., Gilbert E.M., Vlaeminck S.E., Joss A., Horn H. and Van Loosdrecht M.C.M. (2014). Full-scale partial nitritation/anammox experiences – An application survey. *Water Research*, 55, 292-303.

Langeveld J., Nopens I., Schilperoort R., Benedetti L., De Klein J., Amerlinck J. and Weijers S. (2013). On data requirements for calibration of integrated models for urban water systems. *Water Science and Technology*, 68(3), 728-736.

Laurent J., Samstag R.W., Ducoste J.M., Griborio A., Nopens I., Batstone D.J., Wicks J.D., Saunders S. and Potier O. (2014). A protocol for the use of computational fluid dynamics as a supportive tool for wastewater treatment plant modelling. *Water Science and Technology*, 70(10), 1575-1584.

Le Moullec Y., Gentric C., Potier O. and Leclerc J.P. (2010). CFD simulation of the hydrodynamics and reactions in an activated sludge channel reactor of wastewater treatment. *Chemical Engineering Science*, 65(1), 492-498.

Le Moullec Y., Potier O., Gentric C. and Leclerc J.P. (2008). Flow field and residence time distribution simulation of a cross-flow gas–liquid wastewater treatment reactor using CFD. *Chemical Engineering Science*, 63(9), 2436-2449.

Liu R., Hao X., Chen Q. and Li J. (2019). Research advances of Tetrasphaera in enhanced biological phosphorus removal: A review. *Water Research*, 166, 115003.

Lopez-Vazquez C.M., Mithaiwala M., Moussa M.S., Van Loosdrecht M.C.M. and Brdjanovic D. (2013). Coupling ASM3 and ADM1 for wastewater treatment process characterization and biogas production in a developing country: case-study Surat, India. *J Water Sanitation Hygiene Development*, 3(1), 12-25.

Lopez-Vazquez C.M., Oehmen A., Hooijmans C.M., Brdjanovic D., Gijzen H.J., Yuan Z. and Van Loosdrecht M.C.M. (2009). Modelling the PAO–GAO competition: effects of carbon source, pH and temperature. *Water Research*, 43(2), 450-462.

Lu H., Wang J., Li S., Chen G.H., Van Loosdrecht M.C.M. and Ekama G.A. (2009). Steady-state model based evaluation of sulfate reduction, autotrophic denitrification and nitrification integrated (SANI) process. *Water Research*, 43(14), 3613-3621.

Makinia J., Rosenwinkel K. and Phan L. (2006). Modification of ASM3 for the determination of biomass adsorption/storage capacity in bulking sludge control. *Water Science and Technology*, 53(3), 91-99.

Mampaey K.E., Spérandio M., Van Loosdrecht M.C.M. and Volcke E.I. (2019). Dynamic simulation of N_2O emissions from a full-scale partial nitritation reactor. *Biochemical Engineering Journal*, 152, 107356.

Marais G.v.R. and Ekama G.A. (1976). The activated sludge process Part 1 – Steady-state behaviour. *Water SA*, 2(4), 164-200.

Maruejouls T., Vanrolleghem P.A., Pelletier G. and Lessard P. (2012). A phenomenological retention tank model using settling velocity distributions. *Water Research*, 46(20), 6857-6867.

Maurer M. and Boller M. (1999). Modeling of phosphorus precipitation in wastewater treatment plants with enhanced biological phosphorus removal. *Water Science and Technology*, 39(1), 147-163.

Maurer M., Abramovich D., Siegrist H. and Gujer W. (1999). Kinetics of biologically induced phosphorus precipitation in waste-water treatment. *Water Research*, 33, 484-493.

Mbamba C.K., Flores-Alsina X., Batstone D.J. and Tait S. (2016). Validation of a plant-wide phosphorus modelling approach with minerals precipitation in a full-scale WWTP. *Water Research*, 100, 169-183.

Mbamba C.K., Lindblom E., Flores-Alsina X., Tait S., Anderson S., Saagi R., Batstone D.J., Gernaey K.V., and Jeppsson U. (2019). Plant-wide model-based analysis of iron dosage strategies for chemical phosphorus removal in wastewater treatment systems. *Water Research*, 155, 12-25.

Meijer S.C.F. (2004). Theoretical and practical aspects of modelling activated sludge processes. PhD thesis. Delft University of Technology, Delft, The Netherlands. ISBN: 9090180273.

Meijer S.C.F. and Brdjanovic D. (2012). A practical guide to activated sludge modelling. Lecture notes. IHE Delft Institute of Water Education, Delft, The Netherlands. ISBN: 9073445264.

Meijer S.C.F., Van der Spoel H., Susanti S., Heijnen J.J. and Van Loosdrecht M.C.M. (2002). Error diagnostics and data reconciliation for activated sludge modelling using mass balances. *Water Science and Technology*, 25(6), 145-156.

Meijer S.C.F., Van Loosdrecht M.C.M. and Heijnen J.J. (2001). Metabolic modeling of full-scale biological nitrogen and phosphorus removing WWTPs. *Water Research*, 35, 2711-2723.

Meister M. and Rauch W. (2016). Wastewater treatment modelling with smoothed particle hydrodynamics. *Environmental Modelling & Software*, 75, 206-211.

Monod J. (1949). The growth of bacterial cultures. *Annual review of microbiology*, 3(1), 371-394.

Murnleitner E., Kuba T., Van Loosdrecht M.C.M. and Heijnen J.J. (1997). An integrated metabolic model for the aerobic and denitrifying biological phosphorus removal. *Biotechnology and Bioengineering*, 54(5), 434-450.

Ni B.-J., Yuan Z., Chandran K., Vanrolleghem P.A. and Murthy S. (2013). Evaluating four mathematical models for nitrous oxide production by autotrophic ammonia-oxidizing bacteria. *Biotechnology and Bioengineering*, 110(1), 153–163.

Nielsen P.H., Mielczarek A.T., Kragelund C., Nielsen J.L., Saunders A.M., Kong Y., Hansen A.A. and Vollertsen J. (2010). A conceptual ecosystem model of microbial communities in enhanced biological phosphorus removal plants. *Water Research*, 44(17), 5070-5088.

Nopens I., Batstone D.J., Copp J.B., Jeppsson U., Volcke E., Alex J. and Vanrolleghem P.A. (2009). An ASM/ADM model interface for dynamic plant-wide simulation. *Water Research*, 43(7), 1913-1923.

Nopens I., Batstone D.J., Griborio A., Samstag R., Wicklein E. and Wicks J. (2012). Computational Fluid Dynamics (CFD): What is Good CFD-Modeling Practice and What Can Be the Added Value of CFD Models to WWTP Modeling?. Proceedings of the Water Environment Federation, 2012(7), 7400-7405.

Nopens I., Arnaldos M., Belia E., Jeppsson U., Kinnear D., Lessard P., Murthy S., O'Shaughnessy M., Rieger L., Vanrolleghem P. and Weijers S. (2014). Maximising the benefits of activated sludge modelling. *Water 21*, October 2014, 31-33.

Oehmen A., Carvalho G., Lopez-Vazquez C.M., Van Loosdrecht M.C.M. and Reis M.A.M. (2010). Incorporating microbial ecology into the metabolic modelling of polyphosphate accumulating organisms and glycogen accumulating organisms. *Water Research*, 44(17), 4992-5004.

Oehmen A., Lemos P.C., Carvalho G., Yuan Z., Keller J., Blackall L.L. and Reis M.A.M. (2007). Advances in enhanced biological phosphorus removal: from micro to macro scale. *Water Research*, 41(11), 2271-2300.

Onnis-Hayden A., Srinivasan V., Tooker N.B., Li G., Wang D., Barnard J.L., Bott C., Dombrowski P., Schauer P., Menniti A., Shaw A., Stinson B., Stevens G., Dunlap P., Takács I., McQuarrie I., Phillips H., Lambrecht A., Analla H., Russell A. and Gu A.Z. (2019). Survey of full-scale sidestream enhanced biological phosphorus removal (S2EBPR) systems and comparison with conventional EBPRs in North America: Process stability, kinetics, and microbial populations. *Water Environmental Research*, 1-15.

Petersen B., Vanrolleghem P.A., Gernaey K. and Henze M. (2002). Evaluation of an ASM1 model calibration procedure on a municipal-industrial wastewater treatment plant. *J Hydroinformatics*, 4, 15-38.

Petersen E.E. (1965). Chemical reaction analysis. Prentice-Hall, Englegood Cliffs, NJ, USA. ISBN: 0131287281.

Picioreanu C., Kreft J.U. and Van Loosdrecht M.C.M. (2004). Particle-based multidimensional multispecies biofilm model. *Applied and Environmental Microbiology,* 70(5), 3024–3040.

Plósz B.G., Nopens I., Rieger L., Griborio A., De Clercq J., Vanrolleghem P.A., Daigger G.T., Takács I., Wicks J. and Ekama G.A. (2012). A critical review on clarifier modelling: state-of-the-art and characterization practices. In: Proceedings 3rd IWA/WEF Wastewater Modelling Seminar. 26-28 February 2012, Mont-Sainte-Anne, Québec, Canada.

Price R.K. and Vojinovic Z. (2010). Urban hydroinformatics: data, models and decision support for integrated urban water management. IWA Publishing. London, UK. ISBN: 1780401361.

Radu A.I., Vrouwenvelder J.S., Van Loosdrecht M.C.M. and Picioreanu C. (2010). Modelling the effect of biofilm formation on reverse osmosis performance: Flux, feed channel pressure drop and solute passage. *J Membrane Science*, 365(1), 1-15.

Rahaman M.S. and Mavinic D.S. (2009). Recovering nutrients from wastewater treatment plants through struvite crystallization: CFD modelling of the hydrodynamics of UBC MAP fluidized-bed crystallizer. *Water Science and Technology*, 59(10), 1887-1892.

Rauch W., Bertrand-Krajewski J.L., Krebs P., Mar O., Schilling W., Schutze M. and Vanrolleghem P.A. (2002). Deterministic modelling of integrated urban drainage systems. *Water Science and Technology*, 45(3), 81-94.

Regmi P., Stewart H., Amerlinck Y., Arnell M., García P.J., Johnson B., Maere T., Miletić I., Miller M., Rieger L., Samstag R., Santoro D., Schraa O., Snowling S., Takács I., Torfs E., Van Loosdrecht M. C. M., Vanrolleghem P.A., Villez K., Volcke E. I. P., Weijers S., Grau P., Jimenez J. and Rosso D. (2019). The future of WRRF modelling–outlook and challenges. *Water Science and Technology*, 79(1), 3-14.

Rehman U., Audenaert W., Amerlinck Y., Maere T., Arnaldos M. and Nopens I. (2017). How well-mixed is well mixed? Hydrodynamic-biokinetic model integration in an aerated tank of a full-scale water resource recovery facility. *Water Science and Technology*, 76(8), 1950-1965.

Rieger L., Gillot S., Langergraber G., Ohtsuki T., Shaw A., Takács I. and Winkler S. (2012). Guidelines for using activated sludge models. IWA Task Group on Good Modelling Practice. Scientific and Technical Report No. 22. IWA publishing, London, UK. ISBN: 9781843391746.

Rieger L., Koch G., Kühni M., Gujer W. and Siegrist H. (2001). The EAWAG Bio-P module for activated sludge model No. 3. *Water Research*, 35(16), 3887–3903.

Roeleveld P.J. and Van Loosdrecht M.C.M. (2002). Experience with guidelines for wastewater characterization in the Netherlands. *Water Science and Technology*, 45, 77-87.

Rubio-Rincón F.J., Welles L., Lopez-Vazquez C.M., Nierychlo M., Abbas B., Geleijnse M., Nielsen P.H., Van Loosdrecht M.C.M. and Brdjanovic D. (2017). Long-term effects of sulphide on the enhanced biological removal of phosphorus: the symbiotic role of Thiothrix caldifontis. *Water Research*, 116, 53-64.

Sahlstedt K., Aurola A.M. and Fred T. (2004). Practical modelling of a large activated sludge DN-process with ASM3. *Water Intelligence Online*. http://test.iwaponline.com/wio/2004/08/pdf/wio200408017.pdf

Samstag R.W., Ducoste J.J., Griborio A., Nopens I., Batstone D.J., Wicks J.D., Saunders S., Wicklein E.A., Kenny G. and Laurent J. (2016). CFD for wastewater treatment: an overview. *Water Science and Technology*, 74(3), 549-563.

Sanchez A., Medina N., Vojinovic Z. and Price R. (2013). An integrated cellular automata evolutionary-based approach for evaluating future scenarios and the expansion of urban drainage networks. *J Hydroinformatics*, 16(2), 319–340.

Satoh H., Mino T. and Matsuo T. (1994). Deterioration of enhanced biological phosphorus removal by the domination of microorganisms without polyphosphate accumulation. *Water Science and Technology*, 30(6), 203-211.

Saunders A.M., Oehmen A., Blackall L.L., Yuan Z. and Keller J. (2003). The effect of GAO (glycogen accumulating organisms) on anaerobic carbon requirements in full-scale Australian EBPR (enhanced biological phosphorus removal) plants. *Water Science and Technology*, 47(11), 37-43.

Sharma K.R., Yuan Z., De Haas D., Hamilton G., Corrie S. and Keller J. (2008). Dynamics and dynamic modelling of H_2S production in sewer systems. *Water Research*, 42 (10–11), 2527–2538.

Sin G. and Vanrolleghem P.A. (2006). Evolution of an ASM 2 d-like model structure due to operational changes of an SBR process. *Water Science and Technology*, 53(12), 237-245.

Smolders G.J.F., Bulstra D.J., Jacobs R., Van Loosdrecht M.C.M. and Heijnen J.J. (1995c). A metabolic model of the biological phosphorus removal process: II. Validation during start-up conditions. *Biotechnology and Bioengineering*, 48(3), 234-245.

Smolders G.L.F., Klop J.M., Van Loosdrecht M.C.M. and Heijnen J.J. (1995b). A metabolic model for the biological phosphorus removal process: Effect of the sludge retention time. *Biotechnology and Bioengineering*, 48, 222-233.

Smolders G.L.F., Van der Meij J., Van Loosdrecht M.C.M. and Heijnen J.J. (1994b). Stoichiometric model of the aerobic metabolism of the biological phosphorus removal process. *Biotechnology and Bioengineering*, 44(7), 837-848.

Smolders G.L.F., Van der Meij J., Van Loosdrecht M.C.M. and Heijnen J.J. (1995a). Structured metabolic model for anaerobic and aerobic stoichiometry and kinetics of the biological phosphorus removal process. *Biotechnology and Bioengineering*, 47, 277-287.

Smolders G.L.F., Van der Meij J., Van Loosdrecht M.C.M. and Heijnen J.J. (1994a). Model of the anaerobic metabolism of the biological phosphorous removal process: stoichiometry and pH influence. *Biotechnology and Bioengineering*, 43(6), 461-470.

Smolders GJF (1995). A metabolic model of the biological phosphorus removal: stoichiometry, kinetics and dynamic behaviour. PhD dissertation. Delft University of Technology. Delft, the Netherlands.

Solon K., Flores-Alsina X., Mbamba C. K., Ikumi D., Volcke E. I. P., Vaneeckhaute C., Vanrolleghem P.A., Batstone D.J., Gernaey K.V. and Jeppsson U. (2017). Plant-wide modelling of phosphorus transformations in wastewater treatment systems:

impacts of control and operational strategies. *Water Research*, 113, 97-110.

Strande L., Ronteltap M. and Brdjanovic D. (2014) Faecal Sludge Management. IWA Publishing, London, UK.

Takács I., Vanrolleghem P., Wett B. and Murthy S. (2007). Elemental balance-based methodology to establish reaction stoichiometry in environmental modelling. *Water Science and Technology*, 56(9), 37-41.

Torfs E., Martí M.C., Locatelli F., Balemans S., Bürger R., Diehl S., Laurent J., Vanrolleghem P.A., François P. and Nopens, I. (2016). Concentration-driven models revisited: towards a unified framework to model settling tanks in water resource recovery facilities. *Water Science and Technology*, 75(3), 539-551.

Torfs E., Vlasschaert P., Amerlinck Y., Bürger R., Diehl S., Farås S. and Nopens I. (2013). Towards improved 1-D settler modelling: calibration of the Bürger model and case study. In: Proceedings 86th Annual Water Environment Federation Technical Exhibition and Conference (WEFTEC 2013). Water Environment Federation (WEF). 5-9 October 2013, Chicago, IL, USA.

Van Haandel A.C., Ekama G.A. and Marais G.v.R. (1981). The activated sludge process Part 3 - Single sludge denitrification. *Water Research*, 15(10) 1135-1152.

Van Loosdrecht M. C. M., Lopez-Vazquez C. M., Meijer S. C. F., Hooijmans C. M., and Brdjanovic, D. (2015). Twenty-five years of ASM1: past, present and future of wastewater treatment modelling. *Journal of Hydroinformatics*, 17(5), 697-718.

Van Loosdrecht M.C.M. and Brdjanovic D. (2014). Anticipating the next century of wastewater treatment. *Science*, 344(6191), 1452-1453.

Van Loosdrecht M.C.M., Hooijmans C.M., Brdjanovic D. and Heijnen J.J. (1997). Biological phosphorus removal processes: a mini review. *Applied Microbiology and Biotechnology*, 48, 289-296.

Van Veldhuizen H.M., Van Loosdrecht M.C.M. and Heijnen J.J. (1999). Modelling biological phosphorus and nitrogen removal in a full scale activated sludge process. *Water Research*, 33(16), 3459–3468.

Vanhooren H., Meirlaen .J, Amerlinck Y., Claeys F., Vangheluwe H. and Vanrolleghem P.A. (2003). WEST: modelling biological wastewater treatment. *J Hydroinformatics*, 5(1), 27-50.

Vanrolleghem P., Flores-Alsina X., Guo L., Solon K., Ikumi D.S., Batstone D., Brouckaert C., Takács I., Grau P., Ekama G.A., Jeppsson U. and Gernaey K. (2014). Towards BSM2-GPS-X: A plant-wide benchmark simulation model not only for carbon and nitrogen, but also for greenhouse gases (G), phosphorus (P), sulphur (S) and micropollutants (X), all within the fence of WWTPs/WRRFs. In: Proceedings 4th IWA/WEF Wastewater Treatment Modelling Seminar (WWTmod2014). 30 March-2 April 2014, Spa, Belgium.

Vanrolleghem P.A., Rosen C., Zaher U., Copp J., Benedetti L., Ayesa E. and Jeppsson U. (2005). Continuity-based interfacing of models for wastewater systems described by Petersen matrices. *Water Science and Technology*, 52(1-2), 493-500.

Vanrolleghem P.A., Spanjers H., Petersen B., Ginestet P. and Takács I. (1999). Estimating (combinations of) Activated Sludge Model No. 1 parameters and components by respirometry. *Water Science and Technology*, 39(1), 195-214.

Varga E., Hauduc H., Barnard J., Dunlap P., Jimenez J., Menniti A., Schauer P., Lopez Vazquez C.M., Gu A.Z., Sperandio M. and Takács I. (2018). Recent advances in bio-P modelling – a new approach verified by full-scale observations. *Water Science and Technology*, 78 (10), 2119–2130.

Velkushanova K., Strande L., Ronteltap M., Kottatep T., Brdjanovic D. and Buckley C. (2020). Methods for Faecal Sludge Analysis. IWA Publishing, London, UK.

Vojinovic Z. and Abbott M.B. (2011). Flood risk and social justice: from hydroinformatics of quantities to hydroinformatics of qualities. IWA Publishing, London, UK. ISBN: 1843393875.

Vojinovic Z., Sahlu S., Torres A.S., Seyoum S.D., Anvarifar F., Matungulu H., Barreto W., Savic D. and Kapelan Z. (2014). Multi-objective rehabilitation of urban drainage systems under uncertainties. *J. Hydroinformatics*, 16(5), 1044–1061.

Volcke E.I., Van Loosdrecht M.C.M. and Vanrolleghem P.A. (2006). Continuity-based model interfacing for plant-wide simulation: A general approach. *Water Research*, 40(15), 2817-2828.

Von Sperling M. and Chernicharo C.A.L. (2005). Biological wastewater treatment in warm climate regions. IWA Publishing, London, UK. ISBN: 1 84339 002 7.

Wang J., Lu H., Chen G.H., Lau G.N., Tsang W.L. and Van Loosdrecht M.C.M. (2009). A novel sulfate reduction, autotrophic denitrification, nitrification integrated (SANI) process for saline wastewater treatment. *Water Research*, 43(9), 2363-2372.

Warner A.P.C., Ekama G.A. and Marais G.v.R. (1986). The activated sludge process Part 4 - Application of the general kinetic model to anoxic-aerobic digestion of waste activated sludge. *Water Research*, 20(8) 943-958.

Wentzel M. C., and Ekama G. A. (1997). Principles in the modelling of biological wastewater treatment plants. Microbial Community Analysis: The Key to the Design of Biological Wastewater Treatment Systems, *IWA Scientific and Technical Report No. 5*, 73-82.

Wentzel M.C., Dold P.L., Ekama G.A. and Marais G.v.R. (1989b). Enhanced polyphosphate organism cultures in activated sludge systems. Part III: Kinetic model. *Water SA*, 15(2), 89-102.

Wentzel M.C., Ekama G.A., Loewenthal R.E., Dold P.L. and Marais G.v.R. (1989a). Enhanced polyphosphate organism cultures in activated sludge systems. Part II: Experimental behaviour. *Water SA*, 15(2), 71-88.

Wentzel M.C., Loewenthal R.E., Ekama G.A. and Marais G.v.R. (1988). Enhanced polyphosphate organism cultures in activated sludge systems. Part I: Enhanced culture development. *Water SA*, 14(2), 81-92.

Wett B., Omari A., Podmirseg S.M., Han M., Akintayo O., Gómez Brandón M., Murthy S., Bott C., Hell M., Takács I., Nyhuis G. and O'Shaughnesy M. (2013). Going for mainstream deammonification from bench-to full-scale for maximized resource efficiency. *Water Science and Technology*, 68(2), 283-289.

Wicklein E., Batstone D.J., Ducoste J., Laurent J., Griborio A., Wicks J., Saunders S., Samstag R., Potier O. and Nopens I. (2016). Good modelling practice in applying computational fluid dynamics for WWTP modelling. *Water Science and Technology*, 73(5), 969-982.

Wild D., Schulthess R.V. and Gujer W. (1994). Synthesis of denitrification enzymes in activated sludge; modelling with structured biomass. *Water Science and Technology*, 30(6), 113-122.

Wyffels S., Boeckx P., Pynaert K., Zhang D., Van Cleemput O., Chen G. and Verstraete, W. (2004). Nitrogen removal from sludge reject water by a two-stage oxygen-limited autotrophic nitrification denitrification process. *Water Science and Technology*, 49(5-6), 57-64.

Yuan Z., Oehmen A. and Ingildsen P. (2002). Control of nitrate recirculation flow in predenitrification systems. *Water Science and Technology*, 45(4), 29-36.

Zakaria F., Garcia H.A., Hooijmans C.M. and Brdjanovic D. (2015). Decision support system for the provision of emergency sanitation. *Science Total Environment*, 512-513, 645-658.

Websites of model-relevant task groups from the International Water Association

- Task Group on Green House Gas (GHG):
  http://www.iwa-network.org/task/task-group-on-green-house-gas
- Task Group on Good Modelling Practices (GMP) - Guidelines for use of activated sludge models:
  http://www.iwa-network.org/task/good-modelling-practice-gmp-guidelines-for-use-of-activated-sludge-models
- Task Group on Benchmarking of Control Strategies for Wastewater Treatment Plants (BSM):
  http://www.iwa-network.org/task/benchmarking-of-control-strategies-for-wastewater-treatment-plants
  http://www.benchmarkwwtp.org/
- Task Group on Generalized Physicochemical Framework (GPCF):
  http://www.iwa-network.org/task/generalized-physicochemical-framework

References for software simulators (websites)

- AQUASIM: http://www.eawag.ch/en/department/siam/software
- ASIM: http://www.asim.eawag.ch/
- BioWin: https://envirosim.com/getbw
- GPS-X: http://www.hydromantis.com/GPS-X.html
- HEC-CRAS: http://www.hec.usace.army.mil/software/hec-ras/
- Mike Urban: http://www.mikebydhi.com/products/mike-urban
- STOAT: http://www.wrcplc.co.uk/software-development
- SUMO: http://www.dynamita.com/
- WEST: http://www.mikebydhi.com/products/

NOMENCLATURE

Symbol[1)]	Symbol[2)]	Description	Unit
A	A	Surface area	m^2
b_A	b_{ANO}	Specific rate of endogenous mass loss of nitrifying organisms	1/d
b_H	b_{OHO}	Specific rate of endogenous mass loss of ordinary heterotrophic organisms (OHOs)	1/d
F/M	F/M	Food to microorganism ratio or load factor (LF)	gCOD/gVSS.d
f_H	$f_{XE,OHO}$	Unbiodegradable fraction of OHOs	mgCOD/mgCOD
f_N	f_n	Nitrogen content of VSS	mgN/mgVSS
FSA	FSA	Free and saline ammonia	mgN/l
K	K	Half saturation constant	-
k_H	K_h	Maximum specific hydrolysis rate of SBCOD by OHOs under aerobic conditions	mgCOD/mgCOD.d
K_I		Half saturation constant for inhibition compound	mg/l
K_l		External transfer coefficient	m/h
K_N	K_{NHx}	Half saturation constant for growth of organisms with nitrogen (FSA)	mgN/l
$K_{N,A}$	K_{ANO}	Half saturation constant for growth of nitrifiers with nitrogen (FSA)	mgN/l
$K_{N,H}$	$K_{OHO,NHx}$	Half saturation constant for growth of OHOs with nitrogen (FSA)	mgN/l
K_O	KO	Half saturation constant for dissolved oxygen	mgO_2/l
$K_{O,A}$	$K_{ANO,O2}$	Half saturation constant for nitrifiers for dissolved oxygen	mgO_2/l
$K_{O,H}$	$K_{OHO,O2}$	Half saturation constant for OHOs for dissolved oxygen	mgO_2/l
K_S	KS	Half saturation concentration for soluble organics utilization	mgCOD/l
K_x	Kx	Half saturation concentration for utilization of SBCOD by OHOs	mgCOD/mgCOD.d
q		Specific conversion rate	l/h
Q_{in}	Q_i	Influent flow rate	m^3/h
Q_{out}	Q_e	Effluent flow rate	m^3/h
r_i		Observed transformation rate for process i	$ML^{-3}T^{-1}$
S	S	Soluble concentration in bulk liquid	mgCOD/l
S_{HCO}		Bicarbonate concentration	mg/l
S_I	S_U	Soluble unbiodegradable COD concentration	mgCOD/l
S_{in}	S_i	Influent substrate concentration	mgCOD/l
S_{KI}		Inhibition compound concentration	mg/l
S_{max}		Saturation concentration	$gCOD/m^3$
S_N		Nitrogen concentration (ammonia or nitrate)	mgN/l
S_{NH}	S_{NHx}	Free and saline ammonia concentration	mgFSA-N/l
S_{NO}	S_{NO3}	Nitrate concentration	$mgNO_3$-N/l
S_O	S_{O2}	Dissolved oxygen concentration	mgO_2/l
S_{out}	S_e	Effluent substrate concentration	mgCOD/l
S_S	S_S	Soluble readily biodegradable (RB)COD concentration	mgCOD/l
t	t	Time	h
V	V_R	Reactor volume	m^3
$v_{j,i}$		General stoichiometry term in model matrix for component i in process j	
X	X	Biomass concentration	$gCOD/m^3$

X_A	X_{ANO}	Nitrifier biomass concentration	mgCOD/l
X_H	X_{OHO}	Ordinary heterotrophic (OHO) biomass concentration	mgCOD/l
X_I	XI	Unbiodegradable particulate organics from influent wastewater	mgCOD/l
X_S/X_H	XS/X_{OHO}	SBCOD/OHO concentration ratio	mgCOD/mgCOD
X_S	XS	Slowly biodegradable (SB)COD concentration	mgCOD/l
$X_{STO,S}$		Intra-cellularly stored organic concentration	mgCOD/l
X_{TSS}	X_{TSS}	TSS concentration in reactor	mgTSS/l
Y_H	Y_{OHO}	Yield of OHOs	mg COD/mgCOD

[1] Symbols used in this chapter are the original symbols reported as such in the source literature. Consequently, no attempt has been made by the authors of this chapter to align the symbols with the symbols used in chapters 3, 4, 5, 6, 7 and 17. [2] Instead, an additional column with relevant symbols from these chapters was added to this table, for comparison.

Abbreviation	Description
ADM	Anaerobic digestion model
ASM	Activated sludge model
BOD	Biological oxygen demand
COD	Chemical oxygen demand
CSTR	Complete stirred tank reactor
DO	Dissolved oxygen
DR	Death regeneration
EA	Electron acceptor
EBPR	Enhanced biological phosphorus removal
ER	Endogenous respiration
GAO	Glycogen accumulating organism
IWA	International Water Association
OUR	Oxygen utilization rate
OHO	Ordinary heterotrophic organism
PAO	Phosphorus accumulating organism
PHA	Polyhydroxyalkanoate
RBCOD	Readily biodegradable COD
SBCOD	Slowly biodegradable COD
SRT	Sludge retention time
TKN	Total Kjeldahl nitrogen
TSS	Total suspended solids
TUDP	Delft University of Technology EBPR model
VFA	Volatile fatty acid

Greek symbol[1)]	Greek symbol[2)]	Description	
α		Represents a stoichiometric formula	-
η		Reduction factor for utilization of SBCOD under anoxic conditions	-
μ	μ	Specific growth rate of organisms	1/d
μ_A^{max}	$\mu_{ANO,max}$	Maximum specific growth rate of nitrifiers	1/d
μ_H	μ_{OHO}	Specific growth rate of OHOs	1/d
μ_H^{max}	$\mu_{OHO,max}$	Maximum specific growth rate of OHOs	1/d
μ^{max}	μ_{max}	Maximum specific growth rate of organisms	1/d
ρ_j		Kinetic rate of process j	$ML^{-3}T^{-1}$

[1)] Greek symbols used in this chapter are the original symbols reported as such in the source literature. No attempt has been made by the authors of this chapter to align the symbols with the symbols used in chapters 3, 4, 5, 6, 7 and 17. [2)] Instead, an additional column with relevant symbols from these chapters was added to this table, for comparison.

doi: 10.2166/9781789060362_0667

15

Process control

Gustaf Olsson and Pernille Ingildsen

15.1 DRIVING FORCES AND MOTIVATIONS FOR CONTROL

Several driving forces motivate the control of wastewater treatment systems. Some of them are *demand pull* forces and others are *technology push* forces:

- Saving capital costs by optimising operations to reduce the need for additional capacity or to apply more load to a given system.
- Minimizing operational costs for energy, chemicals and sludge management, despite load changes and other disturbances.
- Keeping consistent operation to maintain effluent water quality.
- Understanding fluctuating processes, such as aeration, sludge thickening, and sludge generation.
- Detecting disturbances and giving early warning to counteract sudden events such as rain or toxicity.
- Gaining transparency into what is happening in the plant for trouble-shooting and post-analysis on detrimental events.

None of these aspects were fully understood when instrumentation, control and automation (ICA) attracted the attention of the water and wastewater industry in the early 1970s (Olsson *et al.*, 1973). Since then the technical development of new processes, sensor and instrumentation technology, computer performance, communication technology and the Internet of Things, detection methods, control theory and artificial intelligence has made ICA widespread in all kinds of water operations (Olsson, 2012). The emphasis is not only on more efficient wastewater treatment but on many aspects of early warning systems, plant monitoring, and operator guidance. There has also been major development towards understanding customer expectations, data-collection tools and the handling of 'big' data.

Too often ICA is implemented to improve efficiency or reduce costs but only as a second step for existing plants. Instead, the coupling of design and operation, known as control-integrated design, needs to be improved. Inflexible or under-dimensioned designs cannot be improved by control.

15.1.1 ICA system features

The ideal ICA is founded on some fundamental building blocks:

- The team (people-ware),
- Hardware (sensors, actuators),
- Communication-ware (information systems, hardware, communication protocols etc.),
- Software (the usual elements: PLC, SCADA, displays, and control algorithms).

Any water operation can be symbolized by three dimensions. Mass flows form one dimension: each component can be tracked by its mass balance in the system. Energy, the second dimension, describes the flow of electrical, chemical, and thermal energy. Energy is not only consumed, but the incoming water contains thermal and chemical energy that can be recovered as heat, biogas or nutrients. Energy balances in the wastewater system, in particular the possibility of the 'zero-energy' plant, are receiving increasing attention.

The ICA perspective represents the third dimension, the information flow. The fundamental concept of feedback can form the basis of the understanding of smart systems. Ingildsen and Olsson (2016) called it the 'MAD' approach: measure (M), analyse (A) and decide (D). This feedback concept can be practised at all levels, from simple component control to management and strategic decisions. The principle is illustrated in Figure 15.1.

To measure is to know: we need adequate data in space and in time. Measurement – the 'M' - includes not only signals from sensors and instruments, but also laboratory analysis results, observations and human communications. It comprises daily and monthly management reports and similar information that will form the basis for long-term decisions. Data must be checked, analysed, understood and interpreted – the 'A' – in order to become reliable information. The decision or control – the 'D' – does not need to be automatic as in a typical feedback controller. It can be manual in terms of decisions about plant operation or about more business-oriented operations. Finally, any decision will be turned into action using an 'actuator'.

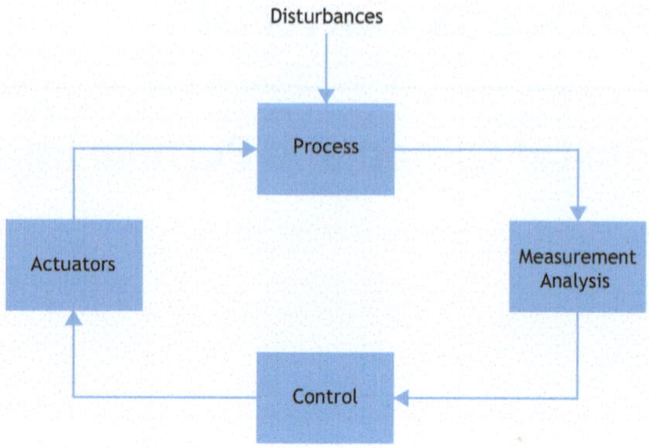

Figure 15.1 The feedback principle.

The word *smart* is used nowadays as a characteristic of devices, people and phenomena. It may mean clever, sharp or intelligent. Sometimes it is a synonym for elegant, stylish or neat. However, a smart water system is not the same as an elegant, glamorous or fashionable system. Instead it operates to fulfil the requirements of producing an acceptable product while keeping the energy and resource requirements at a minimum. Such a system should respond quickly and adequately to disturbances and recover rapidly after a major upset. The operation of the system must be sufficiently transparent so that people involved in the operation have enough information to make rational decisions. The goal is to integrate the complete urban water cycle: water supply, water distribution, urban drainage, wastewater treatment as well as the customer or demand side. The aim of our operations should be to adapt to, protect and preserve the natural water systems we depend on and operate within. Ultimately, our human water consumption should have no detrimental effects on nature.

A water operation can only be smart if people in all positions - not only the utility employees - get involved, so that they can see their respective role in the whole plant or system operation. It is crucial that the water users are engaged as well. For too long water utilities and operations have been invisible to the users, who take the perfect delivery of water and treatment of wastewater for granted. It should be recognized that most of the energy related to the water cycle is used in the home, not within utility operations. Therefore, passing information to the user is crucial. If customers can measure their water consumption and obtain information on how it relates to neighbours or what is 'normal', it can make them more aware of ways to be less wasteful.

The focus of control in most water systems today is on unit processes. In addition, the control of combined sewer overflow (CSO) for the benefit of receiving water quality has also been receiving increasing attention (Ahm *et al.*, 2016). CSO overflow structures are not primarily designed for measurement but to effectively discharge excess water. By using models and estimation it is possible to quantify the CSO discharge so that adequate mass balance calculations can be accomplished. CSO flow rates can be derived by using what are termed soft sensors, where the flow rates are estimated based on level measurements in the system.

Widening the perspective from single processes to a plant-wide view and further on to the entire urban water cycle is necessary in order to handle the increasing complexity of urban water systems. This transition requires system-thinking, where the multitude of couplings between processes and individual controllers are considered (Beck, 2005). Rodriguez-Roda *et al.* (2002) introduced a new way of thinking by describing the development and implementation of a knowledge-based supervisory system to operate a wastewater treatment plant. Their system structured the operation into data gathering, diagnosis and decision support and integrated the control by agent-based reasoning modules. Integrated control is further described in Section 15.10.

15.1.2 Driving forces

There are several driving forces to implement ICA in water systems. Some are demand pulls and others are technology pushes, as described in some detail in Olsson *et al.* (2014).

Key driving forces are more stringent regulatory requirements, financial considerations, and implementation efficiency. Others are growing urbanization and population which imply increasing water use and wastewater load. Climate change and associated extreme weather conditions further add to the operational challenges of urban water systems. However, although water quality is certainly a driver in plant design, it is not typically the main driver for ICA and is unfortunately not usually integrated with operational requirements. Also, despite the need for ICA being recognized decades ago (Vanrolleghem *et al.*, 1996a), it has only been implemented as a second step for existing plants to improve efficiency or reduce costs, even though inflexible or under-dimensioned designs cannot be improved by control.

Process development has become increasingly sophisticated due to effluent requirements, and this has added to the demand for ICA. Bioreactors have more zones; anaerobic, anoxic and aerobic. More recirculation flows can be subject to control. Air supply systems are more sophisticated whereby aeration zones can be controlled separately, pressure losses can be minimized by variable pressure control, and compressors are supplied with variable speed control. Chemicals and external carbon can be added for control purposes.

The importance of actuators should also not be overlooked. In the last few decades there has been a revolution in the development of power electronics. Power electronic devices such as IBGT (Insulated-Gate Bipolar Transistors) are now generally available for currents up to 1,200 A and voltages up to 3,000 V with switching frequencies of more than 1 MHz. This makes frequency control of electric motors both affordable and reliable, from mW scale to MW drives. Variable speed control has a large influence on wastewater treatment operations for flow-rate control as well as for air-flow control. This has a profound impact on both control action quality and on energy efficiency of various operations. All the major electric motor and pump vendors provide a large quantity of information on the internet. There are several short lectures available on www.youtube.com. Search for 'electric drive systems'.

Energy is the single largest operating expense in water operations, so it makes economic sense to reduce these costs where possible through adequate control. The vision of zero or even positive-energy plants has already been realized in some cases (for example, Nowak *et al.,* 2011).

There has been a substantial technology push influencing ICA in water systems. Instrumentation and actuator development has been significant, contributing to making control more successful. Computing power is no longer any obstacle. The real challenge is to convert big data into useful information. The education level of operators and process engineers has been raised considerably over recent decades.

15.1.3 Outline of the chapter

The rest of the chapter is outlined as follows. Disturbances in wastewater treatment systems are significant and they are the reason for control, as explained in Section 15.2. The role of control is further described in Section 15.3. Instrumentation is the basis for advanced operation and its role in monitoring and control is discussed in Section 15.4. Wastewater systems are dynamic systems, implying that any correction needs time to be noticed in the system, as described in Section 15.5. To manipulate any system we need actuators, to translate decisions into mechanical actions, such as motors, pumps, compressors and valves, which is covered in Section 15.6. The following two sections, 15.7 and 15.8, are devoted to basic principles of control and some typical applications in wastewater treatment. Energy is closely related to water and wastewater treatment, and energy and other operating costs are discussed in Section 15.9. A wastewater treatment plant consists of many unit processes and their interaction has to be taken into consideration for more advanced control, which is in Section 15.10.

For the interested reader a comprehensive description of control in wastewater treatment systems is available in the textbook by Olsson and Newell (1999). An updated state-of-the-art description of control issues in wastewater systems is found in Olsson *et al.* (2005). In the more recent book by Ingildsen and Olsson (2016) the feedback principle is demonstrated not only for biological wastewater treatment but for all levels of water operation. In this chapter we limit the discussion to control systems in biological wastewater treatment.

15.2 DISTURBANCES IN WASTEWATER TREATMENT SYSTEMS

One of the incentives for control is the appearance of disturbances either about to enter or already within a plant. The impact of these disturbances should be compensated, but it is even better if the disturbances can be attenuated or best of all eliminated before they hit the plant. Compared to most other process industries, there are many external disturbances that a wastewater treatment plant can be subject to. The wastewater influent typically varies substantially both in terms of concentration of pollutants, composition and flow rate, with timescales ranging from hours to months. Discrete events such as rainstorms, toxic spills and peak loads also occur from time to time. Some disturbances are created within the plant due to process design, poor operation, inadequate equipment or process failures. As a result, a treatment plant is hardly ever in steady state but is subject to transient behaviour all the time.

Consistent performance must be maintained in the presence of these disturbances. The traditional method of reducing their effect has been to design plants with large volumes to attenuate large load disturbances. However, this solution incurs large capital costs. Online control systems, which have been demonstrated to cope well with most of these variations, are much more cost-effective and thus an attractive alternative. Eliminating disturbances is indeed one of the major incentives for introducing online process control to wastewater treatment systems.

Many disturbances are related to the plant influent flow. The influent changes both in terms of flow rate, concentration and composition, as illustrated in Figure 15.2. Any of these changes must be compensated for. If the result of the disturbance is measured within the plant, such as a change in the dissolved oxygen level, a rising sludge blanket, or a varying suspended solids concentration, the measured information is fed back to a controller that will activate a pump, a valve, or a compressor, so that the influence on the plant behaviour is minimized.

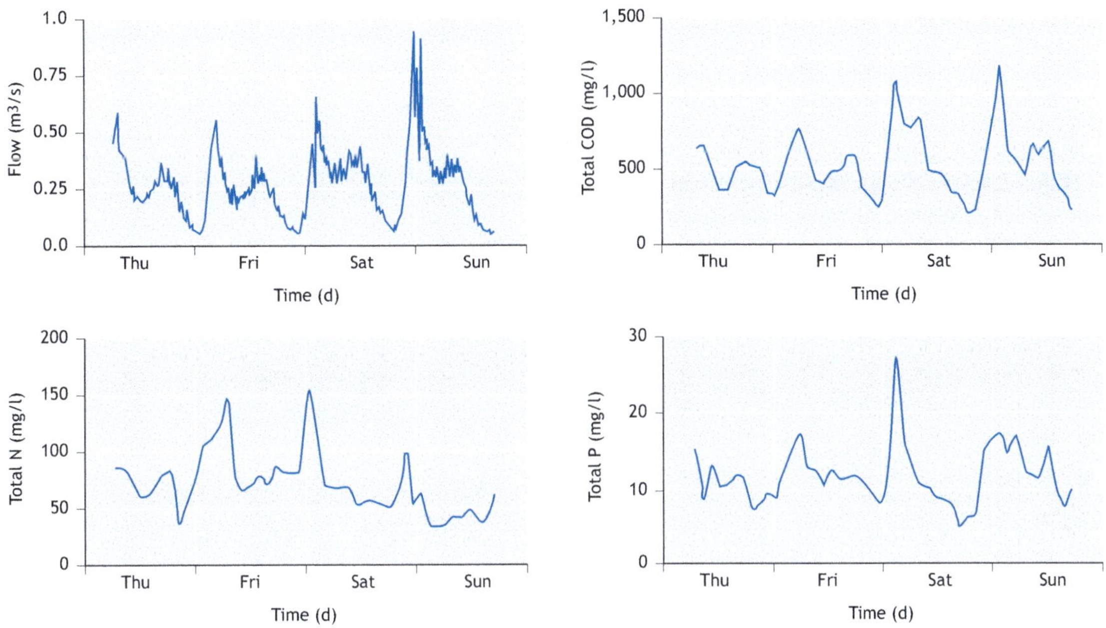

Figure 15.2 Typical dry weather diurnal variations in a municipality with mostly domestic wastewater. The data show variations from Thursday through Sunday (note the peak in P which is on Saturday).

Sometimes a load change can be measured upstream, before it has entered the plant. Then the information can be *fed forward* to prepare the plant. For example, the aeration can be increased before a load increase hits the plant. Another example is when the return sludge pumping can be increased to lower the sludge blanket in order to make the settler ready for an expected increase in the hydraulic load.

However, unfortunately, many disturbances are created within a plant due to inadequate operation. Often this is due to a lack of understanding of how various parts of a plant interact. Figure 15.3 shows such an example. The influent flow is pumped via three on-off pumps. This creates sudden changes of the flow rate, which will deteriorate the behaviour of both the clarification and the settling in the secondary sedimentation unit.

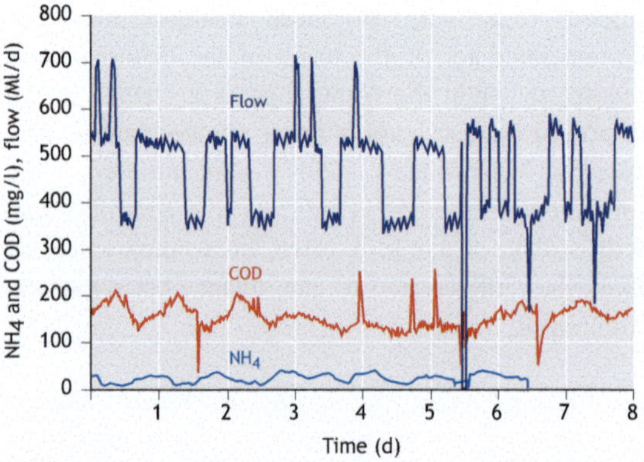

Figure 15.3 Influent variations in a large wastewater treatment plant having only on-off primary pumps, resulting in undesired sudden flow variations into the plant.

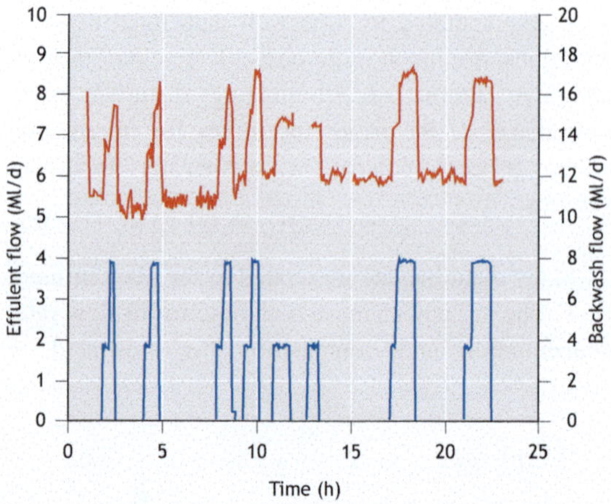

Figure 15.4 Filter backwashing (lower line) and its impact on the plant influent flow rate (upper line) and plant operation.

Filter backwashing can sometimes create huge operational problems. At one medium-sized plant the backwashing increased the influent flow rate by almost 50%, as illustrated in Figure 15.4.

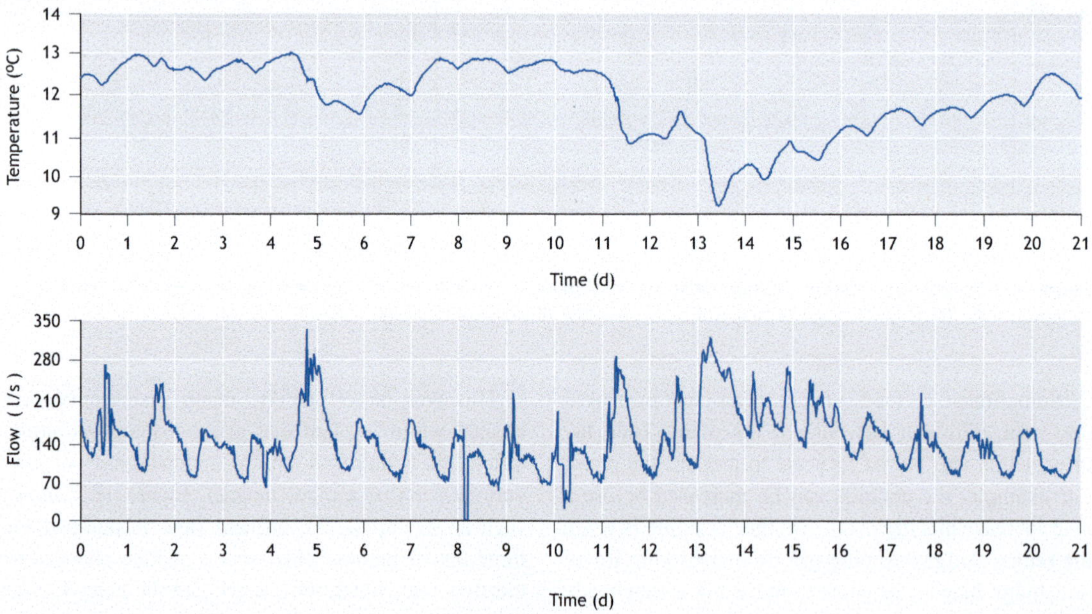

Figure 15.5 Influent flow rate variations during three weeks in the winter. The lower figure shows the daily variations and some rainy periods. The upper figure shows how the temperature decreases as a result of cold rain.

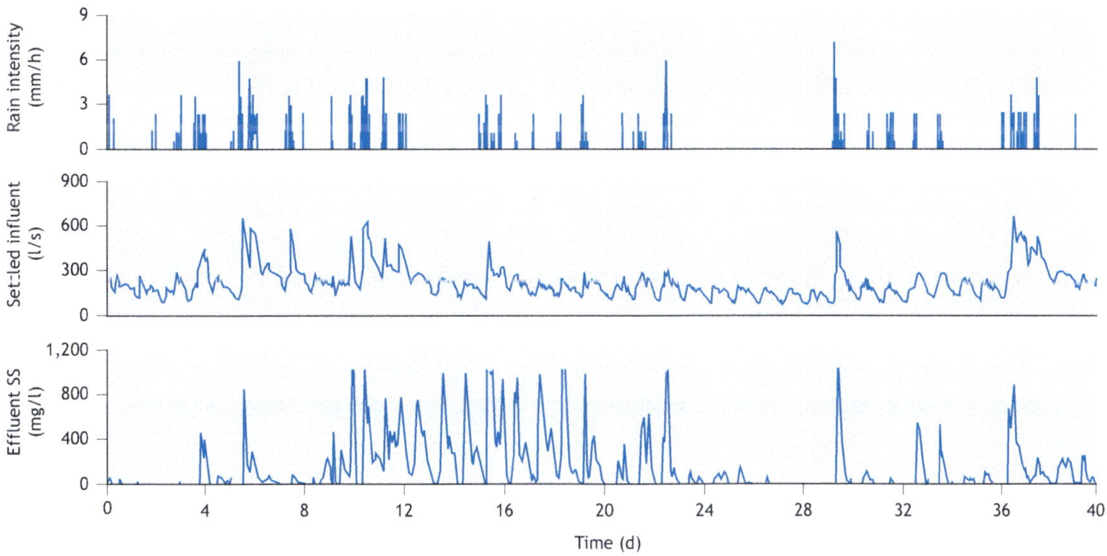

Figure 15.6 The relationship between large hydraulic disturbances and effluent quality. The upper figure shows the rain intensity during approximately 40 days and the middle figure is the corresponding influent flow rate to the municipal treatment plant. The lower figure shows the suspended solids concentration after the secondary settler; it clearly demonstrates that the clarifier is running close to its maximum capacity and is failing during large hydraulic peaks, resulting in large effluent suspended solids concentration values.

The nutrient removal plant had an anaerobic reactor as the first step. This reactor was hit not only by a large flow rate but also by oxygen-rich water, inhibiting the anaerobic reactions. The water propagated into the succeeding anoxic zone, still with some oxygen left. Obviously, the biological reactions suffered considerably and the effluent quality was unsatisfactory.

It transpired that pumping had to be performed differently, and the problem was readily solved, once the disturbance pattern was understood. Instead of pumping the backwash flow directly to the plant influent it was temporarily stored in an available volume and pumped gradually back to the plant input. This attenuated the flow rate change.

In a cold country the temperature of the influent water may change rapidly as a result of rain. Figure 15.5 is a data record over three weeks which shows how heavy rains during the winter will influence the water temperature, resulting in lower microbial activity as well as extra load for the clarifier and the settler. High flow rates will have a significant impact on the clarifier performance. This is illustrated in Figure 15.6.

If sludge supernatant is recycled to the plant influent during a high load, then the nitrogen load to the plant may be very large, as depicted in Figure 15.7. The figure shows how the oxygen uptake rate increases significantly as supernatant is recycled within the plant. It is crucial to identify the sources of the disturbances in order to obtain a high-performance operation of a plant. Then the control system can be structured so that these disturbances can be attenuated or even avoided.

Disturbances also arise from the shift of bacterial populations and changes in their microbial and physical properties. For example, it is not uncommon that a treatment system suffers from sludge settleability problems due to an outbreak of filamentous bacteria. The operations imposed by online control systems may themselves be a cause of a bacterial population shift. These disturbances must be properly dealt with in the control system design and evaluation. Further internal disturbances may be generated due to inadequate or inappropriate operations including human error, unsuitable or malfunctioning actuators, and/or sensor breakdowns, all of which may potentially cause major operational problems. Sudden flow shocks as a result of turning pumps on or off (without any variable speed control) or sudden backwashing of filters occur in many plants as well. Many of the internal disturbances can be avoided (or their impacts minimized) through introducing online control systems, particularly early warning systems.

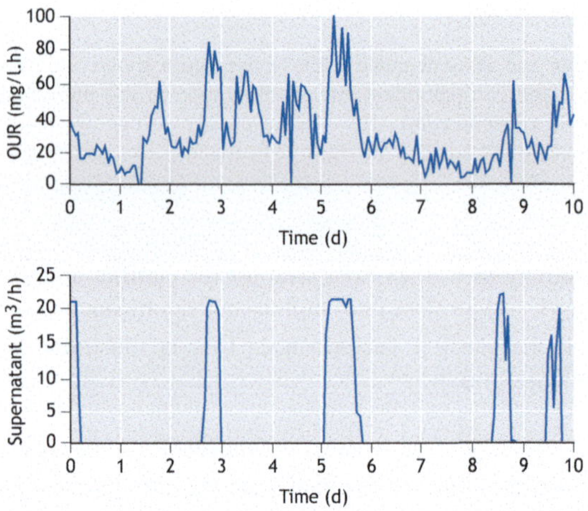

Figure 15.7 The effect of supernatant recycling in a plant during a 10-day period. The lower curve shows the supernatant flow rate (which is not very high but has a high concentration). The upper figure shows the oxygen uptake rate in the aerator (adapted from M.K. Nielsen, Denmark).

15.3 THE ROLE OF CONTROL AND AUTOMATION

ICA in wastewater treatment systems has come a long way and it is now an established and recognized area of technology in the profession. Several factors combined have made this progress possible:

- *Instrumentation technology* – to measure is to know – is today so much more mature. Complex instruments such as online in-situ nutrient sensors are now regularly used in the field. Instrumentation is no longer considered the major obstacle for wastewater treatment control.
- *Actuators* have improved over the years. Today variable speed drives in pumps and compressors are a proven technology and they allow better controllability and flexible operation of plants.
- *Computing power* can be considered almost free.
- *Data collection* is no longer a major obstacle. Software packages are available for data acquisition and plant supervision. The benefits of SCADA and process control systems are no longer questioned. Data analysis and monitoring software is commercially available. The obstacle is not the products but the competence to take full advantage of the available algorithms and software.
- *Control theory* and *automation technology* offer powerful tools. Benchmarking of different control methods is becoming recognized and tools for evaluating control strategy performance are available, such as costs, robustness and 'performance images'.
- Advanced *dynamical models* of many unit processes have been developed and there are commercial *simulators* available to condense the knowledge of plant dynamics.
- *Operators and process engineers* are today much more educated in instrumentation, computers and control understanding. Naturally, with new water quality challenges the need for more education is apparent all the time.
- There are obvious *incentives* for ICA, not the least from an economic point of view. Plants are also becoming increasingly complex which necessitates automation and control.

The development towards process/plant-wide control approaches is still in its infancy, but their implementation is gaining momentum, which is further discussed in Section 15.10. ICA has been accepted as a standard element of wastewater treatment systems. Utilities are now highly dependent on ICA to minimize the resources needed to effectively operate facilities. However, despite almost universal acceptance, there is still a large opportunity to further apply ICA. Probably the biggest obstacle for better control is lack of process flexibility in many plants. Plant design and operation still need to be systematically integrated.

15.3.1 Setting the priorities

Any plant operator must set the priorities for an effective operation, and efficient performance must rely on functioning equipment. All the links in the chain must be working in order to get a satisfactory operational system. Hardware includes not only instrumentation, communication and computers but also actuators, such as compressors, pumps, motors and valves. Communication systems are becoming increasingly important in plant control systems. The software relies not only on proper control algorithms, but also databases, communication systems, data acquisition, and human-friendly displays. Most important of all: people. No control system can be presented to operators who have not been able to influence its design. So much of the performance is built on trust. Any well-intended and functioning control system can be (and has been!) a total failure if the operating people do not trust it. Therefore, people involvement and education is a crucial part of a successful system. So, what are the priorities?

- *Keep the plant running.* Before we even consider controlling the water quality it must be ensured that the plant equipment is working adequately. This includes meters of electric motor speeds, acoustic noise measurements, and indicators that can ensure that pumps and compressors are running. Simple level measurements or hydraulic measurement sensors can confirm that tank levels are within acceptable limits. Gas flow measurements indicate methane production in a digester, and conductivity sensors can detect changes in the influent composition. The whole idea of keeping the plant running is *not* to obtain accurate information of the plant state, but to ensure reliable basic operation. Most of the equipment control actions are traditional process control loops (Figure 15.1), such as air pressure control, liquid level control, and flow rate control.
- *Satisfy the effluent requirements.* It is not enough to keep the physical parameters correct. Other variables that are directly related to the effluent quality must be controlled. This is realized at this level. It involves manipulating the variables of different unit processes, such as dosage control for chemical precipitation, air flow rate control to keep the dissolved oxygen (DO) level satisfactory, and return sludge control or sludge retention time (SRT) control to manage the sludge content.
- *Minimize the cost.* In each unit process the control scheme may be more elaborate. One example is DO control, where the DO setpoint is variable, not only along the aeration basin, but also variable in time (see Section 15.8). The goal at this level is to optimize the unit process operation. All of this depends on suitable sensors and instruments. The cost can be influenced by decreasing the energy demand (for aeration or for mixing), or lowering the cost for dosage chemicals in phosphorous precipitation or in centrifuge operation. The cost is also related to the personnel. Many plants today are satisfactorily operated un-manned during evenings, nights and weekends.
- *Integrate the plant operation.* The purpose of integrating control is also to satisfy the effluent requirement at minimum cost. By coordinating several processes it is possible to decrease the impact of disturbances entering into or within the plant (compare figures 15.4 and 15.7). The combined operation of the processes may make it possible to optimally use the available volumes and the sludge for the best operation.

Present standard computer hardware and software and the growing availability of reliable real-time measurements (properly validated) for an increasing range of different parameters are enabling advanced closed-loop process control on wastewater treatment plants (WWTPs), resulting in increased operating safety and better operational economy. However, these benefits can be limited by the design of the WWTPs themselves, because design has not always been made with controllability in mind.

15.4 INSTRUMENTATION AND MONITORING

'To measure is to know'. Measured or observed phenomena form the basis for all feedback. In water systems it is apparent that flow rates and a multitude of concentrations and quality parameters are the foundation for all operations. The instrumentation must be robust, easy to maintain and cost-effective. This is even more important in an unmanned process.

15.4.1 Sensors and instruments

For a long time instrumentation (here we use the common terms *measuring instruments* or *instrumentation* for sensors, analysers and other measuring instruments) was considered the bottleneck for control and automation in water systems. This has been considerably improved during recent decades. An important development of nutrient sensors has taken place from automated laboratory *ex-situ* analysers that had to be protected from the measured system through filtration units to *in-situ* sensors that can be placed directly in the liquid to be monitored. There is also an interesting development towards increasingly 'smart' sensors with multiple heads, which can be placed anywhere in the processes. Below is a list of just some of the key measurements in water systems. Further details on sensor properties are found both at vendor websites (search for 'water and wastewater sensors' and similar keywords) and in Ingildsen and Olsson (2016).

Measurements form the backbone of the operation in water systems:

- *Physical measurements:* flow rates, water levels, and air flow rates and air pressures.
- *Water quality measurements*: dissolved oxygen (DO), total suspended solids (TSS), ammonia, nitrite and nitrate, phosphate, organic content as chemical oxygen demand (COD) or biochemical oxygen demand (BOD), sludge level, and respirometry.

The introduction of a sensor in a plant or system requires not only confidence in the equipment. If the value of the measurement is not understood, then it is too easy to lose interest in the sensor and consequently the performance of the measurement will gradually decrease because of lack of attention or maintenance. Too many high-quality instruments have failed because of this lack of connection with the purpose of the information. Increased confidence in instrumentation is now driven by the fact that clear definitions of performance characteristics and standardized tests for instrumentation have become available (ISO 15839, 2003).

Standardisation of instrumentation specifications makes it possible to specify, compare and select the most adequate instrumentation, not only in technical terms but also in economic terms through calculation of the cost of ownership (Table 15.1). The investment costs for the device itself are often a minor part of the costs during the lifetime of the instrumentation. However, it is emphasized that a simple cost-benefit analysis is often insufficient to evaluate the benefit of instrumentation. The value of a sensor depends on how intelligently it is used. It is required that operators have a certain level of understanding of their process to run the plant with reasonable confidence.

Measurements from the instrumentation should be available 24 hours a day and 7 days a week. Information needs to be properly extracted from the measured data. Thus, instrumentation must always be combined with adequate data screening, measurement processing, and extraction of features from the measurements, as discussed in the next section.

Table 15.1 Items (and examples) included in the instrumentation cost-of-ownership calculation.

Instrumentation	Cost of the instrumentation itself
Conditioning	Cost of rig, building, pumps, pipes, pretreatment
Installation	Time costs for project and skilled workers
Integration	Time costs for programming of SCADA, control loops
Consumables	Costs of chemicals, power, etc.
Maintenance	Cost of service contract and time costs for calibration, cleaning
Spare parts	Cost of spare parts

There are many interesting sensor developments taking place in companies and research laboratories around the world. This will of course increase the opportunities to make plant operation smarter in many senses. Some significant developments are (Ingildsen and Olsson, 2016):

- *Adding processor power to sensors:* making the sensors more complex by adding processor power to the primary sensor will enhance the monitoring capability, see below. The Internet of Things (IoT) adds another layer of 'smartness' with increasing communication facilities from the sensors directly to the measurement analysis and control decision.
- *Micropollutants:* detection and measurement of micropollutants in water is an increasing challenge. The first step to be able to remove these contaminants is of course to be able to detect them and to measure their concentration.
- *Spectroscopy* offers the ability to measure the whole spectrum online and can give a much more complete picture of water quality than the measurement of only specific wavelengths.
- *Wireless communication* is part of the IoT development and makes it possible to achieve field measurements more easily from distant locations, thus increasing the probability of successfully utilizing early warning systems.

15.4.2 Monitoring

To track the current process operational state via the instrumentation is called *monitoring*. Monitoring can give transparency into what is happening in the plant. Without it, the processes are like a black box. Even reliable instrumentation can fail during operation, which can have serious consequences if the instrumentation is used in closed loop control. Therefore, real-time data validation is needed before using measurements for control purposes. A consistent monitoring of the product quality will help us to avoid problems growing too large, as illustrated in Figure 15.8.

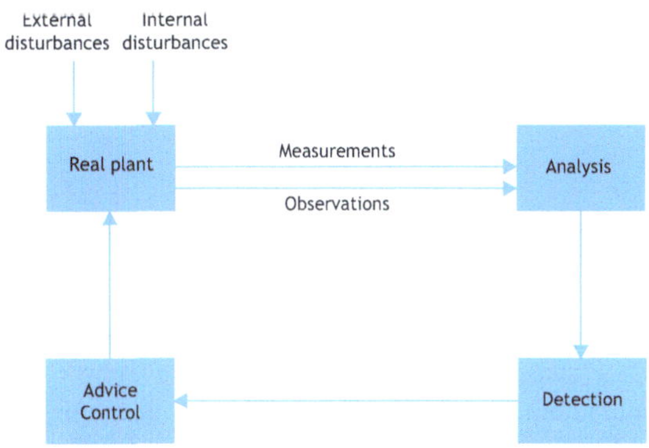

Figure 15.8 A monitoring system is based on analysis of the measurements and observations. A fault or failure can often be detected automatically. Based on the result, some advice or operator support can be given, and a control action can take place.

There is a huge data flow from the processes in a sophisticated treatment plant. More instrumentation and new instrumentation development will further provide more data. However, it should be emphasized that more data is not valuable unless we succeed in translating data into actionable information.

Unlike humans, computers are infinitely attentive and can detect abnormal patterns in plant data. The early detection and isolation of faults in biological processes are very effective because they allow corrective action to be taken well before the situation becomes unfavourable. Slow changes in biological properties are not easily observed by humans and may gradually grow until they become a serious operational problem.

All data analysis may not be in real-time. There is also value in long-term analysis. There is a limit to any human being's ability to take in data. However, supported by computers we may understand more than by just looking at data, but this requires the mastering of relevant analysis tools as well as a deep-seated understanding of what is happening in reality.

Some examples of basic monitoring are described below. Figure 15.9 illustrates daily variations (during some 3 weeks) of influent flow rate. Some significant peaks in the flow rate are obvious. The line indicates the mean value and the $\pm 2\sigma$ and $\pm 3\sigma$ deviations from the mean. It is obvious that deviations larger than 3σ ought to be observed carefully and suitable operations must be implemented.

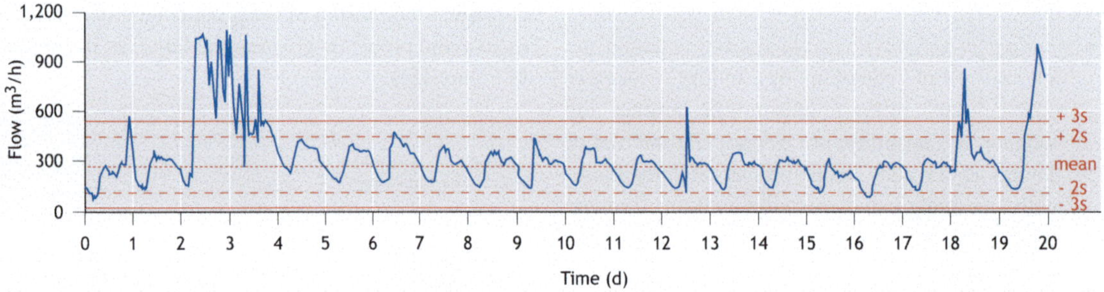

Figure 15.9 Influent flow rate variations during a three-week period. The mean value and the $\pm 2\sigma$ and $\pm 3\sigma$ limits are indicated.

Figure 15.10 illustrates what happens when there is a sensor failure. The upper figure shows the measurement signal and the observant operator can notice a change in the signal character at approximately time 900. By filtering the signal, however, the changes can be made more obvious. A high pass filter essentially shows the variability (or rate of change) of the signal. The filtered signal is displayed in the lower figure and reveals a significant change in the noise character of the signal, thus exposing a sensor problem. Obviously, it is possible to spot the change with the naked eye, but that requires attention by someone. By transforming the noise into a separate signal, it is possible to set up automatic alarms and hence react to the change in a timely manner, rather than realising this too late.

Any monitoring system must determine whether the acquired data are meaningful and correct, so data screening is essential. As a minimum it should include normal range comparisons (high and low limits), rate of change, and variance. Filtering techniques are essential to remove noise while retaining the essential signal information. High pass filters can detect sudden or fast changes.

By rapidly detecting deviations from 'normal' it is possible to minimize the cost of abnormal behaviour. Monitoring as a basis for early warning systems in wastewater treatment is described in literally hundreds of papers. Overviews are found in Irizar *et al.* (2008), Hamouda *et al.* (2009), and Olsson *et al.* (2014). Yuan *et al.* (2019) have reviewed detection methods for water supply, water distribution, sewer systems and wastewater treatment.

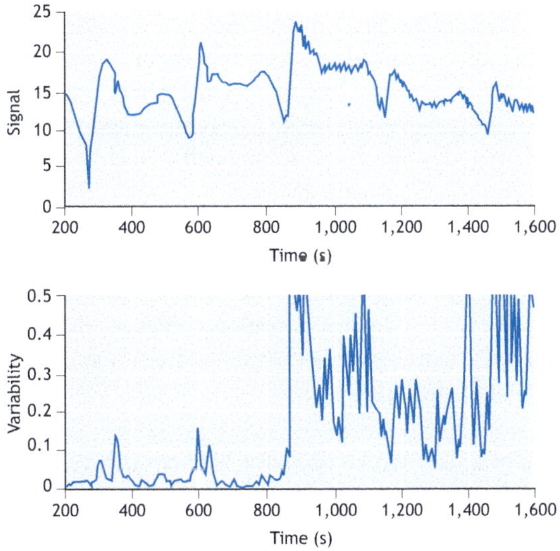

Figure 15.10 Detection of a sensor problem. The upper figure shows the sensor signal. The noise character will change after time 900, indicating a sensor problem. The lower figure shows the variability of the signal. When the variability exceeds a threshold (for example, 0.15) the monitoring system can give an automatic alarm.

Process knowledge is an essential condition for adequate diagnosis. A primary observation may lead to a whole chain of examinations of the problem, as in medical diagnosis. For example, a high turbidity of the effluent may be a primary warning. The reason may be a too high influent flow rate, poor sludge settleability, or a high sludge blanket. By back-chaining and asking questions (automatically or manually), the primary cause of the problem may be tracked down.

By combining information from multiple sensors and mathematical models, other operational parameters that are not directly measurable can be estimated: these are known as soft sensors, and they allow powerful monitoring tools to be built up. An early application of soft sensors to check the performance of instrumentation is described by Lumley (2002). Leonhardt *et al.* (2012) used a soft sensor to estimate stormwater overflow. Racault *et al.* (2011) demonstrated how oxygen transfer rate is estimated from mass balances. By measuring DO concentrations combined with models it is possible to estimate the oxygen uptake rate, and thus the reaction rate. An excellent overview of soft sensors in biological wastewater treatment is presented by Haimi *et al.* (2013).

Multivariate analysis is a method to detect changes in patterns and correlations in large data sets, including more than two variables. By using this analysis, it is possible to detect situations that call for human intervention. It is a viable alternative to manual examination of data; despite many hours of observation, it is still not possible to track all the changes that computer algorithms can. Hence this is a help to understand when to deploy the analysis team.

Multivariate analysis had been used for many years in the chemical process industry and was only introduced into the wastewater industry in the late 1990s (Rosen and Olsson, 1998). The most well-known method to reduce the dimensionality of a data cloud is Principal Component Analysis (PCA). This technique is simple in the sense that the data can readily be projected onto a smaller dimension. PCA has been used in sequencing batch reactors for monitoring (Villez *et al.*, 2008) and as a basis for control of the phase length (Villez *et al.*, 2010). Ruiz *et al.* (2011) used PCA for fault detection.

However, PCA methods are insufficient to deal with highly variable data, such as influent flow rates and compositions (Rosen *et al.*, 2003). Wastewater treatment plants have a wide range of time constants, and it is difficult to look at correlations of data in just one timescale. Various methods to extend the PCA to dynamic data in wastewater treatment systems have been developed (Rosen and Lennox, 2001; Lennox and Rosen, 2002). Flores *et al.* (2007) evaluated selected multivariable statistical techniques for various control strategies for plant-wide wastewater treatment plant control.

Corominas *et al.* (2018) reviewed many commonly used computer-based methods that can transform data into useful information. These include methods based on control charts, mass balances, regression models (multi-linear, partial least squares (PLS)), self-organizing maps (SOM), principal component analysis (PCA), independent component analysis (ICA has another meaning here), artificial neural networks (ANNs), clustering, fuzzy methods, support vector machines (SVMs), and techniques for the recognition of qualitative features in data series. Computer-based tools for information (Environmental Decision Support Systems, EDSS) and knowledge management (ontologies) are also discussed. It is easy to get lost on how to select the best method for data analysis and there is no simple shortcut to how to find the best one. A specific review of machine learning techniques applied to the field of water and wastewater is described by Hadjimichael *et al.* (2016).

Several books and articles on elementary data analysis are freely downloadable from the Internet (search for 'data analysis', 'statistical data analysis', and 'data mining'). There is a lot of freely available as well as commercial software for data analysis. All the methods described here can be executed on some available software. Software such as Excel can be used successfully for elementary data analysis, while more sophisticated software products such as Matlab[1] or SAS[2] contain a multitude of data analysis methods.

15.5 THE IMPORTANCE OF DYNAMICS

It is crucial to be reasonably familiar with dynamics in order to understand control. Basically, this means that the result of a corrective action will take some time; it will never appear instantaneously. Therefore, the timescale of any process change is extremely important. The dynamics of a wastewater treatment system involve a wide range of timescales, from seconds to months. Human perception is not very good at dealing with dynamics that are very fast or very slow. It also requires a lot of training and deep understanding of systems to comprehend processes that work and interact in different time domains. It is too easy to come to the wrong conclusion. Therefore, intelligent control systems that are tuned with the right time constants and gains are important in handling this problem. Additionally, such systems are attentive all around the clock.

The dynamics of the typical process units can be classified as fast, medium, and slow as described in Table 15.2 and this classification influences the type of model that can be developed and also the design of control strategies.

There is a wide difference between the fast and the slow timescales (Table 15.2). This allows the separation of the various control actions into different time domains. In the fast timescale the variables that change very slowly will be considered constant. For example, the DO concentration can change within a fraction of an hour. In this timescale the biomass concentration can be considered constant. Conversely, in the slow timescale, for example when we aim to control the total sludge inventory, then the DO concentration can be considered to be changing instantaneously.

This separation in time makes the control problem less complex. It allows control actions to be compartmentalized into fast, medium and slow timescales and we can consider them and often analyse them separately.

A continuous flow wastewater treatment system is in a transient state most of the time, because the influent flow rate and its concentration and composition are hardly ever constant. The control system aims to take care of these changes. Online

[1] http://se.mathworks.com/
[2] http://www.sas.com

measurements and corrective actions should bring the various process variables to their desired values. By contrast, in a sequential batch reactor the system is purposefully in a transient state. An oxidation phase will continue until the oxidation is completed, and then a reduction phase (such as denitrification) will take over and will finish when the reduction of nitrate has been completed. A sequential batch system is therefore very suitable for dynamic process control.

Table 15.2 Biological nutrient removal process dynamic response times.

Speed	Timescale	Wastewater treatment mechanism
Fast	Minutes-hours	Hydraulics and flow dynamics
		Oxygen mass transfer
		Chemical precipitation
		DO dynamics
		Solids-liquid separation
Medium	Hours-several hours	Concentration dynamics
		Nutrient removal
Slow	Days-months	Biomass growth

Sometimes the time needed for measurements must be taken into consideration. To get a DO reading takes a few seconds. However, this delay is small compared to the typical time for a DO change, which is a fraction of an hour. A respirometer reading will take a longer time, typically half an hour. It is obvious that such a measurement can be used only for slower corrective actions, of the order of hours. It is also important to remember that the measurement value is often corrupted by noise. The noise variations may be quite fast, and the controller must not react to the fast and false variations. Therefore, filtering the signal is crucial.

It is important to remember the dynamic when closing the loop. Controlling the total sludge inventory by the waste sludge flow rate is a very slow process. The rate of change depends on the biomass growth rate and the typical timeframe is of the order of several days. Typically, to change the sludge retention time from 10 to 11 days will take 10-20 days. The sludge retention time is an average value and cannot be calculated on a day-to-day basis. Instead the flow rates and sludge concentrations must be averaged over a longer time, typically weeks.

Sometimes the controllers are tuned to be too 'ambitious'. For example, even if a DO sensor can produce a DO concentration value several times every minute this does not mean that the airflow rate should be changed every minute. Since the typical response time in a full-scale aerator is 15-30 minutes, a change of the airflow every minute will only produce meaningless control actions and wear out the valves. Instead, a control action every 8-12 minutes is adequate and more appropriate. In this timeframe the controller can see some results of the changing air flow.

Modelling for control is not the same as modelling for understanding basic kinetic mechanisms. Consequently, models such as the Activated Sludge Models 1, 2, and 3 (Henze *et al.*, 2000) or the Anaerobic Digestion Model (Batstone *et al.*, 2002) are not meant to be the basis for controller synthesis. Instead, they represent detailed descriptions of the way we understand the mechanisms of the biological processes. The comprehensive models are, however, excellently suited as testbeds for controllers, to understand behaviour in the different time domains. In control implementation, on the other hand, one needs to identify key parameters that are crucial for the operation of the plant. Such parameters can be oxygen uptake rate, respiration rate, and reaction rates for BOD removal, for nitrification or for denitrification. Redox can reflect the progress of the anoxic denitrification process.

The key parameters are calculated from primary measurements. For example, the DO concentration can be used as a basis for the estimation of oxygen uptake rates. Online measurements of ammonia nitrogen or nitrate can be further elaborated to calculate nitrogen removal reaction rates. Consequently, estimation of dynamical parameters is an important part of the modelling that will form the basis for more advanced control (compare Section 15.4.2).

15.6 MANIPULATED VARIABLES AND ACTUATORS

Several variables can be used to manipulate biological wastewater processes. Nevertheless, the possibilities to control the plant in a flexible manner are quite limited. A major problem in many plants is the lack of controllability of pumps or compressors. Hence, designing new plants for flexibility and controllability is an important challenge. This not only requires the use of variable actuators, but also that these are variable in the adequate operational range (Section 15.1.2).

Flexibility and controllability are a key to effective and efficient operation. However, be aware that it is also true that if the controllability is not used wisely, based on sensor inputs and intelligent use of controls, it is of no, little or even negative use.

Section 15.2 gave an example of how on/off pumping can create problems later on in the process. Variable speed control is a proven technology and is one important prerequisite for good control, both for pumping of water and sludge flows and for controlling air flow for DO control.

The manipulated variables can be categorized into the following groups:

- Hydraulic, including sludge inventory variables and recirculation flows,
- Additions of chemicals or carbon sources,
- Air or oxygen supply,
- Pre-treatment of influent wastewater.

There are several other manipulated variables in a plant that are related to the equipment and to basic control loops in the process, such as flow controllers, level controllers etc. They are not included in this discussion.

15.6.1 Hydraulic variables

Most of the manipulated variables change the hydraulic flow patterns through the plant. The different flow rates will influence the retention times in the various units. Moreover, the rate of change is crucial in many parts of the treatment plant, since it influences the clarification and thickening processes. The hydraulic flows also determine the interaction between different unit processes. Thus, the hydraulic manipulated variables can be divided into the four groups:

- Variables controlling the influent flow rate,
- Variables controlling the sludge inventory and its distribution,
- Internal recirculation within the biological process,
- External recycle streams, influencing the interactions between different unit processes.

In this category we will also include the control of the phase length in sequential batch reactors, which is equivalent to controlling the retention time of a continuous unit.

The influent flow rate to the activated sludge system can be manipulated in various ways. From a plant point of view the influent flow rate may be considered an external disturbance that has to be handled by the various control systems. Then we emphasize that the pumping of the influent flow must be smooth and variable speed pumping is recommended (compare Figure 15.3). On the other hand, if there is an equalization basin available or if the sewer network can be used as a buffer volume, then the influent flow rate becomes a control variable. The additional volumes before the plant will allow us to control the influent flow rate to minimize the detrimental consequences of the influent variations.

Many plants are designed with two or more parallel aeration basins. The flow splitting process is crucial if the load is to be evenly distributed. This is often not the case, which results in apparent overloading in some parts of the system. In many plants the flow splitting is carried out by a fixed arrangement of channels, which may not guarantee at all that the real flow is split correctly. If flow splitting has to be guaranteed, then the flow rates must be measured, and the individual flow rates controlled.

Bypassing should be a manipulated variable in the sense that it should never occur, unless it is ordered. It must be compared with the alternative of not performing bypassing and has to be based on some quantitative calculation with a suitable time horizon (compare Section 15.1.1).

All the different modes of influent flow control are simply different ways to increase the control authority. In other words, sewer or equalization or bypass control all contribute to making it easier to obtain a smooth flow rate into the plant. Their goal is disturbance rejection. A smooth variation of the flow rate is crucial for the secondary clarifier operation. It requires not only variable speed pumping at the operating level to avoid disturbances, but an adequate storage capacity in wet wells or upstream tanks to damp disturbances which cannot be avoided. Poor pumping control can deteriorate the plant performance considerably.

The sludge inventory can be controlled primarily by three manipulated variables:

- The waste sludge flow rate,
- The return sludge flow rate,
- Step feed flow rates.

Manipulation of the waste sludge flow rate is used to control the total inventory of sludge in the process. Since the total inventory is a function of the total growth rate of organisms, it is used to control the sludge retention time, or the sludge age. This manipulated variable will influence the system in a timescale of several days or weeks.

Manipulation of the return sludge flow rate is used to distribute the sludge between the aeration basins and the settler units or between the acidogenic and methanogenic reactors in two-stage anaerobic systems. Recycle from the settling stage is an important variable for obtaining the right operating point in the reactors, but seldom useful for the control on an hour-to-hour basis. Some systems are supplied with several feeding points for the return sludge. This has a potential for sludge redistribution for certain loads, such as toxic loading. A combination of different recycle streams may be important. In systems with chemical precipitation, sludge from the secondary settler may be combined with chemical sludge from a post-precipitation settler unit. In this way the floc properties may be influenced, and the chemicals better utilized for phosphorus removal.

By controlling the step feed in an activated sludge plant, the sludge within the aeration basin can be redistributed, given the proper amount of time. Uniquely for step-feed control, one will obtain a contact stabilization structure. Also, the return sludge may be fed back not only to the inlet part of the aeration basin, but into different feeding points along the basin, which is known as step return sludge control. This may prove to be an efficient way of preventing bulking sludge.

Internal or external recirculation flows provide couplings between the different units of the plant. The recycle streams can be considered as controllable disturbances to the reactor-settler system. They must be manipulated so that their detrimental impact is minimized. Some of the recirculation flows have very large flow rates, such as the recirculation of nitrate in a pre-denitrification plant. Having a system with pre-denitrification, it is necessary to recirculate nitrate-rich water from the outlet of the nitrifying aerator. In particular, the oxygen contained in the recirculated water may limit the denitrification rate in the anoxic zone.

Section 15.2 demonstrated that the backwashing flow from a deep-bed filter can create major disturbances and must be manipulated properly. Other streams may have extremely large concentrations, such as supernatants from the sludge treatment, as indicated in Section 15.2. Most of them can be manipulated purposefully to achieve a better plant performance. In a bio-P system there are three types of reactors: anaerobic, anoxic and aerobic. Depending on the design, there are many recirculation patterns in such a plant. In a two-stage anaerobic system the recirculation helps to keep the methanogens washed out of the acidification stage and restores pH buffering capacity to reduce caustic usage.

15.6.2 Chemical addition

Chemicals are added for two different reasons, to achieve chemical precipitation for phosphorus removal, or to form a better settleability of the sludge. For phosphorus removal, ferrous, ferric or aluminium salts are added to obtain chemical precipitation by forming insoluble phosphates. A change in chemical dosage can have quite a fast influence on the floc formation and the settling. In Section 15.8 the control of chemical precipitation is discussed.

In addition to the normal use of chemicals for P removal, chemicals can be added to improve sludge settling properties in the secondary settler. Sometimes chemicals are added to the primary settler to reduce the load to the aerator, although this may lead to insufficient carbon for the nutrient removal.

Polymer addition may be used in emergency situations to avoid major settler failures. On a routine basis it is used for sludge conditioning to improve dewatering properties. In addition, polymers can be used to further enhance the efficiency of the pre-precipitation process.

Caustic addition is applied in two-stage anaerobic processes to control the pH, which can inhibit the methanogenic microorganisms.

15.6.3 Carbon addition

Carbon source addition is sometimes needed in denitrification to obtain an adequate carbon/nitrogen ratio in the system. Too little carbon results in incomplete denitrification, while too much carbon adds a cost for the chemical and its subsequent removal. The timescale of such an operation is related to the retention time of the denitrification. For a pre-denitrification system, carbon is usually supplied via the influent wastewater. However, this may be insufficient during low load periods, so some carbon source needs to be added. In a post-denitrification system a carbon source (such as methanol or ethanol) always has to be added. Then the problem appears of how to adjust the dosage to the carbon need, without extensive measurements.

15.6.4 Air or oxygen supply

Dissolved oxygen (DO) is a key variable in activated sludge operation. From a biological point of view, the choice of a proper DO setpoint is crucial. The dynamics of the DO is such that the DO can be influenced within fractions of an hour. Key factors related to the DO supply are the total air supply, the DO setpoints, and the DO spatial distribution. To obtain the required DO profile one needs individual airflow measurements and feedback control over the valves to each zone of the aerator. DO control will be further discussed in Section 15.8.

The airflow rate is recognized to be of major importance for the whole operation. It is reasonable to assume that a well-functioning DO control system should be available. However, since the energy cost is significant, it is of interest to minimize the air supply. It is well known that insufficient air supply will influence the organism growth, the floc formation and the sludge settling properties. However, once undesirable organisms are formed, it is not always obvious how to get rid of them by only DO control.

15.7 BASIC CONTROL CONCEPTS

The fundamental principle of control is feedback, illustrated by Figure 15.1. The process (for example, an aerator, a chemical dosage system, or an anaerobic reactor) is all the time subject to external disturbances. These are mainly caused by the variations in the influent load, but can also be caused by internal changes, such as recycles, pumping, etc. The current state of the process measured by a sensor forms the basis for a decision. In order to make this decision the goal or purpose must be stated. Having made the decision it then has to be implemented via an actuator, which is typically a motor, a pump, a valve or a compressor.

Control engineering today can offer almost anything in terms of methods and algorithms that the water operation might need. Automatic control is sometimes called the hidden technology. It appears everywhere around us and we do not even think about it. We only notice it when it does *not* work. We are subject to feedback in our daily life. In the human body the nerve cells sense the temperature and the brain controls the muscles to restrict the skin capillaries. Balancing the body requires that we sense the direction via our balancing system. The brain controls the muscles in the feet and legs to keep us upright. While driving a car the driver all the time applies feedback. The eyes watch the speedometer and the road etc. and the brain will compile all that information and decide what to do in the next moment. This is translated to the muscles to turn the steering wheel, to brake or to accelerate. The reason for the constant feedback is that the scene is changing continuously. In other words, the process is subject to disturbances that force us to use feedback.

Even if control is applied in completely different areas, there is a common theory that is independent of the applications. There are literally hundreds of textbooks on control. The textbook by Åström and Murray (2014), available on open access, is recommended, since the authors are world leaders in the development of control theory and engineering.

In other words: control is about how to operate the plant or process towards a defined goal, despite disturbances. With the manipulated and controlled variables identified in the previous step, a proper control configuration has to be defined using an adequate control algorithm. The block diagram in Figure 15.11 is a control engineering representation of the simple feedback controller shown in Figure 15.1. It describes the signals (the information flow) of the control system. Note that the terms 'closed loop control', 'feedback control' or just 'control' are often used as synonyms. This kind of control loop appears in all the local control of levels, pressures, temperatures and flow rates. The controller has two inputs, the measurement (actual) value y and the reference (setpoint) value u_c and one output, the control signal u. In this simple case, however, the controller uses only the difference between the two inputs.

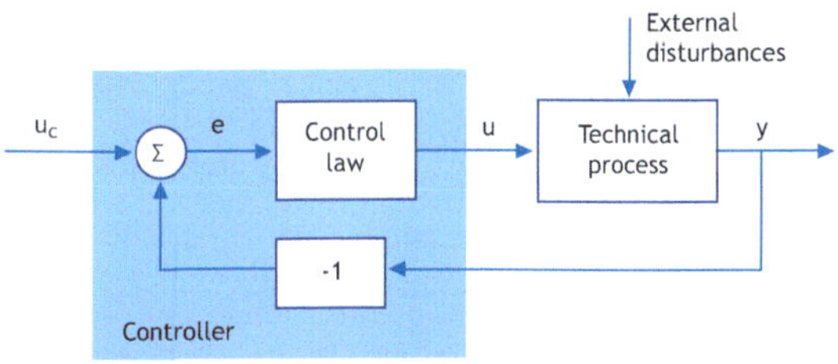

Figure 15.11 The simplest feedback control structure.

The properties of the controller (the controller parameters) can be changed (we usually call this procedure tuning) so that the output of the system gets as close as possible to the setpoint. The controller tries to make the error $e = u_c - y$ as small as possible. It is reasonable to think that the more parameters a complex controller contains, the more degrees of freedom it has. With the help of these parameters this can be changed at will, and the behaviour of the closed loop system can also be changed more arbitrarily.

Note the difference between *open loop* and *closed loop* control. In an open loop controller, the control action is not based on any feedback or measurement, but rather based on time. For example, a compressor providing air to an aerator can be turned on and off at certain times. No measurement of the DO is made and there is no guarantee that the DO concentration will be correct. Such an open loop control is completely different from closed loop control, where the change of aeration is based on true DO measurements.

The design of feedback controllers has attracted considerable attention in the control literature. Many advanced control algorithms based on, for example, dynamic models, neural networks and fuzzy logic have been proposed. Control systems based on simple rules (rule-based control) have also found successful applications. However, little convincing evidence has been presented suggesting that these advanced algorithms produce better unit process control performance in wastewater treatment systems than the conventional PID (proportional-integral-derivative) algorithms, which have been used in most practical process control applications (more than 95% of the controllers in a typical paper and pulp industry are PID controllers). Of course, this assumes that the PID controllers are correctly tuned, which requires technical understanding and experience with PID controllers and dynamic systems.

15.8 EXAMPLES OF FEEDBACK IN WASTEWATER TREATMENT SYSTEMS

The traditional WWTP control is still largely unit-process-oriented. Some examples of state-of-the-art unit process control include:

- DO control with a constant or a variable setpoint as part of the aerator unit process operation. In a nitrifying plant the DO setpoint no longer needs to be constant. It is calculated online based on information of nitrogen-removal performance, monitored with ammonia sensors.
- Aeration phase-length control in alternating plants based on nutrient sensors.
- Nitrate recirculation control in a pre-denitrification plant based on nitrate and DO measurements in the aerator and in the anoxic zone.
- Sludge retention time control based on measurements of effluent ammonia concentration and of estimates of nitrification capacity.
- Return sludge control based on sludge blanket measurements in the settler.
- Aeration tank settling (ATS) as a way of temporarily increasing the plant capacity in storm conditions (Nielsen *et al.,* 1996; Gernaey *et al.,* 2004).
- The control of anaerobic processes aims to stabilize the process and to maximize the biogas production. In most cases the control is based on biogas flow measurements only. Information from the reactor itself, such as pH, is needed for better control. More advanced measurements are under development.
- Dosing control in chemical precipitation control based on local measurements of pH, alkalinity or phosphate concentration.

Comprehensive reviews of these control systems can be found in Olsson *et al.* (2005, 2014), Åmand *et al.* (2013, 2014), Olsson (2012) and in Ingildsen and Olsson (2016).

A further challenge that future ICA systems ought to address is the mitigation of greenhouse gas emissions, such as N_2O, from biological nitrogen removal plants (Mannina *et al.,* 2016). The control of these parameters at levels that minimize N_2O emissions and still allow satisfactory nitrogen removal must be developed and demonstrated.

Example: Dissolved oxygen control

Dissolved oxygen control is of primary importance in the activated sludge process, both in recirculating plants and in alternating or intermittent systems. The control of aeration has been the subject of considerable research since the 1970s, when DO sensors reached a level of robustness and precision suitable for feedback control. Today, the control of DO to a setpoint is a mature technology from a methodological point of view, though in reality it is still suffering from underperformance and even encounters occasional failures due to physical limitations (*e.g.* inadequate capacity of the blowers) and/or hardware malfunctions (*e.g.* breakdown of a DO sensor). We will consider here the control of the DO concentration to a pre-specified setpoint through manipulating the airflow rate, as illustrated by Figure 15.12.

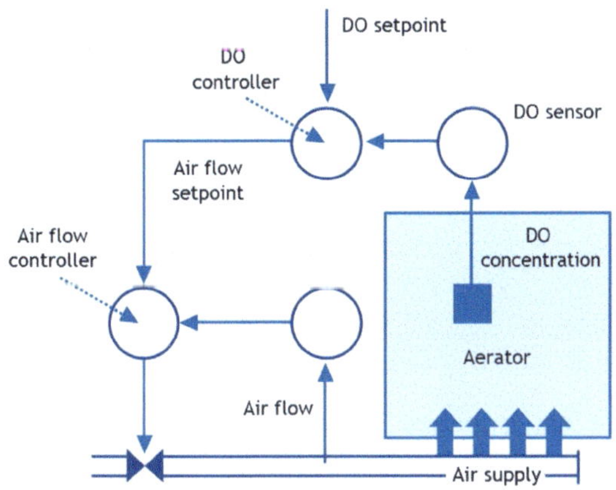

Figure 15.12 Structure of a standard dissolved oxygen control loop.

The DO is measured at one point in the aerator. The concentration is compared with the DO setpoint and the DO controller (the 'master') will calculate the necessary airflow change required to modify the DO concentration towards the desired value. However, the DO controller does not directly manipulate the air valve. Instead the desired airflow is given as a setpoint to a second controller, the airflow controller (the 'slave'). This controller receives the airflow rate measurement and compares it with the desired airflow. This difference will then make the actuator (a compressor or a valve) change the airflow to the required value. The control structure is called a cascaded control loop and is the standard configuration in this kind of system.

There are two important reasons why the DO controller is not directly coupled to the valve. The first reason has to do with the valve characteristics. Usually valves are nonlinear, as in a butterfly valve. A 10% change in the valve signal will produce significantly different responses if the valve is almost closed, in the mid-range or almost fully open. This means that a desired airflow change will require very different valve movements if the valve is almost closed or if it is almost fully opened. If the airflow rate is measured, then the flow controller can produce the required flow rate independent of the valve nonlinearity. Having the closed-loop slave controller in place means that the master controller will see a linear behaviour of the airflow system. The second reason has to do with the commissioning of the control system. The slave controller is tuned while the master controller is set to manual. In this way one can ensure that the response of the airflow system is adequate. Having completed this, the master controller can be put into automatic mode and subsequently tuned.

Example: DO setpoint control based on ammonium measurements

With the development of nutrient sensors, it has been possible to extend the DO control to allow for an online adjustment of the level of oxygen supply. For a recirculating system this means that the appropriate DO setpoint can be determined by online measurements.

An online ammonia analyser is placed near the outlet of the aeration basin. Under ideal conditions the ammonia concentration will decrease along the aerator and reach a low value just before the outlet. If the ammonia concentration is too low, then we may have been too ambitious. If so the effluent quality can be achieved with less air. Consequently, the DO setpoint can be decreased for the last zones of the aerator. Conversely, if the ammonia concentration is too high at the outlet, then we will try to improve the nitrification rate by increasing the DO setpoint so that we can reach the required low ammonia concentration. However, it may not be enough to increase the airflow. The load may simply be too high and the nitrification capacity insufficient for this load. Therefore, the upper value of the DO setpoint should be limited.

Figure 15.13 shows the result of DO control with a variable DO setpoint at the Källby WWTP in Lund, Sweden. It is a 100,000 PE plant of the pre-denitrification type. In one of two parallel identical lines a DO setpoint controller was tested based on ammonium concentrations at the end of the aerated part of the plant. A simple PI controller was used to change the DO setpoint value, based on the ammonia sensor signal at the outlet of the aerator. The DO setpoint value was sent to the DO controller system, as the one shown in Figure 15.12. The resulting controller is expanded to a structure of three controllers working in cascade using the master-slave principle. The controller performance is shown in Figure 15.13. At times with high ammonia concentration in the influent the DO setpoint is raised to its maximum (set to 3 mg/l). At periods with low ammonia load the DO setpoint can be decreased to much smaller values. Here it is limited to stay above 0.5 mg/l. By allowing a variable DO setpoint we will save energy for aeration. During the testing period in this plant, aeration energy savings of 28% were obtained compared to the parallel line where a constant DO setpoint was applied. This corresponds to a significant part of the operating costs and compensates readily for the extra cost of an ammonia analyser.

There are many publications on DO control. Åmand *et al.* (2013, 2014) and the two book chapters Olsson *et al.* (2019a, 2019b) present comprehensive reviews of aeration control.

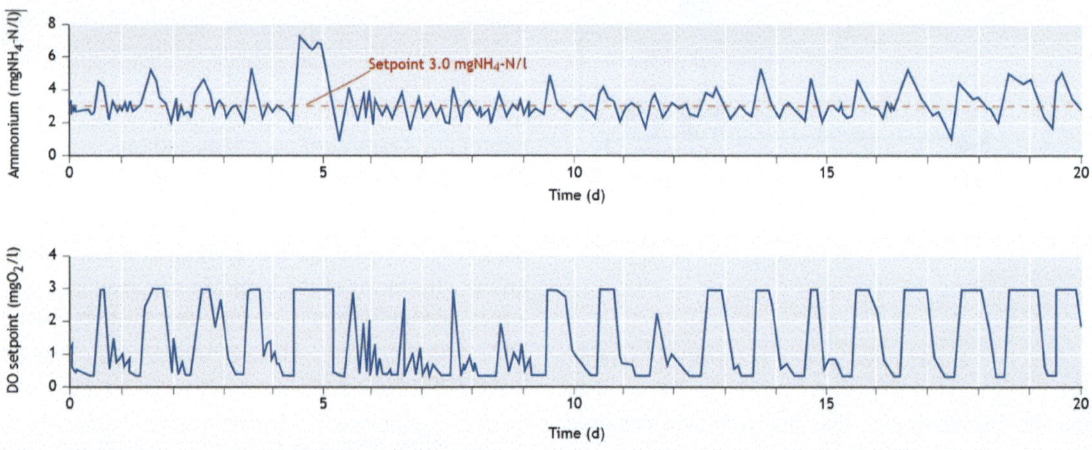

Figure 15.13 DO control with a variable DO setpoint. The upper plot shows the ammonium concentration at the end of the aerated part of the plant. The ammonium setpoint is 3 mg/l of NH4-N. The lower plot shows the DO setpoint in the last aerator zone during the same period of time. The DO setpoint is limited between 0.5 and 3 mg/l (from Ingildsen, 2002).

Example: Chemical precipitation control

In many plants and places phosphorous removal is obtained using chemical precipitation. Chemical precipitation processes are a lot faster than biological reactions. Compared to the timescale of the variations in wastewater flow rate and composition, chemical precipitation can be assumed to occur instantaneously. This represents a useful feature from a control point of view and implies that the disturbance can be quickly dealt with through feedback control. However, the challenging issue is the timely and reliable measurement of the key process variables so that a feedback control system can be established.

Chemical precipitation can be applied either prior to or after the biological treatment step, called *pre-precipitation* and *post-precipitation,* respectively. Chemicals can also be added directly into the aerator, which is known as simultaneous precipitation. Many plants apply a combination of these different types of precipitation. Here we will show that with a phosphate sensor in place, excellent control performance can be achieved using a simple feedback controller, demonstrated here for a post-precipitation process.

Phosphorous is precipitated by post-precipitation at the Källby treatment plant, Lund, Sweden. The line consists of a dosage system, where the precipitation chemicals are introduced into the water stream that runs into a flocculation chamber where soft mixing ensures the build-up of chemical flocs, followed by a settler for sludge removal. The average retention time in the flocculation chamber is 1 hour and the average retention time in the settler is 4.3 hours. The preceding biological lines achieve partial biological phosphorous removal, and chemical precipitation is applied to remove the remaining amount of phosphate which is typically around 2 mgP/l.

Here we will compare two control strategies first implemented by Ingildsen (2002):

1) *Flow proportional dosage:* This is a common strategy but relies on the assumption that the P concentration is constant. This is however *mostly not the case; the assumption of the constant relationship between influent phosphate* load and dosage may not be entirely correct, as factors such as pH may influence the process.
2) *Feedback control*: A feedback loop is applied that controls the dosage toward a certain phosphate setpoint in the effluent. The feedback signal comes from an online phosphate analyser located at the end of the flocculation chamber.

A flow proportional controller was tested for 35 days. The performance in terms of effluent phosphate can be seen in Figure 15.14. Four periods of malfunction are noticed in the effluent phosphate concentration (days 9, 10, 11 and 15). Especially the last incident (day 15) is easily detectable, where the effluent phosphate concentration increases drastically. The effluent criterion is 0.5 mg/l and it is obvious that the concentration is often lower than 0.5 mg/l. At other periods it is far higher, so the variability of the phosphate in the effluent is high. Obviously, the dosage is too ambitious at times, and this will be directly reflected in the operating costs.

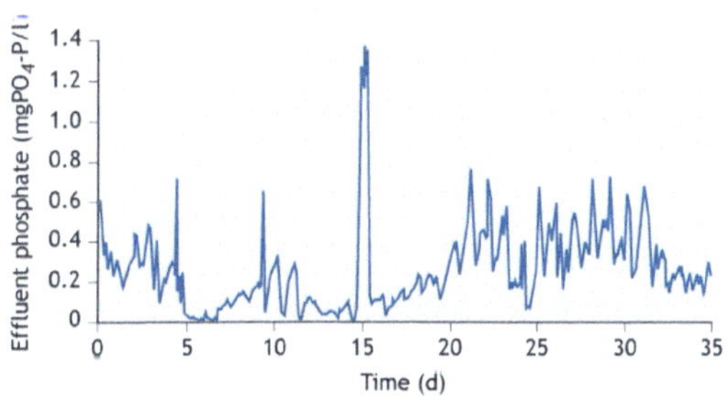

Figure 15.14 The effluent phosphate concentration with control of chemical dosage based on the hydraulic flow rate (from Ingildsen, 2002).

The retention time in the flocculation chamber is short, on average around 1 hour. This is considerably shorter than the time constant of the variation in the influent load of phosphate to the chemical step. Hence, it should be possible to control the phosphate precipitation by means of feedback control based on an *in-situ* phosphate sensor located in the effluent of the flocculation chamber. The influent phosphate concentration to the chemical step varied from about 1 to 3 mg/l, while the target value was 0.5 mg/l.

The performance in terms of effluent phosphate concentration is displayed in Figure 15.15. The setpoint was purposefully changed from 0.5 to 0.4 mg/l PO_4-P on day 23 and back to 0.5 mg/l on day 33. The peak concentration on day 31 is due to a malfunction of the dosage pump. The proposed controller is based on the effluent phosphate concentration while most effluent permits are defined in terms of effluent total phosphorous concentration. At the Källby plant it was verified that the total phosphorous and the orthophosphate concentrations are linearly correlated, with a regression value of 0.96. This means that it is possible to control the process towards a certain phosphate setpoint and be reasonably certain that the total phosphorous concentration will also be in compliance.

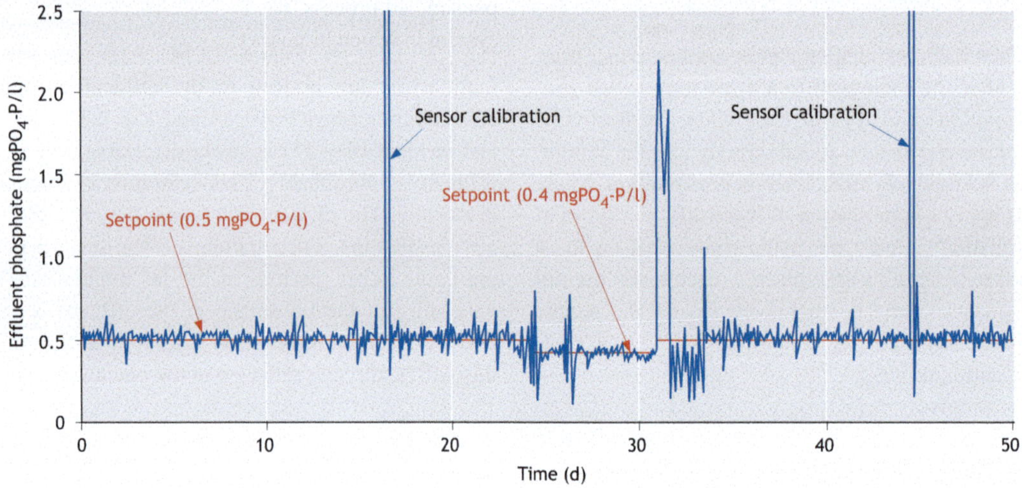

Figure 15.15 The effluent phosphate concentration with control of chemical dosage based on in-situ phosphate measurements (from Ingildsen, 2002).

The feedback controller works well in terms of precision. This can be used to quantify the savings when comparing various strategies. The *in-situ* feedback control with 0.5 mg/l as a setpoint is used as a basis for comparison. We can now compare the chemical dosage for the different strategies based on 90% compliance. If compliance in 90% of the time is accepted, this means that 10 % of the time the dosage may be less than in the feedback controller. This 90% compliance quantification is included to avoid extreme situations. In this case one line of the plant was operated with flow-proportional dosage while the parallel line was operated based on the phosphate measurements. The amount of dosage chemicals using the feedback control was decreased by more than 35% compared with the flow-proportional dosage control. The payback time for a phosphate analyser is consequently short, just a few months.

Example: Anaerobic reactor control

A main difficulty of anaerobic digestion (AD) is that it is often perceived as being unstable during both start-up and steady-state operations. Imbalances in the microbial ecosystem may lead to organic overload, causing severe reduction in degradation capability and washout of the microorganisms, resulting in poor reactor performance, reduced biogas production and even shut-down. The traditional way of avoiding this kind of instability is to operate the process far below its theoretical reactor capacity. In addition, the nature of the influent characteristics involves dynamic variation in both flow rate and composition, which are significant disturbances to the processes. Handling of these disturbances by attenuation or rejection is thus important for stable operation. A more economically viable approach to overcoming the problem is by applying close monitoring and automatic control of the process in order to enhance the operational stability, to attenuate and reject disturbances and to allow the treatment of waste and biogas production at a higher specific rate (Liu, 2003).

It is crucial to characterize the feed to a digester in order to obtain a reliable AD operation. Jinguara and Kamusoko (2017) present a review of feed characterization methods. Today there are analytical tools available to automatically test the biochemical methane (biogas) potential, the anaerobic biodegradability, as well as the specific methanogenic activity. Liu *et al.* (2019) describe the automation potential of feed testing for the non-specialist.

Many anaerobic bioreactors are still operated without close monitoring or control. This is not only because AD involves complicated degradation steps, but also due to lack of proper analytical devices. In fact, sensor technology is the weakest part of the process chain (Liu, 2003; Boe, 2006; Olsson *et al.*, 2005). Close monitoring and control of the AD process firstly requires identification of suitable process parameters, which can give indications of imbalances in the microbial ecosystem and warnings of external disturbances. The activity of the different microbial groups involved in the AD process can be measured indirectly by monitoring the metabolites. In general, it is now possible to analyse pH, alkalinity, biogas flow and composition, VFAs, biodegradable organic matter, dissolved hydrogen, and toxicity online by less expensive sensors and instruments (Liu, 2003).

The timing of measurements is important. Controlling the AD process by regulating the feed based only on the biogas flow rate is not an efficient control, simply because the gas flow rate is very late information for the controller. Any feed rate change will be noticed after a significant time delay. If any variable can be measured in the bioreactor itself, it will provide earlier and more useful information. Measuring the pH also adds important information, as demonstrated by Steyer *et al.* (1999). The alkalinity and hydrogen content in the gas phase have also been used. The feed rate is the typical control variable, since the organic loading rate should be adapted to the treatment capacity of the system in order to obtain an efficient process operation. Jimenez *et al.* (2015) present an overview of instrumentation and control problems in AD.

The start-up of anaerobic reactors is relatively slow due to low net growth of anaerobic biomass and required adaptation to specific wastewater components. Therefore, the start-up is a critical process that relies heavily on the operator skills. Also, after the start-up period the process is vulnerable to disturbances such as temporary overloading, biomass washout and toxicity. With a good control system the anaerobic reactor can – even during the transient start-up mode and at large loading – be operated close to its capacity and still maintain stable operation (Liu *et al.*, 2004).

More advanced control

If a control system cannot perform well enough on measured information by direct feedback of observed variables, then one can seek to incorporate a model of the system in the controller. Such a model forms the basis for more sophisticated predictive control. Consequently, simplified dynamical models that allow all the model parameters to be uniquely updated from available online measurements and with different predictive time horizons will prove useful. Naturally, the timescale of the models must also be related to the timescale in which the controlled variable can influence the process. The effectiveness to manage high process complexity by hierarchical and modular models has been demonstrated in numerous process industry applications. Consequently, it is suggested that future control systems for water and wastewater processes are based on similar principles.

To improve the reliability of a control system a fall-back control is essential. When severe problems occur, for example actuator or sensor failures, the control system should react to this and apply a robust control strategy that may not be optimal but will avoid significant process failures. Once the equipment functionality has been restored the control system can move the process back into a more efficient operating state. For any successful application of control it is also a necessity that the process is flexible enough to allow for a reasonable degree of freedom in terms of manipulation by the control system. Naturally, any new process should be designed for such flexibility rather than having to be subjected to costly redesign in the future. In many situations this is a major bottleneck for successful implementation of control in water and wastewater systems.

15.9 OPERATING COST SAVINGS DUE TO CONTROL

Electrical energy consumption is closely related to advanced wastewater treatment systems. Treatment and transmission of water and wastewater consume large amounts of energy. Estimates indicate that on average wastewater treatment alone requires around 1-3 % of the total electricity energy output of a country, representing a significant fraction of municipal energy bills. The energy utilisation of treatment plants is in the range of between 20 and 45 kWh per head of population per year. Older plants may have even higher demands. This figure does not include the primary pumping of the influent water. In this section we will emphasize how control and automation can reduce the electrical energy requirement. More detailed discussions on energy requirement in wastewater treatment are presented in Olsson (2015, chapters 15-19) and in Capodaglio and Olsson (2020).

DO control has been discussed and it is now obvious that even simple DO control, based on only one DO sensor, will save a lot of electrical energy compared to no control at all. Furthermore, having a time-variable setpoint of the DO concentration will further reduce the energy consumption, as discussed in Section 15.8. There are further possibilities to save energy in the DO control. For example, the air pressure can be minimized. Assume that the plant has two or more parallel aerators. The air system must supply sufficient air to the plant. However, sometimes the pressure can be lowered. This is noticed if the airflow valves are not fully open, which causes a pressure drop across the valves. The idea is to gradually decrease the airflow pressure so that the most open air flow valve becomes almost fully open. This will minimize the pressure drop and further energy savings are possible. Such control methods have been implemented; see *e.g.* Olsson and Newell (1999).

Large pumps, primarily for influent water, are often the most energy-demanding equipment in a plant. In many cases the pumping equipment has not been designed for adequate flow rates. If the pump is over-designed, then it may be operated with a low level of efficiency for small flow rates. In some cases it is profitable to install a special pump for small flow rates. The pump should be designed to work at the most efficient operation point for the most common flow rates (Olsson, 2015, Chapter 16).

Aeration by compressors ought to be continuously variable. However, to control airflow by closing airflow valves will cause considerable energy loss. Variable speed compressors will save energy significantly. Typically, the power requirement for the airflow is proportional to n^3, where n is the rotational speed. This means that only 1/8 of the power is needed to produce half the flow rate. Consequently, the potential for energy savings is considerable.

The cost for chemical precipitation is significant. In Section 15.8 we demonstrated that feedback control can contribute to much lower operating costs.

A wastewater treatment plant in fact should be considered a recovery plant for both nutrients and energy. Considering the energy potential in anaerobic digestion, there is a huge unused potential in most plants. Just one plant (in Sweden) can illustrate this: the plant uses 41 kWh/person.year of electrical energy. At the same time the plant produces biogas corresponding to 72 kWh/person.year. Furthermore, the heat content of the effluent water is taken care of in heat pumps that produce 336 kWh/person/year. The plant is in fact an important energy producer.

As discussed in Section 15.8, AD is a crucial process to obtain a zero-energy plant. In too many plants the AD uses only a fraction of the energy content of the sewage. Major savings happen when adequate instrumentation and control allows the plant to operate at a higher capacity. An implementation of control that avoids a plant extension has a very favourable cost-benefit ratio. By-products from sewage treatment can also provide a valuable source of energy if managed and utilized effectively. In addition, costs of sludge transportation and disposal, which currently place a major burden on the industry, can be reduced.

15.10 INTEGRATION AND PLANT-WIDE CONTROL

Integration aims at minimizing the impact on the receiving water, while ensuring a better resource utilisation. The system resilience is an important factor. This includes its ability to attenuate disturbances but should also reflect its sensitivity to major disturbances or even purposeful and harmful attacks. In the integrated approach the goal is to formulate criteria for the receiving water and its ecological quality while satisfying various economic and technical constraints. It is a major challenge to relate this performance to the plant effluent and possible sewer overflow. We need performance measures of the plant operation that relate effluent quality to the resources that are needed to obtain it, such as energy, chemicals, and other material and operating costs. Models are being developed to find strategies to dynamically find maximum WWTP loading according to continuous monitoring and prediction of the operational state. One example is maximizing the nitrification capacity in the activated sludge process, depending on the load to the system. Some full-scale results are reported by Rosen *et al.* (2004, 2006). Another aspect is storage management (in the sewer system and in retention tanks), not only during storms but also during normal operations. By mixing different types of wastewater to compensate *e.g.* for nutrient deficit or overload, the capacity of the plant can be maximized.

All integration means some compromise. If there were no interactions, then the individual optimization of each sub-process would be the best strategy. An integrated view of the operation will give a better result than if each one of the processes were controlled separately. This is the essence of a multi-criteria index: various performances are weighted and compared with each other. This can be illustrated with some examples:

- The interaction between the aerator and the settler is a classical integration problem, reflected in the compromise that must be made in the control of return sludge flow rate.
- The anoxic zone in a pre-denitrifying plant interacts closely with the nitrifying aerator. Oxygen-rich water is recirculated from the aerobic zone to the anoxic zone. The DO level towards the outlet of the nitrification zone must be a compromise between sufficient nitrification and the denitrification following the nitrification.
- There is interplay between serially linked processes. For example, chemical pre-precipitation in a primary settler will remove not only phosphates but also particulate organic material, saving aeration energy. On the other hand, a pre-denitrification may then obtain too little carbon. Similarly, if the precipitation is combined with a bio-P process the latter may be carbon-limited.

- Recycle streams interconnect various parts of a treatment plant. Supernatants from sludge treatment are most often highly concentrated in nutrients and should be synchronized in time with the plant influent load.
- Backwash water from deep-bed filters is recirculated to the input of the plant. Since the flow rates are often significant a synchronized control of the flow rate to the plant load is necessary.
- The target for the sludge production is not the same in different plants. Sometimes the target is to maximize the methane production, while at other plants the sludge production needs to be minimized.

In the combined sewer and WWTP operation the individual system operations are sometimes in conflict, so the overall goal of minimizing the load to the receiving water must overrule the individual goals (Rauch and Harremoës, 1996a; Schütze *et al.* 1999; Vanrolleghem *et al.*, 1996b). An early approach to integrated control was published by Rauch and Harremoës (1996b).

A plant-wide control system will assume that all the different unit processes are controlled locally. In addition it will consider the interaction between different parts of the plant, for example by computing suitable setpoints for the local controllers. The sewer control system will control the flow rate in the various parts of the sewer system using the information from water level and flow rate sensors, pumping equipment, as well as rain gauges. The coupling between the sewer system and treatment plant control is achieved when the plant influent flow rate can be predicted and manipulated. Typical measurements and control handles for the interacting sewer system and WWTP are listed in Table 15.3.

Table 15.3 The objectives, measurements and control handles for a combined sewer-wastewater treatment system operation.

	Partial aim	Measurements	Control handles
Sewer system	Minimize upstream overflow Utilize basins for most polluted water	Rain levels Flow rates	Pumping stations Adjustable weirs Basins
Wastewater treatment plant	To treat as much wastewater as possible during and after rainfall Reduce hydraulic load and sludge load in secondary sedimentation tanks	Flow rates (inlet, outlet, return sludge, recycles) Suspended solids (aeration tanks and return sludge) Sludge blanket	Return sludge pumping (control of sludge blanket in sec. sedimentation tanks) ATS control (sedimentation in aeration tanks) Primary pumping (bypass before biological section or the total plant)

Following the development of models, researchers together with operators have developed necessary technologies and experience for reliable online monitoring of sewer systems which complements the more established monitoring of treatment plants. Integrated control has now been realized in real life, for example in The Netherlands (Weijers *et al.*, 2012), in Denmark (Grum *et al.*, 2011), and in Germany (Seggelke *et al.*, 2013). Benedetti *et al.* (2013) present a comprehensive review of the integration challenges and solutions.

15.11 CONCLUDING REMARKS

Automation is the method of making a process or a system operate and adapt automatically. Uncertainty in the process or in the environment around the process makes automation both an opportunity and a great challenge. Disturbances are everywhere and are the main reason for control. Another reason is to be able to operate at optimum, and control would be relevant even if there were no disturbances.

Application of automation in wastewater treatment operation can be said to have two primary functions: information acquisition and process control. For the former function, the level of automation is relatively high. Many, often thousands of variables, are today gathered online in the SCADA systems of treatment plants, and data analyses with varying degrees of sophistication are standard components of the treatment operation and quality monitoring. However, the latter function, process control, is less developed and often limited to a few unit process control loops. It should be noted that plant operation becomes sub-optimized with only local controllers. The potential of plant-wide automation is to coordinate the various unit processes so that the overall performance requirements are better fulfilled. Succeeding with this means that automation gets a third purpose beyond information acquisition and process control. This can be called overall optimisation and is realized in the integrated operation of the urban water cycle. Encouraging developments have been reported during the last few years. It is quite apparent that effective operation must rely on functioning equipment. All the links in the chain must be working to obtain a good operational system. The hardware includes not only the instrumentation, but also all the various actuators, such as compressors, pumps, motors and valves. Note that if all the components are not functioning at all times, fall-back strategies must be in place.

Future development will be exploiting the enormous capacity of data distribution that is possible today. Many SCADA systems are also applying the technology from the Internet, which gives an almost unlimited potential for remote data evaluation and decision. The distributed control room is already here. Although there is a limit to how much expertise a treatment plant can afford, given that plant data can be made available anywhere it is possible to utilize specialist competencies wherever they are located. Nevertheless, there are several human and managerial aspects of how to distribute the responsibility and decision-making in various sectors. There is also already commercial software available for decentralized process monitoring and control. Naturally there has to be caution against publicizing sensitive data or against misuse of information and there is also a need to guarantee that data from each individual plant is correctly interpreted.

The increasing incorporation of ICA in water treatment operation is not only driven by the impressive technical development of instrumentation and computer technology, modelling and control, and the progress in automation. It is also motivated by economic and environmental obligations and it turns out to be a necessary and worthwhile investment. It is already proven in several installations that ICA investments have paid off quickly and we will see that ICA will become an increasing part of the total investment.

Mounting evidence has become available demonstrating that the microbial populations and their properties are jointly determined by the wastewater composition and the design and operation of a treatment system. The impact of control systems on the microbial communities has not attracted much attention in the past, and sludge population optimization through online process control is still an emerging concept (Yuan and Blackall, 2002). Fundamental studies to understand how certain microorganisms are selected and how bacterial properties are influenced by particular plant designs and operations are of vital importance and need to be carried out systematically. Modern molecular techniques allowing the identification and quantization of microorganisms present in a system are indispensable tools for these studies. The most rapid fundamental advances will come from the incorporation of detailed micro-scale data into current mathematical models such that these models more closely represent the sludge processes, allowing model-based sludge population optimization. A great deal of effort from both microbiologists and engineers is still required for the practical application of these methods in the context of process control. The close collaboration between microbiologists and engineers cannot be over-emphasized.

ICA is often perceived as the hidden technology. You will only notice it when it does *not* work. The complexity of modern WWTPs is often reflected in ICA systems. Several specialties must cooperate to achieve one system of process technology and automation. The challenge of automation is to comprehend the system aspects from a unit process perspective and to understand the process aspects from a system perspective. This challenge has profound consequences on the profession and on fundamental educational approaches, not the least in civil and environmental engineering curricula. Process specialists and designers should be able to appreciate the implications of ICA and computer and control engineers must understand the process controllability and its constraints. This further emphasizes the multi-disciplinary character of water operations.

The primary aim of developing our utilities into a higher level of smart operation is to ensure not only that the water system is operated more efficiently but also adapting to the environment and the human society in a sustainable fashion. We ought to try to understand our human infrastructure systems as an integral part of the nature we inhabit. We need to sense our environment, our systems and our usage to make sure that the three systems are working in synchronicity. Indeed, we must think in terms of 'partnership with nature' – sustaining, preserving, restoring and learning from nature in all our operations with water. This is not in conflict with the original aim of applying instrumentation, control and automation (ICA) to make operations less expensive. It is a much loftier goal to reach true sustainability and the task is much more complex, requiring a deeper understanding of our infrastructure systems, the natural physical and biological systems they interact with, and our usage.

We should think in terms of feedback at all levels, from the equipment to high-level management and strategic decisions. Also, note that feedback should be used to ensure our continuous learning. When we troubleshoot and implement the wrong solutions occasionally, we should use 'feedback thinking' to find a way to learn from our experiences. The framework is always the same, whereas the measurements, the analysis and the decisions are different. This involves:

- Making use of huge amounts of data,
- Analysing the data to become reliable and useful information,
- Exhausting the possibilities with digital communication,
- Using the potential of control and decision methods at all levels of the operation.

It is also still true many years into the future: water is life and we must treat it wisely. As we said more than two decades ago and it should still be kept in mind (Olsson and Newell, 1998), 'Our societies will need clean water and clean air. Sustainability will not only be a matter of cost. In fact, it is already a matter of survival in some countries. What role will ICA play in this development and how can we meet that challenge?' This is our challenge for years to come.

REFERENCES

Ahm M., Thorndahl S., Nielsen J.E. and Rasmussen M.R. (2016). Estimation of combined sewer overflow discharge: a software sensor approach based on local water level measurements. *Water Science and Technology,* 74(11), 2683-2696.

Åmand L., Olsson G. and Carlsson B. (2013). Aeration control - a review. *Water Science and Technology,* 67(11), 2374-2398.

Åmand L., Laurell C., Stark-Fujii K., Thunberg A. and Carlsson B. (2014). Lessons learnt from evaluating full-scale ammonium feedback control in three large wastewater treatment plants. *Water Science and Technology,* 69(7), 1573-1580.

Åström K.J. and Murray R.M. (2014). Feedback Systems: An Introduction for Scientists and Engineers. The complete book is available free of charge on the web, called FBSwiki http://www.cds.caltech.edu/~murray/amwiki/index.php/Main_Page (accessed 21 January 2020).

Batstone D.J., Keller J., Angelidaki R.I., Kalyuzhnyi S.V., Pavlostathis S.G., Rozzi A., Sanders W.T.M., Siegrist H. and Vavilin V.A. (2002). *Anaerobic Digestion Model No.1.* Scientific and Technical Report No.13, IWA Publishing, London.

Beck M.B (2005). Vulnerability of water quality in intensively developing urban watersheds. *Environmental Modelling & Software*, 20, 381-400.

Benedetti L., Langeveld J., Comeau A., Corominas L., Daigger G., Martin C., Mikkelsen P.S., Vezzaro L., Weijers S. and Vanrolleghem P.A. (2013). Modelling and monitoring of integrated urban wastewater systems: review on status and perspectives. *Water Science and Technology,* 68(6), 1203-1215.

Boe K. (2006). Online monitoring and control of the biogas process. PhD thesis, Danish Technical University (DTU). Available at https://backend.orbit.dtu.dk/ws/portalfiles/portal/127333186/-MR2006_055.pdf (accessed 20 January 2020).

Capodaglio A.G. and Olsson G. (2020). Energy issues in sustainable urban wastewater management: use, demand reduction and recovery in the Urban Water Cycle. *Sustainability 2020,* 12, 266; doi:10.3390/su12010266.

Corominas L., Garrido-Baserba M., Villez K., Olsson G., Cortés U. and Poch M. (2018). Transforming data into knowledge for improved wastewater treatment operation: A critical review of techniques. *Environmental Modelling and Software,* 106, 89-103.

Flores X., Comas J., Roda I.R., Jiménez L. and Gernaey K.V. (2007). Application of multivariable statistical techniques in plant-wide WWTP control strategies analysis. *Water Science and Technology,* 56(6), 75-83; DOI: 10.2166/wst.2007.586.

Gernaey K.V., Nielsen M.K., Thornberg D., Höök B., Munk-Nielsen T., Ingildsen P. and Jørgensen S.B. (2004). First principle suspended solids distribution modelling to support ATS introduction on a recirculating WWTP. *2nd Int. IWA Conference on Automation in Water Quality Monitoring*, Vienna.

Grum M., Thornberg D., Christensen M.L., Shididi S.A. and Thirsing C. (2011). Full-scale real time control demonstration project in Copenhagen's largest urban drainage catchments. In Proceedings of the 12th International Conference on Urban Drainage, Porto Alegre, Portugal.

Hadjimichael A., Comas J. and Corominas L. (2016). Do machine learning methods used in data mining enhance the potential of decision support systems? A review for the urban water sector. *AI Commun.* 29, 747–756. doi:10.3233/AIC-160714.

Haimi H., Mulas M., Corona F. and Vahala R. (2013). Data-derived soft-sensors for biological wastewater treatment plants: an overview. *Environmental Software & Modelling*, 47, 88-107.

Hamouda M.A., Anderson W.B. and Huck P.M. (2009). Decision support systems in water and wastewater treatment process selection and design: a review. *Water Science and Technology,* 60(7), 1757-1770; DOI: 10.2166/wst.2009.538.

Henze M., Gujer W., Mino T. and van Loosdrecht M.C.M. (2000). *Activated Sludge Models: ASM1, ASM2, ASM2d and ASM3.* Scientific and Technical Report No. 9, IWA Publishing, London.

Ingildsen P. (2002). Realising full-scale control in wastewater treatment systems using in situ nutrient sensors. PhD thesis, Dept. of Ind. Electrical Engineering and Automation (IEA), Lund University, Lund. Available at https://www.iea.lth.se/publications/Theses/LTH-IEA-1030.pdf (accessed 20 January 2020).

Ingildsen P. and Olsson G. (2016). *Smart water utilities.* IWA Publications, London.

Irizar I., Alferes J., Larrea L. and Ayesa E. (2008). Standard signal processing using enriched sensor information for WWTP monitoring and control. *Water Science and Technology,* 57(7), 1053-1060; DOI: 10.2166/wst.2008.139.

ISO (2003). *ISO15839:2003 Water Quality – On-line sensors/analysing equipment for water – Specifications and performance tests.* First edition. International Standard Organization.

Jimenez J., Latrille E., Harmand J., Robles Martínez Á., Ferrer Pol J., Gaida D. and Wolf C. (2015). Instrumentation and control of anaerobic digestion processes: a review and some research challenges. *Reviews in Environmental Science and Biotechnology*, 14(4), 615-648. doi:10.1007/s11157-015-9382-6.

Jinguara R.M. and Kamusoko R. (2017). Methods for determination of biomethane potential of feedstock: a review. *Biofuel Research Journal,* 14, 573-586.

Lennox J. and Rosen C. (2002). Adaptive multiscale principal components analysis for online monitoring of wastewater treatment. *Water Science and Technology*, 45(4-5), 227–235.

Leonhardt G., Fach S., Engelhard C., Kinzel H. and Rauch W. (2012). A software-based sensor for combined sewer overflows. *Water Science and Technology,* 66(7), 1475–1482.

Liu J. (2003) Instrumentation, Control and Automation in Anaerobic Digestion. PhD thesis, Dept Biotech., Lund University, Lund. https://lup.lub.lu.se/search/publication/466476 (accessed 20 January 2020).

Liu J., Olsson G. and Mattiasson B. (2004) Control of an Anaerobic Reactor towards Maximum Biogas Production, *Water Science and Technology*, 50(11), 189-198.

Liu J., Strömberg S. and Nistor M. (2019). How standardization and automation of lab equipment could help biogas research. Available at https://www.bioprocesscontrol.com/media/2132/how-standardization-and-automation-of-lab-equipment-could-help-biogas-research.pdf (accessed 20 January 2020).

Lumley D. (2002). On-line instrument confirmation: how can we check that our instruments are working? *Water Science and Technology*, 45(4-5), 469–476.

Mannina G., Ekama G., Caniani D., Cosenza A., Esposito G., Gori R., Garrido-Baserba M., Rosso D. and Olsson G. (2016). Greenhouse gases from wastewater treatment – a review of modelling tools. *Journal Science of the Total Environment*, 551-552, 254-270.

Nielsen M.K., Carstensen J. and Harremöes P. (1996). Combined control of sewer and treatment plant during rainstorm. *Water Science and Tech*nology, 34(3-4), 181-187.

Nowak O., Keil S. and Fimml C. (2011). Examples of energy self-sufficient municipal nutrient removal plants. *Water Science and Technology,* 64(1), 1-6.

Olsson G. (2012). ICA and me – a subjective review. *Water Research,* 46(6), 1585-1624.

Olsson G. (2015). *Water and energy – threats and opportunities*, 2nd edition. IWA Publications, London.

Olsson G., Eklund K., Dahlqvist K.I. and Ulmgren L. (1973). Control Problems in Wastewater Treatment Plants. (Research Report TFRT-3064). Department of Automatic Control, Lund Institute of Technology, Sweden. Available at http://portal.research.lu.se/portal/files/4620667/8726145.pdf (accessed 20 January 2020).

Olsson G. and Newell R.B. (1998). Talking of RAS - Reviewing, Assessing and Speculating), Final address at the 7th IAWQ Symposium on ICA, Brighton, UK, July 1997, *Water Science and Technology,* 37(12), 397-401.

Olsson G. and Newell B. (1999). *Wastewater Treatment Systems. Modelling, Diagnosis and Control.* IWA Publishing, London.

Olsson G., Nielsen M.K., Yuan Z., Lynggaard-Jensen A. and Steyer J.P. (2005). *Instrumentation, Control and Automation in Wastewater Treatment Systems.* Scientific and Technical Report No.15, IWA Publishing, London.

Olsson G., Carlsson B., Comas J., Copp J., Gernaey K.V., Ingildsen P., Jeppsson U., Kim C., Rieger L., Rodríguez-Roda I., Steyer J.-P., Takács I., Vanrolleghem P.A., Vargas Casillas A., Yuan Z. and Åmand L. (2014). Instrumentation, Control and Automation in wastewater – from London 1973 to Narbonne 2013. *Water Science and Technology,* 69(7), 1373-1385. doi: 10.2166/wst.2014.057.

Olsson G., Åmand L., Rieger L. and Carlsson B. (2019a). Aeration control – fundamentals. Chapter 4 in *Aeration, mixing and energy. Bubbles and sparks.* (D. Rosso, editor), IWA Publishing, London.

Olsson G., Åmand L., Rieger L. and Carlsson B. (2019b). Aeration control – implementation. Chapter 5 in *Aeration, mixing and energy. Bubbles and sparks.* (D. Rosso, editor), IWA Publishing, London.

Rauch W. and Harremoës P. (1996a). The importance of the treatment plant performance during rain to acute water pollution. *Water Science and Technology*, 34(3-4), 1-8.

Rauch W. and Harremoës P. (1996b). Minimizing acute river pollution from urban drainage systems by means of integrated real time control. *Proc. 1st Int. Conf. On New/Emerging Concepts for Rivers (RIVERTECH '96)*, 22-26 September, Chicago US.

Racault Y., Stricker A.-E., Husson A. and Gillot S. (2011). Monitoring the variations of the oxygen transfer rate in a full-scale membrane bioreactor using daily mass balances. *Water Science and Technology,* 63(11), 2651–2657.

Rodriguez-Roda I., Sànchez-Marrè M., Comas J., Baeza J., Colprim J., Lafuente J., Cortés U. and Poch M. (2002). A hybrid supervisory system to support WWTP operation: implementation and validation. *Water Science and Technology,* 45(4-5), 289-297.

Rosen C. and Olsson G. (1998). Disturbance detection in wastewater treatment systems. *Water Science and Technology*, 37(12), 197–205.

Rosen C. and Lennox J.A. (2001). Adaptive and multiscale monitoring of wastewater treatment operation. *Water Research*, 35(14), 3402–3410.

Rosen C., Röttorp J. and Jeppsson U. (2003). Multivariate on-line monitoring: challenges and solutions for modern wastewater treatment operation. *Water Science and Technology,* 47(2), 171–179.

Rosen C., Ingildsen P., Guildal T., Nielsen M.K., Jacobsen B.N. and Ønnerth T. (2004). On-line estimation of nitrification and denitrification capacity at Avedøre wastewater treatment plant. *Proc. Int. Conf. Upgrading of Wastewater Treatment Plants (AquaTech2004)*, 30 September-1 October, Amsterdam.

Rosen C., Ingildsen P., Guildal T., Nielsen M.K., Munk-Nielsen T., Jacobsen B.N. and Thomsen H. (2006). Introducing biological phosphorous removal in an alternating plant by means of control – a full scale study. *Water Science and Technology*, 53(4-5), 133-141.

Ruiz M., Sin G., Berjaga X., Colprim J., Puig S. and Colomer J. (2011). Multivariate Principal Component Analysis and Case-Based Reasoning for monitoring, fault detection and diagnosis in a WWTP. *Water Science and Technology*, 64(8), 1661–1667.

Schütze M., Butler D. and Beck M.B. (1999). Optimisation of control strategies for the urban wastewater system – an integrated approach. *Water Science and Technology*, 39(9), 209-216.

Seggelke K., Löwe R., Beeneken T. and Fuchs L. (2013). Implementation of an integrated real-time control system of sewer system and waste water treatment plant in the city of Wilhelmshaven. *Urban Water Journal*, 10(5), 330-341.

Steyer J.P., Buffiére P., Rolland D. and Moletta R. (1999). Advanced control of anaerobic digestion processes through disturbances monitoring. *Water Research,* 33(9), 2059-2068.

Vanrolleghem P., Fronteau C. and Bauwens W. (1996a). Evaluation of design and operation of the sewage transport and treatment system by an EQO/EQS based analysis of the receiving water immission characteristics. *Proc. WEF Conf. Urban Wet Weather Pollution*, 16-19 June, Québec, 14.35-14.46.

Vanrolleghem P.A., Jeppsson U., Carstensen J., Carlsson B. and Olsson G. (1996b). Integration of wastewater treatment plant design and operation - a systematic approach using cost functions. *Water Science and Technology,* 34(3-4), 159–171.

Villez K., Ruiz M., Sin G., Colomer J., Rosen C. and Vanrolleghem P.A. (2008). Combining Multiway Principal Component Analysis (MPCA) and clustering for efficient data mining of historical data sets of SBR processes. *Water Science and Technology*, 57(10), 1659–1666.

Villez K., Rosen C., D'hooge E. and Vanrolleghem P.A. (2010). Online phase length optimization for a sequencing batch reactor by means of the Hotelling's T2 statistic. *Industrial & Engineering Chemistry Research,* 49, 180–188.

Weijers S.R., De Jonge J., Van Zanten O., Benedetti L., Langeveld J., Menkveld H.W. and Van Nieuwenhuijzen A.F. (2012). KALLISTO: cost effective and integrated optimization of the urban wastewater system Eindhoven. *Water Practice and Technology*, 7(2), 1-9.

Yuan Z. and Blackall L. (2002). Sludge Population Optimisation, A New Dimension for the Control of Biological Wastewater Treatment Systems. *Water Research,* 36 (2), 482-490.

Yuan Z., Olsson G., Cardell-Oliver R., Van Schagen K., Marchi A., Deletic A., Urich C., Rauch W., Liu Yanchen and Jiang Guangming (2019). Sweating the assets – the role of instrumentation, control and automation in urban water systems. *Water Research*, 155, 381-402.

NOMENCLATURE

Abbreviation	Description
AD	Anaerobic digestion
ATS	Aeration tank settling
BOD	Biological oxygen demand
DO	Dissolved oxygen
ICA	Instrumentation, control and automation
IWA	International water association
PI	Proportional-integral
PID	Proportional-integral-derivative
SCADA	Supervisory control and data acquisition
SRT	Sludge retention time
VFA	Volatile fatty acid

Figure 15.16 Detail of measuring and control system of activated sludge aeration tank (photo: D. Brdjanovic).

doi: 10.2166/9781789060362_0701

16

Anaerobic wastewater treatment

Jules B. van Lier, Nidal Mahmoud and Grietje Zeeman

16.1 SUSTAINABILITY IN WASTEWATER TREATMENT

16.1.1 Definition and environmental benefits of anaerobic processes

Anaerobic digestion refers to the stabilisation of organic matter without the presence of oxygen; its main end product is biogas, which is composed of mainly methane and carbon dioxide. Anaerobic digestion processes occur in many natural environments where organic material is available and redox potential is low (zero oxygen). This is typically the case in the stomachs of ruminants, in marshes, sediments of lakes and ditches, municipal landfills, and even municipal sewers. Anaerobic treatment refers to a treatment technology that stabilises organic waste, or organic pollutants in wastewater, without the need for aeration. During anaerobic treatment, biodegradable organic compounds are mineralised, leaving inorganic compounds such as NH_4^+, PO_4^{3-}, and HS^- in the solution. Anaerobic treatment can be conducted in technically simple systems, and the process can be applied at any scale and almost anywhere. Moreover the amount of excess sludge produced is very small and well stabilised, and there is even market value when granular anaerobic sludge is produced in the bioreactor. Moreover, during treatment, useful energy in the form of biogas is produced instead of high-grade energy being consumed. Accepting that anaerobic digestion merely removes organic pollutants, there are very few if any serious remaining drawbacks, even with respect to the rate of start-up of the system. Figure 16.1 shows the fate of carbon and energy in both aerobic and anaerobic wastewater treatment (AnWT), assuming that the oxidation of 1 kgCOD requires 0.5-1 kWh of aeration energy, depending on the applied aeration system and reactor type. In contrast to anaerobic treatment, aerobic treatment is generally characterised by high energy costs, and a very large fraction of the waste is converted into another type of waste (sludge). Aerobic treatment in a conventional activated sludge process yields approximately 50% (or even more) newly grown sludge from the COD converted, which requires further treatment, *e.g.* anaerobic digestion, before it is reused, disposed of, or incinerated. The carbon/energy flow principles of aerobic and anaerobic bio-conversion have considerable effect on the set-up of the corresponding wastewater treatment system. Not surprisingly, to date, AnWT has evolved into a competitive wastewater treatment technology. Many different types of organically-polluted wastewaters, even those that were previously believed not to be suitable for AnWT, are now being treated by anaerobic high-rate conversion processes.

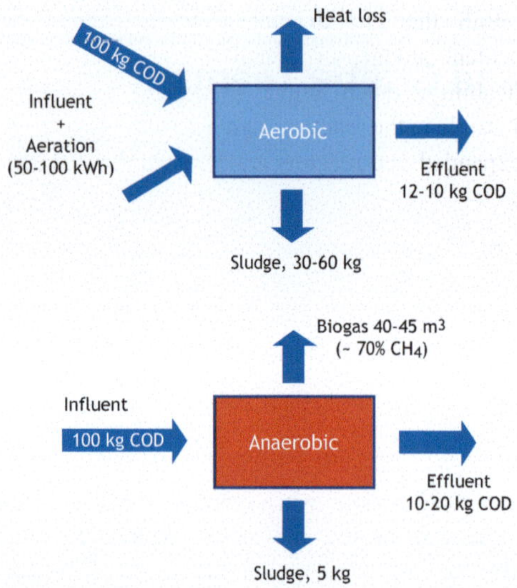

Figure 16.1 Fate of carbon and energy in aerobic (above) and anaerobic (below) wastewater treatment.

Since the 1970s, high-rate anaerobic treatment has been particularly applied to organically-polluted industrial wastewaters that have come from the agro-food sector and the beverage industry (Table 16.1). The shown percentages more or less reflect the present distribution. Currently, in more than 90% of these applications, anaerobic sludge bed technology is applied, for which the presence of granular sludge is of eminent importance. Interestingly, both the number of anaerobic reactors installed, as well the application potential of anaerobic wastewater treatment, is rapidly expanding. The authors estimate that the current number of installed anaerobic high-rate reactors exceeds 4,500; wastewaters are now being treated that were previously not considered suitable for anaerobic treatment, such as chemical wastewaters containing toxic compounds or wastewaters with a complex composition. For the more extreme type of wastewaters, novel high-rate reactor systems have been developed as discussed in section 16.7.2.

Analysing the reasons why AnWT has been selected, the following striking advantages of AnWT over conventional aerobic treatment systems can be given:

- Reduction of excess sludge production by up to 90%.
- up to 90% reduction in space requirement when using expanded sludge bed systems.
- High applicable COD loading rates reaching 20-35 kg COD per m^3 of reactor per day, requiring smaller reactor volumes.

Table 16.1. Worldwide application of anaerobic technology for industrial wastewater treatment. Total number of registered installed reactors = 2,266, according to a survey in January 2007 (adopted from Van Lier, 2008).

Industrial sector	Type of wastewater	Installed reactors[a] (% of total)
Agro-food	Sugar, potato, starch, yeast, pectin, citric acid, canning, confectionary, fruit, vegetable, dairy, bakery	36
Beverage	Beer, malting, soft drink, wine, fruit juice, coffee	29
Alcohol distillery	Cane juice, cane molasses, beet molasses, wine, grain, fruit	10
Pulp & paper	Paper recycling, mechanical pulp, NSSC[b] pulp, sulphite pulp, straw, bagasse	11
Miscellaneous	Chemical, pharmaceutical, sludge liquor, landfill leachate, acid mine water, municipal sewage	14

[a] various types of high-rate anaerobic reactor systems; [b] neutral sulphite semi-chemical.

- No use of fossil fuels for treatment, saving approximately 0.5-1.0 kWh/kgCOD removed, depending on aeration efficiency.Production of 13.8 MJ CH_4energy/kgCOD removed, giving approximately 1.5 kWh electricity (assuming 40% electric conversion efficiency). The value of 13.8 MJ/kgCOD refers to the high heating value (HHV) of dry CH_4. Since biogas is wet, the value of 12.4 MJ/kgCOD should be used for energy recovery calculations, which refers to the low heating value (LHV). The LHV corrects for the evaporation energy losses of the condensed water in biogas.
- Rapid start-up (< 1 week), using granular anaerobic sludge as seed material.
- No or very little use of chemicals.
- Simple technology with high treatment efficiencies.
- Anaerobic sludge can be stored unfed, and reactors can be operated when needed (*e.g.* during agricultural campaigns of 4 months per year in the sugar industry).
- Excess sludge has a market value. In Europe, prices range between 100-200 €/m^3 including transport costs.
- High-rate systems facilitate water recycling in factories which is a step towards closing loops in industrial processes.

Obviously, the exact ranking of the above advantages depends on the local economic and societal conditions. In the Netherlands, excess sludge handling is the cost-determining factor in operating wastewater treatment systems. Since landfill is not an option for excess sewage sludge and biowaste, and prices for incineration can reach €400-500/ton wet sludge, the low sludge production in anaerobic reactors is an immediate economic benefit. The system compactness, another important asset of AnWT, can be illustrated by a full-scale example, where an anaerobic reactor with a 6 m diameter and a height of 25 m suffices to treat up to 25 tons of COD daily. The produced sludge, which is less than 1 ton dry matter per day in this example, is not a waste product, but is marketed as seed sludge for new reactors. Such compactness makes the system suitable for implementation on industrial premises or sometimes even inside the factory buildings where it was produced. This is of particular interest in densely populated areas and for those industries that aim to reclaim process water, using anaerobic treatment as the first treatment step.

The renewed interest in the energy aspects of AnWT directly results from rising energy prices, fossil fuel concerns and the overall concern about global warming. The above 25 tons COD/d of agro-industrial waste(water) can be converted into 7,000 m^3CH_4/d (assuming 80% CH_4 recovery), with an energy equivalent of approximately 250 GJ/d. Working with a modern combined heat power (CHP) gas engine, reaching 40% efficiency, a useful 1.2 MW electric power output can be achieved (Table 16.2). The overall energy recovery could be even higher (reaching up to 60%) if all the excess heat can be used on the industrial premises or in the direct vicinity. Assuming that full aerobic treatment would require approximately 0.5-1 kWh/kgCOD removed, or up to 1 MW installed electric power in the above case, the total energy benefit of using AnWT over the activated sludge process is 2.2 MW. At an energy price of 0.1 €/kWh this equals approximately 5,000 €/d. Apart from the energy itself, the carbon credits that can be obtained by generating renewable energy using AnWT could also become of interest (Table 16.2). For

Table 16.2 Energy output and CO_2 emission reduction applying in anaerobic high-rate wastewater treatment systems.

Loading capacity (kgCOD/m^3.d)	5-40
Energy output (MJ/m^3 reactor installed per d)	50-400
Electric power output (kW/m^3 reactor installed)	0.25-1.9
CO_2 emission reduction (tonCO_2/m^3.y, based on coal-driven power plants)	1.9-14

Assumptions: 80% CH_4 recovery relative to influent COD load and 40% electric conversion efficiency using a modern combined heat power generator. Calculations based on lower heating value of CH_4 (12.4 MJ/kg COD).

an average coal-driven power plant, the generation of 1 MW electricity emits approximately 21 tonCO_2/d, whereas for a natural gas-driven plant it is half that value. At an anticipated stabilised price of 20 €/ton CO_2, the above example industry could earn 500 €/d on carbon credits (based on a coal-powered plant), and no fossil fuels are used for treating the wastewater. Although this amount is negligible in industrialised countries, it could provide a real incentive in developing countries to start treating wastewater using high-rate AnWT, and thereby protecting the local environment.

Table 16.2 gives a summary of the expected energy output as well as the predicted CO_2 emission reduction (if the produced CH_4 is converted to electricity) of an anaerobic reactor, operated at commercially available organic loading rates.

16.2 MICROBIOLOGY OF ANAEROBIC CONVERSIONS

16.2.1 Anaerobic degradation of organic polymers

The anaerobic degradation pathway of organic matter is a multi-step process of series and parallel reactions. This process of organic matter degradation proceeds in four successive stages, namely: (*i*) hydrolysis, (*ii*) acidogenesis, (*iii*) acetogenesis, and (*iv*) methanogenesis. These are discussed below.

Methanogenic bacteria are located at the end of the anaerobic food chain and, partly thanks to their activity, no large quantities of organic matter accumulate in anaerobic environments, where this matter is inaccessible to aerobic organisms. The anaerobic digestion process involves a complex food web, in which organic matter is sequentially degraded by a wide variety of microorganisms. The microbial consortia involved jointly convert complex organic matter and ultimately mineralize it into methane (CH_4), carbon dioxide CO_2, ammonium (NH_3), hydrogen sulphide (H_2S) and water (H_2O).

The anaerobic ecosystem is the result of complex interactions among microorganisms of several different species. The major groupings of bacteria and the reaction they mediate are: (*i*) fermentative bacteria, (*ii*) hydrogen-producing acetogenic bacteria, (*iii*) hydrogen-consuming acetogenic bacteria, (*iv*) carbon dioxide-reducing methanogens, and (*v*) aceticlastic methanogens. The reactions they mediate are presented in Figure 16.2.

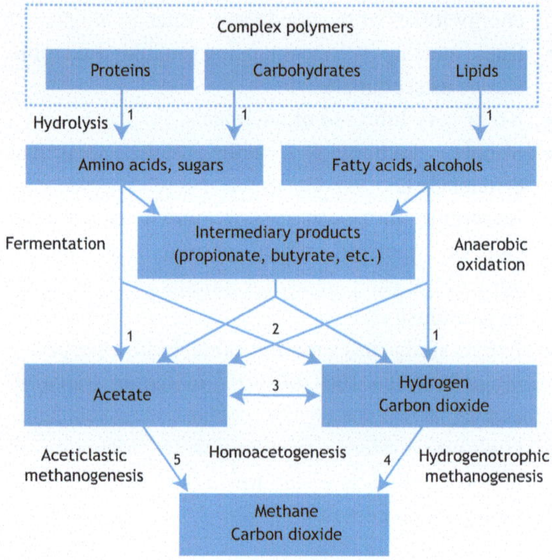

Figure 16.2 Reactive scheme for the anaerobic digestion of polymeric materials. The numbers indicate the bacterial groups involved: 1. Hydrolytic and fermentative bacteria; 2. Acetogenic bacteria; 3. Homo-acetogenic bacteria; 4. Hydrogenotrophic methanogens; 5. Aceticlastic methanogens (after Gujer and Zehnder, 1983).

The digestion process can be subdivided into the following four phases:

1) *Hydrolysis*, where hydrolytic-fermentative bacteria colonise solid substrates as the first step in substrate solubilisation, or where enzymes excreted by fermentative bacteria (referred to as 'exo-enzymes') convert complex, undissolved material into less complex, dissolved compounds which can pass through the cell walls and membranes of the fermentative bacteria.

2) *Acidogenesis*, where the dissolved compounds present in cells of fermentative bacteria are converted into a number of simple compounds which are then excreted. The compounds produced during this phase include volatile fatty acids (VFAs), alcohols, lactic acid, CO_2, H_2, NH_3 and H_2S, as well as new cell material.
3) *Acetogenesis* (intermediary acid production) where digestion products are converted into acetate, hydrogen (H_2) and CO_2, as well as new cell material.
4) *Methanogenesis*, where acetate, hydrogen plus carbonate, formate or methanol are converted into methane, CO_2 and new cell material.

In this global scheme, the following sub-processes can be distinguished (Figure 16.3):

1) Hydrolysis of biopolymers:
   - hydrolysis of proteins
   - hydrolysis of polysaccharides
   - hydrolysis of fats
2) Acidogenesis/fermentation:
   - anaerobic oxidation of amino acids and sugars
   - anaerobic oxidation of higher fatty acids and alcohols
3) Acetogenesis:
   - formation of acetate and H_2 from intermediary products (particularly VFAs)
   - homoacetogenesis: the formation of acetate from H_2 and CO_2
4) Methanogenesis:
   - methane formation from acetate
   - methane formation from hydrogen and carbon dioxide.

In addition to acetate and H_2/CO_2, methanogenesis can also proceed from methylamines, methanol, formate, carbon monoxide, etc., though in much smaller amounts.

Figure 16.2 gives the unidirectional degradation of organic matter to the end products CH_4 and CO_2. The homoacetogenic process illustrates the inter conversion of acetate (the major CH_4 precursor) and H_2/CO_2. In practice, other so-called *back*-reactions also may occur, such as the formation of medium-chain VFAs or alcohols out of acetate and propionate. These *back*-reactions might become of particular importance in the case of malfunctioning or perturbation of the anaerobic reactor or when a specific reaction is deliberately pursued. Under normal AnWT applications, *i.e.* stable reactor performance under mesophilic conditions, acetate is the major precursor of CH_4 (approximately 70% of the COD flux). It is interesting to observe is that there is only COD conversion and no COD destruction. COD removal takes place owing to the fact that the end product of the reaction chain, CH_4, is gaseous and highly insoluble in water.

In the presence of alternative electron acceptors, such as NO_3^- and SO_4^{2-}, other bacterial groups such as denitrifiers and sulphate reducers will also be present in the anaerobic reactor (see Section 16.4).

16.2.1.1 Hydrolysis

Since bacteria are unable to take up particulate organic matter, the first step in anaerobic degradation consists of the hydrolysis of polymers. This process is merely a surface phenomenom in which the polymeric particles are degraded by membrane-bound enzymes of hydrolytic bacteria attached to the solid surfaces, or through the action of excreted exo-enzymes to produce smaller molecules which can cross the cell barrier. During the enzymatic hydrolysis process, proteins are hydrolyzed to amino acids, polysaccharides to simple sugars, and lipids to long chain fatty acids (LCFA). Hydrolysis is considered the rate-limiting step for the overall digestion process and it determines the process and reactor design when (semi-)solid substrates and wastewaters with a high suspended solids (SS)/COD ratio are treated. Moreover, the hydrolysis process is very sensitive to temperature fluctuations. For this reason, the solids retention time (SRT) is the prime design criterion for anaerobic digesters treating (semi-)solid substrates and wastewaters with a high SS/COD ratio, such as distillery slops and low-temperature sewage.

Hydrolysis can be defined as the process in which complex polymeric substrates, particulate or dissolved, are converted into monomeric and dimeric compounds which are readily accessible for acidogenic bacteria. During anaerobic digestion of complex substrates, hydrolysis is usually the first step. With the digestion of biological sludges, such as in the case of waste activated sludge (WAS), hydrolysis is preceded by death and lysis of the biomass. In some cases a preparatory step, *i.e.* physico-chemical pre-treatment or comminution, is applied to enhance the hydrolysis step (Gonzalez *et al.*, 2018). The products of the hydrolysis are the substrates for acidogenic bacteria.

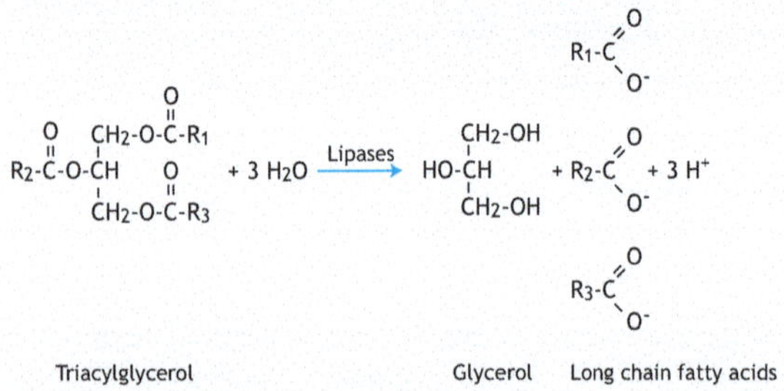

Figure 16.3 The hydrolysis of lipids.

The low surface-volume ratio, or the limited free accessible surface area, of solid substrates largely explains the overall rate limition of hydrolysis in anaerobic digestion (Chandler *et al.*, 1980; Zeeman *et al.*, 1996; Azman *et al.*, 2015). Therefore, the design of sludge and slurry digesters, in which the hydraulic retention time (HRT) equals the SRT, is fully determined by the minimum required SRT. For WAS digestion under mesophilic conditions the minimum SRT is in the range of 20-30 days. Since hydrolytic enzymes are proteins, and thus prone to digestion, there are low concentrations of hydrolytic enzymes in sludge and slurry digesters. However, two-stage or plugflow operation can increase the hydrolytic enzyme concentration and thus accelerate hydrolysis (*e.g.* Ge *et al.*, 2011; Ghasimi *et al.*, 2016). Even in dilute wastewaters such as low temperature domestic sewage, hydrolysis can determine the overall process and thereby also the required reactor design. It must be noted that 45–75% of domestic sewage and 80% of primary sludge consists of suspended matter. The main biopolymers in sewage are proteins, carbohydrates and lipids. A schematic presentation of the hydrolysis of lipids into LCFAs is given in Figure 16.3.

16.2.1.2 Acidogenesis

During the acidogenesis step, the hydrolysis products (amino acids, simple sugars, and LCFAs), which are relatively small soluble compounds, are diffused inside the bacterial cells through the cell membrane and subsequently fermented or anaerobically oxidized. Acidogenesis is a very common reaction and is performed by a large group of hydrolytic and non-hydrolytic microorganisms. Approximately 1% of all known bacteria are (facultative) fermenters. The acidification products consist of a variety of small organic compounds, mainly VFAs, *i.e.* acetate and higher organic acids such as propionate and butyrate, as well as H_2, CO_2, some lactic acids, ethanol and ammonia (Figure 16.2).

Characteristically, neutral compounds such as sugars and proteins are converted into VFAs and carbonic acid, being the main end products. Therefore, fermentative organisms are usually designated as acidifying or acidogenic microorganisms and, accordingly, this process step is denominated acidogenesis. Table 16.3 lists several acidogenic

Table 16.3 Acidogenic reactions with sucrose as the substrate and the corresponding Gibbs free energy change ($\Delta G^{o\prime}$) at 25 °C.

Reactions	$\Delta G^{o\prime}$ (kJ/mol)	Eq.
$C_{12}H_{22}O_{11} + 9H_2O \rightarrow 4CH_3COO- + 4HCO_3- + 8H + 8H_2$	- 457.5	(16.1)
$C_{12}H_{22}O_{11} + 5H_2O \rightarrow 2CH_3CH_2CH_2COO- + 4HCO_3- + 6H + 4H_2$	- 554.1	(16.2)
$C_{12}H_{22}O_{11} + 3H_2O \rightarrow 2CH_3COO- + 2CH_3CH_2COO- + 2HCO_3- + 6H + 2H_2$	- 610.5	(16.3)

reactions starting from sucrose and generating different amounts of VFAs, HCO_3^-, H_2, and H^+. It appears that the type of end products depends on the conditions in the reactor medium. From Table 16.2 it follows that the Gibb free energy change or $\Delta G^{o\prime}$ of the less energetic acidogenic reactions with sucrose as the substrate heavily depends on the prevailing H_2 concentrations. If scavenging organisms such as methanogens effectively remove H_2, acetate will be the main end product. However, if methanogenesis is retarded and H_2 accumulates, more reduced products such as propionate and butyrate are likely to appear and possibly the even more reduced compounds lactate and alcohols. Therefore, effluents of overloaded or perturbed anaerobic reactors (or reactors designed as acidifying reactors in an anaerobic two-step process) often contain these more reduced intermediate products.

Acidogenesis is the most rapid conversion step in the anaerobic food chain. The $\Delta G^{o\prime}$ of acidifying reactions is highest of all the anaerobic conversions, resulting in ten to twentyfold higher bacterial growth rates, and fivefold higher bacterial yields and conversion rates compared to methanogens (Table 16.4). For this reason, anaerobic reactors are subjected to souring, *i.e.* a sudden pH drop, when reactors are overloaded or perturbed by toxic compounds. Once alkalinity is consumed by the produced acids then the pH starts to drop, resulting in a higher concentration of non-dissociated VFAs, leading to a more severe inhibition of methanogens. The latter obviously leads to an even quicker accumulation of VFAs and subsequent pH drop (Figure 16.4). The fact that acidifiers are still active even at a pH as low as 4-5 means the reactor souring to pH 4-5 can and will occur when the methanogenic capacity of the system is exceeded.

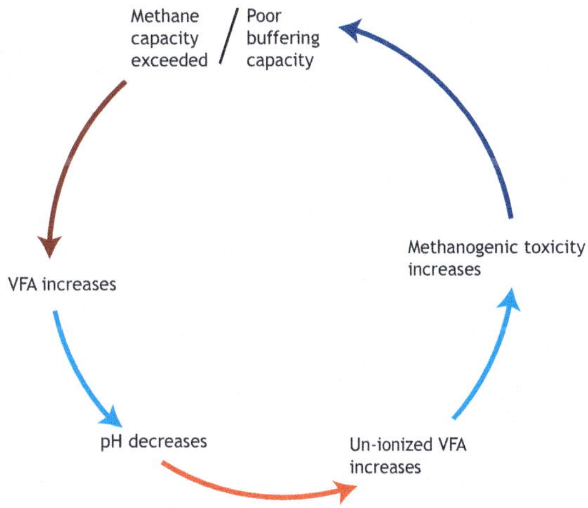

Figure 16.4 Reactor pH drop as a result of methanogenic overloading and accumulating VFAs.

The acidogenic conversion of amino acids generally follows the Stickland reaction, in which an amino acid is de-ammonified by anaerobic oxidation yielding also VFA and H_2, in conjunction with the reductive de-ammonification of other amino acids consuming the produced H_2. From both reactions NH_3 is released and subsequently acts as a proton acceptor, thus leading to a pH increase. In this reaction there is no net proton production and there is no chance of a reactor pH drop.

16.2.1.3 Acetogenesis

The short chain fatty acids (SCFA, also called VFA) other than acetate, which are produced in the acidogenesis step, are further converted to acetate, hydrogen gas and carbon dioxide by the acetogenic bacteria. The most important acetogenic substrates are propionate and butyrate, key intermediates in the

Table 16.4 Averaged kinetic properties of acidifiers and methanogens.

Process	Conversion rate gCOD/gVSS.d	Y gVSS/gCOD	K_s mgCOD/l	μ_m 1/d
Acidogenesis	13	0.15	200	2.00
Methanogenesis	3	0.03	30	0.12
Overall	2	0.03-0.18	-	0.12

anaerobic digestion process. However, lactate, ethanol, methanol and even H_2/CO_2 are also (homo) acetogenically converted into acetate as shown in Figure 16.2 and Table 16.5. LCFAs are converted by specific acetogenic bacteria following β-oxidation in which acetate moieties are split from the aliphatic chain (Table 16.5) (Alves *et al.*, 2009). LCFAs with uneven C atoms also yield propionate as well as acetate. Non-saturated LCFAs such as oleate and linoleate are firstly saturated by H_2 addition prior to the β-oxidation. The acetogenic bacteria are obligate hydrogen producers and their metabolism is inhibited by hydrogen, which immediately follows from the stochiometric conversion reaction, in the same way as for propionate:

$$\Delta G' = \Delta G^{o,} + RT \ln \frac{[\text{acetate}]\cdot[CO_2]\cdot[H_2]^3}{[\text{propionate}]} \qquad (16.4)$$

Studies carried out on acetogenic conversions have elucidated the required narrow associations between the H_2-producing acetogenic bacteria and the H_2-consuming methanogenic bacteria that regulate the H_2 level in their environment. This is of vital importance as these reactions are thermodynamically unfavourable at high H_2 concentrations, indicated by the positive ΔG°' in Table 16.5. From this table it follows that the reactions for ethanol, butyrate, propionate and the LCFAs palmitate will not occur under standard conditions, as the ΔG°' is positive, and thus the bacterial energy yield is negative.

However, under stabilised digestion conditions the hydrogen partial pressure is maintained at an extremely low level. This can be achieved by an effective uptake of the hydrogen by methanogens or sulphate-reducing bacteria. Methanogens have a very high affinity for molecular hydrogen in the anaerobic digester, thus keeping the hydrogen partial pressure to below 10^{-4} atm, which is enough to ensure the actual occurrence of the hydrogen-producing acetogenic reaction (Figure 16.5).

This interdependence means that the degradation of higher fatty acids and alcohols largely depends on the activity of electron-scavenging organisms such as methanogenic archaea. Microbial associations in which an H_2-producing organism can grow only in the presence of an H_2-consuming organism are called syntrophic associations. The coupling of formation and use of H_2 is called interspecies hydrogen transfer. In a properly functioning methane-producing installation, the partial hydrogen pressure will not exceed 10^{-4} atm and is usually between 10^{-4}-10^{-6} atm. At such a low hydrogen concentration, the degradation of ethanol, butyrate or propionate becomes exergonic and will yield energy for the acetogens.

Similar to the other acetogenic substrates, LCFA conversion is highly endergonic and often limits the entire digestion process (Novak and Carlson, 1970). Trials with upflow anaerobic sludge blanket (UASB) reactors have been only partly successful as LCFA tend to absorb to the sludge. The resulting fatty

Table 16.5 Stoichiometry and change of free energy (ΔG°') for some acetogenic reactions, assuming neutral pH, a temperature of 25 °C and a pressure of 1 atm (101 kPa). Water is regarded as a pure liquid, and all soluble compounds have an activity of 1 mole/kg.

Compound	Reaction	ΔG°' (kJ/mol)	Eq.
Lactate	$CH_3CHOHCOO^- + 2H_2O \rightarrow CH_3COO^- + HCO_3^- + H^+ + 2H_2$	–4.2	(16.5)
Ethanol	$CH_3CH_2OH + H_2O \rightarrow CH_3COO^- + H^+ + 2H_2$	+9.6	(16.6)
Butyrate	$CH_3CH_2CH_2COO^- + 2H_2O \rightarrow 2CH_3COO^- + H^+ + 2H_2$	+48.1	(16.7)
Propionate	$CH_3CH_2COO^- + 3H_2O \rightarrow CH_3COO^- + HCO_3^- + H^+ + 3H_2$	+76.1	(16.8)
Methanol	$4CH_3OH + 2\ CO_2 \rightarrow 3CH_3COOH + 2H_2O$	–2.9	(16.9)
Hydrogen-CO_2	$2HCO_3^- + 4H_2 + H^+ \rightarrow CH_3COO^- + 4H_2O$	–70.3	(16.10)
Palmitate	$CH_3\text{-}(CH_2)_{14}\text{-}COO\text{-} + 14H_2O \rightarrow 8CH_3COO\text{-} + 7H\text{+} + 14H_2$	+ 345.6	(16.11)

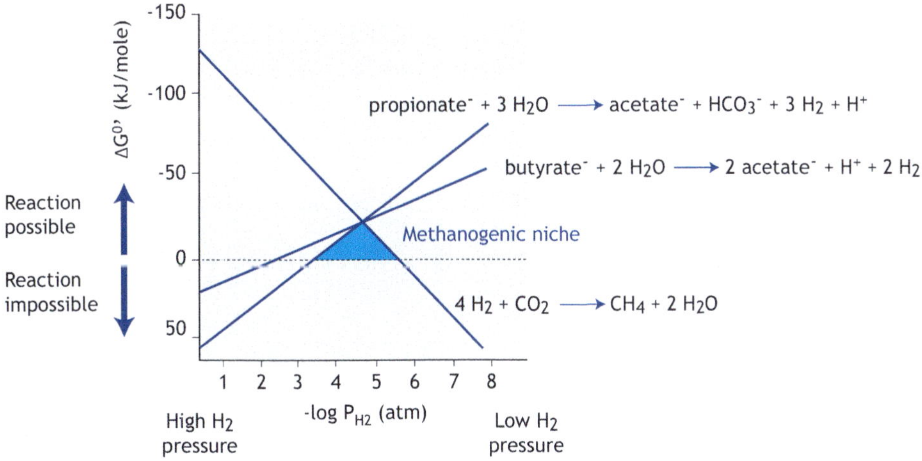

Figure 16.5 Free energy change as a function of the H_2 partial pressure. A negative $\Delta G^{o\prime}$ indicates a possible occurrence of the acetogenic and methanogen reactions referred to in this section.

clumps of biomass showed only little if any methanogenic activity, and the produced biogas in LCFA-laden sludge particles caused sludge flotation and biomass wash-out. Expanded bed reactors, in which the LCFA is more evenly distributed over the available biomass, have been more successful (Rinzema, 1988). The buoyant force of entrapping biogas in LCFA-laden anaerobic sludge can also be used to separate the reactor sludge from the discharging effluent by mounting a floatation device inside the anaerobic reactor. In this way, the effluent is clarified, while at the same time the active methanogenic sludge is retained in the bioreactor. The Dutch contractor Paques developed this anaerobic flotation reactor, the Biopaq®AFR, to convert high concentrations of fats, oil and grease (FOG) into methane (Figure 16.6) (Van Lier *et al.*, 2015).

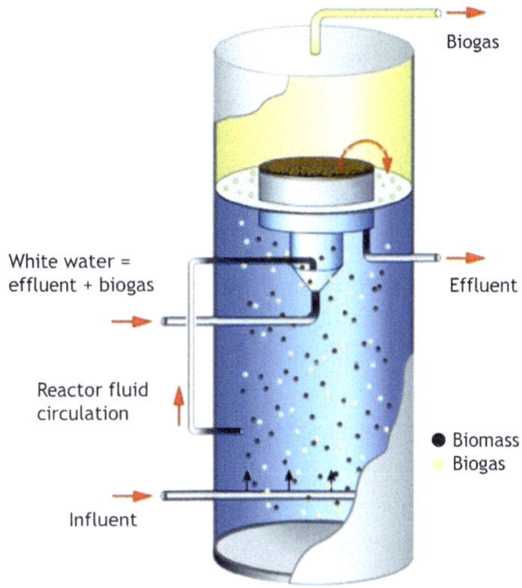

Figure 16.6 Schematic representation of the Biopaq®AFR reactor for the treatment of FOG-rich wastewater, in which sludge separation is based on sludge flotation. These reactors operate with either granular or flocculent sludge.

Other authors have proposed using the absorptive capacity of the sludge and to periodically load the sludge with LCFA after which solid state digestion will convert the absorbed matter to CH_4 (Pereira *et al.*, 2004; Cavaleiro *et al.*, 2015). Such a sequencing bed mode of operation requires multiple reactors to treat a continuous flow of wastewater.

16.2.1.4 Methanogenesis

During the fourth and last stage of anaerobic digestion, a group of methanogenic archaea reduce the available CO_2 to CH_4 using hydrogen as the electron donor, and at the same time acetate is decarboxylated to form CH_4 from the methyl group and CO_2/bicarbonate from the carboxyl group (Figure 16.2). It is during methanogenesis that the influent COD is finally converted to a gaseous form with a low solubility (CH_4) that automatically leaves the reactor system. Methanogens are obligate anaerobes, with a very narrow substrate spectrum. Some can only use certain determined substrates such as acetate, methylamines, methanol, formate, and H_2/CO_2 or CO. For engineering purposes, methanogens are classified into two major groups: the acetate-converting or aceticlastic methanogens and the hydrogen-utilising or hydrogenotrophic methanogens (Table 16.6). Generally speaking, approximately 70% of the produced methane originates from acetate as the main precursor. The remainder mainly originates from H_2 and CO_2. The growth rate of aceticlastic methanogens is very low, resulting in doubling times of several days or even more. The extremely low growth rates explain why anaerobic reactors require a very long start-up time with unadapted seed material and why high sludge concentrations are ideally chosen. Hydrogenotrophic bacteria have a much higher maximum growth rate than acetoclastic bacteria with doubling times of 4 to 12 hours. Because of this feature, and despite the very delicate acetogenic reaction step discussed in the previous section, anaerobic high-rate reactor systems remain remarkably stable under varying conditions.

Table 16.6 lists two types of aceticlastic methanogens with very different kinetic characteristics. Also the morphological characteristics of both methanogenic genera are very different as indicated by Figure 16.7.

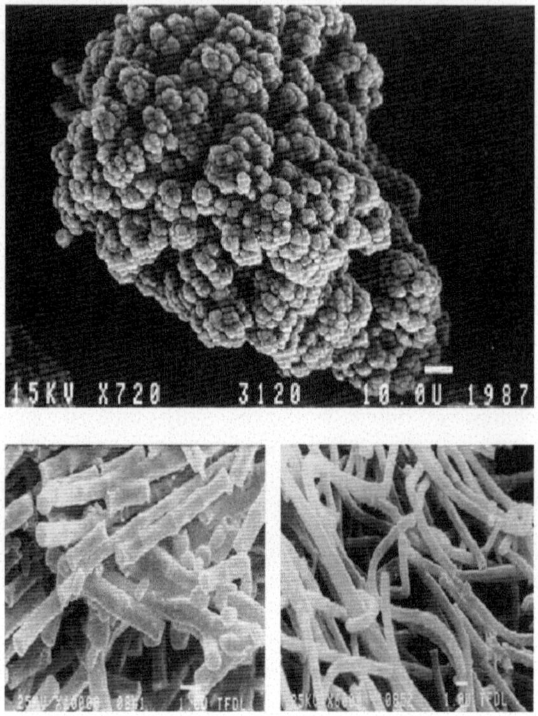

Figure 16.7 Morphology and appearance of the most important acetotrophic methanogens belonging to the genera *Methanosarcina* (above) and *Methanosaeta* (below) (pictures: sub-section Environmental Technology, Wageningen University).

Table 16.6 Most important methanogenic reactions, the corresponding Gibbs free energy change ($\Delta G^{o\prime}$), and some kinetic properties.

Functional step	Reaction	$\Delta G^{o\prime}$ kJ/mol	μ_{max} 1/d	T_d d	K_s mgCOD/l	Eq.
Acetotrophic methanogenesis[1)]	$CH_3COO^- + H_2O \rightarrow CH_4 + HCO_3^-$	–31	0.12[a)] 0.71[b)]	5.8[a)] 1.0[b)]	30[a)] 300[b)]	(16.12)
Hydrogenotrophic methanogenesis	$CO_2 + 4H_2 \rightarrow CH_4 + 2H_2O$	–131	2.85	0.2	0.06	(16.13)

[1)] Two different methanogens with different kinetic properties, belonging to [a]*Methanosaeta* spec. and [b]*Methanosarcina* spec.

Methanosarcina spp. are characterised by a coccoid shape, appearing in small grape-like clumps, and have a relatively wide substrate spectrum as they can convert acetate, H_2/CO_2, methylamines, methanol, formate, etc. They have a relatively high μ_{max} and relatively low substrate affinity. *Methanosaeta* spp. are filamentous, appear in large spaghetti-like conglomerates, can only convert acetate, and are kinetically characterised by a low μ_{max} and a very high substrate affinity. Although the μ_{max} of the latter organism is significantly lower, *Methanosaeta* spp. are the most common acetotrophic methanogens in anaerobic high-rate systems that are operated at high SRTs and low HRTs, such as sludge bed systems and anaerobic filters. The reason for this phenomenon can be attributed to the fact that wastewater treatment systems always aim at the lowest possible effluent concentrations, and substrate concentrations inside anaerobic biofilms or sludge granules approach 'zero' when bulk liquid concentrations are low. Under such conditions, the *Methanosaeta* spp. have a clear kinetic advantage over the *Methanosarcina* spp. (Fig. 16.8).

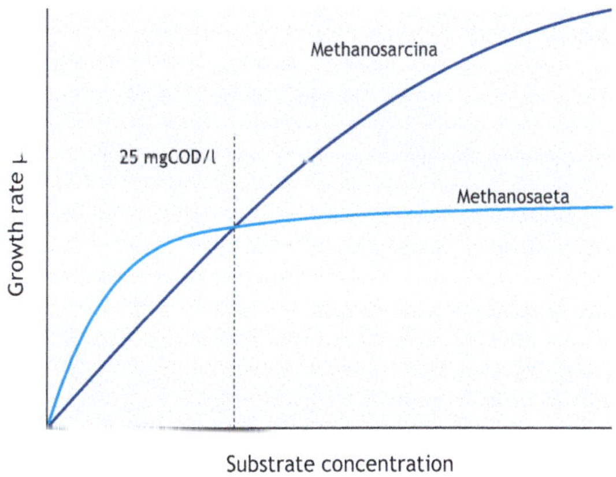

Figure 16.8 Monod growth curves of the acetotrophic methanogens *Methanosarcina* spp. and *Methanosaeta* spp. Both μ_{max} and the Monod half saturation constant (K_s) of both genera are given in Table 16.6.

Once the *Methanosaeta* spp. dominate the sludge bed, a very effective wastewater treatment system is obtained, reaching extremely low effluent acetate concentrations. Considering the inferior kinetic properties at low substrate concentrations and the inferior adherence properties of *Methanosarcina* spp., it is advised to keep the effluent acetate concentrations at a very low level during the first start-up of an anaerobic reactor with unadapted seed material.

16.3 PREDICTING THE CH_4 PRODUCTION

Organic pollution can be classified in different ways, such as based on solubility (soluble and insoluble organic matter) or based on biodegradability. The latter has a direct link to the biochemical conversion potential of organic matter. However, due to the enormous variety of organic compounds in wastewater, it is impractical and generally also impossible to determine these compounds separately. In order to quantify the organic pollution in practice, use is made of the fact that these contaminants can be oxidised by strongly oxidizing agents. In wastewater treatment engineering practice, two standardised tests based on the oxidation of organic material are applied: the biochemical oxygen demand (BOD) test and the chemical oxygen demand (COD) test (Chapter 3). In both tests, the organic material is oxidised and the amount of oxygen consumed stands for the value of the parameter. The BOD test concerns the biochemical amount of oxygen required by the aerobic organisms to oxidize the organic matter. The BOD value therefore is closely related to the biodegradability. For application of anaerobic treatment, it is preferable to use some kind of standardized anaerobic biodegradability test instead of the conventional aerobic BOD test (Spanjers and Vanrolleghem, 2016). In an anaerobic test a sample of the wastewater is exposed to an available amount of anaerobic sludge and the total amount of CH_4 produced after the digestion process has completed is determined and then related to the amount of organic matter present in the wastewater sample. As a certain amount of CH_4 is equivalent to a certain amount of COD, we can then determine the $BOD_{anaerobic}$, or the maximum amount of CH_4 that can be produced per amount of organic matter. The latter test is better known as the biomethane potential, or BMP, test (Holliger *et al.*, 2016).

Since generally not all organic pollutants are biodegradable and also part of the organic substrate will be used for cell synthesis, the BOD value is usually substantially lower than the COD value. The latter is particularly the case for the conventional aerobic BOD test, but much less for the anaerobic BOD test because of the significantly lower growth yield under anaerobic conditions. Efforts to standardise these tests, including ring tests, are currently being made in various laboratories (Holliger *et al.*, 2016).

In the standardized COD test, which generally uses bichromate as the oxidizing agent at an elevated temperature (150 °C), (almost) all the organic compounds are completely oxidised into CO_2 and H_2O. On the other hand, the organically bound nitrogen present in these compounds stays reduced and is converted into NH_3. Similarly, organic matter containing quaternary ammonium salts, such as betaine (trimethyl glycine), also remains reduced and is thus 'invisible' in the COD test.

The total organic carbon (TOC) is another measurement used, but it is a much less useful parameter, since the state of reduction of the carbon is not measured. The organic carbon concentration is measured in the form of carbon dioxide after incineration of the organic material present in a wastewater sample. Correction must be made for inorganic carbon, originally present in the sample. The theoretical value of a pure compound follows from Eq. 16.14:

$$TOC_t = 12n / (12n + a + 16b + 14d) \quad (gTOC/gC_nH_aO_bN_d) \quad (16.14)$$

16.3.1 COD

The COD undoubtedly represents the most important parameter for the concentration of contaminants in wastewater, particularly for industrial wastewaters. This feature in which organic matter is almost completely oxidized makes the COD test very suitable for assessing COD balances. Calculation of the substrate COD and the theoretical quantity of methane produced is presented below.

The COD of an organic compound $C_nH_aO_b$ can easily be calculated on the basis of the chemical oxidation reaction, assuming a complete oxidation:

$$C_nH_aO_b + \frac{1}{4}(4n + 1 - 2b)O_2 \rightarrow nCO_2 + (a/2)H_2O \quad (16.15)$$

Eq. 16.15 shows that 1 mol of organic material demands 1/4·(4n + a – 2b) moles O_2 or 8·(4n + a – 2b) gO_2. Hence the theoretical oxygen demand of organic material can be expressed as:

$$COD_t = 8 \cdot (4n + a - 2b) / (12n + a + 16b) \quad (gCOD/gC_nH_aO_b) \quad (16.16)$$

Obviously, with nitrogen-containing compounds (proteins and amino acids) Eq. 16.16 needs to be corrected for the number of electrons that will stay with N and the total weight of N in the compound.

$$COD_t = 8 \cdot (4n + a - 2b - 3d) / (12n + a + 16b + 14d) \quad (gCOD/gC_nH_aO_bN_d) \quad (16.17)$$

The COD equivalent of an organic compound can be rapidly assesed based on the full oxidation reaction. From the chemical-oxidation equation for acetic acid,

$$CH_3COOH + 2O_2 \rightarrow 2CO_2 + 2H_2O \quad (16.18)$$

follows that 1 mole (60 grams) of acetic acid requires 2 moles (64 grams) of oxygen. This means that 1 gram of acetic acid requires 64/60 (1.067) grams of oxygen, and consequently 1 gram of acetic acid corresponds to 1.067 gram COD.

The ratio between the COD and TOC values is calculated from:

$$\begin{aligned} COD/TOC &= 8 \cdot (4n + a - 2b - 3d) / (12n) \\ &= 8/3 + 2(a - 2b - 3d)/(3n) \end{aligned} \quad (16.19)$$

Table 16.7 Stoichiometric values of COD and TOC per unit mass for different pure organic compounds $C_nH_aO_bN_d$, the COD/TOC values and the mean carbon oxidation state for these compounds, and the estimated CH_4 % in the biogas.

Compound	n	a	b	d	gCOD/ $gC_nH_aO_bN_d$	gTOC/ $gC_nH_aO_bN_d$	COD/ TOC	C-ox. state	CH_4 %
Methane	1	4	0	0	4.00	0.75	5.33	–4.00	100.0
Ethane	2	6	0	0	3.73	0.8	4.67	–3.00	87.5
Methanol	1	4	1	0	1.50	0.38	4.00	–2.00	75.0
Ethanol	2	6	1	0	2.09	0.52	4.00	–2.00	75.0
Cyclohexane	6	12	0	0	3.43	0.86	4.00	–2.00	75.0
Ethylene	2	4	0	0	3.43	0.86	4.00	–2.00	75.0
Palmitic acid	16	32	2	0	3.43	0.75	3.83	–1.75	72.0
Acetone	3	6	1	0	2.21	0.62	3.56	–1.33	67.0
Ethylene glycol	2	6	2	0	1.29	0.39	3.33	–1.00	62.5
Benzene	6	6	0	0	3.08	0.92	3.33	–1.00	62.5
Betaine	5	11	2	1	1.64[a)]	0.51	3.20	–0.80	60.0
Glycerine	3	8	3	0	1.22	0.39	3.11	–0.67	58.0
Phenol	6	6	1	0	2.38	0.77	3.11	–0.67	58.0
Lysine	6	14	2	2	1.53	0.49	3.11	–0.67	58.0
Phenylalanine	9	11	2	1	1.94	0.65	2.96	–0.44	56.0
Insuline	254	377	75	65	1.45	0.53	2.72	–0.08	51.0
Glucose	6	12	6	0	1.07	0.4	2.67	0.00	50.0
Lactic acid	3	6	3	0	1.07	0.4	2.67	0.00	50.0
Acetic acid	2	4	2	0	1.07	0.4	2.67	0.00	50.0
Citric acid	6	8	7	0	0.75	0.38	2.00	1.00	37.5
Glycine	2	5	2	1	0.64	0.32	2.00	1.00	37.5
Formic acid	1	2	2	0	0.35	0.26	1.33	2.00	25.0
Oxalic acid	2	2	4	0	0.18	0.27	0.67	3.00	12.5
Carbondioxide	1	0	2	0	0.00	0.27	0.00	4.00	0.0

[a)] Calculated COD. Theoretical: with standardised bi-chromate COD test no COD will be measured.

Table 16.7 lists the calculated values of COD per unit mass for a number of organic compounds of the type $C_nH_aO_bN_d$. In the case of heavily reduced compounds, for example methane, the COD equivalent is high. By using Eq. 16.17, one calculates for methane (CH_4 *i.e.* n = 1, a = 4, b = 0, d = 0):

$$COD_{CH_4} = 8 \cdot (4 \cdot 1 + 4 - 2 \cdot 0 - 3 \cdot 0) \\ /(12 \cdot 1 + 4 + 16 \cdot 0 + 14 \cdot 0) = 4 \text{ gCOD/gCH}_4 \quad (16.20)$$

The COD/TOC ratio substantially differs for the various compounds, which relates to the differences in the mean oxidation state of the organic carbon (C-ox. state). The C-ox. state can vary from –4, the most reduced state of carbon, as found in CH_4, to +4, the most oxidized as found in CO_2. Figure 16.9 depicts the C-ox. state for a number of compounds. Since during anaerobic digestion, the mean C-ox. state will stay the same, the theoretical CH_4/CO_2 composition of the produced biogas is linearly correlated to the mean C-ox. state of that compound (Table 16.7).

The lower the mean C-ox. state in a compound (*i.e.* the more negative), the more oxygen can be bound by the compound, and thus the higher its COD value and the higher its BMP.

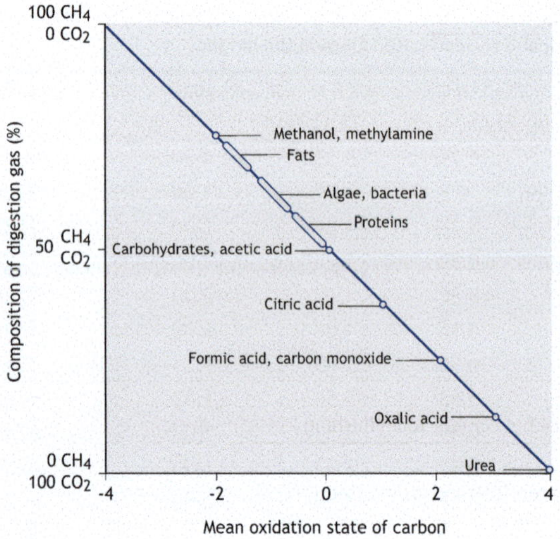

Figure 16.9 Theoretical composition of the biogas produced in relation to the mean oxidation state of the carbon in specific substrates, assuming complete mineralization of the substrate.

How can the C-ox. state be calculated? During the COD test, N remains reduced and organic-N is converted into NH_3-N. Consequently, N takes up 3 electrons. One atom of H provides one electron and one atom of O will take up two electrons, so the mean C-ox. state in a compound $C_nH_aO_bN_d$ follows from:

$$C-\text{ox.state}=(2b-a+3d)/n \tag{16.21}$$

Since the C-ox. state of fully oxidised carbon is +4 (CO_2), the number of electrons made free per atom C in the complete oxidation of $C_nH_aO_bN_d$ amounts to:

$$4-(2b+3d-a)/n=4+(a-2b-3d)/n \tag{16.22}$$

Consequently the number of molecules O_2 required for the oxidation amounts to:

$$n+1/4a-1/2b-3/4d \tag{16.23}$$

Therefore the reaction for complete chemical oxidation of this compound is:

$$\begin{aligned}&C_nH_aO_bN_d+(n+a/4-b/2-3d/4)O_2\\&\rightarrow nCO_2+(a/2-3d/2)H_2O+dNH_3\end{aligned} \tag{16.24}$$

If the compound ($C_nH_aO_bN_d$) is completely biodegradable and is entirely converted by the anaerobic organisms (no sludge yield) into CH_4, CO_2 and NH_3, the theoretical amount of methane gas (and CO_2) produced can be calculated using the Buswell equation:

$$\begin{aligned}&C_nH_aO_bN_d+(n-a/4-b/2+3d/4)H_2O\\&\rightarrow(n/2+a/8-b/4-3d/8)CH_4\\&+(n/2-a/8+b/4+3d/8)CO_2+dNH_3\end{aligned} \tag{16.25}$$

The mean C-ox. state and thus the COD will not change during anaerobic digestion. Therefore, the COD, and not TOC, is generally used to quantify the organics, since COD can then be used for making a mass balance over the anaerobic reactor (see also Section 16.5). For predicting the relative amount of CH_4 in the produced biogas when the exact composition of the organic matter is unknown, the COD/TOC ratio is a very useful tool, as this resembles the mean C-ox. state. In fact, the COD/TOC ratio is linearly correlated with the mean C-ox. state and thus with the expected CH_4 content in the biogas (Figure 16.10).

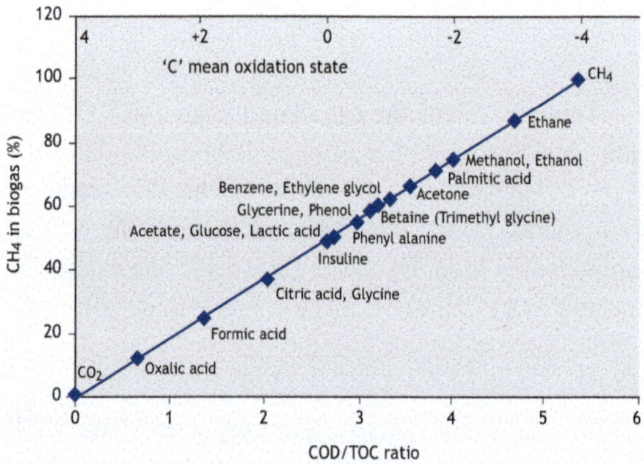

Figure 16.10 Expected CH_4 % in the produced biogas as a function of the COD/TOC ratio: $CH_4\% = 18.75 \cdot COD/TOC$.

In the presence of specific inorganic electron donors such as the oxygenated anions nitrate, sulphate or sulphite, the production of methane will decrease, while the oxygenated anions will be reduced, as *e.g.* follows from reactions:

$$10H + 2H^+ + 2NO_3^- \rightarrow N_2 + 6H_2O \quad (16.26)$$

$$8H + SO_4^{2-} \rightarrow H_2S + 2H_2O + 2OH^- \quad (16.27)$$

For wastewaters containing an excess of organic electron acceptors relative to the amount of nitrate (NO_3^-), nitrite (NO_2^-), sulphate (SO_4^{2-}) or sulphite (SO_3^{2-}), a complete removal of these inorganic electron acceptors can be expected. Whereas the former two will be reduced to N_2 gas, the latter two will end up as H_2S. Since the solubility of H_2S in water considerably exceeds that of CH_4, a substantial lower COD removal from the water phase will be obtained if the wastewater contains sulphate.

The quantity of CO_2 present in the biogas is generally significantly lower than what follows from Buswell's equation or the COD/TOC ratio as depicted in Figure 16.10. This is because of (*i*) the relatively high solubility of CO_2 in water leaving the reactor with the effluent and (*ii*) because part of the CO_2 can become chemically bound to the water phase due to the presence of cations that neutralised the VFA, SO_4^{2-}, and/or NO_3^- in the incoming wastewater and/or due to the formation of ammonia, resulting from the anaerobic conversion of nitrogen containing organic compounds.

16.4 IMPACTS OF ALTERNATIVE ELECTRON ACCEPTORS

16.4.1 Bacterial conversions under anoxic conditions

Anaerobic digesters contain mixed microbial communities. Besides the methanogenic consortia described before, other bacteria are present that can compete with the methanogens for methanogenic substrates (Table 16.8). The listed bacteria have different microbial respiration systems and can use different electron acceptors such as oxygen (O_2) by (facultative) aerobic bacteria, nitrate (NO_3^-) by denitrifiers, sulphate (SO_4^{2-}) or sulphite (SO_3^{2-}) by sulphate-reducing bacteria and iron (Fe^{3+}) by iron reducers. Anoxic means that oxygen in the form of oxygen gas (O_2) is not available as an electron acceptor.

16.4.1.1 Sulphate reduction

In the presence of sulphate, sulphite or thiosulphate, sulphate-reducing bacteria (SRB), which have a much wider substrate spectrum, are able to use several intermediates of the anaerobic mineralisation process (Table 16.8). These bacteria convert sulphate into hydrogen sulphide. Besides the direct methanogenic substrates such as molecular hydrogen (H_2), formate, acetate, methanol and pyruvate, SRB can also use propionate, butyrate, higher and branched fatty acids, lactate, ethanol and higher alcohols, fumarate, succinate, malate, and aromatic compounds (Colleran *et al.*, 1995). Hence, the main intermediary products of the anaerobic degradation process (H_2/CH_3COO^-) can be converted by both SRB, methanogens and/or obligate hydrogen-producing bacteria (OHPB). Because these three groups of bacteria operate under the same environmental conditions (pH, temperature), they will compete for the same substrates. The outcome of this competition depends on the conversion kinetics (see Section 16.10).

If organic material is oxidised via sulphate reduction, 8 electrons can be accepted per molecule of sulphate. Since one molecule of oxygen can only accept 4 electrons, the electron's accepting capacity of 2 moles of O_2 equals that of 1 mole of SO_4^{2-}, which is equivalent to 0.67 g of O_2 per g SO_4^{2-}. This means that for waste streams with a COD/sulphate ratio of 0.67, there is theoretically enough sulphate available to completely remove the organic matter (COD) via sulphate reduction. For COD/sulphate ratios lower than 0.67, the amount of organic matter is insufficient for a complete reduction of the sulphate present and extra substrate then should be added if removal of sulphate is the objective of the treatment. On the

Table 16.8 Stoichiometry and change of free energy $\Delta G^{o'}$ (kJ/mole substrate) of hydrogen and acetate conversion under different conditions.

Reaction	$\Delta G^{o'}$ (kJ/mole substrate)	Eq.
Aerobes		
$H_2 + 0.5O_2 \rightarrow H_2O$	–237	(16.28)
$CH_3COO^- + 2O_2 \rightarrow 2HCO_3^- + H^+$	–844	(16.29)
Denitrifiers		
$H_2 + 0.4NO_3^- + 0.4H^+ \rightarrow 0.2N_2 + 1.2H_2O$	–224	(16.30)
$CH_3COO^- + 1.6NO_3^- + 0.6H^+ \rightarrow 2HCO_3^- + 0.8N_2 + 0.8H_2O$	–792	(16.31)
Fe^{3+}-reducing bacteria		
$H_2 + 2Fe^{3+} \rightarrow 2Fe^{2+} + 2H^+$	–228	(16.32)
$CH_3COO^- + 4Fe^{3+} + 4H_2O \rightarrow 4Fe^{2+} + 5H^+ + 2HCO_3$	–352	(16.33)
Sulphate-reducing bacteria		
$H_2 + 0.25SO_4^{2-} + 0.25H^+ \rightarrow 0.25HS^- + H_2O$	–9.5	(16.34)
$CH_3COO^- + SO_4^{2-} \rightarrow HS^- + 2HCO_3^-$	–48	(16.35)
Methanogens		
$H_2 + 0.25HCO_3^- + 0.25H^+ \rightarrow 0.25CH_4 + 0.75H_2O$	–8.5	(16.36)
$CH_3COO^- + H_2O \rightarrow CH_4 + HCO_3^-$	–31	(16.37)

contrary, for wastewaters with a COD/sulphate ratio exceeding 0.67, the complete removal of the organic matter can only be achieved if, in addition to sulphate reduction, methanogenesis also occurs.

In the presence of sulphate, organic matter is not necessarily degraded less easily, but compared to methane, hydrogen sulphide has the great disadvantage that it dissolves much better in water than methane. This means that, for the same degree of organic waste degradation, a higher quantity of COD will be present in the reactor effluent (as sulphide), when sulphate is in the influent. Sulphide production can also cause the following technical problems during the anaerobic digestion process:

- H_2S is toxic to methanogenic archaea (MA), acetogenic bacteria (AB), and SRB. In the case of methanogenic treatment of waste streams, some of the organic compounds in the wastewater will be used by SRB rather than MA and are, therefore, not converted into methane. This results in a lower methane yield per unit of degraded organic waste and, therefore, negatively affects the overall energy balance of the process. Moreover, the quality of the biogas is reduced, since part of the produced sulphide ends up as H_2S in the biogas. Removal of H_2S from the biogas is therefore usually required.
- The produced sulphide has a bad smell and can cause corrosion problems to pipes, engines and boilers. Thus, the maintenance costs of the installation increase and extra investment is necessary to avoid these problems.
- Part of the sulphide will be present in the effluent of the anaerobic reactor. As mentioned above, this results in a lower overall treatment efficiency of the anaerobic reactor system, as sulphide contributes to the wastewater COD (per mole of sulphide two moles of oxygen are required for a complete oxidation into sulphate). Moreover, sulphide can upset the treatment efficiency of the aerobic post treatment system, *e.g.* algal blooming in lagoons or activated sludge bulking. Thus, an extra post-treatment system to remove the sulphide from the wastewater could be required.

Based on their substrate consumption, SRB can be classified into the following three groups:

1) hydrogen-oxidising SRB (HSRB),
2) acetic acid-oxidising SRB (ASRB), and
3) fatty acid-oxidising SRB (FASRB).

In the last group, two oxidation patterns can be distinguished (16.39, 16.40):

$$CH_3CH_2COOH + 2H_2O \rightarrow CH_3COOH + 3H_2 + CO_2(OHPB) \quad (16.38)$$

$$CH_3CH_2COOH + 0.75SO_4^{2-} \rightarrow CH_3COOH + 0.75S^{2-} + CO_2(FASRB) \quad (16.39)$$

$$CH_3CH_2COOH + 1.75SO_4^{2-} \rightarrow 1.75S^{2-} + 3CO_2 + 3H_2O(FASRB) \quad (16.40)$$

Some SRB are capable of completely oxidising VFA to CO_2 and sulphide as end products. Other SRB lack the tricarboxylic acid cycle and carry out an incomplete oxidation of VFA with acetate and sulphide as the end products. In the latter case, acetate is excreted in the medium. It should be further noted that incomplete oxidation of propionic acid by an SRB yields the same degradation products as the conversion by the OHPB and HSRB.

In addition to the reduction of sulphate, reduction of sulphite and thiosulphate can also occur (Widdel and Hansen, 1992). *Desulfovibrio* strains have been reported to be able to reduce di-, tri- and tetra-thionate (Fitz and Cypionka, 1990). A unique ability of some SRB, *e.g. Desulfovibrio dismutans* and *Desulfobacter curvatus*, is the dismutation of sulphite or thiosulphate (Widdel and Hansen, 1992):

$$4SO_3^{2-} + H^+ \rightarrow 3SO_4^{2-} + HS^- \quad (16.41)$$

$\Delta G^{o'} = -\ 58.9$ kJ/mole SO_3^{2-}

$$S_2O_3^{2-} + H_2O \rightarrow SO_4^{2-} + HS^- + H^+ \quad (16.42)$$

$\Delta G^{o'} = -\ 21.9$ kJ/mole $S_2O_3^{2-}$

The microbial ecology of SRB has been widely studied using different analytical techniques, such as sulphide microelectrodes, nuclear magnetic resonance (NMR) techniques, and genetic analytical tools (Santos *et al.*, 1994; Raskin *et al.*, 1995; Muyzer and Stams, 2008). Some SRB were found to be able to respire oxygen, despite being classified as strictly anaerobic bacteria.

In the absence of an electron-acceptor, SRB are able to grow through a fermentative or acetogenic reaction. Pyruvate, lactate and ethanol are easily fermented by many SRB (Widdel *et al.*, 1988). An interesting feature of SRB is their ability to perform acetogenic oxidation in syntrophy with hydrogenotrophic MA (HMA). This phenomenon has been described for defined co-cultures using lactate, ethanol or propionate (Widdel *et al.*, 1988; Oude Elferink *et al.*, 1994; Wu *et al.*, 1991) as well as in methanogenic bioreactors such as UASB, fluidized-bed and fixed-bed reactors (*e.g.* Wu *et al.*, 1992). In the presence of sulphate, however, these bacteria behave as true SRB and metabolise propionate as electron donors for the reduction of sulphate.

If SO_4^{2-} is present in the wastewater, SO_4^{2-} reduction by SRB cannot be prevented. Several attempts have been made to try to steer the competition to a single-reactor system but have been unsuccessful. On the other hand, several technological solutions are available on the market that lower the H_2S concentration in the anaerobic reactor to minimise the toxicity of MA (Figure 16.11).

16.4.1.2 Denitrification

In general, no denitrification occurs during anaerobic wastewater treatment and digestion. Under anaerobic conditions, organically-bound nitrogen will be converted into ammonium. Denitrification can only be expected if the influent contains nitrate (see Chapter 5) or when nitrified effluent is recycled to the anaerobic reactor (Kassab *et al.*, 2010). The presence of nitrate in methanogenic environments negatively impacts methane formation (Tugtas and Pavlostathis, 2008).

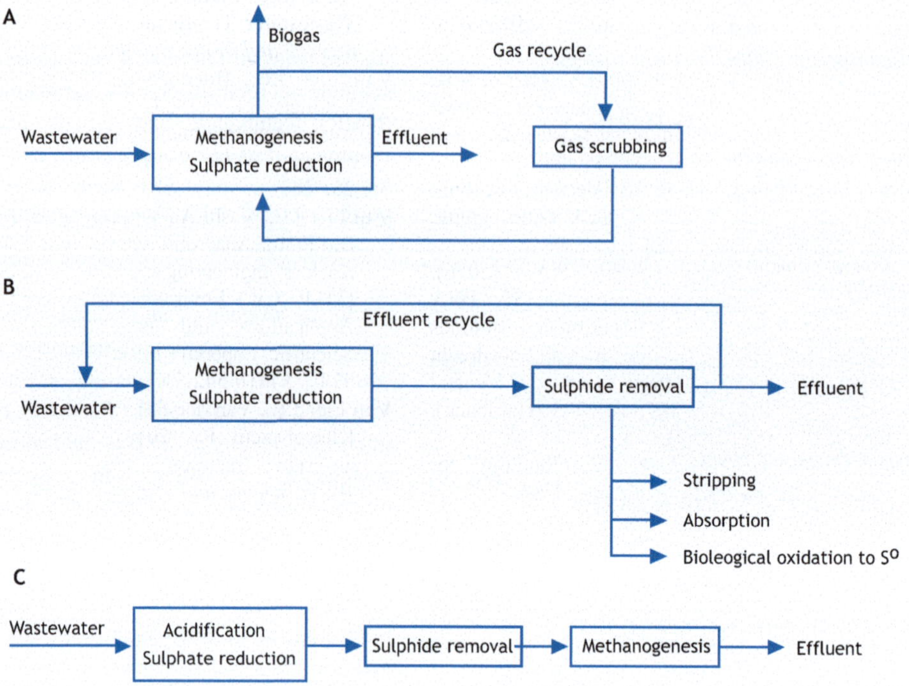

Figure 16.11 Technological solutions to decrease the H_2S concentration in an anaerobic reactor. (A) Enhanced H_2S stripping by biogas recycling and sulphide stripping in the gas line; (B) H_2S removal in a (micro)aerobic post-treatment system and recirculation of the treated effluent to the anaerobic reactor influent for dilution; and (C) combined pre-acidification and sulphate reduction with the sulphide removal step for lowering the S content in the anaerobic reactor. In (C) most of the H_2S will be stripped in the acidification step owing to the low prevailing pH.

Denitrification is mediated by denitrifying microorganisms, *i.e.* chemoheterotrophic bacteria which are capable of oxidising organic matter with nitrate. Nitrate is then converted via nitrite and nitrogen oxide into N_2 gas. Generally, denitrifying microorganisms prefer oxygen as an electron acceptor, as the latter compound yields more energy (Table 16.8). In activated sludge processes, denitrifiers start to use nitrate when O_2 is (almost) depleted and will only occur at a dissolved O_2 concentration of 1 mg/l or below.

Denitrification is a heterotrophic process requiring an electron donor. For oxidising organic matter via denitrification, 5 electrons can be accepted per molecule of NO_3^- (or 3 electrons per molecule of NO_2^-). Therefore, the electron-accepting capacity of 1¼ moles of O_2 equals that of 1 mole of NO_3^-, which is equivalent to 0.65 g of COD per g NO_3^-, or 2.86 g COD per g NO_3^--N.

The stoichiometry of methanol oxidation with nitrate and nitrite occurs according to the following reaction equation:

$$CH_3OH + 2NO_2^- \rightarrow N_2 + CO_3^{2-} + 2H_2O \tag{16.43}$$

$$5CH_3OH + 6NO_3^- \rightarrow 3N_2 + 4HCO_3^- + CO_3^{2-} + 8H_2O \tag{16.44}$$

These reactions show that denitrification will result in a pH increase (carbonate production).

16.5 WORKING WITH THE COD BALANCE

As with any biological system, an anaerobic treatment process must be monitored for relevant parameters, and measurements must be evaluated for adequate operation and control. Section 16.3 discusses the usefulness of the COD as the control parameter for anaerobic systems. The reason for this is that, in contrast to aerobic systems, there is no COD destruction in an anaerobic reactor. During anaerobic treatment the COD is only 're-arranged'. Complex organic compounds are broken down into more simple intermediates and eventually mineralised to CH_4 and CO_2. All the COD that entered the system ends up in the end product CH_4, minus the COD that is incorporated in the new bacterial mass, minus any residual COD in the effluent. Since a perfect mass balance can be made by only using the COD as a parameter, the COD is therefore generally taken as a control tool to operate an anaerobic system:

$$COD_{in} = COD_{out} \tag{16.45}$$

For practical purposes Eq. 16.45 should be expanded to the various outlets of the anaerobic reactor as depicted in Figure 16.12.

In order to identify the fate of COD in an anaerobic reactor, detailed analyses of the gaseous, liquid and solid outlets should be performed (Table 16.9).

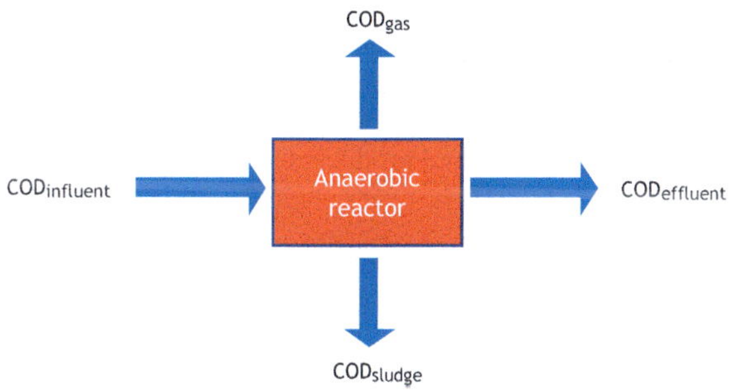

$$COD_{influent} = COD_{effluent} + COD_{gas} + COD_{sludge}$$

Figure 16.12 The COD balance of an anaerobic reactor. By differentiating between the COD fractions of gas, liquid and solids, the missing parameters can be estimated from the more easily measurable parameters.

Based on the basic influent characteristics, *i.e.* flow rate and COD concentrations, and on information on the biodegradability of the COD, the expected CH_4 production rate can be easily estimated. From Section 16.3.1. we can derive that:

$$CH_4 + 2\,O_2 \rightarrow CO_2 + 2H_2O \tag{16.46}$$

which means that 22.4 m^3 CH_4 (STP) requires 2 moles of O_2 (COD), which equals 64 kg COD. Therefore, theoretically, 1 kg COD can be converted into 0.35 m^3 CH_4.

Table 16.9 Various COD fractions and their fate in an anaerobic reactor system. The number of dots indicates the relative importance of the indicated COD fraction in the respective compartment (influent, effluent, sludge, biogas).

COD fraction	Influent	Effluent	Sludge	Biogas
Soluble organic	•••	•••	•	
Soluble inorganic	•	••	•	
Suspended organic	•••	••	••	
Suspended inorganic		•	•	
Colloidal	•	••	•	
Absorbed	•		•••	
Entrapped			•••	
CH_4		•		•••
H_2				•
H_2S	•	••		•
N_2				•
Newly grown biomass		••	••	

Similarly, the theoretical COD equivalent for 1 kg 'bacterial VSS', with an estimated composition of $C_5H_7O_2N$, can be calculated as 1.42 kgCOD/kgVSS. Having both the final products CH_4 and newly grown bacteria expressed as COD, the balance can be made if influent and effluent are properly measured.

Often 'gaps' in the COD balance occur which can be attributed mostly to the 'loss of electrons' when these are channelled to oxidised anions such as SO_4^{2-} and NO_3^-, as explained in Section 16.4. Therefore, in this case, for closing the COD balance either all reduced gases should be taken into account or the concentration of electron acceptors needs to be measured. It should be realised that soluble COD-containing gases such as H_2S, or its inonised form HS^-, will be present in the effluent. In this example, organic COD is converted into inorganic COD of which a pH-dependent fraction will end up in the biogas while the remainder will stay in the effluent.

Another frequently cited cause for a COD gap is the entrapment or accumulation of refractory or slowly biodegradable COD in the sludge bed, sometimes drastically changing the stochiometric value of 1.42 kgCOD/kgVSS. The latter is particularly true during the treatment of FOG-rich or LCFA-containing wastewater. With these substrates, COD removal efficiencies are generally very high, but low CH_4 production rates lead to huge gaps in the balance. In this example, the COD gap indicates potential severe long-term operational problems. Eventually, accumulating fats and/or other inert solids will deteriorate the SMA of the sludge, and FOG/LCFA absorbed to sludge can result in sludge floation and washout. The decrease in active biomass can lead to complete failure of the anaerobic high-rate process.

Operating an anaerobic reactor using the COD balance as a tool to monitor reactor performance gives the operator vital information about the functioning of the system. Adequate action can be undertaken before irreversible deterioration occurs. Also, the impact of alternative electron acceptors on the CH_4 production rate can be easily assessed while, based on the gas production and effluent COD values, an estimate can be made of the amount of newly grown and entrapped biomass.

16.6 IMMOBILISATION AND SLUDGE GRANULATION

The key for modern high-rate biotechnology, whatever systems are considered, is the immobilization of proper bacteria. In fact, the required high sludge retention in anaerobic treatment systems is based on immobilization, which generally leads to the formation of well-balanced bacterial consortia. The presence of these consortia is considered a prerequisite for proper anaerobic process operation, particularly considering the occurrence of various syntrophic conversion reactions in the anaerobic degradation of most organic compounds, the detrimental effect of higher concentrations of specific intermediates, and the major effect of environmental factors such as pH and redox potential. Significant progress in the knowledge of the fundamentals of the immobilisation process has been made since the development and successful implementation of high-rate anaerobic treatment systems in the 1970s (Hulshoff Pol *et al.*, 2004; Van Lier *et al.*, 2015). Immobilisation can occur on inert support material mounted in a fixed matrix in anaerobic filters (AF), which are operated both in upflow and in downflow mode. The matrix can also be free-floating as in moving-bed bioreactors and fluidized-bed (FB) systems. If no inert support material is used, an 'auto-immobilisation' will occur, which is understood as the immobilisation of bacteria on themselves in bacterial conglomerates, or on very fine inert or organic particles present in the wastewater. The bacterial conglomerates will mature in due course and form round-shaped granular sludge.

With respect to immobilization, particularly the phenomenon of anaerobic sludge granulation has puzzled many researchers from very different disciplines. Granulation, in fact, is a completely natural process and will occur in all systems where the basic conditions for its occurrence are met, *i.e.* on mainly soluble substrates and in reactors operated in an upflow manner with hydraulic retention times

(HRT) lower than the bacterial doubling times. Owing to the very low growth rate of the crucial aceticlastic methanogens, particularly under sub-optimal conditions, the latter conditions are easily met. Anaerobic granule formation is mostly observed in anaerobic bioreactors that are operated in upflow mode (Hulshoff Pol *et al.*, 2004). However, successful granulation has also been observed in anaerobic sequencing batch reactors (Sung and Dague, 1995; Wirtz and Dague, 1996). Maybe for the first time since the 1950s, sludge granulation was found to occur in reversed flow Dorr Oliver Clarigesters applied in South Africa. However, this only became apparent by observation of sludge samples taken from such a digester in 1979 (Lettinga, 2014). While studying the start-up and feasibility of anaerobic upflow filters, Young and McCarty (1969) already recognized the ability of anaerobic sludge to form aggregates that settle very well. These granules were as large as 3.1 mm in diameter and settled readily. In AF experiments with potato starch wastewater and methanol solutions conducted in the Netherlands, similar observations were made (Lettinga *et al.*, 1979). However, whereas the interest in AnWT in USA and South Africa diminished, major emphasis was put on developing industrial-scale systems in the Netherlands, where new surface water protection legislation coincided with the world energy crisis of the seventies. As a result, increasing emphasis could be afforded on applied and fundamental research in this field, particularly also on the phenomenon of sludge granulation. This increased worldwide interest occurred in both the engineering and microbiological fields. As a result, insight into the mechanism of the sludge granulation process for anaerobic treatment has been elucidated sufficiently, at least for practical application (see further Hulshoff Pol *et al.*, 2004; Van Lier *et al.*, 2015). Granulation can proceed at all temperatures, *i.e.* under mesophilic, thermophilic and psychrophilic conditions. It is considered of major practical importance to further unravel the fundamentals concerning the growth of mixed balanced granular aggregates, not only from the microbial but also from the process engineering point of view.

16.6.1 Mechanism underlying sludge granulation

In essence, sludge granulation is based on the fact that bacterial retention is imperative when dilution rates exceed the bacterial growth rates. Immobilization further requires the presence of support material and/or specific growth nuclei. The occurrence of granulation can be explained as follows:

1) Proper growth nuclei, *i.e.* inert organic and inorganic bacterial carrier materials as well as bacterial aggregates, are already present in the seed sludge.
2) Finely dispersed matter, including viable bacterial matter, will be retained less and less once the superficial liquid and gas velocities increase, applying dilution rates higher than the bacterial growth rates under the prevailing environmental conditions. As a result, film and/or aggregate formation automatically occurs.
3) The size of the aggregates and/or biofilm thickness are limited, viz. it depends on the intrinsic strength (binding forces and the degree of bacterial intertwinement) and the external forces exerted on the particles/films (shear stress). Therefore, in due course particles/films will fall apart, evolving the next generation. The first generation(s) of aggregates, indicated by Hulshoff Pol *et al.* (1983) as 'filamentous' granules, mainly consist of long multi-cellular rod-shaped bacteria. They are quite voluminous and in fact more flock than granule.
4) Retained secondary growth nuclei will grow in size again, but also in bacterial density. Growth is not restricted to the outskirts, but also takes place inside the aggregates. In due course they will fall apart again, evolving a third generation, etc.
5) The granules will gradually 'age' or 'mature', a process which is characterised by the replacement of the voluminous 'filamentous granules' by dense 'rod' granules.

During the above selection process, both organic and hydraulic loading rates gradually increase, increasing the shear stress inside the system. The latter results in firm and stable sludge aggregates with a high density and a high superficial velocity. Figure 16.13 pictures the course in time of the in-reactor sludge concentrations, expressed as gVSS/l, and the applicable volumetric organic loading rate. The start-up is accomplished when the design loading rate is reached. For mainly soluble wastewaters which are partly acidified, granular sludge will be easily cultivated.

Table 16.10 lists some common characteristics of methanogenic granular sludge.

With respect to the granulation process, essentially there are no principle differences between a UASB reactor, seeded with digested sewage sludge, and an upflow reactor with inert free-floating support material such as the FB reactor, which uses sand particles or pumice as carrier material for the biomass growing in. Indeed granulation can proceed reasonably well in a FB system, provided the reactor is operated with a moderate shear on the particles, *i.e.* in such a mode that biofilms can grow sufficiently in thickness and/or different particles can grow together. Full-scale experiences have shown that complete fluidization is not required and is in fact is detrimental to achieving stable and sufficiently thick biofilms. At present, expanded granular sludge bed (EGSB) reactors are of much more interest for commercial applications than the more expensive FB systems (see also Section 16.7.2.4).

Table 16.10 Proposal for definition and characteristics of good quality granular sludge (photos: Paques B.V.).

Granular sludge examples	Good quality granule characteristics
 Potato wastewater-grown granules	Metabolic activity: Specific methanogenic activity range of granular sludge: 0.1-2.0 kgCOD-CH_4/ kgVSS.d Typical values for industrial wastewater : 0.3- 1.0 kgCOD-CH_4/ kgVSS.d
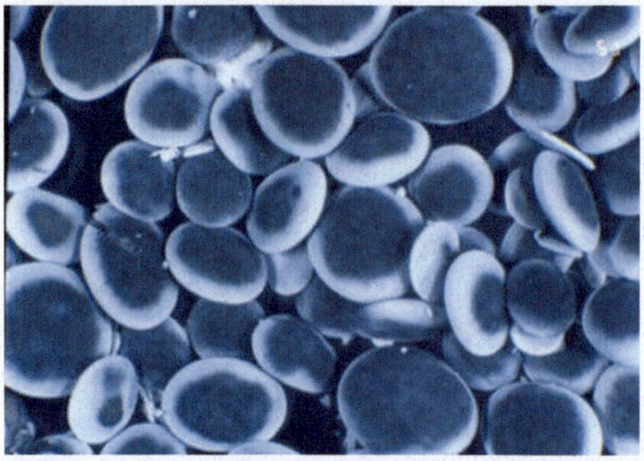 Paper mill wastewater-grown granules	Settleability and other physical properties: • settling velocities: 2-100 m/h, typically: 30-75 m/h • density: 1.0-1.05 g/l • diameter: 0.1-8 mm, typically: 0.15-4 mm • shape: spherical form and well-defined surface • colour: black/grey/white.

Definition: dense spherical-shaped microbial conglomerate, consisting of microorganisms, inert material, and extracellular polymeric substances (EPS), and which is characterised by high levels of metabolic activity and settleability.

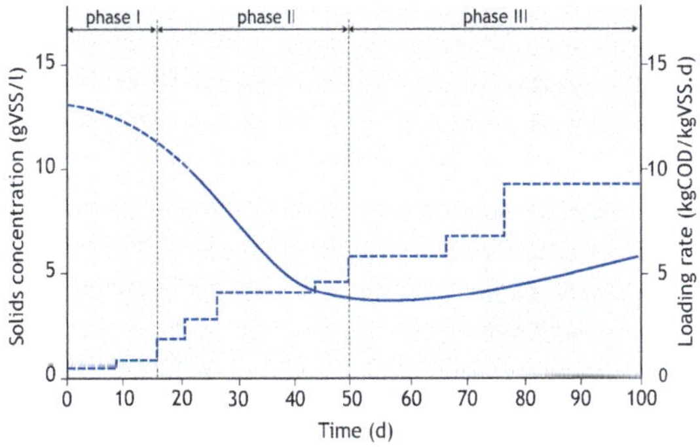

Figure 16.13 Sludge dynamics during the initial start-up of a UASB reactor. Phase I: applied loading rate < 3 kgCOD/m³.d, expansion of the sludge bed and wash-out of colloidal sludge fraction, a flotation layer can occur and the specific methanogenic activity starts to increase. Phase II: heavy sludge wash-out with selection between heavy and light sludge takes place, big increase in loading rate and formation of dense aggregates. Phase III: increase in total sludge concentration, increase in granular sludge quantity, loading rate can be further increased.

16.7 ANAEROBIC REACTOR SYSTEMS

Anaerobic reactors have been in use since the 19th century, when Mouras and Cameron developed the automatic scavenger and the septic tank to reduce the amounts of solids in the sewerage system. Although at a very poor rate, the first anaerobic stabilisation processes occurred in tanks that had been designed for intercepting black-water solids (McCarty, 2001). The first anaerobic reactor was developed in 1905 when Karl Imhoff designed a combined settler-digester tank that treated the entire sewage flow. In the Imhoff tank the solids sediment and are stabilised in a single tank (Imhoff, 1916). The actual controlled digestion of entrapped solids in an ex-situ separate reactor was developed by the Ruhrverband, Essen-Relinghausen in Germany.

At the same time Buswell *et al.* (1932) started to adopt the same technology for treating liquid wastes and industrial wastewater. All these systems can be characterised as low-rate systems since no special features were included in the design to augment the anaerobic catabolic capacity. The process feasibility of these systems was very much dependent on the growth rate of the anaerobic consortia. As a result, reactors were very large and fragile in operation. Also an anaerobic pond can be regarded as a low-loaded anaerobic treatment system. Anaerobic ponds are often constructed in conjunction with facultative and maturation ponds. The applied loading rate to anaerobic ponds ranges between 0.025-0.5 kgCOD/m³.d, while using pond depths of 4 m. The major disadvantage of anaerobic ponds is problems related to odour as these systems easily become overloaded. However, the loss of the energy-rich and very potent greenhouse gas CH_4 into the atmosphere is also a recognised disadvantage.

16.7.1 High-rate anaerobic systems

One of the major successes in the development of anaerobic wastewater treatment was the introduction of high-rate reactors in which biomass retention and liquid retention are uncoupled. In anaerobic processes, the maximum permissible COD load is governed by the amount of viable anaerobic biocatalysts, which are in full contact with the wastewater constituents. In contrast, in aerobic processes, the permissible COD load is determined by the maximum transfer rate of the electron acceptor *i.e.* oxygen. High sludge concentrations in anaerobic high-rate systems are obtained by physical retention and/or immobilisation of anaerobic sludge. High biomass concentrations enable the application of high COD loading rates, while maintaining long SRTs at relatively short HRTs. In the past four to five decades, various high-rate anaerobic treatment configurations have been developed based on different sludge retention mechanisms, such as the anaerobic contact process (ACP), anaerobic filter (AF), upflow anaerobic sludge blanket (UASB) reactor, fluidized-bed (FB) reactor, expanded granular sludge bed (EGSB) reactor, internal circulation (IC) reactor, anaerobic baffled reactor (ABR), membrane-coupled high-rate (UASB/EGSB/FB) reactors, and membrane-coupled CSTR systems. The latter are better known as anaerobic membrane bioreactors (AnMBR) (*e.g.* Dereli *et al.*, 2012).

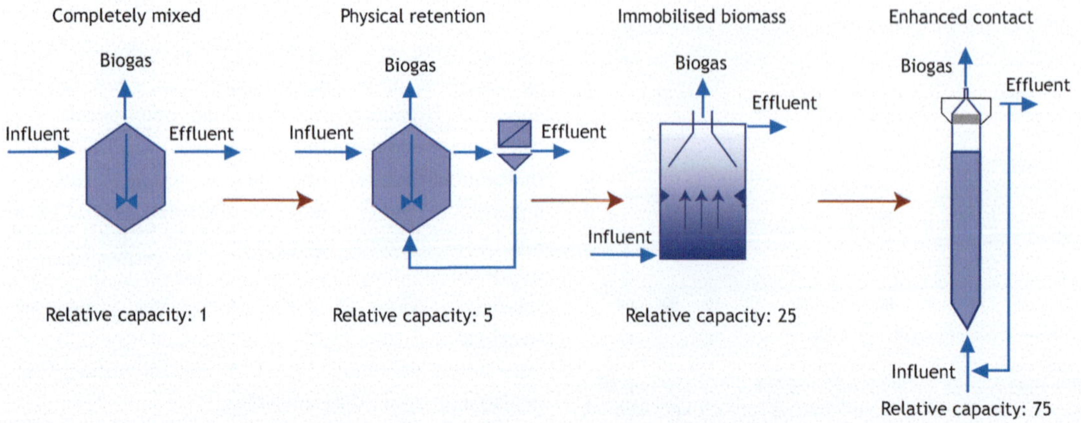

Figure 16.14 Relative loading capacities of different AnWT systems. Maximum applied loading rates under full-scale conditions reach approximately 40 kgCOD/m³.d applying enhanced contact in advanced versions of expanded granular sludge bed systems.

To enable an anaerobic reactor system to accommodate high organic loading rates for treating a specific wastewater, the following conditions should be met:

- High retention of viable sludge in the reactor under operational conditions.
- Sufficient contact between the viable bacterial biomass and wastewater.
- High reaction rates and an absence of serious transport limitations. The latter refers to both substrates and (intermediate) reaction products.
- The viable biomass should be sufficiently adapted and/or acclimatized. Considering the low microbial growth rates under anaerobic conditions, and thus the long SRTs, biomass adaptation can take very long periods of time.
- Prevalence of favourable environmental conditions for all required organisms inside the reactor under all imposed operational conditions, focusing on the rate-limiting steps. It should be emphasized that this condition does not mean that the circumstances should be similar at any location within the reactor and at any point in time. Anaerobic digestion is characterised by a large variety of different organisms (see Section 16.2). The (co-)existence of micro-niches within the system is an absolute prerequisite for full substrate conversion. Depending on the substrate type, anaerobic granules and/or biofilms are characterised by phylogenetic diversities, spatial distributions, and localized activities of bacteria and archaea. For instance, despite relatively high H_2 concentrations in the bulk of the liquid, the H_2 partial pressure in the interior of biofilms and granules is maintained at a very low level to allow even very endergonic acetogenic reactions to proceed *e.g.* the oxidation of propionate.

Figure 16.14 illustrates the development of high-rate reactor systems and the impact of improved sludge retention and enhanced contact on the applicable organic loading rates. Although the first trials by Buswell did not reach loading rates of 1 kgCOD/m³.d, modern AnWT systems are sold on the market with guaranteed loading rates exceeding 40 kgCOD/m³.d.

Thus far, most applications of AnWT can be found as end-of-the-pipe treatment technology for food-processing wastewaters and agro-industrial wastewater. Table 16.1 lists the various industrial sectors where the 2,266 reactors surveyed in 2007 were installed. It should be noted that the current number of commissioned full-scale anaerobic high-rate reactors exceeds 4,000 (Van Lier *et al.*, 2015). In addition to the traditional applications, the number of full-scale anaerobic reactors in the non-food sector has also grown rapidly. Common examples are paper mills and chemical wastewaters, such as those containing formaldehyde, benzaldehydes, terephthalates, etc. (*e.g.* Razo-Flores *et al.*, 2006). It is surprising that chemical wastewaters are also employing full-scale anaerobic reactors since anaerobic technology has not been widely accepted in the chemical industries, owing to general prejudices against biological treatment and anaerobic treatment in particular. With regard to the chemical compounds, it is of interest to mention that certain compounds, such as poly chloro-aromatics and poly nitro-aromatics as well as the azo-dye linkages can only be degraded when a reducing (anaerobic) step is introduced in the treatment line. Anaerobics are then complementary to aerobics for achieving full treatment.

Current high-rate AnWT systems are also able to treat cold and very low-strength wastewaters. In addition to municipal sewage, many industrial wastewaters are discharged at low temperatures, *e.g* beer and malt wastewaters. As such, the application potential of AnWT is still expanding and novel reactor systems are being developed to broaden the field of application.

16.7.2 Single-stage anaerobic reactors

16.7.2.1 The anaerobic contact process (ACP)

Following the historical development of high-rate reactors, the ACP process is the first configuration in which the SRT was uncoupled from the HRT. The reactor biomass concentration was increased by employing a secondary clarifier with return flow, similar to its aerobic homologue (Fig.16.15).

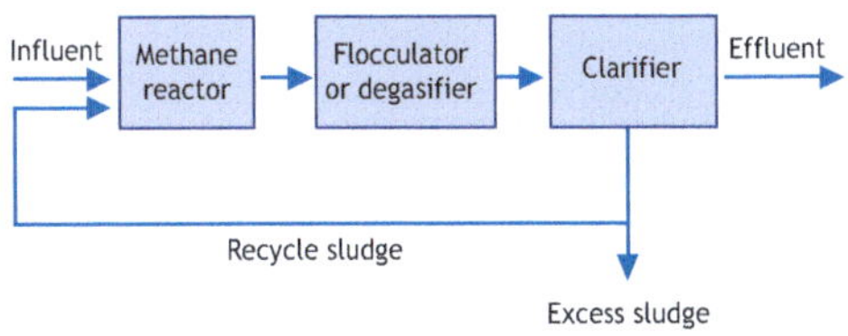

Figure 16.15 The anaerobic contact process, equipped with flocculator or a degasifier unit to enhance sludge sedimentation in the secondary clarifier.

The first ACP process was reported for the treatment of dilute packing house waste that has a COD of approximately 1,300 mg/l (Schroepfer *et al.*, 1955). However, various versions of the first generation of high-rate anaerobic treatment systems for medium strength wastewaters were not very successful. In practice, the main difficulty was a poor separation of the active anaerobic sludge from the treated water in the secondary clarifier. Biogas formation and attachment in the settling tank were the other major problems (Rittmann and McCarty, 2001). The poor sludge separation was attributed to the very intensive agitation applied in the bioreactor, creating very small sludge particles with a poor settleability. In addition, supersaturation of solubilized gases resulted in buoyant upward forces in the clarifier. The idea of intensified mixing was to ensure optimized contact between the sludge and the wastewater. Modern ACP systems apply much milder mixing conditions, and degasifying units are often installed prior to secondary clarification. In fact, modern ACP systems are very effective for concentrated wastewaters with relatively high concentrations of suspended solids. As such, ACP have a consolidated market share within full-scale applied anaerobic high-rate systems (Van Lier, 2008). If well designed, a modern ACP can reach organic loading rates of 10 kgCOD/m^3.d. Nonetheless, ACP effluents require a subsequent treatment step in order to comply with effluent restrictions.

16.7.2.2 Anaerobic filters (AF)

An alternative method of sludge retention was found by applying inert support material on which the anaerobic organisms can adhere in the bioreactor. Whereas the earliest anaerobic filters were already being applied back in the 19th century (McCarty, 2001), the application for industrial wastewater treatment started in the 1960s in the US (Young and McCarty, 1969; Young, 1991). The AF, also called a packed-bed process, has been developed as a biofilm system in which biomass is retained, based on:

- the attachment of a biofilm to the solid (stationary) carrier material,
- entrapment of sludge particles between the interstices of the packing material, and
- the sedimentation and formation of sludge aggregates that have a very high settling rate.

AF technology can be applied in upflow and downflow reactors (Young and Yang, 1989). Various types of synthetic packing materials, as well as natural packing materials such as gravel, coke and bamboo segments, have been investigated in order to be used in AFs. Research results indicate that the shape, size, weight, specific surface area, and porosity of the packing material are important aspects, as well as the surface adherence properties with regard to bacterial attachment. AF systems with the correct support can start rapidly, owing to the efficient adherence of anaerobic organisms to the inert carrier. This ease of starting up the system was the main reason for its popularity in the eighties and nineties. However, UAF systems tend to experience more problems than other systems during their long-term operation. The major disadvantage of the UAF concept is the difficulty in maintaining the required contact between sludge and wastewater, because clogging of the bed easily occurs. This is particularly the case for partly soluble wastewaters. These clogging problems can obviously be (at least partly) overcome by applying a primary settler and/or a pre-acidification step (Seyfried, 1988). However, this requires the construction and operation of additional units. Moreover, in addition to the higher costs, it does not completely eliminate the problem of short-circuiting (clogging of the bed) flows, leading to disappointing treatment efficiencies.

AF technology has been widely applied in the treatment of wastewaters from the beverage, food-processing, pharmaceutical and chemical industries due to its high capability of biosolids retention (Ersahin *et al.*, 2011). Since 1981 approximately 140 full-scale UAF installations have been put into operation for the treatment of various types of wastewater, which is approximately 6% of the total amount of installed high-rate reactors (Figure 16.16).

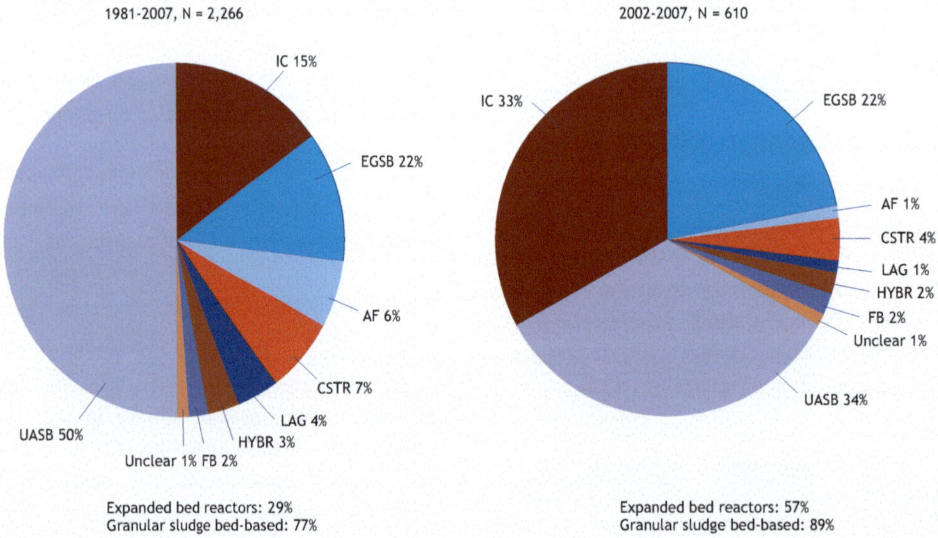

Figure 16.16 Implemented anaerobic technologies for industrial wastewater pictured for the period 1981-2007 (left) and the period 2002-2007 (right). UASB: upflow anaerobic sludge blanket; IC®: internal circulation reactor, a type of EGSB system with biogas-driven hydrodynamics; EGSB: expanded granular sludge bed; AF: anaerobic filter; CSTR: continuous stirred tank reactor; LAG: anaerobic lagoon; HYBR: combined hybrid system with sludge bed at the bottom section and a filter at the top; FB: fluidized bed reactor (Van Lier, 2008).

Experiences with the system are reasonably satisfactory, applying modest to relatively high loading rates of up to 10 kgCOD/m^3.d. The UAF system will remain attractive for treatment of mainly soluble types of wastewater, particularly when the sludge granulation process does not proceed satisfactorily. However, long-term problems related to system clogging and the stability of filter material have caused a decline in the number of full-scale AF systems being installed.

In order to minimise clogging and sludge accumulation in the interstices of the filter material, anaerobic filters are sometimes operated in a downflow mode, which are referred to as downflow fixed film reactors. Various modes of operation and filter material have been investigated but full-scale application is rather disappointing. The limiting factor is the applicable low organic loading rate owing to the limited amount of biomass that can be retained in such a system as it is primarily based on attachment of biomass to the surface of the packing material; in a UAF the majority of the anaerobic activity is found in the non-attached biomass.

16.7.2.3 Anaerobic sludge bed reactors (ASBR)

Anaerobic sludge bed reactors (ASBR) are undoubtedly by far the most popular AnWT systems so far. The sludge retention in such a reactor is based on the formation of easily settling sludge aggregates (flocs or granules), and on the application of an internal gas-liquid-solids separation system (GLSS device).

One of the most remarkable and significant developments in high-rate anaerobic treatment technology is the invention of the UASB reactor by Lettinga and his co-workers in the Netherlands (Lettinga *et al.*, 1976; Lettinga *et al.*, 1980). Sludge retention in this reactor is based on the formation of sludge aggregates (flocs or granules) that settle well, and on the application of a reverse funnel-shaped internal GLSS device. Introduction of UASB technology for industrial wastewater treatment only required 5-6 years to mature from lab-scale in 1971-1972 to pilot reactor in 1976 and full-scale application in 1977-1978 at a sugar mill in the Netherlands (Lettinga, 2014). To our knowledge, this is the shortest recorded development cycle for any new, successful environmental engineering technology so far. Following the first applications, many successful performance results have been reported at lab- and pilot-scale applications using ASBR processes, which has resulted in the establishment of thousands of full-scale reactors worldwide (Nnaji, 2013; Lim and Kim, 2014; Van Lier *et al.*, 2015). ASBRs are undoubtedly the most popular anaerobic wastewater treatment systems so far, having a wide application potential in industrial wastewater treatment. In view of its potential and the fact that almost 90% of newly installed high-rate reactors are sludge bed systems (Van Lier *et al.*, 2015; Figure 16.16), the UASB process will be elaborated in more detail than other systems (Section 16.8).

Figure 16.17 shows a schematic representation of a UASB reactor. Two examples of a full-scale UASB instalation are shown in Figure 16.18.

The first UASB reactors were installed for the treatment of food, beverage and agro-based wastewaters, rapidly followed by applications for paper and board mill effluents in 1983 (Habets and Knelissen, 1985). Most full-scale reactors are used for treating agro-industrial wastewater, but applications for the treatment of wastewaters from chemical industries are also increasing, as discussed below (Van Lier *et al.*, 2015; Rajagopal *et al.*, 2013).

Similar to the UAF system, in a UASB reactor the wastewater moves in an upward mode through the reactor. However, contrary to the AF system, no packing material is present in the UASB reactor. Good sludge settleability, low HRTs, elimination of the packing material cost, high biomass concentrations (up to 80-100 g/l at the bottom), effective solids/liquid separation, and operation at high organic loading rates (OLR) can be achieved by UASB reactor systems (Speece, 1996). The OLR design is typically in the range of 4 to 15 kgCOD/m^3.d (Rittmann and McCarty, 2001). However, one of the major limitations of this process is related to wastewaters that have a high

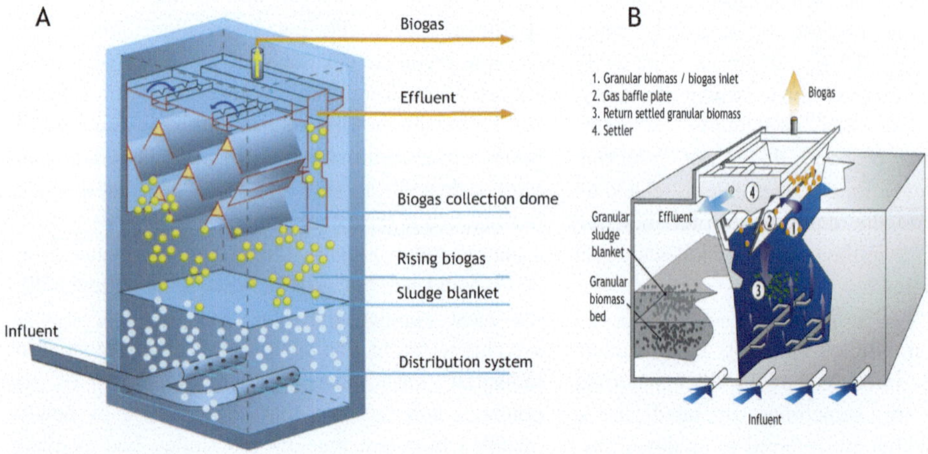

Figure 16.17. UASB reactors of the major anaerobic system manufacturers: (A) Paques B.V. and (B) Biothane B.V.

Figure 16.18 UASB installations for treatment of (A) fruit juice factory wastewater in Bregenz, Germany and (B) dairy wastewater in Indonesia (photos: Paques B.V. and Biothane B.V., respectively).

suspended-solids content, which hampers the development of dense granular sludge (Alphenaar, 1994). The sludge bed reactor concept is based on the following concepts:

1) Anaerobic sludge has or acquires good sedimentation properties, provided the process is operated correctly. Small particles or sludge that settles slowly will be washed out from the system.
2) The required good contact between the sludge and wastewater in UASB reactors is generally accomplished by feeding the wastewater uniformly over the bottom of the reactor. Also, the increased upflow velocity results in a better contact between the sludge particles and the pollutants. At OLRs exceeding 5 kg COD/m^3.d, mixing of sludge and wastewater is brought about by biogas turbulence. Mechanical mixing is not applied in UASB reactors.
3) With wastewaters containing biodegradable inhibitory compounds, the hydrodynamic mixing is additionally achieved by applying a liquid recirculation flow. As a result, a more completely

mixed flow pattern is acquired and stratification of the substrate and intermediate products over the height of the reactor is minimized, thereby minimizing potential inhibition.

4) The washout of active sludge aggregates is prevented by separating the produced biogas using a gas collection dome installed at the top of the reactor. In this way a zone with relatively little turbulence is created in the uppermost part of the reactor, and consequently the reactor is equipped with an built-in secondary clarifier.
5) The gas collection dome acts like a three-phase gas-liquid-solids separator. The GLSS device constitutes an essential part of a UASB reactor and serves to:

   a) Collect, separate and discharge the produced biogas. For a satisfactory performance the gas-liquid surface area within the device should be sufficiently large that gas can easily escape. This is particularly important if scum layers develop.
   b) Minimise liquid turbulences in the settler compartment for enhancement of sludge settling. In order to prevent biogas bubbling to the settling zone at the top, one or more baffles should be installed beneath the aperture between the gas domes as well as between the gas dome and the reactor wall.
   c) Remove sludge particles by a mechanism of sedimentation, flocculation and/or entrapment in a sludge blanket (if present in the settler). The collected sludge can slide back into the digestion compartment, if the sludge bed does not reach into the settler, or can occasionally be discharged together with excess sludge from the digester compartment.
   d) Limit the expansion of the sludge bed in the digester compartment. The system acts as a type of barrier against excessive expansion of the lighter part of the sludge bed. If the sludge bed expands into the settler, the sludge will tend to thicken (because the gas has been separated). This thickened, heavier sludge slides back to the digester compartment.
   e) Reduce or prevent buoying sludge particles from being washed out from the system. For this purpose a scum layer baffle should be installed in front of the effluent weir of the overflow.
   f) Accomplish some polishing of the wastewater with respect to suspended matter.

Some researchers and practitioners suggest replacing the GLSS device with a packed bed in the upper part of the reactor. This is referred to as an upflow hybrid reactor and is a merger of UASB and UAF reactors. In some designs the packing material is mounted only in the settling compartment leaving the GLSS in its original position. Approximately 3% of all anaerobic reactors installed are hybrid reactors (Figure 16.16). In most applications, the majority of organic matter conversion is located in the sludge bed section and enhanced removal of small-sized suspended solids is achieved in the top part. Trials with domestic sewage have showed improved removal of both suspended solids and colloidal matter (Elmitwalli *et al.*, 2002).

For some specific chemical wastewaters, such as purified therephthalic acid (PTA) wastewater (Kleerebezem, 1999a, b), a hybrid reactor shows better performance than a single-stage UASB. However, results have showed that the conversion of therephthalic acid to benzoate is only possible at low concentrations of acetate and benzoate. By applying a hybrid system, these are converted in the sludge bed area, meanwhile therephthalic acid is then converted in the hybrid section, where specific flora is retained for degrading the refractory compound.

The most common disadvantage of hybrid reactors is the deterioration of the filter section after prolonged periods of operation.

16.7.2.4 Anaerobic expanded and fluidized-bed systems (EGSB and FB)

Expanded-bed and fluidized-bed systems are regarded as the second generation of sludge bed reactors, achieving extreme organic volumetric loading rates as well as reaching 30 to 60 kgCOD/m^3.d at lab-scale and 20 to 40 kgCOD/m^3.d at full-scale. The FB process is based on the occurrence of bacterial attachment to mobile carrier particles, which consist, for example, of fine sand (0.1-0.3 mm), basalt, pumice, or plastic. The FB system can be regarded as an advanced anaerobic technology (Li and Sutton, 1981; Heijnen *et al.*, 1990) that can reach high loading rates when operated under defined conditions. High rates of mass transfer resulting from (*i*) liquid turbulence and high flow rate around the particles, (*ii*) less clogging and less short-circuiting due to the occurrence of large pores through bed expansion, and (*iii*) high specific surface area of the carriers due to their small size, make FB reactors highly efficient. However, long-term stable operation appears to be problematic. The system relies on the formation of a more or less uniform (in thickness, density, and strength) attached biofilm and/or particles. In order to maintain a stable situation with respect to the biofilm development, a high degree of pre-acidification is considered necessary and dispersed matter should be absent in the feed (Ehlinger, 1994). Despite that, an even film thickness is very difficult to control and in many situations a segregation of different types of biofilms over the height of the reactor occurs. In full-scale reactors bare carrier particles often segregate from the biofilms leading to operational problems. In order to keep the biofilm particles in the reactor, flow adjustments are necessary after which the support material will start to accumulate in the lower part of the reactor as a kind of stationary bed, whereas light fluffy aggregates (detached biofilms) will be present in the upper part. The latter can only be accomplished when the superficial velocity remains relatively low, which in fact is not the objective of an FB system.

Second-generation FB systems such as the Anaflux system (Holst *et al.,* 1997) rely on bed expansion rather than on bed fluidization. As bed expansion allows a much wider distribution of prevailing biofilms, the system is considerably easier to operate. As in the conventional AF systems, an inert porous carrier material (particles < 0.5 mm, density approximately 2) is used for bacterial attachment in the Anaflux system. The Anaflux reactor also uses a three-phase separator at the top of the reactor, more or less similar to the GLSS device in UASB reactors. When the biofilm layer attached to the media becomes excessively over-developed, and the related (lighter) aggregates subsequently accumulate in the separator device, the material is periodically extracted from the reactor by an external pump in which it is subjected to the application of sufficient shear to remove part of the biofilm. Afterwards both the media and detached biomass are returned to the reactor, and the free biomass is then allowed to be washed out from the system. In this way the density of the media is controlled and a more homogeneous reactor bed is created. Up to 30-90 $kgVSS/m^3_{reactor}$ can be retained in this way and because of the applied high liquid upflow velocities, *i.e.* up 10 m/h, an excellent liquid-biomass contact is accomplished. The system is applicable to wastewaters with a suspended solids concentration < 500 mg/l. Most full-scale anaerobic FB reactors are installed as Anaflux processes. Nonetheless, at present, EGSB-type reactors are of much more interest for full-scale applications than the more expensive FB systems.

The EGSB reactor can be considered to be an upgrade of the conventional UASB reactor. The EGSB system employs granular sludge, which is characterised by good settling characteristics and a high methanogenic activity (see also Table 16.10). As a consequence, the applied OLR and upward flow velocities are distinctly higher in EGSB reactors compared to UASBs. Sludge bed expansion is achieved by the prevailing process conditions. When applying extreme sludge loading rates the settleability will be less owing to the biogas hold-up in the granules. Nonetheless, because of the high sludge settleability, high superficial liquid velocities, *i.e.* exceeding 6 m/h, can also be applied. These high liquid velocities, together with the lifting action of gas evolved in the bed, leads to a (modest) expansion of the sludge bed. In addition, as a result of this, an excellent contact between sludge and wastewater prevails in the system, leading to significantly higher loading potentials compared to conventional UASB installations. In some expanded bed systems, *e.g.* the Biopaq®IC-reactor (Figure 16.20), the superficial flow velocities, resulting from both hydraulic and gas flows, can range between 25-30 m/h, causing an almost complete mixing of the reactor medium with the available biomass.

Excellent results have been obtained with modern full-scale expanded bed installations, such as the Biobed EGSB and Biopaq®IC reactors (Figure 16.19), using various kinds of wastewaters and applying OLRs of 25-35 kg COD/m^3.d. The extreme COD loading rates of EGSB-type systems result in extreme biogas-loading rates following:

$$V_{biogas} = COD_{conc} \cdot \frac{E_{ff-meth}}{100} \cdot \frac{0.35}{F_{meth-biogas}} \cdot \frac{(T+273)}{273} \cdot V_{upw,liquid} \qquad (16.47)$$

In which, $E_{ff\text{-}meth}$ = amount of COD converted to CH_4 or COD efficiency based on CH_4 production, $F_{meth\text{-}biogas}$ = fraction of methane in biogas (*e.g.* 0.6 for 60% CH_4), T = operational temperature of EGSB reactor in °C, $V_{upw,liquid}$ = upward liquid velocity in EGSB reactor.

It must be noted that the actual value of $F_{meth\text{-}biogas}$ will be higher than the theoretical estimate using 18.75/100 · COD/TOC (Figure 16.10), owing to the high solubility of CO_2 in the medium and chemical binding of HCO_3^- to cations such as Na^+, K^+, and NH_4^+ (Section 16.3.1.).

Generally, a biogas loading rate of no more than 2-3 m^3/m^2.h is applied for conventionally designed GLSS devices in UASB reactors. For biogas loading rates exceeding these values, more advanced gas separators are required. EGSB reactors have a high height-diameter ratio, with reactor heights reaching up to 25 meters. Consequently, biogas turbulence accumulates from bottom to top. Since the EGSB systems rely on complete retention of the granular sludge, efficient sludge separation at the top part of the system is of utmost importance. The various contractors supplying EGSB reactors have their own typical features for actively separating the sludge from the liquid and gas flow, applying specifically designed GLSS units. It is clear that, under EGSB conditions, conventionally designed GLSS devices are of no use. Interestingly, by applying an EGSB reactor system, several other types of wastewaters can be treated that cannot be treated using conventional UASB systems. Examples are:

1) Wastewaters containing highly toxic but anaerobically biodegradable compounds. Treatment of these wastewaters requires that external or internal dilution keeps the toxicant concentration to which the biomass is exposed sufficiently low. For example, full-scale reactors have shown stable performance over many years treating wastewaters with high formaldehyde concentrations, reaching values of approximately 10 g/l (Zoutberg and Frankin, 1996).
2) Wastewater containing dyes and other toxic textile auxiliary compounds can be successfully converted into biogas without inhibitory effects on the biomass (Frijters *et al.*, 2006).
3) Cold (< 10°C) and dilute (COD < 1 g/l) wastewaters, *i.e.* when specific gas production is very low and biogas mixing is absent (Rebac *et al.*, 1998). EGSB reactors are characterised by an improved hydraulic mixing, and are thus less dependent on biogas production for proper mixing. As a consequence, and in contrast to UASB systems, all retained sludge is optimally mixed with the incoming wastewater, while small inactive particles are washed out from the system.
4) Wastewaters containing long chain fatty acids (Rinzema, 1988). At low upflow velocities (UASB), LCFAs tend to absorb to the sludge and form inaccessible fatty clumps. At high upflow velocities (EGSB) the substrate is introduced at a lower concentration and is more evenly distributed to the biomass.

A special version of the EGSB concept is the Internal Circulation (IC®) reactor (Vellinga *et al.*, 1986), see also Figure 16.19. In this type of reactor, the produced biogas is separated from the liquid halfway through the reactor by means of a gas/liquid separator device and conveyed upwards through a pipe to a degasifier unit or expansion device. Here, the separated biogas is removed from the system, whereas the sludge-water mixture drops back to the bottom of the reactor via another pipe. In fact, the lifting forces of the collected biogas are used to bring about a recirculation of liquid and granular sludge over the lower part of the reactor, which results in improved contact between the sludge and wastewater. The extent of liquid/sludge recirculation depends on the gas production (Vellinga *et al*, 1986; Pereboom and Vereijken, 1994; Habets *et al*., 1997).

Full-scale examples of IC and EGSB systems are shown in Figure 16.20.

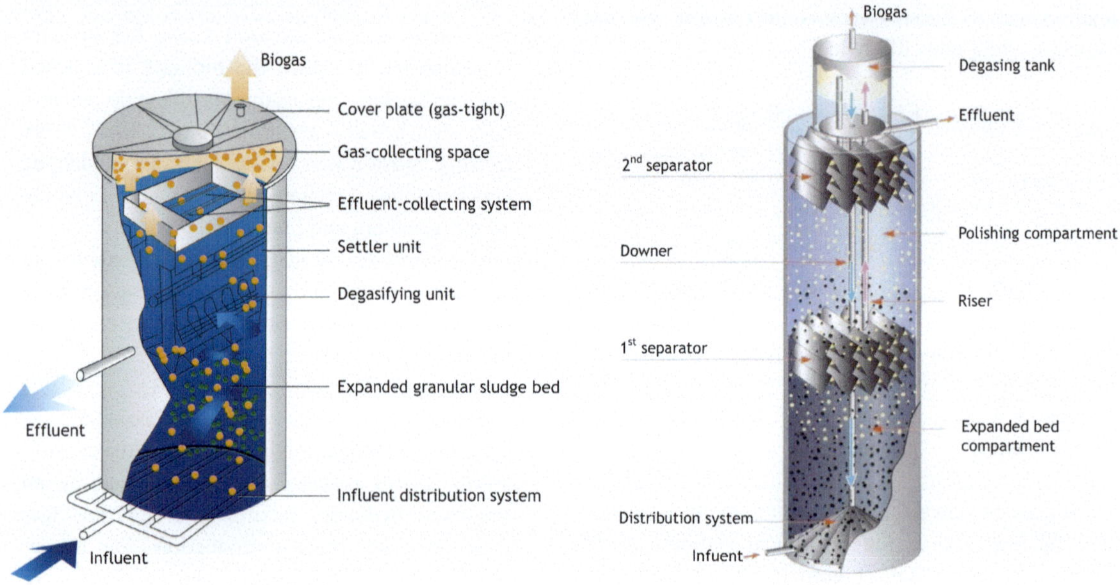

Figure 16.19 EGSB and IC® reactors of the major anaerobic system manufacturers Biothane B.V. (left) and Paques B.V. (right).

Figure 16.20 An EGSB installation for treatment of dairy wastewater in Germany (left) and an IC installation for treatment of brewery wastewater in Den Bosch in the Netherlands (right) (photos: Biothane B.V. and Paques B.V., respectively).

The success of expanded bed reactors in the 1990s has resulted in an increase in the share of sales at the expense of UASB reactors (Figure 16.21).

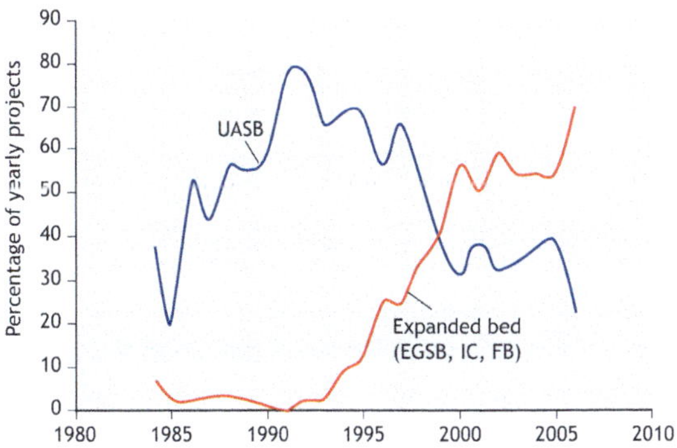

Figure 16.21 Share of UASB and EGSB systems in the full-scale anaerobic treatment systems installed in the period 1984-2007. The EGSB reactors include EGSB, IC®, and FB systems.

16.7.2.5 Advanced sludge liquid separation

Disappointing granule formation restricts contractors in offering EGSB-type anaerobic high-rate reactors to industries for treating their wastewater. Also, for the more complex types of wastewaters, such as those characterized by a high suspended solids (SS) content, expanded bed reactors are not very appropriate. Under the prevailing flow conditions the suspended solids will be washed out from the system, and/or the heavier suspended solids can negatively impact granule formation and granule growth (Alphenaar, 1994). The inability to form granules is also experienced during the treatment of wastewaters that are characterized by COD concentrations exceeding 50 g/l, *e.g.* distillery slops or vinasse. Owing to the high influent COD concentrations, the resulting HRTs are very long, drastically diminishing the hydraulic selective pressure inside the reactor, which is regarded as crucial for sludge granulation (Hulshoff Pol *et al.*, 2004). Novel reactor developments are focusing on system robustness similar to UASB reactors, in which the COD loading potentials should reach the levels of EGSB systems, although the presence of granular sludge cannot be guaranteed. This calls for more enhanced sludge-solids separation devices that can operate under high hydraulic flow conditions.

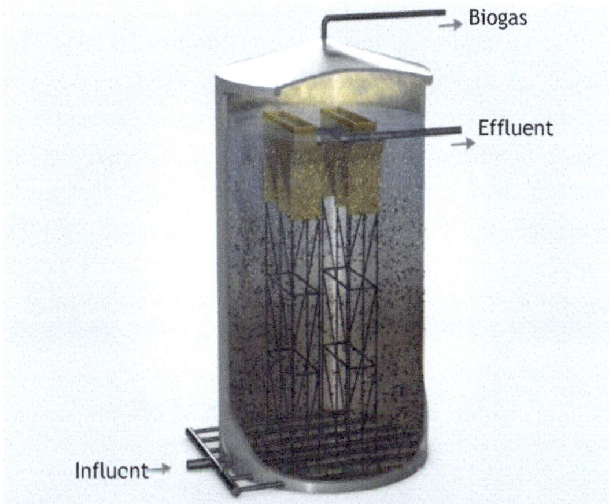

Figure 16.22 The Biopaq®UASBplus reactor for the treatment of concentrated wastewaters, operated with either granular or flocculent sludge.

Physically-enhanced settling can be achieved by mounting a tilted plate settling device for sludge liquid separation in the bioreactor. In fact, Biothane Systems International has already incorporated a tilted plate settler into the GLSS device in their BioBed®EGSB system (Zoutberg and Frankin, 1996). In the past decade, the Dutch contractor Paques has applied this idea to an upflow sludge bed reactor with a high height-diameter ratio, in a system known as the Biopaq®UASBplus (Figure 16.22). Although the UASBplus sludge separator device can also be employed for the retention of anaerobic granules, it is very suitable for anaerobic flocculent sludge, which is prevalent in the case of more concentrated wastewater, such as bioethanol waste(water), *e.g.* vinasse. Approximately one fifth of full-scale UASBplus systems are operational, of which approximately one third of the reactors contain flocks or small aggregates; most UASBplus reactors are installed in China.

More recently the same company has introduced a novel reactor system in which sludge-liquid separation is performed at the bottom part of the high-rate reactor. The reactor, the Biopaq®ICX, carries the treated effluent from the top part to the tilted plate solids separator at the bottom (Figure 16.23). The observed advantages are: (*i*) floating sludge does not wash out; (*ii*) improved sludge-liquid separation, which is attributed to the higher sludge density at the bottom, which results from the increased hydraulic pressure of 1-2 bar; and (*iii*) increased sludge concentrations reaching 70-80 kg SS/m^3, which result in applicable loading rates beyond 40 kg COD/m^3.d.

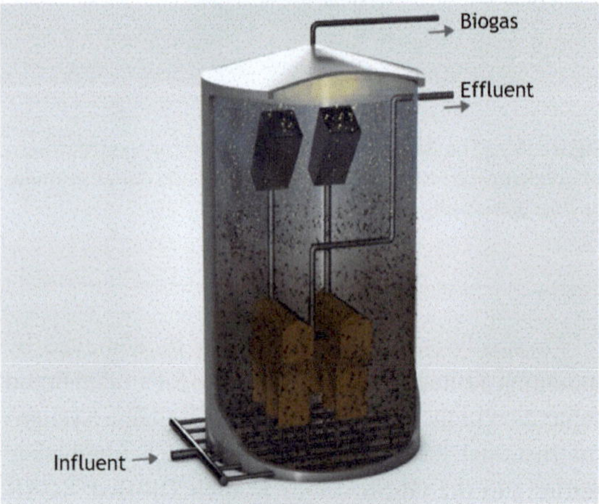

Figure 16.23 The Biopaq®ICX reactor with enhanced biomass separation at the bottom part of the reactor.

16.7.2.6 Other anaerobic high-rate systems

Whereas ACP, UASB and EGSB reactors are based on a mixed/completely-mixed reactor content, various designs have been tested which employ *staging* of the various phases of anaerobic treatment (Van Lier *et al.*, 2001). An extreme example is the two-stage process where the acidification step is completely separated from the methanogenic step (see Section 16.7.2.7). Horizontal staging is obtained in anaerobic baffled reactors (ABR), which is best characterised as a series of serially operated UASB units without GLSS devices (Bachmann *et al.*, 1985; Barber and Stuckey, 1999). Although some larger-scale applications have been made on domestic sewage, the reactor has not been developed further than pilot-scale (Zhu *et al.*, 2015). A problem of concern is the hydrodynamic limitation which constrains the achievable SRT in the system, since the superficial liquid velocity in a baffled system is substantially higher than in a single-step sludge bed reactor. As a result, the sludge mass can slowly move with the liquid flow through the various compartments. Vertically-staged reactors such as the upflow staged sludge bed system (Van Lier *et al.*, 1994, 2001; Guiot *et al.*, 1995; Tagawa *et al.*, 2002) were specifically developed for high temperature treatment. However, although the staged reactor concept showed very promising results on a pilot-scale, so far no full-scale reactors have been developed.

Very interesting possibilities could exist for the anaerobic sequencing batch reactor (ASBR) which consists of a set of anaerobic reactors operated in batch mode using a 'fill and draw' method. A certain amount of the raw wastewater is supplied to the anaerobic reactor, after the supernatant liquid of a previous batch has been discharged. Then a gentle mixing of the reactor contents is started in order to enable the settled viable sludge to contact the wastewater and to eliminate the biodegradable organics. After a sufficient period of reaction time, the sludge is allowed to settle and the supernatant solution is discharged. The next cycle is then started. Granulation proceeds well in an ASBR on dilute wastewaters, and also at lower ambient temperatures (Banik and Dague, 1997). ASBR systems have been shown to be of particular interest for LCFA-containing wastewaters (Alves *et al.*, 2001). During the filling period, LCFAs absorb to the anaerobic sludge after which a gentle digestion period proceeds in which the absorbed sludge is stabilised and completely regenerated to high-active methanogenic biomass.

16.7.2.7 Anaerobic membrane bioreactors

Anaerobic membrane bioreactors (AnMBR) are being increasingly researched, particularly for those applications where the commonly applied sludge bed systems are less successful. AnMBRs combine the advantages of both membrane bioreactors (MBR) and anaerobic technology (Dereli *et al.*, 2012; Ersahin *et al.*, 2014). Operational costs related to energy requirements for gas/liquid recirculation for membrane-fouling control and chemical costs required for membrane cleaning still have a major effect on the economic feasibility of AnMBRs. However, membrane acquisition and/or replacement costs have decreased significantly due to a decline in membrane module costs (Ozgun *et al.*, 2013). Despite these constraints, AnMBRs offer high-quality effluents free of solids and complete retention of biomass, regardless of their settling and/or granulation properties. Furthermore, AnMBRs can be used to retain special microbial communities that can degrade specific pollutants in wastewater. Therefore, AnMBR technology can present an attractive option for treating industrial wastewaters at extreme conditions, such as high salinity, high temperature, high suspended solids concentrations and the presence of toxicity, that hamper granulation and biomass retention or reduce the biological activity (Dereli *et al.*, 2012; Muñoz Sierra *et al.*, 2019). Industrial wastewaters with extreme physicochemical characteristics will probably occur more often in the future as cleaner industrial production processes require a reduction in water consumption, and increased water reuse and resource recovery (Van Lier *et al.*, 2015; Dereli *et al.*, 2012).

Combinations of membranes with different types of high-rate anaerobic reactor configurations such as CSTR, ACP, UASB, EGSB, FB, and hybrid reactors seem possible alternatives for treatment of industrial wastewaters (Ozgun *et al.*, 2013). However, membrane integration eliminates the hydraulic selection pressure required for granulation whereas flocculent biomass with poor immobilization characteristics is retained instead of being washed out. Moreover, by applying cross-flow filtration, the prevailing shear forces will minimize the particle's diameter. Therefore, no granulation is expected in sludge bed reactors coupled to membrane filtration, which would decrease the settleability of the biomass in the long-term operation. Nonetheless, a sequenced approach of a UASB reactor followed by separate membrane modules offers interesting perspectives for full treatment. The preceding UASB provides a pre-elimination of suspended solids by entrapment and biodegradation in the sludge bed, which reduces the suspended solids load to the membrane and thus minimizes membrane fouling related to cake layer formation (Ozgun *et al.*, 2013). Most researched AnMBR systems consist of a CSTR bioreactor coupled to cross-flow membrane modules or a CSTR bioreactor equipped with submerged membranes.

Successful commercial implementation of AnMBR technology started in the early 2000s. In Japan, Kubota realized 13 relatively small-scale plants with flow rates up to 2.5 m^3/h using flat-sheet submerged membranes. The same configuration was picked up at a larger scale in the USA by ADI, where three full-scale systems have been realized since 2008 (Christian *et al.*, 2011; Allison *et al.*, 2013). The year 2008 also saw the construction of the first multi-tube demonstration scale AnMBR treating whey from a cottage cheese producer in the USA. This system utilized Pentair's (formerly Norit) ultrafiltration membranes. Based on this success, Biothane Systems International and Pentair co-developed a low-energy AnMBR system called Memthane. There are now 11 full-scale Memthane plants.

16.7.2.8 Acidifying and hydrolytic reactors

Except for well-stirred tank reactors, so far no specific reactor concepts have been developed for acidogenesis. The process of acidogenesis generally proceeds sufficiently fast in a stirred tank reactor that is concomitantly used as an equalization tank. In practice, there is generally no real need for complete acidogenesis. Moreover, nowadays it is fully understood that joint acidification with methanogenesis is beneficial for granule formation (Verstraete *et al.*, 1996). The occurrence of fermentative substrate conversion by acidifying organisms is indispensable for the production of sufficient exopolymeric substances (EPS) which are required for the formation of a stable granular structure with a high granule strength (Vanderhaegen *et al.*, 1992). Various authors have suggested that the EPS are particularly produced by acidifying organisms, creating a matrix in which all the bacteria and archaea are embedded (*e.g.* Batstone and Keller, 2001). At present, in most full-scale applications, a pre-acidification of maximum 40% is pursued. In addition, the presence of higher concentrations of acidifying organisms in the feed of a methanogenic reactor is relatively detrimental to the stability of the granular sludge bed. This means that the sludge retention of an acidogenic reactor needs to be improved.

Acidifying reactors can be combined with solids entrapment systems, safeguarding the methanogenic reactor from a suspended solids loading that is too high. Trials have been carried out combining primary clarification with anaerobic stabilisation on domestic sewage. However, although Wang (1994) implemented some full-scale systems in China, no large implementations have been implemented so far. Other research on acidifying reactors has focused on the production of VFAs either as a precursor to the production of bio-plastics (Tamis *et al.*, 2018) or for the enhancement of nutrient removal in biological nutrient removal plants for municipal sewage.

16.7.2.9 Current market trends in anaerobic high-rate reactor sales

Many different anaerobic high-rate reactors have been introduced to the market in the past few decades. Figure 16.16 clearly indicates that sludge bed reactors were (and are) by far the most successful. Since the development of UASB technology, a widespread market introduction has occurred. Since the 1990s, expanded sludge bed reactors have taken over the market (Figure 16.21), owing to their cost-efficiency and very high loading potentials. More recently, various other configurations have been introduced to the market, expanding the application potentials of anaerobic high-rate treatment. The two global market leaders in AD technology, *i.e.* Paques B.V. and Biothane-Veolia, are both located in the Netherlands and are cooperating closely with (Dutch) universities in technology development. The reactor sales of both companies are illustrative for AD technology development worldwide. Figure 16.24 clearly shows the sales of Paques B.V., indicating the sequential rise of UASB and IC technology. In the past decade, various novel and previously discussed reactors systems have been developed such as the Biopaq®AFR, Biopaq®UASBplus, Biopaq®ICX, and Biopaq®Ubox.

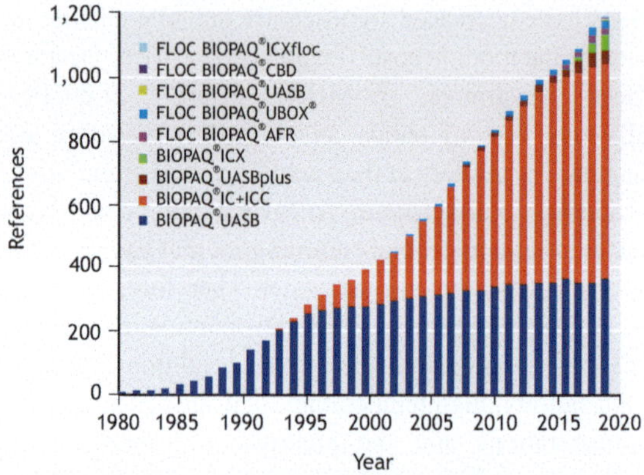

Figure 16.24 Number of references (y-axis) indicating the sales of anaerobic high-rate reactors by Paques B.V. in the period of existence (1981-2018)

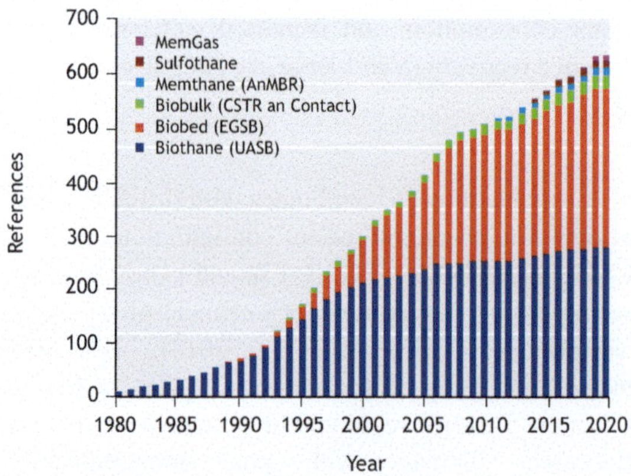

Figure 16.25 Number of references (y-axis) indicating the sales of anaerobic high-rate reactors by Biothane-Veolia in the period of existence (1976-2019), 2019 data not included.

Similar trends in reactor sales can be observed in Figure 16.25 for Veolia-Biothane, which started with the first full-scale UASB reactor constructed at a sugar mill in the Netherlands in 1976. It is worthwhile to mention the advances of Biothane-Veolia in modern BioBulk CSTR technology and the Memthane AnMBR reactor technology.

16.8 UPFLOW ANAEROBIC SLUDGE BLANKET (UASB) REACTOR

16.8.1 Process description

Development and implementation of the UASB reactor (Figure 16.17) has caused the major breakthrough in the acceptance of anaerobic treatment for a wide variety of wastewaters. The success of the UASB reactor can be attributed to its capability to retain a high concentration of sludge, while efficient separation of solids, liquids and the gas phase is attained. The UASB reactor consists of a circular or rectangular tank in which wastewater flows in an upward direction through an activated anaerobic sludge bed which occupies approximately half the volume of the reactor and consists of highly settleable granules or flocs (Figure 16.17). During the passage through the anaerobic sludge, the treatment process takes place by solids entrapment and organic matter conversion into biogas and sludge. The produced biogas bubbles automatically towards the top of the reactor, carrying water and solid particles, *i.e.* biological sludge and residual solids. The biogas bubbles are directed (via baffles) to a gas-liquid surface at the upper part of the reactor, leading to an efficient GLSS. The solid particles drop back to the top of the sludge blanket, while the released gases are captured in an inverted cone or related structure, located at the top of the reactor. Water passes through the apertures between the baffles carrying some solid particles which settle in the settling area because of the drop in upward velocity owing to the increase in the cross-sectional surface area. After settling, the solids slide back to the sludge blanket, while water leaves the settler over overflow weirs.

16.8.2 Design considerations of the UASB reactor

16.8.2.1 Maximum hydraulic surface loading

The methanogenic conversion capacity of UASB reactors, expressed in kgCOD/m^3.d, is directly related to the amount of retained viable biomass and the specific methanogenic activity of the accumulated sludge. In addition to the quantity and quality of the retained sludge, the maximum organic loading potentials also depend on the proper mixing of the sludge with the incoming wastewater. The required sludge retention time (SRT) sets limits on the applicable upward liquid velocities (V_{upw}) as well as on the specific biogas loading resulting from the anaerobic conversion process (Lettinga and Hulshoff Pol, 1991). The design of the UASB reactor combines the features of a high-rate bioreactor with those of an in-built secondary clarifier at the top. Therefore, average V_{upw} in the UASB reactor's cross-sectional area and the clarification section at the top are in the range of 0.5-1.0 m/h. Higher hydraulic loadings can lead to undesirable loss of biomass if flocculent-type sludge accumulates during the reactor operation. The latter can happen, for instance, during the first start-up when the reactor is seeded with non-adapted seed material such as digested sewage sludge or during the anaerobic treatment of domestic sewage. The V_{upw} can be calculated using the average flow and the reactor's cross-sectional area, A (Eq. 16.48).

$$V_{upw} = \frac{Q_{inf}}{A} \qquad \text{(m/h)} \quad (16.48)$$

in which Q_{inf} = the influent flow rate.

With the growth and accumulation of thick flocculent sludge, or granular sludge with a high settleability, much higher hydraulic loadings are admissible in the reactor. High V_{upw} values are applied in expanded bed reactors reaching values up to 8-10 m/h.

Based on the maximum allowable V_{upw}, the minimum surface dimensions can be calculated (Eq. 16.49).

$$A_{min} = \frac{Q_{inf}}{V_{upw,max}} \quad (m^2)\ (16.49)$$

At a given hydraulic retention time (HRT, Θ), the maximum upward velocity determines the H/A ratio, in which H is the reactor height according to Eq. 16.50.

$$\Theta = \frac{A_{min} \cdot H_{max}}{Q} \quad (h)\ (16.50)$$

$$H = V_{upw} \cdot Q \quad (m)\ (16.51)$$

$$V_{reactor} = \Theta \cdot Q \quad (m^3)\ (16.52)$$

For any situation in which the organic loading capacity is not restrictive, Eq. 16.52 gives the volume of the required UASB reactor. The latter is only the case with diluted wastewaters, such as with most domestic wastewaters in the tropical zone of Latin America where COD values are < 1,000 mg/l. Here, the hydraulic load fully determines the accumulating sludge quantity, and the in-reactor methanogenic capacity generally exceeds the applied organic loading rates.

16.8.2.2 Organic loading capacity

In most cases UASB reactors are used for the treatment of more concentrated wastewaters (Table 16.1). The volumetric conversion capacity or organic loading rate (OLR) in kgCOD/m^3 reactor.d is then dependent on the:

- quantity of accumulated biomass, X, in kg volatile suspended solids VSS/m^3 reactor volume,
- specific methanogenic activity (SMA) of the sludge in kgCOD/kgVSS.d,
- the achieved degree of mixing leading to a contact factor between 0 and 1.

The SMA depends on several factors such as:

- temperature,
- presence of inhibitory or toxic compounds,
- biodegradability of the substrate,
- presence of suspended solids (SS) in the influent,
- degree of wastewater pre-acidification.

In conventional UASB reactors the amount of anaerobic sludge generally is in the range 35-40 kgVSS/m^3 reactor volume (settler included). The contact factor depends on the effectiveness and evenness of the feed distribution and the applied organic loading rate with the resulting biogas production largely contributing to the reactor mixing.

Considering the number of unknown factors, a thorough wastewater characterisation is indispensible prior to designing a UASB reactor. In addition, reactor pilot trials are generally performed to achieve a better insight into the growth and development of the anaerobic sludge for a specific wastewater. Based on many pilot trials in the past few decades and the large subsequent amount of full-scale experience, a table of allowable organic loading rates that are dependent on the reactor temperature has been developed (Table 16.11). When the allowable OLR or r_v is known, the required UASB reactor volume can be easily calculated from the influent flow rate and its concentration (Eq. 16.53):

$$V_{reactor} = \frac{C_{inf} \cdot Q_{inf}}{r_v} \quad (16.53)$$

in which r_v = organic loading rate (kgCOD/m^3.d).

A UASB reactor is either hydraulically or organically limited in which the volume of a UASB reactor is calculated by either Eq. 16.52 or 16.53. If the actual situation is not known, the volume is generally calculated based on both considerations after which the largest volume suggested by either equation is taken as the design volume. Figure 16.26 depicts the impact of the wastewater concentration (in kgCOD/m^3) on the required reactor volume.

Table 16.11 Permissible organic loads in single-step UASB reactors for various types of wastewater in relation to the applied operating temperature. The biomass consists of thick flocculant or granular sludge.

Temperature (°C)	Organic loading rate ($kgCOD/m^3.d$)			
	VFA wastewater	Non-VFA wastewater	Wastewater with < 5% SS-COD[a]	Wastewater with 30-40 % SS-COD
15	2 - 4	1.5 - 3	2 - 3	1.5 - 2
20	4 - 6	2 - 4	4 - 6	2 - 3
25	6 - 12	4 - 8	6 - 10	3 - 6
30	10 - 18	8 - 12	10 - 15	6 - 9
35	15 - 24	12 - 18	15 - 20	9 - 14
40	20 - 32	15 - 24	20 - 27	14 - 18

[a] 5% of the influent COD consists of SS-COD, 95% is soluble COD.

Assuming a minimum HRT of 4 h for preventing sludge wash-out, the minimum required reactor volume will be at least 1,000 m^3, irrespective of the concentration of the wastewater. At high influent COD concentrations, it is obvious that the required reactor volume directly depends on the wastewater concentration since the admissible organic loading rate is fixed.

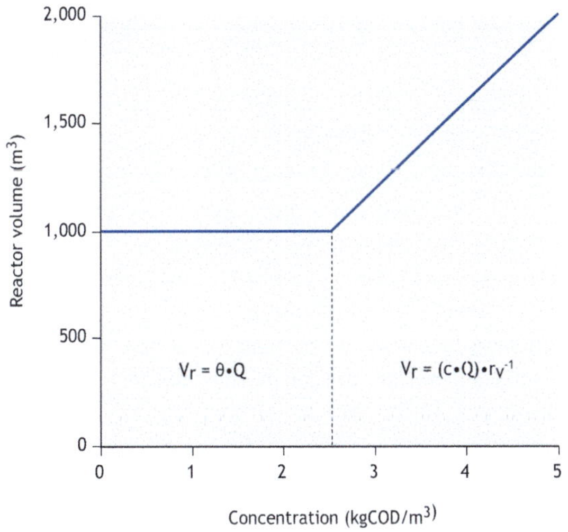

Figure 16.26 Calculating the required UASB reactor volume using the following assumptions: Θ_{min} = 4 h, Q = 250 m^3/h, r_v = 15 $kgCOD/m^3.d$, T = 30 °C. The volume is determined by either the hydraulic or organic loading rate (after Lettinga and Hulshoff Pol, 1991).

Often the great unknown is the maximum hydraulic loading potential or the minimum HRT. It is impossible to give hard numbers since this directly depends on the type of sludge that will be cultivated on that specific wastewater (granular or flocculant). Generally, for UASB reactors, and particularly those operating with non-granular sludge, a maximum upflow velocity of 1 m/h is considered. Figure 16.27 shows the impact on the required reactor volume when upflow velocities of 6 m/h can be tolerated as is the case when good quality granular sludge is cultivated. In the example the same height of the reactor is taken. In effect, reactor volumes can be reduced by a factor of 6.

In addition to liquid velocities, high loaded reactors are also limited by the turbulence brought about by the produced biogas. The biogas upward velocity (V_{biogas}) can be calculated using Eq. 16.47. With conventionally designed GLSS devices the maximum allowable V_{biogas} is between 2 and 3 m/h.

Particularly with reactors characterised by a very high height/diameter ratio, special care is given to the detailed design of the gas/liquid/solids separator as can be viewed in Figure 16.27.

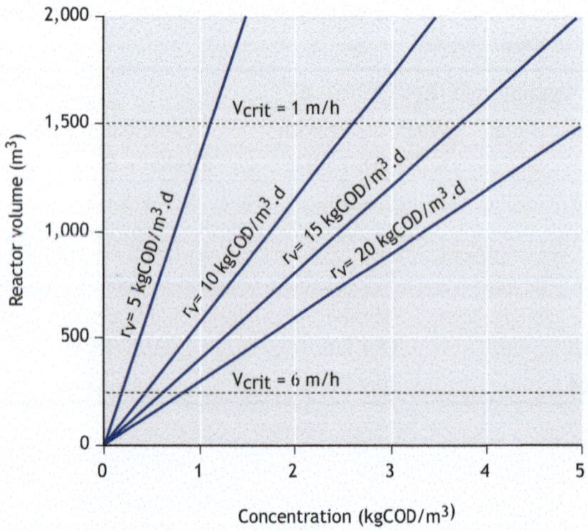

Figure 16.27 Calculating the required UASB reactor volume using the following assumptions: Q = 250 m^3/h, reactor height = 6 m, T = 30 °C. The volume is determined by either the hydraulic or organic loading rate. V_{crit} determines the cut-off level for the minimum required reactor volume based on hydraulic limitations (after Lettinga and Hushoff Pol, 1991).

16.8.2.3 Internal components of the reactor

The most important internal components of the UASB reactor and which require careful consideration are the feed inlet distribution, the effluent outlet, and the GLSS device. Most constructors and contractors apply their own (often patented) design. It goes beyond the purpose of this chapter to address in details the design feature of these internal components, but Section 16.10 gives some general design features of anaerobic sewage treatment reactors.

Of crucial importance are the evenness and density of the feed distribution system, particularly when the UASB system is applied at low loading rates *i.e.* when turbulence brought about by biogas production is limited. Table 16.12 gives some indicative values applicable to UASB reactors operated with either flocculent or granular sludges. Full-scale experiences show that at organic loading rates exceeding 5 kgCOD/m^3.d, biogas-induced reactor turbulence is sufficient for adequate mixing, reducing the resistance to mass transfer rate to an appropriate level. Compared to UASB reactors, the influent distribution systems in EGSB reactors are less critical owing to the relatively small reactor surfaces.

Tentative design guidelines for conventional GLSS devices in UASB reactors are given in Table 16.13. Further design features are explained in detail by *e.g.* Van Haandel and Lettinga (1994) and Chernicharo; Bressani-Ribeiro (2019).

Table 16.12 Required area (m^2) per feed inlet of a UASB reactor, by type of sludge and applied loading rate.

Type of sludge	Loading rate (kgCOD/m^3.d)	Surface area per feed inlet (m^2)
Medium thick flocculant (20-40 kgTS/m^3)	< 1-2	1-2
	> 3	2-5
Dense flocculant (> 40 kgTS/m^3)	< 1	0.5-1
	1-2	1-2
	> 2	2-3
Granular sludge	< 2	0.5
	2-4	1-2
	> 4	> 2

16.8.3 UASB septic tank

The UASB septic tank is a reactor system of particular interest for application in decentralized sanitation concepts. Influents to these reactors can consist of relatively diluted domestic wastewaters or concentrated waste streams, such as separately collected black water. Similar to the UASB reactor configurations, the reactor is operated in an upflow mode and upflow velocities are very low, ranging from approximately 0.01 m/h for black water systems to 0.20 m/h for diluted domestic waters. For concentrated domestic waters, an average upflow velocity of 0.02-0.05 m/h is applied. Because of the low hydraulic loadings, improved solids separation is obtained. In fact, UASB-septic tank systems function as an accumulation and stabilization system for solids and a methanogenic reactor for soluble organic

Table 16.13 Summary of guidelines for the design of the gas-liquid-solids-separator device.

	UASB-GLSS device
1	The angle of the settler sides with the bottom (*i.e.* the inclined wall of the gas collector) should be 45° to 60°.
2	The surface area of the apertures between the gas collectors should be 15-20% of the reactor surface area.
3	The height of the gas collector should be around 1/3 of the reactor height at reactor heights of 5-6 m.
4	To facilitate the release and collection of gas bubbles and to combat scum layer formation, the scum inside the GLS device should be routinely removed.
5	To prevent upflowing gas bubbles from entering the settler compartment, the overlap of the baffles installed beneath the apertures should be 15-20 cm.
6	Generally, scum layer baffles should be installed in front of the effluent weirs.
7	The diameter of the gas exhaust pipes should be sufficient to guarantee the easy removal of the biogas from the gas collection cap, particularly in the event of foaming.
8	In the upper part of the gas cap, anti-foam spray nozzles should be installed in case the treatment of the wastewater is accompanied with heavy foaming.

compounds. The sludge production is very low, *i.e.* the filling period of the reactor is 4-7 years when treating domestic sewage. The UASB-septic tank is a robust and compact system as it can be adequately designed at 2-4 days HRT, which correspond to organic loading rates of 0.23-0.45 (kgCOD/m^3.d), even in Mediterranean countries and temperate-climate countries with a low winter temperature where sewage temperature goes below 13 °C (Al-Jamal and Mahmoud, 2009; Mahmoud and Van Lier, 2011). To enhance mixing, the UASB-septic tank can be optionally equipped with a central 'stirrer' for periodic and gentle movement of the sludge bed.

Optional nutrients and faecal coliform removal need to be addressed in a separate post-treatment step after the UASB-septic tank. In this respect, the very efficient suspended solids removal in the UASB-septic tank is advantageous for the application of simple subsequent processes, such as trickling filters.

16.9 ANAEROBIC PROCESS KINETICS

Bacterial conversion rates, including anaerobic processes, are generally described by Monod kinetics for substrate conversion (see Chapter 2). Anaerobic conversion kinetics, including all the kinetic parameters, were extensively reviewed by Batstone *et al.* (2002) who presented a unified anaerobic digestion model, referred to as ADM1 in analogy with the ASM1 for activated sludge. ADM1 evolved from a number of different anaerobic models which have been presented in the literature in the past decades. For the same convenience as explained in sections 16.3 and 16.5, the ADM1 model also makes use of the COD balance for describing the flow of electrons during the anaerobic conversion process. The large variations in the cited assessed kinetic parameters for the specific conversion reactions are striking, see Table 16.14, after Batstone *et al.* (2002). This means that the process configuration, exact prevailing microbial flora, and actual operation of the system largely determine the applicable kinetic parameters.

So far ADM1 has been a very useful tool for describing existing systems, giving insights into the process dynamics and the impact of changing process parameters such as feed concentration, substrate flow, temperature, etc. on the overall digestion process. Using actual reactor data, the kinetic parameters can be adjusted for realistically predicting the reactor performance on COD removal and CH_4 production. Also, for teaching purposes, ADM1 is a valuable tool giving insight into the importance of specific conversion steps in the entire chain of consecutive reactions. On the other hand, ADM1 is still lacking biofilm kinetics and system hydraulics which can largely determine the actual kinetics in high-rate anaerobic treatment systems.

Table 16.14 Kinetic parameters of main substrates/intermediate products in the anaerobic conversion process (after Batstone *et al.*, 2002). Data from various types of digestion systems. The table presents cited literature review data only if available, otherwise most typical are taken. All substrate and VSS-related weights are expressed as COD equivalents.

Substrate	Uptake rate (kg/kgVSS.d)	μ_{max} (1/d)	Y (kgVSS/kg)	K_s (kg/m^3)	K_d (1/d)
Hydrogen	2-65	0.02-12	0.014-0.183	0.00002-0.0006	0.009
Acetate	3-18	0.05-1.4	0.014-0.076	0.011-0.930	0.004-0.036
Propionate	0.16-0.31	0.004-0.016	0.025-0.05	0.06-1.15	0.01-0.04
Butyrate	5-14	0.35-0.90	0.066	0.012-0.30	0.027
Valerate	15-19	0.86-1.20	0.058-0.063	0.062-0.36	0.01-0.03
LCFA	1.4-37	0.10-1.65	0.045-0.064	0.06-2.0	0.01-0.20
Amino acids	36-107	2.36-16	0.06-0.15	0.05-1.4	0.01-3.2
Monosaccharides	29-125	0.41-21.3	0.01-0.17	0.022-0.63	0.02-3.2

For instance, in a 3-phase system, convective mass transport on a micro- and macro-level, which is induced by the gaseous end products, can have a major effect on the kinetic parameters. The resulting actual system dynamics could fully overrule the model input parameters. Therefore, and so far, as a design tool, ADM1 is of no use and the current challenge is to combine the biological ADM1 model with other hydrodynamic and chemical models in order to create a comprehensive design tool or operation support tool for when operating an anaerobic system in a dynamic environment.

16.10 ANAEROBIC TREATMENT OF DOMESTIC AND MUNICIPAL SEWAGE

Municipal wastewater is in terms of quantity the most abundant type of wastewater on earth. Discharge of non-treated wastewater to surface waters has a huge environmental impact and poses serious health concerns to the population. Minimising both the human health risks and environmental risks have been the main incentives for developing adequate treatment technologies to address these wastewaters (see Chapter 1). However, in many developing countries, financial constraints restrict the application of activated sludge systems and alternatives are being sought. AnWT offers a cost-effective alternative, which was already recognised in the mid-1970s by *e.g.* Lettinga and co-workers. High-rate anaerobic reactors were developed in the 1970s and 1980s for the treatment of high-strength industrial wastewaters with a low concentration of COD_{SS}. It should be noted that domestic sewage and municipal wastewaters are characterised as a very dilute type of wastewater with a high fraction COD_{SS}. In large parts of the world the COD concentrations of municipal sewage is < 1,000 mg/l and often even below 500 mg/l, with temperatures barely reaching 25 °C. According to Figure 16.26, anaerobic treatment of such wastewater is limited by the hydrodynamic constraints in the system rather than the organic conversion capacity. Due to the high fraction of COD_{SS} in the influent, flocculent rather than granular sludge will develop in UASB reactors when treating municipal or domestic wastewater. Hydrolysis of COD_{SS} rather then methanogenesis becomes the rate-limiting step. Moreover, sewage temperatures are often (much) lower than industrial wastewaters. Only under tropical climate conditions will municipal wastewaters reach temperatures ideal for AnWT (Van Haandel and Lettinga, 1994). The first experience with compact/high-rate anaerobic treatment using UASB reactors for sewage treatment was gained in the early 1980s in Cali, Colombia (Van Haandel and Lettinga, 1994). The results obtained from the operation of the 64 m^3 pilot UASB reactor showed the feasibility of the system under the prevailing environmental conditions and sewage characteristics. The initial trials were rapidly followed by full-scale reactors in Colombia,

Brasil and India (Chernicharo *et al.*, 2015). Table 16.15 lists some of the results of these full-scale UASB reactors treating sewage. Since the early 1990s, hundreds of full-scale UASB reactors have been constructed ranging from 50–50,000 m^3 in volume, particularly under (sub)-tropical conditions (Von Sperling and Chernicharo, 2005; Chernicharo *et al.*, 2015; Chernicharo and Bressani-Ribeiro, 2019). Generally, a reduction in the BOD between 70 and 80% is realized, with effluent BOD concentrations of less than 40–50 mg/l. Total removal rates with regard to COD and TSS are approximately 65–75% and sometimes even higher (Chernicharo *et al.*, 2015). In order to comply with local regulations on discharge, the UASB system is generally accompanied by a proper post-treatment system, such as: polishing ponds, trickling filters, activated sludge, physico-chemical treatment, sand filtration, constructed wetlands, etc. (Von Sperling and Chernicharo, 2005).

The UASB reactor and the post-treatment step can be implemented consecutively or in a more integrated set-up. Table 16.16 lists the most important features of high-rate anaerobic sewage treatment. Most of the advantages are in agreement with advantages listed for industrial anaerobic reactors (Section 16.1.1).

However, during the early development of anaerobic sewage treatment, some of the constraints were simply ignored or not taken into consideration in the full-scale design because of financial limitations (*e.g.* Van Lier *et al.*, 2010). This has resulted in negative experiences which is a bad advertisement. Nowadays, measures must be taken to avoid uncontrolled greenhouse gas emissions and non-flaring of captured CH_4 should be prohibited (Chernicharo *et al.*, 2015). For most of the listed constraints technical solutions are available, or at least in development, *e.g.* recovery of the methane from anaerobic effluents (Chernicharo *et al.*, 2017; Heile *et al.*, 2017).

The simplicity of the system also follows from Figure 16.28, which compares the functional units of an activated sludge process with that of an anaerobic high-rate system. The single-step UASB reactor only comprises four functional units:

1) Primary clarifier: removal/entrapment of (non)biodegradable suspended solids from the influent.
2) Biological reactors (secondary treatment): removal of biodegradable organic compounds by converting them into methane.
3) Secondary clarifier: clarifying the treated effluent in the settler zone at the top part of the UASB reactor.
4) Sludge digester: stabilisation (digestion) and improving the dewatering characteristics of the retained sludge.

Table 16.15 Treatment performance of the first full-scale UASB plants treating municipal sewage. COD refers to total COD of the raw wastewater (after Van Haandel and Lettinga, 1994).

Country	Volume (m^3)	Temperature (°C)	HRT (h)	Influent COD (mg/l)	Effluent COD[a] (mg/l)	COD removal (%)
Colombia	64	24-26	4-6	267	110	65
Colombia	6,600	25	5.2	380	150	60-80
Brazil	120	23	4.7-9	315-265	145	50-70
Brazil	67.5	23	7	402	130	74
Brazil	810	30	9.7	563	185	67
India	1,200	20-30	6	563	146	74

[a] Calculated from the influent COD and removal efficiency.

Table 16.16 Main advantages and constraints[a)] of anaerobic sewage treatment in anaerobic high-rate systems.

Advantages

- Substantial savings, reaching 90%, in operational costs as no energy is required for aeration.
- 40-60% reduction in investment cost as fewer treatment units are required.
- If implemented at an appropriate scale, the produced CH_4 is of interest for energy recovery, electricity and/or heat production.
- The technologies do not make use of high-tech equipment, except for the main headwork pumps and fine screens. The treatment system is less dependent on imported technologies.
- The process is robust and can handle periodic high hydraulic and organic loading rates.
- Technologies are compact with average HRTs between 6 and 9 h and are, therefore, suitable for application in urban areas, minimising conveyance costs.
- Small-scale applications allow decentralisation in the treatment, making sewage treatment less dependent on the extent of the sewerage networks.
- The excess sludge production is low, well stabilized and easily dewatered so it does not require extensive post-treatment.
- The valuable nutrients (N and P) are retained in the effluent, giving possibilities for agricultural reuse.

Constraints

- Anaerobic treatment is a partial treatment, requiring post-treatment for meeting the discharge or reuse criteria.
- The produced CH_4 is largely dissolved in the effluent (depending on the influent COD concentration and temperature). So far, no (or only very little) measures have been taken to prevent CH_4 escaping into the atmosphere.
- The collected CH_4 is often not recovered or flared.
- Application of anaerobic treatment of domestic/municipal sewage at low ambient temperatures requires long HRTs and therefore large reactor volumes and/or amended reactors, which limits the feasibility.
- Reduced gases such as H_2S, that are dissolved in the effluent can escape, causing odour problems.

a) Compared to activated sludge processes.

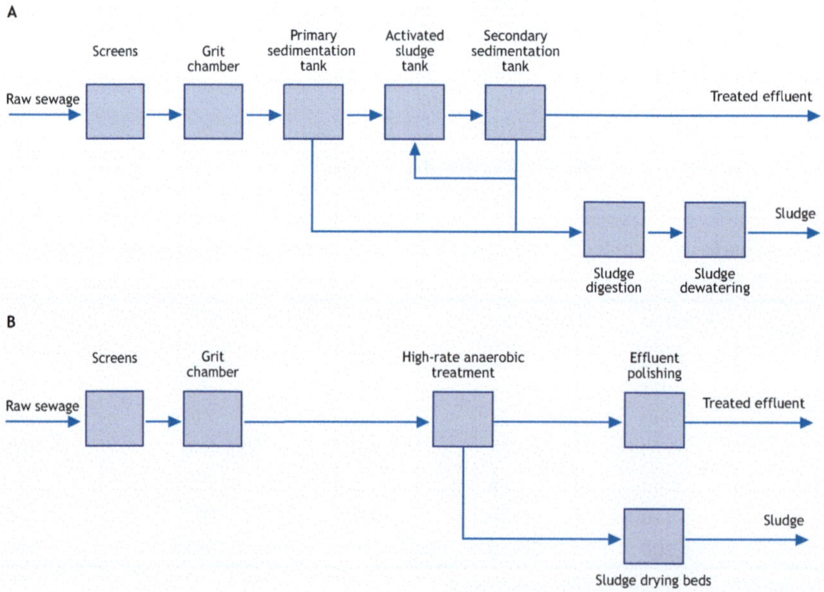

Figure 16.28 Functional units of a sewage treatment plant, comparing activated sludge (A) and UASB technology (B).

Pre-treatment of raw sewage using screens and sand & grit removal are needed for any compact treatment system. Anaerobic sewage treatment generally requires fine screens, of 4-6 mm clear distance between bars, after the coarse screens to minimise operational problems such as clogging of influent distribution. In most cases the fine screen is the most expensive part of the treatment system. The sludge from an anaerobic sewage treatment reactor, when properly designed, is well stabilised owing to the long SRTs and can be dried by applying sludge drying beds. No smell arises from the sludge drying beds.

According to Figure 16.29, the design of a UASB reactor for the treatment of diluted sewage (at temperatures between 20-30 °C) is relatively simple, because the hydraulic criteria limit the loading rate to be applied. Volumetric sizing of a UASB reactor treating sewage with an average COD concentration of 500 mg/l can be calculated using eqs. 16.49-16.53. Applying an average V_{upw}, of 0.7 m/h and an HRT of approximately 8 h (Chernicharo *et al.*, 2015), the reactor height is $0.7 \cdot 8 = 5.6$ m, whereas the reactor volume is $8 \cdot Q$ in m^3/h.

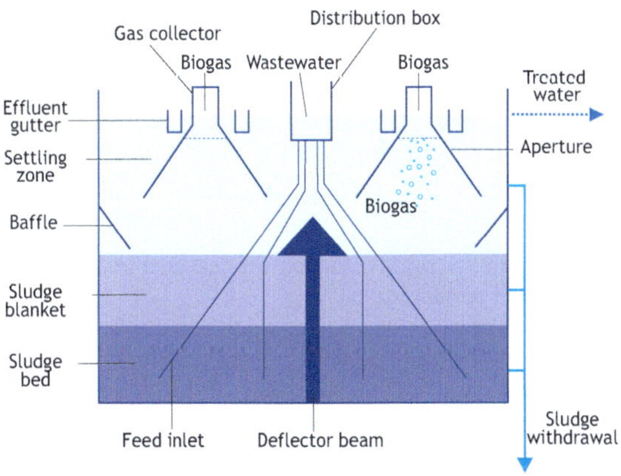

Figure 16.29 Schematic representation of a UASB reactor for treating domestic sewage. The most important design aspects are indicated.

The most critical design aspects are pictured in Figure 16.29 and are well explained by Van Haandel and Lettinga (1994), Von Sperling and Chernicharo (2005) and Chernicharo and Bressani-Ribeiro (2019). Table 16.17 provides some key numbers based on the various full-scale reactors in Latin America.

Table 16.17 Some design criteria for UASB reactors treating sewage in tropical countries.

Parameter	Value
Average HRT	6-8 h
Height	5 m
Feed inlet points	1 inlet per 1-4 m^2
Feed distribution	Each inlet pipe from a separate compartment
Static pressure in feed inlet box	Up to 50 cm
Upflow velocity in aperture	Average daily 4 m/h During 2-4 hrs 8 m/h
Upflow velocity	0.7 m/h

Although domestic sewage is a dilute type of wastewater, it is also characterised as a complex type of wastewater, with a relatively high content of suspended solids, *i.e.* a low $COD_{soluble}/COD_{total}$ ratio and a low temperature. The suspended solids can constitute 50-65% of the total COD. Therefore, total COD conversion is largely limited by hydrolysis of particulate matter and the hydrolyis rate constant at the prevailing temperature determines the SRT to be applied. Zeeman and Lettinga (1999) present the following equation for determining the HRT for treating wastewater with a high fraction of suspended solids (and biomass production << solids accumulation) based on the established SRT, amount of retained sludge in the UASB, fraction of hydrolysed suspended solids, and removal of suspended solids.

$$HRT = \frac{V}{Q} = \frac{SRT \cdot (COD_{SS,0} - COD_{SS,e}) \cdot (1 - \eta_h)}{X} \quad \text{(h)} \quad (16.54)$$

Where η_h = fraction of hydrolysed suspended COD, $COD_{SS,0}$ = influent suspended COD, and $COD_{SS,e}$ = effluent suspended COD.

After calculating the HRT, it is necessary to check that V_{upw} is not higher than the critical V_{upw}. For the treatment of low-concentration domestic/municipal wastewater under mesophilic conditions, the critical V_{upw} will determine the HRT.

When the domestic/municipal sewage temperature is < 20 °C, the SRT rather than the prevailing hydrodynamic conditions will determine the HRT. When the reactor is not well designed (for example, too short HRT at the prevailing temperature), non-digested suspended sludge starts to accumulate in the sludge bed, the hydrolytic and methanogenic capacity of the sludge will gradually decrease, deteriorating both particulate and soluble COD removal, and eventually reactor failure will occur.

As indicated above, the prime design criterion, in addition to the V_{upw} (m/h), is the reactor solids retention time (SRT), which should always be above a minimum value in order to maintain the hydrolytic and methanogenic conversion capacity of the sludge. When treating diluted domestic wastewater under tropical conditions, COD < 1,000 mg/l and t > 20 °C, this condition will always be met, because the hydraulic conditions (V_{upw}) determine the HRT. When considering hydrolyis as the rate-limiting step, which is described by first-order kinetics, the required SRT is determined by the first-order hydrolyis constant and the desired conversion of COD_{ss}.

The actual SRT of a UASB reactor in operation can be calculated using Eq.16.55,

$$SRT = \frac{V \cdot COD_{SS,e}}{Q_e \cdot COD_{SS,e} + Q_{es} \cdot COD_{es}} \qquad (16.55)$$

in which $COD_{SS,e}$ = concentration sludge per m^3 reactor volume ($kgCOD_{SS}/m^3$), V = reactor volume (m^3), Q_e = effluent flow (m^3/d), SS = suspended solids, and es = excess sludge.

Temperature has a major impact on the first-order hydrolysis rates and thus on the minimum-required SRT. The impact of temperature on the required SRT in a UASB reactor treating domestic/municipal wastewater is depicted in Figure 16.30.

Realising the importance of the SRT, it becomes clear that the conventional UASB reactor design for municipal wastewater, as applied in tropical countries, needs to be reconsidered when temperature drops and COD concentrations exceed 1,000 mg/l. In many arid countries with limited water supply, sewage concentrations range between 1,000-2,500 mgCOD/l, *e.g.* the Middle East, Northern Africa, the Arabic peninsula, etc. Furthermore, temperate climates in the Middle East and Northern Africa are characterised by cold winters, particularly in mountainous areas.

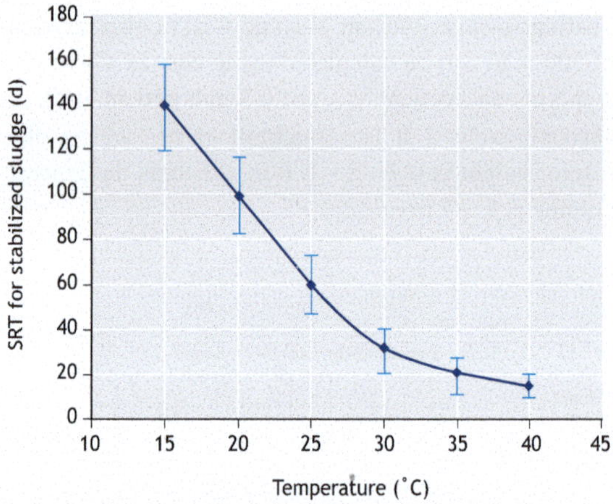

Figure 16.30 Required SRT for domestic sewage treatment as a function of temperature.

Recent experiences in Jordan and Palestine show municipal sewage COD concentrations reaching 2,500 mgCOD/l at TSS/COD ratios of 0.6 (Mahmoud *et al.*, 2003), whereas winter temperatures can drop to 15 °C. When applying the conventional UASB reactor design, the HRT needs to be increased reaching values of 20-24 hours (Hallalsheh *et al.*, 2005). Obviously this will affect the hydrodynamics of the system, requiring changes in influent distribution to prevent short-circuiting. Alternatively, the large suspended solids load can be addressed in separate reactor units such as a primary clarifier or enhanced solids removal in upflow filter systems, coupled to a sludge digester (Elmitwalli, 2000). A novel approach is to link the UASB reactor to a coupled mesophilic digester with sludge exchange (Mahmoud, 2002; Mahmoud *et al.*, 2004; Mahmoud, 2008). With this system, accumulating solids will be digested at higher temperatures, whereas the methanogenic activity in the UASB reactor will be increased by a return digested sludge flow.

The first full-scale UASB reactors in the Middle East region are located in the Nile Delta and Fayoum area, south of Cairo, Egypt. The design is based on the conventional approach taking into account the relatively high strength of the sewage, resulting in a somewhat longer average HRT of 12 h, although HRT drops to 5 h during peak flow rates (Nada *et al.*, 2011). In the Fayoum area the UASB is followed by a conventional trickling filter for post-treating the anaerobic effluent. Long-term results show that effluent specification could meet the local restrictions for BOD and TSS of 60 and 50 mg/l, respectively. However, restrictions for the very stringent COD criterion of 80 mg/l could not be met. The latter is attributed to the very high concentrations of inert organic matter in the influent coming from septic tanks and animal manure. Pilot trials in Amman showed the feasibility of the system as an ideal pre-treatment method for a low-cost reduction in the COD load, while generating energy for post-treatment. Table 16.18 briefly summarises the most important results (Hallalsheh *et al.*, 2005). A community onsite UASB reactor followed by constructed wetlands serving 100 houses in Palestine showed stable process performance with very satisfactory results. The system is operated with zero electricity as both of the combined anaerobic and natural treatment systems do not require electricity, and hydraulic flow is driven by gravity.

However, although the prospects for a full-scale application in Amman looked very promising (Table 16.18), it was decided to change the former pond system into a modern activated sludge plant. With regard to sustainability in domestic sewage treatment this decision is considered a wasted opportunity, particularly since the more concentrated municipal wastewaters are in fact ideal for anaerobic pre-treatment. The recovered energy could then have been beneficially used at the site for extensive treatment up to discharge or reuse standards. Any excess energy could have served as a power supply for *e.g.* irrigation pumps or for settlements in the vicinity of the plant.

Considering the present concern with fossil fuel consumption, related CO_2 emissions and climate change, anaerobic sewage treatment offers a feasible alternative for treating the huge flow of domestic and municipal wastewaters in many parts of the world. However, to prevent any greenhouse gas emissions, the recovery and use of all produced CH_4 should

Table 16.18 UASB pilot reactor trials at the Amman-Zarqa treatment site Khirbet As Samra, Jordan.

Average influent characteristics		Treatment performance (including post-clarification)	
Flow	180,000 m^3/d	COD removal:	up to 80%
COD	1,500 mg/l	BOD removal:	up to 85%
BOD	500-700 mg/l	TSS removal:	up to 80%
TSS	600-700 mg/l	Pathogens:	negligible
NH_4^+-N	70-130 mg/l	Total CH_4 production:	0.25 (winter)-0.44 (summer) $Nm^3CH_4/kgCOD_{removed}$
TKN	90-200 mg/l		
P_{tot}	10-40 mg/l	Assuming recoverable CH_4 of only 0.15 $Nm^3CH_4/kgCOD_{removed}$ gives potential CH_4 production of 27,000 m^3/d, equivalent to a potential power supply of ≈ 5 MW (assuming 40% CHP efficiency).	
T	16-28 °C		

become an intrinsic part of any anaerobic treatment plant design. Owing to its compactness, high-rate anaerobic sewage treatment can be applied in urban areas as well. The latter will lead to huge cost reductions in constructing sewerage networks, pumping stations, and conveyance networks. It must be realised that less than 20% of the produced municipal wastewater in Asia is treated, and in Latin America this value is only 15% (WWAP, 2017). In Africa, the generated wastewater is rarely collected and sewage treatment, with the exception of the Mediterranean area and South Africa, is almost absent. With an increase in basic understanding of the anaerobic process and an increase in the number of full-scale experiences, anaerobic treatment will undoubtedly become one of the prime methods for treating organically-polluted waste(water) streams.

16.11 ANAEROBIC TREATMENT OF BLACK WATER IN NEW SANITATION SYSTEMS

As indicated above, conventionally collected sewage/municipal wastewater is generally very diluted, sometimes even lower than 500 mg COD/l. When, in addition to the strong dilution, low temperatures also prevail (< 20 °C), application of direct anaerobic treatment is not feasible, because of the long HRTs required and, therefore, large reactor volumes and more cumbersome hydraulic profiles. The long HRTs follow from the necessary long SRTs at these low temperatures (see Figure 16.28 and Equation 16.55). Heating municipal sewage for enhancing anaerobic treatment costs too much energy and makes no sense. Based on a COD concentration of 500 mg/l, the maximum biochemical energy recovery via methane is only 0.5 kgCOD/m^3 · 0.7 (COD efficiency) · 12.4 MJ/kg COD (see Section 16.1.1) = 4.3 MJ/m^3, provided that all the removed COD is converted to the energy carrier CH_4. It must be noted that heating water costs 4.2 MJ/m^3.°C, and additional energy losses will be experienced during energy conversion.

Toilet waste (faeces and urine) is originally very high in COD and nutrient concentrations (Kujawa-Roeleveld and Zeeman, 2006). Dilution takes place by mixing with toilet flush water, grey water (shower/bath, kitchen, sink and laundry water) and often rainwater. Conventional sewers for conveying the domestic/municipal wastewater to the wastewater treatment plant are designed for transport of waste with water.

In new sanitation systems, black water (BW) and grey water (GW) are separately collected, transported and treated (Otterpohl *et al.*, 1997; Zeeman, 2012). When toilets are applied that use a small amount of water for flushing (≤1 l per flush), anaerobic treatment of black water at mesophilic conditions (≥ 25 °C) can become a feasible technique (De Graaff *et al.*, 2010). At the moment, vacuum collection (toilet) and vacuum transport to a community on-site anaerobic treatment system is used in full-scale applications. In the Netherlands, Sweden and Belgium UASB-septic tank technology is applied, whereas in Germany the CSTR technology is used for black water treatment.

When applying UASB technology for treatment of (vacuum-collected) black water with a concentration of approximately 10 gCOD/l, shorter HRTs can be applied in comparison with CSTRs. Anaerobic black-water treatment is part of a total community on-site sanitation concept, where nutrients are recovered/removed from the anaerobic effluent and grey water is separately treated, for example in an activated sludge system or constructed wetland. The high nutrient concentration of the black water anaerobic effluent enables more effective and efficient nutrient recovery. A recovery of 0.22 kgP/cap.yr, via struvite precipitation, from anaerobic (vacuum-collected) black water UASB effluent is reported by De Graaff *et al.* (2011a). The latter represents 10% of global artificial phosphorus fertilizer production. Approximately 40% of the influent P is contained in the anaerobic black water sludge.

Though relatively high (approximately 1gN/l), the nitrogen concentration in (vacuum-collected) black water anaerobic effluent is still too low for energy-efficient N recovery with presently available technologies. At the moment N removal is applied via nitritation and anammox processes (Vlaeminck *et al.*,

2009; De Graaff *et al.*, 2011b) in practice. New physicochemical N recovery techniques are under research opting for generating concentrated ammonia streams for reuse (*e.g.* Van Linden *et al.*, 2019, 2020). Fernandes *et al.* (2015) disclosed the possibility of applying a photo-bioreactor for recovering N and P from (vacuum-collected) anaerobic black-water effluent. New sanitation, including anaerobic (vacuum-collected) black-water treatment is now applied at full-scale at four locations in the Netherlands, viz. three office buildings and one housing estate of 250 houses in the city of Sneek (De Graaf and Van Hell, 2014). Novel full-scale applications can be found in Germany for a housing estate of approximately 800 houses, as well as in Belgium and Sweden for 320 apartments (1,800 P.E.) and 430 houses (1,200 P.E.), respectively. Most of the mentioned new sanitation concepts apply co-digestion of (vacuum-collected) black water with kitchen waste.

Recent developments indicate that 'new sanitation' is an attractive option for new housing estates, to recover energy via anaerobic treatment of black water with or without kitchen waste, and subsequent recovery of nutrients for agricultural purposes as schematically depicted in Figure 16.31. In addition to black water treatment, the grey water stream is proposed to be treated to a quality that can be reused for at least soil infiltration (Bisschops *et al.*, 2019). In all 'new sanitation' concepts anaerobic digestion plays a central role in stabilizing organic matter, minimizing (fossil) energy consumption and maximizing biochemical energy recovery in the form of methane gas.

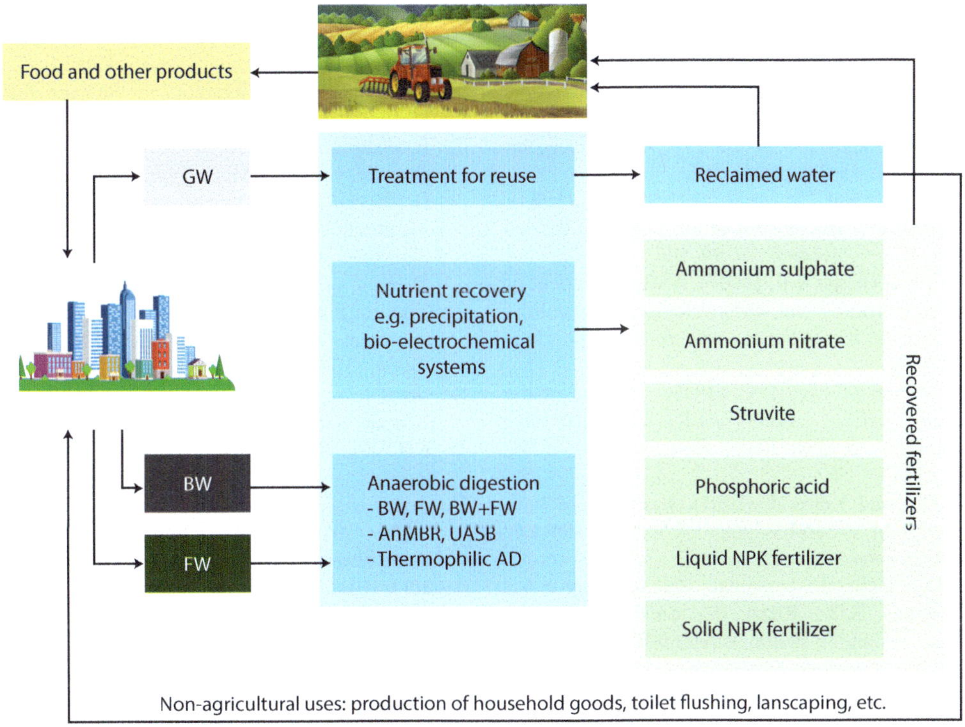

Figure 16.31 Decentralized treatment of source-separated black water (BW) and grey water (GW) with the recovery of nutrients for agricultural purposes (after Bisschops *et al.*, 2019).

REFERENCES

Al-Jamal W. and Mahmoud N. (2009). Community onsite treatment of cold strong sewage in a UASB-septic tank. *Bioresource Technol.*, 100(3), 1061-1068.

Allison M., Grant S., Christian S. and Wilson D. (2013). Full-Scale Operating Experience with USA-based ADI-AnMBR Systems for Food Wastes. *Proceedings of the Water Environment Federation,* 2013(10), 5255-5270.

Alphenaar, P.A. (1994). Anaerobic granular sludge: characterization and factors affecting its functioning. PhD thesis, G. Lettinga (promotor), Wageningen Agricultural University.

Alves M.M., Vieira J.A.M., Álvares Pereira R.M., Pereira M.A. and Mota M. (2001). Effects of lipids and oleic acid on biomass development in anaerobic fixed-bed reactors. Part II: Oleic acid toxicity and biodegradability. *Water Research,* 35(1), 264-270.

Alves M.M., Pereira M.A., Sousa D.Z., Cavaleiro A.J., Picavet M., Smidt H. and Stams A.J.M. (2009). Waste lipids to energy: how to optimize methane production from long-chain fatty acids (LCFA). *Microb. Biotechnol.*, 2(5), 538–550.

Azman S., Khadem A.F., Van Lier J.B., Zeeman G. and Plugge C.M. (2015). Presence and role of anaerobic hydrolytic microbes in conversion of lignocellulosic biomass for biogas production. *Crit. Rev. in Env. Sci. and Technol.*, 45(23), 2523-2564.

Bachmann A., Beard V.L. and McCarty P.L. (1985). Performance-characteristics of the anaerobic baffled reactor. *Water Research,* 19, 99–106.

Banik G.C. and Dague R.R. (1997). ASBR treatment of low strength industrial wastewater at psychrophilic temperatures. *Water Science and Technology,* 36(2-3), 337-344.

Barber W.P. and Stuckey D.C. (1999). The use of the anaerobic baffled reactor (ABR) for wastewater treatment: A review. *Water Research,* 33, 1559–1578.

Batstone D.J. and Keller J. (2001). Variation of bulk properties of anaerobic granules with wastewater type. *Water Research*, 35(7), 1723-1729.

Batstone D.J., Keller J., Angelidaki I., Kalyuzhnyi S.V., Pavlostathis S.G., Rozzi A., Sanders W.T.M., Siegrist H. and Vavilin V.A. (2002). *The IWA Anaerobic digestion Model No 1 (ADM1).* IWA Publishing, Scientific and Technical Report, 13. ISBN 1900222787, p. 77.

Bisschops I., Kjerstadius H., Meulman B. and Van Eekert M. (2019). Integrated nutrient recovery from source-separated domestic wastewaters for application as fertilisers. *Current Opinion in Environmental Sustainability*, 40, 7-13.

Buswell A.M., Boruff C.S. and Wiesman C.K. (1932). Anaerobic Stabilization of Milk Waste. *Industrial and Engineering Chemistry,* 24, 1423-1425.

Cavaleiro A.J., Picavet M.A., Sousa D.Z., Stams A.J.M., Pereira M.A. and Alves M.M. (2015). Anaerobic Digestion of Lipid-Rich Waste. In *Hydrocarbon and Lipid Microbiology Protocols*, Springer Protocols Handbooks, pp. 249–268.

Chandler J.A., Jewell W.J., Gossett J.M., Van Soest P.J. and Robertson J.B. (1980). Predicting methane fermentation biodegradability. *Biotech. Bioeng. Symposium,* 10, 93-107.

Chernicharo C.A.L., Brandt E.M.F., Bressani Ribeiro T., Motafilho C.R. and McAdam E. (2017). Development of a tool for improving the management of gaseous emissions in UASB-based sewage treatment plants. *Wat. Practice and Technol.,* 12(4), pp. 917-926.

Chernicharo C.A.L., Van Lier J.B., Noyola A. and Bressani Ribeiro T. (2015). Anaerobic sewage treatment: state of the art, constraints and challenges. *Reviews in Environmental Science and Biotechnology*, 14 (4), pp. 649-679.

Chernicharo C.A.L. and Bressani-Ribeiro T. (eds.) (2019). Anaerobic Reactors for Sewage Treatment: Design, Construction, and Operation, IWA Publishing, doi: 0.2166/9781780409238_0349.

Christian S., Grant S., McCarty P., Wilson D. and Mills D. (2011). The first two years of full-scale anaerobic membrane bioreactor (AnMBR) operation treating high-strength industrial wastewater. *Water Practice Technology,* 6(2).

Colleran E., Finnegan S. and Lens P. (1995). Anaerobic treatment of sulphate-containing waste streams. *Antonie van Leeuwenhoek,* 67, 29-46.

De Graaf R. and Van Hell A.J. (2014). New Sanitation Noorderhoek, Sneek. P. Hermans. Amersfoort, STOWA (Dutch Foundation for Applied Water Research), 304.

De Graaff M.S., Temmink H., Zeeman G. and Buisman C.J.N. (2010). Anaerobic treatment of concentrated black water in a UASB reactor at a short HRT. *Water,* 2(1), 101-119.

De Graaff M.S., Temmink H., Zeeman G. and. Buisman C.J.N (2011a). Energy and phosphorus recovery from black water. *Water Science and Technology,* 63(11): 2759-2765.

De Graaff M.S., Temmink H., Zeeman G., Van Loosdrecht M.C.M. and Buisman C.J.N. (2011b). Autotrophic nitrogen removal from black water:

Calcium addition as a requirement for settleability. *Water Research,* 45(1): 63-74.

Dereli R.K., Ersahin M.E., Ozgun H., Ozturk I., Jeison D., Van der Zee F. and Van Lier J.B. (2012). Potentials of anaerobic membrane bioreactors to overcome treatment limitations induced by industrial wastewaters. *Bioresource Technology,* 122, 160-170.

Ehlinger F. (1994). Anaerobic biological fluidized beds: Operating experiences in France. *In proceedings of 7th International symposium on anaerobic digestion,* Cape Town, 23-27 January.

Elmitwalli T.A. (2000). Anaerobic Treatment of Domestic Sewage at Low Temperature. Wageningen University.

Elmitwalli T.A., Sklyar V., Zeeman G. and Lettinga G. (2002). Low temperature pre-treatment of domestic sewage in an anaerobic hybrid or an anaerobic filter reactor. *Biores. Tech.,* 82, 233-239.

Ersahin M.E., Ozgun H., Dereli R.K. and Ozturk I. (2011). Anaerobic Treatment of Industrial Effluents: An Overview of Applications. In: Waste Water-Treatment and Reutilization. F.S.G. Einschlag (ed.), InTech, India, 415-456.

Ersahin M.E., Ozgun H., Tao Y. and Van Lier J.B. (2014). Applicability of dynamic membrane technology in anaerobic membrane bioreactors. *Water Research,* 48, 420-429.

Fernandes T.V., Shrestha R., Sui Y.X., Papini G., Zeeman G., Vet L.E.M., Wijffels R.H. and Lamers P. (2015). Closing Domestic Nutrient Cycles Using Microalgae. *Environmental Science & Technology,* 49(20): 12450-12456.

Fitz R.M. and Cypionka H. (1990). Formation of thiosulfate and trithionate during sulfite reduction by washed cells of *Desulfovibrio desulfuricans. Arch. Microbiol.,* 154, 400–406.

Frijters C.T.M.J., Vos R.H., Scheffer G. and Mulder R. (2006). Decolorizing and detoxifying textile wastewater, containing both soluble and insoluble dyes, in a full scale combined anaerobic/aerobic system. *Water Research,* 40(6), 1249-1257.

Ge H., Jensen P.D. and Batstone D.J. (2011). Temperature-phased anaerobic digestion increases apparent hydrolysis rate for waste activated sludge. *Water Research,* 45, 1597-1606.

Ghasimi D.S.M., De Kreuk M., Kyu Maeng S., Zandvoort M.H. and Van Lier J.B. (2016). High-rate thermophilic bio-methanation of the fine sieved fraction from Dutch municipal raw sewage: cost-effective potentials for on-site energy recovery. *Applied Energy,* 165, 569-582.

Gonzalez J.A., Hendriks A.T.W.M., Van Lier J.B. and De Kreuk M.K. (2018). Pre-treatments to enhance the biodegradability of waste activated sludge: elucidating the rate limiting step. *Biotechnology Advances,* 36(5), 1434-1469.

Guiot S.R., Safi B., Frignon J.C., Mercier P., Mullignan C. and Tremblay R. (1995). Performances of a Full-Scale Novel Multiplate Anaerobic Reactor Treating Cheese Whey Effluent. *Biotechnology & Bioengineering,* 45, 398-405.

Gujer W. and Zehnder A.J.B. (1983). Conversion processes in anaerobic digestion. *Water Science and Technology,* 15(8/9), 127-167.

Habets, L.H.A., Knelissen, H.J. (1997). In line biological water regeneration in a zero discharge recycle paper mill. *Wat. Sc. and Technol.* 35(2-3), 41-48.

Halalsheh M., Sawajneh Z., Zu'bi M., Zeeman G., Lier J., Fayyad M. and Lettinga G. (2005). Treatment of strong domestic sewage in a 96 m^3 UASB reactor operated at ambient temperatures: two-stage *versus* single-stage reactor. *Biores. Tech.,* 96, 577-585.

Heijnen J.J., Mulder A., Weltevrede R., Hols P.H. and Van Leeuwen H.L.J.M. (1990). Large-scale anaerobic/aerobic treatment of complex industrial wastewater using immobilized biomass in fluidized bed and air-lift suspension reactors. *Chemical Engineering and Technology,* 13(1), 202-208.

Heile S., Chernicharo C.A.L., Brandt E.M.F. and E.J. McAdam (2017). Dissolved gas separation for engineered anaerobic wastewater systems. *Sep. and Pur. Technol.,* 189, 405-418.

Holliger C., Alves M., Andrade D., Angelidaki I., Astals S., Baier U., Bougrier C., Buffière P., Carballa M., De Wilde V., Ebertseder F., Fernández B., Ficara E., Fotidis I., Frigon J.-C., Fruteau de Laclos H., Ghasimi D.S.M., Hack G., Hartel M., Heerenklage J., Sarvari Horvath I., Jenicek P., Koch K., Krautwald J., Lizasoain J., Liu J., Mosberger L., Nistor M., Oechsner H., Vítor Oliveira J., Paterson M., Pauss A., Pommier S., Porqueddu I., Raposo F., Ribeiro T., Rüsch Pfund F., Strömberg S., Torrijos M., Van Eekert M., Van Lier J.B., Wedwitschka H. and Wierinck I. (2016). Towards a standardization of biomethane potential tests. *Water Science and Technology,* 74(11): 2515-2522.

Holst T.C., Truc A. and Pujol R. (1997). Anaerobic fluidised beds: ten years of industrial experience. *Water Science and Technology,* 36(6-7), 415-422.

Hulshoff Pol L.W., De Zeeuw W.J., Velzeboer C.T.M. and Lettinga G. (1983). Granulation in UASB-reactors. *Water Science and Technology,* 15(8/9), 291-304.

Hulshoff Pol L.W., De Castro Lopes S.I., Lettinga G. and Lens P.N.L. (2004). Anaerobic Sludge Granulation. *Water Research,* 38(6), 1376-1389.

Imhoff K. (1916). Separate sludge digestion improves Imhoff tank operation by keeping sewage fresh. *Engineering Record,* 74: 101-102.

Kassab GH., Halalsheh M., Klapwijk A., Fayyad M. and Van Lier J.B. (2010). Sequential anaerobic-aerobic treatment for domestic wastewater – a review. *Biores. Technol.,* 101, 3299-3310.

Kleerebezem R., Hulshoff Pol L.W. and Lettinga G. (1999b). Anaerobic degradation of phthalate isomers by methanogenic consortia. *Appl. Env. Micr.,* 65(3), 1152.

Kleerebezem R., Hulshoff Pol L.W. and Lettinga G.(1999a). The role of benzoate in anaerobic degradation in terephthalate. *Appl. Env. Micr.,* 65(3), 1161-1167.

Kujawa-Roeleveld K. and Zeeman G. (2006). Anaerobic treatment in decentralised and source-separation-based sanitation concepts. *Reviews in Environmental Science and Biotechnology*, 5,115–139.

Lettinga G. (2014). My anaerobic sustainability story. LeAF Publisher, Wageningen, pp. 200, http://www.leaf-wageningen.nl/en/leaf.htm

Lettinga G. and Hulshoff Pol L.W. (1991). UASB process design for various types of wastewater. *Water Science and Technology,* 24(8), 87-107.

Lettinga G., Van der Sar J. and Van der Ben J. (1976). Anaerobe zuivering van het afvalwater van de bietsuikerindustrie (2). *H2O,* 9, 38-43.

Lettinga G., Van Velsen A.F.M., Hobma S.W., De Zeeuw W.J. and Klapwijk A. (1980). Use of the Upflow Sludge Blanket (USB) reactor concept for biological wastewater treatment. *Biotech. Bioeng.,* 22, 699-734.

Lettinga G., Van Velsen L., De Zeeuw W., and Hobma S.W. (1979). The application of anaerobic digestion to industrial pollution treatment. *Proceedings 1st Int. Symp. on anaerobic digestion.* 167-186, Cardiff, UK.

Lettinga G. and Hulshoff Pol L.W. (1991). UASB process design for various types of wastewater. *Water Science and Technology,* 24(8), 87-107.

Li A. and Sutton P.M. (1981). Dorr Oliver Anitron system, Fluidized Bed technology for methane production from dairy wastes. *Proceedings, Whey Products Institute Annual Meeting*, April, Chicago.

Lim S.J. and Kim T.H. (2014). Applicability and trends of anaerobic granular sludge treatment processes. *Biomass and Bioenergy* 60, 189-202.

Mahmoud N., Zeeman G., Gijzen H. and Lettinga G. (2003). Solids removal in upflow anaerobic reactors. *Biores. Tech.,* 90(1), 1-9.

Mahmoud N., Zeeman G., Gijzen H. and Lettinga G. (2004). Anaerobic sewage treatment in a one-stage UASB reactor and a combined UASB-Digester system. *Water Research,* 38(9), 2348-2358.

Mahmoud N. (2002). Anaerobic Pre-treatment of Sewage Under Low Temperature (15°C) Conditions in an integrated UASB-Digester System. PhD thesis Wageningen University.

Mahmoud N. (2008). High strength sewage treatment in a UASB reactor and an integrated UASB-Digester system. *Bioresource Technol.*, 99(16), 7531-7538.

Mahmoud N. and Van Lier J. B. (2011). Enhancement of an UASB-septic tank performance for decentralized treatment of strong domestic sewage. *Water Sci. Technol.* 64(4), 923–929.

McCarty P.L. (2001). The development of anaerobic treatment and its future. *Water Science and Technology,* 44(8), 149-156.

Muñoz Sierra J., Oosterkamp M., Wang W., Spanjers H. and Van Lier J.B. (2019). Comparative performance of upflow anaerobic sludge blanket reactor and anaerobic membrane bioreactor treating phenolic wastewater: overcoming high salinity. *Chemical Eng. Journal*, 366, 480-490.

Muyzer G. and Stams A.J.M. (2008). The ecology and biotechnology of sulphate-reducing bacteria. *Nature Reviews Microbiol.,* 6(6), 441-454.

Nada T., Moawad A., El-Gohary F.A. and Farid M.N. (2011). Full-scale municipal wastewater treatment by up-flow anaerobic sludge blanket (UASB) in Egypt. *Desal. Wat. Treat.* 30(1-3), 134-145.

Nnaji C.C. (2013). A review of the upflow anaerobic sludge blanket reactor. *Desalination and Water Treatment* 52, 4122-4143.

Novak J.T. and Carlson D. (1970). The kinetics of anaerobic long chain fatty acids degradation. *J. Water Pollut. Control Fed.,* 42 (2), 1932-1943.

Otterpohl R., Grottker M. and Lange J. (1997). Sustainable water and waste management in urban areas. *Water Science & Technology*, 35(9), 121–133.

Oude Elferink S.J.W.H., Visser A., Hulshoff Pol L.W. and Stams A.J.M. (1994). Sulfate reduction in methanogenic bioreactors. *Fems Microbiology Reviews,* 15, 119-136.

Ozgun H., Dereli R.K., Ersahin M.E., Kinaci C., Spanjers H. and Van Lier J.B. (2013). A review of anaerobic membrane bioreactors for municipal wastewater treatment: Integration options, limitations and expectations. *Separation and Purification Technology,* 118, 89-104.

Pereboom, J.H.F. and Vereijken T.L.F.M. (1994). Methenaogenic granule development in full-scale Internal Circulation reactors. *Wat.Sci.Technol.* 30 (8), 9-21

Pereira M.A., Sousa D.Z., Mota M. and Alves M.M. (2004). Mineralization of LCFA associated with anaerobic sludge: kinetics, enhancement of methanogenic activity, and effect of VFA. *Biotech. Bioeng.*, 88 (4), 502-511.

Rajagopal, R., Saady, N.M.C., Torrijos, M., Thanikal, J.V., Hung, Y.T. (2013). Sustainable agro-food industrial wastewater treatment using high rate anaerobic process. *Water* 5, 292-311.

Raskin L., Amann R.I. Poulsen L.K., Rittmann B.E. and Stahl D.A. (1995). Use of ribosomal RNA-based molecular probes for characterisation of complex microbial communities in anaerobic biofilms. *Water Science and Technology,* 31(1), 261-272.

Razo-Flores E., Macarie H. and Morier F. (2006). Application of biological treatment systems for chemical and petrochemical wastewaters. *In: Advanced Biological Treatment Processes for Industrial Wastewaters,* IWA publications, London.

Rebac S., Van Lier J.B., Lens P., Van Cappellen J., Vermeulen M., Stams A.J.M., Swinkels K.T.M. and Lettinga G. (1998). Psychrophilic (6-15 °C) high rate anaerobic traetment of malting wastewater in a two-module expanded granular sludge bed system. *Biotech. Progress,* 14, 856-864.

Rinzema A. (1988) Anaerobic treatment of wastewater with high concentrations of lipids or sulfate. PhD thesis, Wageningen Agricultural University, Wageningen.

Rittmann B.E. and McCarty P.L. (2001). Environmental Biotechnology: Principles and Applications. McGraw-Hill, New York, USA.

Santos H., Fareleira P., Legall J. and Xavier A.V. (1994). In vivo nuclear magnetic resonance in study of physiology of sulfate-reducing bacteria. *Methods in Enzymol.,* 243, 543-558.

Schroepfer G.J., Fullen W.J., Johnson A.S., Ziemke N.R. and Anderson J.J. (1955). The anaerobic contact process as applied to packinghouse wastes. *Sew. and Ind. Waste,* 24, 61.

Seyfried C.F. (1988). Reprints Verfahrenstechnik Abwasserreinigung, GVC - Diskussionstagung, Baden, 17-19 Oktober.

Spanjers H. and Vanrolleghem P.A. (2016). *Respirometry*, chapter 3, pages 133–176. IWA Publishing, London, UK. In Experimental Methods In Wastewater Treatment. Edited by Van Loosdrecht M.C.M., Nielsen P.H., Lopez-Vazquez C.M. and Brdjanovic D.. ISBN: 9781780404745 (Hardback), ISBN: 9781780404752 (eBook).

Speece R.E. (1996). Anaerobic Biotechnology for Industrial Wastewaters, Archae Press, USA.

Sung S. and Dague R.R. (1995). Laboratory studies on the anaerobic sequencing batch reactor. *Water Environment Research*, 67(3), 294-301.

Tagawa T., Takahashi H., Sekiguchi Y., Ohashi A. and Harada H. (2002). Pilot-plant study on anaerobic treatment of a lipid- and protein-rich food industrial wastewater by a thermophilic multi-staged UASB reactor. *Water Science and Technology,* 45 (10), 225-230.

Tamis J., Mulders M., Dijkman H., Van Loosdrecht M.C.M. and Kleerebezem R. (2018). Pilot-Scale Polyhydroxyalkanoate Production from Paper Mill Wastewater: Process Characteristics and Identification of Bottlenecks for Full-Scale Implementation. *J. Env. Eng.,* 144(10) 04018107.

Tugtas A.E. and Pavlostathis S.G. (2008). Inhibitory effects of nitrate reduction on methanogenesis in the presence of different electron donors. *Water Science and Technology*, 57(5), 693-698.

Vanderhaegen B., Ysebaert E., Favere K., Van Wambeke M., Peeters T., Panic V., Vandenlangenbergh V. and Verstraete W. (1992). Acidogenesis in relation to in-reactor granule yield. *Water Science and Technology,* 25, 75-81.

Van Haandel AC. and Lettinga G. (1994). *Anaerobic Sewage Treatment. A Practical Guide for Regions with a Hot Climate.* John Wiley and Sons, New York.

Van Lier J.B. (2008). High-rate anaerobic wastewater treatment: Diversifying from end-of-the-pipe treatment to resource-oriented techniques. *Water Science and Technology,* 57(8), 1137-1148.

Van Lier J.B., Van der Zee F.P., Frijters C.T.M.J. and Ersahin M.E. (2015). Celebrating 40 Years Anaerobic Sludge Bed Reactors for Industrial Wastewater Treatment. *Reviews in Environmental Science and Biotechnology,* 14(4), 681-702.

Van Lier J.B., Boersma F., Debets M.M.W.H. and Lettinga G. (1994). High rate thermophilic wastewater treatment in compartmentalized upflow reactors. *Water Science and Technology*, 30(12), 251-261.

Van Lier J.B., Vashi A., Van der Lubbe J. and Heffernan B. (2010). Anaerobic sewage treatment using UASB reactors: engineering and operational aspects. *In:* Fang H.H.P. (ed.) Environmental Anaerobic technology; Applications and New Developments, World Scientific, Imperial College Press, London, UK, ISBN 978-1-84816-542-7 Chapter 4, pp. 59-89.

Van Lier J.B., Van der Zee F.P., Tan F.P., Rebac S. and Kleerebezem R. (2001). Advances in high-rate anaerobic traetment: staging of reactor systems. *Water Science and Technology*, 44(8), 15-25.

Van Linden N., Bandinu G.L., Vermaas D.A., Spanjers H. and Van Lier J.B. (2020). Bipolar membrane electrodialysis for energetically competitive ammonium removal and dissolved ammonia production. *J. Cleaner Production* (in press).

Van Linden N., Spanjers H. and Van Lier J.B. (2019). Application of dynamic current density for increased concentration factors and reduced energy consumption for concentrating ammonium by electrodialysis. *Water Research*, open access, 163, 114856.

Vellinga S.H.J., Hack P.J.F.M. and Van der Vlugt A.J. (1986). New type 'high rate' anaerobic reactor, first experience on semi-technical scale with a revolutionary and high loaded anaerobic system. *In: Proc. of Anaerobic Treatment, a Grown-up Technology, Aquatech 1986,* 15-19 September, Amsterdam, The Netherlands, 547-562.

Verstraete W., De Beer D., Pena M., Lettinga G. and Lens P. (1996). Anaerobic bioprocessing of organic wastes. *World J. of Microb. and Biotech.,* 12, 221-238.

Vlaeminck S.E., Terada A., Smets B.F., Van der Linden D., Boon N., Verstraete W. and Carballa M. (2009). Nitrogen Removal from Digested Black Water by One-Stage Partial Nitritation and Anammox. *Environmental Science & Technology,* 43(13): 5035-5041.

Von Sperling M. and Chernicharo C.A.L. (2005). *Biological Wastewater Treatment in Warm Climate Regions.* IWA Publishing, London.

Wang K. (1994). Integrated anaerobic and aerobic treatment of sewage. PhD thesis, Department of Environmental Technology, Wageningen University, Wageningen.

WHO/Unicef (2000). Global water supply and sanitation assessment 2000 report. World Health Organisation, United Nations Children's Fund, Geneva, p. 80.

Widdel F. and Hansen T.A. (1992). The dissimilatory sulfate- and sulfur-reducing bacteria in the Prokaryotes, 2nd edn. Springer-Verlag, New York.

Widdel F., Rouviere P.E. and Wolfe R.S. (1988). Classification of secondary alcohol–utilizing methanogens including a new thermophilic isolate. *Arch. Microbiol.*, 150, 477-481.

Wiegant W.M. and De Man A.W.A. (1986). Granulation of biomass in thermophilic anaerobic sludge blanket reactors treating acidified wastewaters. *Biotech. Bioeng.,* 28, 718-727.

Wirtz R.A. and Dague R.R. (1995). Enhancement of granulation and start-up in the anaerobic sequencing batch reactor. *Water Environment Research,* 68(5), 883-892.

Wu W.-M. (1987). Granular sludge in upflow anaerobic sludge blanket (UASB) reactors and its properties. *Water Treat.*, 2, 148-157.

Wu W.-M., Hickey R.F. and Zeikus J.G. (1991). Characterisation of metabolic performance of methanogenic granules treating brewery wastewater: Role of sulfate-reducing bacteria. *Appl. Env. Micr.,* 57, 3438-3449.

Wu W-M., Jain M.K., Conway de Macario E., Thiele J.H. and Zeikus J.G. (1992). Microbial composition and characterisation of prevalent methanogens and acetogens isolated from syntrophic methanogenic granules. *Appl. Microbiol. Biotechnol.,* 38, 282-290.

WWAP (United Nations World Water Assessment Programme) (2017). The United Nations World Water Development Report 2017. Wastewater: The Untapped Resource. Paris, UNESCO.

Young J.C. (1991). Factors affecting the design and performance of anaerobic filters. *Water Science and Technology,* 24(8), 133-156.

Young J.C. and Yang, B.S. (1989). Design considerations for full-scale anaerobic filters. *Journal of Water Pollution Control Federation* 61(9), 1576-1587.

Young J.C. and McCarty P.L. (1969). The anaerobic filter for waste treatment. *J. Water Pollut., Control Fed.,* 41, 160-173.

Zeeman G., Sanders W.T.M., Wang K.Y. and Lettinga G. (1996). Anaerobic treatment of complex wastewater and waste activated sludge- application of an upflow anaerobic removal (UASR) reactor for the removal and pre-hydrolysis of suspended COD. *IAWQ-NVA conference for 'Advanced wastewater treatment',* Amsterdam.

Zeeman G. and Lettinga G. (1999). The role of anaerobic digestion in closing the water and nutrient cycle at community level. *Water Science and Technology,* 39 (5), 187-194.

Zeeman G. (2012). New Sanitation. Sub-department Environmental Technology Wageningen University. Inaugural Speech. https://edepot.wur.nl/246120

Zhu G., Zou R., Jha A.K., Huang X., Liu L. and Liu C. 2015. Recent developments and future perspectives of anaerobic baffled bioreactor for wastewater treatment and energy recovery. *Critical Reviews in Environmental Science and Technology*, 45(12), 1243-1276.

Zoutberg G.R. and Frankin R. (1996). Anaerobic treatment of chemical and brewery wastewater with a new type of anaerobic reactor, the Biobed® EGSB reactor. *Water Science and Technology,* 34(5-6), 375-381.

NOMENCLATURE

Symbol	Description	Unit
A	Reactor cross-sectional area	m^2
A_{min}	Minimum surface area	m^2
$E_{ff\text{-}meth}$	Percentage of COD in kg/m^3 converted to CH_4	%
f_c	Contact factor, between 0 and 1	-
$F_{meth\text{-}biogas}$	Fraction of CH_4 in biogas, generally between 0.6 and 0.9	-
H	Reactor height	m
K_s	Monod half saturation constant	mgCOD/l
Q_{inf}	Influent flow rate	m^3/h
r_v	Organic loading rate	$kgCOD/m^3.d$
T	Temperature	°C
T_d	Doubling time of the biomass	d
V	Reactor volume	m^3
V_{biogas}	Biogas upward velocity	m/h
V_{crit}	Cut-off level for minimum required reactor volume based on hydraulic limitations	m/h
$V_{reactor}$	Volume of the reactor	m^3
V_{upw}	Upward liquid velocity	m/h
$V_{upw,max}$	Maximum allowable upward liquid velocity	m/h
X	Quantity of accumulated biomass	$kgVSS/m^3$ reactor
$X_{reactor}$	Concentration of viable biomass in the reactor	kg/m^3
$\Delta G^{\circ\prime}$	Free energy change	kJ/mol

Abbreviation	Description
AB	Acetogenic bacteria
ABR	Anaerobic baffled reactor
ACP	Anaerobic contact process
ADM1	Anaerobic digestion model
AF	Anaerobic filter
AMBR	Anaerobic membrane bioreactor
AnWT	Anaerobic wastewater treatment
ASBR	Anaerobic sludge bed reactor
ASBR	Anaerobic sequencing batch reactor
ASM1	Activated sludge model no.1
ASRB	Acetic acid oxidising sulphate-reducing bacteria
CHP	Combined heat power
CSTR	Continuous stirred tank reactor
EGSB	Expanded granular sludge bed
EPS	Extracellular polymeric substance

FASRB	Fatty acid-oxidising sulphate-reducing bacteria
FB	Fluidized bed reactor
GLSS	Gas-liquid-solids separation system
HMB	Hydrogenotrophic methanogenic bacteria
HRT	Hydraulic retention time
HSRB	Hydrogen-oxidising sulphate-reducing bacteria
IC	Internal circulation reactor
LCFA	Long chain fatty acid
MB	Methanogenic bacteria
OHPB	Obligate hydrogen-producing bacteria
OLR	Organic loading rate
PTA	Purified therephthalic acid
SCFA	Short chain fatty acid
SMA	Specific methanogenic activity of the sludge
SRB	Sulphate-reducing bacteria
SRT	Sludge retention time
UAF	Upflow anaerobic filter
UASB	Upflow anaerobic sludge blanket
VFA	Volatile fatty acid
VSS	Volatile suspended solids

Greek symbol	Explanation	Unit
μ_{max}	Maximum growth rate	1/d
Θ	Hydraulic retention time (HRT)	h

doi: 10.2166/9781789060362_0757

17

Modelling biofilms

Eberhard Morgenroth

17.1 WHAT ARE BIOFILMS?

Biological treatment processes have the following two conditions in common: (*i*) active microorganisms have to be concentrated within the system, and (*ii*) microorganisms have to be removed from the treated effluent before the water leaves the system. In activated sludge systems microorganisms grow as flocs suspended in water and solid-liquid separation is required to retain biomass within the system (*e.g.* using a settling tank or a membrane). In biofilm reactors microorganisms are immobilized in a dense layer growing attached to a solid surface. Maintaining active biomass in the biofilm reactor does not require a settler. Bacteria in suspension can be washed out with the water flow, but bacteria in biofilms are protected from washout and can grow in locations where their food supply remains abundant. Whether or not a biofilm will develop in a system will depend on the washout of suspended biomass (or the solids retention time). If the rate of washout of suspended bacteria is higher than the growth rate of a particular group of organisms then these organisms will preferentially grow in a biofilm. With low washout rates there is less of an incentive for bacteria to develop a biofilm. Biofilms are composed of bacteria embedded in a matrix of extracellular polymeric substances (EPS) containing polysaccharide, proteins, free nucleic acids, and water (Sutherland, 2001). EPS can be considered as the glue that holds the biofilm in place. Active biomass concentrations inside the biofilm are much higher compared to activated sludge systems. A photo of a biofilm in a moving-bed biofilm reactor is shown in Figure 17.1.

Figure 17.1 Biofilm grown on a suspended biofilm carrier (photo: AnoxKaldnes).

In Figure 17.2B a schematic of the different compartments in a biofilm system are shown, namely: the bulk liquid, boundary layer, biofilm and biofilm support (substratum). Mass transport of substrates and electron acceptors within the biofilm is mostly based on molecular diffusion, which is usually slow compared to substrate removal, resulting in substrate gradients within the biofilm. One consequence of these substrate gradients is that substrate removal in biofilms is often mass transport-limited. This is a drawback for biofilm reactors. On the other hand, substrate gradients also allow for the development of different ecological niches within the biofilm, depending on the local substrate and electron acceptor concentrations. One example is anoxic conditions that can develop inside a biofilm, regardless of aerobic bulk liquid conditions, and allow denitrification to occur inside the biofilm. Understanding the interactions between mass transport and substrate conversion processes is necessary to understand the overall performance of biofilm systems.

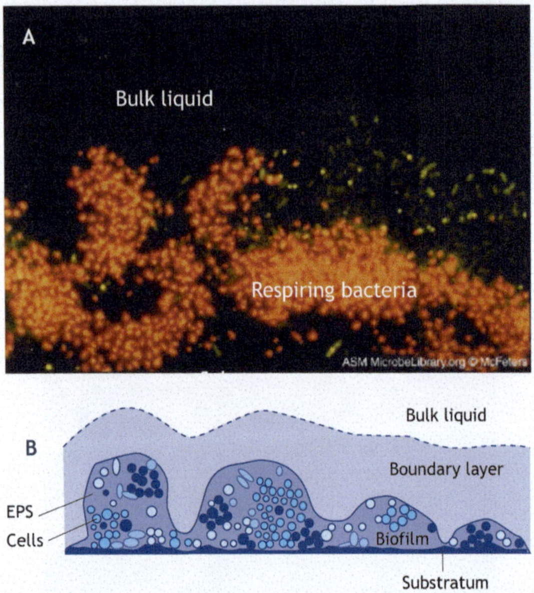

Figure 17.2 Biofilm grown in a flow channel imaged using confocal laser scanning microscopy (A) (photo: Hung *et al.*, 1995; McFeters, 2002), and a schematic representation of the different components of a biofilm system: bulk liquid, boundary layer, biofilm and substratum (B) (adapted from Wanner *et al.*, 2006).

Biofilms can be beneficial, for instance in wastewater treatment, drinking water treatment, soil remediation, and in forming barriers to contain contaminant plumes. On the other hand, biofilms can also be detrimental, for instance in water supply distribution networks, heat exchangers, dental hygiene, biomaterial implants, and on ships' hulls. Biofilms are either beneficial or detrimental because they (*i*) convert compounds available in the bulk liquid which is used in biological wastewater or drinking water treatment for the removal of unwanted compounds, (*ii*) they take up space and interfere with bulk water flow, which in some cases is desired (*e.g.* in bio-barriers) and in other cases is detrimental (*e.g.* in heat exchangers), and (*iii*) they can harbour pathogenic microorganisms that are difficult to remove or inactivate within the biofilm. This chapter is focused on modelling substrate transport and conversion in biofilms, biofilm development, and the overall performance of biofilm reactors.

17.2 MOTIVATION FOR MODELLING BIOFILMS AND HOW TO CHOOSE APPROPRIATE MATHEMATICAL MODELLING APPROACHES

A range of mathematical biofilm models has been developed that vary in terms of the processes considered within the biofilm, the information predicted by the model, and the effort required for solving the model: ranging from simple analytical to complex multi-dimensional numerical models. Before deciding on a particular modelling approach one has to clearly define the modelling objective. The following objectives and questions are relevant for predicting the performance of biofilm reactors and will be addressed in this chapter:

- *Substrate flux as a function of bulk-phase substrate concentration:* How do mass transport limitations and microbial kinetics inside the biofilm influence substrate conversion rates? How do mass transport limitations in the mass-transfer boundary layer influence the availability of substrate within the biofilm? The model should provide the substrate flux into the biofilm (J_{LF}) (quantifying the overall rate of substrate transformation within the biofilm) as a function of bulk-phase substrate concentrations (S_B) (Figure 17.3A).

- *Multi-component diffusion:* How does the local availability of electron donor and electron acceptor and the local presence of inhibitory compounds influence microbial processes? The model should predict the penetration of multiple substrates into the biofilm as a basis for determining the limiting substrate (Figure 17.3B).
- *Distribution of microorganisms:* How does substrate availability influence the distribution of microorganisms within the biofilm and how, in turn, does the distribution of microorganisms influence substrate removal? The model should predict biomass distributions and corresponding substrate removal (Figure 17.3C).
- *Overall reactor performance*: How are local substrate fluxes from the bulk into the biofilm related to the overall reactor performance? The model should integrate local substrate fluxes to predict overall biofilm reactor performance (Figure 17.3D).

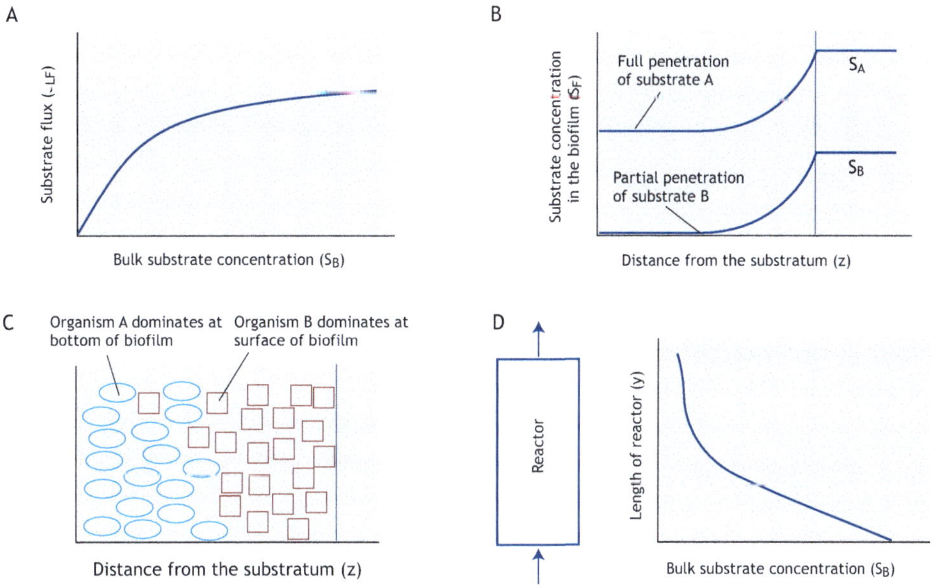

Figure 17.3 Schematic representations of the different questions that mathematical modelling can help to address: (A) How does substrate flux into the biofilm depend on bulk-phase substrate concentrations? (B) For reactions involving multiple substrates (e.g. electron donor and electron acceptor), which substrate will be limiting conversion processes? (C) How will microorganisms be distributed over the thickness of the biofilm and how does this biomass distribution affect conversion processes? (D) How can overall biofilm reactor performance be integrated from local substrate fluxes?

Choosing the right modelling approach requires a balance of the level of detail required to meet the modelling objective and the complexity of the model that one wants to work with. For example, assuming a homogeneous one-dimensional biofilm is in most cases sufficient to evaluate carbon oxidation. However, evaluating the competition between heterotrophic and autotrophic bacteria for substrate and space requires a modelling approach that predicts biomass distributions over the thickness of the biofilm. Analytical solutions are available for one-dimensional biofilms with a homogeneous organism distribution over the thickness of the biofilm with simple first or zero-order rate expressions, assuming Monod kinetics already requires the application of numerical solutions. This chapter will introduce the basic concepts of biofilm models together with analytical approaches to solve simple biofilm models and numerical approaches for more complex biofilm models. Numerical solutions in this chapter were obtained using the free software AQUASIM [1] (Reichert, 1998) or using a commercial simulator (SUMO [2]). Simulation files are available for download[3] so that readers who have the possibility to work with AQUASIM or SUMO can explore these simulations on their own. Numerical solvers for biofilm models are increasingly available in commercial wastewater treatment plant simulators. However, when working with these numerical solvers one must have a solid understanding of basic mechanisms and should regularly perform some calculations by hand using analytical solutions to simplified biofilm models to check the plausibility of the results from more complicated models. The current chapter will combine the discussion of analytical and numerical biofilm models.

17.3 MODELLING APPROACH FOR A BIOFILM ASSUMING A SINGLE LIMITING SUBSTRATE AND NEGLECTING EXTERNAL MASS TRANSFER RESISTANCE

Biofilms, as shown in Figure 17.2, are complex and heterogeneous aggregates. How can these aggregates be described in a simplified mathematical model? What features are relevant and which features can be neglected? Complex and numerically expensive models are available that aim at describing and predicting the multi-dimensional spatial heterogeneous structure of biofilms that is apparent in Figure 17.2. Some applications of such complex models are discussed in Section 17.10, but a detailed analysis and solution of such multi-dimensional modelling approaches is beyond the scope of this chapter; the focus here is on describing the effect of mass-transfer limitations resulting in heterogeneous substrate and biomass distributions in only one dimension. As analysed and discussed in Wanner *et al.* (2006), one-dimensional modelling is sufficient and appropriate to answer most process engineering-related questions. In this approach it is assumed that process rates, biofilm biomass density and composition, and substrate concentrations can be averaged in planes parallel to the substratum. With this simplifying assumption, a biofilm can be described as a one-dimensional structure with reactions and molecular diffusion inside the biofilm and an external mass-transfer boundary layer as shown in Figure 17.4.

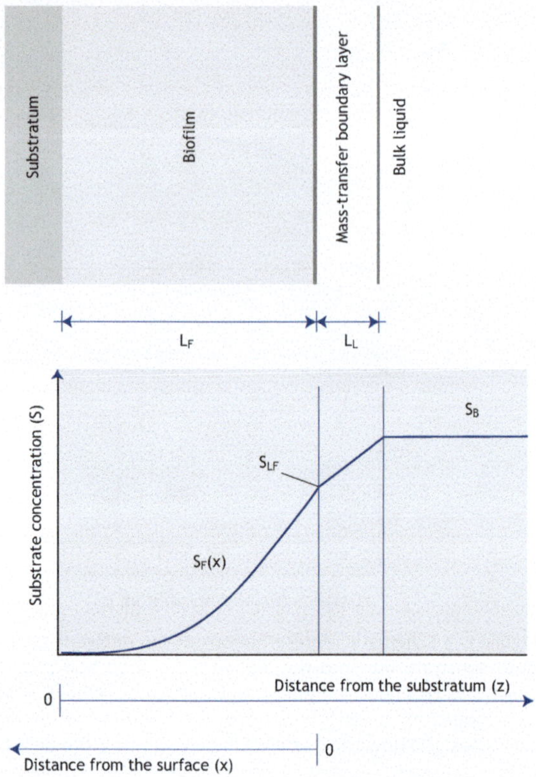

Figure 17.4 Concentration within the biofilm, in the concentration boundary layer, and in the bulk phase. The space coordinate can be measured either from the bottom of the biofilm (z, typically used for numerical simulations) or from the surface of the biofilm (x, simplifies the solution of manual calculations).

[1] www.aquasim.eawag.ch

[2] www.dynamita.com

[3] https://doi.org/10.25678/00021W (AQUASIM files), https://doi.org/10.25678/00022X (SUMO files)

For most models used in this chapter it is assumed that the biofilm biomass density (X_F) and composition are known *a priori* and are constant over the thickness of the biofilm. Using numerical simulations the biomass composition over the thickness of the biofilm can be predicted as a model output (see the example in Section 17.9).

17.3.1 Basic equations

The partial differential equation describing molecular diffusion, substrate utilization inside a biofilm, and dynamic accumulation for a single limiting substrate is given as:

$$\underbrace{\frac{\partial S_F}{\partial t}}_{\text{Accumulation}} = \underbrace{D_F \cdot \frac{\partial^2 S_F}{\partial x^2}}_{\text{Diffusion}} - \underbrace{r_F}_{\text{Reaction}} \tag{17.1}$$

where:

S_F	substrate concentration in the biofilm [M L^{-3}]
x	distance from the biofilm surface [L]
t	time [T]
D_F	diffusion coefficient in the biofilm [L^2 T^{-1}]
r_F	rate of substrate removal per biofilm volume ($r_F > 0$ for removal and $r_F < 0$ for production processes) [M L^{-3} T^{-1}]

Eq.17.1 is based on Fick's second law of diffusion. A detailed derivation of Eq. 17.1 is provided in the sidebar in Section 17.3.4. Different rate equations for the degradation of the limiting substrate within the biofilm can be defined as shown in Table 17.1. Analytical solutions for Eq. 17.1 are available only for first and zero-order rate expressions and assuming steady state. Numerical solutions are required for more complex rate expressions.

Note that in Table 17.1 the rate constants for zero-order ($k_{0,F}$, Eq. 17.2) and first-order ($k_{1,F}$, Eq. 17.3) substrate degradation rates can be related to Monod kinetics (Eq. 17.4) as follows:

$$k_{0,F} \approx \frac{\mu_{max}}{Y} \quad \text{for } S_F >> K_S \tag{17.6}$$

$$k_{1,F} \approx \frac{\mu_{max}}{Y \cdot K_S} \quad \text{for } S_F << K_S \tag{17.7}$$

The general rate expression in Eq. 17.5 refers to more complex systems with multiple processes and components reacting within the biofilm (*e.g.* Table 17.12 for heterotrophic and autotrophic bacteria) with stoichiometric coefficients $\nu_{i,j}$ and process rates ρ_j for a process matrix with compounds (i) and processes (j). The concept of the stoichiometric and kinetic matrix is described in detail in Chapter 14.

Table 17.1 Rate expressions for r_F in Eq. 17.1 [M L^{-3} T^{-1}] where X_F is the concentration of active biomass inside the biofilm [M L^{-3}], $k_{0,F}$ and $k_{1,F}$ are zero and first-order rate constants and μ_{max}, K_C, Y are the maximum growth rate, half-saturation constant, and yield constant, respectively. In Eq. 17.5, $\nu_{i,j}$ and ρ_j are generic stoichiometric coefficients and rates for a process j, respectively.

Rate type	Rate expression	Eq.
Zero-order	$r_F = k_{0,F} \cdot X_F$	(17.2)
First-order	$r_F = k_{1,F} \cdot S_F \cdot X_F$	(17.3)
Substrate utilization assuming Monod growth kinetics	$r_F = \underbrace{\frac{1}{Y}}_{\text{Stoichiometric coefficient }(\nu)} \cdot \underbrace{\mu_{max} \cdot \frac{S_F}{K_S + S_F} \cdot X_F}_{\text{Process rate }(\rho)}$	(17.4)
General rate expression for compound $S_{F,i}$ that is affected by multiple processes (j)	$r_{F,i} = \sum_{j=1}^{n} \nu_{i,j} \cdot \rho_j$	(17.5)

Solving the second-order differential equation (Eq. 17.1) requires two constants that can be derived from the following two boundary conditions:

$$\text{BC1:}\quad \frac{dS_F}{dx} = 0 \quad \text{at } x = L_F \tag{17.8}$$

$$\text{BC2:}\quad S_F = S_{LF} \quad \text{at } x = 0 \tag{17.9}$$

where BC1 (Eq. 17.8) is based on the assumption that there is no flux into the substratum and S_{LF} is the concentration at the surface of the biofilm. The substrate flux at a given location within the biofilm (J(x)) is proportional to the concentration gradient at a given location (x) within the biofilm

$$J_F(x) = -D_F \cdot \frac{dS_F(x)}{dx} \tag{17.10}$$

where D_F is the substrate diffusion coefficient inside the biofilm. Using Eq. 17.10 the flux through the surface of the biofilm (J_{LF}) is calculated as

$$J_{LF} = -D_F \cdot \frac{dS_F}{dx} \quad \text{at } x = 0 \tag{17.11}$$

This flux of substrate, J_{LF}, will be used subsequently in material balances for the overall biofilm reactor.

17.3.2 Solutions of the diffusion-reaction biofilm equation for different rate expressions

17.3.2.1 First-order substrate removal rate within the biofilm

Combining Eq. 17.1 with a first-order rate expression given by Eq. 17.3, and assuming steady state ($\partial S_F/\partial t = 0$) results in the following second-order ordinary differential equation:

$$0 = D_F \cdot \frac{d^2S_F}{dx^2} - k_{1,F} \cdot X_F \cdot S_F \tag{17.12}$$

This second-order differential equation can be solved taking into account the two boundary conditions (eqs. 17.8 and 17.9), resulting in an analytical solution of the substrate concentration within the biofilm assuming a first-order reaction ($S_{F,1}$) of

$$S_{F,1}(x) = \frac{\cosh\left(\frac{L_F - x}{L_{crit}}\right)}{\cosh\left(\frac{L_F}{L_{crit}}\right)} S_{LF} \tag{17.13}$$

where L_{crit} is a characteristic length that is defined as follows,

$$L_{crit} = \sqrt{\frac{D_F}{k_{1,F} \cdot X_F}} \tag{17.14}$$

Biofilms much thicker than L_{crit} will be mass transport-limited (sometimes referred to as deep biofilms) and biofilms much thinner than L_{crit} are fully penetrated (sometimes referred to as shallow biofilms). It is a useful exercise for the reader to differentiate Eq. 17.13 twice and to verify that the result for $S_{F,1}$ does, in fact, satisfy both the original differential equation (Eq. 17.12) and the two boundary conditions (eqs. 17.8 and 17.9). From the concentration profile (Eq. 17.13) the substrate flux into the biofilm assuming first-order substrate removal ($J_{LF,1}$) can be directly calculated using Eq. 17.11:

$$J_{LF,1} = \underbrace{D_F \cdot \frac{\tanh\left(\frac{L_F}{L_{crit}}\right)}{L_{crit}}}_{k_{1,A}} \cdot S_{LF} \tag{17.15}$$

where the terms on the right-hand side of Eq. 17.15 that do not depend on S_{LF} can be summarized in an aggregate rate ($k_{1,A}$). Using this aggregate rate it can be seen that for a given biofilm thickness the substrate flux has a first-order dependency on S_{LF}

$$J_{LF,1} = k_{1,A} S_{LF} \tag{17.16}$$

It should be noted that $k_{1,A}$ is constant only as long as the biofilm thickness, L_F, is constant. For cases where the biofilm thickness varies, it can be seen that $k_{1,A}$ will increase with increasing L_F.

A useful parameter to quantify the influence of mass transport limitations on the substrate flux is the efficiency factor ε. The efficiency factor ε is defined as the ratio of the $J_{LF,1}$ in Eq. 17.15 and a hypothetical substrate flux that assumes that the rate within the biofilm would not be slowed by diffusion:

$$\varepsilon = \frac{J_{LF,1,\text{with diffusion}}}{J_{LF,1,\text{without diffusion resistance}}} \quad (17.17)$$

Assuming a hypothetical flux without diffusion resistance ($= k_{1,F}X_FL_FS_{LF}$) the value of ε can be calculated from the flux with diffusion (Eq. 17.15) as

$$\varepsilon = \frac{\tanh\left(\frac{L_F}{L_{crit}}\right)}{\frac{L_F}{L_{crit}}} \quad (17.18)$$

For low values of L_F/L_{crit} (< 0.4) the biofilm is fully penetrated and the value of $\varepsilon \approx 1$. For thicker biofilms ($L_F/L_{crit} > 4$) the value of ε in Eq. 17.18 decreases with increasing L_F/L_{crit} and substrate conversion in the biofilm will be mass transport-limited, and can be approximated as follows

$$\varepsilon \approx \frac{L_{crit}}{L_F} \text{ for } L_F/L_{crit} > 4 \quad (17.19)$$

Substrate concentration profiles over the thickness of the biofilm and corresponding ε are shown in Figure 17.5 for different values of L_F/L_{crit}.

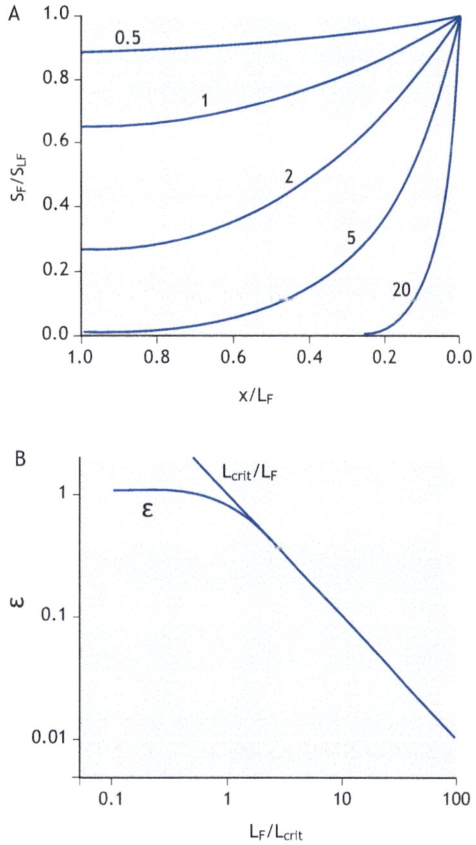

Figure 17.5 Substrate concentrations (S_F/S_{LF}) over the depth of the biofilm (x/L_F) for different values of L_F/L_{crit} (numbers in plot) assuming first-order kinetics inside the biofilm (A). The value of ε according to eqs. 17.18 and 17.19 for a range of L_F/L_{crit} (B).

Example 17.1 First-order substrate removal in biofilm

Question:

Calculate the acetate concentration at the base of a 400 μm thick biofilm assuming an acetate concentration at the surface of the biofilm, S_{LF}, of 3 mgCOD/l, a first-order rate constant, $k_{1,F}$, of 2.4 m^3/gCOD.d, a biofilm biomass density [4] of 10,000 gCOD/m^3, and a diffusion coefficient of

[4] The biofilm biomass density in this example is low compared to suggestions in Section 17.11.1. The value was chosen to make results comparable to the parameters in Table 17.13 and to related examples in Wanner and Gujer, 1985; Wanner and Reichert, 1996; Wanner *et al.*, 2006).

$0.8 \cdot 10^{-4}$ m²/d. Then calculate the acetate flux into this biofilm for bulk-phase concentrations of 3 or 30 mgCOD/l. Discuss the potential problems with your calculations.

Answer:

Step 1: Calculate L_{crit}:

$$L_{crit} = \sqrt{\frac{\left(0.8 \cdot 10^{-4}\ m^2/d\right)}{\left(2.4\ m^3/gCOD.d\right) \cdot \left(10{,}000\ gCOD/m^3\right)}}$$

$$= 58\ \mu m$$

Note that L_{crit} is independent of the substrate concentration at the surface of the biofilm.

Step 2: Calculate L_F/L_{crit}:

$$\frac{L_F}{L_{crit}} = \frac{400\ \mu m}{58\ \mu m} = 6.9$$

With $L_F/L_{crit} > 4$ the biofilm can be considered a deep biofilm that is mass transport-limited.

Step 3: Calculate the substrate concentration at the base of the biofilm, $S_F(x = L_F)$, assuming $S_{LF} = 3$ mgCOD/l

$$S_F(x = L_F) = \frac{\overbrace{\cosh(0)}^{\cosh(0)=1}}{\cosh(400/58)} \cdot 3\ mg/l$$

$$= 0.0061\ mg/l$$

Step 4: Calculate the corresponding substrate flux for $S_{LF} = 3$ mgCOD/l:

$$J_{LF} = 0.8 \cdot 10^{-4}\ m^2/d \cdot \frac{\tanh(400/58)}{58 \cdot 10^{-6}\ m} 3\ g/m^3$$

$$= 4.1\ g/m^2.d$$

The corresponding efficiency factor ε can be calculated from Eq. 17.18 as:

$$\varepsilon = \frac{\tanh\left(\frac{L_F}{L_{crit}}\right)}{\frac{L_F}{L_{crit}}} = \frac{\tanh(400/58)}{400/58} = 0.145$$

Thus, the flux of substrate into the biofilm is 14.5% of the flux that would be expected if the biofilm was fully penetrated with negligible effects of mass-transport limitations. Note that the value of ε is independent of the substrate concentration at the surface of the biofilm.

Step 5: Now calculate the substrate flux for a substrate concentration at the surface of the biofilm of 30 mgCOD/l.

$$J_{LF} = 0.8 \cdot 10 - 4\ m^2/d \cdot \frac{\tanh(400/58)}{58 \cdot 10^{-6}\ m} 30\ g/m^3$$

$$= 41\ g/m^2.d$$

Note that the calculated flux is exactly 10 times the flux assuming $S_{LF} = 3$ mgCOD/l. This flux is unrealistically high. How can this be? The underlying assumptions for all the calculations in this example were (*i*) that substrate removal in the biofilm was first-order and (*ii*) that acetate is the limiting compound. However, while both assumptions seem reasonable for $S_{LF} = 3$ mgCOD/l, it can be expected that neither assumption is satisfied for $S_{LF} = 30$ mgCOD/l. Thus, one has to be very careful when applying solutions developed in this chapter that the underlying assumptions are in fact satisfied. The question of using first-order, zero-order, and Monod-order reaction rates inside the biofilm is discussed in Section 17.3.2.3. The question of multi substrate diffusion is addressed in Section 17.6.

17.3.2.2 Zero-order substrate removal rate within the biofilm

Combining Eq. 17.1 with a zero-order rate expression and assuming steady-state results in the following second-order ordinary differential equation:

$$0 = D_F \cdot \frac{d^2S_F}{dx^2} - \begin{cases} k_{0,F} \cdot X_F & \text{for } S_F > 0 \\ 0 & \text{for } S_F \leq 0 \end{cases} \quad (17.20)$$

The solution to Eq. 17.20 will depend on whether the substrate reaches all the way to the base of the biofilm ($S_F > 0$ for $0 < x < L_F$ or a fully penetrated biofilm) or whether the substrate decreases to zero at some location within the biofilm (partially penetrated biofilm).

Partially penetrated biofilm (β ≤ 1) assuming zero-order substrate removal rates

Solving Eq. 17.20 for a partially penetrated biofilm requires three constants to be determined: two from the integration of the second-order differential equation and a third constant that describes the penetration of substrate into the biofilm where r_F goes to zero. Substrate penetration into the biofilm relative to the biofilm thickness (β) (Figure 17.6) is defined as

$$\beta = \frac{\text{substrate penetration into the biofilm}}{L_F} \quad (17.21)$$

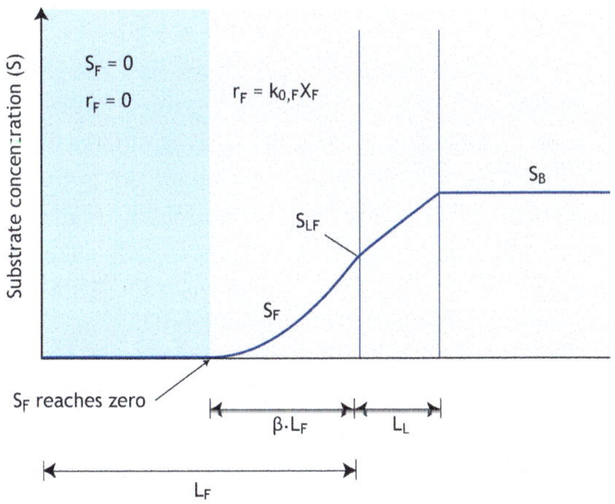

Figure 17.6 Partial penetration of a biofilm with zero-order substrate removal for $x < \beta \cdot L_F$ and an inactive zone without any substrate removal for $x > \beta \cdot L_F$ (shaded in blue) (L_F is the biofilm thickness and L_L is the thickness of the mass-transfer boundary layer).

Three boundary conditions are defined to determine the value of the two integration constants and also the value of β:

$$\text{BC1a:} \quad \frac{dS_F}{dx} = 0 \qquad \text{at } x = \beta \cdot L_F \quad (17.22)$$

$$\text{BC1b:} \quad S_F = 0 \qquad \text{at } x = \beta \cdot L_F \quad (17.23)$$

$$\text{BC2:} \quad S_F = S_{LF} \qquad \text{at } x = 0 \quad (17.9)$$

With these three boundary conditions the integration of Eq. 17.20 provides the following results:

$$S_{F,0,p}(x) = S_{LF} - \left(x \cdot \beta \cdot L_F - \frac{x^2}{2} \right) \frac{k_0 \cdot X_F}{D_F} \quad (17.24a)$$

which can be rearranged to:

$$S_{F,0,p}(x) = S_{LF} \left(1 - \left(\frac{2 \cdot x}{\beta \cdot L_F} - \frac{x^2}{(\beta \cdot L_F)^2} \right) \right) \quad (17.24b)$$

in which:

$$\beta = \sqrt{\frac{2 \cdot S_{LF} \cdot D_F}{L_F^2 \cdot k_0 \cdot X_F}} \quad (17.25)$$

Eq. 17.25 can be rearranged to provide the penetration depth (βL_F):

$$\beta \cdot L_F = \sqrt{\frac{2 \cdot S_{LF} \cdot D_F}{k_0 \cdot X_F}} \quad (17.26)$$

Again, the reader should verify that Eq. 17.24a and Eq. 17.24b do in fact satisfy the differential equation (Eq. 17.20) and the three boundary conditions. The flux into the biofilm assuming zero-order rates in a partially penetrated biofilm ($J_{LF,0,p}$) can be calculated from Eq. 17.24a by calculating the substrate gradient at the surface of the biofilm (Eq.17.11) as:

$$J_{LF,0,p} = \beta \cdot L_F \cdot k_0 \cdot X_F \quad (17.27)$$

Note that β depends on the substrate concentration at the surface of the biofilm (S_{LF}). Substituting β into Eq. 17.27 provides a direct relationship of the flux to the bulk-phase substrate concentration:

$$J_{F,0,p} = \underbrace{\sqrt{2 \cdot D_F \cdot k_0 \cdot X_F}}_{k_{0,p,A}} \sqrt{S_{LF}} \quad (17.28)$$

Collecting all the terms in Eq. 17.27 that are independent of S_{LF} provides a half-order dependence of the substrate flux on S_{LF} with the surface-based reaction rate $k_{0,p,A}$ [$M^{0.5}$ $L^{-0.5}$ T^{-1}]:

$$J_{LF,0,p} = k_{0,p,A} \cdot \sqrt{S_{LF}} \quad (17.29)$$

Fully penetrated biofilm (β ≥ 1) assuming zero-order substrate removal rates

The solution of Eq. 17.20 assuming full penetration with the original boundary conditions (Eq. 17.8 and Eq. 17.9) provides the following solution for the substrate concentration in the biofilm:

$$S_{F,0,f}(x) = S_{LF} - \left(x \cdot L_F - \frac{x^2}{2}\right)\frac{k_0 \cdot X_F}{D_F} \quad (17.30a)$$

which can be rearranged to:

$$S_{F,0,f}(x) = S_{LF}\left(1 - \left(\frac{2 \cdot x}{L_F \cdot \beta^2} - \frac{x^2}{L_F^2 \cdot \beta^2}\right)\right) \quad (17.30b)$$

The flux into the biofilm assuming zero-order rates in a fully penetrated biofilm ($J_{LF,0,f}$) can be calculated from the substrate gradient at the surface of the biofilm as

$$J_{LF,0,f} = L_F \cdot k_0 \cdot X_F \quad (17.31)$$

The ratio of the fluxes for zero-order reactions inside the biofilm for a partially penetrated biofilm (Eq. 17.26) and a fully penetrated biofilm (Eq. 17.30) is β. Concentration profiles inside the biofilm for different values of β are shown in Figure 17.7 for both partially and fully penetrated biofilms.

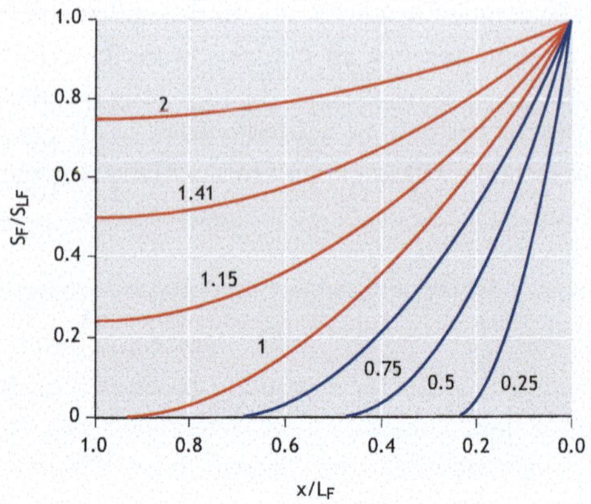

Figure 17.7 Substrate concentrations (S_F/S_{LF}) over the depth of the biofilm (x/L_F) assuming zero-order biofilm reaction rates and different values of β (numbers in plot). Note that different equations need to be used for partial and fully penetrated biofilms: Eq. 17.24 for β < 1 (blue lines) and Eq. 17.29 for β ≥ 1 (red lines).

17.3.2.3 Monod kinetics within the biofilm

More complicated rate expressions for substrate removal within the biofilm (*e.g.* Monod kinetics, Eq. 17.4) in most cases do not allow for analytical solutions of the original differential equation describing diffusion and reaction in a one-dimensional biofilm (Eq. 17.1). However, tools to provide numerical solutions for Eq.17.1 are readily available today and can be used to evaluate steady-state or dynamic conditions. One example is AQUASIM, a computer program for the identification and simulation of aquatic systems (Wanner and Morgenroth, 2004; Wanner and Reichert, 1996). A brief introduction on how to use AQUASIM to model biofilms is provided in the sidebar in Section 17.3.5. The reader can choose other software or commercially available wastewater treatment plant simulators to replicate the results computed in this chapter using AQUASIM. AQUASIM has the ability to simultaneously solve multiple process equations for the diffusion and degradation of soluble substrates, production and utilization of intermediates, and growth, decay, and detachment of different biomass fractions. However, in this section only substrate conversion using Monod kinetics is considered and biomass growth is neglected. This enables a direct comparison of the analytical solutions derived above with numerical modelling results. In later sections (*e.g.* Section 17.8) biofilms are evaluated with multiple processes and components. To clearly identify the processes and components that were included in a simulation, matrix notation is used (as already introduced in Chapter 14). For a single process removing a single substrate, Monod kinetics in Eq. 17.4 is represented in the simple matrix shown in Table 17.2.

Figure 17.8 gives the results for the numerical solution of a steady-state biofilm using Monod kinetics compared to analytical solutions using first and zero-order rate expressions. To make the results from the different simulations comparable, the kinetic parameters for first and zero-order kinetics were derived from the maximum growth rate and half-saturation concentration using Eq. 17.6 and Eq. 17.7. The substrate removal rates within the biofilm are shown in Figure 17.8A. The corresponding substrate fluxes for first-order ($J_{LF,1}$), zero-order partially penetrated biofilm ($J_{LF,0,p}$), zero-order fully penetrated biofilm ($J_{LF,0,f}$), and Monod kinetics ($J_{LF,Monod}$) are shown for 200 and 80 μm thick biofilms in Figure 17.8B and 17.8C, respectively. It can be seen that the local rate within the biofilm assuming Monod kinetics (r_{Monod}) is always smaller compared to either first (r_1) or zero-order (r_0) rates (Figure 17.8A). As a result, substrate fluxes assuming Monod kinetics are always smaller than substrate fluxes assuming first or zero-order rates (Figure 17.8B and 17.8C). Note that solutions for zero-order biofilm kinetics in Figure 17.8 based on Eq. 17.27 are valid only for partially penetrated biofilms ($\beta \leq 1$) and based on Eq. 17.30 are valid only for fully penetrated biofilms ($\beta \geq 1$). The transition from a partially to a fully penetrated biofilm occurs at the intersection of $J_{LF,0,p}$ and $J_{LF,0,f}$ or at $\beta = 1$. Based on the definition of β (Eq. 17.25) the substrate concentration at the surface of the biofilm ($S_{LF,transition}$) resulting in $\beta = 1$ is

$$S_{LF,transition} = \frac{L_F^2 \cdot k_0 \cdot X_F}{2 \cdot D_F} \tag{17.32}$$

Solutions for substrate fluxes for first or zero-order biofilm kinetics can be combined where fluxes assuming first-order kinetics ($J_{LF,1}$) are assumed to apply for lower S_{LF} and fluxes assuming zero-order biofilm kinetics for higher substrate concentrations at the biofilm surface. For the 200 μm thick biofilm, fluxes assuming first-order biofilm kinetics ($J_{LF,1}$, Eq. 17.15) transition to half-order kinetics ($J_{LF,0,p}$, Eq. 17.27) and then zero order kinetics ($J_{LF,0,f}$, Eq. 17.30); for the 80 μm thick biofilm, fluxes assuming first order kinetics transition directly to zero order kinetics. One approach to combining the different analytical flux equations is to simply choose the minimum of the three analytical solutions for where $J_{LF,1}$ and $J_{LF,0,p}$ are a function of S_{LF} and $J_{LF,0,f}$ is independent of S_{LF}:

$$J_{LF}(S_{LF}) = \min(J_{LF,1}(S_{LF}),\ J_{LF,0,p}(S_{LF}),\ J_{LF,0,f}) \tag{17.33}$$

A more sophisticated approach is to use a linear combination of the different substrate fluxes as described by Perez *et al.*, 2005 (with corrections in Gapes *et al.*, 2006). The suitability of simple analytical and more complex numerical solutions is discussed for a range of applications in Wanner *et al.* (2006).

Table 17.2 Matrix of stoichiometry and kinetics for heterotrophic substrate removal, assuming that organic substrate is the limiting compound. Constant biofilm thickness and biofilm biomass density are assumed.

Process name ↓j →i	S_S	Process rate, ρ_j
Heterotrophic substrate removal	$-\frac{1}{Y_{OHO}}$	$\mu_{OHO,max} \cdot \frac{S_S}{K_S + S_S} \cdot X_{OHO}$
Unit	COD	

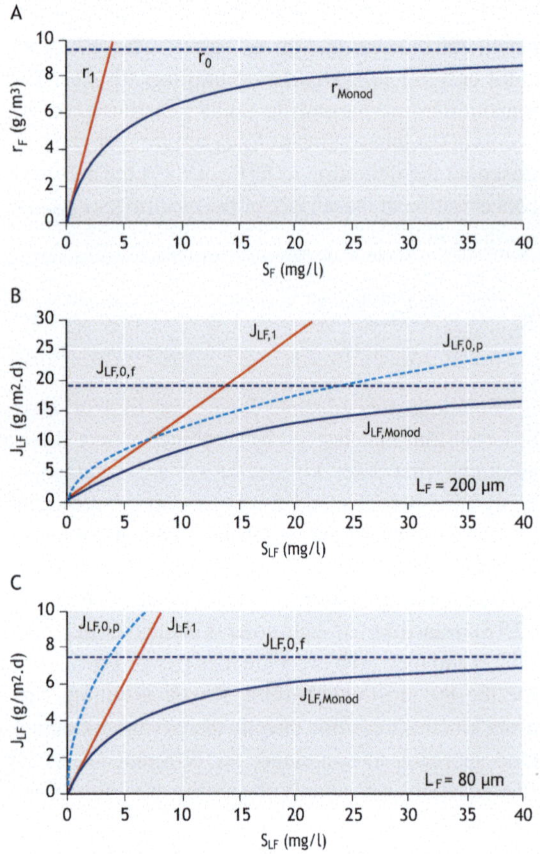

Figure 17.8 First-order, zero-order, and Monod reaction rates within the biofilm (r_F) are shown as a function of local substrate concentrations (S_F) in (A). Corresponding substrate fluxes as a function of the substrate concentration at the biofilm surface (S_{LF}) are calculated for a biofilm thickness of 200 µm (B) and 80 µm (C). (Parameters: μ_{max} = 6/d, K_S = 4 mg/l, Y_{OHO} = 0.63 gCOD/gCOD, D_F = 0.00008 m²/d, X_F = 10,000 g/m³ resulting in $k_{1,F}$ = 2.38 m³/g.d (Eq. 17.7), $k_{0,F}$ = 9.52 1/d (Eq. 17.6), and L_{crit} = 58 µm (Eq. 17.14)).

17.3.3 Summary of analytical solutions for a single limiting substrate

A summary of analytical solutions for substrate profiles over the thickness of the biofilm and corresponding substrate fluxes is provided in Table 17.3. The extent of mass-transfer limitation for first and zero-order (partially penetrated) biofilm kinetics is described using ε and β where both parameters describe the ratio of substrate flux divided by the hypothetical substrate flux, assuming there were no mass-transfer limitations. Thus, $\varepsilon \approx 1$ and $\beta \geq 1$ describe situations where substrate flux is not limited by mass transport into the biofilm. On the other hand, $\varepsilon << 1$ and $\beta << 1$ mean that substrate transport into the biofilm is significantly limiting substrate removal.

17.3.4 Derivation of the reaction diffusion equation (Eq. 17.1) from a mass balance within the biofilm

The reaction diffusion equation in Eq. 17.1 can be derived from a mass balance for the control volume between x and x + Δx as shown in Figure 17.9.

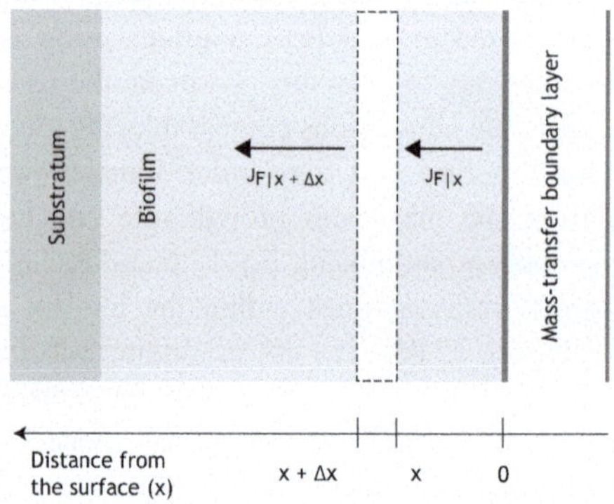

Figure 17.9 Mass balance for a section of the biofilm with a thickness of Δx, substrate fluxes entering and leaving the control volume, and substrate removal within the control volume (Δx A).

$$\underbrace{\Delta x A\frac{\partial S_F}{\partial t}}_{\text{Accumulation}} = \underbrace{AJ_F(x)}_{\text{Input}} - \underbrace{AJ_F(x+\Delta x)}_{\text{Output}} - \underbrace{\Delta x A r_F}_{\substack{\text{Dissappearance}\\ \text{by reaction}}} \quad (17.34)$$

where A_F is the surface area of the biofilm (m²). Using Fick's first law which calculates the substrate flux based on the diffusion coefficient and the local concentration gradient:

Table 17.3. Overview of biofilm kinetics depending on kinetics within the biofilm

Kinetics inside the biofilm	Flux of the limiting substrate (J_S) into the biofilm: …neglecting external mass-transfer resistance	Flux of the limiting substrate (J_S) into the biofilm: …including external mass-transfer resistance (R_L)	Efficiency (ε or β) relative to fully-penetrated biofilm	With
First-order ($k_{1,F}$)	$J_{LF,1} = k_{1,A} S_B$ (17.16)	$J_{LF,1} = k_{1,A} S_B \frac{1}{\underbrace{k_{1,A} R_L + 1}_{\text{Reduction of flux due to external mass-transfer resistance}}}$ (17.51)	$\varepsilon = \frac{\tanh\left(\frac{L_F}{L_{crit}}\right)}{\frac{L_F}{L_{crit}}}$ (17.18) $J_{LF,1} = \varepsilon \cdot k_{1,F} X_F L_F S_{LF}$	$k_{1,A} = D_F \frac{\tanh\left(\frac{L_F}{L_{crit}}\right)}{L_{crit}}$ (17.15) $L_{crit} = \sqrt{\frac{D_F}{k_{1,F} X_F}}$ (17.14)
Zero-order ($k_{0,F}$) partially-penetrated ($\beta \le 1$)	$J_{LF,0,p} = k_{0,p,A} \sqrt{S_B}$ (17.29)	$J_{LF,0,p} = k_{0,p,A} \sqrt{S_B} \left(-\frac{\zeta}{2} + \sqrt{\left(\frac{\zeta}{2}\right)^2 + 1} \right)$ (17.57)	$\beta = \sqrt{\frac{2 D_F S_{LF}}{k_{0,F} X_F L_F^2}}$ (17.25) $J_{LF,0,p} = \beta \cdot k_{0,F} X_F L_F$ (17.27)	$k_{0,p,A} = \sqrt{2 D_F k_{0,F} X_F}$ (17.28) $\zeta = \frac{k_{0,p,A} R_L}{\sqrt{S_B}}$ (17.56)
Zero-order ($k_{0,F}$) fully-penetrated ($\beta \ge 1$)	$J_{LF,0,f} = k_{0,f,A}$ (17.31)	$J_{LF,0,f} = k_{0,f,A}$ (17.31)	No influence of mass-transport limitations for $\beta > 1$	$k_{0,f,A} = k_{0,F} X_F L_F$ (17.31)

$$J_F(x) = -D_F \cdot \frac{\partial S_F(x)}{\partial x} \tag{17.35a}$$

and

$$J_F(x+\Delta x) = -D_F \cdot \frac{\partial S_F(x+\Delta x)}{\partial x} \tag{17.35b}$$

Eq. 17.34 can be modified to:

$$\frac{\partial S_F}{\partial t} = D_F \frac{\frac{\partial S_F(x+\Delta x)}{\partial x} - \frac{\partial S_F(x)}{\partial x}}{\Delta x} - r_F \tag{17.36}$$

By letting Δx approach zero, Eq. 17.36 results in Eq. 17.1:

$$\underbrace{\frac{\partial S_F}{\partial t}}_{\text{Accumulation}} = \underbrace{D_F \cdot \frac{\partial^2 S_F}{\partial x^2}}_{\text{Diffusion}} - \underbrace{r_F}_{\text{Reaction}} \tag{17.1}$$

17.3.5 Overview of AQUASIM

This section provides a brief overview of the equations solved numerically by AQUASIM and some initial comments on how to use AQUASIM to simulate biofilms. More detailed information is available in Wanner and Reichert (1996), Wanner and Morgenroth (2004), and in the AQUASIM manual (Reichert, 1998). AQUASIM is freely available at www.aquasim.eawag.ch.

Underlying equations

AQUASIM evaluates biofilms for a compartment assuming a completely-mixed bulk phase, a mass-transfer boundary layer, and a one-dimensional biofilm. AQUASIM simultaneously solves the corresponding mass balances that are described below. Model inputs include the definition of the initial biofilm characteristics, biofilm detachment kinetics, and a stoichiometric and kinetic matrix following the format of Eq. 17.5. The model performs dynamic simulations and steady-state results are obtained by simulating with constant operating conditions for a sufficient period of time. Model output includes the biofilm characteristics and substrate concentrations within the biofilm and in the bulk water for every time point.

Relevant process parameters

The general mass balance equations for the biofilm compartment are given in Wanner and Reichert (1996) (eqs. 17.22, 17.23 and 17.24) and in the AQUASIM manual. A key difference between the mass balance in Eq. 17.1 and the mass balance in AQUASIM is that it differentiates between a solid fraction (ε_s) and a liquid fraction (ε_l) within the biofilm. In AQUASIM it is assumed that diffusion takes place only in the liquid fraction whereas the general diffusion reaction equation used for analytical calculations (Eq. 17.1) does not differentiate between solid and liquid fractions. In AQUASIM the density of the particulate species is the mass per volume of the solid fraction (Figure 4 in Wanner and Morgenroth, 2004). The overall biofilm biomass density is calculated as the density in the solid fraction multiplied by ε_s.

The mass balance for soluble substrate, taking into account the different definition of substrate concentrations per volume element (and neglecting minor terms), can be given as (detailed mass balance is given in Wanner and Reichert, 1996):

$$\frac{\partial \overbrace{(\varepsilon_\ell \cdot S_F)}^{\text{Mass of } S_F \text{ per total volume}}}{\partial t} = \underbrace{\varepsilon_\ell \cdot D_W}_{\substack{\text{Effective diffusion} \\ \text{coefficient in} \\ \text{the biofilm } (= D_F)}} \frac{\partial^2 S_F}{\partial z^2} + r_F \tag{17.37}$$

where D_W is the diffusion coefficient in water and $\varepsilon_l \cdot D_W$ is the effective diffusion coefficient in the biofilm that corresponds to D_F in Eq. 17.1. The accumulation term also takes into account that soluble components can accumulate in the liquid fraction within the biofilm. In addition, AQUASIM automatically takes into account substrate conversion in the bulk phase based on active biomass in the bulk volume derived from influent biomass, biomass detached from the biofilm, and biomass growing in suspension (Nogueira *et al.*, 2005). Note that the spatial coordinate in AQUASIM, z, is the distance from the substratum rather than the spatial coordinate, x, used in the analytical solutions (Figure 17.4). AQUASIM does not provide the substrate flux as an output but the user can calculate the flux through the biofilm surface from the calculated substrate concentrations based on Eq. 17.10 using $\varepsilon_l \cdot D_W$ as the effective diffusion coefficient:

$$J_{LF} = -\underbrace{\varepsilon_\ell \cdot D_W}_{\substack{\text{Effective diffusion}\\ \text{coefficient in the}\\ \text{biofilm}}} \cdot \frac{dS_F}{dz} \quad \text{at } z = L_F \tag{17.38}$$

In AQUASIM the flux of component S_F can be approximated by replacing the differential (dS_F/dz) with the secant of the substrate concentration ($\Delta S_F/\Delta z$). The value of ΔS_F can be calculated in AQUASIM by using what are referred to as 'probe variables' for different locations, z, within the biofilm and choosing a Δz that is smaller than the grid size that was used to simulate the biofilm:

$$J_{LF} \approx -\varepsilon_\ell \cdot D_W \cdot \frac{S_F(z = L_F) - S_F(z = L_F - \Delta z)}{\Delta z} \tag{17.39}$$

An alternative approach to calculate the flux into the biofilm is based on the change in concentration over the external concentration boundary layer (J_{BL}):

$$J_{BL} = -D_W \cdot \frac{S_B - S_{LF}}{L_L} \tag{17.40}$$

where L_L is the thickness of the concentration boundary layer. AQUASIM calculates the biofilm thickness (L_F) from a balance of growth, decay, attachment, and detachment (Wanner and Reichert, 1996):

$$\underbrace{\frac{dL_F}{dt}}_{\substack{\text{Net change}\\ \text{in biofilm}\\ \text{thickness}}} = \underbrace{u_F(L_F)}_{\text{Growth - decay}} + u_{a,S} - u_{d,S} \tag{17.41}$$

where $u_F(L_F)$ is the net effect of biofilm expansion as the result of growth and decay processes within the biofilm [L T^{-1}], $u_{a,S}$ is the rate of attachment, and $u_{d,S}$ is the rate of detachment. To simulate biofilms with a constant predefined thickness the biofilm detachment velocity $u_{d,S}$ can be set equal to $u_F(L_F)$ (assuming $u_{a,S} = 0$).

17.4 EXAMPLE OF HOW $J_{LF} = F(S_{LF})$ CAN BE USED TO PREDICT BIOFILM REACTOR PERFORMANCE

One motivation for calculating the substrate flux into the biofilm is to estimate the overall performance of a biofilm reactor. This relationship of substrate flux and overall reactor performance is illustrated below using examples of different rate expressions.

The simplest case for a biofilm reactor is the assumption of a completely mixed bulk phase as shown in Figure 17.10.

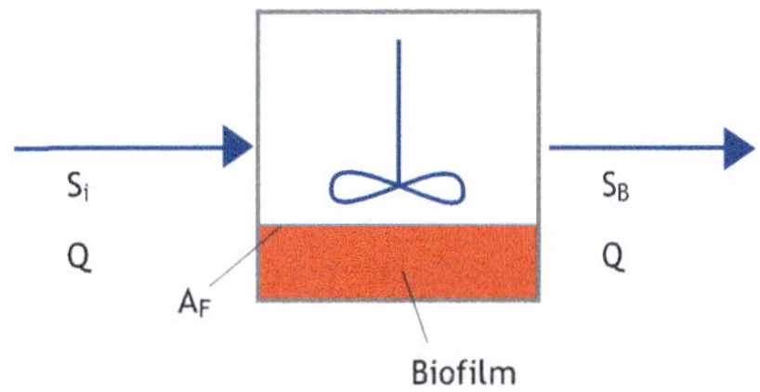

Figure 17.10 Biofilm reactor with completely mixed bulk phase and a biofilm surface area A_F.

For such a system the steady-state mass balance for substrate in the bulk phase is:

$$0 = Q \cdot (S_i - S_B) - J_{LF} \cdot A_F - r_B \cdot V_B \tag{17.42}$$

where:

S_i	influent substrate concentration [M L^{-3}]
A_F	biofilm surface area [L^2]
r_B	substrate conversion in the bulk phase due to suspended biomass [M L^{-3} T^{-1}]
V_B	volume of the bulk phase [L^3]

Assuming that bulk-phase conversion processes are negligible ($r_B \cdot V_B << J_{LF} \cdot A_F$), the material balance in Eq. 17.42 can be used to predict effluent substrate concentration S_B for a given flux of substrate into the biofilm (J_{LF}):

$$S_B = S_i - \frac{J_{LF} \cdot A_F}{Q} \tag{17.43}$$

One problem with using Eq. 17.43 is that J_{LF} is a function of the bulk-phase substrate concentration (S_B). Solving Eq. 17.43 requires simultaneously solving of this equation together with the appropriate equation for the substrate flux (*e.g.* from Table 17.3). Depending on the type of flux equation, Eq. 17.43 can either be solved analytically, by iteration, or can be solved graphically. Note that the mass balance in Eq. 17.42 is limited to soluble substrates. The fate of particulate matter in biofilm reactors is more complex. Particles entering a biofilm reactor can be retained by adsorption and some of the particles can be hydrolyzed resulting in soluble substrate that can be degraded inside the biofilm. See Section 17.7.2 for further discussion of how attachment and detachment are implemented in commercial simulators.

17.4.1 Analytical solution

In the case of a first-order rate expression (Eq. 17.16) an analytical solution for Eq. 17.43 can be derived. The substrate flux into the biofilm assuming first-order kinetics is:

$$J_{LF,1} = k_{1,A} \cdot S_{LF} \tag{17.16}$$

Neglecting external mass-transfer resistance, S_{LF} is equal to S_B. Then $J_{LF,1}$ from Eq. 17.16 can be substituted in Eq. 17.43 resulting in:

$$S_B = \frac{S_i}{\dfrac{k_{1,A} \cdot A_F}{Q} + 1} \tag{17.44}$$

From Eq. 17.44 it can be seen that the effluent substrate concentration is independent of the bulk volume but is determined by the surface area of the biofilm, the influent flow rate, the influent substrate concentration, and the first-order removal rate.

17.4.2 Trial and error or iterative approach

A simple approach to find S_B and the corresponding J_{LF} that solves Eq. 17.43 is to perform iteration. Choose a starting value for S_B, calculate J_{LF} using the appropriate rate equation (*e.g.* Table 17.3), then use Eq. 17.43 to calculate an updated value of the bulk-phase concentration and continue the iteration until S_B and J_{LF} do not vary significantly between iterations. For most flux data this iteration will be numerically stable and provide the unique solution to Eq. 17.43.

17.4.3 Graphical solution

If a graphical representation of the substrate flux is available from calculations, numerical simulations, or from experimental data then the effluent substrate concentration can also be read from this graph directly for given values of Q, A_F, and S_i. The material balance equation in Eq. 17.43 can be rearranged to:

$$J_{LF} = \underbrace{\frac{Q \cdot S_i}{A_F}}_{\text{const.}} - \underbrace{\frac{Q}{A_F}}_{\text{slope}} S_B \tag{17.45}$$

Eq. 17.45 describes a straight line in the flux plot that intersects the y axis ($S_B = 0$) at $J_{LF} = QS_i/A_F$ and the x axis ($J_{LF} = 0$) at $S_B = S_i$, and has a slope of Q/A_F. The intersection between Eq. 17.45 and the plotted substrate flux provides the solution for bulk-phase substrate concentrations and substrate flux that satisfies both the reactor mass-balance equation and also the biofilm mass-balance equation (Figure 17.11). This graphical solution is useful when evaluating measured flux vs. bulk-phase concentration data. In addition, it provides visual insight on how changes of influent substrate concentrations or influent flow rates will affect effluent substrate concentrations.

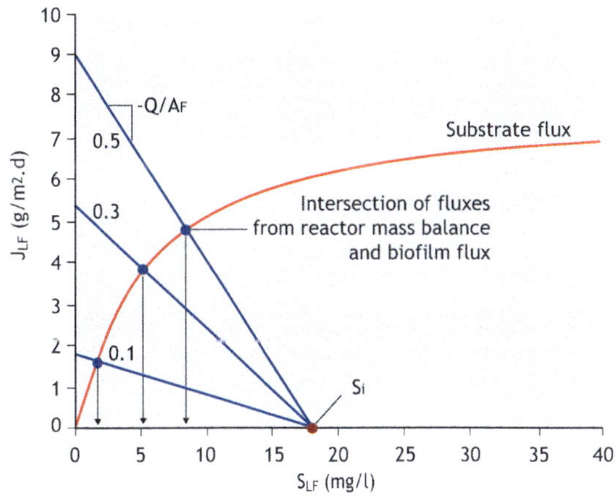

Figure 17.11 Graphical solution of Eq. 17.41 for a given substrate flux (from Figure 17.8C) and for three different values of Q/A_F (numbers in plot) and S_i = 18 mg/l. The arrows indicate the resulting effluent substrate concentrations for the three different values of Q/A_F.

17.4.4 Numerical solution (*e.g.* using AQUASIM)

As described in Section 17.3.5, AQUASIM simultaneously solves mass balances for the processes inside the biofilm and for the bulk phase and provides bulk-phase substrate concentrations as a direct model output.

17.5 EFFECT OF EXTERNAL MASS-TRANSFER RESISTANCE

In Section 17.3, concentration profiles inside the biofilm were calculated assuming that the substrate concentration at the biofilm surface (S_{LF}) is equal to the substrate concentration in the bulk phase (S_B). However, even with vigorous mixing in the bulk phase there will still be a mass-transfer resistance that needs to be considered. The concentration in outside the biofilm increases gradually as shown in Figure 17.12.

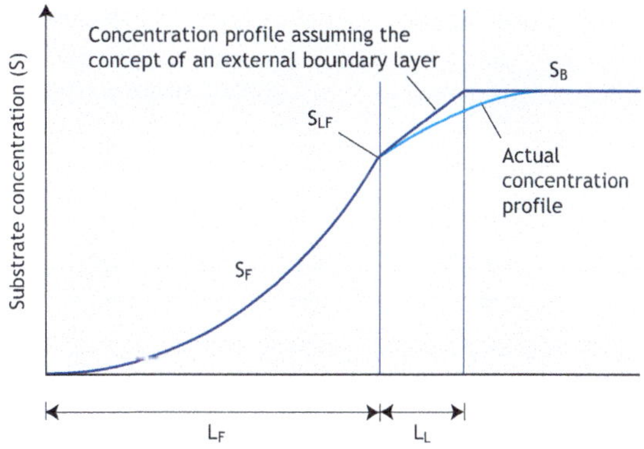

Figure 17.12 Concentration profile outside the biofilm and idealized representation as a concentration boundary layer with $L_L = R_L \cdot D_W$ (Eq. 17.47) (L_F is biofilm thickness and L_L thickness of the mass-transfer boundary layer).

The details of mixing close to the biofilm surface and the actual concentration profile are usually not explicitly modeled. Instead, external mass-transfer is described assuming an external mass-transfer resistance (R_L):

$$J_{BL} = \frac{1}{R_L} \cdot (S_B - S_{LF}) \tag{17.46}$$

where J_{BL} is the substrate flux in the boundary layer towards the biofilm.. It is helpful to visualize R_L by introducing the concept of a concentration boundary layer with an assumed linear concentration profile. The thickness of this concentration boundary layer provides a more intuitive understanding compared to the resistance. Resistance and the thickness of the concentration boundary layer are related as:

$$R_L = \frac{L_L}{D_W} \tag{17.47}$$

where L_L is the thickness of the external mass-transfer boundary layer and D_W is the diffusion coefficient in the water phase.

The substrate flux in the boundary layer (Eq. 17.46) is linked to the substrate flux at the surface of the biofilm (Eq.17.11). This provides an additional equation (boundary condition) that is necessary to calculate the additional unknown value of the substrate concentration at the surface of the biofilm:

$$\text{BC3:} \quad J_{BL} = J_{LF} \tag{17.48}$$

17.5.1 Substrate flux for first-order reaction rate with external boundary layer

The concept of linking substrate flux into the biofilm and an external mass-transfer resistance can be demonstrated by calculating the analytical solution assuming first-order kinetics within the biofilm. Combining the flux through the surface of the biofilm (Eq. 17.16) and the flux through the external boundary layer (Eq. 17.46) results in:

$$J_{LF} = k_{1,A} \cdot S_{LF} = \frac{1}{R_L} \cdot (S_B - S_{LF}) \tag{17.49}$$

This can be solved for S_{LF}:

$$S_{LF} = \frac{S_B}{k_{1,A} \cdot R_L + 1} \tag{17.50}$$

and the corresponding substrate flux can be calculated by substituting Eq. 17.50 into Eq. 17.16:

$$J_{LF} = k_{1,A} \cdot S_B \underbrace{\frac{1}{k_{1,A} \cdot R_L + 1}}_{\text{Reduction of flux due to external mass-transfer resistance}} \tag{17.51}$$

From Eq. 17.51, taking into account the external mass-transfer resistance, it can be seen that the extent of external mass-transfer resistance increases with increasing values of $k_{1,A}$ and R_L. Evaluating Eq. 17.51 for extreme cases results in

$$J_{LF} = \begin{cases} \frac{1}{R_L} \cdot S_B & \text{for } R_L \cdot k_{1,A} >> 1 \\ k_{1,A} \cdot S_B & \text{for } R_L \cdot k_{1,A} << 1 \end{cases} \tag{17.52}$$

For $R_L \cdot k_{1,A} >> 1$ substrate degradation is limited by external mass transfer and the flux in Eq. 17.52 equals the flux in Eq. 17.46 with $S_{LF} = 0$. For $R_L \cdot k_{1,A} << 1$ substrate removal is limited by the degradation and mass transport inside the biofilm resulting in $S_{LF} = S_B$. The value of $R_L \cdot k_{1,A}$ can be used to evaluate different approaches to increase overall substrate conversion. If the external mass-transfer resistance is significant (*e.g.* $R_L \cdot k_{1,A} >> 1$) then J_{LF} can be increased by increasing bulk-phase mixing, which reduces the thickness of the concentration boundary layer (L_L) and therefore reduces R_L. Increasing bulk-phase mixing does not have a direct effect on J_{LF} in the case of $R_L \cdot k_{1,A} << 1$.

17.5.2 Substrate flux for zero-order reaction rate (partially penetrated) with external boundary layer

The influence of external mass-transfer resistance in the case of a partially penetrated biofilm with zero-order reaction rates inside the biofilm can be determined from combining eqs. 17.29 and 17.49:

$$J_{LF} = k_{0,p,A} \cdot \sqrt{S_{LF}} = \frac{1}{R_L} \cdot (S_B - S_{LF}) \tag{17.53}$$

Eq. 17.53 is a quadratic equation in terms of $\sqrt{S_{LF}}$ with the following solution:

$$\sqrt{S_{LF}} = -\frac{k_{0,p,A} \cdot R_L}{2} + \sqrt{\left(\frac{k_{0,p,A} \cdot R_L}{2}\right)^2 + S_B} \tag{17.54}$$

Substituting $\sqrt{S_{LF}}$ from Eq. 17.54 into Eq. 17.29 for the substrate flux into a partially penetrated biofilm with zero-order reaction rates results in

$$J_{LF} = k_{0,p,A} \cdot \left(-\frac{k_{0,p,A} \cdot R_L}{2} + \sqrt{\left(\frac{k_{0,p,A} \cdot R_L}{2}\right)^2 + S_B} \right) \tag{17.55}$$

A dimensionless parameter ζ is defined below (Eq. 17.56) as the ratio of the substrate flux into the biofilm neglecting the external mass-transfer resistance ($S_{LF} = S_B$ in Eq. 17.29) and the substrate flux in the external mass-transfer boundary layer assuming $S_{LF} = 0$:

$$\zeta = \frac{\overbrace{k_{0,p,A} \cdot \sqrt{S_B}}^{\text{Maximum mass-transfer into biofilm}}}{\underbrace{\frac{S_B}{R_L}}_{\text{Maximum external mass-transfer}}} = \frac{k_{0,p,A} \cdot R_L}{\sqrt{S_B}} \qquad (17.56)$$

Substituting ζ from Eq. 17.56 into Eq. 17.55 results in the substrate flux as a function of the bulk-phase substrate concentration (S_D) and the parameter ζ:

$$J_{LF} = k_{0,p,A} \cdot \sqrt{S_B}\left(-\frac{\zeta}{2} + \sqrt{\left(\frac{\zeta}{2}\right)^2 + 1}\right) \qquad (17.57)$$

Evaluating Eq. 17.57 for extreme cases results in

$$J_{LF} = \begin{cases} \frac{1}{R_L} \cdot S_B & \text{for } R_L \cdot k_{0,p,A} >> S_B \text{ or } \zeta >> 1 \\ k_{0,p,A} \cdot \sqrt{S_B} & \text{for } R_L \cdot k_{0,p,A} << S_B \text{ or } \zeta << 1 \end{cases} \qquad (17.58)$$

A biofilm with $\zeta \rightarrow \infty$ is limited by the external mass-transfer resistance while for a biofilm with $\zeta \rightarrow 0$, the external mass transfer can be neglected.

It should be noted that the strict separation of external and internal mass-transfer resistances is to some extent artificial. In the current modelling approach, bulk-phase mixing only influences the external mass-transfer resistance. However, in reality external mixing will have an influence on biofilm development, the biofilm biomass density of the biofilm, and the occurrence of streamers[5]. Lower mixing and lower shear result in fluffy and thicker biofilms while higher shear results in denser biofilms (Van Loosdrecht *et al.*, 1995).

17.5.3 Substrate flux for Monod kinetics inside the biofilm with an external boundary layer

To demonstrate the significance of the external mass transfer resistance, AQUASIM simulations were performed using the same kinetic and biofilm parameters as in Figure 17.8B where substrate fluxes are calculated neglecting external mass-transfer resistances (Figure 17.13).

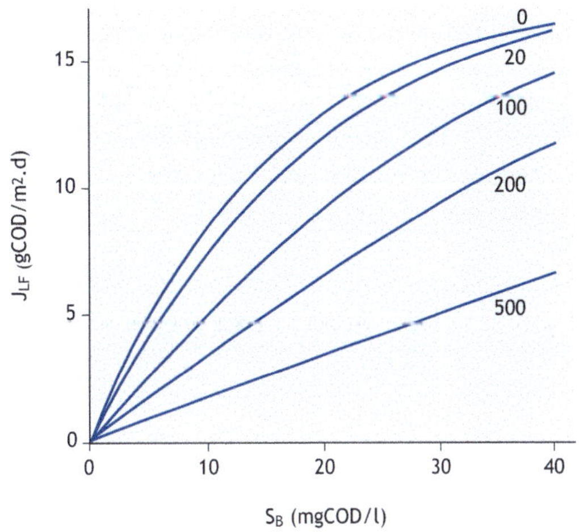

Figure 17.13 Substrate fluxes for different influent substrate concentrations as a function of the external boundary layer thickness (numbers in plot are L_L in μm) for a 200 μm thick biofilm. As in Figure 17.8, substrate removal in the biofilm was modelled using Monod kinetics (Table 17.2) and parameters provided in Table 17.13.

The results shown in Figure 17.13 are based on external mass-transfer boundary layer thicknesses (L_L) ranging from 0 to 500 μm. It can be seen that both with increasing bulk-phase substrate concentration and with a decreasing boundary-layer thickness, the flux increases. A boundary-layer thickness of $L_L = 500$ μm decreases the flux by more than 70%.

[5] Streamers – a technical term for filaments that are attached to the biofilm but that move due to water flow.

17.6 MULTI-COMPONENT DIFFUSION

17.6.1 Two-component diffusion of electron donor and acceptor

Conversion processes in a biofilm usually require the diffusion of an electron donor and an electron acceptor from the bulk phase into the biofilm. Comparing the relative substrate penetration for an electron donor and acceptor into the biofilm enables the compound that is limiting the overall substrate conversion to be identified. Three cases are possible as shown in Figure 17.14:

- Case 1: The electron donor does not fully penetrate the biofilm and substrate conversion is limited by the availability of the electron donor towards the base of the biofilm.
- Case 2: The electron acceptor does not fully penetrate the biofilm and limits overall substrate conversion.
- Case 3: Both electron donor and electron acceptor fully penetrate the biofilm. Substrate conversion is not mass transport-limited.

Section 17.3.2.2 discussed the extent of substrate penetration into the biofilm for a single substrate and assuming zero-order reaction rates in the biofilm based on the value of β (Eq. 17.25). The same approach to calculating substrate penetration can be applied for two-component diffusion where the value of β is calculated separately for the electron donor ($\beta_{e.d.}$) and acceptor ($\beta_{e.a.}$). Based on the values of $\beta_{e.d.}$ and $\beta_{e.d.}$ it is possible to evaluate which of the three cases in Figure 17.14 applies:

$$\text{Case 1: } \beta_{e.d.} < 1 \text{ and } \beta_{e.d.} < \beta_{e.a.} \tag{17.59a}$$

$$\text{Case 2: } \beta_{e.a.} < 1 \text{ and } \beta_{e.d.} > \beta_{e.a.} \tag{17.59b}$$

$$\text{Case 3: } \beta_{e.d.} > 1 \text{ and } \beta_{e.a.} > 1 \tag{17.59c}$$

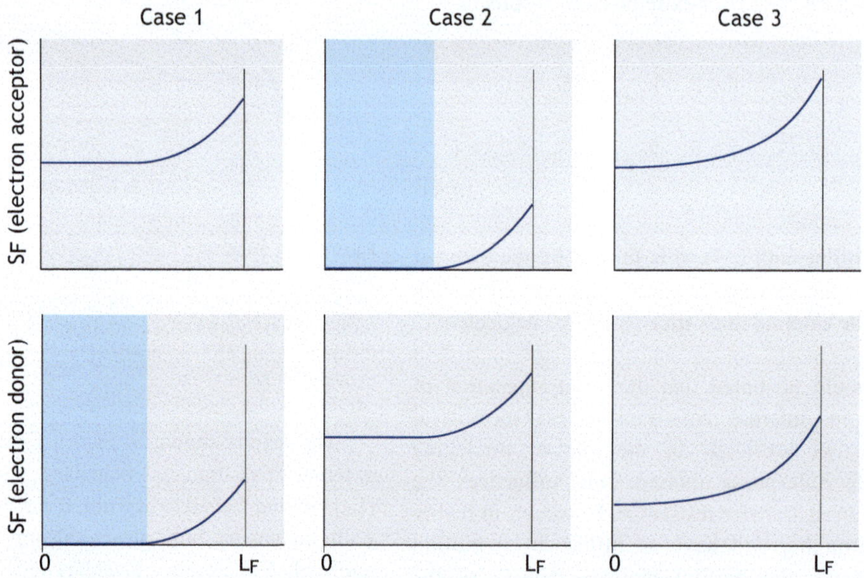

Figure 17.14 Substrate penetration into a biofilm, assuming that the electron donor is limiting (Case 1), that the electron acceptor is limiting (Case 2), or that neither electron donor nor acceptor are limiting (Case 3). No substrate conversion occurs in the section shaded blue of the biofilm due to electron donor or acceptor limitations (L_F is the biofilm thickness).

To differentiate between Case 1 and 2, a simpler way than explicitly evaluating $\beta_{e.d.}$ and $\beta_{e.a}$ can be developed by calculating the ratios of βs rather than comparing their specific values. For a growth process the conversion rates for the electron donor ($r_{0,e.d.}$) and acceptor ($r_{0,e.a.}$) are stoichiometrically linked:

$$r_{0,e.d.} = k_{0,F,e.d.} \cdot X_F \tag{17.60}$$

$$r_{0,e.a.} = (\alpha - Y) \cdot r_{0,e.d} \tag{17.61}$$

in which Y is the biomass yield and α is a stoichiometric factor linking the electron acceptor and donor utilization in the catabolic reaction. For organic substrates the value of $\alpha = 1$ $gO_2/gCOD$, and for nitrification $\alpha = 4.57$ gO_2/gN. Now the values of $\beta_{e.d.}$ and $\beta_{e.a.}$ can be compared by calculating their ratio ($\gamma_{e.d.,e.a.}$):

$$\gamma_{e.d.,e.a.} = \frac{\beta_{e.d.}}{\beta_{e.a.}} \tag{17.62}$$

Substituting the βs in Eq. 17.62 by using the zero-order rate constants from eqs. 17.60 and 17.61 and the definition of β (Eq. 17.25) yields:

$$\gamma_{e.d.,e.a.} = \frac{\sqrt{\dfrac{2 \cdot D_{F,e.d.} \cdot S_{LF,e.d.}}{k_{0,F,e.d.} \cdot X_F \cdot L_F^2}}}{\sqrt{\dfrac{2 \cdot D_{F,e.a.} \cdot S_{LF,e.a.}}{(\alpha - Y) \cdot k_{0,F,e.d.} \cdot X_F \cdot L_F^2}}} \tag{17.63}$$

in which $D_{F,e.d.}$ and $D_{F,e.a.}$ are the diffusion coefficients and $S_{LF,e.d.}$ and $S_{LF,e.a.}$ are substrate concentrations at the biofilm surface for the electron donor and acceptor, respectively. The advantage of calculating the ratios of βs is apparent in Eq. 17.63 as many parameters cancel out and the equation can be simplified to:

$$\gamma_{e.d.,e.a.} = \sqrt{(\alpha - Y) \cdot \frac{D_{F,e.d.} \cdot S_{LF,e.d.}}{D_{F,e.a.} \cdot S_{LF,e.a.}}} \tag{17.64}$$

Using $\gamma_{e.d.,e.a.}$, cases 1 and 2 can be differentiated:

- Case 1 with $\gamma_{e.d.,e.a.} < 1$ or $\dfrac{S_{LF,e.d.}}{S_{LF,e.a.}} < \dfrac{1}{(\alpha - Y)} \cdot \dfrac{D_{F,e.a.}}{D_{F,e.d.}}$

  : The electron donor is potentially limiting inside the biofilm, but the electron acceptor will fully penetrate the biofilm. Note that this assumes that the biofilm conversion is mass transport-limited.
- Case 2 with $\gamma_{e.d.,e.a.} > 1$ or $\dfrac{S_{LF,e.d.}}{S_{LF,e.a.}} > \dfrac{1}{(\alpha - Y)} \cdot \dfrac{D_{F,e.a.}}{D_{F,e.d.}}$

  : The electron acceptor is potentially limiting inside the biofilm, but the electron donor will fully penetrate the biofilm. Note that this assumes that the biofilm conversion is mass transport-limited.

Examples of heterotrophic growth on organic substrate and autotrophic growth on ammonium are provided and conditions for oxygen (*i.e.* electron acceptor) or electron donor-limitation are provided in Table 17.4.

Once it is known whether the electron donor or acceptor are limiting the conversion processes within the biofilm, then substrate flux for the limiting compound can be calculated using the kinetic expressions developed above for a single limiting substrate. The substrate flux for the non-limiting compound can then be calculated based on the overall stoichiometry:

$$J_{LF,e.a.} = (\alpha - Y) \cdot J_{LF,e.d.} \tag{17.65}$$

Example 17.2 Substrate fluxes (J_{LF}) and penetration depths ($\beta \cdot L_F$) for heterotrophic or autotrophic growth or respiration assuming zero-order kinetics

Substrate removal in biofilm reactors is typically mass transport-limited. Substrate flux and the penetration into the biofilm are a function of substrate concentrations at the biofilm surface, reaction rates inside the biofilm, and diffusive mass transport. In Table 17.5 the substrate flux and the penetration depth ($\beta \cdot L_F$) are presented for acetate, ammonium, and oxygen, assuming zero-order substrate removal rates in a partially penetrated biofilm. Comparing penetration depths for electron donors (acetate or

Table 17.4 Calculation of limiting compounds for growth on organic substrate or nitrification (stoichiometric parameters from Table 17.13, assuming $D_F = 0.8 \cdot D_W$, and D_W from Table 17.10).

	Organic substrate removal	Nitrification
Electron donor	Organic substrate	NH_4^+
$D_{F,e.d.}$	$0.8 \cdot 94.1 \cdot 10^{-6}$ m²/d	$0.8 \cdot 169.1 \cdot 10^{-6}$ m²/d
Electron acceptor	O_2	O_2
$D_{F,e.a.}$	$0.8 \cdot 209.1 \cdot 10^{-6}$ m²/d	$0.8 \cdot 209.1 \cdot 10^{-6}$ m²/d
α	1 gO_2/gCOD	4.57 gO_2/gN
Y	0.4 gCOD/gCOD	0.22 gCOD/gN
Electron acceptor-limiting (based on Eq. 17.88)	$\frac{S_{LF,COD}}{S_{LF,O2}} > 3.7\ gCOD/gO_2$	$\frac{S_{LF,NH4}}{S_{LF,O2}} > 0.28\ gN/gO_2$
Oxygen will be limiting (based on Eq. 17.88) for $S_{LF,O2} = 8$ mg/l when $S_{LF,e.d.}$ is higher than →	29.6 mgCOD/l	2.3 mgN/l

Table 17.5 Penetration depth and flux of oxygen and substrate estimated for growth with zero-order biofilm kinetics using eqs. 25 and 27 in Table 17.3 or for respiration only with kinetic parameters in Table 17.13. Fluxes for electron donors marked with (*) are not realistic as the biofilm would be oxygen-limited (Table 17.4).

Heterotrophic growth					
$S_{B,O2}$ (g/m³)	Penetration depth (µm)	$J_{LF,O2}$ (g/m².d)	$S_{B,HAc}$ (gCOD/m³)	Penetration depth (µm)	$J_{LF,HAc}$ (gCOD/m².d)
1	68	4.9	1	35	4.3
3	118	8.5	5	79	9.5
5	153	11.0	15	137	16.5
8	193	13.9	150	434	52.1(*)
Nitrification					
$S_{B,O2}$ (g/m³)	Penetration depth (µm)	$J_{LF,O2}$ (g/m².d)	$S_{B,NH4}$ (gN/m³)	Penetration depth (µm)	$J_{LF,NH4}$ (gN/m².d)
1	42	7.9	1	79	3.4
3	73	13.7	5	177	7.6(*)
5	94	17.7	15	307	13.2(*)
8	119	22.4	70	662	28.6(*)
Heterotrophic respiration in the absence of external substrate			**Autotrophic respiration in the absence of external substrate**		
$S_{B,O2}$ (g/m³)	Penetration depth (µm)	$J_{LF,O2}$ (g/m².d)	$S_{B,O2}$ (gN/m³)	Penetration depth (µm)	$J_{LF,O2}$ (g/m².d)
1	409	0.8	1	818	0.4
3	708	1.4	3	1,417	0.7
5	915	1.8	5	1,829	0.9
8	1,157	2.3	8	2,314	1.2

ammonium) and the electron acceptor demonstrates that for most electron donor concentrations the removal is limited by the availability of oxygen. It also becomes obvious that oxygen penetration into biofilm is only a few hundred micrometers. Therefore thicker biofilms (for typical ranges see Table 17.11) are usually not beneficial for aerobic biofilm processes.

Penetration of oxygen into the biofilm depends on the rate of substrate utilization inside the biofilm. In biofilm systems that are operated as sequencing batch reactors (*e.g.* aerobic granular sludge systems) biofilms are sequentially exposed to high and low substrate bulk-phase substrate concentration. Without substrate, oxygen utilization is reduced to endogenous respiration processes. Table 17.5 provides oxygen penetration depths of more than 1 mm in the case of respiration only. This explains why even large aerobic granules can periodically be fully penetrated by oxygen.

17.6.2 General case of multi-component diffusion

The concept for the application of $\gamma_{e.d.,e.a.}$ can be expanded for more than two compounds. It should be noted that Eq. 17.64 is based only on the assumption of diffusive transport inside the biofilm and on a stoichiometric link between the electron donor and acceptor utilization. Thus, even though the derivation of Eq. 17.64 was based on zero-order kinetics, the concept of $\gamma_{e.d.,e.a}$ can also be applied to other growth kinetics as long as all conversion rates ($r_{F,i}$) are linked to the same overall process rate (ρ) (Gujer and Boller, 1986):

$$r_{F,i} = v_i \cdot \rho \tag{17.66}$$

where v_i is the stoichiometric coefficient for the removal of compound $S_{F,i}$. for a given process (Eq. 17.5). Combined with Eq. 17.1 and assuming steady state this yields

$$\rho = \frac{D_{F,1}}{v_1} \cdot \frac{\partial^2 S_{F,1}}{\partial x^2} = \frac{D_{F,2}}{v_2} \cdot \frac{\partial^2 S_{F,2}}{\partial x^2} = \cdots = \frac{D_{F,i}}{v_i} \cdot \frac{\partial^2 S_{F,i}}{\partial x^2} \tag{17.67}$$

A direct stoichiometric relationship between the fluxes of the different substrates $S_{F,i}$ based on Eq. 17.91 is (Gujer and Boller, 1986):

$$\frac{J_{LF,1}}{v_1} = \frac{J_{LF,2}}{v_2} = \cdots = \frac{J_{LF,i}}{v_i} \tag{17.68}$$

Eq. 17.68 can be used to calculate fluxes of the non-limiting compounds based on the flux of the limiting compound. That means that Eq. 17.68 is a more general form of Eq. 17.89 which provided a link between the electron acceptor and donor fluxes.

The rate-limiting compound can be determined, similar to the approach based on $\gamma_{e.d.,e.a.}$ in Eq. 17.64, by finding the compound with the lowest outcome for the following relation (Andrews, 1988; Wanner *et al.*, 2006):

$$\frac{D_{F,i} \cdot S_{LF,i}}{v_i} \tag{17.69}$$

Once the flux of the limiting compound has been determined, the flux of the other compounds can be calculated directly using Eq. 17.68. Note that a key assumption for Eq. 17.68 is that all the processes in the biofilm are stoichiometrically linked.

17.6.3 Complications for multiple processes inside the biofilm

Note that the derivation of Eq. 17.64 and Eq. 17.65 was based on the assumption that the utilization of the electron acceptor is directly coupled with electron donor utilization and growth-neglecting processes such as endogenous respiration where electron acceptors are utilized even in the absence of an electron donor. Due to this simplification the concept of $\gamma_{e.d.,e.a.}$ needs to be used with caution for $\gamma_{e.d.,e.a.} \approx 1$ and no clear conclusions can be drawn on the limiting substrate. For $\gamma_{e.d.,e.a.} \approx 1$ a multi-substrate model should be used to evaluate the biofilm (Wanner *et al.*, 2006).

17.7 COMBINING GROWTH AND DECAY WITH DETACHMENT

In many biofilm models and in all the calculations in the preceding sections, the user must assume a constant biofilm thickness. Mathematical modelling, however, can predict biofilm development over time and also a steady-state biofilm thickness based on growth, decay, and detachment processes:

$$\underbrace{\frac{dL_F}{dt}}_{\substack{\text{Net change}\\\text{of biofilm}\\\text{thickness}}} = \underbrace{\frac{Y \cdot J_{LF}}{X_F}}_{\text{Growth}} - \underbrace{b_{ina} \cdot L_F}_{\text{Decay}} - \underbrace{u_{d,S}}_{\substack{\text{Surface}\\\text{detachment}\\\text{velocity}}} \tag{17.70}$$

However, while setting up the biofilm thickness balance (Eq. 17.70) is straightforward, finding suitable expressions for the detachment velocity ($u_{d,S}$) is not. An overview of detachment rate expressions is provided in Table 17.6. It can be seen that different researchers have assumed very different detachment mechanisms. There is also no agreement on estimation of values for detachment rate coefficients for the equations presented in Table 17.6 based on reactor type or operation. In practice in biofilm reactor modelling, a constant biofilm thickness is usually assumed (Rittmann *et al*., 2018).

There are different approaches to quantify detachment rates (Morgenroth, 2003). In Eq. 17.70 detachment rates are describe as a constant detachment velocity ($u_{d,S}$) [L T^{-1}]. Other references express detachment as the mass of biofilm removed per area and time ($u_{d,M}$) [M L^{-2} T^{-1}] or as the volume fraction of biofilm detaching per time ($u_{d,V}$) [T^{-1}]. These different detachment rate expressions are related to the detachment velocity as follows:

Table 17.6 Detachment rate expressions (modified from Morgenroth, 2003; Peyton and Characklis, 1993; Tijhuis *et al*., 1995)[1].

Mechanism of detachment related to	Reported detachment rate expression, $u_{d,M}$ [M L^{-2} T^{-1}]	Reference
None specified	0	Kissel *et al*., 1984; Fruhen *et al*., 1991
	Constant biofilm thickness	Wanner and Gujer, 1985
Biofilm thickness	$k_d \cdot (X_F \cdot L_F)^2$	Bryers, 1984; Trulear and Characklis, 1982
	$k_d \cdot X_F \cdot L_F^2$	Wanner and Gujer, 1986
	$k_d \cdot X_F \cdot L_F$	Chang and Rittman, 1987; Kreikenbohm and Stephan, 1985; Rittmann, 1989
Shear	$k_d \cdot X_F \cdot \tau$	Bakke *et al*., 1984
	$k_d \cdot X_F \cdot L_F \cdot \tau^{0.58}$	Rittmann, 1982b
Growth rate or substrate utilization rate	$L_F \cdot (k_d' + k_d'' \cdot \mu)$	Speitel and DiGiano, 1987
	$k_d \cdot r_S \cdot L_F$	Peyton and Characklis, 1993; Robinson *et al*., 1984; Tijhuis *et al*., 1995
Backwashing down to a predefined base thickness	$\begin{cases} k_d' \cdot L_F & \text{normal operation} \\ k_d'' \cdot (L_F - L_{base\ thickness}) & \text{backwashing} \end{cases}$	Lackner *et al*., 2008; Morgenroth and Wilderer, 1999; Rittmann *et al*., 2002

[1] Explanation of symbols: k_d, k_d', k_d'' = detachment rate coefficients, X_F = biofilm biomass density [M L^{-3}], L_F = biofilm thickness [L], $L_{base\ thickness}$ = predefined biofilm thickness after backwashing [L], μ = specific growth rate [T^{-1}], r_S = substrate utilization rate [M L^{-2} T^{-1}], τ = shear stress [M L^{-1} T^{-2}].

$$u_{d,V} = \frac{u_{d,S}}{L_F} \quad (17.71)$$

$$u_{d,M} = u_{d,S} \cdot X_F \quad (17.72)$$

in which X_F is the biofilm biomass density [M L^{-3}] of the biofilm. Most models of biofilm detachment use constant detachment rate coefficients. Dynamic detachment differs from both surface and volume detachment in that detachment is modelled not as a continuous process but as discrete events occurring at certain intervals. An example of dynamic detachment is backwashing of biofilm reactors. The resulting change in overall biofilm thickness can then be calculated assuming a dynamic detachment rate expression:

$$u_{d,S} = \begin{cases} k_d^{'} \cdot L_F & \text{during normal operation} \\ k_d^{''} \cdot \left(L_F - L_{\text{base thickness}}\right) & \text{during backwashing} \end{cases} \quad (17.73)$$

where $u_{d,S}$ can be defined in a way that all biofilm above a predefined base thickness is removed during backwashing (Morgenroth and Wilderer, 1999; Morgenroth, 2003).

17.7.1 Influence of detachment ($u_{d,S}$) on the steady-state biofilm thickness (L_F) and the substrate flux (J_{LF})

The mass balance in Eq. 17.70 can be solved analytically for some selected substrate removal and detachment rate expressions or numerically using AQUASIM.

Example 17.3 Predicting the biofilm thickness assuming half-order substrate flux

Assume the following substrate flux and detachment:

$$J_{LF,0,p} = k_{0,p,A} \cdot \sqrt{S_{LF}} \quad (17.29)$$

$$u_{d,S} = k_d \cdot L_F \quad (17.74)$$

With these definitions, Eq. 17.70 for steady-state conditions results in:

$$0 = \frac{Y \cdot k_{0,p,A} \cdot \sqrt{S_{LF}}}{X_F} - b_{ina} \cdot L_F - k_d \cdot L_F \quad (17.75)$$

which can be solved for L_F:

$$L_F = \frac{Y \cdot k_{0,p,A} \cdot \sqrt{S_{LF}}}{X_F \cdot (b_{ina} + k_d)} \quad (17.76)$$

Thus, both an increase of decay (b_{ina}) and of detachment (k_d) results in a decreased biofilm thickness. Increased removal rates ($k_{0,p,A}$) and surface concentrations (S_{LF}) will result in thicker biofilms.

Example 17.4 Predicting the biofilm thickness assuming zero-order substrate flux

An analytical solution for the biofilm thickness can also be calculated for a zero-order substrate removal rate

$$J_{LF} = k_0 \cdot X_F \cdot L_F \quad (17.31)$$

but assuming a different detachment rate expression

$$u_{d,S} = k_d \cdot L_F^2 \quad (17.77)$$

Note that choosing the detachment rate in Eq. 17.77 instead of Eq. 17.74 was purely for convenience to simplify the analytical solution. Inserting into Eq. 17.70 yields:

$$0 = \frac{Y \cdot k_0 \cdot X_F \cdot L_F}{X_F} - b_{ina} \cdot L_F - k_d \cdot L_F^2 \quad (17.78)$$

which can be solved for L_F:

$$L_F = \frac{Y \cdot k_0 - b_{ina}}{k_d} \quad (17.79)$$

Again, the biofilm thickness decreases with increasing values of decay (b_{ina}) and of detachment (k_d) and with decreasing substrate-removal rates (k_0).

Example 17.5 Predicting the biofilm thickness using a numerical solution assuming Monod kinetics

Analytical solutions for the biofilm thickness are available only for selected combinations of biofilm growth and detachment rate expressions. Using AQUASIM we can evaluate the influence of different detachment rates assuming Monod kinetics within the biofilm. Simulations were performed for bulk-phase substrate concentrations ranging from 0.5 to 100 mg/l and for detachment kinetics of:

$$u_{d,S} = k_d \cdot L_F^2 \qquad (17.80)$$

with a detachment rate coefficient k_d ranging from 0 to 100,000 1/m.d. In Figure 17.15 substrate fluxes and biofilm thicknesses are shown for different bulk-phase substrate concentrations; each line represents a different value of k_d. It can be seen that with increasing values of the detachment rate coefficient, k_d, both the substrate flux and the biofilm thickness decrease while both values increase with increasing bulk-phase substrate concentrations. This interdependence of substrate flux and biofilm thickness can also be seen in Figure 17.16.

Substrate fluxes were determined for different fixed biofilm thicknesses and each line in Figure 17.16 represents a different bulk-phase concentration. For very thin biofilms increasing the biofilm thickness will increase substrate flux: substrate conversions in very thin biofilms are biomass-limited. For relatively thick biofilms the biofilm thickness has only a limited influence on the substrate fluxes. Increasing the biofilm thickness increases the substrate flux only if bulk-phase substrate concentrations are sufficiently high that substrate can penetrate to the base of the biofilm.

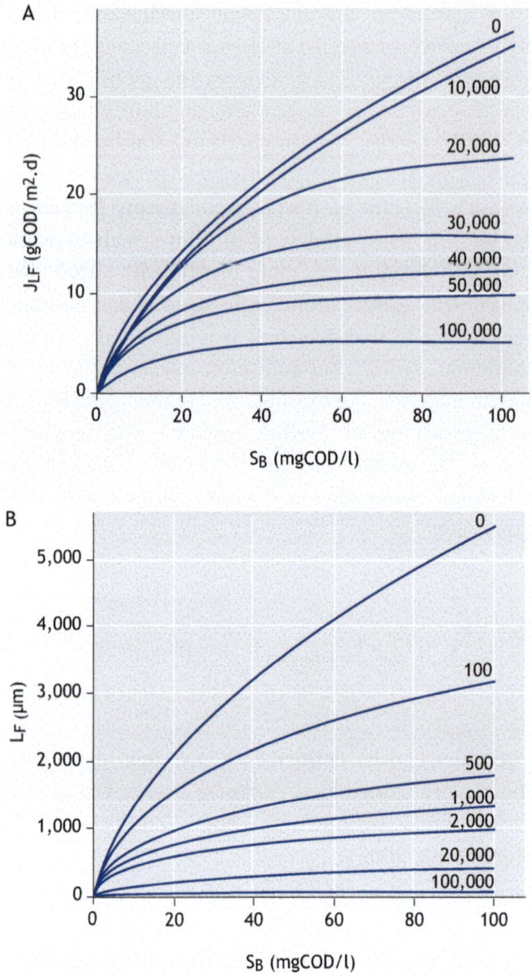

Figure 17.15 Influence of bulk-phase substrate concentrations and the detachment rate coefficient on substrate flux (A) and biofilm thickness (B). Numbers in the plot are the value of k_d in Eq. 17.60. The unit for k_d is 1/m.d. The biofilm model includes substrate removal and growth using the parameters provided in Table 17.13.

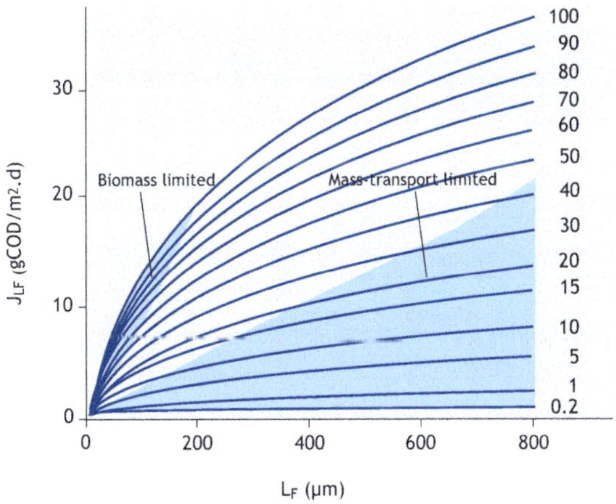

Figure 17.16 Influence of biofilm thickness on substrate flux for different bulk-phase substrate concentration (the numbers in the plot are in mg/l). Two areas are marked in blue: thin biofilms can be biomass-limited where increasing the biofilm thickness results in a significant increase of the substrate flux. For thick biofilms an increase in biofilm thickness does not significantly increase the substrate flux as substrate removal in the biofilm is mass transport-limited and lower regions of the biofilm will not be active. Substrate removal in the biofilm was modelled using Monod kinetics using parameters provided in Table 17.13.

17.7.2 Attachment and fate of particles

Wastewater contains significant amounts of particulate organic carbon which is too large to diffuse into a biofilm. The simple diffusion-reaction equation (Eq. 17.1) and the corresponding derivations in previous sections summarized in Table 17.3 do not apply for such non-diffusible substrates. Practical biofilm reactor modelling requires explicit modelling of particle attachment, hydrolysis, and detachment processes.

Mechanisms for particle attachment in biofilm reactors depend on particle characteristics, local biofilm morphology, and overall reactor hydrodynamics. However, rates and dynamics of particle retention, biofilm attachment, and overall degradation are not well understood. Most fundamental research has neglected the fate of particulate substrate and instead focused on soluble substrates and microscale observations. Detailed full-scale observations evaluating particle retention are rare (Janning *et al.*, 1998). Particle retention is mostly a physical process depending on particle size and surface characteristics. In addition, it has been demonstrated that protozoa and metazoans can be actively involved in removing particles from the bulk phase (De Kreuk *et al.*, 2010).

Different commercial simulators use different approaches to model attachment and detachment of particles where mechanisms and parameters of these approaches are difficult to compare (Boltz *et al.*, 2010). Attachment is typically implemented as a process transferring particles from the bulk water into the top layer of a discretized biofilm with a first-order rate depending on the bulk-phase concentration of the particulate compound (Figure 17.17). Particles in this top layer can be hydrolysed, producing soluble compounds that are transported by diffusion. In addition, most simulators include some spreading mechanism allowing the slow transport of particles into deeper layers of the biofilm. The process of spreading takes into account that biofilms are more heterogeneous than is considered in the simplified 1D representation, allowing particles to reach also deeper layers. Particles can be transferred back into the bulk phase by detachment. Introducing transport terms for particulate matter and biomass between layers results in a form of limited back-mixing. Kinetic parameters for attachment, spreading, and detachment are usually determined by fitting the overall biofilm reactor performance, and parameter values will depend on the type of particles and reactor configuration. With different commercial simulators using different approaches to model attachment, spreading, and detachment, the numeric values of kinetic parameters cannot be directly compared (Boltz *et al.*, 2010). A further complication is that overall fate of particles in a simulation is linked with other biofilm model features such as the thickness of the top layer (and therefore the number of layers) and the biofilm density. However, while there are significant uncertainties in the mechanisms and kinetic parameters, realistic predictions of biofilm reactors require the fate of particles to be modelled.

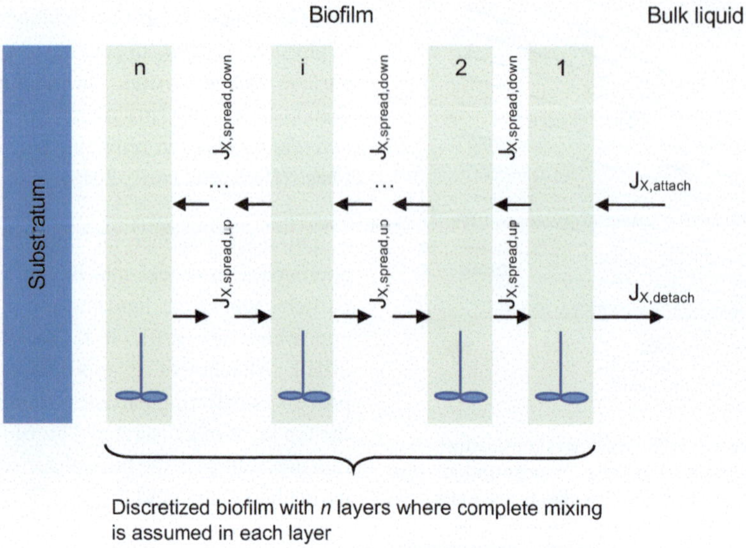

Figure 17.17 Discretization of a biofilm in *n* layers resulting in a layer thickness of L_F/n. Arrows indicate the mass transport of particles from the bulk to the first layer in the biofilm ($J_{X,attach}$), detachment from the first layer of the biofilm ($J_{X,detach}$), and a certain amount of back-mixing of biomass and particulate compounds between layers ($J_{X,spread,up}$, $J_{X,spread,down}$). Rate expressions vary significantly between commercial simulators (Boltz *et al.*, 2010). The attachment flux ($J_{X,attach}$) is typically proportional to the bulk-phase concentration of this component. In most simulators detachment flux ($J_{X,detach}$) is a function of the maximum total solids concentration in the first layer. The net accumulation of solids due to biomass growth, decay or particle hydrolysis) is balanced by detachment (when assuming constant biofilm thickness) or volume expansion (when assuming constant density).

17.8 BIOFILM REACTOR MODELLING IN PRACTICE

Practical biofilm reactor modelling builds on the principles developed in the previous sections. In the previous sections, modelling approaches and analytical solutions often assumed a completely-mixed bulk phase within one process and one compound-limiting overall conversion. However, in practice, in a biofilm reactor the bulk phase is usually not completely mixed and different limiting factors apply for different regions in the reactor. Such a reactor can again be simplified by separating different regions into different sections where each section is modelled as a biofilm compartment with a completely-mixed bulk phase (Figure 17.18). This general approach can be applied independent of the type of reactor where individual biofilm compartments can correspond to physically separated tanks (*e.g.* in an MBBR or in a RBC) or they can simply be a discretization of a plug-flow type reactor (*e.g.* trickling filter or biologically-activated filter). This section is organized as a collection of examples followed by a structured step-wise approach to practical biofilm reactor modelling.

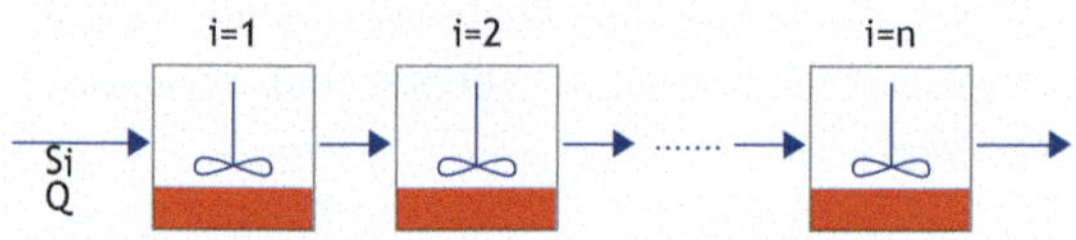

Figure 17.18 Mixing conditions in many biofilm reactors can be approximated as multiple biofilm compartments in series.

17.8.1 Collection of examples

Example 17.6 Limiting substrate changes over the length of a biofilm reactor

This first example evaluates expected ammonium concentrations along the length of a nitrifying biofilm reactor. Bulk-phase mixing is approximated as the plug flow. The reactor is aerated resulting in constant oxygen concentrations along the length of the reactor. As shown in Table 17.4, oxygen is the limiting substrate if ammonium concentrations are higher than 0.28 gN/gO_2. For bulk-phase oxygen concentrations of 8 mgO_2/l this corresponds to ammonium concentrations of 2.3 mgN/l. Thus, close to the inlet, nitrifying biofilm reactors are often oxygen rather than ammonium-limited. For a constant bulk-phase oxygen concentration the corresponding ammonium flux into the biofilm is also constant, as shown in Figure 17.19.

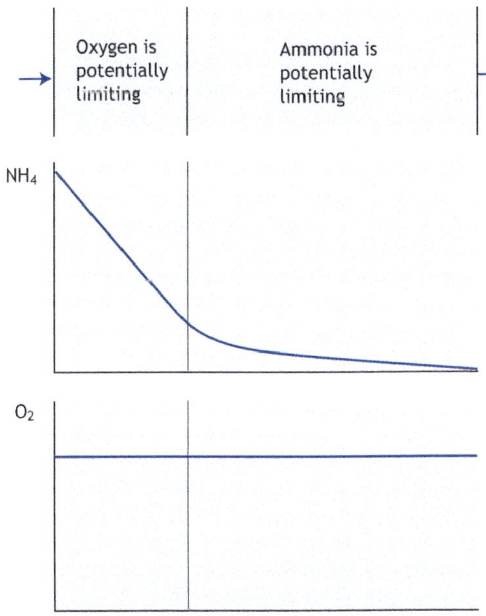

Figure 17.19 Nitrifying biofilm reactor with aeration resulting in constant bulk-phase oxygen concentrations. Depending on bulk-phase ammonium concentrations, the limiting compound will shift along the length of the reactor.

Once bulk-phase ammonium concentrations have been decreased to below 0.28 gN/gO_2 the limiting compound will be ammonium. The ammonium flux into the biofilm and observed removal will then become a function of bulk-phase ammonium concentrations.

Example 17.7 Competition by different groups of organisms for substrate and space within the biofilm

As shown in the previous example, different conversion processes can occur in different regions of the biofilm, which, in turn, can have an influence on the microbial communities which establish themselves in different layers of a biofilm. A classic example of how the availability of substrates and microbial growth rates influences microbial competition and substrate removal is the competition between heterotrophic and autotrophic bacteria in a biofilm for oxygen (Wanner and Gujer, 1985, 1986; Wanner and Reichert, 1996). Predicting the distribution of heterotrophic and autotrophic bacteria in different layers of the biofilm requires the modelling of local growth processes within the biofilm and can only be solved using numerical modelling. Some selected results from Wanner and Gujer (1985) are presented below.

Model prediction for oxygen and biomass profiles over the thickness of the biofilm are shown in Figure 17.20 for four different bulk-phase COD concentrations. A general observation from these simulations is that the faster-growing heterotrophic bacteria tend to overgrow the slower-growing autotrophic bacteria. This has two implications: (*i*) heterotrophic growth and COD removal are only to a very limited extent influenced by autotrophic growth, and (*ii*) autotrophic growth and ammonium oxidation are significantly influenced by the extent of oxygen utilization by the heterotrophic biomass because autotrophic bacteria rely on oxygen passing through the layer of heterotrophic bacteria.

From the biomass distributions in Figure 17.20 it can be seen that heterotrophic and autotrophic bacteria can only coexist with bulk-phase COD concentrations lower than 30 mg/l. This can be explained by the concept of dual substrate limitation (Section 17.8). For bulk-phase COD concentrations of 30 mg/l, the heterotrophic biofilm can be expected to be oxygen-limited (compare Table 17.4) and no oxygen will be available for autotrophic growth below the heterotrophic layer. Thus, coexistence of heterotrophic and autotrophic bacteria is possible only if the oxidation of organic substrate is COD rather than oxygen-limited. Substrate fluxes for a range of bulk-phase COD and ammonium concentrations are shown in Figure 17.21.

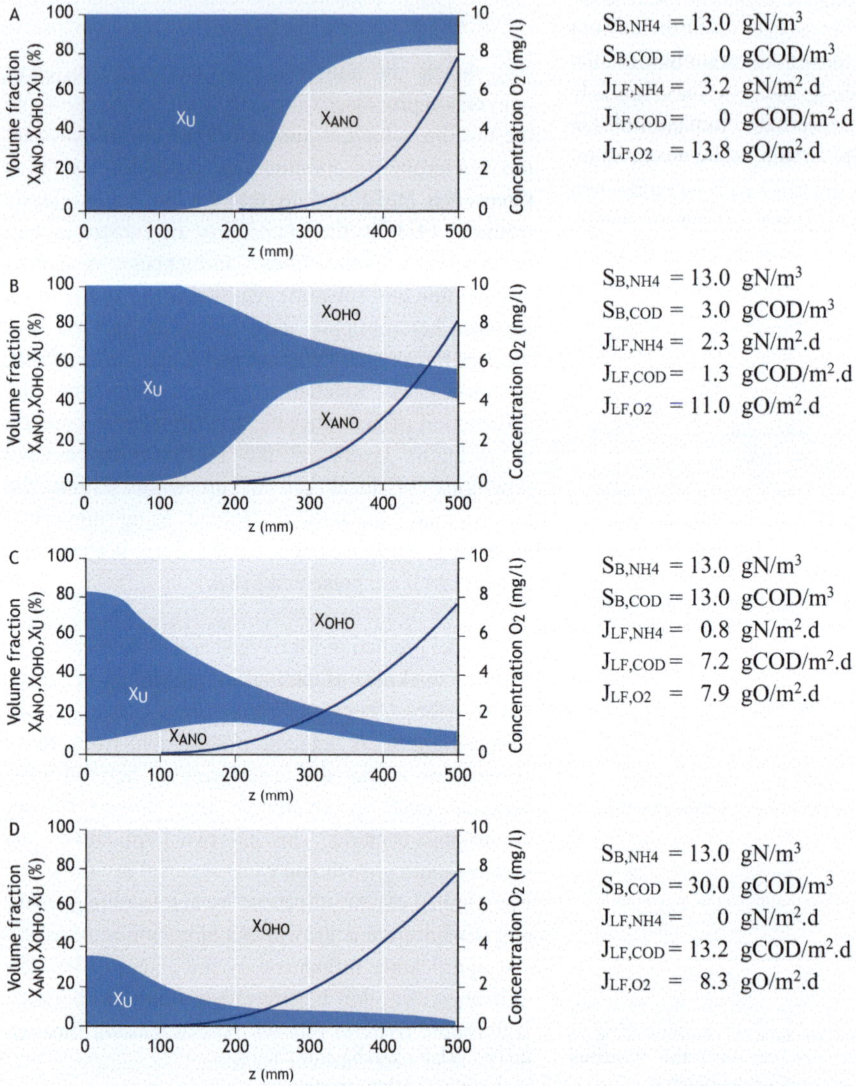

Figure 17.20 Concentration of oxygen over the depth of a biofilm and corresponding distribution of heterotrophic bacteria (X_{OHO}), autotrophic bacteria (X_{ANO}), and inert biomass (X_U) for different bulk-phase organic substrate concentrations of 0 (A), 3 (B), 13 (C), and 30 mg COD/l (D). The bulk-phase oxygen and ammonia concentrations are 8 mg O_2/l and 13 mg N/l, respectively, for all cases (from Wanner and Gujer, 1985).

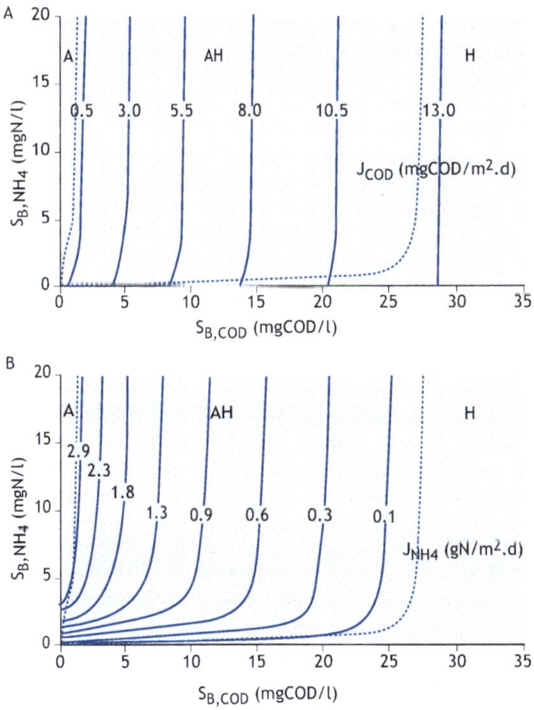

Figure 17.21 Flux of organic substrate (A) and ammonia (B) are both mainly determined by the bulk-phase concentration of the organic substrate for bulk-phase oxygen concentrations of 8 mg/l (Wanner and Gujer, 1985). Dashed lines in the figures separate different regions with only heterotrophic growth (H), only autotrophic growth (A), or coexistence of heterotrophic and autotrophic growth (AH).

It can be seen that the flux of COD into the biofilm increases with increasing bulk-phase COD concentrations until COD removal is oxygen-limited at bulk-phase COD concentrations of 28 mgCOD/l. It is interesting to note that the flux of ammonium into the biofilm is also controlled by bulk-phase COD concentrations rather than by bulk-phase ammonium concentrations (at least for bulk-phase ammonium concentrations higher than 5 mgN/l). This major influence of bulk-phase COD concentrations on ammonium fluxes is due to the fact that the amount of oxygen passing through the heterotrophic layer will decrease with increasing bulk-phase COD concentrations, leaving less and less oxygen for the autotrophic bacteria. Whether a biofilm that is oxidizing organic carbon will be oxygen or organic carbon-limited can be determined using Eq. 17.64.

As shown in Figure 17.22, the value of $\gamma_{e.d.,e.a.}$ in Eq. 17.64 can be directly correlated with the oxygen within the biofilm available for nitrification and the observed nitrification rate.

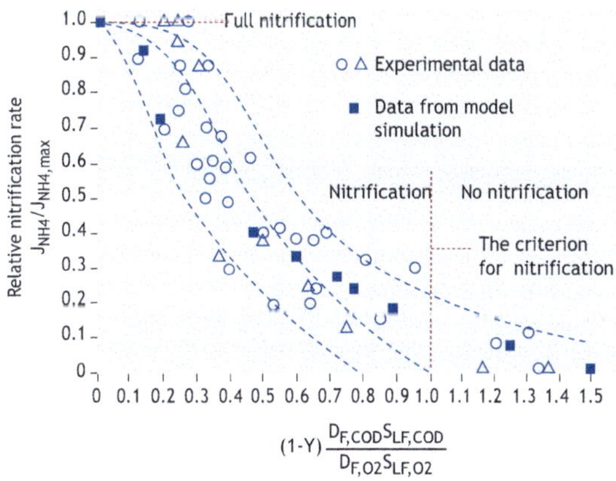

Figure 17.22 Experimental data demonstrating a reduced nitrification rate with increased bulk-phase organic substrate concentrations in a biofilter. The x axis corresponds to $\gamma_{e.d.,e.a.}$ in Eq. 17.72 (Henze *et al.*, 2002).

Example 17.8 How is heterotrophic/autotrophic competition considered in design guidelines?

The competition between heterotrophic and autotrophic bacteria in the previous example translates directly into a design approach and the design curves presented in Figure 18.12. In the previous example, (Figure 17.21) it was demonstrated that nitrification and the corresponding ammonia flux are directly dependent on bulk-phase COD and heterotrophic growth. Microbial competition and substrate fluxes are further influenced by bulk-phase oxygen concentrations. The design curves in Figure 18.12A take into account the influence of bulk-phase organic substrate, oxygen, and ammonia as the basis to predict the ammonia flux. While bulk-phase oxygen and ammonia are explicitly included, organic loading is used as a proxy of bulk-phase COD concentrations. As shown in Section 17.4, bulk-phase organic substrate concentrations are directly linked to influent organic substrate loadings (*e.g.* Eq. 17.44).

Figure 18.12B in Chapter 18 provides ammonia fluxes for a situation with low organic loadings (e.g. tertiary nitrification). The reader should compare the increase of ammonia fluxes to the point where oxygen becomes limiting with half-order kinetics for the partially penetrated biofilm followed by zero-order kinetics in Figure 17.8.

The design curves in Figure 18.12 are in line with the theoretical principles of competition and multi-component diffusion, but the numerical values in the figure are in fact determined from practical reactor operation. Application of these design curves is further discussed in Chapter 18.

Example 17.9 What are the effects of heterotrophic/autotrophic competition in a plug-flow biofilm reactor?

Numerical modelling can be used to evaluate factors limiting the reactor operation and to evaluate strategies to improve performance. This example evaluates the performance of a biofilm reactor operated for carbon and ammonia oxidation (Table 17.7). Mixing conditions in this reactor are described as eight stages where each stage could represent a compartment in a moving-bed biofilm reactor (MBBR) or a section of a biologically-aerated filter (BAF) (Figure 17.23A)[6]. The goal is to achieve effluent ammonia concentrations of less than 1 mgN/l. You are concerned that the reactor is overloaded and you are evaluating four cases (Table 17.8).

Reference case

In the reference case the effluent ammonia concentration significantly exceeds the target value of 1 mgN/l (Table 17.8). This is not surprising as the organic surface loading of the overall system of 9 gCOD/m^2.d is much higher compared to design recommendations (see Section 18.1.1.6). The process performance can be further evaluated based on bulk-phase COD and NH_4^+ concentrations (Figure 17.24A). The initial sections of the biofilm reactor only oxidize organic carbon and the nitrifiers are outcompeted. The observed ammonia removal in the initial sections is due to the nitrogen requirement for cell synthesis. Nitrification starts only in the third section as is evident from the increase in bulk-phase nitrate concentrations.

Reduced surface loading

One approach to improve the reactor performance is to reduce the organic surface loading by increasing the reactor volume (maintaining the specific surface area) or increasing the specific surface area in the given reactor volume (in an MBBR this can be achieved by adding more carriers). Decreasing the surface loading from 9 to 7.5 gCOD/m^2.d increases nitrification and allows the target effluent ammonia concentrations to be met (Figure 17.24B).

Increased aeration

Performance of an overloaded biofilm reactor can be improved by increasing the bulk-phase oxygen concentrations. Increasing the bulk-phase oxygen concentrations from 4 to 6 mgO_2/l allows the effluent ammonia concentrations to meet the target effluent ammonia concentrations (Figure 17.24C). The advantage of increasing bulk-phase oxygen concentrations is that this change in operation can be made during operation if there is sufficient aeration capacity, whereas reducing surface loading would require retrofitting the plant. Increasing aeration is not only used to improve the performance of overloaded biofilm reactors but also for the operator to respond to variable influent loading.

Sludge recirculation

Some types of biofilm reactors (*e.g.* MBBR) can be operated with or without sludge retention (Figure 17.23 and 18.11). An MBBR with sludge retention is referred to as an integrated fixed-film activated sludge (IFAS) process (see Chapter 18 for details).

[6] The SUMO file used for these simulations is available at https://doi.org/10.25678/00022X.

Introducing sludge retention even with a very low bulk-phase solids retention time of 1 d allows the heterotrophic bacteria to accumulate in the bulk phase and to degrade organic carbon. Bulk-phase organic carbon oxidation reduces the organic surface loading of the biofilm and improves nitrification so that effluent target ammonia concentrations are achieved (Figure 17.24D).

The purpose of this example is to demonstrate the power of numerical modelling to evaluate different scenarios in terms of overall reactor operation. The discussion focused on bulk-phase COD and NH_4^+ concentrations. However, the model also provides oxygen concentration profiles (Figure 17.25) in different sections of the reactor.

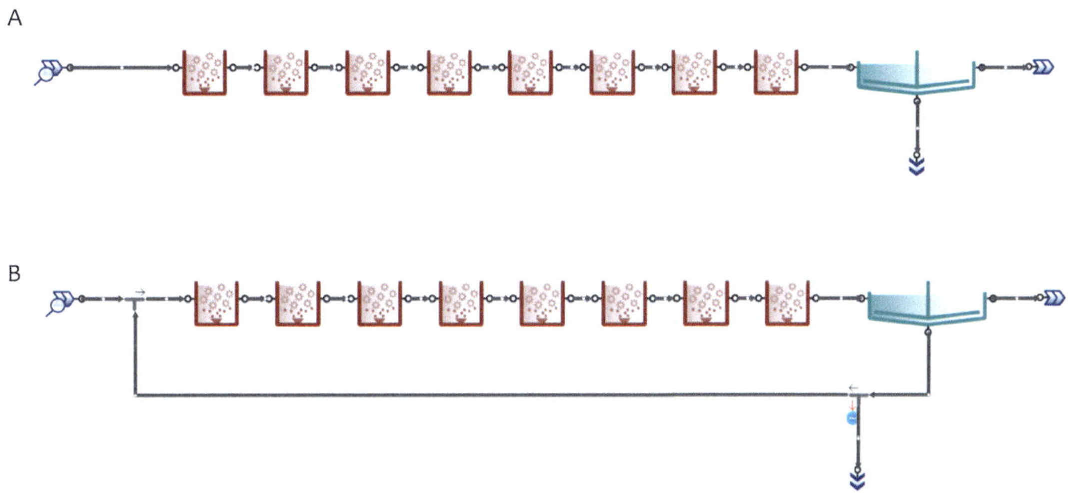

Figure 17.23 A biofilm reactor is modelled as eight biofilm compartments in series without biomass recycle (A) and with biomass recycle (B).

Table 17.7 Operating conditions for the biofilm reactor (base case).

Parameter	Symbol	Value
Influent COD concentration	COD_i	300 mgCOD/l
Influent NH_4^+ concentration	S_{NH4}	45 mgN/l
Influent flow rate	Q_i	24,000 m^3/d
Overall reactor volume	V_R	3,200 m^3 (= 8·400 m^3)
Specific biofilm surface area	a_F	250 m^2/m^3
Total biofilm surface area	A_F	800,000 m^2 (= 8·100,000 m^2)
Biofilm thickness	L_F	400 µm
Boundary layer thickness	L_L	30 µm
Number of layers in a biofilm	n	6
Bulk-phase oxygen concentration	$S_{B,O2}$	4 mgO$_2$/l
COD surface loading	$B_{A,COD}$	9 gCOD/m^2.d
NH_4^+ surface loading	$B_{A,NH4}$	1.4 gN/m^2.d

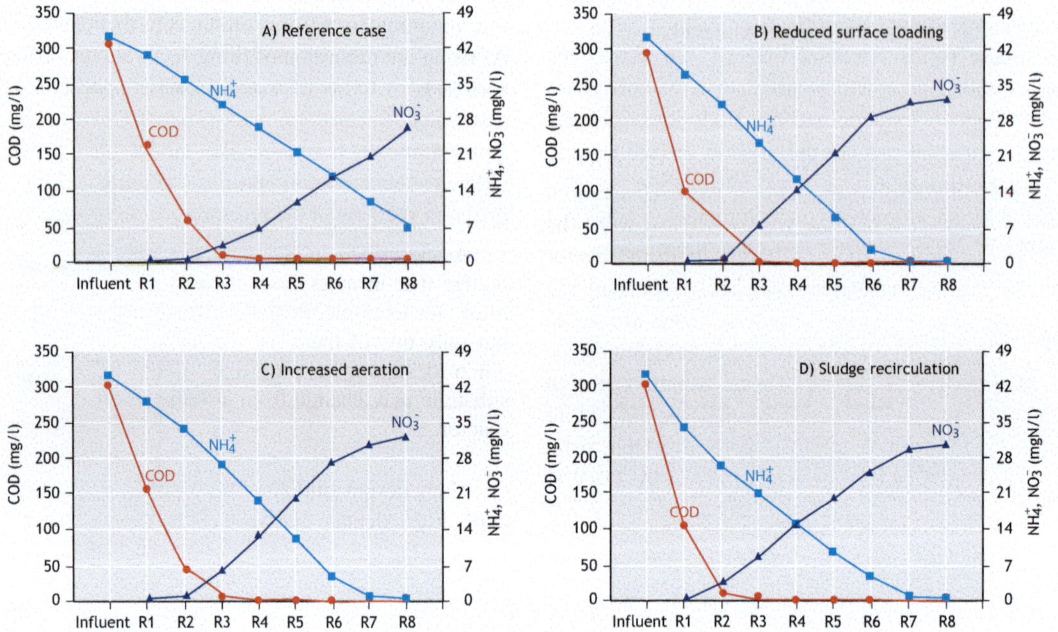

Figure 17.24 Bulk-phase concentrations of COD, ammonia, and nitrate for the four scenarios described in Table 17.8.

Table 17.8 Operating conditions and effluent ammonia concentrations for the evaluated scenarios.

Scenario	Organic surface loading[1)] ($gCOD/m^2.d$)	Bulk phase O_2 (mgO_2/l)	SRT in bulk phase[2)] (d)	Effluent NH_4^+ (mgN/l)
Reference case	9.0	4.0	-0.1	6.7
Reduced surface loading	7.5	4.0	-0.1	0.1
Increased aeration	9.0	6.0	-0.1	0.1
Sludge recirculation	9.0	4.0	1.0	0.2

[(1)] The organic surface loading does not consider bulk-phase removal of organics in the scenario with sludge recirculation.
[(2)] The solids retention time for the system with sludge recirculation (Figure 17.23B) is calculated as the mass of suspended solids in the bulk phase divided by the rate of solids removal from the system. For a biofilm reactor without sludge recycle (Figure 17.23A) the hydraulic retention time is provided.

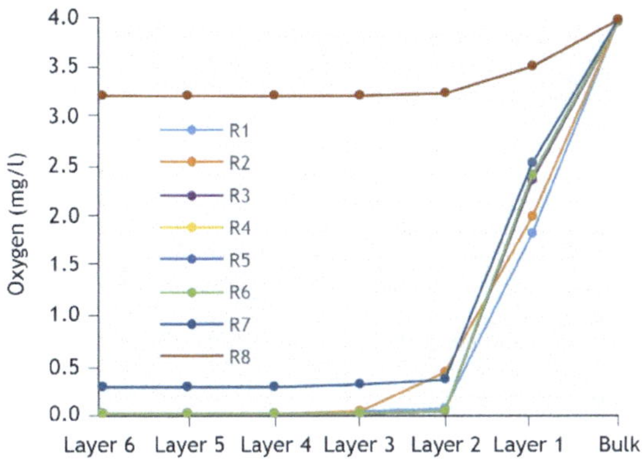

Figure 17.25 Oxygen concentration profiles in the bulk and in the six layers of the biofilm (x axis) for the eight biofilm compartments (different colour curves).

17.8.2 Step-by-step approach to evaluating biofilm reactors

The previous examples demonstrate how biofilm reactor can be dissected into different sections where limiting factors can be different in individual sections. It is beneficial for the engineer to develop a conceptual understanding of the limiting factors and underlying mechanisms. This conceptual understanding is necessary for applying simplified analytical solutions but is equally important when using readily available numerical models. When using numerical models a conceptual understanding is often not required to get the numbers right, but it is essential in order to ask the right questions and to interpret the modelling results. A stepwise conceptual approach is presented in Figure 17.26 and provides a sequence of individual decisions and corresponding modelling approaches. A first step is to determine whether the biofilm reactor can be described using a single process (*e.g.* only carbon oxidation or only nitrification) or whether multiple processes and multiple types of bacteria must be considered (Step 1). In the case of a single process the next question is whether the electron donor or electron acceptor are limiting (Step 2). For a single process and a single compound (*i.e.* either the electron donor or acceptor), analytical solutions, *e.g.* from Table 17.3, can be applied. An alternative approach is to use a numerical model that explicitly takes into account multi-component diffusion. Complex systems with multiple processes can be dissected into different sections where each section is dominated by a single process. For example, a biofilm reactor that is used for combined carbon oxidation and nitrification can be considered as two sections; mainly carbon oxidation in the first section and mainly nitrification in the second section (Step 5). As an alternative, microbial competition as discussed in examples 17.8 and 17.9 can be used to describe interactions. And, again, numerical modelling that takes into account multiple substrates, processes, and biomass fractions can be used to describe the overall system. The final decision is whether or not to include bulk-phase processes (Step 4). Bulk-phase conversion is usually neglected in pure biofilm processes. However, bulk-phase processes are essential in hybrid systems that combine biofilms with activated sludge (*i.e.* integrated fixed-film activated sludge or IFAS). Again, subsystems can be defined that can be solved using simplified analytical solutions or the overall system can be evaluated using numerical modelling that takes into account the processes, organisms, and interactions.

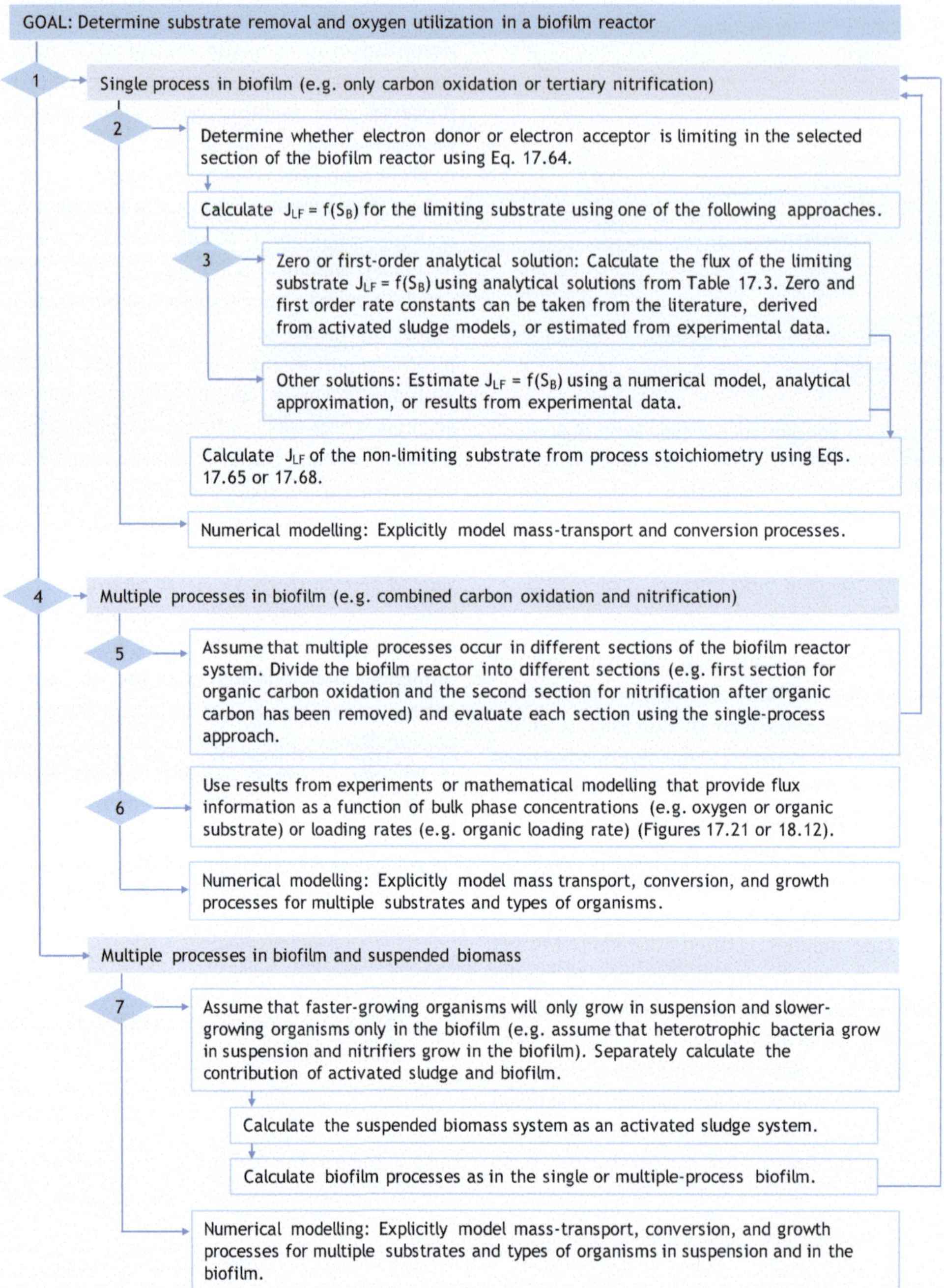

Figure 17.26 Stepwise approach to modelling biofilm reactors. Decision points for steps (numbered) are discussed in the text.

17.9 DERIVED PARAMETERS

17.9.1 Solids retention time

For activated sludge treatment plants, the sludge (solids) retention time (SRT) is a key parameter in their design and operation. The solids retention time can be used to calculate effluent substrate concentrations, the amount of biomass in the system, and the net yield. However, to what extent can this concept of a solids retention time be translated into biofilm systems? In an activated sludge system, solids removal is a stochastic process where random flocs are removed from the system in sludge wastage and in the secondary effluent. The solid retention time then represents the average time a particle remains in the system. However, biomass removal in a biofilm is not a random process as detachment preferentially removes particles from the surface of the biofilm while particles at the base of the biofilm are more protected from detachment. Thus, the concept of a solids retention time cannot be directly applied to biofilms.

For biofilms with multiple groups of organisms competing for substrate and space within the biofilm, the different retention times resulting from preferential detachment at the biofilm surface have a significant influence on microbial competition and calculating an average solids retention time is not useful (Morgenroth, 2003; Morgenroth and Wilderer, 2000). However, for a homogeneous biofilm with only a single type of organism, calculating solid retention times provides a useful comparison with growth conditions in activated sludge systems. The average solids retention time in biofilm systems is defined as:

$$SRT = \frac{\text{average mass of biofilm}}{\text{average rate of biofilm detachment}} = \frac{L_F \cdot X_F}{u_{d,S} \cdot X_F} \tag{17.81}$$

where the right hand side of Eq. 17.81 is equal to the volumetric removal rate due to biofilm detachment ($u_{d,V}$, Eq. 17.71). The definition of SRT can be combined with the mass balance for the overall biofilm development (Eq. 17.70). Assuming steady state, Eq. 17.70 can be rearranged to:

$$u_{d,S} = \frac{Y \cdot J_{LF}}{X_F} - b_{ina} \cdot L_F \tag{17.82}$$

which can be substituted into Eq. 17.81 resulting in:

$$SRT = \frac{L_F}{\dfrac{Y \cdot J_{LF}}{X_F} - b_{ina} \cdot L_F} = \frac{1}{\dfrac{Y \cdot J_{LF}}{X_F \cdot L_F} - b_{ina}} \tag{17.83}$$

From Eq. 17.83 it can be seen that SRT increases with decreasing substrate flux and increasing biofilm thickness. Indirectly, SRT is influenced by bulk-phase concentrations (decreasing bulk-phase concentrations results in decreased substrate fluxes) and biofilm detachment (decreased detachment results in thicker biofilms). It should again be noted that Eq. 17.83 is based on the assumption of a steady state where all biomass growth is balanced by detachment.

The reader should also note that SRT has to be interpreted differently in activated sludge and in biofilm systems. In activated sludge systems, every floc has the same probability of being removed during sludge wastage. However, in biofilms, the probability for removal through detachment is significantly higher at the surface of the biofilm compared to the base of the biofilm. The value of SRT calculated from Eq. 17.83 will provide the overall average solids retention time while it is also possible to calculate solids retention times for different locations within the biofilm (Morgenroth, 2003; Morgenroth and Wilderer, 2000). Localized solids retention times at the base of the biofilm will be higher than the overall average, providing an ecological niche for slower-growing bacteria. One example of slower-growing bacteria developing preferentially towards the base of the biofilm is nitrifying bacteria, as discussed in the example in Section 17.8. In addition, it has also been suggested that the base of the biofilm provides an ecological niche for specialized organisms responsible for the degradation of slowly-biodegradable xenobiotic compounds.

Example 17.10 Solids retention time for a biofilm assuming a Monod-order growth rate

Simulations were performed using AQUASIM to calculate the SRT for different bulk-phase substrate concentrations and biofilm thicknesses based on simulated detachment rates ($u_{d,S}$) and using Eq. 17.81. In Figure 17.27 it can be seen that SRT increases with decreasing bulk-phase substrate concentrations. The different lines in Figure 17.16 represent different biofilm thicknesses where SRT are higher for larger biofilm thicknesses. The effects of bulk-phase substrate concentrations, biofilm thickness, and detachment rate coefficients are similar to those covered by the general Eq. 17.81 above.

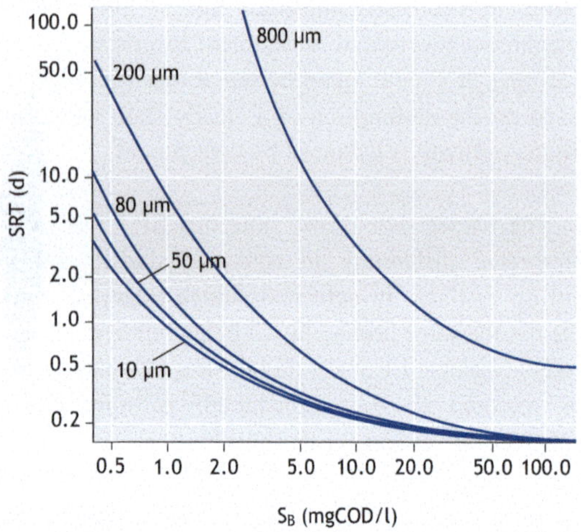

Figure 17.27 Influence of bulk-phase substrate concentrations and biofilm thickness (lines in plot) on the SRT calculated using Eq. 17.81, assuming a constant thickness biofilm and including the processes of growth, decay, and detachment. Substrate removal and growth in the biofilm was modelled using Monod kinetics (Table 17.2) and the parameters provided in Table 17.13.

17.9.2 Lowest effluent substrate concentration supporting biomass growth (S_{min})

Effluent substrate concentrations in a biofilm reactor depend on the availability of a sufficient amount of biomass, on the biofilm surface area, and on the reactor operating conditions. Effluent substrate concentrations for a CSTR-type (continuous stirred-type reactor) biofilm reactor were discussed in Section 17.4. In Section 17.4, however, a fixed biofilm thickness was always provided as the model input regardless of biofilm detachment and decay processes. The question addressed in the following discussion is the lowest effluent concentration that can be achieved in a CSTR-type biofilm reactor, assuming the following conditions: (*i*) biofilm detachment is negligible ($u_{d,S} = 0$), (*ii*) growth in the biofilm can be described using Monod kinetics, (*iii*) decay is described using a first-order decay ($b_{ina}X_F$), and (*iv*) influent flow rates are extremely low resulting in a very thin biofilm, where mass-transfer limitations can be neglected. Neglecting detachment and assuming steady state, Eq. 17.70 can be simplified to:

$$0 = \frac{Y \cdot J_{LF}}{X_F} - b_{ina} \cdot L_F \tag{17.84}$$

The substrate flux into a biofilm neglecting mass-transport limitations and assuming Monod is:

$$J_{LF} = \frac{1}{Y} \cdot \mu_{max} \cdot \frac{S_{LF}}{K_S + S_{LF}} X_F \cdot L_F \tag{17.85}$$

Combining Eq. 17.84 and Eq. 17.85 results in the minimum substrate concentration supporting microbial growth in the biofilm (S_{min}):

$$S_{min} = \frac{K_S \cdot b_{ina}}{Y \cdot \mu_{max} - b_{ina}} \tag{17.86}$$

The concept of S_{min} was shown experimentally (Rittmann and McCarty, 1980), has practical implications for the removal of contaminants to very low concentrations (Rittmann, 1982a), and the value of S_{min} can be used as a scaling factor for the derivation of pseudo-analytical solutions (Rittmann and McCarty, 2001; Saez and Rittman, 1992; Wanner *et al.*, 2006).

17.9.3 Characteristic times and non-dimensional numbers to describe biofilm dynamics

The various processes in biofilms occur at very different timescales. Biomass growth occurs at timescales in the order of hours to days while substrate diffusion and hydrodynamic processes are in the order of seconds to minutes (Esener *et al.*, 1983; Gujer and Wanner, 1990; Kissel *et al.*, 1984; Picioreanu *et al.*, 2000). This concept of characteristic times is useful when evaluating how fast a system approaches steady state, as a basis for defining non-dimensional parameters, and when implementing numerical solutions to model biofilms.

The concept of characteristic times can be explained using a simple example. For a first-order reaction rate, the mass-balance equation for the degradation of a substrate S in a batch process is:

$$\frac{dS}{dt} = -k_1 \cdot S \tag{17.87}$$

where S is the substrate concentration, M L^{-3}, k_1 is the first-order reaction rate, T^{-1}, and t = time, T. Solving Eq. 17.87 with $S(t = 0) = S_0$ yields:

$$\frac{S(t)}{S_0} = e^{-k_1 \cdot t} \tag{17.88}$$

Based on Eq. 17.88 the characteristic time for a first-order reaction rate ($\tau_{reaction,1}$) can be defined as:

$$\tau_{reaction,1} = \frac{1}{k_1} \tag{17.89}$$

Note that for a first-order reaction rate this selection of the characteristic time results in $S(t{=}\tau) = \exp(-1) \cdot S_0 = 36.8\% \cdot S_0$. However, there is nothing special about the value of 36.8% (there are others); it is simply mathematically convenient (Clark, 1996). For example, the characteristic time for a zero-order reaction rate $\tau_{reaction,0}$ corresponds to $S(t{=}\tau) = 0$ in a batch reaction. The timescale is simply a measure of how fast a process will proceed. In Table 17.9 characteristic times for processes of importance in the description of biofilms are summarized. A range of typical values is provided in Figure 17.28.

Table 17.9 Characteristic times for relevant processes in a biofilm (based on Clark, 1996; Esener *et al.*, 1983; Gujer and Wanner, 1990; Kissel *et al.*, 1984; Picioreanu *et al.*, 2000).

Process	Characteristic time
Advection	$\tau_{convection} = \frac{L}{u}$
Retention in completely mixed reactor	$\tau_{HRT} = \frac{V}{Q}$
Diffusion (mass)	$\tau_{diffusion,mass} = \frac{L^2}{D}$
Diffusion (viscous)	$\tau_{diffusion,viscous} = \frac{L^2}{\nu}$
Reaction	$\tau_{reaction} = \frac{S}{r_S}$ (general)
	$\tau_{reaction,0} = \frac{S}{k_0}$ (zero-order)
	$\tau_{reaction,1} = \frac{1}{k_1}$ (first-order)
	$\tau_{reaction,Monod} = \frac{Y}{\mu_{max}} \cdot \frac{(K_S + S)}{X_{OHO}}$ (Monod)
	$\tau_{growth} = \frac{1}{\mu_{max}}$ (maximum growth rate)
	$\tau_{growth}^* = \frac{S_{LF}}{\mu_{max} \cdot X_{LF}}$ (actual growth rate)
	$\tau_{decay} = \frac{1}{b_{decay}}$ (decay)
Detachment	$\tau_{detachment} = \frac{L_F}{u_{d,S}} = \frac{1}{u_{d,V}}$

L = characteristic distance [L]; u = velocity [L T^{-1}]; D = diffusion coefficient [L^2 T^{-1}]; ν = viscosity [L^2 T^{-1}]; V = volume of reactor [L^3]; Q = flow to the reactor [L^3 T^{-1}]; k_0, k_1 = zero and first-order volumetric reaction rates M L^{-3} T^{-1} and T^{-1}, respectively); b_{decay} = decay coefficient [T^{-1}]; μ, μ_{max} = growth rate, maximum growth rate [T^{-1}]; Y = yield coefficient [M M^{-1}]; S_{LF}, X_F = substrate, biomass concentration [M L^{-3}]; K_S = Monod half-saturation constant, [M L^{-3}].

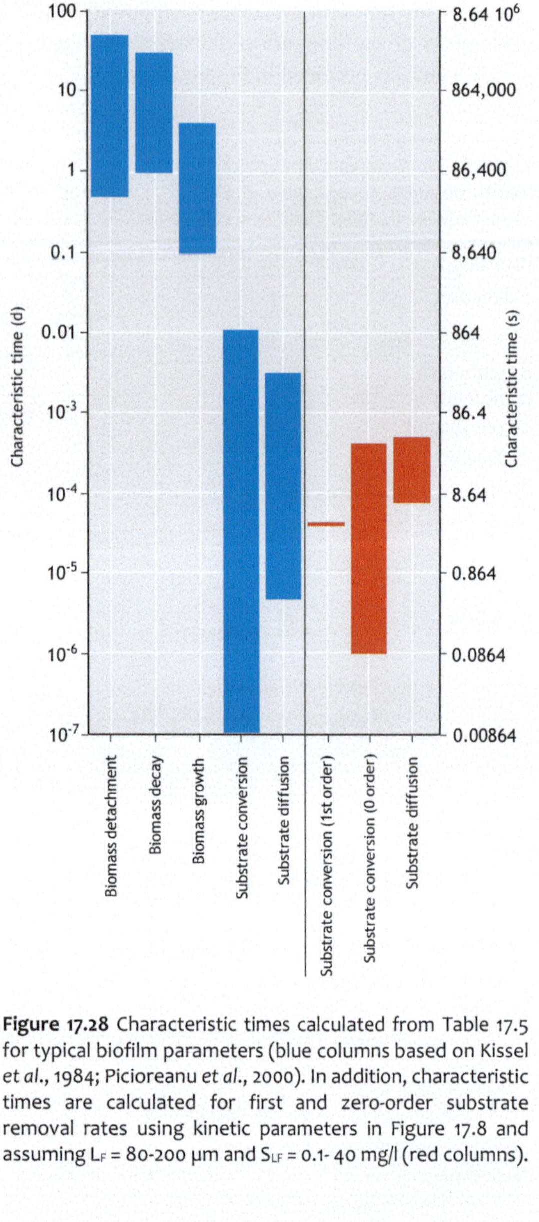

Figure 17.28 Characteristic times calculated from Table 17.5 for typical biofilm parameters (blue columns based on Kissel *et al.*, 1984; Picioreanu *et al.*, 2000). In addition, characteristic times are calculated for first and zero-order substrate removal rates using kinetic parameters in Figure 17.8 and assuming L_F = 80-200 μm and S_{LF} = 0.1- 40 mg/l (red columns).

As can be seen for the process of growth, there are different ways to define characteristic times based on the maximum growth rate (τ_{growth}) or an approximation of the actual growth rate in the system (τ_{growth}^*). However, without strict rules for the definition of characteristic times it is the responsibility of the user to understand the basis of how and why characteristic times or non-dimensional numbers (Section 17.9.3.2) are derived for order-of-magnitude information and to use his/her good judgment.

17.9.3.1. Application of characteristic times to estimate response times

When running a biofilm model, one has to define a timescale of interest (τ_0). All processes with much shorter timescales (*i.e.* faster processes) than τ_0 can be assumed to be at a pseudo steady state. Processes with much longer timescales (*i.e.* slower processes) can be described as 'frozen' in time (Picioreanu *et al*., 2000; Wanner *et al*., 2006), Thus, for experiments lasting a few hours, the substrate concentration profiles can be assumed to be at steady-state as $\tau_{diffusion} \ll \tau_0$. However, although changes in the biofilm thickness can be neglected as $\tau_{growth} \gg \tau_0$, when biofilm development is evaluated over periods of weeks then biofilm growth has to be considered explicitly, as τ_{growth} is of the same order as τ_0. An overview of characteristic times for biofilm systems in general and for the example in Figure 17.8 is provided in Figure 17.28.

Crank (1975) demonstrated that for diffusion of soluble substrate in a biofilm (or a plane sheet) the time required to approach steady state ($\tau_{steady\text{-}state}$) is given by:

$$\tau_{steady\text{-}state} = 0.45 \cdot \tau_{diffusion,mass} \qquad (17.90)$$

Assuming a diffusion coefficient of $D_F = 8 \cdot 10^{-5}$ m²/d (for acetate) and L_F = 500 μm we find using Eq. 17.72 that the time required to approach steady state is 4.5 min. Note that this time to reach steady state in the biofilm is independent of whether diffusion is limiting the substrate removal in the biofilm or not. Timescales for approaching steady state for microorganisms within the biofilm (τ_{growth} or τ_{growth}^*) are much longer ranging from days to weeks. Competition within a biofilm is an even slower process as it relies on the growth rate differential for organism A (μ_A) and organism B (μ_B) within the biofilm ($\tau_{\Delta\text{-}growth} = 1/(\mu_A - \mu_B)$).

17.9.3.2 Non-dimensional numbers:

Damköhler number (Da^{II}), Thiele modulus (Φ), and Growth number (G)

Comparing the values of characteristic times for coupled processes allows an estimation of which processes are rate-limiting and which processes can be neglected. One example is the coupling of the reaction and diffusion of the substrate in a biofilm. Assume, for example, that for a given biofilm the characteristic time for substrate conversion ($\tau_{reaction}$) is short (*i.e.* the reaction is very fast) while the characteristic time for diffusion ($\tau_{diffusion}$) is long (*i.e.* diffusion is slow). If diffusion is slower than reaction then the biofilm can be assumed to be diffusion-limited. For the case of zero-order substrate conversion in the biofilm ($r_F = k_{0,F}X_F$, Eq. 17.1) then the ratio of $\tau_{diffusion}$ and $\tau_{reaction}$ in the biofilm is defined as:

$$\frac{\tau_{diffusion}}{\tau_{reaction}} = \frac{L_F^2}{D_F} \cdot \frac{k_0 \cdot X_F}{S_{LF}} \quad (17.91)$$

In the chemical engineering literature this ratio of characteristic times in Eq. 17.91 is referred to as the second Damköhler number (Da^{II}) (Boucher and Alves, 1959):

$$Da^{II} = \frac{L_F^2}{D_F} \cdot \frac{k_0 \cdot X_F}{S_{LF}} \quad (17.92)$$

For a diffusion-limited (deep) biofilm $Da^{II} >> 1$; for a reaction-limited (shallow) biofilm $Da^{II} << 1$. By comparing Da^{II} with the explicit solution of a biofilm with zero-order kinetics (Table 17.3), it can be seen that Da^{II} is directly related to the penetration of substrate into the biofilm (β):

$$\beta = \sqrt{\frac{2}{Da^{II}}} \quad (17.93)$$

Based on Eq. 17.93 a diffusion-limited biofilm with $Da^{II} >> 1$ corresponds to $\beta << 1$. This makes sense as a diffusion-limited biofilm is only partially penetrated. Note that using characteristic times we were able to evaluate whether diffusion or reaction was the limiting process without explicitly solving detailed differential equations. The second Damköhler number is related to another non-dimensional number, the Thiele modulus (Φ) that is defined as:

$$\Phi = \sqrt{Da^{II}} = \frac{\sqrt{2}}{\beta} \quad (17.94)$$

Picioreanu *et al.* (1998) introduced the Growth number (G) in their study evaluating the influence of local substrate diffusion and growth rates on the development of the multi-dimensional biofilm structure. The definition of G is:

$$G = \frac{\tau_{diffusion}}{\tau_{growth}^{*}} \quad (17.95)$$

where τ_{growth}^{*} is defined as (Table 17.9):

$$\tau_{growth}^{*} = \frac{S_{LF}}{\mu_{max} \cdot X_{LF}} \quad (17.96)$$

Combining Eq. 17.95 with Eq. 17.96 and the definition of $\tau_{diffusion}$ in Table 17.9 results in:

$$G = \frac{L_F^2}{D_F} \cdot \frac{\mu_{max} \cdot X_{LF}}{S_{LF}} \quad (17.97)$$

Note that G is identical to Da^{II} except for using the maximum growth rate (μ_{max}) instead of the zero-order substrate removal rate ($k_{0,F}$). Picioreanu *et al.* (1998) were able to demonstrate using multi-dimensional mathematical modelling that for substrate-limited growth in the biofilm and for large values of the G number (*e.g.* $G > 20$), a porous biofilm develops with many channels and voids. For growth-limited biofilms and for small values of the G number (*e.g.* $G < 7$), compact and dense biofilms develop.

Two lessons can be learned from the different non-dimensional numbers introduced above (Da^{II}, β, Φ, and G). First, there is no unique method to define these non-dimensional numbers as all four numbers relate substrate diffusion with substrate metabolism in a biofilm but each of the four numbers is defined slightly differently. Second, non-dimensional numbers can serve multiple purposes from predicting substrate diffusion into a biofilm, to differentiating diffusion and reaction-limited regimes, and predicting biofilm structure.

Biot number (Bi)

Another example of a non-dimensional number that can be derived from ratios of characteristic times is the Biot number (Bi):

$$Bi = \frac{\tau_{diffusion,internal}}{\tau_{diffusion,external}} = \frac{L_F^2}{D_F} \cdot \frac{D_W}{L_L^2} \qquad (17.98)$$

where $\tau_{diffusion,internal}$ (= L_F^2/D_F) is the characteristic time for diffusion inside the biofilm while $\tau_{diffusion,external}$ (= L_L^2/D_W) is the characteristic time for diffusion in the external concentration boundary layer. If diffusion within the biofilm is much faster than in the concentration boundary layer (*i.e.* Bi << 1) then, for a diffusion-limited biofilm, we can expect that external mass-transfer resistance is limiting the overall conversion processes within the biofilm.

Peclet number (Pe)

The Peclet number (Pe) can be used to compare the characteristic time for diffusion with the characteristic time for advection:

$$Pe = \frac{\tau_{diffusion}}{\tau_{advection}} = \frac{L^2}{D} \cdot \frac{u}{L} \qquad (17.99)$$

The Peclet number is often applied in reactor studies to evaluate the extent of axial dispersion (diffusion in the direction of the water flow) relative to the transport with the water flowing through the reactor. With Pe >> 1, diffusive transport is small compared to overall advection. When applying the Peclet number to biofilm reactors what characteristic length, L, should be used in Eq. 17.99? When comparing advective and diffusive transport along the length of the reactor, the characteristic length would be the length of the overall reactor. Another application of the Peclet number is to evaluate the relative importance of diffusion and advection in a porous biofilm that allows some water flow through the biofilm matrix (Libicki *et al.*, 1988). In this case the characteristic length would be the thickness of the biofilm. This example of two very different length scales in Eq. 17.99 demonstrates that there are always choices to make in defining characteristic times and non-dimensional numbers. No strict rules can be defined but making appropriate choices must be based on a thorough understanding of the underlying principles. Much abuse of non-dimensional parameters is based on a lack of understanding of how the processes that are being compared in terms of characteristic times are related to each other.

17.10 HOW DOES 2D/3D STRUCTURE INFLUENCE BIOFILM PERFORMANCE?

If you have ever taken a closer look at biofilms in reactors or natural systems then you will be aware that biofilms often look quite heterogeneous. You could ask yourself whether the one-dimensional biofilm models described in the previous sections will be suitable to describe real biofilms. All the previous sections have assumed that concentration gradients parallel to the substratum are much lower than concentration gradients perpendicular to the substratum, resulting in a one-dimensional diffusion reaction equation (Eq. 17.1).

In the last 25 years, multi-dimensional biofilm models have been developed to predict the formation of the heterogeneous biofilm structure depending on local substrate availability. In turn, these models predict the influence of this heterogeneous structure on the microbial ecology within the biofilm and subsequently on substrate fluxes. Whether or not a simplified 1D model is suitable or whether a multi-dimensional model needs to be applied will depend on the specific questions to be answered using the mathematical model. A systematic evaluation of different modelling approaches and their suitability for a range of different applications is provided in Wanner *et al.* (2006).

If the focus is on predicting substrate fluxes for processes such as aerobic carbon oxidation, nitrification, and denitrification then 1D models are an excellent choice. These 1D models allow mass-transport limitations to be predicted for a limiting substrate, multi-component diffusion, and microbial competition based on layering the faster-growing heterotrophic and slower-growing autotrophic biomass.

Situations where using a multi-dimensional model is justified include (*i*) complex substratum geometry (*e.g.* micro-scale studies of biofilm development in porous media) and for biofilm spatial distribution (*e.g.* 'patchy' biofilm growth with isolated colonies); (*ii*) complex ecological questions that are influenced by local aggregation of different groups of microorganisms and local mass transport. Examples of such ecological questions include modelling of the anaerobic food web in anaerobic granular sludge (Batstone *et al*., 2002). Another ecological question is the formation of different surface structures in biofilms and granular sludge, where the surface morphology can have, for example, significant influence on the external mass-transfer resistance (Picioreanu *et al*., 2000). Multi-dimensional models are typically used to answer specific questions that are used to gain a better understanding of interactions within the biofilm, and in some cases, these interactions within the biofilm have a significant impact on the overall biofilm reactor performance. In these cases the results and mechanisms from such multi-dimensional models then also need to be taken into account for practical biofilm reactor applications.

Example 17.11 2D distribution of hydrogen-producing and utilizing microorganisms in microbial aggregates

Anaerobic digestion is a multi-step process carried out by a variety of microorganisms that live in a symbiotic relationship. Local concentrations of substrates and intermediates depend on the spatial distribution of these microorganisms. One crucial aspect is the interspecies hydrogen transfer between hydrogen-producing and hydrogen-utilizing microorganisms. Batstone *et al.* (2006) applied a 2D mathematical model to evaluate the development of the spatial organization of microorganisms within anaerobic granules. Spatial distributions of organisms in an actual granule are compared to model simulations in Figure 17.29.

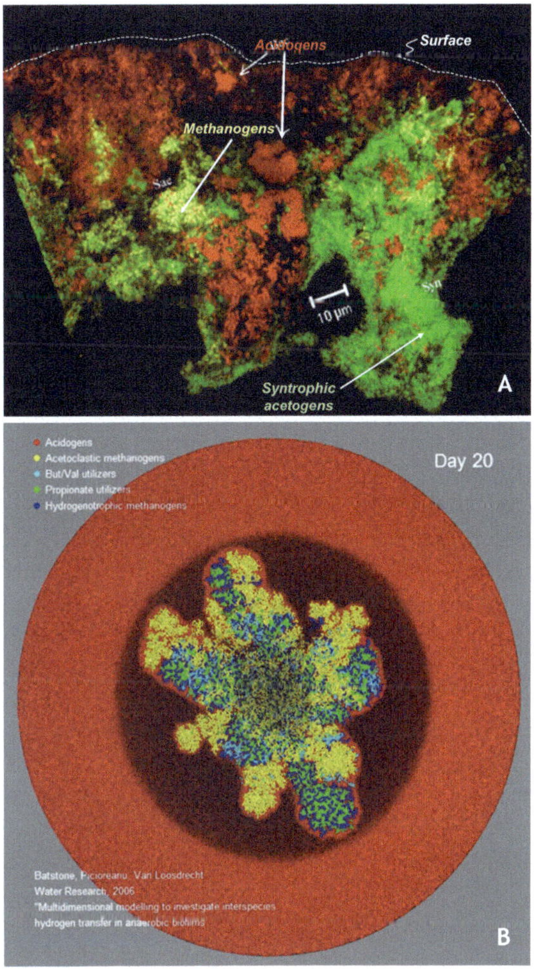

Figure 17.29 Spatial distribution of acidogenic bacteria (red) and methanogens (green) in an anaerobic granule image using fluorescence in-situ hybridization (A). Modelled distribution of acetogens (red), butyrate/valerate utilizers (light blue), propionate utilizers (green), and methanogens (yellow or blue) (B). Images: Batstone *et al.* (2006).

Batstone *et al.* (2006) were able to confirm measured syntrophic co-location of hydrogen producers and utilizers with their modelling results. Microorganism distribution in their model is based on local growth combined with random spreading of biomass without assuming any microbiological-based mechanisms (*e.g.* through chemotaxis). Evaluating potential mechanisms resulting in specific spatial patterns requires a multi-dimensional model. Such a cell-to-cell neighbourhood is not defined in 1D models where biomass distributions are averaged in planes parallel to the substratum.

17.11 MODEL PARAMETERS

Choosing appropriate values for model parameters is essential for obtaining reliable model predictions. Parameters in biofilm models can be classified into three categories: (*i*) microbial parameters that are independent of the form of growth: suspended culture or biofilm (*e.g.* Y, μ_{max}, etc.), (*ii*) mass transport-related parameters that are system-independent (*e.g.* D_W), and (*iii*) biofilm parameters that depend on the type of reactor and reactor operation (*e.g.* L_F, L_L, X_F, etc.). For the application of mathematical models for practical biofilm systems, only limited measurements are available. Direct in-situ measurements of biofilm parameters and representative sampling of biofilms is difficult, and biofilms are not homogeneously distributed over the entire system. Thus, it is usually not possible to identify all system-dependent model parameters based on measurements and therefore parameters need to be assumed from literature information. A discussion of some relevant model parameters is presented below.

17.11.1 Biofilm biomass density (X_F)

Biomass densities in biofilm reactors are significantly higher compared with activated sludge systems. Concentrations of biomass in the biofilm (X_F) are typically in the range of X_F = 10 to 60 gVSS/l. Rittmann *et al.* (2018) provide typical values of X_F for different types of conversion processes:

X_F = 20 to 30 gVSS/l for carbon oxidation
= 40 to 60 gVSS/l or nitrification, tertiary denitrification with methanol, or denitrification on membranes with H_2.

A direct comparison of biomass concentrations in a biofilm reactor with biofilm biomass densities in activated sludge systems ranging from 2 to 6 gVSS/l depends on the relative amounts of biofilm, water, air, and support medium, which can vary significantly between different types of biofilm systems. The value of X_F depends on growth conditions. Biofilms grown under increased shear typically result in denser and stronger biofilms. The biofilm biomass density can be readily measured from the biofilm volume and solids measurements of the biomass as described, for example, in Horn and Hempel (1997). A note for AQUASIM users: in AQUASIM, the biofilm is described as consisting of different phases. In this case, the density of the solid biofilm phase can be calculated from the overall biofilm biomass density divided by the total volume fraction of the solids ($= \Sigma\varepsilon_{S,i}$).

17.11.2 Diffusion coefficients (D_W, D_F)

Diffusion coefficients in water (D_W) are available for most compounds in the literature (Lide, 2008; Perry and Green, 2008). These references also provide approximations for compounds that are not listed, based on properties such as molecular weight. Diffusion coefficients within the biofilm (D_F) are smaller than diffusion coefficients in water depending to some extent on the biofilm biomass density and also the properties of the compound. A reduction factor of 80% is typically used (Horn and Morgenroth, 2006; Stewart, 1998):

$$D_F = 0.8 \cdot D_W \qquad (17.100)$$

The reduction of diffusion coefficients within biofilms cannot be readily measured within biofilm reactors but the approximation in Eq. 17.100 is appropriate in most cases. An overview of values for D_W is provided in Table 17.10.

Table 17.10 Diffusion coefficients in water (D_W) (for further references see Lide, 2008; Perry and Green, 2008; Stewart, 2003).

Compound	D_W (m^2/d)
Oxygen, O_2	$209.1 \cdot 10^{-6}$ (a)
Carbon dioxide, CO_2	$165.0 \cdot 10^{-6}$ (a)
Hydrogen, H_2	$441.5 \cdot 10^{-6}$ (a)
Hydrogen sulphide, H_2S	$117.5 \cdot 10^{-6}$ (a)
Hydrogen sulphide ion, HS^-	$149.6 \cdot 10^{-6}$ (a)
Methane, CH_4	$128.7 \cdot 10^{-6}$ (a)
Nitrogen, N_2	$172.8 \cdot 10^{-6}$ (a)
Nitrous oxide, N_2O	$222.0 \cdot 10^{-6}$ (a)
Ammonium, NH_4^+	$169.1 \cdot 10^{-6}$ (a)
Nitrite, NO_2^-	$165.2 \cdot 10^{-6}$ (a)
Nitrate, NO_3^-	$164.3 \cdot 10^{-6}$ (a)
Perchlorate, ClO_4^-	$154.8 \cdot 10^{-6}$ (a)
Chlorate, ClO_3^-	$148.6 \cdot 10^{-6}$ (a)
Chloride, Cl^-	$175.6 \cdot 10^{-6}$ (a)
$H_2PO_4^-$	$76.0 \cdot 10^{-6}$ (b)
HPO_4^{2-}	$65.7 \cdot 10^{-6}$ (b)
Hydroxide, OH^-	$455.6 \cdot 10^{-6}$ (a)
Hydron, H^+	$804.5 \cdot 10^{-6}$ (a)
Hydrogencarbonate, HCO_3^-	$102.4 \cdot 10^{-6}$ (a)
Carbonate, CO_3^{2-}	$79.7 \cdot 10^{-6}$ (a)
Acetate, CH_3COO^-	$94.1 \cdot 10^{-6}$ (a)
Glucose, $C_6H_{12}O_6$	$57.9 \cdot 10^{-6}$ (a)
Ethanol	$107.1 \cdot 10^{-6}$ (a)
Toluene	$73.4 \cdot 10^{-6}$ (a)
Methanol	$110.6 \cdot 10^{-6}$ (a)
Benzene	$88.1 \cdot 10^{-6}$ (a)
Soluble BOD (3-30 kDa)	$9.7 \cdot 10^{-6}$ (c)
Soluble BOD (30-50 kDa)	$7.3 \cdot 10^{-6}$ (c)
Soluble BOD (50-100 kDa)	$5.6 \cdot 10^{-6}$ (c)
Soluble BOD (100-500 kDa)	$4.3 \cdot 10^{-6}$ (c)
Soluble BOD (500-1,000 kDa)	$2.6 \cdot 10^{-6}$ (c)

(a) For 25 °C (Lide, 2008)

(b) For 25 °C (Stewart, 2003)

(c) Estimated for 20 °C using the Einstein relation linking molecular weight to D_W (Logan *et al.*, 1987; Polson, 1950)

17.11.3 External mass transfer (L_L, R_L)

The external mass-transfer boundary layer thickness (L_L) or the external mass-transfer resistance ($R_L = L_L/D_W$) depends on the reactor configuration, operating conditions, and biofilm characteristics.

L_L = 30 – 100 µm (most reactors)
= 10 – 300 µm (range)

with lower values for higher mixing intensity and smoother biofilms. Predicted biofilm reactor performance is significantly influenced by the value of L_L (Boltz *et al.*, 2011). In biofilm reactor modeling in practice the value of L_L is often calibrated based on observed plant performance (Rittmann *et al.*, 2018).

Dedicated experimental approaches are available to estimate external mass-transfer resistances, depending on the operating conditions in MBBR demonstrating a direct relationship between mixing intensity and the value of L_L (Melcer and Schuler, 2014; Nogueira *et al.*, 2015).

For well-defined hydrodynamic conditions the external mass-transfer resistance can also be estimated using empirical correlations from the chemical engineering literature. The non-dimensional Sherwood number (Sh) is a function of R_L, D_W, and a characteristic length (*e.g.* the diameter of a biofilm support particle) (d_p):

$$Sh = \frac{d_p}{R_L \cdot D_W} \qquad (17.101)$$

The Sherwood number can be expressed as a function of the non-dimensional Reynolds number ($Re = Ud_p/\nu$) and the Schmidt number ($Sc = \nu/D_i$)

$$Sh = A + B \cdot Re^m \cdot Sc^n \qquad (17.102)$$

where U is the free-stream velocity around the support particle and ν is the kinematic viscosity. The Reynolds number depends on flow and geometry, while the Schmidt number only depends on fluid properties. The parameters A, B, m, and n are, in most cases, determined empirically from experimental data for a specific system. For example, for the simple case of fluid flow around a rigid spherical particle A = 2, B = 0.95, m = 1/2, n = 1/3 (for 100 < Re < 700 and 1,200 < Sc < 1,525, Garner and Suckling, 1958).

For three-phase suspensions with non-ideal flow, Nicolella *et al.* (1998) approximated the Sherwood number using Eq. 17.96 with A = 2, B = 0.265, m = 0.241, n = 1/3, and assuming an alternative expression for the Reynolds number based on Kolmogoroff's theory of turbulence:

$$Re = \frac{\varepsilon_{dissipation} \cdot d_p^4}{\nu^3} \tag{17.103}$$

where $\varepsilon_{dissipation}$ is the rate of energy dissipation per unit mass.

For packed-bed reactors Wilson and Geankoplis (1966) provided the following approximation of the Sherwood number that is applicable for bed porosities ($\varepsilon_{porosity}$) between 0.35 and 0.75:

$$Sh = \begin{cases} \dfrac{1.09}{\varepsilon_{porosity}} \cdot Sc^{1/3} & \text{for} \begin{cases} 0.0016 < Re_{pb} < 55 \\ 165 < Sc < 70{,}600 \end{cases} \\ \dfrac{0.25}{\varepsilon_{porosity}} \cdot Re_{pb}^{0.69} \cdot Sc^{1/3} & \text{for} \begin{cases} 55 < Re_{pb} < 1{,}500 \\ 165 < Sc < 10{,}690 \end{cases} \end{cases} \tag{17.104}$$

The Reynolds number in Eq. 17.104 is defined as $Re_{pb} = U_{sup} \cdot d_p / \nu$ where U_{sup} is the superficial bed velocity, which is the total flow through the bed divided by the cross-sectional area (Clark, 1996).

More correlations for Sh can be found for example in tables 5.20 to 5.24 in Perry and Green (2008) or in Kissel (1986). It should be noted that many of the correlations to estimate Sh were derived for relatively simple chemical engineering applications, and they should be applied to biofilm systems with caution.

The parameters A, B, m, and n depend on the geometry of the biofilm support medium and are valid only for a defined range of Re and Sc. Extrapolating from these conditions can yield erroneous results. For complex geometries, Eq. 17.102 might not be sufficient and other forms of Sh = *f* (Re, Sc) can be used to fit experimental data. Most correlations for Sh were derived for rigid particles, but the elastic and heterogeneous nature of the biofilm can influence external mass transfer. Different authors have suggested that R_L will either increase (Nicolella *et al.*, 2000) or decrease (Horn and Hempel, 2001) as a result of a biofilm covering the support medium.

Modelling of external mass-transfer resistance in this chapter is limited to the mass transfer from the bulk phase to the biofilm surface. Many biofilm reactors are, however, three-phase systems (solid/biofilm, liquid, gas phases) where the mass-transfer rate of, for example, oxygen from the gas phase into the bulk phase can limit the overall process performance. Mass transfer is also important from the bulk to the gas phase to remove metabolic products such as N_2 or CO_2 from the bulk phase. Increasing mixing (*e.g.* by increasing water flow) and increasing gas flow will influence both types of mass transfer: from the bulk to the biofilm surface and from the gas phase to the liquid phase.

17.11.4 Biofilm thickness (L_F) and biofilm detachment ($u_{d,S}$, $u_{d,V}$, $u_{d,M}$)

The biofilm thickness (L_F) and biofilm detachment rates ($u_{d,S}$, $u_{d,V}$, $u_{d,M}$) are directly related for a given substrate flux (Eq. 17.70). Biofilm simulations can be performed fixing either the biofilm thickness or the detachment rate. For different types of biofilm reactors a typical range of expected thicknesses are available (Table 17.11). Estimating a typical biofilm thickness (and biofilm detachment) in full-scale biofilm reactors is often difficult as biofilms are heterogeneously distributed over the support medium and over the overall reactor. In practical applications representative sampling and measurement are also often difficult. In biofilm reactors with relatively thick biofilms this uncertainty in estimating L_F is not a problem as for thick biofilms, substrate flux into the biofilm is not sensitive to changes in L_F. If lower layers of the biofilm are substrate-limited (*i.e.* $L_F >> L_{crit}$ for first-order kinetics or $\beta < 1$ for zero-order kinetics) then an increase in biofilm thickness only increases the amount of substrate-limited biomass at the base of the biofilm. In many commercial simulators, the biofilm thickness is fixed as a model input.

Table 17.11 Typical biofilm thickness (L_F) for different reactor types for carbon oxidation or nitrification (based on Rittmann *et al.*, 2018)[(a)].

Reactor type	L_F, µm	
	Carbon oxidation	Nitrification
Trickling filter	500 (top) 100 (low loaded bottom)	100-200 (top) 20-40 (low loaded bottom)
Rotating biological contactor	500 (first section) 100 (low loaded last section)	100-200 (first section) 20-40 (low loaded last section)
Biological aerated filter (dense media)	100 (before backwashing) 40-60 (after backwashing)	20-40
Biological aerated filter (floating media)	100-200	80-120
Moving bed biofilm reactors (MBBR)	500	100-200
Fluidized bed reactor, air lift reactor	100-200	10-20

17.11.5 Caution when using parameters from other types of models

Parameter estimation for complex biofilm systems is difficult and in many cases kinetic and stoichiometric parameters are taken from the literature. This should be done with caution (Wanner *et al.*, 2006). One example of misuse is to directly apply half-saturation constants from activated sludge models for biofilm simulations. The half-saturation constants (*e.g.* K_S, K_{O2}, etc) in the Monod Equation (*e.g.* Eq. 17.4) can be regarded as fundamental microbial parameters that should not depend on system configurations. In wastewater treatment, however, half-saturation constants are often estimated based on experiments with floccular biomass, and activated sludge models are often used for this purpose. The model structure of these activated sludge models does not take into account mass-transfer limitations into flocs, which results in observed half-saturation constants that are artificially high. In biofilms, mass-transfer limitations are explicitly modelled so therefore the true value of the half-saturation constant needs to be applied. Thus, the value of half-saturation constants in biofilm models should be lower than values in activated sludge models (*e.g.* $K_{S,biofilm} = 0.1 \cdot K_{S,activated\ sludge}$).

17.12 MODELLING TOOLS

The purpose of Chapter 17 and Chapter 18 is to focus mainly on basic principles of biofilms and biofilm reactors. But how can the resulting models be implemented in the engineering practice or in research? The answer to this question depends, first, on the objective for wanting to use a model. The question of appropriate model selection in the design of biofilm reactors is addressed in Chapter 18. A comprehensive discussion of different types of models and appropriate use of these models is provided in Wanner *et al.* (2006). Once a model has been selected, the next question is how to implement the chosen model. The following paragraphs provide a brief discussion of the commonly used approaches:

- *Analytical solutions:* In this chapter a range of analytical solutions has been presented. One great benefit of analytical solutions is that they provide the user with a very direct understanding of how different parameters influence modelling results and they are very important for teaching the concept of mass transport in biofilms. For many practical biofilm problems where the user is mainly interested in the substrate flux of a single limiting compound, analytical solutions provide fast and accurate results. However, for complex problems analytical solutions are either not feasible or become very complex.

- *Pseudo-analytical solutions:* Pseudo-analytical solutions are based on numerical solutions of Monod kinetics for a single limiting compound in a homogeneous biofilm. Solutions can be conveniently implemented in a spreadsheet (Rittmann and McCarty, 2001; Saez and Rittmann, 1992). However, the application of pseudo-analytical solutions is, like analytical solutions, limited to relatively simple systems.

- *Numerical solutions (1D, homogeneous biomass distribution):* Numerical solutions are used to calculate substrate concentration profiles for multiple substrates that are degraded or produced inside the biofilm. Biomass fractions are homogeneously distributed over the thickness of the biofilm. A single biomass fraction can be assumed *a priori* as in Section 17.2. Alternatively the density of different biomass fractions inside the biofilm can be determined at each time point from the balance from growth, decay, and detachment (Boltz *et al.*, 2008; Rauch *et al.*, 1999). While changes in biomass fractions over time are considered, it is assumed that all biomass fractions are always homogeneously distributed over the thickness of the biofilm which greatly simplifies the numerical solution. This approach of 1D biofilms with homogeneous biomass distribution is implemented in some commercial wastewater treatment plant simulators.

- *Numerical solutions (1D, heterogeneous biomass distribution):* This approach takes into account concentration gradients for both soluble substrate and biomass fractions as the result of growth, decay, and detachment processes over time. This approach applied in Section 17.8 when evaluating the competition between heterotrophic and autotrophic bacteria in a biofilm. Many commercial wastewater treatment plant simulators have included numerical solutions for 1D biofilms with heterogeneous biomass distributions. While basic modelling features will be the same across commercial simulators (*e.g.* diffusion and degradation of soluble substrates), it is the responsibility of the user to understand how processes such as particle attachment, detachment, and spreading are implemented in a commercial simulator (Boltz *et al.*, 2010; Rittman *et al.*, 2018).

- *Numerical solutions (2D, 3D):* The application of multi-dimensional models is discussed in Section 17.10. The application of these models is currently limited to research. The availability of faster personal computers means that the simulation of multi-dimensional models can be relatively quick. However, regardless of computational speed, the implementation and application of these models and also the interpretation of the multi-dimensional results is complex. Good biofilm reactor modelling practice (Rittmann *et al.*, 2018) recommends that 1D models are the appropriate level of complexity for most biofilm reactor questions.

Table 17.12 Matrix of stoichiometry and kinetic expressions for heterotrophic and autotrophic biofilm processes (adapted from Wanner *et al.*, 2006). Note that nitrogen for cell synthesis is not included so that observed nitrogen removal can be directly related to autotrophic growth.

↓j →i	1	2	3	4	5	6	
Process name	X_{OHO}	X_{ANO}	X_U	S_S	S_{NH4}	S_{O2}	Process rate, ρ_j
1: Heterotrophic growth	1			$-\frac{1}{Y_{OHO}}$		$\frac{-(1-Y_{OHO})}{Y_{OHO}}$	$\mu_{OHO,max}\cdot\frac{S_S}{K_S+S_S}\cdot\frac{S_{O2}}{K_{OHO,O2}+S_{O2}}\cdot X_{OHO}$
2: Heterotrophic inactivation	−1		1				$b_{OHO,ina}\cdot X_{OHO}$
3: Heterotrophic respiration	−1					−1	$b_{OHO,res}\cdot\frac{S_{O2}}{K_{OHO,O2}+S_{O2}}\cdot X_{OHO}$
4: Autotrophic growth		1			$-\frac{1}{Y_{ANO}}$	$\frac{-(4.57-Y_{ANO})}{Y_{ANO}}$	$\mu_{ANO,max}\cdot\frac{S_{NH4}}{K_{ANO}+S_{NH4}}\cdot\frac{S_{O2}}{K_{ANO,O2}+S_{O2}}\cdot X_{ANO}$
5: Autotrophic inactivation		−1	1				$b_{ANO,ina}\cdot X_{ANO}$
6: Autotrophic respiration		−1				−1	$b_{ANO,res}\cdot\frac{S_{O2}}{K_{ANO,O2}+S_{O2}}\cdot X_{ANO}$
Units	COD	COD	COD	COD	N	− COD	

$r_{i,V}=\sum_j \nu_{i,j}\cdot\rho_j$ (with stoichiometric matrix $\nu_{i,j}$ and vector of process rates, ρ_j)

Table 17.13 Kinetic and stoichiometric parameters (adapted from Wanner and Gujer, 1985; Wanner and Reichert, 1996) and zero-order rate constants based on eqs. 17.6 and 17.7)

Heterotrophic growth			Nitrification		
α_{OHO}	1.0	$gO_2/gCOD$	α_{ANO}	4.57	gO_2/gN
$\mu_{OHO,max}$	4.8	1/d	$\mu_{ANO,max}$	0.95	1/d
Y_{OHO}	0.4	gCOD/gCOD	Y_{ANO}	0.22	gCOD/gN
K_S	5.0	$gCOD/m^3$	K_{ANO}	1.00	gN/m^3
$K_{OHO,O2}$	0.1	gO_2/m^3	$K_{ANO,O2}$	0.10	gO_2/m^3
$b_{OHO,ina}$	0.1	1/d	$b_{ANO,ina}$	0.10	1/d
$b_{OHO,res}$	0.2	1/d	$b_{ANO,res}$	0.05	1/d
$k_{0,COD,OHO}$	12.0	gCOD/gCOD.d	$k_{0,NH4,ANO}$	4.30	gN/gCOD.d
$k_{0,O2,OHO}$	7.2	gO_2/gCOD.d	$k_{0,O2,ANO}$	18.80	gO_2/gCOD.d
X_{OHO}	10,000.0	$gCOD/m^3$	X_{ANO}	10,000.00	$gCOD/m^3$

REFERENCES

Andrews G. (1988). Effectiveness Factors for Bioparticles with Monod Kinetics. *Chemical Engineering Journal And The Biochemical Engineering Journal*, 37(2), B31-B37.

Bakke R., Trulear M.G., Robinson J.A. and Characklis W.G. (1984). Activity of *Pseudomonas aeruginosa* in biofilms: Steady state. *Biotechnology and Bioengineering*, 26, 1418-1424.

Batstone D.J., Keller J., Angelidaki I., Kalyuzhnyi S.V., Pavlostathis S.G., Rozzi A., Sanders W.T.M., Siegrist H. and Vavilin V.A. (2002). *Anaerobic Digestion Model No.1 (ADM1)*. (ed.) Scientific and Technical Report No. 13. IWA Task Group for Mathematical Modelling of Anaerobic Digestion Processes (ed.) IWA Publishing, London.

Batstone D.J., Picioreanu C. and Van Loosdrecht M.C.M. (2006). Multidimensional modelling to investigate interspecies hydrogen transfer in anaerobic biofilms. *Water Research*, 40(16), 3099-3108.

Boltz, J. P., Morgenroth, E., Brockmann, D., Bott, C., Gellner, W. J. and Vanrolleghem, P. A. 2011. Systematic evaluation of biofilm models for engineering practice: components and critical assumptions. *Water Science and Technology*, 64(4), 930–944

Boltz J.P., Morgenroth E. and Sen D. (2010). Mathematical Modelling of Biofilms and Biofilm Reactors for Engineering Design. *Water Science and Technology* 62(8), 1821-1836.

Boltz J.P., Johnson B.R., Daigger G.T. and Sandino J. (2008). Modeling integrated fixed film activated sludge (IFAS) and moving bed biofilm reactor (MBBR) systems I: mathematical treatment and model development. *Water Environmental Research*, 81(6), 555-575.

Boucher D.F. and Alves G.E. (1959). Dimensionless numbers. *Chemical Engineering Progress*, 55(9), 55-64.

Bryers J.D. (1984). Biofilm formation and chemostat dynamics: Pure and mixed culture considerations. *Biotechnology and Bioengineering*, 26, 948-958.

Chang H.T. and Rittmann B.E. (1987). Mathematical modeling of biofilm on activated carbon. *Environmental Science and Technology*, 21(3), 273-280.

Clark M.M. (1996). *Transport modeling for environmental engineers and scientists.* Wiley, New York.

Crank J. (1975). *The Mathematics of Diffusion*. (ed.) 2nd edition. Oxford University Press, London.

De Kreuk M.K., Kishida N., Tsuneda S. and Van Loosdrecht M.C.M. (2010). Behavior of polymeric substrates in an aerobic granular sludge system. *Water Research,* 44(20), 5929-5938.

Esener A.A., Roels J.A. and Kossen N.W.F. (1983). Theory and Applications of Unstructured Growth-Models - Kinetic and Energetic Aspects. *Biotechnology and Bioenginering*, 25(12), 2803-2841.

Fruhen M., Christan E., Gujer W. and Wanner O. (1991). Significance of spatial distribution of microbial species in mixed culture biofilms. *Water Science and Technology*, 23(7-9), 1365-1374.

Gapes D., Perez J., Picioreanu C. and Van Loosdrecht M.C.M. (2006). Corrigendum to 'Modeling biofilm and floc diffusion processes based on analytical solution of reaction-diffusion equations' [Water Res. 39 (2005). 1311-1323]. *Water Research*, 40(15), 2997-2998.

Garner F.H. and Suckling R.D. (1958). Mass transfer from a soluble solid sphere. *Aiche Journal*, 4(1), 114-124.

Gujer W. and Boller M. (1986). Design of A Nitrifying Tertiary Trickling Filter Based on Theoretical Concepts. *Water Research*, 20(11), 1353-1362.

Gujer W. and Wanner O. (1990). Modeling mixed population biofilms. In: Marshall K.C. (ed.), *Biofilms*, (11), pp. 397-443. New York Wiley.

Henze M. (2000). *The Activated Sludge Models (1, 2, 2d, and 3)*, IWA Scientific & Technical Report, IWA Publishing, London, U.K.

Henze M., Harremoës P., Jansen J.L.C. and Arvin E. (2002). *Wastewater Treatment* (eds.) 3rd ed. Springer, Berlin.

Horn H. and Hempel D.C. (1997). Substrate utilization and mass transfer in an autotrophic biofilm system: Experimental results and numerical simulation. *Biotechnology and Bioengineering*, 53(4), 363-371.

Horn H. and Hempel D.C. (2001). Simulation of substrate conversion and mass transport in biofilm systems. *Chemical Engineering & Technology*, 24(12), A225-A228.

Horn H. and Morgenroth E. (2006). Transport of oxygen sodium chloride, and sodium nitrate in biofilms. *Chemical Engineering Science*, 61(5), 1347-1356.

Hung C., Yu F., McFeters G. and Stewart P. (1995). Nonuniform spatial patterns of respiratory activity within biofilms during disinfection. *Applied and Environmental Microbiology*, 61(6), 2252-2256.

Janning K.F., Le Tallec X. and Harremoes P. (1998). Hydrolysis of organic wastewater particles in laboratory scale and pilot scale biofilm reactors under anoxic and aerobic conditions. *Water Science and Technology*. 38(8-9), 179-188.

Jansen J.L.C., Arvin E., Henze M. and Harremoës P. (2019). *Wastewater Treatment: Biological and Chemical Processes*, Polyteknisk.

Jansen J.L.C and Harremoes P. (1984). Removal of soluble substrates in fixed films. *Water Science and Technology,* 17(2-3), 1-14.

Kissel J.C. (1986). Modeling mass-transfer in biological waste-water treatment processes. *Water Science and Technology*, 18(6), 35-45.

Kissel J.C., McCarty P.L. and Street R.L. (1984). Numerical simulation of mixed-culture biofilm. *Journal of Environmental Engineering - ASCE*, 110(2), 393-411.

Kreikenbohm R. and Stephan W. (1985). Application of a two-compartment model to the wall growth of *Pelobacter acidigallici* under continuous culture conditions. *Biotechnology and Bioengineering*, 27, 296-301.

Lackner S., Terada A. and Smets B.F. (2008). Heterotrophic activity compromises autotrophic nitrogen removal in membrane-aerated biofilms: Results of a modeling study. *Water Research*, 42(4-5), 1102-1112.

Libicki S.B., Salmon P.M. and Robertson C.R. (1988). Effective diffusive permeability of a nonreacting solute in microbial cell aggregates. *Biotechnology and Bioengineering*, 32(1), 68-85.

Lide D.R. (2008). *CRC Handbook of Chemistry and Physics.* Taylor and Francis Group, London.

Logan B.E., Hermanowicz S.H. and Parker D.S. (1987). A Fundamental Model for Trickling Filter Process Design. *Journal of the Water Pollution Control Federation*, 59(12), 1029-1042.

McFeters G. (2002). Biofilm of Two Bacteria. ASM MicrobeLibrary.org.

Melcer H. and Schuler A.J. (2014). *Mass Transfer Characteristics of Floating Media in MBBR and IFAS Fixed-Film Systems*, IWA Publishing.

Morgenroth E. (2003). Detachment - an often overlooked phenomenon in biofilm research and modeling, In: *Biofilms in wastewater treatment.* Wuertz S., Wilderer P.A. and Bishop P.L. (eds.), pp. 264-290, IWA Publishing.

Morgenroth E. (2003). Detachment - an often overlooked phenomenon in biofilm research and modeling. In: Wuertz S., Wilderer P.A. and Bishop P.L. (eds.), *Biofilms in Wastewater Treatment*, pp. 264-290. IWA Publishing.

Morgenroth E. and Wilderer P.A. (1999). Controlled biomass removal - The key parameter to achieve enhanced biological phosphorus removal in biofilm systems. *Water Science and Technology*, 39(7), 33-40.

Morgenroth E. and Wilderer P.A. (2000). Influence of detachment mechanisms on competition in biofilms. *Water Research*, 34(2), 417-426.

Nicolella C., Van Loosdrecht M.C.M. and Heijnen J.J. (1998). Mass transfer and reaction in a biofilm airlift suspension reactor. *Chemical Engineering Science*, 53(15), 2743-2753.

Nicolella C., Van Loosdrecht M.C.M. and Heijnen J.J. (2000) Wastewater treatment with particulate biofilm reactors. *Journal of Biotechnology*, 80(1), 1-33.

Nogueira R., Elenter D., Brito A., Melo L.F., Wagner M. and Morgenroth E. (2005). Evaluating heterotrophic growth in a nitrifying biofilm reactor using fluorescence in situ hybridization and mathematical modeling. *Water Science and Technology*, 52(7), 135-141.

Nogueira B.L., Perez J., Van Loosdrecht M.C.M., Secchi A.R., Dezotti M. and Biscaia E.C. (2015). Determination of the external mass transfer coefficient and influence of mixing intensity in moving bed biofilm reactors for wastewater treatment. *Water Research,* 80, 90-98.

Perez J., Picioreanu C. and Van Loosdrecht M.C.M. (2005). Modeling biofilm and floc diffusion processes based on analytical solution of reaction diffusion equations. *Water Research,* 39(7), 1311-1323.

Perry R.H. and Green D. (2008). *Perry's Chemical Engineers' Handbook* (eds.) 8th ed. McGraw-Hill.

Peyton B.M. and Characklis W.G. (1993). A Statistical Analysis of the Effect of Substrate Utilization and Shear-Stress on the Kinetics of Biofilm Detachment. *Biotechnology and Bioengineering*, 41(7), 728-735.

Picioreanu C., Van Loosdrecht M.C.M. and Heijnen J.J. (1998). Mathematical modeling of biofilm structure with a hybrid differential discrete cellular automaton approach. *Biotechnology and Bioengineering*, 58(1), 101-116.

Picioreanu C., Van Loosdrecht M.C.M. and Heijnen J.J. (2000). Effect of diffusive and convective substrate transport on biofilm structure formation: A two-dimensional modeling study. *Biotechnology and Bioengineering*, 69(5), 504-515.

Picioreanu C., Vrouwenvelder J.S. and Van Loosdrecht M.C.M. (2009). Three-dimensional modeling of biofouling and fluid dynamics in feed spacer channels of membrane devices. *Journal of Membrane Science*, 345(1-2), 340-354.

Polson A. (1950). Some Aspects of Diffusion in Solution and a Definition of a Colloidal Particle. *The Journal of Physical Chemistry*, 54(5), 649-652.

Rauch W., Vanhooren H. and Vanrolleghem P.A. (1999). A simplified mixed-culture biofilm model. *Water Research*, 33(9), 2148-2162.

Reichert P. (1998). *Aquasim 2.0 - User manual. Computer program for the identification and simulation of aquatic systems,* Swiss Federal Institute for Environmental Science and Technology (EAWAG). CH 8600 Dübendorf, Switzerland.

Rittmann B.E. (1982a). Comparative Performance of Biofilm Reactor Types. *Biotechnology and Bioengineering*, 24, 1341-1370.

Rittmann B.E. (1982b). The effect of shear stress on biofilm loss rate. *Biotechnology and Bioengineering*, 24, 501-506.

Rittmann B.E. (1989). Detachment from biofilms. In: Characklis W.G. and Wilderer P.A. (eds.), *Structure and function of biofilms*, pp. 49-58. New York Wiley.

Rittmann B.E. and McCarty P.L. (1980). Evaluation of Steady-State-Biofilm Kinetics. *Biotechnology and Bioengineering*, 22, 2359-2373.

Rittmann B.E. and McCarty P.L. (2020). *Environmental Biotechnology: Principles and Applications.* 2nd edition. McGraw-Hill, New York.

Rittmann B.E., Stilwell D. and Ohashi A. (2002). The transient-state, multiple-species biofilm model for biofiltration processes. *Water Research*, 36(9), 2342-2356.

Rittmann B.E., Boltz J.P., Brockmann D., Daigger G.T., Morgenroth E., Sorensen K.H., Takacs I., Van Loosdrecht M.C.M. and Vanrolleghem P.A. (2018). A framework for good biofilm reactor modeling practice (GBRMP). *Water Science & Technology,* 77(4), 1149-1164.

Robinson J.A., Trulear M.G. and Characklis W.G. (1984). Cellular reproduction and extracellular polymer formation by *Pseudomonas aeruginosa* in continuous culture. *Biotechnology and Bioenginering*, 26, 1409-1417.

Saez P.B. and Rittmann B.E. (1992). Communication to the Editor: Accurate Pseudoanalytical Solution for Steady-State Biofilms. *Biotechnology and Bioengineering*, 39, 790-793.

Speitel G.E. and DiGiano F.A. (1987). Biofilm shearing under dynamic conditions. *Journal of Environmental Engineering - ASCE*, 113(3), 464-475.

Stewart P.S. (1998). A review of experimental measurements of effective diffusive permeabilities and effective diffusion coefficients in biofilms. *Biotechnology and Bioengineering*, 59(3), 261-272.

Stewart P.S. (2003). Diffusion in biofilms. *Journal of Bacteriology*, 18(5), 1485-1491.

Sutherland I.W. (2001). The biofilm matrix - an immobilized but dynamic microbial environment. *Trends in Microbiology*, 9(5), 222-227.

Tijhuis L., Van Loosdrecht M.C.M. and Heijnen J.J. (1995). Dynamics of biofilm detachment in biofilm airlift suspension reactors. *Biotechnology and Bioengineering*, 45(6), 481-487.

Trulear M.G. and Characklis W.G. (1982). Dynamics of biofilm processes. *Journal of the Water Pollution Control Federation*, 54(9), 1288-1301.

Van Loosdrecht M.C.M., Eikelboom D., Gjaltema A., Mulder A., Tijhuis L. and Heijnen J.J. (1995). Biofilm structures. *Water Science and Technology*, 32(8), 35-43.

Wanner O., Eberl H.J., Morgenroth E., Noguera D.R., Picioreanu C., Rittmann B.E. and van Loosdrecht M.C.M. (2006). *Mathematical Modeling of Biofilms* IWA Publishing, London, UK. Series: Scientific and Technical Report Series Report No. 18.

Wanner O. and Gujer W. (1985). Competition in biofilms. *Water Science and Technology*, 17(2-3), 27-44.

Wanner O. and Gujer W. (1986). A multispecies biofilm model. *Biotechnology and Bioengineering*, 28, 314-328.

Wanner O. and Morgenroth E. (2004). Biofilm modeling with AQUASIM. *Water Science and Technology*, 49(11-12), 137-144.

Wanner O. and Reichert P. (1996). Mathematical-modeling of mixed-culture biofilms. *Biotechnology and Bioengineering*, 49(2), 172-184.

Wanner O. and Reichert P. (1996). Mathematical-modeling of mixed-culture biofilms. *Biotechnology and Bioengineering,* 49(2), 172-184.

Wilson E.J. and Geankoplis C.J. (1966). Liquid Mass Transfer at Very Low Reynolds Numbers in Packed Beds. *Industrial & Engineering Chemistry Fundamentals*, 5(1), 9-14.

NOMENCLATURE

Symbol	Description	Dimension	Example units
A_F	Surface area of the biofilm	L^2	m^2
a_F	Specific surface area of the biofilm = A_F/V_R	L^2/L^3	m^2/m^3
Bi	Biot number	-	-
$b_{ANO,ina}$	Rate of inactivation for autotrophic bacteria	T^{-1}	1/d
$b_{ANO,res}$	Rate of endogenous respiration for autotrophic bacteria	T^{-1}	1/d
$b_{OHO,ina}$	Rate of inactivation for heterotrophic bacteria	T^{-1}	1/d
$b_{OHO,res}$	Rate of endogenous respiration for heterotrophic bacteria	T^{-1}	1/d
Da^{II}	Second Damköhler number	-	-
D_F	Diffusion coefficient in the biofilm	$L^2\ T^{-1}$	m^2/d
D_W	Diffusion coefficient in water	$L^2\ T^{-1}$	m^2/d
G	Growth number	-	-
J	Substrate[(1)] flux	$M\ L^{-2}\ T^{-1}$	$g/m^2.d$
J_F	Substrate[(1)] flux inside the biofilm	$M\ L^{-2}\ T^{-1}$	$g/m^2.d$
J_{LF}	Substrate[(1)] flux at the biofilm surface	$M\ L^{-2}\ T^{-1}$	$g/m^2.d$
$k_{0,F}$	Zero-order substrate[(1)] removal rate inside the biofilm	T^{-1}	1/d
$k_{0,f,A}$	Zero-order substrate[(1)] removal rate per biofilm surface for a fully penetrated biofilm	$M\ L^{-2}\ T^{-1}$	$g/m^2.d$
$k_{0,p,A}$	Zero-order substrate[(1)] removal rate per biofilm surface for a partially penetrated biofilm	$M^{0.5}\ L^{-0.5}\ T^{-1}$	$g^{0.5}/m^{0.5}.d$
$k_{1,A}$	First-order substrate[(1)] removal rate per biofilm surface	$L\ T^{-1}$	m/d
$k_{1,F}$	First-order substrate[(1)] removal rate inside the biofilm	$L^3\ M^{-1}\ T^{-1}$	$m^3/g.d$
k_d	Biofilm detachment rate coefficient	(3)	(3)
$K_{ANO,O2}$	Half-saturation constant for S_{O2} for autotrophic bacteria	$M\ L^{-3}$	mg O_2/l
K_{NH4}	Half-saturation constant for S_{NH4}	$M\ L^{-3}$	mg N/l
$K_{OHO,O2}$	Half-saturation constant for S_{O2} for heterotrophic bacteria	$M\ L^{-3}$	mg O_2/l
K_S	Half-saturation constant for S_S	$M\ L^{-3}$	mg COD/l
L_F	Biofilm thickness	L	µm
L_L	External mass-transfer boundary layer thickness	L	µm
n	Number of layers in a biofilm	-	-
Pe	Peclet number	-	-
Q	Flow rate	$L^3\ T^{-1}$	m^3/d
Q_i	Influent flow rate	$L^3\ T^{-1}$	m^3/d
r_F	Substrate[(1)] conversion rate inside the biofilm	$M\ L^{-3}\ T^{-1}$	$g/m^3.d$
R_L	External mass-transfer resistance	$T\ L^{-1}$	d/m
Re	Reynolds number (= $U.d_p/\nu$)	-	-
S	Substrate[(1)] concentration in the bulk liquid where substrate is a generic term that can relate to any of the rate-limiting compounds such as organic substrate, NH_4^+, NO_3^-, or O_2.	$M\ L^{-3}$	mgCOD/l or mgN/l or mgO_2/l
S_B	Soluble substrate[(1)] concentration in the bulk water	$M\ L^{-3}$	mg/l
Sc	Schmidt number (= ν/D_i)	-	-
S_F	Soluble substrate concentration[(1)] inside the biofilm	$M\ L^{-3}$	mg/l

$S_{F,0,f}$	Analytical solution for substrate[1] concentration inside the biofilm assuming zero-order kinetics inside the biofilm and full penetration	$M\ L^{-3}$	mg/l
$S_{F,0,p}$	Analytical solution for substrate[1] concentration inside the biofilm assuming zero-order kinetics inside the biofilm and partial penetration	$M\ L^{-3}$	mg/l
$S_{F,1}$	Analytical solution for substrate[1] concentration inside the biofilm assuming first-order kinetics inside the biofilm	$M\ L^{-3}$	mg/l
$S_{F,NH4}$	Concentration of ammonia inside the biofilm	$M\ L^{-3}$	mgN/l
$S_{F,O2}$	Concentration of oxygen inside the biofilm	$M\ L^{-3}$	mgO_2/l
$S_{F,COD}$	Concentration of organic substrate inside the biofilm	$M\ L^{-3}$	mgCOD/l
S_i	Soluble substrate [1] concentration in the influent	$M\ L^{-3}$	mg/l
S_{LF}	Soluble substrate[1] concentration at the biofilm surface	$M\ L^{-3}$	mg/l
S_{min}	Minimum substrate[1] concentration supporting microbial growth in a biofilm	$M\ L^{-3}$	mg/l
S_{NH4}	Concentration of ammonium	$M\ L^{-3}$	mgN/l
S_{O2}	Concentration of oxygen	$M\ L^{-3}$	mgO_2/l
S_S	Concentration of organic substrate	$M\ L^{-3}$	mgCOD/l
U	Characteristic flow velocity	$L\ T^{-1}$	m/d
$u_{d,M}$	Biofilm detachment rate in terms of mass removed per area and time (= $u_{d,S} \cdot X_F$)	$M\ L^{-2}\ T^{-1}$	$g/m^2.d$
$u_{d,S}$	Biofilm detachment velocity	$L\ T^{-1}$	m/d
$u_{d,VS}$	Biofilm detachment volumetric removal rate (= $u_{d,S}/L_F$)	T^{-1}	1/d
V_R	Volume of reactor	L^3	m^3
x	Distance into the biofilm measured from the surface of the biofilm	L	m
X_{ANO}	Density of autotrophic bacteria inside the biofilm	$M\ L^{-3}$	$kgCOD/m^3$
X_F	Biofilm biomass [2] density inside the biofilm	$M\ L^{-3}$	$kgCOD/m^3$
X_{OHO}	Density of heterotrophic bacteria inside the biofilm	$M\ L^{-3}$	$kgCOD/m^3$
X_U	Density of unbiodegradable organic matter inside the biofilm	$M\ L^{-3}$	$kgCOD/m^3$
y	Distance along the length of the reactor	L	m
Y	Yield coefficient for X_F[2] growing on the generic substrate[1]	$M\ M^{-1}$	g/g
Y_{ANO}	Yield for autotrophic growth on S_{NH4}	$M\ M^{-1}$	gCOD/gN
Y_{OHO}	Yield for heterotrophic growth on S_S	$M\ M^{-1}$	gCOD/gCOD
z	Distance from the substratum	L	m

Subscript	Description
0	Zero order
0	At time zero
0,f	Zero-order fully penetrated biofilm
0,p	Zero-order partially penetrated biofilm
1	First order
A	Per biofilm surface
ANO	Autotrophic nitrifying organisms
B	In the bulk phase
e.a.	Electron acceptor
e.d.	Electron donor
F	In the biofilm
i	Influent
LF	At the biofilm surface
NH4	Ammonium
O2	Oxygen
OHO	Ordinary heterotrophic organisms
S	Organic substrate
W	In water

Greek symbol	Description	Unit
β	Substrate[(1)] penetration assuming zero-order rates inside the biofilm	-
$\beta_{e.a.}$	Penetration of the electron acceptor assuming zero-order rates inside the biofilm	-
$\beta_{e.d.}$	Penetration of the electron donor assuming zero-order rates inside the biofilm	-
$\gamma_{e.d.,e.a.}$	Penetration of electron donor relative to the penetration of the corresponding electron acceptor ($= \beta_{e.d.}/\beta_{e.a.}$)	-
ε	Efficiency factor assuming first-order rates inside the biofilm	-
ε_l	In AQUASIM: liquid-volume fraction inside the biofilm	-
ε_s	In AQUASIM: solids-volume fraction inside the biofilm	-
ζ	Ratio of the substrate flux into the biofilm neglecting the external mass-transfer resistance and the substrate flux in the external mass-transfer boundary layer assuming $S_{LF} = 0$	-
μ_{max}	Maximum growth rate	T^{-1}
ν	Stoichiometric coefficient	
ν	Kinematic viscosity	$L^2\ T^{-1}$
τ	Characteristic time (see Table 17.9)	T
Φ	Thiele modulus	-

(1) Note that in most of Chapter 17 the type of the limiting substrate and the units are not specified. Examples of possible substrates are electron donors such as organic substrate ($S_{F,COD}$), ammonia ($S_{F,NH4}$), or electron acceptors such as oxygen ($S_{F,O2}$) or nitrate ($S_{F,NO3}$). Units for the substrate must be consistent with units in kinetic and stoichiometric constants.

(2) The type of biomass is not specified. The generic active biomass converts the generic substrate S_F. Examples of possible biomass types are heterotrophic bacteria (X_{OHO}) and autotrophic bacteria (X_{ANO}).

(3) The units of the detachment rate coefficient depend on the chosen detachment rate expression (see Table 17.6).

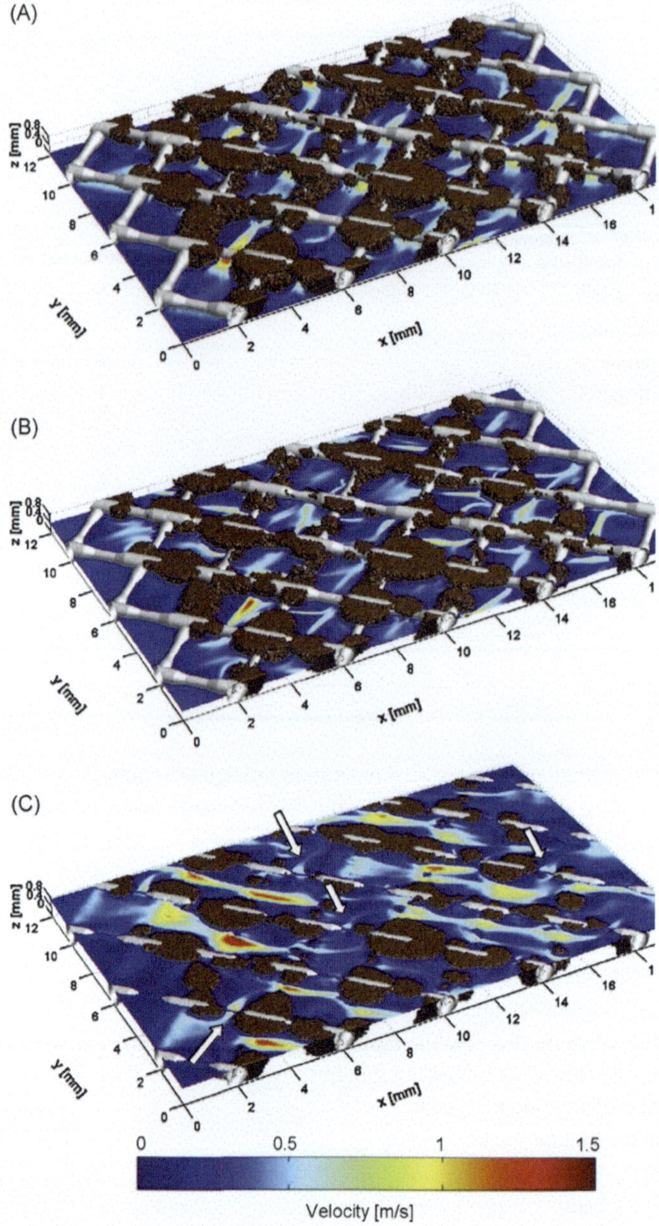

Figure 17.30 Example of biofilm modelling in a drinking water treatment treatment plant: a three-dimensional simulation of flow and biofilm formation (biofouling) in nanofiltration and reverse osmosis membrane modules. Flow velocity profiles are shown in two-dimensional sections at different heights in the feed channel of a spiral-wound membrane device, at (A) z = 78 μm, (B) z = 390 μm and (C) z = 702 μm. Colours show the biofilm (brown volumes), the spacer (white crossing filaments), and the flow velocity from high values (red) to low values (blue). The flow is from left to right. New biomass attached where shear was less than 10 Pa. Arrows in (C) point to channels with flow perpendicular to the main flow direction in that section. (For interpretation of the references to colour in this figure legend, the reader is referred to the web version of the article of Picioreanu *et al.*, 2009).

doi: 10.2166/9781789060362_0813

18

Biofilm reactors

Kim Helleshøj Sørensen and Eberhard Morgenroth

18.1 BIOFILM REACTORS

Biofilm reactors can achieve treatment objectives that are similar for activated sludge systems; examples are organic matter removal, nitrification, denitrification, and chemical or biological phosphorus removal. The same types of microorganisms are involved and these microorganisms have to be exposed to the same local environmental conditions in terms of availability of electron donor, electron acceptor, pH, and temperature. However, some other factors are different in biofilm reactors compared to activated sludge systems: conversion processes in biofilm reactors are typically mass transport-limited so that only bacteria in the outer layers of the biofilm contribute to the overall substrate removal. These mass transport limitations have implications for the design and operation of biofilm reactors as well as for the microbial ecology within the biofilm. Microbial competition is not only based on the availability of substrate in the bulk phase but also on the location of the different groups of bacteria within the biofilm. Bacteria closer to the surface of the biofilm have the advantage of a more direct access to substrates in the bulk phase. On the other hand, bacteria located further away from the surface of the biofilm are better protected from detachment and from toxic events. The fact that the bacteria are better protected against detachment also allows very slow growing specialized species to stay in the deeper layer of the biofilm, resulting in that a biofilm system typically has a superior performance when degrading complex compounds such as micropollutants, as can be seen in Figure 18.1.

Another clear difference between most biofilm-based technologies and activated sludge is that biofilm-based technology allows a clear separation of the bacteria for a specific task by confining them in a specific zone; in contrast, in an activated sludge system all the bacteria pass through all the zones. This also enables the process conditions in each zone to be optimized without having to compromise, which again results in the bacteria attaining their maximum growth rate. Added to this is the fact that the bacteria are fixed to a substratum, which also allows a higher biomass concentration in the reactor to be achieved, which often leads to a very compact design that takes up a small amount of surface area for biofilm reactors compared to activated sludge systems.

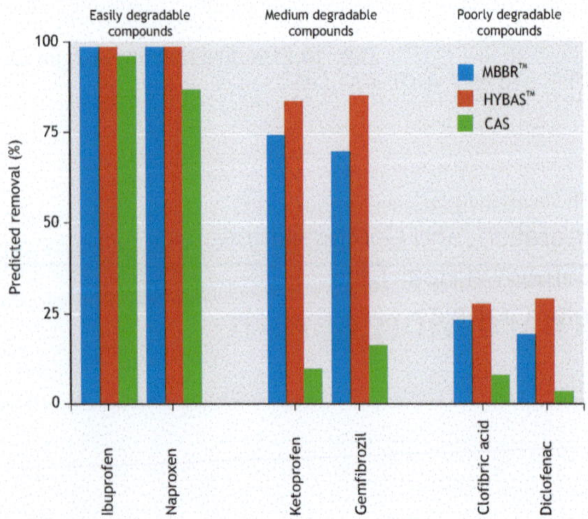

Figure 18.1 Removal of pharmaceuticals in different reactor types (source: The Danish Environmental Protection Agency/MERMISS 2018).

A conceptual overview of different types of bioreactors is provided in Figure 18.2. However, regardless of their differences, all of these biofilm reactors have to meet the following requirements: (*i*) the retention of microorganisms is based on the attachment of biomass to the surface of the support medium (substratum) rather than using solid liquid separation and biomass recycle, (*ii*) water containing the polluting compounds is brought into contact with the biofilm, and local mixing conditions and turbulence will determine the effective mass transport from the bulk water to the biofilm surface, (*iii*) biofilm growth has to be balanced with biofilm detachment to avoid clogging of the reactor while retaining sufficient active biomass in a stable biofilm, and (*iv*) if needed, an electron donor, electron acceptors, nutrients, and/or alkalinity can be added to the system. For example, oxygen can be provided by aeration for aerobic systems.

This chapter provides an overview of some basic types of biofilm treatment processes and design approaches based on the principles introduced in Chapter 17.

18.1.1 Types of reactors

Biofilm reactors can be grouped into three basic categories: (*i*) non-submerged or partially submerged

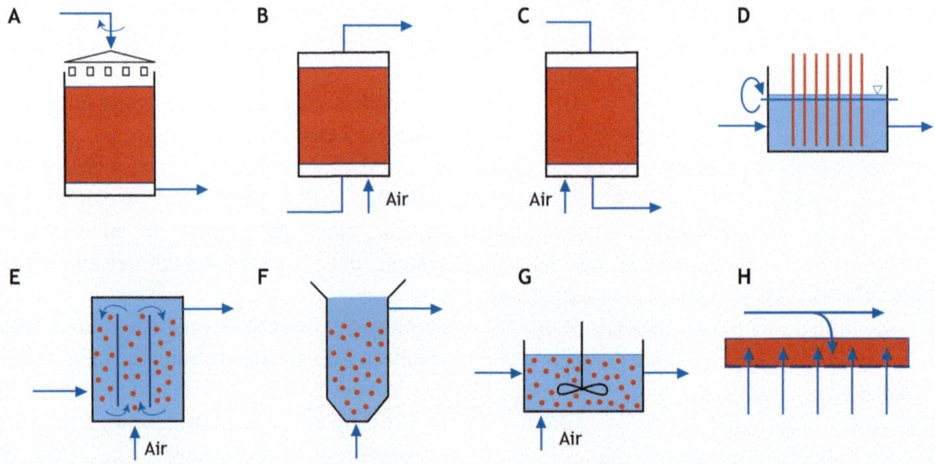

Figure 18.2 Types of biofilm reactors: (A) trickling filter (TF), (B) submerged fixed-bed biofilm reactor operated as upflow or (C) downflow, (D) rotating biological contactors (RBC), and (E) suspended biofilm reactor including an air lift reactor, (F) fluidized-bed reactor, (G) moving-bed biofilm reactor (MBBR) and aerobic granular sludge (AGS), and (H) membrane-attached biofilm reactors (modified from Wanner *et al.*, 2006).

systems including trickling filters and rotating biological contactors (Figure 18.2A and Figure 18.2D), (*ii*) submerged fixed-bed biofilm reactors (Figure 18.2B and Figure 18.2C), and (*iii*) different types of fluidized-bed reactors (Figure 18.2E-G). In addition, there is the most recent biofilm reactor technology where biofilms grow on membranes where the substrate can be supplied both from the water phase and by diffusion through the membrane (Figure 18.2H). The key differences between the different types of reactors are the specific surface area (Table 18.1), mechanisms for removing excess biomass, and gas transfer.

18.1.1.1 Trickling filters

Trickling filters are the oldest form of biofilm reactors, they have been applied since the early 1900s and are still in common use today. The biofilm support is stationary and consists of either of 5-20 cm large rocks, a structured plastic packing material or random plastic media (Figure 18.3).

The height of trickling filters ranges from 1 to 3 m using stones or 4 to 12 m using plastic as a support medium. Influent wastewater is distributed over the filter and then trickles down the packing material. The support medium in trickling filters is chosen to provide sufficiently large pore spaces to allow air to ventilate through the trickling filter regardless of either biofilm growth or the water trickling down the filter. The use of large packing material helps to avoid clogging of the filter medium but also results in specific biofilm surface areas (a_F) of 50 to 200 m^2/m^3 that are relatively small compared to other types of biofilm reactors (Table 18.1). Wastewater is distributed via rotary arms at the top and it then trickles down the filter. Water exits from the bottom of the filter, solids are removed in a settler, and some of the effluent is recirculated to ensure suitable hydraulic loading of the trickling filter (Figure 18.4).

Recirculation flow rates (Q_R) typically range from 0.5 to 4 times the influent flow but can be as high as 10 times the influent flow for strong industrial wastewaters (WEF 2018). Ventilation in trickling filters is the only oxygen source supply and is typically due to natural convection. Hence this often becomes the limiting factor for mass transfer, as already identified by Schroeder and Tchobanoglous,

Table 18.1 Specific carrier surface area for different types of media and biofilm reactors.

Type of reactor	Carrier material	Size of material, mm	Specific carrier surface area (a_F), m^2/m^3	Reference
Trickling filter	Rock	40-80	50-100	ATV-DVWK, 2001
	Plastic	-	100-200	ATV-DVWK, 2001
Rotating biological contactor (RBC)	Plastic	-	100-200	ATV-DVWK, 2001
Moving-bed biofilm reactor (MBBR)	Plastic (K_1) (60% fill volume)	7-9	300	Rusten *et al.*, 2006
	Plastic (K_5) (60% fill volume)	3.5-25	480	Dezotti *et al.*, 2018
Submerged biofilter	Porous clay	1.3-8	1,000-1,400	ATV, 1997
	Porous slate	2-8	1,200-1,400	ATV, 1997
	Polystyrene	3-6	1,100	ATV, 1997
	Anthracite	2.5-3.5	1,900	ATV, 1997
	Quartz sand	0.7-2.2	3,000	ATV, 1997
	Basalt	1.4-2.2	3,600	ATV, 1997
Granular sludge	-	-	200-2,000	
Fluidized bed	Sand or basalt	0.2-0.8	3,000-4,000	Nicolella *et al.*, 2000

1976. The natural ventilation is driven by the temperature differences between water and air, leading to many trickling filters having poor performance during periods with small differences in temperature. To overcome this limitation and to reduce the mass transfer limitations, modern trickling filter design can apply forced ventilation. This ventilation will typically have a head loss of less than 1 mWC, and even with forced ventilation the trickling filter remains one of the technologies that have the lowest running costs for aeration.

Figure 18.3 Trickling filters using (A) structured plastic, (B) random plastic or (C) rock as the biofilm support medium (photos: WesTech Engineering Inc. and Raschig USA Inc.).

Trickling filters are mainly used for the oxidation of organic carbon and ammonia. Soluble substrates that diffuse into the biofilm can be efficiently converted but particle removal and bio-flocculation is less efficient (Parker and Newman, 2006). In industrial application the trickling filter is often used as a pretreatment of wastewater with high organic loadings before activated sludge treatment. This is equivalent to the BAS (Biofilm Activated Sludge) - technology used with MBBR (Malmqvist *et al.*, 2004). Trickling filters have also been applied for denitrification, but preventing the convection of air through the reactor has shown to be almost impossible in practice and hence leads to low efficiency.

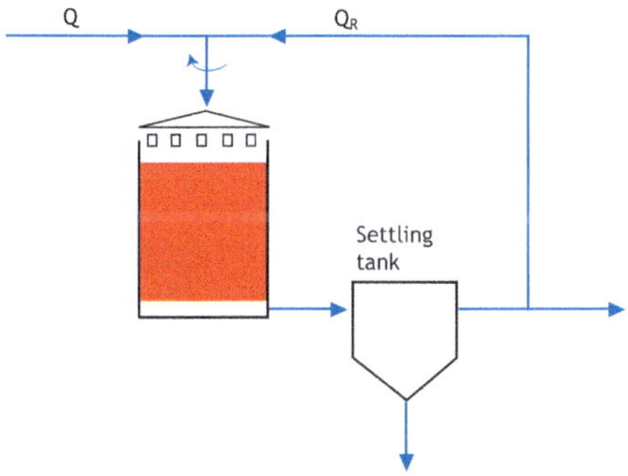

Figure 18.4 Schematic of a trickling filter with recycling of clarified effluent.

18.1.1.2 Rotating biological contactors

Rotating biological contactors (RBCs) use lightweight plastic disks that are mounted on a rotating horizontal shaft and that are partially submerged in water. RBCs were first introduced in the 1960s and can be advantageous due to their low energy demand, small footprint and simple operation. The rotation of the disks provides both aeration (when the biofilm is out of the water) and shear to control biofilm growth (when the biofilm moves through the water). An example of an RBC using corrugated plastic media is provided in Figure 18.5. After having almost disappeared in the 1990s due to many cases of failure with broken shafts or media, due to undersizing of the mechanical elements or organic overloading, RBCs have seen a return to the market as a package solution for small communities all over the globe. This is thanks to a redesign of the shafts and biofilm support to better withstand the weight of the biofilm and also to the utilization of conservative design rules to avoid excessive biofilm growth. Furthermore, frequency controlled motors on the shaft allow them to work periodically with a faster rotational speed for additional biofilm control.

Figure 18.5 (A) Rotating biological contactors using corrugated plastic media and (B) package plant RBC from Veolia PMT on the way to the Fiji Islands.

18.1.1.3 Submerged fixed-bed biofilm reactors

Since the 1980s a range of submerged biofilm reactor technologies has been developed using a small size (0.7 8.0 mm) granular medium that is completely submerged in water. The smaller sized medium results in larger specific surface areas (1,000-3,600 m^2/m^3) compared to trickling filters and RBCs (Table 18.1). The corresponding smaller pore spaces also mean that the biofilm thickness must be more effectively controlled to avoid clogging of the filter. The smaller filter medium in fixed-bed reactors can allow biological conversion processes to be combined with depth filtration, retaining suspended solids, and thus allowing the absence of any downstream treatment for solids removal. Removal of excess biofilm is typically achieved by regular backwashing of the filter where air and treated water are introduced into the reactor to temporarily expand the filter bed and to remove detached biomass and entrapped particulate matter. Backwashing is performed when the headloss across the reactor exceeds a critical value (2.0-2.5 mWC) or after fixed time periods (typically in the order of 24-48 h). Due to the periodic backwashing, several filter cells will be needed to ensure continuous treatment of the influent. Submerged biofilm reactors that are specifically designed for combined biological processes and solids removal are called biological aerated filters (BAF) The absence of a separate solids separation makes BAF one of the most compact treatment technologies, taking up a small area. For carbon removal or for carbon removal and nitrification, one or two stages are used. Several process stages will normally be needed for complete nitrogen removal although examples of complete carbon and nitrogen removal in one stage do exist (Thauré *et al.*, 2007). Biological phosphorous removal is difficult to realize using a BAF and due to the additional solids production related to chemical phosphorous precipitation, the dosing of coagulant is normally done in the primary sedimentation upstream of the BAF. In contrast, submerged aerated filters (SAF) use coarser media requiring no backwashing and are mainly designed for biological oxidation. In a SAF, solids removal has to be carried out in a separate clarifier or filter (WEF 2018). In submerged fixed bed biofilm reactors, oxygen has to be provided by introducing air at the bottom of the filter (Figure 18.6).

Oxygen transfer occurs throughout the filter bed as air bubbles rise to the top of the reactor. Although medium or coarse bubble diffused air is utilized to avoid clogging of the aeration system by biofilm growth, aeration efficiencies close to those of fine bubble diffusers in activated sludge can be achieved (Stenstrom *et al.*, 2008). Different types of submerged biofilm reactors are available that are operated with water introduced at the bottom (upflow) (Figure 18.6A and C) or the top (downflow) (Figure 18.6B) of the reactor. Packing material can either be heavier than water and is supported with an underdrain nozzle floor below the packing material (Figure 18.6A and B) or can be lighter than water and is supported with a ceiling plate with nozzles above the packing material (Figure 18.6C). Photos of packing material in fixed-bed biofilm reactors are shown in Figure 18.7A and B.

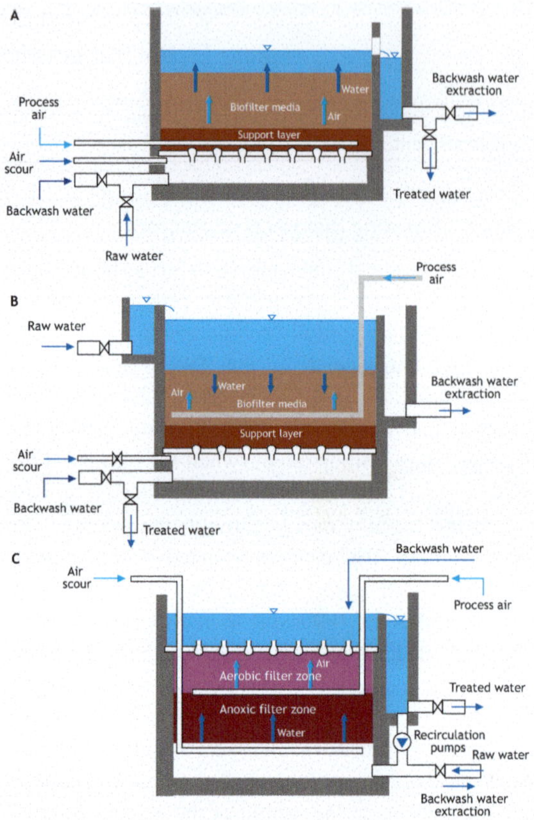

Figure 18.6 Submerged biofilm reactors: (A) upflow, dense media (Biofor®), (B) downflow, dense media (Biocarbone®) and (C) upflow, floating media (Biostyr®) (modified from ATV, 1997; Tschui, 1994).

Figure 18.6 shows schematics of submerged biofilm reactors with details of how air and water are introduced into the reactor during normal operation and during backwashing (scouring). Most fixed-bed biofilm reactors are operated as continuous flow reactors, but they can also be operated as sequencing batch biofilm reactors (SBBR) where the reactor is filled with wastewater at the beginning of a cycle, wastewater is recirculated through the reactor during a react phase, and clean water is discharged at the end of the cycle. One motivation for operating fixed-bed biofilm reactors as SBBR is to enable enhanced biological phosphorus removal (Morgenroth and Wilderer, 1999).

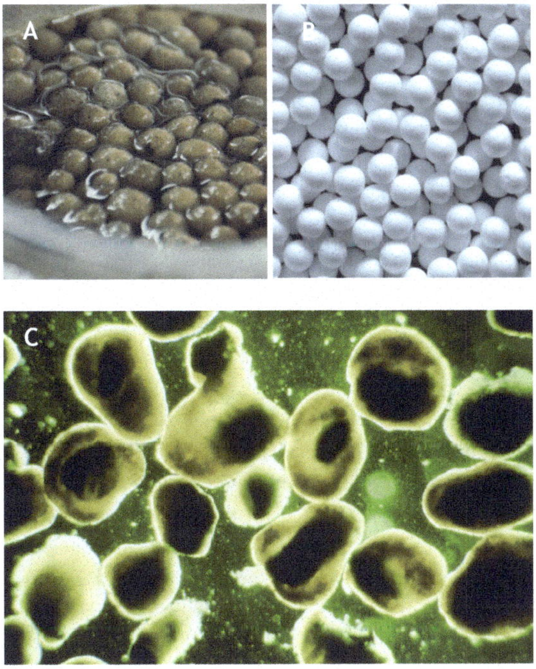

Figure 18.7 (A) Biofor® and (B) BioStyr® support media in submerged fixed-bed biofilm reactors. (C). Sand or basalt can be used as support medium in fluidized reactors. Diameter of the support medium in (A) and (B) is 4 mm and in (C) is 1 mm (photos: E. Morgenroth, Veolia, and M.C.M. van Loosdrecht, respectively).

18.1.1.4 Fluidized and expanded-bed biofilm reactors

In fluidized-bed reactors the support medium is kept in suspension by introducing water or air at the bottom of the reactor resulting in large upflow water velocities. Water upflow velocities range from 10-50 m/h (Boltz *et al.*, 2009; Nicolella *et al.*, 2000). Expanded bed reactors are similar to fluidized-bed reactors but they are operated with smaller upflow velocities resulting in an incomplete fluidization of the biofilm support medium. This continuous agitation enables even smaller filter media to be used compared to submerged biofilm reactors with even larger surface to volume ratios (Table 18.1).

In conventional fluidized-bed reactors the required upflow velocities can be achieved independent of influent flow rates by recycling treated water (Figure 18.2F). Operation of conventional fluidized-bed reactors requires a careful adjustment of upflow water velocities. If the upflow velocities are too low then the filter medium will settle to the bottom of the reactor. If the upflow velocities are too high then the filter medium will be washed out of the reactor. A second problem is the stratification of biofilm particles according to their settling velocity. Particles with a more porous biofilm settle more slowly and accumulate at the top of the bed where they experience less shear. As a consequence the biofilm becomes more and more fluffy and particles at the top of the reactor start to wash out. Both the need for a defined upflow velocity and the inherent instability of the biofilms in the system limit the application of fluidized and expanded-bed systems to low growth systems such as anaerobic wastewater treatment or nitrification.

In reactors utilising air lift a complete suspension of particles in the reactor is achieved by introducing air into the bottom of the reactor as shown in Figure 18.8A. Since all the particles in the air lift reactor experience a similar shear rate, the control of the biofilm is easier than in fluidized-bed reactors and the reactors are successfully used for COD removal and nitrification. It is essential for efficient aeration that the reactor is designed in such a manner that the bubbles also circulate over the down-comer of the reactor (Van Benthum *et al.*, 1999). On top of the reactor is a three-phase separation unit separating the gas, liquid and particles.

All the types of fluidized-bed reactors are sensitive to the hydraulic design; therefore this reactor type is mainly applied in industry where influent flow rates are more constant compared to municipal wastewater.

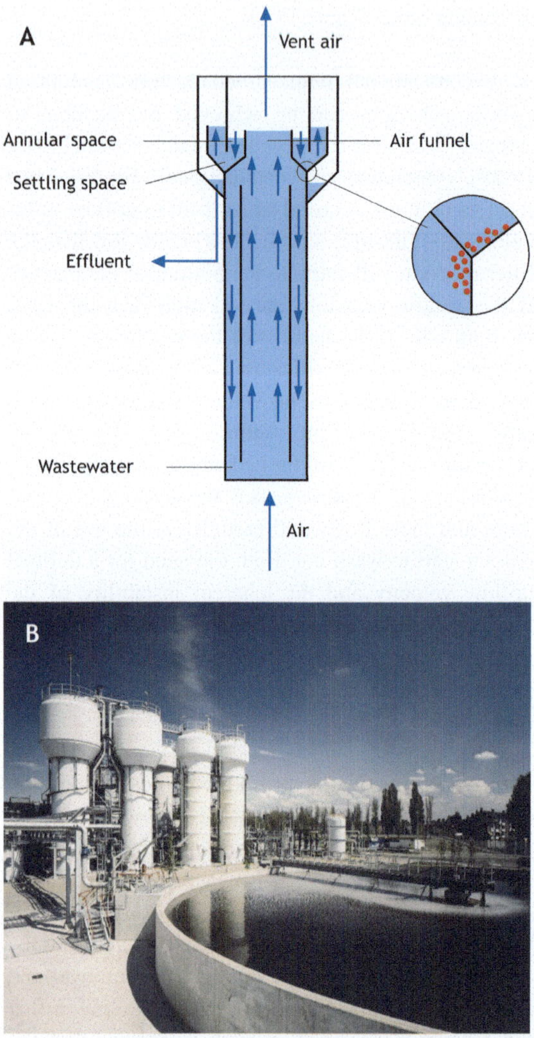

Figure 18.8 (A) Schematic and (B) photo of an airlift reactor (two towers on the left) with an integrated settler at the top of the reactor, at fermentation company DSM in Delft, The Netherlands (photo: J. Blom).

18.1.1.5 Granular sludge reactors

Granular biofilms can be grown also without a support medium (Hulshoff Pol *et al.*, 1982). Even though granular sludge might not fit the strict definition in Chapter 17 of microorganisms growing on a solid support surface, granules share many features with biofilm systems. The morphology, density, and size or granular sludge is, as in biofilm systems, directly influenced by shear forces and corresponding detachment in the reactor (Liu and Tay, 2002; Tay *et al.*, 2006; Van Loosdrecht *et al.*, 1995). In particular the structure of the granule is similar to that of a biofilm, 'fixed' and not subjected to disruption/flocculation, as in activated sludge flocs, meaning that gradients of microbial populations exist as in biofilm systems. A distinction between conventional activated sludge and granular sludge is that during settling of the granules no thickening occurs whereas for activated sludge aggregation and thickening are an important settling characteristic (Chapter 11). The definition of granulation has therefore been proposed as when the SVI after 5 minutes settling is similar to that after 30 minutes settling in a standard SVI test. A typical SVI for granular sludge after 5 minutes settling is 40-60 ml/g. Granulation is observed both in aerobic and in anaerobic reactors where the formation of larger and faster settling microbial aggregates provides an ecological advantage when the reactor is operated in a way where smaller flocs are washed out of the system. Upflow anaerobic sludge blanket reactors (UASB) are a widely used technology to achieve granulation under anaerobic conditions (Chapter 16). One approach that is commonly used to achieve aerobic granulation is to operate a sequencing batch reactor with very short settling times (Beun *et al.*, 1999; Morgenroth *et al.*, 1997). Depending on reactor operation the size of granules can range from a few hundred micrometers up to a few millimetres (Figure 18.9) (Liu and Tay, 2002). Aerobic granular sludge that is formed by slow-growing bacteria is more stable than when fast-growing bacteria are present (Van Loosdrecht *et al.*, 1995, Chapter 11).

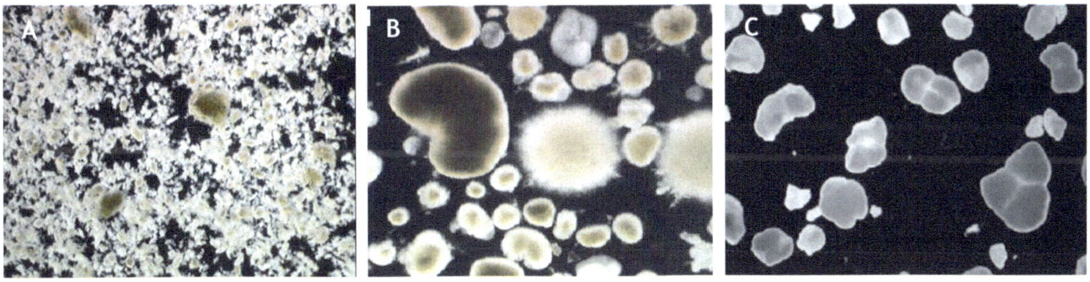

Figure 18.9 Development of aerobic granular sludge in a sequencing batch airlift reactor started up with conventional activated sludge after (A) 4, (B) 13, and (C) 87 days of reactor operation (photos: M.C.M. van Loosdrecht).

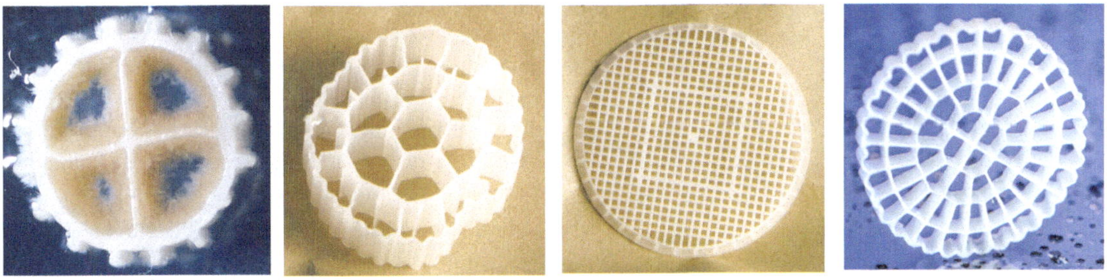

Figure 18.10 MBBR plastic media used in moving-bed biofilm reactors (MBBR) (photos: AnoxKaldnes).

18.1.1.6 Moving-bed biofilm reactors

Moving-bed biofilm reactors (MBBR) use a biofilm support medium with a density close to water so that it can be kept in suspension with minimum mixing energy provided by aeration or mechanical mixing (Ødegaard, 2006). Biofilm support media are manufactured in different shapes and are sufficiently large so that suspended support media can be retained in the reactor by screens or wire wedges (Figure 18.10).

MBBRs can be operated without or with sludge recycles. Without the recycle of biomass (Figure 18.11A), the MBBR biomass retention in the system is in general limited to the biofilm retained on the support medium. A system with biomass recycle retains both biofilm and suspended biomass and this type of reactor is discussed further in the next section (Figure 18.11B).

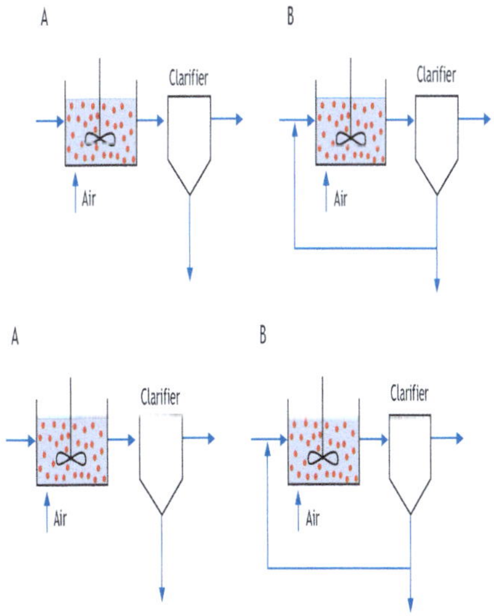

Figure 18.11 MBBR systems that can be configured (A) without or (B) with recycle of suspended biomass.

One of the advantages of MBBR is the very low hydraulic head loss in the reactor that allows a specific reactor to be designated to each specific process step, for example, pre-denitrification, carbon removal and nitrification. This again leads to higher removal rates since specialized biofilm and process conditions can be obtained in each reactor making the MBBR one of technologies with the the smallest footprint. Another advantage of MBBR using media giving a protected surface for the biofilm growth is the ability to achieve stable nitrification down to very low wastewater temperatures, as reported by Delatolla *et al.* (2012).

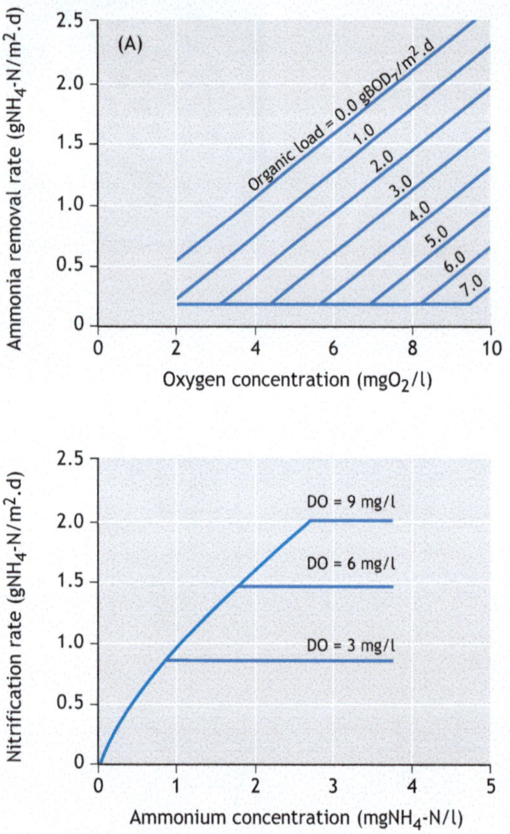

Figure 18.12 Influence of BOD_7 and oxygen concentration on nitrogen removal rates (Ødegaard, 1999).

However, a drawback of MBBR technology is that the fact that the media moves with the water which will increase the thickness of the stagnant laminar water layer surrounding the biofilm, resulting in increased diffusion limitation of the substrate and oxygen. As a result, especially nitrification reactors need to work at a relative high oxygen set point, as illustrated in Figure 18.12.

In MBBR reactors the preferred aeration method is an aeration grid with a medium-size bubble. This is chosen to avoid any maintenance of the aeration grid due to clogging or biofilm growth, since this would necessitate removing the media from the reactor. The low oxygen transfer efficiency of the medium bubble aeration is partly compensated for by a high alpha factor. Further the media and the biomass sitting on it can positively influence the transfer. Collivignarelli *et al.* (2019) go so far as to conclude that 'The fine bubble side aeration system compared to the coarse ones did not show significant advantages in terms of oxygen transfer efficiency'.

18.1.1.7 Hybrid biofilm/activated sludge systems

The introduction of a biofilm support medium can be used to enhance the performance of activated sludge systems. These systems are referred to as hybrid systems or as integrated fixed-film activated sludge systems (IFAS). A hybrid system with both biofilm and activated sludge is shown in Figure 18.13. Biofilm packing material has to be selected in a way that it will not be clogged by the suspended activated sludge in the reactor. Packing material includes suspended media (as in the MBBR) or fixed packing material including plastic strings, structured PVC packing, or submerged rotating biological contactors (Tchobanoglous *et al.,* 2008). In general, the slower growing bacteria will preferentially accumulate in the biofilm. In this way a high-load- or over loaded plants (with short SRT) can be upgraded to nitrification (Van Benthum *et al.,* 1997). Hybrid systems are also proposed in anaerobic wastewater treatment where methanogens grow as biofilm whereas acidogenic microorganisms are present in a flocculated sludge blanket.

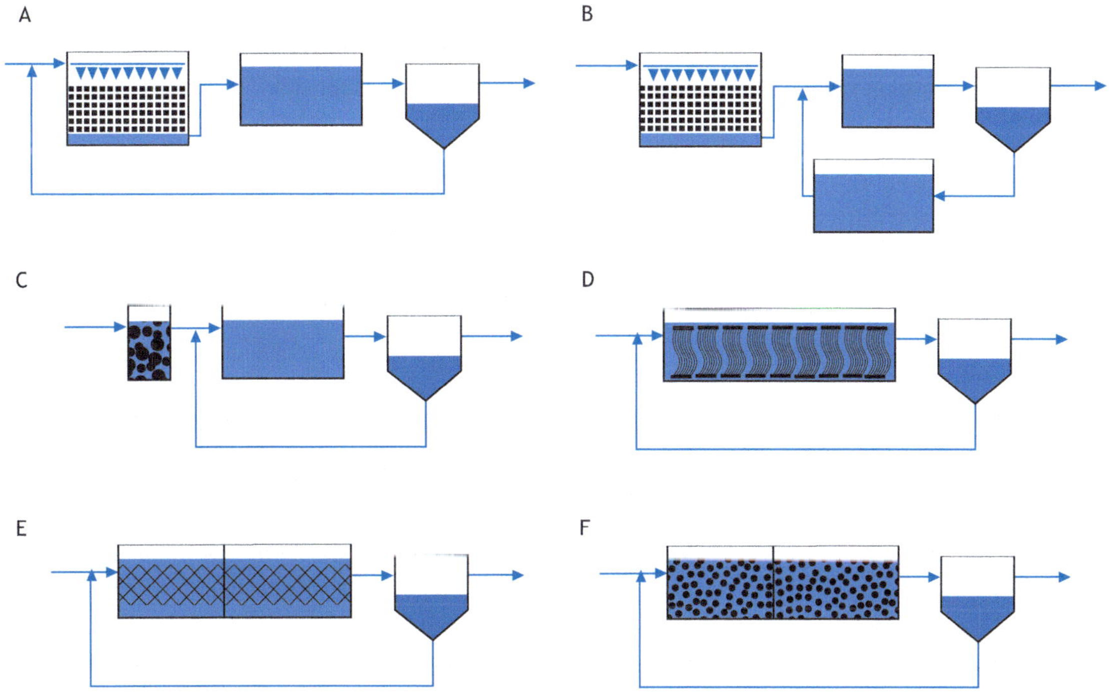

Figure 18.13 Different hybrid systems: (A) ABF, (B) TF/SC, (C) BAS, (D) fixed media IFAS, (E) ringlace IFAS and (F) MBBR IFAS (Ødegaard *et al.*, 2014).

Over time MBBR IFAS systems have become the most widespread hybrid systems used. They are used for a variety of reasons, including: enhanced nitrification; enhanced removal of carbon, nitrogen and biological phosphorous; improved settling and operational stability; a reduced footprint, etc. (Christensson and Welander, 2004). The most frequently used objective is, however, to enhance nitrification, with the goal to be able to fully nitrify at a lower aerobic sludge age of the suspended biomass than in conventional activated sludge systems. This results in a reduced footprint for both nitrifying, nitrogen removal, and biological nutrient removal (BNR) plants. The side effects are improved biomass settling and operational stability. Therefore IFAS systems are frequently used for upgrading of conventional activated sludge (CAS) plants (Ødegaard *et al.*, 2014). Also the lower applied SRT for nitrifying IFAS systems results in a lower oxygen consumption due to reduced endogenous respiration. This again also results in a less stabilized sludge with higher methane production potential if taken to a digester.

18.1.1.8 Membrane-attached biofilm reactors

Biofilms can be grown on gas permeable membranes allowing for mass transfer of substrates both from the surface as well as the base of the biofilm (Figure 18.13).

If the contaminant is an oxidized compound (*e.g.* nitrate or perchlorate) then an electron donor (*e.g.* hydrogen gas) can be provided to the base of the biofilm through the membrane (Nerenberg and Rittmann, 2004). Alternatively, if the contaminant is a reduced compound (*e.g.* ammonium) then an electron acceptor (*e.g.* oxygen) can be provided through the membrane (Terada *et al.*, 2007). The membrane-attached biofilm reactor (MABR) is typically operated in hybrid form, where the membranes support a nitrifying biofilm and suspended heterotrophs carry out the denitrification. This achieves total nitrogen removal. In order to maximize MABR nitrification fluxes, it is critical to understand the underlying biofilm behaviour. This behaviour differs from conventional biofilm processes due to the counter-diffusion of the electron donor (ammonia) and acceptor (oxygen). Furthermore, partial nitrification or partial nitrification by anammox may occur if ammonia-oxidizing bacteria outcompete nitrite-oxidizing bacteria. This can further enhance the cost-effectiveness of the MABR process (Pérez-Calleja *et al.* (2019). Two different commercial reactor configurations exist for the MABR. The first is on a configuration much like the one known from MBR; the membrane is made as a hollow fibre where a gas is supplied in the lumen and biofilm grows on the outside. The fibres are mounted on cassettes that can be submerged in an activated sludge tank to form a hybrid system (Figure 18.13). The second configuration is inspired from the traditional reverse osmosis (RO) membrane configuration with a spiral-wound membrane placed in a tube; water flows on one side of the membrane and air is supplied from the other side of the membrane (Figure 18.14).

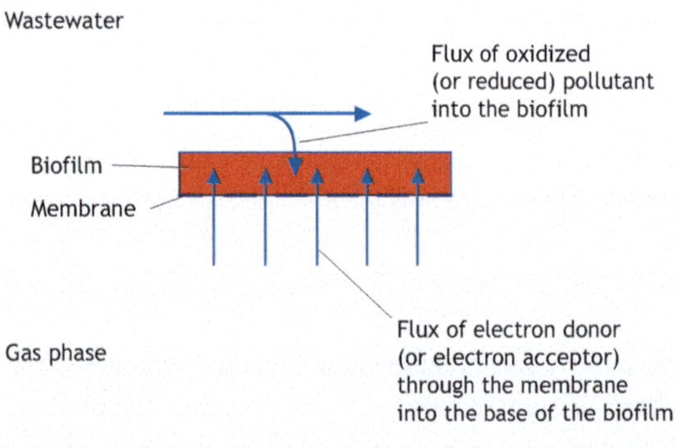

Figure 18.14 Schematic of mass transport into a membrane-attached biofilm.

The configuration of a hybrid reactor set-up makes MABR ideal for upgrading an existing activated sludge plant into an IFAS MABR configuration. Another advantage is the superior oxygen transfer efficiency because the oxygen transfer through the membrane is bubble-free. The IFAS configuration also allows biological phosphorous removal to take place in a MABR system.

Figure 18.15 MABR module (photo: OxyMem).

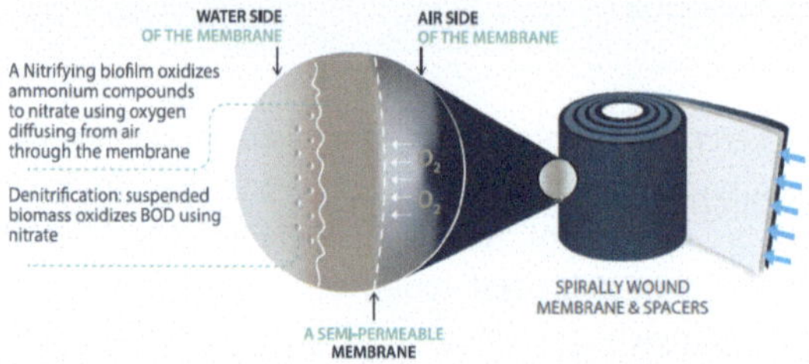

Figure 18.16 MABR module (illustration: Fluence).

18.1.2 Choosing from different biofilm support material options

Carrier media are produced from different materials and come in various shapes and sizes. The larger the carrier material and the corresponding void spaces in the reactor, the smaller the risk of clogging and channelling through biofilm accumulation. Typical examples of biofilm reactors with large void spaces are trickling filters and RBC. The downside of large void spaces is that the specific surface areas are relatively small. Smaller carrier material requires either regular backwashing or continuous fluidization of the carrier material to prevent clogging. The corresponding specific surface areas of the smaller media can be a larger order of magnitude compared to trickling filters and RBC. The specific surface area (a_F) (L^{-1}) is defined as:

$$a_F = \frac{A_F}{V_R} \tag{18.1}$$

where A_F is the effective surface area of the biofilm [L^2] and V_R is the volume of the biofilm reactor containing the filter material [L^3]. Note that A_F is typically smaller than the total surface area of the filter material due to incomplete biofilm coverage. Depending on the type of reactor, V_R can be smaller than the total volume of the biofilm reactor taking into account clear water zones above and below the filter media that are not included in V_R. An overview of specific surface areas and selected properties of different carrier materials is provided in Table 18.1.

Factors other than specific surface area that also affect the choice of biofilm support material include cost, density (floating material or material that is heavier than water), attrition resistance, and suitability for biofilm attachment (Lazarova and Manem, 2000). For sustainability any possibilities to recycle used support material must also be considered.

18.2 DESIGN PARAMETERS

Biofilm reactors are based on a range of different support media, mixing conditions, types of aeration, and methods for biofilm removal. However, the detailed design of these different biofilm reactor systems will be system-specific and is beyond the scope of this book. Nevertheless, there are some general design principles that are common among the different reactor systems. These common design principles will be discussed in this section. More detailed design information is partly available from the technical literature (*e.g.* ATV, 1997; ATV-DVWK (2001); Grady *et al.*, 1999; Tchobanoglous *et al.*, 2013; WEF, 2018) and partly from companies which is therefore often proprietary.

18.2.1 Substrate flux and loading rates

Substrate removal in biofilm reactors is almost always mass-transport-limited. As a result, the extent of substrate removal in a reactor is not determined by the total amount of biomass in the system, but by the available biofilm surface area (A_F) and the substrate flux into the biofilm (J_{LF}). Using the approaches developed in Chapter 17 the substrate flux can be calculated based on effluent substrate concentrations, external mass transfer resistance, and mass transport and reactions in the biofilm. The necessary biofilm surface area for a biofilm reactor with a completely mixed bulk phase (*e.g.* a moving-bed biofilm reactor) can be calculated as:

$$A_F = \frac{Q \cdot (S_i - S_B)}{J_{LF}} \tag{18.2}$$

assuming that substrate removal in the bulk phase is negligible. For reactors with more complex mixing conditions (*e.g.* plug-flow conditions in fixed-bed reactors) the reactor can be modelled as biofilm compartments in series (Figure 17.19). The required A_F for the overall reactor is the sum of the biofilm surfaces areas in the individual compartments. The minimum reactor volume (V_R) can be calculated from A_F using the definition of the specific biofilm surface area (a_F):

$$V_R = \frac{A_F}{a_F} \tag{18.3}$$

As described in Section 18.3 it is important to realize that in most cases, two or more components must, for example, actually diffuse into the biofilm, but only one of them will be limiting, meaning that when designing a plug-flow system the limiting substrate may change from influent to effluent of the reactor.

Many design guidelines for biofilm reactors do not explicitly calculate the surface area based on local substrate fluxes within the reactor, but are based on design loadings that are determined empirically for a given reactor's system. Design loadings can be expressed as surface loadings (B_A), [M L^{-2} T^{-1}]:

$$A_F = \frac{Q \cdot S_i}{B_A} \tag{18.4}$$

or as volumetric loadings (B_V), [M $L^{-3}T^{-1}$]:

$$V_R = \frac{Q \cdot S_i}{B_V} \tag{18.5}$$

B_A and B_V are directly related by the specific surface area (a_F) (Eq. 18.1):

$$B_V = B_A \cdot a_F \tag{18.6}$$

For a reactor with a completely mixed bulk phase, Eq. 18.2 and Eq. 18.3 can be combined providing the following relationship between substrate fluxes and surface loading rates:

$$J_{LF} = B_A - \underbrace{\frac{Q \cdot S_B}{A_F}}_{\text{Small for small values of } S_B} \tag{18.7}$$

For low levels of effluent substrate concentrations, the substrate flux and the design surface loading rates are virtually identical. Approaches for choosing appropriate design fluxes or loading rates are provided in Section 18.3.

18.2.2 Hydraulic loading

The mixing conditions and hydraulic loading of a biofilm reactor have an influence on the concentration gradients along the overall reactor, external mass transfer resistance, and also exposure of the biofilm to shear. The hydraulic loading (q_A) [L T^{-1}], in some systems also referred to as the filter velocity, is defined as:

$$q_A = \frac{Q_i + Q_R}{A_R} \tag{18.8}$$

where A_R is the cross-sectional area of the biofilm reactor in the flow direction [L^2]; for example, for a circular trickling filter, with radius r, $A_R = r^2 \cdot \pi$. The flow rate in Eq. 18.8 is the sum of the wastewater flow rate influent to the treatment plant (Q_i) and a recirculation flow (Q_R) (*e.g.* in Figure 18.3). Typical values for hydraulic loading of biofilm reactors are presented in Table 18.2. For trickling filters, the filter velocity must be seen as a minimum velocity that will

Table 18.2 Typical ranges of filter velocities (q_A, Eq. 18.8) for different types of biofilm reactors. Note that the filter velocities are heavily dependent on pretreatment, the air-water mixture (for submerged biofilters), backwashing frequency, and treatment objective.

Type of reactor	Carrier material	Filter velocity (q_A), m/h	Reference
Trickling filter	Rock	0.4-1.0	ATV, 1997
	Plastic	0.6-1.8	ATV, 1997
UASB	None	1-5	Nicolella *et al.*, 2000
Submerged biofilter	Porous clay	2-6 (max 10) (organics removal) 10 (nitrification) 14 (denitrification)	ATV, 1997; Pujol *et al.*, 1994
	Porous slate	2-5 (max 10)	ATV, 1997
	Polystyrene	2-6 (max 10)	ATV, 1997
	Quartz sand	5-15	ATV, 1997
	Anthracite	5-15	ATV, 1997
Fluidized bed	Sand or basalt	20-40	Nicolella *et al.*, 2000

guarantee a thorough distribution of the influent water and reduce the mass transfer limitation. In UASB and fluidized-bed systems the filter has a minimum value to ensure that bed expansion will take place which will minimize mass transfer limitations and a maximum velocity which will avoid loss of the biofilm support material. For submerged fixed-bed biofilm reactors with biofilm support that have a density higher than water and an upward flow direction, the maximum filtration velocity will avoid bed expansion; for reactors with biofilm support that have a density lower than water, a high filter velocity will lead to a sharp increase in head loss. For MBBR systems the filter velocity can be compared to the flux towards the sieve wall and here a too-high flux will lead to uneven distribution of the biofilm support since the media will be pushed towards the sieve wall.

18.3 HOW TO DETERMINE MAXIMUM DESIGN FLUXES OR DESIGN LOADING RATES

18.3.1 Model-based estimation of the maximum substrate flux

Biofilm reactors can be designed based on the desired effluent quality of soluble substrates using flux calculations from Chapter 17. The necessary level of complexity will depend on whether the pollutant is a limiting substrate and how mass transfer limitations influence microbial competition within the biofilm. Simple design approaches based on analytical solutions are usually sufficient for design purposes. However, for some specific questions more complex numerical models can be useful. It is important to realize that in most cases the influent will not only contain soluble substrate, but also particulate substrate. Depending on the reactor, more or less of this suspended substrate will be hydrolysed so that the total soluble substrate to eliminate will be increased. Four different levels of complexity can be differentiated, and the first three levels are shown in Figure 18.17.

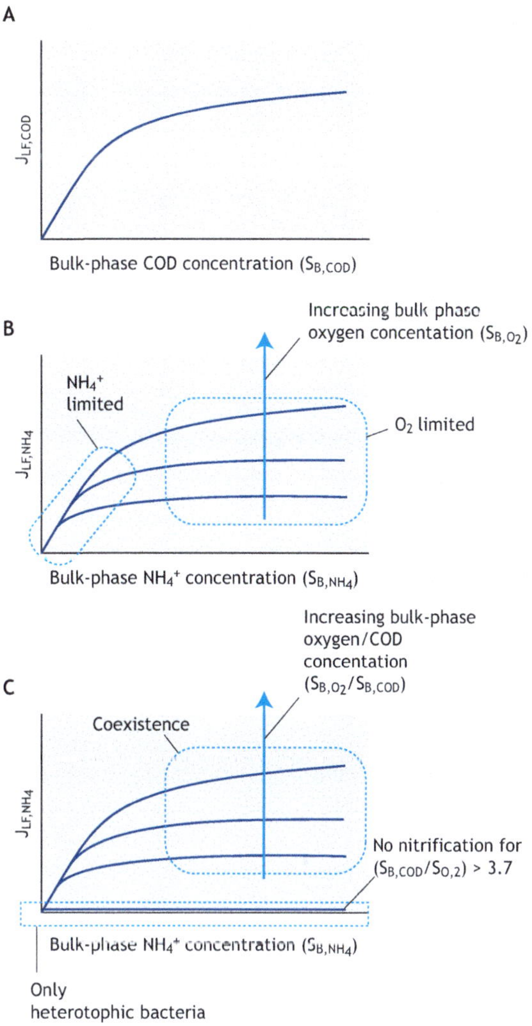

Figure 18.17 Examples of the three different levels of complexity as the basis for the design: (A) level 1 design: organic substrate removal where the organic substrate removal is rate-limiting, and nitrification where ammonia oxidation is ammonia -limited; (B) level 2 design: nitrification where ammonia oxidation can be oxygen limited (see also Figure 17.20 towards the inlet of the reactor), and (C) level 3 design: the ammonia flux is mostly determined by the bulk-phase BOD concentration (see also Figure 17.22).

18.3.1.1 Level 1 design: the compound of interest is the rate-limiting substrate

If the pollutant to be removed in the biofilm reactor is the rate-limiting compound then the design can be based on fluxes estimated for a single limiting substrate (Section 17.3). Criteria for determining what compound is rate-limiting were developed in Section 17.8. Examples of a level-1 design are carbon removal with low bulk-phase COD concentrations or nitrification with very low ammonia concentrations (Figure 18.11A).

18.3.1.2 Level 2 design: removal of the compound of interest is limited by the corresponding electron donor/acceptor

Multi-component diffusion has to be considered in systems where the pollutant is not the rate-limiting compound. In this case, the flux of the limiting compound has to be determined first. Then the flux of the pollutant can be calculated based on stoichiometry (Eqs. 17.89 and 17.92). An example of a level-2 design is the process of nitrification that is typically oxygen rather than ammonia-limited (Figure 18.17B). In this case ammonia flux is limited by the flux of oxygen into the biofilm. The oxygen flux depends on the bulk phase oxygen concentration which in turn depends on the type or aeration and oxygen transfer rates.

18.3.1.3 Level 3 design: removal of the compound of interest is limited by growth processes and by microbial competition within the biofilm for substrate and space

The previous two levels of design assumed that organisms are homogeneously distributed over the thickness of the biofilm and that substrate removal is limited by substrate diffusion into the biofilm. However, mass transport limitations can produce local ecological niches that can result in a heterogeneous distribution of different groups of microorganisms over the thickness of the biofilm. An example where competition for substrate and space significantly influences biofilm reactor performance is the combined oxidation of ammonia and organic substrate. As discussed in Section 17.9, the nitrifying bacteria tend to be covered by faster growing heterotrophic bacteria. As a result the flux of ammonia into the biofilm is controlled by the relative amount of oxygen ($S_{B,O2}$) and organic substrate ($S_{B,COD}$) in the bulk phase (Figure 18.17C). There is a threshold for $S_{B,COD}/S_{B,O2}$ above which nitrifying bacteria are outcompeted and nitrification can no longer take place.

18.3.1.4 Level 4 design: detailed modelling of concentration profiles, heterogeneous biofilm structure and design for dynamic environmental conditions

Microbial growth rates are not only influenced by electron donor and acceptor concentrations but also by factors such as pH, temperature, and the availability of nutrients and a suitable carbon source (*e.g.* CO_2 for autotrophic growth). Numerical simulations are required to take these complex interactions into account. Numerical solutions for heterogeneous 1-D biofilms are readily available using AQUASIM and many commercial wastewater treatment simulators. The advantages of using them are that (*i*) a-priori assumptions of a limiting substrate or of biomass distribution are not necessary, (*ii*) complex interactions within the biofilm can easily be implemented, and (*iii*) bulk-phase processes ($r_B \cdot V_B$ in Eq. 17.42) are automatically taken into account. Furthermore, numerical modelling allows to analyse the consequences of the typical dynamic behaviour of the influent to a treatment plant. However, the disadvantages of numerical solutions are that (*i*) it can be difficult to keep an overview and (*ii*) a complete understanding of what factors are in fact dominating the system performance can only be recognized by a user who understands both the model and the technology being modelled. Therefore, it is always recommended to be flexible and to combine different levels of design. Simple hand calculations are very useful for initial designs and to evaluate the plausibility of numerical simulations. When carrying out complex biofilm modelling it is further recommended to follow the steps given in *A framework for good biofilm reactor modelling practice* (Rittmann *et al.*, 2018).

18.3.2 Empirical maximum loading rates

In engineering practice, biofilm reactors are often designed based on loading rates (B_A or B_V) that are based on empirical observations of maximum loading rates resulting in satisfactory effluent concentrations. However, the specific characteristics of these satisfactory effluent concentrations are only vaguely defined in many publications and guidelines. Design values for surface loadings (B_A) and volumetric loadings (B_V) are in principle directly related (Eq. 18.5), assuming the specific surface area (q_A). Nevertheless, in systems such as submerged biofilters the specific surface area is not well defined and listed values for B_V provide an aggregate rate that includes substrate flux into the biofilm colonizing the support medium and also substrate removal by suspended biomass in the system (*i.e.* $r_B \cdot V_B$ in Eq. 17.42).

Typical design loadings for carbon oxidation and nitrification are provided in Table 18.3 and for denitrification in Table 18.4. The reader should compare these empirical design loadings with substrate fluxes calculated in Table 17.7 for oxygen, organic substrate, and ammonia. These recommended maximum loading rates should be used with caution as they are dependent on the wastewater characteristics, temperature, reactor operation, and desired treatment objectives. Also, as always when using recommended design values, the reader is strongly encouraged to review the specific conditions and references associated with the design values (ATV, 1997; Grady *et al.*, 1999; Tchobanoglous *et al.*, 2013; WEF, 2018).

18.3.3 Design examples

In the follow examples

Example 18.1 Organic substrate removal (level 1 design)

Assignment:

An MBBR should be designed to treat the following wastewater to a target effluent concentration for the biodegradable organic matter of 10 mgCOD/l as soluble COD. Calculate the volume of the reactor and the hydraulic retention time. You can assume that the MBBR can be modelled as a completely mixed reactor and that the oxygen concentration in the reactor is 8 mg/l.

Wastewater characteristics:
Q_i = 150 m^3/d
L_F = 200 μm
COD_i = 300 mgCOD/l
Specific surface area of the biofilm support medium
a_F = 300 m^2/m^3

Answer:

Step 1: Bulk phase organic substrate concentrations in the completely mixed reactor will be identical to the target effluent concentrations of 10 mg COD/l. Check whether oxygen or organic substrate are limiting using Eq. 17.64 or Table 17.4:

$$\frac{S_{LF,COD}}{S_{LF,O2}} = \frac{10}{8}\ gCOD/gO_2$$
$$= 1.25\ gCOD/gO_2 < 3.5\ gCOD/gO_2 \qquad (18.9)$$
$$\Rightarrow \text{Organic substrate is rate-limiting}$$

Step 2: Find the substrate flux assuming a bulk-phase substrate concentration of 10 mg COD/l from Figure 17.8.

$J_{LF,COD}$ = 12.3 gCOD/m^2.d
(assuming zero order partial penetration)

$J_{LF,COD}$ = 8.5 gCOD/m^2.d
(assuming Monod order)

Step 3: Calculate the necessary surface area.

$$A_F = \frac{Q_i \cdot (COD_i - S_B)}{J_{LF}}$$
$$= 3{,}537\ m^2 \text{(zero order partial penetration)} \qquad (18.10)$$
$$= 5{,}118\ m^2 \text{(Monod order)}$$

Table 18.3 Design surface loading rates (B_A) and volumetric loading rates (B_V) for BOD oxidation, combined BOD and ammonia oxidation, or tertiary nitrification. The values apply for the treatment of municipal wastewater to achieve significant removal (e.g. effluent concentration < 10 mg/l for BOD and < 3 mgN/l for ammonia) at normal temperatures (10-15 °C). Note that these values depend on the specific pre-treatment and wastewater composition.

Reactor type	Carrier material	BOD loading		Ammonia loading		Reference
		B_A ($gBOD/m^2.d$)	B_V ($kgBOD/m^3.d$)	B_A ($gN/m^2.d$)	B_V ($kgN/m^3.d$)	
BOD oxidation						
Trickling filter	Rock	4	0.4 (1)	-	-	ATV, 1997
	Plastic	4	0.4-0.8 (1)	-	-	ATV, 1997
Rotating biological contactor (RBC)	Plastic	8-20 (2)	-	-	-	Tchobanoglous *et al.*, 2003
Submerged biofilter	Porous clay (Biofor)	-	10	-	-	ATV, 1997
	Porous slate (Biocarbone)	-	10	-	-	ATV, 1997
	Polystyrene (Biostyr)	-	8	-	-	ATV, 1997
MBBR		5-15 (2)	-	-	-	WEF and ASCE, 1998
Combined BOD and ammonia oxidation						
Trickling filter	Rock	2	0.2 (1) (3)	-	-	ATV, 1997
	Plastic	2	0.2-0.4 (1) (3)	-	-	ATV, 1997
Rotating biological contactor (RBC)	Plastic	5-16	-	0.75-1.5	-	Tchobanoglous *et al.*, 2003
MBBR		4	-	0.8	-	Odegaard, 2006
Tertiary nitrification						
Trickling filter	Rock	-	-	0.5-2.5	0.05-0.25 (1)	Tchobanoglous *et al.*, 2013
	Plastic	-	-	0.5-2.5	0.05-0.5 (1)	Tchobanoglous *et al.*, 2013
Rotating biological contactor (RBC)	Plastic	1-2	-	1.5	-	Tchobanoglous *et al.*, 2013
Submerged biofilter	Porous clay (Biofor)	-	-	-	1.2	ATV, 1997
	Porous slate (Biocarbone)	-	-	-	0.7	ATV, 1997
	Polystyrene (Biostyr)	-	-	-	1.5	ATV, 1997

(1) Surface loadings (B_A) for trickling filters are converted into volumetric loadings (B_V) using Eq. 18.5 and assuming typical specific surface areas q_A of 100 m^2/m^3 for trickling filters using rock and 100-200 m^2/m^3 using plastic as support medium.
(2) BOD loadings > 10 $gBOD/m^2.d$ typically result in low removal efficiencies (*e.g.* BOD removal < 80%).
(3) In ATV (1997) combined BOD and ammonia oxidation is based solely on BOD loadings and assumes typical municipal wastewater composition in terms of BOD/TKN ratios.

Table 18.4 Design surface loading rates (B_A) and volumetric loading rates (B_V) for denitrification. The values apply for the treatment of municipal wastewater to achieve significant removal (> 90 %) at normal temperatures (10-15 °C). Note that these values depend on the treatment objectives, specific pre-treatment, wastewater composition, and the amount and type of added external carbon source.

Type of reactor	Carrier material	Nitrate loading		Reference
		B_A ($gN/m^2.d$)	B_V ($kgN/m^3.d$)	
Denitrification				
Submerged biofilter	Porous clay (Biofor)	-	2	ATV, 1997
	Porous slate (Biocarbone)		0.7	ATV, 1997
	Polystyrene (Biostyr)	-	1.2-1.5	ATV, 1997
	Quartz sand	-	1.5-3	
	Anthracite	-	1.5-3	
MBBR	K1	2.5-3 (1) 1-5-2 (2)	-	Aspegren *et al.*, 1998

(1) Using ethanol as electron donor.

(2) Using methanol as electron donor.

Step 4: Calculate the volume of the reactor and HRT.

$$V_R = \frac{A_F}{a_F} = \frac{3{,}537\ m^2}{300\ m^2/m^3} = 11.8\ m^3$$

(zero order partial penetration)

$$V_R = \frac{A_F}{a_F} = \frac{5{,}118\ m^2}{300\ m^2/m^3} = 17.1\ m^3 \qquad (18.11)$$

(Monod order)

Step 5: How would the substrate flux change now, taking into account an external mass transfer boundary layer of 200 μm? Substrate fluxes assuming different boundary layer thicknesses are given in Figure 17.13:

$J_{LF,COD} = 4$ g $COD/m^2.d$ (assuming Monod order and $L_L = 200$ μm)

The corresponding HRTs are 1.9 and 2.7 h assuming zero order partial penetration or Monod order, respectively.

Step 6: What would be the oxygen flux corresponding to a substrate flux of 4 $gCOD/m^2.d$?

The flux of different components involved in the same process is described by Eq. 17.92 given again below:

$$\frac{J_{LF,1}}{\nu_1} = \frac{J_{LF,2}}{\nu_2} = \dots = \frac{J_{LF,i}}{\nu_i} \qquad (18.12)$$

The stoichiometric coefficient for organic substrate is $\nu_S = 1/Y$ and for oxygen $\nu_{O2} = (1-Y)/Y$. Thus, the flux of oxygen ($J_{LF,O2}$) can be calculated from:

$$J_{LF,O2} = \frac{\nu_{O2}}{\nu_S} \cdot J_{LF,COD} = (1-Y) \cdot J_{LF,COD} \qquad (18.13)$$

Assuming an organic substrate flux of 4 $gCOD/m^2.d$ and a yield coefficient of 0.4 gCOD/gCOD, the flux of oxygen into the biofilm would be 2.4 $gO_2/m^2.d$.

Example 18.2 Nitrification (level 2 design)

An RBC should be designed for tertiary nitrification with a target effluent concentration of 5 $mgNH_4$-N/l. The bulk phase of the reactor can be assumed to be completely mixed with a dissolved oxygen concentration of 8 mg/l.

Wastewater characteristics:
$Q_i = 150\ m^3/d$
$L_F = 200\ \mu m$
$S_{NH4,i} = 40\ mgNH_4\text{-}N/l$
Specific surface area of the biofilm support medium
$a_F = 300\ m^2/m^3$

Step 1: Evaluate whether oxygen or ammonia will be rate-limiting. This could be done using Eq. 17.64, or directly using the penetration depths for ammonia and oxygen in Table 17.5. From Table 17.5 the penetration depths for ammonia concentrations of 5 mg N/l and oxygen concentrations of 8 mg/l are:

Ammonia penetration depth = 177 μm
Oxygen penetration depth = 120 μm

Thus, for the given bulk phase concentrations the availability of oxygen inside the biofilm will be limiting ammonia removal.

Step 2: The flux of oxygen can directly be taken from Table 17.5 to be:

$J_{LF,O2} = 22.4\ gO_2/m^2.d$

Step 3: The flux of ammonia needs to be calculated from the flux of oxygen using Eq. 17.89:

$$J_{LF,NH4} = \frac{v_{NH4}}{v_{O2}} \cdot J_{LF,O2} = \frac{\frac{1}{Y_{ANO}}}{\frac{4.57 - Y_{ANO}}{Y_{ANO}}} \cdot J_{LF,O2} = \frac{1}{4.57 - 0.22} 22.5 = 5.17\ g\ N/m^2.d \tag{18.14}$$

Note that this flux of 5.17 gN/(m².d) is smaller than the value provided in Table 17.5 for a bulk-phase ammonia concentration of 5 mgN/l. This is due to the fact that the ammonia fluxes in Table 17.5 are calculated assuming that no oxygen limitation exists.

Step 4: The fluxes in Table 17.5 do not take into account external mass transfer resistance Section 5 of Chapter 17 discussed the effect of the external mass transfer resistance and the oxygen flux could be explicitly calculated. In many cases, however, design values for ammonia oxidation are available based on measured fluxes in similar systems as summarized in Table 18.3. Ammonia fluxes for nitrification are in the range from 1 to 3 gN/m².d. Thus, for the current system a design value of 2.5 gN/m².d could be chosen.

Thus, steps 1 to 4 would not have been necessary and we could have directly used Table 18.3. A danger in using such design values without additional calculations is that it is often not apparent from such recommended design values what factors are limiting the removal. From steps 2 to 4 it is apparent that nitrification is oxygen-limited and that substrate fluxes are determined by the penetration of oxygen into the biofilm. Using the penetration depths provides insight into the desired biofilm characteristics. Also, by doing the explicit calculations it is immediately clear under what conditions the biofilm will be ammonia-limited rather than oxygen-limited.

Step 5: Calculate the necessary surface area and volume following the previous example.

Example 18.3 Combined organic substrate removal and nitrification (level 3 design)

A trickling filter using plastic media should be designed for combined carbon oxidation and nitrification.

Wastewater characteristics:
$Q_i = 150\ m^3/d$
$COD_{b,i} = 200\ mgBOD/l$
$S_{NH4,i} = 40\ mgNH_4\text{-}N/l$

Typical design parameters should be assumed for the design.

Approach 1: Detailed modelling of combined carbon oxidation and nitrification was discussed in Section 17.9, demonstrating that a condition for nitrification to occur is to have sufficiently low bulk-phase BOD concentrations for given oxygen concentrations. Carbon oxidation and nitrification can be evaluated by explicitly modelling substrate fluxes and the heterogeneous biofilm structure where BOD is oxidized first followed by nitrification (Wanner and Gujer, 1985). Simulations can be performed using biofilm modules available in commonly used treatment plant simulator software or the AQUASIM software that was introduced in Chapter 17.

Approach 2: As was shown in Figure 17.25, there will be three regions in a trickling filter operated for carbon oxidation and nitrification: (*i*) only carbon oxidation, (*ii*) combined carbon oxidation and nitrification, and (*iii*) only nitrification (first oxygen-limited and then ammonia-limited). A simplified design approach is to neglect the region of combined carbon oxidation and nitrification and to separately calculate dimensions for the other two regions:

$$V_{R,total} = V_{R,COD} + V_{R,NH4} \quad (18.15)$$

where $V_{R,total}$ is the overall volume of the reactor and $V_{R,COD}$ and $V_{R,NH4}$ are the reactor volumes for carbon oxidation in the upper part and nitrification in the lower part of the trickling filter. Using typical loading rates from Table 18.3 of 0.6 kgBOD/m^3.d (for BOD oxidation) and 0.1 kgN/m^3.d (for tertiary nitrification) we can estimate the overall volume as:

$$V_{R,COD} = \frac{Q_i \cdot COD_{b,i}}{B_{V,COD}} = \frac{(150\ m^3/d)(200\ gBOD/m^3)}{0.6\ kgBOD/m^3.d} = 50\ m^3 \quad (18.16)$$

$$V_{R,NH4} = \frac{Q_i \cdot S_{NH4,i}}{B_{V,NH4}} = \frac{(150\ m^3/d)(40\ gN/m^3)}{0.1\ kgN/m^3.d} = 60\ m^3 \quad (18.17)$$

Thus Eq. 18.15 gives:

$$V_{R,total} = 50\ m^3 + 60\ m^3 = 110\ m^3$$

Therefore, an overall volume of 110 m^3 should provide both reliable organic substrate oxidation and nitrification. The designer has the choice of either designing one large reactor or two separate reactors with solid liquid removal between the first carbon oxidizing reactor and the second nitrifying reactor.

18.4 OTHER DESIGN CONSIDERATIONS

This chapter provides an overview of different biofilm reactor technologies and is aimed at highlighting common features when designing these systems. Further biofilm design must also address conditions such as aeration, flow distribution, biofilm control and solids removal.

18.4.1 Aeration

For aerobic systems a sufficient supply of oxygen must be provided. Trickling filters and RBC typically rely on natural convection for aeration. If needed, aeration can be enhanced by forced ventilation or submerged air diffusers in trickling filters and RBC, respectively. Submerged biofilm reactors rely entirely on forced aeration. For fixed-bed reactors, air is introduced into the filter material through a grid that helps to ensure equal distribution of air over the cross section of the fixed bed reactor. As air bubbles move up through the filter material they rapidly coalesce resulting in larger bubbles, but due to increased gas hold-up caused by the combination of biofilm support and biofilm blocking the way for the bubbles, the aeration efficiency in BAFs can be compared to that found for a fine bubble diffuser in activated sludge (Stenstrom *et al.,* 2008). In suspended biofilm reactors, aeration often serves the dual purpose of providing oxygen and also energy input for mixing. Usually coarse-to-medium bubble aeration systems are preferred for biofilm reactors since these have less risk of clogging/fouling related to biofilm growth.

18.4.2 Flow distribution

For fixed-bed and fluidized-bed biofilm reactors a homogeneous distribution of the influent water flow over the cross section of the reactor is critical for effective treatment. Water distribution influences both local substrate loading rates and also shear forces acting on the biofilm. In fixed-bed reactors, inhomogeneous flow distribution can result in channelling, reduced substrate removal, and clogging of the filter material.

18.4.3 Biofilm control

Effective reactor operation must retain a sufficiently thick biofilm to allow for substrate removal, and at the same time prevent the accumulation of too much biofilm in order to avoid clogging. A thick biofilm will also lead to the risk of anaerobic conditions on the inside of the biofilm, which again can have an unwanted biofilm sloughing as a consequence. Shear forces in trickling filters are a function of the hydraulic loading (q_A, [L T^{-1}], Eq. 18.7), the number of arms on the rotating distributor (a), and the distributor rotational speed in revolutions per time (n, 1/T). These different factors affecting shear forces and detachment in trickling filters are combined in the flushing force (SK, from *Spülkraft* in German):

$$SK = \frac{q_A}{a \cdot n} \tag{18.18}$$

Typical values for SK range from 4 to 8 mm/arm (ATV-DVWK, 2001). Modern trickling filters are often equipped with an electrically driven distributer arm that allows the SK value to be controlled. Equally, RBC with PLCs and a frequency-controlled drive speed can be programmed to carry out a daily or weekly biofilm control by increasing the rotational speeds for some minutes.

In suspended biofilm reactors, high detachment rates and thin biofilms are the result of high shear and abrasion rates. Too-high carbon surface loading can though lead to such a high biofilm growth that the biofilm will not be sufficiently controlled and this becomes a limiting factor.

In fixed-bed biofilm reactors, backwashing must be performed at regular intervals to remove both excess biofilm and also suspended solids that can accumulate in the pore space of the filter medium. In nitrifying BAFs this also lowers the risk f the nitrifying biomass being grown over by heterotrophs.

18.4.4 Solids removal

Biomass removed from biofilm reactors has to be separated from water using settlers or other methods for solid-liquid separation. There are significant differences in the characteristics of biomass in terms of particle size and settleability differs between the different systems and different loading conditions. Different types of biofilm reactors vary significantly in how they remove particulate matter from the influent wastewater. For example, submerged fixed-bed reactors can be operated as a true filter while trickling filters or RBCs may achieve only limited particle removal (Parker and Newman, 2006). An example of the multitude of possible solids removal methods after MBBR is given in Figure 18.18.

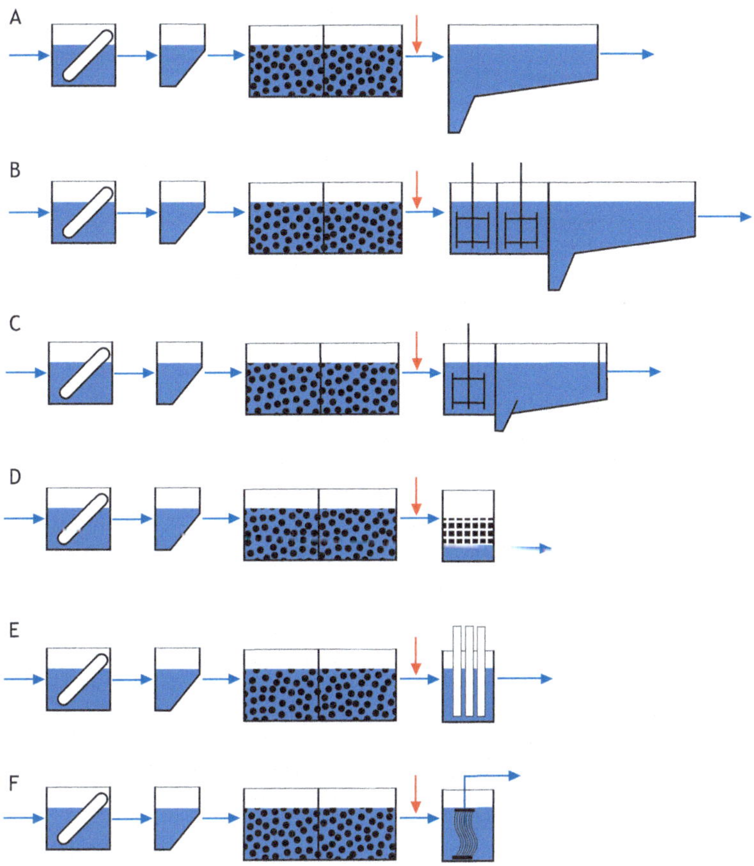

Figure 18.18 MBBR combined with biomass separation alternatives: (A) settling, (B) coagulation-settling (also Actiflo®), (C) flotation, (D) media filtration, (E) micro-screening (*i.e.* disc filtration) and (F) membrane filtration (Ødegaard, 2010).

REFERENCES

Aspegren H., Nyberg U., Andersson B., Gotthardsson S. and Jansen J.C. (1998). Post denitrification in a moving bed biofilm reactor process. *Water Science and Technology*, 38(1), 31-38.

ATV (1997). *Biologische und weitergehende Abwasserreinigung* (in German). 4th edition. Ernst & Sohn, Berlin.

ATV-DVWK (2001). *Bemessung von Tropfkörpern und Rotationstauchkörpern* (in German). DWA, Hennef. Vol. ATV-DVWK-A 281.

Beun J.J., Hendricks A., Van Loosdrecht M.C.M., Morgenroth E., Wilderer P.A. and Heijnen J.J. (1999). Aerobic granulation in a sequencing batch reactor. *Water Research*, 33(10), 2283-2290.

Boltz J.P., Goodwin S.J., Rippon D. and Daigger G.T. (2008). A review of operational control strategies for snails and other macro-fauna infestations in trickling filters. *Water Practice*, 2(4), 1-16.

Boltz J.P., Morgenroth E., de Barbadillo C., Dempsey M.J., McQuarrie J., Ghylin T., Harrison J. and Nerenberg R. (2009). Biofilm Reactor Technology and Design. In: *Design of Municipal Wastewater Treatment Plants*, Volume 2, 5th edition. WEF Manual of Practice No. 8, ASCE Manuals and

Reports on Engineering Practice No. 76, McGraw Hill, New York, USA.

Collivignarelli M.C., Abbà A. and Bertanza G. (2019). Oxygen transfer improvement in MBBR process. *Environmental Science and Pollution Research,* 26, 10727–10737.

Christensson M. and Welander T. (2004). Treatment of municipal wastewater in a hybrid process using a new suspended carrier with large surface area. *Water Science & Technology,* 49(11-12), 207-214.

De Kreuk M.K. and Van Loosdrecht M.C.M. (2004). Selection of slow growing organisms as a means for improving aerobic granular sludge stability. *Water Science and Technology*, 49(11-12), 9-17.

Dezotti M., Lippel G. and Bassin J.P. (2018). Advanced Biological Processes for Wastewater Treatment, Emerging, Consolidated Technologies and Introduction to Molecular Techniques. *Springer Verlag 2018.*

Delatolla R., Tufenkji N., Comeau Y, Gadbois A., Lamarre D. and Berk D. (2012). Effects of long exposure to low temperatures on nitrifying biofilm and biomass in wastewater treatment. *Water Environmental Research,* 84(4):328-38.

Grady C.P.L., Daigger G.T. and Lim H.C. (1999). *Biological wastewater treatment.* 2nd edition. Marcel Dekker, New York.

Hulshoff Pol L., Dolfing J., Dezeeuw W. and Lettinga G. (1982). Cultivation of well adapted pelletized methanogenic sludge. *Biotechnology Letters*, 4(5), 329-332.

Lazarova V. and Manem J. (2000). Innovative biofilm treatment technologies for water and wastewater treatment. In: Bryers J.D. (ed.), *Biofilms II: Process Analysis and Application*, (6), 159-206. Wiley-Liss, Inc.

Liu Y. and Tay J.H. (2002). The essential role of hydrodynamic shear force in the formation of biofilm and granular sludge. *Water Research*, 36(7), 1653-1665.

Malmqvist A., Berggren B., Sjölin C., Welander T., Heuts L., Fransén A. and Ling D. (2004). Full scale implementation of the nutrient limited BAS process at Södra Cell Värö. *Water Science Technology,* 50(3), 123-130.

Morgenroth E., Sherden T., Van Loosdrecht M.C.M., Heijnen J.J. and Wilderer P.A. (1997). Aerobic granular sludge in a sequencing batch reactor. *Water Research*, 31(12), 3191-3194.

Morgenroth E. and Wilderer P.A. (1999). Controlled biomass removal - The key parameter to achieve enhanced biological phosphorus removal in biofilm systems. *Water Science and Technology*, 39(7), 33-40.

Nerenberg R. and Rittmann B.E. (2004). Hydrogen-based, hollow-fiber membrane biofilm reactor for reduction of perchlorate and other oxidized contaminants. *Water Science and Technology,* 49(11-12), 223-230.

Nicolella C., Van Loosdrecht M.C.M. and Heijnen J.J. (2000). Wastewater treatment with particulate biofilm reactors. *Journal of Biotechnology*, 80(1), 1-33.

Ødegaard H. (1999). The Moving Bed Biofilm Reactor. *Water Environmental Engineering and Reuse of Water.* Hokkaido Press, pp. 250-305.

Ødegaard H. (2006). Innovations in wastewater treatment: the moving bed biofilm process. *Water Science and Technology*, 53(9), 17-33.

Ødegaard H. (2010). Separation of biomass from moving bed biofilm reactors (MBBRs). Proceedings at WEF/IWA conference on biofilm reactor technology, Portland, USA, August 2010.

Ødegaard H., Christenson M. and Sørensen K.H. (2014). Chapter 15 Hybrid systems, from *Activated Sludge - 100 Years and Counting.* IWA Publishing, London.

Parker D.S. and Newman J.A. (2006). New process design procedure for dealing with variable trickling filter effluent suspended solids. *Journal of Environmental Engineering*, 132(7), 758-763.

Pérez-Calleja P., Clements E., Grotz L. and Nerenberg R. (2019). Enhancing Nitrification Fluxes in MABRs: Modeling and experimental approaches. *Proceedings of the Water Environment Federation, WEFTEC*

Pujol R., Hamon M., Kandel X. and Lemmel H. (1994). Biofilters: Flexible, reliable biological reactors. *Water Science and Technology*, 29(10-11), 33-38.

Rittmann B., Boltz J.P., Brockmann D., Daigger G.T, Morgenroth E., Sørensen K.H., Takács I., Van Loosdrecht M.C.M. and Vanrolleghem P.A. (2018). A framework for good biofilm reactor modeling practice (GBRMP), *Water Science and Technology,* 77(5), 1149-1164.

Rusten B., Eikbrokk B., Ulgenes Y. and Lygren E. (2006). Design and operations of the Kaldnes moving bed biofilm reactors. *Aquacultural Engineering,* 34(3), 322-331.

Schroeder E.D. and Tchobanoglous G. (1976). Mass Transfer Limitations on Trickling Filter Design. *Journal (Water Pollution Control Federation),* 48(4), 771-777.

Stenstrom M.K., Rosso D., Melcer H, Appleton R., Occiano V., Langworthy A. and Wong P. (2008). Oxygen Transfer in a Full-Depth Biological Aerated

Filter. *Water Environmental Research,* 80(7), 663-671.

The Danish Environmental Protection Agency/MERMISS (2018). Environmentally friendly treatment of highly potent pharmaceuticals in hospital wastewater - Mermiss. *The Danish Environmental Protection Agency.*

Tay J.H., Tay S.T.L., Liu Y., Show K.Y. and Ivanov V. (2006). *Biogranulation Technologies for Wastewater Treatment: Microbial Granules,* Elsevier, Amsterdam. Series: Waste Management Series, Volume 6.

Tchobanoglous G., Stensel H.D., Tsuchihashi R. and Burton F.L. (2013). *Wastewater Engineering: Treatment and Resource Recovery, 5th edition.* McGraw-Hill Education, New York.

Terada A., Lackner S., Tsuneda S. and Smets B.F. (2007). Redox-stratification controlled biofilm (ReSCoBi) for completely autotrophic nitrogen removal: The effect of co- *versus* counter-diffusion on reactor performance. *Biotechnology and Bioengineering,* 97(1), 40-51.

Thauré D., Girodet P., Le Tallec X. and Sørensen K.H. (2007). Application de la nitrification dénitrification simultanée au Biostyr. *Eau, l'Industrie, les Nuisance* 302, 63 66.

Tschui M. (1994). *Submerse Festbettreaktoren* Skriptum der EAWAG (Dübendorf) für den Studiengang Umwelt- und Kulturingenieure an der ETH Zürich, Vertieferblock, WS 94/95 (unpublished).

Van Benthum W.A.J., Van der Lans R.M., Van Loosdrecht M.C.M. and Heijnen J.J. (1999). Bubble recirculation regimes in an internal-loop airlift reactor. *Chemical Engineering Science*, 54(18), 3995-4006.

Van Benthum W.A.J., Van Loosdrecht M.C.M. and Heijnen J.J. (1997). Process design for nitrogen removal using nitrifying biofilm and denitrifying suspended growth in a biofilm airlift suspension reactor. *Water Science and Technology*, 36(1), 119-128.

Van Loosdrecht M.C.M., Eikelboom D., Gjaltema A., Mulder A., Tijhuis L. and Heijnen J.J. (1995). Biofilm structures. *Water Science and Technology*, 32(8), 35-43.

Wanner O., Eberl H.J., Morgenroth E., Noguera D.R., Picioreanu C., Rittmann B.E. and Van Loosdrecht M.C.M. (2006). *Mathematical Modeling of Biofilms.* IWA Publishing, London, UK. Series: Scientific and Technical Report Series Report No. 18.

Wanner O. and Gujer W. (1985). Competition in biofilms. *Water Science and Technology*, 17(2-3), 27-44.

WEF (2018). Design of Water Resource Recovery Facilities, MOP 8. WEF Press (Water Environment Federation and McGraw Hill Education).

NOMENCLATURE

Symbol	Description	Unit
a	Number of arms of the rotating distributor in a trickling filter	-
A_F	Surface area of the biofilm	m^2
a_F	Specific surface area of the biofilm = A_F/V_R	m^2/m^3
A_R	Cross-sectional area of the biofilm reactor in the flow direction	m^2
B_A	Surface specific loading rate	$g/m^2.d$
Bi	Biot number	-
B_V	Volume specific loading rate	$g/m^3.d$
COD_i	Concentration of total COD in the influent	mgCOD/l
$COD_{b,i}$	Concentration of biodegradable COD in the influent	mgCOD/l
h	Height of reactor with plug flow characteristics in the bulk phase	m
J	Substrate$^{(1)}$ flux	$g/m^2.d$
J_F	Substrate$^{(1)}$ flux inside the biofilm	$g/m^2.d$

J_{LF}	Substrate[(1)] flux at the biofilm surface	$g/m^2.d$
L_L	Mass-transfer boundary layer	m
S_B	Soluble substrate [(1)] concentration in the bulk water	mg/l
S_F	Soluble substrate concentration[(1)] inside the biofilm	mg/l
$S_{F,NH4}$	Concentration of ammonia inside the biofilm	mgN/l
$S_{F,O2}$	Concentration of oxygen inside the biofilm	mgO_2/l
$S_{F,S}$	Concentration of organic substrate inside the biofilm	mgCOD/l
S_i	Soluble substrate [(1)] concentration in the influent	mg/l
S_{LF}	Soluble substrate[(1)] concentration at the biofilm surface	mg/l
S_{NH4}	Concentration of ammonium	mgN/l
$S_{NH4,i}$	Concentration of ammonium in the influent	mgN/l
S_{O2}	Concentration of oxygen	mgO_2/l
S_S	Concentration of organic substrate	mgCOD/l
V_B	Volume of bulk-phase	m^3
V_R	Volume of reactor	m^3
Y_{ANO}	Yield for autotrophic growth on S_{NH4}	gCOD/gN

(1) Note that in most of Chapter 18 the type of the limiting substrate and the units are not specified. Examples for possible substrates are electron donors such as organic substrate ($S_{F,COD}$), ammonia ($S_{F,NH4}$), or electron acceptors such as oxygen ($S_{F,O2}$) or nitrate ($S_{F,NO3}$). Units for the substrate must be consistent with units in kinetic and stoichiometric constants.

(2) The type of biomass is not specified. The generic active biomass converts the generic substrate S_F. Examples of possible biomass types are heterotrophic bacteria (X_{OHO}) and autotrophic bacteria (X_{ANO}).

Subscript	Description
0	Zero order
0	At time zero
1	First order
A	Per biofilm surface area
ANO	Autotrophic bacteria
B	In the bulk phase
COD	Organic substrate
F	In the biofilm
OHO	Heterotrophic bacteria
i	Influent
LF	At the biofilm surface
NH4	Ammonium
O2	Oxygen
W	In water

Abbreviation	Description
BAF	Biological aerated filter
IFAS	Integrated fixed-film activated sludge system
MBBR	Moving-bed biofilm reactor
RBC	Rotating biological contactor
SAF	Submerged aerated filter

SBBR	Sequencing batch biofilm reactor
SK	Spülkraft (German for flushing force)
SRT	Sludge retention time
UASB	Upflow anaerobic sludge blanket

Greek symbol	Description	Unit
ν	Stoichiometric coefficient	-

IWA Publishing's authorised EU representative for General Product Safety Regulations is Diane D'Arras, 15 rue Duret, 75116 Paris, France, e-mail: safety@iwap.co.uk.

Printed and bound by CPI Group (UK) Ltd, Croydon, CR0 4YY
14/05/2026
02110309-0001